PHYSICAL CHEMISTRY

UNDERGRADUATE CHEMISTRY

A Series of Textbooks

edited by
J. J. Lagowski
Department of Chemistry
The University of Texas at Austin

Volume 1: Modern Inorganic Chemistry, *J. J. Lagowski*

Volume 2: Modern Chemical Analysis and Instrumentation,
Harold F. Walton and Jorge Reyes

Volume 3: Problems in Chemistry, Revised and Expanded,
Henry O. Daley, Jr. and Robert F. O'Malley

Volume 4: Principles of Colloid and Surface Chemistry,
Paul C. Hiemenz

Volume 5: Principles of Solution and Solubility,
Kōzō Shinoda, translated in collaboration with Paul Becher

Volume 6: Physical Chemistry: A Step-by-Step Approach
Marwin K. Kemp

PHYSICAL CHEMISTRY

A Step-by-Step Approach

MARWIN K. KEMP

Department of Chemistry
The University of Tulsa
Tulsa, Oklahoma

MARCEL DEKKER, INC. NEW YORK AND BASEL

Library of Congress Cataloging in Publication Data

Kemp, Marwin K.
Physical chemistry: a step-by-step approach.

(Undergraduate chemistry ; v. 6)
Includes bibliographies and index.
1. Chemistry, Physical and theoretical. I. Title.
QD453.2.K46 541'.3 79-791
ISBN 0-8247-6640-7

MARCEL DEKKER, INC.
270 Madison Avenue, New York, New York 10016

Current printing (last digit):
10 9 8 7 6 5 4 3

PRINTED IN THE UNITED STATES OF AMERICA

Dedicated to Linda

Preface

TO THE INSTRUCTOR

This text is a different approach to physical chemistry. It is written to provide flexibility in the format of the course in which it is used. In a traditional lecture course, the text, with its objectives, examples, problems with solutions, and self-tests, provides a framework that will guide the students in their studies. The instructor can "flesh out" this framework by providing additional applications and delving into some of the derivations more deeply (as well as clarifying any confusing points).

The textbook has been designed for use in nontraditional formats as well. It is written in such a way that the instructor can tailor the course to fit various disciplines that might be represented in the class. By selecting appropriate objectives, you can construct a course that fits the needs of any discipline, whether it is chemistry, life science, engineering, physics, or earth science. In self-paced or individualized instruction formats, the text is designed to allow students to progress at their own rates, as well as work on topics closely related to their field of specialization. The role of the instructor is no longer that of lecturer. In an individualized (or self-paced) course, the instructor becomes a guide, a resource person, a counselor, and an examiner. His contact with the class is on a one-to-one basis, allowing individual assistance and guidance beyond that normally found in a lecture-style course. More is said about this method in the Appendix A: Notes to the Instructor.

For much too long, we, as teachers of chemistry, have designed and taught courses as though every student in our classes was enamored with the subject. It is time for us to consider the needs of those in our classes. If we are honest with ourselves, we must admit that little of what is learned in a class will be remembered. If that is accepted, then our principal aim must be to provide students with the skills required to learn what they need to know, when they need to know it, on their own. That is one of the objectives of this book.

There are those who say physical chemistry should delve deeply into such areas as thermodynamics, kinetics, quantum mechanics, and statistical mechanics. My opinion is that an introduction to these topics should be given, then students should be directed to more detailed study in those subjects of most use to them in their particular disciplines. The topics listed above are, many times, overemphasized at the expense of some very important subjects, such as colloids and polymers, which are of great practical importance. This book has been designed to allow some flexibility on the part of the instructor in choosing topics for study. For example, students majoring in petroleum engineering, who have a strong background in thermodynamics, should be directed to study in more fruitful areas; whereas, chemistry majors need a thorough introduction to thermodynamics. Appendix A lists some possible combinations of topics for different disciplines in courses using the individualized instruction format. In lecture-style courses, it should be possible to lecture on a basic core of material pertinent to all disciplines, then assign additional topics relevant to individual disciplines. These additional topics can be thought of as enriching topics. Such a procedure avoids forcing every student into the same mold.

The Objectives sections as written vary from very specific to rather general. They are meant as a guide to study, not as a detailed listing of everything that is to be learned. Some instructors will want to be much more specific, others more general. Such variation is easily accomplished by providing the students with substitute objectives, and, occasionally, additional readings.

It is impossible to produce a book of practical length and include sufficient detail on every topic to satisfy every (any) instructor. Each instructor will have to supplement the basic text with additional readings in some areas. However, the framework of a self-study program is included which allows an instructor to design study plans for individuals without an inordinate investment of time.

TO THE STUDENT

This book is written in a format suitable for individualized (self-paced) study, as well as for lecture-style courses. The design is to allow you, along with the instructor, to choose topics for study that are most directly related to your major. This text, when used in a self-paced format, is designed for you to work at your own pace, within reasonable limits as defined by your instructor. You will have a tendency to procrastinate: fight it. It is time to develop the self-discipline that will be required when you enter into employment upon graduation. To become successful you must learn to budget your time and to motivate yourself to do what needs to be done. No one will "hold your hand" in this course, but there is ample assistance when you require it. All you need to do is ask whenever you have difficulty.

Each chapter is divided into sections, and each section or unit contains the following:

1. Unit Objective, a statement of what you are to know when you have finished the unit.
2. A body of reading material which discusses the area of physical chemistry under consideration.
3. Suggested problems, with solutions given in Appendix C.
4. Self-Tests, which may be checked by the instructor or may be used as additional problems.
5. Supplementary reading: Additional Reading list at the end of each section gives detailed reference to sources for additional information. Strongly recommended sources are noted with an asterisk.

You should read the objectives for a section before doing anything else. It has been included to assist you in your study. By keeping the objectives in mind as you read the section, you will be directing your study toward the most important aspects of that section. Studying with an objective in mind is much more effective than without; you know what you are looking for. Next, read fairly quickly through the section to get the trend. Then, go back, reread, and study the derivations and the examples. It is a good idea in some cases to see if you can work through the examples before looking at the solutions. This will show whether you understand the concepts involved. After your careful second (or third) reading, you are ready to try the problems.

You will be tempted to read a problem and say; Aw, I can work that. Don't bet on it. At least sketch out the solution before consulting the solutions in Appendix C. Finally, if required by the instructor, work out the self-test. If you can do it correctly, you are ready to take an examination on that section. Do not try to take short cuts. Learning physical chemistry, like other technical subjects, takes several readings and a large investment of time in solving problems.

A vast body of information about physical chemistry is summarized in this text. All topics are incomplete by necessity. I could not write it all, even if I knew it, which I don't, and you could not learn it all in a reasonable length of time, even if you needed it. However, the basics are here. You will learn basic principles, some applications, and most importantly, I hope you learn an approach to problem solving that involves the thinking process. I do not intend the problems in this book to be of the "plug and grind" type, even though many can be solved that way. Think about your approach to solving each. The method used is more important than the actual solution in many cases.

My principal objective in writing this book was to help you learn on your own while you learn a little physical chemistry. The factual information will be soon forgotten (although quickly recoverable). But, if you develop your ability to learn what you need to know, on your own, you will have gained an invaluable tool that will be useful from now on. A background of classical physics, integral and differential calculus and general chemistry is assumed. If you need a brief review of the mathematics necessary for physical chemistry, the excellent paperback *Mathematical Preparation for Physical Chemistry,* by Farrington Daniels (McGraw-Hill, New York, 1956) is highly recommended. Also, *Applied Mathematics for Physical Chemistry*, by J. R. Barrante (Prentice-Hall, Englewood Cliffs, N. J., 1974) will be useful.

ACKNOWLEDGMENTS

Many people have contributed to the preparation of this book. To the many reviewers, most of whom are unknown to me, I extend my thanks. Their criticisms have been most helpful in clarifying topics and arranging material in a more nearly logical fashion. Especially, I acknowledge the detailed reading and critique by Dr. J. M. Pochan and his wife Dar at Xerox Corporation. Also, I must mention several classes of long-suffering students, who endured various inconveniences and numerous errata sheets as the manuscript was forged into shape. Finally, I want to thank my wife Linda, who interpreted my handwriting through several drafts, and converted it into legible form. On more than one occasion her patience, diligence, and support saved this work from abandonment. If the final version bears any resemblance to all other versions, there will be errors present. I welcome comments from users of the book pointing out those errors. Also, I welcome comments from instructors and students on strengths and weaknesses in the text, and ways of using it.

Marwin K. Kemp

Contents

Tables

Chapter 12

Chapter 14

Examples

Chapter 7

Chapter 8

Chapter 9

Chapter 10

PHYSICAL CHEMISTRY

1

Units, Unit Analysis, and Data Treatment in Physical Chemistry

OBJECTIVES

You shall be able to

1. Estimate errors in results for given errors in experimental quantities
2. Estimate allowed errors in experimental quantities for a given error in a result
3. Perform a dimensional (or unit) analysis given an equation and the units on the variables involved
4. Linearize an equation and properly plot it to find the slope and intercept

1-1.1 INTRODUCTION

Physical chemistry is both a theoretical and an experimental science. In both theory and experiment there comes a time when problems have to be solved, quantities calculated, empirical relationships evaluated, etc. In any calculations a definite advantage is enjoyed by the person who works systematically. One important aid in establishing a system is to apply unit analysis to each problem.

Every experimental quantity has units, even if they happen to be no-unit. The use of the units associated with each quantity in an expression to be solved serves as a good road map in the solution of that expression. If the insertion of units on all variables and constants leads to the desired units on the final answer then there is a good probability, barring mathematical error, that the final answer will be correct. However, if the final unit is incorrect there is no way, other than by chance, that the final answer can be correct.

Another aid in the solution of any mathematical expression is neatness. Neatness serves at least two purposes. First, it shows forethought in one's work. Second, it allows one's steps to be checked with a minimum of labor. It is most difficult to trace one's steps back through a problem when margins and corners of the paper have to be consulted for pertinent intermediate answers. The habit of being neat and orderly both in problem solving and in written discourse is one that is formed only with difficulty. However, the rewards far outweigh the effort.

Before stepping into the main body of physical chemistry, let us borrow a couple of equations to be developed later to show how unit analysis can be used to avoid common pitfalls. In Chap. 2, we will derive an equation for the average speed $\bar{c}$ of a gas molecule. The average speed of a molecule will be shown to be given by

$$\bar{c} = \left(\frac{8RT}{\pi M}\right)^{1/2} \qquad (1\text{-}1.1.1)$$

where R is the gas constant, T is the absolute temperature, and M is the molecular weight of the gas molecule. Before we can solve Eq. (1-1.1.1) we need a few comments on units.

1-1.2 UNITS

The units in this course will, in general, be given according to SI (International System) conventions. There will be occasions on which non-SI units will be necessary simply because the vast majority of data in a given area are in some other units. Care will be taken to ensure that proper conversion factors are given. An important point to be noted is the method of finding a unit for a given quantity. For example, consider the gas constant R which is given in SI units as follows:

$$R = 8.314 \text{ J mol}^{-1} \text{ K}^{-1}$$

where J is Joule, an energy unit, mol is mole, the amount of material containing Avagadro's number (6.023×10^{23}) of particles, and K is the Kelvin degree. It may be necessary at times to break J into its basic units. To do that it is a definite aid to know the definition of energy. You should recognize that

$$\text{Energy} = \text{force} \times \text{distance}$$

But,

$$\text{Force} = \text{mass} \times \text{acceleration}$$

Thus, in SI units

$$\text{Energy} = \text{mass} \times \text{acceleration} \times \text{distance}$$
$$= \text{kg} \times \text{m s}^{-2} \times \text{m}$$
$$= \text{kg m}^2 \text{ s}^{-2} = \text{J}$$

The above is given to point out that it is essential to know the basic units of any derived unit. Table B-1 in the Appendix lists some of the constants with units that will be essential for this course. Table B-2 lists some commonly used conversion factors.

1-1.3 UNIT ANALYSIS

Now return to the problem presented in Eq. (1-1.1.1):

$$\bar{c} = \left(\frac{8RT}{\pi M}\right)^{1/2} \tag{1-1.1.1}$$

From Table B-1 we know that $R = 8.314 \text{ J mol}^{-1} \text{ K}^{-1}$. Assume the gas we are considering is nitrogen N_2 with a molecular weight M of $28 \text{ g mol}^{-1} = 0.028 \text{ kg mol}^{-1}$. Also, choose the temperature T as 25°C = 298.1 K. Solving Eq. (1-1.1.1):

$$\bar{c} = \left[\frac{(8)(8.314 \text{ J mol}^{-1} \text{ K}^{-1})(298.1 \text{ K})}{(3.1416)(0.028 \text{ kg mol}^{-1})}\right]^{1/2} = \left[\frac{(8)(8.314)(298.1)}{(3.1416)(0.028)} \frac{\text{J}}{\text{kg}}\right]^{1/2}$$

$$= \left(225400 \frac{\text{kg m}^2\text{s}^{-2}}{\text{kg}}\right)^{1/2} = \left(225400 \text{ m}^2 \text{ s}^{-2}\right)^{1/2} = 474.8 \text{ m s}^{-1}$$

$$= \underline{\underline{470 \text{ m s}^{-1}}}$$

which gives an acceptable velocity unit of meters per second. Assume that you had chosen R in $\text{cal mol}^{-1} \text{ K}^{-1}$ instead. Equation (1-1.1.1) would then give

$$\bar{c} = \left[\frac{(8)(1.987 \text{ cal mol}^{-1} \text{ K}^{-1})(298.1 \text{ K})}{(3.1416)(0.028 \text{ kg mol}^{-1})}\right]^{1/2} = (53870 \text{ cal kg}^{-1})^{1/2}$$

$$= \underline{\underline{230 \text{ (cal kg}^{-1})^{1/2}}}$$

which is a nonsense unit. The unit $(\text{cal kg}^{-1})^{1/2}$ can be converted to the proper unit, but as it stands it is not recognizable as a velocity unit.

To reinterate, if the units on a problem work out, there is a chance of obtaining the right answer. If the units are incorrect there is no hope of getting a correct numerical value.

1-1.4 ERROR ANALYSIS

The determination of a physical property of a system usually involves the combination of several independent physical measurements. For example, the determination of the molecular weight of a solid by measuring the freezing point depression of the solvent in a solution involves the measurement of the weight of solute, the weight of solvent, and the depression of the freezing point of the solvent. This can be seen from the formula (from Chap. 6) for an ideal solution

$$M = \frac{1000\ K_f g}{G\theta} \tag{1-1.4.1}$$

where g is the weight of solute, G is the weight of solvent, θ is the depression in freezing point, and K_f is the molal freezing point depression constant of the solvent. Do not worry about the form of the equation. The treatment of the variables is the important thing here. None of the experimental quantities will be exact, and hence the molecular weight will be inexact. We will want to find the relation between the errors in the experimental quantities g, G, and θ and the error in the molecular weight. Usually the errors in g, G, and θ are not equal and hence do not contribute equally to the error in the molecular weight.

TYPES OF ERRORS. Before deriving some equations related to the estimation of error, we need to identify the various types of errors that can be present in any experimental determination. Three types of errors are systematic, random, and erratic. The definition of each of these will be given along with some examples. Then, equations will be developed that can be used to determine the magnitude of the error in a final result, given the expected errors in each of the variables.

Systematic errors are those that vary in a regular and predictable manner, if we are aware of their presence. Examples are

1. Use of uncalibrated weights or meters (ammeter, voltmeter, etc.)
2. Failure to apply a stem correction to a temperature measured on a liquid-in-glass thermometer
3. Use of an erroneous physical constant
4. Use of an uncorrected barometer reading
5. Approximations made in the derivation of the equations used to relate variables

In theory, all systematic errors could be eliminated. However, in many cases, their presence may not be known or suspected. This makes systematic errors very insidious. Repeated measurements with the same apparatus will not show these errors.

There are ways to test a procedure or apparatus for systematic errors. In general, the presence of systematic errors leads to results that are consistently too large or too small compared to known values. This fact allows one check on the procedure or apparatus being used. A *standard* (known material with known properties) can be run and the results compared to the "true" value. Another check is obtained if the parameter is measured with a different apparatus or procedure to check the result.

Once the presence of systematic errors has been verified, it is necessary to analyze the whole experiment in an attempt to eliminate them. Even the most careful analysis may fail to reveal some error sources.

Random errors are those that vary irregularly from measurement to measurement. Thus, repeated measurements will yield an estimate of the magnitude of these errors. Examples of random errors are

1. Response time: *personal errors* that occur in starting and stopping a timer, or determining the position of the meniscus in a burette, etc.
2. Ill-defined magnitude: the diameter of a wire, for example, may vary from position to position, the earth's magnetic field varies with time and position, etc.
3. "Occasional" bias: bias associated with given instruments on a given day, the room conditions (humidity, pressure, temperature), and the mental and physical state of the operator. All these may vary from day to day leading to variability in repeated measurements.

Random errors can be treated statistically as we shall see below.

The third type of errors are those called *erratic*. Examples are

1. A reading mistake
2. Calculation error
3. Line voltage change

These events usually lead to a value that is not consistent with those normally obtained. However, just because a value does not "fit" with others is not sufficient reason to eliminate it. Other measured values may be affected to a lesser degree. There are statistical tests which determine if a value can be eliminated. These can be found in Lyon, *Dealing with Data* (see Sec. 1-1.12, Ref. 3).

One additional distinction needs to be made before developing equations for estimating errors. This concerns the difference between accuracy and precision. *Precision* refers to the degree of repeatability of a result or to the freedom from random error. A value may be very precise (little random error) but very inaccurate (large systematic error). *Accuracy* refers to the closeness of a result to an accepted true value. An accurate result is one with little systematic error.

With these concepts in mind, let us turn our attention to estimating the errors in a result given estimates of the errors in variables used to calculate that result.

ESTIMATION OF MAXIMUM AND RELATIVE ERRORS. We will be concerned with two problems here:

1. Given the errors of several directly measured quantities, how can we determine the maximum and relative errors in a quantity derived from the measured quantities?
2. Given a prescribed error in the derived quantity, how can we specify the allowable errors in the measured quantities?

Consider z, which is calculated from the experimentally determined quantities x and y. From differential calculus we can write

$$dz = \left(\frac{\partial z}{\partial x}\right) dx + \left(\frac{\partial z}{\partial y}\right) dy \tag{1-1.4.2}$$

This gives the differential change in z corresponding to the differential changes in x and y. The differentials can be replaced by finite increments Δz, Δx, and Δy, to give to the first approximation

$$\Delta z = \left(\frac{\partial z}{\partial x}\right) \Delta x + \left(\frac{\partial z}{\partial y}\right) \Delta y \tag{1-1.4.3}$$

This approximation is reasonably good since the experimental quantities are usually accurate to within a few percent, so Δx and Δy are small. Δz approximates the finite change in z resulting from finite changes Δy and Δx. When this equation is applied to error analysis, it is possible that the error in y will partially cancel the error in x. (If Δy and Δx have opposite signs this is true.) However, in general, we never assume this. Instead, we assume all errors add, and the *maximum error* is calculated. This overestimates the error and gives an upper limit to it. We write then

$$|\Delta z| = \left|\frac{\partial z}{\partial x}\right| |\Delta x| + \left|\frac{\partial z}{\partial y}\right| |\Delta y| \tag{1-1.4.4}$$

where || indicate absolute values. This can be illustrated by working with Eq. (1-1.4.1) given above. If Eq. (1-1.4.2) is applied to (1-1.4.1) we get

$$dM = \left(\frac{\partial M}{\partial K_f}\right) dK_f + \left(\frac{\partial M}{\partial g}\right) dg + \left(\frac{\partial M}{\partial G}\right) dG + \left(\frac{\partial M}{\partial \theta}\right) d\theta \tag{1-1.4.5}$$

where (evaluate the derivatives to show the following results)

$$\left(\frac{\partial M}{\partial K_f}\right) = \frac{1000\ g}{G\theta} = \frac{M}{K_f}$$

$$\left(\frac{\partial M}{\partial g}\right) = \frac{1000\ K_f}{G\theta} = \frac{M}{g}$$

$$\left(\frac{\partial M}{\partial G}\right) = -\frac{1000\ K_f g}{G^2\theta} = -\frac{M}{G}$$

$$\left(\frac{\partial M}{\partial \theta}\right) = -\frac{1000\ K_f g}{G\theta^2} = -\frac{M}{\theta}$$

Substituting into (1-1.4.5) and taking absolute values gives the *maximum error in M*.

$$\Delta M = \frac{\Delta K_f}{K_f} M + \frac{\Delta g}{g} M + \frac{\Delta G}{G} M + \frac{\Delta\theta}{\theta} M$$

or

$$\Delta M = M \left(\frac{\Delta K_f}{K_f} + \frac{\Delta g}{g} + \frac{\Delta G}{G} + \frac{\Delta\theta}{\theta}\right) \tag{1-1.4.6}$$

The *relative error* in M will be given by $\Delta M/M$ which is easily obtained from (1-1.4.6).

A more convenient approach when working with products and quotients is to write the natural logarithm of both sides. For Eq. (1-1.4.1) this becomes

$$\ln M = \ln 1000 + \ln K_f + \ln g - \ln G - \ln \theta \qquad (1\text{-}1.4.7)$$

(Note: If sums and differences are involved in the initial equation, taking logarithms may not simplify the problem as much as when only products and quotients are involved.)

Differentiating gives

$$\frac{dM}{M} = \frac{dK_f}{K_f} + \frac{dg}{g} - \frac{dG}{G} - \frac{d\theta}{\theta} \qquad (1\text{-}1.4.8)$$

Taking the absolute values, changing d to Δ, and multiplying by M gives Eq. (1-1.4.6) again.

If we know the errors in experimental quantities we can calculate the error in the molecular weight. Remember, this calculation does not take into account any errors due to deviations from ideal solution behavior which was assumed in deriving Eq. (1-1.4.1). Such deviation is a *systematic error* that would have to be estimated by other means.

1-1.4.1 EXAMPLE

Find the maximum error and relative error in molecular weight if 6.02 ± 0.003 grams of solid is added to 210.3 ± 0.8 g of benzene causing a depression in the freezing point of 0.665 ± 0.007 degrees. Assume K_f is exact and equal to $5.12° \text{ molal}^{-1}$.

Solution

We need to apply Eq. (1-1.4.6). Since K_f is a constant, $\Delta K_f = 0$ and we have

$$\frac{\Delta M}{M} = \frac{\Delta g}{g} + \frac{\Delta G}{G} + \frac{\Delta\theta}{\theta}$$

From the problem statement $\Delta g = 0.003$ (not 0.006), $\Delta G = 0.8$, and $\Delta\theta = 0.007$. Thus the relative error is

$$\frac{\Delta M}{M} = \frac{0.003}{6.02} + \frac{0.8}{210.3} + \frac{0.007}{0.665}$$

$$= 0.00050 + 0.00380 + 0.01053$$

$$\boxed{\frac{\Delta M}{M} = 0.015}$$

To get ΔM we must have a value of M. This is obtained by substitution in Eq. (1-1.4.1) of the *average values* of the variables.

$$M = \frac{1000\ K_f g}{G\theta} = \frac{1000(5.12° \text{ molal}^{-1})(6.02 \text{ g})}{(210.3 \text{ g})(0.665°)}$$

(You do not have sufficient information to do a unit analysis on this problem. Accept the fact for now that the units work out to give g mol^{-1}.)

$$M = 220 \text{ g mol}^{-1}$$

Thus the maximum error in M is

$$\Delta M = (220 \text{ g mol}^{-1})(0.015) = \underline{\underline{3.3 \text{ g mol}^{-1}}}$$

Equation (1-1.4.6) can also be used to solve the second problem mentioned above: How accurately must the experimental quantities be determined to yield a given precision in the molecular weight? This is very important in the design and improvement of experiments. A unique solution is not possible since there are many errors which will add up to a given value of ΔM. As a starting point we sometimes assume that each of the measured quantities contribute equally to the error in the function. In designing an experiment, one objective is to make errors in all the variables approximately equal. This condition is known as the *principle of equal effects*. If all the errors are equal, Eq. (1-1.4.6) can be rewritten as

$$\frac{\Delta M}{M} = \frac{4\Delta K_f}{K_f} = \frac{4\Delta G}{G} = \frac{4\Delta g}{g} = \frac{4\Delta\theta}{\theta} \tag{1-1.4.9}$$

In practice we can frequently design an experiment so that one of the quantities has a negligible error. This, of course, allows a larger error in the other measurements for a given error in the result. In Eq. (1-1.4.9) K_f is a constant which can be assumed error-free. With $\Delta K_f = 0$, Eq. (1-1.4.9) becomes

$$\frac{\Delta M}{M} = \frac{3\Delta G}{G} = \frac{3\Delta g}{g} = \frac{3\Delta\theta}{\theta} \tag{1-1.4.10}$$

1-1.4.2 EXAMPLE

Assuming the principles of equal effects, determine the percent errors and actual errors allowed in g, G, and θ if the error in M is 4%. (Use the values of Example 1-1.4.1 for g, G, and θ.)

Solution

Equation (1-1.4.10) can be applied immediately:

$$\frac{\Delta M}{M} = 0.04 = \frac{3\Delta g}{g} = \frac{3\Delta G}{G} = \frac{3\Delta\theta}{\theta}$$

$$\frac{\Delta g}{g} = \frac{\Delta G}{G} = \frac{\Delta\theta}{\theta} = \frac{0.04}{3} = 0.0133$$

$$\Delta g = 0.0133\ g = 0.0133(6.02) = 0.08 \text{ grams}$$

$$\Delta G = 0.0133\ G = 0.0133(210.3) = 2.80 \text{ grams}$$

$$\Delta\theta = 0.0133\ \theta = 0.0133(0.665) = 0.009^\circ$$

These are all attainable limits of precision with modern experimental equipment.

If we have an equation such as

$$x = \frac{vy^2}{z^3} \tag{1-1.4.11}$$

where v, y, and z are variables, we can easily derive an equation for the relative error in x. Show that

$$\frac{\Delta x}{x} = \frac{\Delta v}{v} + \frac{2\Delta y}{y} + \frac{3\Delta z}{z} \tag{1-1.4.12}$$

If we wanted to apply the principle of equal effects, we would say that each term contributes equally to the error in x. Thus

$$\frac{\Delta v}{v} = \frac{2\Delta y}{y} = \frac{3\Delta z}{z} \tag{1-1.4.13}$$

This leads immediately to the result (show it)

$$\frac{\Delta x}{x} = \frac{3\Delta v}{v} = 3\left(\frac{2\Delta y}{y}\right) = 3\left(\frac{3\Delta z}{z}\right) \tag{1-1.4.14}$$

Other equations can be treated in a similar manner.

ESTIMATION OF STANDARD ERROR. The methods described above do not allow for the possibility of cancellation of some of the errors. If we make a number of measurements, there is a good chance that some cancellation will occur. It is not reasonable to assume that all deviations will be in the same direction. If cancellation occurs, the error estimated by the above procedures will be too large. Another method for estimating limits of error is called the *vectoral method*. The vectoral method leads to an estimated error that is smaller than the error given by the previous methods; this estimated error is called the *standard error*.

In the vectoral method Eq. (1-1.4.3) is squared. Any cross-terms that occur are ignored. The assumption is that some of the cross-terms will be positive and some negative. Applying this procedure to Eq. (1-1.4.3) leads to

$$\Delta z^2 = \left(\frac{\partial z}{\partial x}\right)^2 \Delta x^2 + \left(\frac{\partial z}{\partial y}\right)^2 \Delta y^2 \tag{1-1.4.15}$$

The standard error in z, given the symbol S_z, is

$$S_z = \Delta z = \left[\left(\frac{\partial z}{\partial x}\right)^2 \Delta x^2 + \left(\frac{\partial z}{\partial y}\right)^2 \Delta y^2\right]^{1/2} \tag{1-1.4.16}$$

The equation is readily extended to more variables. This is probably a more realistic estimate of the error in an experiment than the maximum error. In any event, when you report an error estimate be sure to label it as maximum, standard, or whatever type you are reporting.

STANDARD DEVIATION. Let us say you are repeatedly measuring a flow time to find t that will be used in an equation. One way of finding Δt that goes into the estimation of maximum or standard error is to find the *standard deviation* of your repeated measurements. The standard deviation for a large number of measurements (which contain only random error) is defined as

$$\sigma_t = \left[\frac{1}{n-1}\left\{\sum_{i=1}^{n} |\bar{t} - t_i|^2\right\}\right]^{1/2} \tag{1-1.4.17}$$

where n = the number of measurements, $\bar{t}$ is the average value of t, and t_i is an individual measurement. This equation is strictly valid for large values of n. For small numbers of measurements, a typical situation in physical chemistry, the standard deviation is well approximated by

$$\sigma_t \approx \frac{w}{\sqrt{n}} \tag{1-1.4.18}$$

where w is the range of values ($t_{highest} - t_{lowest} = w$) and n is the number of measurements (n = 3 to 12).

For a normal error distribution, 68.3% of errors will be within $\pm 1\sigma$ of the average value. To estimate Δt in the above case to be used in determining the maximum or standard error you might set $\Delta t = 1\sigma$ or $\Delta t = 2\sigma$, etc., depending on how your data are to be used. *The important thing is to state in your results how the error is estimated.*

1-1.4.3 EXAMPLE

Determine the standard deviation for the following times in a viscosity experiment: t = 231.7, 231.9, 232.2, 231.6, 231.8, 232.0, and 232.3 s

Solution

The average value of $\bar{t}$ is 231.929. The deviations are

$$|\bar{t} - t_1|^2 = 231.929 - 231.7^2 = |0.229|^2 = 0.052$$
$$|\bar{t} - t_2|^2 = |0.029|^2 = 0.001$$
$$|\bar{t} - t_3|^2 = |-0.271|^2 = 0.074$$
$$|\bar{t} - t_4|^2 = |0.329|^2 = 0.108$$
$$|\bar{t} - t_5|^2 = |0.129|^2 = 0.017$$
$$|\bar{t} - t_6|^2 = |-0.071|^2 = 0.005$$
$$|\bar{t} - t_7|^2 = |-0.371|^2 = 0.138$$

$$\sum |\bar{t} - t_i|^2 = 0.395$$

$$\sigma_t = \left[\frac{1}{(7-1)}(0.395)\right]^{\frac{1}{2}} = 0.256 = 0.26$$

Therefore, $\bar{t} = 231.93 \pm 0.26$ s.

Using the approximate relation

$$\sigma_t = \frac{w}{\sqrt{n}} = \frac{231.3 - 231.6}{\sqrt{7}} = \frac{0.70}{2.646} = \underline{0.265}$$

The fact that these are so close is not significant. The times given are not random experimental data. They are illustrative only.

All the above procedures give error estimates. Remember, however, *the assignment of experimental error in a measurement is usually a matter of judgement.* The final error calculated depends to a great extent on whether you are an optimist or a pessimist. In all experimental work, the best you can do is make an educated guess for the uncertainty of the result. That educated guess is far superior to a number with no estimated uncertainty attached to it. A result without an estimated error is not altogether useless. Such a result does not, however, provide others with an assessment of the quality of the experiment or the degree to which the data can be trusted.

1-1.5 SIGNIFICANT FIGURES AND ROUNDING

There are numerous conventions that can be used to determine how many digits should be retained in a final or intermediate result. Perhaps the most useful is the most general rule. This *general rule* states that enough figures should be retained to ensure that the computational error E_c is less than or equal to 5% of the estimated experimental error E_e, or $E_c \leq E_e/20$.

Before illustrating that and giving a couple of "rules of thumb," perhaps it is wise to review significant figures. This is done in Table 1-1.5.1.

Now, back to our problem of determining how many figures to retain. Let us say we had a final answer of 8.6368 ±0.23 for an experimental result. We should round the answer to 8.64 ±0.23. We have retained two uncertain digits. The right-most digit (the second uncertain figure) is called a *guarding figure*. It is retained to help eliminate accumulation of errors if the number is used in further calculations. Note that we cannot round to 8.6 without violating the general rule. The possible range of values for 8.6 is 8.55 to 8.65. This means our computational error is ±0.05. Compared to the experimental error we have

$$\frac{E_c}{E_e} = \frac{0.05}{0.23} \approx \frac{1}{5} = 20\%$$

This violates our general rule.

Table 1-1.5.1
Review of Significant Figures

Number	Significant figures	Comments
2.01	3	
0.00201	3	Leading zeros after a decimal are not significant.
0.20100	5	Trailing zeros after a decimal are significant.
20100	3	Trailing zeros are not significant unless the decimal point is written (see next example).
20100.	5	If the decimal point is written, all figures to its left are significant.
500	1	If we wanted to indicate two significant figures, we would have to write 5.0×10^2. We could not write 500, since that would give three. Note that the 0 in 5.0 is significant.

In intermediate calculations, two guarding figures (two digits to the right of the first uncertain digit) should be retained. With the availability of pocket calculators, these rules are not as critical as they were when slide rules and logarithm tables had to be used. It is easy to retain two or three additional digits. It is, however, a waste of effort to enter an eight-figure number if only the first two are significant. *Your final answer should be rounded to conform to the general rule given above, or to the rule that the final answer has no more digits than the smallest number of any number used in the calculation.* If you look at the example in Sec. 1-1.3, you can see that the result was rounded to two significant figures to conform to the rule just given. The 0.028 kg mol^{-1} used in the calculation has only two significant figures. This limits the final result to two.

1-1.6 TABLE READING

There are many tables contained in this text. It is necessary for you to be able to read these properly if you are to use the data contained in them correctly. The following two ways of labeling the columns in tables will be used. Note the difference in the way these are interpreted.

Assume we have a column in a table that is as follows:

$d \times 10^6$
5.00

What is the value of d? The column heading tells us that $d \times 10^6 = 5.00$, the table entry. Thus $d = 5.00 \times 10^{-6}$ *not* $5.00 \times 10^{+6}$! Care must be taken in interpreting tables and graphs. The same value of d would be represented by

$d\ (\times 10^{-6})$
5.00

1-1.7 PLOTTING LINEAR EQUATIONS

The procedure given below will be used many times in the following chapters. So, make sure you understand it. Many equations will have to be put into linear form before they can be applied to experimental data. You will have many occasions to do just that during your studies.

The general form for the equation of a straight line is

$$y = mx + b$$

where y and x are variables and m and b are constants. The constants m and b can be found from a graph of y vs. x (y is the ordinate and x the abscissa). The slope of such a plot is m, and the y intercept (x = 0) is the value of b. Figure 1-1.7.1 illustrates this.

Linear equations of this type can be solved to give m and b from values of x and y by methods other than plotting. Almost any computer you might have occasion to use will have a *least squares* subroutine. This computer program "fits" the best "line" through given data points by minimizing the sum of the squares of the deviations of the experimental points from the calculated line. However, before using such a subroutine, you must arrange your equation into a linear form.

Not all equations are in the proper form for plotting as a linear equation. However, a little manipulation is sufficient in many cases to produce a linear form. Consider the function

$$y = bx^a$$

This is obviously not linear. Take the logarithm of both sides (ln = natural logarithm, log = base-10 logarithm)

$$\begin{aligned} \ln y &= \ln bx^a = \ln b + \ln x^a \\ &= \ln b + a \ln x \end{aligned}$$

This is now in a linear form. A plot of ln y vs. ln x yields a line with slope a

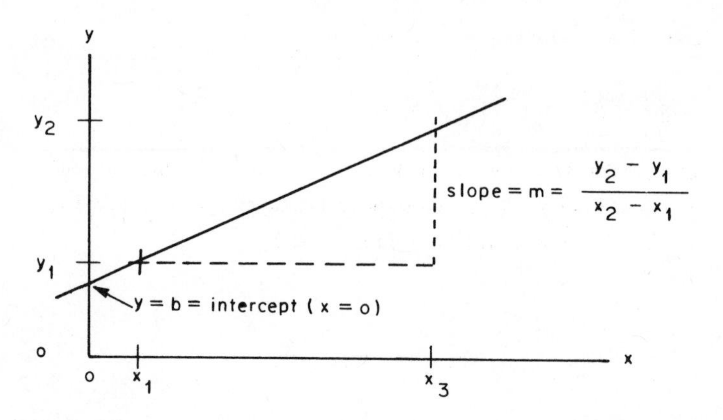

FIG. 1-1.7.1 A plot of y vs. x for the equation y = mx + b.

and an intercept of ln y = ln b (ln x = 0).

Other equations will require different manipulations to obtain a linear form. In all cases the following rules apply to plots you make:

1. Use convenient, easily interpreted scales (each division should equal 1, 2, or 5 units, not some odd division such as 3 or 8).
2. Use as much of the paper for the plot as practical while maintaining rule (1). The following examples illustrate good and poor form:

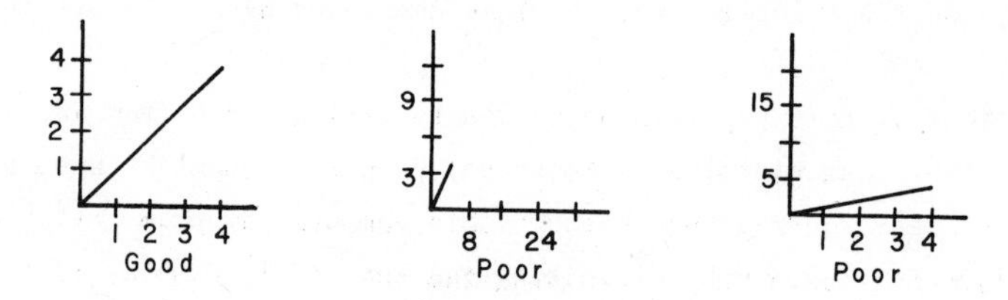

3. Choose variables that give as nearly as possible a straight line.

Further details on estimating errors from plots as well as a more detailed treatment of all the topics presented above can be found in Lyon (see Sec. 1-1.12, Ref. 3). It is worth your while to read through this book, especially before doing any laboratory work.

In our study of physical chemistry we will encounter many models that form the basis for theories. A particular model may bear no resemblance to what actually exists in nature. Truth is not a prerequsite for a model: utility is. Even if a model does not resemble nature in any sense, it it leads to accurate predictions and suggests fruitful questions, it serves a purpose. The point is that you should view all models with skepticism. Accept them for what they are: aids in investigating the nature of physical phenomena.

1-1.8 SOURCES OF DATA AND INFORMATION ON EXPERIMENTAL TECHNIQUES

The following incomplete list of references gives sources for standard chemical and physical data. Also, some sources of information on experimental techniques are provided.

A. *GENERAL PHYSICAL AND CHEMICAL PROPERTIES.*

1. *Journal of Chemical and Engineering Data.* Published by the American Chemical Society. Data on specific systems, quarterly; annual index.
2. *Physikalisch-Chemische Tabellen,* reproduced by Edwards Brothers, Inc., Ann Arbor, Michigan, 1943. Physical and chemical properties tabulated for numerous properties and systems; eight volumes (in German).

   See also, *Landolt--Boernstein Tables,* 6th ec., Springer-Verlag, 1950. An English-German new series is in preparation.
3. *Solubilities of Inorganic and Organic Compounds*, ed. H. Stephen and T. Stephen, MacMillan, New York, 1963-1964; four volumes.
4. *Thermophysical Properties of High Temperature Solid Materials*, ed. Y. S. Touloukian, MacMillan, New York, 1967; six volumes.
5. (a) *Thermophysical Properties of Matter*, ed. Y. S. Touloukian, IFI/Plenum, New York, 1970; thirteen volumes.

   (b) *Thermophysical Properties Research Literature Guide*, Y. S. Touloukian, et. al., ed., Plenum, New York, 1973; several volumes.
6. *Handbook of Thermophysical Properties of Solid Materials*, A. Goldsmith, T. E. Waterman, and H. J. Hirschhorn, MacMillan, New York, 1961.
7. *Oxidation Potentials*, W. M. Latimer, 2nd ed., Prentice-Hall, Englewood Cliffs, N. J., 1952.
8. *Selected Constants: Oxidation-Reduction Potentials of Inorganic Substances in Aqueous Solutions*, G. Charlot, et. al., Butterworth, London, 1972.
9. *Handbook of Chemistry and Physics*, Chemical Rubber Company, Cleveland, updated annually.
10. *Physical Properties of Chemical Compounds*, R. R. Dreishach, American Chemical Society, Washington D. C.; three volumes between 1955 and 1961.
11. *Physico-Chemical Constants of Binary Compounds*, J. Timmermans, Interscience, New York, 1959; four volumes.
12. *Physical Constants of Hydrocarbons*, G. Egloff, American Chemical Society Monograph Series, Reinhold, New York, 1953; five volumes.
13. *International Critical Tables*, ed. E. W. Washburn, McGraw-Hill, New York, 1929; seven volumes.
14. *Chemist's Companion: A Handbook of Practical Data, Techniques and References*, A. J. Gordon and R. A. Ford, Wiley-Interscience, New York, 1972.
15. *Tables of Interatomic Distance and Configuration in Molecules and Ions*, L. E. Sutton. Special Publication Nos 11 and 18, The Chemical Society, London, 1958 and 1965.
16. *The National Standard Reference Data System (NSRDS)* administered by the National Bureau of Standards has published over 50 volumes of quantitative data. Each volume contains a complete list of prior titles. Some examples are:

a. NSRDS-NBS-2. *Thermal Properties of Aqueous Univalent Electrolytes*, 1965.

b. NSRDS-NBS-8. *Thermal Conductivity of Selected Materials*, 1966.

c. NSRDS-NBS-33. *Electrolytic Conductance and the Conductances of the Halogen Acids in Water*, 1970.

d. NSRDS-NBS-36. *Critical Micelle Concentrations of Aqueous Surfactant Systems*, 1971.

Of particular interest is NSRDS-NBS-55, which is a bibliography of NSRDS publications, NBS publications, and selected *Journal of Physical and Chemical Reference Data* articles.

17. *PVT Relationships of Organic Compounds*, R. R. Dreishach, Handbook Publisher, Cleveland, 1952.

18. *Thermophysical Properties Research Literature Retrieval Guide*, 2nd ed., ed., Y. S. Touloukian, Plenum, New York, 1967. A guide to literature, not a compilation.

19. *The Virial Coefficients of Gases, A Critical Compilation*, J. H. Dymond and E. B. Smith, Oxford University Press, 1969.

B. THERMODYNAMIC DATA.

1. *Journal of Chemical Thermodynamics*, Academic, New York. Thermodynamic properties of specific systems. No annual index of key words. Annual contents list. Somewhat difficult to use.

2. *Selected Values of Chemical Thermodynamics Properties*, Circular of the National Bureau of Standards 500, F. O. Rossini, D. D. Wagman, W. H. Evans, S. Levine, and I. Jaffe, U. S. Government Printing Office, Washington, D. C., 1952.

3. *Vapor-Liquid Equilibrium Data*, J. C. Chu, S. L. Wang, S. L. Levy, and R. Paul, J. W. Edwards, Inc., Ann Arbor, Michigan, 1956.

4. *Thermodynamic and Transport Properties of Gases, Liquids and Solids*, ed. Y. S. Touloukian, American Society of Mechanical Engineers, New York, McGraw-Hill, New York, 1959.

5. *The Chemical Thermodynamics of Organic Compounds*, D. R. Stull, E. F. Westrum, Jr., and G. C. Sinke, John Wiley and Sons, New York, 1969.

6. *JANAF Thermochemical Tables*, 2nd ed. NSRDS-NBS-37(1970). (See Note A-16.)

7. *Values of Properties of Hydrocarbons and Related Compounds*, American Petroleum Institute (API) Project, 44 Tables. Thermodynamics Research Center (TRC), Dept. of Chemistry, Texas A and M University, College Station, Texas, 1955. Loose-leaf format. A definitive compilation which includes spectral as well as physical-thermodynamic data.

8. *Selected Values of Properties of Chemical Compounds*. Manufacturing Chemists Association (MCA) Research Project tables. Also produced by TRC. Emphasizes organic compounds other than hydrocarbons.

9. *Thermodynamic Properties of the Elements*, D. F. Stull and G. C. Sinke, American Chemical Society, 1956. No. 18 in the *Advances in Chemistry* series.

10. *Free Energy of Formation of Binary Compounds: An Atlas of Charts for High-Temperature Chemical Calculations*, T. B. Reed, MIT, Cambridge, Mass., 1971.

11. *Theroetical Mean Activity Coefficients of Strong Electrolytes in Aqueous Solutions from 0 to* $100^{\circ}C$, W. J. Hamer, NSRDS-NBS-24. (See Note A-16.)

12. *Vapor-Liquid Equilibrium Data at Normal Pressures*, E. Hala, et. al., Pergamon, New York, 1968.

13. *Thermodynamic Functions of Gases*, ed. F. Din, Butterworth, London, 1962. Multi-volume treatment of industrially important gases.

14. *Data of Geochemistry*, 6th ed., Chapter L. Phase-Equilibrium Relations of the Common Rock-Forming Oxides Except Water. Geological Survey Professional Paper 440-L, G. W. Morey, U. S. Government Printing Office, Washington, D. C., 1964.

15. *Azeotropic Data*, L. H. Horsley, American Chemical Society, 1952 to 1973; three volumes.

16. *Thermodynamic Properties and Reduced Correlations for Gases*, L. Canjar and F. Manning, Gulf Publishing Co., Houston, 1967.

17. *Thermodynamic Tables in S.I. Units*, R. W. Haywood, Cambridge University Press, New York, 1972.

C. GENERAL EXPERIMENTAL TECHNIQUES.

1. *Physical Methods of Organic Chemistry*, ed. A. Weissberger, Interscience, New York, 1960; ten volumes. Reviews theory and techniques for application of numerous chemical and physical methods. An excellent source.

2. *Physical-Chemical Methods*, J. Reilly and W. N. Rae, Methuen, London, 1943, several volumes.

3. *Precision Measurements and Calibration*, National Bureau of Standards Special Publication 300, 1970.

4. *Statistics and Experimental Design in Engineering and the Physical Sciences*, N. L. Johnson and F. C. Leone, Wiley, New York, 1964; two volumes.

5. *An Introduction to Scientific Research*, E. B. Wilson, Jr., McGraw-Hill, New York, 1952.

6. *Experimental Method: A Guide to the Art of Experiments for Students of Science and Engineering*, W. G. Wood and D. G. Martin, Humanities Press, N. J., 1974.

D. SPECTROSCOPIC DATA.

1. *Structure Reports*, International Union of Pure and Applied Chemistry, Osthoek, Scheltema, and Holkema, Utrecht. A continuing series from 1913.

2. *Index of Vibrational Spectra of Inorganic and Organometallic Compounds*, N. N. Greenwood, E. J. F. Ross, and B. P. Straughan, CRC Press, Cleveland, 1972.

3. *Molecular Spectra and Molecular Structure*, G. Herzberg, Van Nostrand Company, Inc.

   Volume I. *Spectra of Diatomic Molecules*, 2nd ed., 1950. Extensive tables of vibration, rotation and electronic parameters. Extensive reference list.
   Volume II. *Infrared and Raman Spectra of Polyatomic Molecules*, 1945.
   Volume III. *Electronic Spectra and Electronic Structures of Polyatomic Molecules*, 1966.

4. (a) *Spectral Data, 1956-1957*, H. Wheeler and L. Kaplan, Wiley-Interscience, New York, 1966 (*Organic Electronic Spectral Data*, Vol. 3.).

   (b) *Spectral Data, 1953-1955*, H. E. Ungnade, Wiley-Interscience, New York, 1960 (*Organic Electronic Spectral Data*, Vol. 2.).

   (c) *Spectral Data, 1946-1952*, M. J. Kamlet, Wiley-Interscience, New York, 1960 (*Organic Electronic Spectral Data*, Vol. 1.).

5. *Spectroscopic Data Relative to Diatomic Molecules*, S. Bourcier, Pergamon, New York, 1971. (Tables of Constants and Numerical Data Series: Vol. 17; contents of Vol. 1-16 unknown.)

6. *Microwave Spectroscopy*, C. H. Townes and A. L. Schawlow, McGraw-Hill, New York, 1955.

7. NSRDS-NBS (see Note A-16.); Numbers 3 (six sections), 5, 6, 11, 14, 17, 22, 26, 31, 34, 35, and 39.

8. NBS Monograph 70, five volumes, P. F. Wacker, et. al., 1964 to 1968.

E. KINETIC DATA.

1. The Interpretation of Rate Data, J. F. Bannett, in *Investigation of Rates and Mechanisms of Reactions*, Part I, ed. S. L. Friss, E. S. Lewis, and A. Weissberger, 2nd ed., Wiley-Interscience, New York, 1961.
2. *Thermochemical Kinetics, Methods for the Estimation of Thermochemical Data and Rate Parameters*, S. W. Benson, Wiley, New York, 1968.
3. (a) NSRDS-NBS 9. *Tables of Bimolecular Gas Reactions*, 1967.

   (b) NSRDS-NBS 20. *Gas Phase Reaction*, 1968. Kinetics of Neutral Oxygen Species.

   (c) NSRDS-NBS 21. *Kinetic Data on Gas Phase Unimolecular Reactions*, 1970.

   (d) NSRDS-NBS 43, 43 Supplement. *Selected Specific Rates of Reactions of Transients from Water in Aqueous Solution*; Vol. I. The Hydrated Electron, 1973, 1975.

   (e) NSRDS-NBS 46. *Reactivity of the Hydroxyl Radical in Aqueous Solution*, 1973.

   (f) NSRDS-NBS 51. *Selected Specific Rates of Reactions of Transients from Water in Aqueous Solution II*, 1975, Hydrogen Atom.

F. MATHEMATICAL AND COMPUTER TECHNIQUES AND DATA TREATMENT.

1. *Computer Programs for Chemistry*, D. F. Detar, et. al., W. A. Benjamin, New York, 1969; three volumes. Numerous programs for data reduction and plotting listed.
2. *Mathematical Techniques in Chemistry*, J. B. Dence, John Wiley and Sons, New York, 1975
3. *A Statistical Manual for Chemists*, E. L. Bauer, Academic, New York, 1971. Written as a manual for working chemists; elementary in content and method.
4. *Data Reduction and Error Analysis for the Physical Sciences*, P. R. Bevington, McGraw-Hill, New York, 1969.
5. *Mathematical Preparation for Physical Chemistry*, F. Daniels, McGraw-Hill, New York, 1956.
6. *Dealing with Data*, A. J. Lyon, Pergamon, New York, 1970.
7. *The Handling of Chemical Data*, P. D. Lark, B. R. Craven, and R. C. L. Bosworth, Pergamon, New York, 1968.

1-1.9 PROBLEMS

1. If $y = 4.35x^3 - 6.55x^2z + 1.22xz^2 + z^3$, find the error in y if $x = 4.23 \pm 0.07$ and $z = 1.65 \pm 0.02$. Give both absolute and relative error.
2. If $y = u^2v^3/(w^{\frac{1}{2}}x^{\frac{1}{2}}z^{\frac{1}{4}})$ derive an equation for the relative error in y for given errors in all the other variables. If the following percent errors are known, find the percent error in y: u - 1.3%, v - 0.6%, w - 2.2%, x - 0.33%, and z - 1.02%.
3. For the problem (2) repeat the calculation using the vectoral method and compare the results.

4. Assuming the principle of equal effects, determine the relative errors allowed in u, v, w, x, and z in Problem (2), if y is to be determined to an accuracy of 0.49%.

5. Put the equation $y = y_m P/(b + P)$ (where y_m and b are constants and y and P are variables) into linear form and indicate how the slope is related to y_m and the intercept to b.

6. How many significant figures are there in each of the following numbers? 45010, 54.900, 0.000034, 5.004, 780, and 33.1.

7. The viscosity of a gas may be found from the following equation:

$$\eta = \frac{m\bar{c}}{2\sqrt{2}\pi\sigma^2}$$

Find the (relative) error in η for the following errors in the variables: m - 3.2%, $\bar{c}$ - 1.23%, and σ - 4.81%.

8. Find the mean and standard deviation for the following set of numbers: 2.223, 2.224, 2.258, 2.198, 2.189, 2.245, and 2.001.

9. The freezing point depression for a solution can be found from the following formula:

$$\Delta T_f = \frac{RT_f^2 m}{\Delta H_f \, n_A}$$

If the errors in T_f and ΔH_f are 1.03% and the errors in n_A and m are 1.45%, find the relative error in ΔT_f.

1-1.10 SELF-TEST

1. From Eq. (1-1.4.8) derive the equation for the error in M by means of the vectoral method. Use the data from Example 1-1.4.1 and the equation you just derived to find the error in M. Compare your results to those of Example 1-1.4.1.

2. In a heat of solution experiment, the heat capacity is determined by measuring the temperature change θ that accompanies a given electrical power input according to the following formula:

$$C_p = \frac{I^2 Rt}{\theta}$$

where I is the current, R is the resistance of the heater involved, t is the time of heating, and θ is the temperature change. Derive an equation relating errors in I, R, t, and θ to a final error in C_p.

3. Assuming the principle of equal effects, determine the relative error in each of the variables that could be tolerated and still have an error of only 1% in C_p for Problem (2).

4. For each of the following write a linear form and indicate how the constants can be evaluated from a graph:

   a. $y = ae^{bx}$

   b. $\frac{a}{x} + b = y$

   c. $\frac{1}{y} = bx^a$

1-1.11 ADDITIONAL READING

1. R. Bevington, *Data Reduction and Error Analysis for the Physical Sciences*, McGraw-Hill, New York, 1969.

2. E. L. Bauer, *A Statistical Manual for Chemists*, Academic, New York, 1971.

3.* A. J. Lyon, *Dealing with Data*, Pergamon, New York, 1970.

4. P. D. Lark, B. R. Craven, and R. C. L. Bosworth, *The Handling of Chemical Data*, Pergamon, New York, 1968.

2

Properties of Gases and the Kinetic-Molecular Theory

SYMBOLS AND DEFINITIONS

T	Absolute (Kelvin) temperature (K)
P	Pressure (atm; 1 atm = 760 mm Hg = 760 torr = 101,325 N m^{-2})
V	Volume (dm^3 = liters = 10^3 cm^3 = 10^{-3} m^3)
R	Gas constant (see Table 2-1.3.1 for values)
ρ	Density (g cm^{-3}, kg dm^{-3}, etc.)
n	Number of moles
N	Avogadro's number (6.023×10^{23})
M	Molecular weight (g mol^{-1} = 10^{-3} kg mol^{-1})
E	Energy (J)
m	Mass of a molecule (g, kg)
u,v,w	Velocity components (m s^{-1})
p	Momentum (kg m s^{-1})
c	Speed (m s^{-1})
α	Thermal expansion coefficient (K^{-1})
β	Isothermal compressibility (atm^{-1})
P_r	Reduced pressure
V_r	Reduced volume
T_r	Reduced temperature
$\bar{q}$	Average quantity
Z_1	Collision number (s^{-1})
Z_{11}	Collision per unit time per unit volume (m^{-3} s^{-1})
λ	Mean free path (m)
σ	Collision diameter [m = 10^{-10} Angstroms (Å)]
η	Coefficient of viscosity (kg m^{-1} s^{-1})
c	Concentration (mol dm^{-3}, molecules dm^{-3}, mol m^{-3})
Q	Permeation (mol or molecules)
J	Flux (mol m^{-2} s^{-1})
A	Area (m^2)
D	Diffusion coefficient (m^2 s^{-1})

The material to be studied in this chapter is concerned with the properties of gases. Our question is: Why study gas properties? There are at least two reasons. First, the experimental and theoretical treatment of gas properties is much simpler than either the solid state or the liquid state. Thus, the behavior of gases provides a good topic with which to "get your feet wet" before moving into more complicated areas of study. Also, the study of gases is important from a practical standpoint. Almost any industrial processing plant handles gases. An engineer designing or working in such a plant must be able to predict the behavior of gases in flows, in reactors, or in storage. Theoretical considerations are important since many situations arise in which needed data are incomplete and properties of gases must be estimated from available data. A theoretical base allows at least intelligent guesses to be made. This is another point to be emphasized: seldom will all the data needed for a design or process be available. Existing data will have to be extrapolated or interpolated to fit the conditions at hand.

Below, empirical relationships derived for gases on the basis of experimental observations will be presented first. Then, we will investigate a theory that attempts to correlate these experimental observations and produce a set of equations that agree with the empirical equations. This procedure is common in science. Seldom do we find theory leading experiment. Theory, in general, initially lags behind experiment until a model is constructed. If the model, which is a conceptual device, is useful it will lead to theoretical expressions which agree with experiments. Also, if the model is good, new experiments will be suggested. From that point, theory, model, and experiment move forward, each requiring modification of the other as new information and insights are gained.

The equations of state (mathematical expressions describing behavior) for gases presented below range from simple (ideal gas equation) to complex. As is usually the case in science, the simple equations are strictly applicable to ideal cases only. Real behavior requires more complex expressions. The various equations below will be used repeatedly in the chapters on thermodynamics. Thus it is imperative that you learn to manipulate these equations to obtain a form for calculating whatever variable may be missing. Do not let complicated mathematical expressions disturb you. Approach each problem in a systematic fashion and *think* about what you are doing, and your difficulties will be minimized.

2-1 IDEAL GASES

OBJECTIVES

You shall be able to apply the equation of state for an ideal gas to predict the P, V, and T behavior of that gas under given conditions.

2-1.1 BOYLE'S LAW

As is the case in much of scientific endeavor, the discovery of the relationship between volume and pressure in a gas had to await the development of an appropriate measuring device. In this case the necessary instrument was the mercury barometer, invented by Evangelista Torricelli in 1643. Figure 2-1.1.1 is a sketch of a Torricellian barometer. A second necessary invention was that of a vacuum pump, first described by Boyle in 1660, in the first edition of *New Experiments, Physico-Mechanical, Touching on the Spring of Air, and Its Effects*. Boyle discovered that when air

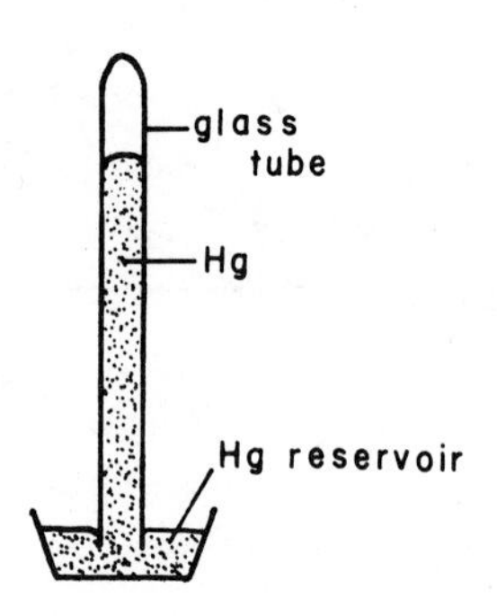

FIG. 2-1.1.1 A simple barometer.

was pumped from around the reservoir of a barometer, the height of the mercury column decreased. He concluded that air pressure was holding the mercury in the tube.

Immediately a controversy arose over the concept of a vacuum at the top of the tube. Such a controversy is not an unusual state of affairs in science. Any new idea usually generates a storm of controversy if it is in opposition to currently accepted views. [For an interesting description of this and other struggles in science see J. B. Conant, *Science and Common Sense* (Sec. 2-1.7, Ref. 1).]

Robert Boyle modified his procedure by forming a J-shaped tube with the short leg closed at the end. By pouring mercury in the long tube (see Fig. 2-1.1.2) the pressure on the air in the closed end can be varied. Some of Boyle's original observations are found in Table 2-1.1.1 (see also Thorpe, Sec. 2-1.7, Ref. 2).

This is a good opportunity to apply the material in Sec. 1-1.7 on plotting equations. Before continuing, see if you can find a function of V that gives a linear plot with P. Figure 2-1.1.3 shows two trials. Plot (a) is obviously nonlinear, while plot (b) fits a straight line nicely. These observations led Boyle to conclude that

$$P \propto \frac{1}{V} \quad \text{or} \quad P = \frac{a}{V} \qquad (2\text{-}1.1.1)$$

or

$$PV = a \qquad (2\text{-}1.1.2)$$

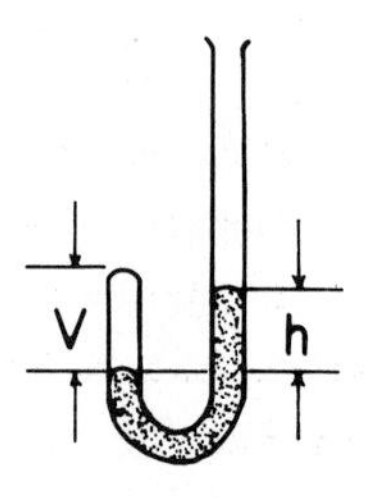

FIG. 2-1.1.2 Boyle's J-tube.

Table 2-1.1.1
Some of Boyle's Observations on the Condensation of Air[d]

V (in.)[a]	h (in.)	Pressure + atmospheric pressure[b]	Pressure[c]
12	00	+ 29-1/8 = 29-2/16	29-2/16
10	6-3/16	+ 29-1/8 = 35-5/16	35
8	15-1/16	+ 29-1/8 = 44-3/16	43-11/16
6	29-11/16	+ 29-1/8 = 58-13/16	58-4/16
5	41-9/16	+ 29-1/8 = 70-11/16	70
4	58-2/16	+ 29-1/8 = 87-14/16	87-6/16
3	88-7/16	+ 29-1/8 = 116-9/16	116-4/8

[a]Boyle measured the length of the air space. If the tube is uniform this is proportional to volume.

[b]Atmospheric pressure was 29-1/8 in. Hg.

[c]Predicted by the reciprocal law.

[d]From Sir Edward Thorpe, *Essays in Historical Chemistry*, MacMillan, 1923, Chap. 1.

where a is a constant. Equation (2-1.1.2), which is called Boyle's law, holds at constant temperature, which was the approximate condition of Boyle's experiment. Stated verbally, Boyle's law is: *At constant temperature the volume of a given sample of gas varies inversely as the pressure.*

2-1.2 GAY-LUSSAC'S LAW

The relationship between the volume of a gas and temperature was delayed more than a century after the publication of Boyle's law. This was due to the difficulty found in quantitatively defining the terms "hot" and "cold". Boyle had observed that the flame of a candle destroyed his inverse relationship between pressure and volume.

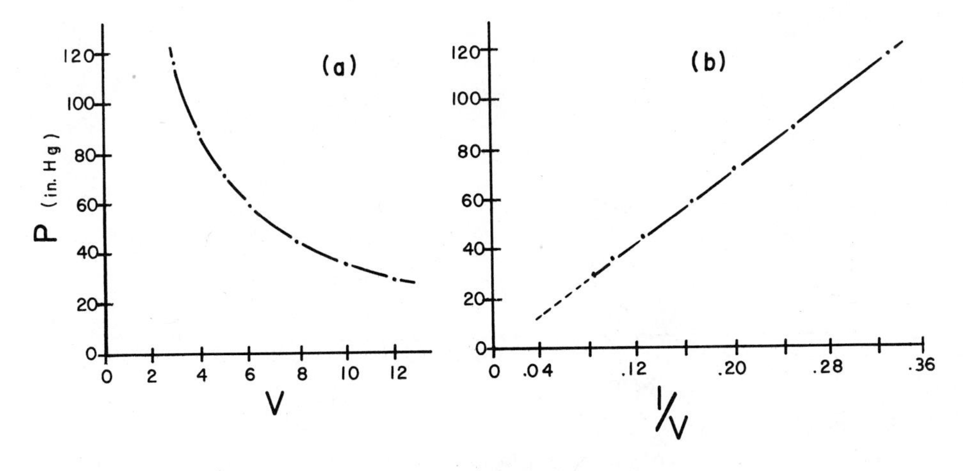

FIG. 2-1.1.3 Plots of: (a) P vs. V and (b) P vs. 1/V.

However, with no quantitative measure of "temperature" or "degrees of heat" available, he could find no quantitative relationship.

Near the end of the eighteenth century, the use of the expansion of a liquid in a glass tube came to be accepted as a way of measuring temperature. After agreement was reached in defining the freezing point of water as 0° and the boiling point as 100°, a scale was available with which to measure temperature. This scale is called Celsius or Centigrade and is given a symbol °C.

Charles in 1787 and Gay-Lussac in 1808 showed that volume varies linearly with temperature if the pressure is kept constant. Figure 2-1.2.1 shows this variation with temperature if the pressure is held at 1 atm.

Detailed studies of many gases show that V is approximately linear in t and that the variation can be expressed as a function of some *absolute temperature* T. The variation of V with this absolute temperature can be written

$$V \propto T \quad \text{or} \quad V = bT \tag{2-1.2.1}$$

or

$$\frac{V}{T} = b \tag{2-1.2.2}$$

where b is a constant. Stated verbally: *The volume of a given mass of gas varies directly as the absolute temperature if the pressure remains constant.*

The T in Eq. (2-1.2.2) refers to an absolute temperature which is given by

$$T = t + \text{constant} \tag{2-1.2.3}$$

where t is in °C. We can find the constant in Eq. (2-1.2.3) if we write Gay-Lussac's relationship in a slightly different form. Let V = volume of the gas at any T and V_0 = volume of the gas at T = constant (i.e., t = 0°C). Then, if the volume varies

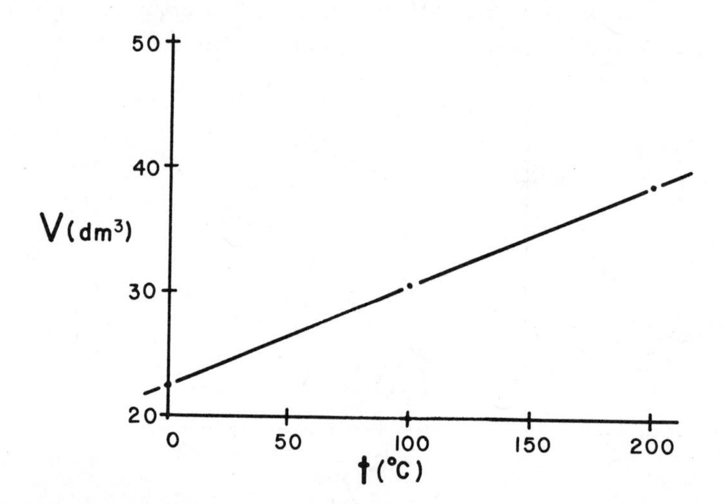

FIG. 2-1.2.1 Variation of volume with temperature at a pressure of 1 atm.

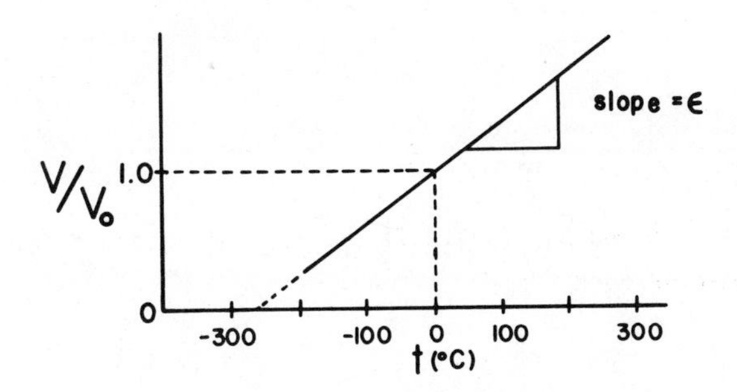

FIG. 2-1.2.2 Plot of V/V_0 vs. t for an ideal gas.

linearly with T (and therefore t),

$$V = V_0(1 + \varepsilon t) \qquad \text{or} \qquad \frac{V}{V_0} = 1 + \varepsilon t \tag{2-1.2.4}$$

If we plot V/V_0 vs. t, we obtain a line with a slope of ε, where ε is the coefficient of thermal expansion. Such a plot is shown in Fig. 2-1.2.2. ε is observed to be 1/273.15. Hence, when $V/V_0 = 0$ from Eq. (2-1.2.4)

$$1 = -\varepsilon t = -\varepsilon(T - \text{constant})$$

or

$$\text{Constant} - T = \frac{1}{\varepsilon} = 273.15$$

From Eq. (2-1.2.1) we know V goes to zero as T goes to zero. Thus at $V/V_0 = 0$, $T = 0$, and

$$\text{Constant} = 273.15$$

We can write

$$T = t + \text{constant} = t(^\circ\text{C}) + 273.15$$

Then the ice point (0°C) becomes T = 273.15 K, and absolute zero (T = 0) is t = -273.15°C.

To avoid experimental difficulties encountered in accurately measuring the ice point, the Tenth Conference of the International Committee on Weights and Measures in 1954, defined a temperature scale based on the triple point of water (i.e., liquid, solid, and vapor are in equilibrium). *The temperature of this point was arbitrarily chosen as 273.16 K, which fixes the ice point at 273.15 K.*

2-1.3 COMBINED GAS LAWS AND AVOGADRO'S HYPOTHESIS

The gas laws given in the preceding sections can be combined into one expression. Combine

$$PV = a \qquad \text{at constant } T \tag{2-1.1.2}$$

and $V = bT$ at constant P (2-1.2.1)

to give

$$\frac{PV}{T} = \text{constant} \equiv c \qquad (2\text{-}1.3.1)$$

It is readily apparent that Eq. (2-1.1.2) and Eq. (2-1.2.1) are special cases of Eq. (2-1.3.1). For example, if T is constant then

$$\frac{PV}{T} = c \qquad (2\text{-}1.3.1)$$

yields $PV = cT = \text{constant}$

as given in Eq. (2-1.1.2)

We must now evaluate the constant in Eq. (2-1.3.1). To do this we make use of Avogadro's hypothesis, first reported in 1811, which takes into account the theoretical concept of molecules and a mole. Avogadro stated that *equal volumes of different gases at the same temperature and pressure contain the same number of molecules.* In terms of the quantity of material containing $N = 6.023 \times 10^{23}$ molecules (Avogadro's number, which is defined as 1 mol), this may be stated: *The same volume is occupied by 1 mol of every gas at a given temperature and pressure.* This hypothesis allows us to write Eq. (2-1.3.1) as

$$\frac{PV}{T} = n(\text{constant})' \qquad (2\text{-}1.3.2)$$

where n is the number of moles of gas and (constant)′ is the constant for 1 mol of material. We designate (constant)′ by the symbol R which is known as the gas constant. Hence, we can write

$$\frac{PV}{T} = nR \quad \text{or} \quad PV = nRT \qquad (2\text{-}1.3.3)$$

This is called the *ideal gas equation*. R in Eq. (2-1.3.3) can be found experimentally by measuring the variables P, V, and T for a given quantity of gas.

One mole of a material is that weight of material that contains 6.023×10^{23} molecules, atoms, particles, or whatever. For molecules (or atoms) this weight is just the molecular weight (or atomic weight) in grams. For example, the molecular weight of oxygen O_2 is 32.000 g = 0.032 kg. If we are given, say, 10.0 g of O_2 we can find the number of moles by

$$n = \frac{10.0 \text{ g}}{32.0 \text{ g mol}^{-1}} = 0.313 \text{ mol}$$

It is found that at 0°C (273.15 K), 1 mol of an ideal gas occupies 22.414 dm^3 (1 dm^3 = 1 liter = 10^{-3} m^3) at 1 atm (101,325 N m^{-2} = 760 mm Hg = 760 torr). Thus, on rearranging Eq. (2-1.3.3), we find

$$R = \frac{PV}{nT} = \frac{(1 \text{ atm})(22.414 \text{ dm}^3)}{(1 \text{ mol})(273.15 \text{ K})} = 0.082058 \text{ dm}^3 \text{ atm mol}^{-1} \text{ K}^{-1}$$

or $$R = \frac{(101{,}325\ \mathrm{N\ m^{-2}})(0.022414\ \mathrm{m^3})}{(1\ \mathrm{mol})(273.15\ \mathrm{K})} = 8.3143\ \mathrm{N\ m^{-2}\ m^3\ mol^{-1}\ K^{-1}}$$

$$= 8.3143\ \mathrm{kg\ m\ s^{-1}\ m^{-2}\ m^3\ mol^{-1}\ K^{-1}}$$

$$= 8.3143\ \mathrm{kg\ m^2\ s^{-1}\ mol^{-1}\ K^{-1}}$$

$$= 8.3143\ \mathrm{J\ mol^{-1}\ K^{-1}}$$

Values of R in various units are given in Table 2-1.3.1. You should prove for yourself that the entries in the table are correct.

2-1.4 EXAMPLE

An ideal gas at 25.00°C and at 0.931 atm occupies 23.01 $\mathrm{dm^3}$. Find (1) the number of moles and (2) the volume occupied at STP (standard temperature and pressure: 0.00°C and 1 atm).

Solution

(1) First let us find the number of moles. Rearranging Eq. (2-1.3.3) gives

$$n = \frac{PV}{RT} = \frac{(0.931\ \mathrm{atm})(23.01\ \mathrm{dm^3})}{R(298.15\ \mathrm{K})}$$

Obviously we need R in $\mathrm{dm^3\ atm\ mol^{-1}\ K^{-1}}$ to get the proper units. Then

$$n = \frac{(0.931\ \mathrm{atm})(23.01\ \mathrm{dm^3})}{(0.082058\ \mathrm{dm^3\ atm\ mol^{-1}\ K^{-1}})(298.15\ \mathrm{K})}$$

$$= 0.87561\ \mathrm{mol}$$

But there are at most three significant figures. Hence

$$n = 0.876\ \mathrm{mol}$$

(2) There are several ways to approach the solution to this problem. Assume we do not have the number of moles given (but we do know that the number is constant). Applying Eq. (2-1.3.3) to both initial ($P_1V_1T_1$) and final ($P_2V_2T_2$) conditions, we can write

$$\frac{P_1V_1}{T_1} = nR = \frac{P_2V_2}{T_2}$$

or $$\frac{P_1V_1}{T_1} = \frac{P_2V_2}{T_2} \qquad (2\text{-}1.4.1)$$

List the known quantities:

$P_1 = 0.931$ atm $\quad P_2 = 1.00$ atm

$T_1 = 298.15$ K $\quad T_2 = 273.15$ K

Table 2-1.3.1
Value of the Ideal Gas Constant R in Various Units

Unit	Value
$J\ K^{-1}\ mol^{-1}$	8.3143
$cal\ K^{-1}\ mol^{-1}$	1.9872
$cm^3\ atm\ K^{-1}\ mol^{-1}$	82.058
$dm^3\ atm\ K^{-1}\ mol^{-1}$	0.082058
$dm^3\ torr\ K^{-1}\ mol^{-1}$ [a]	62.364

[a] 1 torr = 1 mm Hg = 1/760 atm = 133.3 N m^{-2}.

$$V_1 = 23.01\ dm^3 \qquad V_2 = ?$$

Since V_2 is the only unknown, rearrange (2-1.4.1) to give

$$V_2 = \frac{P_1V_1T_2}{P_2T_1} = \frac{(0.931\ atm)(23.01\ dm^3)(273.15\ K)}{(1.00\ atm)(298.15\ K)}$$

$$= \underline{\underline{19.6\ dm^3}} \quad (19.6\ liters)$$

You should look at the problem and ask: Is the answer reasonable? To see if it is, look at how the variables change. P increases, which will give a lower final volume. T decreases, which also lowers the volume, which it is.

Below are some problems dealing with the ideal gas law. The following sections are concerned with real (nonideal) gases.

2-1.5 PROBLEMS

1. The molecular weight of O_2 is 32.00 g. Find the volume of 16.31 g of O_2 at 0.966 atm and 23.2°C.

2. Find the number of moles per liter and the number of molecules per cubic centimeter of an ideal gas at 290 K at (a) 1 atm and (b) 10^{-8} atm. Also give the volume per molecule for part (b).

3. Density ρ is defined as mass per volume (g cm^{-3}, g dm^{-3}, or kg m^{-3}). Use the ideal gas law and the fact that n = w/M, where w is the weight of a gas, n is the number of moles, and M is the molecular weight, to define the density of an ideal gas ρ = w/V in terms of P, T, w, M, etc.

4. Use the result of Problem (3) to find the density of O_2 at STP [standard temperature (0°C) and pressure (1 atm)]; M = 32.00 g mol^{-1}.

5. What gas would have a density of 1.63 g dm^{-3} at 25°C and 1 atm? What would be the density of this gas at STP?

6. What pressure is required to give a gas with a molecular weight of 0.065 kg mol^{-1} a volume of 20.456 dm^3 if there are 0.0567 kg present at 304.5 K?
7. What is the density of the gas in Problem (6)?
8. Under what conditions would the gas in Problem (6) contain 2.034×10^{22} molecules per cubic decimeter if the temperature is held at 304.5 K?

2-1.6 SELF-TEST

1. Calculate the pressure exerted by 31.0 g of nitrogen in a closed 2.0-dm^3 flask at 22°C. (Give P in atm, N m^{-2}, and torr.)
2. Derive an equation relating the density of a given amount of an ideal gas at one temperature and pressure (P_0, T_0, ρ_0) to the density at another temperature and pressure (P_1, T_1, ρ_1)
3. Find the density of a gas that has a molecular weight of 0.1233 kg mol^{-1} at 307.8 K and 10.345 atm.
4. What volume is occupied by 0.2346 kg of the gas under the conditions given in Problem (3)?
5. How many molecules per cubic meter are there for the gas in Problem (3)?

2-1.7 ADDITIONAL READING

1.* J. B. Conant, *Science and Common Sense*, Yale University Press, New Haven, Conn., 1951.

2.* E. Thorpe, "Essays in Historical Chemistry," McMillan, London, 1923; Chapter 1.

2-2 NONIDEAL GASES

OBJECTIVES

You shall be able to

1. State the postulates of the kinetic-molecular model of a gas.
2. Use an equation of state of a gas (ideal or nonideal) to calculate P, V, T, ρ, etc., for that gas, given appropriate data.
3. State the law of corresponding states.
4. Use compressibility factors and appropriate charts to predict the P, V, and T behavior of a gas, or given the required data, find compressibility factors.

2-2.1 THE KINETIC MOLECULAR MODEL OF A GAS

Section 2-1 was concerned only with an empirical treatment of ideal gases. This section will give a rather brief treatment of a simple *model* that attempts to explain the "why" of gas behavior. As pointed out in the introduction to this chapter, a model is a set of assumptions about the characteristics of a system. The validity of the model is measured by its agreement with experiment.

We encountered a portion of the model of a gas when we assumed that gases were composed of molecules (Avagadro's hypothesis). Other assumptions of the *kinetic molecular model* are

1. A gas is made up of a large number of molecules which are small compared to the size of the container and the distance between them.
2. Gas molecules are in continuous random motion.
3. All collisions of molecules (with each other and with the walls of the container) are perfectly elastic.

From physics we know that for a particle of mass m, the kinetic energy is given by

$$E = \frac{1}{2}mu^2 \qquad (2\text{-}2.1.1)$$

where u^2 is the square of the velocity of the molecule.

For N particles the total energy will be the sum of individual energies. Thus

$$E = \frac{1}{2}Nmu^2 \qquad (2\text{-}2.1.2)$$

The derivation of the ideal gas law from the postulates given above is a useful exercise. It shows how a set of assumptions can be converted into mathematical form. Before considering the derivation, a couple of terms need to be defined. The quantity u^2 in (2-2.1.1) can be broken up into components:

$$u^2 = u_x^2 + u_y^2 + u_z^2 \qquad (2\text{-}2.1.3)$$

where x, y, and z denote direction.

We need to use the average of the square of the velocities $\overline{u^2}$ if we are to deal with a large number of molecules (or particles). Also, if the motion is random [Assumption (2)] the components of velocity are all equal (the molecule cannot tell x from y from z), and

$$\overline{u_x^2} = \overline{u_y^2} = \overline{u_z^2} \qquad (2\text{-}2.1.4)$$

From Eq. (2-2.1.3) you will notice that

$$\overline{u^2} = 3\overline{u_x^2} = 3\overline{u_y^2} = 3\overline{u_z^2} \qquad (2\text{-}2.1.5)$$

Now consider a particle contained in a cubic box and moving in only one direction, say x, with a velocity u_x, as shown in Fig. 2-2.1.1. The momentum p of such a particle is

$$p = mu_x \qquad (2\text{-}2.1.6)$$

If one particle strikes face A, after a perfectly elastic collision [Assumption (3)], the velocity will be $-u_x$ and the momentum

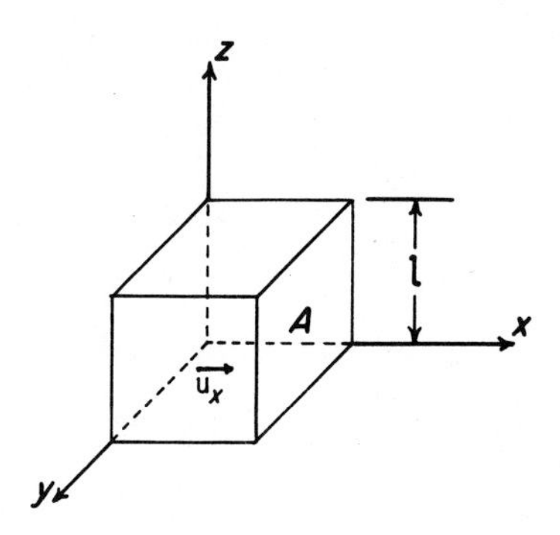

FIG. 2-2.1.1 A cubic box of length ℓ containing a particle moving in the x direction.

$$p' = m(-u_x) \tag{2-2.1.7}$$

The change in momentum Δp is

$$\Delta p = p - p' = 2mu_x \tag{2-2.1.8}$$

How many times will the particle strike face A per second? This will be given by $u_x/2\ell$ (units are m s^{-1}/m = s^{-1}; ℓ = length). Thus the momentum change per second is $(\Delta p)(u_x/2\ell)$ which is a force f.

$$f = 2mu_x \frac{u_x}{2\ell} = \frac{mu_x^2}{\ell} \tag{2-2.1.9}$$

But force/area is defined as pressure P, and the area A is given by $A = \ell^2$.

$$P = \frac{f}{A} = \frac{mu_x^2}{\ell \cdot \ell^2} = \frac{mu_x^2}{V} \tag{2-2.1.10}$$

where V is the volume of the box (ℓ^3). Thus we have

$$PV = mu_x^2 \tag{2-2.1.11}$$

For a collection of N molecules, the average of the squares of the velocities will be $\overline{u_x^2}$ and you can see that

$$PV = Nm\overline{u_x^2} \tag{2-2.1.12}$$

From Eq. (2-2.1.5) we can write this for three dimensions (since the directions are independent):

$$PV = \frac{Nm\overline{u^2}}{3} \tag{2-2.1.13}$$

Finally, rewriting Eq. (2-2.1.2) in terms of averages,

$$\overline{E} = \frac{Nm\overline{u^2}}{2} \tag{2-2.1.14}$$

and combining with Eq. (2-2.1.13), we see that

$$PV = \frac{2\overline{E}}{3} \tag{2-2.1.15}$$

If we choose 1 mol of particles (n = 1, N = Avagadro's number) and combine Eq. (2-2.1.15) with Eq. (2-1.3.3) we find

$$PV = \frac{2\overline{E}}{3} = RT \tag{2-2.1.16}$$

Note that by combining an empirical equation, Eq. (2-1.3.3) and a theoretical expression, Eq. (2-2.1.15), we deduce that the average kinetic energy of 1 mol of particles is

$$\overline{E} = \frac{3RT}{2} \tag{2-2.1.17}$$

We will return to the results of this section after a discussion of the equations used to describe the behavior of nonideal (real) gases.

2-2.2 THE EQUATION OF STATE

An equation of state defines the behavior of a system in terms of a minimum set of variables. For gases (as well as for thermodynamic functions treated in Chaps. 3, 4, and 5), only two variables are necessary to describe their behavior. Thus we can write

$$\begin{aligned} T &= f(P, V) && \text{(a)} \\ P &= g(V, T) && \text{(b)} \\ V &= h(P, T) && \text{(c)} \end{aligned} \tag{2-2.2.1}$$

where f, g, and h are functions. This indicates that if we know any two variables for a gas we can determine the third, if the function is known.

Another property of a function of two independent variables which will be very useful in Chaps. 3, 4, and 5 is that the differential change in the dependent variable may be written as follows:

$$\begin{aligned} dT &= \left(\frac{\partial T}{\partial P}\right)_V dP + \left(\frac{\partial T}{\partial V}\right)_P dV && \text{(a)} \\ dP &= \left(\frac{\partial P}{\partial V}\right)_T dV + \left(\frac{\partial P}{\partial T}\right)_V dT && \text{(b)} \\ dV &= \left(\frac{\partial V}{\partial P}\right)_T dP + \left(\frac{\partial V}{\partial T}\right)_P dT && \text{(c)} \end{aligned} \tag{2-2.2.2}$$

The subscript indicates the variable that is assumed constant for purposes of taking the derivative. (It may not actually be constant, but we take the derivative as though it were.)

Each of the partial derivatives above represent slopes from P, V, and T curves. [For more details on these see Moore (Sec. 2-2.12, Ref. 1).] Some of the partial derivatives above are of particular interest. For example

$$\alpha = \frac{1}{V}\left(\frac{\partial V}{\partial T}\right)_P \qquad (2\text{-}2.2.3)$$

This is the *isobaric* (constant pressure) change in volume for a change in temperature. When $(\partial V/\partial T)_P$ is multiplied by 1/V, the thermal expansivity is obtained.

Perhaps a few comments on partial derivatives would be useful here. The expression $(\partial V/\partial T)_P$ states that the derivative of V is taken with respect to T as though P were constant (whether it is or not). Look at the ideal gas law in which V has been written as a function of P and T:

$$V = \frac{nRT}{P} \qquad (2\text{-}2.2.4)$$

Taking $(\partial V/\partial T)_P$ gives

$$\left(\frac{\partial V}{\partial T}\right)_P = \left[\frac{\partial(nRT/P)}{\partial T}\right]_P = \frac{nR}{P}\left(\frac{\partial T}{\partial T}\right) = \frac{nR}{P}$$

This can be substituted into Eq. (2-2.2.3) to give (for an ideal gas)

$$\alpha = \frac{1}{V}\frac{nR}{P} = \frac{nR}{PV} = \frac{1}{T} \qquad (2\text{-}2.2.5)$$

For a nonideal gas, the appropriate equation of state (which will be discussed below) would have to be used.

Another useful form (at constant temperature) is

$$\beta = -\frac{1}{V}\left(\frac{\partial V}{\partial P}\right)_T \qquad (2\text{-}2.2.6)$$

which is the *isothermal* (constant temperature) volume change for a given pressure change. It is called the *isothermal compressibility*. Both α and β are important quantities for engineering design.

Finally, we can combine parts of Eq. (2-2.2.2) to get some useful relationships between variables (a procedure which will be seen again in thermodynamics). Rearrange Eq. (2-2.2.2)(c) as follows:

$$dP = \frac{dV}{(\partial V/\partial P)_T} - \frac{(\partial V/\partial T)_P\, dT}{(\partial V/\partial P)_T} \qquad (2\text{-}2.2.7)$$

Using Eq. (2-2.2.5) with Eq. (2-2.2.3) and Eq. (2-2.2.6)

$$dP = -\frac{1}{\beta V}\,dV - \frac{\alpha V}{-\beta V}\,dT$$

$$= -\frac{1}{\beta V}\,dV + \frac{\alpha}{\beta}\,dT \qquad (2\text{-}2.2.8)$$

Compare Eq. (2-2.2.8) with Eq. (2-2.2.2)(b). Since the coefficients of dV and dT in the two must be equal, we can see that

$$\left(\frac{\partial P}{\partial V}\right)_T = -\frac{1}{\beta V} \tag{2-2.2.9}$$

and

$$\left(\frac{\partial P}{\partial T}\right)_V = \frac{\alpha}{\beta} \tag{2-2.2.10}$$

Eq. (2-2.2.9) tells us nothing new: it is the same as Eq. (2-2.2.6). However, Eq. (2-2.2.10) relates the *isochoric* (constant volume) change of pressure with a given change in temperature to α and β which were previously defined.

Perhaps the above exercise shows the utility of manipulating equations. We can determine $(\partial P/\partial T)_V$ by measuring α and β. This avoids having to perform an experiment to measure $(\partial P/\partial T)_V$ itself. Finally, you will note that the above equations apply equally as well for liquids and solids. Experimentally, a quantity such as $(\partial V/\partial T)_P$ is determined by measuring V vs. T at a given pressure, plotting, then taking the slope of the resulting line.

After a few more definitions, several empirical and semi-empirical equations of state for gases will be given. Empirical implies that no theory is involved: experimental data are fitted to determine the constants. Semi-empirical means that the constants in the equation are based on theory but are measured experimentally. Three two-constant equations of state will be compared, and one equation of state containing five constants will be presented.

2-2.3 COMPRESSIBILITY FACTORS AND THE LAW OF CORRESPONDING STATES

A useful quantity in finding P, V, and T relationships is the compressibility factor Z. It is defined by

$$Z = \frac{PV}{RT} \tag{2-2.3.1}$$

You should immediately be able to predict that, for an ideal gas, Z = 1 at all P, V, and T values. This is not the case for real gases, as seen in Fig. 2-2.3.1. Nonideal behavior results from the failure of gases to follow our simple model. We assumed that there were no intermolecular attractions. This cannot possibly be true or we would never find chemicals in a liquid or solid state. We know that any gas will change to a liquid given high enough P and low enough T. Figure 2-2.3.2 shows several *isotherms* (constant temperature curves) on a PV plot for a generalized real gas. Notice the high temperature curve looks like a typical ideal gas curve ($P \propto 1/V$). Deviations from this shape become apparent as the temperature is lowered and approaches the *critical temperature* T_c. The critical temperature is defined as the highest temperature at which a liquid can exist as a distinct phase or region. A material can become a liquid only below T_c. Above T_c, the liquid and vapor are indistinguish-

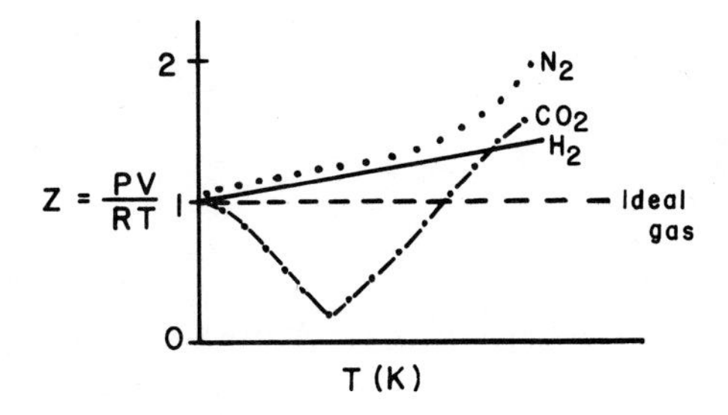

FIG. 2-2.3.1 Compressibility factors as a function of T for 1 mol of several gases at 273 K.

able and the term fluid is used instead of liquid or vapor. Below T_c, there are values of P and V at low temperature where liquid and vapor are both present as distinct regions. This area is shown in Fig. 2-2.3.2 and is outlined by the dotted dome. Any P, V, and T value that falls beneath this dome gives a liquid-vapor region (both are present). The P and V just at the top of the dome (point A) are called the *critical pressure* P_c and the *critical volume* V_c. Point A is a point of inflection so

$$\left(\frac{\partial P}{\partial V}\right)_{T_c} = 0 \qquad \left(\frac{\partial^2 P}{\partial V^2}\right)_{T_c} = 0 \tag{2-2.3.2}$$

The conditions in Eq. (2-2.3.2) serve to define the critical state.

You might guess that, since every gas becomes a liquid, then perhaps there is some important relationship involving P_c, V_c, and T_c. This approach leads to the *law of corresponding states*. Before giving a statement of it let us define three new variables:

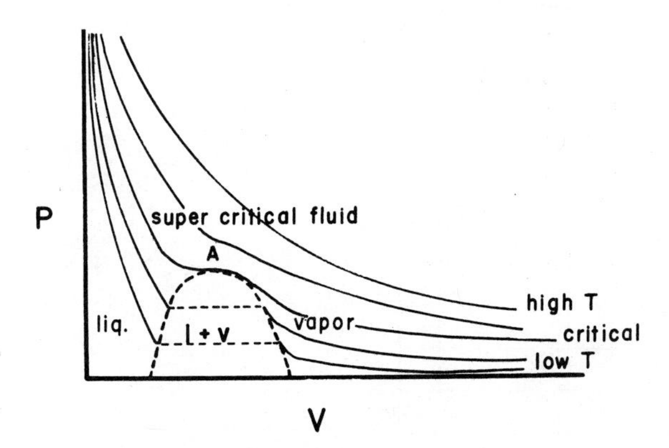

FIG. 2-2.3.2 Typical PV behavior of a real gas at several temperatures (ℓ = liquid, v = vapor).

$$\text{Reduced pressure:} \qquad P_r = \frac{P}{P_c}$$

$$\text{Reduced temperature:} \qquad T_r = \frac{T}{T_c} \tag{2-2.3.3}$$

$$\text{Reduced volume:} \qquad V_r = \frac{V}{V_c}$$

We now state the *law of corresponding states: All gases deviate from ideality in a way that depends only on the reduced pressure and reduced temperature.* This is just a statement of our feeling that the behavior of a gas should depend on how far its variables (P, V, T) are from the critical point (P_c, V_c, T_c).

The validity of this statement is shown in Fig. 2-2.3.3 in which the compressibility factor is plotted against P_r for several T_r and several gases. The close agreement between many different types of gases as shown below is, indeed, satisfying. You should note that both polar (H_2O) and nonpolar (CH_4, CO_2, etc.) gases agree equally well. The lines drawn are "best fit" curves for the experimental points determined for several gases.

2-2.4 THE VAN DER WAALS EQUATION OF STATE FOR REAL GASES

In 1873, van der Waals proposed an equation of state for real gases. In this equation he attempted to eliminate two of the assumptions in the kinetic molecular theory. A corrective term was added to the pressure to take into account the attractive interactions between molecules. This term reduces the true pressure below what it might have been. Also, a volume correction is included to compensate for the container volume taken up by the molecules themselves. The form of the equation is given as

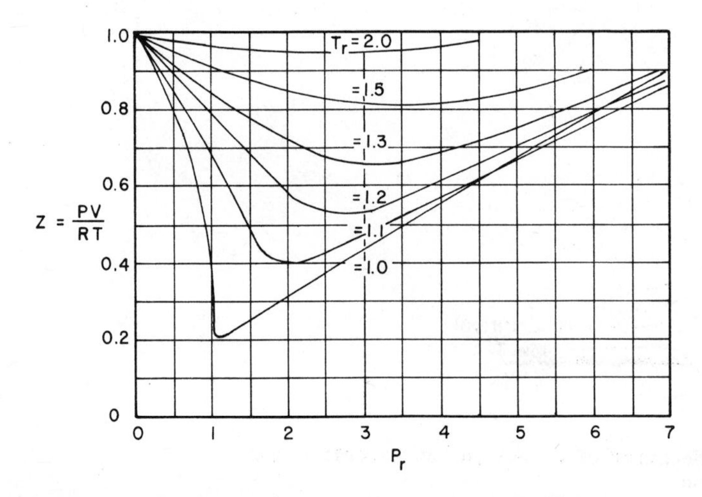

FIG. 2-2.3.3 The compressibility factor as a function of P_r and T_r for several gases. [From Goug-Jen Su, *Ind. Eng. Chem.*, *38*:803 (1946). (See Sec. 2-2.12, Ref. 2.)]

$$\left(P + \frac{an^2}{V^2}\right)(V - nb) = nRT \qquad (2\text{-}2.4.1)$$

in which a and b are empirical constants. They can be related to the critical constants by applying the conditions given in Eq. (2-2.3.2). At T_c we can show (and you should show it) for 1 mol

$$\frac{dP}{dV} = -\frac{RT_c}{(\overline{V}_c - b)^2} + \frac{2a}{\overline{V}_c^{\,3}} = 0 \qquad (2\text{-}2.4.2)$$

and

$$\frac{d^2P}{dV^2} = \frac{2RT_c}{(\overline{V}_c - b)^3} - \frac{6a}{\overline{V}_c^{\,4}} = 0 \qquad (2\text{-}2.4.3)$$

Also, from (2-2.4.1),

$$P_c = \frac{RT_c}{\overline{V}_c - nb} - \frac{a}{\overline{V}_c^{\,2}} \qquad (2\text{-}2.4.4)$$

You should solve these to show

$$b = \frac{\overline{V}_c}{3}$$

$$a = 3P_c\overline{V}_c^{\,2} \qquad (2\text{-}2.4.5)$$

$$R = \frac{8P_c\overline{V}_c}{3T_c}$$

where the bar indicates the value for 1 mol. These can be rearranged:

$$\overline{V}_c = 3b$$

$$P_c = \frac{a}{27b^2} \qquad (2\text{-}2.4.6)$$

$$T_c = \frac{8a}{27bR}$$

or

$$\overline{V} = 3b\overline{V}_r$$

$$P = \frac{a}{27b^2}P_r \qquad (2\text{-}2.4.7)$$

$$T = \frac{8a}{27bR}T_r$$

Using these relationships, the van der Waals equation can be written in terms of P_r, V_r, and T_r as

$$\left(P_r + \frac{3}{\overline{V}_r^{\,2}}\right)\left(\overline{V}_r - \frac{1}{3}\right) = \frac{8}{3}T_r \qquad (2\text{-}2.4.8)$$

This form will be used later (Sec. 2-2.6) in comparing equations of state involving only two constants.

2-2.5 OTHER EQUATIONS OF STATE FOR REAL GASES

Several empirical equations with two constants have been proposed. Only two will be given here. You may be able to see the similarity of the Redlich-Kwong equation to the van der Waal's equation. No such similarity is apparent in Dieterici's equation.

DIETERICI'S EQUATION. Dieterici's equation can be written as

$$P = \left(\frac{nRT}{V - nb}\right) \exp\left(-\frac{an}{VRT}\right) \tag{2-2.5.1}$$

Applying the conditions of Eq. (2-2.3.2) as done in the previous section for the van der Waals equation results in the following (if you are mathematically inclined you will find the derivation a satisfying exercise):

$$a = 2\overline{V}_c RT_c \qquad b = \frac{\overline{V}_c}{2} \tag{2-2.5.2}$$

and

$$P_c = \frac{2RT_c}{\overline{V}_c} e^2 \tag{2-2.5.3}$$

Finally, in terms of P_r, V_r, and T_r:

$$P_r \exp\left(\frac{2}{V_r T_r}\right)\left(V_r - \frac{1}{2}\right) = \frac{e^2}{2} T_r \tag{2-2.5.4}$$

REDLICH-KWONG EQUATION. The last equation of the two-constant type to be considered is the Redlich-Kwong equation (see Sec. 2-2.12, Ref. 3). For 1 mol, this is

$$P = \frac{RT}{V - b} - \frac{a}{T^{1/2}\overline{V}(\overline{V} + b)} \tag{2-2.5.5}$$

where a and b can be found from

$$a = \frac{0.4275R^2 T_c^{2.5}}{P_c}$$

$$b = \frac{0.0866RT_c}{P_c} = 0.260\ \overline{V}_c \tag{2-2.5.6}$$

Equation (2-2.5.5) may be written in terms of reduced variables as

$$P_r = \frac{11.5T_r}{3.85V_r - 1} - \frac{14.8}{T_r^{1/2} V_r (3.85V_r + 1)} \tag{2-2.5.7}$$

2-2.6 COMPARISONS OF EQUATIONS OF STATE WITH TWO CONSTANTS

The compressibility factors as given in Table 2-2.6.1 for each equation of state are plotted as a function of P_r for several T_r in Figs. 2-2.6.1 through 2-2.6.6.

Table 2-2.6.2 contains critical-point data and values of Redlich-Kwong and van der Waals constants. For the Dieterici constants, Eq. (2-2.5.2) can be used. Equation (2-2.4.5) can be used for the van der Waals constants. Actually, R is assumed to be a universal constant and the "best fit" values of a and b are listed in Table 2-2.6.2.

The figures are self-explanatory. However, one point is worth emphasizing: The Redlich-Kwong equation is far superior to other two-constant equations for gases. This should be apparent from the close agreement of compressibility factors calculated from this equation and those determined experimentally. If critical constants are available, the Redlich-Kwong equation is the preferred equation.

2-2.7 EQUATIONS OF STATE FOR REAL GASES -- BEATTIE-BRIDGEMAN

The Beattie-Bridgeman equation (see Sec. 2-2.12, Ref. 4, 1928) contains five empirical constants in addition to R. It has the form (for 1 mol)

$$P\overline{V}^2 = RT\left(1 - \frac{c}{\overline{V}T^3}\right)\left(\overline{V} + B_0 - \frac{bB_0}{\overline{V}}\right) - A_0\left(1 - \frac{a}{\overline{V}}\right) \tag{2-2.7.1}$$

Table 2-2.6.1
Equations of State in Reduced Form

Equation of state	Equation of state (reduced form)	Compressibility factor (reduced form)
van der Waals		
$P = \frac{RT}{\overline{V} - b} - \frac{a}{\overline{V}^2}$	$P_r = \frac{8T_r}{3\overline{V}_r - 1} - \frac{3}{\overline{V}_r^2}$	$Z = \frac{3}{8}\frac{P_r\overline{V}_r}{T_r} = 0.375\frac{P_r\overline{V}_r}{T_r}$
Redlich-Kwong		
$P = \frac{RT}{\overline{V} - b} - \frac{a}{T^{1/2}\overline{V}(\overline{V} + b)}$	$P_r = \frac{11.5T_r}{3.85\overline{V}_r - 1} - \frac{14.8}{T_r^{1/2}\overline{V}_r(3.85\overline{V}_r + 1)}$	$Z = 0.333\frac{P_r\overline{V}_r}{T_r}$
Dieterici		
$P = \frac{RT\exp(-a/\overline{V}RT)}{\overline{V} - b}$	$P_r = \frac{T_r\exp(2 - 2/T_r\overline{V}_r)}{2\overline{V}_r - 1}$	$Z = \frac{2P_r\overline{V}_r}{T_r e^2} = 0.271\frac{P_r\overline{V}_r}{T_r}$

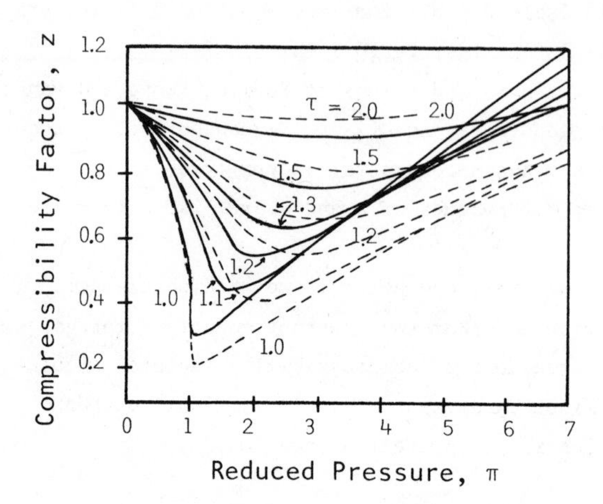

FIG. 2-2.6.1 Comparison of the experimental Z values (dotted lines) with the Z values calculated from the van der Waals equation of state (solid lines). [From J. B. Ott, J. R. Goates, and H. T. Hall, Jr., *J. Chem. Educ.*, *48*:515-517 (1971).] Reprinted by permission.

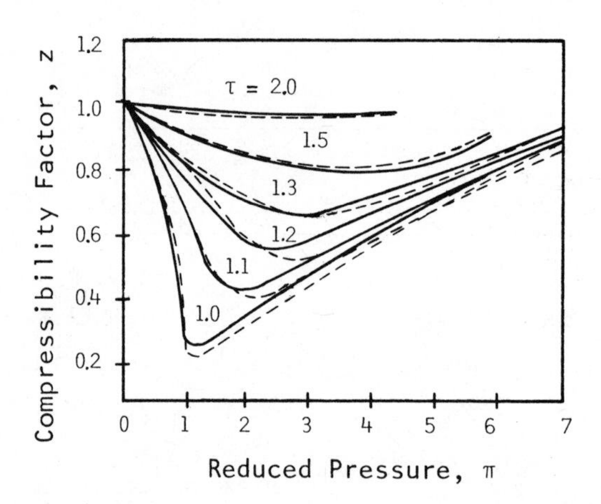

FIG. 2-2.6.2 Comparison of the experimental Z values (dotted lines) with the Z values calculated from the Redlich-Kwong equation of state (solid lines). [From J. B. Ott, J. R. Goates, and H. T. Hall, Jr., *J. Chem. Educ.*, *48*:515-517 (1971).] Reprinted by permission.

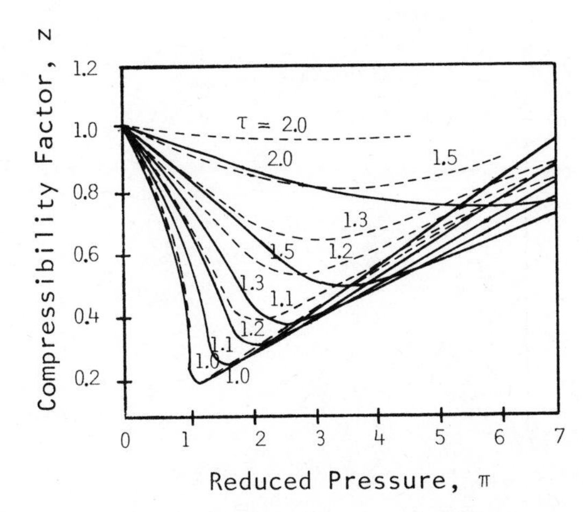

FIG. 2-2.6.3 Comparison of the experimental Z values (dotted lines) calculated from the Dieterici equation of state (solid lines). [From J. B. Ott, J. R. Goates, and H. T. Hall, Jr., *J. Chem. Educ.*, *48*: 515-517 (1971).] Reprinted by permission.

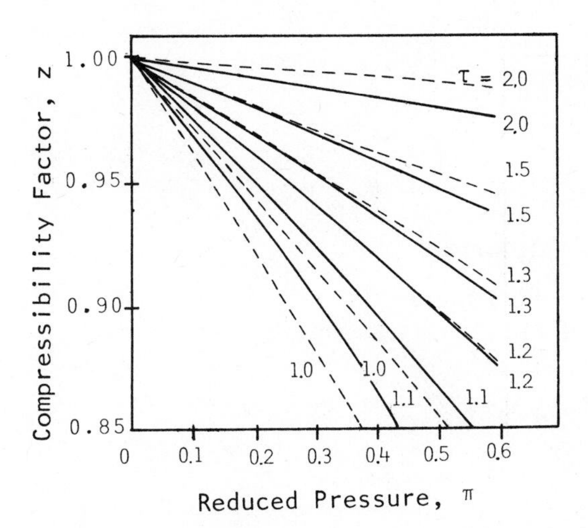

FIG. 2-2.6.4 Comparison of the experimental Z values (dotted lines) in the low pressure region with the z values calculated from the van der Waals equation of state (solid lines). [From M. K. Kemp, R. E. Thompson, and D. J. Zigrang, *J. Chem. Educ.*, *52*:802 (1975).] Reprinted by permission.

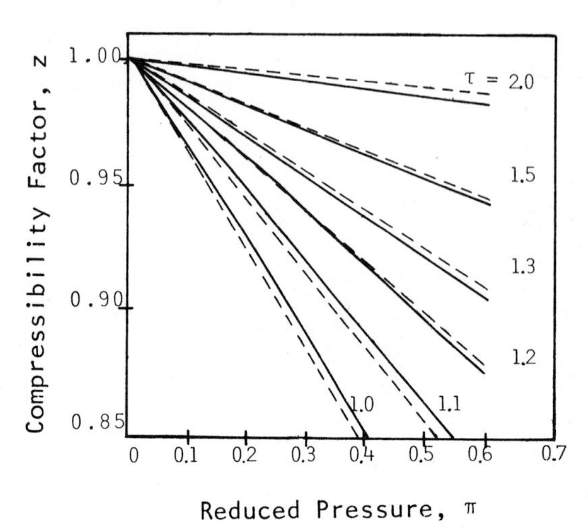

FIG. 2-2.6.5 Comparison of the experimental Z values (dotted lines) in the low pressure region with the Z values calculated from the Redlich-Kwong equation of state (solid lines). [From M. K. Kemp, R. E. Thompson, and D. J. Zigrang, *J. Chem. Educ.*, *52*:802 (1975).] Reprinted by permission.

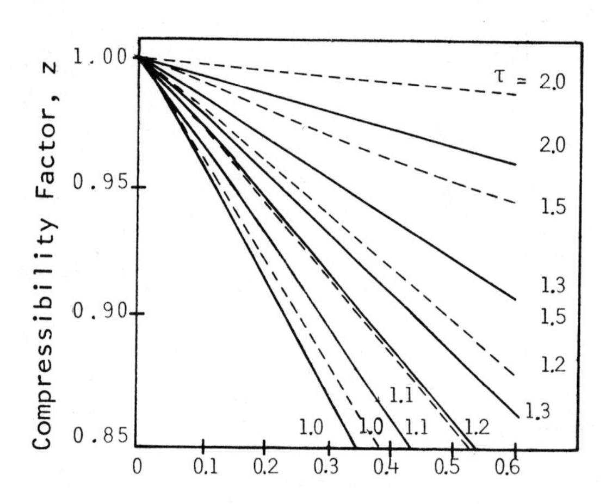

FIG. 2-2.6.6 Comparison of the experimental Z values (dotted lines) in the low pressure region with the Z values calculated from the Dieterici equation of state. [From M. K. Kemp, R. E. Thompson, and D. J. Zigrang, *J. Chem. Educ.*, *52*:802 (1975).] Reprinted by permission.

Table 2-2.6.2
"Best Fit" Values for van der Waals and Redlich-Kwong Constants Calculated from the Listed Values of the Critical Constants.[a]

T_c	P_c	Atom or molecule	Redlich-Kwong		van der Waals	
			a	b	a	b
(K)	(atm)		(atm $K^{1/2}$ dm^6 mol^{-2})	(dm^3 mol^{-1})	(atm dm^6 mol^{-2})	(dm^3 mol^{-1})
5.3	2.26	He	0.0824	0.0167	0.0341	0.0237
33.3	12.8	H_2	1.439	0.0185	0.246	0.0266
126.1	33.5	N_2	15.34	0.0268	1.35	0.0386
134.0	34.6	CO	17.29	0.0275	1.46	0.0392
154.3	49.7	O_2	17.12	0.0221	1.36	0.0319
304.2	72.8	CO_2	63.81	0.0297	3.60	0.0428
405.6	112.2	NH_3	85.00	0.0257	4.19	0.0373
647.2	217.7	H_2O	140.9	0.0211	5.47	0.0305

[a]R is assumed a constant.

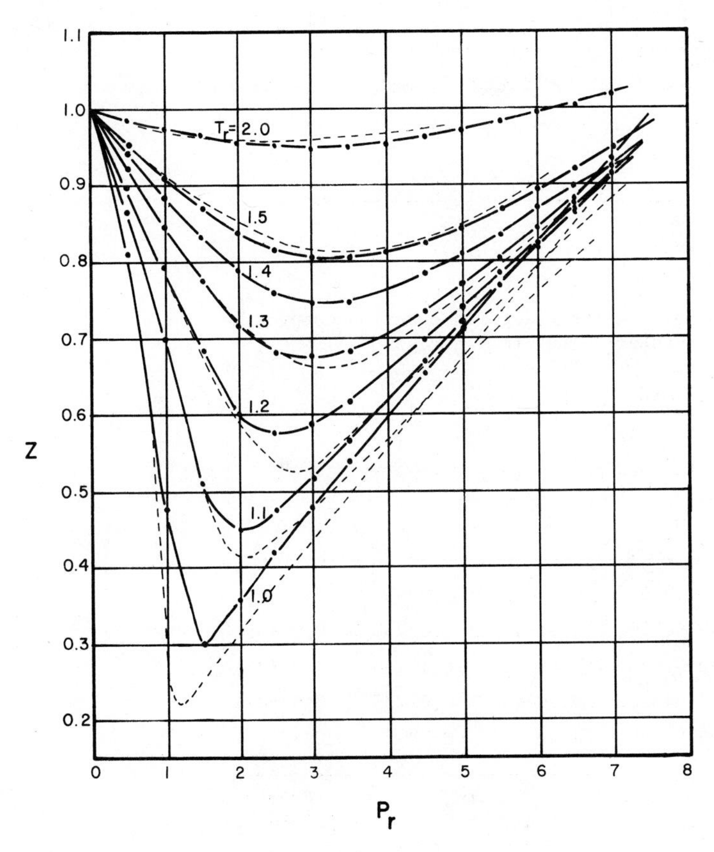

FIG. 2-2.7.1 Comparison of Redlich-Kwong equation compressibility factor with experimental values for several gases: nitrogen, helium, carbon dioxide, hydrogen, air, methane, neon, argon, oxygen, and diethyl ether.

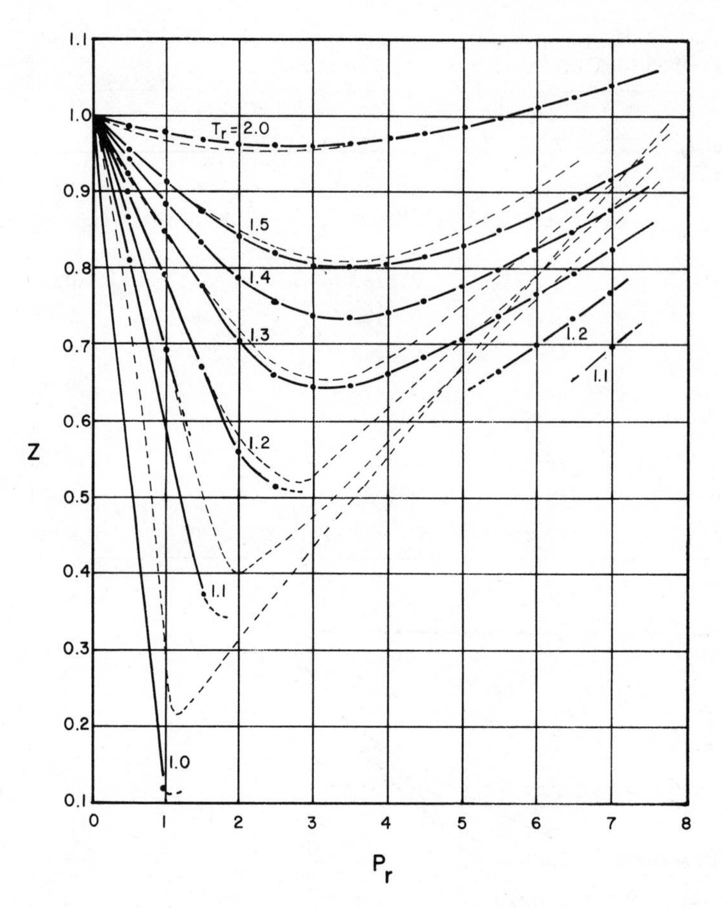

FIG. 2-2.7.2 Comparison of Beattie-Bridgeman equation compressibility factor with experimental values for nitrogen.

Table 2-2.7.1
Values of the Beattie-Bridgeman Constants for Several Gases[a]

	He	H_2	N_2	CO_2	NH_4
A_0	0.0216	0.1975	1.3445	5.0065	2.2769
a	0.05984	-0.00506	0.02617	0.07132	0.01855
B_0	0.01400	0.02096	0.05046	0.10476	0.05587
b	0.0	-0.04359	-0.00691	0.07235	-0.01587
$c \times 10^{-4}$	0.0040	0.0504	4.20	66.00	12.83

[a]Units assumed are: V in $dm^3\ mol^{-1}$, P in atm, T in K. [From J. O. Beattie and O. C. Bridgeman, *J. Am. Chem. Soc., 50*:3136 (1928).]

2-2.8 COMPARISON OF REDLICH-KWONG AND BEATTIE-BRIDGEMAN EQUATIONS

Figures 2-2.7.1 and 2-2.7.2 compare compressibility factors calculated from the Redlich-Kwong and the Beattie-Bridgeman equations with experimental values. In the case of the Redlich-Kwong equation, all the gases tested (the ones listed in Fig. 2-2.3.3) give the same result. However, that is not true for the Beattie-Bridgeman equation. Values shown in Fig. 2-2.7.2 are for N_2 which gives the best fit of all the gases tested. All others deviate more than N_2 from the experimental Z values. It should be clear that the Redlich-Kwong equation more nearly matches experimental Z values. As stated before, it is the preferred equation.

There are other empirical equations of state, such as the virial equation which gives Z as a power series in V:

$$\frac{PV}{nRT} = \frac{Z}{n} = 1 + \frac{nB}{V} + \frac{n^2C}{V^2} + \cdots \tag{2-2.8.1}$$

where B, C, and so on are constants. As many constants as necessary may be included. With the availability of large computers, such equations are solvable and do find use. These power series equations are very useful as curve-fitting equations. With computer capability to do the curve fitting and a reliable set of PVT data, functions can be found to accurately describe the PVT behavior of a gas over a wide range of conditions.

Equation (2-2.8.1) is of theoretical importance, but a discussion of this importance is beyond the scope of this book. (See Sec. 2-2.12, Ref. 5.)

This completes our introduction to the kinetic molecular model and equations of state for nonideal gases. The kinetic molecular theory will be used in the following section to predict many of the properties of gases. You will be using some of the nonideal equations of state when we treat the thermodynamics of gaseous systems in Chaps. 3, 4, and 5. Hence, it is important to gain facility in their use. The examples and problems below are presented for that purpose.

2-2.9 EXAMPLES

Calculate the pressure of 4.00 mol CO_2 contained in a 6.25-dm^3 flask at 298.15 K. Use the Dieterici, van der Waals, Redlich-Kwong, and Beattie-Bridgeman equations; compare the results.

Solution

Dieterici

$a = 4.62 \text{ atm dm}^6 \text{ mol}^{-2}$

$b = 0.0463 \text{ dm}^3 \text{ mol}^{-1}$

$$P = \frac{RT\ \exp(-a/\overline{V}RT)}{V - b}$$

$$\overline{V} = \frac{dm^3}{mol} = \frac{6.25\ dm^3}{4.00\ mol} = 1.563\ dm^3\ mol^{-1}$$

$$P = \frac{(0.0821\ dm^3\ atm\ mol^{-1}\ K^{-1})(298\ K)}{(1.563\ dm^3\ mol^{-1} - 0.0463\ dm^3\ mol^{-1})}$$

$$\exp\left[\frac{-\ 4.62\ atm\ dm^6\ mol^{-2}}{(1.563\ dm^3\ mol^{-1})(0.0821\ dm^3\ atm\ mol^{-1}\ K^{-1})(298\ K)}\right]$$

$$= \frac{24.46\ dm^3\ atm\ mol^{-1}}{1.517\ dm^3\ mol^{-1}}\exp(-0.121) = (16.13\ atm)(0.886)$$

$$= \underline{\underline{14.3\ atm}}$$

van der Waals

$a = 3.6\ atm\ dm^6\ mol^{-2}$

$b = 0.0428\ dm^3\ mol^{-1}$

$$P = \frac{RT}{\overline{V} - b} - \frac{a}{\overline{V}^2}$$

$$P = \frac{(0.0821\ dm^3\ atm\ mol^{-1}\ K^{-1})(298\ K)}{1.563\ dm^3\ mol^{-1} - 0.0428\ dm^3\ mol^{-1}} - \frac{3.60\ atm\ dm^6\ mol^{-2}}{(1.563\ dm^3\ mol^{-1})^2}$$

$$= \frac{24.46\ atm}{1.520} - \frac{3.60\ atm}{2.443\ atm} = 16.09\ atm - 1.47\ atm$$

$$= \underline{\underline{14.6\ atm}}$$

Beattie-Bridgeman

$$P = \frac{RT}{\overline{V}^2}\left(1 - \frac{c}{\overline{V}T^3}\right)\left(\overline{V} + B_0 - \frac{bB_0}{\overline{V}}\right) - \frac{A_0}{\overline{V}^2}\left(1 - \frac{a}{\overline{V}}\right)$$

From Table 2-2.7.2 (note units listed in table):

$A_0 = 5.0065$ $a = 0.07132$

$B_0 = 0.10476$ $b = 0.07235$

$c = 66 \times 10^4$

$$P = \frac{(0.0821)(298.15)}{(1.563)^2}\left[1 - \frac{66 \times 10^4}{(1.563)(298.15)^3}\right]$$

$$\times \left[1.563 + 0.10476 - \frac{(0.07235)(0.10476)}{1.563}\right] - \frac{5.0065}{1.563^2}\left(1 - \frac{0.07132}{1.563}\right)$$

$$= 10.02\ (0.984)(1.663) - \frac{4.778}{(1.563)^2}$$

$$= \underline{\underline{14.4\ \text{atm}}}$$

Redlich-Kwong

$$a = \frac{0.4275R^2T_c^{\ 2.5}}{P_c}$$

$$= \frac{0.4275\ (0.08206)^2(304.2)^{2.5}}{(72.8)}$$

$$a = 63.8$$

$$b = \frac{0.0866RT_c}{P_c}$$

$$= \frac{0.0866\ (0.0821)(304.2)}{(72.8)}$$

$$b = 0.0297$$

$$P = \frac{RT}{\overline{V} - b} - \frac{a}{T^{\frac{1}{2}}\overline{V}(\overline{V} + b)}$$

$$= \frac{(0.0821)(298)}{1.563 - 0.0297} - \frac{63.8}{(298^{\frac{1}{2}})(1.563)(1.563 + 0.0297)}$$

$$= 15.96 - 1.48$$

$$= \underline{\underline{14.5\ \text{atm}}}$$

As an exercise you should show that the units do work out. For this gas, under these conditions, the results are all essentially the same.

2-2.10 PROBLEMS

1. The compressibility factor of H_2O at 776 K is found to be 0.60. From Fig. 2-2.3.3 estimate the pressure of H_2O.

2. A mole of a certain gas occupies a volume of 27.0 dm^3 at 1 atm and 273 K. If the equation of state for this gas is $PV = nRT(1 + bP)$, where b is a constant, find the volume at 2 atm and 295 K.

3. Calculate the temperature of a van der Waals gas which has $a = 1.35\ dm^6\ atm\ mol^{-2}$, $b = 0.0322\ dm^3\ mol^{-1}$ if at 1 atm, 2 mol of the gas occupies 51.5 liters.

4. Calculate by the ideal gas, the Redlich-Kwong, and the Beattie-Bridgeman equations the pressure of 10 mol ethane in a 4.86 dm^3 flask at 300 K. (For ethane, A_0 = 5.880, a = 0.05861, B_0 = 0.09400, b = 0.01915, and c = 90.0×10^4.) Compare with the observed value of 34.0 atm (T_c = 305.4 K, P_c = 48.2 atm).

5. For N_2 the constants in the Beattie-Bridgeman equation are $A_0 = 1.3445$, $a = 0.02617$, $B_0 = 0.05046$, $b = -0.00691$, and $c = 4.2 \times 10^4$. Determine the pressure exerted by 1 mol of N_2 at 275 K if the volume is 11.2 dm^3.

6. Given that $T_c = 405.6$ K and $P_c = 112.2$ atm for NH_3, at what temperature will 1 mol occupy 7.6 dm^3 at 10.1 atm? (Assume the Redlich-Kwong equation applies.)

A CHALLENGE FOR YOU.

7. Estimate the volume of 3 mol of CO_2 gas at 313.15 K and 200 atm from the Beattie-Bridgeman equation. This problem may involve a series of approximations. Try it.

2-2.11 SELF-TEST

1. State the postulates of the kinetic molecular theory.

2. Calculate the molar density (moles per cubic decimeter) of 10 g of O_2 at 10 atm and 300 K from the van der Waals equation.

3. Compare the pressure calculated for 2 mol CO_2 confined to a 3.02-dm^3 vessel at 185 K, using the Redlich-Kwong equation and the Beattie-Bridgeman equation.

4. Plot the compressibility factor of N_2 as a function of temperature at 100 atm, from 100 K to 250 K. (Hint: Use Table 2-2.6.2 and Fig. 2-2.3.3.)

2-2.12 ADDITIONAL READING

1.* W. J. Moore, *Physical Chemistry*, 4th ed., Prentice-Hall, Englewood Cliffs, N. J., 1972, pp. 150-152.

2.* Gouq-Jen Su, *Ind. Eng. Chem. 38*:803 (1946).

3.* O. Redlich and J. N. S. Kwong, *Chem. Rev., 44*:233 (1949).

4.* J. O. Beattie and O. C. Bridgeman, *J. Am. Chem. Soc., 50*:3136 (1928) and *49*:1665 (1927).

5.* S. M. Blinder, *Advanced Physical Chemistry*, MacMillan, New York, 1969, p. 479.

6. See Sect. 1-1.9A for references to data compilations for gases.

2-3 VELOCITY AND SPEED DISTRIBUTION

OBJECTIVES

You shall be able to calculate any of the following for a gas, given appropriate equations and data: velocity, speed, averages of velocities and speeds, mean free path, collision diameter and/or collision number, and viscosity.

2-3.1 INTRODUCTION

In this section you will encounter the derivation of a set of equations that gives the distribution of molecular velocities in a gas. Equations for average properties of gaseous systems (such as average velocity, speed, etc.) are also derived. The derivations are good examples of how theoretical considerations (the postulates of the kinetic molecular theory) can be turned into practical equations. View these derivations in that light, not as something that must be struggled through.

The utility of the equations presented should be obvious by the time you have finished this section. One word of caution in using the results: The final equations for flow properties and mean free path are strictly valid *only* for ideal gases. Any nonideality will require some modification of the equations.

2-3.2 DISTRIBUTION OF VELOCITIES AND SPEEDS

To avoid subscripts the velocity components of a gas will be defined as follows:

$$\begin{aligned} u &= u_x \qquad -\infty \leq u \leq \infty \\ v &= u_y \qquad -\infty \leq v \leq \infty \\ w &= u_z \qquad -\infty \leq w \leq \infty \end{aligned} \tag{2-3.2.1}$$

The velocities are vector quantities, meaning they have both magnitude and direction. Any arbitrary velocity can be written in terms of the velocity components in (2-3.2.1) as shown in Fig. 2-3.2.1. This is an important point. Any velocity can be written in terms of three component velocities defined in mutually perpendicular axes. There will be other occasions on which vector quantities will be resolved into components.

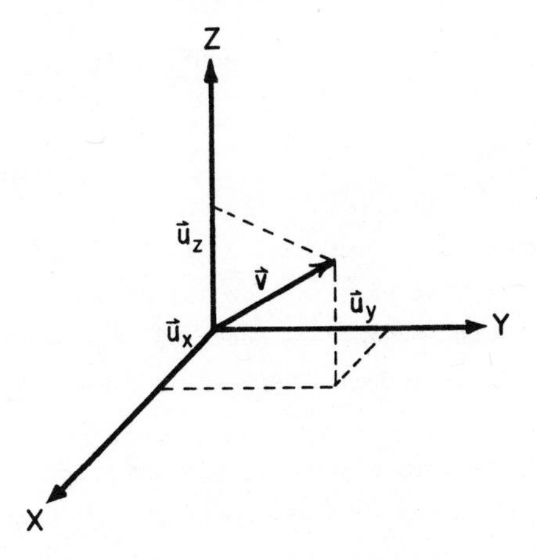

FIG. 2-3.2.1 The three components of a velocity.

A nonvector (scalar) quantity we will use is the speed c. The scalar c has no direction and has only positive values for its magnitude. It is related to the velocity components in (2-3.2.1) by

$$c^2 = u^2 + v^2 + w^2$$

or

$$c = (u^2 + v^2 + w^2)^{\frac{1}{2}} \qquad 0 \leq c \leq \infty \tag{2-3.2.2}$$

We also need to define the velocity increments (these are given by du, dv, and dw) and a speed increment given by dc.

Our first aim is to find a function which will tell us how molecules are distributed with respect to velocity. How many molecules have given values of velocity, within a small range of velocities? This will become clearer as we proceed. The function we are interested in will be called f(u, v, w), the *distribution function*.

Instead of using the number of molecules with each velocity, we will use the *fraction* of molecules with each velocity. This fraction is defined as follows:

$$\frac{dN_{uvw}}{N} = f(u, v, w)\ du\ dv\ dw \tag{2-3.2.3}$$

which is the fraction of molecules in the system with

u between u and u + du
v between v and v + dv
w between w and w + dw

Equation (2-3.2.3) gives, then, the fraction of molecules with velocities u, v, and w within the increments of du, dv, and dw.

Given that dN(u, v, w)/N is a fraction, you should be able to solve Eq. (2-3.2.4)

$$\sum_{u=-\infty}^{\infty} \sum_{v=-\infty}^{\infty} \sum_{w=-\infty}^{\infty} f(u, v, w)\ du\ dv\ dw = ? \tag{2-3.2.4}$$

If we take du, dv, and dw as infinitesimally small (velocities are continuous), we can write the sums as integrals to give

$$\int_{-\infty}^{\infty} \int_{-\infty}^{\infty} \int_{-\infty}^{\infty} f(u, v, w)\ du\ dv\ dw = ? \tag{2-3.2.5}$$

Since f(u, v, w) du dv dw indicates the fraction of molecules with a given u, v, and w, if we sum over all possible velocities, then we must have included every molecule (each molecule must have velocity components between $-\infty$ and ∞). Hence, (2-3.2.4) or (2-3.2.5) must yield a value of 1. Every molecule is included between velocities of $-\infty$ and ∞.

The next question, of course, is: What is the form of f(u, v, w)? To find this, we can follow the arguments of Maxwell. The first assumption is that f depends only on the kinetic energy, E_k.

$$f(u, v, w) = f(E_k) \tag{2-3.2.6}$$

But, $E_k = \frac{1}{2}m(u^2 + v^2 + w^2)$, so

$$f(u, v, w) = f(u^2 + v^2 + w^2) \qquad (2\text{-}3.2.7)$$

which indicates that f is a function of the sum of the squares of the velocity components.

A second assumption is that the velocity components are independent. Thus

$$f(u, v, w) = f(u)\, f(v)\, f(w) \qquad (2\text{-}3.2.8)$$

At this point we have two restrictions given in (2-3.2.7) and (2-3.2.8). It would be a useful exercise for you to stop at this point and try to find a function that will fit these restrictions.

One function that will satisfy the above restrictions is an exponential form:

$$f(u, v, w) = h \exp\left[-\frac{bm}{2}(u^2 + v^2 + w^2)\right]$$

$$= h\left[\exp\left(-\frac{bmu^2}{2}\right)\right]\left[\exp\left(-\frac{bmv^2}{2}\right)\right]\left[\exp\left(-\frac{bmw^2}{2}\right)\right] \qquad (2\text{-}3.2.9)$$

In (2-3.2.9) you will find two constants b and h which must be evaluated. This is a general result. For each restriction applied to a solution, a constant is included which must be evaluated. The constant h can be evaluated very easily by applying (2-3.2.5)

$$\int\int\int_{-\infty}^{\infty} h \exp\left[-\frac{bm}{2}(u^2 + v^2 + w^2)\right] du\, dv\, dw = 1 \qquad (2\text{-}3.2.10)$$

To solve, consider a one-dimensional case first:

$$\int_{-\infty}^{\infty} h' \exp\left(-\frac{bmu^2}{2}\right) du = 1 \qquad (2\text{-}3.2.11)$$

To evaluate (2-3.2.11), put it into standard integral form by making the following substitution

$$a = \frac{bm}{2}$$

Then $\qquad h' \int_{-\infty}^{\infty} e^{-au^2}\, du = 1$

This can be evaluated from the standard integral forms given in Appendix D.

$$-\int_{-\infty}^{\infty} e^{-au^2}\, du = \left(\frac{\pi}{a}\right)^{1/2} = \left(\frac{\pi}{bm/2}\right)^{1/2} = \left(\frac{2\pi}{bm}\right)^{1/2}$$

therefore $h'\left(\frac{2\pi}{bm}\right)^{1/2} = 1$

or

$$h' = \left(\frac{bm}{2\pi}\right)^{1/2} \qquad (2\text{-}3.2.12)$$

Since the velocities are independent you should be able to justify the statement that

$$h' = h^{1/3}$$

Hence
$$h = \left(\frac{bm}{2\pi}\right)^{3/2} \tag{2-3.2.13}$$

and
$$f(u, v, w) = \left(\frac{bm}{2\pi}\right)^{3/2} \exp\left[-\frac{bm}{2}(u^2 + v^2 + w^2)\right] \tag{2-3.2.14}$$

Equation (2-3.2.14) is known as Maxwell's velocity distribution function. It is not of use until we evaluate the constant b. The constant b can be evaluated from statistical mechanical equations for the translational energy of an ideal gas. Suffice it at present simply to state the result [details can be found in Wall (Sec. 2-3.13, Ref. 1)]:

$$b = \frac{1}{kT} \tag{2-3.2.15}$$

where k is Boltzmann's constant ($k = 1.38 \times 10^{-23}$ J molecule^{-1} K^{-1}) and T is the absolute temperature. Combining (2-3.2.3), (2-3.2.14), and (2-3.2.15):

$$\frac{dN_{uvw}}{N} = \left(\frac{m}{2\pi kT}\right)^{3/2} \exp\left[\frac{-m(u^2 + v^2 + w^2)}{2kT}\right] du\ dv\ dw \tag{2-3.2.16}$$

This is the *Maxwell-Boltzmann velocity distribution function.*

Remember dN_{uvw}/N gives the fraction of molecules in a system with velocity components u, v, and w within the increments du, dv, and dw.

The corresponding function for speed is given by [you might try to derive this from (2-3.2.16)]

$$\frac{dN_c}{N} = 4\pi\left(\frac{m}{2\pi kT}\right)^{3/2} \exp\left(-\frac{mc^2}{2kT}\right) c^2\ dc \tag{2-3.2.17}$$

2-3.3 GRAPHICAL PRESENTATION OF RESULTS

Figures 2-3.3.1 and 2-3.3.2 are graphical representations of the distribution functions for He and N_2 at two different temperatures. Note the effect of both mass and temperature on the distribution of molecular speeds and velocities. An increase in mass makes the distribution more narrow, as does a decrease in temperature.

Note that the velocity distribution is symmetrical about $u_x = 0$ while the speed distribution starts at c = 0; there are no negative speeds. These differences are due to the fact that u is a vector and c is a scalar.

A final comment should suffice. The area under the distribution curve for a given molecule is constant even if T changes. (This should make sense, as the number of molecules is constant.) For example, the distribution function for N_2 at 298 K is

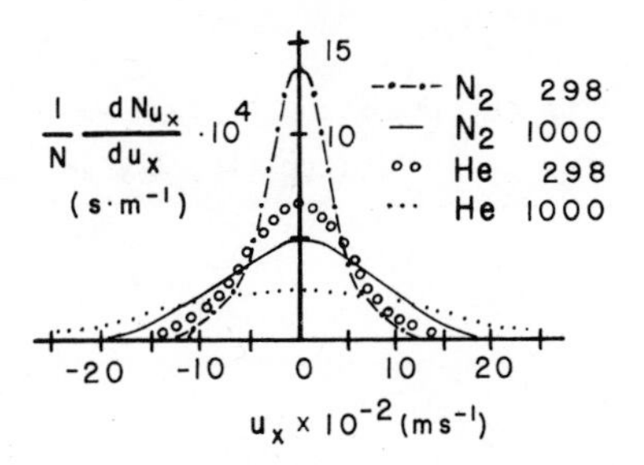

FIG. 2-3.3.1 One dimensional velocity distribution for He and N_2 at 298 and 1000 K.

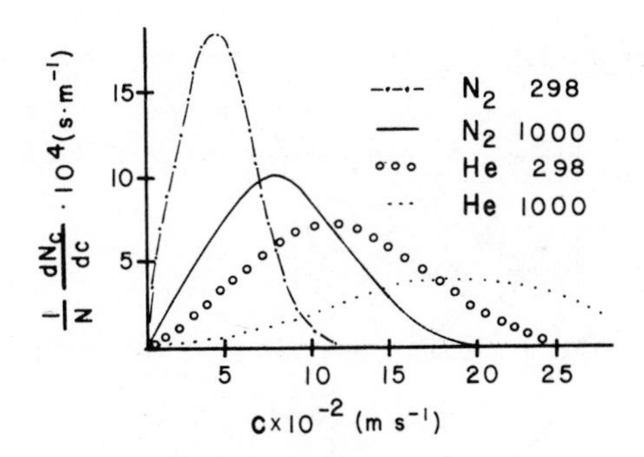

FIG. 2-3.3.2 Distribution of speeds for He and N_2 at 298 and 1000 K.

very different from that at 1000 K. However, the total area is a constant. At 1000 K the function is smaller for any given value of u, but it is much broader than the same function at 298 K.

2-3.4 SOME AVERAGE QUANTITIES

We will now apply the Maxwell-Boltzmann velocity distribution function to find average values for some quantities. Assume we have any variable q which depends in some way on one or more of the velocity components u, v, and w. The average value of that variable (denoted by $\bar{q}$) is given by

$$\bar{q} = \int\int_{-\infty}^{\infty}\int q\ f(u, v, w)\ du\ dv\ dw \qquad (2\text{-}3.4.1)$$

Stated verbally, Eq. (2-3.4.1) says that *the average value of a variable is given by the integral over all space of that variable times the distribution function.* (This definition is a general one, not limited to velocities.)

2-3.4.1 EXAMPLE

To make the meaning of Eq. (2-3.4.1) clear, let us determine the average value of v^2 and v to show the important result that $\overline{v^2} \neq (\bar{v})^2$. (Of course, the results for u and w are the same as for v.)

Solution

From Eq. (2-3.4.1) we can write immediately that

$$\overline{v^2} = \iiint_{-\infty}^{\infty} v^2 f(u, v, w)\, du\, dv\, dw$$

$$= \iiint_{-\infty}^{\infty} v^2 \left(\frac{m}{2\pi kT}\right)^{3/2} \exp\left[\frac{-m(u^2 + v^2 + w^2)}{2kT}\right] du\, dv\, dw \qquad (2\text{-}3.4.2)$$

Since u, v, and w are independent, break (2-3.4.2) into three separate integrals:

$$\overline{v^2} = \int_{-\infty}^{\infty} \left(\frac{m}{2\pi kT}\right)^{1/2} \exp\left(-\frac{mu^2}{2kT}\right) du \int_{-\infty}^{\infty} \left(\frac{m}{2\pi kT}\right)^{1/2} v^2 \exp\left(-\frac{mv^2}{2kT}\right) dv$$

$$\int_{-\infty}^{\infty} \left(\frac{m}{2\pi kT}\right)^{1/2} \exp\left(-\frac{mw^2}{2kT}\right) dw$$

From (2-3.2.11) you should conclude that the first and last integral terms are equal to 1. Thus

$$\overline{v^2} = 1 \int_{-\infty}^{\infty} \left(\frac{m}{2\pi kT}\right)^{1/2} v^2 \exp\left(-\frac{mv^2}{2kT}\right) dv\ 1$$

$$= \left(\frac{m}{2\pi kT}\right)^{1/2} \int_{-\infty}^{\infty} v^2 e^{-av^2} \cdot dv \qquad a = \frac{m}{2kT}$$

From Appendix D the integral can be evaluated immediately.

$$\overline{v^2} = \left(\frac{m}{2\pi kT}\right)^{1/2} \frac{1}{2a}\left(\frac{\pi}{a}\right)^{1/2}$$

$$= \left(\frac{m}{2\pi kT}\right)^{1/2} \frac{1}{2m/2kT} \left(\frac{\pi}{m/2kT}\right)^{1/2}$$

$$= \frac{kT}{m} = \frac{RT}{M} \qquad (2\text{-}3.4.3)$$

(Note: R = kN and M = mN)

By definition

$$v_{rms} = \left(\overline{v^2}\right)^{1/2}$$

Now evaluate $\bar{v}$.

$$\bar{v} = \iiint_{-\infty}^{\infty} v \left(\frac{m}{2\pi kT}\right)^{3/2} \exp\left[\frac{-m(u^2 + v^2 + w^2)}{2kT}\right] du\, dv\, dw$$

$$= \left(\frac{m}{2\pi kT}\right)^{\frac{1}{2}} \int_{-\infty}^{\infty} v \exp\left(-\frac{mv^2}{2kT}\right) dv \qquad (2\text{-}3.4.4)$$

This can be evaluated by noting that the argument of the integral is an *odd* function. An odd function, when integrated from $-\infty$ to ∞, yields a value of 0. It would be a useful exercise to solve the integral explicitly to show that the value is 0.

$$\bar{v} = 0$$

$$\overline{v^2} = \frac{kT}{m} \qquad (2\text{-}3.4.5)$$

and, as we stated,

$$\overline{v^2} \neq (\bar{v})^2$$

It is important to be specific when talking about averages. You can gather from what has been said previously that the results for u and w are the same.

2-3.4.2 EXAMPLE

Below the expressions for $\overline{c^2}$ and $\bar{c}$ are set up. It is left as an exercise for you to prove the results. Applying (2-3.4.1) to (2-3.2.17), we can write (note the limits used for c):

Solution

$$\overline{c^2} = \int_0^{\infty} c^2\, 4\pi \left(\frac{m}{2\pi kT}\right)^{3/2} \exp\left(-\frac{mc^2}{2kT}\right) c^2\, dc$$

$$= 4\pi \left(\frac{m}{2\pi kT}\right)^{3/2} \int_0^{\infty} c^4 \exp\left(-\frac{mc^2}{2kT}\right) dc \qquad (2\text{-}3.4.6)$$

and, $$\bar{c} = 4\pi \left(\frac{m}{2\pi kT}\right)^{3/2} \int_0^{\infty} c^3 \exp\left(-\frac{mc^2}{2kT}\right) dc \qquad (2\text{-}3.4.7)$$

You should use the integral forms in Appendix D to show that

$$\overline{c^2} = \frac{3kT}{m} = \frac{3RT}{m} \qquad (2\text{-}3.4.8)$$

$$\bar{c} = \left(\frac{8kT}{\pi m}\right)^{\frac{1}{2}} = \left(\frac{8RT}{\pi M}\right)^{\frac{1}{2}} \qquad (2\text{-}3.4.9)$$

$$c_{rms} = (\overline{c^2})^{\frac{1}{2}} = \left(\frac{3kT}{m}\right)^{\frac{1}{2}} = \left(\frac{3RT}{M}\right)^{\frac{1}{2}} \qquad (2\text{-}3.4.10)$$

Note that the average of the squares of the speeds [Eq.(2-3.4.8)] is the same as the average of the squares of the three-dimensional velocities. Thus, if we want to calculate the average kinetic energy of translation we need only evaluate

$$\bar{E}_k(\text{tr}) = \frac{1}{2}\, m\overline{c^2} = \frac{3}{2}\, kT \qquad (2\text{-}3.4.11)$$

as given previously. It is also easy to show (do so) that

$$E_{k_x}(\text{tr}) = E_{k_y}(\text{tr}) = E_{k_z}(\text{tr}) = \frac{1}{2}\, kT \qquad (2\text{-}3.4.12)$$

Each direction of translational motion contributes $\frac{1}{2}kT$ to the kinetic energy and each of these three directions corresponds to a *degree of freedom* which will be considered in more depth later.

2-3.5 THE MOST PROBABLE VELOCITY AND SPEED

If we want to predict the most probable value of a variable that depends on one of the velocity components, we apply the same procedure as in finding a maximum value of any function. To find v_{mp} (the most probable value of v) we do the following: set $[\partial f(u, v, w)]/\partial v = 0$ and solve for v. The result is $v = 0$ (show that this is true).

Similarly, for c_{mp} [from Eq. (2-3.2.17)], set

$$\frac{\partial f(c)}{\partial c} = \frac{\partial (dN_c/Ndc)}{\partial c} = 0$$

Solve for c to show

$$c_{mp} = \left(\frac{2kT}{m}\right)^{1/2} = \left(\frac{2RT}{M}\right)^{1/2} \qquad (2\text{-}3.5.1)$$

2-3.6 SOME NUMERICAL RESULTS

Here, all the averages and most probable values for He at 298 K are given ($M = 4.00\ \text{g mol}^{-1} = 0.00400\ \text{kg mol}^{-1}$).

$$\bar{u} = \bar{v} = \bar{w} = 0$$

(The distribution function is symmetrical about u, v, w = 0; as many positive values as negative values are expected. This leads to an average of zero.)

$$\overline{u^2} = \overline{v^2} = \overline{w^2} = \frac{RT}{M} = \frac{(8.31\ \text{J mol}^{-1}\ \text{K}^{-1})(298\ \text{K})}{(0.00400\ \text{kg mol}^{-1})}$$

$$= \frac{(8.31)(298)}{0.004}\ \frac{\text{kg m}^2\ \text{s}^{-2}}{\text{kg}}$$

$$= \underline{\underline{6.19 \times 10^5\ \text{m}^2\ \text{s}^{-2}}}$$

$$u_{rms} = v_{rms} = w_{rms} = \overline{(u^2)}^{1/2} = \underline{\underline{7.87 \times 10^2 \text{ m s}^{-1}}}$$

$$\overline{c^2} = \frac{3RT}{M} = \frac{(3)(8.31)(298)}{(0.004)} \text{ m}^2 \text{ s}^{-2} = \underline{\underline{1.86 \times 10^6 \text{ m}^2 \text{ s}^{-2}}}$$

$$c_{rms} = \overline{(c^2)}^{1/2} = \underline{\underline{1.36 \times 10^3 \text{ m s}^{-1}}}$$

$$\bar{c} = \left(\frac{8RT}{\pi M}\right)^{1/2} = \underline{\underline{1.26 \times 10^3 \text{ m s}^{-1}}}$$

(c is a scalar and can have only positive values; thus $\bar{c} > 0$.)

$$c_{mp} = \left(\frac{2RT}{M}\right)^{1/2} = \underline{\underline{1.11 \times 10^3 \text{ m s}^{-1}}}$$

Note that the following holds:

$$c_{rms}:\bar{c}:c_{mp} = 1.00:0.92:0.82$$

2-3.7 MOLECULAR COLLISIONS

Initially molecules will be treated as hard spheres. Of course they are not, but the results derived here are sufficiently exact for a large number of uses. In our later study of chemical kinetics, the importance of knowing the number of collisions per second will become clear. How do we proceed to find the number of collisions a molecule undergoes? First, assume each molecule has an "effective" or "collision" diameter of σ. A collision occurs whenever molecules approach each other closely enough to repel each other with a significant force. In Fig. 2-3.7.1 several molecules are shown. Center your attention on molecule A. As A travels through space, each molecule that has its center within the cylinder of radius σ (= the collision diameter of A) will collide with A. (B will collide with A, but C will not.) Let the length of the cylinder be $\bar{c}$ (where $\bar{c}$ is the average speed of A) which is the distance molecule A will travel in 1 s. You should conclude that any molecule whose center is in the cylinder of radius σ and length $\bar{c}$ will collide with A in a second. The volume of this cylinder is $\pi\sigma^2\bar{c}$. The number of molecules in a unit volume is given by the ideal gas equation. Designate this number as N^*. Thus the number of

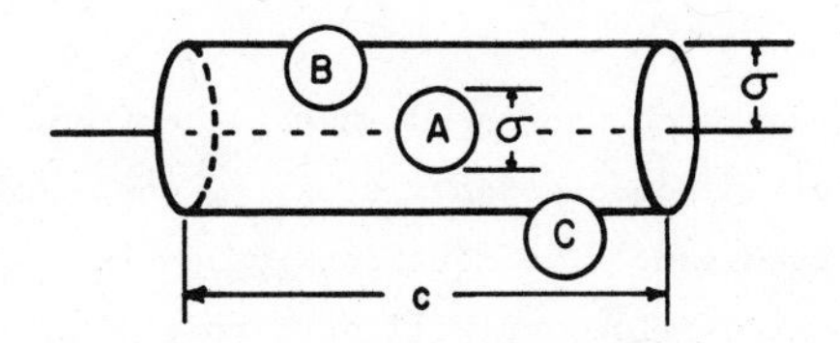

FIG. 2-3.7.1 Molecular collisions.

collisions per second (the *collision number* or *collision frequency*) is

$$Z_1 = \pi\sigma^2\bar{c}N^* \tag{2-3.7.1}$$

This has to be modified slightly to take into account the relative motions of A with respect to the other molecules, giving rise to a factor of $\sqrt{2}$ [see Moore (Sec. 2-5.6, Ref. 3).] Thus,

$$Z_1 = \sqrt{2}\pi\sigma^2\bar{c}N^* \tag{2-3.7.2}$$

To find the number of collisions per unit time per unit volume, we must recognize that each of the N^* molecules in the volume undergoes Z_1 collisions (we multiply by ½ to eliminate counting each collision twice: think about that for a moment). Thus, the collisions per second per unit volume is

$$Z_{11} = \frac{1}{2}Z_1N^* = \frac{\sqrt{2}}{2}\pi\sigma^2\bar{c}(N^*)^2 \tag{2-3.7.3}$$

2-3.8 MEAN FREE PATH

An important property of gases is the distance travelled between collisions. This is called the *mean free path* and is designated by the symbol λ. The derivation of this quantity is very straightforward. The number of collisions of a molecule in unit time is $Z_1 = \sqrt{2}\pi\sigma^2\bar{c}N^*$ from (2-3.7.2). Look at the units to find the form we need.

$$\lambda(\text{m}) \qquad Z_1(\text{s}^{-1}) \qquad \bar{c}(\text{m s}^{-1})$$

It follows directly that

$$\lambda(\text{m}) = \frac{\bar{c}\ (\text{m s}^{-1})}{Z_1\ (\text{s}^{-1})}$$

Thus

$$\lambda = \frac{\bar{c}}{Z_1} = \frac{1}{\sqrt{2}\pi\sigma^2N^*} \tag{2-3.8.1}$$

Notice that λ, Z_1, and Z_{11} depend on the collision diameter, σ. Our next task is to find one additional relation that will allow us to determine σ. This involves viscosity, a quantity that can be measured experimentally.

2-3.9 GAS VISCOSITY

The kinetic theory of gas viscosity will be the approach presented in this section. If you are interested in flow properties and their dependence on viscosity, you should consult Sec. 14-1.5 where liquid viscosity is discussed. The equations given in that section fit gases as well. To follow the derivation below you will have to employ your imagination a bit.

First, suppose we can divide a gas sample into layers. Also, suppose a velocity gradient exists between these layers (one layer is moving faster than the other). For convenience, let the distance separating them be λ, the mean free path. Finally, imagine a unit area on two adjacent layers. Figure 2-3.9.1 is included for those who have trouble visualizing this system (it also gives symbols for things to be discussed).

Molecules are not limited to movement in the planes of the layers. So occasionally one will traverse the distance λ and end up (or down) in another layer. A molecule from A will tend to slow down layer B, if it jumps from A to B (you should be able to tell why). Conversely, a molecule from B will tend to speed up A if it makes the jump from B to A. (If you have ever stepped off a moving vehicle you experienced this effect when your foot touched the ground.) There is a transport of momentum (mass × velocity) whenever a molecule goes from A to B or B to A in one of its mean free paths between collisions (that is why λ was cleverly chosen as the separation between layers).

The net result of numerous exchanges between layers brought about by random thermal motion is an increase in the average velocity of A and a decrease in the average velocity of B. The momentum transport (identified by the term "drag" in fluid flow) tends to counteract the velocity gradient (brought on by shearing forces of some type).

The remainder of the derivation is a simplified approach to a complex problem. The result is of sufficient importance to justify a derivation, albeit a not too exact one.

If a molecule of mass m moves from B to A, it carries with it excess momentum $m\ \Delta v$, which is $m\lambda(dv/dx)$. The number of molecules transversing λ per unit time (up and down) is $\frac{1}{2}N^*\bar{c}$. The factor $\frac{1}{2}$ comes from the fact that in each layer only half the molecules are moving up (or down) while the other half are moving in the opposite direction. Hence, the total momentum transport per second per unit area is

$$f = \frac{1}{2}\,(N^*\bar{c})\left(m\lambda\,\frac{dv}{dx}\right) \qquad (2\text{-}3.9.1)$$

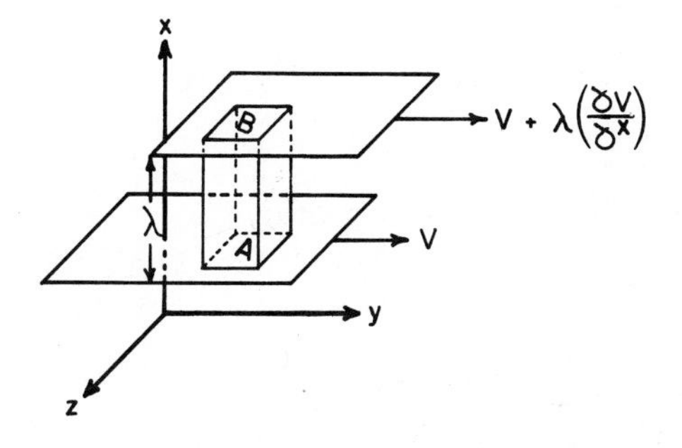

FIG. 2-3.9.1 Model for kinetic theory of gas viscosity.

(Note that this momentum transport per second per unit area is a rate of change of momentum and is, therefore, a force. You might want to check the units to see that force units actually result.)

From fluid mechanics we can find that the force resisting flow in fluids is given by (see Sec. 14-1.5)

$$f = \eta \frac{dv}{dx} \tag{2-3.9.2}$$

where η is called the *coefficient of viscosity*. Its units are kg m^{-1} s^{-1} (in CGS units, the unit is 1 poise = g cm^{-1} s^{-1}).

Equating (2-3.9.1) and (2-3.9.2) gives

$$\eta \frac{dv}{dx} = \frac{1}{2} m\lambda\bar{c}N^* \frac{dv}{dx}$$

or

$$\eta = \frac{1}{2} m\lambda\bar{c}N^* \tag{2-3.9.3}$$

Finally, combine (2-3.8.1) with (2-3.9.3) to get

$$\eta = \frac{1}{m} \frac{1}{\sqrt{2}\pi\sigma^2 N^*} \bar{c}N^* = \frac{m\bar{c}}{2\sqrt{2}\pi\sigma^2} \tag{2-3.9.4}$$

Equation (2-3.9.4) shows some interesting properties of viscosity. Note first there is no dependence on gas density, a rather surprising result on first thought. However, it is reasonable if you consider the model carefully and the effect of density on both the separation of layers λ and the number of molecules jumping from one layer to the next. Making the result reasonable has been left as an exercise for you.

Second, the only dependence on T is through the term $\bar{c}$ which varies as $T^{1/2}$. Therefore, we expect η to vary as $T^{1/2}$.

Experimentally, the first conclusion holds for almost-ideal and ideal gases. The actual dependence of η on T is a little greater than $T^{1/2}$, due to the effect of more energetic collisions (higher T) on the collision diameter, σ. The molecules are not hard spheres, and σ decreases slightly with T. (Can you imagine why this should be so?)

Equation (2-3.9.4) allows a determination of σ by measuring the viscosity of a gas. Once this is done, λ, Z_1, and Z_{11} can be found.

2-3.10 NUMERICAL VALUES OF σ, λ, Z_1, and Z_{11}

The viscosity of Ar is, according to Table 2-3.10.1 (at 298 K, 1 atm),

$$\eta = 22.7 \times 10^{-6} \text{ kg m}^{-1} \text{ s}^{-1}$$

$$N^* = \frac{N}{V} = \frac{N}{RT/P} = \frac{(6.023 \times 10^{23} \text{ molecules mol}^{-1})(1 \text{ atm})}{(0.0821 \text{ dm}^3 \text{ atm mol}^{-1} \text{ K}^{-1})(298 \text{ K})}$$

$$= 2.46 \times 10^{22} \text{ molecules dm}^{-3}$$

$$= \underline{\underline{2.46 \times 10^{25} \text{ molecules m}^{-3}}} \quad (40.8 \text{ mol m}^{-3} = 0.0408 \text{ mol dm}^{-3})$$

$$\bar{c} = \left(\frac{8RT}{\pi M}\right)^{1/2} = \left[\frac{(8)(8.31 \text{ J mol}^{-1} \text{ K}^{-1})(298 \text{ K})}{(3.14)(0.040 \text{ kg mol}^{-1})}\right]^{1/2}$$

$$= \underline{\underline{397 \text{ m s}^{-1}}} \quad (1430 \text{ km hr}^{-1} = 890 \text{ mile hr}^{-1})$$

$$m = \frac{M}{N} = \frac{0.040 \text{ kg mol}^{-1}}{6.023 \times 10^{23} \text{ molecules mol}^{-1}} = \underline{\underline{6.64 \times 10^{-26} \text{ kg mol}^{-1}}}$$

$$\sigma = \left(\frac{\bar{c}m}{2\sqrt{2}\pi\eta}\right)^{1/2} \quad \text{from Eq. (2-3.9.4)}$$

$$= \left[\frac{(397 \text{ m s}^{-1})(6.64 \times 10^{-26} \text{ kg})}{2\sqrt{2}(3.14)(22.7 \times 10^{-6} \text{ kg m}^{-1} \text{ s}^{-1})}\right]^{1/2}$$

$$= (1.31 \times 10^{-19} \text{ m}^2)^{1/2} = \underline{\underline{3.62 \times 10^{-10} \text{ m}}} \quad [3.62 \text{ angstrom } (\text{Å})]$$

Bond lengths in molecules are on the order of 1 to 2 Å. Molecule is a unit that is often ignored in calculations.

$$\lambda = \frac{1}{\sqrt{2}\pi\sigma^2 N^*}$$

$$= \frac{1}{\sqrt{2}\,(3.14)(3.62 \times 10^{-10} \text{ m})^2\,(2.46 \times 10^{25} \text{ m}^{-3})}$$

$$= \frac{1}{1.37 \times 10^7 \text{ m}^{-1}} = \underline{\underline{6.99 \times 10^{-8} \text{ m}}} \quad (699 \text{ Å})$$

$$Z_1 = \sqrt{2}\pi\sigma^2\bar{c}N^*$$

$$= \sqrt{2}\,(3.14)(3.62 \times 10^{-10} \text{ m})^2\,(397 \text{ m s}^{-1})(2.46 \times 10^{25} \text{ m}^{-3})$$

$$= \underline{\underline{5.68 \times 10^9 \text{ s}^{-1}}}$$

(The number of collisions each molecule suffers in each second.)

$$Z_{11} = \frac{1}{2}\sqrt{2}\pi\sigma^2\bar{c}(N^*)^2 = \frac{1}{2}N^* Z_1$$

$$= \frac{1}{2}\,(2.46 \times 10^{25} \text{ m}^{-3})(5.68 \times 10^9 \text{ s}^{-1})$$

$$= \underline{\underline{6.99 \times 10^{34} \text{ m}^{-3} \text{ s}^{-1}}}$$

(The total number of collisions each second in each cubic meter.)

Values of some kinetic molecular gas properties are given in Table 2-3.10.1.

Table 2-3.10.1
Values of Kinetic Molecular Gas Properties at 25°C, and 1 atm

Molecule	$\eta \times 10^6$ (kg m^{-1} s^{-1})	$\sigma \times 10^{10}$ (m)	$\lambda \times 10^8$ (m)	$Z_1 \times 10^{-9}$ (Collision s^{-1})	$Z_{11} \times 10^{-34}$ (collision m^{-3} s^{-1})
O_2	20.8	3.57	7.14	6.1	7.5
CO_2	15.0	4.56	4.41	8.6	10.6
N_2	17.8	3.74	6.50	7.3	9.0
H_2	9.0	2.73	12.3	14.4	17.7
HI	17.2	3.50	7.46	3.0	3.7
He	19.7	2.18	19.0	6.6	8.1
Ar	22.7	3.54	7.31	5.2	6.4

It is appropriate for you to exercise your new skill in calculating gas properties. Problems in the following section offer you that opportunity.

2-3.11 PROBLEMS

1. Calculate all the kinetic molecular gas properties of H_2 from the value of $\eta = 9.0 \times 10^{-6}$ kg m^{-1} s^{-1} at 298.15 K and 1 atm. Compare your results to those listed in Table 2-3.10.1.
2. Assume interstellar space contains four H atoms per cubic meter. Determine whether an H atom released by our sun during a solar flare will collide with another H atom before reaching our moon. Assume $\sigma = 1 \times 10^{-10}$ m, sun to moon distance = 1.56×10^8 km, and the diameter of the moon is 3600 km.
3. Estimate the number of O_2-O_2 collisions and N_2-N_2 collisions per second at 0°C and 0.95 atm pressure in our atmosphere.
4. What pressure would be necessary in a vacuum system to insure a distance of 10^5 m between collisions of He atoms?
5. Calculate the average speed and root mean square velocity for CO_2 gas at 325 K and 10^6 N m^{-2} pressure.
6. Calculate the density at 0.80 atm of a gas which has a molecular weight of 60 if the root mean square velocity is 340 m s^{-1}.
7. For a given gas, $\bar{c}$ = 380 m s^{-1} and $\lambda = 7.309 \times 10^{-8}$ m. Determine the value of Z_1.
8. Find the mean free path for N_2 (M = 0.028 kg mol^{-1}) at 2.2×10^{-4} atm and 100 K if $\sigma = 3.74 \times 10^{-10}$ m.

9. At 298 K and 1.00 atm the collision diameter of HI ($M = 0.128$ kg mol^{-1}) is 5.55 $\times 10^{-10}$ m. Determine the coefficient of viscosity.

10. Assuming the collision diameter is independent of temperature and P is constant, find the value of Z_{11} at 600 K if the value is 9.54×10^{34} m^{-3} s^{-1} at 300 K.

2-3.12 SELF-TEST

1. Calculate the average and root mean square speeds of Kr ($M = 84$ g mol^{-1}) at 326 K.

2. Estimate the collisions per second for a gas whose molecular weight is 20 g mol^{-1} if 3 mol are confined to a volume of 35 liters at 100 K. Also find the collisions per second per unit volume for this case ($\sigma = 2 \times 10^{-10}$ m).

3. Calculate the viscosity of a gas whose collision diateter is 8.2×10^{-10} m. From this information and the fact that $\lambda = 6.2 \times 10^{-8}$ m, find the density of the gas (in molecules m^{-3}). ($M = 0.084$ kg mol^{-1}, $T = 298$ K.)

2-3.13 ADDITIONAL READING

1.* F. T. Wall, *Chemical Thermodynamics*, W. H. Freeman and Company, San Francisco, 1958, Chapter 11.

2-4 AVERAGE VALUES

OBJECTIVES

You shall be able to use a given Maxwell-Boltzmann velocity distribution to find

1. The average value for velocity, speed, and/or kinetic energy for a system of gas molecules
2. The most probable value for velocity, speed, etc.
3. The fraction of molecules which have velocities, speeds, etc., above (or below) a given value

There is no additional reading material for this section. You are required to apply the definitions found in Sec. 2-3 for solving problems of the type in this section. Sections 2-3.1 through 2-3.5 specifically apply to this section.

2-4.1 PROBLEMS

1. Assume for a strange, new gas, scientists observe that the distribution function for velocity (one-dimensional) is

$$\frac{dN_u}{N} = c' \exp\left(-\frac{3mu^2}{2kT}\right)$$

where c′ is a constant. Find $\overline{u^2}$ and u_{rms} at 300 K if the molecular weight is 36 g mol^{-1}.

2-4.2 SELF-TEST

1. Use the distribution function and conditions given in Sec. 2-4.1, Problem (1), to find a value for u_{mp}.

2-5 EFFUSION AND DIFFUSION

OBJECTIVES

You shall be able to estimate rates of effusion and/or diffusion for gases.

2-5.1 INTRODUCTION

Two types of flow will concern us in this section. These are *diffusion* and *effusion* which involve flow through a porous wall or tiny orifice. In the case of diffusion, an appreciable amount of material is lost through the porous wall of the container (or through a membrane). In effusive flow, on the other hand, so little material is lost that none of the properties of the gas within the container is changed appreciably. Effusion is a molecular flow and is important in molecular beam studies.

Diffusion is important in many systems in which concentration gradients exist. Diffusion occurs any time there is a concentration gradient whether a membrane or porous wall exists or not. Many reaction rates are partially diffusion controlled, hence it is necessary to be able to estimate diffusion rates if the rate of a given reaction is to be estimated in many cases. Many processes in biological systems depend on diffusion for the transport of materials. Controlled diffusion is also important in the construction of transistors and other solid state electronic devices. Another significant diffusion process is that of O_2, through the membranes in the lungs into the bloodstream.

2-5.2 EFFUSION

As mentioned, effusion is a type of molecular flow through a tiny orifice in which very little material is lost in unit time relative to the total amount in the system. Effusive flow occurs when the molecules in the container have no knowledge of the orifice, that is, there is no dynamic flow. The number of molecules escaping depends only on the number randomly striking the orifice, as shown in Fig. 2-5.2.1. This

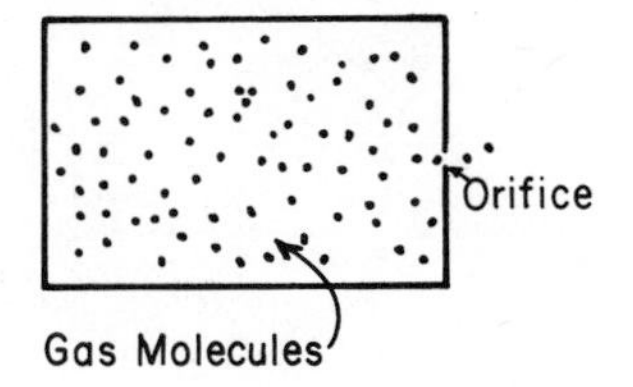

FIG. 2-5.2.1 Effusion of molecules through a tiny orifice. Escape through the hole depends on random collisions.

number is related to the average speed of the molecules within the container.

Experimentally it is difficult to measure absolute rates of effusion through well-characterized orifices. Hence, relative effusion rates are usually determined. In 1829, Graham found that at a given temperature and pressure drop the rates of effusion of gases are inversely proportional to the square root of their densities.

$$\frac{\text{Rate}_1}{\text{Rate}_2} = \left(\frac{\rho_2}{\rho_1}\right)^{1/2} \tag{2-5.2.1}$$

From the ideal gas law

$$\rho = \frac{MP}{RT}$$

Thus

$$\frac{\text{Rate}_1}{\text{Rate}_2} = \left(\frac{M_2}{M_1}\right)^{1/2} \tag{2-5.2.2}$$

This equation shows the rate of effusion of a gas to be inversely proportional to the square root of its molecular weight. (From this fact you should be able to see why vacuum systems are usually leak-tested with helium.)

2-5.3 DIFFUSION OF GASES

A far more important form of movement of molecules through porous barriers is *diffusion*. The driving force for diffusion is the random thermal motion of the molecules. A barrier is not a necessity for diffusion. Diffusion will occur any time there is a concentration difference between regions of a system.

FICK'S LAWS OF DIFFUSION. Consider a cross-section within a gas of area A, in the yz plane (Fig. 2-5.3.1). Assume that a concentration (pressure) difference exists only along the x direction. Diffusion will tend to eliminate this concentration gradient by moving some of the molecules from the higher to the lower concentration region.

To formulate a mathematical description of what happens, let us focus our attention on the quantity Q, which is the amount of gas which crosses the area A in the x

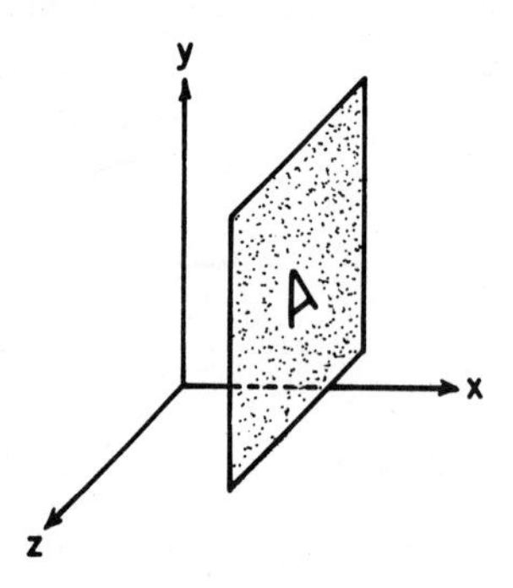

FIG. 2-5.3.1 A cross-section of area A in the yz plane within a gas.

direction in Fig. 2-5.3.1. Now, if $\partial c/\partial x$ is negative (concentration c is greater for smaller x), then Q will be positive. You should agree that Q is proportional to A and proportional to the time (if nothing else changes) since the rate should be constant in the absence of any other changes. If we define a flux J as the transport of material across unit area in unit time, then

$$Q = AJt \tag{2-5.3.1}$$

We need only worry about determining the flux and how it varies with different parameters to be able to describe the diffusion across A.

If $\partial c/\partial x$ equals zero, there should be no flux. Thus we expect J to be proportional to $\partial c/\partial x$, and experimentally this is found to be true. We can, therefore, write

$$J = -D \frac{\partial c}{\partial x} \tag{2-5.3.2}$$

D is the diffusion coefficient. This, or the result of combining (2-5.3.2) and (2-5.3.1) to give

$$Q = -DA\left(\frac{\partial c}{\partial x}\right)t \tag{2-5.3.3}$$

is known as *Fick's first law of diffusion*. This law gives us a means of determining the flux or the permeation Q through any region in a system. D depends, to a small degree, on T and P (and other system parameters in solutions), but we will ignore these at present.

Let us expand Fig. 2-5.3.1 to include another area A separated by dx from that shown. Draw such a figure before continuing. We want to determine the change within the volume defined due to movement into or out of it. The flux at A_x will be slightly different from that at A_{x+dx}. Thus we will have a flux gradient $\partial J/\partial x$ across dx. Finally, assume A is a unit area so that the volume defined is A dx = dx. J_x will represent the flux at A_x, and J_{x+dx} is that at A_{x+dx}. The amount of material entering A_x in time dt is J dt and that leaving A_{x+dx} is $[J + (\partial J/\partial x)\, dx]\, dt$. The net change in the amount in the volume dx (dx is a volume since A is a unit area) will be the difference which is $-\partial J/\partial x \, dx \, dt$. Since the volume is dx, we find the change in concentration c by dividing the amount $(-\partial J/\partial x \, dx \, dt)$ by dx. The result is

$$dc = -\left(\frac{\partial J}{\partial x}\right) dt \qquad (2\text{-}5.3.4)$$

We can also find dc by

$$dc = \left(\frac{\partial c}{\partial t}\right) dt$$

Equating these two will give

$$\frac{\partial c}{\partial t} = -\frac{\partial J}{\partial x} \qquad (2\text{-}5.3.5)$$

This implies that the concentration varies with time as the flux varies with distance.

If we take $\partial/\partial x$ of Eq. (2-5.3.2) we get

$$\frac{\partial J}{\partial x} = -D\frac{\partial^2 c}{\partial x^2} \qquad (2\text{-}5.3.6)$$

where we have used A as a unit area and assumed D is constant. The last two expressions, when combined, give *Fick's second law*:

$$\frac{\partial c}{\partial t} = D\frac{\partial^2 c}{\partial x^2} \qquad (2\text{-}5.3.7)$$

The change in concentration at a given point with time is related to the diffusion coefficient times the slope of the concentration gradient [$\partial c/\partial x$ = concentration gradient: $(\partial/\partial x)(\partial c/\partial x)$ = slope of concentration gradient].

BROWNIAN MOTION. Random thermal motion of molecules can lead to diffusion if a concentration gradient exists. Brownian motion is the zig-zag motion of molecules that results from random collisions between molecules within a system. This motion exists in gases, but obviously cannot be seen. However, for very small particles (just visible under a microscope), Brownian motion can be observed. This motion can be described mathematically, and the resulting equations have the same form as the equation describing random errors in measurements. This might have been expected since molecular bombardment is a random process. Such a development is beyond the scope of this book. The concept is illustrated in the following derivation which follows that found in K. J. Mysel's *Introduction to Colloid Chemistry* (see Sec. 2-5.6, Ref. 1). A further description of Brownian motion is found in that reference.

Consider the system shown in Fig. 2-5.3.2. In this figure the initial concentration gradient is found by noting that there are n particles at level zero, $n - \Delta n$ at ℓ above zero and $n + \Delta n$ at ℓ below. The gradient will be

$$\frac{(n - \Delta n) - (n + \Delta n)}{2\ell} = -\frac{\Delta n}{\ell}$$

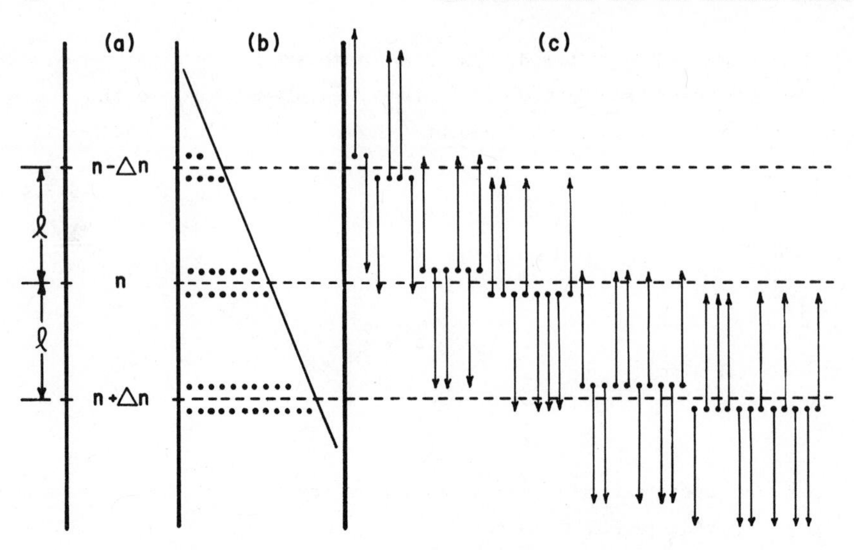

FIG. 2-5.3.2 Schematic diagram for the derivation of Einstein's law of Brownian motion. (a) Concentration at distance ℓ, the average distance travelled in time t. (b) Average concentration and position at t = 0. (c) The result of random motion of each particle after each has travelled ℓ in one of two possible directions. Note that a net permeation in the upward (decreasing concentration) direction results.

$$\frac{\partial c}{\partial x} = -\frac{\Delta n}{\ell} \tag{2-5.3.8}$$

To calculate the permeation Q on the basis of random motion, consider part (c) of the figure. Assume all molecules in time t move a distance ℓ up *or* down. If the motion is random, half will be up and half down. The number of particles below the zero level and within ℓ before movement is given by the volume times the average concentration. The volume is ℓ (if we assume unit dimensions for the other two directions). The average concentration is $n + \Delta n/2$. Thus the number of particles is volume times concentration: $\ell(n + \Delta n/2)$. The number above zero within ℓ is $\ell(n - \Delta n/2)$.

It was mentioned above that half the particles move up and half move down. So half the particles above zero, $[\frac{1}{2}\ell(n - \Delta n/2)]$, and half below zero, $[\frac{1}{2}\ell(n + \Delta n/2)]$, cross zero. The permeation Q will be the difference, or

$$Q = \frac{1}{2}\ell\left(n + \frac{\Delta n}{2}\right) - \frac{1}{2}\ell\left(n - \frac{\Delta n}{2}\right) = \frac{\ell\,\Delta n}{2} \tag{2-5.3.9}$$

Fick's law for this system states that the number crossing the zero level upward will give the permeation Q. So, from (2-5.3.8):

$$Q = -D\left(\frac{\partial c}{\partial x}\right)t = \frac{Dt\,\Delta n}{\ell} \tag{2-5.3.10}$$

Equating (2-5.3.9) and (2-5.3.10) gives

$$Dt\frac{\Delta n}{\ell} = \frac{\ell \Delta n}{2}$$

or

$$D = \frac{\ell^2}{2t} \qquad (2\text{-}5.3.11)$$

This is *Einstein's law of Brownian motion.* The equation relates the average distance travelled in a given time to the diffusion coefficient. Note that ℓ is not the same as λ, the mean free path, studied in Sec. 2-3. λ is the distance between collisions, while ℓ is the distance moved as a result of a number of collisions.

Hopefully, you can see the significance of this result. If we can determine the distance (average) travelled in time t, we can get diffusion coefficients or vice versa. The result is good for gases, liquids, and solids. You will see it again.

DIFFUSION COEFFICIENTS FOR GASES. The following result is given without proof. For a derivation see Sheehan (Sec. 2-5.6, Ref. 2). The self-diffusion coefficient of a gas depends on the root mean square velocity and the mean free path as follows:

$$D = 0.543 c_{rms} \lambda \qquad (2\text{-}5.3.12)$$

The self-diffusion coefficient is the coefficient for a gas diffusing into itself. This diffusion is due to a concentration difference between regions of the system, such as across a porous membrane. The self-diffusion coefficient does not necessarily have the same value as the coefficient of that gas diffusing into another gas. For more detail see Moore (Sec. 2-5.6, Ref. 3).

Now that you have finished this section, reread the derivation of Fick's laws of diffusion to see the correlations.

2-5.3.1 EXAMPLE

From data given previously for He (Sec. 2-3), find the self-diffusion coefficient at 298.15 K and 1 atm.

Solution

From Eq. (2-3.4.10)

$$c_{rms} = \left(\frac{3RT}{M}\right)^{1/2} = \left[\frac{(3)(8.31\ \text{kg m}^2\ \text{s}^{-2}\ \text{mol}^{-1}\ \text{K}^{-1})(298.15\ \text{K})}{0.004\ \text{kg mol}^{-1}}\right]^{1/2}$$

$$= \underline{\underline{1.36 \times 10^3\ \text{m s}^{-1}}}$$

Table 2-3.10.1 gives $\sigma = \underline{\underline{2.18 \times 10^{-10}\ \text{m}}}$

$$N^* = \frac{NP}{RT} = \frac{(6.023 \times 10^{23}\ \text{molecules mol}^{-1})(1\ \text{atm})}{(0.0821\ \text{dm}^3\ \text{atm mol}^{-1}\ \text{K}^{-1})(298.15\ \text{K}^{-1})}$$

$$= \underline{\underline{2.46 \times 10^{22}\ \text{molecules dm}^{-3}}}$$

From Eq. (2-3.8.1):

$$\lambda = \frac{1}{\sqrt{2}\pi\sigma^2 N^*} = \frac{1}{\sqrt{2}\pi(2.18 \times 10^{-10}\ \text{m})^2\ (2.46 \times 10^{22}\ \text{molecules dm}^{-3})}$$

$$= (1.92 \times 10^{-4}\ \text{dm}^3\ \text{m}^{-2})\left(\frac{\text{m}}{10\ \text{dm}}\right)^3$$

$$= \underline{\underline{1.92 \times 10^{-7}\ \text{m}}}$$

which agrees with Table 2-3.10.1. (The practice probably did you good.) Then

$$D = 0.543 c_{rms}\lambda = (0.543)(1.36 \times 10^3\ \text{m s}^{-1})(1.92 \times 10^{-7}\ \text{m})$$

$$= \underline{\underline{1.42 \times 10^{-4}\ \text{m}^2\ \text{s}^{-1}}}$$

(The units will make more sense if you do a unit analysis on Eq. (2-5.3.3) with the knowledge that Q is in molecules or moles depending on the units chosen for $\partial c/\partial x$.)

2-5.3.2 EXAMPLE

Find the average distance a molecule of He will diffuse (self-diffusion) in 30 s using the conditions of Example 2-5.3.1.

Solution

$$D = 1.42 \times 10^{-4}\ \text{m}^2\ \text{s}^{-1}$$

$$D = \frac{\ell^2}{2t} \qquad (2\text{-}5.3.11)$$

Thus $\ell = (2Dt)^{1/2}$

$$= [(1.42 \times 10^{-4}\ \text{m}^2\ \text{s}^{-1})(30\ \text{s})]^{1/2}$$

$$= (8.5 \times 10^{-3}\ \text{m}^2)^{1/2}$$

$$= \underline{\underline{9.2 \times 10^{-2}\ \text{m}}}$$

Note the difference in the two values. The mean distance travelled between collisions is $\lambda = 1.92 \times 10^{-7}$ m, while the average distance each molecule travels from its initial location is $\ell = 9.2 \times 10^{-2}$ m.

2-5.3.3 EXAMPLE

Find the time required for 1.0 g He to diffuse across a porous plug 1.0 mm thick and 1.0 cm in diameter if the pressure on one side is 1 atm and that on the other is 10^{-5} atm; T = 298.15 K. Assume that the D found in the previous example is valid and that the plug has a porosity of 25%.

Solution

The permeation is given by Fick's first law [which is Eq. (2-5.3.3)]:

$$Q = -DA\left(\frac{\partial c}{\partial x}\right) t \qquad (2\text{-}5.3.3)$$

$$D = 1.42 \times 10^{-4}\ m^2\ s^{-1}$$

$$A = \pi r^2 = \pi(0.5\ cm)^2 = (7.85 \times 10^{-1}\ cm^2)\left(\frac{1m}{100\ cm}\right)^2$$

$$= \underline{\underline{7.85 \times 10^{-5}\ m^2}}$$

But the porosity is only 25%, so

$$A_{eff} = 0.25\ A = \underline{\underline{1.96 \times 10^{-5}\ m^2}}$$

We now need the concentration on each side of the plug.

$$N_1^* = \frac{P_1}{RT} = \frac{1\ atm}{(0.0821\ dm^3\ atm\ mol^{-1}\ K^{-1})(298.15\ K)}$$

$$= 4.08 \times 10^{-2}\ mol\ dm^{-3} = \underline{\underline{40.8\ mol\ m^{-3}}}$$

$$N_2^* = \frac{P_2}{RT} = 10^{-5}\ N_1^* = \underline{\underline{4.08 \times 10^{-4}\ mol\ m^{-3}}}$$

We will estimate the term $\partial c/\partial x$ by assuming the concentration gradient is linear. Thus

$$\frac{\Delta c}{\Delta x} = \frac{4.08 \times 10^{-4}\ mol\ m^{-3} - 40.8\ mol\ m^{-3}}{1.0\ mm}$$

$$= -\frac{40.8\ mol\ m^{-3}}{1 \times 10^{-3}\ m} = \underline{\underline{-\ 4.08 \times 10^4\ mol\ m^{-4}}}$$

Finally,

$$Q = \frac{1.0\ g}{4.0\ g\ mol^{-1}} = \underline{\underline{0.25\ mol}}$$

Now find t:

$$Q = -DA_{eff}\left(\frac{\Delta c}{\Delta x}\right) t$$

$$t = \frac{Q}{-DA_{eff}(\Delta c/\Delta x)}$$

$$= -\frac{0.25\ mol}{(1.4 \times 10^{-4}\ m^2\ s^{-1})(1.96 \times 10^{-5}\ m^2)(-4.08 \times 10^4\ mol\ m^{-4})}$$

$$= \underline{\underline{2.2 \times 10^3\ s}}$$

2-5.4 PROBLEMS

1. Estimate the rate of effusion of methane (CH_4) through a pin hole in a pipeline operating at 75 psig (14.7 psi = 1 atm) if a leak test with helium under the same conditions reveals that 0.55 dm^3 He escapes per minute. (Aside: If the gas sells for $1.25 per thousand cubic feet at STP, is the loss significant?)

2. The diffusion equations derived previously apply equally well to solutions. If the diffusion coefficient of serum albumin is 6.1×10^{-7} cm^2 s^{-1}, how long will it take (on the average) for a serum albumin molecule to diffuse 1 cm along a given direction?

3. Estimate the time required for 1 g sucrose ($C_6H_{12}O_6$) to diffuse from a solution of 100 g dm^{-3} into pure water through a porous glass disk 3 mm thick and 2 cm in diameter. The disk is about 33% open space.

4. How is effusion related to density? If the rate of effusion for a gas at 406 K and 2.23 atm is 1.45×10^{-10} mol min^{-1}, what will the rate be at 203 K and 2.23 atm from the same container?

5. For a gas with a diffusion coefficient of 8.2×10^{-11} m^2 s^{-1}, the permeation was found to be 2.5×10^{-5} mol m^{-2} s^{-1} across a certain barrier. Find the concentration gradient across that barrier.

2-5.5 SELF-TEST

1. A given biological polymer molecule (a natural macromolecule) requires about 10 hr. to diffuse 1 mm in a given direction. Estimate the diffusion coefficient.

2. What is the mean free path implied by a diffusion coefficient of 1.85×10^{-4} m^2 s^{-1} for a material with a molecular weight of 40 g mol^{-1} at 273 K ($\sigma = 2 \times 10^{-10}$ m)?

3. A gas with a molecular weight of 35 requires 87 s to effuse through a tiny hole. Another gas requires 145 s for the same amount to effuse through the same hole. What is the molecular weight of this second gas?

4. Determine the flux and permeation in 30 s for a gas across a boundary with an area of 0.034 m^2 under a concentration gradient of -4.3×10^{-2} mol m^{-4} if the gas has a diffusion coefficient of 3.5×10^{-10} m^2 s^{-1}.

2-5.6 ADDITIONAL READING

1.* K. J. Mysels, *An Introduction to Colloid Chemistry*, Interscience, New York, 1967, pp. 111-114. Reprinted by permission of John Wiley and Sons, Inc.

2.* W. F. Sheehan, *Physical Chemistry*, 2nd ed., Allyn and Bacon, Boston, 1970, Sec. 1.9.

3.* W. J. Moore, *Physical Chemistry*, 4th ed., Prentice-Hall, Englewood Cliffs, N. J., 1972, pp. 159-164.

2-6 SUMMARY

You have seen a progression from the simple ideal gas laws to more complicated (and better) equations of state. Sufficient material has been included for you to calculate properties of most gases under ordinary conditions. You should not hesitate to apply the Redlich-Kwong equation if you need P, V, and T relations for a given gas. It is not overly complex and, as you have seen, gives much better results than any of the other equations listed (including the much more complicated Beattie-Bridgeman equation). We will have occasion to use these gas laws extensively in our study of thermodynamics.

A simple model has been used to predict flow behaviors of gases. Also, the average speeds and velocities (and, therefore, the average kinetic energies) of various gases have been described. These concepts are important in any process where gases interact, as in chemical reactions. All the equations derived were for ideal systems. Real systems are more complicated, naturally. However, you have the background needed to study real systems when the occasion demands it.

Finally, we worked out a series of equations describing effusion and diffusion. The diffusion equations will be used again when the behavior of systems of large molecules (or particles) are studied: Colloids (Chap. 11) and Polymers (Chap. 12). As mentioned in the introduction to Sec. 2-5 diffusion is of paramount importance in problems of all types. These range from diffusing impurities into silicon to make transistors, to the diffusion of materials within biological systems.

Only an introduction has been given, as is the case in every chapter. When you meet specific problems, you will have to read material related to that problem. You should have the background to do that after successful completion of the appropriate sections in this text.

3

The First Law of Thermodynamics

SYMBOLS AND DEFINITIONS

E	Internal energy (J; dm^3 atm, 101 J = 1 dm^3 atm)
Q	Heat absorbed *by* system (J; dm^3 atm)
W	Work done *on* system (J; dm^3 atm)
H	Enthalpy = E + PV (J; dm^3 atm)
$\bar{C}_p$	Heat capacity at constant pressure (J mol^{-1} K^{-1})

$\bar{C}_V$	Heat capacity at constant volume (J mol^{-1} K^{-1})
(-)	Molar quantities
μ	Joule-Thomson coefficient (K atm^{-1})
γ	Heat capacity ratio = C_P/C_V
ΔH_f°	Heats of formation (kJ mol^{-1})

Thermodynamics is that portion of science concerned with the interrelationships among heat, mechanical energy, and potential energy. You will observe that thermodynamics is concerned with a system as a whole. No reference is made to the nature of the building blocks (or molecules) that comprise the system. Thermodynamics is concerned with *macroscopic* changes. It is independent of any model concerning the constituents that make up the system. For example, if we calculate the changes that occur in thermodynamic variables upon expanding a gas, we do not care whether we think the gas is made up of molecules or goblins. The result of our calculation is independent of what we choose to use as our model. In a later chapter on statistical thermodynamics (Chap. 9), we will combine our model of the atomic and molecular nature of matter to predict theoretically *microscopic* (atomic and molecular) model.

If you scan the contents of this book, you will observe that thermodynamics occupies more pages than any other subject. This is due both to its importance and to the conceptual difficulties of thermodynamics compared to other areas of study. The easiest way to grasp the meaning of thermodynamics is to work problems -- many problems. Below, several examples will be given. Follow them carefully. At times you may become confused as to what formula applies where. Do not dismay: almost everyone is in the same boat. The remedy is to go back to the basic definitions, apply all the restrictions given in the problem, then derive (or choose) the formula to solve it. You cannot (repeat, cannot) hope to solve thermodynamic problems by memorizing equations. There are too many possible combinations of conditions. Make a determined effort to obtain the equation needed from the definitions and from basic equations.

Thermodynamics is the logical formalization of everyday observations. The laws are simple in statement, but the ramifications are tremendous. The value of having the three laws of thermodynamics lies in the fact that our experimental observations are *systematized*.* One set of experimental measurements allows us to draw conclusions about the system far beyond our initial measurement. This is the purpose of formulating any law.

---

* The following "laws" governing everyday activities have been given. You will recognize some facts of life in them. Their significance will be more apparent after you have learned the laws of thermodynamics.

1. The first law says you can not win; the best you can do is break even.
2. The second law says you can break even only at absolute zero.
3. The third law says you can never reach absolute zero.

There is no need to emphasize the importance of thermodynamics. Almost any aspect of modern life is influenced by applications of thermodynamic principles: power generation, transportation, refining, chemical processing, etc. The aim of the chapters on thermodynamics is to introduce you to the subject and to indicate its utility. You can take the background you obtain here and delve into areas of interest in the future, as the need arises.

For those who like history, the beginning of Chap. 2 in the text *Physical Chemistry* by W. J. Moore (see Sec. 3-1.8, Ref. 1) will give some historical background to the development of thermodynamics.

3-1 AN INTRODUCTION

OBJECTIVES

You shall be able to

1. Define or discuss briefly any of the following: the law of conservation of energy; q, w, ΔE, and ΔH; the first law of thermodynamics; C_p and C_v
2. Calculate q, w, ΔE, and/or ΔH in the expansion or compression of an ideal gas

3-1.1 DEFINITIONS

Thermodynamics is a very precise subject. Terms and symbols are specific in their meanings. Practice using these in the proper way, and you will find a little less difficulty in working with thermodynamic equations. Section 3-1.2, Notes on Notation and Defining Process Paths in Thermodynamics, is very important.

We will deal with two types of energy in this portion of our study: *thermal* and *mechanical*. There are other types of energy (such as gravitational, magnetic, etc.), but these will not be considered here. At this point you will probably expect to find a definition of energy: "Energy is . . . ." Well, you will not. In fact, you can look in any physical chemistry text under "Energy, definition of," and you will find no such entry. Why not? Before considering that point further we will relate an anecdote taken from a book by H. C. van Ness, *Understanding Thermodynamics* (see Sec. 3-1.8, Ref. 2).

First, we will set the stage for this story. We have a boy and his mother, and the boy's room in the northwest corner of a country house. There are two windows in the room. One faces West (call this window W) and the other North (which we will call Q -- surprise!). The boy's name is Johnny. Outside window W is a small pond, and outside Q is a tree containing a family of squirrels. Figure 3-1.1.1 is given as an aid in setting the stage.

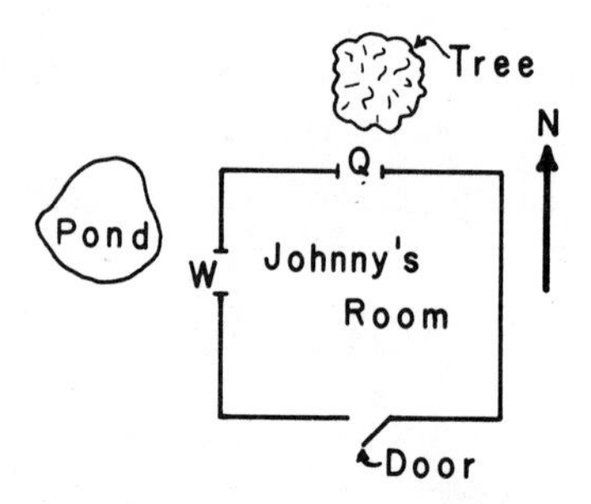

FIG. 3-1.1.1 Johnny's room.

Johnny always plays in his room. One day he asks his mother for some blocks, but his mother has none. Instead, she gives Johnny 25 sugar cubes and admonishes him not to eat any; she will count them every day to check up on him.

Johnny plays with these for days. Every day that week, when Mother checks, there are 25 cubes on the floor. However, on Monday when Mother checks she sees only 21 sugar cubes. She is about to take appropriate action when Johnny points to an old box. Mother starts to open it when Johnny protests, "No! No! Please don't."

Well, Mother is very understanding (and shrewd). She puts the box down, and the next day when she sees all 25 cubes on the floor she takes the box and weighs it. She also weighs a standard sugar cube and then returns both to Johnny's room. The following are the weights:

Box: 121 g
Sugar cube: 4.1 g

She then writes a formula that allows her to calculate the number of sugar cubes without counting those in the box. Her formula is

$$\text{No. on floor} + (\text{wt of box} - 121)/4.1 = 25 \tag{3-1.1.1}$$

On Wednesday she goes into the room and finds only 17 sugar cubes. She weighs the box and finds that it weighs 153.8 g. So she uses her formula

$$17 + (153.8 - 121)/4.1 = 17 + 32.8/4.1 = 17 + 8 = 25$$

and Johnny is spared any punishment. On Friday she checks the room. This time she finds 14 sugar cubes and a box weighing 157.9 g. Checking her formula:

$$14 + (157.9 - 121)/4.1 = 23$$

Johnny is about to get his. However, he manages to struggle free and point out of window W to the pond. Now, how is Mother going to tell if Johnny has eaten the cubes?

Well, Mother was once a nurse and knows how to measure sucrosity (sugar content). So she measures the pond's sucrosity, drops in a couple of cubes, stirs the pond well, and measures the sucrosity again to get k in the expression

$$\text{No. in pond} = k \text{ (sucrosity of pond)}$$

She then modifies her equation to read

$$\text{No. on floor} + \frac{\text{wt. of box} - 121}{4.1} + k \text{ (sucrosity of pond)} = 25 \qquad (3\text{-}1.1.2)$$

This works fine, but she soon gets tired of counting and decides to work only with changes in the various terms. She figures that all the changes have to add to zero. For example, if one disappears from the floor it has to show up in the box or in the pond. So,

$$\Delta(\text{no. on floor}) + \frac{\Delta(\text{wt. of box})}{4.1} + k \text{ } (\Delta \text{ sucrosity}) = 0 \qquad (3\text{-}1.1.3)$$

where Δ = change in.

This formula works fine until Thursday of the following week when (3-1.1.3) yields a value of -1. This time Johnny points to the window Q. Mother looks out Q and sees several squirrels running around. How can she possibly keep track of the number of sugar cubes the squirrels eat?

Fortunately her husband is an electrical engineer. He rigs up sensors on Q and W. Every time a sugar cube passes by the sensors a count is registered. So Mother can now write

$$\Delta(\text{no. on floor}) + \frac{\Delta(\text{wt. of box})}{4.1} + \text{no. passing Q} + \text{no. passing W} = 0 \qquad (3\text{-}1.1.4)$$

To simplify this she writes it as follows:

$$\Delta(\text{no. on floor}) + \frac{\Delta(\text{wt. of box})}{4.1} = -Q - W \qquad (3\text{-}1.1.5)$$

where Q and W are not changes, they are numbers or counts.

What Mother has accomplished up to this point is to define a few terms. She has defined the room as her *system* (we define a system as that part of the universe which we wish to study for a particular problem). Everything outside the room is defined as *surroundings* (we make the same definition: everything outside our system is surroundings). And finally, the walls of the room are the *boundary*. (We define boundary as the dividing line between system and surroundings. It may be a real boundary such as a beaker wall or an imaginary boundary.)

Notice that Mother does not care what happens to the surroundings. She concentrates on the system and what happens at the boundary. This gives her a problem that is not too complex to solve. If she had to worry about the squirrels, she could never keep track of what Johnny does with his sugar cubes.

About now, Johnny has fed the last cube to the squirrels. He asks for more. This time Mother dumps a handful of jelly beans in Johnny's box. She does not even count them this time, but she does instruct Johnny to keep the beans off the floor. Mother has faith in her formula

$$\frac{\Delta(\text{wt. of box})}{a} = -Q - W \qquad (3\text{-}1.1.6)$$

To determine a she merely weighs the box, sets Q and W to zero, and waits a few days. She can read Q and W, then reweigh the box and apply (3-1.1.6) to find a, the weight of one jelly bean. If a jelly bean leaves the box it must pass Q or W (if Johnny does not eat it, which is not allowed). This might be called

The Law of the Conservation of Lumps

Equation (3-1.1.6) will always work as long as Johnny does not eat a jelly bean or crush one on the floor. To bring us closer to where we want to go, let us rewrite (3-1.1.6):

$$f(\text{wt. of box}) = -Q - W \qquad (3\text{-}1.1.7)$$

where f(wt. of box) = wt. of box/a.

We now leave Johnny and his mother and start thinking about energy. We will write a formula similar to (3-1.1.7):

$$(\text{energy of system}) = -(\text{energy out by } Q) - (\text{energy out by } W) \qquad (3\text{-}1.1.8)$$

which can be called the "law of conservation of energy." But, what is energy? We have no definition. We do not have the faintest notion of what it is. We only say that whatever it is, it is conserved.

At this point we state that the energy (whatever it is) of a system should depend in some way on the variables of the system. In that case we write

$$\text{energy of system} = E(T, P, \text{etc.}) \qquad (3\text{-}1.1.9)$$

where this implies that E is a function of T, P, etc. So we rewrite (3-1.1.8) as

$$\Delta[E(T, P, \text{etc.})] = -Q - W$$

or

$$\Delta E = -Q - W \qquad (3\text{-}1.1.10)$$

where Q and W are the energies lost by system through windows Q and W, respectively.

Experience has shown that two ways a system has of losing or gaining energy is by losing or gaining *heat*, or the performance of work by (or on) the system. *Work* is defined as the product of the force component in a given direction and the distance a mass of matter is moved. *Heat* is defined as transfer of energy from one system to another (or to the surroundings) due to a temperature difference. We have now identified Q and W in (3-1.1.10). However, we need to be a little more explicit.

Simply by convention, Q is defined as the *heat gained by the system* and W is defined as the *work done on the system*. In both cases, a positive Q and W (heat gained *by* system and work done *on* system) will increase the internal energy of the system, making ΔE positive. We must, to be consistent, rewrite (3-1.1.10) as

$$\Delta[E(T, P, \text{etc.})] = Q + W \qquad (3\text{-}1.1.11)$$

We could chose any convention we like for Q and W, and as long as we are consistent, we would have no difficulty. Our convention is expressed in (3-1.1.11).

The energy expression E(T, P, etc.) includes all possible contributions to the energy of the system. It will include the energy due to the temperature of the system E(T), kinetic energy due to motion $\frac{1}{2}mv^2$, potential energy resulting from elevation mgz, energy due to the mass $\frac{1}{2}mc^2$, and any other energy terms that are possible. Normally we will be concerned only with systems at rest and those that do not change elevation or mass, so we do not have to worry about those terms. We always consider changes in the energy. Thus, if the elevation *does not change* and the velocity and mass *do not change*, they will not contribute to ΔE.

Another note on Q and W: W includes changes in energy due to changes in elevation, electrical potential, velocity, etc. Q is essentially the change in energy that is left over after all other kinds of energy have been accounted for. In other words, it is not necessary that Q be defined in terms of temperature as was done previously.

Have you noticed that we always talk about energy changes and conservation of energy? We still have not defined energy, and we will not. Thermodynamics tells us nothing about absolute energy values, only about *changes* in energy. The equation that expresses the first law is:

$$\boxed{\Delta E = Q + W} \qquad (3\text{-}1.1.12)$$

ΔE Depends on changes of system variables (T, P, etc.)

Q = Heat absorbed by system

W = Work done on system

This equation is a *statement of the first law of thermodynamics: the law of conservation of energy*.

We continue now with some more definitions. We can define three types of systems:

1. A *closed* system cannot exchange matter with the surroundings but energy can be exchanged.
2. An *open* system can exchange matter and energy with the surroundings.
3. An *isolated* system cannot exchange matter or energy with the surroundings.

We, just like Johnny's mother, ignore the surroundings. By worrying only about what happens in the system and what crosses the boundary we can define a problem of manageable size. This would not be possible if we had to consider everything in the surroundings (the universe).

We will refer to the *state of a system* many times in our study of thermodynamics. The state of a system is defined when a certain number of properties of the system are specified. The number that must be specified depends on how accurately we wish to detail the state or condition of the system. Normally, *intensive properties* are used in describing the state of a system. Intensive properties are those that do not depend on the amount of material present in the system. Examples of intensive properties are density, pressure, and temperature. In each of these, we do not have to say how much material is being considered. The pressure, for example, can be the same (and will be the same, at equilibrium) in two portions of a system containing different amounts of material.

There are other properties, called *extensive properties*, that do depend on the amount of material. Examples are mass and volume. The volume of two regions of a system will be different if the amounts of material in the regions is different. Perhaps Fig. 3-1.1.2 will help clarify the concepts of extensive and intensive variables.

These properties (intensive and extensive) can be used to define functions of state or *state functions*. The state functions (E is an example) can, in turn, be used to specify the condition (or, redundantly, the state) of the system. We will define and use several state functions. Each will be defined in terms of some or all of the following: T, P, V, number of moles of each substance, and other variables.

In thermodynamics the word process is used extensively. A *process* is just a certain way of carrying out a given assignment. If we want to get a system from *State A* to *State B*, we do so by defining some process or *path*. More will be said about this in the following section. Two general types of processes are of interest: reversible and irreversible. A *reversible* process is carried out in such a way that the internal and external intensive variables are the same within an infinitesimal. In other words, an infinitesimal change in an intensive variable is sufficient to stop or reverse the process. For example, let us suppose we want to expand a gas *isothermally* (constant temperature) and *reversibly*. The following must be satisfied:

$$T_{sys} = T_{surr}$$

$$P_{sys} + dP > P_{surr}$$

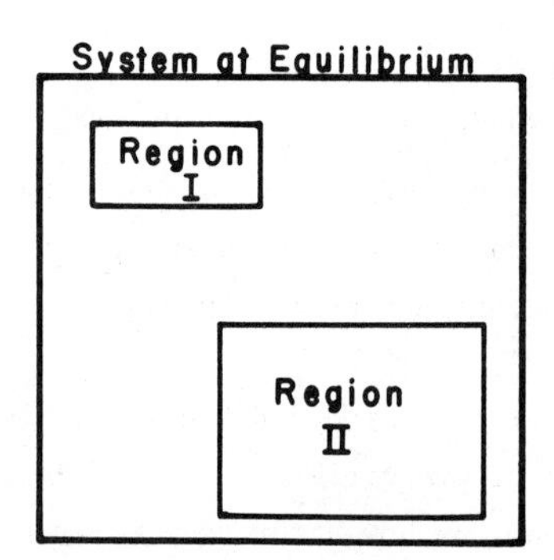

FIG. 3-1.1.2 Illustration of intensive $P_I = P_{II}$, $T_I = T_{II}$, $\rho_I = \rho_{II}$ (density), and $n_I = n_{II}$ (refractive index) and extensive ($V_I \neq V_{II}$ and $m_I \neq m_{II}$) variables.

If P_{surr} is increased by dP, then

$$P_{sys} + dP = P_{surr} + dP$$

and the expansion stops. Reversible processes are *equilibrium* processes. The term equilibrium implies that the intensive variables of a system are independent of time and there is no net flow of matter or energy across the boundary. Other definitions will be required, but they will be given as the need arises.

3-1.2 NOTES ON NOTATION AND DEFINING PROCESS PATHS IN THERMODYNAMICS

In this section an attempt will be made to point out some of the errors students usually make. Perhaps this way you can avoid some common pitfalls. Remember, previously we stated that the definitions in thermodynamics are precise; so are the symbols used.

We have agreed that a state function depends only on the variables of the system. A state function value does not depend on history. This is fortunate. Otherwise we would have to know everything that has happened to a system before we could describe its present state. As it is, we only need to specify a certain number of variables such as T and P. It follows from what we have presented above that if we allow a system to go from State 1 to another condition, State 2, the change in any state function is independent of how we carry out the change.

Energy (and enthalpy H to be defined later) is a state function. This means that

$$\Delta E = \int_1^2 dE = E_2 - E_1 \tag{3-1.2.1}$$

There are several points of interest in (3-1.2.1). First, note that $\Delta E = \int_1^2 dE$. ΔE is read as the change in E in going from State 1 to State 2. It has a value. On the other hand, dE is an infinitesimal change in E. It will not have a value, but will be expressed as some function of dT, dP, etc. *A common mistake made by students is to be very careless in using ΔE and dE interchangeably. They are not interchangeable.*

The second point of importance is that dE is an exact differential. Let us look at some properties of exact differentials (dT, dP, dE, and dV are all exact differentials), then consider dE again. If we have a function $f = f(x_1, x_2, \ldots, x_i, \ldots)$ of n independent variables, the *total differential df* is defined as follows:

$$df = \sum_{i=1}^{n} \left(\frac{\partial f}{\partial x_i}\right)_{x_1, x_2} dx_i \tag{3-1.2.2}$$

Consider a function of two variables (x, y) as an example:

$$df = \left(\frac{\partial f}{\partial x}\right)_y dx + \left(\frac{\partial f}{\partial y}\right)_x dy \tag{3-1.2.3}$$

where the subscript means that the indicated variable is held constant for the purpose of taking the derivative. A property of (3-1.2.3) that will be useful later is the *Euler reciprocity relation*. This is expressed in (3-1.2.4):

$$\frac{\partial}{\partial y}\left(\frac{\partial f}{\partial x}\right)_y = \frac{\partial}{\partial x}\left(\frac{\partial f}{\partial y}\right)_x \qquad (3\text{-}1.2.4)$$

The order of differentiation is not important for exact differentials. Both (3-1.2.3) and (3-1.2.4) will be very useful in later discussions.

Now, let us go back to considering an exact differential dE. *Since dE is exact we can state that if we change a system from condition 1 through a number of steps back to condition 1, the cyclic integral of dE = 0:*

$$\oint dE = 0 \qquad (3\text{-}1.2.5)$$

Our next task is to clarify how we can go about defining a process path for calculating the changes ΔE, Q, W, etc. for a given change. In Fig. 3-1.2.1 we have two states. We will define two paths to get from State 1 to State 2 as shown. Let us consider these paths separately.

PATH I.

Step 1: $T_1P_1V_1 \rightarrow T_1P_2V_x$ isothermal (constant T) change from P_1 to P_2

Step 2: $T_1P_2V_x \rightarrow T_2P_2V_2$ isobaric (constant P) change from T_1 to T_2 (note that an intermediate volume V_x is given: Why?)

Total: $\Delta E_T{}^I = \Delta E_I{}^{(1)} + \Delta E_I{}^{(2)}$

$$Q_T{}^I = Q_I{}^{(1)} + Q_I{}^{(2)}$$

$$W_T{}^I = W_I{}^{(1)} + W_I{}^{(2)}$$

To calculate $\Delta E_I{}^{(1)}$, $\Delta E_I{}^{(2)}$, $Q_I{}^{(1)}$, etc., we would need formulas we have not derived yet. However, we can see some relationships.

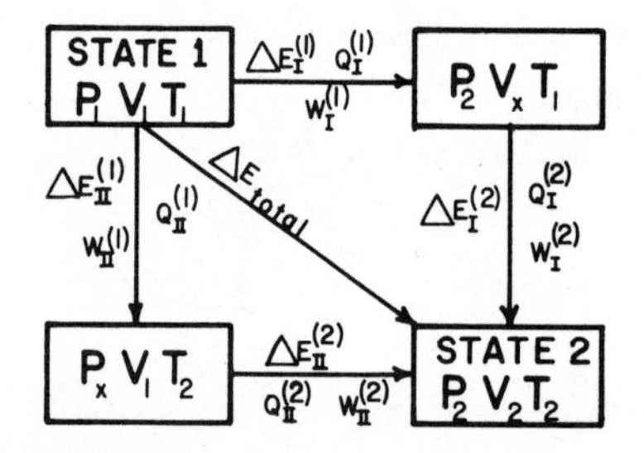

FIG. 3-1.2.1 Thermodynamic paths between State 1 and State 2 for an ideal gas.

PATH II.

Step 1: $T_1P_1V_1 \rightarrow T_2P_xV_1$ isochoric (constant V) change from T_1 to T_2

Step 2: $T_2P_xV_1 \rightarrow T_2P_2V_2$ isothermal change from V_1 to V_2

Total: $\Delta E_T^{\ II} = \Delta E_{II}^{\ (1)} + \Delta E_{II}^{\ (2)}$

$$Q_T^{\ II} = Q_{II}^{\ (1)} + Q_{II}^{\ (2)}$$

$$W_T^{\ II} = W_{II}^{\ (1)} + W_{II}^{\ (2)}$$

Since E is a state function, we can say immediately

$$\Delta E^{I}_{total} = \Delta E^{II}_{total} = \Delta E_{total}$$

But, Q and W are not state functions. Thus,

$$Q^{I}_{total} \neq Q^{II}_{total} \quad \text{and} \quad W^{I}_{total} \neq W^{II}_{total}$$

The important point here is that we can pick any path that is convenient for calculating changes in state functions such as ΔE. However, Q and W will depend on the path we choose.

In any problem that you encounter, it will aid you greatly to outline a path as done in Fig. 3-1.2.1. Such a figure quickly shows what is known and what is unknown about a system. Many of the common errors made in calculating the changes in thermodynamic quantities can be avoided by making a process outline similar to that given above. Foresight and organization will be very effective in your solution of various problems.

The next section will present a few equations for calculating changes that occur in some representative processes.

3-1.3 FIRST LAW CALCULATIONS OF WORK FOR IDEAL GASES

Some basic definitions have been given. We have defined ΔE in terms of Q and W but have said nothing about how to find these two quantities. Below W is defined for the expansion or compression of a gas, then Q is considered.

You should recall that mechanical work is defined as force times distance. Or in differential terms:

$$W = \int_{l_1}^{l_2} f_{opp}\, dl \qquad (3\text{-}1.3.1)$$

The subscript opp is used to indicate the opposing force, since that is the force that determines the work. Consider Fig. 3-1.3.1 which shows a gas enclosed in a cylinder that has one end closed with a massless, frictionless piston (an idealized

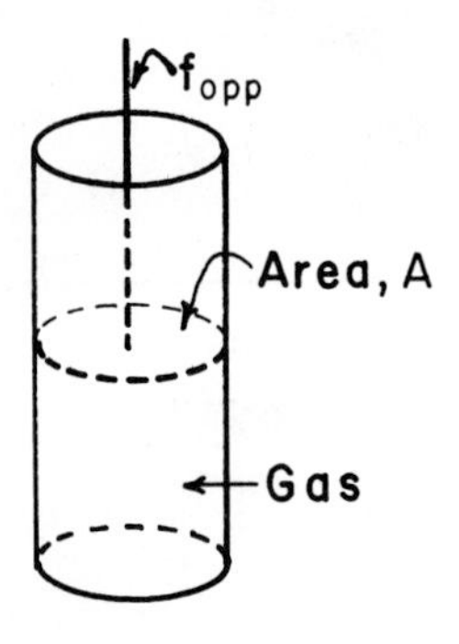

FIG. 3-1.3.1 A cylinder of an ideal gas with a frictionless, massless piston containing the expansion.

system). If we assume the gas is expanding, the opposing force will be f_{opp}. Rewrite (3-1.3.1):

$$W = \int_{Al_1}^{Al_2} \frac{f_{opp}}{A} A\ dl = \int_{V_1}^{V_2} P_{opp}\ dV \qquad (3\text{-}1.3.2)$$

This formula has been multiplied and divided by A which leaves the value unchanged. The term f/A is the opposing pressure, and A dl is the volume change experienced for a movement dl of the piston. For *any gas* we can write

$$W = -\int_{V_1}^{V_2} P_{opp}\ dV \qquad (3\text{-}1.3.3)$$

(The negative sign in (3-1.3.3) makes work done *on* the system positive. Previous expressions were in terms of work done *by* the system.) Notice the term in the integral is P_{opp} not P_{ext} (why?). The reason for this is that on expansion P_{opp} can be represented by P_{ext}; however, on compression P_{opp} will be determined by P_{int}. We can now write:

For expansion,

$$W = -\int_{V_1}^{V_2} P_{ext}\ dV \qquad (3\text{-}1.3.4)$$

For compression,

$$W = -\int_{V_1}^{V_2} P_{int}\ dV = -\int_{V_1}^{V_2} P_{sys}\ dV \qquad (3\text{-}1.3.5)$$

For a particular case we need an expression of P as a function of V before we can do the integral. We will idealize a true expansion and assume we carry out the expansion either reversibly or against a constant external pressure. We will always assume compression is done reversibly, since it is impossible (at constant T) to carry out a compression at constant internal (system) pressure. (Why?) Can you think of a way of compressing a gas while maintaining the system pressure constant?

CONSTANT EXTERNAL PRESSURE EXPANSION. Let the external (opposing) pressure be constant. Then from (3-1.3.4), for expansion:

$$W = -\int_{V_1}^{V_2} P_{ext}\, dV = -P_{ext}\int_{V_1}^{V_2} dV$$

$$= -P_{ext}(V_2 - V_1)$$

$$= -P_{ext}\,\Delta V \qquad (3\text{-}1.3.6)$$

There are several points to note here. P_{ext} is a constant, so it can be taken out of the integral. And V is a state function so $\int_{V_1}^{V_2} dV$ can be evaluated without reference to how we get from V_1 to V_2. So we can write immediately that $\int_{V_1}^{V_2} dV = V_2 - V_1$. If the gas is expanding, $V_2 > V_1$ and $W < 0$, which means negative work has been done on the system (or work is being done by the system).

3-1.3.1 EXAMPLE

One mole of an ideal gas expands against an external pressure of 1.12 atm. The initial volume of the gas is 23.17 dm^3, and the final volume is 35.22 dm^3. What is the work done on the system in this process? (Assume $P_{sys} = P_{ext}$ at the end of the process, and that T is constant.)

Solution

$$W = -\int_{V_1}^{V_2} P_{ext}\, dV = -P_{ext}\int_{V_1}^{V_2} dV = -P_{ext}\,(V_2 - V_1)$$

$$= -1.12 \text{ atm } (35.22 - 23.17) \text{ dm}^3$$

$$= -1.12 \text{ atm } (12.05 \text{ dm}^3) = \underline{\underline{-13.50 \text{ dm}^3 \text{ atm}}}$$

The answer is given, but the units do not tell us much, so let us make a conversion to SI energy units. It is good practice anyway.

$$(1 \text{ dm}^3 \text{ atm})\left(\frac{101{,}325 \text{ N m}^{-2}}{1 \text{ atm}}\right)\left(\frac{10^{-1} \text{ m}}{1 \text{ dm}}\right)^3 = 101.325 \text{ N m}$$

$$= 101.325 \text{ J}$$

Thus 1 dm^3 atm (= 1 liter atm) = 101.325 J

Our answer should be

$$W = -1367.8 \text{ J} = \underline{\underline{-1370 \text{ J}}}$$

(Only three significant figures.)

(Note: Later you will have occasion to add various terms to determine total energy changes, Q or W. Before individual terms can be added, they must be in the same units. A common error is to add one term expressed, say in J mol^{-1}, to another in dm^3 atm mol^{-1}. This is not correct.)

REVERSIBLE EXPANSION. Remember from Sec. 3-1.2 that in a reversible process the internal and external variables (intensive variables such as T and P) are the same within an infinitesimal. Thus $P_{ext} = P_{int} = P_{sys}$. We can then write from (3-1.3.4), for expansion:

$$W = -\int_{V_1}^{V_2} P_{ext}\, dV = -\int_{V_1}^{V_2} P_{sys}\, dV$$

But we cannot solve this unless we know an expression for P_{sys}. We cannot take P_{sys} out of the integral because it is not a constant. For an ideal gas we know

$$P_{sys} = \frac{nRT}{V_{sys}}$$

so

$$W = -\int_{V_1}^{V_2} P_{sys}\, dV = -\int_{V_1}^{V_2} \frac{nRT}{V_{sys}}\, dV$$

$$= -nR \int_{V_1}^{V_2} \frac{T}{V}\, dV \qquad (3\text{-}1.3.7)$$

We drop the subscript sys on T and V since these variables always refer to the system unless specifically indicated otherwise. We still cannot solve (3-1.3.7) unless we know something about T, since T will be a function of V. However, if we state that the process is isothermal, we can then write

$$W = -nRT \int_{V_1}^{V_2} \frac{dV}{V} = -nRT \ln \frac{V_2}{V_1} \qquad (3\text{-}1.3.8)$$

This is good only for a reversible isothermal expansion of an ideal gas. Equation (3-1.3.8) is an example of an equation that is for a *specific* process. Do not try to memorize it. It is appropriate *only* for a *reversible, isothermal expansion* of an *ideal gas*. Any problem you have will likely have different conditions. Always start with $W = -\int P\, dV$ and insert the *appropriate* expression for P to obtain an equation for evaluating W. Remember, any equation you may have memorized is probably not appropriate for the problem you are trying to solve.

3-1.3.2 EXAMPLE

Let us redo the previous example. You should prove that T = 480 K. We reversibly expand 1 mol of an ideal gas from 23.17 dm^3 to 35.22 dm^3.

Solution

From (3-1.3.8):

$$W = -\ nRT\ \ln \frac{V_2}{V_1} = (1\ \text{mol})(8.31\ \text{J mol}^{-1}\ \text{K}^{-1})(480\ \text{K})\left(\ln \frac{35.22}{23.17}\right)$$

$$= \underline{\underline{-1670\ \text{J}}}$$

Now compare the results. *In the constant pressure process the work on the surroundings is 1370 J, and in the reversible process the result is 1670 J.* This is an example of a general result. A reversible process will yield the maximum work possible on the surroundings. This should make sense because in a reversible expansion the system is expanding against the *maximum* opposing pressure possible. Thus, the maximum amount of work is being done.

In fact, any reversible process will always give the maximum work possible in going from one state to another. Any irreversibilities due to friction or other factors will lead to work that is less than W_{max}.

Before doing other examples we need to define some more terms and derive some relationships that will be needed. This will be done in the following section.

3-1.4 RELATIONSHIPS INVOLVING ENERGY, ENTHALPY, AND HEAT CAPACITY

ENERGY. We have already defined $\Delta E = Q + W$. For this to be useful we have to find expressions for Q and W. In all that follows the assumption is that only PV work (expansion or compression) is involved unless specifically stated otherwise. In that case

$$W = -\int_{V_1}^{V_2} P_{opp}\ dV$$

or

$$w = -P_{opp}\ dV \tag{3-1.4.1}$$

(q and w will represent infinitesimal changes in heat and work; Q and W represent finite changes). Remember, in work expressions P refers to the opposing pressure. The opp subscript will now be dropped.

Equation (3-1.4.1) leads to

$$dE = q + w = q - P\ dV \tag{3-1.4.2}$$

In general, E is expressed as a function of T and V. Thus, from (3-1.2.3) we can write

$$dE = \left(\frac{\partial E}{\partial T}\right)_V dT + \left(\frac{\partial E}{\partial V}\right)_T dV \qquad (3\text{-}1.4.3)$$

We will now state, without proof (proof will be found in Chap. 4, Sec. 4-2.2), that

$$\left(\frac{\partial E}{\partial V}\right)_T = T\left(\frac{\partial P}{\partial T}\right)_V - P \qquad (3\text{-}1.4.4)$$

Then

$$\boxed{dE = \left(\frac{\partial E}{\partial T}\right)_V dT + \left[T\left(\frac{\partial P}{\partial T}\right)_V - P\right] dV} \qquad (3\text{-}1.4.5)$$

This equation is a general equation of state for any system involving only PV work. Next $(\partial E/\partial T)_V$ will be defined, and an equation which is useful for calculating ΔE will be found.

HEAT CAPACITY AT CONSTANT VOLUME. The heat capacity for any system is defined as the heat absorbed to produce a 1° change in temperature.

$$C = \frac{q}{dT}$$

At constant volume, (3-1.4.2) and (3-1.4.5) become

$$dE_V = q_V \qquad \text{and} \qquad dE_V = \left(\frac{\partial E}{\partial T}\right)_V dT$$

Equating these two, we have

$$\left(\frac{q}{dT}\right)_V = \left(\frac{\partial E}{\partial T}\right)_V$$

or

$$C_V = \left(\frac{\partial E}{\partial T}\right)_V \qquad (3\text{-}1.4.6)$$

C_V is the heat capacity of a system at constant volume. This is the heat absorbed to raise the temperature 1° when the system is confined to a constant volume. Heat capacity can be defined in terms of moles or mass. We will normally work with moles. The unit on C_V will tell you which is being used. Now combine (3-1.4.6) and (3-1.4.5) to get a very useful *general* expression.

$$\boxed{dE = n\bar{C}_V\, dT + \left[T\left(\frac{\partial P}{\partial T}\right)_V - P\right] dV} \qquad (3\text{-}1.4.7)$$

where $\bar{C}_V$ is the molar heat capacity.

Values of $\bar{C}_v$ are tabulated for many substances, so (3-1.4.7) gives us a means of finding ΔE if we know an equation of state for the substance (P as a function of T and V). The following example will show how the bracketed [] term in (3-1.4.7) is evaluated for one equation of state.

3-1.4.1 EXAMPLE

Find the change in the energy of 2 mol of an ideal gas with a heat capacity (at constant volume) of 3/2R that changes from $T_1 = 298$ K and $V_1 = 24.2\ dm^3$ to $T_2 = 352$ K and $V_2 = 41.6\ dm^3$.

Solution

From (3-1.4.7):

$$dE = n\bar{C}_v\ dT + \left[T\left(\frac{\partial P}{\partial T}\right)_v - P\right] dV$$

But, P = nRT/V for an ideal gas, so

$$\left(\frac{\partial P}{\partial T}\right)_v = \frac{nR}{V}$$

(Note that the subscript v indicates that V is held constant for purposes of taking the derivative. It does not imply that v is constant for the process under consideration.)

$$\text{Then } dE = n\bar{C}_v\ dT + \left[T\left(\frac{nR}{V}\right) - P\right] dV$$

$$= n\bar{C}_v\ dT + \left[T\left(\frac{nR}{V}\right) - \frac{nRT}{V}\right]^{\to 0} dV$$

$$= n\bar{C}_v\ dT \qquad (3\text{-}1.4.8)$$

We have the important result that the *energy of an ideal gas depends only on the temperature* and not on its volume.

Continuing with the solution:

$$\int_1^2 dE = \Delta E = \int_{T_1}^{T_2} n\bar{C}_v\ dT = \int_{T_1}^{T_2} n\frac{3}{2}R\ dT$$

$$= \frac{3}{2}\ nR(T_2 - T_1)$$

$$= \frac{3}{2}(2\ \text{mol})(8.31\ \text{J mol}^{-1}\ \text{K}^{-1})(352 - 298)\text{K}$$

$$= 1346\ \text{J}$$

$$= \underline{\underline{1350\ \text{J}}}$$

This positive value of ΔE means that the internal energy of the system has increased. In this case, the increase is from heat flow.

ENTHALPY AND HEAT CAPACITY AT CONSTANT PRESSURE. From the previous development it should be clear that it is easy to determine ΔE for constant volume processes. Most common reactions are, however, carried out at constant pressure. The determination of ΔE is a little more difficult under these conditions. So, we define a new state function, *enthalpy*, that is very useful in constant pressure situations. Enthalpy is defined as

$$\boxed{H = E + PV} \tag{3-1.4.9}$$

The enthalpy is equal to the internal energy plus the pressure-volume product of the system. This pressure and volume are those for the *system*, not the surroundings. Enthalpy is a state function as it is defined in terms of other state functions, E, P, and V. H is usually written as a function of T and P. Thus

$$dH = \left(\frac{\partial H}{\partial T}\right)_P dT + \left(\frac{\partial H}{\partial P}\right)_T dP \tag{3-1.4.10}$$

Again, without proof at this time, (see Sec. 4-2.2)

$$\left(\frac{\partial H}{\partial P}\right)_T = V - T\left(\frac{\partial V}{\partial T}\right)_P \tag{3-1.4.11}$$

so

$$\boxed{dH = \left(\frac{\partial H}{\partial T}\right)_P dT + \left[V - T\left(\frac{\partial V}{\partial T}\right)_P\right] dP} \tag{3-1.4.12}$$

Now, let us develop the expression for heat capacity at constant pressure.

$$C_P = \left(\frac{q}{dT}\right)_P$$

From (3-1.4.9),

$$dH = dE + P\,dV + V\,dP$$

At constant pressure

$$dH = dE + P\,dV$$

Substituting for dE:

$$dH = q_P - P\,dV + P\,dV$$

$$= q_P \tag{3-1.4.13}$$

And at constant pressure (3-1.4.12) becomes

$$dH = \left(\frac{\partial H}{\partial T}\right)_P dT \tag{3-1.4.14}$$

Equating these last two expressions:

$$q = \left(\frac{\partial H}{\partial T}\right)_P dT$$

or
$$C_P = \left(\frac{q}{dT}\right)_P = \left(\frac{\partial H}{\partial T}\right)_P \qquad (3\text{-}1.4.15)$$

This allows us to rewrite (3-1.4.12):

$$\boxed{dH = n\bar{C}_P\, dT + \left[V - T\left(\frac{\partial V}{\partial T}\right)_P\right] dP} \qquad (3\text{-}1.4.16)$$

where $\bar{C}_P$ is the molar heat capacity at constant pressure. Note that dH depends on changes in T and changes in P. As for energy, we must have an equation of state before the bracketed expression can be evaluated.

At constant pressure this reduces to the simple expression

$$dH = n\bar{C}_P\, dT \qquad (3\text{-}1.4.17)$$

Before considering the example, try to prove that Eq. (3-1.4.17) results from (3-1.4.16) for an ideal gas, even if P is not constant.

3-1.4.2 EXAMPLE

Consider the conditions given in Example 3-1.4.1. Note that $\bar{C}_P = (5/2)R$ for reasons that will be discussed shortly. $T_1 = 298$ K and $T_2 = 352$ K.

Solution

From Eq. (3-1.4.16):

$$dH = n\bar{C}_P\, dT + \left[V - T\left(\frac{\partial V}{\partial T}\right)_P\right] dP$$

For an ideal gas V = nRT/P, hence

$$\left(\frac{\partial V}{\partial T}\right)_P = \frac{nR}{P}$$

As before, the subscript P indicates that the derivative is taken as though P were constant, whether it is or not. Substitution gives

$$\left[V - T\left(\frac{\partial V}{\partial T}\right)_P\right] = \left[\frac{nRT}{P} - \frac{TnR}{P}\right] = 0$$

We have the result that $dH = n\bar{C}_P\, dT$ for an ideal gas. *The enthalpy of an ideal gas depends only on the temperature.* For a real gas the bracketed term in dH may not be (usually is not) zero. To complete the problem:

$$\Delta H = \int_1^2 dH = \int_{T_1}^{T_2} n\bar{C}_P\, dT = \int_{T_1}^{T_2} n\frac{5}{2}R\, dT$$

$$= n\frac{5}{2}R \int_{T_1}^{T_2} dT = n\frac{5}{2}R\ (352 - 298)$$

$$= \frac{5}{2}(2\ \text{mol})(8.31\ \text{J mol}^{-1}\ \text{K}^{-1})(54\ \text{K})$$

$$= \underline{\underline{2240\ \text{J}}}$$

Note that $\Delta H > \Delta E$ for this process. The difference is due to the PV term in ΔH. Recall that $dH = dE + d(PV)$. So ΔH will always be greater than ΔE for an expansion. (Unless it is isothermal, in which case $\Delta H = \Delta E$ for an ideal gas. Why? What are ΔH and ΔE for an isothermal expansion?) From the definition of C_P and C_V you should see that $C_P > C_V$ for a gaseous substance. Can you prove that for an ideal gas?

3-1.5 MORE EXAMPLES

Up to this point we have introduced the terms Q, W, E, H, C_P, and C_V. It is now time to work some problems involving these terms.

3-1.5.1 EXAMPLE

In thermodynamics it is important to keep in mind the conditions that must be met for a particular equation to be used. In each of the following give the conditions that are required for the equation to be valid.

1. $\Delta E = Q - P\ \Delta V$
2. $\Delta E = nC_V\ \Delta T$

Solution

1. ΔE is defined as Q + W, so for PV work only:

$$dE = q + w = q - P\ dV$$

$$\int dE = Q - \int P\ dV$$

$$\Delta E = Q - \int P\ dV$$

which will give

$$\Delta E = Q - P\ \Delta V$$

only if P is constant. Otherwise P must be expressed as a function of V and the integral done explicitly. Our two conditions are (1) only PV work is involved, and (2) P is constant.

2. From (3-1.4.7), if only PV work is involved,

$$dE = n\bar{C}_V\ dT + \left[T\left(\frac{\partial P}{\partial T}\right)_V - P\right] dV$$

This reduces to $dE = n\bar{C}_V\, dT$ for constant V or for an ideal gas. So $\Delta E = \int n\bar{C}_V\, dT$. This can be written as $\Delta E = n\bar{C}_V \int dT = n\bar{C}_V\, \Delta T$ only if $\bar{C}_V$ is a constant (not a function of temperature). The conditions are then: (1) PV work only, (2) constant V or ideal gas, and (3) $\bar{C}_V$ is constant.

3-1.5.2 EXAMPLE

Calculate the Q, W, ΔE, and ΔH for the isothermal reversible compression of 1.21 mol of an ideal gas from 10 dm^3 to 1 dm^3 at 300 K.

Solution

Work

By definition $W = -\int_{V_1}^{V_2} P_{opp}\, dV$, and since the process is reversible, $P_{opp} = P_{sys}$. Also, $P_{sys} = nRT/V$ for ideal gases. Thus $W = -\int_{V_1}^{V_2} (nRT/V)\, dV = -nRT \int_{V_1}^{V_2} (dV/V)$ (we can remove T from the integral *only* because it is constant for this process).

$$W = nRT \ln \frac{V_2}{V_1}$$

$$= -(1.21\ \text{mol})(8.31\ \text{J mol}^{-1}\ \text{K}^{-1})(300\ \text{K}) \ln \frac{1\ \text{dm}^3}{10\ \text{dm}^3}$$

$$= \underline{\underline{+\ 6950\ \text{J}}}$$

The plus sign means that work is being done on the system, as expected for a compression process.

ΔE

For an ideal gas we concluded previously that $\Delta E = \int n\bar{C}_V\, dT$. For an isothermal process, $dT = 0$; therefore,

$$\underline{\underline{\Delta E = 0}}$$

ΔH

Similarly, we showed for an ideal gas that $\Delta H = \int n\bar{C}_P\, dT$. Thus, for an isothermal process,

$$\underline{\underline{\Delta H = 0}}$$

Heat

From the definition $\Delta E = Q + W$ and the fact that $\Delta E = 0$ we conclude $Q = -W$. Thus $\underline{\underline{Q = -6950\ \text{J}}}$. This implies that the system gives off 6950 J of heat (absorbs − 6950 J). Since we have done 6950 J of work on the system, the system *must* lose 6950 J of heat to keep the internal energy constant.

3-1.5.3 EXAMPLE

Calculate the change in energy and enthalpy of 12.0 g H_2O if it is heated from 10°C to 70°C at a constant pressure of 1 atm. $C_P(H_2O_\ell) = 4.18\ J\ g^{-1}\ K^{-1}$, $\rho(H_2O_\ell, 10°C) = 0.9997\ g\ cm^{-3} = 999.7\ g\ dm^{-3}$, and $\rho(H_2O_\ell, 70°C) = 0.9778\ g\ cm^{-3} = 977.8\ g\ dm^{-3}$.

Solution

$\underline{\Delta H}$

For constant pressure, $\Delta H = \int C_P\ dT = C_P\ \Delta T$ (if C_P is constant).

$$\Delta H = (12.0\ g)(4.18\ J\ g^{-1}\ K^{-1})(343 - 283\ K)$$

$$= \underline{\underline{3010\ J}}$$

$\underline{\Delta E}$

$\Delta E = \Delta H - \Delta(PV) = \Delta H - P\ \Delta V$ for constant pressure.

$$\Delta E = \Delta H - P(V_{343} - V_{283})$$

We have to find the volume at each temperature.

$$V_{343} = \frac{12\ g}{\rho_{343}} = \frac{12\ g}{977.8\ g\ dm^{-3}} = 0.01227\ dm^3$$

$$V_{283} = \frac{12\ g}{\rho_{283}} = \frac{12\ g}{999.7\ g\ dm^{-3}} = 0.01200\ dm^3$$

(We must keep more than three significant figures here for obvious reasons.)

$$\Delta E = \Delta H - P(V_{343} - V_{283})$$

$$= 3010\ J - 1\ atm(0.01227 - 0.01200)\ dm^3$$

$$= 3010\ J - (0.00027\ dm^3\ atm)\left(\frac{101\ J}{1\ dm^3\ atm}\right)$$

$$= 3010\ J - 0.027\ J = \underline{\underline{3010\ J}}$$

For all practical purposes $\Delta E = \Delta H$. This means $C_P \approx C_V$ for condensed phases (liquids and solids). If you do the same type of calculations for a gas (and you will, below) you will find ΔE and ΔH differ considerably.

3-1.6 PROBLEMS

1. Give all the conditions which must be met for the following to be valid:

   1. $dH = dE + V\ dP$
   2. $\Delta E = \int C_V\ dT$

2. Find Q, W, ΔE, and ΔH for the compression of 3 mol of an ideal gas reversibly from a volume of 100 dm^3 to 2.5 dm^3 at a constant T of 300 K.

3. An ideal gas produces 475 J work in a reversible expansion in an isothermal process; P_1 = 1 atm and P_2 = 0.1 atm. Find Q, ΔH, and ΔE for this expansion.

4. Find the change in ΔH for an isobaric heating of 10 g of an ideal gas with a molecular weight of 40 g mol^{-1} from T_1 = 105 K to T_2 = 330 K if $\bar{C}_P = (5/2)R$; P_1 = 0.75 atm.

5. For the system in Problem (4) find ΔE if $\bar{C}_V = (3/2)R$ and the process is carried out isochorically (constant volume).

6. For 1.72 mol of an ideal gas, 720 J of work are used to compress it from a volume of 33.2 dm^3 and from an initial temperature of 307 K. If $\bar{C}_P = \bar{C}_V + R = 7R/2$, determine the final temperature (isobaric at P_1).

7. Find Q, W, ΔE, and ΔH for the isothermal expansion of 1.31 mol of an ideal gas against a constant external pressure of 1.05 atm. T = 276 K, P_1 = 4.62 atm, and P_2 = 2.21 atm. $\bar{C}_P = \bar{C}_V + R = 7R/2$.

8. Given $dE = n\bar{C}_V\,dT + [T(\partial P/\partial T)_V - P]\,dV$ and the general definition of dH, derive the equation needed to calculate the value of ΔH for a constant external pressure expansion of an ideal gas from V_1T_1 to V_2T_2.

3-1.7 SELF-TEST

1. State verbally the law of conservation of energy.

2. Give a mathematical and verbal definition of the first law of thermodynamics.

3. Derive from basic definitions the equation needed to calculate the enthalpy change for an isothermal change from P_1 to P_2 for an ideal gas.

4. Calculate Q, W, ΔE, and ΔH for the isobaric heating of 0.56 mol of an ideal gas ($\bar{C}_P = 7R/2$) at 2.1 atm from 270 K to 480 K.

5. Repeat the calculations of Question (4) for a constant volume heating; $\bar{C}_V = 5R/2$.

3-1.8 ADDITIONAL READING

1.* W. J. Moore, *Physical Chemistry*, 4th ed., Prentice-Hall, Englewood Cliffs, N. J., 1972, Chap. 2, Sec. 1-14.

2.* H. C. van Ness, *Understanding Thermodynamics*, McGraw-Hill, New York, 1969. Used with permission of McGraw-Hill Book Company.

3-2 THERMODYNAMIC PROCESSES

OBJECTIVES

You shall be able to

1. Calculate the changes in E and H for constant volume and constant pressure heating of materials if C_P and C_V are given as a function of temperature
2. Calculate ΔH, ΔE, Q, and W and the final T, P, or V for adiabatic processes
3. Find ΔE, ΔH, Q, and W for processes involving changes of state and/or for processes involving nonideal gases
4. Estimate Joule-Thomson coefficients from the van der Waals or Beattie-Bridgeman equations.

3-2.1 INTRODUCTION

In Sec. 3-1 you were introduced to the first law quantities Q, W, ΔE, ΔH, C_P, and C_V. Here we will develop some of these quantities more fully, and you will encounter more difficult problems. Some of the equations of state for nonideal gases that were presented in Chap. 2 will be utilized here in our calculations.

Again, it should be emphasized that the best procedure for attacking any of the thermodynamic problems presented here is to start from the basic definitions. Apply all restrictions indicated in the problem to develop the equation that will solve the problem. You cannot, with any reliability, choose an equation from all that have been (and will be) presented that will fit the conditions of the problem. A continuing aim is to develop your ability to think your way through a solution. Almost everyone can plug numbers into an equation and grind out an answer, but considerable skill is required to develop the equations needed for a particular problem. Nurture that skill: it will serve you well in the future.

3-2.2 HEAT CAPACITIES

$\bar{C}_P - \bar{C}_V$. The derivation of the relationship for $C_P - C_V$ is straightforward. The final expression is simple for an ideal gas but somewhat more complicated for real systems. To find $C_P - C_V$ we will apply some of the definitions given previously. This will be a good review of many of the definitions we have had up to this point.

$$C_P - C_V = \left(\frac{\partial H}{\partial T}\right)_P - \left(\frac{\partial E}{\partial T}\right)_V$$

$$= \left[\frac{\partial(E + PV)}{\partial T}\right]_P - \left(\frac{\partial E}{\partial T}\right)_V$$

$$= \left(\frac{\partial E}{\partial T}\right)_P + \left[\frac{\partial(PV)}{\partial T}\right]_P - \left(\frac{\partial E}{\partial T}\right)_V \qquad (3\text{-}2.2.1)$$

We need to find $(\partial E/\partial T)_P$. From (3-1.4.3) we know that

$$dE = \left(\frac{\partial E}{\partial T}\right)_V dT + \left(\frac{\partial E}{\partial V}\right)_T dV \qquad (3\text{-}1.4.3)$$

Divide by dT at constant P to get

$$\left(\frac{\partial E}{\partial T}\right)_P = \left(\frac{\partial E}{\partial T}\right)_V + \left(\frac{\partial E}{\partial V}\right)_T \left(\frac{\partial V}{\partial T}\right)_P \qquad (3\text{-}2.2.2)$$

Substitute into (3-2.2.1):

$$C_P - C_V = \left(\frac{\partial E}{\partial T}\right)_V + \left(\frac{\partial E}{\partial V}\right)_T \left(\frac{\partial V}{\partial T}\right)_P + \left[\frac{\partial (PV)}{\partial T}\right]_P - \left(\frac{\partial E}{\partial T}\right)_V$$

$$= \left(\frac{\partial E}{\partial V}\right)_T \left(\frac{\partial V}{\partial T}\right)_P + P\left(\frac{\partial V}{\partial T}\right)_P$$

Thus

$$C_P - C_V = \left[\left(\frac{\partial E}{\partial V}\right)_T + P\right] \left(\frac{\partial V}{\partial T}\right)_P \qquad (3\text{-}2.2.3)$$

To use this equation we would need an equation of state to evaluate $(\partial E/\partial V)_T$ and $(\partial V/\partial T)_P$. For an ideal gas this is easy to do as shown below.

3-2.2.1 EXAMPLE

From equation (3-2.2.3) evaluate $C_P - C_V$ for an ideal gas.

Solution

$$C_P - C_V = \left[\left(\frac{\partial E}{\partial V}\right)_T + P\right] \left(\frac{\partial V}{\partial T}\right)_P \qquad (3\text{-}2.2.3)$$

For an ideal gas, $V = nRT/P$ and $(\partial E/\partial V)_T = 0$, since the energy of an ideal gas is independent of volume (it depends only on T).

$$C_P - C_V = P\left[\frac{\partial (nRT/P)}{\partial T}\right]_P$$

$$= P\,\frac{nR}{P}\left(\frac{\partial T}{\partial T}\right) = nR$$

Or we can write for 1 mol of an ideal gas:

$$\bar{C}_P - \bar{C}_V = R$$

You should now see why, when $\bar{C}_V$ in Example 3-1.4.1 was given as 3R/2, $\bar{C}_P$ in Example 3-1.4.2 was stated to be 5R/2

HEAT CAPACITY AS A FUNCTION OF TEMPERATURE. In all the previous problems, we have assumed that $\bar{C}_P$ and $\bar{C}_V$ are constants. This is, in general, not true. Both $\bar{C}_P$ and $\bar{C}_V$ depend on temperature. Usually $\bar{C}_P$ is presented as a function of T in one of the following two ways.

$$\bar{C}_P = a + bT + cT^{-2} \tag{3-2.2.5}$$

or
$$\bar{C}_P = a' + b'T + c'T^2 \tag{3-2.2.6}$$

We will use the former convention. It is important, though, for you to note which form is being used whenever you look up the coefficients in tables. Obviously, if you look up a′, b′, and c′ and apply (3-2.2.5) you are not going to get a good result.

Table 3-2.2.1 lists values of a, b, and c for gases and condensed phases for several materials. Since most processes in which we will be interested will take place at constant pressure, tables of $\bar{C}_V$ will not be given.

3-2.2.2 EXAMPLE

Determine the heat capacity (C_P) of NH_3(g) at 298 K and 500 K.

Solution

$$\bar{C}_P = 29.75 + 25.10 \times 10^{-3}\ (298) - 1.55 \times 10^5\ (298)^{-2}$$
$$= \underline{\underline{35.48\ \text{J mol}^{-1}\ \text{K}^{-1}}} \quad (\text{at } 298\ \text{K})$$

$$\bar{C}_P = 29.75 + 25.10 \times 10^{-3}\ (500) - 1.55 \times 10^5\ (500)^{-2}$$
$$= \underline{\underline{41.68\ \text{J mol}^{-1}\ \text{K}^{-1}}} \quad (\text{at } 500\ \text{K})$$

3-2.2.3 EXAMPLE

Calculate the enthalpy change in 24 g of N_2 if it is heated from 300 K to 1500 K at constant pressure. $\bar{C}_P = 28.6 + 3.8 \times 10^{-3}\ T - 0.50 \times 10^5\ T^{-2}$ (J mol^{-1} K^{-1}).

Solution

From Eq. (3-1.4.12):

$$dH = n\bar{C}_P\ dT + \left[V - T\left(\frac{\partial V}{\partial T}\right)_P\right] dP \tag{3-1.4.12}$$

we can state at constant P (since dP = 0)

$$dH = n\bar{C}_P\ dT$$

or
$$\Delta H = \int_1^2 dH = n\int_{T_1}^{T_2} \bar{C}_P\ dT$$

$$= n\int_{T_1}^{T_2} (28.6 + 3.8 \times 10^{-3}\ T - 0.50 \times 10^5\ T^{-2})\ dT$$

$$= n\left[28.6(T_2 - T_1) + \frac{3.8 \times 10^{-3}}{2}(T_2^{\ 2} - T_1^{\ 2}) + 0.50 \times 10^5\left(\frac{1}{T_2} - \frac{1}{T_1}\right)\right]$$

Table 3-2.2.1
Molar Heat Capacities of Selected Materials [$\bar{C}_p = a + bT + cT^{-2}$ (J mol^{-1} K^{-1})]

Substance	a	b × 10^3	c × 10^{-5}
Gases[a]			
He, Ne, Ar, Kr, Xe	20.79	0	0
H_2	27.28	3.26	0.50
O_2	29.96	4.18	-1.67
N_2	28.58	3.76	-0.50
CO	28.41	4.10	-0.46
Cl_2	37.03	0.67	-2.84
I_2	37.40	0.59	-0.71
CO_2	44.22	8.79	-8.62
H_2O	30.54	10.29	0
NH_3	29.75	25.10	-1.55
CH_4	23.64	47.86	-1.92
SO_2	38.60	9.25	5.02
Liquids[b]			
I_2	80.33	0	0
H_2O	75.48	0	0
Solids[c]			
C(graphite)	16.86	4.77	-8.54
Al	20.67	12.38	0
Cu	22.63	6.28	0
I_2	40.12	49.79	0
S	22.00	-0.418	1.50

[a] Applicable from 298 K to 2000 K.
[b] Applicable from melting point to boiling point.
[c] Applicable from 298 K to melting point or 2000 K.

$$= \frac{24\text{ g}}{28\text{ g}}\text{ mol}\left[28.6\ (1500 - 300) + 1.9 \times 10^{-3}\ (1500^2 - 300^2) + 0.50 \times 10^5 \left(\frac{1}{1500} - \frac{1}{300}\right)\right]\text{ J mol}^{-1}$$

$$= 0.86\text{ mol }(34320 + 4104 - 133)\text{ J mol}^{-1}$$

$$= 32930\text{ J} = \underline{\underline{33000\text{ J}}}$$

3-2.2.4 EXAMPLE

If we can say $\bar{C}_p - \bar{C}_v = R$ for N_2, find ΔE for the process in Example 3-2.2.3.

Solution

From (3-1.4.7):

$$dE = n\bar{C}_v\ dT + \left[T\left(\frac{\partial P}{\partial T}\right)_V - P\right] dV \qquad (3\text{-}1.4.7)$$

At constant volume this becomes

$$dE = n\bar{C}_V \, dT$$

Therefore

$$E = \int_1^2 \Delta E = \int_{T_1}^{T_2} n\bar{C}_V \, dT = n\int_{T_1}^{T_2} (\bar{C}_P - R) \, dT$$

$$= n\int_{T_1}^{T_2} \bar{C}_P \, dT - n\int_{T_1}^{T_2} R \, dT = \Delta H - nR\int_{T_1}^{T_2} dT$$

$$= 33000 \text{ J} - [(0.86 \text{ mol})(8.314 \text{ J mol}^{-1} \text{ K}^{-1})(1500 - 300 \text{ K})]$$

$$= 33000 \text{ J} - 8580 \text{ J} = 24420 \text{ J}$$

$$= \underline{\underline{24000 \text{ J}}}$$

The difference in these two results is the work expended in heating at constant pressure. At constant volume there is no work involved. (Why?)

3-2.3 THE JOULE-THOMSON EXPERIMENT

One very important characteristic of a gas is the effect on its temperature when it is expanded or compressed under conditions in which there is no heat flow *($Q = 0$, an adiabatic process)*. The existence of such a phenomenon is necessary for the liquification of some gases by expansion. It also plays a role in the behavior of compression-expansion type air conditioners.

Initial experiments by Joule in an attempt to observe the temperature change when a gas was expanded into an evacuated chamber, showed no temperature change. Since $Q = 0$ (the system was insulated to prevent heat loss), and $W = 0$ (expansion into a vacuum has no opposing force and, therefore, can do no work), ΔE must equal 0. This led to Joule's conclusion that E was independent of volume. Such is the case for ideal gases, as presented earlier.

Subsequent, refined experiments by Joule and Thomson yielded small temperature changes upon expansion of real gases. Their experiment is illustrated in Fig. 3-2.3.1. Allow a volume V_1 of a gas at a constant pressure P to be forced through a porous plug that allows a pressure difference to be maintained. Volume V_2 is swept out on the other side of the plug by the piston opposed by the constant pressure P_2. We insulate the system and make all components using poor heat conductors to assure $Q = 0$ (adiabatic).

The work done in this case, $\text{work}_1 - \text{work}_2 = P_1V_1 - P_2V_2$. Also, $\Delta H = Q + W + \Delta(PV) = Q + W + (P_2V_2 - P_1V_1)$. But $Q = 0$ and $W = P_1V_1 - P_2V_2$, therefore, $\Delta H = 0 + (P_1V_1 - P_2V_2) + (P_2V_2 - P_1V_1) = 0$. *Enthalpy is constant in this process.* Since $H = H(T, P)$, and

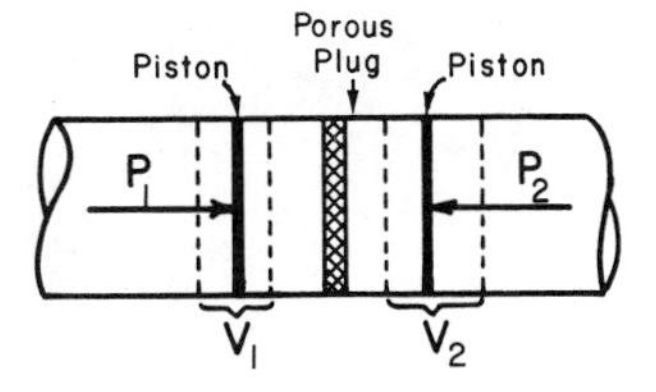

FIG. 3-2.3.1 The Joule-Thomson experiment.

$$dH = \left(\frac{\partial H}{\partial T}\right)_P dT + \left(\frac{\partial H}{\partial P}\right)_T dP \tag{3-1.4.10}$$

We see that

$$dH = 0 = \left(\frac{\partial H}{\partial T}\right)_P dT + \left(\frac{\partial H}{\partial P}\right)_T dP$$

or

$$\left(\frac{\partial H}{\partial P}\right)_T = -\left(\frac{\partial H}{\partial T}\right)_P \left(\frac{\partial T}{\partial P}\right)_H \tag{3-2.3.1}$$

But

$$\left(\frac{\partial H}{\partial T}\right)_P = C_P$$

Thus

$$\left(\frac{\partial H}{\partial P}\right)_T = -\left(\frac{\partial T}{\partial P}\right)_H C_P \tag{3-2.3.2}$$

At this point we define the Joule-Thomson coefficient as

$$\mu = \left(\frac{\partial T}{\partial P}\right)_H \tag{3-2.3.3}$$

The Joule-Thomson coefficient is useful in determining the temperature change of a gas on expansion. It is important in the process of liquification of gases by expansion; see Sheehan (Sec. 3-2.8, Ref. 1).

Equation (3-2.3.2) can be rewritten as

$$\left(\frac{\partial H}{\partial P}\right)_T = -\mu C_P \tag{3-2.3.4}$$

This equation gives the enthalpy change as a function of pressure at constant temperature.

Again we have found a useful relationship. It would be difficult to find $(\partial H/\partial P)_T$ experimentally, but μ, $(\partial T/\partial P)_H$, and C_P can be determined without too much difficulty. The quantity $(\partial H/\partial P)_T$ will be useful later in determining changes in temperature, enthalpy, etc., for various processes. Perhaps, by now you are convinced of the utility of manipulating thermodynamic quantities. In many cases it saves a lot of experimental work.

We can rewrite (3-1.4.10):

$$dH = n\bar{C}_P\, dT - \mu C_P\, dP \tag{3-2.3.5}$$

Comparison with (3-1.4.12):

$$dH = n\bar{C}_P\, dT + \left[V - T\left(\frac{\partial V}{\partial T}\right)_P\right] dP \tag{3-1.4.12}$$

allows us to write

$$\mu = -\frac{1}{C_P}\left[V - T\left(\frac{\partial V}{\partial T}\right)_P\right] \tag{3-2.3.6}$$

So, given an equation of state for a gas, we can calculate the Joule-Thomson coefficient. Values of μ for selected gases are shown in Fig. 3-2.3.1 in which μ as a function of T and P is given. It is readily apparent that μ can depend strongly on T and P.

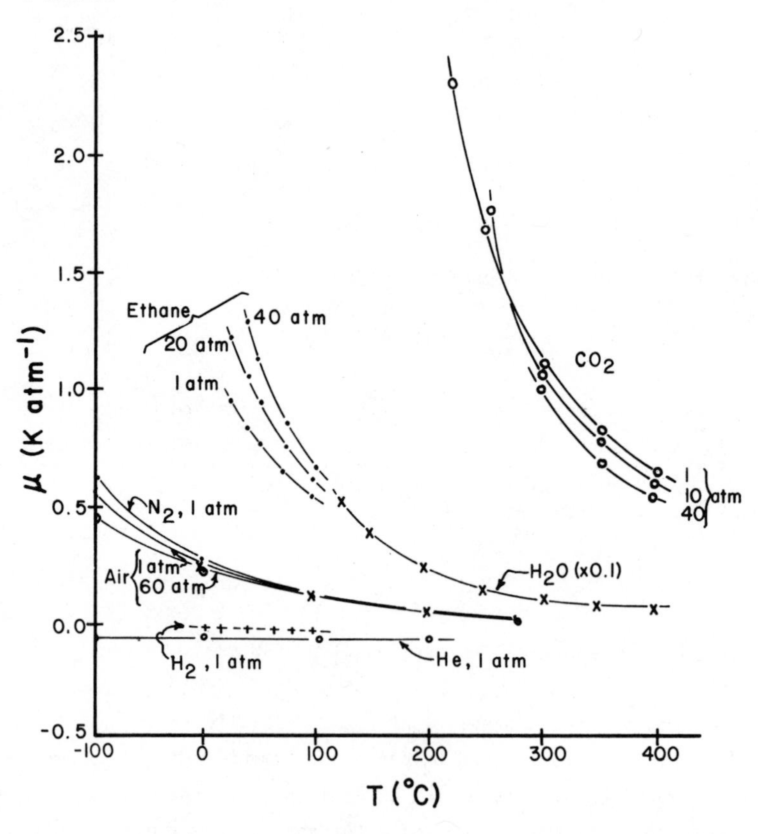

FIG. 3-2.3.2 Plot of μ vs. T for several gases. (Data from W. F. Sheehan, *Physical Chemistry*, 2nd ed., Allyn and Bacon, Boston, 1970, p. 95; and from E. W. Washburn, ed., *International Critical Tables of Numerical Data*, McGraw-Hill, New York, 1926, Vol. V, pp. 144-146.)

3-2.3.1 EXAMPLE

Predict the temperature lowering expected when N_2 at 273 K is expanded across a nozzle from a pressure of 10 atm to 1.0 atm in an insulated system.

Solution

From Fig. 3-2.3.2 we estimate $\mu = 0.21\ \text{K atm}^{-1}$. And $\mu = (\partial T/\partial P)_H$ from (3-2.3.3). If we are operating at constant enthalpy we can write

$$dT = \mu\ dP$$

$$\text{Thus } \Delta T = \int dT = \int \mu\ dP = \mu\ \Delta P$$

$$= +0.21\ \text{K atm}^{-1}\ (1.0 - 10.0)$$

$$= \underline{\underline{-1.9\ \text{K}}}$$

For N_2 we observe about a 2° drop in temperature for a pressure drop of 9.0 atm. This is under ideal conditions. The above assumed μ was independent of pressure when doing the integral. The assumption is valid in this case (as you can tell from the plot in Fig. 3-2.3.2), but not always.

CALCULATION OF μ. We will now apply Eq. (3-2.3.6) to find μ from the van der Waals and the Beattie-Bridgeman equations.

The van der Waals equation can be written for 1 mol:

$$\left(P + \frac{a}{V^2}\right)(V - b) = RT \tag{3-2.3.7}$$

$$PV - bP + \frac{a}{V} - \frac{ab}{V^2} = RT$$

Neglecting second order terms (ab) and substituting P/RT for 1/V (if a is small this introduces very little error) yields on rearranging:

$$PV = RT - \frac{aP}{RT} + bP \tag{3-2.3.8}$$

This can be differentiated to give (after division by P)

$$\left(\frac{\partial V}{\partial T}\right)_P = \frac{R}{P} + \frac{a}{RT^2} \tag{3-2.3.9}$$

Write (3-2.3.8) as

$$R = \frac{(V + a/RT - b)P}{T}$$

and substitute into (3-2.3.9) to get

$$\left(\frac{\partial V}{\partial T}\right)_P = \frac{V - b}{T} + \frac{2a}{RT^2} \tag{3-2.3.10}$$

This gives, upon substitution into (3-2.3.6)

$$\mu_{vdW} = \frac{2a/RT - b}{C_P} \tag{3-2.3.11}$$

In a similar but more complicated fashion, the Beattie-Bridgeman equation gives

$$\mu_{BB} = \frac{1}{C_P}\left\{- B_0 + \frac{2A_0}{RT} + \frac{4c}{T^3} + \left[\frac{2B_0 b}{RT} - \frac{3A_0 a}{(RT)^2} + \frac{5B_0 c}{RT^4}\right]P\right\} \tag{3-2.3.12}$$

3-2.3.2 EXAMPLE

Calculate μ_{BB} for N_2 at 1.0 atm and compare with the values estimated from Fig. 3-2.3.2.

Solution

Table 2-2.7.1 yields the following constants:
A_0 = 1.344, a = 0.0262, B_0 = 0.0505, b = -0.0069, and c = 4.2×10^4. C_P calculated at 298 K from Table 3-2.2.1 = 29.1 J mol^{-1} K^{-1} = 0.288 dm^3 atm mol^{-1} K^{-1}.
Substitution into (3-2.3.12) gives

$$\mu_{BB} = \frac{1}{0.288}\left\{- 0.0505 + \frac{2(1.344)}{(0.0821)(298)} + \frac{4(4.2 \times 10^4)}{(298)^3} + \left[\frac{2(0.0505)(-0.0069)}{(0.0821)(298)} - \frac{3(1.344)(0.0262)}{(0.0821)(298)^2} + \frac{5(0.0505)(4.2 \times 10^4)}{(0.0821)(298)^4}\right]10\right\}$$

$$= \frac{6.38 \times 10^{-2}}{0.288}$$

$$= \underline{\underline{0.222 \text{ K atm}^{-1}}}$$

This value compares with an observed value of approximately 0.21 K atm^{-1}. Do the same calculations with van der Waals equation and compare the results.

3-2.4 ADIABATIC PROCESSES

IDEAL GASES. The previous section dealt with an adiabatic process, one that is carried out with no heat exchange between the system and surroundings. What we will do in the following is apply some of the skills and equations developed so far to find a relationship between T and V, T and P, and P and V for adiabatic processes.

For an adiabatic process we know q = 0, so dE = w. If only PV work is involved, we know from (3-1.4.1) and (3-1.4.7) that

$$n\bar{C}_V\, dT + \left[T\left(\frac{\partial P}{\partial T}\right)_V - P\right] dV = -P\, dV \tag{3-2.4.1}$$

For an ideal gas, we have shown previously that the energy is independent of volume. In that case, (3-2.4.1) reduces to

$$n\bar{C}_V \, dT = -P \, dV \tag{3-2.4.2}$$

Replacing P by nRT/V, which can be done only if the process is carried out reversibly, we obtain

$$n\bar{C}_V \, dT = -\frac{nRT \, dV}{V} \tag{3-2.4.3}$$

This cannot be integrated directly, since T is a function of V, but by dividing through by T we obtain

$$\frac{\bar{C}_V}{T} \, dT = -\frac{R \, dV}{V} \tag{3-2.4.4}$$

Equation (3-2.4.4) can be integrated as follows (if $\bar{C}_V$ is a constant, as it will be for an ideal gas):

$$\int_{T_1}^{T_2} \frac{\bar{C}_V}{T} \, dT = -\int_{V_1}^{V_2} \frac{R \, dV}{V}$$

$$\bar{C}_V \ln \frac{T_2}{T_1} = -R \ln \frac{V_2}{V_1}$$

or $$\frac{\bar{C}_V}{R} \ln \frac{T_2}{T_1} = -\ln \frac{V_2}{V_1} = \ln \frac{V_1}{V_2}$$

Take the exponential of each side to obtain

$$\left(\frac{T_2}{T_1}\right)^{\bar{C}_V/R} = \frac{V_1}{V_2} \tag{3-2.4.6}$$

or $$V_1 T_1^{\bar{C}_V/R} = V_2 T_2^{\bar{C}_V/R} \tag{3-2.4.7}$$

These two equations relate the temperature change to a given volume change (or vice versa). Now suppose a particular problem was specified in terms of P and V changes. Equation (3-2.4.6) can be modified by use of the ideal gas equation T = PV/nR.

$$\left(\frac{T_2}{T_1}\right)^{\bar{C}_V/R} = \left(\frac{P_2V_2/nR}{P_1V_1/nR}\right)^{\bar{C}_V/R} = \frac{V_1}{V_2}$$

$$\left(\frac{P_2V_2}{P_1V_1}\right)^{\bar{C}_V/R} = \frac{V_1}{V_2}$$

$$\left(\frac{P_2}{P_1}\right)^{\bar{C}_V/R} = \frac{V_1}{V_2}\left(\frac{V_1}{V_2}\right)^{\bar{C}_V/R} = \left(\frac{V_1}{V_2}\right)^{\bar{C}_V/R + 1} = \left(\frac{V_1}{V_2}\right)^{(C_V + R)/R}$$

But $\bar{C}_v + R = \bar{C}_p$ for an ideal gas:

$$\left(\frac{P_2}{P_1}\right)^{\bar{C}_v/R} = \left(\frac{V_1}{V_2}\right)^{\bar{C}_p/R} \tag{3-2.4.8}$$

Take the $R/\bar{C}_v$ power of both sides:

$$\left(\frac{P_2}{P_1}\right)^{(\bar{C}_v/R)(R/\bar{C}_v)} = \left(\frac{V_1}{V_2}\right)^{(\bar{C}_p/R)(R/\bar{C}_v)}$$

$$\frac{P_2}{P_1} = \left(\frac{V_1}{V_2}\right)^{\bar{C}_p/\bar{C}_v} = \left(\frac{V_1}{V_2}\right)^{\gamma} \tag{3-2.4.9}$$

or $$P_1V_1^{\gamma} = P_2V_2^{\gamma} \tag{3-2.4.10}$$

where γ is defined as $\bar{C}_p/\bar{C}_v$.

Obviously, other combinations are possible by making appropriate substitutions.

3-2.4.1 EXAMPLE

Three moles of an ideal gas with $\bar{C}_v = 20.8\ \mathrm{J\ mol^{-1}\ K^{-1}}$ are expanded reversibly and adiabatically from an initial pressure of 15 atm and temperature of 365 K to a final pressure of 1 atm. Find (1) the final temperature of the gas, (2) the work done on the gas, and (3) the energy change of the gas.

Solution

1. We need a relationship between P and T to solve this problem in the most straightforward manner. From (3-2.4.6):

$$\left(\frac{T_2}{T_1}\right)^{\bar{C}_v/R} = \frac{V_1}{V_2}$$

Substituting $V = nRT/P$:

$$\left(\frac{T_2}{T_1}\right)^{\bar{C}_v/R} = \frac{nRT_1/P_1}{nRT_2/P_2} = \frac{T_1P_2}{T_2P_1}$$

$$\left(\frac{T_2}{T_1}\right)\left(\frac{T_2}{T_1}\right)^{\bar{C}_v/R} = \left(\frac{T_2}{T_1}\right)^{\bar{C}_v/R + 1} = \left(\frac{T_2}{T_1}\right)^{(\bar{C}_v + R)/R} = \left(\frac{P_2}{P_1}\right)$$

or $$\frac{P_2}{P_1} = \left(\frac{T_2}{T_1}\right)^{(\bar{C}_V + R)/R} = \left(\frac{T_2}{T_1}\right)^{\bar{C}_P/R} \qquad (3\text{-}2.4.11)$$

We are looking for T_2, so rearrange (3-2.4.11):

$$T_2 = T_1 \left(\frac{P_2}{P_1}\right)^{R/(\bar{C}_V + R)}$$

$$= (365)\left(\frac{1}{15}\right)^{8.314/(20.8 + 8.314)}$$

$$= (365)\left(\frac{1}{15}\right)^{0.286}$$

$$= (365)(0.461)$$

$$= \underline{\underline{168 \text{ K}}}$$

2. and 3. Since $Q = 0$, $\Delta E = W$, and $\Delta E = \int n\bar{C}_V \, dT = n\bar{C}_V \, \Delta T$,

$$\Delta E = W = (3 \text{ mol})(20.8 \text{ J mol}^{-1} \text{ K}^{-1})(168 - 365) \text{ K}$$

$$= -12290 \text{ J}$$

$$= \underline{\underline{-12300 \text{ J}}}$$

Which indicates the system *did* 12,300 J of work on the surroundings.

3-2.4.2 EXAMPLE

Two moles of an ideal gas are compressed reversibly and adiabatically from T_1 = 200 K and P_1 = 1 atm, by the expenditure of 900 J of work. If $\bar{C}_V = (3/2)R$ J mol^{-1} K^{-1}, find the final temperature and pressure of the gas. Also, find the energy change for the system.

Solution

Stop for a moment before writing any equations and think about the process. We have done 900 J of work *on* the system and $Q = 0$. Thus

$$\Delta E = W = +900 \text{ J}$$

Since $\Delta E = \int n\bar{C}_V \, dT = n\bar{C}_V \, \Delta T = 900$ J:

$$\Delta T = \frac{900 \text{ J}}{n\bar{C}_V} = \frac{900 \text{ J}}{(2 \text{ mol})(3/2)(8.314 \text{ J mol}^{-1} \text{ K}^{-1})}$$

$$= \underline{\underline{36.1 \text{ K}}}$$

$$T_2 = 36.1 + T_1 = 36.1 + 200 = 236.1 = \underline{\underline{236 \text{ K}}}$$

All that remains is to find P_2, which can be done by using (3-2.4.11)

$$P_2 = P_1 \left(\frac{T_2}{T_1}\right)^{(\bar{C}_V + R)/R} = (1 \text{ atm}) \left(\frac{236}{200}\right)^{[(3/2)R + R]/R}$$

$$= (1 \text{ atm}) \left(\frac{236}{200}\right)^{5/2} = (1 \text{ atm})(1.51)$$

$$= \underline{\underline{1.5 \text{ atm}}}$$

So we have

$\Delta E = W = 900$ J

$T_2 = 236$ K

$P_2 = 1.5$ atm

3-2.4.3 EXAMPLE

Reconsider Example 3-2.4.1. Assume the expansion is carried out by allowing the gas to expand irreversibly into the atmosphere (1 atm). Find all the quantities previously sought.

Solution

Here is another case in which you cannot just plug into formulas. This is an irreversible process, so all the equations in this section are inappropriate. We can find work since this is a constant pressure expansion.

$$W = -\int_{V_1}^{V_2} P_{opp}\, dV = -P_{opp}\, \Delta V = -P_{opp}(V_2 - V_1)$$

$$= -(1 \text{ atm}) \left(\frac{nRT_2}{P_2} - \frac{nRT_1}{P_1}\right)$$

We know $P_1 = 15$ atm, $P_2 = 1$ atm, and $T_1 = 365$ K, but we do not know T_2. However, $Q = 0$ (adiabatic).

$\Delta E = W$

So $\quad n\bar{C}_V(T_2 - T_1) = -(1 \text{ atm}) \left(\frac{nRT_2}{P_2} - \frac{nRT_1}{P_1}\right)$

$$(3 \text{ mol})(20.8 \text{ J mol}^{-1} \text{ K}^{-1})T_2 - (3 \text{ mol})(20.8 \text{ J mol}^{-1} \text{ K}^{-1})(265 \text{ K})$$

$$= -1 \text{ atm} \left[\frac{(3 \text{ mol})(8.314 \text{ J mol}^{-1} \text{ K}^{-1})T_2}{1 \text{ atm}} - \frac{(3 \text{ mol})(8.314 \text{ J mol}^{-1} \text{ K}^{-1})(265 \text{ K})}{15 \text{ atm}}\right]$$

$$(62.4\ \text{J K}^{-1})T_2 - 16536\ \text{J} = -(24.9\ \text{J K}^{-1})T_2 + 441\ \text{J}$$

$$(62.4\ \text{J K}^{-1} + 24.9\ \text{J K}^{-1})T_2 = 441\ \text{J} + 16536\ \text{J}$$

$$T_2 = \frac{16977\ \text{J}}{87.3\ \text{J K}^{-1}}$$

$$= \underline{\underline{194\ \text{K}}}$$

$$\Delta E = W = n\bar{C}_V(T_2 - T_1)$$

$$= (3\ \text{mol})(20.8\ \text{J mol}^{-1}\ \text{K}^{-1})(194 - 265)$$

$$= \underline{\underline{-4430\ \text{J}}}$$

You should note at least two things here. One, the amount of work done *by* the gas (+ 4430 J) is considerably less than the reversible work possible (12300 J). Also, the temperature drops only to 194 K, compared to 168 K in Example 3-2.4.1.

REAL GASES. We have already discussed one aspect of adiabatic processes for real gases. In Sec. 3-2.3, the behavior of a real gas on expansion in a constant enthalpy process was presented. Below, a set of equations will be derived for a van der Waals gas similar to those given for ideal gases above. Other equations of state could be used, but this one will illustrate the technique. The van der Waals gas gives equations that are more complex than those encountered with the ideal gas. The procedures that follow should be studied closely. They are applicable to any equation of state.

We are given

$$\left(P + \frac{an^2}{V^2}\right)(V - nb) = nRT$$

$$dE = n\bar{C}_V\ dT + \left[T\left(\frac{\partial P}{\partial T}\right)_V - P\right]\ dV$$

With Q = 0 for a reversible adiabatic process,

$$dE = w$$

$$n\bar{C}_V\ dT + \left[T\left(\frac{\partial P}{\partial T}\right)_V - P\right]\ dV = -\ P\ dV$$

Since $$P = \frac{nRT}{V - nb} - \frac{an^2}{V^2}$$

we can find

$$\left(\frac{\partial P}{\partial T}\right)_V = \frac{nR}{V - nb}$$

Using the fact that dE = w along with this last equation yields

$$n\bar{C}_V\ dT + \left[\frac{TnR}{V - nb} - \left(\frac{nRT}{V - nb} - \frac{an^2}{V^2}\right)\right] dV = -\left(\frac{nRT}{V - nb} - \frac{an^2}{V^2}\right) dV$$

$$n\bar{C}_V\ dT + \frac{an^2}{V^2}\ dV = -\frac{nRT\ dV}{V - nb} + \frac{an^2}{V^2}\ dV$$

$$n\bar{C}_V\ dT = -\frac{nRT}{V - nb}\ dV$$

$$\frac{\bar{C}_V}{RT}\ dT = -\frac{dV}{V - nb}$$

It is easy to integrate the above equation (*if $\bar{C}_V$ is constant*).

$$\int_{T_1}^{T_2} \frac{\bar{C}_V}{RT}\ dT = -\int_{V_1}^{V_2} \frac{dV}{V - nb}$$

$$\frac{\bar{C}_V}{R} \ln \frac{T_2}{T_1} = -\ln \frac{V_2 - nb}{V_1 - nb} = +\ln \frac{V_1 - nb}{V_2 - nb} \qquad (3\text{-}2.4.12)$$

or

$$\left(\frac{T_2}{T_1}\right)^{\bar{C}_V/R} = +\frac{V_1 - nb}{V_2 - nb} \qquad (3\text{-}2.4.13)$$

To arrive at (3-2.4.13) we have assumed that $\bar{C}_V$ is independent of temperature. Usually this is not the case, so we return to (3-2.4.12) and assume $\bar{C}_V = a + bT + cT^{-2}$. Then we get

$$\int_{T_1}^{T_2} \frac{a + bT + cT^{-2}}{RT}\ dT = -\int_{V_1}^{V_2} \frac{dV}{V - nb}$$

$$\frac{a}{R} \ln \frac{T_2}{T_1} + \frac{b}{R}(T_2 - T_1) - \frac{c}{2R}(T_2^{-2} - T_1^{-2}) = -\ln \frac{V_2 - nb}{V_1 - nb} \qquad (3\text{-}2.4.14)$$

This last equation is rather complicated. For a more complex equation of state, the result would be even more involved. However, in principle we can relate the volume change to the temperature change for any gas. As you will see below, we can manipulate Eq. (3-2.4.14) to obtain the missing variable (T or V, depending on which is known) by a series of approximations. The availability of computers that can do a series of approximations quickly allows for rather rapid solution of very complex expressions.

3-2.4.4 EXAMPLE

Assume we expand 1 mol of a van der Waals gas reversibly and adiabatically from 23 dm^3 to 46 dm^3 with an initial temperature of 298 K. If $\bar{C}_V$ = 20.3

$+ 3.77 \times 10^{-3}\ T - 0.50 \times 10^{5}\ T^{-2}$ (J mol^{-1} K^{-1}), find the final temperature of the gas, the energy change, and the work involved; $a = 1.39$ dm^{6} atm mol^{-2} and $b = 0.039$ dm^{3} mol^{-1}.

Solution

The process is adiabatic, so $Q = 0$ and $\Delta E = W$. From (3-1.4.7)

$$\Delta E = \int_{T_1}^{T_2} \bar{C}_V\, dT + \int \left[T\left(\frac{\partial P}{\partial T}\right)_V - P\right] dV$$

We need to find T_2 before evaluating ΔE.

For 1 mol of a van der Waals gas, Eq. (3-2.4.14) applies:

$$a \ln \frac{T_2}{T_1} + b(T_2 - T_1) - \frac{c}{2}(T_2^{-2} - T_1^{-2}) = -R \ln \frac{V_2 - b}{V_1 - b}$$

This gives

$$20.3 \ln \frac{T_2}{298} + (3.77 \times 10^{-3})(T_2 - 298) + \frac{0.5 \times 10^{5}}{2}(T_2^{-2} - 298^{-2})$$

$$= -8.314 \ln \frac{46 - 0.04}{23 - 0.04} = -8.314 \ln 2.0 = -5.77$$

$$20.3 \ln T_2 - 115.6 + 3.77 \times 10^{-3}\ T_2 - 1.12 + (0.25 \times 10^{5})\ T_2^{-2} - 0.28 = -5.77$$

or $\quad 20.3 \ln T_2 + 3.77 \times 10^{-3}\ T_2 + 0.25 \times 10^{5}\ T_2^{-2} = 111.2$

How do we possibly solve this equation? The easiest way is to make an approximation to the solution. Note that the temperature terms on the left are small except for the first one. So we rearrange to get

$$20.3 \ln T_2 = 111.2 - 3.77 \times 10^{-3}\ T_2 - 0.25 \times 10^{5}\ T_2^{-2}$$

$$\approx 111.2$$

$$\ln T_2 \approx 5.48$$

$$T_2 \approx 240$$

Now substitute T_2 into the above equation on the right side and recalculate T_2 on the left.

$$20.3 \ln T_2 \approx 111.2 - 3.77 \times 10^{-3}\ (240) - 0.25 \times 10^{5}\ (240)^{-2}$$

$$\approx 110$$

$$\ln T_2 \approx 5.41$$

$$T_2 \approx 224$$

Iterate once more to get

$$20.3 \ln T_2 = 111.2 - 3.77 \times 10^{-3}\ (224) - 0.25 \times 10^{5}\ (224)^{-2}$$

$$= 109.9$$

$$\ln T_2 = 5.41$$

$$T_2 = \underline{\underline{224}}$$

This may not be the simplest method of finding T_2, but it is a useful procedure on many problems. Small terms can be ignored to find an approximate solution. This solution is then used in the small terms to find a better solution. Iteration is continued until two successive solutions agree. You should be aware that this does not always work. There are cases in which convergence is too slow to use the technique.

Next we need to find ΔE.

$$\Delta E = \int_{T_1}^{T_2} C_V\ dT + \int_{V_1}^{V_2} \left[T\left(\frac{\partial P}{\partial T}\right)_V - P\right] dV$$

For 1 mol of a van der Waals gas

$$\left(P + \frac{a}{V^2}\right)(V - b) = RT$$

$$P = \frac{RT}{V - b} - \frac{a}{V^2}$$

$$\left(\frac{\partial P}{\partial T}\right)_V = \frac{R}{V - b}$$

$$\left[T\left(\frac{\partial P}{\partial T}\right)_V - P\right] = \frac{TR}{V - b} - \frac{RT}{V - b} + \frac{a}{V^2} = \frac{a}{V^2}$$

Therefore

$$\Delta E = \int_{T_1}^{T_2} [20.3 + 3.77 \times 10^{-3}\ T - 0.50 \times 10^{5}\ T^{-2}]\ dT\ (J) + \int_{V_1}^{V_2} \frac{a\ dV}{V^2}$$

$$= [20.3(224 - 298) + \frac{3.77 \times 10^{-3}}{2} (224^2 - 298^2)$$

$$+ 0.50 \times 10^{5}\ (224^{-1} - 298^{-1})]\ (J)$$

$$- (1\ \text{mol})^2\ (1.39\ \text{dm}^6\ \text{atm mol}^{-2})\left(\frac{1}{23\ \text{dm}^3} - \frac{1}{46\ \text{dm}^3}\right)$$

$$= -\ 1520\ \text{J} - 0.0302\ \text{dm}^3\ \text{atm}\left(\frac{101\ \text{J}}{1\ \text{dm}^3\ \text{atm}}\right)$$

$$= -\ 1520 - 3.05 = -\ 1523\ \text{J} = \underline{\underline{-1520\ \text{J}}}$$

It is clear that the energy change due to the volume change is very small. This is usually the case, but it is not an assumption that can always be made.

3-2.5 PROCESSES INVOLVING PHASE CHANGES

A process involving a phase change is one in which the substance under consideration changes from one physical state (gas, liquid, or solid) to another physical state. At this time, we will limit our consideration to the changes listed in Table 3-2.5.1. There are other possible changes, such as one solid phase to another. These will be treated in Chap. 13, Phase Equilibria.

For reasons that will not be given until the discussion of phase equilibria, the temperature of a pure material is constant if two phases (two physical states) coexist. It follows that all the processes shown in Table 3-2.5.1 will be isothermal. Some of you might be tempted to say that $\Delta E = \Delta H = 0$ if the temperature is constant. *That is not the case at all.* A large amount of energy is required to change a liquid to a gas. (You have boiled water before, so you know that.) This will lead to a discontinuity in the E (or H) vs. T plot for a substance. Consider Fig. 3-2.5.1 for a plot of E. vs. T for an ideal substance. There is no scale, since we cannot calculate absolute values of E (at least not until we get to Chap. 9, Statistical Thermodynamics).

3-2.5.1 EXAMPLE

To illustrate the procedure, ΔE and ΔH will be calculated for the conversion of $H_2O(s)$ at 200 K to $H_2O(g)$ at 450 K, at constant P. We will simplify the problem considerably by assuming the following:

1. C_V Values are independent of temperature.
2. $H_2O(g)$ behaves as an ideal gas.

Neither assumption is accurate, but this is for illustration only.

Table 3-2.5.1
Phase Changes Possible in a System

State 1	State 2	Name
Solid	liquid	Fusion (melting)
Solid	gas	Sublimation
Liquid	gas	Vaporization
Gas	liquid	Condensation
Gas	solid	Condensation

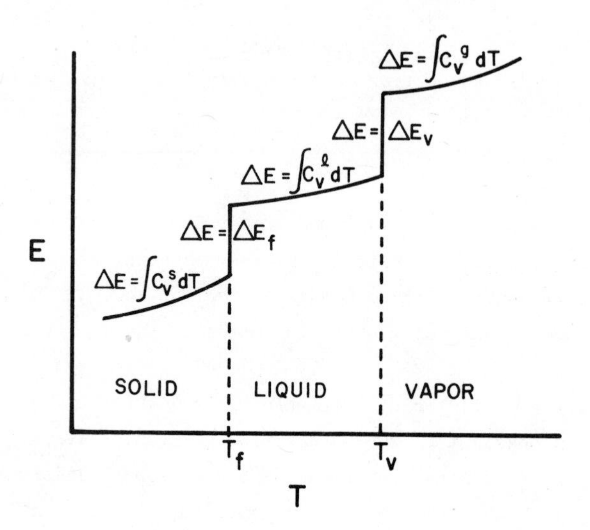

FIG. 3-2.5.1 Idealized plot of E vs. T for a pure substance.

Solution

The data needed to work this problem are given:

$$\bar{C}_P\ (s) = 37.7\ \mathrm{J\ mol^{-1}\ K^{-1}}$$
$$\bar{C}_P\ (\ell) = 75.5\ \mathrm{J\ mol^{-1}\ K^{-1}}$$
$$\bar{C}_P\ (g) = 30.5\ \mathrm{J\ mol^{-1}\ K^{-1}}$$
$$\Delta\bar{H}_{fus} = 6020\ \mathrm{J\ mol^{-1}} \qquad \text{fus = fusion}$$
$$\Delta\bar{H}_{vap} = 40660\ \mathrm{J\ mol^{-1}} \qquad \text{vap = vaporization}$$

Before actual calculation, outline the steps involved.

$H_2O(s)$ 200 K → (ΔE_1, ΔH_1) → $H_2O(s)$ 273 K → (ΔE_2, ΔH_2) → $H_2O(\ell)$ 273 K → (ΔE_3, ΔH_3) → $H_2O(\ell)$ 373 K → (ΔE_4, ΔH_4) → $H_2O(g)$ 373 K → (ΔE_5, ΔH_5) → $H_2O(g)$ 400 K

$$\Delta E_T = \sum_{i=1}^{5} \Delta E_i \qquad \Delta H_T = \sum_{i=1}^{5} \Delta H_i$$

Step 1. From Example 3-1.5.3 we can see for condensed phases $\Delta E \approx \Delta H$.

$$\Delta H_1 = \Delta E_1 = \int_{T_1}^{T_2} \bar{C}_P(s)\ dT$$

$$= (37.7\ \mathrm{J\ mol^{-1}\ K^{-1}})(273 - 200\ \mathrm{K})$$

$$= \underline{\underline{2750 \text{ J mol}^{-1}}}$$

Step 2. $\Delta H_2 = \Delta H_{fus} = \underline{\underline{6020 \text{ J mol}^{-1}}}$. Since the volume change in going from solid to liquid is very small

$$\Delta E_2 \approx \Delta H_2$$

Step 3. As in Step (1), $\Delta H \approx \Delta E$.

$$\Delta H_3 = \Delta E_3 = \int_{273}^{373} C_P(\ell)\, dT$$

$$= (75.5 \text{ J mol}^{-1} \text{ K}^{-1})(373 - 273 \text{ K})$$

$$= \underline{\underline{7550 \text{ J mol}^{-1}}}$$

Step 4. $\Delta H_4 = \Delta H_{vap} = 40660 \text{ J mol}^{-1}$

$$\Delta E_4 = \Delta H_4 - \Delta(PV) = \Delta H_4 - P\,\Delta V$$

$$= \Delta H_4 - P(V_g - V_\ell)$$

$$V_g \approx \frac{nRT}{P}$$

$$\approx \frac{(0.0821 \text{ dm}^3 \text{ atm mol}^{-1} \text{ K}^{-1})(373 \text{ K})}{1 \text{ atm}}$$

$$\approx \underline{30.6 \text{ dm}^3 \text{ mol}^{-1}}$$

$$V_\ell = \frac{m_\ell}{\rho_\ell} \qquad \rho_\ell \approx 1 \text{ g cm}^{-3} = 1000 \text{ g dm}^{-3}$$

$$= \frac{18 \text{ g mol}^{-1}}{1000 \text{ g dm}^{-3}}$$

$$= \underline{0.018 \text{ dm}^3 \text{ mol}^{-1}}$$

Thus $V_g - V_\ell \approx V_g \approx 30.6 \text{ dm}^3 \text{ mol}^{-1}$

Therefore

$$\Delta E_4 = 40660 \text{ J mol}^{-1} - (1 \text{ atm})(30.6 \text{ dm}^3 \text{ mol}^{-1})\left(\frac{101 \text{ J}}{1 \text{ dm}^3}\right)$$

$$= \underline{\underline{37570 \text{ J mol}^{-1}}}$$

Step 5.

$$\Delta H_5 = \int_{373}^{400} \bar{C}_P(g)\, dT = (30.5 \text{ J mol}^{-1} \text{ K}^{-1})(400 - 373 \text{ K})$$

$$= \underline{\underline{825 \text{ J mol}^{-1}}}$$

$$\Delta E_5 = \int_{373}^{400} \bar{C}_V(s)\ dT \approx \int_{373}^{400} (\bar{C}_P - R)\ dT$$

$$= (22.2\ \text{J mol}^{-1}\ \text{K}^{-1})(400 - 373\ \text{K})$$

$$= \underline{\underline{600\ \text{J mol}^{-1}}}$$

$$\Delta H_T = 2750 + 6020 + 7550 + 40660 + 825$$

$$= \underline{\underline{57800\ \text{J mol}^{-1}}}$$

$$\Delta E_T = 2750 + 6020 + 7550 + 37570 + 600$$

$$= \underline{\underline{54500\ \text{J mol}^{-1}}}$$

A useful exercise now is to make a plot similar to Fig. 3-2.5.1. Since you do not know absolute values of E and H at any T, arbitrarily assign these values as E_0 and H_0 for $H_2O(s)$ at 200 K. Make such a plot with both H and E on the same plot and to the same scale.

3-2.6 PROBLEMS

1. Five moles of an ideal gas ($\bar{C}_P$ = 7R/2) are heated at constant volume from 120 K to 240 K. Find ΔE, ΔH, Q, and W.
2. One mole of an ideal gas ($\bar{C}_P$ = 5R/2) is expanded adiabatically from 600 K at 7.00 atm to a final pressure of 1.00 atm. Find ΔE and ΔH for this change if the expansion is
   a. Reversible
   b. Irreversible against a constant pressure of 1 atm
3. Find ΔH for the constant pressure heating of 2 mol Cu(s) from 25°C and 1 atm to 150°C and 1 atm. What would you estimate ΔE to be? What additional information would be needed to find W exactly?
4. How much heat is required to vaporize 37.2 g of water?
5. Estimate W for the reversible, isothermal expansion of 1 mol N_2 from 10 dm^3 to 30 dm^3 at 310 K using the Beattie-Bridgeman equation.
6. The densities of H_2O are as follows:

$$\rho(s,\ 0^\circ C) = 0.917\ \text{kg dm}^{-3}$$
$$\rho(\ell,\ 0^\circ C) = 0.9998\ \text{kg dm}^{-3}$$
$$\rho(\ell,\ 100^\circ C) = 0.9584\ \text{kg dm}^{-3}$$
$$\rho(g,\ 100^\circ C) = 0.000596\ \text{kg dm}^{-3}$$

   Find $\Delta H_{fus} - \Delta E_{fus}$ at 0°C and $\Delta H_{vap} - \Delta E_{vap}$ at 100°C, 1 atm.
7. Three moles of CO are expanded reversibly and adiabatically from T_1 = 400 K and

P_1 = 1.0 atm to P_2 = 0.1 atm. Calculate the final temperature and ΔE for this process ($\bar{C}_P = 7R/2$ and $\bar{C}_V = 5R/2$, if CO behaves

   a. Ideally
   b. As a van der Waals gas

8. Estimate the Joule-Thomson coefficient for CO_2 at 10 atm and 300 K using the van der Waals and Beattie-Bridgeman equations. Compare with Fig. 3-2.3.2.

9. a. Calculate ΔE and ΔH for the heating of 27 g NH_3 from 295 K to 376 K at a constant pressure of 1 atm. Assume $\bar{C}_V = \bar{C}_P - R$, and assume the Redlich-Kwong equation applies. See Table 2-2.6.2 for the appropriate critical constants.
   b. Calculate ΔE for the isothermal compression (T = 280 K) of 27 g NH_3 from 40.1 dm^3 to 30.6 dm^3 if the Redlich-Kwong equation applies.

10. For H_2O, $\bar{C}_P = 75.5$ J mol^{-1} K^{-1}, $\rho(0°C) = 0.9998$ kg dm^{-3}, and $\rho(99°C) = 0.9584$ kg dm^{-3}. Find ΔE and Q for the isobaric heating of 1 mol H_2O at 1.0 atm from 0°C to 99°C.

11. For Cl_2, $\bar{C}_P = 34.0$ J mol^{-1} K^{-1} at 300 K. If this value is constant and $\bar{C}_P = \bar{C}_V - R$, find the final T and P if 0.76 mol of Cl_2 is adiabatically, reversibly expanded from $V_1 = 32.72$ dm^3 and $T_1 = 290$ K to $V_2 = 41.6$ dm^3. Outline the method you would use to find ΔE.

3-2.7 SELF-TEST

1. If CH_4 (methane) can be assumed to behave ideally find Q, ΔE, W, T_{fin}, and V_{fin} (fin = final) for the adiabatic, reversible expansion of 2 mol from 400 K and 2 atm to 0.2 atm, given that γ for CH_4 is 1.31.

2. Find ΔH for the constant pressure heating of 1.36 mol of I_2 vapor from 350 K to 1100 K.

3. Calculate ΔH, ΔE, Q, and W for the reversible, isothermal expansion of 1.5 mol of He at 120 K from P_1 = 5 atm to P_2 = 1 atm, assuming the van der Waals equation applies.

4. Repeat Problem (3) using the Beattie-Bridgeman equation. (Hint: The problem can be solved more easily if the constants are inserted before starting your calculations. Some terms are small enough to be neglected.)

3-2.8 ADDITIONAL READING

1.* W. F. Sheehan, *Physical Chemistry*, 2nd ed., Allyn and Bacon, Boston, 1970, Sec. 2-14.

2.* E. W. Washburn, ed., *International Critical Tables of Numerical Data*, McGraw-Hill, New York, 1926.

3. See Sec. 1-1.9B for references on thermodynamic data compilations.

3-3 THERMOCHEMISTRY

OBJECTIVES

You shall be able to

1. Describe an experimental method used for determining heats of reaction
2. Calculate ΔH and/or ΔE of reaction by applying Hess' law of heat summation and: tables of heats of combustion and/or tables of heats of formations
3.* Calculate ΔH for a reaction at a given T is the heat capacities of reactants and products as a function of T are given and ΔH is known at some T

3-3.1 INTRODUCTION

In this section both the experimental methods of calorimetry and the results will be discussed. A very important tool for determining heats of reactions will be developed. This is Hess' law of heat summation. The importance of this law lies in the fact that for the vast majority of reactions we cannot conveniently (or at all) measure the heats involved. Hess' law allows us to combine reaction equations for which ΔH values are known to get the reaction equation in which we are interested, and at the same time find ΔH for that reaction without any experimental work.

The ability to predict ΔH for reactions that cannot be characterized experimentally is, obviously, a great aid to the engineer designing a chemical processing system, or to the bench chemist who needs to predict what will occur under certain reaction conditions. Finally, we will determine how the heat of reaction varies with temperature. As we will observe, ΔH may be very strongly dependent on temperature. In fact, there are cases in which the sign of ΔH will change if the temperature is changed sufficiently.

3-3.2 CALORIMETRY

Calorimetry is an experimental portion of thermodynamics concerned with precise (and accurate) determination of heat capacities, heats of combustion and reactions, and heats of fusion. Calorimeters measure the amount of heat generated or absorbed by reactions or phase changes. These measurements are used to obtain ΔH (if constant P is maintained) or ΔE (if constant V is used), for reactions and phase changes. Calorimetry is one of the most precise of the experimental sciences, and considerable effort has gone into making calorimeters as accurate as physically possible.

Most calibrations of calorimeters are done by using electrical means. Electrical energies can be measured with exceptional accuracy and with relative ease. A resistance heater inserted in a calorimeter can be used to deliver a precise amount of energy. The heater resistance, voltage across the heater, and the time of electrical current flow are all that is necessary to accurately determine the heat input into the calorimeter. The current is given by

$$I = \frac{V_h}{R_h}$$

where V_h is the voltage across the heater and R_h is the heater resistance. The energy dissipated is

$$E = \int_{t_1}^{t_2} I^2 R_h \, dt = I^2 R_h (t_2 - t_1)$$

if R_h is independent of temperature, which is usually the case over the small temperature changes experienced in a calorimeter.

The changes in the calorimeter (T, liquid level in a dilatometer, etc.) for a given heat input serve as a calibration for changes brought about due to a sample in the calorimeter.

ISOTHERMAL CALORIMETERS. The first calorimeters used were those in which the heat of a reaction was measured by weighing the amount of water produced by the melting of ice during a reaction. This technique was not very accurate. However, an adaptation of the technique, in which the volume change of ice on melting is observed has been used to obtain very good measurements of heat capacities. A sketch of such a calorimeter is given in Fig. 3-3.2.1. Tube B is filled with air-free water and mercury and is immersed in a mixture of ice and water in a Dewar flask. Some of the water in B is frozen (after all the liquid in B has reached 0°C) by placing a freezing mixture into tube A. Then, any other material that is put into A (that releases heat on reaction) will melt some of the ice formed in B, causing the mercury level at D to change. This causes a weight change of the container D. The system is calibrated either by electrical means or by using a material with a known heat capacity

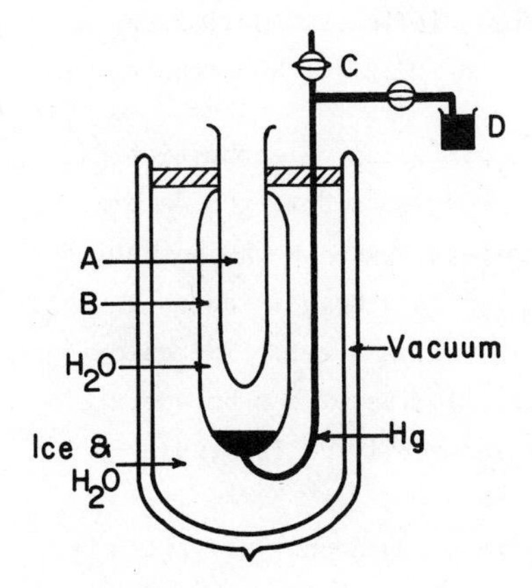

FIG. 3-3.2.1 Ice calorimeter.

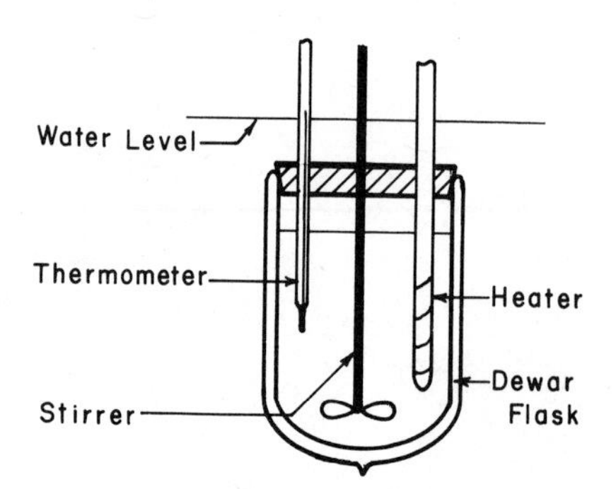

FIG. 3-3.2.2 Simple calorimeter for determining heat capacities of liquids.

and a known temperature. Precision ice calorimeters have been built that attain an accuracy of 0.2% or better.

ADIABATIC CALORIMETERS. Heat capacities are usually measured in an adiabatic calorimeter. In such a calorimeter, every precaution is taken to eliminate heat exchange with the surroundings. In many cases, the container temperature is monitored and the jacket temperature continuously changed to maintain as near zero differential as possible. A simple adiabatic calorimeter for determining heat capacities is shown in Fig. 3-3.2.2. The calorimeter shown uses a Dewar flask to eliminate heat losses. The water surrounding the calorimeter is kept at a constant temperature near that of the inside of the calorimeter, or the temperature is maintained at that of the inside. The liquid in the flask is stirred at a constant rate by the propeller stirrer. Heat is added by passing current through the heater, and the temperature is monitored. Given a certain quantity of liquid, the electrical energy added, and the temperature change, heat capacity can be found. This simple device gives results that can be accurate to about 1%. Much more elaborate setups are accurate to 0.01% or better.

Heat capacity measurements for gases are much more difficult. Solids can be handled in a calorimeter similar to that shown in Fig. 3-3.2.2, without the stirrer.

BOMB CALORIMETERS. Due to the interest in the application of combustion data to the evaluation of heats of formation (see the following section), the determination of heats of combustion has long occupied an important place in thermochemistry. Heats of combustion are usually determined by combustion in a bomb in an excess of oxygen. A schematic of a bomb calorimeter is shown in Fig. 3-3.2.3. The calorimeter is surrounded by a jacket and immersed in water. The calorimeter can be operated adiabatically or isothermally. The bomb calorimeter is usually calibrated by combusting benzoic acid as a standard.

The sample to be combusted is placed in the platinum cup A and is ignited by the electrical heater B after the bomb has been flushed and filled with oxygen to

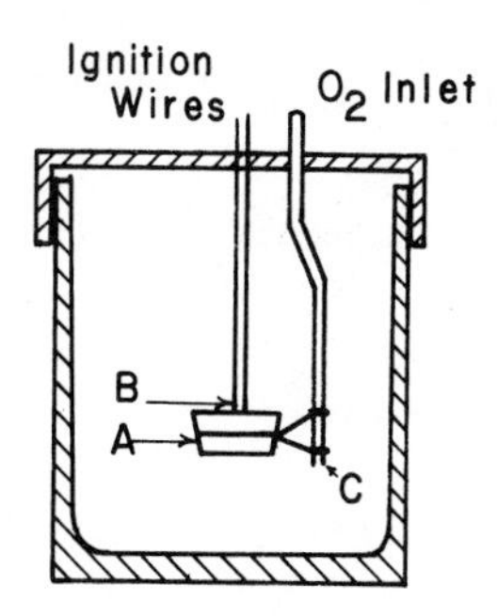

FIG. 3-3.2.3 Bomb for combustion calorimeter.

about 30 atm pressure. In the case of liquids, the liquid is enclosed in thin-walled glass ampoules to prevent evaporation of the sample before combustion.

The bomb calorimeter is a constant volume device. Hence ΔE values are obtained. These can be converted to ΔH by formulas you have seen in previous sections.

The following sections will show how the experimental data obtained by various calorimetric methods are used. In all cases, extreme care is taken to avoid heat leaks. For a complete description of some of the factors that must be taken into account, see Sturtevant.*

3-3.3 HEATS OF REACTION

The methods of measuring heats of reactions have been briefly described above. Now it is time to describe what is being measured. In any chemical reaction the following occurs:

$$\text{Reactants} \rightarrow \text{products}$$

This reaction may release energy (ΔH and ΔE negative), in which case it is called an *exothermic* process. (Before continuing, justify to yourself that an exothermic process should have a negative ΔH and ΔE.) If energy is absorbed (ΔH and ΔE positive), then the reaction is *endothermic*.

ΔH (REACTION) AND ΔE (REACTION). The enthalpy change or energy change in a reaction is determined by

$$\Delta H = H_{prod} - H_{react}$$

or $$\Delta E = E_{prod} - E_{react}$$

Consider a particular reaction as an example.

---

*For a complete treatment of calorimetry see "Calorimetry" by J. M. Sturtevant (Sec. 3-3.10, Ref. 1).

$$C_2H_4(g) + 3O_2(g) \rightarrow 2CO_2(g) + 2H_2O(\ell) \tag{3-3.3.1}$$
Ethylene

Before we can do anything else, we must make certain the reaction equation is balanced. The reason for this is obvious from the following equation:

$$\Delta H = 2\bar{H}_{CO_2(g)} + 2\bar{H}_{H_2O(\ell)} - 3\bar{H}_{O_2(g)} - \bar{H}_{C_2H_4(g)} \tag{3-3.3.2}$$

Note that the coefficients in the stoichiometric equation are required in the calculation of the enthalpy change. The same is true for ΔE.

$$\Delta E = 2\bar{E}_{CO_2(g)} + 2\bar{E}_{H_2O(\ell)} - 3\bar{E}_{O_2(g)} - 3\bar{E}_{C_2H_4(g)} \tag{3-3.3.3}$$

ΔH and ΔE are, of course, related.

$$\Delta H = \Delta E + \Delta(PV) \tag{3-3.3.4}$$

If the volume of liquids and solids are ignored (they are very small in comparison to the volume of any gas in the reaction), and if the gases behave ideally, we can write

$$\Delta H = \Delta E + \Delta(nRT) \tag{3-3.3.5}$$

If the reaction is carried out isothermally, (3-3.3.5) becomes

$$\Delta H = \Delta E + RT\,\Delta n$$

$$= \Delta E + RT(n_{gp} - n_{gr}) \tag{3-3.3.6}$$

(where gp = gaseous products and gr = gaseous reactants).

$$n_{gp} = 2$$

$$n_{gr} = 4$$

$$\Delta n = 2 - 4 = -2$$

Thus $$\Delta H = \Delta E + RT(-2)$$

$$= \Delta E - 2RT \tag{3-3.3.7}$$

Equation (3-3.3.6) indicates that for an exothermic process more heat is released at constant pressure (ΔH) than at constant volume (ΔE) if Δn is negative (meaning a decrease in the number of moles of gas). The opposite is true if Δn is positive.

ΔH and ΔE are both path independent. It is immaterial how the reaction is carried out as long as the initial and final states are brought back to a particular condition. There may be many steps not shown in the stoichiometry of the reaction. This has no effect. The ΔH or ΔE depends only on the products and reactants of the written equation and on their final and initial states, respectively. There may be a catalyst added to promote the reaction. This will not affect ΔH or ΔE. These considerations are embodied in *Hess' law of heat summation*, which will be considered shortly.

For the reaction given in (3-3.3.1), it is found experimentally that $\Delta H = -1411$ kJ mol^{-1} (ethylene) at 298 K. Be careful when you write $\Delta H = x$ kJ mol^{-1}. Designate per mole of what substance. You have to know what substance is your base compound (see Table 3-3.3.1). From (3-3.3.7) we know immediately that

$$\begin{aligned}\Delta E &= \Delta H + 2RT \\ &= -1411 + (2\ \text{mol})(8.314\ \text{J mol}^{-1}\ \text{K}^{-1})(298\ \text{K})\left(\frac{1\ \text{kJ}}{10^3\ \text{J}}\right) \\ &= (-1411 + 4.95)\ \text{kJ} \\ &= \underline{\underline{-1406\ \text{kJ}}}\end{aligned}$$

(for the reaction as written). (Note: Much data in the literature are in terms of calories. Since 1 cal = 4.184 J, ΔE = (-1406 kJ)(1 k cal/4.184 kJ) = 336.0 kcal. The conversion from cal to J is a useful one to remember. Also, R = 1.9872 cal mol^{-1} K^{-1} = 8.314 J mol^{-1} K^{-1}.)

You can see that the reaction produces more heat at constant P (ΔH) than at constant V (ΔE). The volume decreases on reaction at constant pressure since the number of moles has decreased. For a volume decrease at constant T, heat must be lost to the surroundings. Hence, ΔH is more negative than ΔE for this reaction.

3-3.3.1 EXAMPLE

The heat of combustion [heat of complete reaction with O_2 to produce only $H_2O(\ell)$ and $CO_2(g)$] of ethanol (ℓ) at constant pressure and 298 K is -1367 kJ mol^{-1} (ethanol). What is the heat of combustion at constant volume and 298 K? (Assume the gases behave ideally.)

Solution

First you *must* write a balanced reaction.

$$\underset{\text{Ethanol}}{C_2H_5OH(\ell)} + 3O_2(g) \rightarrow 2CO_2(g) + 3H_2O(\ell)$$

$$\Delta H = \Delta E + RT\ \Delta n$$

$$\begin{aligned}\Delta E &= \Delta H - RT\ \Delta n \\ &= \Delta H - RT(2\ \text{mol product gas} - 3\ \text{mol reactant gas}) \\ &= \Delta H + RT(1\ \text{mol}) \\ &= -1367\ \text{kJ} + (1\ \text{mol})(8.314\ \text{J mol}^{-1}\ \text{K}^{-1})(298\ \text{K})\left(\frac{10^{-3}\ \text{kJ}}{\text{J}}\right) \\ &= \underline{\underline{-1364\ \text{kJ}}} = \underline{\underline{326.0\ \text{kcal}}}\end{aligned}$$

for the reaction as written.

If the gases are not ideal, we evaluate $\Delta(PV)$ in (3-3.3.4) using an appropriate equation of state. Otherwise the procedures are the same as illustrated.

Table 3-3.3.1
Heats of Combustion ΔH at 298 K[a]

Substance	ΔH_{comb}
$H_2(g)$	-285.84
C(graphite)	-393.51
CO(g)	-282.99
$CH_4(g)$, Methane	-890.36
$C_2H_6(g)$, Ethane	-1559.90
$C_3H_8(g)$, Propane	-2220.1
n-$C_4H_{10}(g)$, n-Butane	-2878.5
i-$C_4H_{10}(g)$, Isobutane	-2871.6
n-$C_7H_{16}(g)$, n-Heptane	-4811.2
C_2H_4, Ethylene	-1411.0
$C_2H_2(g)$, Acetylene	-1299.6
$C_6H_6(g)$, Benzene	-3301.5
$C_2H_5OH(\ell)$, Ethanol	-1367.0
CH_3COOH, Acetic acid	-872.4

[a] In kJ mol^{-1} of reactant. Products are solely $H_2O(\ell)$ and $CO_2(g)$; 4.184 kJ = 1 kcal.

You should spend a few moments to see what trends are present in Table 3-3.3.1. For example, in the alkane series (methane, ethane, propane, etc.), what is the effect on ΔH_{comb} of each additional $-CH_2-$ unit. Calculate the difference [e.g., $\Delta H_{comb}(C_2H_6) - \Delta H_{comb}(CH_4)$] between adjacent alkanes to see if it is constant.

HESS' LAW. In the previous section Hess' law was mentioned. The essence of this law was stated above. Only initial and final states must be considered. It does not matter if intermediate reactions are involved, if temperatures change during the process, or if catalysts are used. Only the initial and final conditions determine the heat of the reaction. The importance of this should be self-evident. We can take heats of reactions for known reactions and use them to predict heats of reactions for those that have never been studied. For example, suppose we want to know the heat of the reaction

$$3C_2H_2(g) \rightarrow C_6H_6$$

Acetylene Benzene

From heats of combustion of acetylene and benzene given in Table 3-3.3.1 we can write

1. $C_2H_2(g) + \frac{5}{2}O_2(g) \rightarrow 2CO_2(g) + H_2O(\ell)$

$\Delta H = -1299.6$ kJ mol^{-1}

$$C_6H_6(g) + \frac{15}{2}O_2(g) \rightarrow 6CO_2(g) + 3H_2O(\ell)$$

$\Delta H = -3301.5$ kJ mol^{-1}

(The values given for ΔH are for the reaction as written.)

Another point should be stressed. Chemical equations and the corresponding ΔH values can be treated like mathematical equations. For example, let us multiply equation (1) by 3 and equation (2) by - 1. This gives

$$3\left[C_2H_2(g) + \frac{5}{2}O_2(g) \rightarrow 2CO_2(g) + H_2O(\ell) \qquad \Delta H = -1299.6 \text{ kJ}\right]$$

$$-1\left[C_6H_6(g) + \frac{15}{2}O_2(g) \rightarrow 6CO_2(g) + 3H_2O(\ell) \qquad \Delta H = -3301.5 \text{ kJ}\right]$$

$$1.' \quad 3C_2H_2(g) + \frac{15}{2}O_2(g) \rightarrow 6CO_2 + 3H_2O(\ell) \qquad \Delta H = -3898.8 \text{ kJ}$$

$$2.' \quad -C_6H_6(g) - \frac{15}{2}O_2(g) \rightarrow 6CO_2 - 3H_2O(\ell) \qquad \Delta H = +3301.5 \text{ kJ}$$

$$1.' + 2.' \quad 3C_2H_2(g) - C_6H_6(g) \rightarrow 0 \qquad \Delta H = -597.3 \text{ kJ}$$

or

$$3C_2H_2(g) \rightarrow C_6H_6(g) \qquad \Delta H = -597.3 \text{ kJ}$$

The main point here is that whatever you do to the equation, you must also do to ΔH. If the reaction is reversed, the sign of ΔH must be changed. The above procedures work for any reaction. They are not limited to heats of combustion.

3-3.4 STANDARD HEATS OF FORMATION

SOLIDS, LIQUIDS, AND GASES. In order to tabulate enthalpy information for compounds and elements not in their most stable states, use is made of a definition of a *standard state* for each element. This gives rise to values called the *standard heats of formation. The standard state of each element is the physical state and stable form found at 1 atm pressure.* Normally a temperature of 25°C is chosen. The enthalpy of formation (designated by ΔH_f^0) of each element in its most stable form is arbitrarily assigned a value of zero.

When this is done, the combustion data of the previous section and other reaction data can be summarized and tabulated conveniently. For example, the heat of combustion of graphite is given at 25°C and 1 atm.

$$C(\text{graphite}) + O_2(g) \rightarrow CO_2(g) \qquad \Delta H = -393.51 \text{ kJ mol}^{-1} \quad (C)$$

We know

$$\Delta H = \bar{H}(CO_2, g) - \bar{H}(C, \text{graphite}) - \bar{H}(O_2, g)$$

We rewrite this as

$$\Delta H = \Delta\bar{H}_f^0\ (CO_2,\ g) - \Delta\bar{H}_f^0\ (C,\ \text{graphite}) - \Delta\bar{H}_f^0\ (O_2,\ g)$$

where ΔH_f^0 is defined as the enthalpy of formation. But C(graphite) and O_2(g) are in their standard states. So

$$\Delta\bar{H}_f^0\ (C,\ \text{graphite}) = \Delta\bar{H}_f^0\ (O_2,\ g) = 0$$

or $$\Delta H = \Delta\bar{H}_f^0\ (CO_2,\ g) = 393.51\ \text{kJ mol}^{-1}(-94.05\ \text{kcal mol}^{-1})$$

The standard heat of formation of a compound is the enthalpy of reaction of elements in their standard states to form the compound at 1 atm.

The example below shows how heats of combustion can be used to derive heats of formation. Thereafter, heats of formation will be used to calculate enthalpy changes for reactions.

3-3.4.1 EXAMPLE

Try to find $\Delta H_f(CH_4)$ from the data in Table 3-3.3.1) The reaction we are interested in is

$$2H_2(g) + C(\text{graphite}) \rightarrow CH_4(g) \qquad \Delta H_f^0 = ?$$

Solution

From Table 3-3.3.1:

1. $H_2(g) + \frac{1}{2}O_2(g) \rightarrow H_2O(\ell)$ $\qquad \Delta\bar{H}_1 = -285.84\ \text{kJ mol}^{-1}\ (H_2)$
2. $C(\text{graphite}) + O_2(g) \rightarrow CO_2(g)$ $\qquad \Delta\bar{H}_2 = -393.51\ \text{kJ mol}^{-1}\ (C)$
3. $CH_4(g) + 2O_2(g) \rightarrow CO_2(g) + 2H_2O(\ell)$ $\qquad \Delta\bar{H}_3 = -890.36\ \text{kJ mol}^{-1}\ (CH_4)$

If we take [2 × (1)] + [1 × (2)] - [1 × (3)] we should get the desired result. (Note we are multiplying by the number of moles involved so we are left only kilojoules as a unit.)

$$2H_2(g) + O_2(g) \rightarrow 2H_2O(\ell) \qquad \Delta H_1 = -571.68\ \text{kJ}$$

$$C(\text{graphite}) + O_2(g) \rightarrow CO_2(g) \qquad \Delta H_2 = -393.51\ \text{kJ}$$

$$-CH_4(g) - 2O_2(g) \rightarrow -CO_2(g) - 2H_2O(\ell) \qquad \Delta H_3 = +890.36\ \text{kJ}$$

$$2H_2(g) + C(\text{graphite}) - CH_4(g) \rightarrow 0 \qquad \Delta H = -74.83\ \text{kJ}$$

Finally,

$$2H_2(g) + C(\text{graphite}) \rightarrow CH_4(g) \qquad \Delta H_f^0\ (CH_4) = -74.83\ \text{kJ}\quad (-17.88\ \text{kcal})$$

Values of ΔH_f^0 for several compounds are given in Table 3-3.4.1. These are taken from Rossini, et. al., *Selected Values of Chemical Thermodynamic Properties*, (see Sec. 3-3.10, Ref. 2).

We now have the data necessary to calculate the enthalpy change for any reaction (real or hypothetical) that we care to write. We use ΔH_f^0 values as follows:

Reactants → products

$$\Delta H_{reaction} = \Delta H_f^0 \text{ (products)} - \Delta H_f^0 \text{ (reactants)}$$

Let us show that this works for Example 3-3.3.1.

$$3C_2H_2(g) \rightarrow C_6H_6(g)$$

We can write the equation for the formation of each of these from the elements as follows:

1. $2C(\text{graphite}) + H_2(g) \rightarrow C_2H_2(g) \qquad \Delta\bar{H}_f^0 = 226.75 \text{ kJ mol}^{-1}$
2. $6C(\text{graphite}) + 3H_2(g) \rightarrow C_6H_6(g) \qquad \Delta\bar{H}_f^0 = 82.93 \text{ kJ mol}^{-1}$

$(-3 \times 1) \quad -6C(\text{graphite}) - 3H_2(g) \rightarrow -3C_2H_2(g) \qquad -3 \times \Delta\bar{H}_f^0 = -680.25 \text{ kJ}$

$2. - (3 \times 1.) \quad 0 \rightarrow C_6H_6(g) - 3C_2H_2(g) \qquad \Delta H = -597.32 \text{ kJ}$

or $\quad 3C_2H_2(g) \rightarrow C_6H_6(g) \qquad \underline{\underline{\Delta H = -597.32 \text{ kJ}}} \quad (-142.76 \text{ kcal})$

For a general reaction:

$$aA + bB \rightarrow cC + dD$$

$$\Delta H^0 = c\Delta\bar{H}_f^0 \text{ (C)} + d\Delta\bar{H}_f^0 \text{ (D)} - a\Delta\bar{H}_f^0 \text{ (A)} - b\Delta\bar{H}_f^0 \text{ (B)}$$

The superscript indicates standard conditions.

3-3.4.2 EXAMPLE

Calculate the enthalpy change for the reaction

$$SO_2(g) + \tfrac{1}{2}O_2(g) \rightarrow SO_3(g)$$

Solution

From above we know

$$\Delta H^0 = 1 \cdot \Delta\bar{H}_f^0 \text{ (}SO_3\text{)} - 1 \cdot \Delta\bar{H}_f^0 \text{ (}SO_2\text{)} - \tfrac{1}{2}\Delta\bar{H}_f^0 \text{ (}O_2\text{)}$$

And from Table 3-3.4.1:

$$\Delta H^0 = (-395.2 \text{ kJ mol}^{-1})(1 \text{ mol}) - (-296.9 \text{ kJ mol}^{-1})(1 \text{ mol}) - \tfrac{1}{2}(0)$$

$$\Delta H^0 = \underline{\underline{-98.30 \text{ kJ}}}$$

Table 3-3.4.1
Standard Heats of Formation $\Delta\bar{H}_f^0$ of Compounds and Ions at 25°C (kJ mol^{-1})

Substance	$\Delta\bar{H}_f^0$	Substance	$\Delta\bar{H}_f^0$	Substance	$\Delta\bar{H}_f^0$
Ag^+(aq)	1361	C_2H_2(g)	226.75	NO(g)	90.37
Ag_2O(s)	-30.57	C_2H_4(g)	52.283	NO_2(g)	33.84
AgCl(s)	-127.03	C_2H_6(g)	-84.667	NO_3^-(aq)	-206.57
AgBr(s)	-99.50	C_3H_8(g)	-103.85	NH_3(g)	-46.19
Ag_2CO_3(s)	-506.14	n-C_4H_{10}(g)	-126.1	NH_3(aq)	-80.83
Ba^{2+}(aq)	-538.35	iso-C_4H_{10}(g)	-134.5	NH_4^+(aq)	-132.8
$BaCl_2$(s)	-860.06	C_6H_6(g)	82.93	NH_4OH(aq)	-366.7
$BaCl_2 \cdot H_2O$(s)	-1165	Ca^{2+}(aq)	-542.96	Na^+(aq)	-239.66
$BaCl_2 \cdot 2H_2O$(s)	-1461.7	CuF_2(s)	-1215	NaOH(s)	-426.73
$BaSO_4$(s)	-1465	Cl^-(aq)	-167.46	NaCl(s)	-411
Br^-(aq)	-120.9	HCl(g)	-92.31	NaBr(s)	-359.95
Br_2(g)	30.7	F^-(aq)	-329.1	Na_2SO_4(s)	-1384.5
HBr(g)	-36.2	HF(aq)	-269	$Na_2SO_4 \cdot 10H_2O$(s)	-4324.1
C(diamond)	1.896	H^+(aq)	0.000	Na_2CO_3(s)	-1130.9
CO(g)	-110.52	OH^-(aq)	-229.94	$NaHCO_3$(s)	-947.7
CO_2(g)	-393.51	H_2O(g)	-241.83	S(g)	222.8
CO_3^{2-}(aq)	-676.26	H_2O(ℓ)	-285.84	SO_2(g)	-296.9
CH_4(g)	-74.848	I_2(g)	62.24	SO_3(g)	-395.2
HCO_3^-(aq)	-691.11	HI(g)	25.9	SO_4^{2-}(aq)	-907.51
CH_3OH(g)	-201.2	Fe_2O_3(s)	-822.15	H_2S(s)	-20.15
CH_3NO_2(ℓ)	-89.04	Fe_3O_4(s)	-1117.1	H_2SO_4(ℓ)	-811.32

IONS IN SOLUTION. Up to this point we have considered only pure materials. Many reactions take place in solution, and we must be able to find and use *heats of formation of ions in solution.* Consider the dissolving of HCl(g) in a large excess of H_2O. We can write

$$\text{HCl(g)} + \text{excess } H_2O \rightarrow H^+\text{(aq)} + Cl^-\text{(aq)}$$

The (aq) label means the ion is in a large amount of water. Notice below that we treat any effects due to the water in which the solutes are dissolved as though they were contributions due to the solutes themselves. ΔH for this reaction can be written

$$\Delta H = \Delta\bar{H}_f^0\ (H^+,\ aq) + \Delta\bar{H}_f^0\ (Cl^-,\ aq) - \Delta\bar{H}_f^0\ (HCl,\ g)$$

ΔH for this reaction is found to be -75.15 kJ, and from Table 3-3.4.1 $\Delta\bar{H}_f^0$(HCl, g) = -92.31 kJ mol^{-1}. Thus,

$$\Delta\bar{H}_f^0(H^+, aq) + \Delta\bar{H}_f^0(Cl^-, aq) = \Delta H + 1 \cdot \Delta\bar{H}_f^0(HCl, g)$$
$$= -75.15 - 92.31$$
$$= -167.46 \text{ kJ}$$

We can find numerous relations similar to this for a large number of ions; however, there is no way to determine ΔH for a single ion independently. Thus the H^+ ion in solution is taken as a reference ion, and $\Delta\bar{H}_f^0$ is arbitrarily set to zero:

$$\Delta\bar{H}_f^0(H^+, aq) \equiv 0 \tag{3-3.4.1}$$

With this convention you should see immediately that $\Delta\bar{H}_f^0(Cl^-, aq) = -167.46$ kJ mol^{-1}. The other entries in Table 3-3.4.1 are obtained in a similar manner.

We have overlooked one detail (or made an implicit assumption) in the above. We have assumed that the ions act independently in solution. This is found experimentally to be the case for very dilute solutions. If the solutions are not very dilute there will be heats of mixing of the solution, which will be discussed further in Chap. 6, Thermodynamics of Solution. In this chapter, only cases in which our assumption is valid will be considered.

3-3.4.3 EXAMPLE

Given that $\Delta\bar{H}_f^0(NaCl, s)$ is -411.00 kJ mol^{-1}, and the heat of solution of NaCl(s) is 3.88 kJ mol^{-1}, use the value for $\Delta\bar{H}_f^0(Cl^-, aq)$ to predict $\Delta\bar{H}_f^0(Na^+, aq)$ and compare with the values given in Table 3-3.4.1.

Solution

From above,

$$\Delta\bar{H}_f^0(Cl^-, aq) = -167.46 \text{ kJ mol}^{-1}$$

The reaction is

$$NaCl(s) + \text{excess } H_2O \rightarrow Na^+(aq) + Cl^-(aq) \qquad \Delta H = +3.88 \text{ kJ}$$

$$\Delta H = 3.88 \text{ kJ} = 1\cdot\Delta\bar{H}_f^0(Na^+, aq) + 1\cdot\Delta\bar{H}_f^0(Cl^-, aq) - 1\cdot\Delta\bar{H}_f^0(NaCl, s)$$

$$\Delta\bar{H}_f^0(Na^+, aq) = 3.88 \text{ kJ mol}^{-1} - 1\cdot\Delta\bar{H}_f^0(Cl^-, aq) + 1\cdot\Delta\bar{H}_f^0(NaCl, s)$$
$$= 3.88 - (-167.46) + (-411.00) \text{ kJ}$$
$$= \underline{-239.66 \text{ kJ}}$$
$$= \underline{\underline{-239.66 \text{ kJ mol}^{-1}}} \qquad (57.28 \text{ kcal mol}^{-1})$$

(since the above reaction is for 1 mol Na^+). The table value is the same.

3-3.5 HEATS OF REACTION TEMPERATURE DEPENDENCE

In this section, thus far, we have assumed that whatever reaction we are studying is being carried out isothermally. The temperature might change during the course of the reaction, but the final temperature of the products is brought back to the initial temperature of the reactants, 25°C. Our task in this section will be to use heat capacity data to predict the heat of a reaction at any temperature. The reaction will still be carried out isothermally, but the temperature of reaction will be other than 25°C. We will assume constant pressure throughout.

We know, from previous definitions, that

$$\Delta H = H_{prod} - H_{react} \tag{3-3.5.1}$$

Take the partial derivative of this equation with respect to temperature at constant pressure to find

$$\left(\frac{\partial\ \Delta H}{\partial T}\right)_P = \left(\frac{\partial H_{prod}}{\partial T}\right)_P - \left(\frac{\partial H_{react}}{\partial T}\right)_P \tag{3-3.5.2}$$

The right side can be rewritten by applying the definition of heat capacity at constant pressure

$$\left(\frac{\partial\ \Delta H}{\partial T}\right)_P = [C_P(\text{prod}) - C_P(\text{react})] \tag{3-3.5.3}$$

The heat capacity terms as written in (3-3.5.3) are taken to imply the heat capacity of all the products minus the heat capacity of all the reactants.

Equation (3-3.5.3) can be integrated as follows (if P is constant):

$$\int_1^2 d\ \Delta H = \int_{T_1}^{T_2} [C_P(\text{prod}) - C_P(\text{react})]\ dT$$

$$\Delta H_{T2} - \Delta H_{T1} = \int_{T_1}^{T_2} \Delta C_P\ dT \tag{3-3.5.4}$$

where ΔC_P is the difference between $C_P(\text{prod})$ and $C_P(\text{react})$.

Consider the hypothetical reaction

$$xX + yY \rightarrow zZ$$

$$\Delta H_{T2} - \Delta H_{T1} = \int_{T_1}^{T_2} \Delta C_P\ dT$$

$$= \int_{T_1}^{T_2} [z\bar{C}_P(Z) - x\bar{C}_P(X) - y\bar{C}_P(Y)]\ dT \tag{3-3.5.5}$$

If we know $\bar{C}_P(Z)$, $\bar{C}_P(X)$, and $\bar{C}_P(Y)$ as a function of T (and we do; see Sec. 3-2.2) we can determine ΔH_{T2} if we also know ΔH_{T1}. We can find ΔH_{T1} if we choose

T1 = 25°C, the standard condition, from tables similar to Table 3-3.4.1. Thus we are now able to find ΔH for a reaction at any temperature *within the range of validity of our heat capacity data.*

Let us draw a process diagram (Fig. 3-3.5.1) which will, perhaps, help clarify the origin of (3-3.5.5). Before continuing write ΔH_{T2} in terms of ΔH_1, ΔH_{T1}, and ΔH_2. You should see that

$$\Delta H_1 = \int_{T_2}^{T_1} [x\bar{C}_p(X) + y\bar{C}_p(Y)]\ dT = -\int_{T_1}^{T_2} [x\bar{C}_p(X) + y\bar{C}_p(Y)]\ dT$$

(the heat required or released in changing the temperature of the reactants from $T_2 \to T_1$).

$$\Delta H_{T1} = \Delta H_{T1} \qquad \text{heat of reaction at } T_1$$

$$\Delta H_2 = \int_{T_1}^{T_2} z\bar{C}_p(Z)\ dT \qquad \text{heat gained or lost on changing the temperature of the products from } T_1 \to T_2$$

From the equation you should have written already

$$\Delta H_{T2} = \Delta H_1 + \Delta H_{T1} + \Delta H_2$$

We see immediately

$$\Delta H_{T2} = -\int_{T_1}^{T_2} [x\bar{C}_p(X) + y\bar{C}_p(Y)]\ dT + \Delta H_{T1} + \int_{T_1}^{T_2} z\bar{C}_p(Z)\ dT$$

which can be rearranged to give (3-3.5.5).

In Sec. 3-2.2 we were given that $\bar{C}_p = a + bT + cT^{-2}$. Equation (3-3.5.5) can now be written as

$$\Delta H_{T2} - \Delta H_{T1} = \int_{T_1}^{T_2} \Delta C_p\ dT$$

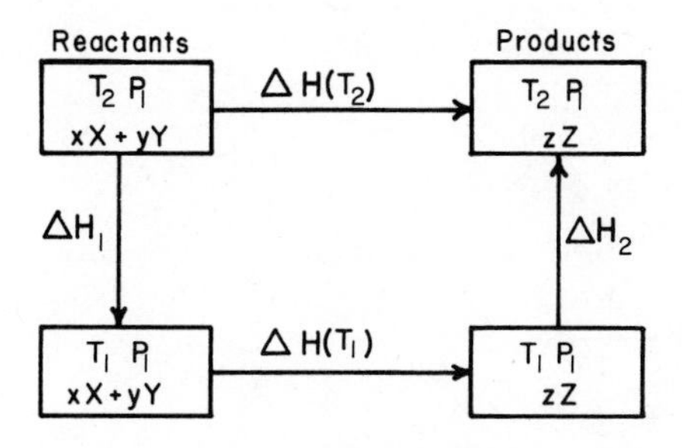

FIG. 3-3.5.1 Process diagram showing how ΔH_{T2} may be found from ΔH_{T1} and heat capacity data.

$$= \int_{T_1}^{T_2} \left[(za_Z - xa_X - ya_Y) + (zb_Z - xb_X - yb_Y)T + (zc_Z - xc_X - yc_Y)T^{-2}\right] dT$$

or
$$\Delta H_{T2} - \Delta H_{T1} = \int_{T_1}^{T_2} \left[\Delta a + \Delta bT + \Delta cT^{-2}\right] dT \quad (3\text{-}3.5.6)$$

Integration yields:

$$\Delta H_{T2} - \Delta H_{T1} = \Delta a\ (T_2 - T_1) + \frac{\Delta b}{2}\ (T_2{}^2 - T_1{}^2) - \Delta c \left(\frac{1}{T_2} - \frac{1}{T_1}\right) \quad (3\text{-}3.5.7)$$

Equation (3-3.5.7) is the one we need. For any particular reaction we must know ΔH_{T1} and Δa, Δb, and Δc, all of which can be found in tables.

Note: A common error made by students is to evaluate ΔC_p in (3-3.5.4) at a given T, say T_2, then assume ΔC_p is constant to integrate as follows:

$$\Delta H_{T2} - \Delta H_{T1} = \int_{T_1}^{T_2} \Delta C_p(T_2)\ dT = \Delta C_p(T_2) \int_{T_1}^{T_2} dT$$

$$= \Delta C_p(T_2)(T_2 - T_1)$$

This is not correct. The integral *must* be done before values of Δa, Δb, and Δc are inserted. In other words, (3-3.5.7) *must* be used.

As stated previously, ΔH_{T1} is usually taken as ΔH^0_{298}. Thus (3-3.5.7) can be rewritten as

$$\Delta H_{T2} = \Delta H^0_{298} + \Delta a\ (T_2 - 298) + \frac{\Delta b}{2} \left[T_2{}^2 - (298)^2\right] - \Delta c \left(\frac{1}{T_2} - \frac{1}{298}\right) \quad (3\text{-}3.5.8)$$

3-3.5.1 EXAMPLE

Find ΔH_{1000} for $2H_2(g) + O_2(g) \rightarrow 2H_2O(g)$ from the following data; $\bar{C}_p = a + bT + cT^{-2}$ (J mol^{-1}) and $\Delta\bar{H}^0_{298} = \Delta\bar{H}^0_f = -241.83$ J mol^{-1}.

Item	$2H_2$	O_2	$2H_2O$	Δ (Item)
ΔH^0_f	0	0	2(-241.83)	-483,660
a	2(27.3)	30.0	2(30.5)	-23.6
$b \times 10^3$	2(3.3)	4.2	2(10.3)	9.8
$c \times 10^{-5}$	2(0.50)	-1.7	0	0.67

Solution

First, find an equation for ΔH at any T_2, then put in the value of 1000. Such a form will be useful in a later chapter.

$$\Delta H_{T2} = \Delta H^0_{298} + \int_{298}^{T_2} \Delta C_p \, dT$$

$$= -483{,}660 + \int_{298}^{T_2} (-23.6 + 9.8 \times 10^{-3} T + 0.67 \times 10^5 T^{-2}) \, dT$$

$$= -483{,}660 - 23.6(T_2 - 298) + \frac{9.8}{2} \times 10^{-3}(T_2{}^2 - 298^2)$$
$$- 0.67 \times 10^5 \left(\frac{1}{T_2} - \frac{1}{298}\right)$$

$$= -483{,}660 + 23.6(298) - 4.9 \times 10^{-3}(298)^2 + 0.67 \times 10^5\left(\frac{1}{298}\right) - 23.6T_2$$
$$+ 4.9 \times 10^{-3} T_2{}^2 - 0.67 \times 10^5\left(\frac{1}{T_2}\right)$$

$$= -476{,}840 - 23.6T_2 + 4.9 \times 10^{-3} T_2{}^2 - 0.67 \times 10^5\left(\frac{1}{T_2}\right) (J) \qquad (3\text{-}3.5.9)$$

To find ΔH at any T we only have to substitute that T into (3-3.5.9). At 1000 K this becomes

$$\Delta H_{1000} = -476{,}840 - 23.6(1000) + 4.9 \times 10^{-3}(1000)^2 - 0.67 \times 10^5\left(\frac{1}{1000}\right)$$
$$= -495{,}610 \text{ J} = \underline{\underline{-496{,}000 \text{ J}}}$$

3-3.6 BOND ENERGIES

The last portion of this section will indicate how thermodynamic data can be used to estimate bond energies. Once bond energies are available, heats of reaction can be estimated for many types of reactions even in the absence of ΔH^0_f data. From our investigation of bond energies we can obtain important information about the nature of bonding in molecules.

It is easy to estimate bond energies for diatomic molecules. Since a diatomic molecule, such as O_2, contains two atoms and only one bond, all we need to do is measure the *dissociation energy* of O_2, which is the energy required for the reaction:

1. $O_2(g) \rightarrow 2O(g) \qquad \Delta H_{diss} = 495.0$ kJ

Values of ΔH_{diss} for several diatomics are found in Table 3-3.6.1.

For more complex molecules, only estimates can be obtained. For example, let us find the O-H bond energy in H_2O. First, for H_2O: The dissociation energy of H_2 can be measured.

2. $H_2(g) \rightarrow 2H(g) \qquad \Delta H_{diss} = 436.0$ kJ

From calorimetric data

3. $H_2 + \frac{1}{2} O_2 \rightarrow H_2O \qquad \Delta H = -241.8$ kJ

If we combine (1), (2), and (3) as follows: (3) - ½(1) - (2) we get

Table 3-3.6.1
Dissociation Energies for Several Diatomic Molecules (kJ)

Molecule	ΔH_{diss}	Molecule	ΔH_{diss}
O_2	495.0	S_2	427.0
H_2	436.0	N_2	941.8
OH	424.7	NO	628.0
F_2	158.0	C_2	602.0
HF	563.2	CO	1073.0
Cl_2	243.0	CH	340.0
HCl	431.8	Li_2	110.0
Br_2	193.0	Na_2	75.3
HBr	366.0	K_2	51.0

$$H_2(g) + \tfrac{1}{2}O_2 \rightarrow H_2O(g) \qquad \Delta H = -241.8 \text{ kJ}$$

$$-\tfrac{1}{2}O_2(g) \rightarrow -O(g) \qquad \Delta H = -247.0 \text{ kJ}$$

$$-H_2(g) \rightarrow -2H(g) \qquad \Delta H = -436.0 \text{ kJ}$$

$$0 \rightarrow H_2O(g) - O(g) - 2H(g) \qquad \Delta H = -924.8 \text{ kJ}$$

or

$$2H(g) + O(g) \rightarrow H_2O(g) \qquad \Delta H = -924.8 \text{ kJ}$$

The molecule H_2O contains two O-H bonds which we assume are identical. Thus the O-H bond energy is found to be 462.4 kJ.

3-3.6.1 EXAMPLE

Find the C-H bond energy given ΔH_f^0 for CH_4, ΔH_{sub} (sublimation) for graphite, and whatever diatomic dissociation enthalpies you may need.

Solution

We want to find ΔH for

1. $C(g) + 4H(g) \rightarrow CH_4(g)$

We apply Hess' law to do so:

2. $C(\text{graphite}) + 2H_2(g) \rightarrow CH_4(g) \qquad \Delta H_f^0 = -74.85 \text{ kJ}$

For ΔH_{sub} an accepted value is

3. $C(\text{graphite}) \rightarrow C(g) \qquad \Delta H_{sub} = 713.0 \text{ kJ}$

Finally from Table 3-3.6.1:

4. $H_2(g) \rightarrow 2H(g)$ $\Delta H_{diss} = 436.0$ kJ

Combine these as follows: (2) - 2(4) - (3)

$$C(graphite) + 2H_2(g) \rightarrow CH_4 \qquad \Delta H = -74.85 \text{ kJ}$$

$$-2H_2(g) \rightarrow -4H(g) \qquad \Delta H = -872.0 \text{ kJ}$$

$$-C(graphite) \rightarrow -C(g) \qquad \Delta H = -713.0 \text{ kJ}$$

$$0 \rightarrow CH_4(g) - 4H(g) - C(g)$$

or

$$C(g) + 4H(g) \rightarrow CH_4(g) \qquad \Delta H = -1660 \text{ kJ}$$

Thus the C-H bond energy is -1660 kJ/4 = -415 kJ. This value can be compared to the value given in Table 3-3.6.2.

The data in Table 3-3.6.2 are subject to considerable uncertainty since assumptions must be made for more complex molecules. For example, in CH_3-CH_3 (ethane) the assumption is made that the C-H bond energy is the same as in CH_4. This is not precisely true. In fact, the bond energy of a C-H bond in CH_4 is not the same as in CH_3, CH_2, etc. The loss of an H will affect the bond energies of the other C-H bonds. The values in Table 3-3.6.2 are "best guesses," as consistent as possible with all available data. If calorimetric data are available, it is far more precise and, therefore, more desirable. However, in many cases the calorimetric data are not available, and bond energy tabulations offer at least a reasonable estimate of a particular heat of reaction.

3-3.6.2 EXAMPLE

Estimate ΔH for the reaction from bond or dissociation energy data

$$C(graphite) + \frac{1}{2} O_2(g) \rightarrow CO(g)$$

Solution

We need the following equations:

	Reaction	ΔH (kJ)
1.	$C(graphite) \rightarrow C(g)$	713.0
2.	$O_2(g) \rightarrow 2O(g)$	495.0
3.	$CO(g) \rightarrow C(g) + O(g)$	1073.0

Combine as follows: (1) + ½(2) - (3)

$$C(graphite) \rightarrow C(g) \qquad 713.0$$

$$\frac{1}{2}O_2(g) \rightarrow O(g) \qquad 248.0$$

Table 3-3.6.2
Bond Energies[a]

Single bonds	Energy	Multiple bonds	Energy
H-H	436.0	C=C	615.0
H-F	563.2	C≡C	812.0
H-Cl	431.8	C≡O	1073.0
C-H	414.0	N=N	418.0
C-C	350.0	N≡N	946.0
C-Cl	330.0	O_2	495.0
N-H	390.0		
O-H	464.0		
O-O	138.0		
F-F	153.0		
Cl-Cl	243.0		
Cl-F	254.0		

[a] These are averages from a number of sets of available data (see Sec. 3-3.10, Ref. 3).

$$-CO(g) \rightarrow -C(g) - O(g) \qquad -1073.0 \text{ kJ}$$

$$C(\text{graphite}) + \tfrac{1}{2}O_2(g) - CO(g) \rightarrow 0$$

or

$$C(\text{graphite}) + \tfrac{1}{2}O_2(g) \rightarrow CO(g) \qquad \Delta H = -112 \text{ kJ}$$

This compares with a value of -111 kJ from Table 3-3.4.1. Agreement will not be as good for more complex cases.

3-3.7 ENRICHMENT -- ADIABATIC FLAME TEMPERATURE

The following outlines the method for finding the hypothetical temperature which would result if a compound were burned adiabatically. Actual temperatures are of course lower, due to heat losses and incomplete combustion. Consider the reaction

$$CH_4(g) + 2O_2(g) \rightarrow CO_2(g) + 2H_2O(g)$$

(H_2O is gaseous since the products will be at high temperature.)

Outline our process in a block diagram:

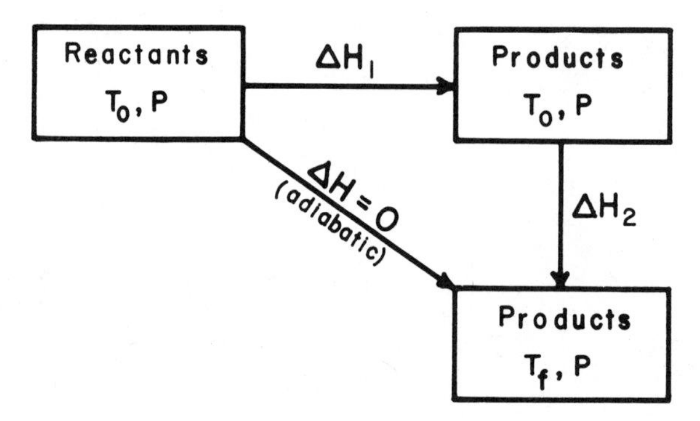

You should see immediately that

$$\Delta H_1 = \Delta H_f^0 \text{ (prod)} - \Delta H_f^0 \text{ (react)}$$

$$\Delta H_2 = \int_{T_0}^{T_f} C_p(\text{prod})\, dT$$

And $\Delta H_2 = -\Delta H_1$

It is easy to evaluate ΔH_1. You should show from data in Table 3-3.4.1 that

$$\Delta H_1 = -802 \text{ kJ}$$

Now as a first guess, assume C_p values can be represented by the constant term in Table 3-2.2.1:

$$C_p(H_2O) = 30.5$$

$$C_p(N_2) = 28.6 \quad \text{for reasons you will see later}$$

$$C_p(CO_2) = 44.2$$

Thus $C_p(2H_2O + CO_2) \approx 105 \text{ J K}^{-1}$

$$\Delta H_2 = \int_{T_0}^{T_f} C_p(\text{prod})\, dT \approx C_p \int_{T_0}^{T_f} dT = C_p(T_f - T_0) = -\Delta H_1$$

$$T_f - T_0 \approx \frac{-\Delta H_1}{C_p} = \frac{802 \text{ kJ}}{105 \text{ J K}^{-1}} \times \frac{10^3 \text{ J}}{\text{kJ}} \approx 7630 \text{ K}$$

Therefore, $T_f \approx 8000$ K

This is obviously a very rough guess. A more accurate guess would result if C_p as a function of T were used, except that these values of C_p are not usually valid above 2000 K. Also, molecules would begin to dissociate at these extreme conditions.

Now look at the effect of N_2. The above assumed pure O_2. For each mole of O_2 from air, 4 mol of N_2 are present which must be heated along with the products.

$$C_p(2H_2O + CO_2 + 8N_2) = 334 \text{ J K}^{-1} \qquad \text{why } 8N_2?$$

$$T_f - T_0 \approx \frac{802 \times 10^3 \text{ J}}{334 \text{ J K}^{-1}} = 2400 \text{ K}$$

$$T_f \approx 2700 \text{ K}$$

This is the maximum temperature possible if methane is burned in air. We are assuming

1. C_p values constant

2. No heat loss
3. No absorption of heat in dissociating the products

All are poor assumptions, but very rough comparisons of fuels can be made this way. The CH_4 air flame temperature is about 1100°C ≈ 1400 K in practice. Our predicted value is not very good.

3-3.8 PROBLEMS

1. From heats of combustion find ΔH for the reaction

$$C_2H_2 + 2H_2 \rightarrow C_2H_6$$

2. Repeat the calculation in Problem (1) using ΔH_f^0 values and compare the results.

3. From whatever data that are applicable find the heat of reaction for the oxidation of ethanol to acetic acid:

$$CH_3\text{-}CH_2OH + O_2 \rightarrow CH_3COOH + H_2O$$

4. Find the heat of reaction of $S(g) + O_2(g) \rightarrow SO_2(g)$ at 600 K.

5. Ca^{2+} ions can be removed from hard water by bubbling $CO_2(g)$ through the solution, resulting in a $CaCO_3$ precipitate. $\Delta H_f^0 = -1206.0$ g kJ mol^{-1}. Write a balanced equation for this reaction and find ΔH(reaction).

6. Calculate the heat of reaction of $H_2(g) + I_2(g) \rightarrow 2HI(g)$ at 520 K. Assume C_p (HI) = 29.2 J mol^{-1} K^{-1}.

7. The heats of combustion of H_2, C, and butane, C_4H_{10}, are -285.84, -393.51, and -2877.1 kJ mol^{-1}, respectively. Find the heat of formation of butane.

8. Find ΔH_f^0 (H_2S) at 300 K, given

$$C_p\ (H_2) = 27.28 + 3.26 \times 10^{-3}\ T\ (J\ mol^{-1}\ K^{-1})$$
$$C_p\ (S,\ g) = 22.2 - 0.42 \times 10^{-3}\ T\ (J\ mol^{-1}\ K^{-1})$$
$$C_p\ (H_2S) = 32.68 + 12.38 \times 10^{-3}\ T\ (J\ mol^{-1}\ K^{-1})$$

ΔH_f^0 (H_2S) at 450 K = -22.131 kJ (mol $H_2S)^{-1}$.

9. For the reaction: $CH_4(g) + 2O_2(g) \rightarrow CO_2(g) + 2H_2O(\ell)$

$$\Delta H_f^0\ (CH_4) = -74.81\ kJ\ mol^{-1};$$
$$\Delta H_f^0\ (H_2O) = -285.8\ kJ\ mol^{-1};$$
$$\Delta H_f^0\ (CO_2) = -393.51\ kJ\ mol^{-1}$$

Find ΔH_r^0 at 298 K.

10. Given: $\Delta H_f^0(Br_2,\ g)$ = 30.7 kJ mol^{-1}, and $\Delta H_f^0(HBr,\ g)$ = -36.2 kJ mol^{-1}, find ΔH_r^0 for $H_2(g) + Br_2(g) \rightarrow 2HBr(g)$.

3-3.9 SELF-TEST

1. Find the heat of reaction of C(graphite) + $2H_2 \rightarrow CH_4$ at 1000 K.

2. Which of the following two reactions is more favorable energetically?

$$CO_2(g) + H_2O(\ell) \rightarrow HCO_3^-(aq) + H^+(aq)$$

or $$CO_2(g) + H_2O(\ell) \rightarrow CO_3^{2-}(aq) + 2H^+(aq)$$

(In each case assume that excess water is available to hydrate the ions formed by the reaction given.)

3. Given the following data for C_p (J mol^{-1} K^{-1}) = $a' + b'T + c'T^2$, find the heat of reaction of $N_2(g) + O_2(g) \rightarrow 2NO(g)$ at 600 K.

Substance	a′	b′ × 10^3	c′ × 10^7
N_2	26.98	5.91	-3.38
O_2	25.50	13.61	-42.56
NO	29.37	-1.55	106.50

4. Describe (one page) at least one experimental method used in determining calorimetric data. You are expected to do more than repeat what is in the text. List your references.

3-3.10 ADDITIONAL READING

1.* J. M. Sturtevant "Calorimetry", in *Physical Methods of Organic Chemistry*, 2nd ed., Vol. I, Part 1, Chap. XIV, Interscience, New York, 1949.

2.* F. D. Rossini, D. D. Wagman, W. H. Evans, S. Levine, and I. Jaffee, *Selected Values of Chemical Thermodynamic Properties*, Washington, D. C., Circular 500 of the National Bureau of Standards, 1952.

3.* For additional data tabulations and discussions of experimental methods see Sec. 1-1.9B and C.

3-4 ENRICHMENT -- A RUBBER BAND DEVELOPMENT TO STRETCH YOUR IMAGINATION

Throughout this chapter we have considered only PV work for gases. Thermodynamics is not limited to PV work, as we will illustrate by deriving some equations for a rubber band.

ENERGY, HEAT CAPACITY AT CONSTANT LENGTH. When a rubber band is stretched it exerts a restoring force f which is a function of the length L and the temperature T. If we assume E = E(T, L) we can write immediately

$$dE = \left(\frac{\partial E}{\partial T}\right)_L dT + \left(\frac{\partial E}{\partial L}\right)_T dL \tag{3-4.1}$$

Also, the work done on the rubber band is force × distance

$$w = f\ dL \tag{3-4.2}$$

Then

$$q = dE - w = \left(\frac{\partial E}{\partial T}\right)_L dT + \left[\left(\frac{\partial E}{\partial L}\right)_T - f\right] dL \tag{3-4.3}$$

The generalized heat capacity is

$$\left(\frac{q}{dT}\right) = C = \left(\frac{\partial E}{\partial T}\right)_L + \left[\left(\frac{\partial E}{\partial L}\right)_T - f\right] \frac{dL}{dT} \tag{3-4.4}$$

It should be clear that the heat capacity at constant L is

$$C_L = \left(\frac{\partial E}{\partial T}\right)_L \tag{3-4.5}$$

ENTHALPY, HEAT CAPACITY AT CONSTANT FORCE. We can define a counterpart to enthalpy for the rubber band:

$$H = E - fL \tag{3-4.6}$$

$$H = H(T, f) \tag{3-4.7}$$

$$\begin{aligned} dH &= dE - f\ dL - L\ df \\ &= q + w - f\ dL - L\ df \\ &= q + f\ dL - f\ dL - L\ df \\ &= q - L\ df \end{aligned} \tag{3-4.8}$$

Also, $$dH = \left(\frac{\partial H}{\partial T}\right)_f dT + \left(\frac{\partial H}{\partial f}\right)_T df \tag{3-4.9}$$

Combine (3-4.8) and (3-4.9):

$$q = \left(\frac{\partial H}{\partial T}\right)_f dT + \left[\left(\frac{\partial H}{\partial f}\right)_T + L\right] df \tag{3-4.10}$$

The heat capacity at constant f is then

$$\left(\frac{q}{\partial T}\right)_f = C_f = \left(\frac{\partial H}{\partial T}\right)_f \tag{3-4.11}$$

AN IDEAL RUBBER BAND -- REVERSIBLE ADIABATIC EXTENSION. An ideal rubber band has an equation of state of the form $f = T\phi(L)$ where $\phi(L)$ depends only on the length L. Also, $(\partial E/\partial L)_T = 0$ for an ideal rubber band, which means that $E = E(T)$. (This is similar to the case for an ideal gas.)

An adiabatic extension indicates stretching a band with $q = 0$. Thus

$$dE = w$$

For an ideal band with q = 0, Eq. (3-4.3) reduces to

$$0 = \left(\frac{\partial E}{\partial T}\right)_L dT + \left[\left(\frac{\partial E}{\partial L}\right)_T - f\right] dL$$

$$= \left(\frac{\partial E}{\partial T}\right)_L dT - f\ dL \qquad (3\text{-}4.12)$$

or applying (3-4.5):

$$C_L\ dT = f\ dL = T\phi\ dL \qquad (3\text{-}4.13)$$

Divide by T:

$$\frac{C_L\ dT}{T} = \phi(L)\ dL \qquad (3\text{-}4.14)$$

This is the counterpart to Eq. (3-2.4.4). For an ideal system, C_L will be a constant and $\phi(L)$ will increase with L. In this case we can integrate (3-4.14):

$$\int_{T_1}^{T_2} \frac{C_L}{T}\ dT = \int_{L_1}^{L_2} \phi(L)\ dL \qquad (3\text{-}4.15)$$

$$C_L\ \ln \frac{T_2}{T_1} = \int_{L_1}^{L_2} \phi(L)\ dL \qquad (3\text{-}4.16)$$

We cannot actually solve (3-4.16) without a form for $\phi(L)$, but we know if $\phi(L)$ increases with L, then if $L_2 > L_1$, the integral on the right side will be positive. This leads to an interesting result: For an adiabatic extension of a rubber band, the temperature will rise.

You can test this by pulling a large rubber band rapidly while allowing the band to touch your lip. To approach an adiabatic process you must do it rapidly. You will feel the band heat up. On contraction the rubber band will cool.

3-5 SUMMARY

You have now completed your introduction to the first law of thermodynamics. Perhaps most important for you to learn are the basic definitions and how to manipulate these definitions to give the equations appropriate for whatever problem you may encounter. As has been emphasized, it is impossible to recall the proper equation for the numerous problems contained in this chapter. You must have the ability to derive the equation you need.

You have probably noted a lack of real situations. This is due to the fact that this chapter is mainly for laying the groundwork for the remainder of our study of thermodynamics. Hopefully, you have thought of some applications of concepts contained here as you were reading. It is particularly easy to see the importance of the equation developed in the section on thermochemistry. Any engineer who must design or control synthetic processes must be able to predict heat flows caused by the reactions involved.

The last section on the rubber band serves two purposes. First, it predicts behavior you can check easily. Grab a rubber band, put it next to your lip, and stretch the band. Does it get hot or cold? Second, the exercise shows that thermodynamics is applicable to more than the expansion and compression of gases. In the general introduction, the statement was made that thermodynamics is not concerned with whether a system is made up of molecules or goblins. It is applicable to any system. We usually treat gases because we have equations of state that can easily (?) be solved to find P, V, and T behavior, but, thermodynamics is in no way limited to any one system type.

4

The Second and Third Laws of Thermodynamics

SYMBOLS AND DEFINITIONS

$\bar{E}$ Internal energy (J; dm^3 atm)
$\bar{H}$ E + PV = enthalpy (J; dm^3 atm)
$\bar{C}_V$ Heat capacity at constant V (J mol^{-1} K^{-1})
$\bar{C}_P$ Heat capacity at constant P (J mol^{-1} K^{-1})
T_h Temperature of hot reservoir
T_ℓ Temperature of cold reservoir
W Work done *on* system (J; dm^3 atm)
Q Heat absorbed *by* system (J dm^3 atm)
ε Efficiency of a process or cycle

γ Heat capacity ratio C_p/C_v
S Entropy = $\int q_{rev}/T$
S_0 Entropy at 298.15 K, 1 atm (J mol^{-1} K^{-1})

We have just completed your introduction to the first law of thermodynamics, the law of conservation of energy. We found a relationship among heat, work, and changes in the stored energy of a system. The first law does not limit the extent of any energy conversion. Neither does the first law give any indication of whether a particular process is possible, other than limiting processes to those in which the total energy is conserved.

However, experience has shown that at least one type of energy conversion, heat into work, cannot be carried out completely. The first law does not rule out such a conversion, but it does not cccur. It is of significance that Joule found the mechanical equivalent of thermal energy from converting work into heat, not heat into work. Our study of the second law will show that energy behavior is not symmetrical. Work can be completely converted to heat, but heat cannot be completely converted to work without external help.

In Sec. 4-1, several examples of impossible processes will be noted. Reversibility and irreversibility will be defined, and general statements of the second law of thermodynamics will be given. Hopefully, by the end of Sec. 4-1 you will have a general idea of what the second law is. Then, various thermodynamic cycles for the conversion of heat to work will be described. The relation of these idealized cycles to familiar engines will be discussed.

In Sec. 4-2, the concept of entropy will be introduced and its significance will be discussed. Entropy changes for various processes will be calculated. A procedure for finding absolute entropy values will be introduced in Sec. 4-3. These will be useful in Chap. 5, where a new function is defined that gives us an indication of where equilibrium lies in a process or reaction.

Throughout this chapter, you should remember that the entropy function is one that will be used later in determining how processes should go. Entropy changes a alone, like energy changes, are not sufficient to predict the direction of a reaction or process. The two must be combined. All the mathematical manipulations in this chapter either prepare you for utilizing free energy equations in Chap. 5, or are given to show how useful relationships among variables can be obtained. Thus, the material in this chapter can be viewed as the development of basic ideas that will be applied later (in Chap. 5 and 6, particularly).

4-1 GENERAL STATEMENTS OF THE SECOND LAW AND A DISCUSSION OF THERMODYNAMIC CYCLES

OBJECTIVES

You shall be able to

1. Discuss the importance of the second law and give examples showing processes that violate it
2. Give two general statements of the second law
3. Discuss the meaning and significance of the terms reversible and irreversible
4. Discuss the Carnot, Otto, and Diesel cycles, pointing out the major steps in each cycle
5. Apply the Carnot cycle to any system, given an equation of state for the working medium
6. Calculate the maximum efficiency of a Carnot engine

4-1.1 LIMITS ON THE CONVERSION OF HEAT ENERGY TO MECHANICAL ENERGY

POSSIBLE AND IMPOSSIBLE PROCESSES. A few examples in this section will illustrate some of the limits on the conversion of thermal energy to mechanical energy. Some of the processes are real, some imaginary. None of the hypothetical processes described violate the first law. The real processes demonstrate limits imposed by the second law.

First, look at a steam power plant as a real example. A steam power plant uses about 250 J of energy stored in fuel (oil, coal, or natural gas) to produce 100 J of work. This means about 150 J are lost to the surroundings as some form of energy other than work. Your question should be: Is this necessary?

Consider two steel blocks at different temperatures contained in a perfectly insulated box. What happens as time goes by? Our experience tells us that the higher temperature block will lose heat to the lower temperature block until the two temperatures are the same. The first law tells us that the heat lost by the first block equals the heat gained by the second block. However, nothing in the first law prevents the transfer of heat from the cooler block to the warmer one. You should now ask: Why does heat flow from a higher temperature object to one of lower temperature, but the reverse will not occur (with no external assistance)?

Why are you not able to extract heat from a block of ice to warm your hands on a cold day? The first law does not eliminate that possibility. We just know from experience that it does not work.

When a block of wood falls to the floor, several energy changes occur. The potential energy of the block is converted to kinetic energy during the fall. Upon impact, the kinetic energy is dissipated as heat which distributes itself throughout the block, the floor, and the room. Why are we not able to extract some heat from the floor to raise the block? The first law would not be violated.

These questions lead to the formulation of the second law of thermodynamics.

THE SECOND LAW OF THERMODYNAMICS. The second law will be stated in two forms. Each statement, like that of the first law, can be treated as a postulate; it cannot be derived. The second law leads to the introduction of a new function, entropy, which allows it to be applied to chemical systems. This will be introduced in Sec. 4-2.

The first statement of the second law to be given was formulated by Clausius in 1850: *It is impossible for any device to operate in such a manner that it produces no effect other than the transfer of heat from one body to another body at a higher temperature.*

You might say: But a refrigerator does just that; it takes heat from the cold interior and ejects heat to the warm surroundings. If that is your feeling, you have overlooked the words "produces no effect other than ... ." A refrigerator operates only if energy is input from some source other than the two bodies between which the heat is flowing. Planck interpreted the statement by Clausius as follows: "As Clausius repeatedly and expressly pointed out, this principle does not merely say that heat does not flow directly from a cold to a hot body -- that is self-evident, and is a condition of the definition of temperature -- but it expressly states that heat can in no way and by no process be transported from a colder to a warmer body without leaving further changes."

A second statement, the Kelvin-Planck statement, is: *It is impossible for any device to operate in a cycle and produce work while exchanging heat only with bodies at a single fixed temperature.* This can be reworded as: It is impossible for any device operating in a cycle to absorb heat from a single reservoir and produce an equivalent amount of work. (Reservoir refers to an *energy reservoir* from which a finite amount of energy can be withdrawn without changing its temperature.)

CHEMICAL SIGNIFICANCE. The above are general statements which summarize our knowledge that natural processes tend to go to equilibrium. It does not look that way at first glance. The second law is, however, concerned with the equilibrium state, and the tendency of processes to occur spontaneously. Work cannot be extracted from a system at equilibrium. There must be a path for spontaneous change to occur before useful work is obtained. We will formulate deductions from the second law that can be applied to chemical problems after a short discussion of impossible engines and the meaning of the terms *reversible* and *irreversible*. More will be said about equilibrium and spontaneous processes as we develop the second law.

4-1.2 PERPETUAL MOTION MACHINES

There are three kinds of perpetual motion machine. A perpetual motion machine of the first kind operates in a cycle and produces a greater net work output than the net heat input. Even though several machines of this type have been proposed and some patented, the first law is violated, since for a cycle $\Delta E = 0$ and $Q = -W$. No machine of this kind has ever been sucessfully operated.

A perpetual motion machine of the second kind operates in a cycle producing work and exchanging heat with bodies (reservoirs) at a fixed temperature (i.e., both reservoirs are at the same temperature). This does not violate the first law since no energy is created, but does violate the second law. Even though no energy is created, this type of machine would be invaluable, since the almost limitless supply of energy available in the oceans and atmosphere would be tapped. Imagine a ship that could pump warm water from the ocean, extract energy in the form of work to drive the ship, then dump the cold water (or ice cubes) overboard. The second law tells us it will not work.

A third type of perpetual motion machine is a device which, once set into motion will continue in motion without slowing down. This violates neither the first nor second law. All that must be done is to eliminate all loss from friction. We will not consider the third type further. There is no way to extract useful work from such a machine; hence, it is of little interest here. (If you were a navigator, dependent on a gyroscope, then you would have a great interest in this type of machine, however.)

4-1.3 REVERSIBILITY AND IRREVERSIBILITY

The second law is useful in pointing out those processes that are impossible. Of the possible processes, some can be carried out in such a way that all changes in *both* the system and the surroundings can be undone: everything can be returned to the initial state. Other possible processes are such that it is impossible to return *both* the system and the surroundings to the initial state. The former is called a *reversible* processes, the latter *irreversible*.

In chemical processes, we limit ourselves further. We restrict the term *reversible* to those processes which proceed through a series of infinitesimal changes in controlling variables, and both system and surroundings can be returned to the initial states by reversing the process through infinitesimal changes. In other words, a reversible process is one that occurs by means of a series of equilibrium states. An infinitesimal change in controlling variables allows the system and surroundings to proceed to the next equilibrium state. The process can be reversed by an infinitesimal change in the controlling variables. Such a process would take infinite time.

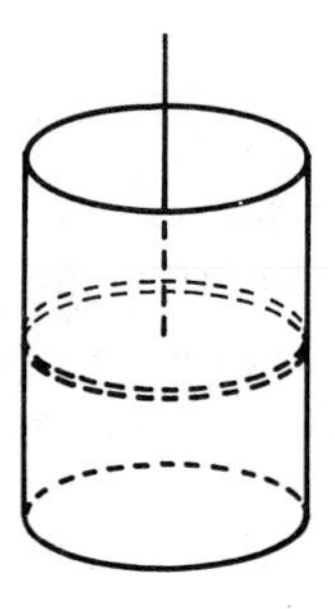

FIG. 4-1.3.1 A cylinder of gas supporting an external pressure. The piston is weightless and frictionless.

However, time plays no role in equilibrium thermodynamics. You will not find the variable "time" in any thermodynamic equation presented.

Consider the system shown in Fig. 4-1.3.1, which consists of a cylinder with a weightless, frictionless piston. Assume the system is at equilibrium, which means

$$P_{ext} = P_{int} \qquad T_{ext} = T_{int}$$

If P_{ext} is allowed to increase an infinitesimal amount dP, you should predict from your study in Chap. 3 that the gas will be compressed until $P'_{int} = P'_{ext}$. If the walls of the cylinder are good heat conductors (which we will assume), then T_{int} will remain constant. The above compression can be reversed by allowing P'_{ext} to be decreased by dP. When equilibrium is again reached, we find that our system and surroundings are in their initial states.

Irreversible processes are given labels such as *real*, *spontaneous*, *natural*, or *actual*. All these terms are meant to indicate that a process will proceed on its own if left alone. The concept of spontaneity will arise again after our definition of entropy in Sec. 4-2. Chapter 5 will also be concerned with the definition of a function that determines whether a given process or reaction should be spontaneous.

4-1.4 THE CARNOT CYCLE

Perhaps the most well known of the reversible cyclic processes is that described by Sadi Carnot in 1824. The cycle is not the basis for a practical device, but it offers a good system for analysis. One complete cycle of an engine operating on a Carnot cycle has the net effect of using heat from a hot reservoir, producing work, and delivering some heat to a cold reservoir.

Usually this analysis is done using an ideal gas. Here, two different working media, a van der Waals gas and a rubber band will be used. You should look up the Carnot cycle for an ideal gas as the working fluid in another physical chemistry text to familiarize yourself with all the steps before reading on. The development below for the van der Waals gas is much more complicated than for an ideal gas. It is presented to show that the results usually derived for an ideal gas are general. The development using a rubber band is for fun. The rubber band example does, however, emphasize that thermodynamics applies to systems other than gaseous ones. Specifically, it shows that the Carnot cycle can be applied to systems other than gases.

THE CARNOT CYCLE USING A VAN DER WAALS GAS. Before presenting the cycle we need some equations from Chaps. 2 and 3. The general equation for an energy change is

$$dE = n\bar{C}_V\ dT + \left[T\left(\frac{\partial P}{\partial T}\right)_V - P\right] dV \tag{3-1.4.7}$$

van der Waals equation is

$$\left(P + \frac{an^2}{\bar{V}^2}\right)(V - nb) = nRT$$

or, for 1 mol

$$P = \frac{RT}{V - b} - \frac{a}{V^2} \tag{2-2.4.1}$$

$$\left(\frac{\partial P}{\partial T}\right)_V = \frac{R}{V - b} \tag{4-1.4.1}$$

Equation (3-1.4.7) becomes (show this)

$$dE = \bar{C}_V\ dT + \frac{a}{V^2}\ dV \tag{4-1.4.2}$$

Consider Fig. 4-1.4.1 and 4-1.4.2 during the following description of the cycle. All steps are carried out reversibly. You should verify each result shown, since some steps have been left out for brevity. The following strokes refer to those given in Fig. 4-1.4.2.

Stroke 1. The gas is expanded isothermally at a temperature of T_h from V_1 to V_2. The heat absorbed is Q_h and the work done on the system is W_1 (which will be negative: Why?).

$$\Delta E_1 = \int_1^2 dE = \int_{T_h}^{T_h} \bar{C}_V\ dT + \int_{V_1}^{V_2} \frac{a}{V^2}\ dV = -a\left(\frac{1}{V_2} - \frac{1}{V_1}\right) \tag{4-1.4.3}$$

$$W_1 = -\int_{V_1}^{V_2} P\ dV = -\int_{V_1}^{V_2} \left(\frac{RT}{V - b} - \frac{a}{V^2}\right) dV$$

$$= -RT_h \ln \frac{V_2 - b}{V_1 - b} - a\left(\frac{1}{V_2} - \frac{1}{V_1}\right) \tag{4-1.4.4}$$

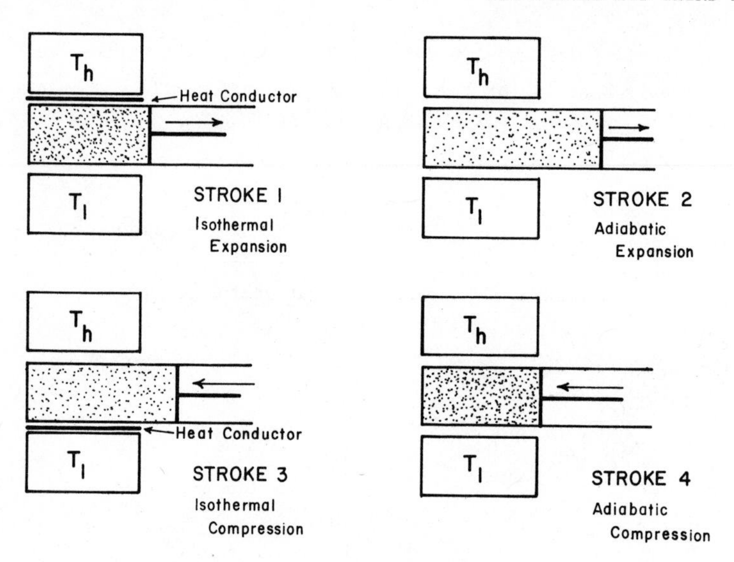

FIG. 4-1.4.1 The four strokes of a Carnot engine. (A heat conductor is added whenever heat is lost or gained; otherwise, assume no heat transfer.

$$Q_1 = \Delta E_1 - W_1 = RT_h \ln \frac{V_2 - b}{V_1 - b} \tag{4-1.4.5}$$

Stroke 2. The gas expands adiabatically from a volume V_2 to V_3. The temperature drops from T_h to T_ℓ. Work W_2 is done on the gas (again, it will be negative). No heat is transferred.

$$Q_2 = 0 \tag{4-1.4.6}$$

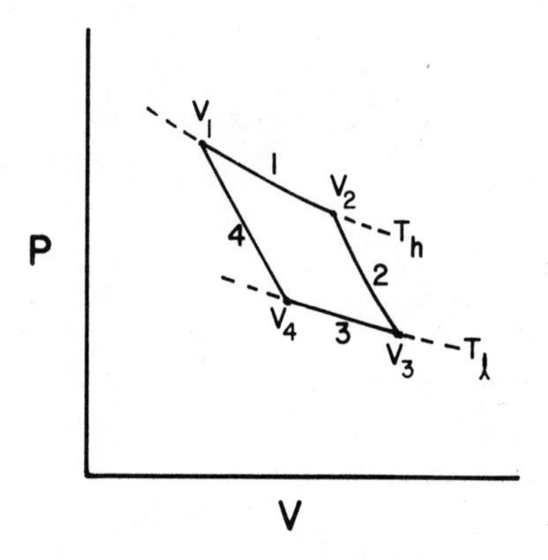

FIG. 4-1.4.2 Plot of the four strokes of a Carnot cycle on a PV diagram (not to scale).

$$\Delta E_2 = W_2 = \int_{T_h}^{T_\ell} \bar{C}_v \, dT + \int_{V_2}^{V_3} \frac{a}{V^2} \, dV \qquad (4\text{-}1.4.7)$$

You should refer to Sec. 3-2.4, under Real Gases, to find a discussion related to this part of the problem. Equation (3-2.4.12) gives one equation we will need:

$$\frac{\bar{C}_v}{R} \ln \frac{T_\ell}{T_h} = -\ln \frac{V_3 - b}{V_2 - b} \qquad (4\text{-}1.4.8)$$

This relates the temperature change $T_h \to T_\ell$ to the volume change $V_2 \to V_3$. We will need this later.

Stroke 3. The gas is compressed isothermally at T_ℓ from volume V_3 to V_4. W_3 is the work done on the gas (positive this time), and Q_1 heat flows from the gas to the cold reservoir.

The equations from Stroke (1) apply with appropriate substitutions.

$$\Delta E_3 = -a\left(\frac{1}{V_4} - \frac{1}{V_3}\right) \qquad (4\text{-}1.4.9)$$

$$W_3 = -RT_\ell \ln \frac{V_4 - b}{V_3 - b} - a\left(\frac{1}{V_4} - \frac{1}{V_3}\right) \qquad (4\text{-}1.4.10)$$

$$Q_3 = RT_\ell \ln \frac{V_4 - b}{V_3 - b} \qquad (4\text{-}1.4.11)$$

Stroke 4. The gas is compressed adiabatically from V_4 to V_1 with a temperature change from T_ℓ to T_h; $Q_4 = 0$ and W_4 is the work done on the gas. As in Stroke (2):

$$Q_4 = 0$$

$$\Delta E_4 = W_4 = \int_{T_\ell}^{T_h} \bar{C}_v \, dT + \int_{V_4}^{V_1} \frac{a}{V^2} \, dV \qquad (4\text{-}1.4.12)$$

We will need Eq. (3-2.4.12) again:

$$\frac{\bar{C}_v}{R} \ln \frac{T_h}{T_\ell} = -\ln \frac{V_1 - b}{V_4 - b} \qquad (4\text{-}1.4.13)$$

We have to add all the changes to see what the total cycle gives.

$$\Delta E_{total} = 0 = \Delta E_1 + \Delta E_2 + \Delta E_3 + \Delta E_4$$

$$0 = -a\left(\frac{1}{V_2} - \frac{1}{V_1}\right) + \int_{T_\ell}^{T_h} \bar{C}_v \, dT - \frac{a}{V_3} + \frac{a}{V_2} - a\left(\frac{1}{V_4} - \frac{1}{V_3}\right)$$

$$+ \int_{T_h}^{T_\ell} \bar{C}_v \, dT - \frac{a}{V_1} + \frac{a}{V_4}$$

$$= -\frac{a}{V_2} + \frac{a}{V_1} + \bar{C}_v(T_h - T_\ell) - \frac{a}{V_3} + \frac{a}{V_2} - \frac{a}{V_4} + \frac{a}{V_3}$$

$$+ \bar{C}_v(T_\ell - T_h) - \frac{a}{V_1} + \frac{a}{V_4} \qquad (4\text{-}1.4.14)$$

This, indeed, does add to zero, as expected for a cyclic process.

$$W_{total} = W_1 + W_2 + W_3 + W_4$$

$$= -RT_h \ln \frac{V_2 - b}{V_1 - b} - a\left(\frac{1}{V_2} - \frac{1}{V_1}\right) + \int_{T_h}^{T_\ell} \bar{C}_v \, dT - \frac{a}{V_3} + \frac{a}{V_2}$$

$$- RT_\ell \ln \frac{V_4 - b}{V_3 - b} - a\left(\frac{1}{V_4} - \frac{1}{V_3}\right) + \int_{T_\ell}^{T_h} \bar{C}_v \, dT - \frac{a}{V_1} + \frac{a}{V_4}$$

$$= -RT_h \ln \frac{V_2 - b}{V_1 - b} - RT_\ell \ln \frac{V_4 - b}{V_3 - b} \qquad (4\text{-}1.4.15)$$

Since $\Delta E = 0$, $Q_{total} = -W$. Thus,

$$Q_{total} = Q_1 + Q_2 + Q_3 + Q_4 = -W$$

$$= RT_h \ln \frac{V_2 - b}{V_1 - b} + 0 + RT_\ell \ln \frac{V_4 - b}{V_3 - b} + 0$$

$$= RT_h \ln \frac{V_2 - b}{V_1 - b} + RT_\ell \ln \frac{V_4 - b}{V_3 - b} \qquad (4\text{-}1.4.16)$$

We now have the equations required to calculate all the variables in this cycle. A particular set of conditions will be given in one of the problems as the end of this section to allow you to calculate all the changes that take place.

EFFICIENCY OF AN ENGINE. Efficiency is defined as the net work produced for a given consumption of heat from the hot reservoir.

$$\varepsilon = -\frac{W}{Q_h} = -\left(\frac{-RT_h \ln \frac{V_2 - b}{V_1 - b} - RT_\ell \ln \frac{V_4 - b}{V_3 - b}}{RT_h \ln \frac{V_2 - b}{V_1 - b}}\right)$$

$$= \frac{+T_h \ln \frac{V_2 - b}{V_1 - b} + T_\ell \ln \frac{V_4 - b}{V_3 - b}}{T_h \ln \frac{V_2 - b}{V_1 - b}} \qquad (4\text{-}1.4.17)$$

We can simplify this by use of (4-1.4.8) and (4-1.4.13):

$$-\ln\frac{V_3-b}{V_2-b}=\frac{\bar{C}_v}{R}\ln\frac{T_\ell}{T_h}=-\frac{\bar{C}_v}{R}\ln\frac{T_h}{T_\ell}=-\left(-\ln\frac{V_1-b}{V_4-b}\right)$$

$$\ln\frac{V_2-b}{V_3-b}=\ln\frac{V_1-b}{V_4-b} \tag{4-1.4.18}$$

$$\ln(V_2-b)-\ln(V_3-b)=\ln(V_1-b)-\ln(V_4-b)$$

$$\ln(V_2-b)-\ln(V_1-b)=\ln(V_3-b)-\ln(V_4-b)$$

$$\ln\frac{V_2-b}{V_1-b}=\ln\frac{V_3-b}{V_4-b}=-\ln\frac{V_4-b}{V_3-b} \tag{4-1.4.19}$$

Applying (4-1.4.19) to (4-1.4.17) yields

$$\frac{T_h\ln\frac{V_2-b}{V_1-b}-T_\ell\ln\frac{V_2-b}{V_1-b}}{T_h\ln\frac{V_2-b}{V_1-b}}=\frac{T_h-T_\ell}{T_h}$$

$$\varepsilon=-\frac{W}{Q_h}=\frac{T_h-T_\ell}{T_h} \tag{4-1.4.20}$$

The result is the same as that found for an ideal gas. Any gas will produce the same result. Now, let us apply the Carnot cycle to a rubber band.

4-1.5 THE CARNOT CYCLE INVOLVING A RUBBER BAND

The first step is to write the equations used in Sec. 3-4. For an ideal rubber band, the energy depends only on T.

$$dE=\left(\frac{\partial E}{\partial T}\right)_L dT=C_L\ dT \tag{4-1.5.1}$$

$$w=f\ dL=T\ \phi(L)\ dL \tag{4-1.5.2}$$

From (4-1.5.1) and (4-1.5.2)

$$q=dE-w=C_L\ dT-T\ \phi(L)\ dL \tag{4-1.5.3}$$

$$\frac{C_L\ dT}{T}=\phi(L)\ dL \tag{3-4.13}$$

The cycle for a rubber band is similar to that for a gas. Figure 4-1.5.1 shows a schematic representation of the cycle.

Step 1. Reversible isothermal contraction from $L_1 \rightarrow L_2$ at T_h. If T is constant, dE = 0 and q = -w. So q = -f dL. If we allow heat to be absorbed q > 0 and -f dL > 0 or f dL < 0, which implies that $L_2 < L_1$. That is unimportant since it does

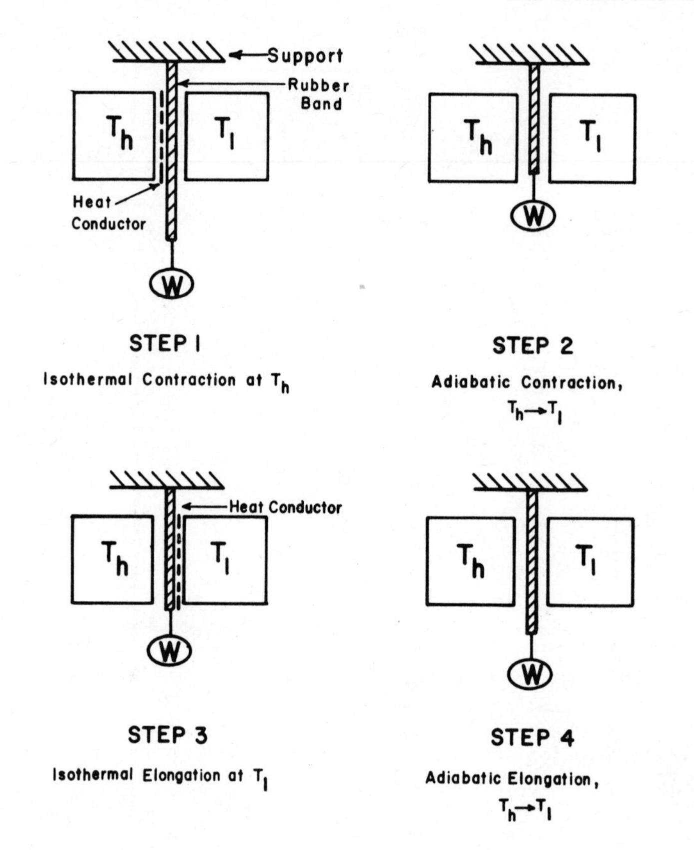

FIG. 4-1.5.1 The Carnot cycle for an engine using a rubber band as the working medium.

not matter whether the band extends or contracts on heating. Either way, dL changes, leading to the performance of work. We have the following changes [$\int \phi(L)\ dL \equiv \phi'(L)$ where $\phi'(L)$ is some function of L; see Sec. 3-4.]:

$$\Delta E_1 = 0$$

$$Q_1 = +\int_{T_h}^{T_h} C_v\ dT - T_h \int_{L_1}^{L_2} \phi(L)\ dL = -T_h[\phi'(L_2) - \phi'(L_1)] \tag{4-1.5.5}$$

$$W_1 = -Q_1 = T_h \int_{L_1}^{L_2} \phi(L)\ dL = T_h[\phi'(L_2) - \phi'(L_1)] \tag{4-1.5.6}$$

Step 2. Adiabatic reversible contraction from $L_2 \rightarrow L_3$. Temperature changes from $T_h \rightarrow T_\ell$:

$$\Delta E_2 = \int_2^3 dE = \int_{T_h}^{T_\ell} C_L\ dT \tag{4-1.5.7}$$

$$Q_2 = 0 \tag{4-1.5.8}$$

$$W_2 = \Delta E_2 = \int_{T_h}^{T_\ell} C_L \, dT \tag{4-1.5.9}$$

Later we will need the relationship from (3-4.13):

$$\frac{C_L \, dT}{T} = \phi(L) \, dL \tag{3-4.13}$$

$$\int_{T_h}^{T_\ell} \frac{C_L \, dT}{T} = \int_{L_2}^{L_3} \phi(L) \, dL \tag{4-1.5.10}$$

Integrating yields

$$C_L \ln \frac{T_\ell}{T_h} = \phi'(L_3) - \phi'(L_2) \tag{4-1.5.11}$$

Step 3. Isothermal expansion from $L_3 \to L_4$ at T_ℓ. The band loses heat to the low temperature reservoir and becomes more elongated.

$$\Delta E_3 = 0 \tag{4-1.5.12}$$

$$Q_3 = \int_{T_\ell}^{T_\ell} C_L \, dT - \int_{L_3}^{L_4} T_\ell \phi(L) \, dL = -T_\ell [\phi'(L_4) - \phi'(L_3)] \tag{4-1.5.13}$$

$$W_3 = -Q_3 = T_\ell \int_{L_3}^{L_4} \phi(L) \, dL = T_\ell [\phi'(L_4) - \phi'(L_3)] \tag{4-1.5.14}$$

Step 4. Adiabatic elongation from T_ℓ to T_h, $L_4 \to L_1$. As in Step (2):

$$\Delta E_4 = \int_{T_\ell}^{T_h} C_L \, dT \tag{4-1.5.15}$$

$$Q_4 = 0 \tag{4-1.5.16}$$

$$W_4 = \Delta E_4 = \int_{T_h}^{T_\ell} C_L \, dT \tag{4-1.5.17}$$

Also similar to (4-1.5.11):

$$C_L \ln \frac{T_h}{T_\ell} = \phi'(L_1) - \phi'(L_4) \tag{4-1.5.18}$$

Add all the changes:

$$\Delta E_{total} = 0 = \Delta E_1 + \Delta E_2 + \Delta E_3 + \Delta E_4$$

$$0 = 0 + \int_{T_h}^{T_\ell} C_L \, dT + 0 + \int_{T_\ell}^{T_h} C_L \, dT$$

$$= 0 \qquad (4\text{-}1.5.19)$$

$$Q_{total} = Q_1 + Q_2 + Q_3 + Q_4$$

$$= -T_h[\phi'(L_2) - \phi'(L_1)] + 0 - T_\ell[\phi'(L_4) - \phi'(L_3)] + 0$$

$$= -T_h[\phi'(L_2) - \phi'(L_1)] - T_\ell[\phi'(L_4) - \phi'(L_3)] \qquad (4\text{-}1.5.20)$$

$$W_{total} = W_1 + W_2 + W_3 + W_4$$

$$= T_h[\phi'(L_2) - \phi'(L_1)] + \int_{T_h}^{T_\ell} C_L \, dT + T_\ell[\phi'(L_4) - \phi'(L_3)] + \int_{T_\ell}^{T_h} C_L \, dT$$

$$= T_h[\phi'(L_2) - \phi'(L_1)] + T_\ell[\phi'(L_4) - \phi'(L_3)] \qquad (4\text{-}1.5.21)$$

All we need is a form of the function $\phi(L)$ to find all the quantities for the Carnot cycle for a rubber band. An equation for the efficiency is given below.

THE EFFICIENCY OF A CARNOT ENGINE USING A RUBBER BAND AS THE WORKING MEDIUM. As before, the definition of efficiency is

$$\varepsilon = -\frac{W}{Q_h}$$

$$\varepsilon = -\frac{T_h[\phi'(L_2) - \phi'(L_1)] + T_\ell[\phi'(L_4) - \phi'(L_3)]}{-T_h[\phi'(L_2) - \phi'(L_1)]} \qquad (4\text{-}1.4.22)$$

Use (4-1.5.11) and (4-1.5.18) to relate the ϕ's:

$$C_L \ln \frac{T_\ell}{T_h} = \phi'(L_3) - \phi'(L_2) = -C_L \ln \frac{T_h}{T_\ell} = -[\phi'(L_1) - \phi'(L_4)]$$

$$\phi'(L_3) - \phi'(L_2) = -\phi'(L_1) + \phi'(L_4)$$

Rearrange:

$$\phi'(L_4) - \phi'(L_3) = -\phi'(L_2) + \phi'(L_1)$$

or

$$[\phi'(L_4) - \phi'(L_3)] = -[\phi'(L_2) - \phi'(L_1)] \qquad (4\text{-}1.5.23)$$

(Note: We do not have to know the form of $\phi(L)$ or $\phi'(L)$ to establish these results. Thus the formula we are about to obtain is general in nature.)

Substitute (4-1.5.23) into (4-1.5.22):

$$\varepsilon = \frac{T_h[\phi'(L_2) - \phi'(L_1)] - T_\ell[\phi'(L_2) - \phi'(L_1)]}{T_h[\phi'(L_2) - \phi'(L_1)]}$$

$$= \frac{T_h - T_\ell}{T_h} \qquad (4\text{-}1.5.24)$$

This is the same result we obtained earlier for a van der Waals gas. *It is a general result for reversible Carnot engines.* The working material does not alter the relationship. As you would probably guess, the closer T_h is to T_ℓ, the smaller the efficiency. If T_ℓ could be made zero, we would always have an efficiency of 1, no matter what the value of T_h. Unfortunately, $T_\ell = 0$ is not possible.

A final note on Carnot engines: The efficiency calculated is for a *reversible* process. Any irreversibility will be reflected in a loss of efficiency. A reversible process yields the maximum efficiency allowed by thermodynamics. This point should be clear by now. Reversible processes yield the maximum work possible and the maximum efficiency.

4-1.6 OTHER CYCLES

The Carnot cycle leads to an expression that can be used to calculate the *maximum* efficiency of any engine operating reversibly in a cycle. It is, however, not commonly used as the basis of actual engines. Below, two additional cycles are discussed that are used in engines you can readily recognize. The first is the Otto cycle on which the automobile internal combustion engine is based. The second is the Diesel cycle which is the basis of the Diesel engine.

OTTO CYCLE. Figure 4-1.6.1 gives the strokes in the Otto cycle and plots of P vs. V and T vs. S for an ideal Otto cycle. The steps in the cycle are given (the letters correspond to those of the figure).

- (i)-(a) Intake stroke: cylinder fills with fuel and air
- (a)-(b) Isentropic compression
- (b)-(c) Addition of heat at constant volume: ignition of fuel
- (c)-(d) Isentropic expansion
- (d)-(e) Constant volume rejection of heat: through the cylinder walls to circulating coolant
- (e)-(i) Exhaust stroke

The efficiency ε of this cycle using an ideal gas approximation is given by (no proof; see a general text on internal combustion engines, Sec. 4-1.10, Ref. 1)

$$\varepsilon = 1 - \left(\frac{V_b}{V_a}\right)^{\gamma - 1} \qquad (4\text{-}1.6.1)$$

where V_b/V_a is the compression ratio and γ is the heat capacity ratio. Note that $\gamma - 1$ will be less than one. Therefore, the larger V_b/V_a, the larger ε.

Figure 4-1.6.2 shows how ε varies with the compression ratio for an ideal gas. Modern internal combustion engines operate with a compression ratio of about 8.5. The maximum efficiency is about 56%. Actual realizable efficiency is about 27%. The

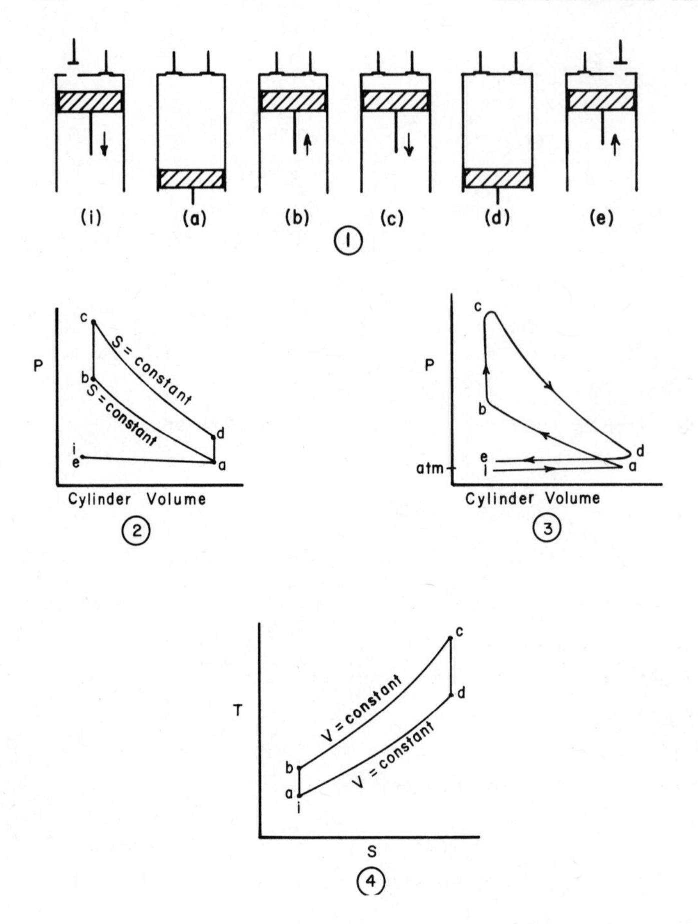

FIG. 4-1.6.1 The Otto cycle. (1) Strokes in the cycle. (2) Idealized PV plot. (3) More nearly real PV diagram. (4) ST diagram.

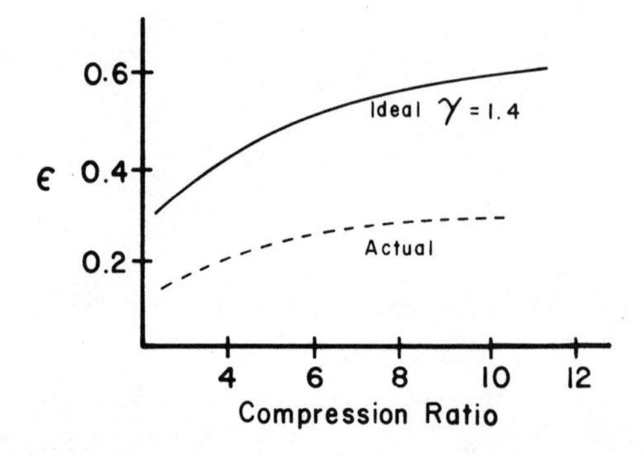

FIG. 4-1.6.2 Efficiency vs. compression ratio for ideal and real Otto cycle.

difference is due to nonideal behavior of the gases, heat losses in the cycle, and incomplete combustion.

You might wonder why higher compression ratios are not used, since these would result in higher efficiencies. If the compression ratio is increased too much, preignition takes place. The temperature rises so much during compression that spontaneous combustion occurs before the spark is given. The result is "knocking" in the engine which can lead to damage.

DIESEL CYCLE. Figure 4-1.6.3 gives the pressure-volume plot for the Diesel cycle. The following steps are present:

(a)-(b) Reversible adiabatic compression
(b)-(c) Heat addition at constant volume: combustion
(c)-(d) Reversible adiabatic expansion
(d)-(a) Heat rejected at constant V

These steps are essentially the same as for the Otto cycle. The difference is that in step (a)-(b) air is compressed, then the fuel is injected. The compression is sufficient to raise the temperature to the point that the fuel ignites when it is injected (no spark is needed).

The efficiency expression for this cycle is more complicated than for the Otto cycle. The exact form will not concern us here. Suffice it to say that the Diesel cycle is less efficient than the Otto cycle at the same compression ratio. However, preignition is not a problem with the Diesel (ignition occurs when the fuel is injected: there is no spark to ignite the mixture), so compression ratios can be made higher than in the Otto cycle. Practical Diesel engines achieve an efficiency of about 42%. This is much better than the Otto cycle.

Considering the relative efficiencies, it is surprising that more vehicles (especially passenger cars) are not powered by Diesel engines. The Diesel is hard to start in cold weather, and is generally heavier than an equivalent power spark-ignition engine. Otherwise, the Diesel seems to have many advantages over a spark-ignition engine.

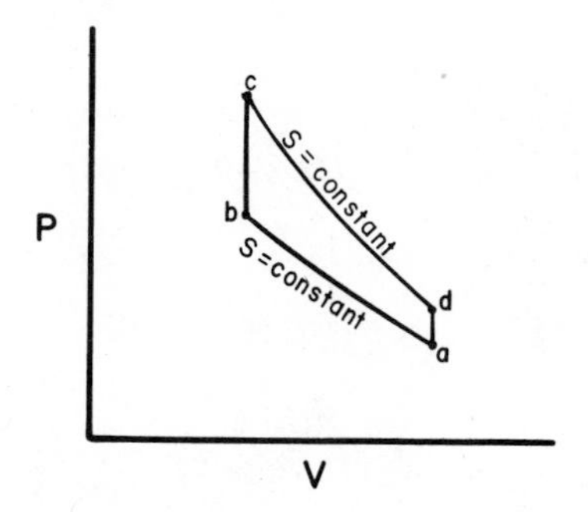

FIG. 4-1.6.3 Idealized PV diagram for the Diesel cycle.

4-1.7 EXAMPLE

Determine Q_h, Q_ℓ, and the efficiency of a Carnot engine using 1 mol N_2 as the working fluid between two reservoirs at 900 K and 220 K. For N_2: a = 1.39 dm^6 atm mol^{-2} and b = 0.0391 dm^3 mol^{-1}. Assume V_1 = 2 dm^3, V_2 = 4 dm^3, and $\bar{C}_V$ = 22.8 J mol^{-1} K^{-1}. (Also, assume van der Waals behavior for N_2.)

Solution

All we need do is apply the equation given previously from (4-1.4.5) (for 1 mol):

$$Q_h = RT_h \ln \frac{V_2 - b}{V_1 - b} = (8.314 \text{ J K}^{-1})(900 \text{ K}) \ln \frac{4 \text{ dm}^3 - 0.0391 \text{ dm}^3}{2 \text{ dm}^3 - 0.0391 \text{ dm}^3}$$

$$= (8.314 \text{ J})(900) \ln 2.02 = \underline{\underline{5.3 \text{ kJ}}}$$

From (4-1.4.11):

$$Q_\ell = RT_\ell \ln \frac{V_4 - b}{V_3 - b}$$

But Eq. (4-1.4.18) indicates

$$\ln \frac{V_2 - b}{V_3 - b} = \ln \frac{V_1 - b}{V_4 - b}$$

or rearranging:

$$\ln \frac{V_4 - b}{V_3 - b} = \ln \frac{V_1 - b}{V_2 - b} = -\ln \frac{V_2 - b}{V_1 - b} = -\ln 2.02$$

Then $Q_\ell = (8.314 \text{ J})(220)(-\ln 2.02) = \underline{\underline{-1.3 \text{ kJ}}}$

Finally the efficiency is given by

$$\varepsilon = \frac{T_h - T_\ell}{T_h} = \underline{\underline{0.76}}$$

Notice also

$$\varepsilon = \frac{Q_h + Q_\ell}{Q_h} = \frac{5.3 - 1.3}{5.3} = \underline{\underline{0.75}}$$

which agrees within rounding errors.

4-1.8 PROBLEMS

1. Analyze the system given in Fig. 4-1.3.1 for changes that will occur if T_{int} is increased by dT and then decreased by dT. Assume an ideal gas in the cylinder.

State what other assumptions you make.

2. Determine all variables for a Carnot cycle which uses 2 mol He as the working fluid between reservoirs at 1100 K and 300 K. The van der Waals constants are given in Chap. 2. Assume $V_1 = 0.1\ dm^3$, $V_3 = 1.2\ dm^3$, and $\bar{C}_V = (3/2)R$.
3. Give one of the two general statements of the second law of thermodynamics and explain it in your own words.
4. A power plant with steam of 400°C uses a lake with 15°C water as its cooling water source. What fraction of the heat generated by the power plant is wasted in the lake?

4-1.9 SELF-TEST

1. Discuss the significance of the second law in terms of possible and impossible processes.
2. Apply the Carnot cycle to a gas that follows the equation of state

$$\left(P + \frac{a}{V^2}\right)V = RT$$

to derive the efficiency expression.

3. Proposals have been made to use the thermal gradient in the ocean (the surface is warmer than the depths) to generate power. Discuss how this might be done and why the second law is not violated.

4-1.10 ADDITIONAL READING

1.* J. A. Polson, *Internal Combustion Engines*, John Wiley and Sons, New York, 1942.
2. H. A. Bent, *The Second Law*, Oxford University Press, New York, 1965, Chaps. 3 and 4.

4-2 ENTROPY

OBJECTIVES

You shall be able to

1. State the mathematical definition of dS_{sys} (system) and $đS_{surr}$ (surroundings)
2. Calculate entropy changes for any process, given the equation of state for the material used in the process

4-2.1 ENTROPY

In Sec. 4-1 various types of engines that could be operated in reversible cycles were discussed. Here use will be made of a Carnot cycle to introduce the definition of entropy. Once the entropy function is defined, equations will be derived to calculate entropy changes. Some attention will be given to the relationship of entropy to spontaneity in a process. The real significance of this new function will not be evident until Chap. 5, when it will be combined with the energy function to define another function that will give us a means of determining where equilibrium lies in a reaction or process.

You should work several problems in this chapter to gain facility in using the equations involving entropy. The procedures developed here will be very useful in the following section and the following chapter.

If you skipped Sec. 4-1, go back at this point and read Sec. 4-1.4 so you will be familiar with the operation of a Carnot cycle. The equations used need not be studied in great detail, but you need to find the definition of ε, the efficiency. ε is defined as the work done by the engine (- W), divided by the heat absorbed at the high temperature reservoir.

$$\varepsilon = -\frac{W}{Q_h} = \frac{Q_h + Q_\ell}{Q_h} \qquad (4\text{-}2.1.1)$$

We also showed previously

$$\varepsilon = \frac{T_h - T_\ell}{T_h} \qquad (4\text{-}1.4.20)$$

Combine these two equations to get

$$\frac{Q_h + Q_\ell}{Q_h} = \frac{T_h - T_\ell}{T_h} \qquad (4\text{-}2.1.2)$$

which can be rearranged to give (show it)

$$\frac{Q_h}{T_h} = -\frac{Q_\ell}{T_\ell} \qquad (4\text{-}2.1.3)$$

or

$$\frac{Q_h}{T_h} + \frac{Q_\ell}{T_\ell} = 0 \quad \text{or} \quad \sum\frac{Q}{T} = 0 \qquad (4\text{-}2.1.4)$$

Remember that the Carnot cycle is a reversible cycle. Equation (4-2.1.4) can be extended to any reversible cycle. We can imagine any reversible cycle involving some working medium as being made up of a large number of Carnot cycles. Figure 4-2.1.1 illustrates this point. The outermost parts of the individual Carnot cycles approximate the overall cycle. The portions of the Carnot cycles inside the general process boundary cancel each other, since each is traced in both a forward and a

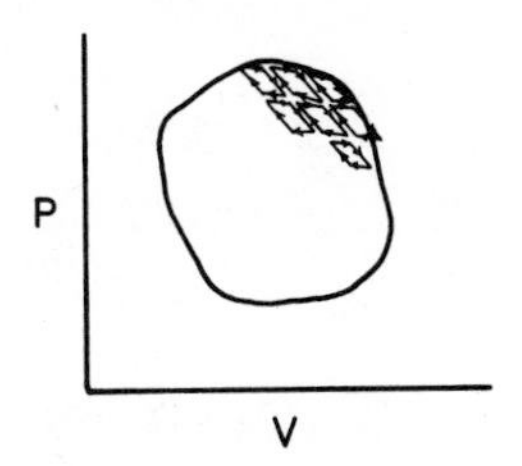

FIG. 4-2.1.1 A reversible cyclic process approximated by Carnot cycles.

reverse direction. (Spend a moment looking at the interior of the general cycle to prove to yourself that this is true.)

We can imagine making the Carnot cycles in Fig. 4-2.1.1 smaller and smaller to approximate the area enclosed more precisely. The net effect of performing all these Carnot cycles will be the same as performing the general process. We can apply Eq. (4-2.1.4) to these Carnot cycles. Thus, for this set of Carnot cycles

$$\sum \frac{Q}{T} = 0 \tag{4-2.1.4}$$

But, we stated previously that the sum of all the individual Carnot processes yielded the same result as the general process.

We conclude, then, that a *general*, *reversible* process obeys Eq. (4-2.1.4). And since we have reduced the process to infinitesimal steps, the summation in Eq. (4-2.1.4) may be replaced by an integral to give

$$\oint \frac{q_{rev}}{T} = 0 \tag{4-2.1.5}$$

As before, the symbol $\oint$ indicates that integration is carried out over the whole cycle.

Equation (4-2.1.5) is a general result for any reversible process. No proof will be given for that statement. Note that Eq. (4-2.1.5) has the same properties as E and H. The integral over the cycle is zero. This is a property of a state function.

On the basis of these considerations we define a new quantity, *entropy*, as follows:

$$dS_{sys} = \frac{q_{rev}}{T} \tag{4-2.1.6}$$

We are defining the entropy changes in a system in terms of the reversible heat absorption and the temperature. We can immediately write [since S is a state function; see Eqs. (4-2.1.5) and (4-2.1.6)]

$$\Delta S_{sys} = \int_a^b \frac{q_{rev}}{T} = \int_a^b dS = S_b - S_a \qquad (4\text{-}2.1.7)$$

The new function we have defined is independent of path.

You have had experience in calculating $\int q_{rev}$ in Chap. 3. Thus, Eq. (4-2.1.7) is sufficient to allow us to calculate the entropy changes in a system for a given process. Before deriving the equations needed to do that, a couple of observations are in order.

First, from examination of Eq. (4-2.1.7) you should conclude that, even though we defined dS_{sys} in terms of q_{rev}/T, we will find the same value for ΔS_{sys} in going from State A to State B no matter how we actually carry out the process. Equation (4-2.1.7) gives us directions on how to calculate ΔS_{sys} for any process. We simply imagine a convenient reversible path from A to B, then calculate ΔS for that path. We know that the ΔS we calculate is the same as for the actual path.

Second, we have not yet said anything about the surroundings. We do not define a state function to describe entropy changes in the surroundings due to difficulties in describing all the changes that might occur there. Instead we define

$$đS_{surr} = -\frac{q_{act}}{T} \qquad (4\text{-}2.1.8)$$

where q_{act} signifies the actual heat absorbed in a process. We can, of course, write

$$\Delta S_{surr} = \int_a^b đS_{surr} = -\int_a^b \frac{q_{act}}{T} \qquad (4\text{-}2.1.9)$$

ΔS_{surr} written in this way is path dependent, since q_{act} is.

One more statement is required, then we can deduce some very important conclusions. Let us consider two paths from State A to State B, one reversible, one adiabatic (and irreversible and spontaneous). See Fig. 4-2.1.2. You should see immediately that

$$q_1 = q_{rev}$$

$$q_2 = 0$$

We are interested, thus far, in processes which absorb heat ($q > 0$) to produce work. So, assume $q_1 > 0$. Obviously, $q_1 > q_2$ for this particular process. In fact, for all processes of the type in which heat is converted to work,

$$q_{rev} \geq q_{act} \qquad (4\text{-}2.1.10)$$

This equation allows us to find some interesting relationships.

The total entropy change in a process ΔS_{total} will be equal to the sum of the entropy change for the system ΔS_{sys} and the entropy change for the surroundings ΔS_{surr}:

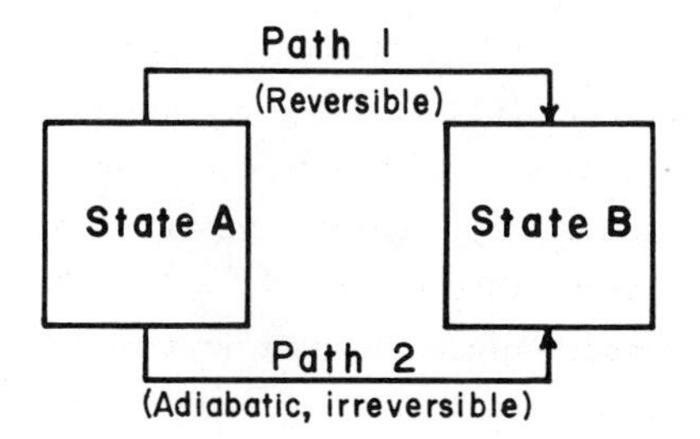

FIG. 4-2.1.2 Two paths from State A to State B.

$$\Delta S_{total} = \Delta S_{sys} + \Delta S_{surr}$$

$$= \int \frac{q_{rev}}{T} - \int \frac{q_{act}}{T} \quad (4\text{-}2.1.11)$$

For a reversible process, $q_{rev} = q_{act}$, so we have

$$\Delta S_{total}(\text{rev}) = \int \frac{q_{rev}}{T} - \int \frac{q_{rev}}{T} = 0$$

The total entropy change of the universe (system and surroundings) for a reversible process is zero.

For an irreversible (spontaneous) process, $q_{rev} > q_{act}$. Then

$$\Delta S_{total}(\text{spontaneous}) = \int \frac{q_{rev}}{T} - \int \frac{q_{act}}{T} > 0$$

The entropy change of the universe for an irreversible process is greater than zero.

You will recall that a reversible process is one that is carried out by means of infinitesimal changes in system variables and is therefore always at equilibrium (within an infinitesimal). For such a process no net entropy changes occur in the universe. For a spontaneous process, however, we have a net increase in the entropy of the universe. This increase in entropy for spontaneous processes is one of the forces that drives a system toward equilibrium.

We can summarize the above by the following three statements for a given process *in which* $\Delta E = 0$.

1. If $\Delta S_{total} > 0$, the process will tend to proceed spontaneously in the direction written.
2. If $\Delta S_{total} = 0$, the system is at equilibrium and no spontaneous processes will occur.
3. If $\Delta S_{total} < 0$, the process will tend to proceed spontaneously in the reverse direction.

For cases in which $\Delta E \neq 0$, we must wait until the definition of a new function in the next chapter to describe the system's behavior.

At this point you are probably asking: What is this thing called entropy? That is a good question, and one that is exceedingly difficult to answer. Entropy, up to now, is just a function whose change gives us a measure of the tendency to spontaneity. Conceptually, it can be related to disorder or randomness within a system. If State B is more random (has more disorder) than State A, then $S_B > S_A$. This will be pursued further in Chap. 9, Statistical Thermodynamics. A point that might be of interest is that the entropy of the universe is increasing. Heat spontaneously moves from high-temperature portions to those of lower temperatures. As you saw above, we can do work by using reservoirs of different temperature to operate an engine. What will happen if heat is ever uniformly distributed in the universe?

Just knowing the definition of entropy, and that changes in it are related to spontaneity, is sufficient for our purposes. You really do not know much more about energy or work: the terms are just more familiar.

4-2.2 GENERAL EQUATIONS FOR CALCULATING ENTROPY CHANGES

The equations below are derived for gases, but the final results are good for any system. If you know an equation of state you should be able to calculate entropy changes for any process.

Reversible processes are assumed in the following, but since entropy is a state function, the final results will apply to any system. Note the word *system*. The equations do not apply to the surroundings unless the process is actually carried out reversibly.

T AND V AS VARIABLES. We write S as S(T, V) in this case. As usual, it takes at least two variables to specify a thermodynamic function.

$$dS = \left(\frac{\partial S}{\partial T}\right)_V dT + \left(\frac{\partial S}{\partial V}\right)_T dV \qquad (4\text{-}2.2.1)$$

We can write

$$dE = q + w \qquad (3\text{-}1.4.2)$$

If only PV work is involved and if the process is reversible,

$$w = -P\ dV \qquad q_{rev} = T\ dS \qquad (4\text{-}2.2.2)$$

Thus

$$dE = T\ dS - P\ dV \qquad (4\text{-}2.2.3)$$

But from Eq. (3-1.4.3):

$$dE = \left(\frac{\partial E}{\partial T}\right)_V dT + \left(\frac{\partial E}{\partial V}\right)_T dV \qquad (3\text{-}1.4.3)$$

Equating (4-2.2.3) and (3-1.4.3) yields

$$T\,dS - P\,dV = \left(\frac{\partial E}{\partial T}\right)_V dT + \left(\frac{\partial E}{\partial V}\right)_T dV \qquad (4\text{-}2.2.4)$$

Rearranged, this becomes

$$dS = \frac{1}{T}\left(\frac{\partial E}{\partial T}\right)_V dT + \frac{1}{T}\left[\left(\frac{\partial E}{\partial V}\right)_T + P\right] dV \qquad (4\text{-}2.2.5)$$

Equating the coefficients of dT and dV in (4-2.2.5) and (4-2.2.1) gives

$$\left(\frac{\partial S}{\partial T}\right)_V = \frac{1}{T}\left(\frac{\partial E}{\partial T}\right)_V = \frac{n\bar{C}_V}{T} \qquad (4\text{-}2.2.6)$$

and

$$\left(\frac{\partial S}{\partial V}\right)_T = \frac{1}{T}\left[\left(\frac{\partial E}{\partial V}\right)_T + P\right] \qquad (4\text{-}2.2.7)$$

Rearrange these last two equations,

$$\left(\frac{\partial E}{\partial T}\right)_V = T\left(\frac{\partial S}{\partial T}\right)_V \qquad (4\text{-}2.2.8)$$

$$\left(\frac{\partial E}{\partial V}\right)_T = T\left(\frac{\partial S}{\partial V}\right)_T - P \qquad (4\text{-}2.2.9)$$

Take $(\partial/\partial V)_T$ of Eq. (4-2.2.8), and then $(\partial/\partial T)_V$ of Eq. (4-2.2.9). (You should show that the following equations result.)

$$\left(\frac{\partial^2 E}{\partial V\,\partial T}\right) = T\left(\frac{\partial^2 S}{\partial V\,\partial T}\right) \qquad (4\text{-}2.2.10)$$

$$\left(\frac{\partial^2 E}{\partial T\,\partial V}\right) = T\left(\frac{\partial^2 S}{\partial T\,\partial V}\right) + \left(\frac{\partial S}{\partial V}\right)_T - \left(\frac{\partial P}{\partial T}\right)_V \qquad (4\text{-}2.2.11)$$

We can write $(\partial^2 E/\partial V\,\partial T) = (\partial^2 E/\partial T\,\partial V)$, since the order of differentiation is immaterial for state functions. We then equate the right sides of Eqs. (4-2.2.10) and (4-2.2.11)

$$T\left(\frac{\partial^2 S}{\partial T\,\partial V}\right) + \left(\frac{\partial S}{\partial V}\right)_T - \left(\frac{\partial P}{\partial T}\right)_V = T\left(\frac{\partial^2 S}{\partial V\,\partial T}\right) \qquad (4\text{-}2.2.12)$$

As in the case of E

$$\left(\frac{\partial^2 S}{\partial T\,\partial V}\right) = \left(\frac{\partial^2 S}{\partial V\,\partial T}\right)$$

So Eq. (4-2.2.12) can be simplified to

$$\left(\frac{\partial S}{\partial V}\right)_T = \left(\frac{\partial P}{\partial T}\right)_V \qquad (4\text{-}2.2.13)$$

This is an important equation since it relates the dependence of entropy on volume to a dependence of P on T. We can, given an equation of state, evaluate $(\partial P/\partial T)_V$, which gives us $(\partial S/\partial V)_T$. If you substitute (4-2.2.13) and (4-2.2.6) into (4-2.2.1) you should get (show it)

$$dS = \frac{n\bar{C}_V}{T}\,dT + \left(\frac{\partial P}{\partial T}\right)_V dV \tag{4-2.2.14}$$

All we need is an equation of state, and we can evaluate ΔS for a given process.

4-2.2.1 EXAMPLE

Calculate the entropy change for the heating of 2 mol argon gas at constant pressure from 100 K to 200 K. C_V = 12.4 J mol^{-1} K^{-1}. Assume Ar behaves ideally.

Solution

$$dS = \frac{n\bar{C}_V}{T} + \left(\frac{\partial P}{\partial T}\right)_V dV$$

P = nRT/V. Therefore,

$$\left(\frac{\partial P}{\partial T}\right)_V = \frac{nR}{V}$$

(Note that V is not actually constant in the process, but we take the derivative as though it is.) For an ideal gas we can write the above equation as

$$dS = \frac{n\bar{C}_V}{T}\,dT + \frac{nR}{V}\,dV \tag{4-2.2.15}$$

Take the integral to find ΔS.

$$\Delta S = \int dS = \int_{100}^{200} \frac{n\bar{C}_V}{T}\,dT + \int_{V_1}^{V_2} \frac{nR}{V}\,dV$$

The quantities C_V, n, and R are constants, so they may be taken out of the integral.

$$\Delta S = n\bar{C}_V \int_{100}^{200} \frac{dT}{T} + nR \int_{V_1}^{V_2} \frac{dV}{V}$$

$$= n\bar{C}_V \ln \frac{200}{100} + nR \ln \frac{V_2}{V_1}$$

But $V_2 = nRT_2/P_2$, $V_1 = nRT_1/P_1$, and $P_1 = P_2$. Thus $V_2/V_1 = T_2/T_1$. Substitution gives

$$\Delta S = n\bar{C}_V \ln \frac{200}{100} + nR \ln \frac{T_2}{T_1}$$

$$= n(\bar{C}_V + R)\ \ln \frac{200}{100}$$

$$= 2\ \text{mol}\ (12.4\ \text{J mol}^{-1}\ \text{K}^{-1} + 8.314\ \text{J mol}^{-1}\ \text{K}^{-1})\ \ln \frac{200}{100}$$

$$= \underline{\underline{28.7\ \text{J K}^{-1}}} \qquad (6.86\ \text{cal K}^{-1})$$

Note the units on S are J K^{-1} (or cal K^{-1}).

T AND P AS VARIABLES. Here we will go through the same steps as previously, only this time we will use T and P as the variables. If S = S(T, P),

$$dS = \left(\frac{\partial S}{\partial T}\right)_P dT + \left(\frac{\partial S}{\partial P}\right)_T dP \qquad (4\text{-}2.2.16)$$

From the definition of enthalpy, H, we can write

$$H = E + PV \qquad (3\text{-}1.4.9)$$

$$dH = dE + P\ dV + V\ dP = q + V\ dP$$

For a reversible process, q_{rev} = T dS, thus

$$T\ dS = dH - V\ dP \qquad (4\text{-}2.2.17)$$

But H = H(T, P), from which we get

$$dH = \left(\frac{\partial H}{\partial T}\right)_P dT + \left(\frac{\partial H}{\partial P}\right)_T dP \qquad (3\text{-}1.4.10)$$

Combine (4-2.2.17) and (3-1.4.10):

$$dS = \frac{1}{T}\left(\frac{\partial H}{\partial T}\right)_P dT + \frac{1}{T}\left[\left(\frac{\partial H}{\partial P}\right)_T - V\right] dP \qquad (4\text{-}2.2.18)$$

And from (4-2.2.16):

$$dS = \left(\frac{\partial S}{\partial T}\right)_P dT + \left(\frac{\partial S}{\partial P}\right)_T dP \qquad (4\text{-}2.2.16)$$

Equating coefficients of dT and dP in (4-2.2.16) and (4-2.2.18) we can write

$$\left(\frac{\partial S}{\partial T}\right)_P = \frac{1}{T}\left(\frac{\partial H}{\partial T}\right)_P = \frac{n\bar{C}_P}{T} \qquad (4\text{-}2.2.19)$$

and

$$\left(\frac{\partial S}{\partial P}\right)_T = \frac{1}{T}\left[\left(\frac{\partial H}{\partial P}\right)_T - V\right]$$

or

$$T\left(\frac{\partial S}{\partial P}\right)_T = \left(\frac{\partial H}{\partial P}\right)_T - V \qquad (4\text{-}2.2.20)$$

As before, take $(\partial/\partial P)_T$ of Eq. (4-2.2.19) and $(\partial/\partial T)_P$ of (4-2.2.20) to yield

$$\left(\frac{\partial^2 S}{\partial P\ \partial T}\right) = \frac{1}{T}\left(\frac{\partial^2 H}{\partial P\ \partial T}\right) \qquad (4\text{-}2.2.21)$$

and $$T\left(\frac{\partial^2 S}{\partial T\ \partial P}\right) + \left(\frac{\partial S}{\partial P}\right)_T = \left(\frac{\partial^2 H}{\partial T\ \partial P}\right) - \left(\frac{\partial V}{\partial T}\right)_P \qquad (4\text{-}2.2.22)$$

Combine these last two equations and note that $(\partial^2 H/\partial P\ \partial T) = (\partial^2 H/\partial T\ \partial P)$, (why?), to find (do it)

$$\left(\frac{\partial S}{\partial P}\right)_T = -\left(\frac{\partial V}{\partial T}\right)_P \qquad (4\text{-}2.2.23)$$

Again, we have found a very useful function. Now, substitute (4-2.2.23) and (4-2.2.19) into Eq. (4-2.2.16)

$$\boxed{dS = \frac{n\bar{C}_P}{T}\,dT - \left(\frac{\partial V}{\partial T}\right)_P dP} \qquad (4\text{-}2.2.24)$$

This is a second general equation for calculating entropy changes within a system. Whether we choose to use (4-2.2.24) or (4.2.2.14) is strictly a matter of convenience. For a process carried out at constant pressure, (4-2.2.24) is convenient, since the last term goes to zero. Similarly, Eq. (4-2.2.14) is useful for constant volume processes.

4-2.2.2 EXAMPLE

Recalculate the entropy change for the heating of Ar(g) (2 mol) from 100 K to 200 K, at constant volume. $C_P = 20.7\ \text{J mol}^{-1}\ \text{K}^{-1}$. Compare with the results of Example 4-2.2.1.

Solution

$$dS = \frac{n\bar{C}_P}{T}\,dT - \left(\frac{\partial V}{\partial T}\right)_P dP$$

$$= \frac{n\bar{C}_P}{T}\,dT - \frac{nR}{P}\,dP \qquad \text{prove this}$$

$$\Delta S = \int_{100}^{200} \frac{n\bar{C}_P}{T}\,dT - \int_{P_1}^{P_2} \frac{nR}{P}\,dP$$

$$= n\bar{C}_P \ln\frac{200}{100} - nR\ln\frac{P_2}{P_1}$$

$$= n\bar{C}_P \ln\frac{200}{100} - nR\ln\frac{T_2}{T_1}$$

$$= n(\bar{C}_P - R)\ln\frac{200}{100}$$

$$= 2\ \text{mol}\ (20.7 - 8.3)\ \text{J mol}^{-1}\ \text{K}^{-1} \ln\frac{200}{100}$$

$$= \underline{\underline{17.2 \text{ J K}^{-1}}}$$

Note two things. $\Delta S_V < \Delta S_P$. This can be rationalized as follows. At constant P, the volume increases, giving the argon molecules more freedom of motion. The larger the volume, the more random the system at a given T. Since S_{init} (initial) is the same for both processes, and S_{fin} (final) (constant P) > S_{fin} (constant V), we find $\Delta S_P > \Delta S_V$.

Also, note that Example 4-2.2.1 is solved much more easily with Eq. (4-2.2.24) than with Eq. (4-2.2.14). The converse is true for Example 4-2.2.2.

4-2.2.3 EXAMPLE

Calculate the entropy change for the system and surroundings for the constant pressure and constant volume heating of 3 mol of a gas which obeys the equation of state

$$P(V - nb) = nRT$$

from 150 K to 300 K. Given $b = 0.039 \text{ dm}^3 \text{ mol}^{-1}$, $\bar{C}_P = 28.6 + 3.77 \times 10^{-3} T - 0.50 \times 10^5 T^{-2}$ (J mol^{-1} K^{-1}), and $\bar{C}_V = \bar{C}_P - R$.

Solution

At Constant V

Apply Eq. (4-2.2.14):

$$dS = \frac{n\bar{C}_V}{T} dT + \left(\frac{\partial P}{\partial T}\right)_V dV \qquad (4\text{-}2.2.14)$$

which becomes at constant V

$$dS = \frac{n\bar{C}_V}{T} dT$$

$$\Delta S_V = \int dS = 3 \text{ mol} \int_{150}^{300} \left[\frac{28.6 + 3.77 \times 10^{-3} T - 0.50 \times 10^5 T^{-2} - R}{T}\right] dT$$

$$= 3 \text{ mol} \left[(28.6 - R) \ln T + 3.77 \times 10^{-3} T + \tfrac{1}{2}(0.50 \times 10^5 T^{-2}\right]_{150}^{300} (\text{J mol}^{-1} \text{ K}^{-1})$$

$$= 3\left[(28.6 - 8.31) \ln \frac{300}{150} + 3.77 \times 10^{-3}(300 - 150) + \tfrac{1}{2}(0.50 \times 10^5)\left(\frac{1}{300^2} - \frac{1}{150^2}\right)\right] \text{ J K}^{-1}$$

$$= 3(13.8) \text{ J K}^{-1}$$

$$= \underline{\underline{41.4 \text{ J K}^{-1}}} \qquad (9.89 \text{ cal K}^{-1})$$

At Constant P

Applying Eq. (4-2.2.24):

$$dS = \frac{n\bar{C}_P}{T}\,dT + \left(\frac{\partial V}{\partial T}\right)_P dP \qquad (4\text{-}2.2.24)$$

At constant P this becomes

$$dS = \frac{n\bar{C}_P}{T}\,dT = \frac{n(\bar{C}_V + R)}{T}\,dT = dS_V + \frac{nR}{T}\,dT$$

$$\Delta S = \Delta S_V + \int_{150}^{300} \frac{R}{T}\,dT = 41.4 \text{ J K}^{-1} + (3)(8.31) \ln \frac{300}{150}$$

$$= \underline{58.7 \text{ J K}^{-1}} \qquad (14.0 \text{ cal K}^{-1})$$

Note: We cannot find ΔS_{surr} since the path of the process is not sufficiently well defined. We do know, however, that $-\Delta S_{surr}$ must be less than or equal to 58.7 J K^{-1}. For a spontaneous process, $\Delta S(\text{spont}) = \Delta S_{sys} + \Delta S_{surr} > 0$ for which $-\Delta S_{surr} < \Delta S_{sys} = 58.7$ J K^{-1}. If the process were reversible $-\Delta S_{surr} = \Delta S_{sys} = 58.7$ J K^{-1}.

ENTROPY CHANGES FOR PHASE CHANGES: AN EXAMPLE. This section outlines the method for calculating change in entropy when a substance changes from one phase to another (solid to liquid, liquid to vapor, etc.). Let us take our argon examples (Examples 4-2.2.1 and 4-2.2.2) and calculate the change in entropy for the following process at constant P = 1 atm, 1 mol.

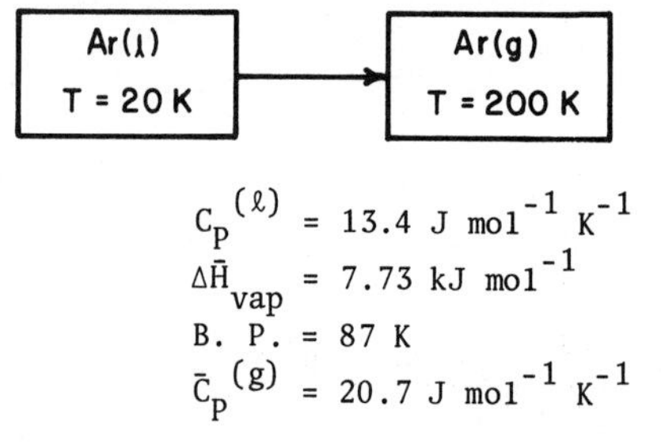

$C_P^{(\ell)} = 13.4$ J mol^{-1} K^{-1}
$\Delta\bar{H}_{vap} = 7.73$ kJ mol^{-1}
B. P. = 87 K
$\bar{C}_P^{(g)} = 20.7$ J mol^{-1} K^{-1}

As usual, the first step is to outline a process: we assume reversibility to ease calculation.

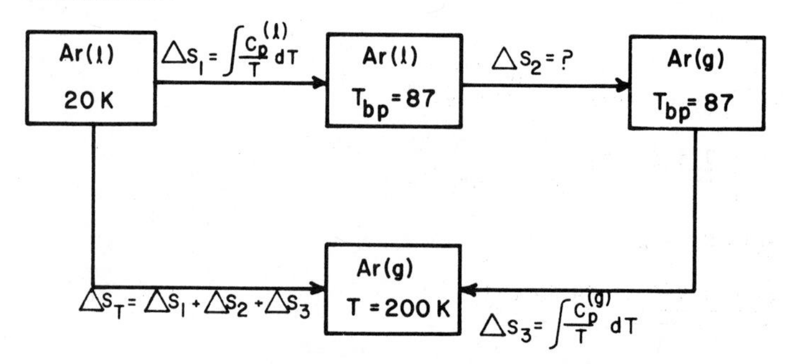

You should already know how to calculate ΔS_1 and ΔS_3 from previous examples. For ΔS_2, we need to find q_{rev}. For a phase change at constant pressure, we know q_{rev} = dH, which for vaporization gives $q_{rev} = dH_{vap}$.

$$\Delta S_2 = \int_{\ell}^{v} \frac{q_{rev}}{T} = \int_{\ell}^{v} \frac{dH_{vap}}{T} = \frac{1}{T}\int_{\ell}^{v} dH_{vap} \qquad \text{T is constant}$$

$$= \frac{\Delta H_{vap}}{T} = \frac{n\Delta \bar{H}_{vap}}{T} \tag{4-2.2.25}$$

For argon, this is 7.73×10^3 J mol^{-1}

$$\Delta S_2 = \frac{(7730 \text{ J mol}^{-1})(1 \text{ mol})}{87 \text{ K}}$$

$$= \underline{\underline{88.9 \text{ J K}^{-1}}}$$

For the other two steps (we remove $\bar{C}_P$ from the integral only because it is constant)

$$\Delta S_1 = \int_{20}^{87} \frac{n\bar{C}_P^{(\ell)}}{T} dT = n\bar{C}_P^{(\ell)} \int_{20}^{87} \frac{dT}{T}$$

$$= \bar{C}_P^{(\ell)} \ln \frac{T_2}{T_1}$$

$$= (1 \text{ mol})(13.4 \text{ J mol}^{-1} \text{ K}^{-1})\left(\ln \frac{87}{20}\right)$$

$$= \underline{\underline{19.7 \text{ J K}^{-1}}}$$

$$\Delta S_3 = \int_{87}^{200} \frac{n\bar{C}_P}{T} dT$$

$$= (1 \text{ mol})(20.7 \text{ J mol}^{-1} \text{ K}^{-1})\left(\ln \frac{200}{87}\right)$$

$$= \underline{\underline{17.2 \text{ J K}^{-1}}}$$

$$\Delta S_T = \Delta S_1 + \Delta S_2 + \Delta S_3 = \underline{\underline{126 \text{ J K}^{-1}}} \qquad (30.1 \text{ cal K}^{-1})$$

4-2.3 PROBLEMS

1. Two moles of an ideal gas are heated from 300 K to 400 K, with a corresponding volume increase from 46.2×10^{-3} m^3 to 57.4×10^{-3} m^3. Calculate the increase in entropy of the system if $\bar{C}_P = 20.8$ J mol^{-1} K^{-1}.

2. Repeat the calculation for the conditions in Problem (1) if the gas follows the van der Waals equation written in the following form:

$$\left(P + \frac{an^2}{V^2}\right)(V - nb) = nR'T$$

$a = 8.1 \times 10^{-3}$ N m^4 mol^{-2}, $b = 3.0 \times 10^{-5}$ m^3 mol^{-1}, $R' = 6.5$ J mol^{-1}, and $\bar{C}_P = 28.58 + 3.76 \times 10^{-3}$ T (J mol^{-1} K^{-1}). Assume $C_P = C_V + R'$

3. A refrigerator is placed in a perfectly insulated room. The door of the refrigerator is left open. After thermal equilibrium has been established, the refrigerator is plugged into an electrical outlet and turned on. What happens to the temperature of the room as time passes? Explain.

4. Calculate the increase in entropy of 3 mol H_2 as they change from 300 K and 1 atm to 1000 K and 3 atm. (Look up $\bar{C}_P$ in Chap. 3.)

5. For graphite we may write

$$C_P = 1.265 + 14.008 \times 10^{-3}\ T - 103.31 \times 10^{-7}\ T^2$$

$$C_V = C_P - 0.002\ R$$

Find ΔS_{sys} for the constant pressure and constant volume heating of graphite from 100 K to 225 K. Why is the difference so small?

6. A sample of 0.051 kg CO_2 (M = 0.044 kg mol^{-1}) undergoes a change from $T_1 = 400$ K and $V_1 = 71.0$ dm^3 to $T_2 = 350$ K and $V_2 = 93.0$ dm^3. If CO_2 behaves ideally and $C_P = 4.3R = C_V + R$, find ΔS_{sys}.

7. The Bertholet equation of state is $P = RT/(\bar{V} - b) - a/T\bar{V}^2$. For a given gas, $a = 3.82 \times 10^4$ atm dm^6 K mol^{-2} and $b = 0.019$ dm^3 mol^{-1}. Find ΔS for the isothermal (T = 225 K) expansion from $V_1 = 26.2$ dm^3 to $V_2 = 54.3$ dm^3.

8. Derive the equation necessary to find ΔS for the change from T_1V_1 to T_2V_2 for 1 mol of a gas that follows the Beattie-Bridgeman equation. Do the integrals and leave in terms of T_1, T_2, V_1, and V_2.

4-2.4 SELF-TEST

1. A gas is expanded from $T_1P_1V_1$ to $T_2P_2V_2$. Outline a feasible process for this expansion in which no adiabatic steps are involved, and write the equations you would use to find W, Q, ΔE, and ΔS. If you use more than one step, indicate how you would determine values of T, P, and V at each step.

2. Calculate the entropy change for both the system and surroundings for the reversible expansion of 2 mol of a gas from 2 atm to 1 atm if $T_1 = 100$ K and $C_V = 16.7$ J mol^{-1} K^{-1}, $V_2 = 12.6$ dm^3.

   a. If the gas is ideal

   b. If the gas follows van der Waals equation with $a = 1.46$ dm^6 atm mol^{-2} and $b = 0.0392$ dm^3 mol^{-1}.

3. One mole of steam is *compressed* reversibly to liquid water at the boiling point, 373 K. $\Delta H_{vap} = 2255$ J/g (1 atm). Calculate Q, W, ΔE, ΔH, and ΔS.

4. For a Redlich-Kwong gas (CO; P_c = 34.6 atm, T_c = 134.0 K) determine the entropy change for isothermal compression of 1 mol from P_1 = 2 atm and T_1 = 280 K to P_2 = 4 atm.

4-2.5 ADDITIONAL READING

1. J. A. Polson, *Internal Combustion Engines*, 2nd ed., John Wiley and Sons, New York, 1942, Chap. 4 and 5.
2. H. A. Bent, *The Second Law*, Oxford University Press, New York, 1965, Chaps. 7 through 12.

4-3 THE THIRD LAW AND ABSOLUTE ENTROPIES

OBJECTIVES

You shall be able to

1. State the third law of thermodynamics
2. Calculate absolute entropies from heat capacity data
3. Calculate entropy changes for reactions and phase transitions

4-3.1 A STATEMENT OF THE THIRD LAW OF THERMODYNAMICS

Perhaps the best way to begin this section is to state that *the absolute zero of temperature is unattainable*. This statement can be used to form the basis of the third law which will be given shortly. The statement that the attainment of T = 0 is not possible grows out of both experimental attempts to reach absolute zero and theoretical considerations of the conditions necessary to do so.

An interesting question for you to ponder is: Does all motion cease at absolute zero temperature? On the basis of our study so far you would probably say yes. However, we will see in Chap. 7 that there is always some motion (vibration) no matter how low the temperature becomes.

Scientists have expended considerable effort in achieving temperatures closer and closer to T = 0. Their objective is to study molecular phenomena at these low temperatures. Superconductivity is an example of a physical phenomenon which requires low temperatures to become operative. The measurement of absolute entropies, as we shall see, requires observations of heat capacity behavior at very low temperatures. A description of some experimental methods will be given before our statement of the third law. This will be brief. A more detailed account may be found in Moore (see Sec. 4-3.6, Ref. 1).

The Joule-Thomson effect has been used to obtain temperatures to about 4 K. If a gas is below its inversion temperature (the temperature at which a gas cools on expansion; see Sec. 3-2), it can be cooled by expansion sufficiently to liquify a portion of the gas that is expanding (see Sheehan for a calculation of the amount of liquid formed for a given amount of gas; Sec. 4-3.6, Ref. 2). Such a process can be used to produce liquid nitrogen, which boils at 77 K, from air. (The oxygen produced boils at 90 K and can be separated by fractional distillation.)

If the liquid nitrogen is used to cool hydrogen gas (inversion T = 193 K), then liquid hydrogen can be obtained by expansion. The boiling point of hydrogen is 22 K. This hydrogen is then used to lower the temperature of helium to below its inversion temperature of 100 K; the helium can then be liquified by expansion. Liquid helium boils at 4 K, and with large pumps removing vapor as it forms, the temperature can be lowered to 0.84 K. Pumping vapor from above a liquid results in a lowering of the temperature of the liquid. To become a vapor molecule, heat must be added to a liquid molecule. In an insulated system (Q = 0) the only source of heat is the liquid itself. The heat is added to the liquid molecule to produce a vapor molecule at the expense of the internal energy of liquid. This results in a decrease in temperature within the liquid. This temperature is about the lower limit for liquid helium due to the tremendous quantity of vapor that must be removed to lower the temperature any further.

To attain temperatures below about 1 K another technique is required. This is the *adiabatic demagnetization of paramagnetic* materials.

A paramagnetic material is one which has unpaired electrons (e.g., O_2, CoF_6^{3-}). When such a material is placed in a magnetic field the unpaired spins, which behave as though they were small magnets, line up in the direction of the applied external magnetic field. Under these conditions the material is *magnetized.*

From what we mentioned briefly in Sec. 4-2.1 about entropy and randomness, you should conclude that such a *mangetized* state has low entropy (it is an *ordered* system; high order means low entropy). If the magnetic field is removed, then the unpaired spins will lose their ordering leading to a rise in entropy. It can do this only by extracting heat from the material itself since the system is isolated from the surroundings (q = 0). This, in turn, drops the temperature of the sample.

This technique has led to temperatures as low as 10^{-5} K. Perhaps you can see why absolute zero cannot be attained (at least in this manner). As the sample temperature drops closer and closer to zero there is less and less thermal energy available to cause the spins to become unaligned. So the more closely zero is approached, the less effective adiabatic demagnetization is in lowering temperature. It would take an infinite number of cycles to lower the temperature to absolute zero.

In this cooling process, use has been made of the entropy difference between the material itself and the unpaired spins. We conclude that this entropy difference goes to zero as T goes to zero. Otherwise, we could continue to lower the temperature of the material by successive application of the steps outlined above. This can be generalized to the statement: *The entropies of all perfect crystalline materials are the same at absolute zero.* The condition "perfect crystalline" is required to eliminate the possibility of disorder in the solid. If disorder remains, there will be a residual entropy at 0 K.(Sec. 9-2 will give the reasons for this behavior). Most substances do not form perfect crystalline solids. We can still use the "perfect crystal" as our reference point even though the substance is actually imperfect.

Expanding our statement above: *The entropy of all perfectly crystalline substances at absolute zero are the same and we take that value to be zero.* This implies, then, that at any temperature above zero, every substance will have a finite positive entropy. [For a good discussion of the effects of disorder and impurities in a substance see Sheehan (see Sec. 4-3.6, Ref. 2).] We are now in a position to calculate absolute entropies from heat capacity data.

4-3.2 ABSOLUTE ENTROPIES

Now that we have a common reference point for the entropies of all perfect crystalline solids, we can calculate absolute entropies from heat capacity data.

For a constant pressure process

$$\Delta S = S(T, P) - S(0, P)$$

$$= \int_0^T \left(\frac{\partial S}{\partial T}\right)_P dT = \int_0^T \frac{C_P}{T} dT \qquad (4\text{-}3.2.1)$$

If the substance is a pure, perfect crystal,

$$S(0, P) = 0$$

so

$$S(T, P) = \int_0^T \frac{C_P}{T} dT = \int_0^T C_P \, d \ln T \qquad (4\text{-}3.2.2)$$

If the substance is not a perfect crystal, it will have a finite positive value, so (4-3.2.2) becomes

$$S(T, P) = S_0 + \int_0^T C_P \, d \ln T \qquad (4\text{-}3.2.3)$$

How heat capacity varies with temperature will not be discussed. Instead, some empirical plots of C_P/T vs. T and C_P vs. ln T will be presented to show how absolute entropies can be calculated. Figure 4-3.2.1 shows both types of plots for Cl_2. These are taken in part from Example 4.6 in Sheehan (Sec. 4-3.6, Ref. 2).

To obtain S(T, P) we need to graphically integrate the area under either of the curves in Fig. 4-3.2.1. To that value, we must add the entropy contribution due to any phase changes that have occurred. For this case we have the following process.*

---

* We have insufficient data in the figure to find the change in entropy from 0 K to 20 K. This would have to be determined by other means.

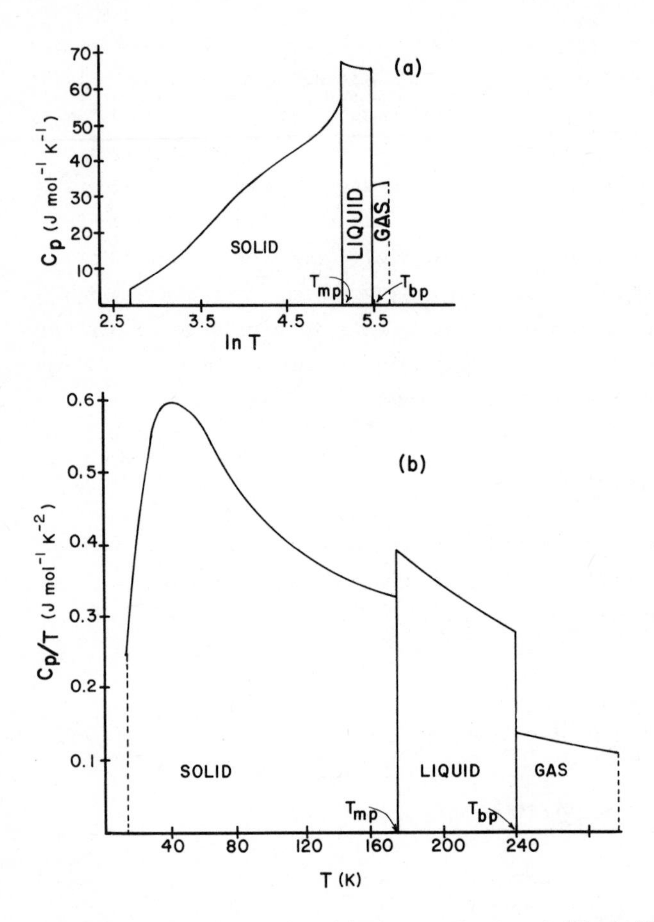

FIG. 4-3.2.1 Plots of: (a) C_p vs. ln T and (b) C_p/T vs. T. (Taken in part from W. F. Sheehan, *Physical Chemistry*, 2nd ed., Allyn and Bacon, Boston, 1970.)

$$\begin{aligned} S(T_f, P) = S_0 &+ S_{0\to 20} \\ &+ \int_{20}^{T_{mp}} \frac{C_P{}^{(s)}}{T}\, dT \quad \left(\text{or} \int_{20}^{T_{mp}} C_P{}^{(s)}\, d\ln T\right) + \frac{\Delta H_{fus}}{T_{mp}} \\ &+ \int_{T_{mp}}^{T_{bp}} \frac{C_P{}^{(\ell)}}{T}\, dT \quad \left(\text{or} \int_{T_{mp}}^{T_{bp}} C_P{}^{(\ell)}\, d\ln T\right) + \frac{\Delta H_{vap}}{T_{bp}} \\ &+ \int_{T_{bp}}^{T_f} \frac{C_P{}^{(g)}}{T}\, dT \quad \left(\text{or} \int_{T_{bp}}^{T_f} C_P{}^{(g)}\, d\ln T\right) \end{aligned} \qquad (4\text{-}3.2.4)$$

$$= \text{area under curve (a) [or area under curve (b)]}$$

$$+ \frac{\Delta H_{fus}}{T_{mp}} + \frac{\Delta H_{vap}}{T_{bp}} + S_0 + \Delta S_{0\to 20}$$

where mp = melting point and bp = boiling point.

The graphical integration can be done by a planimeter, or by tracing the curve on a uniform sheet of paper, cutting out the area under the curve, and weighing (thus referencing the result to a standard weight per unit area), or by applying Simpson's rule. You will have the opportunity to determine the area under such a curve in the problems section (Sec. 4-3.4).

If the process ends at some temperature below the vaporization temperature, we simply terminate the integration at that point and include only the transitions that have actually occurred. If S_0 is nonzero we must include it. Table 4-3.2.1 contains selected entropy values at 298.15 K.

Table 4-3.2.1
Standard Entropy Values (Substances in the Standard State at 298.15 K)

Substance	S_0 ($J\ K^{-1}\ mol^{-1}$)	Substance	S_0 ($J\ K^{-1}\ mol^{-1}$)
Gases			
H_2	130.59	H_2O	188.72
N_2	191.50	NH_3	192.50
O_2	205.1	SO_2	248.5
Cl_2	223.0	CH_4	186.2
HCl	186.6	C_2H_2(acetylene)	200.8
CO	197.5	C_2H_4(ethylene)	219.5
CO_2	213.7	C_2H_6(ethane)	229.5
Liquids			
Hg	76.02	C_2H_5OH(ethanol)	161
Br_2	152	C_6H_6(benzene)	173
H_2O	70.00	C_7H_9(toluene)	220
CH_3OH(methanol)	127	C_6H_8(n-hexane)	296
Solids			
C(diamond)	2.44	Fe	27.2
C(graphite)	5.694	I_2	116.1
S(rhombic)	31.9	NaCl	72.38
S(monoclinic)	32.6	$CuSO_4$	113
Ag	42.72	AgCl	96.23
Cu	33.3	AgBr	107.1

4-3.2.1 EXAMPLE

The data used to draw Fig. 4-4.2.1 are given below. Use these and the following data to find $S^0_{298}(Cl_2)$.

$$\Delta H_f = 6.406 \text{ kJ mol}^{-1} \quad \text{at } T_f = 172.12 \text{ K}$$

$$\Delta H_v = 20.41 \text{ kJ mol}^{-1} \quad \text{at } T_v = 239.05 \text{ K}$$

$$S_{15}(Cl_2) = 1.26 \text{ J mol}^{-1} \text{ K}^{-1}$$

$$C_p^{(g)}(Cl_2) = 37.0 + 0.67 \times 10^{-3}\, T - 2.85 \times 10^5\, T^{-2}$$

T(K)	$\bar{C}_p$ [a]	C_p/T [b]	ln T	T(K)	$\bar{C}_p$	C_p/T	ln T
15	3.7	0.25	2.71	150	51.04	0.340	5.01
20	7.74	0.387	3.00	160	53.05	0.332	5.08
25	12.1	0.484	5.22	170	55.10	0.324	5.14
30	16.7	0.557	3.40	172.12	55.52	0.323	5.15
35	20.8	0.595	3.56	172.12	67.07	0.390	5.15
40	24.0	0.600	3.69	180	67.03	0.372	5.19
45	26.7	0.593	3.81	190	66.90	0.352	5.25
50	29.2	0.584	3.91	200	66.73	0.334	5.30
60	33.5	0.558	4.09	210	66.48	0.317	5.35
70	36.3	0.519	4.25	220	66.28	0.301	5.39
80	38.6	0.483	4.38	230	65.98	0.287	5.44
90	40.6	0.451	4.50	239.05	65.69	0.275	5.48
100	42.26	0.423	4.60	240	32.21	0.134	5.48
110	43.81	0.398	4.70	260	32.95	0.127	5.56
120	45.48	0.379	4.79	280	33.55	0.120	5.64
130	47.24	0.363	4.87	290	33.81	0.117	5.67
140	49.08	0.351	4.94				

[a] In J mol^{-1} K^{-1}.
[b] In J mol^{-1} K^{-2}.

Solution

The results are given below. You should expand one of the two plots given in Fig. 4-3.2.1 and actually do the graphical integration yourself. You will have to in Sec. 4-3.5, Self-Test, so the practice will be useful.

We have not addressed the subject of how to determine C_p at low temperatures. Accept the given value of S_{15}.

$$\Delta S_{0\to 15} = 1.26 \text{ J mol}^{-1} \text{ K}^{-1}$$

$$\Delta S_{15\to 172.12} = 70.17 \text{ J mol}^{-1} \text{ K}^{-1} \qquad \text{graphical integration}$$

$$\Delta S_f = \frac{\Delta H_f}{T_f} = \frac{6.406 \text{ J mol}^{-1}}{172.12 \text{ K}} = 37.22 \text{ J mol}^{-1} \text{ K}^{-1}$$

$$\Delta S_{172.12\to 239.05} = 22.26 \text{ J mol}^{-1} \text{ K}^{-1} \qquad \text{graphical integration}$$

$$\Delta S_V = \frac{\Delta H_V}{T_V} = \frac{20.410}{239.05} = 85.38 \text{ J mol}^{-1}\text{ K}^{-1}$$

$$\Delta S_{239.05\to 298.15} = 7.57 \text{ J mol}^{-1}\text{ K}^{-1} \qquad \text{graphical or actual integration}$$

$$\Delta S_{total} = S^0_{298.15}(Cl_2) = 223.9 \text{ J mol}^{-1}\text{ K}^{-1}$$

The value in Table 4-3.2.1 is 223.0. The small difference comes from using different sets of data in the calculation of the two values.

4-3.3 ENTROPY CHANGES IN REACTIONS

Since we have found a method of finding absolute entropies, we can tabulate these results. Table 4-3.2.1 contains these values. To find entropy change in a reaction we use the same procedures as for ΔE or ΔH.

$$\Delta S_{reaction} = S_{prod} - S_{react} \tag{4-3.3.1}$$

Perhaps an example will suffice to present the method.

4-3.3.1 EXAMPLE

Find the entropy change at 298.15 K for the reaction

$$C_2H_2(g) + 2H_2(g) \rightarrow C_2H_6(g)$$

Solution

$$\Delta S^0 = S^0_{prod} - S^0_{react}$$

$$= 1\cdot S^0_{C_2H_6} - 2S^0_{H_2} - 1\cdot S^0_{C_2H_2}$$

$$= (1 \text{ mol})(229.5 \text{ J mol}^{-1}\text{ K}^{-1}) - (2 \text{ mol})(130.59 \text{ J mol}^{-1}\text{ K}^{-1}) - (1 \text{ mol})(200.8 \text{ J mol}^{-1}\text{ K}^{-1})$$

$$= \underline{\underline{-232.48 \text{ J K}^{-1}}}$$

This reaction would not tend to go spontaneously (on the basis of entropy) since $\Delta S < 0$. The next chapter deals with a new function that considers both ΔE and ΔS to determine whether a process or reaction should tend to go spontaneously.

4-3.4 PROBLEMS

1. Find the entropy change at 298.15 K for the reaction

$$N_2(g) + 3H_2(g) \rightarrow 2NH_3$$

2. Evaluate the entropy change for the formation of methanol (CH_3OH) from the elements at 25°C.

3. From the following data for Cd find the absolute entropy at 298.15 K.

C_p (J mol^{-1} K^{-1})	T (K)	C_p (J mol^{-1} K^{-1})	T (K)
0.0009	1	19.44	70
0.0037	2	20.59	80
0.010	3	22.11	100
0.024	4	23.09	120
0.15	6	23.78	140
0.90	10	24.26	160
2.81	15	24.62	180
5.17	20	24.92	200
7.54	25	25.18	220
9.65	30	25.41	240
13.21	40	25.60	260
15.91	50	25.82	280
17.92	60	26.04	298.15

4. Calculate the entropy change for the transition $H_2O(\ell) \rightarrow H_2O(g)$ at 298.15 K. Justify the sign of ΔS in terms of randomness and disorder.

5. From heat capacity data given in Chap. 3, entropy data in this section, and Eq. (4-3.2.24), find the entropy change at 1 atm for the transition $H_2O(\ell) \rightarrow H_2O(g)$ at 360 K for 1 mol.

6. Find ΔS for the transition $CHCl_3(\ell) \rightarrow CHCl_3(g)$ at the normal boiling point of 334 K if $\Delta H_f^0(\ell) = -131.8$ kJ mol^{-1} and $\Delta H_f^0(g) = -100.0$ kJ mol^{-1}.

7. Given:

	S^0 (J mol^{-1} K^{-1})	C_p (J mol^{-1} K^{-1})
C(s)	5.694	$16.86 + 4.77 \times 10^{-3}\,T - 8.54 \times 10^5\,T^{-2}$
$O_2(g)$	205.1	$29.96 + 4.18 \times 10^{-3}\,T - 1.67 \times 10^5\,T^{-2}$
CO(g)	197.5	$28.41 + 4.10 \times 10^{-3}\,T - 0.46 \times 10^5\,T^{-2}$

Find ΔS^0 for

$$C(s) + \tfrac{1}{2}\,O_2(g) \rightarrow CO(g) \quad \text{at 500 K.}$$

4-3.5 SELF-TEST

1. Evaluate the entropy changes in the reactions below

   a. $C_2H_2(g) + H_2(g) \rightarrow C_2H_4(g)$

b. $C_2H_2(g) + 2H_2(g) \rightarrow C_2H_6(g)$

c. $S(\text{monoclinic}) + O_2(g) \rightarrow SO_2(g)$

2. Calculate the entropy increase in raising the temperature of 2 mol of $SO_2(g)$ from 300 K to 900 K.

3. From the following data evaluate the entropy of platinum at 100 K.

C_P (J mol^{-1} K^{-1})	T (K)	C_P (J mol^{-1} K^{-1})	T (K)
0.0068	1	2.67	25
0.014	2	4.14	30
0.0238	3	7.4	40
0.0363	4	11.	50
0.072	6	13.3	60
0.13	8	15.4	70
0.218	10	17.2	80
0.64	15	18.3	90
1.4	20	19.5	100

4. The absolute entropy of $Br_2(\ell)$ is 152.3 J mol^{-1} K^{-1} and that for $Br_2(g)$ is 245.34 J mol^{-1} K^{-1}. Find ΔS for the vaporization at 298.15 K. Does this agree with the value found from a ΔH_v of 30.71 kJ mol^{-1} and a boiling point of 58°C?

4-3.6 ADDITIONAL READING

1.* W. J. Moore, *Physical Chemistry*, 4th ed., Prentice-Hall, Englewood Cliffs, New Jersey, 1972, Chap. 3, Sec. 22.

2.* W. F. Sheehan, *Physical Chemistry*, 2nd ed., Allyn and Bacon, Boston, 1970, Sec. 2.14, 4.15.

4-4 SUMMARY

By now you should be able to describe several impossible processes and why they are impossible. Also, the terms reversible and irreversible should be a permanent part of your vocabulary. The entropy concept developed here will be of utmost significance in Chap. 5, where a new function is defined to determine where equilibrium lies in a process or reaction. The relationship between entropy and disorder will be more clearly deliniated in Chap. 9, Statistical Thermodynamics.

This chapter, like Chap. 3, is concerned mainly with the definitions of terms and with gaining facility in manipulating thermodynamic equations. Application of the concepts in Chaps. 3 and 4 are found in Chaps. 5 and 6. You are now prepared for those applications.

5

Free Energy and Chemical Equilibrium

SYMBOLS AND DEFINITIONS

$\bar{G}$	Gibbs free energy function = H - TS (J mol^{-1} K^{-1})
$\bar{A}$	Helmholtz free energy function = E - TS (J mol^{-1} K^{-1})
L_p	Quotient of product pressure over reactant pressure
K^p	Reaction equilibrium constant in terms of fugacity
μ^p	Chemical potential

Thus far, our study of thermodynamics has introduced two new concepts: energy and entropy. A summary of the general properties of these functions can be found in the famous maxim of Clausius: "The energy of the universe is constant; the entropy of the universe always tends toward a maximum." It was indicated in Chap. 4 that a new function would be defined involving E and S which will tell where equilibrium lies in a reaction or process.

We know two tendencies in a system: (1) In a system of constant entropy, the energy of the system will tend toward a minimum, and (2) In a system of constant energy, the tendency is toward maximum entropy. The true equilibrium in a reaction or process will be achieved by balancing these tendencies: minimum energy and maximum entropy. As you might suspect, our new functions will involve both energy and entropy. Two functions will be defined. One, the *Gibbs free energy function* G, is useful at constant pressure; the other, the *Helmholtz free energy function* A, is useful at constant volume.

Various relationships will be derived in the following, and conditions for equilibrium in systems will be defined. This allows us to use thermodynamic data to predict equilibrium concentrations in chemical reactions. Such an ability is of obvious utility in designing experiments and processes involving chemical reactions.

5-1 THE FREE ENERGY FUNCTIONS

OBJECTIVES

You shall be able to

1. Give a mathematical definition of the Gibbs and the Helmholtz free energy functions and state the conditions for equilibrium
2. Predict the direction of a reaction or process from values of ΔG and ΔA
3. Apply restrictions given in a problem to find the proper equation for finding ΔG and ΔA.
4. Find simple relationships among various thermodynamic variables

5-1.1 THE GIBBS AND HELMHOLTZ FUNCTIONS

You may recall that we invented a function, enthalpy, when we began considering constant pressure processes. The reason for this was to eliminate the volume dependence of energy that was troublesome in constant pressure processes. We will use similar "tricks" to define our energy functions.

Let us first review our previous definitions. For a process involving only PV work, we have

$$dE = q + w = q - P\,dV \qquad (5\text{-}1.1.1)$$

Then, we define $H = E + PV$ from which we obtain

$$d(H - PV) = q - P\ dV$$

$$dH - P\ dV - V\ dP = q - P\ dV$$

$$dH = q + V\ dP \qquad (5\text{-}1.1.2)$$

which, for a constant pressure process, reduces to the very useful form

$$dH = q_p \qquad (5\text{-}1.1.3)$$

Now return to (5-1.1.1). Since $q_{rev} = T\ dS$ (for the system), for a *reversible* process we can write (5-1.1.1) as

$$dE = q_{rev} - P\ dV = T\ dS - P\ dV \qquad (5\text{-}1.1.4)$$

(dS will always refer to the system unless specified otherwise).

In order to eliminate the $T\ dS$ term in (5-1.1.4), add $d(-TS)$ to each side and simplify:

$$dE + d(-TS) = T\ dS - P\ dV + d(-TS)$$

$$d(E - TS) = T\ dS - P\ dV - T\ dS - S\ dT$$

$$d(E - TS) = -P\ dV - S\ dT \qquad (5\text{-}1.1.5)$$

We now have a function $E - TS$ which depends on temperature and volume. We define this as the *Helmholtz function A*. Thus

$A = E - TS$	general	(5-1.1.6)
$dA = -S\ dT - P\ dV$	reversible process, PV work only	

A is a state function since E, T, and S are state functions.

For completeness (and to define conditions for equilibrium in a later section), let us see what relations hold for a process which is not reversible. (The following also hold for a reversible process.)

$$dA = d(E - TS) = dE - T\ dS - S\ dT$$

$$= q - P\ dV - T\ dS - S\ dT \qquad (5\text{-}1.1.7)$$

Of course, when the process is reversible, $q = T\ dS$ and (5-1.1.6) is obtained. We will use (5-1.1.7) later.

GIBBS FUNCTION G. For convenience in describing a process at constant pressure, we defined enthalpy which can be written as in Eq. (5-1.1.2):

$$dH = q + V\ dP \qquad (5\text{-}1.1.2)$$

If the process is reversible, we can write

$$dH = q_{rev} + V\ dP = T\ dS + V\ dP \qquad (5\text{-}1.1.8)$$

As before, add d(-TS) to each side

$$dH + d(-TS) = T\,dS + V\,dP + d(-TS)$$

$$d(H - TS) = T\,dS + V\,dP - T\,dS - S\,dT$$

$$d(H - TS) = -S\,dT + V\,dP \qquad (5\text{-}1.1.9)$$

We call the new state function H - TS, the *Gibbs free energy function G*

$$G = H - TS \qquad \text{general}$$

$$dG = -S\,dT + V\,dP \qquad \text{reversible process, PV work only} \qquad (5\text{-}1.1.10)$$

If the process is not reversible, we have the more general form which reduces to (5-1.1.10) for reversible processes.

$$dG = d(H - TS) = dH - d(TS)$$

$$= d(E + PV) - d(TS)$$

$$= dE + P\,dV + V\,dP - T\,dS - S\,dT$$

$$= q - P\,dV + P\,dV + V\,dP - T\,dS - S\,dT$$

$$= q + V\,dP - T\,dS - S\,dT \qquad (5\text{-}1.1.11)^*$$

5-1.2 CONDITIONS FOR EQUILIBRIUM

We have defined four functions so far. Let us write the general forms for all these, to see what we can say about equilibrium (assuming only PV work at this time):

$$dE = q - P\,dV \qquad (a)$$

$$dH = q + V\,dP \qquad (b)$$

$$dA = q - P\,dV - T\,dS - S\,dT \qquad (c) \qquad (5\text{-}1.2.1)$$

$$dG = q + V\,dP - T\,dS - S\,dT \qquad (d)$$

You should remember that we stated (in Chap. 4) that $q \leq T\,dS$ (the equal sign holds for equilibrium) for any real process. We can then rewrite the equations in (5-1.2.1) for any real process as

* There is some confusion of symbols and terms used for the Gibbs and Helmholtz functions. The Helmholtz function is sometimes called the work function and has been given the symbols A and F. We will use A for the Helmholtz function. The Gibbs function has been given symbols of G and F (again!) and is usually referred to as free energy. The symbol G and the term free energy will be used. When you pick up any other book or look at tabulated data, be sure you know what function is being described.

$$dE \leq T\,dS - P\,dV \qquad (a)$$

$$dH \leq T\,dS + V\,dP \qquad (b)$$

$$\begin{aligned} dA &\leq T\,dS - P\,dV - T\,dS - S\,dT \\ &\leq -P\,dV - S\,dT \qquad (c) \end{aligned} \qquad (5\text{-}1.2.2)$$

$$\begin{aligned} dG &\leq T\,dS + V\,dP - T\,dS - S\,dT \\ &\leq V\,dP - S\,dT \qquad (d) \end{aligned}$$

The inequality holds for any real (spontaneous, irreversible) process. Thus, if a system is left alone (isolated), each function will tend to a minimum. Any natural process that occurs must result in a lowering of these four functions. Once equilibrium is reached we find the following conditions from (5-1.2.2):

$$dE = T\,dS - P\,dV \qquad (a)$$

$$dH = T\,dS + V\,dP \qquad (b)$$

$$dA = -P\,dV - S\,dT \qquad (c) \qquad (5\text{-}1.2.3)$$

$$dG = V\,dP - S\,dT \qquad (d)$$

Look specifically at Eq. (5-1.2.2d) and (5-1.2.3d). If we have a system at constant T and P we have:

For a spontaneous process

$$dG < 0 \quad \text{or} \quad \Delta G < 0 \qquad (5\text{-}1.2.4)$$

For an equilibrium process

$$dG = 0 \quad \text{or} \quad \Delta G = 0 \qquad (5\text{-}1.2.5)$$

A similar treatment can be given to the other functions to find the conditions for equilibrium or spontaneity when other variables are held constant.

A final note on spontaneity before we consider what G and A actually signify. We can conclude from the above discussion that:

- $\Delta G < 0$ the process or reaction tends to be spontaneous in the direction written
- $\Delta G = 0$ the system is at equilibrium
- $\Delta G > 0$ the process will tend to be spontaneous in the opposite direction from what we have written

The use of the term "tends to" will be explained later.

5-1.3 FURTHER CONSIDERATION OF THE PROPERTIES OF G AND A

HELMHOLTZ FUNCTION A. We can write for any process

$$dE = q + w$$

and since $q \leq T\,dS$,

$$dE \leq T\,dS + w \tag{5-1.3.1}$$

For *constant T*,

$$T\,dS = d(TS)$$

and

$$dE - d(TS) = d(E - TS) \leq w \tag{5-1.3.2}$$

From the definition of A = E - TS

$$dA \leq w \tag{5-1.3.3}$$

We have defined w as the work done *on* the system. Thus, the work done *by* the system will be -w. We can rewrite (5-1.3.3) as

$$-dA \geq -w \quad \text{or} \quad -\Delta A \geq -W \tag{5-1.3.4}$$

This implies that the work done *by* the system in a real process is less than the negative of the change in the Helmholtz function. For a reversible change,

$$-dA = -w \quad \text{or} \quad -\Delta A = -W \tag{5-1.3.5}$$

The maximum work possible *by* a system is equal to the negative of the change in the Helmholtz function. Any irreversibility in the process yields a work output lower than this maximum. For these reasons, A is sometimes called the *work function* of the system. You should remember that a decrease in A represents the maximum work that can be done *by* a system for that process. Any real process will yield less work.

GIBBS FUNCTION G. Consider now a system at constant T and P. Also, let us define a new term W′, which will represent all work done *on* the system other than PV work. Then

$$dE = q + w + w' \tag{5-1.3.6}$$

$$dE = q - P\,dV + w'$$

or, since P is constant,

$$P\,dV = d(PV)$$

$$dE = q - d(PV) + w' \tag{5-1.3.7}$$

If T is constant, we can write

$$q \leq T\,dS = d(TS)$$

so

$$dE \leq d(TS) - d(PV) + w' \tag{5-1.3.8}$$

or

$$d(E + PV - TS) \leq w'$$

$$d(H - TS) \leq w'$$

and, from the definition of G,

$$dG \leq w' \tag{5-1.3.9}$$

The term $-W'$ will be the work (other than PV work) done *by* the system. So rearrange Eq. (5-1.3.9) to

$$-dG \geq -w' \quad \text{or} \quad -\Delta G \geq -W' \tag{5-1.3.10}$$

Equation (5-1.3.10) shows that the work done *by* the system at constant T and P will be less than the negative of the change in the Gibbs free energy function for any real process. We conclude:

For a spontaneous process

$$-\Delta G > -W' \tag{5-1.3.11}$$

For a reversible process

$$-\Delta G = -W'$$

The maximum work available beyond PV work at *constant T and P* is given for a particular process by the negative of the change in Gibbs free energy. Perhaps you can see why the words "free energy" are used. Any real process will yield less work (other than PV work) than the decrease in G.

5-1.4 FUNDAMENTAL EQUATIONS

As in Chaps. 3 and 4, you cannot hope to remember all the equations necessary to solve problems for all possible conditions. The best procedure for each problem is to go to the appropriate, basic definitions, then apply whatever restraints (such as constant T, adiabatic, reversible or what have you) given in the problem to obtain the equation you need.

ENERGY AND WORK.

[By definition]

$$w = -P_{opp}\ dV \qquad \text{(a)}$$

[By definition]

$$dE = q + w \qquad \text{(b)}$$

[From E = E(T, V)]

$$dE = \left(\frac{\partial E}{\partial T}\right)_V dT + \left(\frac{\partial E}{\partial V}\right)_T dV \qquad \text{(c)} \tag{5-1.4.1}$$

[From definition $C_V = (\partial E/\partial T)_V$; $(\partial E/\partial V)_T$ is derived in Sec. 5-1.5]

$$dE = n\bar{C}_V\ dT + \left[T\left(\frac{\partial P}{\partial T}\right)_V - P\right] dV \qquad \text{(d)}$$

[Derived from (b) for reversible process]

$$dE = T\ dS - P\ dV \qquad \text{(e)}$$

ENTHALPY.

[By definition]

$$dH = dE + d(PV) \qquad (a)$$

[From H = H(T, P)]

$$dH = \left(\frac{\partial H}{\partial T}\right)_P dT + \left(\frac{\partial H}{\partial P}\right)_T dP \qquad (b) \qquad (5\text{-}1.4.2)$$

[From definition $C_P = (\partial H/\partial T)_P$; $(\partial H/\partial P)_T$ is derived in Sec. 5-1.5]

$$dH = C_P\, dT + \left[V - T\left(\frac{\partial V}{\partial T}\right)_P\right] dP \qquad (c)$$

[Derived from (a) for reversible process]

$$dH = T\, dS + V\, dP \qquad (d)$$

ENTROPY.

[By definition]

$$dS_{sys} = \frac{q_{rev}}{T}$$

$$dS_{surr} = \frac{-q_{act}}{T} \qquad (a)$$

[From S = S(T, P); see Sec. 5-1.5]

$$dS = \left(\frac{\partial S}{\partial T}\right)_P dT + \left(\frac{\partial S}{\partial P}\right)_T dP$$

$$= \frac{C_P}{T} dT - \left(\frac{\partial V}{\partial T}\right)_P dP \qquad (b) \qquad (5\text{-}1.4.3)$$

[From S = S(T, V); see Sec. 5-1.5]

$$dS = \left(\frac{\partial S}{\partial T}\right)_V dT + \left(\frac{\partial S}{\partial V}\right)_T dV$$

$$= \frac{C_V}{T} dT + \left(\frac{\partial P}{\partial T}\right)_V dV \qquad (c)$$

HELMHOLTZ FUNCTION.

[By definition]

$$dA = dE - d(TS) \qquad (a)$$

[From A = A(T, V)]

$$dA = \left(\frac{\partial A}{\partial T}\right)_V dT + \left(\frac{\partial A}{\partial V}\right)_T dV \qquad (b) \qquad (5\text{-}1.4.4)$$

[Derived from (a) for reversible process]

$$dA = -S\,dT - P\,dV \qquad (c)$$

GIBBS FREE ENERGY FUNCTION

[By definition]

$$dG = dH - d(TS) \qquad (a)$$

[From G = G(T, P)]

$$dG = \left(\frac{\partial G}{\partial T}\right)_P dT + \left(\frac{\partial G}{\partial P}\right)_T dP \qquad (b) \qquad (5\text{-}1.4.5)$$

[Derived from (a) for a reversible process]

$$dG = -S\,dT + V\,dP \qquad (c)$$

The above are sufficient to find changes in any of the thermodynamic functions for any given process (at least any process we have considered up to now).

For example, find the Gibbs free energy change for the reversible, isothermal expansion of an ideal gas:

$$\begin{aligned} dG &= dH - d(TS) \qquad (5\text{-}1.4.5a) \\ &= d(E + PV) - T\,dS \\ &= dE + d\,PV - T\,dS \\ &= q - P\,dV + P\,dV + V\,dP - T\,dS \\ &= q + V\,dP - T\,dS \end{aligned}$$

If reversible,

$$\begin{aligned} dG &= T\,dS + V\,dP - T\,dS \\ &= V\,dP \end{aligned}$$

$$\Delta G = \int_{P_1}^{P_2} V\,dP \qquad (5\text{-}1.4.6)$$

[We could have gotten the result directly from (5-1.4.5c).] All we need to solve the problem is an equation of state (to relate V and P) and the initial and final conditions.

5-1.5 SOME OTHER RELATIONSHIPS

For a function of two variables f = f(x, y), we may write the following:

$$df = \frac{\partial f}{\partial x}\,dx + \frac{\partial f}{\partial y}\,dy \qquad (5\text{-}1.5.1)$$

If df is an exact differential, we know that

$$\frac{\partial}{\partial y}\left(\frac{\partial f}{\partial x}\right) = \frac{\partial^2 f}{\partial y\ \partial x} = \frac{\partial^2 f}{\partial x\ \partial y} = \frac{\partial}{\partial x}\left(\frac{\partial f}{\partial y}\right) \tag{5-1.5.2}$$

since the order of differentiation is unimportant. This cross-derivative rule is very helpful. For example, consider Eq. (5-1.4.4c)

$$dA = -S\ dT - P\ dV \tag{5-1.4.4c}$$

Applying (5-1.5.2) yields the *Maxwell relation*.

$$\left(\frac{\partial S}{\partial V}\right)_T = \left(\frac{\partial P}{\partial T}\right)_V \tag{5-1.5.3}$$

and from

$$dG = -S\ dT + V\ dP \tag{5-1.4.5c}$$

we get

$$-\left(\frac{\partial S}{\partial P}\right)_T = \left(\frac{\partial V}{\partial T}\right)_P \tag{5-1.5.4}$$

These two relationships allow us to evaluate entropy changes with respect to V and P from an equation of state of the system.

In Chap. 3 the equations given in (5-1.4.1d) and (5-1.4.2d) were presented without proof. The derivation of these is now possible.

dE. At equilibrium (or for reversible processes)

$$dE = T\ dS - P\ dV \tag{5-1.4.1e}$$

If we hold T constant, we can write

$$\left(\frac{\partial E}{\partial V}\right)_T = T\left(\frac{\partial S}{\partial V}\right)_T - P \tag{5-1.5.5}$$

But we just showed in Eq. (5-1.5.3) that

$$\left(\frac{\partial S}{\partial V}\right)_T = \left(\frac{\partial P}{\partial T}\right)_V \tag{5-1.5.3}$$

Thus $$\left(\frac{\partial E}{\partial V}\right)_T = T\left(\frac{\partial P}{\partial T}\right)_V - P \tag{5-1.5.6}$$

Substitution of this result in (5-1.4.1c) gives (5-1.4.1d) as stated earlier.

dH. At equilibrium (or for reversible processes)

$$dH = T\ dS + V\ dP \tag{5-1.4.2d}$$

Again, if T is held constnnt

$$\left(\frac{\partial H}{\partial P}\right)_T = T\left(\frac{\partial S}{\partial P}\right)_T + V \tag{5-1.5.7}$$

But $$\left(\frac{\partial S}{\partial P}\right)_T = -\left(\frac{\partial V}{\partial T}\right)_P \tag{5-1.5.4}$$

Thus $$\left(\frac{\partial H}{\partial P}\right)_T = V - T\left(\frac{\partial V}{\partial T}\right)_P \tag{5-1.5.8}$$

Substitution of this into Eq. (5-1.4.2b) gives (5-1.4.2c).

dA. From (5-1.4.4b) and (5-1.4.4c):

$$dA = \left(\frac{\partial A}{\partial T}\right)_V dT + \left(\frac{\partial A}{\partial V}\right)_T dV \tag{5-1.4.4b}$$

$$dA = -S\ dT - P\ dV \tag{5-1.4.4c}$$

since the coefficients of dT and dV must be equal, we can immediately write:

$$\left(\frac{\partial A}{\partial T}\right)_V = -S \tag{5-1.5.9}$$

$$\left(\frac{\partial A}{\partial V}\right)_T = -P \tag{5-1.5.10}$$

We will make use of these relationships later.

dG. From (5-1.4.5b) and (5-1.4.5c):

$$dG = \left(\frac{\partial G}{\partial T}\right)_P dT + \left(\frac{\partial G}{\partial P}\right)_T dP \tag{5-1.4.5b}$$

$$dG = -S\ dT + V\ dP \tag{5-1.4.5c}$$

we find

$$\left(\frac{\partial G}{\partial T}\right)_P = -S \tag{5-1.5.11}$$

$$\left(\frac{\partial G}{\partial P}\right)_T = V \tag{5-1.5.12}$$

Notice that (5-1.5.12) is merely the relationship we obtained in (5-1.4.6). These last two equations will be used extensively in the next two sections.

Several other relationships among thermodynamic functions could be developed. However, you should realize by now that mathematical manipulations can yield new relations that allow us to evaluate easily some functions that cannot be evaluated directly.

5-1.6 PROBLEMS

1. For a given process, A → B at a constant T of 302 K: $\Delta H = -102$ kJ and $\Delta S = -330$ J K^{-1}. Is this process spontaneous? Explain.

2. Let us define a new function called "Kemp's useless energy function:"

$$K = H + TS$$

where $K = K(T, S)$.

a. Derive a general equation for dK for a process involving only PV work.
b. Add the additional assumptions that the process is reversible and isobaric to derive functions for

$$\left(\frac{\partial K}{\partial S}\right)_T \quad \text{and} \quad \left(\frac{\partial K}{\partial T}\right)_S$$

3. Indicate whether each of the following are true or false, and explain. (State the conditions necessary to make the statement true if it can be made true.)

a. $dG = P\,dV - S\,dT$
b. $dA = -P\,dV - S\,dT$
c. $G = H - TS$ for an irreversible process

4. Derive the expression for ΔA for a reversible, isothermal process given

$$A = E - TS \quad \text{and} \quad A = A(T, V).$$

5. From the cross-derivatives in $dH = T\,dS + V\,dP$, find $(\partial T/\partial P)_S$.

6. (a) Define the Helmholtz free energy function A.
(b) If $A = A(T, V)$, show for a reversible process involving only PV work that $(\partial A/\partial T)_V = -S$ and $(\partial A/\partial V)_T = -P$.
(c) If some other work (say, W_{other}) in addition to PV work is involved, derive the expression for dA for a reversible process.

7. Given that a function $B = B(T, P)$ and that $(\partial J/\partial T)_P = (\partial L/\partial P)_T$, fill in the blanks in the following expression:

$$dB = \_\_\_\_\_\_\_\_ dT + \_\_\_\_\_\_\_\_ dP.$$

5-1.7 SELF-TEST

1. For a particular process, ΔA was found to be -10.7 J. What is the maximum work that the system can do by this process?

2. For the process $D \rightarrow E$, ΔG was found to be +36.1 kJ. Indicate the direction expected for this process if it is allowed to proceed without interference. If ΔH is negative and independent of temperature, how will a temperature change affect ΔG (assume ΔS is also independent of T)?

3. Define another function "Kemp's second useless function"

$$K' = E + TS$$

a. Derive a general equation for dK′ for a process involving only PV work.

b. Apply the additional condition that volume is constant and the process is reversible. Then find $(\partial K'/\partial T)_S$ and $(\partial K'/\partial S)_T$.

4. Under what conditions will the following be true?

a. $A = E - TS$

b. $dA = -P\ dV$

c. $dG = -S\ dT$

5-2 FREE ENERGY AS A FUNCTION OF TEMPERATURE

OBJECTIVES

You shall be able to

1. Find ΔG^0 for a reaction if ΔG_f^0, or ΔH_f^0 and ΔS^0 data are given
2. Find ΔG or ΔA for any process given appropriate data
3. Find ΔG^0 at any temperature from ΔG^0 at some temperature and heat capacity data or from ΔH and ΔS at some temperature and heat capacity data (at constant pressure)

5-2.1 INTRODUCTION

The previous section was concerned with the definitions and mathematical properties of G and A. We now turn to calculating ΔG and ΔA for particular processes. Most attention here will be centered on the Gibbs free energy function since it is so useful for constant pressure processes. Constant pressure is the normal condition of chemical reactions. Some experience with the Helmholtz function will be gained in working the suggested problems.

For particular processes or reactions, we can calculate ΔG^0 [the superscript implies standard conditions (1 atm, normally 25°C)] from ΔH_f^0 and ΔS^0. Then we can predict whether that process should be spontaneous. *You must bear in mind that even if we find a large negative value for* ΔG^0*, the process or reaction may not actually take place at a finite rate.* Rate is a term that belongs to kinetics, not to thermodynamics. Time is not a variable that enters into our study of thermodynamics. The reaction of $H_2 + O_2$ yielding H_2O has a large negative ΔG. But, a mixture of H_2 and O_2 will not react to any appreciable degree in any reasonable length of time unless a spark or a catalyst is added to the system. Then the reaction may be extremely rapid. Nothing in thermodynamics can predict this kind of behavior. Thermodynamics only tells us that the reaction of $H_2 + O_2$ is energetically (in terms of free energy) favorable. Other information must come from other sources. On the other hand, if we find $\Delta G > 0$ for a given process, it *cannot* and *will not* proceed as written.

5-2.2 STANDARD FREE ENERGIES (GIBBS)

For a process or reaction at constant T, we can write immediately from the definition of G,

$$dG = dH - d(TS) = dH - T\,dS \tag{5-2.2.1}$$

Or if we integrate from State 1 to State 2,

$$\Delta G = \Delta H - T\,\Delta S \tag{5-2.2.2}$$

We will be concerned here with only two types of processes: phase changes and chemical reactions.

PHASE CHANGES. A *phase change* involves the transformation from one physical state to another, such as from liquid to gas or from one solid form to another. For a phase change for a pure substance we found in Chap. 4 that

$$\Delta S = \frac{\Delta H}{T} \tag{4-3.2.25}$$

The corresponding ΔG value is

$$\Delta G = \Delta H - T\,\frac{\Delta H}{T} = 0 \tag{5-2.2.3}$$

or

$$\Delta G = G_2(T, P) - G_1(T, P) = 0 \tag{5-2.2.4}$$

This important result will be used in the chapter on phase equilibria (Chap. 13). It states that at a constant pressure (temperature must be constant for reasons found in Chap. 13), the free energies of the two phases are equal. This, in turn, implies that a phase change is an equilibrium process.

CHEMICAL REACTIONS. Since we have already tabulated standard enthalpies of formation (Table 3-3.4.1) and entropies of formation (Table 4-3.2.1), we can employ (5-2.2.2) to find ΔG_f^0 values immediately.

$$\Delta G_f^0 = \Delta H_f^0 - T\,\Delta S_f^0 \tag{5-2.2.5}$$

[Recall that $\Delta S_f^0 = S^0(\text{compound}) - S^0(\text{elements comprising compound})$.]

Table 5-2.2.1 contains some selected values taken from Rossini, *Selected Values of Chemical Thermodynamic Properties*, (see Sec. 5-2.6, Ref. 1). To find ΔG^0 for a reaction we proceed as follows.

For a hypothetical reaction,

$$xX + yY \rightarrow zZ$$

$$\Delta G^0_{\text{reaction}} = z\,\Delta G_f^0\,(Z) - x\,\Delta G_f^0\,(X) - y\,\Delta G_f^0\,(Y)$$

Alternatively, you can find ΔH_r^0 and ΔS_r^0 then apply (5-2.2.5). (The subscript r stands for *reaction*.)

Table 5-2.2.1
Selected Values of Free Energies of Formation of Compounds and Ions at 25°C and 1 atm[a] (kJ mol^{-1})

Substance	G_f^0	Substance	G_f^0
$Ag^+(aq)$	77.111	$OH^-(aq)$	-157.30
$Ag_2O(s)$	-10.82	$H_2O(g)$	-228.596
$AgCl(s)$	-109.72	$H_2O(\ell)$	-237.192
$AgBr(s)$	-95.939	$I_2(g)$	19.4
$Ba^{2+}(aq)$	-560.7	$HI(g)$	1.30
$BaCl_2(s)$	-810.9	$Fe_2O_3(s)$	-741.0
$BaSO_4(s)$	-1353.	$Fe_3O_4(s)$	-1014.
$Br^-(aq)$	-102.82	$NO(g)$	86.688
$Br_2(g)$	3.14	$NO_2(g)$	51.840
$HBr(g)$	-53.22	$NO_3^-(aq)$	-110.5
C(diamond	2.866	$NH_3(g)$	-16.64
$CO(g)$	-137.27	$NH_3(aq)$	-26.61
$CO_2(g)$	-394.38	$NH_4^+(aq)$	-79.50
$CO_3^{2-}(aq)$	-528.10	$Na^+(aq)$	-261.87
$CH_4(g)$	-50.79	$NaOH(s)$	-377.0
$HCO_3^-(aq)$	-587.06	$NaCl(s)$	-384.03
$CH_3OH(g)$	-161.9	$NaBr(s)$	-347.6
$C_2H_2(g)$	209.0	$Na_2SO_4(s)$	-1266.8
$C_2H_4(g)$	68.124	$Na_2SO_4 \cdot 10H_2O(s)$	-3643.97
$C_2H_6(g)$	-32.89	$Na_2CO_3(s)$	-1048.
$C_3H_8(g)$	-23.49	$NaHCO_3(s)$	-851.9
$C_6H_6(g)$	129.66	$S(g)$	182.3
$Ca^{2+}(aq)$	-533.04	$SO_2(g)$	-300.4
$Cl^-(aq)$	-131.17	$SO_3(g)$	-370.4
$F^-(aq)$	-276.5	$SO_4^{2-}(aq)$	-741.99
$HF(g)$	-271.	$H_2S(g)$	-33.02
$H^+(aq)$	0.00	$H_2SO_4(aq)$	-741.99

[a] Selected values from F. A. Rossini, et.al., eds. *Selected Values of Chemical Thermodynamic Properties*, U. S. National Bureau of Standards, Circular 500, Washington, D. C., 1952.

$$\Delta G_r^0 = \Delta H_r^0 - 298.15\Delta S_r^0$$

The following section will deal with finding ΔG at temperatures and pressures other than standard conditions.

5-2.2.1 EXAMPLE

Find the free energy of reaction for

$$C_2H_2(g) + 2H_2(g) \rightarrow C_2C_6(g)$$

from ΔG^0_f data and from ΔH^0_f and S^0 data. Compare the results.

Solution

ΔG^0_f Data

From Table 5-2.2.1,

$$\Delta G^0_f(C_2H_6) = -32.89 \text{ kJ mol}^{-1}$$

$$\Delta G^0_f(H_2) = 0.00$$

$$\Delta G^0_f(C_2H_2) = 209 \text{ kJ mol}^{-1}$$

$$\Delta G^0_r = (1 \text{ mol})(-32.89 \text{ kJ mol}^{-1}) - (2 \text{ mol})(0.00) - (1 \text{ mol})(209 \text{ kJ mol}^{-1})$$

$$= -241.89 \text{ kJ}$$

$$= \underline{\underline{-242 \text{ kJ}}} \qquad (-57.8 \text{ kcal})$$

ΔH^0_f and S^0 Data

From Tables 3-3.4.1 and 4-3.2.1,

	ΔH^0_f (kJ mol^{-1})	S^0 (J mol^{-1} K^{-1})
C_2H_6	-84.67	229.49
H_2	0.00	130.59
C_2H_2	226.75	200.82

$$\Delta H^0_r = (1 \text{ mol})(-84.67 \text{ kJ mol}^{-1}) - (2 \text{ mol})(0.00) - (1 \text{ mol})(226.75 \text{ kJ mol}^{-1})$$

$$= -311.42 \text{ kJ}$$

$$= \underline{\underline{-311.4 \text{ kJ}}}$$

$$\Delta S^0_r = (1 \text{ mol})(229.49 \text{ J mol}^{-1} \text{ K}^{-1}) - (2 \text{ mol})(130.59 \text{ J mol}^{-1} \text{ K}^{-1})$$

$$- (1 \text{ mol})(200.82 \text{ J mol}^{-1} \text{ K}^{-1})$$

$$= \underline{\underline{-232.52 \text{ J K}^{-1}}}$$

$$\Delta G^0_r = \Delta H^0_r - (298.15 \text{ K})\Delta S^0_r$$

$$= -\ 311{,}420\ \text{J} - (298.15\ \text{K})(-232.52\ \text{J K}^{-1})$$

$$= -\ 311{,}420\ \text{J} + 69325\ \text{J}$$

$$= -\ 242095\ \text{J} = \underline{\underline{-242\ \text{kJ}}}$$

This agrees with the previous value as expected.

5-2.3 FREE ENERGY AS A FUNCTION OF TEMPERATURE

Isobaric conditions are assumed for all processes considered in this section. And, as usual, only PV work will be considered.

Recall that for a reversible process

$$dG = -S\ dT + V\ dP \tag{5-1.4.5c}$$

which reduces, under isobaric conditions, to

$$dG = -S\ dT \tag{5-2.3.1}$$

For changes in G (ΔG):

$$\frac{d\ \Delta G}{dT} = -\Delta S \tag{5-2.3.2}$$

[Note that Eq. (5-2.3.2) is valid even if P is not constant, if the partial derivative is written: $(\partial\ \Delta G/\partial T)_P = -\Delta S$.]

You should recall that $\Delta G = \Delta H - T\ \Delta S$ at any given temperature. Thus

$$\frac{d\ \Delta G}{dT} = -\Delta S = \frac{\Delta G - \Delta H}{T} \tag{5-2.3.3}$$

or

$$\frac{d\ \Delta G}{dT} - \frac{\Delta G}{T} = -\frac{\Delta H}{T} \tag{5-2.3.4}$$

Verify at this point that

$$T\frac{d}{dT}\left(\frac{\Delta G}{T}\right) = \frac{d\ \Delta G}{dT} - \frac{\Delta G}{T} \tag{5-2.3.5}$$

Which on substitution into (5-2.3.4) yields

$$\frac{d}{dT}\left(\frac{\Delta G}{T}\right) = -\frac{\Delta H}{T^2} \tag{5-2.3.6}$$

or

$$d\left(\frac{\Delta G}{T}\right) = -\frac{\Delta H}{T^2}\ dT \tag{5-2.3.7}$$

Integrate this last equation from State 1 to State 2:

$$\int_1^2 d\left(\frac{\Delta G}{T}\right) = -\int_{T_1}^{T_2} \frac{\Delta H}{T^2}\ dT \tag{5-2.3.8}$$

If ΔH is not temperature-dependent, then (5-2.3.8) can be solved easily. However,

ΔH usually *is* a function of T, so we have to find ΔH (T) to put into the integral on the right side of (5-2.3.8).

Go back to Sec. 3-3.5 to review how to find ΔH as a function of T. The results are stated here for the case of a reaction. If we are interested only in one material, the same equations apply, except ΔC_p will be replaced by C_p.

For a reaction [see Eq. (3-3.5.6)],

$$\Delta H\ (T) = \Delta H\ (298) + \int_{298}^{T} \Delta C_p\ dT$$

If we integrate the last term of the above equation, we will get a form as follows

$$\Delta H\ (T) = \Delta H\ (298) + f(T)\Big|_{298}^{T} \qquad (5\text{-}2.3.9)$$

where f(T) is the form found on integrating ΔC_p dT (see Sec. 3-3.5 for more detail). We can write

$$\Delta H\ (T) = \Delta H\ (298) + f(T) - f(298) \qquad (5\text{-}2.3.10)$$

But both ΔH (298) and f(298) are constants. We can combine these into a new constant, *const*. Thus

$$\Delta H\ (T) = \text{const} + f(T) \qquad (5\text{-}2.3.11)$$

This can be substituted directly into (5-2.3.8) to give

$$\frac{\Delta G_2}{T_2} - \frac{\Delta G_1}{T_1} = -\int_{T_1}^{T_2} \frac{\Delta H\ (T)}{T^2}\ dT$$

$$= -\int_{T_1}^{T_2} \frac{\text{const} + f(T)}{T^2}\ dT$$

or

$$\frac{\Delta G_2}{T_2} = \frac{\Delta G_1}{T_1} - \int_{T_1}^{T_2} \frac{\text{const}}{T^2} - \int_{T_1}^{T_2} \frac{f(T)}{T^2}\ dT \qquad (5\text{-}3.2.12)$$

where f(T) and const are defined as above.

Note: Another way to find ΔG (T_2) is to find ΔH (T_2) as done in Sec. 3-3.5 and ΔS (T_2) as in Sec. 4-3.2, then apply the equation $\Delta G\ (T_2) = \Delta H\ (T_2) - T_2\ \Delta S\ (T_2)$. Both methods will be shown in the following example for a reaction.

5-2.3.1 EXAMPLE

Find ΔG^0 (1000) for the reaction

$$2H_2(g) + O_2(g) \rightarrow 2H_2O(g)$$

The data below are to be used. C_p is given as

$$C_p = a + bT + cT^{-2} \ (\text{J mol}^{-1} \text{K}^{-1})$$

Item	$2H_2$	O_2	$2H_2O$	Δ(Item)
ΔH_f^0	0	0	2(-241,830)	-483,660 (J)
ΔG_f^0	0	0	2(-228,500)	-457,000 (J)
S^0	2(130.5)	205	2(189)	-88 (J K^{-1})
a	2(27.28)	29.96	2(30.54)	-23.44
b × 10^3	2(3.26)	4.18	2(10.29)	9.88
c × 10^{-5}	2(0.50)	-1.67	0	+0.67

Solution

Method I

Applying Eq. (5-2.3.9)

$$\Delta H^0 (T') = \Delta H^0 (298) + \int_{298}^{T'} \Delta C_p \, dT$$

$$= -483{,}660 + \int_{298}^{T'} (-23.4 + 9.88 \times 10^{-3} T + 0.67 \times 10^5 T^{-2}) \, dT$$

$$= -483{,}660 - 23.44(T' - 298) + \frac{9.88 \times 10^{-3}}{2}(T'^2 - 298^2)$$

$$- 0.67 \times 10^5 (T'^{-1} - 298^{-1})$$

$$= -483{,}660 + (23.44)(298) - 4.94 \times 10^{-3}(298^2) + 0.67 \times 10^5 (298^{-1})$$

$$- (23.44)T' + (4.94 \times 10^{-3})T'^2 - (0.67 \times 10^5)T'^{-1}$$

$$\Delta H^0 (T')^* = -476{,}889 - 23.44T' + 4.94 \times 10^{-3} T'^2 - 0.67 \times 10^5 T'^{-1}$$

This is ready to substitute into Eq. (5-2.3.12).

$$\frac{\Delta G_2^0}{T_2} = \frac{\Delta G_1^0}{T_1} - \int_{298}^{T_2} \frac{-476{,}889}{T'^2} dT' - \int_{298}^{T_2} \frac{-23.44 \, T'}{T'^2} dT'$$

$$- \int_{298}^{T_2} \frac{4.94 \times 10^{-3} \, T'^2}{T'^2} dT' - \int_{298}^{T_2} \frac{-0.67 \times 10^5 \, T'^{-1}}{T'^2} dT'$$

---

* You cannot evaluate ΔH (1000) at this point if you plan to use Eq. (5-2.3.9). You must leave ΔH as a function of T.

$$= \frac{-457{,}000}{298} - 476{,}889\left(\frac{1}{T_2} - \frac{1}{298}\right) + 23.44 \ln \frac{T_2}{298} - 4.94 \times 10^{-3}(T_2 - 298)$$

$$- \frac{0.67 \times 10^5}{2}\left(\frac{1}{T_2^{\ 2}} - \frac{1}{298^2}\right)$$

$$= \frac{-457{,}000}{298} + \frac{476{,}889}{298} - 23.44 \ln 298 + 4.94 \times 10^{-3}(298) + \frac{67{,}000}{2}\frac{1}{298^2}$$

$$- \frac{476{,}889}{T_2^{\ 2}} + 23.44 \ln T_2 - 4.94 \times 10^{-3}\ T_2 - \frac{67{,}000}{2T_2^{\ 2}}$$

(Note: All units have been kept in Joules.)

$$\frac{\Delta G_2^0}{T_2} = -64.95 - \frac{476{,}889}{T_2} + 23.44 \ln T_2 - 4.94 \times 10^{-3}\ T_2 - \frac{67{,}000}{2T_2^{\ 2}}$$

The last equation is useful for finding ΔG_2^0 for any T_2 within the range for which the C_p function is valid. Substitute T_2 = 1000 K to solve this problem.

$$\frac{\Delta G^0(1000)}{1000} = -\ 64.4 - 476.9 + 161.9 - 4.9 - 0.033$$
$$= -384.9$$

$$\underline{\underline{\Delta G^0(1000) = -385{,}000\ \text{J}}}$$

Method II

Find ΔH^0 (1000) and ΔS^0 (1000), then find ΔG^0 (1000) from (5-2.2.2).

$$\Delta H^0\ (1000) = -476{,}889 - 23.44(1000) + 4.94 \times 10^{-3}(1000^2)$$
$$- 0.67 \times 10^5(1000^{-1})$$
$$= \underline{\underline{-495{,}500\ \text{J}}}$$

From Sec. 4-3.2,

$$\Delta S^0\ (1000) = \Delta S^0\ (298) + \int_{298}^{1000} \frac{\Delta C_p}{T}\, dT$$

$$= -88 + \int_{298}^{1000} \frac{-23.44}{T}\, dT + \int_{298}^{1000} \frac{9.88 \times 10^{-3}\ T}{T}\, dT$$

$$+ \int_{298}^{1000} \frac{0.67 \times 10^5}{T}\frac{1}{T^2}\, dT$$

$$= -88 - 23.44 \ln \frac{1000}{298} + 9.88 \times 10^{-3}(1000 - 298)$$

$$- \frac{67,000}{2} \left(\frac{1}{1000^2} - \frac{1}{298^2}\right)$$

$$= \underline{-109.1 \text{ J K}^{-1}}$$

$$\Delta G^0 (1000) = \Delta H^0 (1000) - (1000)\, \Delta S^0 (1000)$$

$$= -495,500 - 1000(-109.1)$$

$$= \underline{-386,000 \text{ J}}$$

This compares with -385,000 J obtained with Method 1. Since we expect errors to arise in the third decimal place due to the accuracy of some of our data (why?), the results are in agreement.

Refer to Secs. 3-3 and 4-3 for a discussion of calorimetric studies which are used to evaluate values of S^0. The precision of the heat capacity data used to find S^0 values in general limit S^0 to three significant figures.

5-2.4 PROBLEMS

1. From appropriate tabulations find ΔG^0 for the reaction:

   $$C(\text{graphite}) + \tfrac{1}{2}O_2(g) \rightarrow CO(g)$$

   Use two methods and compare the results.

2. Find ΔG^0 for the heating of 3 mol Cl_2 from 306 K to 750 K (1 atm).

3. Find ΔG^0 (1100) for the reaction given in Problem (1).

4. From Table 5-2.2.1 find ΔG^0 for each of the following reactions:

   a. $Ba^{2+}(aq) + 2Cl^-(aq) \rightarrow BaCl_2(s)$

   b. $Ba^{2+}(aq) + SO_4^{2-}(aq) \rightarrow BaSO_4(s)$

   c. $N_2(g) + 3H_2(g) \rightarrow 2NH_3(g)$

5. Find ΔG for the vaporization of water at 298.15 K and at 373 K. Discuss the results in terms of equilibrium at 1 atm.

5-2.5 SELF-TEST

1. Find ΔG_r^0 at 298.15 K and 598 K for

   $$N_2(g) + 3H_2(g) \rightarrow 2NH_3(g)$$

   by two different methods and compare the results.

2. Evaluate ΔG^0_{298} for the reactions

$$H_2(g) + F_2(g) \rightarrow 2HF(g)$$

$$HF(g) + \text{excess } H_2O \rightarrow H^+(aq) + F^-(aq)$$

What is ΔG^0 for the net reaction

$$\frac{1}{2}H_2(g) + \frac{1}{2}F_2(g) \rightarrow H^+(aq) + F^-(aq)$$

5-2.6 ADDITIONAL READING

1.* F. A. Rossini, et. al., (eds.), *Selected Values of Chemical Thermodynamic Properties*, Circular 500, U. S. National Bureau of Standards, Washington, D. C., 1952.

2. See Sec. 1-1.9B for additional references to data compilations.

5-3 CHEMICAL EQUILIBRIUM

OBJECTIVES

You shall be able to

1. Find ΔG at any pressure given an equation of state and ΔG at some other pressure (isothermal conditions)
2. Determine the equilibrium constant for a reaction if ΔG^0 is known, or vice versa
3. Determine the equilibrium constant for a reaction as a function of temperature
4. Define fugacity and chemical potential, and evaluate fugacities for gases from given equations of state.

5-3.1 INTRODUCTION

So far we have looked at the definitions of ΔG and ΔA and have shown how these functions can be used to define equilibrium. In this section you will be shown how to use ΔG values to find the equilibrium point for a reaction as well as demonstrate how to determine ΔG as a function of pressure. The material in the last section will show how the equilibrium constant for a reaction varies with temperature. A factor to correct for nonideality of gases will be introduced, and a partial molar free energy (the chemical potential) will be defined.

5-3.2 THE PRESSURE DEPENDENCE OF ΔG

First, let us review the procedure for finding an appropriate equation for a problem. Consider a process that is reversible and involves PV work only. Write G as a function of T and P:

$$G = G(T, P)$$

So
$$dG = \left(\frac{\partial G}{\partial T}\right)_P dT + \left(\frac{\partial G}{\partial P}\right)_T dP \quad (5\text{-}3.2.1)$$

From the definition of G,

$$G = H - TS \quad (5\text{-}1.1.10)$$

$$dG = dH - T\ dS - S\ dT$$

$$= dE + d(PV) - T\ dS - S\ dT$$

$$= q - P\ dV + P\ dV + V\ dP - T\ dS - S\ dT$$

But $q_{rev} = T\ dS$

$$dG = T\ dS + V\ dP - T\ dS - S\ dT$$

$$= V\ dP - S\ dT \quad (5\text{-}1.2.3d)$$

The above is just an application of definitions. Now, if you compare (5-3.2.1) and (5-1.2.3d) you see

$$\left(\frac{\partial G}{\partial P}\right)_T = V \quad (5\text{-}3.2.2)$$

(found in Sec. 5-1.4). The only purpose of going through this again is to refresh your memory.

If we state that a particular process is isothermal, Eq. (5-3.2.2) becomes

$$dG = V\ dP \quad (5\text{-}3.2.3)$$

The conditions for (5-3.2.3) are: (1) reversible process (2) *PV work only*, and (3) *isothermal process*.

IDEAL GAS. All we need is an equation of state relating P and V, and we can find how G varies with pressure. If PV = nRT, we can write V = nRT/P.

$$dG = \frac{nRT}{P}\ dP \quad (5\text{-}3.2.4)$$

or
$$\Delta G = G_{P_2} - G_{P_1} = \int_1^2 dG = \int_{P_1}^{P_2} \frac{nRT}{P}\ dP \quad (5\text{-}3.2.5)$$

T is constant so this can be written as

$$\Delta G = nRT \int_{P_1}^{P_2} \frac{dP}{P} = nRT \ln \frac{P_2}{P_1} \quad (5\text{-}3.2.6)$$

which shows how G varies with P for a single, ideal gas.

P_1 is chosen as 1 atm which is defined as the standard state. Whence,

$$G - G^0 = nRT \ln P \tag{5-3.2.7}$$

or in terms of 1 mol

$$\bar{G} - \bar{G}^0 = RT \ln P \tag{5-3.2.8}$$

Note that P must be expressed in atmospheres since P^0 is assumed to be 1 atm.

REACTIONS OF IDEAL GASES. Equation (5-3.2.8) is good for one ideal gas. Consider the reaction of ideal gases as follows:

$$aA(g) + bB(g) \rightarrow cC(g) \qquad T = 298.15 \text{ K}$$

We can write immediately [using (5-3.2.7) or (5-3.2.8)]:

$$G_A = a\bar{G}_A = a\bar{G}_A^0 + aRT \ln P_A = a\bar{G}_A^0 + RT \ln P_A^a$$

$$G_B = b\bar{G}_B = b\bar{G}_B^0 + bRT \ln P_A = b\bar{G}_B^0 + RT \ln P_B^b$$

$$G_C = c\bar{G}_C = c\bar{G}_C^0 + cRT \ln P_C = c\bar{G}_C^0 + RT \ln P_C^c$$

But

$$\Delta G_r = c\bar{G}_C - b\bar{G}_B - a\bar{G}_A$$

$$= c\bar{G}_C^0 - b\bar{G}_B^0 - a\bar{G}_A^0 + RT \ln P_C^c - RT \ln P_B^b - RT \ln P_A^a$$

$$= \Delta G_r^0 + RT \ln \frac{P_C^c}{P_B^b P_A^a} \tag{5-3.2.9}$$

You should be able to extend this to other cases. ΔG_r^0 can be found from tabulations such as Table 5-2.2.1.

5-3.2.1 EXAMPLE

Find ΔG_r for the reaction (at 25°C)

$$\underset{\text{Acetylene}}{C_2H_2(g)} + 2H_2(g) \qquad \underset{\text{Ethane}}{C_2H_6(g)}$$

if the reactants and products are each at 100 atm pressure.

Solution

From Eq. (5-3.2.9),

$$\Delta G_r = \Delta G_r^0 + RT \ln \frac{P_{C_2H_6}}{P_{H_2}^2 P_{C_2H_2}}$$

$$\Delta G_r^0 = \Delta G_f^0(C_2H_6) - 2\Delta G_f^0(H_2) - \Delta G_f^0(C_2H_2)$$
$$= (-32.89 - 0 - 209)\ \text{kJ}$$
$$= -241\ \text{kJ} \qquad \text{from Table 5-2.2.1}$$

$$\Delta G_r = -241\ \text{kJ} + (8.31\ \text{J K}^{-1})(298.15)\ \ln \frac{100}{(100^2)(100)}$$

[Note the unit mol^{-1} on R has been eliminated since we took the number of moles into account to get Eq. (5-3.2.9)].

$$\Delta G_r = -241\ \text{kJ} - 22820\ \text{J}$$
$$= (-241 - 22.8)\ \text{kJ}$$
$$= \underline{\underline{-264\ \text{kJ}}}$$

ΔG is more negative for higher pressures, thus the reaction has a greater tendency to be spontaneous. We still use the word tendency since the reaction may not occur unless special circumstances hold.

Do you remember *Le Chatelier's principle* from your introductory chemistry? This principle states that a system responds to a stress in such a manner as to relieve that stress as much as possible. We have just shown that above. Increasing the pressure on the system tends to push the reaction toward $C_2H_6(g)$. This lowers the number of moles of gas, and hence tends to relieve some of the pressure stress by lowering the volume of the system.

NONIDEAL GASES -- FUGACITY. There are two ways to treat nonideal gases. One is to solve (5-3.2.3) explicitly for a nonideal state. This will be done shortly. The second is to use a "fudge factor" to make Eq. (5-3.2.8) apply even to nonideal systems. Why do this? Well, Eq. (5-3.2.8) has a very simple and appealing form. In order to avoid complicated expressions for $\bar{G}$ as a function of P (which you will see shortly), we insert a factor for P that makes the equation valid even for nonideal systems.

This factor is called the *fugacity* and is defined as

$$f = \gamma(P)\ P \tag{5-3.2.10}$$

where $\gamma(P)$ is a *fugacity coefficient* which is pressure dependent. Values of $\gamma(P)$ can be determined and tabulated (see Barrow, *Physical Chemistry*, Sec. 5-3.8, Ref. 1), then Eq. (5-3.2.9) can be written

$$\bar{G} = \bar{G}^0 + RT\ \ln f$$
$$= \bar{G}^0 + RT\ \ln \gamma P \tag{5-3.2.11}$$

The fugacity relates to the free energy of a real gas in the same way that pressure does for an ideal gas.

Consideration of the molar free energy difference between two pressures allows us to write

$$\bar{G}_2 - \bar{G}_1 = RT \ln \frac{f_2}{f_1} \tag{5-3.2.12}$$

with appropriate substitutions (see Wall, Sec. 5-3.8, Ref. 2) we find

$$RT \ln f = \ln P + \int_{P^*}^{P} \left(\frac{\bar{V}}{RT} - \frac{1}{P}\right) dP \tag{5-3.2.13}$$

or relative to the standard state of $P^* = 0$,

$$RT \ln \frac{f}{f^*} = P\bar{V} - P^*V^* - \int_{V^*}^{V} P\, dV \tag{5-3.2.14}$$

where f^*, P^*, and V^* refer to values as $P^* \to 0$. All that is needed is an equation of state for a gas to evaluate the fugacity of that gas. Note that if the ideal gas law is used, ln (f/P) = 0, which means f/P = 1, as expected. Fugacities are of use in calculating gas properties in flow or pressure systems where nonideality occurs. The following example shows how an equation of state for a gas can be used to find fugacities.

5-3.2.2 EXAMPLE

Derive a general equation for the fugacity of a gas that follows the Redlich-Kwong equation (see Sec. 2-2.6). This equation will be used in one of the problems [(7), work through this one carefully. It requires a number of useful manipulations.]

Solution

$$RT \ln \frac{f}{f^*} = P\bar{V} - P^*V^* - \int_{V^*}^{V} P\, dV$$

But

$$P = \frac{RT}{\bar{V} - b} - \frac{a}{(T^{1/2}\bar{V})(\bar{V} + b)}$$

$$RT \ln \frac{f}{f^*} = \frac{RT\bar{V}}{\bar{V} - b} - \frac{a\bar{V}}{(T^{1/2}\bar{V})(\bar{V} + b)} - \frac{RT\bar{V}^*}{\bar{V}^* - b} + \underbrace{\frac{a\bar{V}^*}{(T^{1/2}\bar{V}^*)(\bar{V}^* + b)}}_{\to 0}$$

$$- \int_{V^*}^{\bar{V}} \frac{RT}{\bar{V} - b} dV + \int_{V^*}^{\bar{V}} \frac{a\, dV}{(T^{1/2}\bar{V})(\bar{V} + b)}$$

Note that as $P^* \to 0$, $V^* \to \infty$, so the fourth term on the right side goes to zero. Also, $V^* - b \simeq V^* \approx RT/P^*$ (why?), $f^* \to P^*$, $1/(\bar{V}^* + b) \to 0$, and $V^*/(V^* + b) \to 1$.

Performing the integral and simplifying the third term gives

$$RT \ln f - RT \ln f^* = \frac{RT\bar{V}}{\bar{V} - b} - \frac{a\bar{V}}{(T^{\frac{1}{2}}\bar{V})(\bar{V} + b)} - RT - RT \ln \frac{\bar{V} - b}{\bar{V}^* - b} \quad (RT/P^*)$$

$$- \frac{a}{bT^{\frac{1}{2}}} \ln \left(\frac{\bar{V} + b}{\bar{V}}\right)\left(\frac{\bar{V}^*}{\bar{V}^* + b}\right) \quad 1$$

$$= \frac{RT\bar{V}}{\bar{V} - b} - \frac{a\bar{V}}{(T^{\frac{1}{2}}\bar{V})(\bar{V} + b)} - RT - RT \ln \frac{(\bar{V} - b)P^*}{RT}$$

$$- \frac{a}{bT^{\frac{1}{2}}} \ln \frac{\bar{V} + b}{\bar{V}}$$

Continuing to simplify

$$RT \ln f - RT \ln f^* = \frac{RT\bar{V}}{\bar{V} - b} - \frac{a\bar{V}}{(T^{\frac{1}{2}}\bar{V})(\bar{V} + b)} - RT - RT \ln \frac{\bar{V} - b}{RT} - RT \ln P^*$$

$$- \frac{a}{bT^{\frac{1}{2}}} \ln \frac{\bar{V} + b}{\bar{V}}$$

$$\text{But, } \bar{V} \left|\frac{RT}{\bar{V} - b} + \frac{a}{(T^{\frac{1}{2}}\bar{V})(\bar{V} + b)}\right| = \bar{V}P$$

so the first two terms on the right give $\bar{V}P$. Thus,

$$RT \ln f = P\bar{V} - RT - RT \ln \frac{\bar{V} - b}{RT} - \frac{a}{bT^{\frac{1}{2}}} \ln \frac{\bar{V} + b}{\bar{V}}$$

Given appropriate conditions this can be evaluated to yield the desired fugacity.

NONIDEAL GASES -- EXPLICIT TREATMENT. The Beattie-Bridgeman equation will be applied at this point. Equation (2-2.7.2) gives, on rearranging, for 1 mol,

$$P = \frac{RT}{\bar{V}} \left(1 - \frac{C}{\bar{V}T^3}\right)\left(1 + \frac{B_0}{\bar{V}} - \frac{bB_0}{\bar{V}^2}\right) - \frac{A_0}{\bar{V}^2}\left(1 - \frac{a}{\bar{V}}\right) \quad (5\text{-}3.2.15)$$

Apply Eq. (5-3.2.3):

$$dG = V\ dP \quad (5\text{-}3.2.3)$$

Here you have two choices. Substitute for V or for dP. It should be obvious that it is easier to find dP than to find V as a function of P. Collect powers of $\bar{V}$ in (5-3.2.15):

$$P = \frac{RT}{\bar{V}} + \frac{RT}{\bar{V}^2}\left(B_0 - \frac{c}{T^3} - \frac{A_0}{RT}\right) + \frac{RT}{\bar{V}^3}\left(\frac{aA_0}{RT} - \frac{cB_0}{T^3} - B_0 b\right) + \frac{RT}{\bar{V}^4}\frac{cB_0 b}{T^3} \quad (5\text{-}3.2.16)$$

If T is constant,

$$dP = \left[-\frac{RT}{\bar{V}^2} - \frac{2RT}{\bar{V}^3}\left(B_0 - \frac{c}{T^3} - \frac{A_0}{RT}\right) - \frac{3RT}{\bar{V}^4}\left(\frac{aA_0}{RT} - \frac{cB_0}{T^3} - B_0 b\right) - \frac{4RT}{\bar{V}^5}\frac{cB_0 b}{T^3}\right] dV \tag{5-3.2.17}$$

Substitute into (5-3.2.3) and integrate:

$$\Delta G = \int_1^2 dG = -\int_{V_1}^{V_2} \left[\frac{RT}{-\bar{V}} + \frac{2RT}{\bar{V}^2}\left(B_0 - \frac{c}{T^3} - \frac{A_0}{RT}\right) + \frac{3RT}{\bar{V}^3}\left(\frac{aA_0}{RT} - \frac{cB_0}{T^3} - B_0 b\right) + \frac{4RT}{\bar{V}^4}\left(\frac{cB_0 b}{T^3}\right)\right] dV \tag{5-3.2.18}$$

Which gives, on integration,

$$\Delta\bar{G} = -RT \ln \frac{\bar{V}_2}{\bar{V}_1} + 2RT\left(B_0 - \frac{c}{T^3} - \frac{A_0}{RT}\right)\left(\frac{1}{\bar{V}_2} - \frac{1}{\bar{V}_1}\right) + \frac{3}{2}RT\left(\frac{aA_0}{RT} - \frac{cB_0}{T^3} - B_0 b\right)\left(\frac{1}{\bar{V}_2^{\,2}} - \frac{1}{\bar{V}_1^{\,2}}\right) + \frac{4}{3}RT\,\frac{cB_0 b}{T^3}\left(\frac{1}{\bar{V}_2^{\,3}} - \frac{1}{\bar{V}_1^{\,3}}\right) \tag{5-3.2.19}$$

Other equations of state can be treated in a similar fashion as you will discover in the problem section.

5-3.2.3 EXAMPLE

Use both the Beattie-Bridgeman and ideal gas equations to calculate the change in free energy at 300 K for 10 mol of ethane for the pressure change associated with an increase in volume from 4.86 dm^3 to 9.72 dm^3.

$A_0 = 5.880$ $B_0 = 0.0940$

$a = 0.05861$ $b = 0.01915$

$c = 90.0 \times 10^4$

Note, if $V = 4.86\ dm^3$ then $\bar{V} = 0.486\ dm^3$

Solution

Ideal Gas

$$\Delta\bar{G} = \int_{P_1}^{P_2} \bar{V}\, dP = \int_{V_1}^{V_2} \bar{V}\, d\left(\frac{RT}{\bar{V}}\right) = -RT\int_{V_1}^{V_2} \frac{dV}{\bar{V}} = -RT \ln \frac{\bar{V}_2}{\bar{V}_1}$$

$$= -(8.314\ J\ mol^{-1}\ K^{-1})(300\ K) \ln \frac{0.972}{0.486}$$

$$= -1730\ J\ mol^{-1}$$

$$\Delta G = 10\ (\Delta\bar{G}) = \underline{\underline{-17{,}300\ J}}$$

Note the pressure change is

$$P_1 = \frac{RT}{\bar{V}_1} = \frac{(0.0821)(300)}{0.486\ dm^3} = \underline{\underline{50.7\ atm}}$$

$$P_2 = \frac{RT}{\bar{V}_2} = \frac{(0.0821)(300)}{0.972\ dm^3} = \underline{\underline{25.3\ atm}}$$

Beattie-Bridgeman

Evaluate the constants (all constants assume atm and dm^3 as units)

$$B_0 - \frac{c}{T^3} - \frac{A_0}{RT} = 0.0940 - \frac{90 \times 10^4}{(300)^3} - \frac{5.880}{(0.0821)(300)}$$

$$= \underline{-0.178}$$

$$\frac{aA_0}{RT} - \frac{cB_0}{T^3} - B_0 b = \frac{(0.05861)(5.880)}{(0.0821)(300)} - \frac{(90 \times 10^4)(0.0940)}{(300)^3} - (0.0940)(0.01915)$$

$$= \underline{0.00906}$$

$$\frac{cB_0 b}{T^3} = \frac{(90.0 \times 10^4)(0.0940)(0.01915)}{(300)^3}$$

$$= \underline{6 \times 10^{-5}}$$

Now evaluate the pressure just to compare with the ideal gas pressure. From (5-3.2.12),

$$P_1 = RT \left| \frac{1}{\bar{V}} + \frac{1}{\bar{V}^2}(-0.178) + \frac{1}{\bar{V}^3}(0.00906) + \frac{1}{\bar{V}^4}(6 \times 10^{-5}) \right|$$

$$= (0.0821)(300) \left[\frac{1}{0.486} - \frac{0.178}{(0.486)^2} + \frac{0.00906}{(0.486)^3} + \frac{6 \times 10^{-5}}{(0.486)^4} \right]$$

$$= (24.63)(1.384)$$

$$= \underline{\underline{34.1\ atm}}$$

(The observed value is 34.0 atm. The ideal gas law does not fit very well as you can tell.)

$$P_2 = (0.0821)(300) \left[\frac{1}{0.972} - \frac{0.178}{(0.972)^2} + \frac{0.00906}{(0.972)^3} + \frac{6 \times 10^{-5}}{(0.972)^4} \right]$$

$$= (24.63)(0.8503)$$

$$= \underline{\underline{20.9\ atm}}$$

Now back to the original problem. From (5-3.2.16),

$$\Delta G_{BB} = RT\left[-\ln\frac{\bar{V}_2}{\bar{V}_1} + 2(-0.178)\left(\frac{1}{\bar{V}_2} - \frac{1}{\bar{V}_1}\right) + \frac{3}{2}(0.00906)\left(\frac{1}{\bar{V}_2^{\,2}} - \frac{1}{\bar{V}_1^{\,2}}\right) + \frac{4}{3}(6\times10^{-5})\left(\frac{1}{\bar{V}_2^{\,3}} - \frac{1}{\bar{V}_1^{\,3}}\right)\right]$$

$$= RT\left[-\ln\frac{0.972}{0.486} + 2(-0.178)\left(\frac{1}{0.972} - \frac{1}{0.486}\right) + \frac{3}{2}(0.00906)\left(\frac{1}{0.972^2} - \frac{1}{0.486^2}\right) + \frac{4}{3}(6\times10^{-5})\left(\frac{1}{0.972^3} - \frac{1}{0.486^3}\right)\right]$$

$$= RT(-0.3700) = -923 \text{ J mol}^{-1}$$

$$= 10\ \Delta\bar{G} = \underline{\underline{-9230 \text{ J}}}$$

You should notice the very large discrepancy between the two values

$$\Delta G_{ideal} = \underline{\underline{-17{,}300 \text{ J}}}$$

$$\Delta G_{BB} = \underline{\underline{-9{,}230 \text{ J}}}$$

Since the Beattie-Bridgeman equation predicts the pressure accurately, we can feel fairly certain that the value of ΔG is good.

5-3.3 THE EQUILIBRIUM CONSTANT FOR A REACTION

Return to Eq. (5-3.2.9) which applies to any reaction that involves only ideal gases:

$$\Delta G_r = \Delta G_r^0 + RT \ln \frac{P_C^c}{P_B^b\ P_A^a} \tag{5-3.2.9}$$

Let us define the quotient in the ln expression as

$$L_P = \frac{P_C^c}{P_B^b\ P_A^a} \tag{5-3.3.1}$$

Equation (5-3.2.9) can be written very compactly as

$$\Delta G_r = \Delta G_r^0 + RT \ln L_P \tag{5-3.3.2}$$

This equation applies to any reaction involving ideal gases. If nonideal gases are involved, then the pressures in (5-3.3.1) must be replaced by fugacities as indicated in (5-3.2.11). Such a case will not be considered here.

At this point stop for a moment and consider what happens to ΔG_r when equilibrium is reached. Refer to Eq. (5-1.2.5) and note that dG = 0 for an equilibrium process. From this we can conclude that at equilibrium $\Delta G_r = 0$, which means

$$\Delta G_r^0 = -RT \ln L_e$$

The subscript on L designates the pressure quotient at equilibrium. This is usually given the symbol K_p. Thus at equilibrium

$$\Delta G_r^0 = -RT \ln K_p \qquad (5\text{-}3.3.3)$$

The equilibrium constant K_p for a reaction of ideal gases can be found if ΔG_r^0 is known. ΔG_r^0 can, of course, be found from tables of ΔG_f^0 values. For nonideal gases, we write an equation analogous to (5-3.3.3):

$$\Delta G_r^0 = -RT \ln K_f \qquad (5\text{-}3.3.4)$$

where K_f is the equilibrium constant in terms of fugacities instead of pressures. In both cases ΔG_r^0 is a definite number at a given temperature, so K_p (or K_f) will be a constant at that temperature.

5-3.3.1 EXAMPLE

Calculate the equilibrium constant for the reaction

$$C_2H_2(g) + 2H_2(g) \rightarrow C_2H_6(g)$$

at 298.15 K, 1 atm.

Solution

ΔG_r^0 can be found in Example 5-2.2.1:

$$\Delta G_r^0 = -242 \text{ kJ}$$

Apply Eq. (5-3.3.3):

$$\ln K_p = -\frac{\Delta G_r^0}{RT} = \frac{+\ 242{,}000 \text{ J}}{(8.314 \text{ J mol}^{-1} \text{ K}^{-1})(298.15 \text{ K})}$$

$$= \underline{97.6 \text{ mol}^{-1}}$$

$$K_p = \frac{P_{C_2H_2}}{P_{H_2}^2\, P_{C_2H_2}} = e^{+97.6}$$

$$= \underline{\underline{2.4 \times 10^{42}}}$$

Such a large equilibrium constant indicates the reaction tends to go to products almost completely. Note that all pressure must be in atmospheres since we defined our standard state as 1 atm. Also, note that the position of equilibrium is predicted, not the rate of attainment of that equilibrium.

5-3.4 TEMPERATURE DEPENDENCE OF THE EQUILIBRIUM CONSTANT

From the previous section you can see that we can write the equilibrium constant as

$$\ln K_p = -\frac{\Delta G_r^0}{RT} \qquad (5\text{-}3.4.1)$$

This gives, upon differentiation,

$$\frac{d \ln K_p}{dT} = -\frac{1}{R}\frac{d\,(\Delta G^0/T)}{dT} \qquad (5\text{-}3.4.2)$$

You should recognize the term $[d(\Delta G^0/T)/dT]$, since we used such an expression in Sec. 5-2 to find ΔG as a function of T. Look at Eq. (5-2.3.7) to see that (5-3.4.2) can be written

$$\frac{d \ln K_p}{dT} = \frac{\Delta H^0}{RT^2} \qquad (5\text{-}3.4.3)$$

Equation (5-3.4.3) gives one form of the equation we need to determine K_p as a function of temperature. If ΔH^0 is positive (endothermic), then an increase in temperature leads to a larger value of K_p. This is consistent with Le Chatelier's principle (see Example 5-3.2.1). Adding heat to an endothermic reaction shifts the equilibrium toward the products.

A more convenient form for graphical presentation of data can be obtained from (5-3.4.3). You should show that this equation can be written as

$$\frac{d \ln K_p}{d\,(1/T)} = -\frac{\Delta H^0}{R} \qquad (5\text{-}3.4.4)$$

This indicates that a plot of $\ln K_p$ vs. $1/T$ will yield a straight line with a slope of $-\Delta H^0/R$ if ΔH^0 is constant. If ΔH^0 is temperature dependent (and it is if very large temperature ranges are considered), then the plot will not be quite linear.

The utility of Eq. (5-3.4.4) should be obvious. If ΔH^0 is known (and can be assumed constant), and if K_p is known at T_1, then K_p can be calculated for T_2. On the other hand, if K_p is known at two temperatures, ΔH^0 can be found. Again, if ΔH^0 is not constant, an average ΔH^0 is found if two K_p values are used.

You should have no trouble finding an equation for determining K_p if ΔH^0 is a function of temperature. Tables of heat capacity data are sufficient to do the job. You will have an opportunity to do this in one of the exercises.

5-3.4.1 EXAMPLE

For the reaction

$$2H_2(g) + O_2(g) \rightarrow 2H_2O(g)$$

we found in Example 5-2.3.1 that $\Delta G^0_{298.15} = -457$ kJ and $\Delta G^0_{1000} = -385$ kJ. (The superscript zero indicates 1 atm pressure, the standard state.) Find K_p at 298.15 K and at 1000 K and determine the average value of ΔH. Compare this to the values found in Example 5-2.3.1.

Solution

$$\ln K_p(298.15) = -\frac{(-457{,}000 \text{ J})}{(8.314 \text{ J mol}^{-1} \text{ K}^{-1})(298.15 \text{ K})}$$

$$K_p(298.15) = 1.17 \times 10^{80}$$

Similarly,

$$K_p(1000) = 1.29 \times 10^{20}$$

Note that K_p decreases with increasing T. What does this tell us about the sign of ΔH?

Now apply Eq. (5-3.4.4):

$$d \ln K_p = -\frac{\Delta H^0}{R} d\frac{1}{T}$$

(ΔH^0 is assumed constant). Integrating this yields

$$\int_1^2 d \ln K_p = -\frac{\Delta H^0}{R} d\frac{1}{T}$$

$$\ln \frac{K_{p(2)}}{K_{p(1)}} = -\frac{\Delta H^0}{R}\left(\frac{1}{T_2} - \frac{1}{T_1}\right) \qquad (5\text{-}3.4.5)$$

For this particular problem.

$$\ln \frac{1.29 \times 10^{20}}{1.17 \times 10^{80}} = -\frac{\Delta H}{8.314 \text{ J mol}^{-1} \text{ K}^{-1}}\left(\frac{1}{1000 \text{ K}} - \frac{1}{298 \text{ K}}\right)$$

$$-\Delta H^0 = \frac{(8.314 \text{ J mol}^{-1})(-1.38 \times 10^2)}{(-2.36 \times 10^{-3})}$$

$$= 4.86 \times 10^5 \text{ J}$$

$$\Delta H^0 = \underline{\underline{-486{,}000 \text{ J}}}$$

The unit mol^{-1} has been dropped because the K_p given refers to the reaction written which involves 2 mol of the product water. Now compare with the values given in Example 5-2.3.1:

$$\Delta H^0_{298.15} = -484{,}000 \text{ J}$$

$$\Delta H^0_{1000} = -496{,}000 \text{ J}$$

The value obtained here is between these two values, as expected.

5-3.5 THE CHEMICAL POTENTIAL μ

This section will simply introduce the quantity called the chemical potential. Use will be made of it later. The chemical potential μ is useful when we are considering mixtures of materials. Thus far we have assumed that only pure materials are involved. In this case

$$G = G(T, P) \tag{5-3.5.1}$$

as stated in Sec. 5-1. However, if several materials are present in a mixture, G will depend on n_1, n_2, ... (the number of moles of each of the i substances present) as well as on T and P. We then must write

$$G = G(T, P, n_1, n_2, \ldots) \tag{5-3.5.2}$$

You should be able to write dG immediately

$$dG = \left(\frac{\partial G}{\partial T}\right)_{P,n_i} dT + \left(\frac{\partial G}{\partial P}\right)_{T,n_i} dP + \left(\frac{\partial G}{\partial n_i}\right)_{T,P,n_j} dn_i + \left(\frac{\partial G}{\partial n_z}\right)_{T,P,n_j} dn_z \cdots \tag{5-3.5.3}$$

Here the subscript n_i on the derivative implies that the number of moles of all substances is constant. The subscript n_j implies that the number of moles of every substance except the one involved in the derivative is constant.

We define the chemical potential μ as

$$\mu_i = \left(\frac{\partial G}{\partial n_i}\right)_{T,P,n_j} \tag{5-3.5.4}$$

which allows (5-3.4.3) to be written

$$dG = -S\ dT + V\ dP + \mu_1 dn_1 + \mu_2\ dn_2 + \cdots$$

$$= -S\ dT + V\ dP + \sum_i \mu_i\ dn_i \tag{5-3.5.5}$$

(Use has been made of the fact that $(\partial G/\partial T)_{P,ni} = -S$ and $(\partial G/\partial P)_{T,ni} = V$ as noted in Sec. 5-1.)

The partial molar free energy of a substance μ_i denotes the free energy change (*per mole of i added*) of a system, if an *infinitesimal amount of i* is added and the system is at constant T, P, and n_j. The amount of i added can only be infinitesimal, thus the composition of the system is not changed.

Use will be made of this term in future sections when we consider the behavior of mixtures of gases and solutions. Right now you should learn only the definition of μ.

5-3.6 PROBLEMS

1. From the results of Problems (1) and (3) in Sec. 5-2.4, evaluate K_p at 298.15 and 1100 K. Then estimate a value for ΔH^0. Compare this value with that found for ΔH^0 (298.15).
2. What pressure of water is required for equilibrium at 1 atm and 25°C if $H_2O(g)$ behaves ideally?
3. Derive the equation for the isothermal pressure dependence of ΔG for a van der Waals gas.
4. Find ΔG for the isothermal compression of 0.015 kg $CCl_4(g)$ from a pressure of 0.2 atm to 2 atm at 306 K if
   a. CCl_4 behaves ideally
   b. $CCl_4(g)$ behaves as a van der Waals gas with a = 19.6 dm^3 atm mol^{-2} and b = 0.127 dm^3
5. If the pressure of $C_2H_2(g)$ and $H_2(g)$ in the reaction

   $$C_2H_2(g) + H_2(g) \rightarrow C_2H_4(g)$$

   are maintained at 10^{-2} atm, what will be the pressure of $C_2H_4(g)$ at equilibrium at 298.15 K?
6. For the reaction

   $$H_2(g) + \tfrac{1}{2}O_2(g) \rightarrow H_2O(g)$$

   a. Derive an expression for K_p as a function of T from heat capacity data
   b. Evaluate $K_p(298.15)$ and $K_p(1000)$.
7. Given for NH_3, T_c = 405.6 K and P_c = 112.2 atm, evaluate the fugacity at P_r = 2 and T_r = 1.3 using the results of Example 5-3.2.2. Compare fugacity and pressure.
8. For the reaction $C(s) + 2H_2(g) \rightarrow CH_4(g)$, ΔG_r^0 = -50.79 kJ (at 298 K). If $K_p(900)$ = 4.8×10^{17}, what is the average ΔH^0 for this reaction over this temperature range?
9. (a) Write an expression for the equilibrium constant in Problem (8). (b) Dis-

cuss briefly whether an increase in the *total pressure* on the system in Problem (8) will lead to a larger or smaller value of K_p. Give reasons for your conclusion.

10. Derive the equation for the variation of G with pressure for a gas that follows Berthelot's equation

$$P = \left(\frac{RT}{\bar{V} - b} - \frac{a}{T\bar{V}^2}\right)$$

5-3.7 SELF-TEST

1. Define fugacity and chemical potential.
2. Assuming ideal behavior, find ΔG (10 atm) for the reaction

$$C_2H_2(g) + 2H_2(g) \rightarrow C_2H_6(g)$$

(Assume each reactant and product is maintained at a pressure of 10 atm.)

3. Find K_p for the reaction in Question (2) if equilibrium is attained at 298.15 K.
4. For the reaction in Question (2) at equilibrium at 298.15 K, what is the pressure of H_2 if $P(C_2H_6)$ = 100 atm and $P(C_2H_2) = 10^{-16}$ atm? How many H_2 molecules are there per cubic decimeter in this equilibrium mixture?
5. If ΔH^0 for the following reaction is constant at the value observed at 298.15 K, find K_p at 298.15 K and 330 K.

$$C(g) + 2H_2(g) + \tfrac{1}{2}O_2(g) \rightarrow CH_3OH(g)$$

5-3.8 ADDITIONAL READING

1.* G. Barrow, *Physical Chemistry*, 3rd ed., McGraw-Hill, New York, 1973, Sec. 9-5, 9-6.

2.* F. T. Wall, *Chemical Thermodynamics*, 3rd ed., W. H. Freeman and Company, San Francisco, 1958.

5-4 SUMMARY

This chapter completes the introduction to thermodynamics. Chapter 6 will make extensive use of the material in the last three chapters to find properties of solutions. Other chapters will apply some of the results just derived. Even before considering additional applications you have an extensive array of useful skills to use in practical applications.

You are now able to predict the direction of a given chemical reaction or process for ideal or real gaseous systems. You are able to determine how a change in pressure or temperature will affect the equilibrium point of a reaction. Such ability is of use if you ever need to design or optimize a system involving chemical reactions. You can also predict the heat generated or absorbed by a reaction, an important quantity in designing a reaction vessel and controlling feed rates in a reaction.

Nonideal equations of state have been used for many calculations. This was done to ensure that you would not assume that ideal behavior must be used in a process. Few processes even approximate ideal behavior. Even though useful guesses can be obtained from the ideal equations of state for gases, little credence should be put into the validity of the final result. You should no longer fear attacking a problem with a complicated equation of state.

6

Thermodynamics of Solutions

SYMBOLS AND DEFINITIONS

μ	Chemical potential or partial molar free energy (J mol^{-1})
G	Gibbs free energy function (J)
S	Entropy (J K^{-1})
$(\bar{\ })$	Indicates a molar quantity, e.g., $\bar{G}$ = molar free energy
x_i	Mole fraction of substance i (unitless)
μ_i	Chemical potential of substance i (J mol^{-1})
μ_i^0	Standard chemical potential of substance i
n_A	Number of moles of solvent
n_B	Number of moles of solute
P_i^0	Vapor pressure of pure substance i (atm, torr = mm Hg)
K_i	Henry's law constant (atm)
c_i	Concentration of substance i (M - moles (dm^3 solution)$^{-1}$, m - moles (kg solvent)$^{-1}$)
a_i	Activity of substance i
γ_i	Activity coefficient of substance i
ΔT_{bp}	Boiling point elevation (K)
K_{bp}	Boiling point elevation constant (ebullioscopic constant, K $molal^{-1}$)
ΔT_{fp}	Freezing point depression(K)
K_f	Freezing point depression constant (K $molal^{-1}$)
Π	Osmotic pressure (atm)
$\underline{i}$	van't Hoff's factor
ν	Total number of ions produced on dissociation
α	Degree of dissociation
L	Electrical conductance (ohm^{-1})
κ	Specific conductance (ohm^{-1} m^{-1})
Λ	Equivalent conductance (m^2 ohm^{-1} $equivalent^{-1}$)
Λ^0	Limiting equivalent conductance
λ^0	Limiting equivalent ionic conductance
$t_{\pm}$	Transference numbers
K	Equilibrium Constant
$a_{\pm}$	Mean ionic activity (concentration unit)
$\gamma_{\pm}$	Mean ionic activity coefficient (unitless)
$c_{\pm}$	Mean concentration (concentration unit)
$\underline{\mu}$	Ionic strength (concentration unit)
A, B	Debye-Hückel constants
z_+, z_-	Ionic charges
K_{th}	Thermodynamic equilibrium constant (activities)
K_{diss}	Dissociation equilibrium constant (concentrations)

K_{sp} Solubility product (concentration)
ϕ Electrochemical potential (volts)
$\mathcal{E}^0$ Standard electrode potential (volts)
$\mathcal{F}$ Faraday constant (96,500 coulombs)

This will be a rather long chapter, but a very important one. The thermodynamic principles presented in Chaps. 3 through 5 will be applied to mixtures of materials in this chapter. Previously, except for a brief digression at the end of Chap. 5, only pure materials have been considered. We will remedy that in short order by studying mixtures of gases in gases, and solids, liquids, and gases in liquids. Solids in solids will not be considered.

As usual, ideal cases will be considered first, then nonideality will be introduced. An important technique that you should master will be developed: For nonideal systems, a certain property will be measured at convenient concentrations. Extrapolation back to zero concentration will be performed to find the ideal value of that property. The assumption is made that the interactions leading to nonideality will disappear in the limit of infinite dilution (zero concentration).

No effort will be made to present rigorously the theoretical basis for all the concepts introduced in this chapter. If you learn the concepts and learn to apply them properly, you will have fulfilled the objectives of this chapter. It is, as always, important to remember what restrictions apply to given equations.

A preview of what is in store is probably beneficial at this time. Ideal mixtures (gases, liquids) and deviations from ideality will be presented first. Then *colligative properties* will be considered. Colligative properties are those that depend only on the number of particles (molecules, atoms, ions, etc.) present in solution and not on the nature of these particles. Examples are freezing point depression and vapor pressure lowering.

Next, the behavior of electrolytes (solutes that dissociate to some extent in solution) will be considered. This will lead to another "fudge factor," the activity coefficient. Since concentration cannot be used directly to calculate the thermodynamic properties in nonideal solutions, we use the same procedure as introduced in Chap. 5 where fugacity is defined. The reason for this is to maintain the simplicity found in the equations for ideal systems.

Finally, electrochemical cells are discussed and methods are outlined for determining solubilities and dissociation constants of electrolytes. An important application of electrochemistry will be discussed. This application is in fuel cells, devices for converting chemical energy into electrical energy.

6-1 THE CHEMICAL POTENTIAL AND SOLUTIONS OF NONELECTROLYTES

OBJECTIVES

You shall be able to

1. Define equilibrium in terms of the chemical potential μ
2. Calculate the chemical potential of a material given its mole fraction
3. Calculate the entropy, enthalpy, and free energy of mixing for ideal solutions
4. State and apply Raoult's and Henry's laws
5. Discuss deviations from Raoult's law in terms of solvent-solvent, solute-solvent, and solute-solute interactions

6-1.1 THE CHEMICAL POTENTIAL AND EQUILIBRIUM

In Chap. 5 the term chemical potential, which is given the symbol μ, was briefly introduced. Refer to Chap. 5 to verify that for a reversible process involving only PV work

$$dG = -S\ dT + V\ dP + \sum_i \mu_i\ dn_i \qquad (6\text{-}1.1.1)$$

where dn_i is the change in the number of moles of the ith constituent in the process, and μ_i is the chemical potential or partial molar free energy of the ith constituent. μ is defined by

$$\mu_i = \left(\frac{\partial G}{\partial n_i}\right)_{T,P,n_{j \neq i}} \qquad (6\text{-}1.1.2)$$

What is the significance of μ? Suppose that μ_i is different in two portions of a system, A and B. That is, $\mu_i{}^A \neq \mu_i{}^B$. Also suppose that T, P, and n_j $(j \neq i)$ are constant throughout the system. What happens if dn_i moles of i are transferred from region A to region B? Look at Eq. (6-1.1.1) which reduces under these conditions to

$$dG = \mu_i\ dn_i \qquad (6\text{-}1.1.3)$$

if T, P, and n_j are constant. You should see that

$$dG^A = \mu_i{}^A\ dn_i{}^A = \mu_i{}^A\ (-\ dn_i)$$

(dn_i is transferred from A to B, therefore $dn_i{}^A = -\ dn_i$)

and $$dG^B = \mu_i{}^B\ dn_i{}^B = \mu_i{}^B\ (dn_i)$$

This gives an overall change in dG of

$$dG = dG^A + dG^B = (\mu_i^B - \mu_i^A)\ dn_i \tag{6-1.1.4}$$

Remember we have stated that dn_i is the transfer from A → B. (This makes dn_i^A negative as indicated above.) If $\mu_i^A > \mu_i^B$, then, from (6-1.1.4) you should deduce that $dG < 0$. A negative dG indicates a spontaneous process. Thus, material will tend to flow spontaneously from a region of higher chemical potential to one of lower chemical potential at constant T, P, and n_j.

As you might predict this tendency will continue until $\mu_i^A = \mu_i^B$ (dG = 0). This gives us a statement of the equilibrium condition in a system of more than one constituent:

At equilibrium (T, P, and n_j constant throughout),

$$\mu_i^A = \mu_i^B \tag{6-1.1.5}$$

An analogy can be made to electricity. Just as electricity flows spontaneously from a region of high electrical potential to one lower in electrical potential, matter will tend to flow from a region of high chemical potential to one of lower chemical potential.

A final statement is given without proof.

$$G = \sum_i n_i\ \mu_i \tag{6-1.1.6}$$

This equation gives as a way to find the free energy of a mixture of materials if we know the number of moles of each (n_i) and the chemical potential (μ_i) of each. The term μ_i is called the *partial molar free energy* as well as the *chemical potential* of substance i.

6-1.2 PROPERTIES OF IDEAL MIXTURES

THE CHEMICAL POTENTIAL OF AN IDEAL GAS MIXTURE. From the results in Chap. 5 we can write immediately, for a pure ideal gas,

$$\mu = \mu^0(T) + RT\ \ln P \tag{6-1.2.1}$$

where $\mu^0(T)$ is the standard state (P = 1 atm) value of μ at a given temperature. Now what happens to μ if we introduce this material into a mixture? Figure 6-1.2.1 will be useful in helping us decide. Assume we have two materials, A and B, both of which behave ideally. The compartment on the right contains pure A (at P_A) and the left compartment contains A and B (at P_A and P_B). If the membrane is porous to A, then $P_A(\text{pure}) = P_A(\text{mixture})$. (Do you agree with that statement?) Also, from the previous section, we can write *at equilibrium* for this system

$$\mu_A(\text{pure}) = \mu_A(\text{mixture})$$

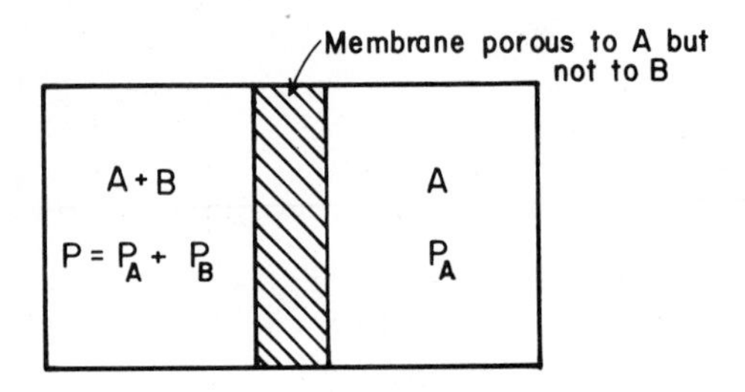

FIG. 6-1.2.1 Figure to deduce the chemical potential of A in a mixture.

And [using Eq. (6-1.2.1)]

$$\mu_A(\text{pure}) = \mu_A^0(\text{pure}) + RT \ln P_A$$

We can readily extend this argument to mixtures of several gases to give

$$\mu_i = \mu_i^0(T) + RT \ln P_i \qquad (6\text{-}1.2.2)$$

in which P_i represents the partial pressure of substance i in the mixture of gases.

If we substitute the relationship

$$P_i = x_i P \qquad (6\text{-}1.2.3)$$

where $x_i = n_i/\sum n_j$ = mole fraction of substance i in the mixture and P = total pressure, into (6-1.2.2) the following is readily obtained:

$$\mu_i = \mu_i^0(T) + RT \ln x_i P$$

$$= \mu_i^0(T) + RT \ln P + RT \ln x_i$$

$$= \mu_i^0(T, P) + RT \ln x_i \qquad (6\text{-}1.2.4)$$

You can tell from the definition of the mole fraction x_i, which is always less than or equal to 1, that $\mu_i \leq \mu_i^0(T, P)$. You should also recognize that $\mu_i^0(T, P)$ is the chemical potential of *pure i* at a given T and P. The important result is that the chemical potential of a substance in a mixture is always less than the chemical potential of the pure material.

6-1.2.1 EXAMPLE

Find the potential change associated with introducing 1 mol pure H_2 into 3 mol pure N_2 if the total pressure and temperature are held constant (T = 298 K).

Solution

$$\Delta\mu_{H_2} = \mu_{H_2} - \mu_{H_2}^0(T, P) = RT \ln x_{H_2}$$

$$\Delta\mu_{N_2} = \mu_{N_2} - \mu^0_{N_2}(T, P) = RT \ln x_{N_2}$$

$$x_{H_2} = \frac{n_{H_2}}{n_{H_2} + n_{N_2}} = \frac{1 \text{ mol}}{1 \text{ mol} + 3 \text{ mol}}$$

$$= \underline{0.25}$$

$$x_{N_2} = 1 - x_{H_2}$$

$$= \underline{0.75}$$

$$\Delta\mu_{total} = \Delta\mu_{H_2} + \Delta\mu_{N_2}$$

$$= RT(\ln x_{H_2} + \ln x_{N_2})$$

$$= (8.314 \text{ J mol}^{-1} \text{ K}^{-1})(298 \text{ K})(\ln 0.25 + \ln 0.75)$$

$$= -4147 \text{ J mol}^{-1}$$

$$= \underline{\underline{-4100 \text{ J mol}^{-1}}}$$

This means that -4100 J *per mole of mixture* is the decrease in the chemical potential on mixing these two "ideal" gases.

A STATEMENT OF RAOULT'S LAW FOR LIQUID SOLUTIONS. A brief statement of Raoult's law will be given, then we will return to the task of formulating equations for the chemical potential of an ideal solution. After that, more detail on Raoult's law and deviations from it will be given.

Raoult's law can be stated very simply mathematically:

$$P_A = x_A^{\ell} P_A^0 \tag{6-1.2.5}$$

P_A is the vapor pressure of A above a solution in which A has a mole fraction x_A^{ℓ}. P_A^0 is the vapor pressure of pure liquid A at the same temperature and total pressure. This equation holds for an ideal solution. It tells us that the vapor pressure of constituent A varies linearly with its mole fraction in the solution. You will see more about this later.

THE CHEMICAL POTENTIAL OF AN IDEAL SOLUTION. Assume we have an ideal solution of *solvent A* with mole fraction x_A^{ℓ} and *solute B* with mole fraction x_B^{ℓ}. What will the chemical potential be for A in this solution?

First, we need to determine the vapor pressure of A above the solution. This is easily done. Apply (6-1.2.5):

$$P_A = x_A^{\ell} P_A^0 \tag{6-1.2.5}$$

The chemical potential of A in the vapor will, according to (6-1.2.2), be

$$\mu_A^v = \mu_A^{0,v}(T) + RT \ln P_A \tag{6-1.2.6}$$

What will the relationship be between μ_A^v and μ_A^ℓ at equilibrium? By the definition of equilibrium these two will be equal so $\mu_A^\ell = \mu_A^v$. Then we can write

$$\begin{aligned} \mu_A^\ell = \mu_A^v &= \mu_A^{0,v}(T) + RT \ln P_A \\ &= \mu_A^{0,v}(T) + RT \ln x_A^\ell P_A^0 \\ &= \mu_A^{0,v}(T) + RT \ln P_A^0 + RT \ln x_A^\ell \end{aligned} \tag{6-1.2.7}$$

But $\mu_A^{0,v}(T) + RT \ln P_A^0$ is just equal to $\mu_A^{0,v}(T, P^0)$, the chemical potential of pure A in the vapor at the normal vapor pressure of A. Again, by the definition of equilibrium $\mu_A^{0,\ell}(T, P^0) = \mu_A^{0,v}(T, P^0)$. Thus

$$\mu_A = \mu_A^{0,\ell}(T,P^0) + RT \ln x_A^\ell \tag{6-1.2.8}$$

which can be shortened (since the chemical potential is the same in ℓ and v at equilibrium) to

$$\mu_A = \mu_A^0 + RT \ln x_A^\ell \tag{6-1.2.9}$$

A similar equation will apply to each substance in the system.

FREE ENERGY OF MIXING. Let us continue with an ideal solution of B in A. Assume we start with pure A and pure B with chemical potentials of μ_A^0 and μ_B^0, respectively, and form an ideal solution with mole fractions x_B^ℓ and x_A^ℓ ($x_B^\ell + x_A^\ell = 1$, of course). From here on the superscript ℓ will be dropped: *x refers to the liquid unless otherwise specified.* How do we find the free energy of mixing? Look at Fig. 6-1.2.2 for a hint.

Initial conditions:

$$G_{total}^i = n_A\mu_A^0 + n_B\mu_B^0$$

Final conditions:

$$G_{total}^f = n_A\mu_A + n_B\mu_B$$

(Note the use of G_{total} instead of μ_{total}. μ is a *partial molar free energy* and should be used only as such. G will be the free energy for the whole system, no matter how many moles are present.)

Change:

$$\Delta G_{mix} = G^f - G^i$$

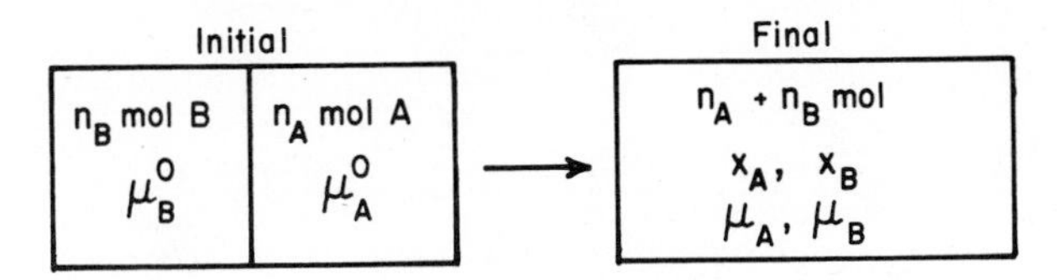

FIG. 6-1.2.2 Schematic diagram of the mixing of n_A mol A and n_B mol B to form an ideal solution.

$$= n_A\mu_A + n_B\mu_B - n_A\mu_A^0 - n_B\mu_B^0$$

$$= n_A(\mu_A - \mu_A^0) + n_B(\mu_B - \mu_B^0) \qquad (6\text{-}1.2.10)$$

But according to (6-1.2.9) this is

$$\Delta G_{mix} = n_A RT \ln x_A + n_B RT \ln x_B \qquad (6\text{-}1.2.11)$$

Divide by the total moles $(n_A + n_B)$ to get $\Delta\bar{G}$ (*per mole of solution*):

$$\Delta\bar{G}_{mix} = \frac{\Delta G}{n_A + n_B} = \left(\frac{n_A}{n_A + n_B}\right) RT \ln x_A + \left(\frac{n_B}{n_A + n_B}\right) RT \ln x_B$$

$$\boxed{\Delta\bar{G}_{mix} = x_A RT \ln x_A + x_B RT \ln x_B} \qquad \text{ideal} \qquad (6\text{-}1.2.12)$$

This gives the free energy of mixing for 1 mol of an ideal solution containing two materials. If more than two materials are involved this equation can be readily extended. Also, look at Eq. (6-1.2.4). If we go through the same arguments for gases as has been done for liquids in this section, we get an identical result. So, for ideal solutions of gases in gases or gases, liquids, and solids in liquids, we have the general result (for two materials A and B):

$$\Delta\bar{G}_{mix} = x_A RT \ln x_A + x_B RT \ln x_B \qquad (6\text{-}1.2.13)$$

ENTHALPY OF MIXING. This is easy to state. For an ideal solution

$$\boxed{\Delta\bar{H}_{mix} = 0} \qquad \text{ideal} \qquad (6\text{-}1.2.14)$$

The reasons for this are given in Sec. 6-1.3 where Raoult's law and deviations from it are treated further.

ENTROPY OF MIXING. You should be able to find the entropy of mixing from what has been given above. Since $\Delta G = \Delta H - T\,\Delta S$ for an isothermal process,

$$\Delta\bar{G}_{mix} = \Delta\bar{H}_{mix} - T\,\Delta\bar{S}_{mix} = -T\,\Delta\bar{S}_{mix}$$

or

$$\boxed{\Delta\bar{S}_{mix} = -\frac{\Delta\bar{G}_{mix}}{T} = -x_A R \ln x_A - x_B R \ln x_B} \qquad \text{ideal} \qquad (6\text{-}1.2.15)$$

This is valid for the formation of *1 mol of an ideal solution* containing two materials.

6-1.2.2 EXAMPLE

Find the free energy and entropy of mixing if 0.270 kg of a sugar with a molecular weight of 0.358 kg mol^{-1} is dissolved in 0.800 kg water at 295 K, assuming an ideal solution results.

Solution

You need to find the mole fractions of sugar and water.

$$n_w = \frac{0.800\ kg}{0.0180\ kg\ mol^{-1}} = \underline{\underline{44.4\ mol\ H_2O}}$$

$$n_s = \frac{0.270\ kg}{0.358\ kg\ mol^{-1}} = \underline{\underline{0.754\ mol\ sugar}}$$

$$n_{total} = n_w + n_s = \underline{\underline{45.2\ mol}}$$

$$x_w = \frac{44.4\ mol\ H_2O}{45.2\ mol} = \underline{\underline{0.983}}$$

$$x_s = \frac{0.754\ mol\ sugar}{45.2\ mol} = \underline{\underline{0.017}}$$

(Note: In any solution $\sum x_i = 1$ from the definition of mole fraction.)

From Eq. (6-1.2.13),

$$\Delta\bar{G}_{mix} = x_w RT \ln x_w + x_s RT \ln x_s \qquad (6\text{-}1.2.13)$$

$$= (8.314\ J\ mol^{-1}\ K^{-1})(295\ K)(0.983 \ln 0.983 + 0.017 \ln 0.017)$$

$$= (2453\ J\ mol^{-1})(-0.0861)$$

$$= \underline{\underline{-211\ J\ mol^{-1}}}$$

Again, it needs to be emphasized that this is *per mole of solution*. For the amount of solution we have

$$\Delta\bar{G}_{mix} = (n_w + n_s)\,\Delta\bar{G}_{mix} = 45.2\ mol(-211\ J\ mol^{-1})$$

$$= \underline{\underline{-9540\ J}}$$

To find $\Delta\bar{S}_{mix}$:

$$\Delta\bar{S}_{mix} = -\frac{\Delta\bar{G}_{mix}}{T} = -\frac{-211\ \text{J mol}^{-1}}{295\ \text{K}}$$

$$= \underline{\underline{0.716\ \text{J mol}^{-1}\ \text{K}^{-1}}}$$

6-1.2.3 EXAMPLE

Plot the free energy and entropy of mixing of an ideal binary (two-component) solution at 300 K.

Solution

The actual calculations are not given. Feel free to verify the numbers in the plot shown in Fig. 6-1.2.3 (in fact, you should verify some of the values). Use will be made of the results in the next section.

6-1.3 THE VAPOR PRESSURE OF IDEAL SOLUTIONS

RAOULT'S LAW. Consider a binary system that forms an ideal solution: one which obeys Raoult's law. Assume both substances are volatile. According to (6-1.2.5),

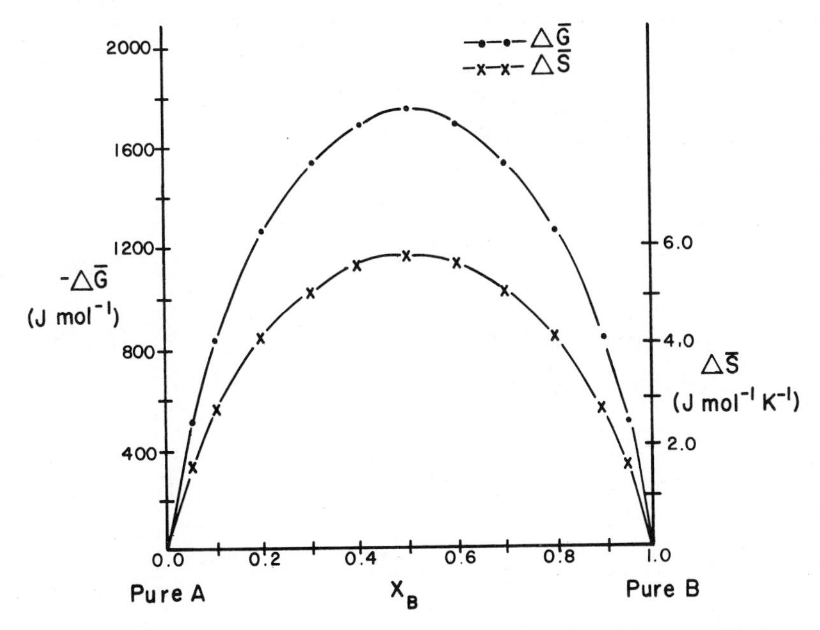

FIG. 6-1.2.3 Plot of ΔG_{mix} and ΔS_{mix} for an ideal binary solution.

$$P_1 = x_1^{\ell} P_1^0 \qquad P_2 = x_2^{\ell} P_2^0 = (1 - x_1^{\ell})P_2^0 \tag{6-1.3.1}$$

The last equation is valid since $x_1 + x_2 = 1$ in a binary system. The total pressure over a binary solution will be

$$\begin{aligned} P_T &= P_1 + P_2 \\ &= x_1^{\ell} P_1^0 + (1 - x_1^{\ell})P_2^0 \\ &= P_2^0 + (P_1^0 - P_2^0)x_1^{\ell} \end{aligned} \tag{6-1.3.2}$$

This shows that P_T is linear with x_1^{ℓ} as shown in Fig. 6-1.3.1(a).

Now let us express the total pressure in terms of x_1^{v}, the mole fraction of one of the substances in the vapor. For a gas, the mole fraction of a given substance is

$$x_1^{v} = \frac{P_1}{P_T} \tag{6-1.3.3}$$

Using (6-1.3.1) and (6-1.3.2) this becomes (verify each of the following steps)

$$x_1^{v} = \frac{x_1^{\ell} P_1^0}{P_2^0 + (P_1^0 - P_2^0)x_1^{\ell}} \tag{6-1.3.4}$$

Solve for x_1^{ℓ} to give

$$x_1^{\ell} = \frac{x_1^{v} P_2^0}{P_1^0 + (P_2^0 - P_1^0)x_1^{v}} \tag{6-1.3.5}$$

This can be substituted into (6-1.3.2) to give the total pressure in terms of the mole fraction of substance 1 in the vapor

$$P_T = \frac{P_1^0 P_2^0}{P_1^0 + (P_2^0 - P_1^0)x_1^{v}} \tag{6-1.3.6}$$

Figure 6-1.3.1(a) has a plot of P_T as a function of the mole fraction of substance 1 in the liquid, while Fig. 6-1.3.1(b) is the same plot for the mole fraction of 1 in the vapor. More detail will be provided about Fig. 6-1.3.1(b) in Chap. 13, Phase Equilibria.

Consider Fig. 6-1.3.1(a) in greater detail. The lines for the vapor pressure of substances 1 and 2, as well as the total pressure, have been drawn. These are found from Eq. (6-1.3.1). As is evident, the pressures for an ideal solution are easily calculated.

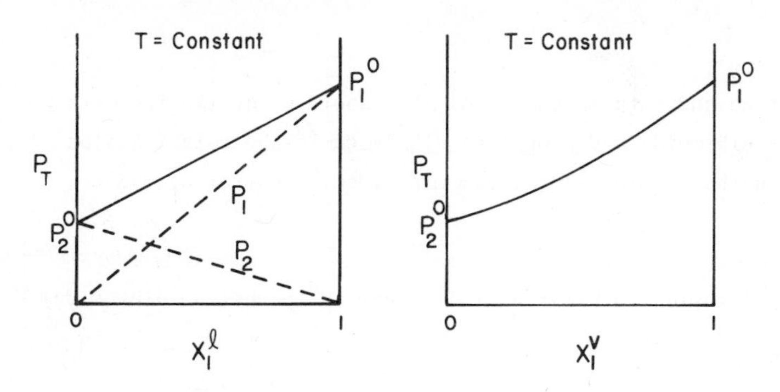

FIG. 6-1.3.1 Vapor pressure as a function of composition.

INTERACTIONS IN SOLUTIONS. In an ideal solution the interactions of solute with solute, solvent with solute, and solvent with solvent are all the same. This leads to the statement that $\Delta H_{mix} = 0$. To see how this statement follows the first consider the binary system of solvent A and solute B. Before mixing we have the situation depicted in Fig. 6-1.3.2(a). Figure 6-1.3.2(b) is representative of the interactions present after mixing. We can write the solution reaction schematically as (..... indicates an interaction)

$$\underset{\Delta H_{AA}}{A.....A} + \underset{\Delta H_{BB}}{B.....B} \rightarrow 2(\underset{\Delta H_{AB}}{A.....B}) \qquad (6\text{-}1.3.7)$$

Before mixing we have A-A and B-B interactions in the separate beakers, after mixing we have A-B interactions. For each A-A and B-B pair broken on dissolution, two A-B interactions are formed. It takes ΔH_{AA} to separate A-A in the solvent and ΔH_{BB} to separate B-B in the solute. But $2\Delta H_{AB}$ is released when the 2A and 2B molecules interact. Then if $\Delta H_{BB} = \Delta H_{AA} = \Delta H_{AB}$ (which is the assumption for an ideal solution), ΔH_{mix} will equal zero as stated above.

Before turning to a study of nonideal (real) solutions, note one example of a real system that exhibits almost ideal behavior. This is the ethylene dibromide--propylene dibromide system depicted in Fig. 6-1.3.3.

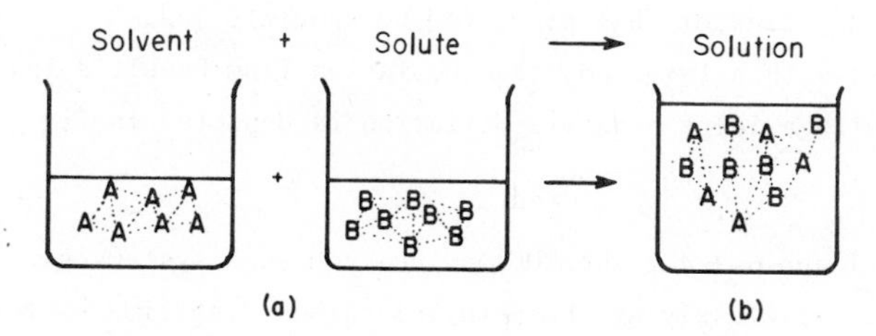

FIG. 6-1.3.2 Interaction present (a) before and (b) after mixing.

6-1.3.1 EXAMPLE

Predict the composition and total pressure of the vapor in equilibrium with a solution of ethylene dibromide and propylene dibromide if the mole fraction of ethylene dibromide in the liquid is 0.6, assuming ideal behavior at 358 K.

Solution

From Fig. 6-1.3.3, P^0(ethylene dibromide) ≈ 172 torr; P^0(propylene dibromide) ≈ 128 torr. From Eq. (6-1.3.1),

$$P_{ed} = x^{\ell}_{ed} P^0_{ed} = (0.6)(172 \text{ mm Hg})$$

$$= \underline{\underline{103 \text{ mm Hg}}}$$

$$P_{pd} = x^{\ell}_{pd} P^0_{pd} = (0.4)(128 \text{ mm Hg})$$

$$= \underline{\underline{51 \text{ mm Hg}}}$$

$$P_{total} = P_{ed} + P_{pd} = \underline{\underline{154 \text{ mm Hg}}}$$

$$x^v_{ed} = \frac{P_{ed}}{P_{total}} = \frac{103 \text{ mm Hg}}{154 \text{ mm Hg}} = \underline{\underline{0.67}}$$

$$x^v_{pd} = \frac{P_{pd}}{P_{total}} = \frac{51 \text{ mm Hg}}{154 \text{ mm Hg}} = \underline{\underline{0.33}}$$

Look at Fig. 6-1.3.3 to check the answers.

6-1.4 THE VAPOR PRESSURES AND INTERACTIONS IN NONIDEAL SYSTEMS

SOME VAPOR PRESSURE DIAGRAMS. The vapor pressure diagrams for some nonideal systems will be given first. Then, Henry's law will be introduced. Finally, an attempt will be made to explain the observed deviations from ideality in terms of interactions in the solutions.

The system depicted in Fig. 6-1.4.1 shows only a slight deviation from Raoult's law. It should be evident from the figure that the total vapor pressure and the vapor pressure of each component are greater than predicted by Raoult's law.

Figure 6-1.4.2 shows a system with a large positive deviation from Raoult's law ($P_{total} > P_{Raoult}$). A solution with a large negative deviation is depicted in Fig. 6-1.4.3.

HENRY'S LAW. Figures 6-1.4.1 and 6-1.4.2 should convince you some systems are far from ideal. Now let us look more closely at the methylal-carbon disulfide system.

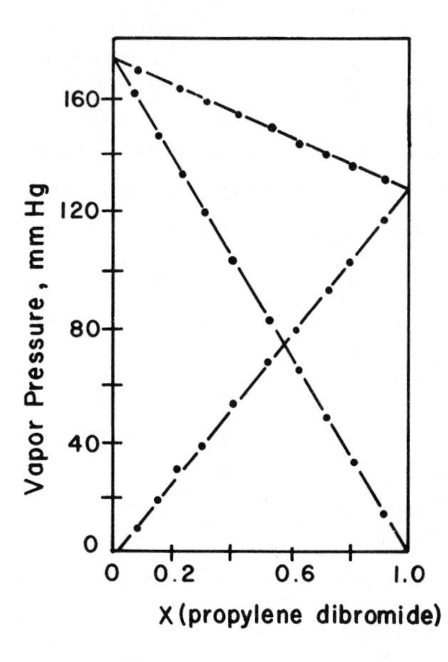

FIG. 6-1.3.3 Vapor pressure of ethylene dibromide-propylene dibromide system at 358.21 K (760 mm Hg ≡ 1 atm). Reprinted with permission of Macmillan Publishing Co., Inc. from *Principles of Physical Chemistry*, 4th ed., by S. H. Maron and C. F. Prutton. Copyright 1965 by Macmillan Publishing Co., Inc.

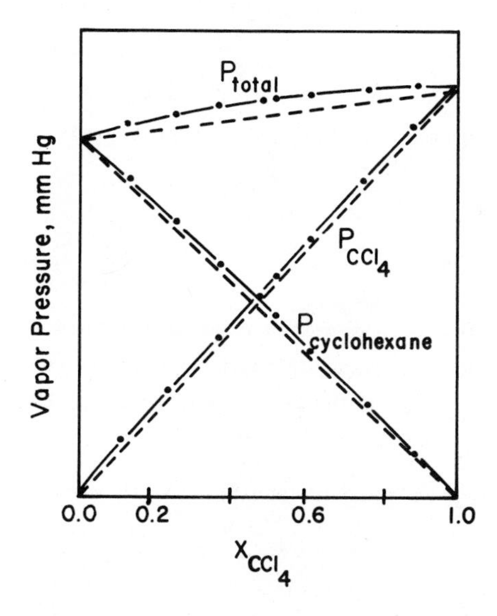

FIG. 6-1.4.1 Vapor pressures of the cyclohexane-carbon tetrachloride system at 313.1 K (—•—, real; ---, ideal). Reprinted with Permission of Macmillan Publishing Co., Inc. from *Principles of Physical Chemistry*, 4th ed., by S. H. Maron and C. F. Prutton. Copyright 1965 by Macmillan Publishing Co., Inc.

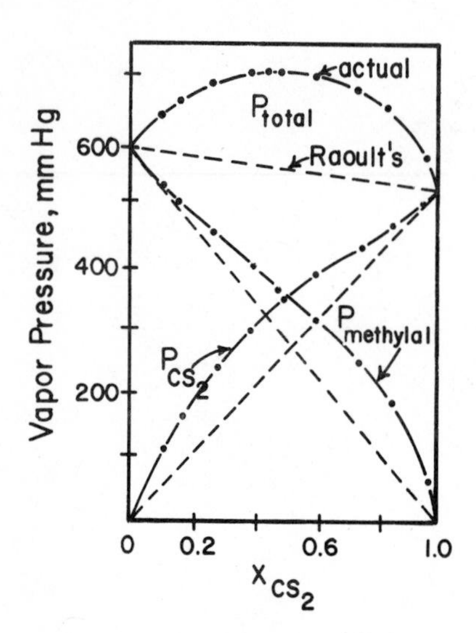

FIG. 6-1.4.2 The vapor pressure of the methylal-carbon disulfide system at 308.4 K. An example of positive deviation from Raoult's law. Reprinted with permission of Macmillan Publishing Co., Inc., from *Principles of Physical Chemistry*, 4th ed., by S. H. Maron and C. F. Prutton. Copyright 1965 by Macmillan Publishing Co., Inc.

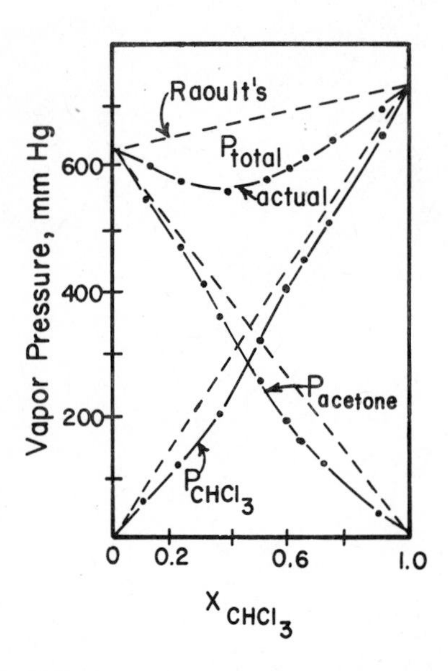

FIG. 6-1.4.3 The vapor pressure of the acetone-chloroform system at 328.3 K. An example of negative deviation from Raoult's law. Reprinted with permission of Macmillan Publishing Co., Inc. from *Principles of Physical Chemistry*, 4th ed., by S. H. Maron and C. F. Prutton. Copyright 1965 by Macmillan Publishing Co., Inc.

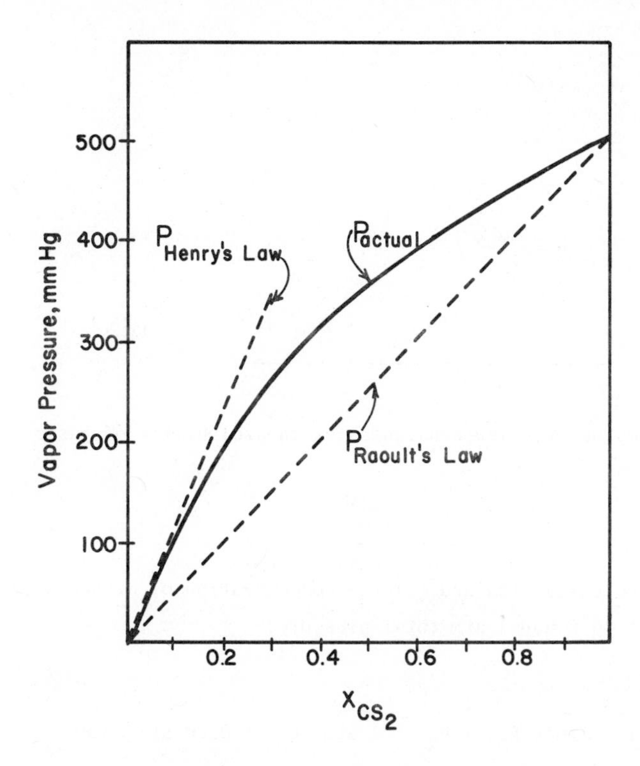

FIG. 6-1.4.4 Vapor pressures of carbon disulfide in the methylal-carbon disulfide system at 308.4 K.

The vapor pressure of carbon disulfide as a function of mole fraction is replotted in Fig. 6-1.4.4 on a larger scale. You will see that two dotted lines are included on the plot. The lower one is the vapor pressure line for CS_2 as predicted by Raoult's law. The upper line is the vapor pressure line predicted by Henry's law, which is defined below. In the region near $x_{CS_2} = 0$ (CS_2 present in low concentration) the vapor pressure curve for CS_2 is almost linear. (The slope of this curve is different than predicted for Raoult's law.) Since the curve is almost linear for small x_{CS_2}, we can write

$$P_{CS_2} = K_{CS_2} x_{CS_2} \tag{6-1.4.1}$$

where K_{CS_2} is a constant. *Equation (6-1.4.1) is a statement of Henry's law.*

Henry's law applies to the *solute* at low concentrations. As an exercise, answer this question: For an ideal solution of volatile components what is K in Henry's law expression? To answer that you should relate Henry's law to Raoult's law. (They are, in fact, the same for an ideal solution, so $K_B = P_B^0$ for solute B in an ideal solution.)

Table 6-1.4.1
Selected Henry's Law Constants for Gases in Water ($K \times 10^{-4}$ atm)

Gas	0°C	20°C	40°C	60°C	80°C
H_2	5.81	6.85	7.63	7.63	7.52
N_2	5.38	7.58	10.0	11.5	--
O_2	2.51	3.88	5.43	6.37	6.94
C_2H_4	0.49	0.99	1.62	--	--

Table 6-1.4.1 contains some Henry's law constants for the solubility of gases in water at various temperatures.

6-1.4.1 EXAMPLE

Find (1) the equilibrium mole fraction and (2) the concentration of N_2 and O_2 in water exposed to air at 40°C and 1 atm total pressure.

Solution

1. To start you must realize that $P_{O_2} + P_{N_2} = 1$ atm, $P_{O_2} = 0.20$ atm and $P_{N_2} = 0.80$ atm.

Henry's law applies to a gas in a mixture if the *partial pressure* of that gas is used. From Eq. (6-1.4.1):

$$x_{O_2} = \frac{P_{O_2}}{K_{O_2}} = \frac{0.20 \text{ atm}}{5.43 \times 10^4 \text{ atm}} = \underline{\underline{3.68 \times 10^{-6}}}$$

$$x_{N_2} = \frac{P_{N_2}}{K_{N_2}} = \frac{0.80 \text{ atm}}{10.0 \times 10^4 \text{ atm}} = \underline{\underline{8.00 \times 10^{-6}}}$$

2. The easy part is done. Now we need to find the concentration. This will not be difficult since the mole fractions of O_2 and N_2 are very small.

Choose 1 mol of water as reference. (One mole of water = one mole of solution for all practical purposes in this problem.) Since the mole fraction of O_2 is 3.68×10^{-6}, this tells us that there are 3.68×10^{-6} mol of O_2 in 1 mol of water. Similarly there are 8.00×10^{-6} mol of N_2 per mole of water.

(*Note:* Assume we had a system with $x_B = 0.01$ and $x_C = 0.02$. Finding the concentrations of B and C in this case is more difficult than before. Pick 1 mol of solution as basis. Then $x_W + x_B + x_C = 1$ which gives $x_W = 0.97$. The concentration of B, c_B for example, will be given by mol B/dm^3 solution. If the solution is ideal, 1 dm^3 solution = 1 dm^3 water, and if the density of the solution is equal to 1, then 1 dm^3 water = 1 kg water. So $c_B \approx$ mol B/kg water. We know

$$c_B = \frac{0.01 \text{ mol B/mol total}}{(0.97 \text{ mol w/mol total})(0.018 \text{ kg w/1 mol w})}$$

$$= \frac{0.01 \text{ mol B}}{0.0175 \text{ kg w}} = 0.573 \frac{\text{mol B}}{\text{kg w}} = 0.573 \frac{\text{mol B}}{\text{dm}^3 \text{ w}}$$

$$= \underline{\underline{0.573 \text{ M}}}$$

A similar procedure holds for C. Why go to all this trouble? Simply to show that we make assumptions almost any time we solve a problem. State those assumptions so you can check the validity of your answers.)

Back to the original problem. We have 8.00×10^{-6} mol N_2 and 3.68×10^{-6} mol H_2O. Assuming ρ_{H_2O} = 1 (you can look up the exact value at 40°C if you like like), we can write

$$c_{O_2} = \frac{3.68 \times 10^{-6} \text{ mol } O_2}{(1 \text{ mol } H_2O)(0.018 \text{ kg } H_2O/1 \text{ mol } H_2O)(1 \text{ dm}^3 \text{ } H_2O/1 \text{ kg } H_2O)}$$

$$= 2.0 \times 10^{-4} \frac{\text{mol}}{\text{dm}^3 \text{ } H_2O}$$

$$= 2.0 \times 10^{-4} \frac{\text{mol}}{\text{dm}^3 \text{ soln}}$$

$$= \underline{\underline{2.0 \times 10^{-4} \text{ M}}}$$

$$c_{N_2} = \frac{8.00 \times 10^{-6} \text{ mol } N_2}{(1 \text{ mol } H_2O)(0.018 \text{ kg } H_2O/1 \text{ mol } H_2O)(1 \text{ dm}^3 \text{ } H_2O/1 \text{ kg } H_2O)}$$

$$= \underline{\underline{4.4 \times 10^{-4} \text{ M}}}$$

ACTIVITIES AND ACTIVITY COEFFICIENTS. The form of Eq. (6-1.2.13) and of Eq. (6-1.2.15) is so simple and easy to use we would like to retain that form even for nonideal solutions. We do this by introducing a term called *activity a*. The first thing we must do is redefine our standard state. Instead of using P = 1 atm let us use $P = P^0$, the normal vapor pressure of the material. Equation (6-1.2.6) then becomes for component A

$$\mu_A^v = \mu_A^{0,v}(T) + RT \ln \frac{P_A}{P_A^0} \qquad (6\text{-}1.4.2)$$

Since $\quad \mu_A^v = \mu_A^\ell$

we can write immediately

$$\mu_A^{\ell} = \mu_A^{0,v}(T) + RT \ln \frac{P_A}{P_A^0} \tag{6-1.4.3}$$

Defining the activity a_A

$$a_A = \frac{P_A}{P_A^0} \tag{6-1.4.4}$$

allows (6-1.4.3) to be written (remember $\mu_A^{0,v} = \mu_A^{0,\ell} \equiv \mu_A^0$)

$$\mu_A^{\ell} = \mu_A^0 + RT \ln a_A \tag{6-1.4.5}$$

Following the arguments used in deriving Eq. (6-1.2.12) you should be able to show (try it) that the free energy of mixing of a solution of n_A moles of A and n_B moles of B is

$$\Delta\bar{G}_{mix} = x_A\ RT \ln a_A + x_B\ RT \ln a_B \tag{6-1.4.6}$$

This equation is very important.

A little more discussion of a is in order. From the definition

$$a_A = \frac{P_A}{P_A^0} \tag{6-1.4.4}$$

you should conclude for an ideal solution [see Eq. (6-1.3.1)],

$$a_A(\text{ideal}) = x_A^{\ell} \tag{6-1.4.7}$$

For a very dilute solution we might expect the activity to approach the mole fraction. (Solutions usually become more nearly ideal as they become more dilute in solute.) To reflect that, we define an *activity coefficient*

$$a_A = \gamma_A\ x_A^{\ell} \tag{6-1.4.8}$$

If you like, you can view γ as a "fudge factor." It is included to allow the simple form of the free energy and entropy of mixing formulas to be retained. γ is not even a constant, as can be seen in Table 6-1.4.2.

The value of the procedure should be obvious. Once γ has been empirically determined for a material, then the simple equations can be used by replacing x_A by a_A ($= \gamma_A x_A$) in the appropriate places. Values of γ can be measured, as suggested by Eqs. (6-1.4.4) and (6-1.4.8), be measuring the vapor pressure of A as a function of concentration.

Table 6-1.4.2
Activities and Activity Coefficients of Solutions of Chloroform in Acetone (Assuming the Vapor Is an Ideal Gas)

$x_{acetone}$	$a_{acetone}$	$\gamma_{acetone}$
1.00	1.00	1.00
0.94	0.94	1.00
0.88	0.87	0.99
0.73	0.70	0.96
0.63	0.57	0.90
0.51	0.42	0.82

6-1.4.2 EXAMPLE

Find the mole fractions in the liquid and the activity and activity coefficient of water for a solution if 0.122 kg of a nonvolatile solute (molecular weight = 0.241 kg mol^{-1}) is dissolved in 0.920 kg water at 293 K. The vapor pressure of pure H_2O is 0.02308 atm, while that of the solution is 0.02239 atm.

Solution

$$x_{H_2O} = \frac{(0.920\text{ kg})/(0.018\text{ kg mol}^{-1})}{(0.920\text{ kg})/(0.018\text{ kg mol}^{-1}) + (0.122\text{ kg})/(0.241\text{ kg mol}^{-1})}$$

$$= \frac{51.1\text{ mol}}{51.1\text{ mol} + 0.506\text{ mol}}$$

$$= \underline{\underline{0.990}}$$

$$x_B = 1 - x_{H_2O} = \underline{\underline{0.010}}$$

$$a_{H_2O} = \frac{P_{H_2O}}{P^0_{H_2O}} = \frac{0.02239}{0.02308} = \underline{\underline{0.970}}$$

$$a_{H_2O} = \gamma_{H_2O}\, x_{H_2O}$$

$$\gamma_{H_2O} = \frac{a_{H_2O}}{x_{H_2O}} = \frac{0.970}{0.990} = \underline{\underline{0.980}}$$

INTERACTIONS IN NONIDEAL SOLUTIONS. Refer to Sec. 6-1.3 [specifically Eq. (6-1.3.7)] for the discussion of interactions in ideal solutions. First consider *negative deviation* from Raoult's law.

Negative Deviations from Raoult's Law. Assume that solute-solvent interactions are strong compared to solute-solute and solvent-solvent interactions (A.....B > A.....A and B.....B). What is ΔH_{AB}, relative to ΔH_{BB} and ΔH_{AA}? $\Delta H_{AB} > \Delta H_{BB}$ and ΔH_{AA}. Remember that $2\Delta H_{AB}$ is released and $\Delta H_{AA} + \Delta H_{BB}$ absorbed on solution for each A-A and B-B pair broken. Thus, hopefully, you will conclude that ΔH_{mix} will be negative: heat will be released on dissolution. This is in opposition to the ideal case in which $\Delta H_{mix} = 0$.

What about ΔS?

$$\Delta S = 2S_{AB} - S_{AA} - S_{BB} \tag{6-1.4.9}$$

If all interactions are equal, S_{AB} will be a maximum since the final solution will have no ordering. In other words, the ideal solution has a maximum disorder. Thus, $\Delta S_{ideal} = \Delta S_{max}$. If A.....B interactions are strong, there will be some ordering in the solution which lowers S_{AB} without changing S_{AA} and S_{BB}. Ordering of any type lowers ΔS_{mix}. We conclude $\Delta S_{mix} < \Delta S_{ideal}$.

What happens to ΔG? You know at constant T:

$$\Delta G_{mix} = \Delta H_{mix} - T\,\Delta S_{mix} \qquad \text{see Eq. (6-1.2.15)}$$

ΔH_{mix} is negative, tending to make ΔG_{mix} more negative than ideal. $T\,\Delta S_{mix}$ is less positive than $T\,\Delta S_{ideal}$, which tends to make $-T\,\Delta S_{mix}$ and ΔG_{mix} less negative. The final value of G_{mix} depends on the relative size of ΔH_{mix} and $T\,\Delta S_{mix}$. No general statement can be made about the direction of change expected for ΔG_{mix} for the case of strong A.....B interactions.

Finally, what happens to P_A? In the absence of a difference in interactions among A-A, A-B, and B-B, Raoult's law tells us that the vapor pressure of A is proportional to the mole fraction of A. If, however, the A.....B interactions are strong, A will not be as free to vaporize. Its interaction with B will tend to hold both in the solution. Thus

$$P_A < P_A(\text{Raoult}) = x_A P_A^0 \tag{6-1.4.10}$$

To summarize: *For strong A-B interactions*

$$\begin{aligned} &\Delta H_{mix} < 0 \\ &\Delta S_{mix} < \Delta S_{ideal} \\ &\Delta G_{mix} \quad \text{depends on system} \\ &P_A < P_{ideal} = x_A P_A^0 \end{aligned} \tag{6-1.4.11}$$

An example of such a system is given in Fig. 6-1.4.5. The vapor pressure diagram is in Fig. 6-1.4.3.

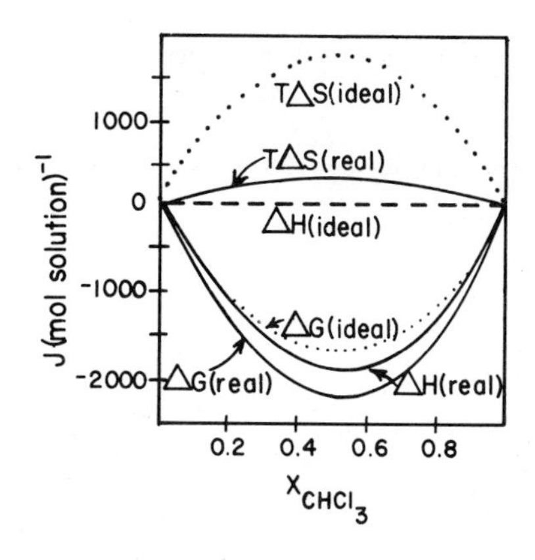

FIG. 6-1.4.5 Thermodynamic functions for 1 mol of solution of chloroform and acetone at 25°C, from *Physical Chemistry*, 3rd ed., G. M. Barrow, 1973, McGraw-Hill Book Co., with permission of McGraw-Hill Book Company.

Positive Deviations from Raoult's Law. Next, let us consider the case in which either solute-solute (B-B) or solvent-solvent (A-A) interactions are strong compared to solvent-solute (A-B). For purposes of description assume A-A is a stronger interaction than A-B or B-B. If that is so, $\Delta H_{AA} > \Delta H_{BB}$ and ΔH_{AB}. This makes $\Delta H_{mix} > 0$ (it takes more heat to break an A.....A interaction than we get back from an A.....B interaction). Since A will tend to stay associated with A there will be some ordering in the solution, so $S_{AB} < S_{AB}(\text{ideal})$. Examination of (6-1.4.9) should convince you then that $\Delta S_{mix} < \Delta S_{ideal}$.

You should now be able to describe ΔG_{mix} compared to ΔG_{ideal}. ΔH_{mix} is positive, which will tend to make $\Delta G_{mix} > 0$. $T\,\Delta S_{mix} < T\,\Delta S_{ideal}$ which will make ΔG_{mix} less negative than the ideal case. So $\Delta G_{mix} > \Delta G_{ideal}$ (i.e., less negative). This is shown in Fig. 6-1.4.6.

Finally we need to see how P_A is affected. If A associates with A more than with B (which will be true if either A-A or B-B is stronger than A-B), then A will be more loosely attached to the solution and will tend to vaporize more readily giving $P_A > P_A(\text{Raoult's})$

A summary of the effects if A-A or B-B is stronger than A-B:

$$\begin{aligned} &\Delta H_{mix} > 0 \\ &\Delta S_{mix} < \Delta S_{ideal} \\ &\Delta G_{mix} > \Delta G_{ideal} \\ &P_A > P_{ideal} = x_A P_A^0 \end{aligned} \qquad (6\text{-}1.4.12)$$

Figure 6-1.4.6 shows such a case.

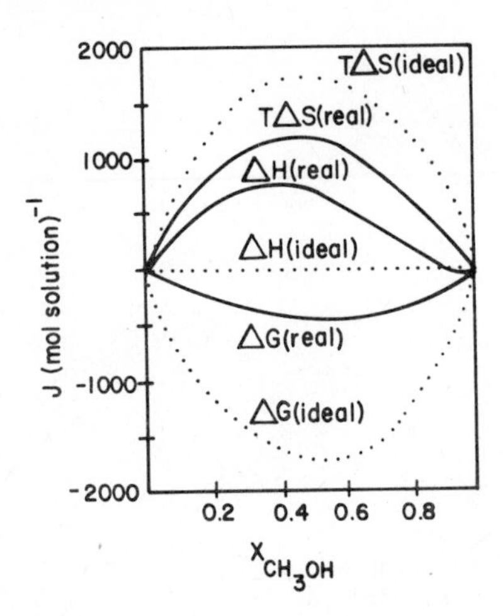

FIG. 6-1.4.6 Thermodynamic functions for solutions of carbon tetrachloride-methanol at 25°C, from *Physical Chemistry*, 3rd ed., G. M. Barrow, 1973, McGraw-Hill Book Co., with permission of McGraw-Hill Book Company.

Perhaps a brief physical description will show why the effects found in Figs. 6-1.4.5 and 6-1.4.6 are observed. In the case of acetone-chloroform (Fig. 6-1.4.5), we expect the following A-B interaction (hydrogen bonding, see Sec. 14-1).

$$Cl_3C - H \text{----} O = C(CH_3)_2$$

There is a strong interaction (a hydrogen bond) between the hydrogen (which carries a partial positive charge) on the chloroform with the oxygen (which carries a partial negative charge) on the acetone. This is a strong solvent-solute interaction which leads to negative deviation from Raoult's law.

In Fig. 6-1.4.6 we have a system in which nonpolar carbon tetrachloride and polar methanol are involved. The dominant interaction here is

$$H_3C - O(H) \text{----} H - O - CH_3$$

The hydrogen on the oxygen of one methanol carries a partial positive charge and interacts strongly with the partially negative charge of the oxygen on another methanol. This strong A-A (or B-B depending on your point of view) interaction leads to positive deviations from Raoult's law.

Immiscible Liquids. Very strong solvent-solvent or solute-solute interactions can lead to the situation in which the two liquids are not soluble in each other. The materials separate into two phases (layers). There is a very strong positive deviation from Raoult's law in this case. In fact, the vapor pressure of each material is just P^0. The total pressure in the limit of zero miscibility is $P_A^0 + P_B^0$: the liquids exhibit their normal vapor pressure as though the other liquid was not present.

This has practical consequences in steam distillation (can you describe how steam distillation works?). Also, a lqiuid spill (such as mercury) can still be hazardous even if covered with water or another immiscible liquid. The spilled liquid still exhibits its normal vapor pressure and can be dangerous if the vapor of the liquid is toxic.

6-1.5 VARIATION OF VAPOR PRESSURE WITH TOTAL PRESSURE

An interesting effect is the variation of the total pressure of a pure liquid under the pressure of an inert gas. In opposition to what one would suppose, the vapor pressure is greater in the presence of an inert gas than in its absence. Below a derivation is given for this phenomenon based on an article by N. O. Smith (see Sec. 6-1.8, Ref. 1).

Consider the closed system in Fig. 6-1.5.1 containing liquid A in equilibrium with its vapor and with an *inert, insoluble gas B*. At equilibrium, you know already that $\mu_A^\ell = \mu_A^v$. If an infinitesimal displacement from equilibrium occurs, $d\mu_A^\ell = d\mu_A^v$. Assume also that the vapor is ideal, so

$$\mu_A^v = \mu_A^0 + RT \ln P_A \tag{6-1.5.1}$$

A(vap) B(gas)
A(liq)

FIG. 6-1.5.1 A liquid in equilibrium with its vapor and an inert, insoluble gas.

But μ_A^v is a function of T and P and the mole fraction of A in the vapor; $\mu_A^v = \mu_A^v(T, P, x_A^v)$. We can write $d\mu_A^v$ using the chain rule:

$$d\mu_A^v = \left(\frac{\partial \mu_A^v}{\partial T}\right)_{P,x_A} dT + \left(\frac{\partial \mu_A^v}{\partial P}\right)_{T,x_A} dP + \left(\frac{\partial \mu_A^v}{\partial x_A}\right)_{T,P} dx_A^v \tag{6-1.5.2}$$

Also, $\mu_A^\ell = \mu_A^\ell(T, P)$ since $x_A^\ell = 1$.

$$d\mu_A^\ell = \left(\frac{\partial \mu_A^\ell}{\partial T}\right)_P dT + \left(\frac{\partial \mu_A^\ell}{\partial P}\right)_T dP \tag{6-1.5.3}$$

From Chap. 5 we can find

$$\left(\frac{\partial \mu_A^\ell}{\partial P}\right)_T = \bar{V}_A^\ell \tag{6-1.5.4}$$

and $$\left(\frac{\partial \mu_A^v}{\partial P}\right)_{T,x} = \bar{V}_A^v = \frac{RT}{P} \quad \text{if ideal}$$

Assume T is constant. Equation (6-1.5.3) and (6-1.5.4) can be combined at constant T to give

$$d\mu_A^\ell = V_A^\ell \, dP \tag{6-1.5.5}$$

Take the derivative of (6-1.5.1) with respect to x_A at constant T,P:

$$\left(\frac{\partial \mu_A^v}{\partial x_A^v}\right)_{T,P} = RT \left(\frac{\partial \ln P_A}{\partial x_A^v}\right)_{T,P} \tag{6-1.5.6}$$

But for an ideal system $P_A = x_A^v P$, so (6-1.5.6) reduces to (show it)

$$\left(\frac{\partial \mu_A^v}{\partial x_A^v}\right)_{T,P} = \frac{RT}{x_A^v} \tag{6-1.5.7}$$

Combine (6-1.5.2), (6-1.5.4) and (6-1.5.7) to yield (show it):

$$d\mu_A^v = \frac{RT}{P} dP + \frac{RT}{x_A^v} dx_A^v \tag{6-1.5.8}$$

Equating (6-1.5.5) and (6-1.5.8) gives ($d\mu_A^v = d\mu_A^\ell$):

$$\bar{V}_A^\ell \, dP = \frac{RT}{P} dP + \frac{RT}{x_A^v} dx_A^v$$

$$= RT \, d \ln P + RT \, d \ln x_A^v \tag{6-1.5.9}$$

But remember $P_A = x_A{}^V P$, so from that fact and (6-1.5.9),

$$\bar{V}_A{}^{\ell}\, dP = RT\, d\ln P_A = \frac{RT}{P_A}\, dP_A \tag{6-1.5.10}$$

Rearranging:

$$\frac{dP_A}{dP} = \frac{\bar{V}_A{}^{\ell} P_A}{RT} \quad \text{or} \quad d\ln P_A = \frac{\bar{V}_A{}^{\ell}\, dP}{RT} \tag{6-1.5.11}$$

You should see that an increase in P (the total pressure) leads to an increase in the vapor pressure.

If the inert gas is soluble, then (6-1.5.11) must be modified slightly to give

$$\frac{dP_A}{dP} = \frac{\bar{V}_A{}^{\ell} P_A}{RT} - \frac{P_A}{K_B} \tag{6-1.5.12}$$

where K_B is the Henry's law constant for B dissolved in A. The last term is usually negligible.

The article by Smith (Sec. 6-1.8, Ref. 1) presents the following data for water at 50°C.

Gas B	P (atm)	Water vapor (mg/liter)	
		Measured	Predicted
None	equil.	83	83
H_2	200	95	94
H_2	400	102	108
N_2	200	126	94
N_2	400	148	108

The chief causes of failure of the predicted values to match the measured values are probably due to assumptions made in Eq. (6-1.5.1) and to the assumption that $\bar{V}_A{}^V = RT/P$ in (6-1.5.4). Also, P_A should be replaced by the fugacity f_A defined in Chap. 5 to give more precise results.

An interesting problem that requires combinations of concepts from several chapters is to assume a Redlich-Kwong gas to find f_A and $\bar{V}_A{}^V$ then compare the results. Several modifications in the derivation would be necessary. If you can do that, you know you understand a considerable amount about gas properties and thermodynamics.

6-1.6 PROBLEMS

1. Assume you have a system of three phases (A, B, and C) with two substances (1 and 2) distributed among these phases. At equilibrium you observe that

   $$\mu_1(A) = 0.85\mu_2(B)$$

   What will be the relationship between $\mu_1(B)$ and $\mu_2(C)$?

2. Ten grams of a material whose molecular weight is 0.60 kg mol^{-1} is dissolved in 0.090 kg of water. Calculate ΔG_{mix}, ΔS_{mix}, and P_{H_2O} at 100°C assuming the solution behaves ideally. What is the activity coefficient if P_{H_2O} is observed to be 0.87 atm?

3. Assuming ideal behavior, find $\Delta\bar{V}_{mix}$, $\Delta\bar{H}_{mix}$, $\Delta\bar{G}_{mix}$, and $\Delta\bar{S}_{mix}$ when an ideal solution is formed from 0.61 mol of substance A and 1.33 mol of substance B.

4. The chemical potential of pure water is - 285.84 kJ mol^{-1}. What is the chemical potential of water in a solution containing 0.1 mol solute per 0.120 kg H_2O at 298.15 K and 1 atm total pressure?

5. What are the activity and activity coefficient of acetone and chloroform if in a solution of the two with x_{CHCl_3} = 0.6 the observed partial pressures are P_{CHCl_3} = 0.553 atm and $P_{acetone}$ = 0.261 atm at 330 K? ($P^0_{CHCl_3}$ = 0.586 atm and $P^0_{acetone}$ = 0.336 atm.)

6. For the problem in (5) determine ΔG_{mix} and compare with ΔG_{mix} (ideal). From ΔG_{mix} above can you tell whether this represents positive or negative deviation from Raoult's law. What does the vapor pressure data tell you about the direction of deviation from Raoult's law?

7. Two liquids (A and B) form an ideal solution. At 320 K the total vapor pressure of a solution I containing 3 mol A and 1 mol B is 400 mm Hg. Addition of 2 mol B results in a solution II that has a vapor pressure of 460 mm Hg.

   a. Find x_A and x_B for both solution I and solution II.
   b. Determine P^0_A and P^0_B.
   c. Predict the vapor pressure of a solution III that results from the addition of 3 mol B to solution II.
   d. Find ΔG_{mix} for solutions I and II.
   e. What is x_B in the *vapor* for solution I?

8. 100 g $H_2O(\ell)$ are placed in a steel flask (V = 5 dm^3) and pressurized with O_2 gas to a pressure of 2.5 × 10^3 atm. The Henry's law constant for O_2 in water at 293 K is 3.88 × 10^4 atm.

   a. Determine the mole fraction and concentration of O_2 in the water.

b. The normal vapor pressure of water at 293 K is 20 torr = 20 mm Hg. Determine the vapor pressure above the water for the solution.

c. If the observed vapor pressure of water is 18.36 mm Hg, give the activity coefficient of water in this solution.

9. A mixture of 0.035 kg of a nonpolar solute B (M_B = 0.086 kg mol^{-1} and P_B^0 = 300 mm Hg) and 0.095 kg of a polar solvent A (M_A = 0.037 kg mol^{-1} and P_A^0 = 400 mm Hg) is made. Assume ideal behavior.

a. Find the values of x_B and x_A.

b. Which material probably has the higher boiling point?

c. Find the entropy of mixing at 300 K.

d. Calculate the total vapor pressure above the solution.

6-1.7 SELF-TEST

1. Given the structure of acetone ($CH_3\text{-}\overset{\displaystyle O}{\overset{\|}{C}}\text{-}CH_3$) and carbon disulfide (S=C=S), discuss the relative strength of A-A, A-B, and B-B interactions and indicate whether positive or negative deviation from Raoult's law is expected.

2. Benzene and toluene form nearly ideal solutions. At 300 K, $P^0_{toluene}$ = 0.0422 atm and $P^0_{benzene}$ = 0.136 atm.

a. Compute the vapor pressure of each if the mole fraction of toluene is 0.45.

b. Find the mole fraction of benzene in the vapor.

3. At 275 K the vapor pressure of pure liquid A is 0.291 atm and that of B is 0.197 atm. If 3 mol of each are mixed, the total vapor pressure is found to be 0.221 atm and the vapor contains a mole fraction of A of 0.52. Assume the vapor behaves ideally (the liquid is nonideal), to find

a. a_A and a_B in the solution

b. γ_A and γ_B in the solution

c. ΔG_{mix} for the solution

6-1.8 ADDITIONAL READING

1.* N. O. Smith, *J. Chem. Educ.*, *40*:317, 1963.

6-2 COLLIGATIVE PROPERTIES: NONELECTROLYTES

OBJECTIVES

You shall be able to

1. Calculate each of the colligative properties (osmotic pressure, vapor pres-

sure of solution, freezing point depression, and boiling point elevation) for nonelectrolytes given appropriate data and equations

2. Use the Clausius-Clapeyron equation to find ΔH_{vap} or vapor pressures

6-2.1 INTRODUCTION

Colligative properties are those properties of solutions that depend on the number of particles (ions, atoms, or molecules) in the solution. There is no dependence on the nature of the particles. They can be charged or noncharged, small or large, etc. The physical basis for these colligative properties will be presented, then the equations necessary to calculate then will be derived.

Some colligative properties such as the osmotic pressure and vapor pressure lowering are very important for the determination of molecular weights of polymers and other macromolecules. Also, measurement of the colligative properties for electrolytes (materials that partially or wholly dissociate into ions in solution) gives a means of determining the amount of dissociation. You will learn more about that in Sec. 6-3.

Keep in mind that we are considering nonvolatile solutes. Volatile solutes were considered in the previous section. Equations will be derived only for ideal solutions. Comments on nonideality will be made where appropriate.

6-2.2 VAPOR PRESSURE LOWERING

From the previous section we can write that the vapor pressure of a solvent containing a *nonvolatile* solute as (if the solution is ideal)

$$P_A = x_A P_A^0 \qquad (6\text{-}1.2.5)$$

or, since $x_A = 1 - x_B$

$$P_A = (1 - x_B)P_A^0 = P_A^0 - x_B P_A^0$$

whence,

$$\frac{P_A^0 - P_A}{P_A^0} = x_B \qquad (6\text{-}2.2.1)$$

Finally, since $P_A^0 - P_A = \Delta P$ = vapor pressure lowering,

$$\boxed{\frac{\Delta P}{P_A^0} = x_B} \qquad (6\text{-}2.2.2)$$

This shows that the vapor pressure lowering depends only on the mole fraction of the nonvolatile solute. No dependence is shown on the nature of the solute (we stipulate a nonelectrolyte; electrolytes will be considered in Sec. 6-3). This makes it a *colligative property*. Molecular weights can be found by this method. Before continuing see if you can relate the molecular weight of the solute B to ΔP. It is done below so you can check your results.

$$x_B = \frac{\text{mol B}}{\text{mol B} + \text{mol A}} = \frac{n_B}{n_B + n_A} = \frac{m_B/M_B}{m_B/M_B + m_A/M_A} \qquad (6\text{-}2.2.3)$$

where m_A and m_B are the weights of A and B, respectively, and M_A and M_B are the molecular weights of A and B, respectively. By equating (6-2.2.3) and (6-2.2.2) and simplifying considerably (check the result)

$$M_B = \frac{M_A m_B}{m_A} \frac{P_A^0 - \Delta P}{\Delta P} = \frac{M_A m_B}{m_A} \frac{P_A^0 - (P_A^0 - P_A)}{\Delta P}$$

$$\boxed{M_B = \frac{M_A m_B}{m_A} \frac{P_A}{\Delta P}} \qquad (6\text{-}2.2.4)$$

By measuring the vapor pressure of a solution of a known weight of solute in a known weight of a solvent we can find the molecular weight of the solute (of course, we must know P_A^0). Obviously, the equation can be used to predict P_A if M_B is known.

The following example gives an introduction to a useful technique in treating colligative properties. Work through it carefully. More will be said about the vapor pressure method of determining molecular weights when polymers are discussed in Chap. 12.

6-2.2.1 EXAMPLE

From the following data on mannitol solutions in water predict the molecular weight of mannitol (20°C, P^0 = 0.02304 atm).

kg Mannitol per kg H_2O	$\Delta P \times 10^5$ (atm)	$P \times 10^2$ (atm)	$\frac{M_A}{m_A}$ (mol^{-1})	$M_B = \frac{M_A m_B}{m_A} \frac{P_A}{\Delta P}$ (kg mol^{-1})
0.01792	4.04	2.30	0.0180	0.1836
0.03600	8.08	2.30	0.0180	0.1845
0.05394	12.1	2.292	0.0180	0.1839
0.08993	20.21	2.284	0.0180	0.1829
0.1263	28.45	2.276	0.0180	0.1819

Solution

Figure 6-2.2.1 is a plot of the result. There are several important points to consider.

1. The technique used here is to plot the colligative property of interest (or some function of it) then extrapolate the plot to zero concentration to get a limiting value of that property. This procedure allows us to use ideal equations for real solutions. The solution should approximate ideality as the concentration goes to zero, since solute-solute interactions should disappear as the solution becomes very dilute. Thus the extrapolation to zero concentration should yield a valid result.

2. Before the extrapolation can be done, some idea of the accuracy of the data must be available. For the two lowest concentrations the error expected in ΔP is in the third digit. Since ΔP has the largest error, ignore the error contribution from the other terms. From Chap. 1, in which error analysis is discussed, we can write that

$$\frac{\Delta M_B}{M_B} \approx \frac{\Delta\ (\Delta P)}{\Delta P} \quad \text{or} \quad \Delta M_B = \frac{\Delta\ (\Delta P)}{\Delta P} M_B$$

For $\Delta P = (4.04 \pm 0.01) \times 10^{-5}$:

$$\Delta M_B = \frac{0.01 \times 10^{-5}}{4.04 \times 10^{-5}} 0.184 = 0.46 \times 10^{-3}$$

and for $\Delta P = (8.08 \pm 0.01) \times 10^{-5}$:

$$\Delta M_B = \frac{0.01 \times 10^{-5}}{8.08 \times 10^{-5}} 0.184 = 0.23 \times 10^{-3}$$

For the other values, errors from other variables would have to be included. "Error bars" have been added to Fig. 6-2.2.1 to the lowest two concentration values. The extrapolation is a "best guess" using these error limits and the fact that we expect M_B to approach a limiting value at low concentration. That is, the M_B vs. m_B (or C_B) curve should approach M_B(ideal) as M_B goes to zero; the curve should "level out" (become horizontal).

3. The limiting value obtained is 0.184 kg mol^{-1}. The true value is 0.18211 kg mol^{-1} (182.11 g mol^{-1}). This gives an error of 1% which is not bad for this particular technique.

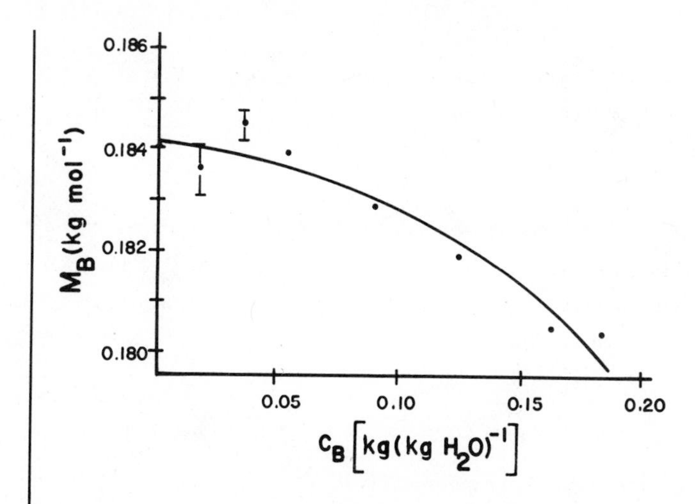

FIG. 6-2.2.1 A plot of the calculated molecular weight of mannitol vs. mannitol concentration at 20°C from vapor pressure data.

6-2.3 BOILING POINT ELEVATION

THE CLAUSIUS-CLAPEYRON EQUATION. Before a discussion can be given of the effect of solutes on the boiling point of a solution we need to know how the vapor pressure of a pure material varies with temperature. You get the opportunity to refresh your memory further on the use of thermodynamic variables in this section. Consider the transition of pure substance A from liquid to vapor:

$$A_\ell \rightarrow A_v$$

$$\Delta\bar{G} = \mu_v - \mu_\ell \tag{6-2.3.1}$$

Suppose we start with the system at equilibrium (meaning $\Delta\bar{G} = 0$). Now allow the pressure on the system to change by dP. A temperature change of dT must occur to reestablish equilibrium. We have, then, the following initial and final relations for dG (note that $dG_\ell = dG_v$ since $\Delta G = 0$ before and after the change in T and P):

$$d\bar{G}_\ell = -\bar{S}_\ell\, dT + \bar{V}_\ell\, dP$$

$$d\bar{G}_v = -\bar{S}_v\, dT + \bar{V}_v\, dP$$

so

$$-\bar{S}_v\, dT + \bar{V}_v\, dP = -\bar{S}_\ell\, dT + \bar{V}_\ell\, dP$$

$$\frac{dP}{dT} = \frac{\bar{S}_v - \bar{S}_\ell}{\bar{V}_v - \bar{V}_\ell} = \frac{\Delta\bar{S}_{vap}}{\Delta\bar{V}_{vap}} \tag{6-2.3.2}$$

where the subscript *vap* indicates vaporization.

Since $\Delta G = 0$ and T is essentially constant, you should recall that $\Delta S = \Delta H/T$. Equation (6-2.3.2) becomes upon substitution

$$\frac{dP}{dT} = \frac{\Delta\bar{H}_{vap}}{T\,\Delta\bar{V}_{vap}} \tag{6-2.3.3}$$

This is known as the *Clapeyron equation*. More discussion about it will be found in Chap. 14. By making several assumptions we can reduce (6-2.3.3) to a simpler form (again more detail will be given in Chap. 14, Sec. 14-1).

Assumptions:

1. $\bar{V}_\ell \ll \bar{V}_v$ thus $\Delta\bar{V}_{vap} \approx \bar{V}_v$

2. $\bar{V}_v = RT/P$ the vapor behaves ideally

Applying these to Eq. (6-2.3.3) yields (you should show the intermediate steps)

$$\frac{dP}{dT} = \frac{\Delta\bar{H}_{vap}\,P}{RT^2} \tag{6-2.3.4}$$

which may be written (again, prove it)

$$\frac{d\ln P}{dT} = \frac{\Delta\bar{H}_{vap}}{RT^2} \quad \text{or} \quad \frac{d\ln P}{d(1/T)} = -\frac{\Delta H_{vap}}{R} \tag{6-2.3.5}$$

Equation (6-2.3.5) is known as the *Clausius-Clapeyron equation.*

We can proceed with our discussion of boiling point elevation. We have succeeded in relating a change in vapor pressure to a change in T.

6-2.3.1 EXAMPLE

Calculate the heat of vaporization of butane at 277 K and 323 K from the following data from the Chemical Rubber Company *Handbook of Chemistry and Physics* (see Sec. 6-2.8, Ref. 1).

Vapor Pressure 10^{-4} (N m^{-2})	T (K)	ln P	1/T (K^{-1}) × 10^3
9.90	272.04	11.50	3.676
12.15	277.59	11.71	3.602
14.90	283.15	11.91	3.532
18.13	288.71	12.11	3.464
21.76	294.26	12.29	3.398
25.87	299.82	12.46	3.335
30.67	305.37	12.63	3.275
35.97	310.93	12.79	3.216
41.85	316.48	12.94	3.160
48.30	322.04	13.10	3.105
56.06	327.59	13.24	3.053
63.80	333.15	13.37	3.002

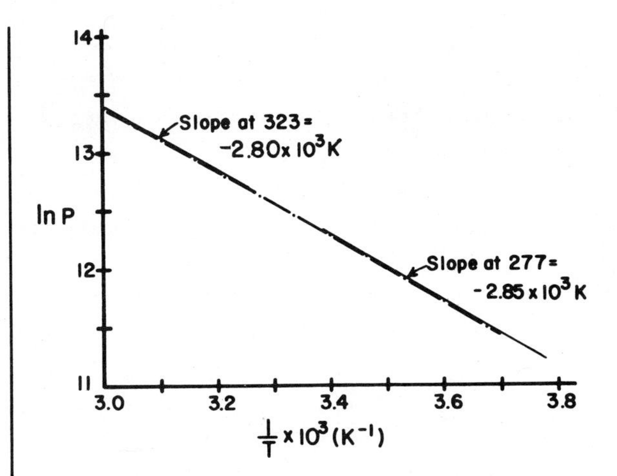

FIG. 6-2.3.1 Plot of ln P vs. 1/T for butane.

Solution

Equation (6-2.3.5) tells us we need a plot of ln P vs. 1/T. The slope of such a plot will be $-\Delta H_{vap}/R$. Figure 6-2.3.1 is such a plot. Take the slope at 323 K and 277 K (the solid lines in the figure).

At 323 K:

$$\text{Slope} = -2.80 \times 10^3 \text{ K}$$

$$\Delta\bar{H}_{vap}(323) = -R \times \text{slope}$$
$$= (8.314 \text{ J mol}^{-1} \text{ K}^{-1})(2.62 \times 10^3 \text{ K})$$
$$= \underline{\underline{2.33 \times 10^4 \text{ J mol}^{-1}}} \qquad (\text{CRC value} = 2.43 \times 10^4 \text{ J mol}^{-1})$$

At 277 K:

$$\text{Slope} = -2.85 \times 10^3 \text{ K}$$

$$\Delta\bar{H}_{vap}(277) = -R \times \text{slope}$$
$$= (8.31)(2.85 \times 10^3) \text{ J mol}^{-1}$$
$$= \underline{\underline{2.37 \times 10^4 \text{ J mol}^{-1}}} \qquad (\text{CRC value} = 2.43 \times 10^4 \text{ J mol}^{-1})$$

Note that the graphical values are 2 to 4% too low. There is some error in graphically determining the slope from Fig. 6-2.3.1. Another error is in the assumption that the vapor behaves ideally. For materials with lower vapor pressures the ideal behavior assumption is more likely to hold.

BOILING POINT ELEVATION. Look at Fig. 6-2.3.2 which shows the vapor pressure behavior of a pure liquid and a solution as a function of temperature. Note that the vapor pressure of the solution is less than that of the pure material, as was shown in the previous section.

In Fig. 6-2.3.2 the symbols have the following meanings: P_{ext} = ambient pressure, T_{bp}(solvent), and T_{bp}(solution) = normal boiling point of solvent and solution, respectively, as a pressure of P_{ext}. dP(A) = vapor pressure lowering due to addition of dx_B mol of B to a solution containing 1 mol of solvent A. dP(T) is the change in the vapor pressure of the solution due to a temperature change from T_{bp}(solvent) to T_{bp}(solution).

You can look at the process as follows: Adding dx_B mol of solute B to a solution of 1 mol of A lowers the vapor pressure by dP(A). To get the *solution* to vaporize we must raise the temperature by $dT_b = T_{bp}(\text{solution}) - T_{bp}(\text{solvent})$ which increases the vapor pressure by dP(T) = dP(A).

All we need to do to find dT_b is to find equations for dP(T) and dP(A) and equate the two.

From Eq. (6-1.2.5) we know for an ideal solution with mole fraction x_A (= 1 - x_B) the vapor pressure is

$$P_A = x_A P_A^0 \qquad (6\text{-}1.2.5)$$

The vapor pressure lowering due to dx_B moles of B is

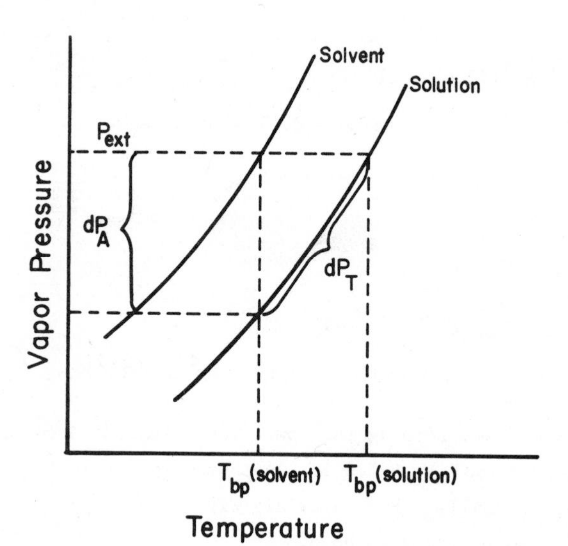

FIG. 6-2.3.2 The vapor pressure of solvent and solution as a function of temperature.

$$\begin{aligned} dP_A &= P_A^0 \, dx_A \\ &= P_A^0 \, d(1 - x_B) \\ &= -P_A^0 \, dx_B \end{aligned} \qquad (6\text{-}2.3.6)$$

For small changes in P_A,

$$\frac{dP_A}{P_A^0} \approx \frac{dP_A}{P_A} = d(\ln P_A) = -dx_B \qquad (6\text{-}2.3.7)$$

Thus, dx_B moles of solute added to 1 mol of solvent lowers the vapor pressure as follows:

$$d \ln P_A = -dx_B \qquad (6\text{-}2.3.8)$$

The effect of a temperature change was derived previously and found to be, at T_{bp} see Eq. (6-2.3.5) ,

$$d \ln P_A = \frac{\Delta\bar{H}_{vap}}{RT_{bp}^2} \, dT_b \qquad (6\text{-}2.3.9)$$

We want the two effects to sum to zero, so we add these last two equations:

$$\frac{\Delta\bar{H}_{vap}}{RT_{bp}^2} \, dT_b - dx_B = 0 \qquad (6\text{-}2.3.10)$$

or

$$dT_b = \frac{RT_{bp}^2}{\Delta\bar{H}_{vap}} \, dx_B \qquad (6\text{-}2.3.11)$$

We need to make a few assumptions to simplify (6-2.3.11). First, assume only small temperature changes (so that $T_{bp} \approx$ constant); second, assume for these small changes $\Delta\bar{H}_{vap}$ = constant.

What is the effect of adding B to change the solution from $x_B = 0$ to $x_B = x_B$? T will change from T_{bp}(solvent) to T_{bp}(solution) so

$$\int_{T_{bp}(\text{solvent})}^{T_{bp}(\text{solution})} dT_B = T_{bp} = \int_0^{x_B} \frac{RT_{bp}^2}{\Delta\bar{H}_{vap}} \, dx_B$$

$$\approx \frac{RT_{bp}^2}{\Delta\bar{H}_{vap}} \int_0^{x_B} dx_B \qquad (6\text{-}2.3.12)$$

The last step can be carried out only because we assumed T_{bp} constant and $\Delta\bar{H}_{vap} \approx$

constant. Finishing the integral gives

$$T_{bp} \approx \frac{RT_{bp}^2}{\Delta\bar{H}_{vap}} x_B \qquad (6\text{-}2.3.13)$$

or

$$\boxed{\Delta T_{bp} = K'_{bp} x_B} \qquad (6\text{-}2.3.14)$$

where $RT_{bp}^2/\Delta\bar{H}_{vap}$ has been taken as a constant K'_{bp}.

Usually Eq. (6-2.3.14) is used with molality, which is defined as moles of solute per kilogram solvent. Let m = moles of solute and n_A = moles solvent in 1000 g solvent. Then

$$x_B = \frac{m}{n_A + m} \approx \frac{m}{n_A} \qquad m \ll n_A$$

This, when substituted into Eq. (6-2.3.14), gives

$$\Delta T_{bp} = K'_{bp} x_B = K'_{bp} \frac{m}{n_A} = K_{bp} m \qquad (6\text{-}2.3.15)$$

The constnat K_{bp} is called the *ebullioscopic constant* or the *boiling point elevation constant*.

Table 6-2.3.1 contains some values observed from measurements of boiling point elevations. K_{bp} can be calculated from

$$\boxed{K_{bp} = \frac{RT_{bp}^2}{\Delta\bar{H}_{vap} n_A}} \qquad (6\text{-}2.3.16)$$

This is accurate only for dilute solutions or those that behave almost ideally. For nonideal cases we can apply procedures similar to those found in Example 6-2.2.1 in which extrapolation of results to zero concentration was carried out to get an estimate of the ideal values.

6-2.3.2 EXAMPLE

1.00 g of a sample is dissolved in 0.100 kg chloroform. This chloroform solution boils at 335.25 K at 1 atm pressure. Determine the molecular weight of the solute.

Solution

Apply Eq. (6-2.3.15) to find the molality of the solution.

Table 6-2.3.1
Values of Ebullioscopic Constants for Various Solvents (Based on Molalities)[a]

Solvent	T_{bp} (K)	K_B (K m^{-1})
Acetic acid	391.65	3.07
Acetone	329.4	1.71
Aniline	457.47	3.52
Benzene	353.3	2.53
Bromobenzene	429.0	6.26
Carbon tetrachloride	349.90	5.03
Chloroform	334.4	3.63
Ethanol	351.7	1.22
Methanol	338.11	0.83
Nitrobenzene	484.0	5.24
Toluene	383.8	3.33
Water	373.15	0.512

[a] From the CRC *Handbook of Chemistry and Physics* (Sec. 6-2.8, Ref. 1).

$$m = \frac{\Delta T_{bp}}{K_{bp}} = \frac{335.25\ \text{K} - 334.4\ \text{K}}{3.63\ \text{K}\ m^{-1}}$$

$$= \underline{\underline{0.234\ m}} = 0.234\ \text{mol solute (kg solvent)}^{-1}$$

Next, find the weight of solute per kilogram solvent.

$$\frac{0.001\ \text{kg solute}}{0.1\ \text{kg solvent}} = \underline{\underline{0.0100\ \text{kg solute (kg solvent)}^{-1}}}$$

Equate these two results.

$$0.0100\ \text{kg solute (kg solvent)}^{-1} = 0.234\ \text{mol solute (kg solvent)}^{-1}$$

Then $M_B = \dfrac{0.010\ \text{kg}}{0.234\ \text{mol}} = \underline{\underline{0.0427\ \text{kg mol}^{-1}}}$ (42.7 g mol^{-1})

6-2.4 FREEZING POINT DEPRESSION

In general we observe a depression of the freezing point of a solvent when a solute is added. You have made use of this phenomenon in the winter if you have ever added salt to thaw the ice on a sidewalk or driveway. We now investigate this phenomenon a little more closely and derive a quantitative expression to calculate such a change in the freezing point.

It is possible to follow arguments similar to those in Sec. 6-2.3 on the boiling point elevation. However, it will be more instructive to use a thermodynamic treatment to show that there is more than one way to approach a given problem.

Let us assume we have a solvent with freezing point T_f and a heat of fusion ΔH_f. If we add x_B (= 1 - x_A) moles of a solute per mole of solution, what happens to the freezing point? The freezing point changes in such a manner to make $\mu_A^s = \mu_A^\ell$, the condition for equilibrium. If the mole fraction of solvent is x_A you know from Eq. (6-1.2.8)

$$\mu_A^\ell = \mu_A^{0,\ell} + RT \ln x_A \tag{6-2.4.1}$$

And, at equilibrium $\mu_A^s = \mu_A^\ell$, so

$$\mu_A^s = \mu_A^{0,\ell} + RT \ln x_A \tag{6-2.4.2}$$

Remember at this point that μ_A is the partial molar free energy. From Eq. (5-2.3.6) it should be evident that

$$\frac{\partial(\bar{G}/T)}{\partial T} = \frac{\partial(\mu/T)}{\partial T} = -\frac{\bar{H}}{T^2} \tag{6-2.4.3}$$

Rearrange (6-2.4.2) to

$$\frac{\mu_A^s - \mu_A^{0,\ell}}{RT} = \ln x_A \tag{6-2.4.4}$$

and differentiate, $(\partial/\partial T)$, to give (prove it)

$$\frac{\bar{H}_A^{0,\ell} - \bar{H}_A^s}{RT^2} = \frac{d \ln x_A}{dT} \tag{6-2.4.5}$$

The numerator on the left side is just $\Delta\bar{H}_f$ (the heat of fusion):

$$\frac{\Delta\bar{H}_f}{RT^2} = \frac{d \ln x_A}{dT} \tag{6-2.4.6}$$

Integrate this equation from pure A (x_A = 1) to x_A.

$$\int_{T_f^0}^{T_f} \frac{\Delta\bar{H}_f}{RT^2}\, dT = \int_1^{x_A} d \ln x_A \tag{6-2.4.7}$$

If $\Delta\bar{H}_f$ is constant over small T changes, Eq. (6-2.4.7) can be evaluated immediately:

$$\frac{\Delta\bar{H}_f}{R}\int_{T_f^0}^{T_f} \frac{dT}{T^2} = -\frac{\Delta\bar{H}_f}{R}\left(\frac{1}{T_f} - \frac{1}{T_f^0}\right) = \ln x_A \tag{6-2.4.8}$$

or, on rearrangement,

$$\frac{\Delta \bar{H}_f}{R}\left(\frac{T_f - T_f^0}{T_f T_f^0}\right) = \ln x_A = \ln (1 - x_B) \tag{6-2.4.9}$$

Let us define $T_f^0 - T_f = \Delta T_f$, the *freezing point depression*. If ΔT_f is small, $T_f T_f^0 \approx (T_f^0)^2$. Then

$$\frac{\Delta \bar{H}_f \, \Delta T_f}{R(T_f^0)^2} = -\ln(1 - x_B) \tag{6-2.4.10}$$

If $\ln(1 - x_B)$ is expanded as a power series we obtain

$$-\ln (1 - x_B) = x_B + \frac{1}{2}x_B^2 + \frac{1}{3}x_B^3 + \cdots$$

When x_B is small, higher order terms can be neglected and the ln approximated by ($x_B \ll 1$)

$$-\ln (1 - x_B) \approx x_B \tag{6-2.4.11}$$

This allows Eq. (6-2.4.10) to be written

$$\frac{\Delta \bar{H}_f \, \Delta T_f}{R(T_f^0)^2} \approx x_B \tag{6-2.4.12}$$

or

$$\Delta T_f \approx \frac{R(T_f^0)^2}{\Delta \bar{H}_f} x_B \tag{6-2.4.13}$$

Applying the procedure used to obtain Eq. (6-2.3.15) from Eq. (6-2.3.14) we can write (you should supply the missing steps)

$$\boxed{\Delta T_f = \frac{R(T_f^0)^2}{\Delta \bar{H}_f \, n_A} m = K_f \, m} \tag{6-2.4.14}$$

where K_f is the *molal freezing point depression constant*, and m is the molality of the solute in the solution [moles solute (kg solvent)$^{-1}$].

It should be obvious that these equations are valid only for very low molalities. Some values of K_f are given in Table 6-2.4.1.

6-2.4.1 EXAMPLE

What weight of methanol must be added to 0.100 kg of water to lower water's freezing point by 0.35 K?

Table 6-2.4.1
Values of K_f for Various Solvents

Solvent	T_f (K)	K_f (K m^{-1})
Acetic acid	289.9	3.90
Benzene	278.7	5.12
Bromoform	281.0	14.4
Camphor	451.6	37.7
Cyclohexane	279.7	20.0
Naphthalene	353.4	6.9
Water	273.2	1.86

Solution

Find the molality. From Eq. (6-2.4.14) and data in Table 6-2.4.1,

$$m = \frac{\Delta T_f}{K_f} = \frac{0.35 \text{ K}}{1.86 \text{ K m}^{-1}} = 0.19 \text{ m}$$

$$= \underline{\underline{0.19 \text{ mol methanol (kg } H_2O)^{-1}}}$$

You should be able to find the weight of methanol in 0.100 kg water.

$$\frac{0.19 \text{ mol } CH_3OH}{1 \text{ kg } H_2O} \frac{32.04 \text{ g } CH_3OH}{1 \text{ mol } CH_3OH} 0.100 \text{ kg } H_2O = \underline{\underline{0.61 \text{ g } CH_3OH}}$$

6-2.5 OSMOTIC PRESSURE

Osmotic pressure is a phenomenon observed when a solution is separated from a more dilute solution or pure solvent by a membrane that is permeable to the solvent but not to the solute. Solvent will tend to move from the pure solvent into the solution due to a difference in chemical potential between the two. Solvent will continue to be transferred to the solution until the concentrations are equalized or until sufficient pressure has been developed on the solution side to balance the chemical potentials. This will be investigated below.

Think of the ways in which osmotic processes are important. Within the body there are many membranes which selectively pass some molecules while rejecting others. Root tips in plants selectively allow substances to pass. In each of these cases, processes other than osmosis are operable, but osmosis is also involved. The osmotic effect can be reversed to yield fresh water from mineralized water in a process called reverse osmosis.

Figure 6-2.5.1 is a schematic representation of a cell in which osmosis can take place. Consider the two effects represented in this figure. Solvent will tend to flow from pure solvent to solution since there is a potential difference between the two sides (assume T = constant):

$$\mu_A(\text{solution}) = \mu_A^0 + RT \ln x_A$$

$$\mu_A(\text{solvent}) = \mu_A^0$$

$$\Delta\mu_A = RT \ln x_A = \Delta\bar{G}_A (x_A) \tag{6-2.5.1}$$

$\Delta\bar{G}_A (x_A)$ means the free energy difference between solvent and solution of mole fraction x_A. The question we want to answer is: What pressure is required on the solution just to counteract the chemical potential difference so that there is no net solvent flow?

The effect of pressure on the solution is given by Eq. (5-1.5.12):

$$\left(\frac{\partial \bar{G}_A}{\partial P}\right)_{T,n_j} = \bar{V}_A \tag{5-1.5.12}$$

or at constant T and n_j this may be written as

$$d\bar{G}_A = \bar{V}_A \, dP \tag{6-2.5.2}$$

If the molar volume of A is assumed to be constant, as it will be for small changes in x_A, we can integrate this last equation to get the free energy change due to applying pressure to the solution side.

$$\Delta\bar{G}_A (P) = \int_{P_1=}^{P_2} \bar{V}_A \, dP = \bar{V}_A (P_2 - P_1) \tag{6-2.5.3}$$

At equilibrium $\Delta G_{\text{total}} = 0$ so we can write

$$\Delta\bar{G}_A (x_A) + \Delta\bar{G}_A (P) = 0 \tag{6-2.5.4}$$

or

$$\bar{V}_A (P_2 - P_1) = -RT \ln x_A \tag{6-2.5.5}$$

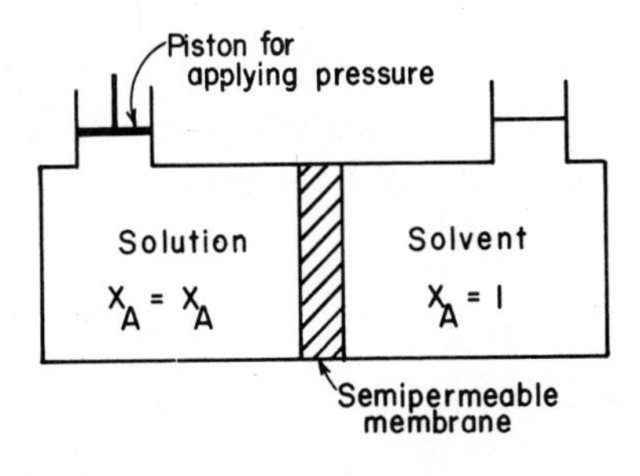

FIG. 6-2.5.1 A representation of a cell in which osmosis can take place.

Define $(P_2 - P_1)$, the pressure required to stop solvent flow into the solution, as Π, *the osmotic pressure.*

$$\Pi\bar{V}_A = -RT \ln x_A = -RT \ln (1 - x_B) \qquad (6\text{-}2.5.6)$$

Applying the approximation given in Eq. (6-2.4.11) gives the result

$$\boxed{\Pi\bar{V}_A = RT\, x_B} \qquad (6\text{-}2.5.7)$$

for $x_B << 1$. This is the equation we will need. Note that $\bar{V}_A$ is the *molar volume* of the solvent. It relates the pressure necessary to prevent solvent flow through a membrane into a solution with a mole fraction of solute of x_B.

Equation (6-2.5.7) will now be rearranged into a more useful form. First, recall for small x_B that $n_B << n_A$, so

$$x_B = \frac{n_B}{n_A + n_B} \approx \frac{n_B}{n_A} \qquad (6\text{-}2.5.8)$$

This allows Eq. (6-2.5.7) to be written

$$\Pi\bar{V}_A = RT\frac{n_B}{n_A}$$

or

$$\Pi n_A \bar{V}_A = n_B RT \qquad (6\text{-}2.5.9)$$

But $n_A\bar{V}_A = V_A$ = the volume occupied by n_A mol of solvent.

$$\Pi = \frac{n_B}{V_A} RT \qquad (6\text{-}2.5.10)$$

Finally, for dilute solutions, $V_A \approx$ volume of the solution. This gives

$$\boxed{\Pi = \frac{n_B}{V} RT = MRT} \qquad (6\text{-}2.5.11)$$

where M is the molarity of the solution. Another way of expressing Eq. (6-2.5.7) or Eq. (6-2.5.11) is sometimes convenient. Equation (6-2.5.11) may be written

$$\Pi = MRT = \frac{n_B}{V} RT$$

$$= \frac{m_B/M_B}{V} RT \qquad (6\text{-}2.5.12)$$

where m_B = mass of solute dissolved in a volume V of solution and M_B is the molecular weight of B. Depending on how data are presented, either Eq. (6-2.5.11) or Eq.

(6-2.5.12) may be used to find Π, M, M_B, etc.

You should keep in mind that the approximations made from Eq. (6-2.5.2) through Eq. (6-2.5.11) lead to less and less accuracy. The final equations are good only for dilute solutions. However, we can overcome this difficulty in many cases by proper data handling. Example 6-2.5.1 uses the technique first introduced in Example 6-2.2.1. Review that example if necessary.

6-2.5.1 EXAMPLE

Given the following data for sucrose solutions, find the molecular weight M_B of the solute (T = 20°C).

$\frac{m_B}{V}\left(\frac{g}{dm^3}\right)$	Π (atm)	$\frac{m_B/V}{\Pi}\left(\frac{g\ dm^{-3}}{atm}\right)$
33.5	2.59	13.0
65.7	5.06	13.0
96.5	7.61	12.7
155	12.75	12.2
209	18.13	11.5
259	23.72	10.9

Solution

In the last column the quantity $(m_B/V)/\Pi$ in Eq. (6-2.5.12) is tabulated. To get a limiting value for M_B, take

$$\lim_{\frac{m_B}{V} \to 0} \frac{m_B/V}{\Pi}$$

You should see from Eq. (6-2.5.12) that $M_B = (RT/\Pi)(m_B/V)$. Thus, if we get a limiting value of $(m_B/V)/\Pi$ we can calculate a limiting value of M_B. The plot in Fig. 6-2.5.2 allows the determination of the limit. The limit is about 13.0 g dm^{-3} atm^{-1}. Thus

$$M_B = RT\left(\frac{m_B/V}{\Pi}\right)_0$$

$$= 0.0821\ \frac{dm^3\ atm}{mol\ K}\ (293.15\ K)(13.0\ g\ dm^{-3}\ atm^{-1})$$

$$= \underline{\underline{313\ g\ mol^{-1}}} = \underline{\underline{0.313\ kg\ mol^{-1}}}$$

The true value is 0.342 kg mol^{-1}. The error is approximately 8% which is not very good. To get a better value more data are needed at lower concentrations to aid in the extrapolation process. (If the last point is ignored and the

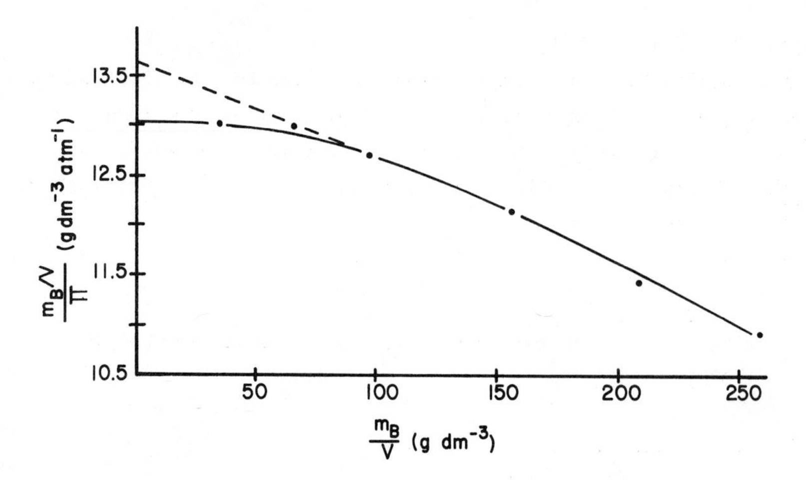

FIG. 6-2.5.2 Plot of $(m_B/V)/\Pi$ vs. m_B/V to obtain a limiting molecular weight of the solute.

dotted extrapolation is made, then a molecular weight of 0.328 kg mol^{-1} is obtained, a 4% error. However, we would expect the slope of the curve to approach zero as m_B approaches zero and the solution becomes more nearly ideal. Thus the curve drawn is probably the best curve we can draw with the given data.)

More will be said about osmotic pressure measurements in the chapter on colloids and polymers. Osmotic pressure measurements are used extensively to determine the molecular weights of macromolecules.

6-2.6 PROBLEMS

1. 10 g of sucrose (M_B = 0.342 kg mol^{-1}) is dissolved in 350 ml of H_2O at 298 K (ρ_{H_2O} = 0.997).
   a. Find x_{H_2O} and P_{H_2O} ($P^0_{H_2O}$ = 0.03126 atm).
   b. Assume the observed vapor pressure is 0.0296 atm for this solution; find γ_{H_2O}.
   c. Calculate Π, T_{bp}, and T_{fp} for this solution.
2. Assuming ideal behavior at 293 K for a solution of diethyl ether and a nonvolatile solute, find the molecular weight of the solute if 0.0100 kg dissolved in 0.100 kg ether lowers the vapor pressure from 0.582 atm to 0.560 atm.
3. An aqueous solution contains 3% urea by weight (M = 0.060 kg mol^{-1}) and 4% glycerol by weight (M = 0.092 kg mol^{-1}). Find the freezing point and boiling point for this solution.

4. A solution of a nonvolatile solute in benzene boils (at 1 atm) at 354.7 K. If $P^0_{benzene}$ at 26.1°C = 0.131 atm and if $\rho^{26°C}_{benzene}$ = 0.875 kg dm^{-3}, find

    a. $P_{solution}$ at 26.1°C.

    b. The osmotic pressure of the solution at 26.1°C.

    c. The freezing point of the solution.

5. From the following vapor pressure data estimate the heat of vaporization of benzene at 278.6 K and at 333 K.

T (K)	236.45	261.65	280.75	299.25	333.75	353.75
P (atm) × 10^3	1.31	13.1	52.6	131.6	526.3	1000

6. Use the results of Problem (5) to calculate K_{bp} for benzene and compare with the value in Table 6-2.3.1.

7. The heat of vaporization of ethyl ether is 2.91 × 10^4 J mol^{-1}. The vapor pressure at 283 K is 3.95 × 10^4 N m^{-2}. What do you predict for the vapor pressure at 275 K?

8. A solution of 2.58 g phenol in 0.100 kg bromoform (M = 0.253 kg mol^{-1}, T_f = 280.95 K, and K_f = 14.4 K molal^{-1}) freezes at 278.58 K

    a. Determine the molecular weight of phenol.

    b. Find the boiling point of a solution of 2.21 g phenol in 0.090 kg chloroform (M_c = 0.119 kg mol^{-1}, T_b = 334.5 K, and K_B = 3.63 K molal^{-1}).

6-2.7 SELF-TEST

1. The heat of vaporization of butane at 322 K is 1.94 × 10^4 J mol^{-1}. The vapor pressure at this temperature is 4.88 × 10^5 N m^{-2}. Assuming ΔH_{vap} is constant, find the vapor pressure at 311 K.

2. A 1.00 g sample of a nonvolatile solute was dissolved in 0.100 kg acetic acid. The boiling point of this solution at 1 atm was found to be 392.0 K.

    a. Find the molecular weight of the solute.

    b. Determine the freezing point of this solution.

    c. Calculate the osmotic pressure of this solution at 293 K ($\rho_{acetic\ acid}$ = 1.0491 kg dm^{-3}).

    d. If the vapor pressure of acetic acid is 5.33 × 10^4 N m^{-2} at 316.15 K, find the vapor pressure of the solution at that temperature.

3. What weight of methanol must be added per gram of ice to deice a windshield at 10°F?

6-2.8 ADDITIONAL READING

1.* R. C. Weast, ed., *Handbook of Chemistry and Physics*, The Chemical Rubber Company, Cleveland.

6-3 COLLIGATIVE PROPERTIES: ELECTROLYTES

OBJECTIVES

You shall be able to

1. Define, discuss, or describe

   Weak and strong electrolytes
   The Arrhenius theory of dissociation
   van't Hoff's $\underline{i}$ factor
   Equivalent and specific conductance, and equivalent ionic conductance
   Kohlrausch's law
   Transference numbers

2. Calculate the

   Colligative properties of weak and strong electrolytes
   Equivalent conductances from equivalent ionic conductances or from concentration and the dissociation constant
   Degree of dissociation from conductance data and/or van't Hoff's $\underline{i}$ factor
   Dissociation constant from the degree of dissociation for any given electrolyte

6-3.1 INTRODUCTION

In this section a number of new terms that are important in solution chemistry will be introduced. Previously we have only discussed nonelectrolytes, substances that do not dissociate into charged species in solution. However, many substances form ions when they are dissolved. You should note immediately that this behavior will have a dramatic effect on some solution properties, such as colligative properties, since these are dependent on the number of particles in the solution and not on their nature. For example, the electrical conductance of a solution will be greatly affected by the presence of charged species.

Very little theoretical detail is contained in this or the following sections. In this section you should learn the language used to describe electrolytes in solution and gain some facility in calculating solution parameters. However, if you ever work in area involving solutions do not think what you have learned here is sufficient. You will have to consult more detailed treatments before you can have sufficient understanding to work effectively (and accurately).

6-3.2 THE ARRHENIUS THEORY

Prior to 1887, when Arrhenius proposed his theory of dissociation, the electrical conductance of some solutions was impossible to explain adequately. It probably seems strange to you that such an idea as ions in solution was so difficult to formulate. However, before Arrhenius, chemists were confronted with the fact that it took a tremendous amount of energy to dissociate a molecule such as HCl in the gas phase. So it was difficult to imagine HCl coming apart easily in solution to form ions. It is easy for us to "explain" such behavior on the basis of solvent-solute interactions, but people in the nineteenth century did not have these concepts with which to work.

Arrhenius, however, did suggest that it is possible for some materials in solution to break up to an appreciable extent to form ions. His theory was based to a large degree on the work of Julius Thomson on the heats of neutralization of acids and bases. Thomson found that strong acids and strong bases all produced about 58 kJ per equivalent on neutralization at 298.15 K. Arrhenius explained this on the basis of having free OH^- and H^+ ions in solution which react to form water. The reaction would be the same for any strong acid or strong base and should give the same value for the heat of neutralization, as is observed.

Conductance data were also used by Arrhenius to support his theory; these data will be presented shortly. Additional support came from van't Hoff, who observed for many materials the osmotic pressure equation, Eq. (6-2.5.13), had to be modified to read

$$\Pi = \underline{i}MRT \qquad (6\text{-}3.2.1)$$

where M is the molarity of the solution and $\underline{i}$ is van't Hoff's $\underline{i}$ factor. This factor is often found to be 1, 2, 3, or more. Such a factor can be explained on the basis of a solute dissociating into two, three, or more ions, as we will see later.

In many cases $\underline{i}$ is found to be greater than 1 (it is seldom less than 1; why?) but not an integer. This too can be explained on the basis of partial dissociation of the electrolyte. This leads us to a definition of weak and strong electrolytes.

A *strong electrolyte* is one which dissociates completely to ions (hence $\underline{i}$ will not be equal to 1). For example, consider a general electrolyte A_xB_y (z_+, z_- are the ionic charges)

$$A_xB_y \rightarrow xA^{z+} + yB^{z-} \qquad (6\text{-}3.2.2)$$

The total number of ions produced is

$$\nu = x + y \qquad (6\text{-}3.2.3)$$

which will be an integer.

A *weak electrolyte* is defined as an electrolyte that only partially dissociates to produce ions in solution. An equilibrium between the ions and the electrolyte is established. Again, look at a general molecule which partially dissociates (that is what the double arrow $\rightleftarrows$ signifies).

$$A_xB_y \rightleftarrows xA^{Z+} + yB^{Z-} \tag{6-3.2.4}$$

A term called the *degree of dissociation* will be designated as α. Perhaps the best way to define α is to say that if the molality of A_xB_y initially is m, then $m\alpha$ represents the amount of A_xB_y that dissociates. α varies from *0* for a nonelectrolyte to *1* for a completely dissociated electrolyte. For the above equation, if $m\alpha$ is the amount of A_xB_y that dissociates, how much A^{Z+} and B^{Z-} will be produced? The answer is in the following.

6-3.3 VAN'T HOFF'S i FACTOR

By using the stoichiometry of Eq. (6-3.2.4) we can determine the concentration of A^{Z+} and B^{Z-}.

	A_xB_y $\rightleftarrows$	xA^{Z+} +	yB^{Z-}	
Init. conc.	m	0	0	(6-3.2.4)
Conc. after $m\alpha$ dissociates	$m - m\alpha$	$xm\alpha$	$ym\alpha$	

The total concentration of particles is the sum of the individual concentrations given above:

$$\text{Total conc.} = m - m\alpha + xm\alpha + ym\alpha$$

$$= m(1 - \alpha + x\alpha + y\alpha) \tag{6-3.3.1}$$

The total number of particles *per* A_xB_y will be given by dividing the total concentration by m; this will be van't Hoff's i factor.

$$\underline{i} = \frac{\text{total conc.}}{\text{initial conc.}} = \frac{m(1 - \alpha + x\alpha + y\alpha)}{m} \tag{6-3.3.2}$$

If we make use of Eq. (6-3.2.3) to define the number of particles that *would* be produced *if* A_xB_y dissociated completely we write

$$\nu = x + y \tag{6-3.2.3}$$

Then (6-3.3.2) can be written

$$\underline{i} = 1 - \alpha + \alpha\nu \tag{6-3.3.3}$$

$$\underline{i} = 1 - \alpha(1 - \nu) \tag{6-3.3.4}$$

which can be rearranged (do it!) to

$$\alpha = \frac{\underline{i} - 1}{\nu - 1} \tag{6-3.3.5}$$

This relationship allows us to determine α if $\underline{i}$ is known. From what has been given previously, how would you go about determining $\underline{i}$?

We could use colligative properties as suggested by Eq. (6-3.2.1). Extending Eq. (6-3.2.1) to other colligative properties allows us to write immediately

$$\Delta T_{bp} = (\underline{i}m)K_{bp} \tag{6-3.3.6a}$$

$$\Delta T_{fp} = (\underline{i}m)K_{fp} \tag{6-3.3.6b}$$

$$\Pi = \underline{i}MRT \tag{6-3.3.6c}$$

By measuring any one of these colligative properties we can find $\underline{i}$. At this point you might be inclined to say, "So what?" The next section, in which equilibria in solutions are considered, may convince you of the utility of finding $\underline{i}$ and thereby α.

6-3.3.1 EXAMPLE

A 0.010 molal (m) solution of HCl freezes at 273.114 K. Calculate van't Hoff's $\underline{i}$ factor and α, the degree of dissociation for HCl at this concentration.

Solution

$$\Delta T_{fp} = 273.15 - 273.114 = 0.036 \text{ K}$$

From (6-3.3.6b),

$$\underline{i} = \frac{\Delta T_{fp}}{K_{fp}m} = \frac{0.036 \text{ K}}{(1.86 \text{ K m}^{-1})(0.010 \text{ m})} = \underline{\underline{1.9}}$$

We know that ν is 2 for HCl (two particles formed if dissociation were complete)

$$\nu = 2$$

Then, from (6-3.3.5),

$$\alpha = \frac{\underline{i} - 1}{\nu - 1} = \underline{\underline{0.9}}$$

This means that HCl in a 0.02 m solution is 90% dissociated (or so it seems; it may be that HCl just appears to be 90% dissociated due to nonideality of the solution.)

6-3.4 CONDUCTANCE MEASUREMENTS

The resistance (or its reciprocal, conductance) of a solution will depend on the concentration and nature of the charged species in solution. This fact makes conductance measurements very useful for determining the amount of dissociation a particular electrolyte undergoes in a solution.

Conductance is usually determined by measuring the resistance of a solution with a cell which has platinum electrodes in a fixed configuration. Conductance is just the reciprocal of resistance. Figure 6-3.4.1 shows a sketch of a typical conductance cell. The conductance L of such a cell depends on the area of the electrodes A, the distance between them ℓ, and the *specific conductance* κ. κ depends on the concentration of the solution. It is the conductance that would be observed for a given solution in an imaginary cell of unit dimensions. L is then given by

$$L = \kappa \frac{A}{\ell} = \kappa\,(\text{cell constant}) \tag{6-3.4.1}$$

where A/ℓ is the dimension of a particular cell and is designated as a cell constant. In general, A and ℓ are difficult to measure precisely so the cell constant is determined by using a solution of known κ (which, of course, has to be determined in a cell in which A and ℓ are very acurately known). The standard is KCl. Specific conductance of different concentrations of KCl solution are given in Table 6-3.4.1. Once the cell constant is known, the specific conductance of any other electrolyte solution can be found from a conductance measurement.

6-3.4.1 EXAMPLE

In a conductance cell at 25°C, a KCl solution which had a concentration of 0.010 mol dm^{-3} gave a conductance of 6.6667×10^{-3} ohm^{-1}. In the same cell a solution of HCl with a concentration of 0.010 mol dm^{-3} gave a conductance of 1.946×10^{-2} ohm^{-1}. Find the specific conductance of the HCl solution.

Solution

We can use the data for KCl to determine the cell constant since $\kappa = 0.14088$ $ohm^{-1}\ m^{-1}$ for 0.010 M KCl at 298.15 K

$$\text{Constant} = \frac{L_{KCl}}{\kappa}$$

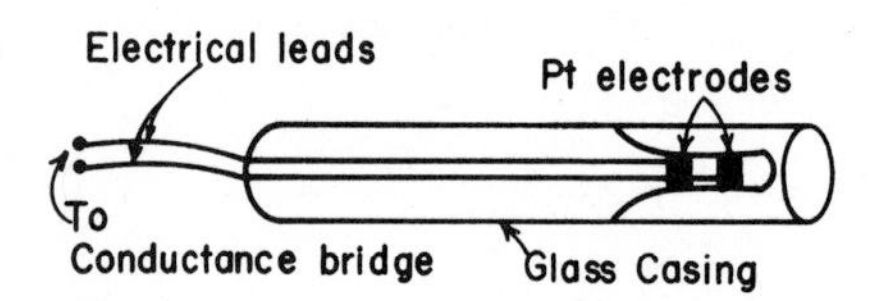

FIG. 6-3.4.1 Typical conductance cell.

Table 6-3.4.1
Specific Conductance of KCl Solutions at Different Temperatures[a]

Conc. (mol dm^{-3})	Conc. $\frac{g\ KCl}{kg\ H_2O}$	Conc. $\frac{g\ KCl}{kg\ solution}$	κ ($ohm^{-1}\ m^{-1}$) 273.15 (K)	291.15 (K)	298.15 (K)
0.01	0.74625	0.74526	0.077364	0.122052	0.14087
0.10	7.47896	7.41913	0.71379	1.11667	1.28560
1.00	76.6276	71.1352	6.5176	9.7838	11.1342

[a] From G. Jones and B. C. Bradshaw, *J. Am. Chem. Soc.* *55*:1799 (1933).

$$= \frac{6.6667 \times 10^{-3}\ ohm^{-1}}{0.14088\ ohm^{-1}\ m^{-1}}$$

$$= \underline{\underline{4.732 \times 10^{-2}\ m}}$$

Then the specific conductance of HCl is

$$\kappa = \frac{L_{HCl}}{constant}$$

$$= \frac{1.946 \times 10^{-2}\ ohm^{-1}}{4.732 \times 10^{-2}\ m}$$

$$= \underline{\underline{0.4112\ ohm^{-1}\ m^{-1}}}$$

We have not yet shown how conductance data has anything to do with the degree of dissociation. We will get there yet. The next step in the process is to define a quantity that eliminates the concentration implicitly included in κ. This quantity is called the equivalent conductance and is defined as

$$\boxed{\Lambda = \frac{10^{-3}\ \kappa}{c}} \qquad (6\text{-}3.4.2)$$

where c will be in *equivalents per cubic decimeter*, the units on κ are $ohm^{-1}\ m^{-1}$ and the 10^{-3} converts the cubic decimeters to cubic meters so Λ has units of

$$\frac{(m^3\ dm^{-3})(ohm^{-1}\ m^{-1})}{equiv.\ dm^{-3}} = equiv.^{-1}\ m^2\ ohm^{-1} \qquad (6\text{-}3.4.3)$$

An equivalent is that amount of material that produces or uses Avagadro's number of electrons in a reaction, or produces or uses Avagadro's number of protons or hydroxl ions in a neutralization. An equivalent of material cannot be specified unless a reaction is written.

6-3.4.2 EXAMPLE

From the results and data in Example 6-3.4.1 find Λ for a 0.01000 M solution of HCl at 298.15 K.

Solution

You must recognize that for HCl molarity is equal to normality.* Thus c = 0.010 mol dm^{-3} = 0.010 equiv. dm^{-3}.

From the previous example κ = 0.4112 ohm^{-1} m^{-1}. Thus

$$\Lambda = \frac{(10^{-3}\ m^3\ dm^{-3})(0.4112\ ohm^{-1}\ m^{-1})}{0.01000\ equiv.\ dm^{-3}}$$

$$= \underline{\underline{0.04112\ equiv.^{-1}\ m^2\ ohm^{-1}}}$$

Λ can be thought of as the conductance of a cell with electrodes 1 m apart and sufficiently large to contain 1 equiv. of electrolyte. From this definition, answer the following question: What would be the effect of concentration on Λ for an electrolyte that is 100% dissociated at all concentrations? The answer is that Λ would be constant.

Experimentally, however, Λ is found to *decrease as concentration increases.* This decrease in Λ suggests that the degree of dissociation decreases as concentration increases (does it not?). The quantitative results will be discussed after considering the experimental data in Table 6-3.4.2 for some weak and strong electrolytes. These data show the variation of Λ with c.

Figure 6-3.4.3 indicates clearly that all the electrolytes plotted except acetic acid (CH_3COOH) are strong electrolytes. Λ approaches a limiting value (called Λ^0) as concentration approaches zero. This limiting value can be found from extrapolation of the Λ vs. $\sqrt{c}$ curve to zero concentration ($\sqrt{c}$ is used instead of C to make the lines more nearly linear). However, that cannot be done for acetic acid. Extrapolation of such a rapidly changing curve is highly inaccurate. Some other means must be found to find Λ^0 for weak electrolytes. This brings us to Kohlrausch's law, after which we finally find out how conductance data can give us information about degrees of dissociation.

KOHLRAUSCH'S LAW. On the basis of conductivity data such as those given above, Kohlrausch proposed his *law of independent migration of ions.* The law states that at infinite dilution (where all electrolytes are completely dissociated and all interionic effects disappear) each ion migrates independently of its co-ion. The contribution of the ion to the total conductance depends only on its own nature and is

---

* One normal = 1 equiv. solute (dm^3 solution)$^{-1}$.

Table 6-3.4.2
Equivalent Conductances (Λ) for Some Weak and Strong Electrolytes in Water at 25°C[a]

c	Λ (ohm^{-1} m^2 $equiv.^{-1}$)							
(equiv. dm^{-3})	$BaCl_2$	HCl	KCl	NH_4Cl	NaCl	NaOH	$NaOOCCH_3$	CH_3COOH
0.0000	0.013998	0.042616	0.014986	0.014970	0.012645	0.02478	0.00910	0.03907
0.0005	0.013596	0.042274	0.014781	--	0.012450	0.02456	0.00892	0.00677
0.001	0.013434	0.042136	0.014695	0.01468	0.012374	0.02447	0.00885	0.00492
0.005	0.012802	0.041580	0.014395	0.01435	0.012065	0.02408	0.008572	--
0.01	0.012394	0.041200	0.014127	0.014128	0.011851	0.02380	0.008376	0.00163
0.02	0.011909	0.040724	0.013834	0.013833	0.011551	--	0.008124	0.001156
0.05	0.011148	0.039909	0.013337	0.013329	0.011106	--	0.007692	0.000736
0.1	0.010519	0.039132	0.012896	0.012875	0.010674	--	0.007280	0.000520

[a] Values at c = 0 are determined by the methods discussed in the text. (Some data are taken from the Chemical Rubber Company *Handbook of Chemistry and Physics*.)

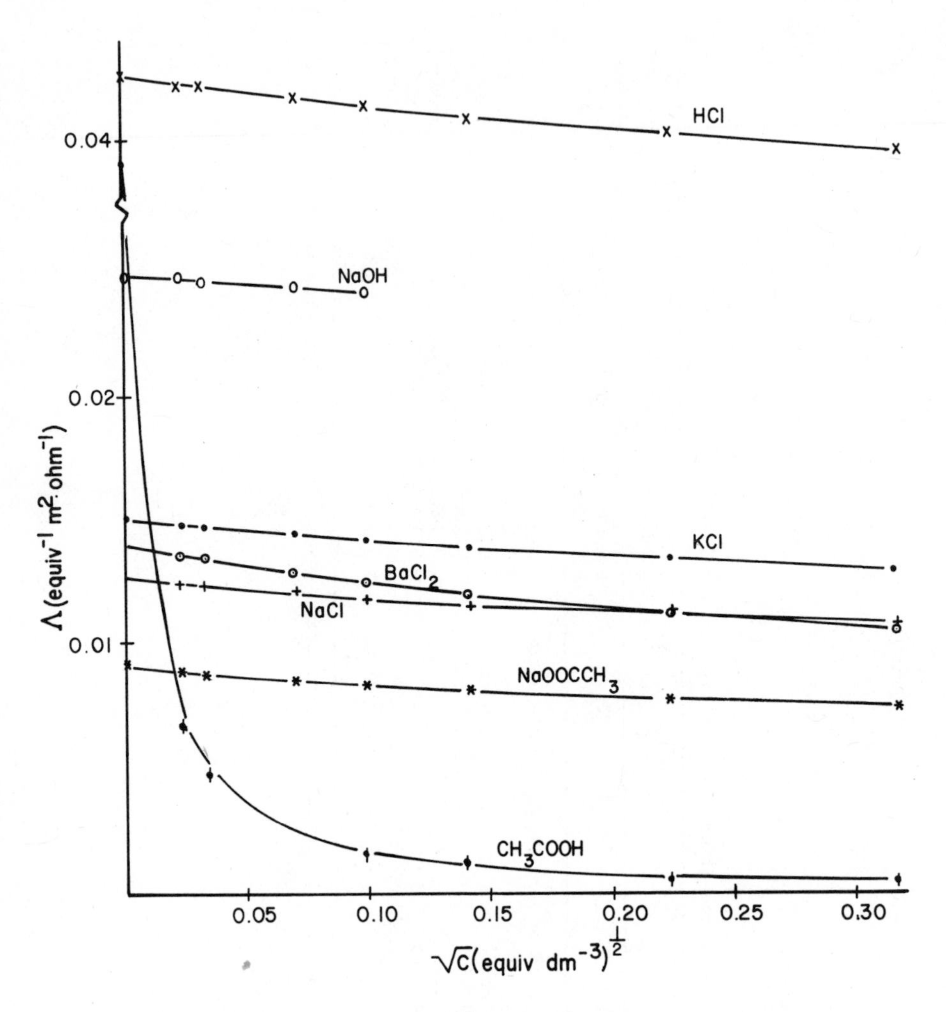

FIG. 6-3.4.2 Variations of Λ with $\sqrt{c}$ for various electrolytes.

independent of the ion or ions with which it was associated before dissociation. This allows us to write immediately for an electrolyte AB:

$$\Lambda^0_{AB} = \lambda^0_{A^+} + \lambda^0_{B^-} \tag{6-3.4.4}$$

where λ^0 are *limiting equivalent conductances of individual ions*. These values cannot be obtained directly from conductance data. Instead another parameter, transference number, is used.

The *transference number* of an ion is the *fraction* of charge carried in a solution of the electrolyte by that ion. No detail is given about how these transference numbers are determined. If you are interested or need to know consult almost any other physical chemistry test. Suffice it to say that for an electrolyte

$$t_+ + t_- = 1 \qquad (6\text{-}3.4.5)$$

where t_+ is the transference number for the positive ion and t_- is that for the negative ion.

Combining transference numbers with conductance data allows us to determine λ^0 values. Some of these are tabulated in Table 6-3.4.3. These are determined by the following formula for an electrolyte AB:

$$\lambda^0_{A+} = t_{A+}\Lambda^0_{AB} \qquad \lambda^0_{B-} = t_{B-}\Lambda^0_{AB} \qquad (6\text{-}3.4.6)$$

You should now wonder of what use are these data. We can use them to find Λ^0 values for weak electrolytes for which extrapolation is inaccurate. For example, consider the case of acetic acid (abbreviate acetic acid as HAc and the acetate ion as Ac^-):

$$\Lambda^0_{HAc} = \lambda^0_{H+} + \lambda^0_{Ac-}$$

$$= 349.82 \times 10^{-4} + 40.9 \times 10^{-4}$$

$$= 390.7 \times 10^{-4} \ (\text{ohm}^{-1}\ \text{m}^2\ \text{equiv.}^{-1})$$

There is another way of obtaining this value. We can add and subtract appropriate Λ^0 values as follows:

$$\Lambda^0_{HAc} = \Lambda^0_{HCl} + \Lambda^0_{NaAc} - \Lambda^0_{NaCl}$$

$$= \lambda^0_{H+} + \lambda^0_{Cl-} + \lambda^0_{Na+} + \lambda^0_{Ac-} - \lambda^0_{Na+} - \lambda^0_{Cl-}$$

$$= \lambda^0_{H+} + \lambda^0_{Ac-}$$

Inserting values from Table 6-3.4.2 gives

Table 6-3.4.3
Equivalent Ionic Conductances at Infinite Dilution (298.15 K)[a]

Cation	$\lambda^0_+ \times 10^4$	Anion	$\lambda^0_- \times 10^4$
H^+	349.82	OH^-	198.0
Li^+	38.69	Cl^-	75.23
Na^+	50.11	Br^-	78.4
K^+	73.52	I^-	76.8
NH_4^+	73.4	CH_3COO^-	40.9
*Ca^{2+}	59.50	*SO_4^{2-}	79.8

[a] In $\text{ohm}^{-1}\ \text{m}^2\ \text{equiv.}^{-1}$; (*) 1 equiv. = ½ mol for these ions.

$$\Lambda^0_{HAc} = 0.042616 + 0.00910 - 0.012645$$
$$= 0.039071$$
$$= 390.71 \times 10^{-4}\ (\text{ohm}^{-1}\ \text{m}^2\ \text{equiv.}^{-1})$$

which is in agreement with our previous results.

Finally, we are ready to consider the second part of the Arrhenius theory, in which the degree of dissociation is related to conductance.

6-3.5 THE DEGREE OF DISSOCIATION AND EQUILIBRIUM IN SOLUTION

ARRHENIUS REVISITED. Arrhenius not only recognized that molecules can dissociate to form ions in solution, he also proposed a formula that can be used to calculate the degree of dissociation for weak or strong electrolytes. Combining two ideas we have already seen leads to Arrhenius' result. The first concept is that the equivalent conductance Λ depends on the number of charged species present to carry the electrical current. The second idea is that, at infinite dilution, all electrolytes are completely dissociated into ions. This led Arrhenius to propose that the comparison of Λ at a given concentration to Λ^0 at infinite dilution should give a measure of the degree of dissociation at the given concentration. Thus, the *degree of dissociation* can be written as

$$\alpha = \frac{\Lambda}{\Lambda^0} \tag{6-3.5.1}$$

You now have an expression that relates α to Λ. We will use this relationship to determine equilibrium constants after an example.

6-3.5.1 EXAMPLE

Calculate α for HCl and HAc at each concentration given in Table 6-3.4.2. The results will be used in Example 6-3.5.2.

Solution

See Table 6-3.5.1. These results should make the difference between weak and strong electrolytes clear.

DISSOCIATION EQUILIBRIA. The results of the last section can be used to determine the equilibrium constant for the dissociation of an electrolyte in solution. Refer to Eq. (6-3.2.4) in which the dissociation of a general electrolyte at concentration c (c is a generalized concentration, it might be molality, molarity, or normality) is written

Table 6-3.5.1
Degrees of Dissociation for HCl and HAc at Various Concentrations

c	$\alpha = \Lambda/\Lambda^0$	
(equiv. dm^{-3})	HCl	HAc
0.0	1.000	1.000
0.0005	0.993	0.173
0.001	0.989	0.126
0.005	0.976	--
0.01	0.967	0.0417
0.02	0.956	0.0296
0.05	0.936	0.0188
0.10	0.918	0.0133

$$A_xB_y \rightleftarrows xA^{z+} + yB^{z-} \qquad (6\text{-}3.2.4)$$

conc. $c(1 - \alpha)$ $\quad cx\alpha \quad cy\alpha$

The equilibrium constant for this dissociation (diss) is (if ideal; Sec. 6-4 treats nonideal cases)

$$K_{diss} = \frac{[A^{z+}]^x [B^{z-}]^y}{[A_x B_y]} \qquad (6\text{-}3.5.2)$$

where the brackets indicate concentration. Substituting the appropriate concentrations (and note the concentration units can be anything we like since K_{diss} acquires whatever units we choose):

$$K_{diss} = \frac{(cx\alpha)^x (cy\alpha)^y}{c(1 - \alpha)} \qquad (6\text{-}3.5.3)$$

or

$$K_{diss} = \frac{(c\alpha)^{(x + y)} x^x y^y}{c(1 - \alpha)} \qquad (6\text{-}3.5.4)$$

It should be apparent that if we know the initial concentration of the electrolyte and α at that concentration, we can find the equilibrium constant *and* the concentration of all ions produced by the dissociation.

6-3.5.2 EXAMPLE

From the results in Example 6-3.5.1 find K_{diss} for HCl and HAc at each concentration given.

Solution

For both HCl and HAc which dissociate into one positive ($z_+ = 1$) and one negative ($z_- = 1$) ion, Eq. (6-3.5.4) reduces to

Table 6-3.5.2
Dissociation Constants for HCl and HAc

c	$K = \frac{c\alpha^2}{1-\alpha}$ (equiv. dm^{-3})	
(equiv. dm^{-3})	HCl	HAc $\times 10^5$
0.0005	0.070	1.81
0.001	0.089	1.82
0.005	0.198	--
0.01	0.283	1.81
0.02	0.415	1.80
0.05	0.684	1.80
0.10	1.03	1.79

$$K_{AB} = \frac{(c\alpha)^{(1+1)}\,(1)^1\,(1)^1}{c(1-\alpha)} = \frac{c^2\alpha^2}{c(1-\alpha)}$$

$$= \frac{c\alpha^2}{(1-\alpha)}$$

(see Table 6-3.5.2.) For illustration let us see how to find K_{HAc} at 0.0500 equiv. dm^{-3}. The results for the other cases are listed below. (Note that we allow c to stand for concentration in any desired units right now.)

$$K_{HAc}(0.0500) = \frac{(0.0500\ \text{equiv. dm}^{-3})(0.0188)^2}{(1 - 0.0188)}$$

$$= \underline{\underline{1.80 \times 10^{-5}\ \text{equiv. dm}^{-3}}}$$

We have now reviewed the basis of the conductance behavior and equilibrium in ideal solutions. The following section is concerned with nonideal solutions. As we shall see, the form of our equations remains unchanged, we just insert appropriate correction factors. It is time for you to do some calculations to help clarify the material just presented.

6-3.6 PROBLEMS

1. Use the values of Λ^0 in Table 6-3.4.2 to predict the value for Λ^0 for NH_4OH. Now use that value and the following Λ values for NH_4OH to find α and K_{diss} at each concentration given.

c (equiv. dm^{-3})	0.001	0.010	0.100
Λ (ohm^{-1} m^2 equiv.$^{-1}$	0.0034	0.00113	0.00036

2. From the following data find Λ^0 for H_2SO_4.

c (equiv. dm^{-3})	0.0005	0.001	0.010	0.100
Λ (ohm^{-1} m^2 $equiv.^{-1}$)	0.04131	0.03995	0.03364	0.02508

3. Use the data from Table 6-3.4.3 to find Λ^0 for H_2SO_4 and compare with the results of Problem (2).

4. (a) Find the cell constant of a cell which gives a conductance reading of 0.120 ohm^{-1} for 0.01 M KCl at 25°C. (b) What conductance would be measured with this cell for a 0.00500 N solution of NH_4Cl at 25°C?

5. A 0.100 m aqueous solution of $CuSO_4$ is observed to freeze at 272.94 K. Calculate van't Hoff's $\underline{i}$ factor and the boiling point of the solution.

6. The boiling point of a 0.500 m aqueous solution of NaCl is 373.62 K at 1 atm pressure. Find van't Hoff's $\underline{i}$ factor and α for NaCl.

7. What would you expect for the osmotic pressures of the solutions in Problems (5) and (6) at 298 K?

8. Calculate the dissociation constant for the $CuSO_4$ solution in Problem (5).

9. Given the following data for the freezing point depression of NH_4Cl in H_2O, graphically find $\underline{i}$, at infinite dilution. Discuss your result and find α for the two concentrations 0.189 and 0.778 mol NH_4Cl/kg H_2O.

$\frac{\text{mol } NH_4Cl}{\text{kg } H_2O}$	0.094	0.284	0.480	0.676
T_f (K)	0.32	0.95	1.59	2.24

10. Given the following, find Λ^0 for NaOH.

Molecule	$CaCl_2$	NaCl	$Ca(OH)_2$	LiCl
Λ^0 (ohm^{-1} m^2 $equiv.^{-1}$)	0.01358	0.012645	0.02579	0.011503

(a) If Λ for NaOH = 0.0238 at 0.0100 equiv. dm^{-3}, what is the value for α and K_{diss}? (b) What is the freezing point of a 0.0100 molal solution of NaOH? (K_f = 1.86 K $molal^{-1}$.)

6-3.7 SELF-TEST

1. The Arrhenius theory is scattered throughout this section. Summarize the important aspects (and supporting evidence) in a half-page discussion.

2. Define or briefly discuss three of the terms from this section.

3. From the conductance data given for acetic acid in aqueous solution, calculate van't Hoff's i factor and the freezing point of a 0.0200 m (≈ 0.0200 equiv. dm^{-3}) solution.

4. Use the data given below to find Λ^0. Then calculate α and K_{diss} at each concentration given for $KHCO_3$ which dissociates into K^+ and HCO_3^-.

c (equiv. dm^{-3})	0.000500	0.00100	0.00500	0.0100	0.100
Λ (ohm^{-1} m^2 $equiv.^{-1}$)	0.01161	0.01153	0.01122	0.01101	0.010311

Is this a weak or strong electrolyte?

6-4 NONIDEAL SOLUTIONS

OBJECTIVES

You shall be able to

1. Define the following for any given electrolyte: ionic strength, mean concentration, mean activity coefficient, mean activity
2. Apply the Debye-Hückel theory to determine activity coefficients and mean activity coefficients from ionic concentrations of electrolytes for dilute solutions
3. Calculate ionic strength, activities, and mean activities from concentration data for electrolytes and the Debye-Hückel theorem
4. Find solubility constants and dissociation constants in terms of activities, concentrations, and activity coefficients for substances that partially dissolve or partially dissociate in solution

6-4.1 INTRODUCTION

In Sec. 6-1.4 some of the interactions in nonideal solutions were mentioned, and a term called activity was introduced. The purpose of using a quantity such as *activity* in the place of *concentration* in nonideal solutions is to retain the form of the ideal equations we have presented for chemical potential and other functions.

A theory formulated by Debye and Hückel will be presented. Actually only the results will be presented. You are referred to other physical chemistry texts for the details on the derivation of this theory.

After defining activities and activity coefficients and giving Debye-Hückel formulas for calculating activity coefficients, use will be made of the activities to reformulate the equations for solubility and dissociation equilibria. Little will be said about how activity coefficients are determined in this section. You saw in Sec. 6-1.4 how vapor pressure data can be used for this purpose for nonelectrolytes. Colligative property measurements can also be used. The following section (6-5) will show how electromotive force measurements in electrochemical cells can be used to obtain activity coefficients.

6-4.2 DEFINITIONS

ACTIVITIES AND ACTIVITY COEFFICIENTS. The first step in our development is to agree on a basic set of definitions involved in finding and using activities and activity coefficients. Consider an equilibrium involving the electrolyte A_xB_y:

$$A_x B_y \rightleftarrows xA^{z+} + yB^{z-} \qquad (6\text{-}4.2.1)$$

The *activity* of each ion is given by

$$a_+ = \gamma_+ c_+$$
$$a_- = \gamma_- c_- \qquad (6\text{-}4.2.2)$$

(*Note*: c_+ and c_- stand for the ionic concentrations. This may be expressed as molarity, molality, or perhaps some other concentration unit. If you have to use activity coefficients in future work be sure to check the headings of the tables to see what concentration unit is being considered. Values of γ will be different for different concentration units. Also, for nonelectrolytes, mole fraction is often used as the "concentration" unit.)

In Eq. (6-4.2.2) γ_+ and γ_- are the *ionic activity coefficients* and a_+ and a_- are the *ionic activities*. We introduce γ to make nonideal solutions correspond to the equations we have derived previously for ideal solutions.

We define the *mean activity* as (for A_xB_y)

$$a^{(x+y)} = a_+^x a_-^y$$

or

$$\boxed{a = (a_+^x a_-^y)^{1/(x+y)}} \qquad (6\text{-}4.2.3)$$

Similarly we define the *mean activity coefficient* and the mean concentration of ions as

$$\gamma_\pm^{(x+y)} = \gamma_+^x \gamma_-^y$$

$$\gamma_\pm = \left(\gamma_+^x \gamma_-^y\right)^{1/(x+y)}$$

and

$$c^{(x+y)} = c_+^x c_-^y$$

or
$$c_{\pm} = (c_+^{x}\, c_-^{y})^{1/(x + y)} \tag{6-4.2.4}$$

Combining Eqs. (6-4.2.3) and (6-4.2.4) gives

$$\begin{aligned} a_{\pm}^{(x + y)} &= a_+^{x}\, a_-^{y} = (\gamma_+ c_+)^{x}\, (\gamma_- c_-)^{y} \\ &= \gamma_+^{x}\, \gamma_-^{y}\, c_+^{x}\, c_-^{y} \\ &= \gamma_{\pm}^{(x + y)}\, c_{\pm}^{(x + y)} \end{aligned} \tag{6-4.2.5}$$

or
$$a_{\pm} = \gamma_{\pm}\, c_{\pm} \tag{6-4.2.6}$$

For illustration purposes, let us assume we have a strong electrolyte that completely dissociates into xA^{z+} and yB^{z-} ions according to (6-4.2.1). If the initial concentration of A_xB_y were c, then

$$c_+ = xc \qquad c_- = yc$$

Applying the relationships in (6-4.2.5) gives

$$\begin{aligned} a_{\pm}^{x + y} &= (\gamma_+ c_+)^{x}\, (\gamma_- c_-)^{y} \\ &= \gamma_+^{x}(xc)^{x}\, \gamma_-^{y}(yc)^{y} \\ &= \gamma_{\pm}^{(x + y)}\, c^{(x + y)}\, x^{x}\, y^{y} \end{aligned} \tag{6-4.2.8}$$

or
$$a_{\pm} = \gamma_{\pm}\, c(x^{x}\, y^{y})^{1/(x + y)} \tag{6-4.2.9}$$

This applies *only* to a *strong* electrolyte.

6-4.2.1 EXAMPLE

Deduce the relationship among mean activities, mean activity coefficients, and concentrations for each of the following, assuming complete dissociation: (1) 0.005 M $BaCl_2$ and (2) 0.01 M $Ca_3(PO_4)_2$.

Solution

$BaCl_2$

It is time to get back into the habit of first writing the stoichiometric equation for the process under consideration. That helps immensely in solving problems.

	$BaCl_2$	→	Ba^{2+}	+	$2Cl^-$
Initial c	0.005		0		0
Final c (after dissociation)	0		0.005		0.01 (= 2 × 0.005)

For $BaCl_2$, x = 1, z+ = +2, y = 2, and z- = -1. So, applying Eq. (6-4.2.5) gives

$$a_{\pm}^{(1+2)} = \gamma_+^{1}\,\gamma_-^{2}\,c_+^{1}\,c_-^{2}$$

$$= \gamma_{\pm}^{3}\,(0.005)^{1}\,(0.01)^{2}$$

$$a_{\pm}^{3} = \gamma_{\pm}^{3}\,(5 \times 10^{-7})$$

or

$$a_{\pm} = (8 \times 10^{-3})\,\gamma_{\pm}$$

We could, of course, use Eq. (6-4.2.9).

$$a_{\pm} = \gamma_{\pm}\,c\,(x^x\,y^y)^{1/(x+y)}$$

$$= \gamma_{\pm}\,(0.005)(1^1\,2^2)^{1/3}$$

$$= (8 \times 10^{-3})\,\gamma_{\pm}$$

The same value results.

$Ca_3(PO_4)_2$

As an exercise, you should show ($z_+ = +2$, $z_- = -3$ in this case)

$$a_{\pm} = (2.6 \times 10^{-2})\,\gamma_{\pm}$$

IONIC STRENGTH. The ionic strength of a solution will be the important parameter in the Debye-Hückel formulas for activity coefficients to be presented next. The *ionic strength* $\underline{\mu}$ is defined as

$$\underline{\mu} = \frac{1}{2}\sum_i c_i z_i^2 \qquad (6\text{-}4.2.10)$$

(Do not confuse this with chemical potential.) In this equation c_i represents the concentration of the ith ion and z_i is its charge.

6-4.2.2 EXAMPLE

Find the ionic strength of each solution in Example 6-4.2.1 and an equal volume mixture of the two.

Solution

$BaCl_2$ (0.005 M)

$$BaCl_2 \rightarrow Ba^{2+} + 2Cl^-$$

$$c_+ = 0.005\ M \qquad z_+ = 2$$

$$c_- = 2(0.005\ M) = 0.010\ M \qquad z_- = -1$$

$$\mu_{BaCl_2} = \frac{1}{2}[(0.005)(2)^2 + (0.010)(-1)^2]$$

$$= \underline{\underline{0.015\ M}}$$

$Ca_3(PO_4)_2$ (0.01 M)

$$Ca_3(PO_4)_2 \rightarrow 3Ca^{2+} + 2PO_4^{3-}$$

$$c_+ = 3(0.01) = 0.03\ M \qquad z_+ = 2$$

$$c_- = 2(0.01) = 0.02\ M \qquad z_- = -3$$

$$\mu = \frac{1}{2}[(0.03)(2)^2 + (0.02)(-3)^2]$$

$$= \underline{\underline{0.15\ M}}$$

Equal Volume Mixture

If we mix equal volumes of the two solutions, the final concentration of each is half the initial concentration if the volumes are additive (prove that). Thus, in the mixture $BaCl_2$ (0.0025 M) and $Ca_3(PO_4)_2$ (0.005 M)

$$c_{Ba^{2+}} = 0.0025\ M \qquad z_{Ba^{2+}} = 2$$

$$c_{Cl^-} = 0.0050\ M \qquad z_{Cl^-} = -1$$

$$c_{Ca^{2+}} = 0.015\ M \qquad z_{Ca^{2+}} = +2$$

$$c_{PO_4^{3-}} = 0.010\ M \qquad z_{PO_4^{3-}} = -3$$

$$\mu = \frac{1}{2}[(0.0025)(2)^2 + (0.005)(1)^2 + (0.015)(2)^2 + (0.010)(-3)^2]$$

$$= \underline{\underline{0.0825\ M}}$$

Table 6-4.2.1
Selected Values of Mean Activity Coefficients at 298.15 K (Based on Molalities)

	m									
Compound	0.001	0.005	0.01	0.05	0.1	0.5	1.0	2.00	3.00	4.00
HCl	0.966	0.920	0.905	0.830	0.796	0.757	0.809	1.009	1.316	1.762
NaCl	0.965	0.927	0.902	0.819	0.778	0.681	0.657	0.668	--	0.783
KCl	0.965	0.927	0.902	0.817	0.769	0.651	0.606	0.576	0.571	0.579
$CaCl_2$	0.888	0.789	0.732	0.584	0.531	0.457	0.509	0.807	--	--
H_2SO_4	0.830	0.639	0.544	0.340	0.265	0.154	0.131	0.124	0.141	0.171
$ZnSO_4$	0.734	0.477	0.387	0.202	0.148	0.063	0.044	0.035	0.041	--
$CdSO_4$	0.697	0.476	0.383	0.199	0.150	0.061	0.041	0.032	--	--

You have seen all the definitions required to deal with the activities of nonideal solutions. As pointed out earlier, activity coefficients can be obtained experimentally. Some of these values are given in Table 6-4.2.1. These values are based on molalities. Members of the same types of electrolyte are grouped in brackets. It would be instructive for you to plot some of these data to see how the activity coefficients change with concentration for each type of electrolyte.

Obviously, values of $\gamma_\pm$ are not available for all electrolytes. It would be useful to have a method to calculate these for those electrolytes for which data are not available. The Debye-Hückel theory presented below allows us to do that (within rather restricted limits).

6-4.3 THE DEBYE-HÜCKEL EQUATION

P. Debye and E. Hückel in 1923 and 1924 theoretically investigated the ion-ion interactions in solution. Their work led to a quantitative equation for calculating activity coefficients of ions in dilute solutions. Only the skeleton of a derivation of their important equation is given below.

The electrostatic interaction between two ions of charges z_1 and z_2 in a solvent with dielectric constant θ separated by r is given by

$$\text{Force} = \frac{1}{\theta}\left(\frac{z_1 z_2}{r^2}\right) \tag{6-4.3.1}$$

In general an ion of one charge will be surrounded by an "atmosphere" of the oppositely charged ions produces a potential ξ_i at the surface of the central ion. Debye and Hückel derived a formula for this potential of the form

$$\xi_i = -\frac{z_i e K}{\theta(1 + k\sigma_i)} \tag{6-4.3.2}$$

where e is the electronic charge, z_i is the charge on the central ion, and σ_i is the ionic diameter. K is given by

$$K = \left(\frac{4\pi e^2 \sum n_i z_i^2}{\theta k T}\right)^{1/2} \tag{6-4.3.3}$$

In this equation n_i is the concentration of the ith ion (per cubic centimeter) and z_i is its charge. The sum is carried over all ions in the solution (k is Boltzmann's constant = R/N).

The units on K are reciprocal length, and it has been interpreted as related to the average thickness of the atmosphere about a given ion.

The electrical free energy at the surface of an ion due to the potential ξ_i produced by the ionic atmosphere is

$$G = -\frac{z_i^2 e^2 K}{2\theta(1 + K\sigma_i)} \tag{6-4.3.4}$$

This free energy is in excess of what the solution would have if no ionic atmosphere were produced. Thermodynamically this free energy can be related to the activity coefficient of the ion in solution

$$G = kT \ln \gamma_i \tag{6-4.3.5}$$

These two equations can be combined to give

$$\ln \gamma_i = -\frac{z_i^2 e^2 K}{2kT\theta(1 + K\sigma_i)} \tag{6-4.3.6}$$

If K is substituted into this equation, along with the relationship

$$c_i = \frac{n_i}{N} 1000$$

to give concentration in molarity, Eq. (6-4.3.6) can be written compactly as

$$\log \gamma_i = -\frac{A z_i^2 (\mu)^{1/2}}{1 + B\sigma_{\pm}(\mu)^{1/2}} \tag{6-4.3.7}$$

(note that the common logarithm is used in this and subsequent equations) where

$$A = \frac{e^3}{2.303(\theta kT)^{3/2}}\left(\frac{2\pi N}{1000}\right)^{1/2}$$

$$= \frac{(1.6022\times10^{-19})^3}{(2.303)\,[illegible]\,(8.85419\times10^{-12})[(1.3807\times10^{-23})(298.15)(78.54)]^{3/2}}\left[\frac{2\pi(6.022\times10^{23})}{1000}\right]$$

$$B = \left(\frac{8\pi Ne^2}{1000\theta kT}\right)^{1/2} \tag{6-4.3.8}$$

This equation is valid for the activity of a single ion. However, we usually consider the mean value of an electrolyte. Equation (6-4.3.8) can be written as

$$\log \gamma_{\pm} = \frac{A\, z_+\, z_- \sqrt{\mu}}{1 + B\sigma_{\pm}\sqrt{\mu}} \tag{6-4.3.9}$$

where $\sigma_{\pm}$ is the average diameter of the (+) and (-) ions and is given in units of $Å = 10^{-10}$ m. Remember z_+ and z_- are the ionic charges, *including sign*.

Values of A and B for water as the solvent are given in Table 6-4.3.1 for different temperatures. However, before Eq. (6-4.3.9) can be used, $\sigma_{\pm}$ must be estimated. That is no easy task. With a good choice of σ, activity coefficients for 1:1 electrolytes (e.g., NaCl) can be predicted up to about c = 0.1 and for 2:1 or 1:2 electrolytes (e.g., $CaCl_2$) up to c = 0.05.

A more useful approximation can be made when $B\sigma_{\pm}\sqrt{\mu}$ is small compared to one. In this case we do not have to try to estimate $\sigma_{\pm}$ since we can neglect the term $B\sigma_{\pm}\sqrt{\mu}$. Equation (6-4.3.9) reduces to the *Debye Hückel limiting law for very dilute solutions (up to* $\sqrt{\mu} = 0.1$*)*:

$$\log \gamma_{\pm} = A\, z_+\, z_-\, \sqrt{\mu} \tag{6-4.3.10}$$

At 25°C, this becomes

$$\log \gamma_{\pm} = 0.5091\, z_+\, z_-\, \sqrt{\mu} \tag{6-4.3.11}$$

6-4.3.1 EXAMPLE

For each type of electrolyte (monovalent 1:1, 1:2, or 2:1, and divalent 1:1) from Table 6-4.2.1 calculate $\gamma_{\pm}$ from the Debye-Hückel limiting law up to m = 0.05 and compare the results with an example of each type.

Table 6-4.3.1
Debye-Hückel Constants A and B for Water as Solvent

T (°C)	A	$B \times 10^{-8}$
0	0.4883	0.3241
15	0.5002	0.3267
25	0.5091	0.3286
40	0.5241	0.3318
50	0.5410	0.3353
70	0.5599	0.3392

Solution

Monovalent 1:1 Electrolyte

$$AB \rightarrow A^{+} + B^{-}$$
$$\quad\quad m \quad m$$

$$\underline{\mu} = \frac{1}{2}\sum c_i z_i^{2} = \frac{1}{2}[m(1)^2 + m(1)^2] = m$$

$$\log \gamma_{\pm}(m) = 0.5091(+1)(-1)\sqrt{m}$$
$$= +0.5091(+1)(-1)\sqrt{m}$$

$$\log \gamma_{\pm}(0.001) = -0.5091\sqrt{0.001} = -0.0161$$
$$\gamma_{\pm}(0.001) = \underline{0.964}$$

$$\log \gamma_{\pm}(0.005) = -0.5091\sqrt{0.005} = -0.3600$$
$$\gamma_{\pm}(0.005) = \underline{0.920}$$

And so on. See Table 6-4.3.2. The numbers in parentheses are $\gamma_{\pm}$ values calculated in a concentration region where the limiting law is not valid. Note that, in general, the agreement between experiment and theory gets worse as the concentration increases.

1:2 or 2:1 Electrolyte

$$A_2B \rightarrow 2A^{+} + B^{2-}$$
$$\quad\quad 2m \quad m$$

or
$$AB_2 \rightarrow A^{2+} + 2B^{-}$$
$$\quad\quad m \quad 2m$$

$$\underline{\mu} = \frac{1}{2}[(2m)(1)^2 + m(-2)^2] = 3m$$

$$\log \gamma_{\pm}(m) = 0.5091(1)(-2)\sqrt{\underline{\mu}}$$
$$= -(2)(0.5091)\sqrt{3m}$$

$$\log \gamma_{\pm}(0.001) = -1.0182\sqrt{0.003} = -0.0558$$
$$\gamma_{\pm}(0.001) = \underline{0.879}$$

$$\log \gamma_{\pm}(0.005) = -1.0182\sqrt{0.015} = -0.125$$
$$\gamma_{\pm}(0.005) = \underline{0.750}$$

And so on.

Divalent 1:1 Electrolyte

$$AB \rightarrow A^{2+} + B^{2-}$$
$$\quad\quad m \quad\quad m$$

$$\underline{\mu} = \frac{1}{2}[(m)(2)^2 + (m)(-2)^2] = 4m$$

$$\log \gamma_\pm(m) = 0.5091(2)(-2)\sqrt{\underline{\mu}}$$
$$= -2.0364\sqrt{4m}$$

$$\log \gamma_\pm(0.001) = -2.0364\sqrt{0.004} = -0.129$$
$$\gamma_\pm(0.001) = \underline{0.743}$$

$$\log \gamma_\pm(0.005) = -2.0364\sqrt{0.020} = -0.288$$
$$\gamma_\pm(0.005) = \underline{0.515}$$

(Note: In the Debye-Hückel equation, $\log \gamma_\pm < 0$ in all cases. If you find $\log \gamma_\pm > 0$ check your calculations.)

6-4.3.2 EXAMPLE

For 0.001 molal $CaCl_2$ find the mean activity.

Solution

$CaCl_2$ produces these ions on dissociation

$$CaCl_2 \rightarrow Ca^{2+} + 2Cl^-$$
$$\quad\quad 0.001 \quad 2(0.001)$$

Table 6-4.3.2
Comparison of Activity Coefficients from the Debye-Hückel Limiting Law to Those Observed Experimentally[a]

Molality		0.001	0.005	0.01	0.05
Monovalent 1:1	$\sqrt{\underline{\mu}}$	0.0316	0.0707	0.100	0.224
	$\gamma_\pm$	0.964	0.920	0.889	(0.769)
	(HCl)*	0.966	0.929	0.905	0.830
2:1 or 1:2	$\sqrt{\underline{\mu}}$	0.055	0.122	0.173	0.387
	$\gamma_\pm$	0.879	(0.750)	(0.666)	(0.403)
	($CaCl_2$)*	0.888	0.789	0.732	0.584
Divalent 1:1	$\sqrt{\underline{\mu}}$	0.0633	0.141	0.200	0.447
	$\gamma_\pm$	0.743	(0.515)	(0.391)	(0.122)
	($ZnSO_4$)*	0.743	0.477	0.387	0.202

[a] (*) See Table 6-4.2.1 for these experimental values. The limiting law is not valid for the numbers in parenthesis.

The calculation of $\gamma_\pm$ for $CaCl_2$ at this concentration is shown in Example 6-4.3.1. Table 6-4.3.2 gives the values of $\gamma_\pm(0.001) = 0.879$.

Applying Eq. (6-4.2.5) gives

$$a_\pm^3 = \gamma_\pm^3 c_+ c_-^2 \tag{6-4.2.5}$$

$$= (0.879)^3 (0.001)(0.002)^2$$

$$= 2.717 \times 10^{-9}$$

$$a_\pm = (2.717 \times 10^{-9})^{1/3} = \underline{\underline{1.40 \times 10^{-3}}}$$

You now have the information you need to calculate mean activity and mean activity coefficients for strong electrolytes. However, the Debye-Hückel theory applies equally well to weak electrolytes. In fact, we can use the Debye-Hückel limiting law to predict activities of the ions produced on dissociation, and thereby find a more precise equilibrium constant for the dissociation.

If you need the activity coefficient of individual ions, the Debye-Hückel equation in slightly modified form can be applied

$$\log \gamma_+ = -0.5091\, z_+^2 \sqrt{\mu} \tag{6-4.3.12}$$

$$\log \gamma_- = -0.5091\, z_-^2 \sqrt{\mu} \tag{6-4.3.13}$$

As an exercise you should show that the $\gamma_\pm$ values calculated from the γ_+ and γ_- values from the above equations are the same as $\gamma_\pm$ calculated directly from (6-4.3.11) for a specific case.

6-4.4 SOLUTION EQUILIBRIA

In Sec. 6-3.5 the equation for the equilibrium constant for a dissociation of a weak electrolyte was written as

$$K_{diss} = \frac{[A^{z+}]^x [B^{z-}]^y}{[A_x B_y]} = \frac{c_A^x c_B^y}{c_{A_x B_y}} \tag{6-3.5.2}$$

This equation is in terms of concentration. A more precise equilibrium constant is the thermodynamic function based on activities. For the general reaction with degree of dissociation α.

$$\begin{array}{llll} & A_x B_y & \rightleftarrows x A^{z+} & + \; y B^{z-} \\ \text{conc.} & c(1-\alpha) & c x \alpha & c y \alpha \end{array} \tag{6-3.2.4}$$

We can write the thermodynamic equilibrium constant in terms of the activities as follows:

$$K_{th} = \frac{a_A^{\ x}\ a_B^{\ y}}{a_{A_xB_y}} \qquad (6\text{-}4.4.1)$$

For practice, before reading on, you should express Eq. (6-4.4.1) in terms of concentrations and mean activity coefficients. It might be helpful to refer to Eq. (6-3.5.4) to see the form of K_{diss} in terms of α and c. It is done below so you can check your work.

$$K_{th} = \frac{a_A^{\ x}\ a_B^{\ y}}{a_{A_xB_y}} = \frac{(\gamma_A\ c_A)^x\ (\gamma_B\ c_B)^y}{c_{A_xB_y}} \qquad (6\text{-}4.4.2)$$

Note that $a_{A_xB_y} \simeq c_{A_xB_y}$ since for a noncharged species $\gamma \simeq 1$. Continuing

$$K_{th} = \frac{\gamma_A^{\ x}\ \gamma_B^{\ y}\ c_A^{\ x}\ c_B^{\ y}}{c_{A_xB_y}} = \gamma_\pm^{(x+y)}\ \frac{c_A^{\ x}\ c_B^{\ y}}{c_{A_xB_y}}$$

$$= \gamma_\pm^{(x+y)}\ K_{diss} \qquad (6\text{-}4.4.3)$$

Equation (6-4.4.3) shows how K_{th} is related to K_{diss} through the activity coefficients of the ions produced. K_{th} is related to ΔG_r^0 (see Chap. 5) so, at a given T, it is constant. K_{diss}, which is not exact, will very likely vary slightly with concentrations.

6-4.4.1 EXAMPLE

From Table 6-3.5.2 the dissociation constant of acetic acid at 0.01 mol dm^{-3} is 1.81×10^{-5} mol dm^{-3}. Estimate K_{th} for acetic acid at this concentration.

Solution

We have to find $\gamma_\pm$ some way. This will involve making some estimates. First, find the concentration of H^+ and Ac^- from Eq. (6-3.5.2), then use these to find $\underline{\mu}$ and $\gamma_\pm$. Equation (6-3.5.2) is

$$K_{diss} = \frac{c_A^{\ x}\ c_B^{\ y}}{c_{A_xB_y}} \qquad (6\text{-}3.5.2)$$

For the degree of dissociation α for acetic acid

$$HAc \rightarrow H^+ + Ac^-$$

conc. $c(1 - \alpha)$ $\quad c\alpha \quad c\alpha$

we know

$$c_{H^+} = c\alpha \qquad c_{Ac^-} = c\alpha \qquad c_{HAc} = c(1 - \alpha) \qquad x = y = 1$$

Thus, from (6-3.5.4),

$$K_{HAc} = \frac{c_{H^-}\ c_{Ac^-}}{c_{HAc}} = \frac{(c\alpha)(c\alpha)}{c(1 - \alpha)}$$

$$= \frac{c\alpha^2}{1 - \alpha}$$

$$= \underline{1.81 \times 10^{-5}}$$

To solve for α, assume $\alpha << 1$. So

$$c\alpha^2 \approx 1.81 \times 10^{-5}$$

$$\alpha^2 = \frac{1.81 \times 10^{-5}}{10^{-2}} = 1.81 \times 10^{-3}$$

$$\alpha = \underline{\underline{0.043}}$$

The assumption $\alpha << 1$ was not too bad. If a more accurate value for α is desired, iterate as follows:

$$\frac{c\alpha^2}{1 - \alpha} \approx \frac{c\alpha^2}{1 - 0.043} \approx 1.81 \times 10^{-5}$$

$$\alpha^2 = \frac{(1.81 \times 10^{-5})(1 - 0.043)}{10^{-2}}$$

$$\alpha = \underline{\underline{0.042}}$$

Very little improvement resulted in this case. However, iteration like this is a very useful technique, as has been emphasized before.

With α known we can find all the concentrations of interest, as well as $\underline{\mu}$ and $\gamma_\pm$.

$$c_{Ac^-} = c_{H^+} = c\alpha = (4.2 \times 10^{-2})(10^{-2}) = 4.2 \times 10^{-4}$$

$$\underline{\mu} = \frac{1}{2} [(4.2 \times 10^{-4})(+1)^2 + (4.2 \times 10^{-4})(-1)^2]$$

$$= \underline{\underline{4.2 \times 10^{-4}}}$$

(Note: $\sqrt{\underline{\mu}} = 0.065$, so the Debye-Hückel limiting law can be used.)

$$\log \gamma_\pm = 0.5091(+1)(-1)(4.2 \times 10^{-3})^{1/2} = -0.0104$$

$$\gamma_\pm = \underline{\underline{0.98}}$$

(This γ is based on concentration in mol dm^{-3} and can only be used with concentration in that set of units.)

Finally, from Eq. (6-4.4.3),

$$K_{th} = \gamma_{\pm}^{2} K_{diss} = (0.98)^2 \ (1.81 \times 10^{-5})$$

$$= \underline{\underline{1.7 \times 10^{-5}}}$$

This is the thermodynamic equilibrium constant for the dissociation of acetic acid.

Other equilibria in solution can be treated in a similar manner, as illustrated for the solubility of a material in the following example.

6-4.4.2 EXAMPLE

The solubility product of lead fluoride at 25°C is approximately 3.4×10^{-8} (based on concentrations in mol dm^{-3}). Find the thermodynamic equilibrium constant.

Solution

You must, of course, know the stoichiometry:

$$\begin{array}{lccc} & PbF_2 \rightleftarrows & Pb^{2+} & + \ 2F^- \\ \text{conc.} & -- & s & 2s \end{array} \qquad (6\text{-}4.4.4)$$

s is defined as the solubility of PbF_2 in mol dm^{-3}. So, $c_{Pb^{2+}} = s$ and $c_{F^-} = 2s$. The solubility product is defined (on the basis of concentration) as

$$K_{sp} = c_{Pb^{2+}} c_{F^-}^{2} = (s)(2s)^2$$

$$= 4s^3 \qquad (6\text{-}4.4.5)$$

The thermodynamic equilibrium constant is written as

$$K_{th} = a_{Pb^{2+}} a_{F^-}^{2} \qquad (6\text{-}4.4.6)$$

(The activity of a pure solid is a constant and is included in K_{th}.)

$$K_{th} = a_{Pb^{2+}} a_{F^-}^{2} = \gamma_+ c_{Pb^{2+}} (\gamma_- c_{F^-})^2$$

$$= \gamma_{\pm}^{3} c_{Pb^{2+}} c_{F^-}^{2}$$

$$= \gamma_{\pm}^{3} K_{sp} \qquad (6\text{-}4.4.7)$$

We are now ready to do the same thing as in Example 6-4.4.1. Try it before reading on.

$$K_{sp} = 4s^3 = 3.4 \times 10^{-8}$$

$$s^3 = \frac{3.4 \times 10^{-8}}{4} = 8.5 \times 10^{-9}$$

$$s = \underline{\underline{2.04 \times 10^{-3} \text{ (mol dm}^{-3})}}$$

From s, find $\underline{\mu}$ and $\gamma_\pm$:

$$\underline{\mu} = \frac{1}{2}(c_{Pb^{2+}} z_{Pb^{2+}}^2 + c_{F^-} z_{F^-}^2)$$

$$= \frac{1}{2}[s(+2)^2 + (2s)(-1)^2]$$

$$= 3s$$

$$= 3(2.04 \times 10^{-3})$$

$$= \underline{\underline{6.12 \times 10^{-3}}}$$

(Note: $\sqrt{\underline{\mu}} = 0.078$, so the Debye-Hückel limiting law can be used.)

$$\log \gamma_\pm = 0.5091(+2)(-1)(6.12 \times 10^{-3})^{\frac{1}{2}}$$

$$= -7.97 \times 10^{-2}$$

$$= \underline{\underline{0.833}}$$

Then $K_{th} = (0.833)^2 K_{sp} = (6.94 \times 10^{-1})(3.4 \times 10^{-8})$

$$= \underline{\underline{2.4 \times 10^{-8}}}$$

You now have had an introduction to calculating and using activities and activity coefficients for solutions. The problems in Sec. 6-4.6 will sharpen your skills.

6-4.5 SUMMARY OF DEFINITIONS AND SYMBOLS

The following list of definitions and symbols is to aid you in keeping track of the large number of new terms found in this section.

a_+, a_- — Activities of positive and negative ions in solution. Used in place of concentrations to account for nonideality.

γ_+, γ_- — Activity coefficients that relate activities to concentration; $a_+ = \gamma_+ c_+$; $a_- = \gamma_- c_-$

$a_\pm, \gamma_\pm, c_\pm$ — Mean ionic activity, mean ionic activity coefficient, and mean ionic concentration. For a molecule A_xB_y that dissociates into x positive ions and y negative ions; all three terms have the same form: $a_\pm = (a_+^x a_-^y)^{1/(x+y)}$

$\underline{\mu}$ — The ionic strength. This gives the effective concentration of all the ions in a particular solution; $\underline{\mu} = \frac{1}{2}\sum^i c_i z_i^2$. z_i is the charge of ion i in the solution.

$\log \gamma_\pm = 0.5091\, z_+ z_- \sqrt{\underline{\mu}}$. The Debye-Hückel limiting law. This relates the mean ionic activity coefficient to the charges on the ions (including signs) and the ionic strength of the *whole* solution. It is limited to solutions with $\sqrt{\underline{\mu}} < 0.1$. Note that log is base 10.

K_{diss} — The dissociation constant of a weak electrolyte in terms of concentration. For A_xB_y, $K_{diss} = (c_A^x c_B^y)/(c_{A_xB_y})$

K_{th} The dissociation constant of a weak electrolyte in terms of activities. For A_xB_y,

$$K_{th} = \frac{a_A^{\ x} a_B^{\ y}}{a_{A_xB_y}} = \frac{(\gamma_A c_A)^x (\gamma_B c_B)^y}{c_{A_xB_y}} \qquad \gamma_{A_xB_y} = 1$$

since A_xB_y is noncharged.

These last two quantities are related by $K_{th} = \gamma_\pm^{x+y} K_{diss}$. Similar equations apply for solubility products. The only difference is that the activity or concentration of the solid material is not expressed explicitly. It is included in the appropriate K.

6-4.6 PROBLEMS

1. Define the mean activity coefficient (by means of a mathematical relationship) for the compound A_2B_3 that completely dissociates in a solution to A^{3+} and B^{2-} in terms of ionic activity coefficients.

2. Find the ionic strength, the mean activity coefficient, and the mean activity of 0.00500 m NaCl.

3. For $K_3Fe(CN)_6$ which dissociates completely to $3K^+ + Fe(CN)_6^{\ 3-}$ the mean activity coefficient of a 0.0100 molal solution is 0.571. Determine the mean molality and the mean activity.

4. A 0.00100 molar solution of $K_3Fe(CN)_6$ [see Problem (3)] has a mean activity coefficient of 0.808. Calculate the mean activity coefficient from the Debye-Hückel limiting law and compare with the observed value.

5. A solution is 0.00300 molar in NaCl and 0.00200 molar in $ZnSO_4$. Calculate the activity coefficient of SO_4^{2-} in this solution.

6. A solution is prepared by mixing 0.0500 dm^3 of 0.00300 M $FeSO_4$ (completely dissociated) with 0.100 dm^3 of 0.00200 M $FeCl_3$ (also completely dissociated). Assume the volumes are additive and calculate the following quantities (also assume the Debye-Hückel limiting law is valid):

   a. All ionic concentrations
   b. The ionic strength of the solution
   c. The mean activity coefficient of each electrolyte
   d. The mean activity of each electrolyte

7. The dissociation constant of a 0.0100 molar solution of propionic acid ($CH_3CH_2COOH \rightleftarrows CH_3CH_2COO^- + H^+$) is $K_{diss} = 1.34 \times 10^{-5}$ at 298 K. Determine the following:

a. The degree of ionization

b. The ionic strength of the solution

c. The mean activity coefficient and the mean activity

d. The thermodynamic equilibrium constant K_{th}.

8. The ionic strength of a solution of $BaCl_2$ is found to be $\underline{\mu}$ = 0.015 molar. If $BaCl_2$ is assumed to dissociate completely, determine the concentration and ionic activity coefficient of Cl^- in this solution.

9. A solution of $CaCl_2$ and $Ca_3(PO_4)_2$ has an ionic strength of 0.0825 molar. Assume both salts dissociate completely. If $[Ca_3(PO_4)_2]$ = 0.00500 M and the total ionic activity coefficient for Ca^{2+} is 0.260, and only if the Debye-Hückel equation applies, determine the concentration of Cl^- in the solution.

6-4.7 SELF-TEST

1. Define or briefly discuss (a) ionic strength, (b) mean concentration, and (c) mean activity coefficient.

2. For a solution of 0.00500 molar $CdCl_2$ which completely dissociates, find the mean activity of $CdCl_2$.

3. A solution is made by mixing 0.0250 dm^3 of a 0.00600 M solution of Na_2SO_4 and 0.0750 dm^3 of a 0.00100 M solution of $FeCl_2$. Assume the Debye-Hückel equation is valid and each electrolyte is completely dissociated. Find the ionic strength of the solution and the mean activity and mean activity coefficient of each electrolyte.

4. (a) Derive an equation relating the solubility of CoF_3 to the mean activity coefficient and the thermodynamic solubility product. (b) If the mean activity coefficient of the saturated solution at 298.2 K is 0.781 and the thermodynamic solubility product is 0.100, find the solubility s.

5. The dissociation constant (based on concentration) for HCN ($HCN \rightleftarrows H^+ + CN^-$) at 298.15 K is 4.93×10^{-10}. Determine α and estimate K_{th} for HCN. Assume c_{HCN} = 0.0100 mol dm^{-3}.

6-4.8 ADDITIONAL READING

1. W. J. Hamer, "Theoretical Mean Activity Coefficients of Strong Electrolytes in Aqueous Solutions from 0 to 100°C," NSRDS-NBS 24 (see Sec. 1-1.19A-16).

6-5 ELECTROMOTIVE FORCE

OBJECTIVES

You shall be able to

1. Write the proper electrochemical cell notation given the chemical reaction that takes place, or vice versa
2. Determine the reversible cell potential (with correct sign) for a given electrochemical cell from a table of standard electrode potentials
3. Use cell potentials to determine any of the following:

    Free energy of the reaction involved
    Entropy and enthalpy of reaction if the cell potential as a function of temperature is given
    Activities and activity coefficients of the electrolytes in the cell
    Equilibrium constant of the reaction
    Solubility products or dissociation constants for partially soluble or partially dissociated electrolytes

4. Discuss the operation and importance of fuel cells

6-5.1 INTRODUCTION

This last section of our study of solution processes will involve electrochemical cells. Electrochemical cells, as the name implies, are devices for the interconversion of electrical and chemical energy. Not all electrochemical cells involve solutions in the liquid state, which has been the topic under consideration in this chapter. However, almost all electrochemical cells involve solutions of some sort.

Aside from their importance as electrical energy storage devices and power sources, electrochemical cells provide a very useful means of determining thermodynamic functions, as we shall see. Activity coefficients, free energies, and entropies can be determined by making electrical measurements on cells.

Fuel cells are becoming more and more important as power sources. In a fuel cell a chemical reaction is carried out in a controlled manner such that the chemical energy of the reaction is converted to electrical energy. In this case the cell is acting as a primary source of power instead of as a storage device.

The cells in which we will be interested and those that yield useful thermodynamic data are reversible cells. The term reversible means the same as in our study of thermodynamics. A cell connected to an external circuit will cause a flow of electricity in that circuit, while a particular chemical reaction is taking place within the cell. There is another condition imposed on actual measurements on electrochemical cells. We make these measurements with devices which compare the cell potential with a known potential in such a manner that no current is drawn from the cell. This insures that the *reversible* potential of the cell is being measured. More on this later.

Below a brief description of the methods of measuring electrochemical cell potentials is given and standard potentials are defined. Then the proper procedure for writing cell notation is reviewed. Some discussion is given to standard (or reference) cells. Finally, the relationship of electrochemical data to thermodynamics is presented.

Very little discussion of the various types of cells will be found (except for fuel cells), even though in practice, this is very important. If you know cell notation and the conventions for assigning signs to the cell potential, you will be able to learn what you need to know about specific types of cells when the occasion arises. Many topics of significance in actual practice will be treated only superficially, if at all. Two of these are overvoltage and polarization, both of which limit the utility of electrochemical processes. For example, the rate of discharge of a battery is limited by *polarization* that results from the buildup of reaction products in the electrode compartments. *Overvoltage* is the excess voltage that must be supplied above the theoretical value to make a given electrode reaction occur. This can be very significant in electrolytic processes that use high current densities. Both problems are significant in industrial processes.

6-5.2 MEASUREMENT OF THE EMF OF CELLS AND STANDARD CELLS

First, a definition is required for the electromotive force (emf) of an electrochemical cell. *Electromotive force* is defined as the difference in potential which causes a current to flow from an electrode of higher potential to one at a lower potential. An electrode is the region in a cell where an oxidation or reduction takes place. An electrode is also the point to which connection to an external circuit is made. It is the source (oxidation) or sink (reduction) for electrons for the external circuit.

Let us look at the most common method of measuring the emf of a cell. The measurement is usually done with a *potentiometer* which allows a direct reading of the emf with no current drain from the cell.

To get the *reversible potential* of a cell it is important to eliminate any current drain since any current will result in a decreased potential reading due to a potential drop across the internal resistance of the cell (think about that for a moment).

POTENTIOMETER. Figure 6-5.2.1 shows the schematic of a *potentiometer*. This device allows the direct reading of the cell emf with essentially no current since it balances the potential of a *working cell* against that of the cell of interest, so a *null* is observed in which no current flows in the cell.

The following procedure is used to determine the potential of an unknown cell. (Let us assume the standard cell has a potential of 1.0181 V for this example.)

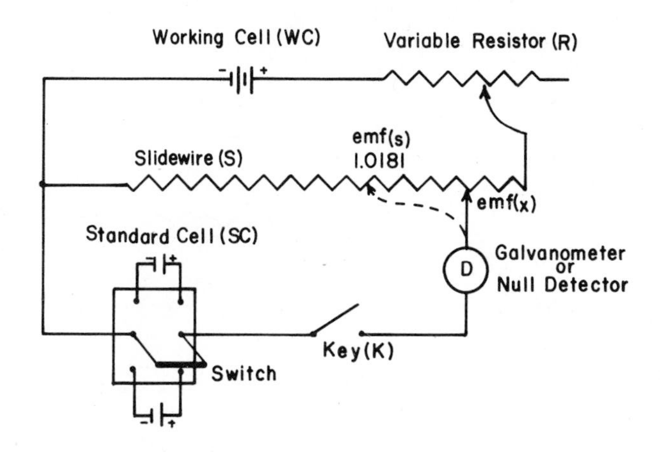

FIG. 6-5.2.1 A schematic of a direct reading potentiometer.

1. The switch is set to the S.C. position and the slidewire contact set at 1.0181 V. The key K is depressed (by tapping) and the variable resistor R is adjusted until no deflection is observed on D. The slidewire S is now calibrated so that the reading on the slidewire corresponds to the actual potential drop being measured.
2. Set the switch to X and adjust S while tapping K until no deflection is observed on D. The reading on S is the same as the reading on X and since the potential on S just matches that across X, we have a null (no current is flowing) and the potential read for X is the reversible emf.

Potentiometers of this general type can be made that can measure small values of emf to a precision of 1×10^{-7} V. This, of course, requires considerable sophistication but the principle is as given above.

STANDARD CELLS. The most common standard cell is the *saturated Weston cell.* This cell has many desirable characteristics. It gives a reproducible voltage, is reversible, and does not suffer permanent damage if current is passed through it. Also, it has a low temperature coefficient meaning the emf varies only slightly with temperature. Figure 6-5.2.2 is a sketch of such a cell.

The reactions involved are:

$$\text{Left:} \quad \tfrac{8}{3}H_2O + Cd(s) + SO_4^{2-} \rightarrow CdSO_4 \cdot \tfrac{8}{3}H_2O(s) + 2e^-$$

$$\text{Right:} \quad Hg_2SO_4(s) + 2e^- \rightarrow 2Hg(\ell) + SO_4^{2-} \qquad (6\text{-}5.2.1)$$

$$\text{Net:} \quad \tfrac{8}{3}H_2O + Cd(s) + Hg_2SO_4(s) \rightarrow 2Hg(\ell) + CdSO_4 \cdot \tfrac{8}{3}H_2O(s)$$

This net reaction occurs when the cell is producing current. If current is being passed through the cell from an external source, the reverse reaction takes place.

The potential of this cell in international volts at t (in °C) is given by

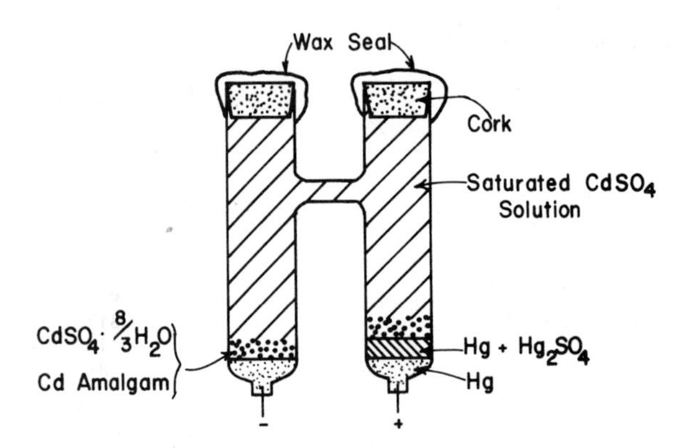

FIG. 6-5.2.2 A saturated Weston cell.

$$\mathcal{E}_t = 1.01830 - 4.06 \times 10^{-5}(t - 20) - 9.5 \times 10^{-7}(t - 20)^2 \tag{6-5.2.2}$$

At 25°C, $\mathcal{E}_{25} = 1.01807$ V.

Two other cells of particular importance are half-cells which are used to measure the potential of other half-cells. A half-cell involves only one-half of an oxidation-reduction reaction. No reaction can take place until the half-cell is combined with another half-cell so that reduction and oxidation are both possible. This will become clearer below when some examples are given.

The *normal hydrogen electrode* is one of the standards used for determining the emfs of other cells. In fact, the normal hydrogen electrode is assigned an emf of *0* arbitrarily and is the primary reference electrode for all of electrochemistry. Figure 6-5.2.3 shows a normal hydrogen electrode. The reaction is $H_2(g) \rightarrow 2H^+(aq) + 2e^-$. As before, this reaction cannot occur unless the electrode is connected to another half-cell. $\mathcal{E}^0_{NHE} = 0$. (The reverse reaction is, of course, also possible.)

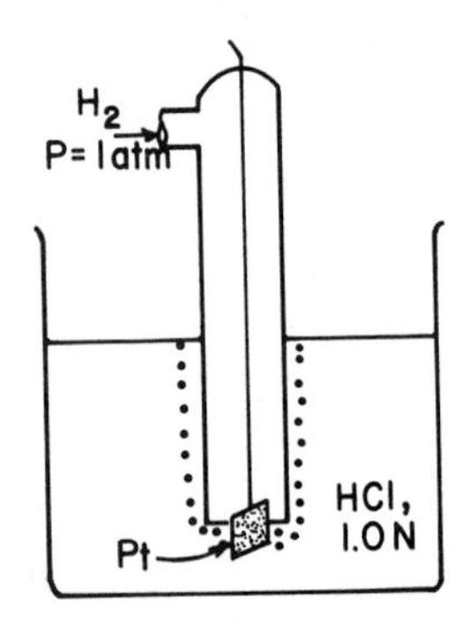

FIG. 6-5.2.3 The normal hudrogen electrode.

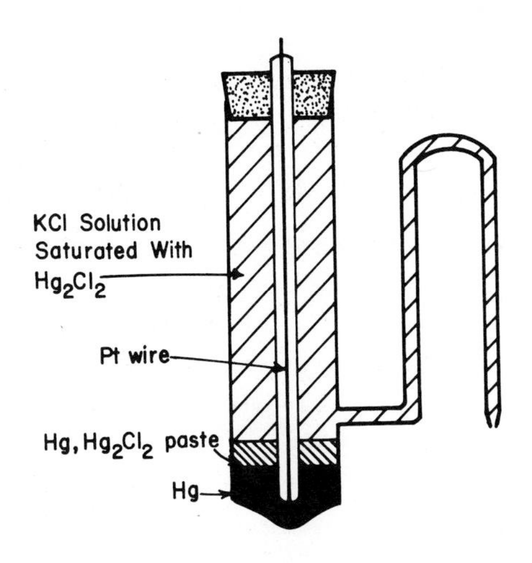

FIG. 6-5.2.4 The calomel electrode (saturated).

A secondary reference electrode is the *calomel electrode*. Its potential is measured by comparison to the NHE as will be seen shortly. Figure 6-5.2.4 shows the sketch of a calomel electrode. The reaction is: $Hg_2Cl_2(s) + 2e^- \rightarrow 2Hg(\ell) + 2Cl^-(c)$ where (c) on the Cl^- indicates the concentration of the chloride. The reaction is reversible, and its potential depends on the concentration of the Cl^- as shown later in Table 6-5.3.1. This half-cell or electrode must be paired via a solution to another electrode before a potential can be developed. Other half-cells can be used as secondary standards, but these will not be considered.

6-5.3 ELECTRODE POTENTIALS, CELL REACTIONS, CELL NOTATION, AND SIGN CONVENTIONS FOR EMF

Before a cell is well characterized, the following must be given: sign of the emf, the reactions taking place, the activities and phases of all the materials present, and the temperature.

The Weston cell will be made an example to define all terms mentioned in the title to this section (see Fig. 6-5.2.2). *This section is most important; read it carefully.*

The potential of the cell when the following reaction takes place is 1.01807 V at 25°C.

$$\frac{8}{3}H_2O + Cd(s) + Hg_2SO_4(s) \rightarrow 2Hg(\ell) + CdSO_4\cdot\frac{8}{3}H_2O(s) \qquad (6\text{-}5.2.1)$$

The cell notation for this cell is

$$Cd(s)\,|\,CdSO_4\cdot\frac{8}{3}H_2O(s)\,|\,CdSO_4(\text{sat'd. sol'n.})\,|\,Hg\text{-}Hg_2SO_4(s)\,|\,Hg(\ell)\,|\,Pt(s) \qquad (6\text{-}5.3.1)$$

Each vertical line in this cell corresponds to the separation between two phases. *The emf of this or any cell is defined as positive if the tendency is for electrons to flow through the external circuit from the electrode written on the left to the electrode written on the right.*

Using the above convention it follows that the *oxidation reaction occurs at the electrode written on the left, and reduction occurs at the electrode written on the right.*

Finally, since electrons flow spontaneously from a negative electrode through the external circuit to the positive electrode, *the electrode written to the left will be negative* (with respect to the external circuit) and *that on the right will be positive.*

(The above is important in determining the direction of a reaction. *If an electrode A is connected to the negative terminal of a potentiometer and electrode B is connected to the positive terminal at balance, then oxidation is occuring at electrode A* (it is the source of electrons for the external circuit) *and reduction is taking place at electrode B.*)

Let us combine our two half-cells from the previous section to illustrate how single electrode potentials can be determined. Figure 6-5.3.1 shows an arrangement that is useful for measuring the emf of these two half-cells. The side arm on the calomel electrode is present to prevent solution mixing in the two electrode compartments.

The potential of this cell is observed to be 0.2800 V at 25°C when the NHE is connected to the negative terminal of the potentiometer. From our convention this indicates oxidation at the NHE and reduction at the calomel electrode. The following reaction occurs with the emf indicated to the right of each reaction.

$$\text{Left} \qquad H_2(1\ \text{atm}) \rightarrow 2H^+(1\ N) + 2e^- \qquad \mathcal{E}^0 = 0 \ \text{(by definition)}$$

$$\text{Right} \qquad Hg_2Cl_2(s) + 2e^- \rightarrow 2Hg(\ell) + 2Cl^-(1\ N) \qquad \mathcal{E}^0 = ? \tag{6-5.3.2}$$

$$\text{Net: } H_2(1\ \text{atm}) + Hg_2Cl_2(s) \rightarrow 2H^+(1\ N) + 2Cl^-(1\ N) + 2Hg(\ell) \qquad \mathcal{E}^0 = 0.2800\ \text{V}$$

The cell notation is:

$$Pt|H_2(1\ \text{atm})|HCl(1\ N)||KCl(1\ N)|Hg\text{-}Hg_2Cl_2(s)|Hg(\ell)|Pt(s) \tag{6-5.3.3}$$

The double vertical bar represents a salt bridge (such as the KCl in the side arm of the calomel electrode) or some membrane used to prevent the mixing of solutions in the electrode compartments.

How do we find the emf of the calomel cell reaction as written? We simply treat emf as we do any other thermodynamic function. *Write the emf for the reaction given, then the overall emf for the reaction that results from summing all reactions is the sum of individual emfs.* For Eq. (6-5.3.2):

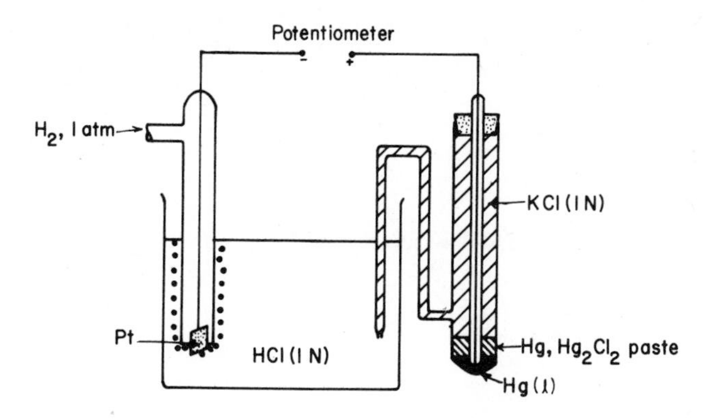

FIG. 6-5.3.1 A cell involving the NHE and a 1 N calomel electrode.

$$\mathcal{E}^0_{overall} = \mathcal{E}^0_{left} + \mathcal{E}^0_{right}$$

$$0.2800 \text{ V} = 0 \text{ V} + \mathcal{E}^0_{right} \tag{6-5.3.4}$$

Then $$\mathcal{E}^0_{right} = \mathcal{E}^0_{red} \text{ (calomel)} = 0.2800 \text{ V}$$

where red means the reduction reaction is considered.

As mentioned previously, the potential of the calomel electrode depends on the concentration of the Cl^- in solution. Table 6-5.3.1 contains some useful information on calomel cells. The emf of each cell is determined by measuring it against the NHE as shown above.

Once the above have been determined it is more convenient to use calomel electrodes as the standard than it is to use the NHE. It is difficult to use an NHE since we must have a cylinder of hydrogen present and care must be taken not to contaminate the platinum electrode. The most common calomel electrode is the SCE, since the KCl solution is easily kept saturated by keeping excess KCl present. In the 0.1 CE and 1 CE care must be taken to keep the concentration of KCl constant.

6-5.4 STANDARD ELECTRODE POTENTIALS

Suppose we want to know the potential of the reaction $AgCl(s) + 1e^- \rightarrow Ag(s) + Cl^-$ $(a = 1)$. The activity is specified as a = 1 for each ionic species present to ensure that the measured emf is the standard value $\mathcal{E}^0$. Shortly you will see how emf varies with the concentration. Right now let us restrict the discussion to cases in which the activity of each substance is 1. This is automatically true for pure liquids and solids (see discussion of activity coefficients in Sec. 6-5.6).

To determine the emf of this reaction we must design a cell in which the above reaction can be coupled with a standard half-cell. Let us, in this case, use the

Table 6-5.3.1
Potentials of Calomel Electrodes

Electrode	Notation	emf $\mathcal{E}$ (at t °C)	Reaction
0.100 N (0.1 CE)	$KCl(0.1) \vert Hg_2Cl_2\text{-}Hg(s) \vert Hg(\ell)$	$0.3338 - 7 \times 10^{-5}(t - 25)$	$Hg_2Cl_2(s) + 2e^- \rightarrow 2Hg(\ell) + 2Cl^-(0.1\ N)$
1 N (1 CE)	$KCl(1) \vert Hg_2Cl_2\text{-}Hg(s) \vert Hg(\ell)$	$0.2800 - 2.4 \times 10^{-4}(t - 25)$	$Hg_2Cl_2(s) + 2e^- \rightarrow 2Hg(\ell) + 2Cl^-(1\ N)$
Saturated calomel (SCE)	$KCl(sat'd) \vert Hg_2Cl_2\text{-}Hg(s) \vert Hg(\ell)$	$0.2415 - 7.6 \times 10^{-4}(t - 25)$	$Hg_2Cl_2(s) + 2e^- \rightarrow 2Hg(\ell) + 2Cl^-(sat'd)$

saturated calomel cell (SCE) as the other electrode to produce the following cell

$$Pt(s)|Hg(\ell)|Hg_2Cl_2\text{-}Hg(s)|KCl(sat'd)||KCl(a = 1)|AgCl(s)|Ag(s) \qquad (6\text{-}5.4.1)$$

The measured potential is 0.0190 V with the *Ag electrode negative*. This means the *silver is being oxidized*, indicating the above cell is written incorrectly. (The above reaction implies that AgCl is being reduced.) So, let us rewrite the cell as

$$Ag(s)|AgCl(s)|KCl(a = 1)||KCl(sat'd)|Hg_2Cl_2\text{-}Hg(s)|Hg(\ell)|Pt(s) \qquad (6\text{-}5.4.2)$$

The reactions involved in (6-5.4.2) are

$$2[Ag(s) + Cl^-(a = 1) \rightarrow AgCl(s) + 1e^-] \qquad \mathcal{E}^0_{Ag,AgCl} = ?$$

$$Hg_2Cl_2(s) + 2e^- \rightarrow 2Hg(\ell) + 2Cl^-(sat'd) \qquad \mathcal{E}^0_{SCE} = 0.2415\ V$$

$$2Ag(s) + Hg_2Cl_2(s) + 2Cl^-(a = 1) \rightarrow 2AgCl(s) + 2Hg(\ell) + 2Cl^-(sat'd)$$

$$\mathcal{E}^0_r = 0.0190\ V$$

where $\mathcal{E}^0_r$ is $\mathcal{E}^0$ for the reaction. (*Note*: We multiply the top reaction by 2 so we can cancel electrons. We *do not* multiply the $\mathcal{E}^0$ value by 2. *$\mathcal{E}^0$ is independent of the amount of material present.*)

We know that

$$\mathcal{E}^0_r = \mathcal{E}^0_{Ag,AgCl} + \mathcal{E}^0_{SCE}$$

$$0.0190\ V = \mathcal{E}^0_{Ag,AgCl} + 0.2415\ V$$

or

$$\mathcal{E}^0_{Ag,AgCl} = 0.0190 - 0.2415$$

$$= -0.2225\ V \qquad (6\text{-}5.4.3)$$

The subscript indicates the direction of reaction.

$$\mathcal{E}^0_{Ag,AgCl} = -0.2225\ V$$

means

$$Ag(s) + Cl^-(a = 1) \rightarrow AgCl(s) + 1e^- \qquad \mathcal{E}^0_{Ag,AgCl} = -0.2225\ V \qquad (6\text{-}5.4.4)$$

You should be able to verify that for

$$AgCl(s) + 1e^- \rightarrow Ag(s) + Cl^-(a = 1) \qquad \mathcal{E}^0_{AgCl,Ag} = +0.2225\ V \qquad (6\text{-}5.4.5)$$

This last value is the *standard reduction potential* or *standard electrode potential* and is the value usually tabulated. Table 6-5.4.1 lists some reduction potentials that we may be using. The following will clarify the use of these tabulated values.

Table 6-5.4.1
Standard (Reduction) Electrode Potentials at 298.15 K[a]

Electrode	Reaction	$\mathscr{E}^0$ (V)
$Ag^+ \| Ag$	$Ag^+ + 1e^- \rightarrow Ag$	+0.7996
$Br^- \| AgBr \| Ag$	$AgBr + e^- \rightarrow Ag + Br^-$	+0.0713
$Cl^- \| AgCl \| Ag$	$AgCl + e^- \rightarrow Ag + I^-$	+0.2225
$I^- \| AgI \| Ag$	$AgI + e^- \rightarrow Ag + I^-$	-0.1519
$Br^- \| Br_2(\ell) \| Pt$	$Br_2(\ell) + 2e^- \rightarrow 2Br^-$	+1.065
$Br^- \| Br_2(aq) \| Pt$	$Br_2(aq) + 2e^- \rightarrow 2Br^-$	+1.087
$Ca^{2+} \| Ca$	$Ca^{2+} + 2e^- \rightarrow Ca$	-2.76
$Cl^-(sat'd) \| Hg_2Cl_2\text{-}Hg(s) \| Hg(\ell)$	$Hg_2Cl_2 + 2e^- \rightarrow 2Hg(\ell) + 2Cl^-$	+0.2415
$Cd^{2+} \| Cd$	$Cd^{2+} + 2e^- \rightarrow Cd$	-0.4026
$Cd^{2+} \| Cd(Hg)$	$Cd^{2+} + 2e^- + Hg(\ell) \rightarrow Cd(Hg)$	-0.3521
$Cl^- \| Cl_2(g) \| Pt$	$Cl_2(g) + 2e^- \rightarrow 2Cl^-$	+1.3583
$Co^{2+} \| Co$	$Co^{2+} + 2e^- \rightarrow Co$	-0.28
$Cu^+, Cu^{2+} \| Pt$	$Cu^{2+} + 1e^- \rightarrow Cu^+$	+ 0.158
$Cu^+ \| Cu$	$Cu^+ + 1e^- \rightarrow Cu$	+0.522
$Cu^{2+} \| Cu$	$Cu^{2+} + 2e^- \rightarrow Cu$	+0.3402
$Fe^{2+}, Fe^{3+} \| Pt$	$Fe^{3+} + 1e^- \rightarrow Fe^{2+}$	+0.770
$Fe^{2+} \| Fe$	$Fe^{2+} + 2e^- \rightarrow Fe$	-0.409
$H_2(1\ atm), H^+(a = 1) \| Pt$	$2H^+(a = 1) + 2e^- \rightarrow H_2$	0.000
$K^+ \| K$	$K^+ + 1e^- \rightarrow K$	-2.924
$Li^+ \| Li$	$Li^+ + 1e^- \rightarrow Li$	-3.045
$Mg^{2+} \| Mg$	$Mg^{2+} + 2e^- \rightarrow Mg$	-2.375
$O_2, H^+ \| Pt$	$O_2 + 4H^+ + 4e^- \rightarrow 2H_2O$	+1.229
$O_2, OH^- \| Pt$	$O_2 + 2H_2O + 4e^- \rightarrow 4OH^-$	+0.401
$Pb^{2+} \| Pb$	$Pb^{2+} + 2e^- \rightarrow Pb$	-0.1263
$Zn^{2+} \| Zn$	$Zn^{2+} + 2e^- \rightarrow Zn$	-0.7628
$Zn^{2+} \| Zn(Hg)$	$Zn^{2+} + 2e^- + Hg \rightarrow Zn(Hg)$	-0.7628

[a] All solution activities equal 1. (From the Chemical Rubber Company *Handbook of Chemistry and Physics*.

6-5.4.1 EXAMPLE

Write the cell notation for the reaction

$$Mg(s) + Pb^{2+}(a = 1) \rightarrow Mg^{2+}(a = 1) + 2Cl^-(a = 1) + Pb(s)$$

and determine $\mathscr{E}^0$ for the reaction.

Solution

The oxidation reaction is

$$Mg \rightarrow Mg^{2+} + 2e^-$$

$$\mathcal{E}^0_{Mg,Mg^{2+}} = -\mathcal{E}^0_{Mg^{2+},Mg}$$

$$= -(-2.375) = \underline{2.375\ V}$$

and the reduction reaction is

$$Pb^{2+} + 2e^- \rightarrow Pb$$

$$\mathcal{E}^0_{Pb^{2+},Pb} = \underline{-0.1263\ V}$$

Add these two to get the correct overall reaction and add the appropriate emf values.

$$\mathcal{E}^0_r = \mathcal{E}^0_{Mg,Mg^{2+}} + \mathcal{E}^0_{Pb^{2+},Pb}$$

$$= 2.375 - 0.1263 = \underline{\underline{2.249\ V}}$$

Note: Some texts use different conventions. In some cases only reduction potentials are written, and the potential of the reaction is given by

$$\mathcal{E}_{reaction} = \mathcal{E}^0_{right\ electrode} - \mathcal{E}^0_{left\ electrode}$$

We are accustomed to adjusting the sign of thermodynamic functions to fit the reactions as written, then summing reactions and corresponding function values. If we write a reduction reaction, we can take a value directly out of the table. If we write an oxidation reaction we have reversed a reduction reaction and must reverse the sign of $\mathcal{E}^0$ found in the table of reduction potentials. Then, we add these potentials. Either procedure gives the same result as long as you consistently use that procedure.

There are other areas of confusion. Some tables of reduction potentials are given in which reactions are written as oxidations. Be careful in using emf values. Make sure you are aware of the conventions being used. They all lead to the same result if applied correctly.

Now to finish the problem. We write the cell notation by writing reduction to the right.

$$Mg(s)|Mg^{2+},Cl^-(a = 1)||Pb^{2+},Cl^-(a = 1)|Pb(s) \qquad \mathcal{E}^0 = 2.249\ V$$

6-5.4.2 EXAMPLE

Given the following electrochemical cell at 25°C, find $\mathscr{E}^0$ and write the overall reaction described by the cell.

$$Ag(s)|AgCl(s),KCl(a = 1)||KBr(a = 1)|Br_2(\ell)|Pt(s)$$

Solution

Reduction (right)

$$Br_2(\ell) + 2e^- \rightarrow 2Br^-(a = 1) \qquad \mathscr{E}^0_{Br_2,Br^-} = 1.065\ V$$

Oxidation (left)

$$2\ Ag(s) + Cl^-(a = 1) \rightarrow AgCl(s) + e^- \qquad \mathscr{E}^0_{Ag,AgCl} = -\mathscr{E}^0_{AgCl,Ag} = -0.2225\ V$$

Net

$$2Ag(s) + Br_2(\ell) + 2Cl^-(a = 1) \rightarrow 2AgCl(s) + 2Br^- \qquad \mathscr{E}^0_r = 0.843\ V$$

Note: Which electrode, in this and the previous example, would be attached to the negative terminal of the potentiometer?

6-5.5 THERMODYNAMIC VALUES FROM EMF

The electrical work performed by the transfer of 1 equiv. of electrons at an emf $\mathscr{E}$ is equal to

$$\text{Electrical work (equiv.)}^{-1} = \mathscr{F}\mathscr{E} \qquad (6\text{-}5.5.1)$$

where $\mathscr{F}$ = 96,490 coulombs (C) equiv.$^{-1}$, the quantity of charge in 1 equiv. of electrons (1 equiv. is just Avagadro's number, 6.023×10^{23}.) For n equiv.,

$$\text{Electrical work} = n\mathscr{F}\mathscr{E}\ (\text{volt} \cdot \text{coulomb} = \text{Joule}) \qquad (6\text{-}5.5.2)$$

This electrical work is performed at the expense of the free energy of the cell. ΔG decreases by $n\mathscr{F}\mathscr{E}$ for n equiv. of electrons transferred. Thus

$$\Delta G = -n\mathscr{F}\,\mathscr{E} \qquad (6\text{-}5.5.3)$$

This equation ties electrochemistry to thermodynamics.

If you refer to Eq. (5-2.3.2) in Chap. 5 you should see how to find ΔS for a reaction if we know $\mathscr{E}$ as a function of T at constant pressure.

$$\frac{d\,\Delta G}{dT} = -\Delta S \qquad (5\text{-}2.3.2)$$

$$\Delta S = -\frac{d\,\Delta G}{dT} = -\,n\mathscr{F}\left(\frac{d\mathscr{E}}{dT}\right) \qquad (6\text{-}5.5.4)$$

or, to be more specific,

$$\Delta S = -n\mathcal{F}\left(\frac{\partial \mathcal{E}}{\partial T}\right)_P \tag{6-5.5.5}$$

Also, from Eq. (5-2.3.6) we can write

$$\Delta H = -T^2\left[\frac{\partial}{\partial T}\left(\frac{\Delta G}{T}\right)\right]_P \tag{6-5.5.6}$$

$$\Delta H = +n\mathcal{F}T^2\left[\frac{\partial}{\partial T}\left(\frac{\mathcal{E}}{T}\right)\right]$$

$$= n\mathcal{F}T^2\left[\frac{T(\partial\mathcal{E}/\partial T) - \mathcal{E}}{T^2}\right]$$

$$= n\mathcal{F}\left[T\left(\frac{\partial \mathcal{E}}{\partial T}\right)_P - \mathcal{E}\right] \tag{6-5.5.7}$$

These equations give us another means of finding values for ΔH, ΔS, and ΔG for reactions for which tabulated values are not available. We do need $\mathcal{E}$ as a function of T, however.

6-5.5.1 EXAMPLE

From the data below find the values for ΔH, ΔS, and ΔG for the reaction at 298.15 K.

$$H_2(g) + 2AgCl(s) \rightarrow 2Ag(s) + 2H^+Cl^-(m = 0.0100)$$

carried out in the cell

$$Pt|H_2(1\ atm)|HCl(0.0100)|AgCl(s)|Ag(s)$$

T (K)	273.15	293.15	298.15	303.15
$\mathcal{E}$ (V)	0.45787	0.46323	0.46412	0.46493

Solution

ΔG

This is easy. There are two electrons involved, so

$$\Delta G = -n\mathcal{F}\mathcal{E} = -(2)(96{,}490\ C)(0.46412\ V)$$

$$= \underline{\underline{8.96 \times 10^4\ J}}$$

ΔH and ΔS

Here we need $(\partial \mathcal{E}/\partial T)_p$. We can get it from a plot of the above data (see following figure). Using $(\partial \mathcal{E}/\partial T)_p = 1.7 \times 10^{-4}$ V K^{-1} in Eq. (6-5.5.5) gives

$$\Delta S = -(2)(96{,}490 \text{ C})(1.7 \times 10^{-4} \text{ V K}^{-1})$$
$$= \underline{\underline{33 \text{ J K}^{-1}}}$$

and from Eq. (6-5.5.7),

$$\Delta H = (2)(96{,}490 \text{ C})[298.15(1.7 \times 10^{-4}) - 0.46412] \text{ V}$$
$$= \underline{\underline{-8.0 \times 10^4 \text{ J}}}$$

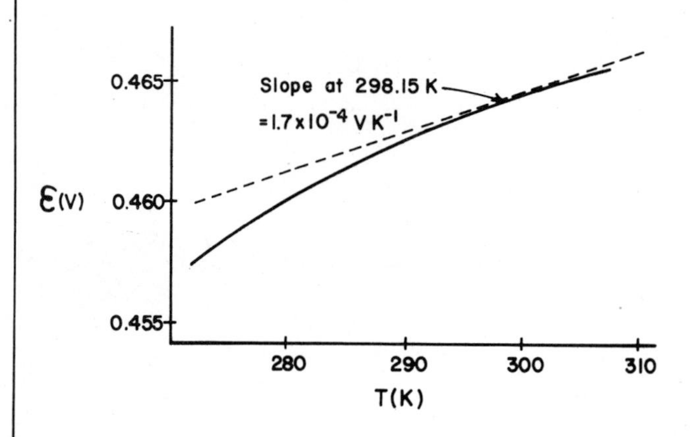

We should keep only two significant figures, since the slope determination on this plot is not very precise.

6-5.6 EMF, ACTIVITIES, AND EQUILIBRIUM CONSTANTS

ACTIVITY COEFFICIENTS. EMF measurements offer one of the best ways of determining activities (and thereby, activity coefficients) for electrolytes. The relationships for a general reaction will be derived, then applied to specific examples.

You should be able to write the free energy change for the following reaction in terms of activities from what you have studied previously in this chapter.

$$aA + bB \rightarrow cC + dD \tag{6-5.6.1}$$

The free energy of each species is given by

$$G_A = a\bar{G}_A = a\bar{G}_A^0 + aRT \ln a_A$$
$$G_B = b\bar{G}_B = b\bar{G}_B^0 + bRT \ln a_B \qquad (6\text{-}5.6.2)$$

And so on. [See discussion following Eq. (6-5.6.5).] And

$$\Delta G_r = G_C + G_D - G_B - G_A$$
$$= c\bar{G}_C^0 + RT \ln a_C^c + d\bar{G}_D^0 + RT \ln a_D^d - a\bar{G}_A^0 - RT \ln a_A^a - b\bar{G}_B^0 - RT \ln a_B^b$$

$$\Delta G_R = \Delta G^0 + RT \ln \frac{a_C^c \, a_D^d}{a_A^a \, a_B^b} \qquad (6\text{-}5.6.3)$$

$$= \Delta G^0 + RT \ln L_r \qquad (6\text{-}5.6.4)$$

where L_r is defined by (6-5.6.3). It is the activity quotient for the reaction (see Sec. 5-3.3 for further details, if needed).

Before reading on, use Eq. (6-5.5.3) along with (6-5.6.4) to find $\mathscr{E}$ in terms of activities. Your result should be:

$$\mathscr{E} = \mathscr{E}^0 - \frac{RT}{n\mathscr{F}} \ln L_r \qquad (6\text{-}5.6.5)$$

This equation is called the *Nernst equation.*

Note: The quotient L_r can be related to activity coefficients and concentrations for materials in solution. Remember we have chosen pure liquids and pure solids and solutions with a = 1 as standard states. Each activity term in L_r is really a ratio of the activity at the given condition to the activity at the standard state. For example, a_A should really be written a_A'/a_A^0 where a_A' is the actual activity. If A is a solid then $a_A = 1$. If B is a solute, $a_B = a_B'/a_B^0 \neq 1$, normally. This distinction is important since we will set a = 1 for pure liquids and solids. We are not saying $a' = 1$. Instead we are recognizing that these materials are in their standard states, so $a_A'/a_A^0 = 1$.

From our previous definitions we can write

$$L_r = \frac{a_C^c \, a_D^d}{a_A^a \, a_B^b} = \frac{(\gamma_C \, c_C)^c \, (\gamma_D \, c_D)^d}{(\gamma_A \, c_A)^a \, (\gamma_B \, c_B)^b} \qquad (6\text{-}5.6.6)$$

where γ is an activity coefficient and c represents concentration terms in the proper units. An example will show how to handle individual terms.

6-5.6.1 EXAMPLE

Set up the expression for $\mathscr{E}$ as a function of concentration for the following cell:

$Pt|H_2(1\ atm)|HCl(m)|AgCl(s)|Ag(s)$

Solution

This corresponds to the reaction

$$H_2(g,\ P = 1\ atm) + 2AgCl(s) \rightarrow 2Ag(s) + 2H^+Cl^-(m)$$

$$L_r = \frac{a_{Ag}^2\ a_{H+}^2\ a_{Cl-}^2}{a_{AgCl}^2\ a_{H_2}}$$

But $a_{Ag} = a_{AgCl} = 1$ since they are in their standard states. Then

$$L_r = \frac{a_{H+}^2\ a_{Cl-}^2}{a_{H_2}} = \frac{(\gamma_{H+}\ m_{H+})^2\ (\gamma_{Cl-}\ m_{Cl-})^2}{\gamma_{H_2}\ P_{H_2}}$$

$$= \frac{\gamma_{H+}^2\ \gamma_{Cl-}^2\ m^4}{\gamma_{H_2}\ P_{H_2}} = \frac{\gamma_{\pm}^4\ m^4}{\gamma_{H_2}\ P_{H_2}} \qquad (6\text{-}5.6.7)$$

Care must be taken here. If the standard state for H_2 has been chosen as 1 atm, then P_{H_2} is used in L_r. If some other standard state is used, then another concentration unit for H_2 will have to be used.

We can now write $\mathcal{E}$ (subscript r indicates reaction).

$$\mathcal{E}_r = \mathcal{E}_r^0 - \frac{RT}{n\mathcal{F}} \ln L_r$$

$$= \mathcal{E}_r^0 - \frac{RT}{2\mathcal{F}} \ln \frac{(\gamma_{\pm}\ m)^4}{\gamma_{H_2}\ P_{H_2}}$$

At 298.15 K,

$$\mathcal{E}_r = \mathcal{E}_r^0 - \frac{0.02569}{2} \ln \frac{(\gamma_{\pm}\ m)^4}{\gamma_{H_2}\ P_{H_2}}$$

$$= \mathcal{E}_r^0 - \frac{0.05915}{2} \log \frac{(\gamma_{\pm}\ m)^4}{\gamma_{H_2}\ P_{H_2}} \qquad (6\text{-}5.6.8)$$

Again, it should be emphasized that you must see what standard states have been chosen before you can set up an equation like the last.

It should be clear how emf data can be used to determine activity coefficients. The values of $\mathcal{E}^0$ can be obtained from tables for half-reactions to get $\mathcal{E}_r^0$. If $\mathcal{E}_r$ is measured, then values of γ can be estimated in many cases (if we carefully choose the cells we use).

6-5.6.2 EXAMPLE

The emf of the cell

$$Zn(s)|ZnCl_2(m = 0.500)||CdSO_4(m = 0.100)|Cd(s)$$

at 298.15 K is found to be 0.325 V. Estimate the ratio of $\gamma_{Zn^{2+}}$ to $\gamma_{Cd^{2+}}$.

Solution

The reaction for this cell (*always* write the reactions involved)

Oxid. $Zn(s) \rightarrow Zn^{2+}(m = 0.500) + 2e^-$ $\qquad \mathcal{E}^0 = 0.7628$

Red. $Cd^{2+}(m = 0.100) + 2e^- \rightarrow Cd(s)$ $\qquad \mathcal{E}^0 = -0.4026$

Net $Zn(s) + Cd^{2+}(m = 0.100) \rightarrow Zn^{2+}(m = 0.500) + Cd(s)$ $\qquad \mathcal{E}_r^0 = 0.3602$

(*Note:* We do not have to include Cl^- and SO_4^{2-}, since neither is consumed or used in the reaction.)

$$\mathcal{E} = \mathcal{E}^0 - \frac{RT}{2\mathcal{F}} \ln \frac{a_{Zn^{2+}}}{a_{Cd^{2+}}}$$

$$= \mathcal{E}^0 - \frac{RT}{2\mathcal{F}} \ln \frac{\gamma_{Zn^{2+}} m_{Zn^{2+}}}{\gamma_{Cd^{2+}} m_{Cd^{2+}}}$$

$$= \mathcal{E}^0 - \frac{RT}{2\mathcal{F}} \ln \frac{m_{Zn^{2+}}}{m_{Cd^{2+}}} - \frac{RT}{2\mathcal{F}} \ln \frac{\gamma_{Zn^{2+}}}{\gamma_{Cd^{2+}}}$$

$\mathcal{E} = 0.325$ V $\qquad \mathcal{E}^0 = 0.3602 \qquad m_{Zn^{2+}} = 0.500$

$m_{Cd^{2+}} = 0.100 \qquad \frac{RT}{2\mathcal{F}}$(at 298.15 K) $= 0.0128$ V

Then

$$0.325 = 0.3602 - 0.0128 \ln \frac{0.500}{0.100} - 0.0128 \ln \frac{\gamma_{Zn^{2+}}}{\gamma_{Cd^{2+}}}$$

$$\frac{0.325 - 0.3602 + 0.0206}{0.0128} = -\ln \frac{\gamma_{Zn^{2+}}}{\gamma_{Cd^{2+}}} = -1.14$$

$$\frac{\gamma_{Zn^{2+}}}{\gamma_{Cd^{2+}}} = \underline{3.13}$$

You will notice that we have not found individual activity coefficients. We cannot from this cell. A judicious choice of cells is required to get individual ionic (or mean ionic) activity coefficients. The next example is a case in which a mean activity coefficient can be found.

6-5.6.3 EXAMPLE

For the cell in Example 6-5.6.1 assume that $a_{H_2} = 1$ (not a bad assumption for $P_{H_2} = 1$ atm). Determine the mean activity coefficient of HCl if the emf of the cell containing 0.01 m HCl is 0.46412 V at 298.15 K.

Solution

In Example 6-5.6.1, Eq. (6-5.6.8) was derived for the cell

$$\text{Pt(s)}|H_2\text{(1 atm)}|\text{HCl(m)}|\text{AgCl(s)}|\text{Ag(s)}$$

$$\mathcal{E}_r = \mathcal{E}_r^0 - \frac{0.02569}{2} \ln \frac{(\gamma_\pm\ m)^4}{a_{H_2}}$$

If $a_{H_2} = 1$,

$$\mathcal{E}_r = \mathcal{E}_r^0 - 0.01284 \ln \gamma_\pm^4 - 0.01284 \ln m^4$$

for a 0.0100 m solution ($\mathcal{E}_r^0 = 0.2225$ V; see Table 6-5.4.1) $\mathcal{E}_r = 0.46412$.

Thus $0.46412 = 0.2225 - 0.0128 \ln (0.0100)^4 - 0.0128 \ln \gamma_\pm^4$

$$\frac{0.46412 - 0.2225 + 4(0.01284) \ln 0.0100}{4(0.01284)} = -\ln \gamma_\pm$$

$$\ln \gamma_\pm = -0.0993$$

$$\gamma_\pm = \underline{\underline{0.905}}$$

The value from Table 6-4.2.1 is 0.905: not bad agreement.

DETERMINATION OF $\mathcal{E}^0$. You may be wondering at this point how to determine $\mathcal{E}^0$ since there will always be concentration terms to consider, so we can only measure $\mathcal{E}$. We employ a technique you have seen before. We find $\mathcal{E}$ as a function of concentration and perform an appropriate extrapolation to m = 0.

For example, look at the equation we used previously for the H_2, AgCl, Ag cell.

$$\mathcal{E}_r = \mathcal{E}_r^0 - 4(0.01284) \ln \gamma_\pm - 4(0.01284) \ln m$$

Rearrange this:

$$\mathcal{E}_r + 4(0.01284) \ln m = \mathcal{E}^0 - 4(0.01284) \ln \gamma_\pm$$

If we could measure $\mathcal{E}$ at a low enough value of m, the $\gamma_\pm$ term would drop out since $\gamma_\pm \to 1$ as $m \to 0$. So we measure $\mathcal{E}_r$ as a function of m, then plot $\mathcal{E}_r + 4(0.01284) \ln m$ vs. $\sqrt{m}$ and extrapolate to m = 0. This gives $\mathcal{E}_r^0(\gamma_\pm \to 1)$. (*Note:* $\sqrt{m}$ is used as the abscissa to make the curve more nearly linear.)

Table 6-5.6.1
$\mathcal{E}_r + 4(0.01284) \ln m$ vs. $\sqrt{m}$ for the Cell $Pt(s)|H_2|H^+Cl^-(m)|AgCl|Ag(s)$ at 298.15 K

m	$\mathcal{E}_r$	$\mathcal{E}_r + 4(0.01284) \ln m$	$\sqrt{m}$
0.003215	0.52053	0.22573	0.05670
0.004488	0.50384	0.22617	0.06699
0.005619	0.49257	0.22644	0.07496
0.007311	0.47948	0.22687	0.08550
0.009138	0.46860	0.22745	0.09559
0.011195	0.45861	0.22789	0.10581

Consider the data in Table 6-5.6.1. A plot of these data is given in Fig. 6-5.6.1. The extrapolated $\mathcal{E}^0$ value is 0.2230 V compared to a true value of 0.2225 V. The error is probably due to the long extrapolation that has been made. More data at a lower concentration would yield a more precise value.

EQUILIBRIUM CONSTANTS. For any reaction we know ΔG at equilibrium. (what is the value?). $\Delta G_r = 0$ is the definition of equilibrium. This means $\mathcal{E}_r = 0$ since $\Delta G_r^0 = n\mathcal{F}\mathcal{E}^0$. So we can write, at equilibrium,

$$\mathcal{E}_r = \mathcal{E}_r^0 - \frac{RT}{n\mathcal{F}} \ln L_r = 0 \qquad (6\text{-}5.6.9)$$

or

$$\mathcal{E}_r^0 = \frac{RT}{n\mathcal{F}} \ln K_{eq} \qquad (6\text{-}5.6.10)$$

where K_{eq} $[= (L_r)_{eq}]$ is the equilibrium value of the activity quotient.

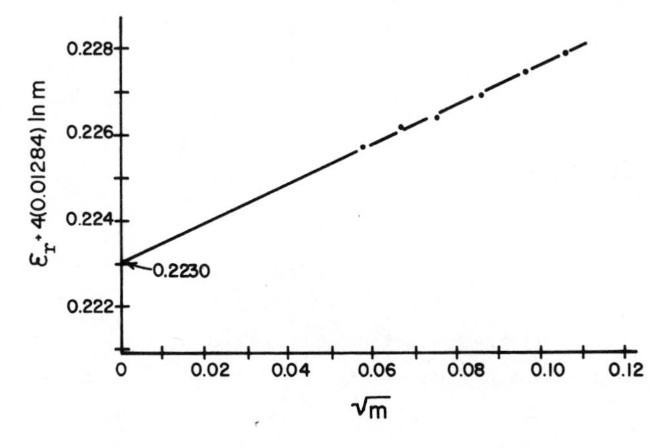

FIG. 6-5.6.1 Plot of $\mathcal{E}_r + 4(0.01284) \ln m$ vs. $\sqrt{m}$ from data in Table 6-5.6.1.

To determine the equilibrium constant for a given reaction, we must determine $\mathcal{E}_r^0$ from tabulated values of $\mathcal{E}^0$ for half cells, then solve for K_{eq} in Eq. (6-5.6.10).

6-5.6.4 EXAMPLE

Find the equilibrium constant at 298.15 K for the reaction indicated by the following cell

$$Pt(s)|H_2(P)|HCl(m)|AgCl(s)|Ag(s)$$

Solution

The reactions are:

Red. $AgCl(s) + 1e^- \rightarrow Ag(s) + Cl^-(m)$ $\qquad \mathcal{E}^0 = 0.2225$

Oxid. $\frac{1}{2}H_2(g) + Cl^-(m) \rightarrow H^+(m) + Cl^-(m)$ $\qquad \mathcal{E}^0 = 0.000$

Net $AgCl(s) + \frac{1}{2}H_2 \rightarrow Ag(s) + H^+(m) + Cl^-(m)$ $\qquad \mathcal{E}^0 = 0.2225$

The value of K_{eq} is found as follows:

$$\ln K_{eq} = \frac{n\mathcal{F}\mathcal{E}^0}{RT}$$

$$= \frac{(1 \text{ equiv./mol})(96{,}490 \text{ C/equiv.})(0.2225 \text{ V})}{(8.31 \text{ C}\cdot\text{V mol}^{-1} \text{ K}^{-1})(298.15 \text{ K})}$$

$$= 8.67$$

$$K_{eq} = \frac{a_{H^+} a_{Cl^-}}{(a_{H_2})^{\frac{1}{2}}} = e^{8.67} = \underline{\underline{5.83 \times 10^4}}$$

This is the equilibrium constant for the reaction involving $1e^-$. If we had written

$$2AgCl(s) + H_2 \rightarrow 2Ag(s) + 2H^+(m) + 2Cl^-(m)$$

$$\ln K'_{eq} = 2(8.67) = 17.3$$

$$K'_{eq} = \frac{a^2_{H^+} a^2_{Cl^-}}{a_{H_2}} = K^2_{eq} = \underline{\underline{3.40 \times 10^7}}$$

The value you determine for an equilibrium constant will depend on the number of electrons involved in the reaction you have written.

We can use the procedures above to find the equilibirum constants for dissociations and solubilities.

6-5.7 SOLUBILITY PRODUCTS AND DISSOCIATION CONSTANTS FROM EMF DATA

SOLUBILITY PRODUCTS. This is just an application of the last part of the previous section where a method was derived to determine equilibrium constants from emf data. Determination of a solubility product is accomplished in the same way. All that must be done is to choose a proper set of cell reactions so the overall reaction is the one of interest.

6-5.7.1 EXAMPLE

Assume we would like the solubility product of AgI. We need to devise a cell reaction where the net reaction is

$$AgI \rightleftarrows Ag^+ + I^-$$

(*Note:* We do not have to be able to actually construct the cell. We are going to look up values of $\mathscr{E}^0$, not measure them.)

Solution

Such a set of reactions is

Oxid.	$Ag(s) \rightarrow Ag^+ + 1e^-$	$\mathscr{E}^0 = -0.7996$
Red.	$AgI(s) + 1e^- \rightarrow Ag(s) + I^-$	$\mathscr{E}^0 = -0.1519$
Net	$AgI(s) \rightarrow Ag^+ + I^-$	$\mathscr{E}^0 = -0.9515$

From (6-5.6.10), after rearranging,

$$\ln K_{eq} = \frac{n\mathscr{F}\mathscr{E}_r^0}{RT}$$

$$= \frac{(1)(96{,}490)(-0.9515)}{(8.314)(298.15)}$$

$$= \underline{-37.04}$$

Then $K_{eq} = \underline{\underline{8.21 \times 10^{-17}}}$

This may be compared to an experimental value of 15×10^{-17}. This last value is based on concentrations, whereas the value of K_{eq} is based on activities. However, these solutions are very dilute so we could expect $\gamma_\pm \approx 1$. The discrepancy may be in $\mathscr{E}^0$ values since a small error in $\mathscr{E}^0$ can make a large error in the exponential. Or the experimental value may be in error. Anyway, the values agree approximately.

6-5.7.2 EXAMPLE

Now let us try an example involving dissociation. Pick a dissociation that has a small K_{eq} so we can assume $\gamma_{\pm} = 1$.

Solution

An appropriate reaction is $H_2O \rightarrow H^+ + OH^-$. A suitable cell is (at 298.15 K):

$$Pt(s)|H_2|H^+(a)||OH^-(a)|H_2|Pt(s)$$

Note that we are ignoring any ions that do not take part in the reaction.

The half-reactions are

Oxid.	$\frac{1}{2} H_2(g) \rightarrow H^+ + 1e^-$	$\mathcal{E}^0 = 0.000$
Red.	$H_2O(\ell) + 1e^- \rightarrow \frac{1}{2} H_2(g) + OH^-$	$\mathcal{E}^0 = -0.82806$
Net	$H_2O(\ell) \rightarrow H^+ + OH^-$	$\mathcal{E}^0 = -0.82806$

$$\mathcal{E} = \mathcal{E}^0 - \frac{RT}{n\mathcal{F}} \ln a_{H^+} a_{OH^-}$$

At equilibrium, $\mathcal{E} = 0$ (as usual), so

$$\mathcal{E}^0 = \frac{0.02569}{1} \ln K_{diss}$$

$$\ln K_{diss} = -\frac{0.82806}{0.02569}$$

$$K_{diss} = \underline{\underline{1.003 \times 10^{-14}}}$$

This compares to the value of 1.008×10^{-14} listed for water. This is in much better agreement than the solubility product found previously. And the better agreement is probably due to better accuracy in the emf data in this case.

6-5.8 MISCELLANEOUS ITEMS

You have now seen how to write electromotive cell notation; how to find emf values for standard conditions; and how to use emf data to find equilibrium constants, activities, and activity coefficients, as well as thermodynamic functions. The following list will remind you of what you have not seen so you will know what you do not know (as important as knowing what you do know). The following topics are of importance for a practical understanding of electrochemical cells. For a good introduction to these topics see S. H. Maron and J. B. Lando, *Fundamentals of Physical Chemistry* (Sec. 6-5.12, Ref. 1). Some important topics that have been omitted are:

Types of electrodes
pH and emf
Junction potentials
Potentiometric titrations
Electrolysis
Commercial cells

6-5.9 FUEL CELLS

Fuel cells are devices for the direct conversion of chemical energy to electrical energy. They differ from what we commonly call a "battery" in that they do not normally act as storage devices. The line of demarcation between a battery and a fuel cell is not always clear, however, in that some fuel cells are *regenerative*, meaning that the products from the reactions within the fuel cell can be converted (chemically, electrically, or thermally) back into the original reactants. The following will not consider regenerative systems (see Crouthamel and Recht, Sec. 6-5.12, Ref. 3). Instead the major concern is on *primary* or *once-through* fuel cells. These cells take reacting materials and produce electrical energy plus chemical products. The products are discarded.

EFFICIENCIES OF FUEL CELLS. Before a brief discussion of the types of fuel cells and some principles of operation, it is perhaps instructive to see the thermodynamic basis for interest in the cells. The major importance for fuel cell research lies in the fact that fuel cells are not limited by the Carnot efficiency. Electrical energy is produced directly from chemical energy without an intermediate heat-producing step. Table 6-5.9.1 gives data on typical energy conversion systems.

This table shows vividly the promise and problems of fuel cell technology. Fuel cell efficiencies are very high compared to common energy converters. However, the cost factor is prohibitive at present. The major source of this high cost is the platinum needed in the catalysts necessary to get chemical reactions to take place at appropriate electrodes. If research can find an inexpensive substitute for the platinum catalyst, then fuel cells will become very common.

Let us investigate the efficiency expression for a fuel cell a little more carefully. After that, a short discussion will be given for two types of fuel cells involving H_2 and O_2. This will give you an idea of what a fuel cell is. One is the type used on space probes. The other uses a unique electrolyte that offers promise for the future.

Table 6-5.9.2 gives some performance characteristics of fuel cells using various fuels as well as efficiencies of some familiar converters for comparison. Perhaps the most striking feature in this table is the very low fuel cost for the methane fuel cell. The overall cost is high but fuel cost is low. In all cases, these costs are based on 1969 prices, so they are no longer accurate. However, relative costs should still be valid.

Table 6-5.9.1
Performance Characteristics of Various Energy Converters[a]

	Efficiency (%)		Power (hp/lb)[c]		Cost ($/hp)	
Energy converter	1969	Future[b]	1969	Future	1969	Future
Gas turbine	28	28	1	1	25	25
Automobile engine	27	27	0.3	0.3	3	3
Diesel engine	42	42	0.2	0.2	3	3
Low-temperature fuel cell (<100°C)	70	90	0.04	1	400	100
High-temperature fuel cell (500-1000°C)	60	90	0.2	2	100	30

[a] From Bockris and Srinivasan (see Sec. 6-5.12, Ref. 2).
[b] Maximum feasible.
[c] 1 hp = 1 horsepower = 746 J s^{-1}; 1 lb = 0.454 kg.

You should recall from Chap. 4 that the Carnot efficiency of a heat engine is

$$\varepsilon = \frac{T_h - T_\ell}{T_h} \tag{6-5.9.1}$$

For a fuel cell the following efficiency is found:

$$\varepsilon_i = \frac{\Delta G}{\Delta H} \tag{6-5.9.2}$$

This definition follows from the fact that ΔG is related to the electrical energy output and ΔH is the heat input (from the reactions that occur). The subscript i stands for intrinsic efficiency. An interesting result follows from (6-5.9.2). From previous definitions show that:

$$\varepsilon_i = \frac{\Delta G}{\Delta H} = 1 - \frac{T\,\Delta S}{\Delta H} \tag{6-5.9.3}$$

Table 6-5.9.2
Costs for Various Energy Converters[a]

Converter	Efficiency (%)	Initial cost ($ kw^{-1})	Fuel costs (¢ kwh^{-1})*	Total cost (¢ kwh^{-1})*
Methane fuel cell	29	385	0.39	5.46
Octane fuel cell	46	150	1.26	3.24
H_2, O_2 fuel cell	50-75	40-120	3.1-4.5	3.6-6.1
Internal combustion engine	27	4	1.49	1.54
Diesel engine	42	4	0.46	0.52

[a] (*) 1969 costs.

For most reactions used in fuel cells, $\Delta H < 0$ (exothermic) and $\Delta S < 0$. But, in a few cases reactions exist in which $\Delta H < 0$ and $\Delta S > 0$. In this case $\varepsilon_i > 1$. This means that the fuel cell absorbs additional heat from the surroundings (in addition to the heat generated in the reaction) to produce electrical energy. Such an occurrence offers interesting possibilities for efficient energy conversion. An example is

$$C + \frac{1}{2} O_2 \rightarrow CO \qquad \begin{array}{l} \varepsilon_i(298\ K) = 1.24 \\ \varepsilon_i(423\ K) = 1.372 \end{array}$$

Practical problems still prevent the widespread use of carbon in fuel cells, but the possibilities are intriguing.

TYPES OF FUEL CELLS.

General. After all the above, you may still not have any idea what a fuel cell is. Figure 6-5.9.1 should eliminate the problem. Look at the fuel side. The fuel (H_2, CH_4, hydrazine, methanol, or another oxidizable material) flows across a porous anode where electrons are released and products formed. The electrons flow through the external circuit to the cathode where oxygen (the universal oxidant due to its ready availability) combines with them to form products. To complete the circuit, ions must travel from anode to cathode or vice versa through the electrolyte. The specific example below will show that step.

The H_2-O_2 Cell Used in Space Applications. Figure 6-5.9.2 shows the fuel cell used in the Gemini space flights. The reactions that occur are:

$$\begin{array}{ll} \text{Anode} & 2H_2 \rightarrow 4H^+ + 4e^- \\ \text{Cathode} & 4H^+ + O_2 + 4e^- \rightarrow 2H_2O \\ \hline \text{Overall} & 2H_2 + O_2 \rightarrow 2H_2O \end{array}$$

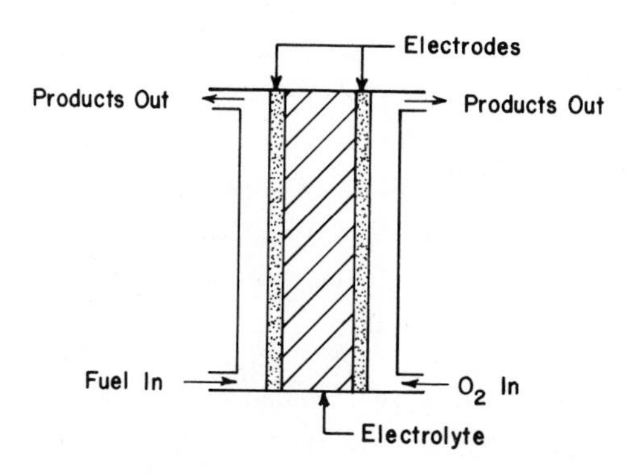

FIG. 6-5.9.1 Schematic of a fuel cell.

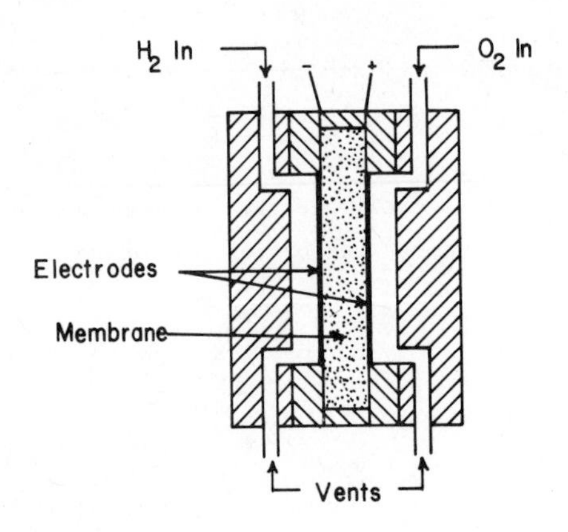

FIG. 6-5.9.2 An H_2, O_2 fuel cell used in space applications.

The hydrogen ions produced at the anode migrate through the electrolyte to the cathode while the electrons generate flow through the external circuit. The electrolyte in this case is an ion exchange material that allows easy passage of protons. At the cathode the hydrogen ions react with oxygen and electrons to form water. This, obviously, is not a description of the actual mechanisms of the reactions; those are much more complex.

The major advantages of this cell are the reliability and the fact that the waste product is water, which can be used. The disadvantages are the high cost of fuel and difficulty of storage and handling very explosive gases.

A Solid Electrolyte Cell. Figure 6-5.9.3 shows a cell that has a unique construction. The electrolyte in this cell is stabilized zirconia (ZrO_2). At elevated temperatures (1000°C) the zirconia is sufficiently porous to oxygen ions to allow it to be used as a membrane. In operation, oxygen flows over (or surrounds) the outside of the cell where a reaction occurs to give O^{2-} ions (using electrons produced on the fuel side). The O^{2-} ions diffuse through the zirconia and react with the fuel (H_2 in many cases) to form oxidation products (H_2O if the fuel is H_2) and electrons. The electrons flow through the external circuit to the oxygen electrode.

A unique feature of this arrangement is that individual cells can be stacked and connected in series to generate a higher voltage. Other advantages are: the cell is very light since the electrolyte is only about 0.05 cm in thickness and the catalyst is easily deposited. Major disadvantages are material problems at elevated temperatures and incomplete reactions at the electrode surface.

KINETIC PROBLEMS IN FUEL CELLS. The actual performance of fuel cells, like that of storage cells, is limited in a number of ways. Below only the briefest mention will be given to the problems that restrict the operation of cells to well below

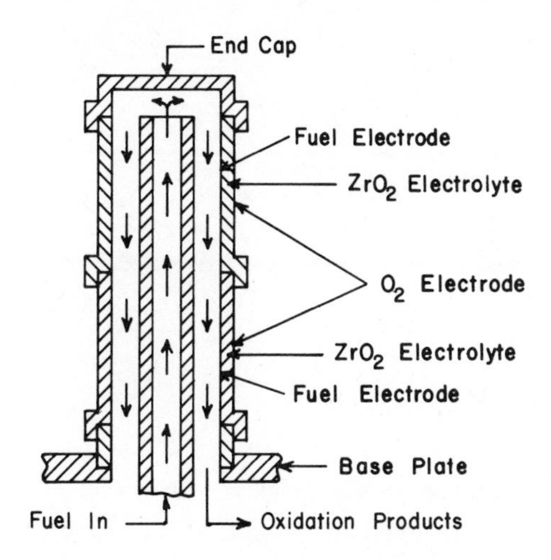

FIG. 6-5.9.3 Solid electrolyte fuel cell.

theoretical capabilities. Polarization within the cell is created by a number of factors and becomes more important as the current density increases. A few effects will now be outlined.

Activation Polarization. Drawing a current from a cell results in a decrease in the cell potential. The process of drawing a current destroys equilibrium and leads to an *overpotential* at the electrode surface. The objective of scientists concerned with fuel cell electrodes is to devise catalysts that lower this overpotential. At present, platinum is the most effective material for an electrode catalyst. The high cost of platinum used in the electrode is the major cost in the production of fuel cells. Figure 6-5.9.4 shows a typical polarization vs. current curve. Note the scale on the right that gives the voltage efficiency. The intrinsic efficiency mentioned earlier is calculated when the voltage efficiency is 100% (zero current). Actual efficiencies are much less.

The overpotential is created by a number of effects:

1. Slow electron transfer of the electrode surface
2. Slow absorption of material on the electrode
3. Slow reactions on the electrode surface

The key to minimizing all these effects is the development of suitable catalyst systems.

Concentration Polarization. A fuel cell generates electricity by the reaction of materials at an electrode. To generate more current, more material must react. At high current densities a concentration gradient may be generated with the area

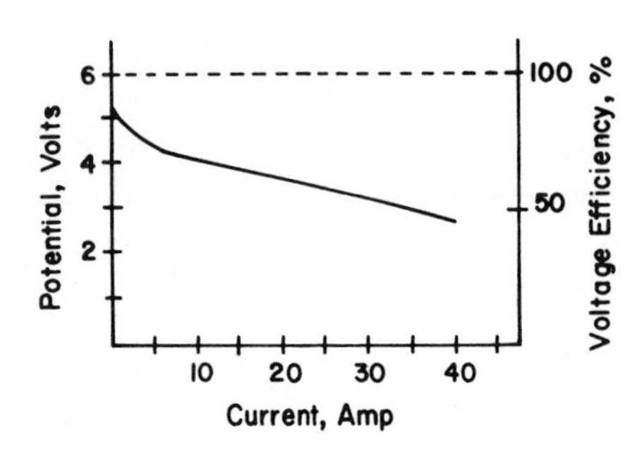

FIG. 6-5.9.4 Potential vs. current for a typical cell system.

immediately surrounding the electrode being depleted of the reacting substance. This leads to a drop in the cell potential and a loss of efficiency. The design of cells requires that careful consideration be given to mass transfer problems. Also, the actual rate of absorption of reactants and desorption of products on the electrode surface may lead to concentration difficulties on the surface.

Ohmic Polarization. The cell itself has some internal resistance. When an attempt is made to draw current, this resistance within the electrolyte may act to limit the current. Power is being dissipated within the cell. This has two detrimental effects. First, this power is lost and cannot be used for useful work. Second, the heat generated can lead to excessive cell temperatures.

COMMENTS. Fuel cells offer a very attractive means for the conversion of fossil fuels into electricity with higher efficiencies than are possible with steam generating stations. However, this promise, to be realized, awaits the development of inexpensive catalyst materials that would allow the conversion to compete economically with present generating methods.

6-5.10 PROBLEMS

1. Write the half-cell notation for the reaction of $Br_2(\ell)$ to form Br^- on a platinum electrode. Also, find $\mathscr{E}^0$ for the half-cell.

2. Write the cell notation and find $\mathscr{E}^0$ for the reaction

$$AgBr(s) + Br^-(m) \rightarrow Ag(s) + Br_2(\ell)$$

3. Write the reaction and find $\mathscr{E}^0$ for each of the following cells.

   a. $Cd|Cd^{2+}(a = 1)||KCl(1\ N)|Hg_2Cl_2(s)|Hg(Pt)$
   b. $Pt|Cl_2(g),Cl^-||Br^-,Br_2(\ell)|Pt$
   c. $Co|Co^{2+}||Cu^{2+},Cu^{1+}|Pt$

d. $Cu|Cu^{2+}||Cu^{1+},Cu^{2+}|Pt$
e. $Pb|Pb^{2+}||Zn^{2+}|Zn$

4. For each cell in Problem (3) find ΔG^0 for the reaction (as written) and find the equilibrium constant.

5. Given

$$Pb|PbSO_4|SO_4^{2-} \qquad \mathcal{E}^0 = -0.3588$$

$$Pb|Pb^{2+} \qquad \mathcal{E}^0 = -0.1263$$

find the solubility product of $PbSO_4$ at 298.15 K.

6. In a cell in which Zn amalgam [Zn(Hg)] is used as electrodes with $ZnCl_2$ as the electrolyte we can write the notation as follows for a Zn^{2+} activity of 0.500:

$$Zn(Hg)(a = x)|Zn^{2+}(a = 0.500)|Zn(Hg)(a = y)$$

For this concentration cell $\mathcal{E}^0 = 0$ (why?). Find the ratio of Zn activities if the measured potential of this cell is 0.062 V at 312 K.

7. In the following cell the potential at 25°C is found to be 0.43783 V. Find $\gamma_\pm$ for HCl.

$$H_2(a = 1)|HCl(m = 0.0171)|AgCl(s)|Ag$$

8. Given a table of emf values, devise a cell that will allow you to find the emf of the reaction

$$HClO(aq) \rightarrow H^+ + Cl^-$$

Determine the potential of the call. Also, find ΔG^0 and the dissociation constant at 298 K. (A good tabulation such as a late edition of the Chemical Rubber Company *Handbook of Chemistry and Physics* will be necessary.)

6-5.11 SELF-TEST

1. For cells of the type

$$H_2(1\ atm)|HBr(m)|AgBr(s)|Ag(s)$$

the emfs at different concentrations are given below. Find $\mathcal{E}^0$ by an appropriate extrapolation and find $\gamma_\pm$ for each molality.

m	0.0003198	0.0004042	0.0008444	0.001355	0.001850
$\mathcal{E}$	0.48569	0.47381	0.43636	0.41243	0.39667

Compare your $\mathcal{E}^0$ value with that calculated from Table 6-5.4.1.

2. (a) Devise and write the proper cell notation for a cell in which the reaction

$$\frac{1}{2}\,Cl_2(g) + Fe^{2+} \rightarrow Cl^- + Fe^{3+}$$

will take place. (b) Find $\mathscr{E}^0$ for your cell. (c) Find ΔG^0 for the reaction. (d) Determine the equilibrium constant for the reaction. (e) If $a_{Cl_2} = 1$, $m_{Fe^{2+}} = 0.00200$, $m_{Fe^{3+}} = 0.00100$, and $m_{Cl^-} = 0.00500$, and if we assume no other ions are present, use the Debye-Hückel theory to estimate activities and find the actual potential of the cell.

6-5.12 ADDITIONAL READING

1.* S. H. Maron and J. B. Lando, *Fundamentals of Physical Chemistry*, Macmillan, New York, 1974, Chap. 14.

2.* C. E. Crouthamel and H. L. Recht, Symposium Chairman, "Regenerative EMF Cells," *Advances in Chemistry Series 64*, American Chemical Society, Washington, D. C., 1967.

3.* J. O'M. Bockris and S. Srinivasan, *Fuel Cells: Their Electro-Chemistry*, McGraw-Hill, New York, 1969.

4. W. M. Latimer, *Oxidation Potentials*, 2nd ed., Prentice-Hall, New York, 1952.

5. G. Charlot, A. Columeau, and M. J. C. Marchon, *Selected Constants: Oxidation-Reduction Potentials of Inorganic Substances in Aqueous Solutions*, Butterworth, London, 1971.

6-6 SUMMARY

You have just completed a fairly extensive introduction to solution chemistry. You now are at least familiar with the terms activity, activity coefficient, mean ionic activity, etc. In the future, if you ever have to do equilibrium constant, vapor pressure, or any other calculations involving solutions, you should be able to estimate the answer much more closely than if you assume ideal behavior. Few solutions are ideal, and it is usually a mistake to solve the problem as though ideality exists. One objective of the chapters on thermodynamics and this chapter on solutions is to get you to the point where you are no longer afraid to include nonideality in your calculations.

You will see many of the concepts developed in this chapter as you continue your study. For example, osmotic pressure and vapor pressure lowering will be seen again in the discussion of molecular weights of polymers and colloids (Chaps. 11 and 12).

Section 6-5.9 was concerned with fuel cells. These devices offer much promise for future power sources since they are not limited by the Carnot efficiency expression. Chemical energy is converted directly to electrical energy without having to go through a heat stage. This offers hope of extremely efficient power sources if the catalyst problem can be solved economically.

This concludes our study of thermodynamics per se. Various functions will show up in the following chapters, however.

7

Introduction to Atomic Molecular Structure: The Quantum Theory

SYMBOLS AND DEFINITIONS

e	Charge of electron, 1.6022×10^{-19} Coulomb (C)
m_e	Mass of electron, 9.1096×10^{-31} kg
m_p	Mass of proton, 1.6726×10^{-27} kg
E_λ	Black body energy distribution of a given wavelength, (J m^{-4})
ε	Energy of a quantum (J)
h	Planck's constant, 6.6262×10^{-34} J s $molecule^{-1}$
m, v	Mass and velocity of particle
ω	Threshold energy for the photoelectric effect
λ	Wavelength (m, Å = 10^{-10} m)
k, n	Hydrogen atom electronic level indices
R_H	Rydberg's constant, 109,677.581 cm^{-1}
c	Speed of light, 2.9979×10^{10} cm s^{-1} = 2.9979×10^{8} m s^{-1}
ν	Frequency of radiation (s^{-1} = Hz)
$\bar{\nu}$	Wave number of radiation (cm^{-1})
m	Mass of α particle (He nucleus), 6.6×10^{-27} kg
Ψ, Φ	Wave functions that describe state of a system
$d\tau$	Volume increment
$\hat{\alpha}$	An operator for position, momentum, energy, etc.
p	Momentum (kg m s^{-1})
a_s	An eigenvalue
$\langle\hat{\alpha}\rangle$	Expectation or average value of a variable
$\hat{\mathcal{H}}$	Hamiltonian energy operator in the Schrödinger equation
E	Total energy of the system (J)
x	Dimension of box for the particle in a box problem
m	Mass of a particle (kg)
p_x	Expectation value for momentum
Δx, Δp_x	Uncertainties in the values of x and p_x
r, θ, ϕ	Variables for the hydrogen atom problem
μ	Reduced mass (kg $molecule^{-1}$)
$\Psi(r, \theta, \phi)$	Total wave function for the hydrogen atom
$R(r)$, $\Theta(\theta)$, $\Phi(\phi)$	One-variable functions for the hydrogen atom
F_j	Force in the j direction (N)
V	Potential energy of a system (J)
T	Kinetic energy of system (J)
L	Lagrangian for system
$\mathcal{H}$	Hamiltonian for system
q_i	Coordinate (m)
$\dot{q}_i$	Velocity (m s^{-1})
p_i	Momentum
I	Moment of inertia (kg m^2)
μ	Reduced mass
m	Quantum number for Φ equation solution
β, z, F, P, G	Variables and functions in the Θ equation solution
ℓ	Quantum number for Θ equation solution
$\alpha, \lambda, \rho, S, F, L$	Variables and functions in R equation solution

n	Principal quantum number, from solution of R equation
E_n	Energy levels for the hydrogen atom (J)
a_0	Bohr radius for the hydrogen atom, 0.529 Å
E'	Variation method energy (J)
Z	Nuclear charge
L	Angular momentum of a particle
L_x, L_y, L_z	Components of angular momentum
m_s	Quantum number for spin
μ_s	Magnetic moment due to spin
M_s	Total spin angular momentum
B	Bohr magneton, 9.273×10^{-24} J T^{-1} = 9.273×10^{-24} kg s^{-1} A^{-1}
$\mu_{z,s}$	z-Axis projection of spin magnetic moment
μ_{orb}	Magnetic moment due to electronic orbital motion
$\mu_{z,orb}$	z-Axis projection of μ_{orb}
H	Magnetic field strength (along z axis)
E_n'	Energy of nth level of the hydrogen atom in the presence of a magnetic field
I.E.	Ionization energies (J)
E.A.	Electron affinity (J)
χ	Electronegativity

Quantum mechanics is the theoretical (and mathematical) study of the structure of matter. Thermodynamics is concerned with *macroscopic properties* and is independent of any model of the building blocks making up a system. Quantum mechanics, on the other hand, deals with *microscopic properties*: the structure of atoms and molecules.

Quantum mechanics is a very large segment of physical chemistry. The problem in writing a chapter such as this becomes one of limiting it to an appropriate degree for undergraduate study. Two extremes could be employed: either a very superficial development could be given or a rigorous mathematical approach could be taken. This chapter will take a middle road and present considerable mathematical detail for some topics, then only briefly indicate how these techniques can be used for other problems.

There are good reasons for studying quantum theory. It is the theoretical basis of all our notions about the make-up of atoms and molecules, the building blocks of nature. Few of you will ever find a direct application of the theories presented here. You will, however, need an understanding of interactions within and between molecules. Quantum mechanics gives a description of the origin of the forces within molecules which lead to those interactions.

The energy level concepts developed here will be used in the following chapter in our study of *spectroscopy*. Spectroscopy is the study of interactions of electromagnetic radiation with molecules. Such study leads to information about the structure of molecules that allows us to predict physical properties.

The energy levels for atoms and molecules will be used in Chap. 9 to form the basis of *statistical thermodynamics*. Statistical thermodynamics allows us to take the energy level distributions in molecules and calculate the absolute values of thermodynamic quantities. Such an approach is useful if data are not available from experimental work for some molecule.

In this chapter, the first section is concerned with experimental and theoretical work that led to the rise and fall of the old quantum theory. This is an historical account of one of the major scientific developments of the twentieth century. Also, in the first section, the postulates of quantum mechanics and a simple example are given.

Section 7-2 contains classical mechanics and mathematical procedures needed for actual quantum mechanical work. The hydrogen atom problem will be solved in detail as an example of the procedures used in a wave mechanics problem. Also, an introduction to approximate methods will be given since exact solutions are not possible in any but the simplest of cases.

Atomic structure is discussed in Sec. 7-3, where the solutions to the equations describing the hydrogen atom are applied to other atoms. The periodic properties of atoms are also given.

Quantum mechanics is a fascinating field since it is concerned with things we cannot see. The methods used are different from any you have encountered in the solution of other problems. You should approach the subject with the anticipation of learning a new way to view nature and nature's building blocks, atoms and molecules.

7-1 HISTORICAL BACKGROUND AND THE POSTULATES OF QUANTUM MECHANICS

OBJECTIVES

You shall be able to

1. Describe several of the important developments that led to the old quantum theory and some of the problems that required the formulation of the new quantum theory
2. Discuss the photoelectric effect and calculate the magnitude of the effect
3. Apply the spectroscopic formula for finding the frequency, energy, or wavelength of transitions for the hydrogen atom
4. Convert from any given set of spectroscopic units to any other
5. State the postulates of quantum mechanics and briefly give the meaning of each
6. Briefly discuss the model of the hydrogen atom
7. Apply the equations associated with the particle in a box problem.

7-1.1 INTRODUCTION

We will very briefly touch on some of the important developments that led to the formulation of the old and new quantum theories. After a tour of the origins of the concepts of atoms, molecules, electrons and nuclei, some of the problems that forced a revision of the classical models that physics had before the beginning of the twentieth century will be pointed out. We can only mention the wave-particle dualism that has been a major source of worry for people concerned with the basic nature of matter and its interactions with radiation.

Several well-known scientists played a role in the unfolding of the quantum theory. Such people as Max Planck, Albert Einstein, Neils Bohr, Louis de Broglie, Erwin Schrodinger, and Werner Heisenberg, among others, created a revolution in classical physics and sought explanations for phenomena that were not explicable using classical concepts.

The developments cannot be as complete as might be desired. The story is a fascinating study of the struggle over a relatively short period of about 30 years to lay the groundwork for the quantum theory satisfactorily. If you have an interest in reading more, there are several books on the history of chemistry that give further detail on some aspects of the development of the theory.

Following the historical discussion, the postulates of quantum mechanics will be given and illustrated with an example. Finally, the postulates will be applied to set up the equations necessary to describe the energy levels in the hydrogen atom.

7-1.2 EARLY IDEAS ABOUT THE STRUCTURE OF MATTER AND LIGHT

ATOMS AND MOLECULES. The first modern attempt to explain the nature of the building blocks that make up matter can be attributed to John Dalton. Dalton, in 1808, postulated that every element consists of indivisible particles called atoms. Over the next several decades the weights of these elements were established, and the periodic relationships among them were worked out. At the same time, Michael Faraday's work on electrolysis was establishing the connection between electrical charge and matter. His work led to the concept of positive and negative portions of atoms and molecules.

In 1897, the studies of J. J. Thomson produced the discovery of free electrons. Thomson studied the discharge between two electrodes in a low-pressure gas. He determined that the rays emitted from the cathode, *cathode rays*, were negatively charged and were particle-like. In 1913, R. A. Millikan performed his well-known "oil-drop" experiment to find the charge on the electron. Oil drops that had been charged by passage through a region of ionizing radiation were suspended in an electrostatic field. By varying the field and observing the rate of fall of the oil drops, Millikan was able to evaluate the charge of a single electron. The value is

$$e = -1.6022 \times 10^{-19}\ \text{C}\ \ \text{(Coulomb)}$$

This value, when combined with the results of Thomson's experiments on cathode rays, yielded the value of the mass of the electron. Thomson could find the value of the ratio of the electronic charge to the mass from observation of the way cathode rays were deflected in an electric field. The resultant mass is

$$m_e = 9.1096 \times 10^{-31}\ \text{kg}$$

Other work with the conduction of gases using the discharge tube led Goldstein, in 1886, to the discovery of positive rays within the tube. These positive particles move in the direction opposite to the negative (cathode) rays. They are found to have almost all the mass of the atoms from which they arise. Studies of the behavior of these rays (or beams) of charged particles in electrical and magnetic fields allowed the identification of *isotopes* of some of the elements. It was observed that the rays were generated in the gas in the tube by bombardment of the gas with the cathode rays. The charge to mass ratio (e/m) of various elements was determined in this apparatus.

Other significant discoveries such as X rays by W. K. Röentgen in 1895, and the discovery of α, β, and γ rays by H. Becquerel in 1896, established the identity of some of the fundamental particles that make up atoms and molecules. Beta rays were found to have the same e/m as the electron, α rays were positively charged, and the γ rays were uncharged. Subsequent work has shown that β rays are simply electrons, α rays are helium nuclei, and γ rays are short wavelength (high-energy) light rays.

During this period the mass of the hydrogen atom was found to be 1836 times the mass of the electron. Hydrogen atoms were found to possess a single positive charge in their normal ionized state. This gives the mass of the hydrogen ion, the proton, as

$$m_p = 1.6726 \times 10^{-27}\ \text{kg}$$

All this activity caused Thomson, in 1906, to propose the following model of the hydrogen atom: The positive charge in the atom is confined to a sphere of definite radius with electrons embedded in this sphere to maintain electroneutrality. This model did not survive long. In 1909, and subsequent years H. Geiger, Marsden, and E. Rutherford performed experiments in which α particles were directed toward thin metal foils. Most of the α particles passed through the foil unchanged. However, some were deflected, and some even scattered back toward the source. From this phenomenon Rutherford concluded that most of the metal foil was empty space. Most of the mass was concentrated in a small space, termed the nucleus, possessing positive charge. Calculations showed that only about 10^{-12} to 10^{-15} of the total space of a material is occupied by nuclei or electrons.

Continued effort produced the notion that the charge within the nucleus is equal to the atomic number of an atom. Also, the number of electrons outside the nucleus is equal to the nuclear charge. This last conclusion was based on study by O. Mosley of the emission of X rays by metallic elements bombarded with high energy electrons. The extra weight of an atom was attributed to uncharged particles called neutrons.

LIGHT. While one group of physical scientists was investigating the nature of the fundamental particles that are present in atoms and molecules, another group was busy trying to explain some anomalous results for light. Classical optics and electromagnetic theory were well known and accepted by the turn of the century. However, there were a few loose ends that did not fit readily into the classical theory. As we have seen, atoms and molecules were being modeled as particles. Light had been fairly well established as wavelike in its behavior.

The wave nature of light was substantiated by the observation that light can be made to form interference patterns and can be made to diffract. These phenomena can be explained on the basis of wave behavior. There was a problem, however, with the failure of classical physics to explain the distribution of radiation observed in a black body.

A *black body* is a material that absorbs all incident radiation. A nearly perfect black body is a hollow sphere that has a small pin hole in it to allow radiation to enter. Any radiation that enters is trapped and, therefore, absorbed. Figure 7-1.2.1 shows the energy distribution in a black body. Theoretical attempts to reproduce this behavior, in 1896 by Wein and in 1900 by Lord Rayleigh, were unsuccessful. Wein derived an equation that fit the data at short wavelengths, and Rayleigh found an equation appropriate to the long wavelengths. The inability to find one mathematical expression that fit all wavelengths was the impetus for Max Planck, in 1900, to propose a startling new concept.

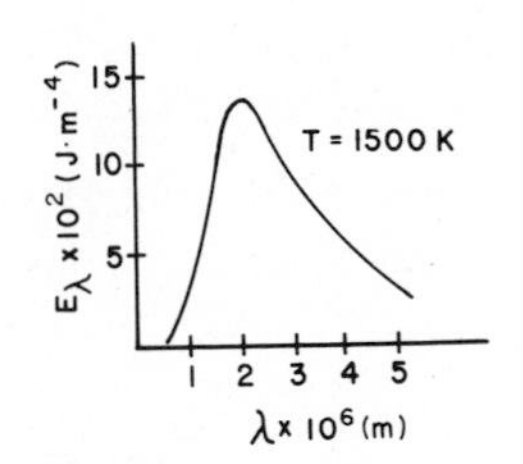

FIG. 7-1.2.1 The distribution of energy in a black body.

The classical approach that had been unsuccessful had assumed that the black body was made up of many tiny oscillators. The total energy of the body was then partitioned among these oscillators according to classical procedures. The results predicted that the energy of radiation would increase without bounds as the wavelength decreased, in marked disagreement with experimental data. Max Planck, in his study of the problem, gave an explanation that was without precedent and one that was justifiable only because it gave the correct answer. Planck postulated that the oscillators making up the body can oscillate only at discrete frequencies, but not at all possible frequencies. Thus, the black body can emit energy only in discrete amounts called *quanta*.

In addition, the energy of these quanta are related to the frequency of the radiation by the Planck equation

$$\varepsilon = h\nu$$

where ε is the energy of the quantum, h is a proportionality constant given by Planck as $h = 6.6262 \times 10^{-34}$ J s molecule^{-1}, and ν is the frequency. (Be careful with the unit molecule. Many times it is omitted.)

This was a startling proposal, since now light had been endowed with some corpuscular (particle) nature. Previously, light had been treated as undulatory (wavelike) and matter as corpuscular (particle-like). (Newton was a strong proponent of a corpuscular theory of light. However, the wave nature of light was used prior to Planck to explain observed phenomena.) Now Planck had suggested that light itself might come in bundles. Such an absurd proposition was immediately rejected by most physicists in spite of the agreement with experiment. However, support was not long in coming from another physical phenomenon.

Albert Einstein was the next proponent of the theory. His study of the *photoelectric effect* required a similar explanation. The photoelectric effect is that observed when light below a certain wavelength is projected onto an active metal surface and electrons are ejected. The kinetic energy of the ejected electrons is related to the frequency of the incidnet light above the threshold frequency. These observations were first made by Heinrich Hertz, in 1887. Hertz also observed that the number of electrons ejected was proportional to the intensity of the beam, and that there is no time lag between the incidence of light and the emission of an electron.

The total energy of the ejected electron is the sum of the energy binding it to the metal (the amount of energy that has to be supplied to free the electron ω) and the kinetic energy due to its velocity. Thus

$$E_t = \frac{1}{2} m_e v^2 + \omega \tag{7-1.2.1}$$

Einstein applied the Planck concept to this problem and resolved the difficulties that had previously surrounded it. He proposed that the light incident on the surface is made up of discrete units or quanta, each with an energy $h\nu$. This packet of energy is totally absorbed by the electron with no allowance for division of the energy between electrons. Einstein's formula becomes

$$h\nu = \frac{1}{2} mv^2 + \omega \qquad (7\text{-}1.2.2)$$

This equation immediately explains many observations. If the frequency of the light is below a threshold value (ν_0: note that $\omega = h\nu_0$), there will be no electrons emitted: there is not enough energy to overcome the binding energy holding the electrons to the metal. Greater intensity just means more quanta are available so more electrons are released, but they all will have the same energy since all the quanta have the same energy.

The fact that the value of h required by Einstein's equation is the same as that found in Planck's equation was powerful support for Planck's theory. From this foundation has grown the whole edifice of the quantum theory that will be investigated in the remainder of this chapter and in the next. We now see that light has some corpuscular properties. Later we will see that matter has some wave nature.

It is time to turn our attention to *spectroscopy*, the study of the wavelengths of radiation emitted from atoms and molecules. This field supplied the majority of the data that were used to formulate and perfect the quantum theory.

7-1.3 SPECTROSCOPY AND ATOMIC STRUCTURE: THE OLD QUANTUM THEORY

THE BOHR ATOM. This discussion will be restricted to the observations made on the hydrogen atom, since it is the simplest possible system and has played a key role in the development of the quantum theory. Spectroscopists were very active during the nineteenth century recording the wavelengths of lines observed in the emission spectra of atoms. These were carefully cataloged, and empirical formulas were found to correlate with the observed wavelengths. The most famous of these formulas is the Rydberg formula for hydrogen which has the form

$$\frac{1}{\lambda} = R_H\left(\frac{1}{k^2} - \frac{1}{n^2}\right) \qquad (7\text{-}1.3.1)$$

where k and n are integers and λ is the wavelength of the radiation. The value of the Rydberg constant is $R_H = 109{,}677.581\ \text{cm}^{-1}$. You probably would guess, as did others, that the spectrum emitted by an atom must somehow be related to its structure. In 1913, Niels Bohr devised a model that was consistent with the data for the hydrogen atom.

The Bohr model is probably familiar to all of you, so it is not necessary to bore you with details. A verbal description of the model and the resulting equation will be presented. (If you want to see a mathematical derivation of the equation look in almost any other physical chemistry text.)

Bohr assumed that the hydrogen atom consisted of a central nucleus with a charge of +e and an electron of charge -e in orbit about the nucleus. The positive and negative charges are attractive and are balanced by the centrifugal force of the electron in its orbit. Such a system is not classically stable since a charge, when accelerated, should radiate energy. Bohr avoided that difficulty mainly by ignoring it. His assumptions were as follows.

1. The electron can move about the nucleus in only certain orbits: a break from classical thought.
2. The orbits thus defined are stable, ignoring classical instability.
3. The electron can move from one orbit to another only by absorption or emission of radiation. The frequency of the transition is given by

   $$\Delta E = h\nu \tag{7-1.3.2}$$

   ΔE is the energy difference, ν the frequency, and h is the now-familiar Planck's constant.

Bohr's next task was to find the stable orbits out of the infinite number of possible orbits. From the additional assumption that the angular momentum is quantized (can have only certain values), Bohr derived the expression for the energy involved in a transition between two stationary (stable) orbits. This assumption is just another way of stating assumption (1). Bohr's formula, in terms of the wave number $\bar{\nu}$ (the *wave number* is the recipricol of the *wave length*) is

$$\boxed{\bar{\nu} = \frac{2\pi^2 m_e e^4}{h^3 c}\left(\frac{1}{k^2} - \frac{1}{n^2}\right)} \tag{7-1.3.3}$$

where m_e = the electronic mass, e = the electronic charge, c = the speed of light, and k and n are integers. If the values of all the constants are inserted, a value of 109,737 cm^{-1} results, which is in excellent agreement with the Rydberg constant of Eq. (7-1.3.1).

The agreement in the above was a triumph for the quantum theory, but the jubilation was short-lived. No such simple formula could be found for any atom with more than one electron. Also, many details of the electronic spectrum of hydrogen could not be predicted. The extension of the quantum theory will be developed after the next historical note (♩).

THE WAVE NATURE OF PARTICLES. Louis de Broglie was the person who first suggested that matter itself might have some wave character. After all, it is only fair. If light can behave like particles in some cases, our sense of fair play tells us that particles should act as waves, on occasion. And sure enough, that is observed to be the case. In 1924 de Broglie proposed, on theoretical grounds, that particles should have a wavelength associated with them. A particle of mass m should have a wavelength of

$$\boxed{\lambda = \frac{h}{mv}} \tag{7-1.3.4}$$

where v is the velocity of the particle, and h is, again, Planck's constant.

Shortly thereafter, two different groups observed the diffraction of electrons from crystals or metal foils. C. Davisson and L. H. Germer of the Bell Telephone Laboratories and G. P. Thomson (son of J. J. Thomson) and A. Reid at the University of Aberdeen observed the diffraction of electrons, a definite wave behavior. This completes the circle. Matter behaves as particles at times and as waves at others. This represents the duality of nature. Such duality was the source of much consternation for a long time, but it is now accepted as a fundamental property of nature.

RISE OF THE NEW QUANTUM MECHANICS. The truth is probably that light and matter are neither waves nor particles but some unexplained entity. What we observe depends on the method we use for observing. Certain types of measurement reveal particle character, while others reveal wave character.

Anyway, the observation of the wave character of electrons quickly produced the development of wave or quantum mechanics. In 1926, Erwin Schrödinger and Werner Heisenberg independently developed a new kind of mechanics: wave mechanics. The formulations of the two are very different, but they yield consistent results. They both assume some wavelike nature for particles. Heisenberg's development is based on matrix algebra that is less familiar to most students, so the Schrödinger treatment is usually presented. We will follow that tradition in Sec. 7-2.

Now is a good time to present several examples showing how the equations and concepts given in this section can be applied. Then the postulates of the new quantum mechanics will be presented.

NOTES ON UNITS. Joule is the appropriate unit for energy. However, in this chapter and the next you will see spectroscopic data expressed as meter (m), per centimeter (cm^{-1}), or per second (s^{-1}). These are not energy units, but we can easily convert from one unit to another by multiplying or dividing by a universal physical constant such as the speed of light ($c = 2.9979 \times 10^{10}$ cm s^{-1} or 2.9979×10^{8} m s^{-1}) or Planck's constant ($h = 6.626 \times 10^{-34}$ J s molecule^{-1}). Use the Bohr relation $\Delta E = h\nu$ and the definitions of wave length λ, wave number $\bar{\nu}$, and frequency ν to show this is true. The wave number will be expressed in per centimeter (cm^{-1}) since this is the traditional unit used by spectroscopists.

$$\nu(s^{-1}) = \frac{c}{\lambda}\left(\frac{m\ s^{-1}}{m}\right)$$

$$\bar{\nu}(cm^{-1}) = \frac{1}{\lambda}\left(\frac{1}{m}\right)\left(\frac{1\ m}{100\ cm}\right)$$

$$\nu(s^{-1}) = \frac{c}{\lambda}\left(\frac{m\ s^{-1}}{m}\right) = c\bar{\nu}$$

Thus

$$\Delta E\ (J) = h\nu\ (J\ s \cdot s^{-1})$$

$$= \frac{h\ c}{\lambda}\left(\frac{J\ s \cdot m\ s^{-1}}{m}\right)$$

$$= h\ c\bar{\nu}\ (J\ s \cdot m\ s^{-1} \cdot m^{-1})$$

Put units on all your conversions and you should have no difficulty. Note also that per second is now expressed as Hz for Hertz. In older literature cps was used for per second (s^{-1}).

7-1.4 EXAMPLES

7-1.4.1 EXAMPLE

Determine the frequency, wavelength, wave number, and energy of several of the possible transitions between energy levels in the hydrogen atom. Hint: Use the equations in Sec. 7-1.3 and the information in the following tables and figures.

The energy levels observed for the hydrogen atom are shown in Fig. 7-1.4.1. The names given correspond to the spectroscopist who discovered the particular series of spectral lines. Figure 7-1.4.2 is a reproduction of what the spectrum of the hydrogen atom for the Balmer series would look like. Also, Table 7-1.4.1 shows the allowed transitions for several series for the hydrogen atom. Instrumentally, the way spectra are observed is as follows:

1. The atoms of the gas under study are excited by some means (thermal, electrical, etc.) and the radiation emitted by the excited atoms is introduced into a spectrograph (see Fig. 7-1.4.3).
2. In the spectrograph, a prism or diffraction grating separates the emission frequencies into a spectrum that is directed toward a film or detector of some sort for recording the intensity of radiation as a function of frequency.

Solution

Use the Rydberg equation

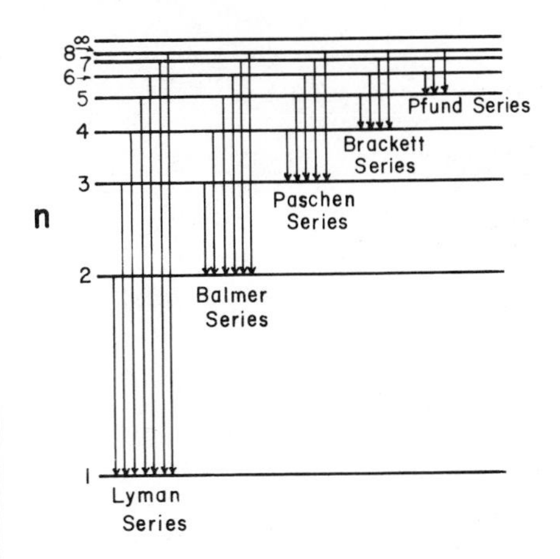

FIG. 7-1.4.1 Energy levels of the hydrogen atom.

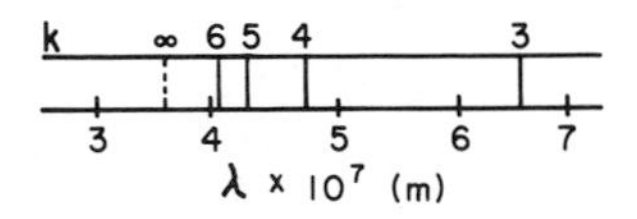

FIG. 7-1.4.2 Emission spectrum of the Balmer series for the hydrogen atom.

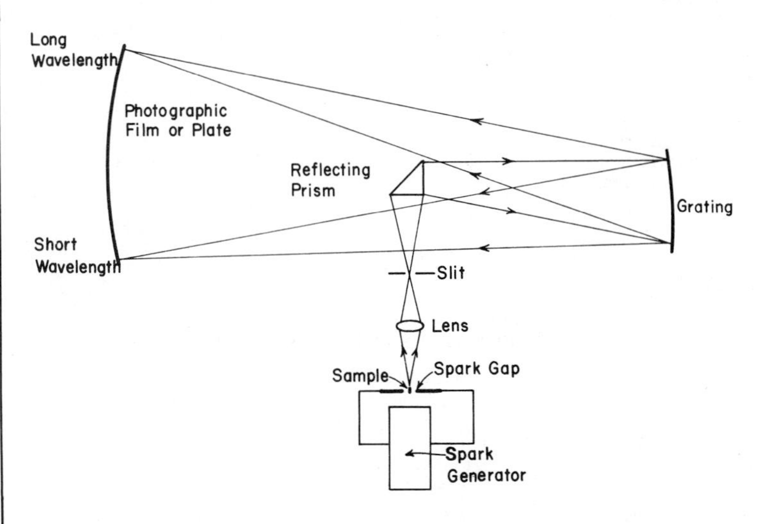

FIG. 7-1.4.3 An emission spectrograph.

Table 7-1.4.1
Spectral Series Observed for Atomic Hydrogen

Series	k	n	Spectral region
Lyman	1	2, 3, 4, ...	Ultraviolet
Balmer	2	3, 4, 5, ...	Visible
Paschen	3	4, 5, 6, ...	Infrared
Brackett	4	5, 6, 7, ...	Infrared

$$\frac{1}{\lambda} = \bar{\nu} = 1.097 \times 10^5 \text{ cm}^{-1} \left(\frac{1}{k^2} - \frac{1}{n^2}\right)$$

For the Lyman series k = 1; choose n = 2 and n = ∞.

n = 2

$$\bar{\nu} = (1.097 \times 10^5 \text{ cm}^{-1})\left(\frac{1}{1^2} - \frac{1}{2^2}\right)$$

$$= \underline{\underline{8.23 \times 10^4 \text{ cm}^{-1}}}$$

$$\lambda = \frac{1}{\bar{\nu}} = 1.22 \times 10^{-5} \text{ cm}$$

$$= \underline{\underline{1.22 \times 10^{-7} \text{ m}}}$$

$$\nu = \frac{c}{\lambda} = \frac{2.998 \times 10^8 \text{ m s}^{-1}}{1.22 \times 10^{-7} \text{ m}}$$

$$= \underline{\underline{2.46 \times 10^{15} \text{ s}^{-1}}}$$

n = ∞

$$\bar{\nu} = \underline{\underline{1.097 \times 10^5 \text{ cm}^{-1}}}$$

$$\lambda = \underline{\underline{9.12 \times 10^{-8} \text{ m}}}$$

$$\nu = \underline{\underline{3.29 \times 10^{16} \text{ s}^{-1}}}$$

For the Balmer series k = 2; choose n = 3 and n = ∞.

n = 3

$$\bar{\nu} = (1.097 \times 10^5 \text{ cm}^{-1})\left(\frac{1}{2^2} - \frac{1}{3^2}\right)$$

$$= \underline{\underline{1.52 \times 10^4 \text{ cm}^{-1}}}$$

$$\lambda = \frac{1}{\bar{\nu}} = 6.56 \times 10^{-5} \text{ cm}$$

$$= \underline{\underline{6.56 \times 10^{-7} \text{ m}}}$$

$$\nu = \underline{\underline{4.57 \times 10^{14} \text{ s}^{-1}}}$$

$n = \infty$

$$\bar{\nu} = (1.097 \times 10^{5} \text{ cm}^{-1})\left(\frac{1}{2^2} - 0\right)$$

$$= \underline{\underline{2.74 \times 10^{4} \text{ cm}^{-1}}}$$

$$\lambda = \underline{\underline{3.65 \times 10^{-7} \text{ m}}}$$

$$\nu = \underline{\underline{8.22 \times 10^{14} \text{ s}^{-1}}}$$

And so on.

7-1.4.2 EXAMPLE

(1) Determine the wavelength associated with an electron accelerated through a potential of 400 volts. (2) Find the wavelength associated with an α particle moving at 1.5×10^{7} m s^{-1}. (3) Calculate the wavelength associated with a snail with a mass of 10^{-2} kg, moving with a velocity of 2×10^{-3} m s^{-1}.

Solution

Note: 1 eV = 1.602×10^{-19} J

1. An electron accelerated to 400 V has a kinetic energy of 400 eV = 6.41×10^{-17} J $= \frac{1}{2} m_e v^2$ $\quad m_e = 9.1 \times 10^{-31}$ kg.

$$v = \left[\frac{2(6.41 \times 10^{-17} \text{ J})}{9.1 \times 10^{-31} \text{ kg}}\right]^{1/2}$$

$$= \underline{1.2 \times 10^{7} \text{ m s}^{-1}}$$

$$\lambda = \frac{h}{mv} = \frac{6.626 \times 10^{-34} \text{ J s}}{(9.1 \times 10^{-31} \text{ kg})(1.2 \times 10^{7} \text{ m s}^{-1})}$$

$$= \underline{\underline{6.1 \times 10^{-11} \text{ m}}}$$

2. $m_{\alpha} = 6.6 \times 10^{-27}$ kg.

$$\lambda = \frac{6.626 \times 10^{-34} \text{ J s}}{(6.6 \times 10^{-27} \text{ kg})(1.5 \times 10^{7} \text{ m s}^{-1})}$$

$$= \underline{\underline{6.7 \times 10^{-15} \text{ m}}}$$

3.

$$\lambda = \frac{6.626 \times 10^{-34} \text{ J s}}{(10^{-2} \text{ kg})(2 \times 10^{-3} \text{ m s}^{-1})}$$

$$= \underline{\underline{3.3 \times 10^{-29} \text{ m}}}$$

Notice that λ is negligibly short for these cases. A lower energy electron (smaller velocity) would have a longer wavelength, on the order of atomic dimensions.

7-1.4.3 EXAMPLE

From the following data collected by Millikan, determine the threshold energy (the binding energy) and the value of Planck's constant.

Electron kinetic energy (J) $\times 10^{19}$	ν (s^{-1}) $\times 10^{-14}$
3.41	9.58
2.56	8.21
1.95	7.40
1.64	6.91
0.75	5.49

Solution

Plot kinetic energy vs. ν. When the kinetic energy is zero we are at ν_0. From examination of Eq. (7-1.2.2) you should conclude that the slope of the plot is h.

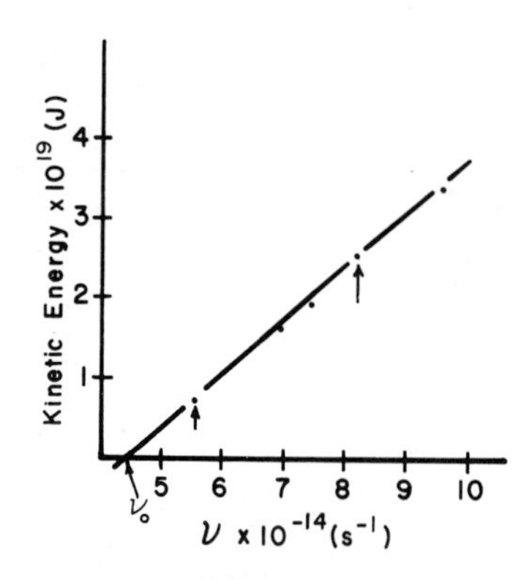

$$\nu_0 = \underline{\underline{4.35 \times 10^{15}\ \text{s}^{-1}}}$$

$$\text{Slope} = h = \frac{(2.56 - 0.75) \times 10^{-19}\ \text{J}}{(8.21 - 5.49) \times 10^{14}\ \text{s}^{-1}}$$

$$= 0.665 \times 10^{-33}\ \text{J s}$$

$$= \underline{\underline{6.65 \times 10^{-34}\ \text{J s}}}$$

This compares with the accepted value of 6.626×10^{-34} J s.

7-1.5 THE POSTULATES OF QUANTUM MECHANICS

The preceeding sections gave the historical background to the development of the quantum theory and quantum mechanics. Here we will go into considerable detail on the postulates that make up the foundation of quantum mechanics. Then, an example to show how practical problems can be solved using the techniques of quantum mechanics will be given.

The problem of the energies allowed for a particle in a box of given dimensions will be given. This may sound like an exercise in futility, but it is not. The energy levels found apply to those allowed for molecules in translational motion. These energy levels will be used in Chap. 9 in making calculations of the absolute energies of molecules. Also, the particle in a box model can be used to explain the electronic properties of some molecules.

Quantum mechanics, just like classical mechanics, is based on a set of unproved (and unprovable) assumptions called *postulates*. The test of the validity of these postulates rests on the degree of agreement between the calculated results and the results obtained in laboratory measurements. The postulates will be stated first, then an example will show how these are applied. Before beginning the postulates it would be a good idea (a very good idea) to read Sec. 7-2.2, where some mathematical preliminaries are given.

POSTULATE I.

a. A function Ψ exists that completely describes any state of the system.
b. Ψ has the property that $\Psi^* \Psi\, d\tau$ is the probability that the variables of the system will be found in the volume increment $d\tau$. (Ψ^* is the complex conjugate of Ψ.)

Part (a) is the basic assumption that we can find some wave function that will completely describe any state of the system we are studying. Without this assumption, there would be no need to try to find equations describing the system behavior at all.

Part (b) is related to the definition of the probability of finding a system in a state with given variables. Note the relationship to the probability function (the distribution function) found in Sec. 2-3. If we want to find the probability of the system having variables within a particular range, we integrate the probability function between those variable limits $\int_a^b \Psi^*\Psi \, d\tau$ = the probability of finding variables between the limits a and b).

There are some rules that are imposed to help in the solution of problems.

1. Ψ is finite and single-valued.
2. The integral $\int \Psi^*\Psi \, d\tau$ exists.
3. Ψ is normalized by requiring that $\int_{\text{all space}} \Psi^*\Psi \, d\tau = 1$.

That is, the integral of $\Psi^*\Psi \, d\tau$ over all space is equal to 1.

POSTULATE II. For every dynamic variable A there is a linear Hermitian operator $\hat{\alpha}$ that follows the rules

a. If A is a coordinate q, then $\hat{\alpha}$ is multiplication by the variable q.

b. If A is momentum (p = mv),

$$\hat{\alpha} = \frac{h}{2\pi i} \frac{\partial}{\partial q}$$

These are rules for finding operators to be used in the Schrödinger equation to be given below. (The Hermitian character is required to assure real values of predicted quantities. There is no need at this point to worry about the definition of Hermitian.)

POSTULATE III. Assume we have a system described by the wave function Ψ_s. If Ψ_s is an eigenfunction of the operator $\hat{\alpha}$, $\hat{\alpha}\Psi_s = a_s\Psi_s$, where a_s is a number. This last equation serves to define the terms known as eigenfunction and eigenvalue. If this equation holds for a given operator $\hat{\alpha}$, then Ψ_s is an *eigenfunction* with *eigenvalue* a_s. The operator $\hat{\alpha}$ corresponds to some observable such as velocity, position, energy, etc.

This postulate ties theoretical developments to experimental data. *If the function Ψ_s describes the system completely, and if it is an eigenfunction of the operator corresponding to an observable, we can predict the exact result we will observe if we measure that observable while the system is in the state described by Ψ_s.*

POSTULATE IV. If a function ϕ which is not an eigenfunction of the operator $\hat{\alpha}$ is used to describe a system, then we can only predict an "expectation value" or "average value" for the observable corresponding to $\hat{\alpha}$. This expectation value is defined as

$$\langle\hat{\alpha}\rangle = \frac{\int \phi^* \hat{\alpha} \phi \, d\tau}{\int \phi^* \phi \, d\tau} \qquad (7\text{-}1.5.1)$$

Measurement of the variable corresponding to $\hat{\alpha}$ will not necessarily give the same result as predicted by $\langle\hat{\alpha}\rangle$. Since ϕ is not the "true" wave function, the values of variables predicted from it will not be "true" values. An approximate wave function will predict results that are in error (usually).

The four postulates given above (or some equivalent statement of the basic assumptions) are sufficient to derive the properties of many quantum mechanical systems. This will be done below.

7-1.6 THE SCHRÖDINGER EQUATION

The best way to approach the development of the Schrödinger wave equation is to accept it as a postulate. Schrödinger developed his wave equation by making an analogy between the wave properties of matter and the wave behavior of a vibrating string. The analogy is not complete in that the final result cannot be derived. Some of the assumptions given in the postulates must be used. For example, after writing the classical Hamiltonian for the motion of a vibrating string, Schrödinger applied Postulate II to convert this Hamiltonian to operator notation. There is no obvious physical reason to believe that this procedure will lead to anything of use. It does, however.

Schrödinger applied operator algebra to write that the total energy of a system is determined by the following equation:

$$\hat{\mathcal{H}} \Psi = E\Psi \qquad (7\text{-}1.6.1)$$

Here, $\hat{\mathcal{H}}$ is the Hamiltonian operator, Ψ is the complete (exact) wave function that describes the system exactly, and E is the total energy of the system. The Hamiltonian operator is given by the following equation:

$$\hat{\mathcal{H}} = -\frac{h^2}{8\pi^2 m} \nabla^2 + V \qquad (7\text{-}1.6.2)$$

V is the potential energy expression which will be a function only of coordinates and

$$\nabla^2 = \frac{\partial^2}{\partial x^2} + \frac{\partial^2}{\partial y^2} + \frac{\partial^2}{\partial z^2} \qquad (7\text{-}1.6.3)$$

So the Schrödinger equation becomes

$$-\frac{h^2}{8\pi^2 m} \nabla^2 \Psi + V\Psi = E\Psi \qquad (7\text{-}1.6.4)$$

The example that follows should help clarify how this simple-looking equation can be used to describe the behavior of some physical systems. At this point you are not supposed to know how (7-1.6.4) is derived. Some hint of that is found in Sec. 7-2. However, you do not have to derive the equation to use it.

7-1.7 THE PARTICLE IN A BOX PROBLEM

The simplest system of physical significance is that of a particle constrained in space to move in a box of given dimensions. We can restrain the particle to a certain region of space by allowing the potential energy to be infinite in any region outside that of interest. If the potential is zero within the defined region (the box), we have a particularly simple system that still has physical applications: translational motion of molecules. First, assume the particle can move in a line only (one dimension). A physical system of this type might be an electron restricted to motion in the bond between two atoms. Figure 7-1.7.1 gives a schematic of the potential energy of such a system.

The potential outside the box of length a is ∞. A general result of quantum mechanics is the obvious one that Ψ is equal to zero in any region where $V = \infty$. You should see that this is reasonable. If the potential energy of a region is infinite, there is no normal particle that will have sufficient energy to exist in this region. If $\Psi = 0$, the second part of Postulate I tells us the probability of finding the particle in that region is zero ($\Psi^*\Psi \, d\tau = 0$ when $\Psi = 0$). Now we need to find the equation to describe the motion of the particle within the box where $V = 0$.

For the region $x = 0$ to $x = a$ we know $V = 0$, so the Schrödinger equation (7-1.6.4) reduces to (for one dimension)

$$-\frac{h^2}{8\pi^2 m}\frac{d^2\Psi}{dx^2} = E\Psi \qquad (7\text{-}1.7.1)$$

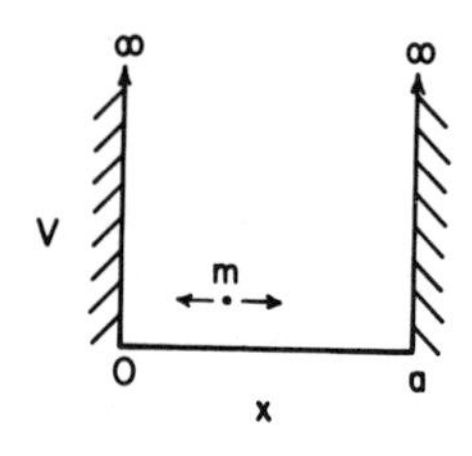

FIG. 7-1.7.1 A potential energy diagram for a particle restricted to motion in one dimension.

Rearranging this produces a second-order differential equation that can be solved by inspection.

$$\frac{d^2\Psi}{dx^2} = -\frac{8\pi^2 mE}{h^2}\Psi \tag{7-1.7.2}$$

We want a function that when differentiated twice gives back the initial function multiplied by a constant. Inspection (guesswork) shows that a solution is

$$\Psi = A \sin \alpha x \tag{7-1.7.3}$$

That this is a solution is proven by substitution into the differential equation. Such a substitution yields (do it)

$$\frac{d^2\Psi}{dx^2} = -\alpha^2 A \sin \alpha x$$

$$= -\alpha^2\Psi \tag{7-1.7.4}$$

If $\alpha^2 = 8\pi^2 mE/h^2$, Eq. (7-1.7.3) is a solution to (7-1.7.2).

Now comes an important step in the solution. To this point, we have placed no restrictions on the allowed values of E in the last equation. However, there are restrictions on the solutions at the edges of the box. $\Psi = 0$ outside the box, and Postulate I requires that the solution be *single valued*, so you should conclude that $\Psi(0) = 0$ and $\Psi(a) = 0$. The fact that $\Psi(0) = 0$ tells us nothing new. But look what happens when we impose the *boundary condition* that $\Psi(a) = 0$.

$$\Psi(a) = A \sin \alpha a = 0 \tag{7-1.7.5}$$

This is satisfied only if $\alpha a = n\pi$, where n is any integer. Thus we must satisfy the condition that

$$\alpha = \frac{n\pi}{a} \qquad n = 1, 2, 3, \ldots \tag{7-1.7.6}$$

But

$$\alpha^2 = \frac{8\pi^2 mE}{h^2} = \frac{n^2\pi^2}{a^2} \tag{7-1.7.7}$$

or

$$E = \frac{n^2h^2}{8ma^2} \qquad n = 1, 2, 3, \ldots \tag{7-1.7.8}$$

n is a quantum number.

The energy can have only discrete values as given in Eq. (7-1.7.8). *This result is important.* In quantum mechanical problems, imposition of boundary conditions on the solutions to the Schrödinger equation results in the appearance of *quantum numbers* that restrict the energies to discrete values. This is the source of the term *quantum mechanics*: the energies of the system are *quantized* (can have only certain values).

The solution is not yet quite complete. We need to evaluate the other constant in Ψ. To do this we can apply the normalization condition of Postulate I. This condition states that the probability of finding a system (the particle in this case) somewhere in space has to be 1. In this example, all space is the dimension of the box since the particle is not allowed outside this region. The complex conjugate of Ψ, which is designated as Ψ^*, is the same as Ψ since it is a real function (no imaginary parts). So,

$$\int_0^a \Psi^*\Psi \, dx = \int_0^a \left(A \sin \frac{n\ x}{a}\right)^2 dx = 1 \qquad (7\text{-}1.7.9)$$

From a table of integrals you should find that

$$A = \left(\frac{2}{a}\right)^{1/2} \qquad (7\text{-}1.7.10)$$

The final results for a particle constrained to move in one dimension within the region 0 to a under a potential of V = 0 are

$$\Psi_n = \left(\frac{2}{a}\right)^{1/2} \sin \frac{n\pi x}{a}$$

$$E_n = \frac{n^2h^2}{8ma^2} \qquad (7\text{-}1.7.11)$$

Several important results can be arrived at by consideration of the solutions to this problem. The following is based on an analysis by M. W. Hanna in *Quantum Mechanics in Chemistry* (see Sec. 7-1.11, Ref. 1). We will not solve the problem of the particle constrained to move in a three-dimensional box. Some discussion of that will be found in Chap. 9 when statistical thermodynamics is developed.

A very useful exercise for you at this point is to plot the form of the wave function for the first three or four energy levels (n = 1, 2, and 3). The abscissa will be the dimension of the box (0 to a). Also, on the same plot, show the form of the function $\Psi^*\Psi$. You will see how the probability function changes with n. Figure 7-1.7.2 shows the results for n = 1, 2, and 3.

There are a few important features you should note. As the mass of the particle is increased, the spacings of the energy levels come closer and closer together. If the mass becomes large enough for the quantity ma^2 to be much larger than h^2, then the energy levels appear to be continuous since they come so close together. *Only when the quantity* ma^2 *is on the order of* h^2 *do quantum mechanical results show up.* Example 7-1.7.1 will give you a feel for the sizes involved.

7-1.7.1 EXAMPLE

Calculate the energy between the first two levels (n = 1 and n = 2) for the following:

1. A 50-g golf ball constrained to a 100-m fairway.
2. An α particle (He nucleus) moving in a 10-m accelerator tube
3. An electron in a 1.5-Å (1.5×10^{-10} m) bond

Compare the results.

Solution

1.
$$E_1 = \frac{(1)^2(6.626 \times 10^{-34}\ \text{J s})^2}{8(0.050\ \text{kg})(100\ \text{m})^2}$$
$$= 1.097 \times 10^{-70}\ \text{J}^2\ \text{s}^2\ \text{kg}^{-1}\ \text{m}^{-2}$$
$$= \underline{\underline{1.097 \times 10^{-70}\ \text{J}}}$$
$$E_2 = \underline{\underline{4.388 \times 10^{-70}\ \text{J}}}$$

2.
$$E_1 = \frac{(1)^2(6.626 \times 10^{-34}\ \text{J s})^2}{8(0.004\ \text{kg}/6.023 \times 10^{23})(10\ \text{m})^2}$$
$$= \underline{\underline{8.26 \times 10^{-44}\ \text{J}}}$$
$$E_2 = \underline{\underline{33.0 \times 10^{-44}\ \text{J}}}$$

3.
$$E_1 = \frac{(1)^2(6.626 \times 10^{-34}\ \text{J s})^2}{(8)(9.1 \times 10^{-31}\ \text{kg})(1.5 \times 10^{-10}\ \text{m})^2}$$
$$= \underline{\underline{2.68 \times 10^{-18}\ \text{J}}}$$
$$E_2 = \underline{\underline{10.7 \times 10^{-18}\ \text{J}}}$$

Only in case (3) are the energy level spacings significant. The others are too small to be of consequence. Note that 2.68×10^{-18} J molecule^{-1} = 1.61×10^{6} J mol^{-1}. At room temperature thermal energy is RT = 2.48×10^{3} J mol^{-1}. So, the energies involved for small particles confined to small spaces are significant.

Careful consideration of the expectation values found by applying Postulate IV to the momentum in the x direction and the square of the momentum in that direction. leads to a statement of Heisenberg's uncertainty principle. (The wave function found is not an eigenfunction of the momentum operator which is defined as $(ih/2\pi)(d/dx)$; show that this statement is true.)

The expectation value for the momentum as defined by Postulate IV is (for a

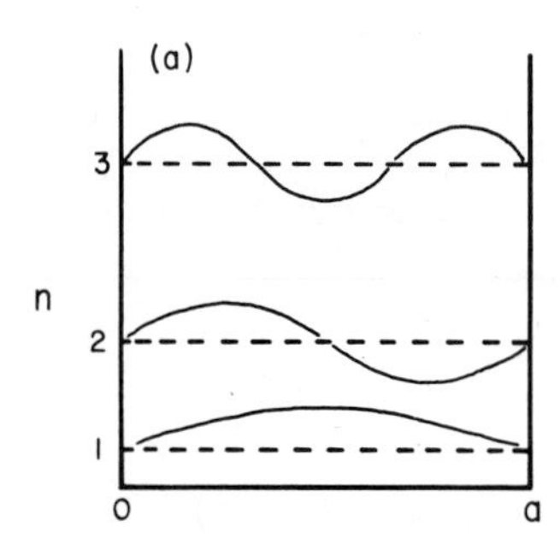

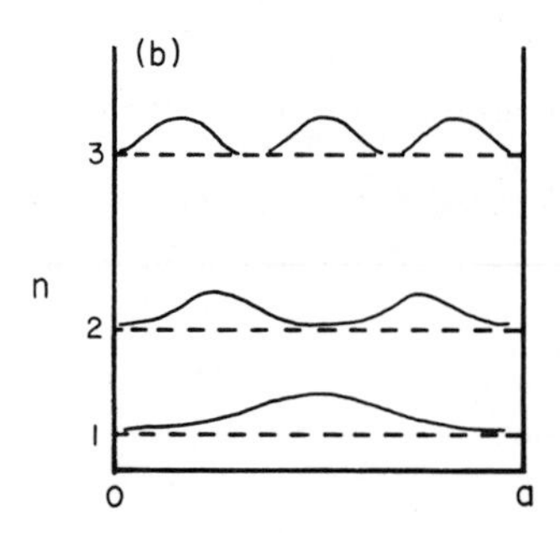

FIG. 7-1.7.2 Plot of (a) Ψ for n = 1, 2, and 3 and (b) Ψ*Ψ for n = 1, 2, and 3.

particle in *State 1* designated by Ψ_1)

$$\langle \hat{p}_x \rangle = \frac{\int_0^a \Psi_1 \hat{p}_x \Psi_1 \, dx}{\int_0^a \Psi_1^2 \, dx}$$

$$= \frac{(2/a) \int_0^a \sin (\pi x/a) [-(ih/2\pi)(d/dx)] \sin (\pi x/a) \, dx}{1}$$

$$= \frac{2}{a} \int_0^a \sin \frac{\pi x}{a} \left(\frac{ih}{2\pi} \frac{\pi}{a} \right) \cos \frac{\pi x}{a} \, dx = 0 \qquad (7\text{-}1.7.12)$$

This indicates that a large number of measurements of $\hat{p}_x$ on identical systems would yield an average value of momentum of zero.

The operator that can be used for the square of the momentum is

$$\hat{p}^2 = -\frac{h^2}{4\pi^2} \frac{d^2}{dx^2} \qquad (7\text{-}1.7.13)$$

You should set up the equation for finding the average value of the square of the momentum as was done in Eq. (7-1.7.12) for the momentum. Solve this before reading on (no peeking!). The first thing you should notice is that the wave function is an eigenfunction of the operator. This means that measurements of this quantity on a large set of identical systems will yield results which will correspond to the eigenvalues.

For State 1 you will find that the average value (for State 1) is given by

$$\hat{p}_x^{\,2} \text{ (State 1)} = \frac{h^2}{4a^2}$$

Comparison with (7-1.7.11) shows that this is equal to

$$p_x^{\,2} = 2mE_1$$

or taking the square root gives

$$p_x = \pm(2mE_1)^{\frac{1}{2}} \qquad (7\text{-}1.7.14)$$

If you compare this result to that found in Eq. (7-1.7.13) you might think you have discovered an inconsistency. The inconsistency is apparent, not real. Since our wave function is an eigenfunction of $p_x^{\,2}$, we will always observe the value of 2mE for $p_x^{\,2}$. Thus, the value of p_x will always be either $+(2mE)^{\frac{1}{2}}$ or $-(2mE)^{\frac{1}{2}}$. (Note: A positive value of p_x indicates motion in one direction, while a negative value is for motion in the opposite direction.) We cannot predict which it will be on any given measurement. A large number of measurements will give an average of $p_x = 0$ as predicted by Eq. (7-1.7.12), since $+(2mE)^{\frac{1}{2}}$ is equally as likely as $-(2mE)^{\frac{1}{2}}$. This leads us to another fundamental concept in quantum mechanics, the *Heisenberg Uncertainty principle*.

The uncertainty in any measurement of p_x is $2(2mE)^{\frac{1}{2}}$, since any one result might be positive or negative. We can also argue that the uncertainty in observing the position of the particle is the size of the box a. The total uncertainty will be the product of these two which gives (for any level)

$$\Delta x \Delta p_x \geq a\ 2(2mE_n)^{\frac{1}{2}} \geq 2a\,\frac{nh}{2a} \geq nh \qquad (7\text{-}1.7.15)$$

For the first state (n = 1) we find the smallest uncertainty (h). This is an important result. The uncertainty in the measurement of both the position and the momentum simultaneously is on the order of Planck's constant. It is impossible to measure both position and momentum at the same time to just any degree of accuracy we might want. The more precisely we determine one of these quantities, the more uncertainty we produce in the other.

For everyday objects this is of no consequence. Only for very small objects does the uncertainty principle limit our measurements. The uncertainty principle is a fundamental property of matter that limits accuracy of measurements on atomic and molecular systems. We shall see the uncertainty principle in operation in our study of spectroscopy (Chap. 8).

One final result which is left for you to prove is the *orthogonality condition*. For proper solutions to the Schrödinger equation we require that (Ψ_i and Ψ_j are two different eigenfunctions of $\mathcal{H}$) [Show that the following are true for some of the solutions given above in Eq. (7-1.7.11).]

$$\int_{\text{all space}} \Psi_i^* \Psi_i \, dx = 1 \qquad \text{the functions are } \textit{normalized}$$

$$\int_{\text{all space}} \Psi_i^* \Psi_j \, dx = 0 \qquad \text{the functions are } \textit{orthogonal}$$

7-1.8 THE HYDROGEN ATOM PROBLEM

The first major triumph of quantum mechanics was the solution of the hydrogen atom problem. Starting with the model of the hydrogen atom as a proton with an electron obeying Coulomb's law, a very classical picture as postulated by Bohr, it is possible to apply Schrödinger's wave equation to obtain allowed energy levels. These levels agree (or, at least, the energies of transitions agree) with the data observed spectroscopically.

Below, we will simply set up the equations for the hydrogen atom problem. Section 7-2 gives much of the mathematical preliminaries necessary to actually set up the equations, as well as a detailed solution of the problem. Section 7-3 gives the final results and applications of those results.

THE MODEL AND THE SCHRÖDINGER EQUATION. Figure 7-1.8.1 shows the model we will use for the H atom. The potential energy of this system is given by classical physics as

$$V = -\frac{e^2}{r} \tag{7-1.8.1}$$

and the Schrödinger equation is

$$-\frac{h^2}{8\pi^2 m_e}\left(\frac{\partial^2}{\partial x_e^2} + \frac{\partial^2}{\partial y_e^2} + \frac{\partial^2}{\partial z_e^2}\right)\Psi - \frac{h^2}{8\pi^2 m_p}\left(\frac{\partial^2}{\partial x_p^2} + \frac{\partial^2}{\partial y_p^2} + \frac{\partial^2}{\partial z_p^2}\right)\Psi - \frac{e^2}{r}\Psi = E\Psi \tag{7-1.8.2}$$

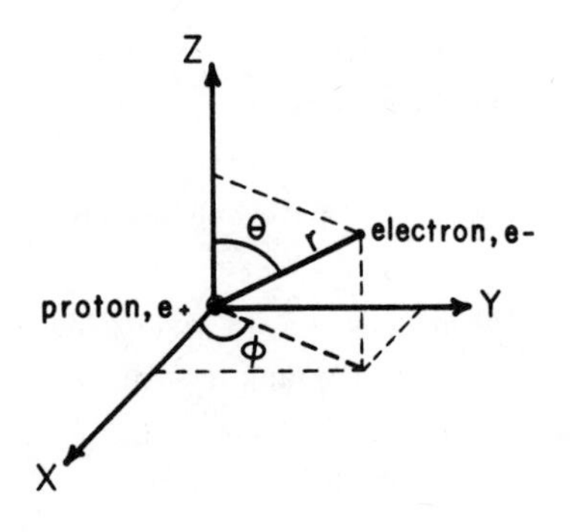

FIG. 7-1.8.1 Model of the hydrogen atom.

or
$$\left(-\frac{h^2}{8\pi^2 m_e}\nabla_e^{\,2} - \frac{h^2}{8\pi^2 m_p}\nabla_p^{\,2} - \frac{e^2}{r}\right)\Psi = E\Psi \qquad (7\text{-}1.8.2)$$

The steps that are necessary to solve this are:

1. Separate the equation into nuclear and electronic parts.
2. Convert to center of mass coordinates and to spherical polar coordinates (using the procedures of Example 7-2.3.3)

Step (2) is by no means easy. Not only are the intermediate steps rather involved, the final equation has to be rearranged somewhat to make the resulting operator Hermitian as required in the postulates. After a large amount of mathematical manipulation the equation given below, Eq. (7-1.8.3), is obtained. You are not expected to fill in the missing steps.

$$-\frac{h^2}{8\pi^2\mu}\left[\frac{1}{r^2}\frac{\partial}{\partial r}\left(r^2\frac{\partial\Psi}{\partial r}\right) + \frac{1}{r^2\sin\theta}\frac{\partial}{\partial\theta}\left(\sin\theta\,\frac{\partial\Psi}{\partial\theta}\right) + \frac{1}{r^2\sin^2\theta}\frac{\partial^2\Psi}{\partial\phi^2}\right] - \frac{e^2}{r}\Psi = E\Psi \qquad (7\text{-}1.8.3)$$

$$\mu = \frac{m_e m_p}{m_e + m_p} = \text{reduced mass}$$

SEPARATION OF VARIABLES. The next step is the separation of the variables. You will note that in Eq. (7-1.8.3) $\Psi(r, \theta, \phi)$ is a function of three variables. What we want to do is find three equations, each a function of one variable, that can be solved. This is a standard procedure for finding solutions to differential equations. You have surely done that in mathematics courses you have had. We start with the assumption that $\Psi(r, \theta, \phi)$ can be written as

$$\Psi(r, \theta, \phi) = R(r)\,\Theta(\theta)\,\Phi(\phi) \qquad (7\text{-}1.8.4)$$

where each of the functions R, Θ, and Φ are dependent on only one variable.

Here is one for you to try. Substitute (7-1.8.4) into (7-1.8.3) and simplify as much as possible to obtain

$$\frac{1}{r^2 R}\frac{d}{dr}\left(r^2\frac{dR}{dr}\right) + \frac{1}{\Phi r^2\sin^2\theta}\frac{d^2\Phi}{d\phi^2} + \frac{1}{\Theta r^2\sin\theta}\frac{d}{d\theta}\left(\sin\theta\,\frac{d\Theta}{d\theta}\right) + \frac{8\pi^2\mu}{h^2}(E - V) = 0 \qquad (7\text{-}1.8.5)$$

Multiply this equation by $r^2\sin^2\theta$ and rearrange to get

$$\frac{\sin^2\theta}{R}\frac{d}{dr}\left(r^2\frac{dR}{dr}\right) + \frac{1}{\Theta}\sin\theta\frac{d}{d\theta}\left(\sin\theta\frac{d\Theta}{d\theta}\right)$$

$$+ \frac{8\pi^2\mu r^2\sin^2\theta}{h^2}(E - V) = -\frac{1}{\Phi}\frac{d^2\Phi}{d\phi^2}$$

$$= f(r, \theta) \qquad (7\text{-}1.8.6)$$

Note that in this last equation we have

$$-\frac{1}{\Phi}\frac{d^2\Phi}{d\phi^2} = f(r, \theta)$$

This means that a function of ϕ only is equal to a function of r and θ. The only way for this to be true for all possible values of r, θ, and ϕ is for both equations to equal a constant. Thus

$$-\frac{1}{\Phi}\frac{d^2\Phi}{d\phi^2} = f(r, \theta) = \text{constant} \equiv m^2$$

$$(7\text{-}1.8.7)$$

or $$\frac{d^2\Phi}{d\phi^2} = -m^2\Phi$$

If you substitute Eq. (7-1.8.7) into Eq. (7-1.8.6) and rearrange (as usual) you can show

$$\frac{1}{R}\frac{d}{dr}\left(r^2\frac{dR}{dr}\right) + \frac{8\pi^2\mu r^2}{h^2}(E - V) = \frac{m^2}{\sin^2\theta} + \frac{1}{\Theta\sin\theta}\frac{d}{d\theta}\left(\sin\theta\frac{d\Theta}{d\theta}\right) \qquad (7\text{-}1.8.8)$$

This time we have a function of r only equal to a function of θ. As before this can be true for all r and θ only if both parts are equal to a constant.

$$f(r) = f(\theta) = \text{constant} \equiv \beta \qquad (7\text{-}1.8.9)$$

We now have the three equations we sought originally. Since we have found three equations, each a function of only one variable, our original assumption that we could do so is valid.

Θ Equation:

$$\frac{1}{\sin\theta}\frac{d}{d\theta}\left(\sin\theta\frac{d\Theta}{d\theta}\right) - \frac{m^2}{\sin^2\theta}\Theta + \beta\Theta = 0 \qquad (7\text{-}1.8.10a)$$

R Equation:

$$\frac{1}{r^2}\frac{d}{dr}\left(r^2\frac{dR}{dr}\right) - \frac{\beta}{r^2}R + \frac{8\pi^2\mu}{h^2}(E - V)R = 0 \qquad (7\text{-}1.8.10b)$$

Φ Equation:

$$\frac{d^2\Phi}{d\phi^2} = -m^2\Phi \tag{7-1.8.10c}$$

This is as far as we will go in this section. The next section continues and gives a solution to these last three equations. The procedures presented here are used (with considerable expansion) to solve all kinds of quantum mechanical problems such as the energy levels of a rotating or vibrating molecule, the binding energy between atoms, and the effect of electric and magnetic fields on energy levels in molecules. We will not do any of these cases explicitly, but the results will be used in the remainder of this chapter (after Sec. 7-2) and in Chaps. 8, 9, and 10.

The point to be made is that it is possible to calculate reliable energy levels (and many, many other properties) from our atomic and molecular model. It is difficult in that exact solutions are not possible, but accurate approximation methods are available to do the job.

7-1.9 PROBLEMS

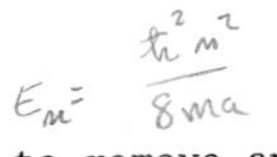

1. The *ionization energy* of an atom is the energy required to remove an electron from a given orbit to $n = \infty$. Determine ionization energies for electrons in the $k = 1$, $k = 2$, and $k = 6$ levels for the hydrogen atom.
2. What is the wavelength associated with a golf ball with $m = 5 \times 10^{-2}$ kg and $v = 35$ m s^{-1}?
3. The kinetic energy of an electron ejected from a metal with a binding energy of 2.8×10^{-19} J is observed to be 2.1×10^{-19} J. What is the wavelength and frequency of the incident radiation?
4. Determine the energy of the first three energy levels for a N_2 molecule constrained to a box 0.10 m on each side (solve the problem for one dimension only: the other two yield the same values). Compare these energies to RT at room temperature.
5. For the particle in a box, find the probability of finding the particle between a/4 and 3a/4 for n = 2.
6. What is the average value of x for the n = 3 level in the particle in a box?
7. Give the frequency and energy of a transition with $\bar{\nu} = 2.06 \times 10^6$ m^{-1}. If $k_1 = 2$, what is k_2?
8. What is the equivalent mass of a photon that has a wavelength of 1.22×10^{-7} m?

7-1.10 SELF-TEST

1. Look up and list the frequencies, wavelengths, and wave numbers for the limits of each of the spectral regions (visible, untraviolet, and infrared).
2. Calculate the frequency, wavelength, and wave number for each of the first four transitions of the Brackett series in the hydrogen atom.
3. Find the mass of a particle that has a wave length of 8.2×10^{-12} m and a velocity of 2.4×10^{3} m s^{-1}.
4. A given metal has a binding energy of 4.00×10^{-19} J. Calculate the kinetic energy of an electron ejected by photons with a wavelength of 3.3×10^{-7} m. What is the threshold wavelength?
5. Use the $\hat{p}^2$ operator in Eq. (7-1.7.13) and the solutions to the particle in a box problem to find the average value of p^2.

7-1.11 ADDITIONAL READING

1.* M. W. Hanna, *Quantum Mechanics in Chemistry*, W. A. Benjamin, New York, 1966, Sec. 3-4.

7-2 MATHEMATICAL CONSIDERATIONS AND SOLUTIONS OF THE HYDROGEN ATOM PROBLEM

OBJECTIVES

You shall be able to

1. Apply the concepts of operator algebra to determine if an operator is linear and to find eigenvalues for eigenfunction equations
2. Write Newton's equations of motion for a given potential function in classical, Lagrangian, and Hamiltonian form
3. Outline the methods used in the solution to the hydrogen atom problem
4. Show whether a function is normalized or whether two functions are orthogonal
5. Outline the procedures used in the variation method

7-2.1 INTRODUCTION

The mathematical groundwork for the remainder of the chapter is laid in this section. Make sure you understand it. The organization of this section follows that found in M. W. Hanna, *Quantum Mechanics in Chemistry* (see Sec. 7-2.8, Ref. 1). If more detail is required, that text is recommended. Some operations with which you may be familiar will be defined. If you are not familiar with these, then get an advanced engineering mathematics book to review.

Following the mathematical review, the solution to the hydrogen atom problem, begun in Sec. 7-1.8, will be completed. Some discussion of approximate methods for solving more complex problems will be given.

7-2.2 MATHEMATICAL PRELIMINARIES

COORDINATE SYSTEMS. We will be concerned with *Cartesian coordinates* and *spherical polar coordinates*. In Cartesian coordinates, three mutually perpendicular axes designated as X, Y, and Z are used. The position of any point is defined by projection of x, y, and z, respectively, onto these axes. Thus, a point is given as P(x, y, z) where the x, y, and z are the projections.

Figure 7-2.2.1 shows the definition of the *spherical polar coordinates* in relation to a Cartesian axis system. The x, y, and z projections are also shown. You should verify that

$$x = r \sin \theta \cos \phi$$
$$y = r \sin \theta \sin \phi \qquad (7\text{-}2.2.1)$$
$$z = r \cos \theta$$

Also you should prove $x^2 + y^2 + z^2 = r^2$.

DETERMINANTS AND VECTORS. It is assumed that you can evaluate determinants. It will be necessary for you to do so in Sec. 7-4, so review if you need to. *Vectors* are symbols that have both *magnitude* and *direction*. *Scalars*, with which you are more familiar, have only *magnitude*. Let us designate a vector by a letter with a $\rightharpoonup$ over it so the symbol $\vec{r}$ implies a vector and r, a scalar. Usually vectors are defined in terms of three mutually perpendicular *unit vectors*, $\vec{i}$, $\vec{j}$, and $\vec{k}$, that lie along the X, Y, and Z axes, respectively. Any vector can be expressed in terms of projections onto these unit vectors. Thus the vector A can be written

$$\vec{A} = A_x\vec{i} + A_y\vec{j} + A_z\vec{k} \qquad (7\text{-}2.2.2)$$

The radius vector of Fig. 7-2.2.1 can be written

$$\vec{r} = x\vec{i} + y\vec{j} + z\vec{k}$$

Some of the rules of vector algebra follow:

Addition and Subtraction. The vector sum $\vec{A} + \vec{B} = \vec{C}$ can be done graphically. You should have done that previously, so it is left as an exercise for you. Pick

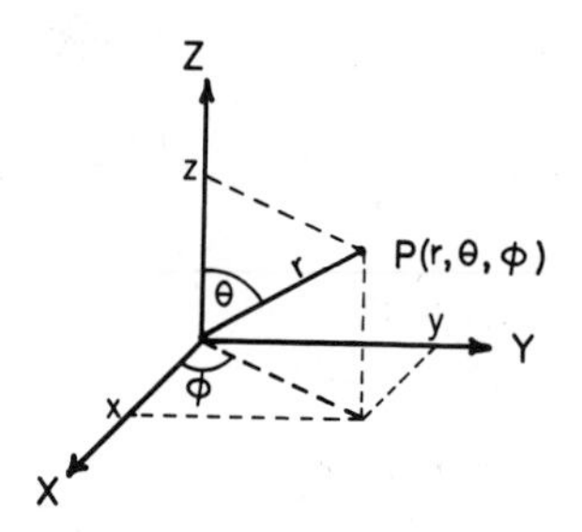

FIG. 7-2.2.1 Definition of spherical polar coordinates in terms of a Cartesian axis system.

two vectors $\vec{A}$ and $\vec{B}$ and graphically find the new vectors $\vec{C}$ and $\vec{D}$ corresponding to the sum and difference of $\vec{A}$ and $\vec{B}$.

If we write $\vec{A}$ and $\vec{B}$ in terms of the components, we can determine $\vec{A} + \vec{B}$ as

$$\begin{aligned}\vec{C} &= \vec{A} + \vec{B}\\ &= A_x\vec{i} + A_y\vec{j} + A_z\vec{k} + B_x\vec{i} + B_y\vec{j} + B_z\vec{k}\\ &= (A_x + B_x)\vec{i} + (A_y + B_y)\vec{j} + (A_z + B_z)\vec{k}\end{aligned} \tag{7-2.2.3}$$

Subtraction is handled in a similar manner.

Magnitude. The magnitude of any vector is given by

$$\|A\| = (A_x^2 + A_y^2 + A_z^2)^{\frac{1}{2}} \tag{7-2.2.4}$$

Scalar (Dot) Product. The *scalar (dot) product* of two vectors results in a number and is defined as

$$\vec{A} \cdot \vec{B} \equiv AB \cos\theta \tag{7-2.2.5}$$

where A and B are the magnitude of $\vec{A}$ and $\vec{B}$, and θ is the angle between them.

An important case is when $\vec{A}\cdot\vec{B} = 0$. If this is true the vectors are said to be *orthogonal*. You will see this word again in a slightly different context.

Cross (Vector) Product. The vector product results in a new vector perpendicular to both the original vectors. The direction of the new vector is defined by the "right hand rule." For $\vec{A} \times \vec{B}$, if you put the bottom edge of your palm on $\vec{A}$ and curl your fingers toward $\vec{B}$, your thumb will point in the direction of the resultant vector. The cross-product is defined by

$$\vec{C} = \vec{A} \times \vec{B} = \vec{n}AB \sin\theta \tag{7-2.2.6}$$

$\vec{n}$ is a unit vector perpendicular to $\vec{A}$ and $\vec{B}$. The other symbols have the same meaning as in Eq. (7-2.2.5). The magnitude of C is AB sin θ.

The vector product is not *commutative*, which means

$$A \times B \neq B \times A \qquad (7\text{-}2.2.7)$$

The right-hand rule should convince you of that.

The components of the cross-product are written conveniently in determinant form

$$A \times B = \begin{vmatrix} \vec{i} & \vec{j} & \vec{k} \\ A_x & A_y & A_z \\ B_x & B_y & B_z \end{vmatrix}$$

$$= (A_yB_z - A_zB_y)\vec{i} - (A_xB_z - B_xA_z)\vec{j} + (A_xB_y - B_xA_y)\vec{k} \qquad (7\text{-}2.2.8)$$

COMPLEX NUMBERS. Complex numbers have two parts: a *real* part and an *imaginary* part. They are written as

$$D = E + iF \qquad (7\text{-}2.2.9)$$

where D, E, and F are numbers and $i = \sqrt{-1}$, the complex number. If i is replaced by - i everywhere it occurs, a *complex conjugate* D* results with D* = E - iF. You should show that

$$DD^* = E^2 + F^2 \qquad (7\text{-}2.2.10)$$

The *magnitude* of D is $\|D\| = (DD^*)^{\frac{1}{2}}$. In adding and subtracting complex numbers the real and imaginary parts are used separately.

OPERATORS. You use operators in any mathematical manipulation. An operator is a symbol that is an instruction to do something. The symbol $(\partial/\partial x)$ says to take the derivative with respect to x of whatever follows the symbol. We will denote operators with the superscript symbol ^. Thus, $\hat{P}$ implies that P is an operator.

Two operators $\hat{P}$ and $\hat{Q}$ *commute* if $\hat{P}\hat{Q} = \hat{Q}\hat{P}$. Assume, for example, that

$$\hat{P} = \frac{d}{dx} \quad \text{and} \quad \hat{Q} = x$$

Also, assume we have a function $f(x) = x^2$. Does $\hat{P}\hat{Q}\ f(x) = \hat{Q}\hat{P}\ f(x)$?

$$\hat{P}\hat{Q}\ f(x) = \frac{d}{dx}(x)(x^2) = \frac{d}{dx}(x^3) = 3x^2$$

(Operators are applied from right to left; $\hat{Q}$ above is performed then $\hat{P}$.)

$$\hat{Q}\hat{P}\ f(x) = (x)\ \frac{d}{dx}(x^2) = x(2x) = 2x^2$$

Thus, $\hat{P}\hat{Q} \neq \hat{Q}\hat{P}$, so $\hat{P}$ and $\hat{Q}$ do not commute.

In the case of vector operators such as ∇

$$\vec{\nabla} = \vec{i}\,\frac{\partial}{\partial x} + \vec{j}\,\frac{\partial}{\partial y} + \vec{k}\,\frac{\partial}{\partial z} \qquad (7\text{-}2.2.11)$$

the components are used separately. For example,

$$\vec{\nabla}\, f(xyz) = \vec{i}\,\frac{\partial}{\partial x}\, f(xyz) + \vec{j}\,\frac{\partial}{\partial y}\, f(xyz) + \vec{k}\,\frac{\partial}{\partial z}\, f(xyz)$$

Finally, an operator is *linear* if

$$\hat{P}(af + bg) = a\hat{P}f + b\hat{P}g \qquad (7\text{-}2.2.12)$$

where a and b are constants and f and g are functions. You should show that $\partial/\partial x$ is linear, but the squaring operator $(\ )^2$ is not.

7-2.2.1 EXAMPLE

Show that the operators $\partial/\partial x$ and multiplication by 2, are linear and commute.

Solution

$\partial/\partial x$ is linear since $(\partial/\partial x)\ (af + bg) = a\ (\partial f/\partial x) + b\ (\partial g/\partial x)$

$2\cdot$ is linear since $2(af + bg) = a2f + b2g$

To see if they commute pick an arbitrary function of x, f(x)

$$\frac{\partial}{\partial x}\,(2)\ f(x) = \frac{\partial(2)}{\partial x}\, f(x) + (2)\,\frac{\partial f(x)}{\partial x}$$

$$= 2\,\frac{\partial}{\partial x}\, f(x)$$

Therefore, $2\cdot$ and $\partial/\partial x$ commute.

7-2.3 NEWTON'S EQUATIONS OF MOTION

LAGRANGIAN FORM. In Cartesian coordinates let us define velocity and acceleration in the x direction as

$$\dot{x} = \frac{dx}{dt} \qquad \ddot{x} = \frac{d^2x}{dt^2} = \frac{d\dot{x}}{dt} \qquad (7\text{-}2.3.1)$$

Other components are defined in a similar manner. Newton's second law relates force to the acceleration and mass of a particle. A particle of mass m under the influence of a force in the x direction F_x will exhibit an acceleration of $\ddot{x}$. Similarly for y and z, so

$$F_x = m\ddot{x}$$
$$F_y = m\ddot{y} \qquad (7\text{-}2.3.2)$$
$$F_z = m\ddot{z}$$

Also, for a conservative system (one in which the sum of the kinetic and potential energies is constant) with a potential V [V is a function of coordinates, V = V(xyz)]

$$F_x = -\frac{\partial V}{\partial x}$$
$$F_y = -\frac{\partial V}{\partial y} \qquad V = V(xyz) \qquad (7\text{-}2.3.3)$$
$$F_z = -\frac{\partial V}{\partial z}$$

Finally, for a system of n particles (with masses m_i) the total kinetic energy T will be (remember for one particle the kinetic energy $\frac{1}{2}mv^2$ for one dimension)

$$T = \frac{1}{2} m_1 (\dot{x}_1^2 + \dot{y}_1^2 + \dot{z}_1^2) + \frac{1}{2} m_2 (\dot{x}_2^2 + \dot{y}_2^2 + \dot{z}_2^2) + \ldots$$
$$= \frac{1}{2} \sum_i m_i (\dot{x}_i^2 + \dot{y}_i^2 + \dot{z}_i^2) \qquad (7\text{-}2.3.4)$$

Now, get your pencil and combine the last three equations to prove that Newton's equations of motion can be written

$$\frac{d}{dt} \frac{\partial T}{\partial \dot{x}_i} + \frac{\partial V}{\partial x_i} = 0$$
$$\frac{d}{dt} \frac{\partial T}{\partial \dot{y}_i} + \frac{\partial V}{\partial y_i} = 0 \qquad (7\text{-}2.3.5)$$
$$\frac{d}{dt} \frac{\partial T}{\partial \dot{z}_i} + \frac{\partial V}{\partial z_i} = 0$$

Next, the Lagrangian function L will be defined. There is nothing mysterious about it. It, like H, G, and A in thermodynamics, is defined for convenience. L is given by

$$L = L(x_i y_i z_i ; \dot{x}_i \dot{y}_i \dot{z}_i)$$
$$= T - V \qquad (7\text{-}2.3.6)$$

It is the difference of the kinetic and potential energies.

You should show that Eq. (7-2.3.5) can now be written [show that the operations below lead to Eq. (7-2.3.5)].

$$\frac{d}{dt}\frac{\partial L}{\partial \dot{x}_i} - \frac{\partial L}{\partial x_i} = 0$$

$$\frac{d}{dt}\frac{\partial L}{\partial \dot{y}_i} - \frac{\partial L}{\partial y_i} = 0 \qquad (7\text{-}2.3.7)$$

$$\frac{d}{dt}\frac{\partial L}{\partial \dot{z}_i} - \frac{\partial L}{\partial z_i} = 0$$

To do this, you must recognize that

$$\frac{\partial V}{\partial \dot{x}_i} = \frac{\partial V}{\partial \dot{y}_i} = \frac{\partial V}{\partial \dot{z}_i} = 0$$

and $$\frac{\partial T}{\partial x_i} = \frac{\partial T}{\partial y_i} = \frac{\partial T}{\partial z_i} = 0$$

Why? Perhaps an example will help.

7-2.3.1 EXAMPLE

Write Newton's equations of motion for a single particle of mass m in a gravitational field defined by $V = mgz$, where z is the height and g is the gravitational acceleration constant. Then write the Lagrangian and apply Eq. (7-2.3.7) to show Newton's equations are obtained.

Solution

$$F_x = m\ddot{x} = -\frac{\partial V}{\partial x} = -\frac{\partial\ (mgz)}{\partial x} = 0$$

$$F_y = m\ddot{y} = -\frac{\partial V}{\partial y} = -\frac{\partial\ (mgz)}{\partial y} = 0$$

$$F_z = m\ddot{z} = -\frac{\partial V}{\partial z} = -\frac{\partial\ (mgz)}{\partial z} = mg$$

So $m\ddot{x} = 0$

$m\ddot{y} = 0$

$m\ddot{z} = mg$

The only force and acceleration are in the z direction as expected. The Lagrangian is

$$L = T - V = \frac{1}{2} m\ (\dot{x}^2 + \dot{y}^2 + \dot{z}^2) - mgz$$

$$\frac{d}{dt}\frac{\partial L}{\partial \dot{x}} - \frac{\partial L}{\partial x} = \frac{d}{dt}\ (m\dot{x}) - 0 = 0 \qquad m\ddot{x} = 0$$

$$\frac{d}{dt}\frac{\partial L}{\partial \dot{y}} - \frac{\partial L}{\partial y} = \frac{d}{dt}(m\dot{y}) - 0 = 0 \qquad m\ddot{y} = 0$$

$$\frac{d}{dt}\frac{\partial L}{\partial \dot{z}} - \frac{\partial L}{\partial z} = \frac{d}{dt}(m\dot{z}) - mg = 0 \qquad m\ddot{z} - mg = 0$$

The results are the same in the two cases.

GENERALIZED COORDINATES. To keep from writing x, y, or z (or r, θ, or φ) let us define general coordinates that can be any-position variable. These will be defined as q_i and the corresponding velocity is $\dot{q}_i = dq_i/dt$. There are 3n of these coordinates and 3n velocities for n particles. Equation (7-2.3.7) can now be written in much more compact form

$$\frac{d}{dt}\frac{\partial L}{\partial \dot{q}_i} - \frac{\partial L}{\partial q_i} = 0 \tag{7-2.3.8}$$

The partial derivatives are taken, holding all other coordinates and velocities constant. The momentum for a particle of mass m is

$$p_i = m\dot{q}_i \tag{7-2.3.9}$$

We can rewrite the Lagrangian in generalized coordinates for a conservative system with particles of mass m

$$L = T - V$$

$$= \sum_i \left[\frac{1}{2} m\dot{q}_i^{\,2} - V(q_i)\right] \tag{7-2.3.10}$$

You should see immediately that

$$p_i = \left(\frac{\partial L}{\partial \dot{q}_i}\right)_{\dot{q}_j, q_i} \tag{7-2.3.11}$$

The subscripts mean all variables except $\dot{q}_i$ are held constant while taking the derivative. These subscripts will not be written anymore: they will be assumed.

THE HAMILTONIAN FUNCTION. We are now ready to present the Hamiltonian function $\mathcal{H}$ that is related to the total energy of the system. This function is used in Sec. 7-3 in deriving the energy levels in the hydrogen atom.

$$\mathcal{H} = \sum_{i=1}^{3N} p_i\dot{q}_i - L \tag{7-2.3.12}$$

We state, without proof, that $\mathcal{H} = T + V$, the total energy of the system. Also, without mathematical proof,

$$\frac{\partial \mathcal{H}}{\partial p_i} = \dot{q}_i$$

and
$$\frac{\partial \mathcal{H}}{\partial q_i} = - \frac{\partial L}{\partial q_i} = - \dot{p}_i \qquad (7\text{-}2.3.13)$$

These are the equations of motion in Hamiltonian form. The following is an example showing how to write the Hamiltonian function for a system.

7-2.3.2 EXAMPLE

For the particle in a gravitational field given in Example 7-2.3.1, find the Hamiltonian function, and from Eq. (7-2.3.13) find the equations of motion.

Solution

From Example 7-2.3.1:

$$L = \frac{1}{2} m (\dot{x}^2 + \dot{y}^2 + \dot{z}^2) - mgz$$

Eq. (7-2.3.11) can be used to find p_i. Let $q_1 = x$, $q_2 = y$, and $q_3 = z$, so we can write immediately

$$L = \frac{1}{2} m \sum_{i=1}^{3} \dot{q}_i^2 - mgq_3$$

From Eq. (7-2.3.11):

$$p_i = \frac{\partial L}{\partial \dot{q}_i} = \frac{1}{2} (2m\dot{q}_i) = m\dot{q}_i$$

Then, from (7-2.3.12),

$$\mathcal{H} = \sum_{i=1}^{3} m\dot{q}_i\dot{q}_i - \frac{1}{2} \sum_{i=1}^{3} m\dot{q}_i^2 + mgq_3$$

$$= \frac{1}{2} m \sum_{i=1}^{3} \dot{q}_i^2 + mgq_3$$

This can be written in terms of momenta ($p_i = m_i\dot{q}_i$):

$$\mathcal{H} = \frac{1}{2m} \sum_i (m\dot{q}_i)^2 + mgq_3$$

$$= \frac{1}{2m} \sum_i p_i^2 + mgq_3$$

Now apply Eq. (7-2.3.13):

$$\frac{\partial \mathcal{H}}{\partial p_i} = \frac{\partial}{\partial p_i}\left(\frac{1}{2m}\sum p_i^{\ 2} + mgq_3\right)$$

$$= \frac{p_i}{m} = \frac{m\dot{q}_i}{m} = \dot{q}_i$$

These three equations (i = 1, 2, 3) tell us nothing new. They just define the momenta.

$$\frac{\partial \mathcal{H}}{\partial q_i} = \frac{\partial}{\partial q_i}\left(\frac{1}{2m}\sum_i p_i^{\ 2} + mgq_3\right)$$

so $\frac{\partial \mathcal{H}}{\partial q_1} = -\dot{p}_1 = 0$ or $m\ddot{q}_1 = 0$ $\quad m\ddot{x} = 0$

$\frac{\partial \mathcal{H}}{\partial q_2} = -\dot{p}_2 = 0$ or $m\ddot{q}_2 = 0$ $\quad m\ddot{y} = 0$

$\frac{\partial \mathcal{H}}{\partial q_3} = -\dot{p}_3 = mg$ or $m\ddot{q}_3 = -mg$ $\quad m\ddot{z} = mg$

These are the same results as obtained previously.

The final thing you need to be able to do is to convert from one coordinate system to another. If we have a variable x_i that is a function of new coordinates $x_i = x_i(q_1, q_2, q_3, \ldots)$, we can evaluate the time derivative using the chain rule.

$$\frac{dx_i}{dt} = \dot{x}_i = \frac{\partial x_i}{\partial q_1}\frac{dq_1}{dt} + \frac{\partial x_i}{\partial q_2}\frac{dq_2}{dt} + \cdots$$

$$= \frac{\partial x_i}{\partial q_1}\dot{q}_1 + \frac{\partial x_i}{\partial q_2}\dot{q}_2 + \cdots$$

$$= \sum_j \frac{\partial x_i}{\partial q_j}\dot{q}_j \qquad (7\text{-}2.3.14)$$

Let us do this by an example.

7-2.3.3 EXAMPLE

Write the Hamiltonian and the equations of motion in terms of the center of mass (x, y) and the angle ϕ for the following system that rotates in the XY plane and can move vertically (along Y) under the influence of gravity. R is the distance between m_1 and m_2. The connecting rod is weightless.

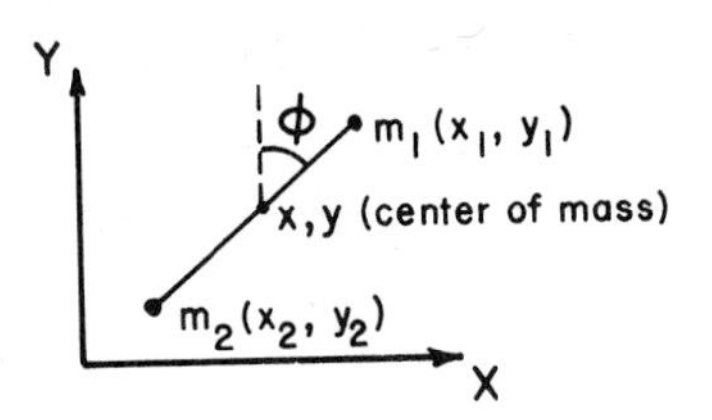

Solution

You should convince yourself that

$$x_1 = x + \frac{m_2}{m_1 + m_2} R \sin \phi \qquad x_2 = x - \frac{m_1}{m_1 + m_2} R \sin \phi$$

$$y_1 = y + \frac{m_2}{m_1 + m_2} R \cos \phi \qquad y_2 = y - \frac{m_1}{m_1 + m_2} R \cos \phi$$

This follows from the definition of the center of mass.

The kinetic energy is given by

$$T = \frac{1}{2} m_1 (\dot{x}_1^2 + \dot{y}_1^2) + \frac{1}{2} m_2 (\dot{x}_2^2 + \dot{y}_2^2)$$

Use Eq. (7-2.3.14) to convert to x, y, and ϕ, the center of mass coordinates. (Fill in the missing steps.)

$$\dot{x}_1 = \frac{\partial x_1}{\partial x} \dot{x} + \frac{\partial x_1}{\partial \phi} \dot{\phi} = \dot{x} + \frac{m_2}{m_1 + m_2} R \cos \phi\dot{\phi}$$

$$\dot{x}_2 = \dot{x} - \frac{m_1}{m_1 + m_2} R \cos \phi\dot{\phi}$$

$$\dot{y}_1 = \dot{y} + \frac{m_2}{m_1 + m_2} R \sin \phi\dot{\phi}$$

$$\dot{y}_2 = \dot{y} - \frac{m_1}{m_1 + m_2} R \sin \phi\dot{\phi}$$

Now define $M = m_1 + m_2$ and the moment of inertia

$$I = \frac{m_1 m_2}{M} R^2 = \mu R^2$$

Substitution into T gives (provide the missing steps; remember $\cos^2 \phi + \sin^2 \phi = 1$)

$$T = \frac{M}{2} \dot{x}^2 + \frac{M}{2} \dot{y}^2 + \frac{I}{2} \dot{\phi}^2$$

$V = Mgy$ the system moves as a mass M under the influence of gravity

The Lagrangian is

$$L = T - V = \frac{M}{2}(\dot{x}^2 + \dot{y}^2) + \frac{I}{2}\dot{\phi}^2 - Mgy$$

Now apply Eq. (7-2.3.11):

$$p_x = \frac{\partial L}{\partial \dot{x}} = M\dot{x}$$

$$p_y = \frac{\partial L}{\partial \dot{y}} = M\dot{y}$$

$$p_\phi = I\dot{\phi}$$

We can write L in terms of momenta:

$$L = \frac{1}{2M}(p_x^{\ 2} + p_y^{\ 2}) + \frac{1}{2I}p_\phi^{\ 2} - Mgy$$

The Hamiltonian from Eq. (7-2.3.12) is

$$\mathcal{H} = p_x\dot{x} + p_y\dot{y} + p_\phi\dot{\phi} - L$$

$$= \frac{1}{M}\left[p_x(M\dot{x}) + p_y(M\dot{y})\right] + \frac{1}{I}\left[p_\phi(I\dot{\phi})\right] - L$$

$$= \frac{1}{M}\left(p_x^{\ 2} + p_y^{\ 2}\right) + \frac{1}{I}p_\phi^{\ 2} - \frac{1}{2M}\left(p_x^{\ 2} + p_y^{\ 2}\right) - \frac{1}{2I}p_\phi^{\ 2} + Mgy$$

$$= \frac{1}{2M}\left(p_x^{\ 2} + p_y^{\ 2}\right) + \frac{1}{2I}p_\phi^{\ 2} + Mgy$$

The equations of motion follow directly from Eq. (7-2.3.13).

$$\frac{\partial \mathcal{H}}{\partial p_x} = \dot{x} = \frac{1}{M}p_x$$

$$\frac{\partial \mathcal{H}}{\partial p_y} = \dot{y} = \frac{1}{M}p_y$$

$$\frac{\partial \mathcal{H}}{\partial p_\phi} = \dot{\phi} = \frac{1}{M}p_\phi$$

The previous three equations are just the definition of momenta and tell us nothing new.

$$\frac{\partial \mathcal{H}}{\partial x} = -\dot{p}_x = 0 : M\ddot{x} = 0$$ no acceleration in the x direction

$$\frac{\partial \mathcal{H}}{\partial y} = -\dot{p}_y = Mg : M\ddot{y} = Mg$$ the system accelerates in the y direction due to g

$$\frac{\partial \mathcal{H}}{\partial \phi} = -\dot{\phi} = 0 : I\ddot{\phi} = 0$$ no acceleration in angular velocity

These are the equations of motion for this rigid rotating rod. You will see it again in Sec. 8-2 when the rotation of molecules is discussed.

The above is an important exercise. We have separated motion into two parts: motion of the center of mass and motion of the system relative to that center. *We can always separate the internal motion of a system from the overall motion of the center of mass.* This leads to considerable simplification in many cases.

7-2.4 SOLUTION OF THE EQUATIONS FOR THE HYDROGEN ATOM

The procedures above were used to obtain the three equations listed for the hydrogen atom in Sec. 7-1.8. Similar procedures are necessary to solve various other quantum mechanical problems, such as the rotation and vibration of molecules.

We now turn to the detailed solution of the equations given in Sec. 7-1.8.

SOLUTION OF THE Φ EQUATION. The easiest equation to solve from the three given in Eq. (7-1.8.10) is the equation involving ϕ.

$$\frac{d^2\Phi}{d\phi^2} = -m^2\Phi \qquad \phi = 0 \text{ to } 2\pi \tag{7-2.4.1}$$

This is solved by inspection yielding

$$\Phi = e^{\pm im\phi} \tag{7-2.4.2}$$

You should substitute this into Eq. (7-2.4.1) to show that it works.

From the postulates we know that Φ must be single valued. Since ϕ goes from 0 to 2π radians, $\Phi(0)$ must equal $\Phi(2\pi)$ if Φ is to be single valued. Thus

$$e^{\pm im(0)} = 1 = e^{\pm im(2\pi)}$$

And $e^{\pm im(2\pi)} = 1$ only if m is an integer.

Again, we have applied restrictions to the solutions found for the Schrödinger equation and have generated a set of quantum numbers. Our boundary conditions can be satisfied only for specific values of m. The solution to the Φ equation is

$$\Phi = N \cdot e^{\pm im\phi} \qquad m = 0, 1, 2, \ldots \tag{7-2.4.3}$$

The N is a constant added to satisfy the condition that Φ must be normalized. That is

$$\int_0^{2\pi} \Phi^*\Phi \, d\phi = 1$$

$$= \int_0^{2\pi} N\cdot e^{\mp im\phi} \, N\cdot e^{\pm im\phi} \, d\phi \tag{7-2.4.4}$$

or $$\int_0^{2\pi} N^2 \, d\phi = 1 \qquad (7\text{-}2.4.4)$$

from which $N = 1/\sqrt{2\pi}$

Thus the final solution for Φ is

$$\Phi = \frac{1}{\sqrt{2\pi}} e^{\pm im\phi} \qquad m = 0, 1, 2, \ldots$$
$$\Phi = \frac{1}{\sqrt{2\pi}} e^{im\phi} \qquad m = 0, \pm 1, \pm 2, \ldots \qquad (7\text{-}2.4.5)$$

SOLUTION OF THE Θ *EQUATION.* The solution to the Θ and R portions of the equation are much more complex than the Φ equation. Both will be presented in some detail. The Θ equation solution will be useful in Chap. 8 when rotational motion of molecules is discussed, and the R equation solution determines the electronic energy levels allowed for the hydrogen atom. These will be discussed in detail in Sec. 7-3. The Θ and Φ equations are important in defining the shapes of the hydrogen atomic orbitals.

The Θ equation from (7-1.8.10) is

$$\frac{1}{\sin\theta}\frac{d}{d\theta}\left(\sin\theta\frac{d\Theta}{d\theta}\right) - \frac{m^2}{\sin^2\theta}\Theta + \beta\Theta = 0 \qquad (7\text{-}2.4.6)$$

Multiply this by $(\sin^2\theta/\Theta)$ to get

$$\frac{\sin\theta}{\Theta}\frac{d}{d\theta}\left(\sin\theta\frac{d\Theta}{d\theta}\right) - m^2 + \beta^2\sin^2\theta = 0 \qquad (7\text{-}2.4.7)$$

Several substitutions will be made to put this into a useful form. The purpose of these substitutions is to put the equation into a form that has been investigated previously by mathematicians, so that the solutions can be written immediately.

The first substitution to be made is $z = \cos\theta$. Since θ varies from 0 to π in spherical polar coordinates, z will vary from 1 to - 1. Make this substitution. Designate the new function as P(z), which will be equal to $\Theta(\theta)$, since all we have done is change variables. Note that for any function of θ, $F(\theta)$,

$$\frac{dF(\theta)}{d\theta} = \frac{dF}{dz}\frac{dz}{d\theta} = -\frac{dF}{dz}\sin\theta$$

Also, $$\frac{d\Theta}{d\theta} = \frac{dP}{d\theta} = \frac{dP}{dz}\frac{dz}{d\theta} = -\frac{dP}{dz}\sin\theta \qquad dz = -\sin\theta\, d\theta \qquad (7\text{-}2.4.8)$$

and $$\sin^2\theta = 1 - \cos^2\theta = 1 - z^2$$

So (prove each step),

$$\frac{\sin\theta}{\Theta}\frac{d\sin\theta(d\Theta/d\theta)}{dz}(-\sin\theta) - m^2 + \beta(1 - z^2) = 0$$

$$-\frac{\sin^2\theta}{\Theta}\frac{d}{dz}\left[\sin\theta\left(-\frac{dP}{dz}\sin\theta\right)\right] - m^2 + \beta(1 - z^2) = 0$$

$$\frac{\sin^2\theta}{\Theta}\frac{d}{dz}\left(\sin^2\theta\,\frac{dP}{dz}\right) - m^2 + \beta(1 - z^2) = 0$$

And, since $P(z) = \Theta(\theta)$,

$$\frac{1 - z^2}{P}\frac{d}{dz}\left[(1 - z^2)\frac{dP}{dz}\right] - m^2 + \beta(1 - z^2) = 0$$

Multiply by $P/(1 - z^2)$ to get

$$\frac{d}{dz}\left[(1 - z^2)\frac{dP}{dz}\right] + \left(\beta - \frac{m^2}{1 - z^2}\right)P(z) = 0 \tag{7-2.4.9}$$

This equation is well known in physics as the associated Legendre equation. The solutions can be written immediately since they have been worked out long ago. However, we will briefly go through the solution.

We do have a problem in Eq. (7-2.4.9). For $z = \pm 1$, singular points arise. These can be eliminated using standard mathematical procedures. Without proof, an acceptable solution is (see Pauling and Wilson, Sec. 7-2.8, Ref. 2)

$$P(z) = (1 - z^2)^{|m|/2}\,G(z) \tag{7-2.4.10}$$

$G(z)$ is a new function which will be represented as a power series. Substituting (7-2.4.10) into (7-2.4.9) results in (after much work; if you like a challenge, give it a try)

$$(1 - z^2)G'' - 2(|m| + 1)zG' + [\beta - |m|(|m| + 1)]G = 0 \tag{7-2.4.11}$$

We assume $G(z)$ is a power series. Also, define $G' = dG/dz$ and $G'' = d^2G/dz^2$. This gives

$$G = \sum_{i=0}^{\infty} a_i z^i$$

$$G' = \sum_{i=1}^{\infty} i a_i z^{i-1} \tag{7-2.4.12}$$

$$G'' = \sum_{i=2}^{\infty} i(i - 1)\,a_i z^{i-2}$$

Substitution of these into (7-2.4.11) will give an equation that is a polynomial in z. This is another challenge for those who are mathematically inclined. A solution to the resulting equation is found if the coefficient of each power of z is set equal to zero. The coefficient of the ith power of z (z^i) in (7-2.4.11) is

$$(i + 1)(i + 2)a_{i + 2} + \{[\beta - |m|(|m| + 1)] - 2i(|m| + 1) - i(i - 1)\} a_i = 0 \tag{7-2.4.13}$$

Rearrangement leads to a relation of the coefficients. Such a relation is called a *recursion formula*. The recursion formula for our equation is given by the following:

$$a_{i + 2} = \frac{(i + |m|)(i + |m| + 1) - \beta}{(i + 1)(i + 2)} a_i \tag{7-2.4.14}$$

All we need to do is pick two coefficients (one with i even, the other with i odd), and we have the solution to the equation.

However, the solution is not valid for $z = + 1$ or $z = - 1$, since we require single-valued, finite functions. To make it valid the series must be terminated at some term. The series can be terminated at the jth term by requiring $a_{j + 2} = 0$. $a_{j + 2}$ can be set equal to zero by making the numerator of Eq. (7-2.4.14) equal to zero:

$$(j + |m|)(j + |m| + 1) = \beta \tag{7-2.4.15}$$

This will terminate either the even or odd series. The other one can be terminated by setting a_0 or $a_1 = 0$.

We are just following established procedures. The solution we have is not valid until we impose some conditions to make it fit the postulates. When we do this we generate quantum numbers. Here I will introduce a new quantum number ℓ defined by

$$\ell = j + |m| \tag{7-2.4.16}$$

which gives immediately that

$$\beta = \ell(\ell + 1) \tag{7-2.4.17}$$

where $\ell = |m|, |m| + 1, |m| + 2, \ldots$ depending on the choice of j. The significance of ℓ will be discussed after the solution of the R equation.

The final normalized solution in terms of Θ is given by

$$\Theta(\theta) = \left[\frac{(2\ell + 1)(\ell + |m|)!}{2(\ell + |m|)!}\right]^{1/2} P_\ell^{|m|}(\cos \theta) \tag{7-2.4.18}$$

As you can see, Θ depends on ℓ and m. Table 7-2.4.1 lists the normalized solutions for $\ell = 0, 1, 2$. The values of m shown are consistent with (7-2.4.16). These solutions will be discussed in Sec. 7-3. What we are doing is not mysterious or devious. We have a set of differential equations that results from Schrödinger's equation. These equations are solved using procedures you very likely have used yourself in advanced applied mathematics courses; no quantum theory applies here.

Table 7-2.4.1
Normalized Associated Legendre Polynomials

ℓ	m	$\Theta(\ell, m)$
0	0	$1/\sqrt{2}$
1	0	$(3/2)^{1/2} \cos\theta$
1	± 1	$(3/4)^{1/2} \sin\theta$
2	0	$(5/8)^{1/2} (3\cos^2\theta - 1)$
2	± 1	$(15/4)^{1/2} \sin\theta \cos\theta$
2	± 2	$(15/16)^{1/2} \sin^2\theta$

The quantum theory applies at each end of the solution. When the problem is set up, quantum mechanical principles are used. After standard mathematical procedures are used to obtain a set of solutions, quantum mechanical considerations are again imposed. The solutions are restricted so that the postulates are not violated. The restrictions give rise to quantum numbers.

Only for specified values of the constants in the equation will valid solutions exist. Later in this section we will see the physical significance of the quantum numbers. It is somewhat amazing that the procedures we are applying lead to numbers that have physical significance. They do, however. As you will see, the quantum numbers we get can be related directly to the periodic chart. So, all our effort is not in vain.

7-2.4.1 EXAMPLE

Show that $\Theta(2, 1)$ is a solution of Eq. (7-2.4.6).

Solution

$\Theta(2, 1) = (15/4)^{1/2} \sin\theta \cos\theta \qquad \beta = \ell(\ell + 1) = 2(2 + 1) = 6$

$$\frac{1}{\sin\theta}\frac{d}{d\theta}\left\{\sin\theta \frac{d}{d\theta}\left[(15/4)^{1/2}\sin\theta\cos\theta\right]\right\} - \frac{1^2}{\sin^2\theta}(15/4)^{1/2}\sin\theta\cos\theta + 6(15/4)^{1/2}\sin\theta\cos\theta \stackrel{?}{=} 0$$

$$\frac{1}{\sin\theta}\frac{d}{d\theta}\left[\sin\theta(\cos^2\theta - \sin^2\theta)\right] - \frac{\cos\theta}{\sin\theta} + 6\sin\theta\cos\theta \stackrel{?}{=} 0$$

$$\frac{d}{d\theta}(\sin\theta\cos^2\theta - \sin^3\theta) - \cos\theta + 6\sin^2\theta\cos\theta \stackrel{?}{=} 0$$

$$\sin\theta(2\cos\theta)(-\sin\theta) + \cos^2\theta(\cos\theta) - 3\sin^2\theta(\cos\theta) - \cos\theta + 6\sin^2\theta\cos\theta \stackrel{?}{=} 0$$

$$- 2 \sin^2 \theta \cos \theta + \cos^3 \theta - 3 \sin^2 \theta \cos \theta - \cos \theta + 6 \sin^2 \theta \cos \theta \stackrel{?}{=} 0$$

$$\sin^2 \theta \cos \theta + \cos^3 \theta - \cos \theta \stackrel{?}{=} 0$$

divide by $\cos \theta$:

$$\sin^2 \theta + \cos^2 \theta - 1 \stackrel{?}{=} 0$$

$$1 - 1 = 0$$

It works. Why not try one of the other functions?

SOLUTION OF THE R EQUATION. We will also go through the solution of this equation in some detail since it illustrates many of the operations common to quantum mechanics. From Eq. (7-1.8.10) and (7-2.4.17),

$$\frac{1}{r^2} \frac{d}{dr} \left(r^2 \frac{dR}{dr} \right) - \frac{\ell(\ell + 1)}{r^2} R + \frac{8\pi^2 \mu}{h^2} (E - V)R = 0 \qquad (7\text{-}2.4.19)$$

The first step is to define some new constants and variables to simplify the appearance of this equation.

$$V = -\frac{e^2}{r} \qquad \alpha^2 = -\frac{8\pi^2 \mu E}{h^2}$$

$$\lambda = \frac{4\pi^2 \mu e^2}{h^2 \alpha} \qquad \rho = 2\alpha r \qquad 0 \le r \le \infty \qquad (7\text{-}2.4.20)$$

Replacing $2\alpha r$ with ρ results in a new function that will be denoted by $S(\rho)$. $S(\rho)$ must equal $R(r)$ since all we have done is change variables.

If you want some practice in mathematical manipulation, substitute the new variables and constants of Eq. (7-2.4.20) into (7-2.4.19) and replace $R(r)$ with $S(\rho)$. The result is

$$\frac{1}{\rho^2} \frac{d}{d\rho} \left(\rho^2 \frac{dS}{d\rho} \right) + \left[-\frac{1}{4} - \frac{\ell(\ell + 1)}{\rho^2} + \frac{\lambda}{\rho} \right] S = 0 \qquad 0 \le \rho \le \infty \qquad (7\text{-}2.4.21)$$

The reasons for these substitutions will become apparent later.

Now, how do we solve Eq. (7-2.4.21)? The initial step is to investigate the solutions as ρ approaches ∞.

What happens as $\rho \to \infty$?. Note that

$$\frac{1}{\rho^2} \frac{d}{d\rho} \left(\rho^2 \frac{dS}{d\rho} \right) = \frac{d^2 S}{d\rho^2} + \frac{1}{\rho} \frac{dS}{d\rho}$$

so we can write

$$\frac{d^2S}{d\rho^2} + \frac{1}{\rho}\frac{dS}{d\rho} + \left[-\frac{1}{4} - \frac{\ell(\ell+1)}{\rho^2} + \frac{\lambda}{\rho}\right] S = 0 \qquad (7\text{-}2.4.22)$$

If $\rho \to \infty$, any term with ρ in the denominator will approach zero. So. Eq. (7-2.4.22) reduces to (for large ρ)

$$\frac{d^2S}{d\rho^2} \simeq \frac{1}{4} S \qquad (7\text{-}2.4.23)$$

The solution to this should be apparent.

$$S = \exp\left(\pm \frac{\rho}{2}\right)$$

However, the solution with the plus sign is not allowed. If it were allowed S would approach ∞ as $\rho \to \infty$. This cannot be since we require our functions to be finite. Thus

$$S = \exp\left(-\frac{\rho}{2}\right) \quad \text{for very large } \rho \qquad (7\text{-}2.4.24)$$

What happens for smaller values of ρ? Assume

$$S(\rho) = \exp\left(-\frac{\rho}{2}\right) F(\rho) \qquad (7\text{-}2.4.25)$$

where $F(\rho)$ is a new function. Designate $F' = \partial F/\partial\rho$ and $F'' = \partial^2 F/\partial\rho^2$. Substitute this into Eq. (7-2.4.21) and simplify to get (do it)

$$F'' + \left(\frac{2}{\rho} - 1\right)F' + \left[\frac{\lambda}{\rho} - \frac{\ell(\ell+1)}{\rho^2} - \frac{1}{\rho}\right]F = 0 \qquad (7\text{-}2.4.26)$$

Immediately, a major problem arises. At $\rho = 0$ we have a singularity. This must be eliminated.

There are standard mathematical procedures for eliminating singularities in differential equations, as mentioned earlier. Application of these procedures gives (without proof)

$$F(\rho) = \rho^{\ell} L(\rho) \qquad (7\text{-}2.4.27)$$

where $L(\rho)$ is a new function.

The next step is to assume $L(\rho)$ is a polynomial in ρ.

$$L(\rho) = \sum_{j=0}^{\infty} a_j \rho^j \qquad (7\text{-}2.4.28)$$

This procedure, which is the same as that used for the Θ equation, is called the *series method* for solving differential equations. We assume the solution is a polynomial and see if we can find one that fits.

Define $L' = \partial L/\partial\rho$ and $L'' = \partial^2 L/\partial\rho^2$, and show that Eq. (7-2.4.26) becomes

$$\rho L'' + [2(\ell + 1) - \rho]L' + (\lambda - \ell - 1)L = 0 \qquad (7\text{-}2.4.29)$$

We will continue with the solution. However, when early workers got to this point, the problem was finished. The solutions to equations of the form given in (7-2.4.29) had been worked out by mathematicians already. This equation is known as the associated Laquerre equation. So, from Eq. (7-2.4.29) the solutions could be written directly. The remainder is presented just to show how it is done.

If $$L = \sum_{j=0}^{\infty} a_j \rho^j$$

then $$L' = \sum_{j=1}^{\infty} j a_j \rho^{j-1}$$

and $$L'' = \sum_{j=2}^{\infty} j(j-1) a_j \rho^{j-2}$$

Thus Eq. (7-2.4.28) becomes (show it: this one is easy)

$$\rho \sum_{j=2}^{\infty} j(j-1) a_j \rho^{j-2} + [2(\ell + 1) - \rho] \sum_{j=1}^{\infty} j a_j \rho^{j-1} + (\lambda - \ell - 1) \sum_{j=0}^{\infty} a_j \rho^j = 0 \qquad (7\text{-}2.4.30)$$

Rearrange to get

$$\sum_j j(j-1) a_j \rho^{j-1} + [2(\ell + 1)] \sum_j a_j j \rho^{j-1} - \rho \sum_j j a_j \rho^{j-1} + (\lambda - \ell - 1) \sum_j a_j \rho^j = 0 \qquad (7\text{-}2.4.31)$$

Continuing the simplification (show that it works)

$$\sum_j j(j-1) a_j \rho^{j-1} + [2(\ell + 1)] \sum_j j a_j \rho^{j-1} - \sum j a_j \rho^j + (\lambda - \ell - 1) \sum_j a_j \rho^j = 0 \qquad (7\text{-}2.4.32)$$

Collect all the coefficients of ρ^{j-1} into one term and all the coefficients of ρ^j into another to get

$$\sum_j [j(j-1) + 2(\ell + 1)j] a_j \rho^{j-1} - \sum_j [j a_j - (\lambda - \ell - 1) a_j] \rho^j$$

or $$\sum_j \{[j(j-1) + 2j(\ell + 1)] a_j \rho^{j-1} - [j - \lambda + \ell + 1)] a_j \rho^j\} = 0 \qquad (7\text{-}2.4.33)$$

A solution for this type of equation is found if the coefficient of each power of ρ equals zero.

In Eq. (7-2.4.33) let us pick the kth term (the coefficient of ρ^k). For the first part of Eq. (7-2.4.33) set $j = k + 1$ to get the coefficient of ρ^k, and for the second term set $j = k$. Make those substitutions to show that the coefficient of ρ^k (which we will set equal to zero) is

$$[(k + 1)(k) + 2(k + 1)(\ell + 1)]a_{k+1} - (k - \lambda + \ell + 1)a_k = 0 \quad (7\text{-}2.4.34)$$

This equation relates the coefficients of our original polynomial (the a_j). The relation which can be found immediately from Eq. (7-2.4.34) is another *recursion formula*.

$$a_{k+1} = -\frac{(\lambda - \ell - 1 - k)}{(k + 1)(k) + 2(k + 1)(\ell + 1)} a_k \quad (7\text{-}2.4.35)$$

The usefulness of this formula should be clear. As before, if we know any coefficient in our original polynomial (or if we choose one), all other coefficients can be found from the recursion formula.

Lest you forget, let us review the solution we have up to this point for the R equation.

$$\begin{aligned} R(r) = S(\rho) &= \exp\left(-\frac{\rho}{2}\right) F(\rho) \\ &= \exp\left(-\frac{\rho}{2}\right) \rho^{\ell} L(\rho) \\ &= \exp\left(-\frac{\rho}{2}\right) \rho^{\ell} \sum_{j=0}^{\infty} a_j\rho^j \\ &= \exp\left(-\frac{\rho}{2}\right) \rho^{\ell} \sum_{k=0}^{\infty} a_k\rho^k \end{aligned} \quad (7\text{-}2.4.36)$$

(j and k are interchangeable since each is only an index.) Equation (7-2.4.35) is the relation that defines all the coefficients of the polynomial in (7-2.4.36), except for one, of course. The above is a general solution. However, it is not adequate in its present form for our problem. We must now insure that the restrictions given in the postulates are met.

It can be shown that the solution given increases without limit as $\rho \to \infty$ unless the series is terminated at some point. We can terminate the series by making sure every term with k greater than some number, say n', is zero. To terminate the series at $k = n'$ we want $a_{k+1} = 0$. This is easily done by setting the numerator of Eq. (7-2.4.35) equal to zero for $k = n'$. (Take a moment and justify that statement to yourself.) Thus we can terminate the series at the n' term if

$$\lambda - \ell - 1 - n' = 0$$

From which

$$\lambda = n' + \ell + 1 \tag{7-2.4.37}$$

Define another constant n as

$$n = n' + \ell + 1 \tag{7-2.4.38}$$

n will be an integer since n' and ℓ are. Before reading on try to determine what values n can have. Note that n' can be any integer ≥ 0, and ℓ is any integer ≥ 0. Thus, n can have any value ≥ 1.

Combining Eq. (7-2.4.37) and (7-2.4.38) yields

$$\lambda = n$$

This is an important result. Rewrite Eq. (7-2.4.20) in which λ is defined.

$$\lambda = \frac{4\pi^2 e^2 \mu}{h^2 \alpha} \qquad \text{and} \qquad \alpha^2 = -\frac{8\pi^2 \mu E}{h^2} \tag{7-2.4.39}$$

Combine Eq. (7-2.4.38) and (7-2.4.39) to find an expression for E, the energy levels of the hydrogen atom.

$$\boxed{E_n = -\frac{2\pi^2 \mu e^4}{h^2 n^2} \qquad n = 1, 2, \ldots} \tag{7-2.4.40}$$

The subscript means the nth level. Remember that μ is the reduced mass, e is the electronic charge, h is Planck's constant, and n is a *quantum number*.

Again we have shown that imposing conditions on the solutions to Schrödinger's equation leads to *quantization* of the results. Terminating the series in the R equation solution has produced a *quantum number* n that determines the energies associated with different solutions.

Note that the energy is determined by n and not by ℓ or m. The potential energy is a function of r only, so the total energy is determined by the solution to the R(r) equation.

Finally, you should compare Eq. (7-2.4.40) with Eq. (7-1.3.3). They are identical (if you remember that $\Delta E = hc\bar{\nu}$ and that (7-2.4.40) refers to only one level. ΔE would be the difference between two energy levels.) The fact that Schrödinger's equation leads to a result identical to that of Bohr is a triumph for the new quantum theory.

After normalization, the solution to the R equation is given by

$$S(\rho) = \left\{\left(\frac{2}{na_0}\right)^3 \frac{(n - \ell - 1)!}{2n[(n + 1)!]}\right\}^{\frac{1}{2}} \exp\left(-\frac{\rho}{2}\right) \rho^{\ell} L_{n+\ell}^{2\ell+1}(\rho) \tag{7-2.4.41}$$

a_0 is the Bohr radius (the radius of the first orbit) of the hydrogen atom given by

$$a_0 = \frac{h^2}{4\pi^2 m_e e^2} = 0.529 \text{ Å}$$

$$= 0.529 \times 10^{-10} \text{ m} \qquad (7\text{-}2.4.42)$$

(For the hydrogen atom $m_e \approx \mu$.) The functions $L(\rho)$ are called *associated Laguerre polynomials* in honor of the man who investigated their properties. Given values for n and ℓ we can look up the proper function that corresponds to $L(\rho)$. The constant term is a normalization constant.

This all looks terribly complex, but it is not so bad for small values of n, as you will see below. Table 7-2.4.2 contains some normalized radial wave functions R(r) for the hydrogen atom. As you can tell, the functions become complicated rapidly as the value of n increases. Note also that R(r) is a function of n and ℓ but not of m. Verify from definitions that $\rho = 2r/na_0$.

7-2.4.2 EXAMPLE

Show that R(1, 0) is a solution of the R equation, (7-2.4.19).

Solution

This is most easily done by substitution: $n = 1$, $\ell = 0$. $\left[\text{const} = 2\left(\frac{1}{a_0}\right)^{3/2}\right]$

$$\frac{1}{r^2}\frac{d}{dr}\left\{r^2\frac{d}{dr}\left[(\text{const})\exp\left(-\frac{r}{a_0}\right)\right]\right\} - \frac{0}{r^2}R(1, 0)$$

$$+ \frac{8\pi^2\mu}{h^2}\left(E + \frac{e^2}{r}\right)\left[(\text{const})\exp\left(-\frac{r}{a_0}\right)\right] \overset{?}{=} 0$$

$$\frac{1}{r^2}\frac{d}{dr}\left\{r^2\left[\text{const}\left(-\frac{1}{a_0}\right)\exp\left(-\frac{r}{a_0}\right)\right]\right\} + \frac{8\pi^2\mu}{h^2}\left(E + \frac{e^2}{r}\right)(\text{const})\exp\left(-\frac{r}{a_0}\right) \overset{?}{=} 0$$

divided by const:

$$\left(-\frac{1}{a_0}\right)\frac{1}{r^2}\frac{d}{dr}\left[r^2\exp\left(-\frac{r}{a_0}\right)\right] + \frac{8\pi^2\mu}{h^2}\left(E + \frac{e^2}{r}\right)\exp\left(-\frac{r}{a_0}\right) \overset{?}{=} 0$$

Note: $\frac{8\pi^2\mu}{h^2} = \frac{2}{a_0 e^2}$ $\quad E = -\frac{e^2}{2a_0}$ $\quad a_0 = 0.529$ Å

$$-\frac{1}{a_0}\left(\frac{1}{r^2}\right)\left[2r\exp\left(-\frac{r}{a_0}\right) - \frac{r^2}{a_0}\exp\left(-\frac{r}{a_0}\right)\right] + \frac{2}{a_0 e^2}\left(-\frac{e^2}{2a_0} + \frac{e^2}{r}\right)\exp\left(-\frac{r}{a_0}\right) \overset{?}{=} 0$$

divided by $\exp(-r/a_0)$:

Table 7-2.4.2
Some Normalized Radial Wave Functions for the Hydrogen Atom: $R(n, \ell)$

$$R(1, 0) = 2\left(\frac{1}{a_0}\right)^{3/2} \exp\left(-\frac{r}{a_0}\right)$$

$$R(2, 0) = \left(\frac{1}{2a_0}\right)^{3/2} \left(2 - \frac{r}{a_0}\right) \exp\left(-\frac{r}{2a_0}\right)$$

$$R(2, 1) = \left(\frac{1}{3}\right)^{1/2} \left(\frac{1}{2a_0}\right)^{3/2} \frac{r}{a_0} \exp\left(-\frac{r}{2a_0}\right)$$

$$R(3, 0) = \frac{2}{27}\left(\frac{1}{3a_0}\right)^{3/2} \left[27 - \frac{18r}{a_0} + 2\left(\frac{r}{a_0}\right)^2\right] \exp\left(-\frac{r}{3a_0}\right)$$

$$R(3, 1) = \frac{1}{81(3)^{\frac{1}{2}}} \left(\frac{2}{a_0}\right)^{3/2} \left(6 - \frac{r}{a_0}\right) \exp\left(-\frac{r}{3a_0}\right)$$

$$-\frac{2}{ra_0} + \frac{1}{a_0^2} - \frac{1}{a_0^2} + \frac{2}{a_0 r} \stackrel{?}{=} 0$$

$$0 = 0$$

As you might guess, the others are more difficult, but they can be done.

7-2.4.3 EXAMPLE

Show that $R(1, 0)$ and $R(2, 0)$ are normalized and orthogonal.

Solution

A function is normalized if $\int_{\text{all space}} \Psi_a^* \Psi_a \, d\tau = 1$, and orthogonal if $\int \Psi_a^* \Psi_b \, d\tau = 0$; for a radial function $d\tau = r^2 \, dr$.

R(1, 0)

$$\int_0^\infty R(1, 0)^* R(1, 0) \, r^2 \, dr = 4\left(\frac{1}{a_0}\right)^3 \int_0^\infty \exp\left(-\frac{r}{a_0}\right) \exp\left(-\frac{r}{a_0}\right) r^2 \, dr$$

$$= 4\left(\frac{1}{a_0}\right)^3 \int_0^\infty r^2 \exp\left(-\frac{2r}{a_0}\right) dr$$

$$= 4\left(\frac{1}{a_0}\right)^3 \left[\frac{2!}{(2/a_0)^{2+1}}\right]$$

$$= 4\left(\frac{1}{a_0}\right)^3 \left(\frac{2a_0^3}{8}\right) = 1$$

Remember:

$$\int_0^\infty x^n e^{-bx}\, dx = \frac{n!}{b^{n+1}}$$

n = 2 here.

R(2, 0)

$$\int_0^\infty R(2, 0)^* R(2, 0)\, r^2\, dr = \frac{1}{8a_0^3}\int_0^\infty \left(2 - \frac{r}{a_0}\right)^2 \exp\left(-\frac{r}{a_0}\right) r^2\, dr$$

$$= \frac{1}{8a_0^3}\int_0^\infty \left(4r^2 - \frac{4r^3}{a_0} + \frac{r^4}{a_0^2}\right) \exp\left(-\frac{r}{a_0}\right) dr$$

$$= \frac{1}{8a_0^3}\left\{4\left[\frac{2!}{(1/a_0)^3}\right] - \frac{4}{a_0}\left[\frac{3!}{(1/a_0)^4}\right] + \frac{1}{a_0^2}\left[\frac{4!}{(1/a_0)^5}\right]\right\}$$

$$= \frac{1}{8a_0^3}\left[8a_0^3 - \frac{4\cdot 3\cdot 2\cdot 1}{a_0} a_0^4 + \frac{4\cdot 3\cdot 2\cdot 1}{a_0^2} a_0^5\right] = 1$$

R(2, 0); R(1, 0)

$$\int_0^\infty R(1, 0)^* R(2, 0)\, d\tau = 4\left(\frac{1}{a_0}\right)^{3/2}\left(\frac{1}{2a_0}\right)^{3/2}\int_0^\infty \exp\left(-\frac{r}{a_0}\right)\left(2 - \frac{r}{a_0}\right) \times \exp\left(-\frac{r}{a_0}\right) r^2\, dr$$

$$= 2\left(\frac{1}{a_0}\right)^{3/2}\left(\frac{1}{2a_0}\right)^{3/2}\int_0^\infty \left[2r^2 \exp\left(-\frac{3r}{2a_0}\right) - \frac{r^3}{a_0} \times \exp\left(-\frac{3r}{2a_0}\right)\right] dr$$

$$= 2\left(\frac{1}{a_0}\right)^{3/2}\left(\frac{1}{2a_0}\right)^{3/2}\left\{2\left[\frac{2!}{(3/2a_0)^3}\right] - \frac{1}{a_0}\left[\frac{3!}{(3/2a_0)^4}\right]\right\}$$

$$= 2\left(\frac{1}{a_0}\right)^{3/2}\left(\frac{1}{2a_0}\right)^{3/2}\left(\frac{2\cdot 2\cdot 2^3}{3^3} a_0^3 - \frac{1}{a_0}\frac{3\cdot 2\cdot 2^4}{3^4} a_0^4\right)$$

$$= 0$$

Therefore these functions are orthogonal and normalized. They are *orthonormal*.

7-2.5 SUMMARY OF THE ALLOWED QUANTUM NUMBERS

The solution to the hydrogen atom is complete. The significance of the results will be given in Sec. 7-3. All we will do here is list the quantum numbers we have found. We start with n and call it the principle quantum number, since it determines the energy of the levels. Then, we apply the other relations found throughout 7-2.4 to find m and ℓ. The relations given in 7-2.4 are

$$m = 0, \pm 1, \pm 2, \pm 3, \ldots$$

$$\ell = |m|, |m| + 1, |m| + 2, \ldots$$

$$n = n' + \ell + 1 = \ell + 1, \ell + 2, \ell + 3, \ldots$$

Using n as the principal quantum number gives

$$n = 1, 2, 3, \ldots$$

$$\ell = 0, 1, 2, \ldots, n - 1$$

$$m = -\ell, -\ell + 1, -\ell + 2, \ldots, -1, 0, 1, \ldots, \ell - 1, \ell$$

These will be used in Sec. 7-3. Section 7-4 extends this treatment to many-electron atoms.

7-2.6 PROBLEMS

1. Write the Hamiltonian and the equations of motion for a rigid rotor in three dimensions, shown below. This problem will be encountered again in our study of rotational (microwave) spectroscopy.

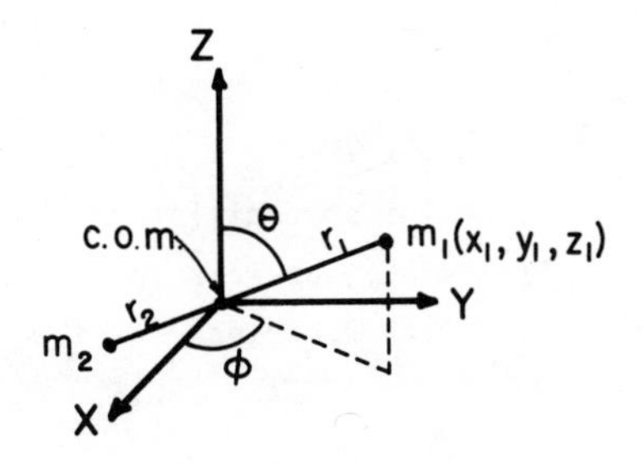

c.o.m.: Origin at center of mass

r, θ, ϕ: Spherical polar coordinates

$$r_1 + r_2 = R$$

$$r_1 = \frac{m_2}{m_1 + m_2} R = \frac{m_2}{M} R$$

$$r_2 = \frac{m_1}{m_1 + m_2} R = \frac{m_1}{M} R$$

The moment of inertia is

$$I = m_1 r_1^2 + m_2 r_2^2$$

$$= \frac{m_1 m_2}{m_1 + m_2} R^2$$

$$= \mu R^2$$

where μ = reduced mass.

Follow the procedures of Example 7-2.3.3. This is a little more complicated but it should be fun to do.

2. Apply Postulate IV to find the "average value" of r for the R(1, 0) and R(2, 1) states. Discuss the results briefly.

3. Show that $\Theta(1, 0)$ and $\Theta(2, 0)$ are orthogonal and normalized. Also show each is a solution to the Schrödinger equation for the hydrogen atom.

4. Derive Eq. (7-2.4.40) from Eq. (7-2.4.39) and $\lambda = n$. Show all steps.

5. Show that R(2, 1) is a solution to the Schrödinger equation for the hydrogen atom.

6. Given that the energy of the hydrogen atom depends only on the value of n, sketch the energy level diagram to scale for the first eight values of n.

7. Given that the most probable value of r can be found by setting $\partial r^2 R^2/\partial r = 0$ and solving for r. Find r_{mp} for R(1, 0) and R(2, 1). Discuss the results briefly.

7-2.7 SELF-TEST

1. Write the Hamiltonian function and the equations of motion for an oscillating particle of mass m moving under a potential of $V = \frac{1}{2} k(x^2 + y^2)$, where k is a constant. $T = \frac{1}{2} m(\dot{x}^2 + \dot{y}^2)$. Put the results into cylindrical coordinates. The final equations of motion will be in terms of r and θ.

2. Outline, in your own words, the procedures used to solve the R(r) equation (one-and-a-half to two pages required).

3. Show that R(2, 1) is normalized and is orthogonal to R(3, 1).

4. Find $\langle r \rangle$ and r_{mp} for R(2, 0). [See Problems (2) and (7).]

7-2.8 ADDITIONAL READING

1.* M. W. Hanna, *Quantum Mechanics in Chemistry*, W. A. Benjamin, New York, 1966.

2.* L. Pauling and E. B. Wilson, Jr., *Introduction to Quantum Mechanics*, McGraw-Hill, New York, 1935.

3. H. Eyring, J. Walter, and G. E. Kimball, *Quantum Chemistry*, John Wiley and Sons, New York, 1964.

7-3 ATOMIC STRUCTURE AND PROPERTIES

OBJECTIVES

You shall be able to

1. Give the allowed quantum number combinations for the hydrogen atom and discuss their significance
2. Sketch the shapes for the first few solutions to the hydrogen atom and sketch probability distributions
3. Define or discuss the terms: Aufbau principle, Pauli exclusion principle, ionization energy, and electron affinity
4. Give the electronic designation of any atom, or, given the designation, identify the atom from a periodic chart

7-3.1 INTRODUCTION

The quantum numbers and equation solutions from the last section will be investigated in more detail here. The purpose of this section is to tie the mathematical results of the last section to properties of atoms. The familiar shapes of the hydrogen orbitals will be given. This section can be studied without having read Sec. 7-2. However, some reference is made to particular aspects of Sec. 7-2.

Some of the considerations involved if more than one electron is present will be discussed. When there is more than one electron, an exact solution to the Schrödinger equation is impossible. The reason for this is that it is no longer possible to separate the Schrödinger equation into one-variable equations. There is a Coulombic repulsion term between any two electrons that precludes separating the variable. Approximate methods must be used.

A new quantum number will be introduced, the *spin quantum number*. After this, the allowed quantum number combinations will be presented along with the *Aufbau principle* and the *Pauli exclusion principle* to show how the periodic chart can be derived from the H atom solutions. Finally, a few comments will be made on atomic spectroscopy and atomic properties.

7-3.2 ALLOWED QUANTUM NUMBERS AND THEIR SIGNIFICANCE

If you look back at Sec. 7-2.5 you will find the relations among the hydrogen quantum numbers. To refresh your memory

$$m = 0, \pm 1, \pm 2, \ldots$$
$$\ell = |m|, |m| + 1, |m| + 2, \ldots \qquad (7\text{-}3.2.1)$$
$$n = n' + \ell + 1 = \ell + 1, \ell + 2, \ell + 3, \ldots$$

Since n is the quantum number that determines the energy of the level associated with a given solution, it is called the *principal quantum number*. So, we assign values to it first, then allow ℓ and m to take whatever values they can consistent with (7-3.2.1). The minimum value of n is 1. Thus,

$$n = 1, 2, 3, \ldots$$

Also, ℓ can have any integer value from 0 up to n - 1. Thus

$$\ell = 0, 1, 2, \ldots, n - 1$$

where ... means all integer values in between.

Finally, from the definition of m we can conclude that m can have any integer value from $-\ell$ through $+\ell$ but cannot exceed $|\ell|$. So,

$$m = -\ell, -\ell + 1, -\ell + 2, \ldots, -1, 0, 1, \ldots, \ell - 1, \ell$$

7-3.2.1 EXAMPLE

Find all the allowed quantum numbers associated with n = 3.

Solution

For n = 3, ℓ = 0, 1, and 2. Also, we have a set of m values for each ℓ.

n = 3 $\quad \ell = 2 \quad$ m = -2, -1, 0, 1, 2

n = 3 $\quad \ell = 1 \quad$ m = -1, 0, 1

n = 3 $\quad \ell = 0 \quad$ m = 0

We find there are nine allowed combinations. Later we will add another quantum number that will increase the number to 18. (Do you know what that quantum number is?)

The interpretation of these numbers is straightforward. n is the principle quantum number which determines the energy of the wave function. It also is valid for any of the hydrogen-like atoms such as He^{+}, Li^{2+}, etc., each of which has only one electron. The only difference in the expression between any of the other one-electron atoms and the hydrogen atom solution is that the quantity Z/a_0 (Z is the nuclear charge) must be used for $1/a_0$ in all the solutions of Table 7-2.4.2 and Z^2e^4 replaces e^4 in the energy expression. Thus, for any one-electron atom, we write

$$E_n = -\frac{Z^2 2\pi^2 \mu e^4}{h^2 n^2} = -\frac{2.6973 \times 10^{-38}\ Z^2}{n^2}\ \frac{\text{kg C}^4}{(\text{J s})^2}\left(\frac{8.9877 \times 10^9\ \text{J m}}{\text{C}^2}\right)^2$$

$$= -2.1788 \times 10^{-18}\ \frac{Z^2}{n^2}\ (\text{J})$$

$$= -3.288 \times 10^{15}\ \frac{Z^2}{n^2}\ (\text{s}^{-1})$$

$$= -109{,}700\ \frac{Z^2}{n^2}\ (\text{cm}^{-1}) \tag{7-3.2.2}$$

and, for example, the R(2, 1) solution would be written

$$R(2,\ 1) = \left(\frac{1}{3}\right)^{\frac{1}{2}}\left(\frac{Z}{2a_0}\right)\left(\frac{Zr}{a_0}\right)\exp\left(-\frac{Zr}{2a_0}\right) \tag{7-3.2.3}$$

[See Table 7-2.4.2 for explanation of the meaning of the symbol R(2, 1).] The energies of each one-electron atom depend on n only. We will see in Sec. 7-4 that the energy of atoms with more than one electron is a function of ℓ as well as n.

The quantum number ℓ is called the *azimuthal quantum number*. It is the quantum number associated with *total angular momentum* of the electron. This should not be surprising since ℓ is from the solution to the Θ equation whose variable is θ, the angular motion of the electron.

Finally, m is the *magnetic quantum number*. It is the total angular momentum along a specified axis. In a free atom there is no specified axis so we do not see the effects of m in the emission spectrum. If, however, we place the atom in an electric or magnetic field we define an axis (usually called the z axis) and the projections of ℓ onto this axis give us the values of m allowed. Perhaps a little more detail is warranted on this topic.

7-3.3 ANGULAR MOMENTUM

Any particle in curvilinear motion has some angular momentum associated with it. Specifically, a particle in circular motion will have an angular momentum as shown in Fig. 7-3.3.1. L is defined as L = r × p (read r cross p). Applying the techniques of Postulate II, we can convert this to operator notation. In Cartesian coordinates the three components are

$$\hat{L}_x = -\frac{ih}{2\pi}\left(y\frac{\partial}{\partial z} - z\frac{\partial}{\partial y}\right)$$

$$\hat{L}_y = -\frac{ih}{2\pi}\left(z\frac{\partial}{\partial x} - x\frac{\partial}{\partial z}\right) \tag{7-3.3.1}$$

$$\hat{L}_z = -\frac{ih}{2\pi}\left(x\frac{\partial}{\partial y} - y\frac{\partial}{\partial x}\right)$$

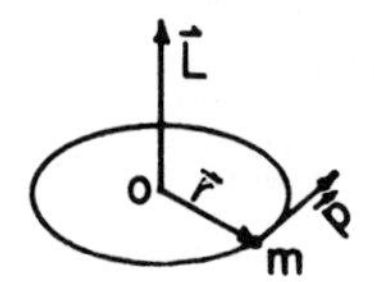

FIG. 7-3.3.1 The angular momentum L associated with a particle of mass m moving about point 0; p is the linear momentum.

These three can be combined ($\hat{L}^2 = \hat{L}_x^{\ 2} + \hat{L}_y^{\ 2} + \hat{L}_z^{\ 2}$) and converted to spherical polar coordinates to give $\hat{L}^2$. The $\hat{L}^2$ that results is identical to the angular part of the $\bar{\nabla}^2$ operator. The details will not be given, but the following important results can be derived.

$$\hat{L}^2\Psi = \ell(\ell + 1) \frac{h^2}{4\pi^2} \Psi \tag{7-3.3.2}$$

$$\hat{L}_z\Psi = m \frac{h}{2\pi} \Psi$$

So, the hydrogen atom wave functions are eigenvalues of both $\hat{L}^2$ and $\hat{L}_z$. Note that the eigenvalues are related to ℓ and m. The magnitude of the total angular momentum is

$$\| L \| = \sqrt{\ell(\ell + 1)} \frac{h}{2\pi} \tag{7-3.3.3}$$

The corresponding values allowed for $\hat{L}_z$ are

$$\| L_z \| = m \frac{h}{2\pi} \tag{7-3.3.4}$$

The values of L_z are simply the projections of L^2 along the axis we have defined as z. These projections are shown in Fig. 7-3.3.2. For the hydrogen atom the values of ℓ and m in no way affect the energy, but for many electron atoms or atoms in an electric or magnetic field, ℓ and m values must be used in the energy expression. [Ψ is not an eigenfunction of $\hat{L}_x$ or $\hat{L}_y$ so we do not have simple relations like (7-3.3.4) for x and y.]

Figure 7-3.3.2 is drawn to imply the angular momentum about the z axis. The cones represent the possible orientation of the angular momentum with respect to the x and y directions.

The important point for you to remember now is that ℓ is related to the total angular momentum and m to the projection on the z axis as defined by a field.

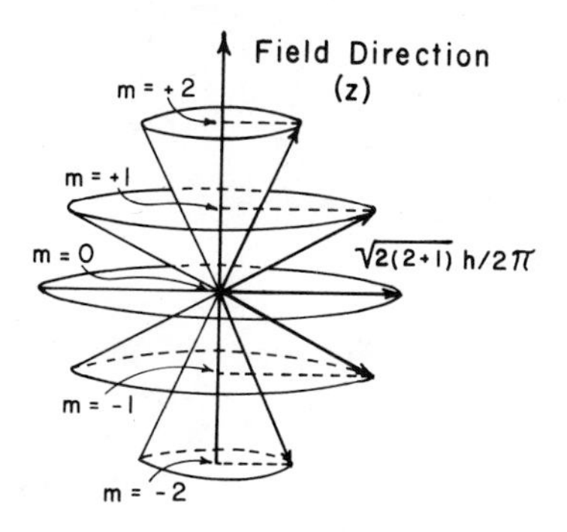

FIG. 7-3.3.2 Allowed projections of the angular momentum defined by ℓ = 2. Axis (z) is defined by a magnetic field.

7-3.4 QUANTUM NUMBERS, THE SHAPES OF THE HYDROGEN ORBITALS, AND PROBABILITY DISTRIBUTIONS

QUANTUM NUMBERS. In the hydrogen atom the principle quantum number n determines the energy of the orbital. The value of ℓ gives the angular momentum of the orbital. Values of ℓ can be associated with the familiar letters s, p, d, f, ... The association is

Value of ℓ	0	1	2	3	4	....
Orbital	s	p	d	f	g	....

The reason there is no order in the letter designations is that they were derived from spectroscopic studies performed before any quantum mechanics was known.

The following summarizes the allowed quantum numbers in terms of notation you have seen before.

n	1	2		3			4			
ℓ	0	0	1	0	1	2	0	1	2	3
Notation	1s	2s	2p	3s	3p	3d	4s	4p	4d	4f

More will be said about this later when many-electron atoms are discussed.

SHAPES AND PROBABILITY DISTRIBUTION FUNCTIONS.

Radial Portion. To designate the shape of an orbital we have to combine the three solutions (R, Θ, and Φ). Only for orbitals with ℓ = 0 will there be no shape

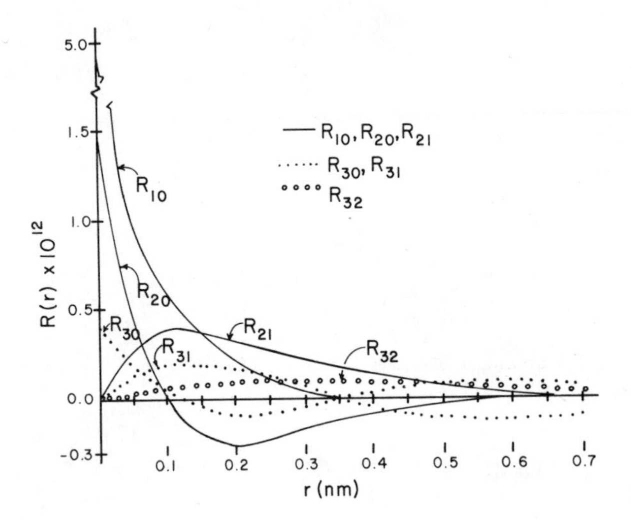

FIG. 7-3.4.1 The radial portion of the hydrogen atom solution.

dependence on θ and ϕ. Figure 7-3.4.1 shows the radial portion of the solution for n = 1, 2, and 3. More important is the square of this function, called the *radial distribution function*. This function (r^2R^2) is related to the probability as defined in Postulate I. Figure 7-3.4.2 shows this function, which gives the probability as finding an electron at a given distance from the nucleus regardless of the direction. We have to include Θ and Φ if we want to get actual electron densities.

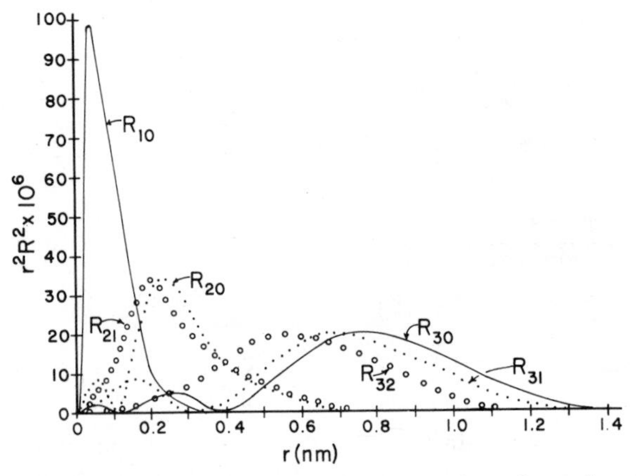

FIG. 7-3.4.2 Radial distribution functions for the H atom.

Angular Dependence. Except for orbitals with $\ell = 0$ it is impossible to draw the angular portions ($\Theta\Phi$) of the H atom solutions. They are imaginary. What is done is to combine solutions with a particular ℓ value to generate real functions (remember, any linear combination of solutions to a differential equation is still a solution) which can be plotted. The only purpose of this is to present a pictorial view to help us visualize the behavior of electrons in given orbitals. Do not think there is any correspondence between the pictures presented below and reality. The shapes drawn below are useful in discussing bonding in certain compounds. Later, we will discuss *hybridization* in which other combinations are required to explain the directional character of bonding. These approaches are not reflections of reality, they are just useful as models.

Anyway, Table 7-3.4.1 shows the combination used to get the familiar p_x, p_y, and p_z orbitals. Figure 7-3.4.3 shows the angular portions of the $\ell = 0$ and $\ell = 1$ orbitals. Also shown are the probability functions which are related to the electron densities. The electron density is obtained by multiplying $(\Theta\Phi)^2$ times R^2. The figures below indicate *only* shape, which is independent of r.

7-3.5 ELECTRON SPIN AND ITS EFFECTS

SPIN. To make the solution of the hydrogen atom fit experimental data it was necessary to postulate that the electron possesses an intrinsic angular momentum.

Table 7-3.4.1
Combination of Φ Functions for $\ell = 1$ to Generate Real Functions[a]

$\Phi(m)$	$\Theta(\ell, m)$	Real combinations
$\Phi(+1) = \frac{1}{\sqrt{2\pi}} e^{i\phi}$	$\Theta(1, 1) = \frac{\sqrt{3}}{2} \sin\theta$	$p_x = \frac{1}{\sqrt{2}}[\Phi(1) + \Phi(-1)]\ \Theta(1, 1) = \left(\frac{3}{4\pi}\right)^{1/2} \sin\theta \cos\phi$
$\Phi(0) = \frac{1}{\sqrt{2\pi}}$	$\Theta(1, 0) = \frac{\sqrt{6}}{2} \cos\theta$	$p_z = \frac{1}{\sqrt{2\pi}}\ \Theta(1, 0) = \left(\frac{3}{4\pi}\right)^{1/2}$
$\Phi(-1) = \frac{1}{\sqrt{2\pi}} e^{-i\phi}$	$\Theta(1, -1) = \frac{\sqrt{3}}{2} \sin\theta$	$p_y = -\frac{i}{\sqrt{2}}[\Phi(1) - \Phi(-1)]\ \Theta(1, -1) = \left(\frac{3}{4\pi}\right)^{1/2} \sin\theta \sin\phi$

[a] $\sin\phi = (e^{i\phi} - e^{-i\phi})/2i$; $\cos\phi = (e^{i\phi} + e^{-i\phi})/2$.

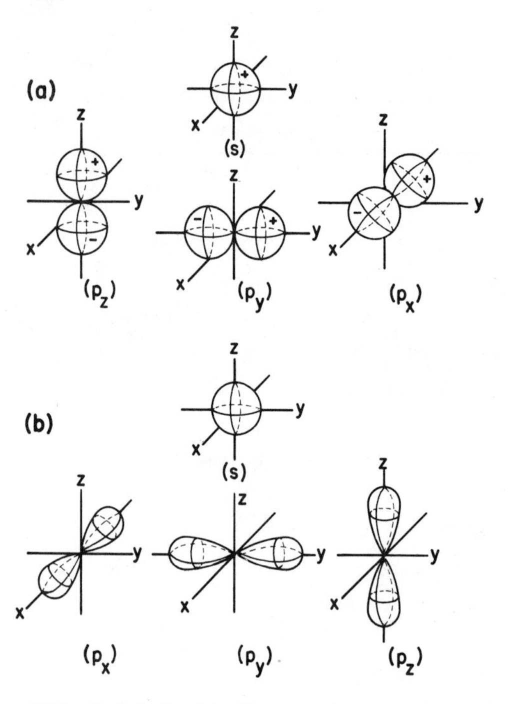

FIG. 7-3.4.3 (a) The angular portion of the H atom solution and (b) the probability function $(\Theta\Phi)^2$, for the H atom for $\ell = 0$ and $\ell = 1$.

The electron appears to behave as a tiny sphere of negative charge spinning on its axis. (Do not take this model seriously: it is hard to justify treating the electron as a wave, yet say that it behaves as a spinning sphere. This is another example of the *wave-particle dualism* alluded to in Sec. 7-1.)

The square of the spin angular momentum is found to be

$$M_s^{\,2} = s(s + 1)\,\frac{h^2}{4\pi^2} \tag{7-3.5.1}$$

with $s = \frac{1}{2}$. As with other angular momenta the spin is quantized, and projections along a defined axis are present. The projections along z are

$$M_z(\text{spin}) = m_s\,\frac{h}{2\pi} \tag{7-3.5.2}$$

with $m_s = +\frac{1}{2}$ or $-\frac{1}{2}$.

We now have four quantum numbers to define a solution to the H atom. These are: n, ℓ, m, and m_s. Contrary to what you might be lead to believe by the above, it is not necessary to postulate the spin quantum number. P. A. M. Dirac has formulated the Schrödinger equation in a form that satisfies relativity theory and has found that all four quantum numbers arise naturally.

MAGNETIC EFFECTS DUE TO ELECTRON MOTION.

Spin Effects. The charged electron spinning on its axis leads to a magnetic moment. The existence of such a moment was the basis of the first observation of the spin property. The magnetic moment due to spin is given by

$$\mu_s = -g\left(\frac{e}{2m_e c}\right) M_{spin} \qquad (7\text{-}3.5.3)$$

where g is a constant, M_{spin} is the total spin angular momentum [see Eq. (7-3.5.1)], and the other variables have their usual meanings. Substituting for M_{spin} gives

$$\mu = -g\left(\frac{eh}{4\pi m_e c}\right)\sqrt{s(s+1)} \qquad (7\text{-}3.5.4)$$

The constant quantity is given the name *Bohr magneton* and has the value

$$\mu_B = \frac{eh}{4\pi mc} = 9.273 \times 10^{-24}\ \text{J T}^{-1}$$
$$= 9.2732 \times 10^{-21}\ \text{erg gauss} \qquad (7\text{-}3.5.5)$$

(T = Tesla = 10^4 gauss = 1 Weber m^{-2} = 1 kg s^{-2} amp^{-1}.)

Thus we can write the total spin magnetic moment as

$$\mu_s = -g\sqrt{s(s+1)}\,\mu_B \qquad (7\text{-}3.5.6)$$

The z component of the magnetic moment is given by

$$\mu_{z,s} = -g\frac{2\pi\mu_B}{h} M_z\ (\text{spin})$$
$$= -gm_s\mu_B \qquad (7\text{-}3.5.7)$$

For an electron the value of g = 2 (approximately) and s = ½, so

$$\mu_s = -2\sqrt{\tfrac{1}{2}(\tfrac{1}{2}+1)}\,\mu_B$$
$$= -\sqrt{3}\,\mu_B$$

and $$\mu_{z,s} = \pm\mu_B \qquad (7\text{-}3.5.8)$$

The projection of the spin magnetic moment along the specified z axis is ± 1 Bohr magneton. You will see the importance of the interaction of spin with a magnetic field in Sec. 8-4 when nuclear magnetic resonance spectroscopy is considered. In that case the spin is nuclear spin, not electron spin, but the concept is the same.

Electron Motion Effects. If ℓ is not zero for an electron, the motion of the electron in its orbit gives rise to an orbital magnetic moment. The motion can be thought of as a current flowing around the atom which generates a magnetic field.

The orbital magnetic moment equations are analogous to the spin equations. Thus [see Eqs. (7-3.3.3) and (7-3.3.4)],

$$\mu_{orb} = -\frac{2\pi}{h}\mu_B \,\|L\| = -\mu_B\sqrt{\ell(\ell+1)} \qquad (7\text{-}3.5.9)$$

and $$\mu_{z,orb} = -\frac{2\pi}{h}\mu_B \,\|L_z\| = -\mu_B\, m$$

You can see that the m levels will be distinguishable if a magnetic field is present (see Fig. 7-3.3.2)

The Zeeman Effect. In studies of atomic spectra, transitions are observed from one electronic energy level to another. In the absence of complications the spectrum could be easily interpreted. As usual, things are not that simple. At least two complicating factors arise. One is the interaction of the electron spin angular momentum with its orbital angular momentum. This interaction (called the spin-orbit coupling) gives rise to several energy levels where only one existed previously. This will not be discussed further here, but more detail can be found in Sec. 7-4. It leads to rather complex emission spectra for many electron atoms.

The second complicating factor is the effect of a magnetic field on the spectrum. If we put a hydrogen atom in a uniform magnetic field with a strength $\vec{H}$ aligned in the z direction (we define the direction), we will note a definite effect on the energy levels possible due to the effects suggested by Eq. (7-3.5.9). Assume we are interested in level n with E_n in the absence of a magnetic field. The magnetic field will lower the energy by $-\mu_z\vec{H}$ where μ_z is the z component (direction of the field) of the magnetic moment.

Thus with the magnetic field present

$$E_n' = E_n - \mu_z\vec{H} \qquad (7\text{-}3.5.10)$$

The magnetic moment is the sum of the spin and orbital contributions ($\mu_z = \mu_{z,s} + \mu_{z,orb}$). This allows us to write immediately $\mu_z = -(m + 2m_s)\mu_B$ and

$$E_n' = E_n + (m + 2m_s)\,\mu_B\vec{H} \qquad (7\text{-}3.5.11)$$

Now consider the n = 1 and n = 2 levels of the hydrogen atom. For the s levels $\ell = 0$ and m = 0, so

$$E_1' = E_1 \pm \mu_B\vec{H} \qquad (7\text{-}3.5.12)$$

This is true for any s level. What happens to the p level ($\ell = 1$) in the n = 2 state? If $\ell = 1$, m can equal 1, 0, and - 1, so there are six combinations of m + $2m_s$, two of which are the same (zero). The $\ell = 1$ state in a magnetic field splits into five levels as shown in Fig. 7-3.5.1.

The allowed transitions are shown. What you should do is show that these transitions yield only *three different values for* ΔE [show set a is the same as set b by applying Eq. (7-3.5.11) to the upper and lower levels, then take the differences].

A word is in order about *allowed transitions*. In any quantum mechanical system there are only certain transitions allowed between energy levels. These are governed by *selection rules*. How these selection rules are derived will be discussed in some detail later. Suffice it to say right now that the only allowed transitions are those where $\Delta m_s = 0$ and $\Delta m = 0, +1$, or -1.

The spectrum of hydrogen in a magnetic field was first observed by Zeeman in 1896. Hence the name the *Zeeman effect*.

7-3.6 PERIODIC PROPERTIES

THE PERIODIC TABLE. One of the fascinating developments in quantum mechanics is the explanation of the structure of the periodic table in terms of the solutions to the hydrogen atom problem. By assuming that every electron in a complex atom can be described by a set of four quantum numbers n, ℓ, m, and m_s, and that the energy level system is similar to that of the hydrogen atom, we can explain the structure of the periodic table. This is a major accomplishment.

It would appear at first that it would be unreasonable to expect the hydrogen atom solutions to be even approximately applicable to many electron atoms. The fact is that they are.

Before any progress can be made in interpreting the periodic table, an additional postulate is required. This is the *Pauli exclusion principle which states that: No two electrons of the same atom can have the same set of four quantum numbers*. A more general statement of the Pauli principle (due to Wolfgang Pauli in 1924) can be made in terms of properties of the solutions of the hydrogen atom. For our purposes the above statement is adequate.

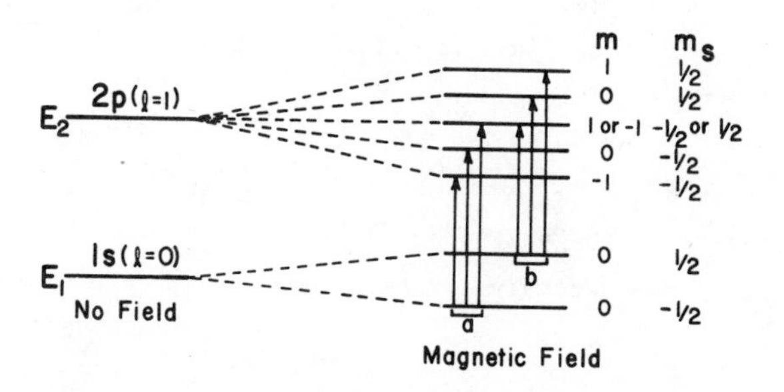

FIG. 7-3.5.1 Splitting of energy levels in a magnetic field and the allowed transitions.

Given that statement and an ordered set of energy levels based on the hydrogen atom solutions to the wave equation, it becomes an easy task to list the quantum numbers allowed for any element. Note that the H atom energy levels are not strictly applicable. Detailed calculations, discussed briefly later, do give the proper order of the energy levels.

To construct the periodic table we apply the *Aufbau principle* (building principle) which can be summarized as:

1. Each electron in a complex atom is described by four quantum numbers n, ℓ, m, and m_s, which are those derived from solution of the hydrogen atom wave equation.
2. The energy level arrangement in complex atoms is similar to that of the hydrogen atom. The electronic levels are arranged in increasing energy order with electrons added one at a time to the lowest available energy level in such a manner that the Pauli exclusion principle is observed. (Section 7-4 will show how these energy levels are obtained.)

With these rules it is easy for us to write the electronic configuration of each element. Before presenting a table showing the designation there are a few comments that might be useful concerning the periodic chart of the elements.

The energy levels with a given n value are assigned to a *shell* with a set of *subshells* defined by the allowed values of ℓ. For example, choose the shell with n = 4. Allowed ℓ values are 0, 1, 2, and 3. These correspond to the s, p, d, and f subshells. The allowed m values are given in Table 7-3.6.1.

The reason for the general form of the periodic chart should be clear. Figure 7-3.6.1 should be consulted at this point. The first two groups correspond to ℓ = 0, the center groups consist of 10 orbitals and fit ℓ = 2, and the last six groups are for the p electron, ℓ = 1. The f groups are below the chart (ℓ = 3) and have 14 available orbitals. To designate the electronic configuration of an element we do not need to memorize them. Just look at the periodic chart and read the designation directly. The only precaution (other than a few exceptions in the order of filling) is to note that the d orbitals in a given period n are designated as n - 1. The 3d orbitals should fill before the 4s according to the H atom solutions. However, calculations show that this is not true. Also, the f orbitals corresponding to n - 2 fill immediately after the ns. The f orbitals are below the chart only to make the chart compact. They fit between ns and (n - 1)d.

A couple of notes of explanation are in order before some examples are given. Each group ends with an "inert" gas which has a filled shell. The configuration of this "inert" filled shell forms the basis for the next group. Thus in Fig. 7-3.6.1, only the outer electrons are shown unless some anomaly exists. Now look at some examples.

← ns → ← (n-1)d → ← np →

Group / Period	Ia	IIa	IIIa	IVa	Va	VIa	VIIa	VIII	VIII	VIII	Ib	IIb	IIIb	IVb	Vb	VIb	VIIb	O
1 1s	1 H $1s^1$	2 He $1s^2$															1 H $1s^1$	2 He $1s^2$
2 2s2p	3 Li $2s^1$	4 Be $2s^2$											5 B $2s^22p^1$	6 C $2s^22p^2$	7 N $2s^22p^3$	8 O $2s^22p^4$	9 F $2s^22p^5$	10 Ne $2s^22p^6$
3 3s3p	11 Na $3s^1$	12 Mg $3s^2$											13 Al $3s^23p^1$	14 Si $3s^23p^2$	15 P $3s^23p^3$	16 S $3s^23p^4$	17 Cl $3s^23p^5$	18 Ar $3s^23p^6$
4 4s3d4p	19 K $4s^1$	20 Ca $4s^2$	21 Sc $4s^23d^1$	22 Ti $4s^23d^2$	23 V $4s^23d^3$	24 Cr $4s^13d^5$	25 Mn $4s^23d^5$	26 Fe $4s^23d^6$	27 Co $4s^23d^7$	28 Ni $4s^23d^8$	29 Cu $4s^13d^{10}$	30 Zn $4s^23d^{10}$	31 Ga $3d^{10}4p^1$	32 Ge $3d^{10}4p^2$	33 As $3d^{10}4p^3$	34 Se $3d^{10}4p^4$	35 Br $3d^{10}4p^5$	36 Kr $3d^{10}4p^6$
5 5s4d5p	37 Rb $5s^1$	38 Sr $5s^2$	39 Y $5s^24d^1$	40 Zr $5s^24d^2$	41 Nb $5s^14d^4$	42 Mo $5s^14d^5$	43 Tc $5s^14d^6$	44 Ru $5s^14d^7$	45 Rh $5s^14d^8$	46 Pd $5s^04d^{10}$	47 Ag $5s^14d^{10}$	48 Cd $5s^24d^{10}$	49 In $4d^{10}5p^1$	50 Sn $4d^{10}5p^2$	51 Sb $4d^{10}5p^3$	52 Te $4d^{10}5p^4$	53 I $4d^{10}5p^5$	54 Xe $4d^{10}5p^6$
6 6s(4f) 5d6p	55 Cs $6s^1$	56 Ba $6s^2$	57 La * $6s^25d^1$	72 Hf $6s^25d^2$	73 Ta $6s^25d^3$	74 W $6s^25d^4$	75 Re $6s^25d^5$	76 Os $6s^25d^6$	77 Ir $6s^05d^9$	78 Pt $6s^15d^9$	79 Au $6s^15d^{10}$	80 Hg $6s^25d^{10}$	81 Tl $5d^{10}6p^1$	82 Pb $5d^{10}6p^2$	83 Bi $5d^{10}6p^3$	84 Po $5d^{10}6p^4$	85 At $5d^{10}6p^5$	86 Rn $5d^{10}6p^6$
7 7s(5f)6d	87 Fr $7s^1$	88 Ra $7s^2$	89 Ac ** $7s^26d^1$															

← (n-2)f →

Series														
* Lanthanide Series	58 Ce $6s^24f^2$	59 Pr $6s^24f^3$	60 Nd $6s^24f^4$	61 Pm $6s^24f^5$	62 Sm $6s^24f^6$	63 Eu $6s^24f^7$	64 Gd $6s^25d^14f^7$	65 Tb $6s^24f^9$	66 Dy $6s^24f^{10}$	67 Ho $6s^24f^{11}$	68 Er $6s^24f^{12}$	69 Tm $6s^24f^{13}$	70 Yb $6s^24f^{14}$	71 Lu $-5d^14f^{14}$
** Actinide Series	90 Th $7s^26d^25f^0$	91 Pa $7s^26d^15f^2$	92 U $-6d^15f^3$	93 Np $-6d^05f^5$	94 Pu $-6d^05f^6$	95 Am $-6d^05f^7$	96 Cm $-6d^15f^7$	97 Bk -	98 Cf -	99 Es -	100 Fm -	101 Md -	102 No -	103 Lw -

FIG. 7-3.6.1 Periodic chart of the elements.

Table 7-3.6.1
Allowed Quantum Number Combinations for n = 4

Subshell	ℓ	m	m_s	Total allowed combinations in subshell
s	0	0	$\pm\frac{1}{2}$	2
		-1	$\pm\frac{1}{2}$	
p	1	0	$\pm\frac{1}{2}$	6
		1	$\pm\frac{1}{2}$	
		-2	$\pm\frac{1}{2}$	
		-1	$\pm\frac{1}{2}$	
d	2	0	$\pm\frac{1}{2}$	10
		1	$\pm\frac{1}{2}$	
		2	$\pm\frac{1}{2}$	
		-3	$\pm\frac{1}{2}$	
		-2	$\pm\frac{1}{2}$	
		-1	$\pm\frac{1}{2}$	
f	3	0	$\pm\frac{1}{2}$	14
		1	$\pm\frac{1}{2}$	
		2	$\pm\frac{1}{2}$	
		3	$\pm\frac{1}{2}$	

$$\text{H: } (1s)^1 \qquad \text{He: } (1s)^2 \equiv [\text{He}]$$

The [He] is defined as the filled n = 1 shell. Similarly

$$\text{Ne: } (1s)^2(2s)^2(2p)^6 = [\text{He}]\ (2s)^2(2p)^6 \equiv [\text{Ne}]$$

$$\text{Cl: } [\text{Ne}]\ (3s)^2(3p)^5$$

$$\text{Mn: } [\text{Ar}]\ (4s)^2(3d)^5$$

There are a few exceptions, due to anomalous behavior, such as Cu in the *transition elements* which has [Ar] $(4s)^1(3d)^{10}$. This should not be surprising, since many approximations are required to derive the ordering at the energy levels. What is surprising is that the system works so well.

The reason for periodic properties of the elements should be evident now. Elements in the same group but different periods have similar outer electron distributions. The filled inner shells have only minor effects on the properties of the elements. Thus, a similar outer electron distribuiton should give similar properties. We will look at some properties below.

ATOMIC RADII. Atomic radius varies appreciably across the periodic chart and down any given group. Comparison of the radii of atoms down any group shows a general increase. This is easily explained on the basis of the filling of shells with larger values of n as the atomic number increases. For example, the last electron in lithium goes into the n = 2 (2s) shell, potassium n = 4 (4s), and cesium n = 6 (6s). If you refer to Fig. 7-3.4.2 you will note that as n increases, the region of maximum probability of finding an electron gets farther and farther from the nucleus. Thus the average "size" of the atom, as defined by region of maximum probability, increases as n increases.

You should realize that the atomic radius is not a well-defined quantity. The electrons in the outer orbital that define the radius are not confined to a narrow region of space but are "smeared out" over a large region as suggested in Fig. 7-3.4.2. *The radius assigned to an atom depends on the method used to measure it or the purpose for which it is being defined.*

Within a given period (for n = 2 and 3), the radius decreases irregularly across the period. For n = 4, the transition elements interrupt this sequence. As the d orbitals are being filled the radius decreases in general.

In each case two trends are present. As electrons are added within a given period the nuclear charge is also being increased. This increased charge tends to pull the electrons toward the nucleus tending to decrease the radius. However, the additional electrons increase electron-electron repulsion, which tends to increase the size of the atom. A balance of these two trends results in the observed behavior.

IONIZATION ENERGIES. The *ionization energy* (I.E.) measures the strength with which an electron is bound to an atom. It is equivalent to the energy required to remove an electron from a given value of n to n = ∞. An atom has several ionization energies, one for each electron present. Table 7-3.6.2 gives some of these values. A plot of first ionization energies is given in Fig. 7-3.6.2 (note the different units used in the table and the figure).

From Table 7-3.6.2 you should note several things.

1. The second ionization energy is greater than the first, and subsequent ones get progressively larger.
2. For similar outer shell configurations (Li, Na, for example) the I.E. gets smaller as the atomic number increases.
3. The "inert" gases have very high I.E. values.
4. The I.E. increases dramatically when a new subshell is involved; note the difference between I.E. (2) and I.E. (3) for Be.

The explanation to the first point is that there is an excess positive charge in the nucleus after an electron is lost. This results in a tighter binding of the remaining electrons. As subsequent electrons are removed, more excess positive charge is present leading to more tightly bound electrons.

The second point is easily explained on the basis of the increase in distance between the nucleus and the electron as new shells are acquired. The third and fourth point are related. There is apparently a marked stability associated with a

Table 7-3.6.2
Ionization Energies of Some Ions

Element	Atomic Number	I.E. $(J\ atom^{-1}) \times 10^{18}$ [a]				
		1	2	3	4	5
H	1	2.18				
He	2	3.94	8.71			
Li	3	0.863	12.1	19.61		
Be	4	1.49	2.92	24.65	34.65	
B	5	1.33	4.03	6.07	41.54	54.23
C	6	1.80	3.91	7.68	10.33	62.49
N	7	2.33	4.74	7.60	12.41	15.6
O	8	2.18	5.63	8.80	12.40	18.10
F	9	2.79	5.60	10.04	13.97	18.21
Ne	10	3.45	6.58	10.3	15.56	
Na	11	0.823	7.58	11.48	15.84	
Mg	12	1.22	2.41	12.48	17.51	

[a] These are usually given in eV: $1\ eV = 1.6021 \times 10^{-19}\ J\ atom^{-1}$

complete shell or subshell. Once the shell or subshell has been broken the trends are the same as noted before.

ELECTRON AFFINITY. The energy that is released when an atom binds an electron to itself is called the *electron affinity* (E.A.). This quantity is difficult to measure or deduce from known data. The only purpose of introducing these at this time is to acquaint you with the concept. These values will be used along with ionization energies in the next chapter to define *electronegativity*. Electronegativity is a very useful qualitative concept that is appropriate for discussions of bonding in molecules. Some accepted values are given in Table 7-3.6.3.

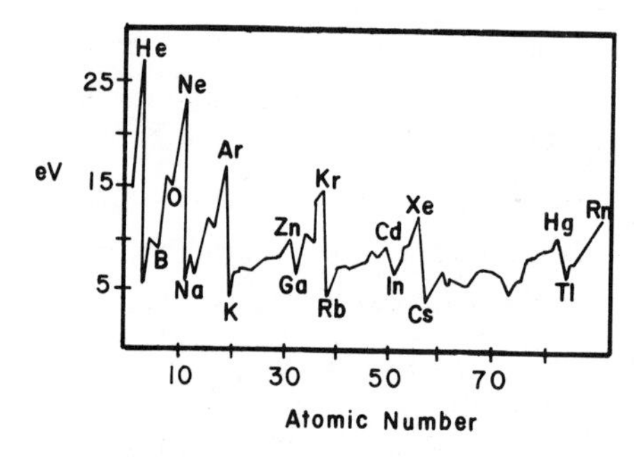

FIG. 7-3.6.2 First ionization energies for several atoms.

Table 7-3.6.3
Some Atomic Electron Affinity Values

Element	H	F	Cl	Br	I	O	S	C
E.A. ($J\ atom^{-1}$) $\times 10^{19}$	1.20	5.52	5.78	5.38	4.90	2.35	3.32	2.00

7-3.7 PROBLEMS

1. Verify the value of the constant in Eq. (7-3.5.5). Note: $1\ C^2 = 8.98 \times 10^9$ J m.

2. In a magnetic field of 2T calculate the splitting of the 1s level in hydrogen due to electron spin.

3. Determine the elements corresponding to the following electronic designations:

   $[Ar]\ (4s)^2(3d)^2 \qquad [Ar]\ (4s)^2(3d)^{10}(4p)^1 \qquad [Kr]\ (5s)^2(4d)^2$

4. Explain the observation that the radius generally decreases across the d orbitals in a period.

5. Assume the energy expression for the hydrogen atom applies except for the inclusion of a Z^2_{eff} term [see Eq. (7-3.2.2)]. Calculate Z_{eff} for the first through fourth ionization energies of boron, B (see Table 7-3.6.2). Explain your results.

6. Plot the electron affinity values for F, Cl, Br, and I. Try to interpret the trends you note.

7. The angular momentum operator about the z axis can be written $\hat{L}_\phi = (ih/2\pi)(\partial/\partial\phi)$. Show that several Φ solutions [see Eq. (7-2.4.5)] are eigenfunctions. What are the eigenvalues?

8. Define or discuss: Pauli exclusion principle, electron affinity, Aufbau principle, and ionization potential.

9. Determine the elements corresponding to

   a. $[Ne]\ 3s^2 3p^3$

   b. $[Ar]\ 4s^2 3d^5$

   c. $[Ar]\ 4s^2 3d^{10} 4p^2$

   d. $[Kr]\ 5s^1$

10. Write the electronic designation for the following elements: (a) Ar, (b) V, (c) Sr, and (d) Si.

7-3.8 SELF-TEST

1. Define the terms in Objective (3).
2. Calculate all the allowed transitions shown in Fig. 7-3.5.1 in Joules, per meter, and per second. Assume $\overline{\mathcal{H}} = 2T$.
3. List all allowed quantum number combinations and corresponding orbital designations for the n = 5 level in the hydrogen atom.
4. List the electronic designations for P, K. Zn, As, and Fe.
5. Explain the observed change between I.E. (2) and I.E. (3) for Be.
6. Sketch the energies (to scale) of the first four levels of the hydrogen atom (n = 1, 2, 3, 4, Z = 1) and the Li^{2+} atom (Z = 3). Comment on the relationship of this plot and the ionization energies given in Table 7-3.6.2.

7-4 APPROXIMATE METHODS AND ATOMIC SPECTRA

OBJECTIVES

You shall be able to

1. Discuss both the variation and perturbation methods for obtaining approximate solutions to the Schrödinger equation for many-electron atoms.
2. Devise term symbols to designate electronic states in atoms
3. Discuss techniques and results of atomic electronic spectroscopy

7-4.1 INTRODUCTION

The previous section dealt with the properties of atoms and the periodic table of the elements. It was mostly qualitative in nature. In this section we will deal with the theoretical basis of the last section by showing how approximate solutions of the Schrödinger equation can be obtained for many-electron atoms.

As we will see, if we add an electron to the hydrogen atom we complicate matters greatly. It is no longer possible to separate the variables as in the hydrogen atom. This is due to interactions between the two electrons. There are various ways to obtain approximate solutions. We will consider two: *the perturbation method* and *the variation method*. The helium atom will be used for illustrative purposes throughout.

Various interactions within atoms lead to rather complex energy level diagrams. They are adequately described by *term symbols*.. The meaning of these and how they are derived will be presented. Transitions among these various energy levels are possible, giving rise to absorption and emission spectra that can be used for identification of the atoms present. Finally, a discussion of practical applications of atomic spectroscopy will be given.

7-4.2 THE PERTURBATION METHOD

The perturbation method assumes that the equations and solutions for a particular system are similar to, but somewhat "perturbed" from, those of a simpler system. There are many cases in which, except for a couple of terms, the equations for a system are the same as for one that has been solved exactly. As an example, consider the He atom. Figure 7-4.2.1 gives the model that will be used. The potential energy of this system is given by

$$V = -\frac{2e^2}{r_1} - \frac{2e^2}{r_2} + \frac{e^2}{r_{12}} \tag{7-4.2.1}$$

The Schrödinger equation, $\hat{\mathcal{H}}\Psi = E\Psi$ becomes

$$\hat{\mathcal{H}} = -\frac{h^2}{8\pi^2 m_1}\nabla_1^2 - \frac{h^2}{8\pi^2 m_2}\nabla_2^2 + V$$

$$\frac{h^2}{8\pi^2 m_1}\nabla_1{}^2\Psi + \frac{h^2}{8\pi^2 m_2}\nabla_2{}^2\Psi + \left(E + \frac{2e^2}{r_1} + \frac{2e^2}{r_2} - \frac{e^2}{r_{12}}\right)\Psi = 0 \tag{7-4.2.2}$$

If it were not for the term e^2/r_{12} in (7-4.2.2) we could write it as

$$\frac{h^2}{8\pi^2 m_1}\nabla_1{}^2\Psi_1 + V_1\Psi_1 = E_1\Psi_1$$

$$\text{and} \quad \frac{h^2}{8\pi^2 m_2}\nabla_2{}^2\Psi_2 + V_2\Psi_2 = E_2\Psi_2 \tag{7-4.2.3}$$

with $E = E_1 + E_2$ and $\Psi = \Psi_1\Psi_2$. This cannot be done due to the e^2/r_{12} term. Thus, a different approach is needed.

In the perturbation method we write the Hamiltonian as

$$\hat{\mathcal{H}} = \hat{\mathcal{H}}^0 + \lambda\hat{\mathcal{H}}' + \lambda^2\hat{\mathcal{H}}'' + \cdots \tag{7-4.2.4}$$

where $\hat{\mathcal{H}}^0$ is the Hamiltonian for the case we can solve exactly ($\mathcal{H}^0$ is the hydrogen atom Hamiltonian for the solution of the helium atom) by the equation

$$\mathcal{H}_n^0\Psi_n^0 = E_n^0\Psi_n^0 \tag{7-4.2.5}$$

E_n^0 are the unperturbed energy levels and Ψ_n^0 the corresponding unperturbed wave

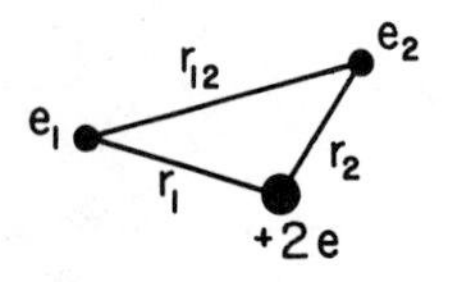

FIG. 7-4.2.1 Model of the helium atom.

functions. $\hat{\mathcal{H}}'$ and $\hat{\mathcal{H}}''$ are correction terms, and λ is a parameter which varies from 0 to 1 depending on the amount of perturbation. The energy is written as

$$E_n = E_n^0 + \lambda E_n' + \lambda^2 E_n'' + \cdots \tag{7-4.2.6}$$

where E_n' and E_n'' are energy correction terms. Also, the wave function is assumed to be given by

$$\Psi_n = \Psi_n^0 + \lambda\Psi_n' + \cdots \tag{7-4.2.7}$$

We are interested only in first-order corrections here (higher order corrections are possible, but are beyond the scope of this book) so all terms beyond λ^1 are ignored. The equation to be solved is the now familiar Schrödinger equation.

$$\hat{\mathcal{H}}_n\Psi_n = E_n\Psi_n \tag{7-4.2.8}$$

Inserting Eqs. (7-4.2.7), (7-4.2.6), and (7-4.2.4) into (7-4.2.8) and rearranging in terms of powers of λ gives

$$(\hat{\mathcal{H}}^0\Psi_n^0 - E_n^0\Psi_n^0) + \lambda(\hat{\mathcal{H}}^0\Psi_n' + \hat{\mathcal{H}}'\Psi_n^0 - E_n^0\Psi_n' - E_n'\Psi_n^0) + \lambda^2(\ldots) + \cdots = 0 \tag{7-4.2.9}$$

As stated, we will ignore coefficients of λ^2 and higher order terms. Setting the coefficients of λ^0 and λ^1 to zero yields two solutions:

$$\hat{\mathcal{H}}^0\Psi_n^0 = E_n^0\Psi_n^0$$

and

$$\hat{\mathcal{H}}^0\Psi_n' - E_n^0\Psi_n' = (E_n' - \hat{\mathcal{H}}')\Psi_n^0 \tag{7-4.2.10}$$

The first equation is simply the unperturbed equation. The second is the first-order perturbation term. We need to find Ψ_n' and E_n' since we know Ψ_n^0, E_n^0, $\hat{\mathcal{H}}^0$, and $\hat{\mathcal{H}}'$. To do so we assume Ψ_n' can be written as a linear combination of unperturbed (hydrogen atom) wave functions.

$$\Psi_n' = \sum_j a_{jn}\Psi_j^0 \tag{7-4.2.11}$$

We know

$$\hat{\mathcal{H}}^0 \Psi_n' = \hat{\mathcal{H}}^0 \sum_j a_{jn} \Psi_j^0$$
$$= \sum_j a_{jn} E_j^0 \Psi_j^0 \qquad (7\text{-}4.2.12)$$

Since the Ψ_j^0 are eigenfunctions of $\hat{\mathcal{H}}^0$. Substitution into (7-4.2.10) gives

$$\sum_j a_{jn}(E_j^0 - E_n^0)\Psi_j^0 = (E_n' - \hat{\mathcal{H}}')\Psi_n^0 \qquad (7\text{-}4.2.13)$$

Let us multiply by $\Psi_n^{0}*$ to get

$$\sum_j a_{jn}(E_j^0 - E_n^0)\Psi_n^0{*}\Psi_j^0 = E_n'\Psi_n^0{*}\Psi_n^0 - \Psi_n^0{*}\hat{\mathcal{H}}'\Psi_n^0 \qquad (7\text{-}4.2.14)$$

Multiplication by $d\tau$ and integrating over all space gives

$$\sum_j a_{jn}(E_j^0 - E_n^0)\underbrace{\int \Psi_n^0{*}\Psi_j^0\, d\tau}_{0} = E_n' \overbrace{\int \Psi_n^0{*}\Psi_n^0\, d\tau}^{1} - \int \Psi_n^0{*}\hat{\mathcal{H}}'\Psi_n^0\, d\tau \qquad (7\text{-}4.2.15)$$

The 0 and 1 arise due to orthonormality of the wave functions and the fact that $E_j^0 - E_n^0 = 0$ when $j = n$. Rearranging gives

$$E_n' = \int \Psi_n^0{*}\hat{\mathcal{H}}'\Psi_n^0\, d\tau \equiv \langle n|\hat{\mathcal{H}}'|n\rangle \qquad (7\text{-}4.2.16)$$

Thus we obtain the result that the first-order correction to the energy of the nth level is just the integral of the perturbation Hamiltonian on the wave function for the nth state.

The coefficients of Eq. (7-4.2.11) can be found (without proof) by

$$a_{mn} = -\frac{\langle m|\hat{\mathcal{H}}'|n\rangle}{E_m^0 - E_n^0} \qquad m \neq n \qquad (7\text{-}4.2.17)$$

The <> notation is a compact method of writing integrals such as (7-4.2.16). You will note that this treatment is limited to nondegenerate cases, since if $E_m^0 = E_n^0$, the coefficient a_{mn} is infinite. Degenerate cases can be handled, but will not be in this treatment.

Now let us see how this works for the helium problem. Helium contains two electrons. If the e^2/r_{12} term were not present, each electron would contribute $E_n^0(1) = E_n^0(2) = Z^2E_H = 4E_H$. Therefore,

$$E_n^0 = E_n^0(1) + E_n^0(2) = 2(4E_H) = 8(E_H) = -108.24 \text{ eV.}$$

The perturbation Hamiltonian is given by $\hat{\mathcal{H}}' = e^2/r_{12}$. If we use the lowest level of the hydrogen atom for our wave function (Ψ_{100}) we find

$$E'_{100,100} = \left\langle 100,100 \left| \frac{e^2}{r_{12}} \right| 100,100 \right\rangle$$

$$= \int \frac{e^2}{r_{12}} \Psi^2_{100,100}\, d\tau \qquad (7\text{-}4.2.18)$$

(the double subscript means each electron is in a 100 state). This gives (no proof)

$$E = -\frac{5}{4} ZE_H = -\frac{5}{2} E_H = 33.82 \text{ eV} \qquad (7\text{-}4.2.19)$$

Thus, $E = E^0 + E' = -74.42$ eV

which is close to the true value of -78.99 eV, an error of 5.8%. The agreement is particularly good considering the large size of the perturbation. Expanding the degree of perturbation to second or higher order leads to more accurate results.

A factor that we ignore can be included to improve the results. This is *electron correlation*. That is, one electron affects the distribution of the others. Such a procedure is involved and will not be considered here.

The procedure outlined here can be (and is) used for many types of perturbation, such as the effect of electron spin on electronic energy levels, and the effect of nuclear spin on molecular rotation. But in many cases the *variation method* is easier to apply. It is discussed in the next section.

7-4.3 THE VARIATION METHOD

One very useful technique for approximating solutions is the *variation method*. We will have occasion to apply this method to molecules in Sec. 8-4, where resonance energies of organic molecules are calculated. Here we will consider only atoms. The first step in this procedure is to multiply Eq. (7-4.2.8) by Ψ^*_n and integrate over all space to get

$$\int \Psi^*_n \hat{\mathcal{H}} \Psi_n \, d\tau = \int \Psi^*_n E'_n \Psi_n \, d\tau$$

$$= E'_n \int \Psi^*_n \Psi_n \, d\tau \qquad (7\text{-}4.3.1)$$

where E' indicates the energy that results from an inexact wave function and $\hat{\mathcal{H}}$ is the complete Hamiltonian for the system.

The last step is legitimate since E' is a constant. This can be rearranged to

$$E'_n = \frac{\int \Psi^*_n \hat{\mathcal{H}} \Psi_n \, d\tau}{\int \Psi^*_n \Psi_n \, d\tau} = \frac{\langle \Psi_n | \hat{\mathcal{H}} | \Psi_n \rangle}{\langle \Psi_n | \Psi_n \rangle} \qquad (7\text{-}4.3.2)$$

Eq. (7-4.3.2) will give a true value for E'_n if Ψ_n is the exact wave function (Postulate III). However, if Ψ_n is only an approximation, the value of E'_n will not be the true value. In fact, the *variation theorem* tells us that for any wave function we choose

$$E_n' \geq E_{true} \tag{7-4.3.3}$$

No matter what function we choose to describe the system, the energy we predict will always be greater than or equal to the true energy. We could prove this, but it would serve no purpose here. What is of importance is the procedure used to find solutions.

The procedure is to choose a function, then minimize the energy associated with that function by varying parameters in the function. Obviously, there are mathematical procedures for doing this. For the helium atom, the simplest assumption we can make is to assume that the repulsion term (e^2/r_{12}) does not exist. Instead, assume each electron moves about a helium nucleus of charge Z = + 2. The lowest energy orbital for the helium would be the product of two hydrogen-like orbitals (those given below are unnormalized).

$$\Psi_1 = \exp\left(-\frac{Zr_1}{a_0}\right)\exp\left(-\frac{Zr_2}{a_0}\right) \tag{7-4.3.4}$$

If this function is used in Eq. (7-4.3.2), an energy of -74.81 eV is found. The true value is -78.99 eV (1 eV = 1.60×10^{-19} J molecule^{-1}). That is not bad for an obviously poor wave function. The next step is to improve the wave function by using some variable parameters. And, the first choice for a variable parameter is Z. Let us assume Z is not fixed at two. We will minimize E_1' with respect to Z to get a better E and an effective value of Z. This should be a reasonable approach since we expect the presence of an extra electron will have a *screening effect* on the nuclear charge felt by the other electron, thereby changing the effective nuclear charge from Z = 2 to Z′. The appropriate trial wave function (normalized) is

$$\Psi_1 = \phi_1\phi_2 = \frac{Z'^3}{\pi a_0^3}\exp\left(-\frac{Z'r_1}{a_0}\right)\exp\left(-\frac{Z'r_2}{a_0}\right) \tag{7-4.3.5}$$

(ϕ_1 and ϕ_2 are individual wave functions for electrons 1 and 2, respectively).

It is perhaps instructive to show the procedure for minimizing E′ in detail at this point. Recall the operator for the helium atom

$$\hat{\mathcal{H}} = -\frac{h^2}{8\pi^2 m_e}(\nabla_1^2 + \nabla_2^2) - Ze^2\left(\frac{1}{r_1} + \frac{1}{r_2}\right) + \frac{e^2}{r_{12}} \tag{7-4.3.6}$$

where Z is the true nuclear charge. Operating $\hat{\mathcal{H}}$ on our wave function Ψ_1 gives

$$\hat{\mathcal{H}}\Psi_1 = \hat{\mathcal{H}}\phi_1\phi_2 = -\frac{h^2}{8\pi^2 m_e}(\nabla_1^{\ 2} + \nabla_2^{\ 2})\phi_1\phi_2 - Ze^2\left(\frac{1}{r_1} + \frac{1}{r_2}\right)\phi_1\phi_2 + \frac{e^2}{r_{12}}\phi_1\phi_2$$

or

$$\hat{\mathcal{H}}\Psi_1 = -\frac{h^2}{8\pi^2 m_e}(\phi_2\nabla_1^{\ 2}\phi_1 + \phi_1\nabla_2^{\ 2}\phi_2) - \frac{Ze^2}{r_1}\phi_1\phi_2$$

$$-\frac{Ze^2}{r_2}\phi_1\phi_2 + \frac{e^2}{r_{12}}\phi_1\phi_2 \qquad (7\text{-}4.3.7)$$

Remember that ϕ_1 and ϕ_2 are hydrogen-like orbitals with Z replaced by Z′. They must satisfy the hydrogen atom-like equation

$$-\frac{h^2}{8\pi^2 m_e}\nabla_1^{\ 2}\phi_1 - \frac{Z'e^2}{r_1}\phi_1 = Z'^2 E_H\phi_1$$

$$-\frac{h^2}{8\pi^2 m_e}\nabla_2^{\ 2}\phi_2 - \frac{Z'e^2}{r_2}\phi_2 = Z'^2 E_H\phi_2 \qquad (7\text{-}4.3.8)$$

Substitution into Eq. (7-4.3.7) and rearrangement of the terms gives (try it)

$$\hat{\mathcal{H}}\Psi_1 = 2Z'^2 E_H\phi_1\phi_2 + (Z' - Z)e^2\left(\frac{1}{r_1} + \frac{1}{r_2}\right)\phi_1\phi_2 + \frac{e^2}{r_{12}}\phi_1\phi_2$$

$$= 2Z'^2 E_H\Psi_1 + (Z' - Z)e^2\left(\frac{1}{r_1} + \frac{1}{r_2}\right)\Psi_1 + \frac{e^2}{r_{12}}\Psi_1 \qquad (7\text{-}4.3.9)$$

Now, multiply by Ψ_1^* and integrate over all space to get E_1' from (7-4.3.2). (Note $\langle\Psi_1|\Psi_1\rangle = 1$ since Ψ_1 is normalized.)

$$E_1' = \langle\Psi_1|\mathcal{H}|\Psi_1\rangle = \int \Psi_1^*\hat{\mathcal{H}}\Psi_1\, d\tau$$

$$= \langle\Psi_1|2Z'^2E_H|\Psi_1\rangle + (Z' - Z)e^2\langle\Psi_1|\frac{1}{r_1} + \frac{1}{r_2}|\Psi_1\rangle + \langle\Psi_1|\frac{e^2}{r_{12}}|\Psi_1\rangle \qquad (7\text{-}4.3.10)$$

The first term is easily evaluated since $2Z'^2E_H$ is a constant

$$2Z'^2 E_H\langle\Psi_1|\Psi_1\rangle = 2Z'^2 E_H \qquad (7\text{-}4.3.11)$$

The second term gives

$$(Z' - Z)e^2\langle\Psi_1|\frac{1}{r_1} + \frac{1}{r_2}|\Psi_1\rangle = (Z' - Z)e^2\langle\phi_1\phi_2|\frac{1}{r_1} + \frac{1}{r_2}|\phi_1\phi_2\rangle$$

$$= (Z' - Z)e^2\langle\phi_1\phi_2|\frac{1}{r_1}|\phi_1\phi_2\rangle$$

$$+ (Z' - Z)e^2\langle\phi_1\phi_2|\frac{1}{r_2}|\phi_1\phi_2\rangle$$

$$= (Z' - Z)e^2\langle\phi_1|\frac{1}{r_1}|\phi_1\rangle + (Z' - Z)e^2\langle\phi_2|\frac{1}{r_2}|\phi_2\rangle$$

(Note in the r_1 term $\langle\phi_2|\phi_2\rangle = 1$ and in the r_2 term $\langle\phi_1|\phi_1\rangle = 1$. Write these in integral notation and it becomes clear.) The two terms in the last equation are identical except for subscripts. So,

$$(Z' - Z)e^2\langle\Psi_1|\frac{1}{r_1} + \frac{1}{r_2}|\Psi_1\rangle = 2(Z' - Z)e^2\langle\phi_1|\frac{1}{r_1}|\phi_1\rangle$$

$$= 2(Z' - Z)e^2 \int \frac{\phi_1^2}{r_1}\, d\tau$$

$$= -4(Z' - Z)Z'\, E_H \qquad (7\text{-}4.3.12)$$

The last term in Eq. (7-4.3.10) has already been evaluated in our treatment of perturbation theory. The results are the same as in Eq. (7-4.2.18) and Eq. (7-4.2.19) except Z must be replaced with Z'. Thus

$$\langle\Psi_1|\frac{e^2}{r_{12}}|\Psi_1\rangle = -\frac{5}{4}Z'\, E_H \qquad (7\text{-}4.3.13)$$

Combining the last three results gives

$$E_1' = \left[2Z'^2 + (Z' - Z)(-4Z') - \frac{5}{4}Z'\right] E_H$$

$$= \left[-2Z'^2 + 4ZZ' - \frac{5}{4}Z'\right] E_H \qquad (7\text{-}4.3.14)$$

Our goal is to find the best value of E', by varying Z'. We can do this by minimizing E_1' with respect to Z'. This is accomplished by setting

$$\frac{\partial E_1'}{\partial Z'} = 0 = \left(-4Z' + 4Z - \frac{5}{4}\right) E_H$$

Solving for Z' gives

$$Z' = Z - \frac{5}{16} \qquad (7\text{-}4.3.15)$$

This value of Z' gives the best energy possible for the chosen wave function. Substitution of Z' back into (7-4.3.14) gives

$$E_1' = \left[-2\left(Z - \frac{5}{16}\right)^2 + 4Z\left(Z - \frac{5}{16}\right) - \frac{5}{4}\left(Z - \frac{5}{16}\right)\right]E_H$$

$$= -77.46 \text{ eV}$$

This value is only 2% in error. This is better than the perturbation theory result.

Inclusion of other parameters leads to a more complex problem, but to a better energy. At this point you might wonder: Why bother with it? If it were just to find the energy, the process would not be worth the effort. The hope is that once a wave function has been found that gives good values for the energy, this function can be used to calculate other properties. It is, however, not obvious (or even always true) that a wave function that gives a good energy will give accurate predictions for other variables.

EXTENSION TO OTHER ATOMS. If we attempt to apply the above treatment to many-electron atoms the problem becomes progressively more difficult as the number of electrons increases. The procedures used are based on the *variation method* derived by Douglas Hartree. Hartree's method is called the *self-consistent field method*. Each electron is assumed to move under the influence of a spherically symmetrical field made up of the nuclear charge and the field due to all other electrons present. This assumption allows the Schrödinger equation to be factored into separate equations for each electron. The result is a set of one-electron wave functions.

These functions are not very accurate. The method has been extended by including effects due to electron spin and *correlation effects* (interactions between electrons themselves other than the average repulsion). These methods are complex and beyond the scope of this treatment.

The major point here is that for even the simplest system (the He atom) approximations must be made. As the number of electrons increase, the effort required to obtain accurate values of the energy and acceptable wave functions also increase. However, with the available high-speed computers, scientists have "solved" rather complex systems. Section 8-4 will give some treatment to similar studies on molecules which, as you might guess, are more difficult to describe theoretically.

7-4.4 ATOMIC ENERGY LEVELS

Atoms can absorb or emit radiation due to transitions between electronic energy levels. The electronic spectrum for multielectron atoms is not at all simple. In fact, a new notation beyond the n, ℓ, m, and m_s quantum numbers is required to satisfactorily label all the energy levels present.

Even for the hydrogen atom the emission spectrum is not simple. Instead of having a single emission line for each $k \rightarrow n$ transition, several lines are observed. This is due to the interaction of the electron spin angular momentum with the orbital angular momentum. Such a case is the *general rule*: when more than one angular momentum is present in an atom or molecule these will usually interact to produce a new angular momentum. You will see that illustrated in Sec. 8-2.2 where the nuclear spin interacts with the rotational angular momentum.

What is needed is a new notation that takes into account the new situation. It is not sufficient to say the transition is from n = 1 to n = 3 in an atom. The appropriate designation is given by *term symbols*. If more than one electron is present the maximum total angular momentum is the sum of the individual angular momenta

$$\vec{L} = \sum_i \vec{\ell}_i \qquad (7\text{-}4.4.1)$$

where $\vec{L}$ is the total angular momentum and $\vec{\ell}_i$ are the individual electron momenta. For the hydrogen atom this is simple since there is only one electron.

For the n = 1 state the only possible ℓ is $\ell = 0$, so $\vec{L} = 0$. The corresponding symbol is S. For n = 2, $\ell = 1$, which gives $\vec{L} = 1$ (symbol P) and $\ell = 0$, which gives L = 0 (symbol S). So, for the hydrogen atom the possible states are (show the states for n = 3)

n = 1 S

n = 2 S, P

n = 3 S, P, D

The spin s of the electron must be included due to the spin-orbit interaction: the interaction of the electron spin with the orbital angular momentum. A new quantum number j is defined as $\ell \pm s$, which is written as a subscript for the term symbol. For a p electron, $\ell = 1$ or 0, so $j = 1 + \frac{1}{2}$ or $1 - \frac{1}{2} = 3/2$ or 1/2 for P and $j = 0 + \frac{1}{2}$ (j cannot be negative) for S. (*Note*: We use lower case letters to designate individual angular momenta and a capital letter for total angular momentum.) We then write

n = 1 $S_{1/2}$

n = 2 $S_{1/2}$, $P_{1/2}$, $P_{3/2}$

n = 3 $S_{1/2}$, $P_{1/2}$, $P_{3/2}$, $D_{3/2}$, $D_{5/2}$

and so on. Each of these term symbols represents a state with slightly different energy. Thus, within each n level there are several sublevels designated by the term symbol.

For heavier atoms (more electrons) the situation is more complicated. One additional symbol is required. If the electrons are paired there is no net spin angular mometnum giving a *singlet state* ($\vec{S} = \sum^i s_i = 0$). This is designated by a superscript 1. If two electrons are unpaired, three projections of the spin are possible since the total spin angular momentum is $\vec{S} = 1$ ($\vec{S} = \sum^i s_i = \frac{1}{2} + \frac{1}{2} = 1$), and it can have projections of 1, 0, or -1. Hence a *triplet state* results.

The *ground state* (lowest energy) for He will have two electrons in the 1s orbital with the spins paired ($\ell_1 = 0$, $\ell_2 = 0 \rightarrow \vec{L} = 0$, $s_1 = \frac{1}{2}$, $s_2 = -\frac{1}{2} \rightarrow \vec{S} = 0$, $\vec{J} = \vec{L} + \vec{S}$) thus the term symbol is

$${}^1S_0$$

In general the symbol will be (χ stands for S, P, D, ...)

$$^{(2\vec{S}+1)}\chi_{\vec{J}}$$

where $\vec{S}$ = total electron spin, $\vec{J} = \vec{L} + \vec{S}, \ldots, \vec{L} - \vec{S}$. [Note, total angular momenta are expressed as $\vec{L}$, $\vec{S}$, or $\vec{J}$. Allowed projections are given as capital letters without the arrow. Thus for $\vec{L} = 2$, L = 2, 1, 0 for D, P, S states, for $\vec{J} = 2$, J = 2, 1, 0 (if $\vec{S} = 0$).]

The lowest excited state would have one electron promoted to the 2s orbital. This would be designated as $1s^1 2s^1$. There are two possible combinations. The electrons could be paired ($\vec{S} = 0$) or unpaired ($\vec{S} = 1$, S = 1, 0, -1). Thus for $1s^1 2s^1$: $\vec{L} = 0$, $\vec{S} = 1 \rightarrow {}^3S_1, {}^3S_0, {}^3S_{-1}$: $\vec{L} = 0$, $\vec{S} = 0 \rightarrow {}^1S_0$. The different J levels (χ_J) are degenerate unless a magnetic field is present.

Figure 7-4.4.1 shows other states involving n = 1 and n = 2. You should verify some of the states given for $1s^1 2p^1$. Figure 7-4.4.2 shows the energy levels of the potassium atom. Note that all levels below n = 4 are ignored. The only states shown are those generated by promoting the outer s electron to excited configurations. Several important transitions are shown. If a magnetic field is present each level shown will be split into a multiplet as shown in Fig. 7-4.4.1 for He. Thus, complicated spectra result.

You may wonder what all this is worth. That is a legitimate concern. Perhaps one experimental technique of analysis based on the above will suffice to show the importance in analytical work. Besides the utility of atomic spectroscopy in analytical work, an accurate energy level diagram is necessary if atoms are to be well

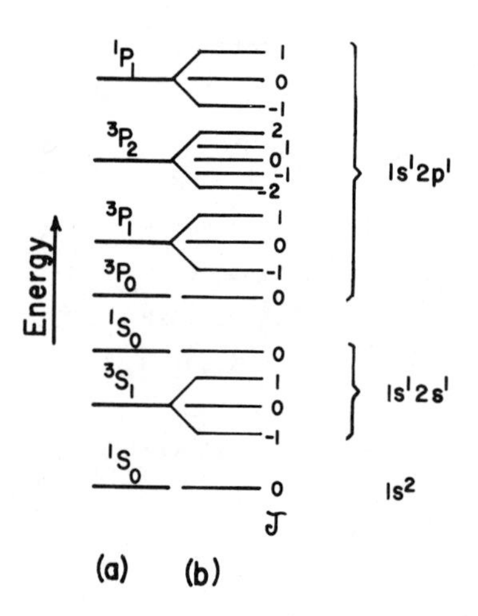

FIG. 7-4.4.1 Energy level diagram (not to scale) showing the terms for $1s^1 2s^2$ and $1s^1 2p^1$ for He.

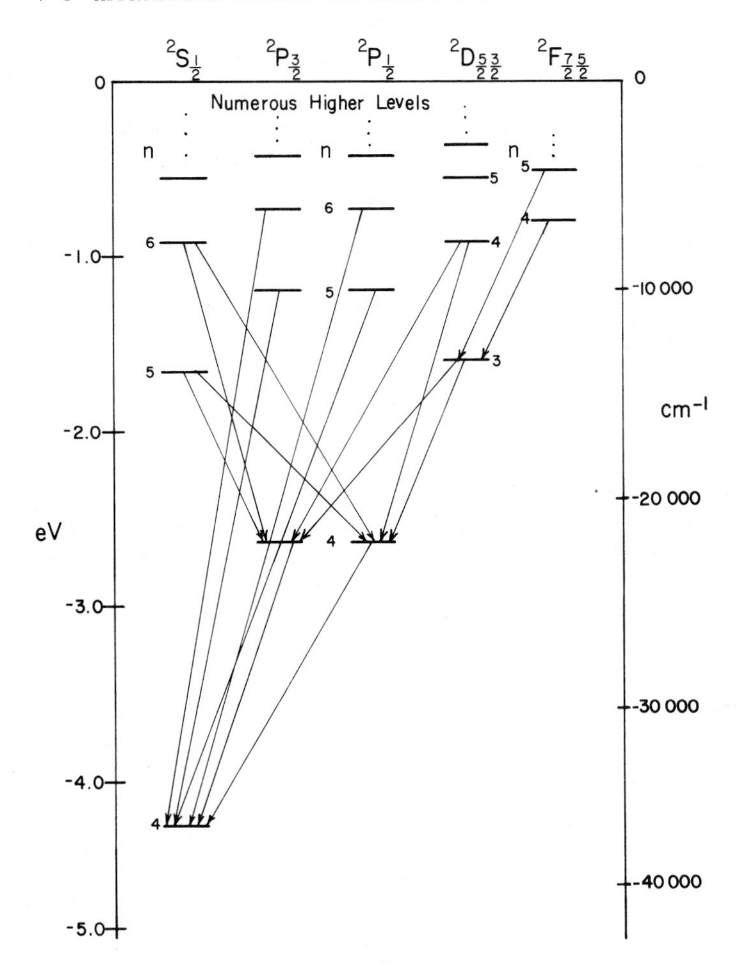

FIG 7-4.4.2 Energy level diagram for K.

characterized, a requirement for predicting whether reactions or energy exchanges will occur between atoms. One of the first lasers (see Sec. 8-4) involved exchange of energy between He and Ne to achieve lasing action. The design would not have been possible if the energy levels of He and Ne had not been well known. Anyway, let us turn our attention to a very precise and useful analytical technique that uses transitions between energy levels in atoms, *atomic absorption spectroscopy*.

7-4.5 ATOMIC ABSORPTION SPECTROSCOPY

In atomic absorption spectroscopy (AAS) a sample is heated to a high temperature (by a flame or other means) to produce atoms. These atoms absorb incident radiation of a specific wavelength. The amount of absorption is then related to the concentration of the sample. Figure 7-4.5.1 shows a schematic of a double-beam instrument. In this device a burner is used to form atoms which are then analyzed. A solution of the sample to be analyzed is aspirated into the flame where the high temperature atomizes it. Radiation from a lamp, which has a cathode made of the element to be analyzed, passes through a monochrometer where a specific atomic transition frequency is selected. This frequency passes through the flame. Atoms of the given element (but not others, since they would absorb at different frequencies) absorb this radiation. The beam coming through the flame is compared to a reference beam that has not passed through the flame. Any changes that have occurred in the flame show up at the detector as a difference signal between the sample and reference beams. This signal is related to the sample absorption.

To use the data, a standardization curve is constructed using solutions of known concentrations. Samples with concentrations of a few parts per million are routinely run. Higher sensitivity can be obtained by some instrumental changes. The flame itself generates "noise" that limits sensitivity. Also, only relatively low-viscosity materials can be run since atomization depends on aspirating the sample into the flame.

These drawbacks have been overcome by using a carbon tube atomizer (this technique is called flameless atomic absorption). In this system the sample is placed in a heated carbon tube (with temperatures up to 3000 K) where atoms are formed thermally. After the atoms are formed, the operation is similar to that described above. The major advantages of this technique are that any type of sample solution (very viscous, very high solid contents, etc.) can be used. Also, solids can be handled in some cases. Extremely small samples can be run (which is not the case with a flame instrument). As little as 0.5 to 50 μl (1 μl = 10^{-6} liters = 10^{-6} dm^3) of solution will do. Absolute detection limits for atoms are 10^{-10} g or less. For example, as little as 10^{-13} g Zn can be detected. (Note: in all cases a different lamp and wavelength is needed for each element analyzed.)

The applications of AAS are too numerous to list. A few will be listed to give you an idea of the range.

1. A 5 to 10 mg sample of brain tissue can be analyzed for metals. In one case about 15 ± 0.5 ppm (parts per million) Cu and 1.5 ± 0.05 ppm Mn were observed.
2. Analysis of a 25-μl sample of decarbonated beer yielded 0.1 ± 0.007 ppm Al.
3. Analysis of Si, Mn, Cr, Ni, Cu, and Mo in stainless steel.
4. The following data were found in one food analysis (ppm).

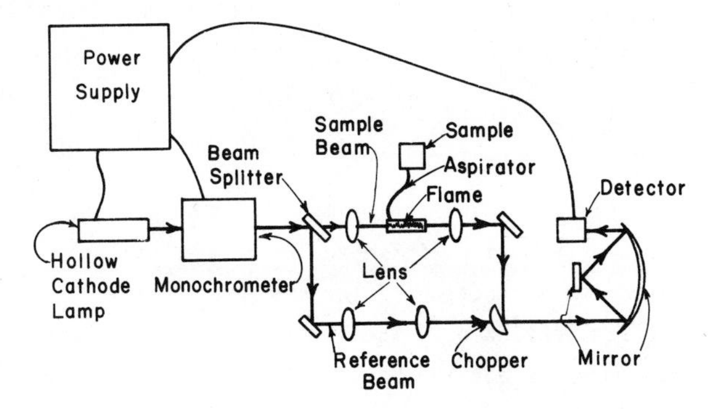

FIG. 7-4.5.1 Double-beam atomic absorption spectrometer.

Food	Na	K	Ca	Mg	Fe	Zn	Cu	Mn	Sn
Canned peaches	60.0	101.0	21.8	94.0	3.0	1.1	0.6	0.4	44.0
Canned corn	262.0	196.0	37.2	409.0	4.7	6.8	0.9	1.3	ND[a]
Tomato soup	10,450	272.0	71.6	199.0	6.0	2.9	0.9	1.5	80.0
Cereal	11,500	80.0	78.0	505.5	41.5	33.0	1.8	9.5	ND
Potatoes	80.0	6,200	12.3	433.0	4.0	3.9	2.1	1.1	ND
White bread	4,950	92.0	175.0	403.0	23.5	29.0	1.5	4.7	ND

[a] Not detected.

5. AAS is used routinely in the determination of metal in water. Some permissible levels in drinking water and approximate detection limits are shown below (ppm; flame technique).

	Permissible level	Detection limit
Ag	0.05	0.001
Ca	60	0.002
Cd	0.01	0.001
Cu	1.0	0.001
Fe	0.3	0.003
Pb	0.05	0.008
Zn	5.0	0.001

There are numerous other applications, but the above should give you an idea of the versatility of the technique.

7-4.6 PROBLEMS

1. Write the term symbols possible for the Li ground state and the $(1s)^2(2p)^1$ excited state. Show how a magnetic field would affect the levels.

2. An atom placed in an electric field will be polarized by the field to produce an induced electric dipole moment (the centers of positive and negative charge are separated: (+ -)). The energy and wave function of the system are affected by this distortion. As usual, the simplest case is the hydrogen atom placed in a uniform electric field $\bar{E}$ in the z direction. The Hamiltonian operator for this case is

$$\hat{\mathcal{H}} = \hat{\mathcal{H}}^0 + e\vec{E}z = \hat{\mathcal{H}}^0 + e\vec{E}r\cos\theta$$

where $\hat{\mathcal{H}}^0$ is the normal Hamiltonian for the H atom. Use first-order perturbation theory to calcuaate the energy of the ground state of an H atom in the electric field $\vec{E}$.

3. Discuss the terms "precision" and "accuracy" as they apply to atomic absorption spectroscopy.

7-4.7 SELF-TEST

1. Write the term symbols for the ground state of Be^+ and the $(1s)^2(3p)^1$ excited state. On your energy level diagram show the splittings caused by a magnetic field.

2. Calculate the energy of the ground state of the hydrogen atom in an electric field $\vec{E}$ using the variation procedure with $\phi = (1 + A\,r\cos\theta)\Psi_{1s}$ as the trial function. Minimize the energy with respect to A and neglect powers of $\vec{E}$ higher than $\bar{E}^2$. Compare with Problem (2).

3. From a text on atomic absorption spectroscopy, give a half-page discussion of the meaning of the terms "sensitivity" and "detection limit".

7-4.8 ADDITIONAL READING

1. L. Pauling and E. B. Wilson, Jr., *Introduction to Quantum Mechanics*, McGraw-Hill New York, 1935.

2. J. C. Davis, Jr., *Advanced Physical Chemistry*, Ronald Press, New York, 1965, Chaps. 6 and 7.

3. W. Slavin, *Atomic Absorption Spectroscopy*, Wiley Interscience, New York, 1968.

7-5 SUMMARY

In this chapter you have encountered a development of quantum mechanics from the earliest years to date. The first section gave historical background outlining the observations that lead to the development of quantum theory. After the postulates, a simple example was given, then the equations for solutions of the energy levels of the hydrogen atom were set up.

Section 7-2 was a mathematical exercise. First, some mathematical preliminaries were given. Then, a rather detailed outline of the solution to the equations for the hydrogen atom was presented. The purpose of this was to show how such problems are approached.

Section 7-3 used the results of Sec. 7-2 to show how the periodic chart can be designed, based on the results for the hydrogen atom. Some detail was given on the shapes of atomic orbitals and electronic distributions. The periodic properties of the elements were discussed briefly to show the relationship between electronic structures and properties.

The final section extended our skills in quantum mechanical techniques. We saw how to apply two different approximation methods to problems that cannot be solved exactly. We will have occasion to use these techniques in Chap. 8. Finally, atomic spectra were discussed. Term symbols, a means of labeling levels in complex atoms, were devised. A very useful and versatile experimental technique, atomic absorption spectroscopy, was discussed.

The purpose of all the above is to show you how problems are approached. It is not intended to make you into a competent theoretician. Fundamental work in quantum mechanics underlies many recent theoretical and practical advances in the science and art of chemistry. To be an educated scientist, you must at least be aware of the world of quantum theory.

8

Spectroscopy

SYMBOLS AND DEFINITIONS

λ	Wavelength (m = 10^{10} angstrom, Å)
ν	Frequency (s^{-1} = Hertz, Hz)
$\bar{\nu}$	Wave number (cm^{-1})
I, I_0	Intensity and initial intensity of radiation
C	Concentration (appropriate units)
ℓ	Path length of radiation in a sample (cm)
R_{km}	Transition moment between two states
Ψ_j	Molecular wave function for the jth state
$\hat{\mu}$	Dipole moment operator
I	Moment of inertia (kg m^2)
h	Planck's constant, 6.6262×10^{-34} J s
μ	Reduced mass (kg $molecule^{-1}$)
r_{12}	Internuclear distance (m, Å)
m_i	Mass of ith particle (kg)
ε_J	Energy of rotational level J (J $molecule^{-1}$)
M	Projection of rotational quantum number onto z axis
A, B, C	Rotational constants about the principal rotational axes (s^{-1}, J)
$\Delta\varepsilon_{J\to J+1}$	Energy of the transition from rotational state J to J + 1
$\hat{\mathcal{H}}$	Hamiltonian operator for a given type of motion
$\bar{k}$	Force constant for vibrational motion (N m^{-1})
q	Displacement from bond equilibrium distance
T, V	Kinetic and potential energies (J)
α, β, ξ	Variables in the solution to the harmonic oscillator problem
H_v	Solution to the harmonic oscillator problem; Hermite polynomials
v	Quantum number for vibrational motion
ν_0	Fundamental vibrational frequency for a given molecular motion (s^{-1}, Hz)
$\Delta\varepsilon_{v\to v+1}$	Energy of the transition from vibrational state v to v + 1
n	Number of atoms in a molecule
D	Centrifugal distortion constant for rotational motion (s^{-1}, cm^{-1}, J)
F	Hyperfine quantum number
$\underline{D}_e$	Dissociation energy of a molecule (J $molecule^{-1}$)
$\bar{\nu}_e$	Equilibrium value of the fundamental vibrational energy (cm^{-1})
$\overline{\nu_e x_e}$	Anharmonicity constant for vibrational motion (cm^{-1})
$\bar{B}_e$	Equilibrium value of the rotational constant (cm^{-1})
α_e	Correction factor relating $\bar{B}_e$ and $\bar{B}_v$
J	Coulomb integral
K	Exchange integral
σ, π, δ	Orbital designations
Σ, Π, Δ	Symbols for electronic states of molecules
g, u	Subscripts denoting symmetric and unsymmetric orbitals
ν''	Observed frequency in a Raman experiment (s^{-1})
ν'	Raman shift (s^{-1})
g_N	Nuclear g factor
μ_N	Magnetic moment of nucleus, 5.0493×10^{-27} J T^{-1}
$\mathcal{H}$	Magnetic field (T)
δ	Chemical shift in nuclear magnetic resonance experiments
ϕ_i	Atomic orbital wave functions used for molecular orbitals
σ, π	Molecular orbital designations
sp, sp^2, sp^3	Hybrid orbital designations
α, β	Variables in the Hückel approximation
E_{loc}	Energies of electrons localized in bonds
E_{mo}	Energies of electrons delocalized into the molecular system
E_{res}	Resonance energy

The last chapter presented the fundamentals of quantum mechanics for atoms. With that basis, we can turn our attention to bonding in molecules, as well as to the various molecular motions possible. The first section will investigate some theoretical approaches to the description of bonding within molecules. The remaining three sections will touch on various experimental methods (along with some necessary mathematical treatments) for deducing molecular parameters. These methods all involve spectroscopic studies. Spectroscopy, the study of the emission or absorption of radiation, is the experimental tool that connects real structures and properties with theoretical investigations and wave equations. There are numerous types of spectroscopy. We have already mentioned investigations of atomic properties by observing transitions among electronic energy levels of atoms (Sec. 7-4). Such electronic transitions occur in the ultraviolet and visible regions of the electromagnetic spectrum. This will be extended to molecules in Sec. 8-4. Vibrational motion in molecules (the oscillation of one nucleus with respect to another) gives rise to quantized energy levels. Transitions between these levels yield absorptions or emissions of radiation in the infrared region. Observed vibrational frequencies can be related to the bond strengths between the nuclei in question. These transitions are extremely useful for identification purposes. Rotational motion is also quantized. Molecules cannot rotate at just any frequency, but are restricted to discrete energies. The transitions between rotational levels give rise to absorptions or emissions in the microwave region. Rotational energies are determined by the moment of inertia of a given molecule. The moment of inertia is, in turn, defined by bond angles and bond lengths. Thus, studies of rotational spectra of molecules lead to accurate structure determinations. The final section in this chapter will deal with other types of spectroscopic data. Nuclear magnetic resonance (NMR) spectroscopy depends on the interaction of the magnetic moment of the hydrogen (or other) nucleus with a magnetic field. This interaction yields extremely valuable data on the electronic environment about the nucleus. NMR is a valuable tool in the identification of molecules. Also, Raman spectroscopy, which also depends on vibrational motion, will be reviewed briefly.

The following table and diagram should assist you in putting the relative magnitudes of the energies in perspective. The table shows the regions of the electromagnetic spectrum which are of concern. The figure shows relative magnitude (*not to scale*) of the first three types of spectroscopy mentioned. NMR involves energy levels much more closely spaced than the rotational levels. The quantum numbers used for the figure are defined in Sec. 8-2. A good general reference for spectroscopy is G. M. Barrow, *Introduction to Molecular Spectroscopy* (see Sec. 8-1.12, Ref. 1).

The Spectral Regions[a]

Unit	Microwave	Infrared	Visible	Ultraviolet
λ (m)	0.30-0.001	3×10^{-5}-2.5×10^{-6}	7×10^{-7}-3×10^{-7}	3×10^{-7}-1×10^{-7}
(Å)		300,000-25,000	7,000-3,000*	3,000-1,000*
$\bar{\nu}$ (cm^{-1})		300-4,000*		
ν (s^{-1})	1×10^9-300×10^9*			

[a] (*) Indicates a commonly used unit. The older literature has a variety of units. You might run across Hz (hertz), s^{-1} (per second), or cps (cycles per second) for frequency; Hz is now standard. Micron (μ) is sometimes used for micrometer (μm = 10^{-6} m) and millimicron (mμ) for nanometer (nm = 10^{-9} m). Also, 1 Å = 10^{-10} m. (Fill in the blanks for practice.)

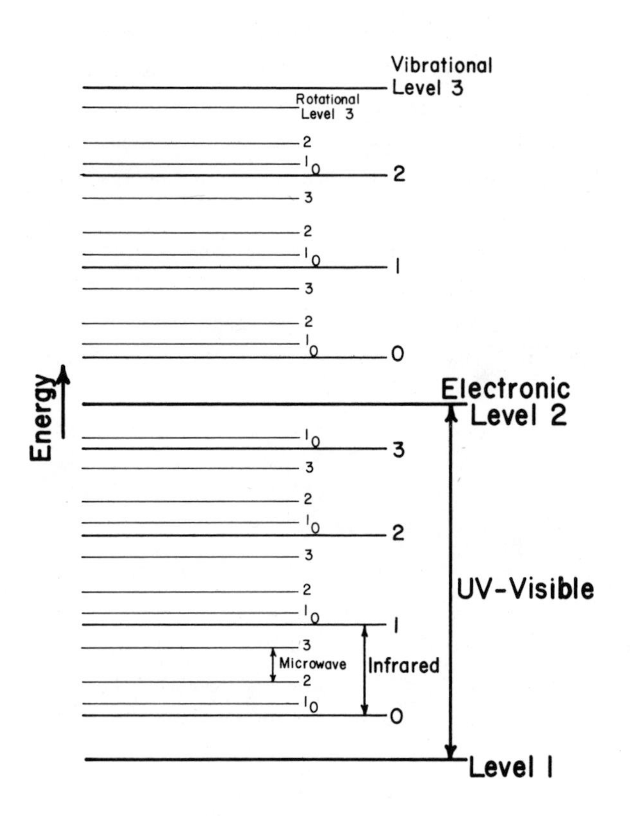

8-1 MOLECULAR STRUCTURE

OBJECTIVES

You shall be able to

1. Define and/or discuss the terms ionic, covalent, polar, electronegativity, bonding and antibonding orbitals, hybrid orbitals, resonance energy, and symmetric and antisymmetric functions
2. Discuss the linear combination of atomic orbitals to form molecular orbitals method (LCAO-MO) and valence-bond description of H_2 and H_2^+
3. Apply molecular orbital diagrams to diatomic molecules to predict bond order, the number of unpaired bonds, and electron distribution in molecular orbital energy levels
4. Form hybrid orbitals with specified directional characteristics from given atomic orbitals
5. Use the Hückel method to estimate *resonance stabilization* in conjugated organic molecules

8-1.1 INTRODUCTION

We now begin our study of the combination of atoms to form molecules. This will not be a very extensive development; however, limiting this topic poses a problem. To gain much of an understanding of the application of quantum mechanics to molecular bonding requires much detail. There is insufficient space to do an adequate job, so only a bare skeleton of procedures will be given. Then, specific topics will be discussed. Molecular orbital theory for simple systems and molecular orbital diagrams for homonuclear and heteronuclear diatomics will be given. The Hückel approach for calculating energies of organic compounds will be presented, introducing the concept of *resonance stabilization*. Finally, directional character in bonding will be represented by a geometric development. We will ignore bonding due to d orbitals a topic of utmost importance to inorganic and organometallic chemistry, catalysis, biochemistry, and other areas.

8-1.2 GENERAL COMMENTS ON BONDING

Up to this point we have considered only atoms. It is time to describe how atoms can combine to produce molecules. There are two extremes possible in bonding: *pure ionic* and *pure covalent bonds*. Pure ionic bonding requires one atom to *completely* lose one or more of its electrons and transfer it (them) to one or more other atoms, forming ions. There are probably no examples of pure ionic bonding, although there are many examples of *ionic compounds*, those with appreciable ionic character. These are discussed in Sec. 14-3. Pure covalent bonding requires two atoms to share two or more electrons equally. This is possible in compounds known as *homonuclear diatomics* such as N_2, Cl_2, O_2, etc. There is no reason to believe one atom of the pair has more attraction for the electrons than the other. The vast majority of compounds lie somewhere between these extremes of ionic and covalent bonding. Before getting into mathematical details, it might be useful to consider this more carefully.

An index has been developed that relates the degree with which an atom *in a compound* attracts bonding electrons to itself. This index is called *electronegativity*. R. S. Mulliken developed a formula for calculating the electronegativity designated by χ from ionization potentials and electron affinities defined in Sec. 7-3.6. Another approach has been taken by Linus Pauling. Pauling calculated electronegativity values by comparing actual bond energies with those calculated, assuming purely covalent bonding. This procedure reflects the deviation from pure covalent character toward ionic character. Pauling's values and those of Mulliken agree very well. Figure 8-1.2.1 shows Pauling's electronegativity values arranged in periodic order. The importance of these values lies in the information they give

			H 2.2				
Li 1.0	Be 1.5	B 2.0		C 2.5	N 3.0	O 3.5	F 4.0
Na 0.9	Mg 1.2	Al 1.5		Si 1.8	P 2.1	S 2.5	Cl 3.0
K 0.8	Ca 1.0	Sc 1.3	Ti-Ga 1.7 ± 0.2	Ge 1.8	As 2.0	Se 2.4	Br 2.8
Rb 0.8	Sr 1.0	Y 1.2	Zr-In 1.9 ± 0.3	Sn 1.8	Sb 1.9	Te 2.1	I 2.5
Cs 0.7	Ba 0.9	La-Lu 1.1	Hf-Tl 1.9 ± 0.4	Pb 1.8	Bi 1.9	Po 2.0	At 2.2
Fr 0.7	Ra 0.9	Ac 1.1	Th → 1.3 →				

FIG. 8-1.2.1 Electronegativity values as calculated by Pauling [see L. Pauling, *The Nature of the Chemical Bond*, 3rd ed., Cornell University Press, Ithaca, New York, 1960 (Sec. 8-1.12, Ref. 2)].

us about the degree of ionic character. The greater the electronegativity difference between two atoms the more ionic character there will be in the compound. The more ionic character there is the more *polar* the molecule will be. This is an important concept in discussing interaction *between* molecules as seen in Sec. 14-1.

8-1.3 MOLECULAR ORBITAL METHODS: THE H_2^+ ION

General Procedure. The simplest molecule imaginable is the H_2^+ ion which has two nuclei (each + 1) and one electron. Even this simple system can be solved only by making an approximation called the *Born-Oppenheimer approximation.* The Born-Oppenheimer approximation states that the electron moves so fast compared to the heavy nuclei that we can assume the nuclei remain fixed. From the study of vibrational motion (infrared spectroscopy) which you will encounter in Sec. 8-2, we know that nuclei move. However, the motion is slow with respect to electronic motion. The following procedures are usually used.

1. Pick appropriate wave functions (with or without some arbitrary parameters in them); we will do that shortly.
2. Pick a value of the internuclear distance.
3. Solve the Schrödinger equation to find the energy.
4. If arbitrary parameters have been included, minimize the energy with respect to these parameters by using the variation theorem (see Sec. 7-4.3).
5. Repeat the above for other internuclear distances to get a plot of E vs. internuclear distance.

Model and Equation. Figure 8-1.3.1 gives the model of the H_2^+ ion. Before reading on see if you can write the Schrödinger equation for this. The Hamiltonian we need is

$$\hat{\mathcal{H}} = -\frac{h^2}{8\pi^2 m_e}\nabla_e^2 - \frac{h^2}{8\pi^2 m_A}\nabla_A^2 - \frac{h^2}{8\pi^2 m_B}\nabla_B^2 - \frac{e^2}{r_A} - \frac{e^2}{r_B} + \frac{e^2}{r_{AB}} \qquad (8\text{-}1.3.1)$$

Using the Born-Oppenheimer approximation, it is possible to separate the nuclear motions from the electronic motions for a given internuclear distance. Thus, the problem of interest (the electronic motion) reduces to a Hamiltonian of the form

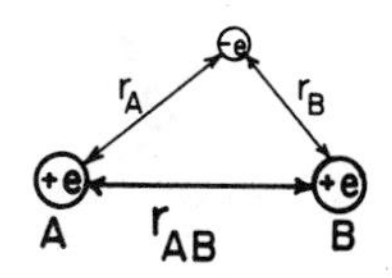

FIG. 8-1.3.1 Model of the H_2^+ ion.

$$\hat{\mathcal{H}} = -\frac{h^2}{8\pi^2 m_e}\nabla_e^2 + \left(-\frac{e^2}{r_A} - \frac{e^2}{r_B} + \frac{e^2}{r_{AB}}\right) \tag{8-1.3.2}$$

The equation we must solve ($\hat{\mathcal{H}}\Psi = E\Psi$) is

$$\hat{\mathcal{H}}\Psi = -\frac{h^2}{8\pi^2 m_e}\nabla_e^2\Psi + \left(-\frac{e^2}{r_A} - \frac{e^2}{r_B} + \frac{e^2}{r_{AB}}\right) = E\Psi \tag{8-1.3.3}$$

By a proper choice of coordinates (confocal elliptical, in this instance), we can cast the equation into a form that is more manageable. We will not do that. Rather, the results will be given (see Davis, Sec. 8-1.12, Ref. 3).

Perhaps the simplest, most reasonable assumption is that the H_2^+ ion wave function is made up of two individual H atom wave functions. Specifically, the 1s wave functions for the hydrogen atom will be used. This is the linear combination of atomic orbitals to form molecular orbitals (LCAO-MO) method. Using this assumption we can write (let $1s_A$ represent the 1s orbital of atom A, etc.)

$$\psi = \frac{1}{N'}(1s_A + 1s_B) \tag{8-1.3.4}$$

where N′ is a normalization constant. Note that $1s = (1/\sqrt{\pi})(1/a_0)^{3/2}\exp(-r/a_0)$ with $a_0 = 0.529$ Å. The energy can be calculated as in previous cases by evaluating

$$E = \frac{\int \psi^*\hat{\mathcal{H}}\psi\, d\tau}{\int \psi^*\psi\, d\tau}$$

$$= \frac{(1/N')^2 \int (1s_A^*\hat{\mathcal{H}}1s_A + 1s_A^*\hat{\mathcal{H}}1s_B + 1s_B^*\hat{\mathcal{H}}1s_A + 1s_B^*\hat{\mathcal{H}}1s_B)\, d\tau}{(1/N')^2 \int (1s_A^*1s_A + 1s_A^*1s_B + 1s_B^*1s_A + 1s_B^*1s_B)\, d\tau}$$

$$= \frac{H_{AA} + H_{AB} + H_{BA} + H_{BB}}{S_{AA} + S_{AB} + S_{BA} + S_{BB}} \tag{8-1.3.5}$$

where $H_{AA} = \int 1s_A^*\mathcal{H}\, 1s_A\, d\tau$, $S_{AB} = \int 1s_A^*1s_B\, d\tau$, etc.

From the symmetry of the H_2^+ ion, as seen in Fig. 8-1.3.1, we know A and B are interchangeable; also, $1s_A$ and $1s_B$ are normalized. Thus,

$$H_{AA} = H_{BB} \qquad H_{AB} = H_{BA} \qquad S_{AB} = S_{BA} \qquad S_{AA} = S_{BB} = 1$$

S_{AB} is called an overlap integral. Its value is related to the "overlap" or shared region in space between atoms A and B. We can now rewrite Eq. (8-1.3.5):

$$E = \frac{2H_{AA} + 2H_{AB}}{2 + 2S_{AB}} = \frac{H_{AA} + H_{AB}}{1 + S_{AB}} \tag{8-1.3.6}$$

Each of these terms can be evaluated by methods beyond the scope of this treatment. For example, by substitution of (8-1.3.2) into the definition of H_{AA},

$$H_{AA} = \int 1s_A^* \hat{\mathcal{H}} \, 1s_A \, d\tau$$

$$= \int 1s_A^* \left(\frac{h^2}{8\pi^2 m_e} \nabla_e^2 - \frac{e^2}{r_A} \right) 1s_A \, d\tau + \int 1s_A^* \left(- \frac{e^2}{r_B} + \frac{e^2}{r_{AB}} \right) 1s_A \, d\tau$$

$$= E_{1s} - \int 1s_A^* \frac{e^2}{r_B} 1s_A \, d\tau + \int 1s_A^* \frac{e^2}{r_{AB}} 1s_A \, d\tau$$

The last term has an operator that does not depend on A (r_{AB} is fixed), so it can be factored out. The second term is defined as the *coulomb integral* and is given the symbol J. Thus, (8-1.3.7) can be written as

$$H_{AA} = E_{1s} + \frac{e^2}{r_{AB}} + J \tag{8-1.3.8}$$

Similarly, H_{AB} can be written as

$$H_{AB} = E_{1s} S_{AB} + \frac{e^2}{r_{AB}} + K \tag{8-1.3.9}$$

where $K = \int 1s_A(-e^2/r_A) \, 1s_B \, d\tau$. K is called the *exchange integral* since it involves electron motion about both nuclei. Both J and K can be evaluated, as can S_{AB}. The final energy expression (in terms of J, K, and S_{AB}), from Eq. (8-1.3.5), is

$$E = E_{1s} + \frac{e^2}{r_{AB}} + \frac{J + K}{1 + S} \tag{8-1.3.10}$$

The exchange integral K is found to be negative, while J is positive. Equation (8-1.3.10) can be solved for several given r_{AB}, yielding corresponding values of E. These can be plotted as shown in Fig. 8-1.3.2. E_s corresponds to Eq. (8-1.3.10). As you can see, a stable molecule is predicted for some values of r_{AB}.

We could just as easily have chosen our trial wave function as

$$\psi' = \frac{1}{N'} (1s_A - 1s_B) \tag{8-1.3.11}$$

The corresponding expression for E is

$$E_a = E_{1s} + \frac{e^2}{r_{AB}} + \frac{J - K}{1 - S} \tag{8-1.3.12}$$

This is also plotted in Fig. 8-1.3.2. The symbols s and a, used as subscripts denote symmetric and antisymmetric. Equation (8-1.3.4) is *symmetric* with respect to exchange of A and B. Equation (8-1.3.11) changes sign under the same exchange and is called *antisymmetric*. We have an important result. When we write a function that allows the electron to be "shared" or "resonate" between the two nuclei, we predict a stable molecule. The results are not very accurate, as you would probably

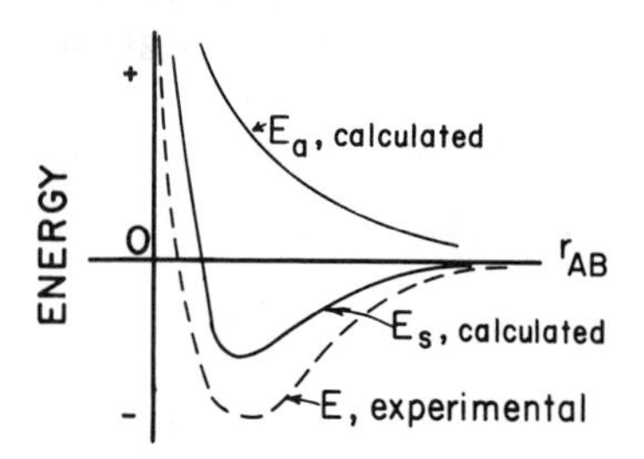

FIG. 8-1.3.2 Calculated and experimental potential energy curves for H_2^+: E_s from Eq. (8-1.3.10) and E_a from Eq. (8-1.3.12).

guess. E_s is calculated as -1.76 eV at r_{AB} = 1.32 Å (the minimum) whereas the experimental value is E = -2.79 eV at r_{AB} = 1.06 Å. Improvements can be made in our trial function. The parameter Z′ can be introduced in our 1s wave function as we did in Sec. 7-4.3. This gives E_s = -2.25 eV and r_{AB} = 1.06 Å for Z′ = 1.228. Further improvement can be obtained by including hydrogen 2p orbital wave functions in addition to the 1s functions. As always, as the wave function improves, the amount of effort required to calculate the result increases rapidly.

Excited (higher energy) states of H_2^+ can be treated in a similar fashion to calculate their energies. Also, the electron distribution for each state can be calculated. The results are given in Fig. 8-1.3.3. Note that all these are approximate solutions. The symbols on the molecular orbitals need some explanation. We will make use of these later, after a discussion of the H_2 molecule. A σ orbital corresponds to a direct overlap of two atomic orbitals, while a π orbital occurs where two p orbitals approach with their orbital axes parallel to each other and perpendicular to the A-B bond axis. The symbol * indicates an antibonding orbital. An antibonding orbital is one that is unstable for all values of r_{AB}. At no internuclear distance is the energy less than the separated atoms. A g or u subscript corresponds to *gerade* (symmetric) and *ungerade* (asymmetric), respectively, with respect to changing all coordinates to their negative (x, y, z → -x, -y, -z). This is equivalent to a reflection of every point in the figure through a central point in the figure (also called *inversion*, see Sec. 14-4). More will be said about these symbols following the H_2 molecule treatment.

8-1.4 MOLECULAR ORBITAL METHODS: THE H_2 MOLECULE

The extension of the discussion given for H_2^+ is straightforward, even though the calculations become very complex. In the H_2 molecule (take a moment and draw a model of H_2 similar to Fig. 8-1.3.1 with another electron added), in addition to electron-nucleus attractions and nucleus-nucleus repulsion, there is the electron-electron

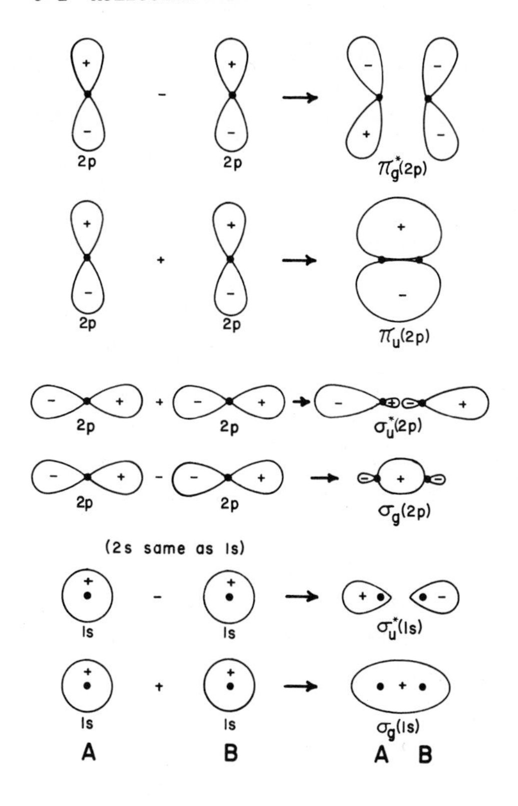

FIG. 8-1.3.3 LCAO molecular orbital wave functions and corresponding electronic probability distributions for the H_2^+ ion (signs represent symmetry of the function).

repulsion term that was so troublesome in the He atom. In fact, an exact solution is impossible, even with the Born-Oppenheimer approximation.

The procedure followed is the same as in the H_2^+ case. It is just more difficult. The trial functions we will use are the same as Eq. (8-1.3.4). Each electron is assigned to an orbital described by ψ. Thus (in all cases below, the electron spin functions are being ignored),

$$\psi(1) = \frac{1}{N'}\,[1s_A(1) + 1s_B(1)]$$
$$\psi(2) = \frac{1}{N'}\,[1s_A(2) + 1s_B(2)] \qquad (8\text{-}1.4.1)$$

To find a function for the Schrödinger equation we must take the product of $\psi(1)$ and $\psi(2)$:

$$\psi_I = \frac{1}{N}\,\psi(1)\ \psi(2) \qquad (8\text{-}1.4.2)$$

where N is a constant for normalization. If this is expanded (do it)

$$\psi_I = \frac{1}{N^2}[1s_A(1)\ 1s_A(2) + 1s_B(1)\ 1s_B(2)] + \frac{1}{N^2}[1s_A(1)\ 1s_B(2) + 1s_A(2)\ 1s_B(1)]$$

$$= \psi_{ionic} + \psi_{cov}$$

where cov indicates covalent. The first two terms are *ionic terms*. Both electrons are on one nucleus (corresponding to $H_A^+ H_B^-$ or $H_A^- H_B^+$; verify that). The second two are *covalent terms* with the electron being shared. The energy calculated from this function is not very good. This is not surprising since it gives both covalent and ionic terms equal weight. If we ignore the ionic terms completely we obtain much better agreement with experimental values. The resulting energy for the covalent function is shown in Fig. 8-1.4.1. An even better result is obtained by putting a variable coefficient λ in the ionic terms and minimizing the energy with respect to it. This procedure results in a value of $\lambda = 0.16$. The energy calculated for this function is also given in Fig. 8-1.4.1 (labeled as $\psi_{cov} + \lambda\psi_{ionic}$).

The wave functions we have chosen lead to a net binding energy (an energy lower than the separated atoms). We could just as easily have chosen our initial wave function as the difference between atomic wave functions.

$$\psi'(1) = \frac{1}{N}[1s_A(1) - 1s_B(1)]$$
$$\psi'(2) = \frac{1}{N}[1s_A(2) - 1s_B(2)] \qquad (8\text{-}1.4.1)$$

and $$\psi_{II} = \frac{1}{N'}\psi(1)\ \psi(2)$$

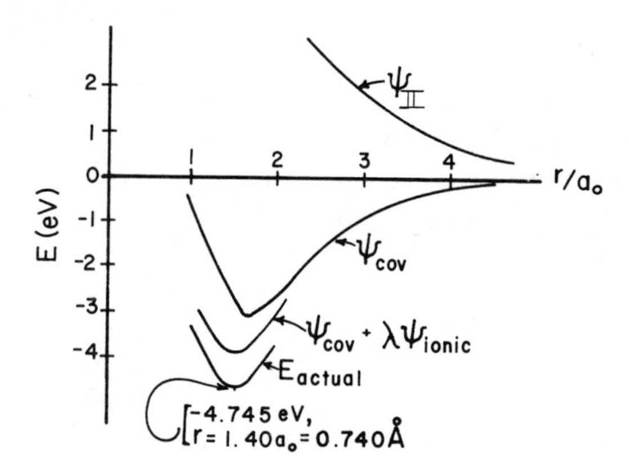

FIG. 8-1.4.1 The potential (binding) energy of H_2 for several approximate wave functions.

As shown in Fig. 8-1.4.1, this wave function always gives energy values larger than the energy of the separated atoms. Thus, a molecule is not stable if it is described by ψ_{II}. This is an *antibonding orbital*, whereas ψ_I represents a *bonding orbital*.

It is possible to improve the results further by including a variable parameter Z' in the 1s wave functions used, then minimizing the energy with respect to it. Also, if excited H_2^+-type molecular orbitals are included, further improvement is realized. In each case, the computational complexity increases. There are multiparametic wave functions that closely approximate the experimental potential energy curve. However, these are not readily applicable to other diatomic molecules. After a brief look at another method of constructing wave functions for H_2 we will return to the LCAO-MO method for diatomic molecules.

8-1.5 THE VALENCE-BOND DESCRIPTION OF H_2

Above, we assumed that molecular orbitals were formed, then electrons were added to make the molecule. Another approach is to assume each atom retains its identity as the molecule is formed. That is, let electron 1 remain on atom A and electron 2 on atom B. A wave function of this type is

$$\psi_I' = 1s_A(1)\ 1s_B(2) \tag{8-1.5.1}$$

This wave function gives an energy minimum of -0.4 eV at about 1 Å compared to the experimental value of -4.745 eV at 0.740 Å. Obviously, this is poor agreement, but it does predict a stable system. You might be able to suggest how to improve the wave function. (Can you?)

One way is to allow electron sharing by letting

$$\psi_{II}' = 1s_A(2)\ 1s_B(1)$$

and combining ψ_I' and ψ_{II}' to give

$$\psi_S' = \frac{1}{N}\left(\psi_I' + \psi_{II}'\right)$$

$$= \frac{1}{N}\,[(1s_A(1)\ 1s_B(2) + 1s_A(2)\ 1s_B(1)] \tag{8-1.5.2}$$

N is a normalization constant. This function (called the Heitler-London or valence bond function) gives an energy minimum of E_S = -3.140 eV at 0.869 Å. The result is much better than using ψ_I' or ψ_{II}' alone. Note one shortcoming of the ψ_S' function. It contains only covalent terms: there are no ionic terms in it. Addition of $\psi_{III}' = 1s_A(1)\ 1s_A(2)$ and $\psi_{IV}' = 1s_B(1)\ 1s_B(2)$ improves the result. However, when we do that we have an expression identical to Eq. (8-1.4.3) derived from LCAO-MO considerations.

8-1.6 ELECTRON SPIN FUNCTIONS

In all the above, we have ignored the spin of the electron. A brief discussion of spin was given in Sec. 7-3.5. However, no detail was given on how to include electron spin wave functions in the atomic wave functions. The spin functions must be included to ensure that the total molecular wave function obeys the *Pauli exclusion principle*. The Pauli principle (previously stated as: "no two electrons on an atom can have the same values for all four quantum numbers") can be stated as: "Every allowable wave function for a system of two or more electrons must be antisymmetric (change sign) for the simultaneous interchange of the position and spin coordinates of any pair of electrons."

We can easily write the spin functions for an electron. Each electron is in one of two possible states of angular momentum. It has either $+\frac{1}{2}(h/2\pi)$ or $-\frac{1}{2}(h/2\pi)$ units of angular momentum with respect to a z axis defined by an external magnetic field (see Sec. 7-3.5 for a discussion of the effect of a magnetic field on electron spin). The wave functions are simple to write. The value of the spin quantum number m_s of $+\frac{1}{2}$ is associated with a wave function designated as α. Conversely, $m_s = -\frac{1}{2}$ is associated with the wave function denoted by β. Remember $m_s = \pm\frac{1}{2}$ are the spin angular momenta (in units of $h/2\pi$) projected on a z axis defined by a magnetic field.

The complete wave function for an atom or molecule must include α or β. For example, the 1s function for the hydrogen atom must be written $\psi_{1s}\alpha$ or $\psi_{1s}\beta$. For a molecule containing two electrons there are several possible combinations of spin functions. These are

$$\alpha(1)\alpha(2) \qquad \beta(1)\beta(2) \qquad [\alpha(1)\beta(2) + \alpha(2)\beta(1)] \tag{8-1.6.1}$$

$$[\alpha(1)\beta(2) - \alpha(2)\beta(1)] \tag{8-1.6.2}$$

The first three functions are symmetric with respect to interchange of electrons 1 and 2 $[\alpha(1)\alpha(2) = \alpha(2)\alpha(1)]$ while the last function is antisymmetric (show that $[\alpha(1)\beta(2) - \alpha(2)\beta(1)] = -[\alpha(2)\beta(1) - \alpha(1)\beta(2)]$). Finally, you may wonder why other functions are not included. They are omitted since any other combination is neither symmetric nor antisymmetric. For example, $\alpha(1)\beta(2)$ does not equal either $\alpha(2)\beta(1)$ or $-\alpha(2)\beta(1)$, and is therefore not a valid wave function.

Now, we are ready to modify our previous molecular orbital wave functions to make them consistent with the Pauli principle. Look at Eq. (8-1.5.2). It is symmetric with respect to interchange of electrons 1 and 2. For the total wave function to be antisymmetric we must multiply ψ_s by the antisymmetric spin function (8-1.6.2). This gives

$$\psi_1 = \frac{1}{N}[1s_A(1)1s_B(2) + 1s_A(2)1s_B(1)][\alpha(1)\beta(2) - \alpha(2)\beta(1)] \tag{8-1.6.3}$$

You should show that this is antisymmetric. We could have chosen

$$\psi_a = \frac{1}{N}\left(\psi_I' - \psi_{II}'\right) = \frac{1}{N}\,[1s_A(1)1s_B(2) - 1s_A(2)1s_B(1)]$$

which is antisymmetric. The total wave function will be ψ_a times a symmetric spin function. Thus,

$$\psi_2 = \frac{1}{N}\,[1s_A(1)1s_B(2) - 1s_A(2)1s_B(1)] \times [\alpha(1)\alpha(2)]$$

$$\text{or} \quad \times [\beta(1)\beta(2)]$$

$$\text{or} \quad \times [\alpha(1)\beta(2) + \alpha(2)\beta(1)]$$

There are three functions with symmetric spin functions and an antisymmetric space function. Such a state is called a *triplet state* (due to the three functions). On the other hand, ψ_1 is described by only one spin function and is called a *singlet state*. For more than two electrons, there are ways of systematically writing appropriate space and spin functions. See Davis for further details (Sec. 8-1.12, Ref. 3).

8-1.7 CORRELATION DIAGRAMS FOR DIATOMIC MOLECULES

In Sec. 7-3 we saw how we used the simple H atom solution to produce the periodic table of the elements. Even though the electronic interactions in multi-electron atoms are complex, the number of electronic states and their fundamental symmetry are the same as calculated for the H atom. A similar approach is useful for molecules. *Correlation diagrams* give the number of electronic states in a diatomic molecule with corresponding quantum numbers for the molecular orbitals. These molecular orbitals are correlated with atomic electronic states for the separated atoms and the combined atoms.

We have already seen, in Fig. 8-1.3.3, how separate atomic orbitals can be merged to form molecular orbitals in H_2^+. This can be carried one step further. We can allow the nuclei to merge into a single nucleus with z = 2, which would correspond to a He^+ atomic orbital. This is not to say the nuclei can actually overlap. The cannot, due to internuclear repulsion. However, we can conveniently ignore these repulsions in generating our correlation diagrams. We see, then, that the two 1s orbitals in H_2^+ combine to form a σ_g molecular orbital. If we allow the nuclei to overlap (r_{AB} = 0), we form an He^+-like atomic orbital which has 1s properties. We denote the combined atomic orbital as $1s\sigma_g$ and the separated atomic orbital as $\sigma_g(1s)$ to indicate the connection between the two. Before considering the actual diagrams, let us work through another example. Choose the 2p orbitals on the separated atoms which combine to form the π_u molecular orbital in Fig. 8-1.3.3. If the two atoms form a united atom we have a He^+-like 2p orbital. Thus, we correlate $\pi_u(2p)$ on the separated atoms with $2p\pi_u$ on the united atoms.

Figure 8 1.7.1 shows the correlation diagrams for homonuclear diatomics, and Fig. 8-1.7.2 gives the diagram for heteronuclear diatomics. The utility of these diagrams should be apparent. Just as the periodic chart gave us the atomic orbital filling sequence, these correlation diagrams give the molecular orbital filling sequences. Figure 8-1.7.3 is readily obtained from Fig. 8-1.7.1. The use of the diagrams is simple. Use the Aufbau principle and add electrons to the lowest available energy level, as long as the Pauli exclusion principle (in slightly different form) is not violated. That is, no two electrons can have the same spin if they are in the same orbital. A few examples will show the techniques.

8-1.7.1 EXAMPLE

Determine the molecular orbital designations for He_2, N_2, and O_2. Give the bond order and the number of unpaired electrons in each.

Solution

He_2

He_2 has four electrons. The first goes into $\sigma(1s)$ with a given spin (see below: the arrow designates a spin direction). The second electron goes into $\sigma(1s)$ with the opposite spin.

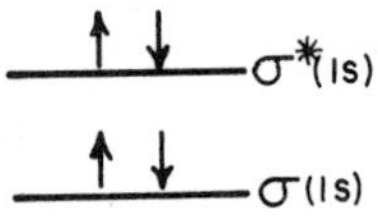

↑ ↓ —— $\sigma^*(1s)$

↑ ↓ —— $\sigma(1s)$

Each orbital can only hold two electrons, so the third and fourth must go into the next higher level, the $\sigma^*(1s)$. Our designation is $\sigma(1s)^2\sigma^*(1s)^2$, and there are no unpaired electrons. The bond order will be given by

$$\frac{\text{No. of bonding electrons - No. of antibonding electrons}}{2} = \frac{N_B - N_{AB}}{2}$$

$$= \frac{2 - 2}{2} = 0$$

There is a bond order of zero and therefore no tendency for He_2 to form a stable compound.

N_2

Figure 8-1.7.3 (b) fits this case. The first eight electrons fill the

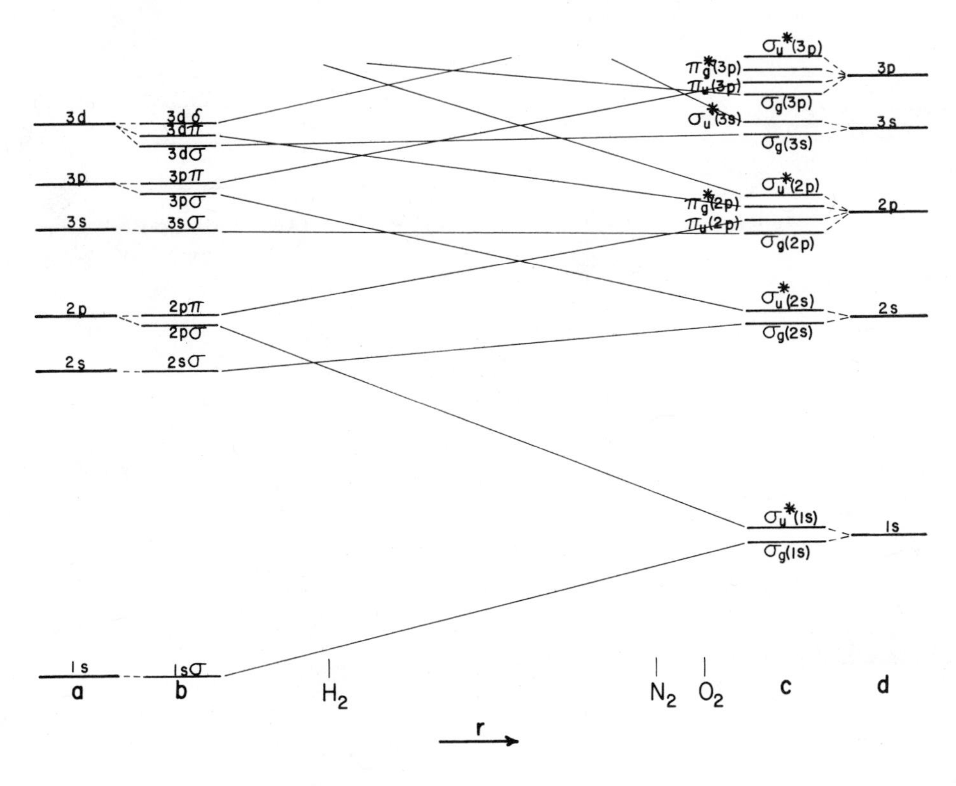

FIG. 8-1.7.1 Correlation diagram for homonuclear diatomic molecules.

σ(1s)σ*(1s)σ(2s)σ*(2s) orbitals completely. This leaves 14 -8 = 6 electrons for the molecular orbitals derived from the p orbitals.

σ*(2p)
π*(2p)
σ(2p)
π(2p)

The ninth electron goes in one π(2p) orbital, and the 10th in the other as shown above. (Electrons will remain unpaired as long as possible.) The 11th and 12th electrons pair up in the π(2p) orbitals, completely filling them. The last two electrons fill the σ(2p). The completed diagram is given below.

σ*(2p)
π*(2p)
σ(2p)
π(2p)

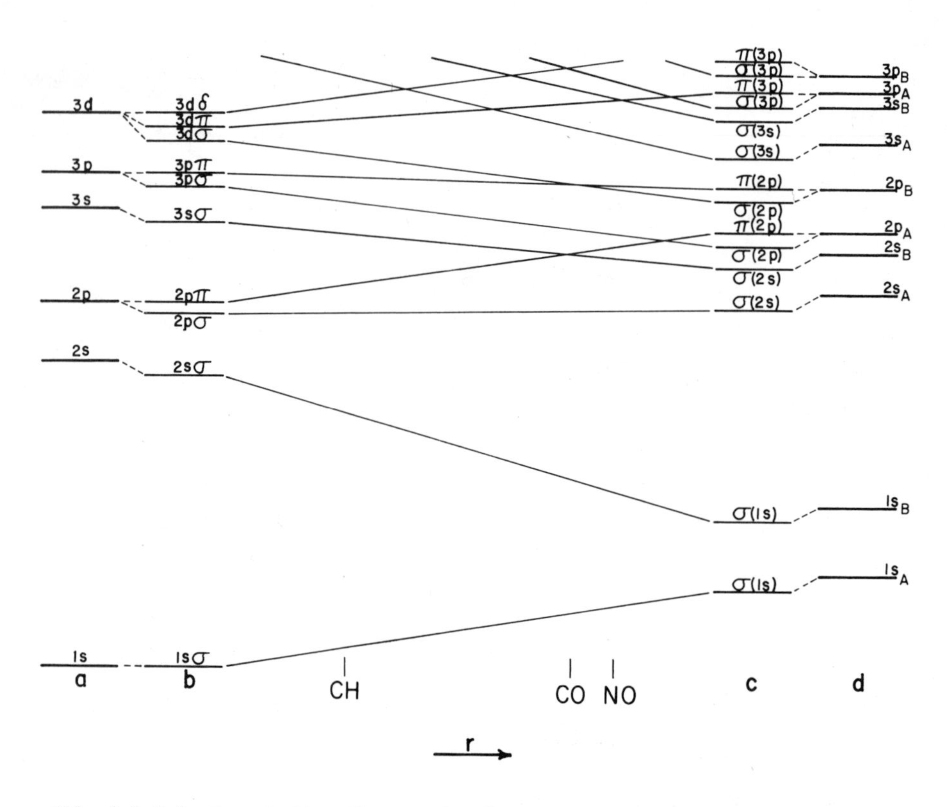

FIG. 8-1.7.2 Correlation diagram for heteronuclear diatomic molecules.

there are no unpaired electrons. Bond order = (10 -4)/2 (remember the levels not shown). The designation is

$$\sigma(1s)^2\sigma^*(1s)^2\sigma(2s)^2\sigma^*(2s)^2\pi^*(2p)^2\pi(2p)^2\sigma(2p)^2$$

O_2

This is left as an exercise for you to do. Use part (a) of Fig. 8-1.7.3. The designation is

$$\sigma(1s)^2\sigma^*(1s)^2\sigma(2s)^2\sigma^*(2s)^2\sigma(2p)^2\pi(2p)^2\pi(2p)^2\pi^*(2p)^1\pi^*(2p)^1$$

Next we need to consider further the shapes of the orbitals formed. The following gives a little more detail about Fig. 8-1.3.3.

Shapes. The easiest way to visualize these orbitals is to assume we bring atomic orbitals together and let them interact. We can do this pictorially without great

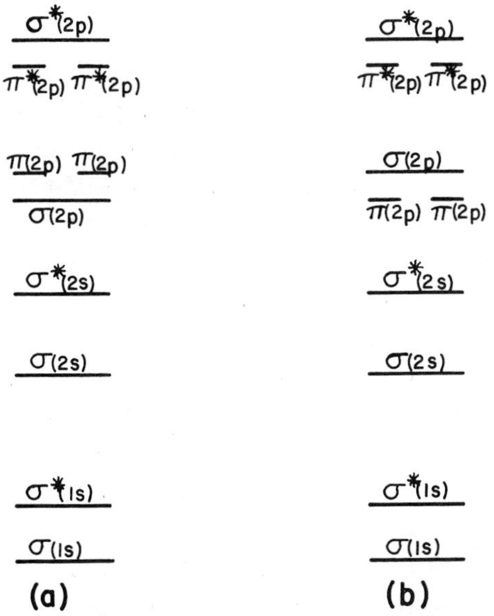

(a)	(b)
$\sigma^*(2p)$	$\sigma^*(2p)$
$\pi^*(2p)\ \pi^*(2p)$	$\pi^*(2p)\ \pi^*(2p)$
$\pi(2p)\ \pi(2p)$	$\sigma(2p)$
$\sigma(2p)$	$\pi(2p)\ \pi(2p)$
$\sigma^*(2s)$	$\sigma^*(2s)$
$\sigma(2s)$	$\sigma(2s)$
$\sigma^*(1s)$	$\sigma^*(1s)$
$\sigma(1s)$	$\sigma(1s)$

FIG. 8-1.7.3 Energy level diagrams for homonuclear diatomic molecular orbitals: (a) for O, F, and Ne, and (b) for Li, Be, B, C, and N. The asterisk (*) indicates antibonding orbitals. The separation between orbitals varies from molecule to molecule.

difficulty. There is one additional point that must be made before doing that. This concerns symmetry. As presented in Sec. 8-1.6, the functions making up atomic and molecular orbitals are *symmetrical* or *antisymmetrical* with respect to a change in sign of all the coordinates. If a function has the property of being *symmetrical* $f(x, y, z) = f(-x, -y, -z)$, it is said to be *gerade* g. If it is *antisymmetrical* $f(x, y, z) = -f(-x, -y, -z)$, it is *ungerade* u. Pictorially this is equivalent to inversion of a figure through a central spherical mirror. Finally, we must consider the sign of the function. Figure 8-1.3.3 is a representation of s and p orbitals. Note the signs included. An s function is always positive. A p function is positive for positive values of the coordinate involved and negative for negative values of the coordinate. If we bring two orbitals together to add them we take the sign of the function into account as shown in Fig. 8-1.3.3. Note in all cases the bonding molecular orbitals end up with electron density built up between the nculei. Such a situation is bonding due to the electron between the two nuclei being attracted to both. In antibonding orbitals electron density is found outside the volume between the nuclei and therefore does not contribute to bonding. In fact, this lack of electron density leads to an increased internuclear repulsion which makes the molecule unstable with respect to the separated atoms if these orbitals are filled.

8-1.8 HYBRIDIZATION

As you know from other studies in chemistry, molecules have definite shapes. This implies there is some directional character to the bonds formed about an atom. There are two ways to approach this subject. One is a long, involved procedure, the other is a geometrical approach. We will take the geometrical route. Let us accept the fact that covalent bonds have directional character and try to formulate *hybrid* orbitals that fit those directions. No theory is involved. The task is one of geometry: given a set of atomic orbitals and a set of bonding orbitals with specified directions, formulate a set of hybrid orbitals to fit the requirements. It sounds baffling, but it is actually easy to do. This section should be fun. Even so, one important concept will be developed. That concept is symmetry. Molecules possess certain symmetry properties (see Sec. 14-4 for a discussion of molecular symmetry), and the functions used to describe these molecules also have symmetry. Symmetry arguments will be used to simplify the construction of hybrid orbitals below. The assumption below is that you have heard of sp, sp^2, and sp^3 hybridization before. These terms imply that we take an s function and combine it with one, two, or three p functions to give two, three, or four hybrid orbitals. As you will see, we choose our directions arbitrarily. After you finish this section you should be able to construct hybrid orbitals that point in any desired direction.

HYBRIDS. Let us assume we want sp^2 orbitals formed that have the following orientation to a defined axis system.

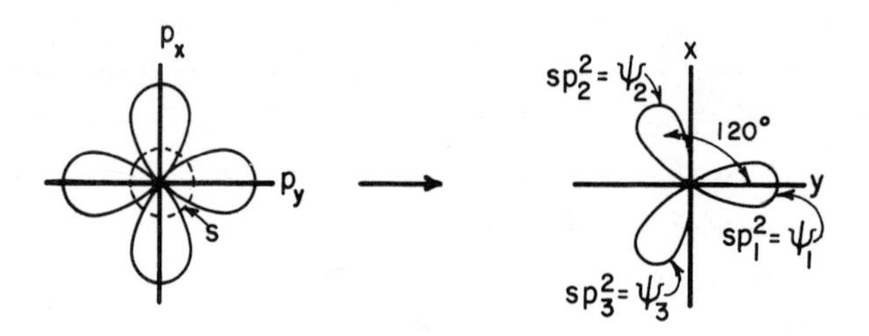

There will be a p function not involved at all perpendicular to the xy plane. The following rules will apply.

1. LCAO-MO approach is used

$$\Psi_j(\text{hybrid}) = \sum_{i=1}^{\text{a.o.}} C_{ji}\,\Phi_i$$

2. To be normalized

$$\sum_j C_{ij}^2 = 1 \qquad \sum_i C_{ji}^2 = 1$$

3. The ϕ_s contribution to all hybrids is the same.
4. An atomic orbital perpendicular to those chosen as hybrids will not contribute to the hybrid.

Now let us assume we are using 2s and 2p functions, so we will leave the 2 out of the designation. Let $\phi_{p_x} \equiv \phi_x$, $\phi_{p_y} \equiv \phi_y$, and $\phi_{p_z} \equiv \phi_z$. Apply Rule (1):

$$\psi_1 = C_{1s}\phi_s + C_{1x}\phi_x + C_{1y}\phi_y + C_{1z}\phi_z$$

$$\psi_2 = C_{2s}\phi_s + C_{2x}\phi_x + C_{2y}\phi_y + C_{2z}\phi_z$$

$$\psi_3 = C_{3s}\phi_s + C_{3x}\phi_x + C_{3y}\phi_y + C_{3z}\phi_z$$

$$\psi_4 = C_{4s}\phi_s + C_{4x}\phi_x + C_{4y}\phi_y + C_{4z}\phi_z$$

We have already stated that ψ_4 is just the ϕ_z function, so $C_{4s} = C_{4x} = C_{4y} = 0$. Also, by Rule (4) $C_{3z} = C_{2z} = C_{1z} = 0$ since ϕ_z is not involved in the hybrids. The set of orbitals are simplified to

$$\psi_1 = C_{1s}\phi_s + C_{1x}\phi_x + C_{1y}\phi_y$$

$$\psi_2 = C_{2s}\phi_s + C_{2x}\phi_x + C_{2y}\phi_y$$

$$\psi_3 = C_{3s}\phi_s + C_{3x}\phi_x + C_{3y}\phi_y$$

$$\psi_4 = \phi_z$$

Apply Rule (3) and Rule (2):

$$C_{1s} = C_{2s} = C_{3s} \quad \text{and} \quad C_{1s}^2 + C_{2s}^2 + C_{3s}^2 = 1$$

You should solve this to show $C_{1s} = 1/\sqrt{3}$. We now have

$$\psi_1 = \frac{1}{\sqrt{3}}\phi_s + C_{1x}\phi_x + C_{1y}\phi_y$$

$$\psi_2 = \frac{1}{\sqrt{3}}\phi_s + C_{2x}\phi_x + C_{2y}\phi_y$$

$$\psi_3 = \frac{1}{\sqrt{3}}\phi_s + C_{3x}\phi_x + C_{3y}\phi_y$$

$$\psi_4 = \phi_z$$

Note that ψ_1 is perpendicular to x, so by Rule (4) p_x will not contribute to $\psi_1(C_{1x} = 0)$. Then ψ_1 can be written

$$\psi_1 = \frac{1}{\sqrt{3}}\phi_s + C_{1y}\phi_y$$

And by Rule (2),

$$\left(\frac{1}{\sqrt{3}}\right)^2 + (C_{1y})^2 = 1$$

which yields (prove it)

$$C_{1y} = \left(\frac{2}{3}\right)^{\frac{1}{2}}$$

Our orbitals are now

$$\psi_1 = \frac{1}{\sqrt{3}}\phi_s \qquad + \left(\frac{2}{3}\right)^{\frac{1}{2}}\phi_y$$

$$\psi_2 = \frac{1}{\sqrt{3}}\phi_s + C_{2x}\phi_x + C_{2y}\phi_y$$

$$\psi_3 = \frac{1}{\sqrt{3}}\phi_s + C_{3x}\phi_x + C_{3y}\phi_y$$

$$\psi_4 = \phi_z$$

Next we need to apply some symmetry considerations. Note the ψ_2 and ψ_3 are both 60°from the y axis. This means the projections of ψ_2 and ψ_3 on y are both the same, so $C_{2y} = C_{3y}$. If we apply Rule (2) to column 3 (the y column) we get

$$\left[\left(\frac{2}{3}\right)^{\frac{1}{2}}\right]^2 + C_{2y}^2 + C_{3y}^2 = \frac{2}{3} + 2C_{2y}^2 = 1 \qquad \text{or} \qquad C_{2y}^2 = \frac{1}{6}$$

Then, $C_{2y} = \pm\ (1/6)^{\frac{1}{2}}$. But, note, these are both negative as seen from the figure. The orbitals are now

$$\psi_1 = \frac{1}{\sqrt{3}}\phi_s \qquad + \left(\frac{2}{3}\right)^{\frac{1}{2}}\phi_y$$

$$\psi_2 = \frac{1}{\sqrt{3}}\phi_s + C_{2x}\phi_x - \left(\frac{1}{6}\right)^{\frac{1}{2}}\phi_y$$

$$\psi_3 = \frac{1}{\sqrt{3}}\phi_s + C_{3x}\phi_x - \left(\frac{1}{6}\right)^{\frac{1}{2}}\phi_y$$

$$\psi_4 = \phi_z$$

There are several ways to get C_{2x} and C_{2y}. Three ways will be shown.

1. From Rule (2), $C_{2x}^2 + C_{3x}^2 = 1$, and since both ψ_2 and ψ_3 are 30° from the x axis, the projection on x will be the same (except for sign). Thus $|C_{2x}| = |C_{3x}|$. This gives $C_{2x} = 1/\sqrt{2}$, $C_{3x} = -1/\sqrt{2}$.

2. Look at ψ_2 alone. The tangent of the angle, 120°, is C_{2x}/C_{2y}. Thus $C_{2x} = C_{2y}\tan(120°) = -(1/6)^{\frac{1}{2}}(-1.732) = (1/6)^{\frac{1}{2}}\sqrt{3} = (1/2)^{\frac{1}{2}}$. As before, $C_{3x} = -C_{2x}$.

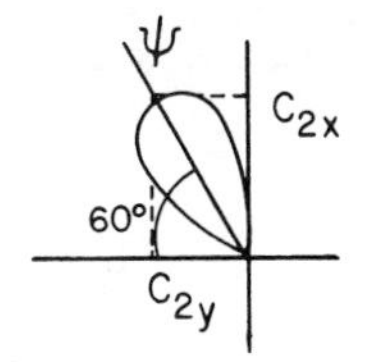

3. We could apply Rule (2) to the second row, getting

$$\left(\frac{1}{\sqrt{3}}\right)^2 + C_{2x}^2 + \left(-\frac{1}{\sqrt{6}}\right)^2 = \frac{1}{3} + C_{2x}^2 + \frac{1}{6} = 1$$

$C_{2x} = 1/\sqrt{2}$ as before.

All three ways were used to show you there are many relationships among the coefficients that can be used. The final orbitals are:

$$\psi_1 = \frac{1}{\sqrt{3}}\phi_s \qquad + \left(\frac{2}{3}\right)^{1/2}\phi_y$$

$$\psi_2 = \frac{1}{\sqrt{3}}\phi_s + \frac{1}{\sqrt{2}}\phi_x - \frac{1}{\sqrt{6}}\phi_y$$

$$\psi_3 = \frac{1}{\sqrt{3}}\phi_s - \frac{1}{\sqrt{2}}\phi_x - \frac{1}{\sqrt{6}}\phi_y$$

$$\psi_4 = \phi_z$$

These functions reproduce the shape of the orbitals we wanted. The functions could be used for calculating properties such as energy, dipole moment (defined in Sec. 14-2), etc. The point of this is to show there is nothing magical about the hybrid orbitals you have seen before. They are constructed to fit known geometry and have no significance apart from their usefulness. Before discussing shapes you should try a couple of constructions.

8-1.8.1 EXAMPLE

Construct an sp hybrid orbital with the following orientation.

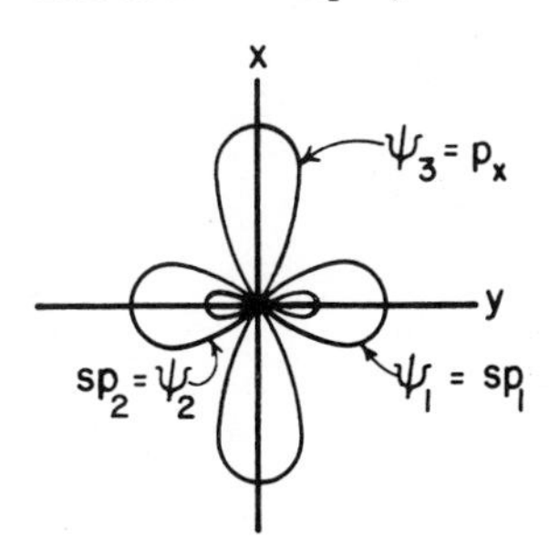

$\psi_4 = p_z \perp$ to xy plane

Do not look below until you have tried it. At least prove the solution fits.

Solution

The solution is

$$\psi_1 = \frac{1}{\sqrt{2}}\phi_s + \frac{1}{\sqrt{2}}\phi_y$$

$$\psi_2 = \frac{1}{\sqrt{2}}\phi_s - \frac{1}{\sqrt{2}}\phi_y$$

$$\psi_3 = \phi_x$$

$$\psi_4 = \phi_z$$

8-1.8.2 EXAMPLE

Construct sp^3 hybrids with the following orientation

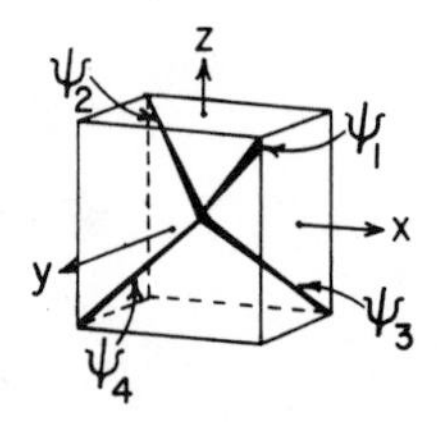

Solution

The solution to this is not given. It is simple, even though it looks difficult. Apply symmetry arguments.

SHAPES.

sp^2. The shapes of the orbitals you may have seen in an organic chemistry course are easily derived from the hybrids. Assume we have two atoms we have assigned sp^2 hybridization. If these come together we get the following (side view).

The overlap of the two sp^2 orbitals gives a σ bond, and the sideways overlap of the p_z orbitals $\perp$ to the sp^2 gives a π bond. Together these give a double bond. The simplest example is ethylene, which we write as

```
H       H
 \     /
  C = C
 /     \
H       H
```

The hydrogens are bonded to the carbon by overlap of the hydrogen s orbital with an sp^2 orbital on carbon (draw it).

sp. In an sp hybrid we have two sp hybrids and two p orbitals perpendicular to these on each hybridized atom.

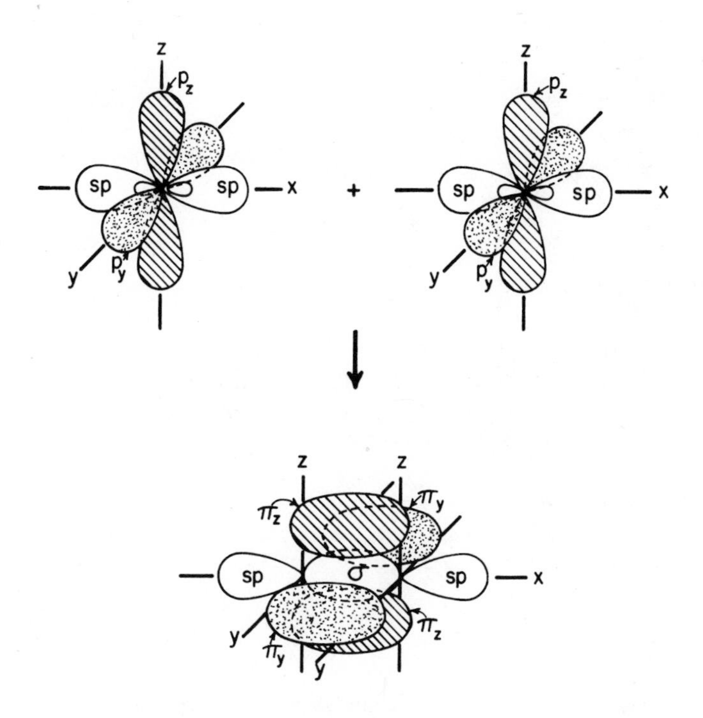

The p_z orbitals form π_z with a lobe above and below the xy plane. The p_y orbitals overlap to form a π_y orbital above and below the xz plane. The sp orbitals pointing inward overlap to form a σ bond. This gives a triple bond. Also, sp hybrids point out from each end to form bonds with other atoms. The simplest example is acetylene, H-C≡C-H.

sp³. If all three p orbitals are used to form a hybrid with an s orbital, sp^3 results. These orbitals have the directional character shown in Example 8-1.8.2. If you like geometry, prove from that figure that the angle between orbitals is 109.5°. This is known as a tetrahedral shape. The simplest example with this geometry is CH_4.

CLOSING COMMENTS. Some texts to into a long discussion of why some molecules such as methane are perfect tetrahedra, and others such as H_2O with an H–O–H angle of about 105° deviate from tetrahedral shape. The results for water are interpreted in terms of deviations from sp^3 hybridization. This is an unnecessary exercise. Molecules take on the shape that leads to the lowest energy for the system. The fact that we can write simple combinations of atomic orbital functions (themselves combinations of solutions found for the hydrogen atom) is fortuitous. All these orbitals are mental constructs that very likely bear no resemblance to reality. They are useful in that they allow us to predict certain molecular properties. The value of the above lies in the fact that we can combine some not too accurate one-electron wave functions into new functions. The new functions (in many cases) can give us accurate energies, electron distributions, and other properties. Thus, even though we are constructing mental orbitals, the results predicted by these functions are real and significant.

8-1.9 HÜCKEL METHOD FOR CALCULATING RESONANCE ENERGY

It is observed that many organic (and other) compounds with certain sequences of double bonds are more stable than would be predicted at first glance. For example, if we measure the heat liberated when 3 mol of cyclohexene are hydrogenated to give 3 mol of cyclohexane we find $\Delta H = -359$ kJ:

$$3C_6H_{10}(g) + 3H_2(g) \rightarrow 3C_6H_{12}(g) \qquad \Delta H = -359 \text{ kJ}$$

Also, if benzene is completely hydrogenated by 3 mol of H_2 we find only 206 kJ liberated:

$$C_6H_6(g) + 3H_2(g) \rightarrow C_6H_{12} \qquad \Delta H = -206 \text{ kJ}$$

Not nearly as much energy is released when the three double bonds of benzene are hydrogenated. The difference of about 150 kJ is attributed to resonance stabilization. Resonance in quantum mechanics occurs when two or more wave functions we have written for a system combine to produce a new wave function that more nearly describes the system. Thus *resonance stabilization* is an artifact. It is a statement of our ignorance about the true wave function. If we could write the true wave function we could predict the true energy directly. Since we cannot usually do that we write several approximate wave functions and state that the true wave function is some combination of all these (resonance). When you draw the various bonding schemes in benzene (alternating double bonds) you are, in essence, writing wave functions, since there is a wave function that corresponds to each "resonance structure" you draw.

We have already seen that we can approximate molecular orbitals by linear combinations of atomic orbitals. For a molecule like LiH we could write

$$\Psi_{LiH} = a_1\phi_{1s(Li)} + a_2\phi_{1s(H)} + a_3\phi_{2s(Li)} + a_4\phi_{2p(Li)} + \cdots$$

Depending on the amount of work we want to do we can include as many terms as we like. We then minimize the energy calculated with respect to each coefficient by setting

$$\frac{\partial E}{\partial a_i} = 0 \tag{8-1.9.1}$$

This obviously involves considerable work, as can be seen from the following. Assume $\psi = \sum_i a_i\phi_i$

$$\hat{\mathcal{H}}\,\psi = E\psi \rightarrow \hat{\mathcal{H}}\sum_i a_i\phi_i = E\sum_i a_i\phi_i \tag{8-1.9.2}$$

Applying the variation theorem,

$$E = \frac{\int \left(\sum_j a_j\phi_j\right)^* \hat{\mathcal{H}} \sum_i a_i\phi_i \; d\tau}{\int \left(\sum_j a_j\phi_j\right)^* \sum_i a_i\phi_i \; d\tau} \tag{8-1.9.3}$$

Choose two real functions with real coefficients for illustration purposes.

$$E = \frac{\int (a_1\phi_1 + a_2\phi_2)\,\hat{\mathcal{H}}\,(a_1\phi_1 + a_2\phi_2)\; d\tau}{\int (a_1\phi_1 + a_2\phi_2)(a_1\phi_1 + a_2\phi_2)\; d\tau}$$

$$= \frac{\int (a_1^2\phi_1\hat{\mathcal{H}}\phi_1 + a_1a_2\phi_1\hat{\mathcal{H}}\phi_2 + a_2a_1\phi_2\hat{\mathcal{H}}\phi_1 + a_2^2\phi_2\hat{\mathcal{H}}\phi_2)\; d\tau}{\int (a_1^2\phi_1\phi_1 + a_1a_2\phi_1\phi_2 + a_2a_1\phi_2\phi_1 + a_2^2\phi_2\phi_2)\; d\tau} \tag{8-1.9.4}$$

To simplify this define

$$H_{ij} \equiv \int \phi_i^*\hat{\mathcal{H}}\phi_j \; d\tau$$

$$S_{ij} \equiv \int \phi_i^*\phi_j \; d\tau \qquad \text{the overlap integral} \tag{8-1.9.5}$$

Equation (8-1.9.4) simplifies (in appearance only) to

$$E = \frac{a_1^2 H_{11} + a_1 a_2 H_{12} + a_2 a_1 H_{21} + a_2^2 H_{22}}{a_1^2 S_{11} + a_1 a_2 S_{12} + a_2 a_1 S_{21} + a_2^2 S_{22}} \qquad (8\text{-}1.9.6)$$

Apply Eq. (8-1.9.1), and after considerable manipulation,

$$\begin{aligned} a_1(H_{11} - S_{11}E) + a_2(H_{12} - S_{12}E) &= 0 \\ a_1(H_{21} - S_{21}E) + a_2(H_{22} - S_{22}E) &= 0 \end{aligned} \qquad (8\text{-}1.9.7)$$

This can be solved by the corresponding *secular determinant*.

$$\begin{vmatrix} H_{11} - S_{11}E & H_{12} - S_{12}E \\ H_{21} - S_{21}E & H_{22} - S_{22}E \end{vmatrix} = 0 \qquad (8\text{-}1.9.8)$$

To this point nothing new has been done. This is just the variation procedure. Now, we need to apply this to molecules with π-electron systems. In this case the functions described above will apply to the π electrons only. The σ-bonded framework is ignored. Equation (8-1.9.8) would apply to ethylene since two functions are involved in forming the π system. Hückel first simplified the problem by the following definitions and assumptions (α and β are not spin functions).

1. $H_{ii} = \alpha$. We are considering at this point π systems in carbon compounds. We would expect all the integrals of the type $\int \phi_i^* \hat{\mathcal{H}} \phi_i \, d\tau$ to be equal.
2. $H_{ij} = \beta$ if i and j are adjacent atoms. $H_{ij} = 0$ if i and j are not adjacent. Here we are assuming that the π bond is between adjacent atoms only. This will possibly be a poor assumption in some cases.
3. $S_{ii} = 1$; the functions on a given atom are normalized.
4. $S_{ij} = 0$; the functions on adjacent and other atoms are orthogonal, sometimes a poor assumption.

For the above case, ethylene, the determinant takes on a very simple form. (Use ε_i to designate the energy of a π system.)

$$\begin{vmatrix} \alpha - \varepsilon_i & \beta \\ \beta & \alpha - \varepsilon_i \end{vmatrix} = 0 \qquad (8\text{-}1.9.9)$$

This can be solved to give

$$(\alpha - \varepsilon_i)^2 - \beta^2 = 0$$

from which $\varepsilon_i = \alpha \pm \beta$. Thus

$$\varepsilon_1 = \alpha + \beta \qquad \varepsilon_2 = \alpha - \beta \qquad (8\text{-}1.9.10)$$

It can be shown that both α and β are less than zero, so

$$\varepsilon_1 < \varepsilon_2$$

Using these values we can solve for the coefficients to get the corresponding wave functions. The details will not be presented, but the results are (if you enjoy mathematics, you might do it yourself).

$$\varepsilon_1 = \alpha + \beta \qquad \psi_1 = \frac{1}{\sqrt{2}}(\phi_1 + \phi_2)$$

$$\varepsilon_2 = \alpha - \beta \qquad \psi_2 = \frac{1}{\sqrt{2}}(\phi_1 - \phi_2)$$

Where ϕ_1 and ϕ_2 are p orbitals from the two carbons.

If we choose α *as the energy of a p electron localized on a carbon, we note that* β *represents the stabilizing energy of the* π *system.* The following diagram (Fig. 8-1.9.1) will apply. The two electrons that occupy the π orbital in ethylene have been shown. (Remember, we ignore all electrons in σ bonds.) The π energy (ε_π) in ethylene will be

$$E_\pi = 2\alpha + 2\beta$$

Note that each electron in an ethylene-type π system contributes $\alpha + \beta$ to the energy. We will need that information in the examples to follow. The question now is what happens in a system such as propylene ($CH_2{=}CH_2{-}CH_3 \leftrightarrow CH_3{-}CH_2{=}CH_2$) or benzene

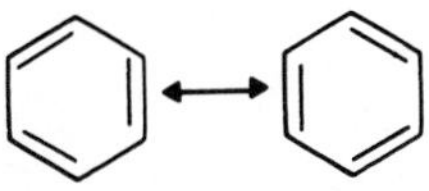

in which resonance is usually involved. The following examples will show what happens.

8-1.9.1 EXAMPLE

Find the resonance stabilization in propylene. Use the model

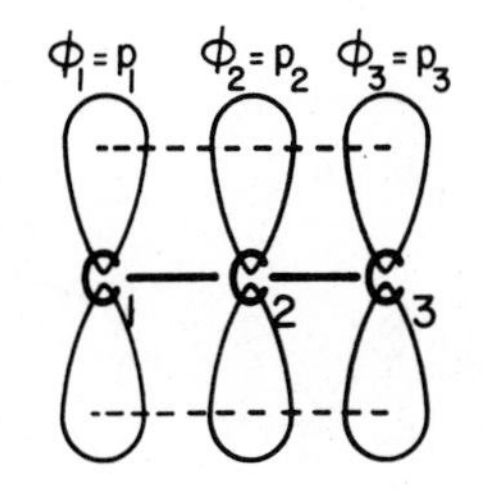

$$\text{———} \quad \varepsilon_2 = \alpha - \beta$$

$$\text{- - - - - - -} \quad \alpha$$

$$\uparrow\downarrow \quad \varepsilon_1 = \alpha + \beta$$

FIG. 8-1.9.1 Energy level system in ethylene.

Start by setting up the secular determinant

$$\begin{vmatrix} \alpha - \varepsilon & \beta & 0 \\ \beta & \alpha - \varepsilon & \beta \\ 0 & \beta & \alpha - \varepsilon \end{vmatrix} = \begin{vmatrix} x & 1 & 0 \\ 1 & x & 1 \\ 0 & 1 & x \end{vmatrix} = 0$$

where $x = (\alpha - \varepsilon)/\beta$. [Before continuing, let us determine the energy if the double bond is localized between C_1 and C_2 or C_3 and C_3. If so, we have two electrons in a π orbital ($\varepsilon = 2\alpha + 2\beta$), and the third is in a p orbital ($\varepsilon = \alpha$). The total localized energy is $E_{loc} = 3\alpha + 2\beta$.] Solve this determinant to get

$$x^3 - 2x = 0$$

$$x = 0 \rightarrow \varepsilon = \alpha$$

$$x = \pm\sqrt{2} \rightarrow \varepsilon = \alpha \pm \sqrt{2}\beta$$

The energy level diagram is

$$\text{———} \quad \varepsilon_3 = \alpha - \sqrt{2}\,\beta \qquad \psi_3$$

$$\uparrow \quad \varepsilon_2 = \alpha \qquad \psi_2$$

$$\uparrow\downarrow \quad \varepsilon_1 = \alpha + \sqrt{2}\,\beta \qquad \psi_1$$

The π system has a total energy of (E_{mo} = energy of molecular orbital system)

$$E_{mo} = 2(\alpha + \sqrt{2}\,\beta) + \alpha$$

$$= 3\alpha + 2\sqrt{2}\,\beta$$

The resonance energy will be the difference between E_{mo} and E_{loc}

$$E_{res} = E_{loc} - E_{mo}$$

$$= 3\alpha + 2\beta - 3\alpha - 2\sqrt{2}\,\beta$$

$$= \underline{\underline{-0.83\ \beta}}$$

The resonance stabilization is about $-0.83\,\beta$. All we need is a value of β. An average value from many experiments (β must be determined experimentally) is about -75 kJ. So, for propylene, the molecule is about (-0.83)(-75 kJ) = 62 kJ (mol^{-1}) more stable than predicted by our simple model. This difference is called resonance energy.

8-1.9.2 EXAMPLE

Estimate the resonance energy of benzene. The model is

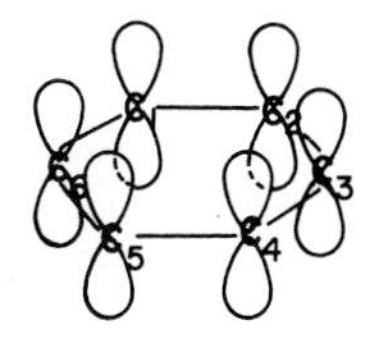

In the absence of resonance we would have six electrons in π orbitals giving an energy of $6\alpha + 6\beta$.

Solution

The determinant can be written immediately by noting the following overlaps: 12, 16; 21, 23; 32, 34; 43, 45; 54, 56; and 65, 61. Define $x = (\alpha - \varepsilon)/\beta$.

$$\begin{vmatrix} x & 1 & 0 & 0 & 0 & 1 \\ 1 & x & 1 & 0 & 0 & 0 \\ 0 & 1 & x & 1 & 0 & 0 \\ 0 & 0 & 1 & x & 1 & 0 \\ 0 & 0 & 0 & 1 & x & 1 \\ 1 & 0 & 0 & 0 & 1 & x \end{vmatrix} = 0$$

Expansion gives (try it)

$$(x + 1)^2(x - 1)^2(x + 2)(x - 2) = 0$$

The roots are

$x = 2$	nondegenerate, one solution
$x = 1$	degenerate, two solutions, two energy levels
$x = -1$	degenerate, two solutions, two energy levels
$x = -2$	nondegenerate, one solution

Substituting for x and arranging in increasing energy (remember β is negative)

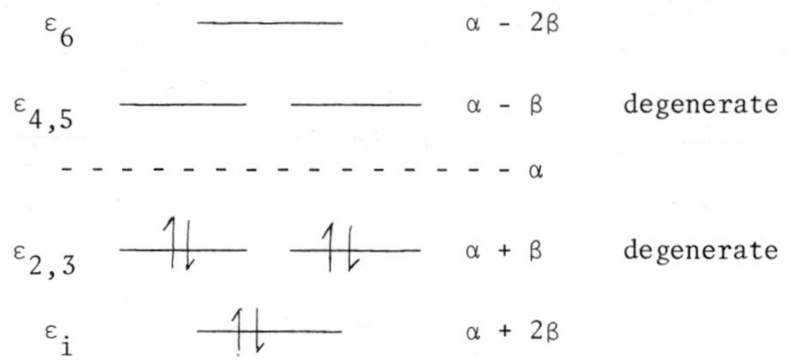

ε_6 ——— $\alpha - 2\beta$

$\varepsilon_{4,5}$ ——— ——— $\alpha - \beta$ degenerate

- - - - - - - - - - - - - - - - α

$\varepsilon_{2,3}$ —⇅— —⇅— $\alpha + \beta$ degenerate

ε_i —⇅— $\alpha + 2\beta$

The six electrons have been put in. E_{mo} is given as

$$E_{mo} = 2(\alpha + 2\beta) + 4(\alpha + \beta) = 6\alpha + 8\beta$$

From before

$$E_{loc} = 6\alpha + 6\beta$$

Thus $E_{res} = E_{loc} - E_{mo}$

$$= -2\beta = -2(-75 \text{ kJ}) = \underline{\underline{150 \text{ kJ}}}$$

Compare this to the experimental value mentioned for hydrogenation experiments. They are almost identical. Lest you are led to believe that the procedure works that well, you must realize that benzene data were used to get the value of β initially.

8-1.9.3 EXAMPLE

From the results in Example 8-1.9.1 estimate the resonance stabilization of the propylene carbonium ion and the corresponding carbanion. (You furnish the details.)

Solution

$H_2C{=}CH\overset{+}{-}CH_2$ $2e^-$ in π, 0 in p; $E_{loc} = 2\alpha + 2\beta$

$$E_{mo} = 2(\alpha + \sqrt{2}\beta)$$

(both e^- in lowest level).

$$E_{res} = E_{loc} - E_{mo} = \underline{\underline{-0.83\ \beta}}$$

$H_2C{=}CH\overset{-}{-}CH_2$ $2e^-$ in π, 2 in p; $E_{loc} = 2\alpha + 2\beta + 2\alpha = 4\alpha + 2\beta$.

$$E_{mo} = 2(\alpha + \sqrt{2}\beta) + 2\alpha \quad \text{see energy level diagram}$$

$$= 4\alpha + 2\sqrt{2}\beta$$

$$E_{res} = E_{loc} - E_{mo} = \underline{\underline{-0.83\ \beta}}$$

The above is sufficient to show how estimates of resonance stabilization can be made. Obviously, the results will not be very good in many cases. We assume β is a constant but it cannot be in something like butadiene (see Problems, Sec. 8-1.10), since the C-C bond lengths are different. However, very quick estimates are available using this technique.

8-1.10 PROBLEMS

1. Draw the molecular orbital diagram for CN^-. Place electrons in the appropriate orbitals and give the bond order, electronic designation, and number of unpaired electrons. Also, draw the shapes of the highest *two occupied* orbitals.
2. Devise the functions necessary to dsscribe the following sp^2 hybrid; $\psi_4 = p_z$ and is perpendicular to the sp^2 hybrids.

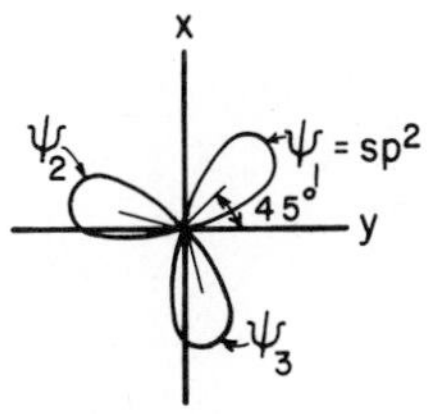

x; ψ_2; $\psi_1 = sp^2$; 45°; y; ψ_3

3. Perform a Hückel calculation to obtain the resonance stabilization for butadiene (CH_2=CH-CH=CH_2).
4. Perform a Hückel calculation for cyclopropene.

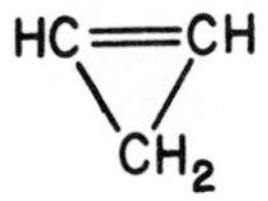

HC=CH; CH_2

5. Construct a molecular orbital diagram for NO. Determine the bond order and number of unpaired electrons. Give the molecular orbital designation. Which is more stable and why, NO or NO^+?
6. Construct sp hybrid orbitals from atomic 2s and 2p orbitals for the following shape.

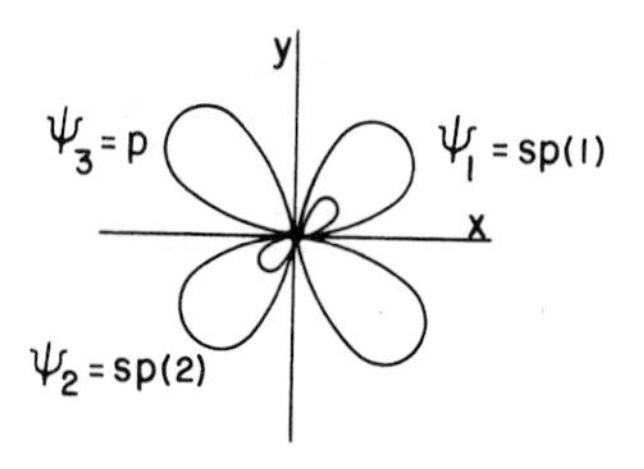

8-1.11 SELF-TEST

1. For O_2^+ draw a molecular orbital diagram and insert the electrons. Give the electron designation, the number of unpaired electrons, and the bond order. Draw the shape of the highest *two occupied* orbitals.

2. Order the following molecules in terms of stability; list the *most stable* first:

   O_2, O_2^+, O_2^{2+}, O_2^-, and O_2^{2-}

3. Devise the functions necessary to describe the following sp^3 hybrid orbitals.

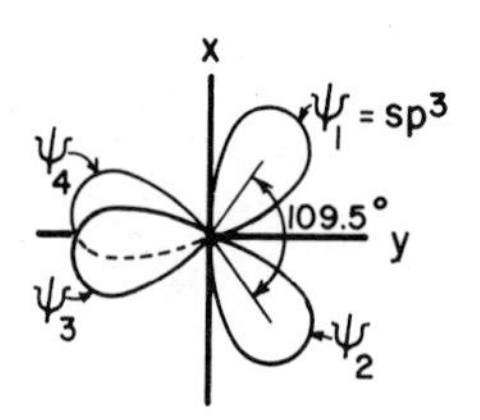

ψ_3 and ψ_4 are in the yz plane

4. Perform a Hückel calculation to estimate the resonance stabilization for cyclobutadiene and compare with the results for Problem (3) of Sec. 8-1.10.

8-1.12 ADDITIONAL READING

1.* G. M. Barrow, *Introduction to Molecular Spectroscopy*, McGraw-Hill, New York, 1962.

2.* L. Pauling, *The Nature of the Chemical Bond*, 3rd ed., Cornell University Press, Ithaca, N. Y., 1960.

3.* J. C. Davis, Jr., *Advanced Physical Chemistry*, Ronald Press, New York, 1965, Chap. 10.

4. A. Liberles, *Introduction to Theoretical Organic Chemistry*, MacMillan, New York, 1968.

8-2 ROTATION AND VIBRATION OF MOLECULES

OBJECTIVES

You shall be able to

1. Discuss the factors affecting the intensity of radiation absorbed in a transition from one energy level to the next
2. Define and/or discuss: transition moment, dipole moment, degeneracy, selection rule, rigid rotor approximation, harmonic oscillator approximation, moment of inertia, normal modes of vibration, degrees of freedom, symmetric and asymmetric top, and tunneling
3. Given appropriate data, calculate the energies (in any desired unit) of rotational and vibrational transitions, or from such energies determine molecular parameters

8-2.1 TRANSITIONS: ABSORPTION OR EMISSION OF ELECTROMAGNETIC RADIATION

All types of spectroscopy depend on the emission or absorption of electromagnetic radiation. Usually the energy (frequency, wave number, etc.) and the intensity of the absorption or emission are both measured in an experiment. Below we will discuss absorption spectra, but with minor changes the discussion applies to emission as well.

Electromagnetic radiation is visualized as consisting of oscillating electric and magnetic fields. Figure 8-2.1.1 shows a schematic of a plane-polarized light beam consistent with this model. Each wave moves with a velocity of light in the direction indicated. If electromagnetic radiation falls on a molecule, it is possible for the oscillating electric field to induce a transition from one quantum state in the molecule to another. This process is called *absorption* if the transition is to a higher energy level and *emission* if to a lower level. Let us assume the molecule occupies a position (such as point b) about which the electric field oscillates (as the wave crests pass point b). If the frequency of the radiation is ν, the electric field at x varies according to the relation

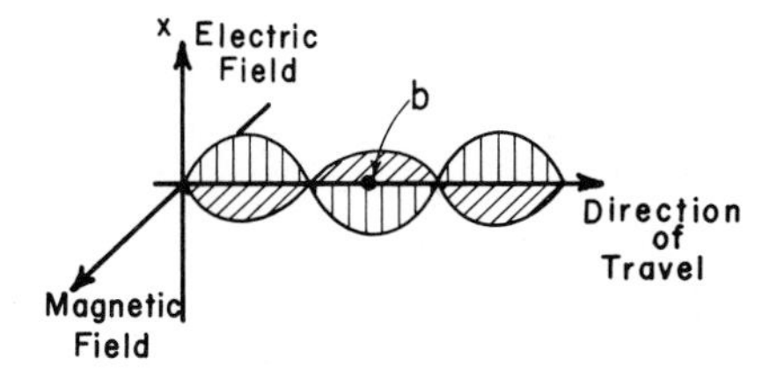

FIG. 8-2.1.1 A beam of plane-polarized radiation. Point b represents a molecule in this beam of radiation.

$$E_x = 2E_x^0 \cos 2\pi\nu t \tag{8-2.1.1}$$

where E_x^0 is the amplitude of the electric field. The strength of an absorption (the absorption strength) depends on several variables. The dependence is characterized by the Lambert-Beer law, which is applicable to many (but not all) experimental determinations of absorption.

$$-\log \frac{I}{I_0} = aC\ell \tag{8-2.1.2}$$

where I_0 is the intensity of the incident radiation and I the intensity after passing through the sample as shown in Fig. 8-2.1.2. C is a concentration term (concentration, pressure, whatever) and a is the absorptivity (in a proper set of units to go with C; if C is in moles per cubic decimeter, then a will be a *molar absorptivity*). Finally, ℓ is the path length, the distance the radiation traveled through the sample.

The variable of interest is a, since that is the property in (8-2.1.2) that is characteristic for a given species at a given frequency. This quantity depends on the transition probability between the two energy levels involved. If wave functions are known for both states, then it should be possible to calculate the theoretical absorption strength. The absorption strength depends on at least two factors:

1. The degree of interaction of the molecule with the electromagnetic wave.
2. The population difference between the initial and final states; population means the number of molecules in the system that are in a given state or energy level.

The second factor will not be considered in detail here. Population distributions are discussed in Chap. 9.

Also, this discussion will be restricted to molecules that interact with the electric field of the electromagnetic radiation. For an interaction to occur there *must* be a change in the *charge distribution* in the molecule when the transition from one energy level to the next occurs. The charge distribution is defined by a quantity called the dipole moment μ which is a measure of the degree to which the positive

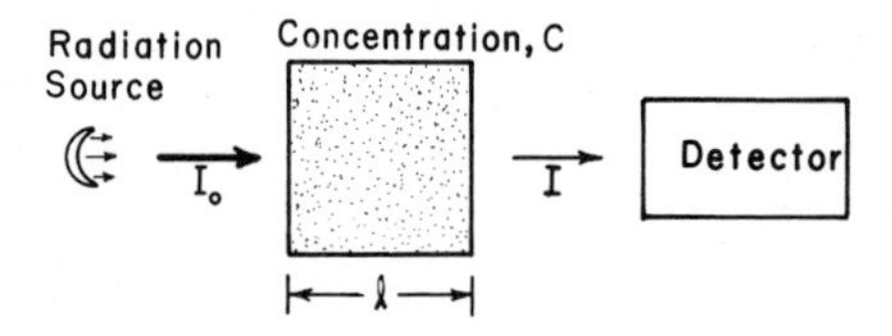

FIG. 8-2.1.2 Absorption of radiation by a sample.

and negative centers in a molecule do not coincide. (You *must* read about the dipole moment, Sec. 14-2.3, before continuing.)

If a molecule has a dipole moment which changes in a transition from one energy state to another, it will be possible for it to interact with the electric field portion of the electromagnetic wave. Such a transition is called an *electric dipole transition*. The actual derivation of probability of a transition requires development of the time-dependent Schrödinger equation. That is beyond the scope of this study. (For details, see Barrow, Sec. 8-2.10, Ref. 1). We can, however, give a little more detail on the interaction.

Assume we are interested only in the x direction. The electric field at point b (see Fig. 8-2.2.1) will act on the x component of the dipole moment of a molecule located there. This produces an energy change of $E_x \mu_x$, which adds to the potential energy of the system. This potential energy change of the system is treated as a perturbation (see Sec. 7-4.2) for purposes of calculating the probability of a transition occuring. (Remember, transitions can occur only if $h\nu = \varepsilon_k - \varepsilon_m$, where k and m refer to two different energy states in the molecule.) It is found that the probability of a transition between two energy levels k and m (when $h\nu = \varepsilon_k - \varepsilon_m$) is dependent on the *energy density* ρ of the radiation and the *transition moment* R_{km}. The radiation density is given by

$$\rho = \frac{6}{4\pi} (E_x^0)^2 \qquad (8\text{-}2.1.3)$$

and the transition moment is defined as

$$R_{km(x)} = \int \psi_k^* \mu_x \psi_m \, d\tau \qquad (8\text{-}2.1.4)$$

where ψ_k and ψ_m are the wave functions for states k and m. The transition moment is, in essence, a measure of the ability of states k and m to be "coupled" or "mixed" by an electric dipole moment. The relation of interest is

$$\text{"Probability of transition (x component)"} = \frac{8\pi^3}{3h^2} R_{km(x)}^2 \rho \qquad (8\text{-}2.1.5)$$

The actual value of R_{km}^2 can be obtained from a detailed study of the experimentally determined absorption intensities. We are not interested in absolute numbers. What we will want to know is whether a given transition is allowed. Equation (8-2.1.4) allows us to determine *selection rules* for each type of spectroscopy. By applying this equation we can find out what states can be reached from a given state. Not all imaginable transitions are possible. If $R_{km} = 0$ for a given k and m, the transition is not allowed. For $R_{km} \neq 0$ the transition is allowed, and its intensity is described (in part) by the value of R_{km}. The selection rules will be presented as each type of spectroscopy is discussed.

8-2.2 ROTATIONAL SPECTROSCOPY: LINEAR MOLECULES

MODEL AND SOLUTIONS. The model for a rotating diatomic molecule is the same as for the rotating hydrogen nucleus and electron. The solution to the equation is also identical, so the work is already done. Assume we have two atoms, masses m_1 and m_2, connected by a weightless rod (the bond) making a rigid system. This whole system will rotate about the center of mass (see Sec. 7-2.3 for the classical Hamiltonian and Sec. 7-1.8 and 7-2.4 for the quantum mechanical development). The approximation involved here is called the *rigid rotor approximation*. We assume that the internuclear distance r_{12} is constant: the nuclei do not vibrate.

The moment of inertia of this system is (relative to the center of mass)

$$I = \sum_i m_i r_i^2 = m_1 r_1^2 + m_2 r_2^2 = \frac{m_1 m_2}{m_1 + m_2} r_{12}^2$$

$$I = \mu r_{12}^2 \qquad \text{where} \qquad \mu = \frac{m_1 m_2}{m_1 + m_2} \tag{8-2.2.1}$$

The resulting Schrödinger equation is identical to the Θ equation given before. The solutions are

$$\frac{8\pi^2 I\varepsilon}{h^2} = \ell(\ell + 1) \tag{8-2.2.2}$$

For rotation the quantum number ℓ is replaced by J to distinguish the solutions from the H atom problem. Also, m is replaced by M. Thus the energy levels will be designated by ε_J

$$\boxed{\varepsilon_J = \frac{h^2}{8\pi^2 I} J(J + 1)} \qquad J = 0, 1, 2, \ldots \tag{8-2.2.3}$$

The allowed wave functions can be written immediately from tables in Sec. 7-2. The rotational wave functions are given by

$$\Psi_{rot}(J, M) = \Theta(J, M)\ \Phi(M) \tag{8-2.2.4}$$

The values allowed for M are 0, ± 1, ± 2, ..., ± J, as before. Define the constant in Eq. (8-2.2.3) as

$$\boxed{B = \frac{h}{8\pi^2 I}} \qquad \text{the rotational constant} \tag{8-2.2.5}$$

This gives B with units per second. For B in Joules we must multiply by h to give

$B = h^2/8\pi^2 I$. ε_J can be written compactly as

$$\boxed{\varepsilon_J = J(J + 1)\ B} \tag{8-2.2.6}$$

Figure 8-2.2.1 shows a schematic of the first few energy levels defined by Eq. (8-2.2.6). Also, see the figure in the introduction to this chapter.

TRANSITIONS. If the rotational wave functions are inserted into Eq. (8-2.1.4) to determine the transition moment, it is found that R_{km} exists only for transitions from J to J $\pm$ 1. That is, the change in J, ΔJ, must not be more than 1, $\Delta J = \pm 1$. Also, it can be shown that $\Delta M = 0, \pm 1$ must also be satisfied. As was the case for the hydrogen atom, projections of the angular momentum J (designated by values of M) onto a defined z axis do not occur unless an external magnetic or electric field is present.

Back to the allowed transitions: let us assume only absorptions, so $\Delta J = +1$.

$$\varepsilon_J = J(J + 1)B \qquad \varepsilon_{J+1} = (J + 1)(J + 1 + 1)B$$

$$\Delta\varepsilon_{J\rightarrow J+1} = \varepsilon_{J+1} - \varepsilon_J$$

$$= \underline{\underline{2(J + 1)\ B}} \tag{8-2.2.7}$$

where J refers to the lower J level involved. The energy change for the first few transitions is shown in Fig. 8-2.2.1. (Remember $\Delta\varepsilon = h\nu$. We measure ν experimentally.)

EXPERIMENTAL DETAILS. The frequency of transitions between rotational levels for most molecules lies in the microwave region of the electromagnetic spectrum

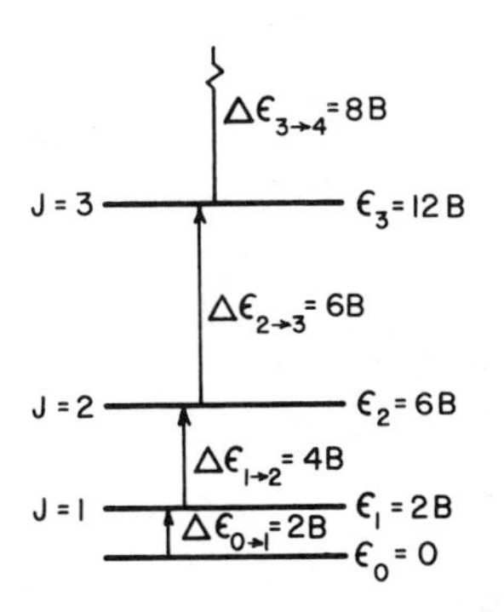

FIG. 8-2.2.1 The first few rotational energy levels of a diatomic molecule with rotational constant B. Also shown are some energies of transitions.

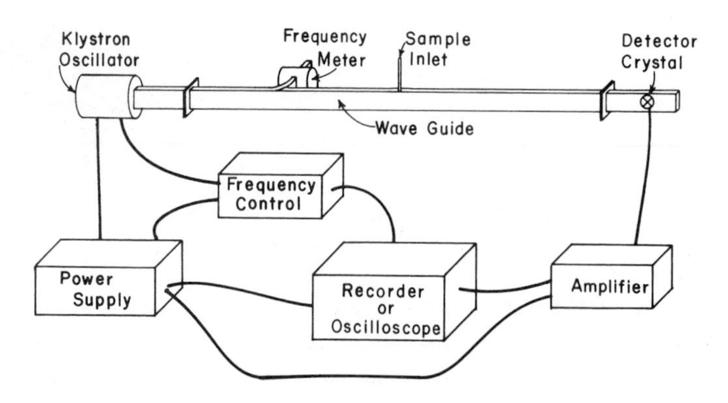

FIG. 8-2.2.2 Simplified sketch of a microwave spectrometer.

(1×10^9 to 300×10^9 s^{-1}, 0.30 to 0.001 m). Experimentally, only molecules in the gas phase can be studied, since rotation is severely limited or eliminated in the liquid and solid states. Figure 8-2.2.2 gives an experimental apparatus that could be used to study rotational transitions in the microwave range. The wave guide is a hollow metal tube of rectangular cross-section which propagates the microwaves and serves as the gas sample holder. Microwave radiation is generated by a Klystron or other oscillator. This radiation is transmitted through the wave guide where it interacts with the sample. The crystal detector is tuned to oscillate in the same range as the microwave frequencies being produced. When microwaves strike the crystal, a signal is generated which is then amplified and displayed on an oscilloscope or strip chart recorder. The frequency is varied by varying the voltage to the Klystron. As the frequency is varied, the signal from the crystal is monitored. When an absorption takes place the amount of microwave radiation reaching the detector decreases, leading to a drop in signal strength. Figure 8-2.2.3 shows a schematic of a spectrum for a diatomic molecule. First, note the equally spaced absorptions and compare with the energy level diagram of Fig. 8-2.2.1. The different intensities are caused by population differences in the various levels, which is one of the two factors mentioned in Sec. 8-2.1 describing transition intensities.

Finally, a note that might interest you. The frequencies of the transitions can easily be determined to $\pm$ 0.100 kHz = 100 s^{-1}. For the $J = 2 \rightarrow 3$ transition, this represents an error of

$$\frac{\Delta\nu}{\nu} = \frac{100\ s^{-1}}{3.6 \times 10^{10}\ s^{-1}} = 2.8 \times 10^{-7}\ \%$$

To state it another way, the uncertainty in the position of an absorption is only about one part in 10^8. Few other experimental techniques approach this precision.

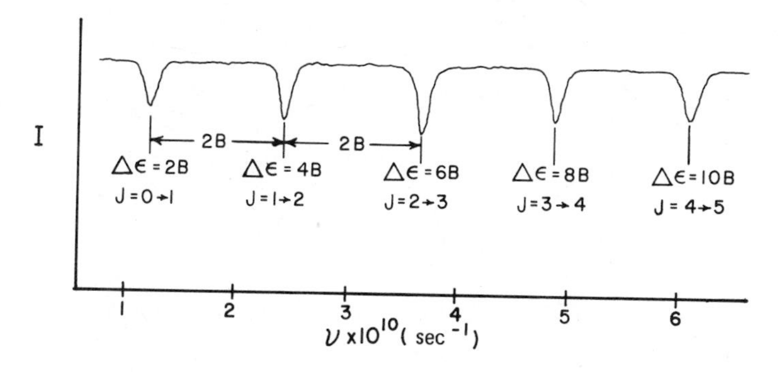

FIG. 8-2.2.3 Schematic representation of a microwave spectrum of $^{16}O^{12}C^{32}S$. Linear molecules follow Eq. (8-2.2.7), as do diatomics. A diatomic molecule is not used, since most diatomics absorb radiation higher in frequency than the microwave.

STRUCTURAL INFORMATION. The major value of microwave spectroscopic studies is the precise structural information that can be obtained from measurements of rotational transitions. Isotopic effects are easily observed as can be seen from the following data.

$$\Delta\varepsilon_{0\to1}\ (^{12}C^{16}O) = 115.2712 \times 10^9\ s^{-1}$$

$$\Delta\varepsilon_{0\to1}\ (^{12}C^{18}O) = 109.7822 \times 10^9\ s^{-1}$$

The superscripts correspond to the atomic weights (in grams). Using these data we can determine the molecular parameters very accurately. From Eq. (8-2.2.7) we can find B if we know $\Delta\varepsilon$. Then from Eq. (8-2.2.5) we obtain the moment of inertia which depends explicitly on the internuclear distance as shown in Eq. (8-2.2.1). Care must be used in interpreting the values. Remember, we have ignored vibration, so the value we obtain is for a particular vibrational state, usually the lowest.

8-2.2.1 EXAMPLE

Determine the internuclear distance for $^{12}C^{16}O$ if $\Delta\varepsilon_{0\to1} = 115.2712 \times 10^9\ s^{-1}$.

Solution

$$\Delta\varepsilon = 2(J + 1)B = 2B = 115.2712 \times 10^9\ s^{-1}$$

$$B = 57.6356 \times 10^9\ s^{-1}$$

But, $B = h/8\pi^2 I$ or $I = h/8\pi^2 B = (6.6262 \times 10^{-34}\ J\ s)/[8\pi^2(57.6356 \times 10^9\ s^{-1})]$

$$I = 1.4561 \times 10^{-46}\ J\ s^2 \qquad (kg\ m^2\ s^{-2}\ s^2 = kg\ m^2)$$

$$I = \mu r^2 = \frac{m_1 m_2}{m_1 + m_2} r^2$$

$$= \underline{\underline{1.4561 \times 10^{-46} \text{ kg m}^2}}$$

The reduced mass (in kg) is

$$\mu = \frac{(0.012)(0.016)}{0.012 + 0.016} \quad \frac{1}{6.023 \times 10^{23}} \text{ kg molecule}^{-1}$$

(*Note*: These masses are exact integers. We observe specific isotopes in microwave spectroscopy.)

$$\mu = 1.138 \times 10^{-26} \text{ kg molecule}^{-1}$$

$$r^2 = \frac{I}{\mu} = \frac{1.456 \times 10^{-46} \text{ kg m}^2}{1.139 \times 10^{-26} \text{ kg molecule}^{-1}}$$

$$= 1.279 \times 10^{-20} \text{ m}^2$$

$$r = \underline{\underline{1.131 \times 10^{-10} \text{ m}}} = \underline{\underline{1.131 \text{ Å}}}$$

More complex linear molecules follow the same equations. The only difference is in the moment of inertia expression for a linear polyatomic molecule which is more complicated than for the diatomic system. Also, it is not possible to determine more than one bond length from the rotational spectrum of one isotopic species. To determine all the parameters in a given molecule, several isotopic substitutions have to be made. Also, the assumption is made that the internuclear distance is independent of the isotopic species present. The following example for O=C=S will show what is involved for a simple case.

8-2.2.2 EXAMPLE

Given the following data (see also Table 8-2.2.1) find r_{CO} and r_{CS} for O=C=S.

$$\Delta\varepsilon_{0\to1} \; (^{16}O^{12}C^{32}S) = 12.163 \times 10^9 \text{ s}^{-1}$$

$$\Delta\varepsilon_{0\to1} \; (^{16}O^{12}C^{34}S) = 11.9664 \times 10^9 \text{ s}^{-1}$$

Also, given that for this linear triatomic molecule

$$I = \frac{m_C m_O r_{CO}^2 + m_C m_S r_{CS}^2 + m_O m_S (r_{CO} + r_{CS})^2}{m_C + m_O + m_S}$$

Solution

First, calculate the moment of inertia for each.

$$I(OC^{32}S) = \frac{(0.012)(0.016)r_{CO}^2 + (0.012)(0.032)r_{CS}^2 + (0.016)(0.032)(r_{CO} + r_{CS})^2}{(0.012 + 0.016 + 0.032)}$$

$$I(OC^{32}S) \times 10^{27} \text{ molecules kg}^{-1} = 5.313r_{CO}^2 + 10.63r_{CS}^2 + 14.17(r_{CO} + r_{CS})^2$$

Also,

$$I(OC^{34}S) \times 10^{27} \text{ molecules kg}^{-1} = 5.142r_{CO}^2 + 10.93r_{CS}^2 + 14.57(r_{CO} + r_{CS})^2$$

From the data given with $I = h/8\pi^2 B$,

$$B(OC^{32}S) = 6.091 \times 10^9 \text{ s}^{-1} \rightarrow I(OC^{32}S) = 1.380 \times 10^{-45} \text{ kg m}^2$$

$$B(OC^{34}S) = 5.9333 \times 10^9 \text{ s}^{-1} \rightarrow I(OC^{34}S) = 1.414 \times 10^{-45} \text{ kg m}^2$$

This gives two simultaneous equations with two unknowns. The solution is not simple but yields

$$r_{CS} = \underline{\underline{1.559 \times 10^{-10} \text{ m}}} \qquad r_{CO} = \underline{\underline{1.161 \times 10^{-10} \text{ m}}}$$

You might substitute these in to show they are solutions. The larger the molecule the more isotopic substitutions that must be made, one for each variable to be determined.

8-2.3 ROTATIONAL SPECTROSCOPY: NONLINEAR MOLECULES

Most molecules do not fit our linear model. These are subdivided into two classes: *symmetric top and asymmetric top*.

SYMMETRIC TOP. A *symmetric top* molecule is one which has two of the three moments of inertia equal. (*Note*: There are three principal moments of inertia in any molecule, designated by I_A, I_B, and I_C. Any rotation can be resolved into projections onto these axes. A linear molecule has $I_A = I_B$ and $I_C = 0$. I_C would be the moment which includes the axis, so this moment does not exist.) In the symmetric top $I_A \neq I_B = I_C$. An example is NH_3. The energy levels in a symmetric top are not as easily calculated as for the linear system. An additional quantum number is required to specify rotation about I_A. The energy levels are given by (per second)

Table 8-2.2.1
Some Bond Lengths for Linear Molecules

Molecule	CO	HCN	OCS	HCl	HBr
Length ($\times 10^{-10}$ m)	1.128	H-C 1.064 C≡N 1.156	O=C 1.161 C=S 1.559	1.275	1.604

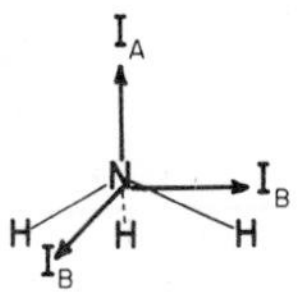

FIG. 8-2.3.1 Moments of inertia in a symmetric top molecule.

$$\varepsilon_{JK} = J(J+1)\frac{h}{8\pi^2 I_B} + \left(\frac{h}{8\pi^2 I_A} - \frac{h}{8\pi^2 I_B}\right) K^2$$

$$\varepsilon_{JK} = J(J+1)B + (A - B)K^2 \qquad (8\text{-}2.3.1)$$

where A and B are the appropriate rotational constants. The selection rules are

$$\Delta J = \pm 1 \qquad \Delta K = 0 \qquad \text{if} \qquad K = 0$$

$$\Delta J = 0, \pm 1 \qquad \Delta K = 0 \qquad \text{if} \qquad K \neq 0$$

What we observe is a spectrum with several peaks, where in the linear case, only one peak was present for each transition of $J \rightarrow J + 1$. In this case, for $J \rightarrow J + 1$, we observe K transitions, making the spectrum more complicated. The data are used as before to obtain molecular parameters. Usually sufficient isotopic substitutions cannot be made to completely characterize the structure of any but the simplest molecules. Some bond angles and lengths must be assumed. Table 8-2.3.1 lists some structural parameters determined by microwave spectroscopy.

ASYMMETRIC TOP. The most difficult case is the one in which $I_A \neq I_B \neq I_C$. This is an *asymmetric top*. No simple expressions can be used to find the energy levels for various J and K combinations. Computational techniques are available to predict energy levels if I_A, I_B, and I_C are known. However, these are beyond the scope of this section (see Townes and Schawlow, Sec. 8-2.10, Ref. 2). Suffice it to say that rather complex molecules have been studied by microwave techniques. The results of microwave work give the most accurate values available for molecular parameters.

Table 8-2.3.1
Some Molecular Parameters for Symmetric Top Molecules

Molecule	Bond angle		Bond distance ($\times 10^{-10}$ m)			
CH_3F	HCH	110°0′	CH	1.109	CF	1.384
CH_3Cl	HCH	110°20′	CH	1.103	CCl	1.782
CH_3Br	HCH	110°48′	CH	1.101	CBr	1.938
CH_3CN	HCH	109°48′	CH	1.092	CC	1.460
NH_3	HNH	107° ± 2°	NH	1.016		

8-2.4 DIPOLE MOMENTS

One of the major uses of rotational spectroscopy is the measurement of molecular dipole moments. If a molecule is put in an electric field, the rotational energies are shifted. The quantum number M designates the projections of J on the axis defined by the applied field. Different values of M will have different energies in the presence of a field, even though the M values are degenerate in the absence of a field. The splitting of the M levels by an electric field is called the Stark effect. The equations for calculating the dipole moment are far from simple, but microwave spectroscopy is the source for the most accurate values of dipole moments for molecules that are found in the gaseous state. (Again, see Townes and Schawlow, Sec. 8-2.10, Ref. 2.)

8-2.5 VIBRATIONAL SPECTROSCOPY: DERIVATIONS OF ENERGY LEVELS

MODEL. Vibrating molecules are usually assumed to behave as two particles with a spring between them (see Fig. 8-2.5.1). The equilibrium extension of the spring may be denoted as x_e. The general coordinate q is just the distance of displacement away from x_e; $q = x - x_e$. (*Note*: In terms of x the kinetic energy is given by

$$T = \frac{1}{2} m_1 \dot{x}_1^2 + \frac{1}{2} m_2 \dot{x}_2^2 \qquad \text{and} \qquad V = \frac{1}{2} \bar{k} (x - x_e)^2$$

You might want to use the procedures of Sec. 7-2 to show that the Hamiltonian (classical) of the system is $\mathcal{H} = \frac{1}{2} \mu \dot{q}^2 + \frac{1}{2} \bar{k} q^2$.) $\bar{k}$ is a spring force constant and μ the reduced mass. Classically this problem is easily solved using procedures from Sec. 7-2 to give ν_0, the classical vibrational frequency.

$$\nu_0 = \frac{1}{2\pi} \left(\frac{\bar{k}}{\mu}\right)^{1/2} \qquad (8\text{-}2.5.1)$$

or

$$\bar{k} = 4\pi^2 \mu \nu_0^2$$

where μ is the reduced mass.

SOLUTION. Let us now sketch the quantum mechanical solution to show the similarities and differences between classical and quantum mechanical results. Using Postulate II from Sec. 7-1 we can write the Schrödinger equation after converting $\mathcal{H}$ to operator form.

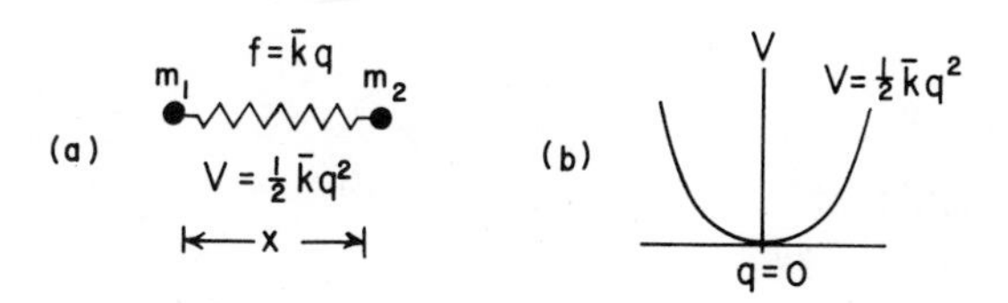

FIG. 8-2.5.1 (a) Model and (b) potential energy function for a harmonic oscillator.

$$\frac{d^2\Psi}{dq^2} + \frac{8\pi^2\mu}{h^2}\left[\varepsilon - V(q)\right]\Psi = 0 \tag{8-2.5.2}$$

V(q) can be written as $2\pi^2\mu\nu_0^2q^2$ by using Eq. (8-2.5.1). Now define two constant terms:

$$\alpha = \frac{8\pi^2\mu\varepsilon}{h^2} \qquad \beta = \frac{4\pi^2\mu\nu_0}{h} \tag{8-2.5.3}$$

Substitution of these into Eq. (8-2.5.2) yields a simple-looking expression.

$$\frac{d^2\Psi}{dq^2} + (\alpha^2 - \beta^2q^2)\Psi = 0 \tag{8-2.5.4}$$

The solution to this equation follows that given for the hydrogen atom. Note that q can vary from $-\infty$ to ∞.

The second substitution is to set $\xi = \sqrt{\beta}\,q$. This can be substituted in Eq. (8-2.5.4) to give

$$\frac{d^2\Psi}{d\xi^2} + \left(\frac{\alpha}{\beta} - \xi^2\right)\Psi = 0 \tag{8-2.5.5}$$

Next, see what happens as $\xi \to \pm\infty$. Equation (8-2.5.5) reduces to

$$\frac{d^2\Psi}{d\xi^2} = \xi^2\Psi \rightarrow \Psi = A\exp\left(\pm\frac{\xi^2}{2}\right) \tag{8-2.5.6}$$

The positive sign cannot be allowed since the solution is not finite for all values of ξ in that case. Now assume that the solution everywhere can be represented as

$$\Psi = f(\xi)\exp\left(-\frac{\xi^2}{2}\right) \tag{8-2.5.7}$$

where $f(\xi)$ is a power series. Substitution into (8-2.5.5) and considerable rearrangement yields

$$\frac{d^2f}{d\xi^2} - 2\xi\frac{df}{d\xi} + \left(\frac{\alpha}{\beta} - 1\right)f = 0 \tag{8-2.5.8}$$

We could write $f(\xi)$ as a power series and solve as we did for the hydrogen atom R equation. It is easier, however, to recognize that Eq. (8-2.5.8) is an equation known as the Hermite equation if the following condition is met:

$$\frac{\alpha}{\beta} - 1 = 2v \qquad v = \text{integer} \tag{8-2.5.9}$$

The solutions $f(\xi)$ are known as Hermite polynomials and are designated as $H_v(\xi)$. Some of the polynomials are shown in Table 8-2.5.1. Perhaps it would be instructive

Table 8-2.5.1
Some Hermite Polynomials

v Even	v Odd
$H_0(\xi) = 1$	$H_1(\xi) = 2\xi$
$H_2(\xi) = 4\xi^2 - 2$	$H_3(\xi) = 8\xi - 12$
$H_4(\xi) = 16\xi^4 - 48\xi^2 + 12$	$H_5(\xi) = 32\xi^5 - 160\xi^3 + 120$

to present the general equation that can be used to generate the polynomials. (Show the equation works for a couple of examples.)

$$H_v(\xi) = \sum_{k=0} \frac{(-1)^k v!(2\xi)^{v-2k}}{(v - 2k)!\ k!} \tag{8-2.5.10}$$

If 2k > n, the factorial in the denominator becomes infinite, which is not allowed. So $k \leq n/2$.

The final solution is

$$\Psi(\xi) = \left(\frac{\beta^{\frac{1}{2}}}{\pi^{\frac{1}{2}} 2^v\ v!}\right) H_v(\xi) \exp\left(-\frac{\xi^2}{2}\right) \tag{8-2.5.11}$$

$$\xi = \sqrt{\beta}\, q$$

The restriction that $(\alpha/\beta - 1) = 2v$ gives

$$\alpha = (2v + 1)\beta \tag{8-2.5.12}$$

On substitution from (8-2.5.3) (do it)

$$\boxed{\varepsilon_v = (v + \tfrac{1}{2})h\nu_0} \qquad v = 0, 1, 2, \ldots \tag{8-2.5.13}$$

The classical expression is obtained with one significant difference. The $(v + \frac{1}{2})$ coefficient shows the energy is quantized. Also, even in the lowest quantum state, molecules still have a vibrational energy of $\frac{1}{2} h\nu_0$. This is called the *zero point vibrational energy*.

8-2.5.1 EXAMPLE

Calculate the ratio of the zero point energies for the C-H and C-D vibrations. Assume the force constants (a property of the bond strength) are the same.

Solution

To find ε_0 for each we need ν_0 which is defined in Eq. (8-2.5.1).

$$\frac{\varepsilon_0\ (C\text{-}D)}{\varepsilon_0\ (C\text{-}H)} = \frac{\nu_0\ (C\text{-}D)}{\nu_0\ (C\text{-}H)} = \frac{(1/2\pi)(\bar{k}/\mu_{CD})^{1/2}}{(1/2\pi)(\bar{k}/\mu_{CH})^{1/2}} = \left(\frac{\mu_{CH}}{\mu_{CD}}\right)^{1/2}$$

$$= \left(\frac{m_C\, m_H}{m_C + m_H}\ \frac{m_C + m_D}{m_C\, m_D}\right)^{1/2} = \left[\frac{m_H(m_C + m_D)}{m_D(m_C + m_H)}\right]^{1/2}$$

$$= \left[\frac{(0.0010)(0.0120 + 0.0020)}{(0.002)(0.0120 + 0.0010)}\right]^{1/2}$$

$$= (0.538)^{1/2} = \underline{\underline{0.734}}$$

$$\varepsilon_0\ (C\text{-}D) = \underline{\underline{0.734\ \varepsilon_0\ (C\text{-}H)}}$$

8-2.6 VIBRATIONAL WAVE FUNCTIONS AND SELECTION RULES

WAVE FUNCTIONS. An interesting contrast can be made between a classical harmonic oscillator and its quantum mechanical counterpart. Figure 8-2.6.1 shows a plot of the wave function amplitudes for different values of ν. The quadratic curve represents the potential energy of the system. A couple of points can be made. First, the wave functions extend beyond the classical potential energy curve. This is not possible in a classical system. There is a finite probability of the system being found outside the potential barrier. Such a possibility leads to an interesting quantum mechanical effect called *tunneling*. An example of tunneling is found in the NH_3 molecule. NH_3 can exist in two stable and one unstable state, as shown in Fig. 8-2.6.2. Note that if the vibrational wave functions overlap (which they do), it is possible for the ammonia to invert even though it does not have enough energy to get *over* the potential barrier. This tunneling effect is widespread and is of utmost importance in solid state electronic design.

Back to our main argument. The second point from Fig. 8-2.6.1 is that the number of *nodes* increases as v increases (a node is a point where a function crosses zero). This is always true. The more nodes a function has the higher the associated energy. Another significant observation can be made if the square of the wave function is plotted. Remember that $\Psi^*\Psi$ is related to the probability of finding a system in a given region of space. Figure 8-2.6.3 is such a plot. Classically, we expect to find the oscillator spending most of its time at the extremes of the oscillation where it slows, stops, and begins to return. This is not observed for small values of the quantum number in the quantum mechanical oscillator. For low values of v the highest probability of the quantum mechanical harmonic oscillator is in the center of its region of motion. As v increases, the probability function begins to look

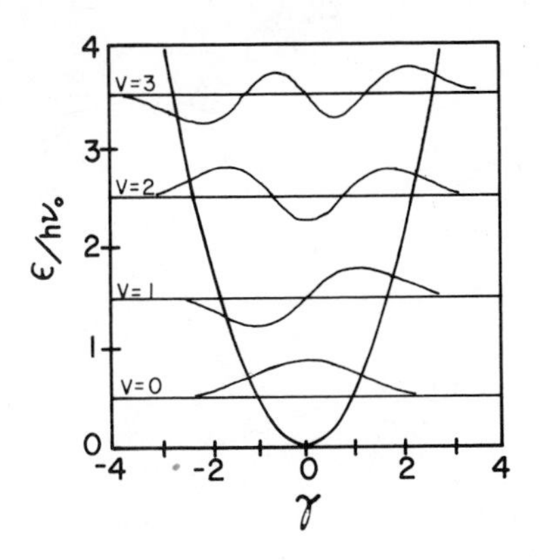

Fig. 8-2.6.1 Wave functions and potential for a harmonic oscillator. The functions are normalized; $q/(h/2\pi\mu\nu_0)^{1/2}$. [From W. J. Moore, *Physical Chemistry*, 4th ed., (C) 1972, pp 544-546, 622-623. Reprinted by permission of Prentice-Hall, Inc., Englewood Cliffs, New Jersey (see Sec. 8-2.10, Ref. 3).]

more like a classical oscillator. At high v the quantum mechanical result merges into the classical result. This is an example of the *correspondence principle*. Quantum mechanical results *correspond* to classical results for large values of the quantum numbers or for large systems.

SELECTION RULES. The transition moment for vibrational transitions involves a dipole moment operator. The molecule does not have to have a permanent dipole moment, but there must be a *change in the dipole moment* if a transition is to have an emission or absorption of radiation. The selection rules that result are

$$\Delta v = \pm 1 \tag{8-2.6.1}$$

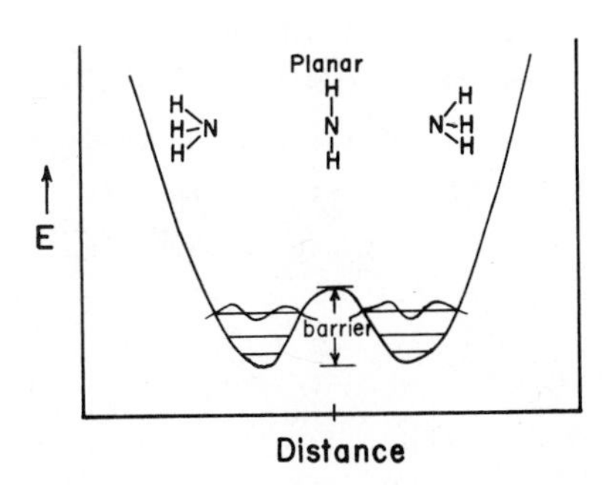

FIG. 8-2.6.2 Potential energy diagram for the inversion of ammonia. Vibrational levesl are shown (not to scale) with one wave function.

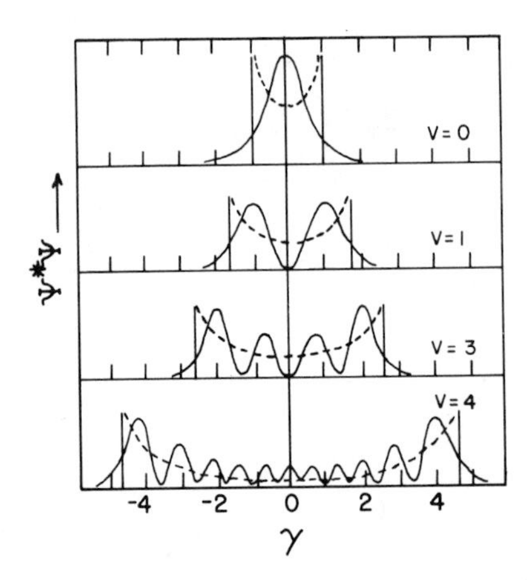

FIG. 8-2.6.3 Sketches of the probability functions for the harmonic oscillator. The dotted curve corresponds to the probability of the classical oscillator. [From W. J. Moore, *Physical Chemistry*, 4th ed., (C) 1972. pp 544-546, 622-623. Reprinted by permission of Prentice-Hall, Inc., Englewood Cliffs, New Jersey (see Sec. 8-2.10, Ref. 3).]

Thus, the energy difference between two adjacent vibrational levels is [from Eq. (8-2.5.13)]

$$\Delta\varepsilon_{v\rightarrow v+1} = h\nu_0 \tag{8-2.6.2}$$

The result is that all vibrational transitions for a given type of motion in a molecule fall at the same frequency (energy). This is not to say only one absorption will be observed for any given molecule. There may be many different types of vibration allowed in a molecule, as detailed in the following section. Table 8-2.6.1 lists some fundamental frequencies and bond force constants for several diatomic molecules. The data for homonuclear diatomics are obtained from Raman spectroscopy (see Sec. 8-4.3), since vibrational transitions cannot be observed directly (no dipole moment change). Each vibrational transition will actually be a *band* of transitions, since within each vibrational level there are numerous rotational levels. Each rotational level has a slightly different energy, so the vibrational *band* is spread over a range of frequencies. High resolution spectroscopy can resolve the different rotational transitions. This is discussed in Sec. 8-3.

EXPERIMENTAL DETAILS. Figure 8-2.6.4 shows a schematic of an infrared (IR) spectrometer used for studying vibrational transitions. A source, usually a glow bar that is electrically heated, emits radiation in the infrared region. This radiation is directed by mirrors to a reference cell and a sample cell. The reference is as nearly like the sample as possible but without sample present. If a solution

Table 8-2.6.1
Fundamental Vibrational Energies and Corresponding Force Constants for Several Diatomics

Molecule	ν_0 (cm^{-1})	$\bar{k}$ ($N\ m^{-1}$) $\times 10^{-2}$
H_2	4159.2	5.2
D_2	2990.3	5.3
HF	3958.4	8.8
HCl	2885.6	4.8
HBr	2559.3	3.8
HI	2230.0	2.9
CO	2143.3	18.7
NO	1876.0	15.5
F_2	0892.0	4.5
O_2	1556.3	11.4
N_2	2330.7	22.6
Li_2	0246.3	1.3

(a)

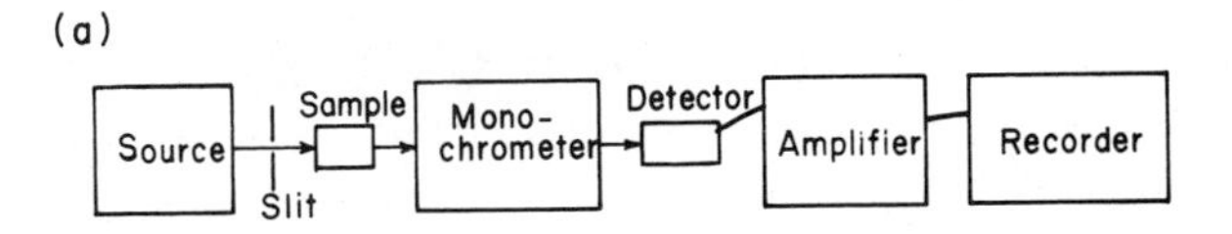

(b)

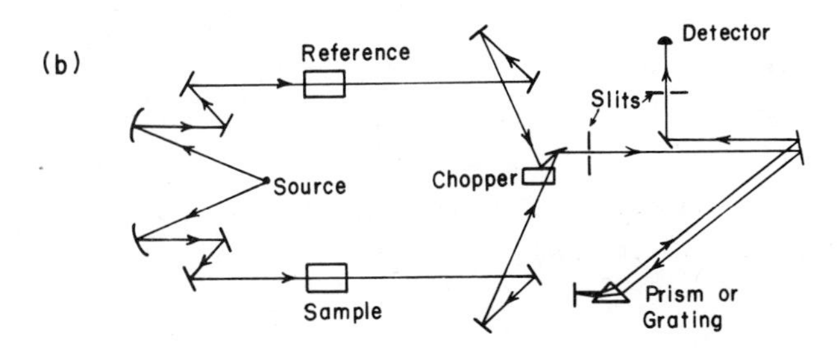

FIG. 8-2.6.4 (a) Block diagram and (b) optical arrangement for an infrared spectrophotometer.

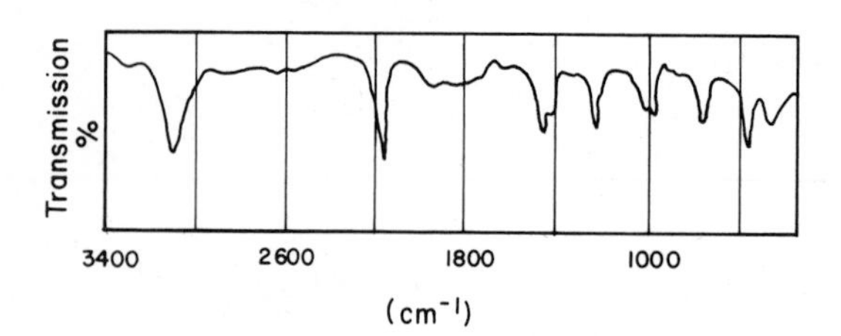

FIG. 8-2.6.5 Infrared spectrum of CH_3I (liquid).

is being studied, the reference is the solvent used. (If KBr pellets are used, the reference is a pellet with no sample present.) The purpose of the reference is to eliminate effects not due to the sample, as explained below. The two IR beams (sample S and reference R) are directed to a chopper which sends the S beam through the monochrometer half the time and the R beam the other half. These beams enter the monochrometer where a particular frequency is selected (by rotating the prism or grating) and sent to the detector. The detector is usually a thermocouple which responds to the heat of the IR beam. The signal received by the detector alternates between the reference and the sample. If the sample is not absorbing, the signals are the same, and the amplifier registers no signal. (The amplifier is a difference amplifier; it only responds if there is a difference between S and R.) When the sample absorbs, there is a difference between S and R that the amplifier will respond to.

In actual operation, the prism or grating is turned slowly to select different frequencies. A strip chart recorder is synchronized with this motion, and the amplifier output is recorded as a function of frequency. Figure 8-2.6.5 shows a typical spectrum in which the percent transmission vs. the wave number is plotted. Zero absorption corresponds to 100% transmission.

COMPOUND IDENTIFICATION. Infrared (vibrational) spectroscopy is a powerful tool in the identification of compounds. Specific groups within molecules absorb in definite regions of the infrared. In many cases, this fact allows one to tell at a glance what groups are present on a molecule. Table 8-2.6.2 lists some of the group frequencies that can be used to identify compounds. The following example shows how these can be used.

8-2.6.1 EXAMPLE

Label each of the absorptions in the following spectrum for CH_3-CH_2-OH.

Table 8-2.6.2
Some Representative Group Stretching Frequencies (cm^{-1})

-O-H	Stretch	3640-3600	(Lower if hydrogen bonding occurs)
-C-H	in C=C-H	3305-3270	
	aromatic ring	3030	
	$-CH_3$ Symmetric	2872	Also, 1450-1100
	$-CH_3$ Asymmetric	2962	
	$-CH_2$ Symmetric	2853	Also, 1450-1100
	$-CH_2$ Asymmetric	2926	
-C≡N	Stretch	2260 2240	
>C=O	Stretch	1900-1580	
>C=C<	Stretch	1680-1630	
$-C\langle^{O}_{O}$	Stretch Symmetric	1400-1300	
	Stretch Asymmetric	1610-1550	

Solution

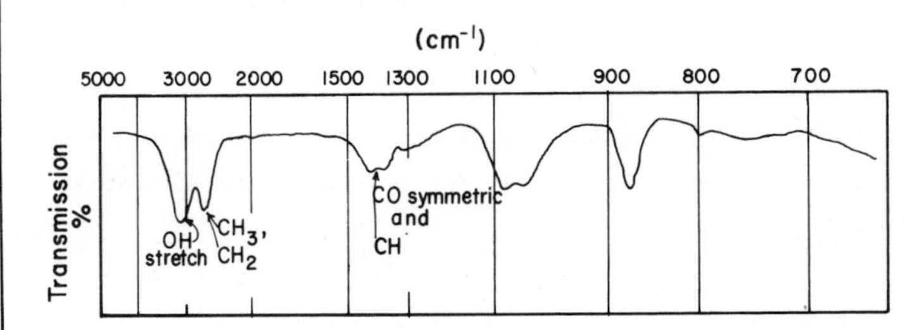

Note the nonlinear scale. Many instruments produce a spectrum that is linear in wavelength, and therefore nonlinear in $\bar{\nu}$. Not all the absorptions can be labeled readily with the information at hand.

Another major use is the comparison of an experimental spectrum with the spectrum of a known compound. This allows us to identify compounds definitely in many cases. Further details of the use of IR spectra for identification are outside the scope of this discussion.

Another use of vibrational spectroscopy is in the measurement of force constants. The force constant of a bond is directly related to the bonding between the atoms involved. This offers a good test of our predicted bond strengths. The above treatment is only approximately correct. Section 8-3 will give a discussion of some correction factors that must be applied.

8-2.7 VIBRATIONAL MODES: DEGREES OF FREEDOM

One question which might arise is: Can we find a way to predict how many absorptions there will be? The answer is yes. All we need to do is calculate the *number of degrees of freedom of vibration* there are. That will give the number of vibrational transitions that *can* occur.

In any molecule with n atoms, the location in space of every atom will be specified by 3n coordinates. Thus there are 3n *degrees of freedom* required to completely specify the position of every atom. It is not necessary that all the degrees of freedom be coordinates. Some can be coordinates and some velocities. There will still be 3n variables required to completely specify the shape or position of a molecule. Usually we reference everything to the *center of mass* of the molecule. If we do that we can describe motion of the *whole* molecule through space with three velocity components. *This uses three degrees of freedom, leaving 3n - 3.*

Rotation relative to the center of mass can be described as components along the principle moments of inertia of the molecule. *For linear systems there are two moments of inertia. For nonlinear systems there are three. Thus, rotation accounts for two degrees of freedom in a linear system and three in a nonlinear system.* This leaves 3n - 5 or 3n - 6 degrees of freedom to describe any other motion. The remaining degrees of freedom are the *normal modes of vibration*. There are *3n - 5* normal modes in a *linear* molecule and *3n - 6* in a *nonlinear* molecule. Normal modes can be thought of as principle axes of vibration. Just as the velocity of the center of mass can be resolved into three components, and rotation can be resolved along two (or three) principle inertial axes, vibration can be resolved into 3n - 5 (or 3n - 6) modes. Any vibration can be described as combinations of these modes.

For some representative molecules we get the following results:

H_2O nonlinear n = 3
vibrational modes 3n - 6 = 9 - 6 = 3

CO linear n = 2
vibrational modes 3n - 5 = 6 - 5 = 1

CO_2 linear n = 3
vibrational modes 3n - 5 = 9 - 5 = 4

In CO_2 only two modes are observed. This is due to the fact that two of the modes are degenerate (same energy) so they appear as one absorption. Also, one mode is infrared inactive. Figures 8-2.7.1 and 8-2.7.2 show the normal modes of vibration for H_2O and CO_2. Figure 8-2.7.3 gives the observed spectrum for water. The vibra-

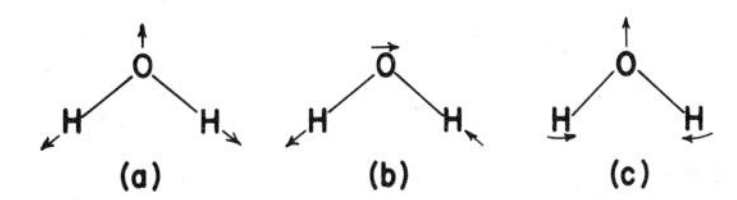

FIG. 8-2.7.1 Normal modes of vibration for H_2O: (a) symmetric stretch, ν_0^i = 3,652 cm^{-1}; (b) asymmetric stretch, ν_0^{ii} = 3,756 cm^{-1}; and (c) scissor bend, ν_0^{iii} = 1,545 cm^{-1}

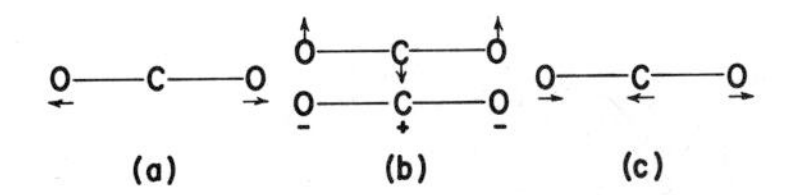

FIG. 8-2.7.2 Normal modes of vibration for CO_2: (a) symmetric stretch, ν_0^i = 1,320 cm^{-1} (infrared inactive); (b) bend (degenerate; motion out of place of the paper up is indicated by +, and down is indicated by -), ν_0^{ii} = 668 cm^{-1}; and (c) asymmetric stretch, ν_0^{iii} = 2,350 cm^{-1}.

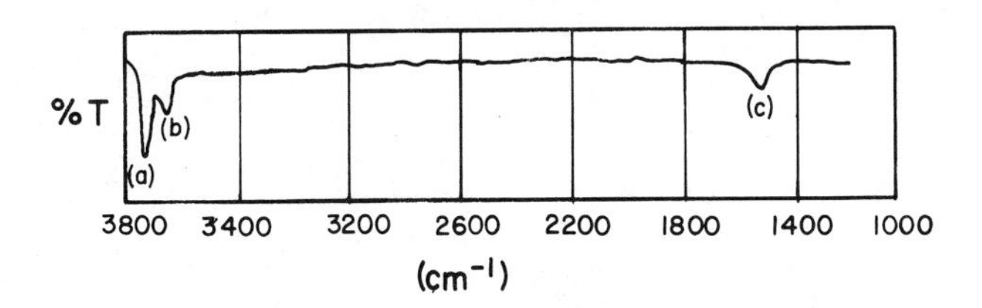

FIG. 8-2.7.3 Spectrum of H_2O (in CCl_4). The letters correspond to the motions depicted in Fig. 8-2.7.1

tional spectrum of each of the molecules shown is not simply two or three absorptions if high sensitivity is used in recording the spectrum. Many combination bands (sums of two modes) and overtones ($\Delta v = \pm 2, \pm 3, \ldots$) are also observed. These are lower in intensity than the fundamentals. Their occurrence represents a breakdown of the harmonic oscillator selection rules which forbid them.

8-2.8 PROBLEMS

1. The internuclear distance for NaCl (g) is 2.36×10^{-10} m. Determine the frequency, energy, and wave number of the first four rotational transitions.
2. Carbon sulfide (CS) has $B = 24.58435 \times 10^9\ s^{-1}$ and $\nu_0 = 1285.1\ cm^{-1}$. Plot on an energy diagram the first three vibrational levels and several rotational levels for each. On the same scale mark an energy difference equal to kT at room temperature (k is Boltzmann's constant = R/N. The quantity kT represents the thermal energy available at a given temperature. You will see more of this in Chap. 9.)
3. An absorption for CO is observed at 2,170 cm^{-1}. Find the force constant for CO.
4. The molecule $^{81}Br^{19}F$ has J = 1 to J = 2 transition at $42.6228 \times 10^9\ s^{-1}$. Find $\Delta\varepsilon_{2\to3}$, $\Delta\varepsilon_{4\to5}$, I, μ, and r for this molecule.
5. For the linear molecule $^{14}N^{14}N^{16}O$, $r_{NN} = 1.126 \times 10^{-10}$ m and $r_{NO} = 1.191 \times 10^{-10}$ m. Determine I and $\Delta\varepsilon_{1\to2}$ for this molecule.
6. The molecule $^{79}Br^{12}C^{14}N$ has three vibrational frequencies, two of which are $\bar{\nu}_1 = 580\ cm^{-1}$ and $\bar{\nu}_3 = 2187\ cm^{-1}$. Determine the force constants for these two motions. State the assumption you had to make to do the calculations.

8-2.9 SELF-TEST

1. For $^{12}C^{32}S$, $B = 24{,}58435 \times 10^9\ s^{-1}$. Assuming r is independent of isotopes present, find r and B for $^{13}C^{32}S$ and $^{12}C^{34}S$.
2. The molecule $^{12}C^{16}O^{32}S$ has three fundamental vibrations with $\bar{\nu}_1 = 859$, $\bar{\nu}_2 = 527$ (degenerate), and $\bar{\nu}_3 = 2079\ cm^{-1}$. Identify the vibrational mode associated with each frequency.
3. Define or discuss five terms from the Objective section.
4. For $^{12}C^{16}O$, $\bar{\nu}_0 = 2170.21\ cm^{-1}$ and for $^{13}C^{16}O$, $\bar{\nu}_0 = 2074.81\ cm^{-1}$. Show whether the bond force constant depends on isotopic substitution. Assume r is constant.

8-2.10 ADDITIONAL READING

1.* G. M. Barrow, *Introduction to Molecular Spectroscopy*, McGraw-Hill, New York, 1962.

2.* C. H. Townes and A. L. Schawlow, *Microwave Spectroscopy*, McGraw-Hill, New York, 1955.

3.* W. J. Moore, *Physical Chemistry*, 4th ed., Prentice-Hall, Englewood Cliffs, N. J., 1972, pp. 622, 623.

4. J. C. Davis, Jr., *Advanced Physical Chemistry*, Ronald Press, New York, 1965.

5. J. E. Wollrab, *Rotational Spectra and Molecular Structure*, Academic, New York, 1967.

8-3 ROTATIONAL-VIBRATIONAL SPECTROSCOPY

OBJECTIVES

You shall be able to

1. Apply appropriate correction terms to rotational and vibrational energy expressions to fit experimental data more closely
2. Interpret a rotation-vibration spectrum and derive appropriate physical parameters from it

8-3.1 INTRODUCTION

At various points in the previous section it was indicated that our simple models (the rigid rotor and the harmonic oscillator) were inadequate. The internuclear distance is not a constant, so the moment of inertia will depend on the vibrational state. Also, the bond length will be affected to some extent by the rate of rotation. Centrifugal effects will tend to increase the bond length as J increases. Finally, interactions between nuclear spin states and rotational states can occur. These complicate the rotational spectrum considerably. The potential energy function for vibration is not a simple harmonic function. Instead, anharmonicity corrections have to be applied to compensate for distortion of the potential from a simple harmonic. This correction becomes more important as v increases and the energy gets closer to the top of the potential well. Finally, in the gas phase it is possible to resolve individual J transitions within a vibrational transition. This gives the rotation-vibration spectrum. Such a resolution is not possible in liquids or solids, since in these phases no well-defined rotational energy levels exist.

This is a good place to reintroduce a concept developed in Sec. 7-1.7. That concept is the Heisenberg uncertainty principle, which states that we cannot simultaneously know the position and momentum of a particle to any desired degree of accuracy. If position is determined precisely, there will be a large uncertainty in the momentum. An alternate form of this principle is

$$\Delta'\varepsilon \, \Delta't \gtrsim h \qquad (8\text{-}3.1.1)$$

which states that the uncertainty, in the energy ($\Delta'\varepsilon$) of a given quantum state, multiplied by the uncertainty in the time ($\Delta't$) the system is in that state, is greater than Planck's constant. (Do not confuse $\Delta'\varepsilon$ here with a transition.)

How does this apply to gas phase and solution rotational motion? Let us take a little digression to get to that. Consider a gas phase system in which we are studying rotational transitions. If microwave energy is put into the system at an appropriate frequency, transitions between rotational states can occur. The molecules that undergo the transitions are not in their most stable states and can "relax" back to their initial states if they undergo collisions. It is observed in microwave spectroscopy that increasing the pressure in a system generally results in a broader absorption. Can you explain that fact on the basis of the previous paragraphs? As the pressure increases, the number of collisions increases, leading to more rapid relaxation. This means that the time spent in the excited rotational states is decreased. That implies $\Delta't$ is decreased, which leads to a larger $\Delta'\varepsilon$. A larger uncertainty in ε (or ν) shows up experimentally as a wider, or broader, absorption peak. A similar argument applies to solutions. In a solution, collisions are very frequent. This gives a small $\Delta't$ for any excited rotational states. The uncertainty in energy becomes very large. Large enough, in fact, that the rotational absorption is so broad that it is unobservable.

8-3.2 ROTATIONAL LEVEL CORRECTIONS

Here we will discuss the effects of centrifugal distortion and what is called *hyperfine splitting*, a splitting of energy levels due to the interaction of rotational motion with other angular momenta. The effect of vibration on the rotational levels will be discussed in the section on rotation-vibration spectra. Only linear molecules or symmetric tops with K = 0 are considered.

CENTRIFUGAL DISTORTION. The effect of centrifugal stretching on rotational levels can be derived. However, it is more convenient here to treat the effect as an empirically derived quantity. You may expect the effect to increase rapidly as J (and the rate of rotation) increases. If so, you are correct. This is reflected in the equation

$$\varepsilon_J = BJ(J + 1) - DJ^2(J + 1)^2 \qquad (8\text{-}3.2.1)$$

D is the *centrifugal distortion constant* which is always much less than B. The effect is significant for high values of J. Table 8-3.2.1 shows the effect of includ-

Table 8-3.2.1
Comparison of Theoretical and Experimental Absorption for HCl

Transition	($\times 10^{12}$ s^{-1})[a]		
$J \rightarrow J + 1$	ν_{obs}	ν_{cal}[b]	ν_{cal}[c]
$3 \rightarrow 4$	2.489	2.480	2.490
$4 \rightarrow 5$	3.121	3.0998	3.1103
$5 \rightarrow 6$	3.7264	3.7198	3.7291
$6 \rightarrow 7$	4.3479	4.3398	4.3464
$7 \rightarrow 8$	4.9618	4.9597	4.9615
$8 \rightarrow 9$	5.5719	5.5797	5.5743
$9 \rightarrow 10$	6.1871	6.1997	6.1847
$10 \rightarrow 11$	6.7902	6.8196	6.7917

[a] Observed (obs) and calculated (cal) values.
[b] $B = 3.100 \times 10^{11}$ s^{-1} and $\nu = 2B(J + 1)$.
[c] $B = 3.1163 \times 10^{11}$ s^{-1}, $D = 1 \times 10^{7}$ s^{-1}, and $\nu = 2B(J + 1) - 4D(J + 1)^3$.

ing D in the calculations of transition energies for HCl. These data are from G. Herzberg, *Spectra of Diatomic Molecules* (see Sec. 8-3.7, Ref. 1). It is clear from Table 8-3.2.1 that the inclusion of a distortion constant allows a much closer fit to the experimental data.

8-3.2.1 EXAMPLE

Prove that $\Delta\varepsilon_{J\rightarrow J+1} = 2B(J + 1) - 4\bar{D}(J + 1)^3$ from Eq. (8-3.2.1)

Solution

$$\varepsilon_J = BJ(J + 1) - DJ^2(J + 1)^2$$

$$\varepsilon_{J+1} = B(J + 1)(J + 2) - D(J + 1)^2(J + 2)^2$$

$$\begin{aligned}\Delta\varepsilon_{J\rightarrow J+1} &= B[(J + 1)(J + 2) - J(J + 1)] - D[(J + 1)^2(J + 2)^2 - J^2(J + 1)^2]\\ &= B[(J + 1)(J + 2 - J)] - D[(J + 1)^2(J + 2)^2 - J^2]\\ &= B[2(J + 1)] - D[(J + 1)^2(J^2 + 4J + 4 - J^2)]\\ &= B[2(+ 1)] - D[(J + 1)^2 4(J + 1)]\end{aligned}$$

$$\boxed{\Delta\varepsilon_{J\rightarrow J+1} = 2B(J + 1) - 4D(J + 1)^3}$$

8 3.2.2 EXAMPLE

For gaseous $^{133}Cs^{79}Br$, $B = 1.08134 \times 10^9\ s^{-1}$, and $D = 2.7 \times 10^2\ s^{-1}$. Find $\Delta\varepsilon_{2\to3}$ for CsBr.

Solution

$$\begin{aligned}\Delta\varepsilon_{2\to3} &= 2B(2 + 1) - 4D(2 + 1)^3 \\ &= 6B - 108D \\ &= 6(1.08134 \times 10^9) - 108(2.7 \times 10^2) \\ &= 6.48804 \times 10^9 - 2.9160 \times 10^4 \\ &= \underline{\underline{6.4880 \times 10^9}}\end{aligned}$$

In this case D is insignificant for low values of J (0.00045% change). You find the percent effect for $J = 10 \to 11$.

8-3.3 VIBRATIONAL LEVEL CORRECTIONS

In Sec. 8-2, the potential energy was treated as a harmonic oscillator function. However, physically this cannot be true since molecules can dissociate. In a harmonic potential this could not occur. (Why not?) Figure 8-3.3.1 shows a Morse potential which more nearly coincides with observed data. Also shown is the harmonic potential we have used previously. Near the bottom of the potential, the harmonic approximation is adequate in some cases but not in others. A better approximation is to apply a Maclaurin series expression about q = 0.

$$V(q) = V_{q=0} + \left(\frac{dV}{dq}\right)_{q=0} q + \frac{1}{2!}\left(\frac{d^2V}{dq^2}\right)_{q=0} q^2 + \frac{1}{3!}\left(\frac{d^3V}{dq^3}\right)_{q=0} q^3 + \cdots \qquad (8\text{-}3.3.1)$$

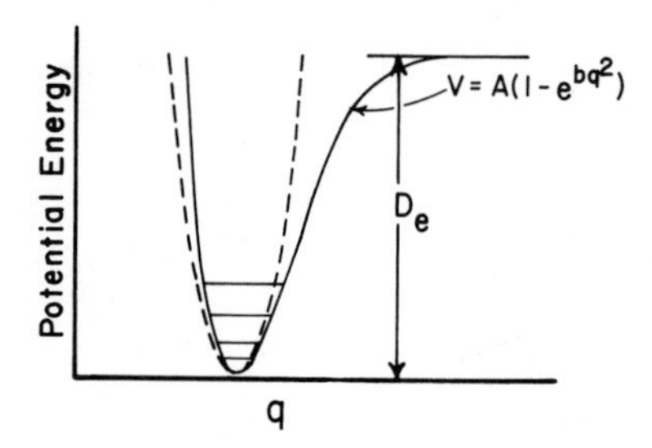

FIG. 8-3.3.1 A Morse potential and a harmonic potential for a diatomic molecule. D_e is the energy required to dissociate the molecule.

$V = 0$ at $q = 0$ is the usual choice, and $(dV/dq)_{q=0} = 0$ since $q = 0$ represents the minimum. The third term is the harmonic approximation. By including the fourth term a better approximation is obtained. This potential can be used in the Schrödinger equation to produce a better energy expression. This is not an easy task. It is usually done by a method known as the *perturbation method*, in which the correction term is assumed to be small and is added to the Hamiltonian (see Sec. 7-4.2). This technique allows the unperturbed wave functions to be used to calculate the energy as follows:

$$\varepsilon_n' = \int \Psi_n^{0*}\hat{\mathcal{H}}'\Psi_n^0 \, d\tau \tag{8-3.3.2}$$

where ε_n' is the correction to the energy, Ψ_n^0 are the wave functions for the unperturbed state (before the correction is applied), and $\hat{\mathcal{H}}'$ is the correction term in the Hamiltonian operator. The result of this treatment is

$$\bar{\varepsilon}_v = \bar{\nu}_e(v + \tfrac{1}{2}) - \bar{\nu}_e x_e(v + \tfrac{1}{2})^2 \tag{8-3.3.3}$$

where $\bar{\nu}_e$ is spacing (per centimeter, the most common unit) of energy levels for the pure harmonic oscillator, and $\bar{\nu}_e x_e$ is the *anharmonicity constant*. It is always small and positive. As v increases you should conclude that the spacings decrease slightly. Before continuing, derive the energy expression for a transition from v to v′. (The selection rule is still approximately $\Delta v = \pm 1$, but overtone bands can occur, although their intensity will be much less than the fundamental.) The result you should have obtained is

$$\boxed{\Delta\bar{\varepsilon}_{v\to v'} = \bar{\nu}_e(v' - v) - \bar{\nu}_e x_e[v'(v' + 1) - v(v + 1)]} \tag{8-3.3.4}$$

If we make all our observations from $v = 0$, Eq. (8-3.3.4) reduces to

$$\Delta\varepsilon_{0\to v'} = \bar{\nu}_e v' - \bar{\nu}_e x_e[v'(v' + 1)] \tag{8-3.3.5}$$

Table 8-3.3.1 shows the effect of the anharmonicity constant on the agreement of observed and calculated spectra of HCl. This table is from G. M. Barrow, *Introduction to Molecular Structure*, (see Sec. 8-3.7, Ref. 2). The table should convince you of the importance of the anharmonicity term. Table 8-3.3.2 lists values of $\bar{\nu}_e$ and $\bar{\nu}_e x_e$ for several molecules. (Note the differences in Table 8-2.6.1.)

8-3.3.1 EXAMPLE

Given the following data find $\bar{\nu}_e$ and $\bar{\nu}_e x_e$ in Eq. (8-3.3.3) for $^{16}O^{16}O$.

$$\Delta\bar{\varepsilon}_{v=0\to1} = 1556.22 \text{ cm}^{-1}$$

Table 8-3.3.1
Observed and Calculated Vibrational Transitions for HCl[a]

Transition	$\bar{\nu}_{obs}$ (cm^{-1})	$\bar{\nu}_{cal}$ (cm^{-1}) Harmonic oscillator	$\bar{\nu}_{cal}$ (cm^{-1}) Anharmonic oscillator
$0 \to 1$	2,885.9	2,885.90[b]	2,885.70
$0 \to 2$	5,668.0	5,771.8	5,668.20
$0 \to 3$	8,347.0	8,657.7	8,347.50
$0 \to 4$	10,923.1	11,543.6	10,923.6
$0 \to 5$	13,396.5	14,429.5	13,396.5

[a] $\bar{\nu}_0$(harmonic oscillator) = 2885.9 cm^{-1}, $\bar{\nu}_e$ = 2988.90 cm^{-1}, and $\bar{\nu}_e x_e$ = 51.60 cm^{-1}.
[b] This must fit since $\bar{\nu}_0$ is determined by the $0 \to 1$.

$$\Delta\bar{\varepsilon}_{v=0\to2} = 3088.28 \text{ cm}^{-1}$$

$$\Delta\bar{\varepsilon}_{v=0\to3} = 4596.21 \text{ cm}^{-1}$$

Solution

$$-\ 2(\Delta\bar{\varepsilon}_{v=0\to1} = \bar{\nu}_e - 2\bar{\nu}_e x_e = 1556.22 \text{ cm}^{-1})$$

$$\Delta\bar{\varepsilon}_{v=0\to2} = 2\bar{\nu}_e - 6\bar{\nu}_e x_e = 3088.28 \text{ cm}^{-1}$$

$$-2\bar{\nu}_e x_e = -24.16$$

therefore,

$$\bar{\nu}_e x_e = \pm 12.08 \text{ cm}^{-1} \quad \text{and} \quad \bar{\nu}_e = 1580.38 \text{ cm}^{-1}$$

Table 8-3.3.2
Values of $\bar{\nu}_e$ and $\bar{\nu}_e x_e$ for Several Molecules

Molecule	$\bar{\nu}_e$ (cm^{-1})	$\bar{\nu}_e x_e$ (cm^{-1})
$^{1}H^{1}H$	4395.2	117.90
$^{1}H^{2}H$	3817.09	94.958
$^{2}H^{2}H$	3118.5	64.10
$^{12}C^{16}O$	2170.21	13.461
$^{14}N^{14}N$	2359.61	14.456

Check:

$$\Delta\bar{\varepsilon}_{v=0\to3} = 3\bar{\nu}_e - 12\bar{\nu}_e x_e = \underline{\underline{4596.18\ \text{cm}^{-1}}}$$

Close enough!

8-3.4 VIBRATION-ROTATION IN MOLECULES

ROTATIONAL CONSTANT. Earlier, it was indicated that the rotational constant is affected by the vibrational state. As v increases the effective internuclear distance is increased. This leads to a moment of inertia that is larger, and therefore a rotational constant that is smaller. The dependence of the rotational constant on v is usually expressed as

$$\bar{B}_v = \bar{B}_e - \bar{\alpha}_e(v + \tfrac{1}{2}) \qquad (8\text{-}3.4.1)$$

where the overbar indicates all quantities are per centimeter. Some values of $\bar{B}_e$ are given in Table 8-3.4.1 along with other appropriate data. Note that r is very dependent on the vibrational level in question. In pure rotational spectroscopy care must be exercised in reporting internuclear distances, since the value obtained is dependent on the vibrational state (as well as the rotational state as shown in Sec. 8-3.2) being studied. Also note that our assumption that r does not change with isotopic substitution appears to be valid for the equilibrium length and the length measured in v = 0.

ROTATION-VIBRATION SPECTRUM. In the gas phase, many vibrational transitions show fine structure due to the rotational levels. Figure 8-3.4.1 illustrates the origin of this fine structure, and Fig. 8-3.4.2 gives an actual rotation-vibration spectrum. The figure summarizes all the important points. Perhaps it would be instructive to write general expressions for the R-branch (J → J + 1) and P-branch

Table 8-3.4.1
$\bar{B}_e$ and Bond Distance for Two Vibrational Levels for Several Molecules

			(10^{-10} m)		
Molecule	$\bar{B}_e$ (cm^{-1})	$\bar{\alpha}_e$ (cm^{-1})	r_e	r_0	r_1
H_2	60.809	2.993	0.7417	0.7505	0.7702
HD	45.655	1.993	0.7414	0.7495	0.7668
D_2	30.429	1.049	0.7416	0.7481	0.7616
HCl	10.5909	0.3019	1.27460	1.2838	1.3028
DCl	5.445	0.1118	1.275	1.282	1.295
CO	1.9314	0.01748	1.1282	1.1307	1.1359

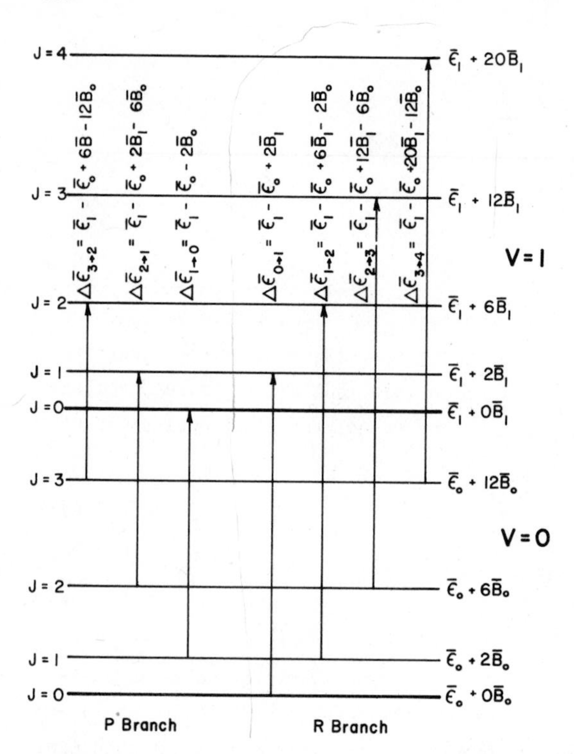

FIG. 8-3.4.1 Rotation-vibration transitions for the v = 0 to v = 1 transition of a linear molecule $\bar{\varepsilon}_0 = (1/2)\bar{\nu}_e - (1/4)\bar{\nu}_e x_e$, and $\bar{\varepsilon}_1 = (3/2)\bar{\nu}_e - (9/4)\bar{\nu}_e x_e$. $\bar{B}_0$ and $\bar{B}_1$ are defined by Eq. (8-3.4.1).

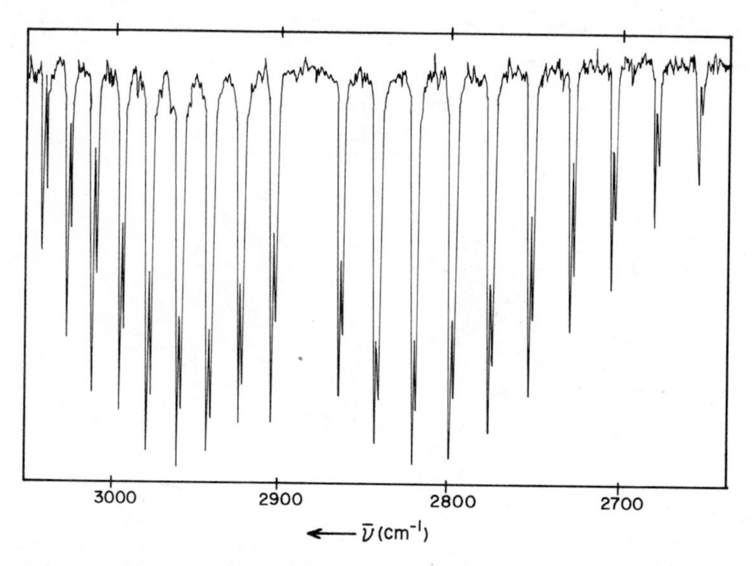

FIG. 8-3.4.2 Rotation-vibration spectrum of HCl (gas phase).

(J → J - 1) transitions.

R branch:

$$\Delta\bar{\varepsilon}_R(v = 0 \rightarrow 1) = \bar{\varepsilon}_1 - \bar{\varepsilon}_0 + (J + 1)(J + 2)\bar{B}_1 - J(J + 1)\bar{B}_0$$

(8-3.4.2)

P branch:

$$\Delta\bar{\varepsilon}_P(v = 0 \rightarrow 1) = \bar{\varepsilon}_1 - \bar{\varepsilon}_0 + (J - 1)(J)\bar{B}_1 - J(J + 1)\bar{B}_0$$

Substituting for $\bar{\varepsilon}_1$ and $\bar{\varepsilon}_0$:

$$\bar{\varepsilon}_1 = \frac{3}{2}\bar{\nu}_e - \frac{9}{4}\bar{\nu}_e x_e$$

$$\bar{\varepsilon}_0 = \frac{1}{2}\bar{\nu}_e - \frac{1}{4}\bar{\nu}_e x_e$$

and $\bar{B}_1$ and $\bar{B}_0$:

$$\bar{B}_1 = \bar{B}_e - \frac{3}{2}\bar{\alpha}_e$$

$$\bar{B}_0 = \bar{B}_e - \frac{1}{2}\bar{\alpha}_e$$

gives the general forms (show all steps), except for the absence of the distortion constant D. You should modify the equation to include it.

R branch J → J+1

$$\Delta\bar{\varepsilon}_R(v = 0 \rightarrow 1) = \bar{\nu}_e - 2\bar{\nu}_e x_e + 2(J + 1)\bar{B}_e - (J + 3)(J + 1)\bar{\alpha}_e$$

P branch J → J-1 (8-3.4.3)

$$\Delta\bar{\varepsilon}_P(v = 0 \rightarrow 1) = \bar{\nu}_e - 2\bar{\nu}_e x_e - 2J\bar{B}_e - J(J - 2)\bar{\alpha}_e$$

Depending on the data given, Eq. (8-3.4.3) or Eq. (8-3.4.2) can be used to calculate the expected vibration-rotation spectrum.

The utility of this procedure lies in the fact that rotational data can be obtained from an infrared spectrometer for many molecules. Each must have sufficient vapor pressure to yield an infrared spectrum for the gas phase. If conditions are ideal the rotational fine structure can be observed, leading to values of $\bar{\nu}_e$, $\bar{B}$, and perhaps even $\bar{\nu}_e x_e$ and $\bar{\alpha}_e$.

8-3.4.1 EXAMPLE

For $^{12}C^{16}O$, $\bar{B}_e = 5.78975 \times 10^{10}\ s^{-1}$ ($1.9314\ cm^{-1}$), $\bar{\alpha}_e = 5.24 \times 10^8\ s^{-1}$ ($0.01748\ cm^{-1}$), and $\bar{\nu}_e = 2170.21\ cm^{-1}$. Calculate the first two R-branch and P-branch transitions.

Solution

R branch

$$\Delta\bar{\varepsilon}_{J=0\to1} = \bar{\nu}_e + 2\bar{B}_e - 3\bar{\alpha}_e$$

$$= 2170.21 + 3.863 - 0.052$$

$$= \underline{\underline{2174.02 \text{ cm}^{-1}}}$$

$$\Delta\bar{\varepsilon}_{J=1\to2} = \bar{\nu}_e + 4\bar{B}_e - 8\bar{\alpha}_e$$

$$= 2170.21 + 7.726 - 0.140$$

$$= \underline{\underline{2177.80 \text{ cm}^{-1}}}$$

P branch

$$\Delta\bar{\varepsilon}_{J=1\to0} = \bar{\nu}_e - 2\bar{B}_e - 0\bar{\alpha}_e$$

$$= 2170.21 - 3.863$$

$$= \underline{\underline{2166.35 \text{ cm}^{-1}}}$$

$$\Delta\bar{\varepsilon}_{J=2\to1} = \bar{\nu}_e - 4\bar{B}_e - 2\bar{\alpha}_e$$

$$= 2170.21 - 7.726 - 0.035$$

$$= \underline{\underline{2162.45 \text{ cm}^{-1}}}$$

Compare with the solution to Problem (1), Sec. 8-3.5 in which $\bar{\nu}_e x_e$ is included.

8-3.5 PROBLEMS

1. Use the data in Tables 8-3.3.2 and 8-3.4.1 to calculate the vibration-rotation spectrum of $^{12}C^{16}O$, including corrections for $\bar{\nu}_e x_e$ and α_e. Find and plot the energies of the first two R- and P-branch transitions. Compare with Example 8-3.4.1.
2. For $^{2}H^{79}Br$, $B_0 = 127.3582 \times 10^9$ s^{-1}, $\alpha_e = 1.258 \times 10^9$ s^{-1}, and $D = 2.8 \times 10^6$ s^{-1}. Find B_e and find $\Delta\varepsilon$ for $J = 3 \to 4$, $J = 10 \to 11$, and $J = 20 \to 21$ in the $v = 0$ state. Calculate the percent error that would result for each if D were ignored.
3. The fundamental and overtone frequencies of the CH stretch in $CHCl_3$ are 3,019, 5,900, 8,700, and 11,315 cm^{-1}. Determine $\bar{\nu}_e$ and $\bar{\nu}_e x_e$.
4. The following data are known for $^{12}C^{16}O$: $B_e = 5.78975 \times 10^{10}$ s^{-1}, $\alpha_e = 5.24 \times 10^8$ s^{-1}, $r_e = 1.128 \times 10^{-10}$ m, $\bar{\nu}_e = 2170.21$ cm^{-1}, $D = 1.834 \times 10^5$ s^{-1}, and

$\bar{\nu}_e x_e = 13.5\ \mathrm{cm}^{-1}$. Derive a general equation relating the frequency of a v, J to v + 1, J + 1 transition for HCl, then determine the frequency of the v = 1, J = 10 → v = 2, J = 11 transition.

8-3.6 SELF-TEST

1. Use the data from Sec. 8-3.5, Problem (2) to find $\Delta\varepsilon_{J=0\to1}$ for v = 0 and $\Delta\varepsilon_{J=0\to1}$ for v = 4.
2. For a given diatomic molecule the following energies were observed for rotation-vibration transition (v = 0 → 1). Assume D = 0 and find all the following as accurately as possible: $\bar{\nu}_e$, $\bar{\nu}_e x_e$, α_e, $\bar{B}_1$, and $\bar{B}_0$ (or $\bar{B}_e$ and α_e). All values are per centimeter: 2978.75, 2961.07, 2942.72, 2923.72, 2904.11, 2863.02, 2841.58, 2819.56, 2796.97, and 2773.82. Also find I and r.

8-3.7 ADDITIONAL READING

1.* G. Herzberg, *Spectra of Diatomic Molecules*, Van Nostrand, Princeton, N. J., 1950.
2.* G. M. Barrow, *Introduction to Molecular Structure*, McGraw-Hill, 1962, p. 45.

8-4 OTHER TYPES OF SPECTROSCOPY

OBJECTIVES

You shall be able to

1. Discuss fully any type of spectroscopy presented in this section
2. Interpret nuclear magnetic resonance (NMR) patterns given or predict NMR patterns for given molecules

8-4.1 INTRODUCTION

The major emphasis up to this point has been on rotational and vibrational spectroscopy. We did learn a little about electronic spectra of atoms in Sec. 7-4. Details for molecules will be contained in this section. There are many other important types of spectroscopy, including Raman, nuclear magnetic resonance (NMR), electron spin resonance (ESR), and mass spectroscopy, in addition to other specialized techniques. Electronic, Raman, and NMR spectroscopy will be discussed in order to indicate what is being measured. Also, the methods used to relate the measurements to molecular or atomic structure will be given. Finally, a short discussion of a very important recent development (20 years ago!), the laser, will be presented.

8-4.2 ELECTRONIC SPECTROSCOPY: DIATOMIC MOLECULES

Molecules can absorb or emit radiation due to transitions between electronic energy levels. Such transitions were mentioned for atoms in Secs. 7-4.4 and 7-4.5. The electronic spectrum of a multi-electron atom is not simple. Even less simple are the spectra of molecules. Instead of having fairly distinct transitions as in atoms, electronic transitions in molecules usually occur as bands. The bands arise because of the vibrational and rotational levels within each electronic level involved. Thus, an electronic transition is not one, but rather multiple transitions among various allowed rotational and vibrational levels. Usually, the band structure is too complex to be interpreted readily. However, the center of the band and its general shape gives information about the electronic level spacing.

We have already seen, in Secs. 8-1.4 and 8-1.7, that the electronic energy levels in diatomic molecules can be adequately described by molecular orbital methods. Figure 8-4.2.1 shows two allowed electronic states for a molecule. Also shown are two (of many) possible transitions between these two levels. A few points need to be made. First, note that the energy minima of the two levels are different. Excited electronic states usually will have different amounts of bonding leading to a different interncular distance. The lines a and b are drawn in accordance with the *Frank-Condon principle*. This principle says the electronic transition occurs so rapidly the nuclei can be assumed fixed. In this event r does not change in the transition. Thus, the transition line is drawn vertically. Transitions are allowed from various vibrational levels in the ground state to vibrational levels in the excited state. The transitions will normally occur from the region in the vibrational state that has the highest probability. For v = 0 this is the center of the level (as shown by a). As v increases, the region of maximum probability moves toward the ends of the vibration (as in b). (See Fig. 8-2.6.3.) Another point is

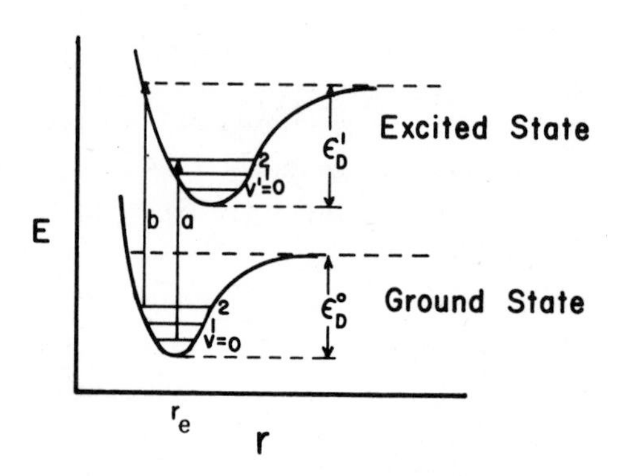

FIG. 8-4.2.1 Transition between two electronic energy levels.

that electronic transitions can lead to dissociation. If transition b occurs, the molecule ends up in the excited state with an energy greater than the dissociation energy (ε_D') of the molecule in the excited state. If this transition occurs the molecule immediately dissociates. From any of the vibrational levels in the ground state, there are several vibrational levels in the excited state that are accessible. Also, rotational levels are present within all the vibrational levels. Thus, any transition from one electronic and vibrational state to an excited electronic state leads to several absorptions. Emission spectra also show rotational-vibrational fine structure. The considerations mentioned above apply to emission spectra as well as absorption spectra. Figure 8-4.2.2 shows only one of several possible series of emissions for N_2. Those shown correspond to $v' \rightarrow v' + 2$. There are other series for which $\Delta v = \ldots, -2, -1, 0, 1, 2, 3, \ldots$. The large number of emissions within each band is due to the rotational levels within each vibrational level.

Thus far we have considered only two electronic levels, but in Secs. 8-1.4 and 8-1.7 we saw that many electronic levels exist. Since transitions can occur between various electronic levels, the spectrum exhibits a large number of bands. A real case is shown in Fig. 8-4.2.3 from G. Herzberg, *Spectra of Diatomic Molecules*, (see Sec. 8-4.9, Ref. 1). The number of electronic states available in addition to all the rotational and vibrational fine structure, leads to a very complicated spectrum.

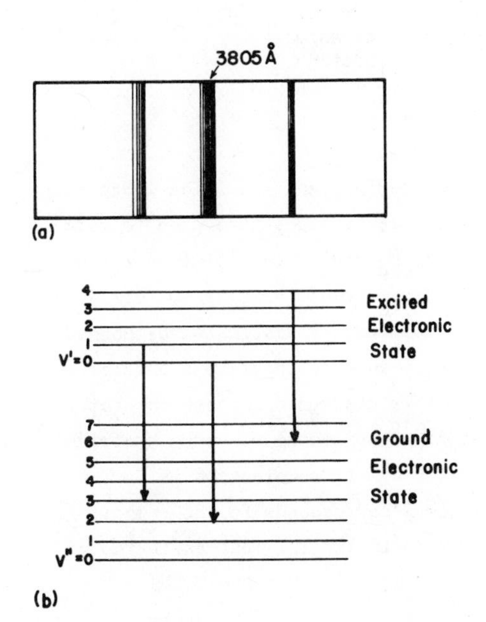

FIG. 8-4.2.2 (a) A small portion of the emission band of N_2 for which $\Delta v = 2$. (b) A schematic showing the vibrational levels involved.

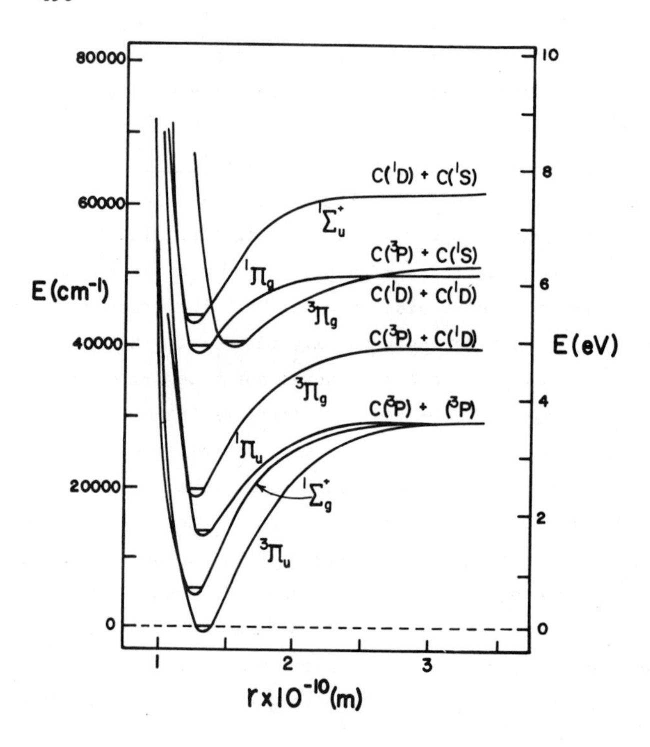

FIG. 8-4.3.2 Potential energy curves for various electronic states of the molecule C_2. [From *Molecular Spectra and Molecular Structure*, Vol. 1., by G. Herzberg (C) 1950, by Litton Educational Publishing, Inc. Reprinted by permission of Van Nostrand Reinhold Company.]

Perhaps some of the symbols used in Fig. 8-4.2.3 need explanation. The letters to the right [such as $C(^3P) + C(^3P)$] give the atomic electronic states used to form the indicated molecular electronic states. The Σ, Π, etc., are explained as follows. The electronic states of molecules are identified by symbols similar to term symbols given for atoms. The electronic arrangement is most easily designated by considering the total orbital (electronic) angular momentum and the total spin angular momentum along the internuclear axis. The letter Λ is used to designate the total orbital angular momentum. Λ is obtained by summing individual electron contributions (labeled λ_i). For an electron in a σ orbital, $\lambda = 0$. If an electron is in a π or δ orbital $\lambda = 1$ or 2, respectively. $\Lambda = 0$, 1, or 2 corresponds to the symbols Σ, Π, or Δ. The spin angular momentum is indicated by a superscript just as in the atomic case. An example is H_2 with a ground state of $(\sigma_{1s})^2$. $\Lambda = 0 + 0 = 0$. $S = 0$. Therefore, the symbol is $^1\Sigma$. More detail is usually given, and the symbol is usually written $^1\Sigma_g^+$. The g stands for *gerade*, meaning even, and designates the net symmetry of the orbital (see Sec. 7-4.4 for a discussion of g and u for atoms).

The plus (or minus) is used to express whether the wave function is symmetric or antisymmetric, respectively, with respect to inversion (see Sec. 8-1.7). Further detail will not be given. (See Davis, Sec. 8-4.9, Ref. 2.)

Several selection rules are applicable. In general,

$$\Delta\Lambda = 0, \pm 1$$

This means that $\Sigma \to \Sigma$, $\Sigma \to \Pi$, $\Pi \to \Sigma$, $\Pi \to \Pi$, $\Pi \to \Delta$, etc., are allowed, but $\Sigma \to \Delta$, $\Phi \to \Pi$, etc., are not. Also, only g → u and u → g are allowed by symmetry considerations. Finally, $\Delta s = 0$ for molecular electronic transitions. This means the spin multiplicity cannot change during a transition. Thus, for a single transition, singlet → singlet or triplet → triplet is allowed, but singlet → triplet or triplet → singlet is not. An interesting example of this is found in some photochemical reactions in the upper atmosphere. By a series of interactions it is possible to produce molecules in excited triplet states. This triplet state cannot "relax" directly to the singlet ground state by emission of radiation. Thus, the molecule is trapped in a triplet state. This plays an important role in its reactivity with other molecules present.

8-4.3 ELECTRONIC SPECTROSCOPY: POLYATOMIC MOLECULES

AREAS OF USE OF ELECTRONIC SPECTROSCOPY. The use of electronic spectroscopy for fundamental theoretical studies in polyatomic molecules is considerably more restricted than in diatomics. This is due mainly to the lack of structure in absorption levels for polyatomics. Most polyatomic molecules are studied as liquids or solids. The absorption band is usually broad and rather structureless in condensed phases. Also, in diatomics, emission spectra are available for a large number of states. In polyatomics, decomposition normally occurs before a sample can be heated to a temperature high enough to yield significant populations in excited states. Thus, few emission bands can be observed. This leads to a dearth of data. For diatomics, theoretical calculations could be made and the results related or checked by experimental results. All but the simplest polyatomics present formidable problems when attempts are made to calculate electronic energy levels. All this is not to say that electronic spectroscopy is useless for polyatomic molecules; far from it.

The absorption of radiation is often associated with a specific group within a molecule. This group normally gives an absorption in a particular spectral region which is almost independent of the rest of the molecule. You have already seen (Sec. 8-2.6) that certain infrared absorptions can be identified with specific bond vibrations (such as the C-H stretch, C=O stretch, C-H bend, etc.), Similar considerations apply for ultraviolet-visible absorptions for electronic transitions in specific groups of atoms. An atomic group and its associated chemical bonds that give rise to an electronic absorption spectrum is called a *chromophore*. Examples are the $\gt C{=}O$ group in aldehydes and ketones, C=C bonds, and NO and NO_2 groups, among others. The absorptions associated with these groups are only slightly sensitive to the composition of the remainder of the molecule. Environmental parameters (pH, polarity of solvent, etc.) have a noticeable effect on the position of the absorption maximum but not enough to obscure the identification of the chromophores present in many cases. The remainder of this section will be concerned, mainly, with some properties of chromophores and how transitions are designated. A few comments will be made about absorptions of aromatic systems. Also, some detail will be given on two phenomena that can occur when a system relaxes from an excited state to a ground state by the emission of radiation. These two are *fluorescence* and *phosphorescence*.

CHROMOPHORES. Before continuing you should review the portions of Sec. 8-1.7 concerning designations of molecular orbitals in diatomics. The terminology developed there will be used here. The visible and ultraviolet absorptions in organic molecules are generated by valence electrons and by nonbonding electrons such as those found isolated on N, O, S, or other atoms. (An example is the lone pair of electrons found on N when it is participating in three bonds.) The spectroscopic designation of an electron in a nonbonding orbital is n. The orbital associated with that electron is also given the symbol n. An electron localized in a bond between two atoms is designated as σ, as is the associated molecular orbital. You have seen this notation used for diatomics. The delocalized electrons from shared p orbitals are called π electrons. Many types of transitions are observed. Some of these are detailed below:

1. *Bonding to antibonding transition:* $\sigma \rightarrow \sigma^*$ or $\pi \rightarrow \pi^*$. Normally, the σ bond is so strong only high energy radiation (vacuum UV, $\lambda \approx 100$ nm or higher) is necessary to excite the $\sigma \rightarrow \sigma^*$ transition. The π bond is considerably weaker so it can usually be observed in the visible or UV.
2. *Nonbonding to antibonding:* $n \rightarrow \sigma^*$ and $n \rightarrow \pi^*$. Again, the $n \rightarrow \sigma^*$ is usually a rather high energy transition (far UV, $\lambda \approx 200$ nm). The $n \rightarrow \pi^*$ transition is forbidden by symmetry, but it is often observed as a weak absorption.

Table 8-4.3.1 gives a few transitions of each type. These are all straightforward cases. Note the small variation of the $\pi \rightarrow \pi^*$ for the C=C in various compounds. For more complicated systems it is difficult to assign all the observed absorptions to particular transitions.

CONJUGATED AND AROMATIC SYSTEMS. Conjugated and aromatic systems give rise to $\pi \rightarrow \pi^*$ transitions. It is observed that a displacement to longer wavelengths occurs as the number of conjugated bonds increase. This can be seen in Table 8-4.3.2. (This can be related to increasing the size of the box in the particle in a box

Table 8-4.3.1
Some n → σ* and π → π* Transitions

Compound	Solvent	Transition type	λ (nm = 10^{-9} m)
CH_3OH	Vapor	n → σ*	184
$(CH_3)_2O$	Vapor	n → σ*	184
$(CH_3)_2S$	Ethanol	n → σ*	210
CH_3I	Heptane	n → σ*	258
	Water	n → σ*	249
$H_2C{=}CH_2$	Vapor	π → π*	162
1-Hexene	Vapor	π → π*	179
cis-2-Butene	Vapor	π → π*	174
Cyclohexene	Vapor	π → π*	176

example in Sec. 7-1.7.) Alkyl groups attached to a benzene ring have only a small effect. The larger effect due to the addition of atoms with nonbonded electrons is attributed to conjugation of these with the π system. If this occurs, a n → π* transition is possible. The effect of the C=C group is due to conjugation. Numerous other examples could be given. Perhaps, however, the above will give you an indication of how UV and visible spectroscopy can be used to interpret molecular structure.

FLUORESCENCE AND PHOSPHORESCENCE. All the preceding was concerned with absorption of radiation. Fluorescence and phosphorescence are two processes by which electronically excited molecules can return to the ground state by emission of radi-

Table 8-4.3.2
Absorptions of Conjugated Polyenes (trans) and Aromatic Systems

Compound	Number of double bonds	Solvent	λ (nm)
1,3 Butadiene	2	Hexane	217
1,3,5 Hexatriene	3	Isooctane	268
1,3,5,7,9,11 Dodecahexene	6	Isooctane	364
1,3,5,7,9,11,13,15, Hexadecaoctene	8	Isooctane	410
Benzene	--	Hydrocarbon	254
Toluene	--	Hydrocarbon	262
Phenol	--	Hydrocarbon	271
Benzoic acid	--	Ethanol	272
Aniline	--	Methanol	280
Styrene	--	Hexane	282
Nitrobenzene	--	Hexane	280

ation. There are nonradiative pathways, in general, but they will not concern us here. Perhaps the most cogent reason for interest in phosphorescence is because of its presence or potential presence in photochemical reactions. Photosynthesis is a photochemical process in which the triplet states characteristic of phosphorescence may be involved.

In both fluorescence and phosphorescence, the emitted radiation has a different wavelength than the exciting radiation. The difference in the two will become apparent shortly. Figure 8-4.3.1 shows the vibrational energy levels in two or three excited electronic states. These are simplified diagrams. In fluorescence, after excitation, a return to the ground state is not always instantaneous. Collisions can cause changes in the vibrational state of the excited molecule. When this occurs, a return to the ground state yields an emission that has a different wavelength than the absorption. This is called *fluorescence*. *Phosphorescence* is similar, except the time delay is much longer. This long delay is apparently due to the formation of a triplet excited state. This can occur if a triplet excited state and a singlet excited state overlap. After the absorption has produced a molecule in an excited singlet state that overlaps a triplet, collisions cause deactivation of the vibrational levels. It is possible for the molecule to undergo an internal conversion to yield a triplet state. If this occurs, the molecule is trapped since triplet → singlet is not allowed. Such transitions do occur, but are generally slower than singlet → singlet transitions. The relatively long lifetime and high reactivity of the unpaired electrons make the triplet state a very important entity in photochemical reactions. As mentioned, photosynthesis likely involves the formation of such a triplet state in one of the early steps.

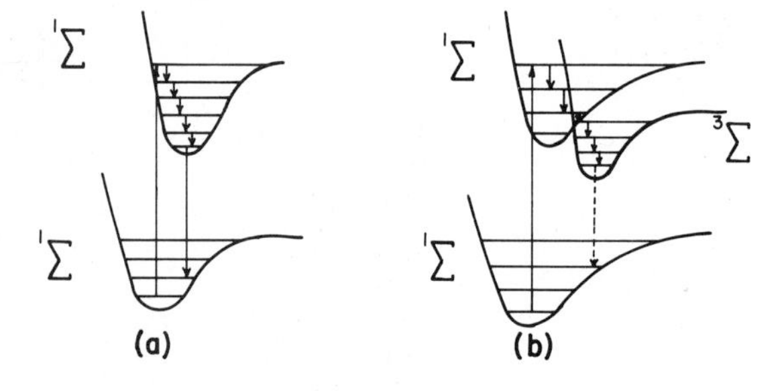

FIG. 8-4.3.1 (a) Absorption and fluorescence. (b) Absorption, internal conversion to a triplet state and phosphorescence.

8-4.4 RAMAN SPECTROSCOPY

Raman spectroscopy complements infrared studies in characterizing vibrational levels. The major difference is that infrared spectroscopy is useful for transitions involving changes in the dipole, while Raman spectra are observed for transitions giving changes in the polarizability of the molecule. (See Sec. 14-2 for definitions and discussion of polarizability.) C. V. Raman (in 1928) was the first person to observe that a small fraction of the light scattered from a sample perpendicular to an incident beam had the wavelength altered. The vast majority of the scattered light is unaltered in wavelength (called *Rayleigh scattering*). If monochromatic (single frequency) radiation is used as the incident beam, the scattered spectrum exhibits a number of lines displaced from the original frequency. The explanation of this phenomenon is the *inelastic scattering* of radiation by molecules. This inelastic scattering results when the radiation interacts with the molecules present and gives up or absorbs energy from them. The energy absorbed or given up by the molecules will result in an emitted beam that is shifted in frequency from the incident beam. The emitted frequency will be

$$\nu'' = \nu \pm \nu' \tag{8-4.4.1}$$

where ν is the incident frequency and ν' is the *Raman shift*. You should note that ν' is independent of ν. ν can be any frequency, not necessarily one that matches some absorption in the molecule. The Raman shift ν' can be used to identify rotational and vibrational transitions. The usefulness of the Raman technique is that rotational transitions in molecules with no dipole moment can be studied, and vibrational transitions which have no dipole moment changes can also be studied, leading to information not available through infrared spectroscopy. (Figure 8-4.4.1 shows a typical spectrum.)

The development of Raman spectroscopy was rather slow due to the lack of an adequate excitation source. This has been overcome in the last few years since the development of reliable, inexpensive lasers (see Sec. 8-4.6). The intense, monochromatic beam generated by a laser is suited perfectly for Raman work. Figure 8-4.4.2 gives a block diagram of a laser Raman apparatus. Other than the ability to detect vibration unobservable with IR, Raman spectroscopy has another major advantage in that water can be used as a solvent. Water absorbs strongly in IR thus it is not a suitable choice as a solvent in IR spectroscopy. However, H_2O is a very weak scatterer making it an ideal solvent for Raman spectroscopy. The ubiquitous nature of water as a solvent in inorganic and biological systems makes the Raman technique extremely important. Some proteins and amino acids have already been studied. It has been found that their spectra are not too complicated to interpret. Thus, it appears that Raman spectroscopy will play a major role in biological work.

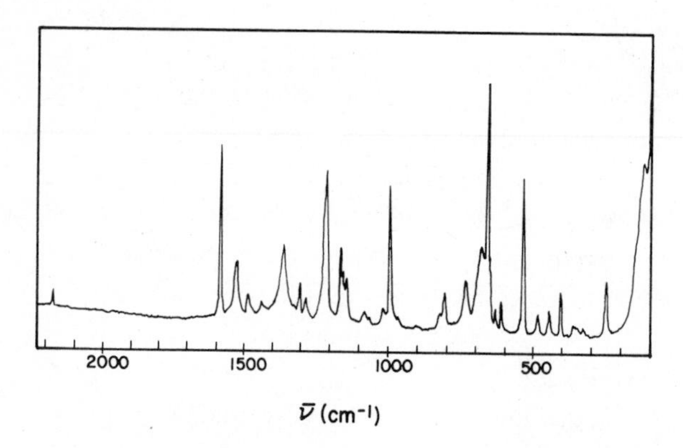

FIG. 8-4.4.1 Laser Raman spectrum for thioacetanilid ($CH_3CSNHC_6H_5$) is measured from the center of the exciting radiation. The large peak at 0 is the Rayleigh scattering of the laser beam.

8-4.5 NUCLEAR MAGNETIC RESONANCE

Nuclear magnetic resonance (NMR) spectroscopy is an invaluable tool in the identification of organic molecules. Also, much theoretical information about molecular structure can be obtained from NMR. However, we will only mention some of these topics since the major use of NMR is in identification. An NMR spectrum is a "fingerprint" for a molecule. Comparison of the spectrum of an unknown with known spectra allows its identification in many cases. The major thrust of this section will be: How can NMR be used to identify compounds?

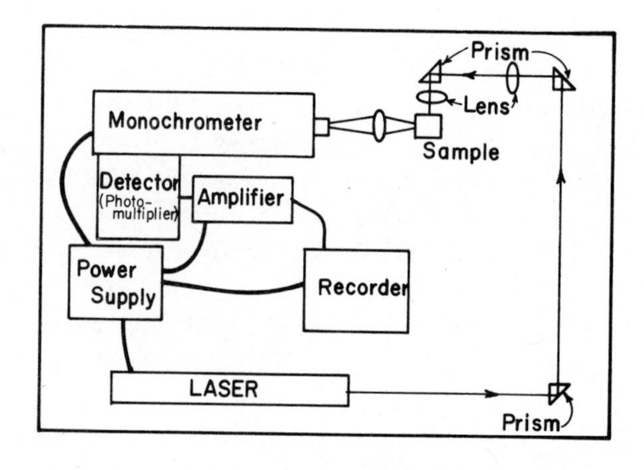

FIG. 8-4.4.2 Laser Raman spectroscopy apparatus.

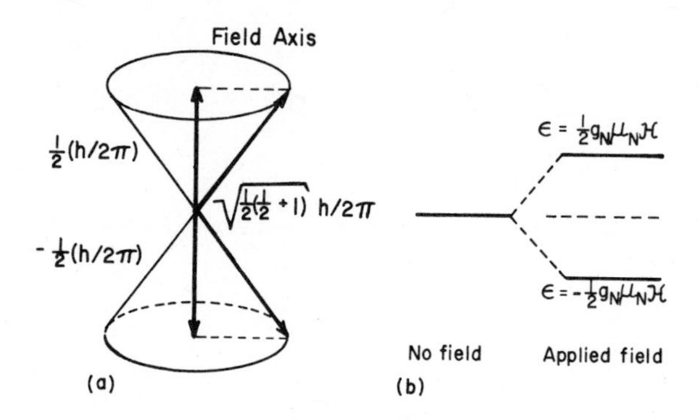

FIG. 8-4.5.1 (a) Angular momenta allowed for a proton along a field axis. (b) The energies of these orientations.

Theory. NMR is most commonly used for studying protons and their environments. There are other nuclei that can be studied, but these are used for specialized research. This discussion is limited to proton NMR. In Sec. 7-3.5 the effect of a magnetic field on an electron (spin ½) was mentioned. The proton (hydrogen nucleus) has a nuclear spin of $I = \frac{1}{2}$. The spin angular momentum is given by

$$\sqrt{I(I + 1)}\ \frac{h}{2\pi}$$

If a magnetic field is applied to define a z axis, the nuclear spin will be quantized on that axis with $\pm\frac{1}{2}\ (h/2\pi)$. Each orientation will have a different energy in the magnetic field. Figure 8-4.5.1 illustrates this effect. In this figure g_N is the nuclear g factor which equals 5.5854 for a proton, and μ_N is the magnetic moment of the nucleus which is equal to 5.0493×10^{-27} J T^{-1} (T = tesla, 1 tesla = 10^4 gauss = 1 weber m^{-1} amp^{-1} = 1 kg s^{-1} amp^{-1}). The energy of the transition between the two states will be

$$\Delta E = g_N\mu_N\mathcal{H} = 2.8202 \times 10^{-26}\ (J)$$

$$\text{or} \qquad \nu = \frac{\Delta E}{h} = 4.2562 \times 10^{7}\ \mathcal{H}\ (s^{-1}) \qquad (8\text{-}4.5.1)$$

NMR spectrometers (see Fig. 8-4.5.2) normally operate with a field of about 1.4 T, so the frequency corresponding to the transition from the lower nuclear state to the upper is about

$$\nu_0 = 60 \times 10^6\ s^{-1}$$

A typical arrangement for a NMR spectrometer is shown in Fig. 8-4.5.2. In an instrument of this type there are permanent magnets with a field of about 1.4 T. The transmitter and receiver operate at a radio frequency of $60 \times 10^6\ s^{-1}$, and this

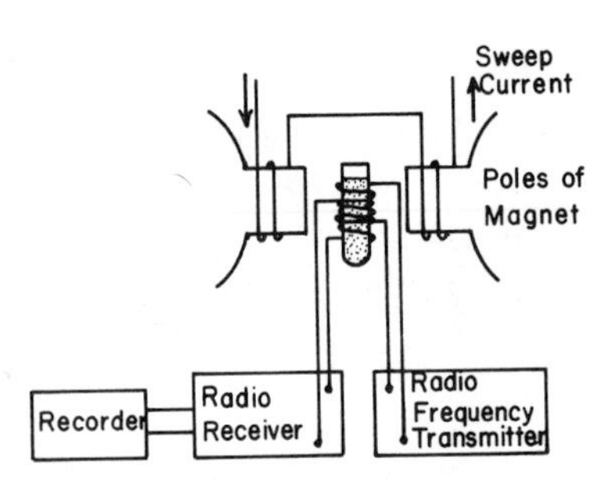

FIG. 8-4.5.2 Schematic of a typical NMR spectrometer.

is not variable. A small sweep field is superimposed on the permanent field. A transition will occur when the nuclear energy levels splitting is just equal to the $60 \times 10^6\ s^{-1}$ radio frequency. The small variable sweep field is adjusted until the 60 MHz ($1\ MHz = 1 \times 10^6\ s^{-1}$) resonant condition is met. When the resonant condition is met some of the energy from the radio transmitter is absorbed by the protons when they make their transition. This results in a decreased signal at the receiver which is amplified and presented as an absorption line on the recorder.

INTERPRETATION OF SPECTRA. An NMR spectrum is a plot of absorption versus sweep field strength. This is usually represented by a term δ defined by

$$\delta = \frac{\mathcal{H}_{reference} - \mathcal{H}_{sample}}{\mathcal{H}_{reference}} \times 10^6 \qquad (8\text{-}4.5.2)$$

δ is a measure of the chemical shift relative to a reference compound. The reference compound is usually tetramethyl silane, $(CH_3)_4$-Si. The *chemical shift* is different for different protons in a molecule. This is due to the effect of the electrons surrounding the proton which influence the *effective magnetic field* of the proton. The external magnetic field is the same for all protons in the sample, but different electron environments give different effective magnetic fields. Figure 8-4.5.3 shows the low resolution NMR spectrum for ethanol. The spectrum can be interpreted in terms of the electron withdrawing ability of the atom attached to the proton. Oxygen withdraws electrons to itself more than carbon. Thus, the proton bonded to the oxygen has its electrons withdrawn toward the electronegative oxygen. This *deshields* the proton. Thus, this proton experiences a higher *effective* field than other protons present so it absorbs at a lower external field. The $-CH_2-$ and $-CH_3$ positions can be explained in a similar manner (do it; remember the CH_2 is also attached to the O). The numbers shown in the figure are the relative integrated intensities of the peaks.

Additional information is available from the *chemical splitting* that occurs. The effective magnetic field is also influenced by the other nuclear spins present.

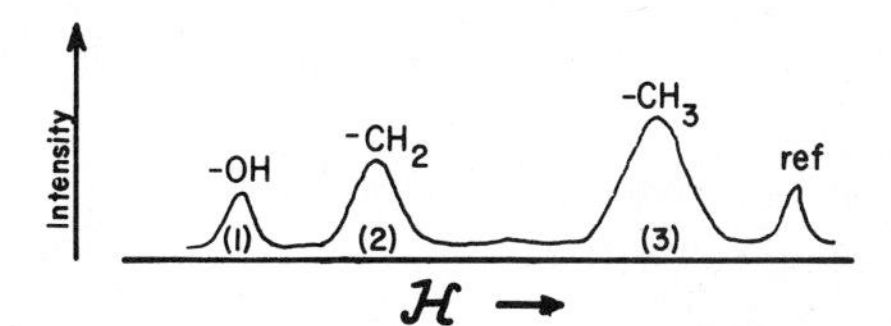

FIG. 8-4.5.3 Low resolution NMR spectrum for C_2H_5OH.

Both carbon and oxygen have spins of zero, so they do not affect the spectrum. However, protons (I = ½) on *adjacent* carbon atoms will affect a given proton. Figure 8-4.5.4 shows the high resolution spectrum for CH_3CH_2OH. Consider the $-CH_2-$ first. The hydrogens on the methyl group will influence the CH_2 proton absorption. Each proton on $-CH_3$ has a spin of ½. The spin on each proton can be aligned parallel or antiparallel with the field. For a large group of molecules, all the orientations in Fig. 8-4.5.5 are possible in the proportions shown. You should study Fig. 8-4.5.5 carefully. It is essential in interpreting the spectra for samples given below. Justify to yourself that the $-CH_3$ will exhibit three peaks with 1:2:1 intensity ratio due to the $-CH_2$ protons.

8-4.5.1 EXAMPLE

Interpret the spectrum for propanol given below by using data from Table 8-4.5.1 and the discussion of Fig. 8-4.5.5.

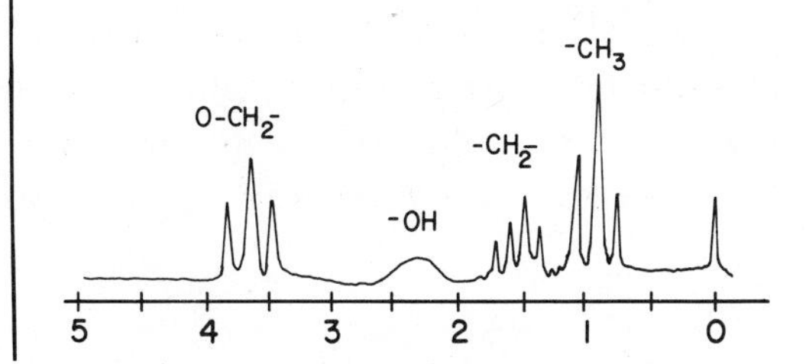

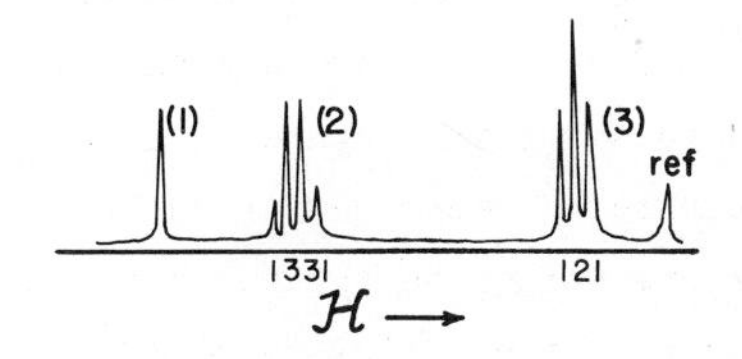

FIG. 8-4.5.4 High resolution NMR spectrum of CH_3CH_2OH.

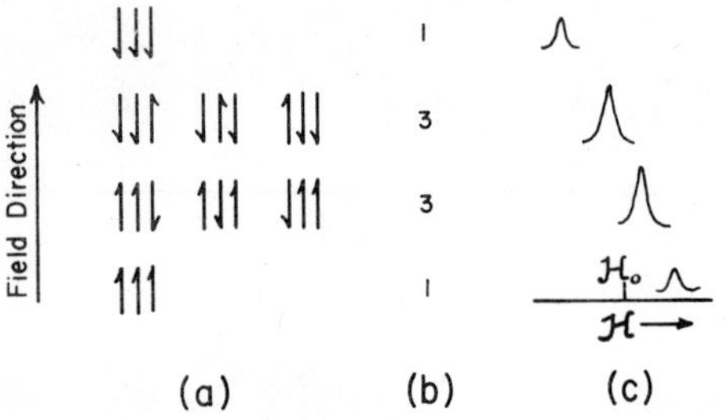

FIG. 8-4.5.5 (a) Allowed orientations of the three proton spins on a methyl group. (b) Statistical distribution of allowed orientation for a large number of methyl groups. (c) Effective field due to given proton spin orientation ($\mathcal{H}_0$ is the absorption in the absence of spin effects).

Solution

The labels above the diagram are determined from Table 8-4.5.1. The splittings shown are easily determined for α-CH_2 and CH_3. The β-CH_2 probably needs some explanation. Note the environment of the β-CH_2.

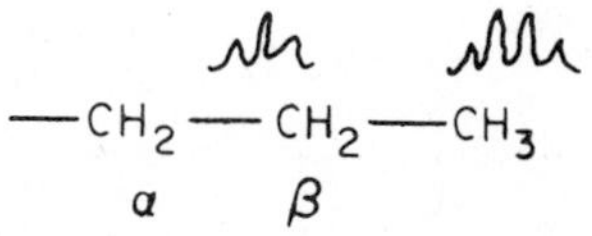

—CH_2—CH_2—CH_3
α β

The patterns above the bonds are the splittings caused by the α-CH_2 and the CH_3. The net effect is a combination of these two patterns leading to a complicated pattern that defies easy or general interpretation. The exact pattern depends on the relative *coupling* of the α protons and the methyl protons with the β protons.

Another concept needs to be developed. That is relaxation. A transition between two levels depends, among other things, on the difference in population in the two levels involved. Any absorption process increases the population of the upper state and decreases that of the lower state. If *relaxation*, the process of returning molecules from excited states to lower energy states, did not occur, the upper level would soon become overpopulated and absorption would cease. There are many ways the excited nuclear state of the proton can relax to the ground state. *Spin-lattice relaxation* is due to interaction with fields created by surrounding nuclei. *Spin-spin relaxation* occurs when one excited spin state transfers energy to some other spin state. The importance of these phenomena is that they allow detailed investigation of the interaction of spins of one molecule with the environment formed by other molecules.

Table 8-4.5.1
Characteristic Values for the Chemical Shift for Hydrogen in Organic Compounds

Type of proton (underlined)		Chemical shift δ
Cyclopropane		0.2
Primary	$RC\underline{H}_3$	0.9
Secondary	$R_2C\underline{H}_2$	1.3
Tertiary	$R_3C\underline{H}$	1.5
Vinylic	$C{=}C{-}\underline{H}$	4.6-5.9
Acetylenic	$C{\equiv}C{-}\underline{H}$	2-3
Aromatic	$Ar{-}\underline{H}$	6-8.5
Benzylic	$Ar{-}C{-}\underline{H}$	2.2-3
Allylic	$C{=}C{-}C\underline{H}_3$	1.7
Fluorides	$\underline{H}C{-}F$	4-4.5
Chlorides	$\underline{H}C{-}Cl$	3-4
Bromides	$\underline{H}C{-}Br$	2.5-4
Iodides	$\underline{H}C{-}I$	2-4
Alcohols	$\underline{H}C{-}OH$	3.4-4
Ethers	$\underline{H}C{-}OR$	3.3-4
Esters	$\underline{H}C{-}COOR$	2-2.2
Acids	$\underline{H}C{-}COOH$	2-2.6
Carbonyl compounds	$\underline{H}C{-}C{=}O$	2-2.7
Aldehydic	$RC\underline{H}O$	9-10
Hydroxylic	$RO\underline{H}$	1-5.5
Phenolic	$ArO\underline{H}$	4-12
Enolic	$C{=}C{-}O\underline{H}$	15-17
Carboxylic	$RCOO\underline{H}$	10.5-12
Amino	$RN\underline{H}_2$	1-5

Another use of NMR is studying the *coupling* (chemical splitting) between nuclei in detail. Careful work allows conclusions to be drawn about the bonding around individual atoms. The coupling can be related to hybridization. Still, the major use is for identification.

8-4.6 LASERS

Lasers are becoming important in so many fields of science, engineering, and technology that you should know the basic principles of operation. The word laser is an acronym for light amplification by stimulated emission of radiation. (For a detailed introductory account, see A. L. Schawlow, *Lasers and Light*, Sec. 8-4.9, Ref. 3.)

The operation of a laser requires a stimulated emission of radiation from a high energy state to a lower state. Figure 8-4.6.1 illustrates this. Stimulated emission leads to a photon with a higher amplitude, but with the same frequency and phase. In order to get this amplification it is necessary somehow to get more molecules in the excited state than in the ground state. This is a *population inversion*. Such an inverted population is not a stable configuration. Usually, the ground state has a higher population (more on this in Chap. 9). Since the system is unstable, stimulated emission is possible. To achieve the population inversion several techniques have been used: *optical pumping* (irradiation with high intensity radiation), *electrical discharges*, and *chemical reactions* leading to products in an excited state. Various types of transitions have been used, such as electronic and vibrational.

The *monochromatic*, *coherent* (photons in phase) radiation from a laser gives it its unique properties of high power density and focusing ability. Lasers are just beginning to be used in communications systems and are in the developmental stage for computer memory devices. Their use in guidance and surveying systems is well established. Section 8-4.4 illustrates a spectroscopic use of lasers.

8-4.7 PROBLEMS

1. If a Hg arc lamp is used to irradiate a sample of CO_2, a Raman spectrum can be observed. Use the data in Fig. 8-2.7.2 to predict the frequencies of Raman emissions if the Hg line at 4358.35×10^{-10} m is used for irradiation.

2. Give the expected NMR spectrum (including chemical splitting) for each of the following:

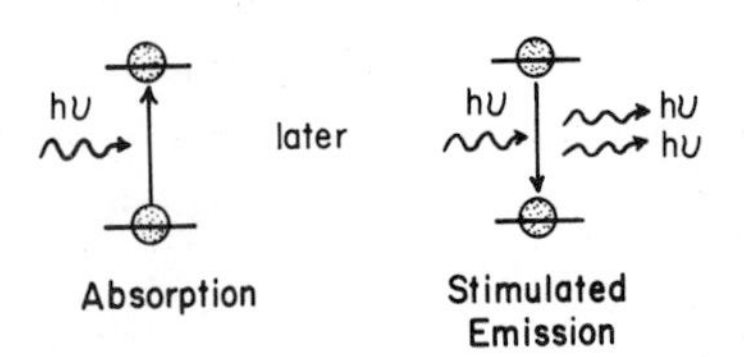

FIG. 8-4.6.1 (a) Absorption of a photon to produce a molecule in an excited state. (b) Interaction of the excited state with a photon to cause emission of a photon.

a. $CH_3-\overset{\overset{O}{\|}}{C}-OH$

b. $CH_3-CH_2-\overset{\overset{O}{\|}}{C}-CH_2-CH_3$

c. $CH_2=CH-CH_2-\overset{\overset{O}{\|}}{C}-H$

d. $CH_3-\underset{CH_3}{\overset{OH}{C}}-CH_3$

e. $CH_3-\overset{\overset{O}{\|}}{C}-CH_2-CH_2-CH_3$

3. Predict what compound would produce the following NMR patterns.

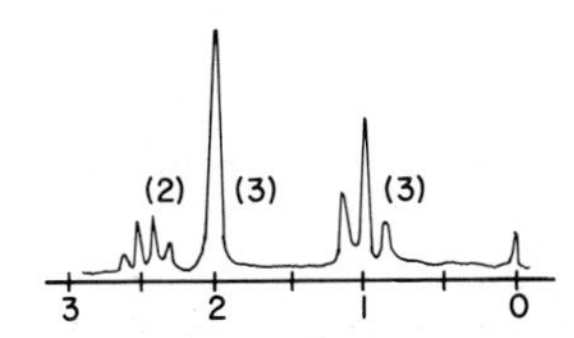

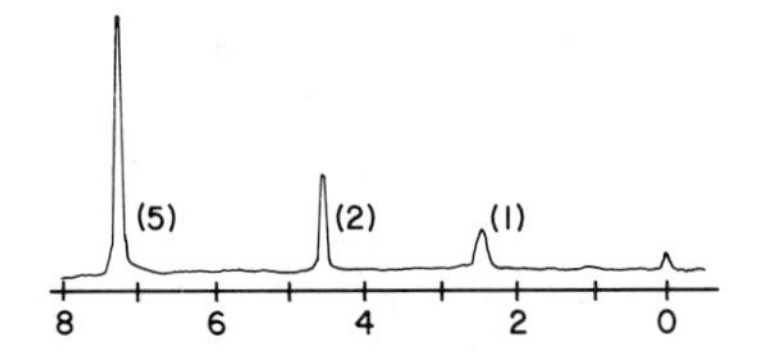

4. From another source, list the differences between fluorescence and phosphorescence. Indicate a use for each of these phenomena.

8-4.8 SELF-TEST

1. A Raman experiment using Hg radiation with a wavelength of 4358.35×10^{-10} m on NH_3 yields four emissions. These are at 26279.5 cm^{-1}, 23894.5 cm^{-1}, 26358.5 cm^{-1}, and 24572.0 cm^{-1}. What are the wave numbers and frequencies of the four fundamental vibrational modes? How many vibrational modes are there supposed to be?

2. Sketch the high resolution NMR spectrum of each of the following.

a. CH_3-OH

b. $(CH_3)_2C{=}C(CH_3)_2$

c. $CH_3\text{-}\overset{\displaystyle O}{\overset{\|}{C}}\text{-}CH_2\text{-}\overset{\displaystyle O}{\overset{\|}{C}}\text{-}CH_2\text{-}CH_3$

d. $HO\text{-}CH_2\text{-}CH_2\text{-}\overset{\displaystyle O}{\overset{\|}{C}}\text{-}OH$

3. Predict the structure of molecule that would give each of the following NMR spectrum.

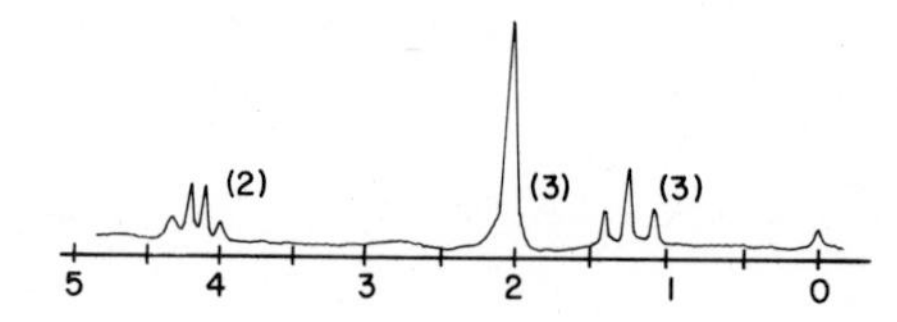

4. Give a half-page discussion of lasers emphasizing one particular use. List your source.

5. Write a one-page summary of electronic spectroscopy; use your own words.

8-4.9 ADDITIONAL READING

1.* G. Herzberg, *Spectra of Diatomic Molecules*, Van Nostrand, Princeton, N. J., 1950.

2.* J. C. Davis, Jr., *Advanced Physical Chemistry*, Ronald Press, New York, 1965, Chap. 10.

3.* A. L. Schawlow, *Lasers and Light*, W. H. Freeman Co., 1969.

4. For additional sources of spectroscopic data, see Sec. 1-1.9D.

9

Statistical Thermodynamics

SYMBOLS AND DEFINITIONS

N	Number of molecules, usually Avagadro's number
$\mathcal{N}$	Number of systems in an ensemble
$A_n{}^n$	Permutations of n distinguishable objects in n boxes
$A_m{}^n$	Permutations of m distinguishable objects in n boxes
$B_m{}^n$	Permutations of m indistinguishable objects in n boxes
q_i	A priori probability of event i

$Q^n_{n_1,n_2\ldots}$	Total a priori probability of n_1 events of type 1, n_2 events of type 2, etc.
$P^n_{n_1,n_2\ldots}$	Total probability of a given set of events
W_k	Thermodynamic probability; the number of ways of arranging a system
P_j	Probability of observing system j in an ensemble of *N* systems
E_j, ε_j	Energy of system j or level j in an ensemble (J mol^{-1}, J molecule^{-1})
n_j	Number of system of type j in an ensemble
α, β	Constants (undetermined multipliers in LaGrange's method)
n_j/N	Fraction of systems of type j in an ensemble
Z	Total partition function of N molecules or *N* systems
z	Partition function for one molecule
g_j	Degeneracy of level or system j
F	Fluctuation
$\overline{\delta F^2}$	Mean square deviation; variance in an ensemble
z_{tr}	Translational partition function for one molecule (unitless)
z_{rot}	Rotational partition function for one molecule (unitless)
z_{vib}	Vibrational partition function for one molecule (unitless)
z_{elec}	Electronic partition function for one molecule (unitless)
D	Dissociation energy from bottom of molecular potential energy well (J mol^{-1}, J molecule^{-1})
D_0	Dissociation energy from v = 0 level (J mol^{-1}, J molecule^{-1})
$\underline{\nu}_i$	Fundamental vibrational frequency (s^{-1})
ν_i	Wave number for a fundamental vibration (cm^{-1})
ξ	Extent of reaction
K_p	Equilibrium constant for a reaction (units vary)
σ	Symmetry factor for rotational partition functions

This chapter will present the elements of statistical mechanics and statistical thermodynamics. Statistical mechanics deals with collections of molecules and their properties. Statistical thermodynamics applies the procedures of statistical mechanics to predict values of thermodynamic variables. It is the objective of statistical mechanics to use the energy levels for different types of molecular motion to predict values for thermodynamic and kinetic variables.

In our study of thermodynamics, we made no reference to the nature of the atoms or molecules in a system. We calculated changes in thermodynamic quantities with no model of the nature of the building blocks comprising the system. We were concerned with macroscopic changes. Such is not the case in statistical mechanics and thermodynamics. We start with assumptions about the structure of atoms and molecules as illustrated in Chaps. 7 and 8. From these assumptions, energy levels for different types of motion (translation, rotation, vibration, electronic, etc.) are obtained by the use of quantum mechanics. These energy levels are then used by statistical mechanics and thermodynamics to predict thermodynamic functions. Thus, statistical mechanics is the bridge between the *microscopic* realm of atoms and molecules and the *macroscopic* realm of thermodynamics.

In this chapter the postulates of statistical mechanics will be given. The procedures for finding thermodynamic and kinetic data will be presented with some examples. Also, the groundwork will be laid for the application of statistical mechanics to chemical kinetics. The actual applications for kinetics will be found in

Chap. 10. Remember, as we go through this chapter, we are attempting to tie quantum mechanics and thermodynamics together, so we can predict changes in thermodynamic functions from molecular data.

9-1 DEFINITIONS AND THE BOLTZMANN DISTRIBUTION FUNCTION

OBJECTIVES

You shall be able to

1. Define or discuss the terms ensemble, canonical ensemble, grand canonical ensemble, microcanonical ensemble, Boltzmann distribution function, partition function, thermodynamic probability, fluctuations
2. Apply permutation and probability results to given situations
3. Calculate relative populations or populations of given energy levels in a system of energy levels
4. Calculate fluctuations in given variables

9-1.1 INTRODUCTION

Like quantum mechanics in Chap. 7, statistical mechanical procedures are based on a set of unproven statements and assumptions. These statements are called postulates. The validity of these postulates, as with all postulates, rests on the accuracy of the results. We will see in subsequent sections that the results agree very well with experiment when all the energy levels in a system are well characterized.

Although statistical mechanics concerns itself with the molecular model of nature, it does not require a detailed account of the behavior of individual molecules. Advantage is taken of the fact that molecules are very numerous. Many properties of a collection of molecules can be predicted accurately, even though little information is available about specific molecules. An analogy can be made to actuarial tables used in the life insurance business. Very precise predictions of the average life expectancy of all people in a given country born in a given year can be made. This can be done despite the fact that the state of health of any specific person is totally unknown. Since we would have a difficult time following the behavior of a specific molecule to get a *time average* of its properties, we use assemblies (collections) of molecules. The collections are called *ensembles*. We find *average values* for properties of these ensembles and postulate that these values are the same as we would determine if we took a long *time average* of an individual molecule's properties. What we will do in this section is define several types of *ensembles*, then present the *postulates of statistical mechanics*. Ensembles are constructs that allow us to apply the postulates. From these we will derive a *distribution function* which will give an equation for predicting the number of molecules in each energy level in a system. The equation that connects the third law of thermodynamics with statistical mechanics will be presented. The relationships for other functions (other than the system energy) will be developed in Sec. 9-2. Finally, a few words will be said about fluctuations (variations of some parameters in the system about the average value of that parameter) and their significance in different systems.

9-1.2 ENSEMBLES

Statistical mechanics allows us to calculate *macroscopic* "mechanical" thermodynamic properties from *microscopic* (molecular) properties. Mechanical properties are those (such as pressure, volume, energy, etc.) that can be defined without explicit reference to temperature. On the other hand, "nonmechanical" thermodynamic variables (such as temperature, entropy, free energies, chemical potential, etc.) will be calculated from mechanical properties by using the procedures of thermodynamics.

What we want to do is find the time average of any given mechanical variable. Within a system there will be fluctuations (small variations) about a mean value. For example, if we accept the fact that molecules are in constant random motion, we have to conclude there is a possibility of small variations of the number of molecules within a small volume element. These variations, called *fluctuations*, are the result of fluctuations of density within the sample. More detail will be given in Sec. 9-1.5. To obtain the mean value of some variable, such as pressure, we need to take the average over a sufficiently long time to smooth out the fluctuations. To do this would require consideration of the behavior of individual molecules in the system over a long period of time, an impossible task for any finite system due to the very large number of molecules. Statistical mechanics offers a solution by considering a large number of "identical" systems which can be used to calculate average values of mechanical properties. Before presenting the postulates that connect the mental construct called an ensemble with the real world of thermodynamics we need to define types of ensembles and develop some useful permutation and probability equations.

An ensemble consists of a mental collection of a large number of systems, each of which is a replica of the system under consideration. For example, let us say we want to study the properties of a *closed, isothermal* system which is a system with N_t (the total number of molecules), V_t (total volume), and T given. A *canonical* ensemble is used to describe such a system. The ensemble is pictured in Fig. 9-1.2.1. We assume we have a large number $\mathcal{N}$ of systems each with N and V in a lattice, all in thermal equilibrium at T. N_t would be the sum of all the individual N, $N_t = \mathcal{N}N$. Also, $V_t = \mathcal{N}V$. The whole ensemble is *isolated*, meaning no energy or materials can cross the barrier. Energy, but not material, can flow between systems.

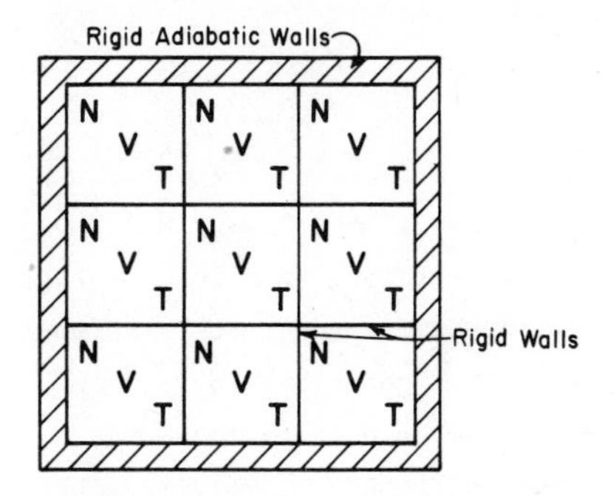

FIG. 9-1.2.1 Representative of a canonical ensemble with *N* system; all with N, V, and T. Walls can transmit energy but not material.

The value of E will not be the same in all systems of the ensemble. There will be *fluctuations* about the *average* of all the systems. Calculating these averages and fluctuations is the subject of Sec. 9-1.4.

A second type of ensemble is more restrictive than the one above. A *microcanonical ensemble* is constructed for an *isolated system* (meaning one with N_t, V_t, and E_t given). The microcanonical ensemble consists of *N* systems each with a given N, V, and E separated by rigid, adiabatic, impermeable walls. In this case, the energy of each system is constant. From a thermodynamic point of view, each system of the ensemble is identical. However, on a molecular level, this is far from true. The total energy of each system of the ensemble is E, but the distribution of molecules within the allowed quantum levels can be very different in different ensembles. There are many combinations of molecules in allowed energy levels that will add to a given total energy E. The size of *N* can vary from a minimum of one system with each allowed distribution of molecules in energy levels (that leads to a total energy E) to $N = \infty$. The fact that *N* can approach ∞ will be useful later. We will make use of both the ensembles defined above. A third type of ensemble is for an *open, isothermal* system in which V_t, T, and μ_t are given, but N and E can vary between systems. The walls between systems are permeable to both heat and matter. We will not consider this system (a grand canonical ensemble) further. See Hill for more details (Sec. 9-1.8, Ref. 1).

The concept of ensembles combined with two postulates (to be given) allows the calculation of average mechanical properties without reference to the behavior of individual molecules with time. The problem becomes one that can be solved instead of one that is impossible to solve. This will be done after an introduction to permutations and probability.

9-1.3 PERMUTATIONS AND PROBABILITY

PERMUTATIONS.

Distinguishable Objects. One of our goals is to derive the Boltzmann distribution function that describes how molecules or systems are arranged in allowed energy levels or states. This will be used to calculate thermodynamic properties. Before deriving the Boltzmann distribution formula it is probably a good idea to refresh your memory about permutations and probability. The number of *permutations* in a system is just the *number of ways* of arranging a system to fit given criteria. Let us assume we have n boxes available and n *different* (distinguishable: we can tell them apart) objects to put in them, one object per box. We have n choices as to where the first object is placed, n - 1 choices for the second, n - 2 for the third, etc. For object n, the last one, there is n - (n - 1) = 1 choice, since all but one box is filled.

The number of permutations (allowed arrangements) is the product of these individual number of choices

$$A_n^{\,n} = n(n - 1)(n - 2) \ldots 2 \cdot 1 = n! \tag{9-1.3.1}$$

where $A_n^{\,n}$ is read "the number of ways to put n objects in n boxes."

That was easy, so let us try m distinguishable objects in n boxes with $n \geq m$. As before, there are n boxes available for the first object, n - 1 for the second, and so on, and n - (m - 1) for the mth object (be sure that last number makes sense). Thus "the number of ways to put m *distinguishable* objects in n boxes, $A_m^{\,n}$," or the permutations, are

$$A_m^{\,n} = n(n - 1)(n - 2) \cdots n - (m - 1) \tag{9-1.3.2}$$

This can be manipulated to a different form by multiplying by an appropriate 1.

$$A_m^{\,n} = n(n - 1) \cdots n - (m - 1) \frac{(n - m)!}{(n - m)!}$$

$$= \frac{n!}{(n - m)!} \tag{9-1.3.3}$$

9-1.3.1 EXAMPLE

How many permutations are there for two distinguishable objects placed in four boxes?

Solution

$$A_2^{\,4} = \frac{4!}{2!} = \frac{4 \cdot 3 \cdot 2 \cdot 1}{2 \cdot 1} = \underline{\underline{12}}$$

1	O	0			7	0	O		
2	O		0		8	0		O	
3	O			0	9	0			O
4		O	0		10		0	O	
5		O		0	11		0		O
6			O	0	12			0	O

FIG. 9-1.3.1 Different arrangement for two objects in four boxes.

Figure 9-1.3.1 shows these permutations.

Indistinguishable (Identical) Objects. Consider Fig. 9-1.3.1 in a little more detail. If the objects are identical (O = 0), notice that 1 is the same as 7, 2 as 8, 3 as 9, etc. In this case there are only six different ways of putting two objects in four boxes. For identical objects we will use the symbol $B_m^{\ n}$ for the permutations.

$$B_2^{\ 4} = \frac{1}{2!} A_2^{\ 4} \tag{9-1.3.4}$$

In general (without proof), for m objects in n boxes we divide by m! to eliminate counting identical arrangements more than once.

$$B_m^{\ n} = \frac{1}{m!} A_m^{\ n} = \frac{n!}{m!(n - m)!} \; . \tag{9-1.3.5}$$

The next step in the development is a little more obscure. Assume we have several *events* that can happen (such as throwing a number on a set of dice or the total number of molecules in a system of known energy levels). Call one such possibility *event of type 1*, another *event of type 2*, etc. Now, out of n total events assume we observe

$$\begin{array}{l} n_1 \text{ events of type 1 (or } n_1 \text{ molecules in energy level 1)} \\ n_2 \text{ events of type 2 (or } n_2 \text{ molecules in energy level 2)} \\ \cdot \quad \cdot \quad \cdot \quad \cdot \quad \cdot \quad \cdot \quad \cdot \\ \cdot \quad \cdot \quad \cdot \quad \cdot \quad \cdot \quad \cdot \quad \cdot \\ \cdot \quad \cdot \quad \cdot \quad \cdot \quad \cdot \quad \cdot \quad \cdot \\ n_r \text{ events of type r (or } n_r \text{ molecules in energy level r)} \end{array} \tag{9-1.3.6}$$

where $n = \sum_{j=1}^{r} n_j$

The allowed permutations [the number of ways of obtaining the results given in (9-1.3.6)] are

$$B^n_{n_1,n_2,\ldots,n_r} = \frac{n!}{n_1!n_2!n_3!\cdots n_r!}$$

$$= \frac{n!}{\prod_{j=1}^{r} (n_j!)} \qquad (9\text{-}1.3.7)$$

where $n_1,..,n_r$ are defined in Eq. (9-1.3.6) and $\prod$ is the product notation meaning the product of the components following the symbol is taken.

9-1.3.2 EXAMPLE

How many permutations are there for throwing 3 ones, 2 twos, 0 threes, 1 four, 1 five, and 3 sixes on a six-sided die in 10 tosses?

Solution

$$B^{10}_{3,2,0,1,1,3} = \frac{10!}{3!2!0!1!1!3!}$$

$$= \frac{10 \cdot 9 \cdot 8 \cdot 7 \cdot \not{6} \cdot 5 \cdot \overset{2}{\not{4}} \cdot \not{3}!}{(\not{3} \cdot \not{2} \cdot 1)(\not{2} \cdot 1)(1)(1)(1)(\not{3}!)}$$

$$= 10 \cdot 9 \cdot 8 \cdot 7 \cdot 5 \cdot 2 = \underline{\underline{50,400}}$$

There are over 50,000 ways of throwing the particular set of numbers given. Two notes: (1) 0! = 1 and (2) *very important*: no order is specified in the above. If an order is specified for the results, the permutations are decreased significantly. For example, if we say 3 ones are thrown, then 2 twos, then a four, then a five, and then 3 sixes, there are no permutations. There is only *one* way to get such a result. So, when we calculate permutations we are *not* specifying any order to the events.

PROBABILITY.

A Priori. For any given event of a group, there is an intrinsic probability of it occurring. For example, if we have a perfect cube with sides numbered from one through six (a die), the probability of any one of the numbers showing on a given toss is equal to that of any other number. Thus the a priori probability is 1/6 for each of the six numbers. For a coin toss, the a priori probability of a head is 1/2. The same is true for a tail. Designate the a priori probability of *event of type i* by the symbol q_i. For the coin toss, $q_H = q_T$. For the die, $q_1 = q_2 = q_3 = q_4 = q_5 = q_6 = 1/6$. Finally, for any set of events

$$\sum_i q_i = 1 \qquad (9\text{-}1.3.8)$$

The total probability that one of the possible events will occur is one. Refer to Eq. (9-1.3.6) now. Let

$$\begin{aligned} q_1 &= \text{a priori probability of } \textit{event of type 1} \\ q_2 &= \text{a priori probability of } \textit{event of type 2} \\ &\vdots \\ q_r &= \text{a priori probability of } \textit{event of type r} \end{aligned} \tag{9-1.3.9}$$

The total a priori probability is

$$Q^n_{n_1,n_2,\ldots,n_r} = q_1^{n_1} \; q_2^{n_2} \cdots q_r^{n_r}$$

$$= \prod_{j=1}^{r} q_j^{n_j} \tag{9-1.3.10}$$

9-1.3.3 EXAMPLE

What is the a priori probability of tossing three heads and five tails in eight coin tosses?

Solution

$$q_H = \frac{1}{2} \qquad q_T = \frac{1}{2}$$

$$Q^8_{3,5} = (\tfrac{1}{2})^3(\tfrac{1}{2})^5 = (\tfrac{1}{2})^8$$

You probably have already noted that if all the individual a priori probabilities are the same for a given number of events, the total a priori probabilities are the same. For eight coin tosses, every possible result has

$$Q^8_{n_i,n_j} = (\tfrac{1}{2})^8$$

Total Probability. Total probability is the fraction of times the given results would be obtained in a large number of trials. The *total probability* of the given set of events described by (9-1.3.6) and (9-1.3.9) is the product of the probability times the permutations.

$$P^n_{n_1,n_2,\ldots,n_r} = Q^n_{n_1,n_2,\ldots,n_r} \cdot B^n_{n_1,n_2,\ldots,n_r}$$

$$= \prod_{j=1}^{r} (q_j)^{n_j} \cdot \frac{n!}{\prod_{j=1}^{r} (n_j!)}$$

$$= n! \prod_{j=1}^{r} \frac{(q_j)^{n_j}}{n_j!} \qquad (9\text{-}1.3.11)$$

Equation (9-1.3.11) is the one we will use in Sec. 9-1.4 to determine the probability of a given distribution of molecules in available energy levels.

9-1.3.4 EXAMPLE

What is the probability of three heads and five tails in eight coin tosses (no order)?

Solution

$$q_H = q_T = \frac{1}{2} \qquad Q^8_{3,5} = \frac{1}{2}^8 = \underline{3.9 \times 10^{-3}}$$

$$B^8_{3,5} = \frac{8!}{3!\ 5!} = \frac{4.032 \times 10^4}{6.120} = \underline{56}$$

$$P^8_{3,5} = 56(3.9 \times 10^{-3}) = 2.19 \times 10^{-1} = \underline{\underline{0.219}}$$

An interesting exercise for you is to calculate the probability of all possible results for, say, 10 tosses of a coin. This is done in Fig. 9-1.3.2 for 10 tosses and 100 tosses. Note that relative probability is plotted vs. fraction of heads. (The curves are idealized. They are actually a series of discontinuous points.) The values for the maximum probabilities are $P^{100}_{50,50} = 0.0796$ and $P^{10}_{5,5} = 0.246$. An important observation can be made from this plot. Note how as N increases, the distribution of results becomes more and more narrow, until for very large N, no set of tosses deviates appreciably from 0.5. This will be important later.

Exercise 9-1.3.1. Calculate the probabilities of each possible result when a pair of dice is thrown repeatedly. Note that the a priori probability of any result is (1/6)(1/6) = 1/36 since there are two dice. Also, you will have to figure permutations without a formula. For example, the number of ways of obtaining seven is six. The following results were obtained by the author's son in 1000 throws.

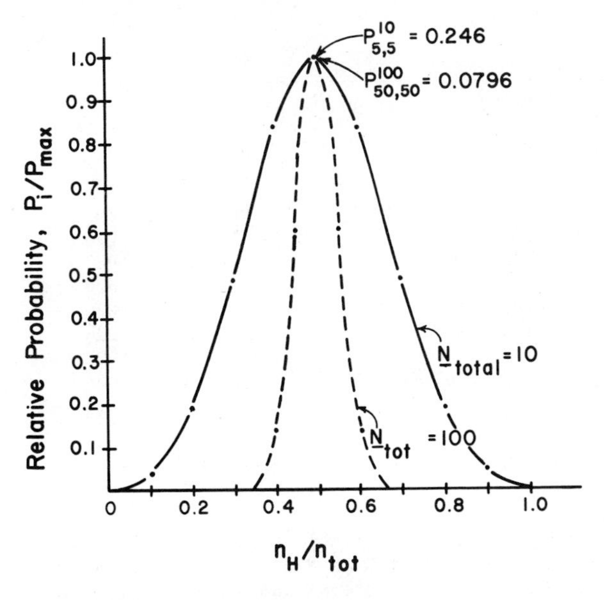

FIG. 9-1.3.2 Relative probabilities for different results for 10 and 100 coin tosses.

Numbers on dice	2	3	4	5	6	7	8	9	10	11	12
Numbers of results	26	37	67	107	129	196	147	121	96	51	23

Do these results agree with your prediction? Table 9-1.3.1 gives results of various numbers of "throws" done by computer. The computer throws the dice by generating random numbers which control the counting of the "numbers" on the dice. Two things are to be noted. These results are from one computer experiment. Another just like it would produce slightly different results. Also, as N increases the results approach the theoretical result, and the distribution becomes more nearly symmetrical. You should have noticed by now that the number of twos equal the number of twelves, etc. For small values of N this is not observed, but is approximately true for large N. For example, in 100,000 throws the computer predicts 2766 twos and 2705 twelves. For 100 throws the predicted result is 1 two and 6 twelves. The purpose of this exercise is two-fold: (1) For fun and (2) to show you that you sometimes have to think. Formulas do not always apply.

9-1.4 THE BOLTZMANN DISTRIBUTION LAW

POSTULATES. There are two basic postulates that are used to derive the results of statistical mechanics:

Table 9-1.3.1
Results of Dice "Throwing" Done by Computer

N	Fraction of trials a given result occurs										
	2	3	4	5	6	7	8	9	10	11	12
100	.01	.08	.09	.11	.17	.13	.12	.11	.08	.04	.06
500	.038	.050	.080	.124	.124	.160	.146	.098	.076	.068	.036
1,000	.034	.049	.087	.130	.129	.141	.148	.099	.080	.066	.037
5,000	.027	.055	.087	.116	.134	.158	.147	.114	.076	.058	.028
10,000	.028	.054	.086	.112	.137	.164	.141	.111	.084	.058	.026
50,000	.028	.054	.085	.114	.138	.167	.140	.110	.082	.055	.026
100,000	.028	.055	.084	.111	.139	.166	.140	.112	.083	.055	.027
Theoretical	.028	.056	.083	.111	.139	.167	.139	.111	.083	.056	.028
Actual trial 1000 tosses	.026	.037	.067	.107	.129	.196	.147	.121	.096	.051	.023

Postulate I: The *time average* of a mechanical variable M in a thermodynamic system is *equal to the ensemble average* of M as $N \to \infty$ if the systems of the ensemble replicate the thermodynamic state of the system being considered.

Postulate II: In a microcanonical ensemble (isolated thermodynamic systems) there is an *equal probability* of every set of *quantum states* consistent with the specified values of N, V, and E.

We discussed the reasoning behind Postulate I in Sec. 9-1.2. Postulate II is the *principle of equal a priori probability*. It is possible to achieve a given total energy in a system by a number of different distributions of molecules in available energy levles. Postulate II states that any of the possible distributions are equally likely. Consider Fig. 9-1.4.1 where three molecules are distributed among four energy levels. This treatment follows that given by T. L. Hill, *Statistical Thermodynamics* (see Sec. 9-1.8, Ref. 1).

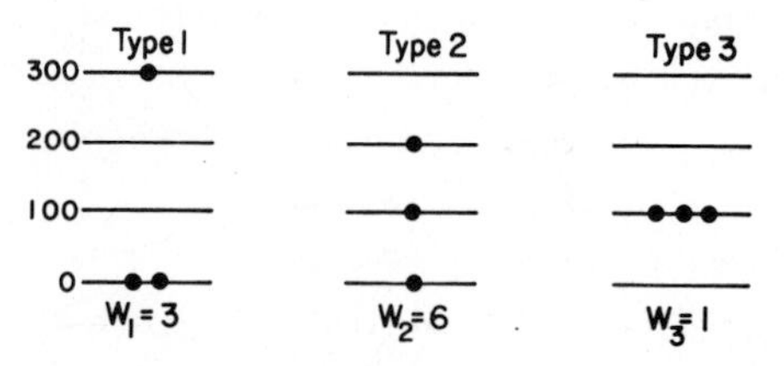

FIG. 9-1.4.1 Distributions of three molecules in four energy levels so that the total energy is $E_t = 300$. W_k is the degeneracy or number of ways of obtaining the kth type of distribution. [W is is the same as B_m^n in Eq. (9-1.3.7). For example, for W_1 there are three molecules; two in level 0, one in level 3: $W_1 = 3!/2!0!0!1! = 3$; $W_2 = 3!/1!1!1!0! = 6$, etc.]

We note there are 10 different distributions which give $E_t = 300$. These are all equally probable according to Postulate II. These ten distributions can be broken into *three* types. The number of ways to get type 1 is three. For type 2 there are six ways ($W_2 = 3!/1!1!1! = 6$). Only one way is possible for type 3. If we say that each distribution is equally likely, then the probability of a given type is just equal to W for that type. This W is the *thermodynamic probability*, the number of ways of obtaining the given results.

For a general system of N molecules with n_1 in energy level 1, n_2 in energy level 2, etc., we can write immediately (from the development in Sec. 9-1.3) the *thermodynamic probability* of such a distribution (call this distribution type k):

$$W_k = \frac{N!}{\prod_j n_j!} \tag{9-1.4.1}$$

Equation (9-1.4.1) results directly from Postulate II (the probability of a distribution is just equal to the number of ways of getting that distribution). This result and Postulate I are sufficient for us to derive the *most probable* distribution (type). Note for large numbers of molecules the most probable distribution is essentially the *only* distribution of importance. (See Fig. 9-1.3.1 and discussion.) So, when we find the most probable distribution we can use that as *the* distribution to calculate average values.

THE DISTRIBUTION FUNCTION. For a canonical ensemble we have $\mathcal{N}$ systems with N_t, V_t, and T given. However, E_j the energy of a given system of the ensemble, may deviate from the average $\bar{E}$. What we want to find is the probability of observing a particular quantum state, E_j. For a given distribution (type) with n_1 systems with energy E_1, n_2 systems with energy E_2, etc., Eq. (9-1.4.1) will apply with N replaced by $\mathcal{N}$. We have already determined that the most probable distribution (type) is essentially the only distribution (type). It is the only distribution that occurs frequently enough to consider. Thus, all we need to do is find the probability of observing E_j in the most probable distributions which will be labeled the k* distribution. The probability of observing E_j in the k* distribution is just

$$P_j = \frac{n_j^*}{\mathcal{N}} \tag{9-1.4.2}$$

where n_j^* is the number of systems in state j in the most probable distribution.

If we find P_j, we can calculate the average of any mechanical variable. The average value of the energy will be given by

$$\bar{E} = \sum_j P_j E_j \tag{9-1.4.3}$$

with similar equations for other variables. The most probable distribution is the one with the largest W_k. All we need to do is find the maximum value of W_k (or in our case, $\ln W_k$). Do you agree that the maximum value of $\ln W_k$ is the same as the maximum value of W_k? Our problem becomes one of finding the distribution with the maximum W_k subject to the conditions that the total energy (E_t) and total number of systems (N) is fixed:

$$E = \sum_j n_j E_j \qquad N = \sum_j n_j \tag{9-1.4.4}$$

Stirling's Approximation. To find W_{k*} we will find the maximum of $\ln W_k$. This can be done by

$$d \ln W_k = 0 \tag{9-1.4.5}$$

But

$$\ln W_k = \ln N! - \ln \left(\prod_j n_j!\right) = \ln N! - \sum_j \ln (n_j!) \tag{9-1.4.6}$$

At this point, we apply Stirling's approximation [For additional details on Stirling's approximation, see S. M. Blinder, *Advanced Physical Chemistry*, (Sec. 9-1.8, Ref. 2).] which states for large X (this is one form that is often used),

$$\ln X! = X \ln X - X \tag{9-1.4.7}$$

9-1.4.1 EXAMPLE

Determine the error in Stirling's approximation for 5! 10! 20! 30! 40! 50! 60! 70! and 100!

Solution

X	ln X!	x ln X - X	Error (%)
5	4.79	3.05	- 36.4
10	15.10	13.03	- 13.8
20	42.34	39.91	- 5.7
30	74.66	72.04	- 3.5
40	110.3	107.6	- 2.5
50	148.5	145.6	- 1.9
60	188.6	185.7	- 1.6
70	230.4	227.4	- 1.3
100	361.4	360.5	- 0.25

As you can see, as X becomes larger the percent error becomes smaller. For very large X the approximation is almost exact.

Apply this to (9-1.4.5) to obtain

$$\ln W_k = N \ln N - N - \sum_j (n_j \ln n_j - n_j)$$

$$= N \ln N - N - \sum_j n_j \ln n_j + \sum_j n_j \qquad (9\text{-}1.4.8)$$

But $\sum_j n_j = N$. Thus, the second and last terms of Eq. (9-1.4.8) cancel, leaving

$$\ln W_k = N \ln N - \sum_j n_j \ln n_j \qquad (9\text{-}1.4.9)$$

Maximum of W_{k*} by LaGrange's Method of Undetermined Multipliers. To find the maximum value of $\ln W_{k*}$ apply (9-1.4.5) and (9-1.4.9):

$$d \ln W_k = d(N \ln N) - d \left[\sum_j (n_j \ln n_j)\right] = 0 \qquad (9\text{-}1.4.10)$$

The first term is zero since N and $\ln N$ are constants. The last term can be written as

$$-\sum_j (n_j^* \, d \ln n_j^* + \ln n_j^* \, dn_j^*) = 0$$

where the asterisk (*) means we are considering the most probable distribution. Changing the sign and taking the derivatives gives

$$\sum_j \left[n_j^* \frac{dn_j^*}{n_j^*} + \ln n_j^* \, dn_j^*\right] = 0$$

The first term is zero ($\sum_j dn_j^* = 0$) since, if a molecule leaves a level (say level 1 making $dn_1^* = -1$) it must go to some other level (say level m making $dn_m^* = +1$). Thus, the sum of all energy level changes is zero. We have, finally,

$$\sum_j (\ln n_j^* \, dn_j^*) = 0 \qquad (9\text{-}1.4.11)$$

The conditions of Eq. (9-1.4.4) can be written (α and β can be introduced since a constant times zero is still zero. The procedure here is a common one for problems that must fit given boundary conditions).

$$dN = 0 = \sum_j dn_j^* = \sum_j \alpha \, dn_j^*$$

$$dE = 0 = \sum_j E_j \, dn_j^* = \sum_j \beta E_j \, dn_j^* \qquad (9\text{-}1.4.12)$$

where α and β are undetermined constants. The equation that must be solved to find the most probable distribution subject to the conditions of (9-1.4.4) is the sum of (9-1.4.11) and (9-1.4.12). By summing these, we obtain one equation which contains all the restrictions applicable to the system.

$$\sum_j (\ln n_j^* + \alpha + \beta E_j)\, dn_j^* = 0 \tag{9-1.4.13}$$

The trivial solution is found if $dn_j^* = 0$. For the solution of interest, the coefficient of each dn_j^* must equal zero.

$$\ln n_j^* + \alpha + \beta E_j = 0$$

or $$n_j^* = e^{-\alpha} e^{-\beta E_j} \tag{9-1.4.14}$$

Recall that $\sum j\ n_j^* = N$. Therefore,

$$N = e^{-\alpha} \sum_j e^{-\beta E_j}$$

$$\text{or} \qquad e^{-\alpha} = \frac{N}{\sum e^{-\beta E_j}} \tag{9-1.4.15}$$

Substituting into (9-1.4.14) gives

$$\frac{n_j^*}{N} = \frac{e^{-\beta E_j}}{\sum_j e^{-\beta E_j}} \tag{9-1.4.16}$$

This is the equation we need. It gives the population of molecules in the energy states for the most probable distribution. Two things still need to be done. One is to find β and the other is to generalize (9-1.4.16) to cases in which several systems may have energy E_j. If this is the case, we include a *degeneracy factor* or *statistical weight* g_j, where g_j is the number of systems with energy E_j. For this case

$$W_k = N! \prod \left(\frac{g_j^{\,n_j}}{n_j} \right)$$

Including this factor gives

$$\frac{n_j^*}{N} = \frac{g_j\, e^{-\beta E_j}}{\sum_j g_j\, e^{-\beta E_j}} \tag{9-1.4.17}$$

Finding β is more difficult. Calculation of the average energy of an ideal gas will yield the value, since we know from Chap. 2 that the average energy of a mole of gas with one degree of translational freedom is $\bar{E} = \frac{1}{2} RT$. (Note: Our ensemble here would be moles of ideal gas with each mole having N_t, V_t, and T.) The result (without derivation) is

$$\beta = \frac{1}{RT} \tag{9-1.4.18}$$

[For details of the derivation, see Blinder, Sec. 17.9 (see Sec. 9-1.8, Ref. 2).]

Thus

$$\frac{n_j^*}{N} = \frac{g_j e^{-(E_j/RT)}}{\sum_j g_j e^{-(E_j/RT)}} \tag{9-1.4.19}$$

The denominator of Eq. (9-1.4.19) is defined as the *partition function* Z.

$$Z = \sum_j g_j e^{-(E_j/RT)} \tag{9-1.4.20}$$

This will be important in Sec. 9-2 when statistical mechanics is related to thermodynamics.

We often want to know the number of systems in a given energy state relative to another state. This is found easily by applying Eq. (9-1.4.19) to both states, say states i and o

$$\frac{n_i^*}{n_o^*} = \frac{n_i^*/N}{n_o^*/N} = \frac{g_i e^{-(E_i/RT)}/Z}{g_o e^{-(E_o/RT)}/Z}$$

$$\frac{n_i^*}{n_o^*} = \frac{g_i}{g_o} \exp\left(-\frac{E_i - E_o}{RT}\right) \tag{9-1.4.21}$$

It is important to note the difference between Eq. (9-1.4.21) and Eq. (9-1.4.19). Equation (9-1.4.19) will give the actual number of systems in a given state, while (9-1.4.21) gives the relative numbers in two states.

9-1.4.2 EXAMPLE

Calculate the population of systems with various energies if the available energy states are singly degenerate and if they are separated by RT. Assume N = 500. (Note: From Example 9-1.4.1, we can be assured of a small error even though N is not infinite.)

Solution

j	E_j	$\exp\left(-\frac{E_j}{RT}\right)$	n_j
0	0	1.0000	316
1	RT	0.3679	116
2	2RT	0.1353	43
3	3RT	0.0498	16
4	4RT	0.0183	6
5	5RT	0.0067	2
6	6RT	0.0025	1
7	7RT	0.0009	0
8	8RT	0.0003	0
9	9RT	0.0001	0
10	10RT	0.0000	0

$\sum = 1.582 = Z$

AVERAGE VALUES. If we want to find an average value for any mechanical variable such as energy E or pressure P we need only apply Eq. (9-1.4.3). Thus

$$\bar{E} = \sum_j P_j E_j \quad \text{and} \quad \bar{P} = \sum_j P_j P_j \tag{9-1.4.22}$$

combining this with Eq. (9-1.4.2) and (9-1.4.19) gives

$$\bar{E} = \frac{\sum_j E_j g_j \, e^{-(E_j/RT)}}{\sum_j g_j \, e^{-(E_j/RT)}}$$

$$= \frac{\sum_j E_j g_j \, e^{-E_j/RT)}}{Z} \tag{9-1.4.23}$$

This will be used in Sec. 9-2.

A MICROCANONICAL ENSEMBLE. The procedures for deriving the Boltzmann distribution for a microcanonical ensemble are similar to those used for the canonical ensembe. A model of the microcanonical ensemble could be N mol of a gas. Each mole contains N molecules in a volume $\bar{V}$ (the molar volume) and will have a total energy E_t. Each molecule in the volume of gas has energy levels of $\varepsilon_1, \varepsilon_2, \ldots, \varepsilon_r$ available to it. There are n_1 molecules in ε_1, n_2 in ε_2, etc. Subject to the conditions that

$$E_t = \sum_i n_i \varepsilon_i$$

$$N = \sum_i n_i \tag{9-1.4.24}$$

The remaining steps parallel those used above. The resulting distribution (including the degeneracy factors) is

$$\boxed{\frac{n_1}{N} = \frac{g_i \, e^{-(\varepsilon_i/kT)}}{\sum_i g_i \, e^{-(\varepsilon_i/kT)}} = \frac{g_i \, e^{-(\varepsilon_i/kT)}}{z}} \qquad (9\text{-}1.4.25)$$

where z is the partition function for molecular energy levels. The ratio of numbers in different levels is (show it)

$$\boxed{\frac{n_i}{n_j} = \frac{g_i}{g_j} \exp - \frac{\varepsilon_i - \varepsilon_j}{kT}} \qquad (9\text{-}1.4.26)$$

These last two quantities are the ones we will use most often, since we are usually interested in the individual energy levels and molecules involved instead of the systems of the canonical ensemble.

9-1.5 FLUCTUATIONS

In our study of thermodynamics, we never considered the possibility that the system under study could undergo changes from the equilibrium state without outside influence. In fact, thermodynamics would predict that when an isolated system has reached its state of equilibrium it has reached its maximum entropy, and no deviations from equilibrium are possible since a decrease in entropy would result. On the other hand, statistical mechanics defines the equilibrium state as an overwhelmingly probable one. Since there are distributions possible that are only slightly removed from the most probable one, it follows that changes from the equilibrium state are possible. They will (in general) be very small. In order to estimate the degree to which fluctuations occur, we must construct an ensemble that fits the system. For calculating thermodynamic variables, we somewhat arbitrarily pick an ensemble for convenience. We cannot do that for studying fluctuations. Only a couple of examples of fluctuation calculations will be given below. The major reasons for giving any are: (1) to show we are justified in ignoring all but the most probable state in most cases, and (2) there are some cases in which fluctuations give rise to observable phenomena not explicable in terms of equilibrium thermodynamics, such as the opalescence exhibited at the critical point of a material.

ENERGY FLUCTUATIONS. Consider first a closed system which has N_t, V_t, and T fixed. Such a system can be represented by a canonical ensemble, as shown previously. If energy fluctuations are to occur, there must be heat transfer between adjacent systems in the ensemble, since N_t and V_t are fixed. These fluctuations will be very small as shown below.

For a general function $\bar{F}$ the mean value is given by

$$\bar{F} = \sum P_i F_i \qquad (9\text{-}1.5.1)$$

A measure of the sharpness of a distribution is the *mean square deviation* which is defined

$$\overline{(F - \bar{F})^2} = \sum_i P_i (F_i - \bar{F})^2$$

$$= \sum_i P_i F_i^2 - 2\bar{F} \sum_i P_i F_i + \bar{F}^2 \sum_i P_i \qquad (9\text{-}1.5.2)$$

You should recall at this point that $\sum^i P_i = 1$. Thus

$$\overline{(F - \bar{F})^2} = \overline{F^2} - 2\bar{F}^2 + \bar{F}^2 = \overline{F^2} - \bar{F}^2 \qquad (9\text{-}1.5.3)$$

(Note the difference between $\overline{F^2}$ and $\bar{F}^2$. Try to explain the difference in your own words.) This shows that the mean square deviation (the *variance*) is the *mean of the square minus the square of the mean*. For simplicity this will be written

$$\overline{\delta F^2} = \overline{F^2} - \bar{F}^2 \qquad (9\text{-}1.5.4)$$

Now let us return to the canonical ensemble and find $\overline{\delta E^2}$. Recall that ($\beta = 1/kT$).

$$\bar{E} = \frac{\sum_j E_j g_j e^{-\beta E_j}}{Z} \qquad (9\text{-}1.5.5)$$

and (by definition)

$$\overline{E^2} = \frac{\sum_j E_j^2 g_j e^{-\beta E_j}}{Z} \qquad (9\text{-}1.5.6)$$

Convince yourself that (for $Z = \sum^j g_j e^{-\beta E_j}$

$$\left(\frac{\partial Z}{\partial \beta}\right)_{V,N} = - \sum_j g_j E_j e^{-\beta E_j} = -\bar{E}Z$$

and

$$\left(\frac{\partial^2 Z}{\partial \beta^2}\right)_{V,N} = - \sum_j g_j E_j^2 e^{-\beta E_j} = \overline{E^2} Z \qquad (9\text{-}1.5.7)$$

(V and N are constant for the canonical ensemble).

With these, (9-1.5.6) can be written (go through these steps)

$$\overline{E^2} = \frac{1}{Z}\left(\frac{\partial^2 Z}{\partial \beta^2}\right)_{V,N} = \frac{1}{Z}\frac{\partial}{\partial \beta}\left(\frac{\partial Z}{\partial \beta}\right)$$

$$= \frac{1}{Z}\frac{\partial}{\partial \beta}(-\bar{E}Z) = -\frac{\bar{E}}{Z}\left(\frac{\partial Z}{\partial \beta}\right) - \frac{Z}{Z}\frac{\partial \bar{E}}{\partial \beta}$$

$$= -\frac{\bar{E}}{Z}(-\bar{E}Z) - \frac{\partial \bar{E}}{\partial \beta} = \bar{E}^2 - \frac{\partial \bar{E}}{\partial \beta} \tag{9-1.5.8}$$

Before continuing, write $\partial\bar{E}/\partial\beta$ in terms of $\partial\bar{E}/\partial T$. The result you should get is

$$-\left(\frac{\partial \bar{E}}{\partial \beta}\right)_V = -\left[\frac{\partial \bar{E}}{\partial(1/kT)}\right]_V = -k\left[\frac{\partial \bar{E}}{\partial(1/T)}\right]$$

$$= kT^2\left(\frac{\partial \bar{E}}{\partial T}\right)_V = kT^2 C_V \tag{9-1.5.9}$$

Combining (9-1.5.9) with (9-1.5.8):

$$\overline{E^2} - \bar{E}^2 = -\frac{\partial \bar{E}}{\partial \beta} = kT^2 C_V$$

or

$$\overline{\delta E^2} = kT^2 C_V \tag{9-1.5.10}$$

One measure of the *fractional fluctuation* is

$$\left(\frac{\overline{\delta E^2}}{\bar{E}^2}\right)^{1/2} = \left(\frac{kT^2 C_V}{\bar{E}^2}\right)^{1/2} \tag{9-1.5.11}$$

To get an order of magnitude for this quantity assume we are considering an ideal monoatomic gas which has $\bar{E} = (3/2)NkT$ and $C_V = (\partial\bar{E}/\partial T)_{V,N} = (3/2)Nk$. This leads to

$$\left(\frac{\overline{\delta E^2}}{\bar{E}^2}\right)^{1/2} = \left[\frac{kT^2\ (3/2)Nk}{(3/2NkT)^2}\right]^{1/2}$$

$$= \left(\frac{2}{3N}\right)^{1/2} = O(N^{-1/2}) \tag{9-1.5.12}$$

where *O means on the order of*. If we consider 1 mol of molecules, then the fractional fluctuation is about 10^{-12}. That means

$$\left(\overline{\delta E^2}\right)^{1/2} \approx 10^{-12}\ \bar{E}\ (1\ \text{mol}) \tag{9-1.5.13}$$

This is an extremely small variation. The same order of magnitude would result for gases which are not monoatomic or ideal. Such considerations indicate that our neglect of all distributions other than the most probable is generally a good assumption.

OTHER FLUCTUATIONS. Other cases will not be considered in detail. Rather, a couple of cases in which particular fluctuations give rise to observable phenomena will be listed.

Concentration (Density) Fluctuations. A grand ensemble is required to describe this situation, since the number of particles is not fixed in a system of given T and V_t. A rather involved analysis yields

$$\frac{\overline{\delta\rho^2}}{\bar{\rho}^2} = \frac{kT\kappa}{V} \tag{9-1.5.14}$$

when κ is the compressibility of the material. The fluctuations in density are small in general [except at the *critical point* and at the *liquid-vapor equilibrium* where κ is infinite; Eq. (9-1.5.14) then does not apply] but can be observed by *light scattering*. Fluctuations in density in a liquid give rise to refractive index fluctuations, which in turn cause light scattering in the media. The opalescence at the critical point is an example of density fluctuations giving rise to observable behavior. As mentioned previously, Eq. (9-1.5.14) does not apply at that point.

Electrical Noise. Within electrical components there is always *thermal noise* present. This *thermal noise* is the result of fluctuating currents and voltages within the material making up the component. This results from random thermal motion of the charges (ions, electrons, etc.) in the material. Thermal noise is of practical importance in that it limits the accuracy attainable for electrical measurements. The mean square thermal emf generated by any circuit element with an effective impedence z in a given frequency increment $d\nu$ is given by

$$d(\overline{\mathcal{E}^2}) = 4zkT\ d\nu$$

An example is the random thermal emf (noise) generated by a 100,000 ohm resistor at 300 K in the frequency range of 100 to 15,000 Hz which equals

$$\mathcal{E}_{rms} = (\overline{\mathcal{E}^2})^{1/2} = 5 \times 10^{-6}\ \text{V}$$

You can probably understand why noise is a limiting factor in the precise measurement of low voltages (or currents).

9-1.6 PROBLEMS

1. You have a weighted six-sided die (numbered one to six) with the following a priori probabilities: $P_1 = 1/6$, $P_2 = 1/5$, $P_3 = 1/7$, $P_4 = 1/12$, and $P_5 = 1/3$. (a) What is P_6? (b) Find the probability of throwing 2 ones, 2 twos, 0 threes, 0 fours, 1 five, and 3 sixes in eight tosses.

2. In Problem (1), assume you have thrown 3 ones in a row. What is the probability of the next throw being a one? What is the probability of 4 fours in four tosses?

3. Given the following set of energy levels, determine the number of distributions that will give a total energy of 500 J, and find W for each if there are 4 molecules present (there is an energy level every 50 J).

```
500 ————————————————
450 ————————————————
     :     :     :
150 ————————————————
100 ————————————————
 50 ————————————————
  0 ————————————————
```

4. Calculate the relative population of the first six levels of a set of energy levels separated by 4×10^{-21} J molecule^{-1} at 298 K. (Assume $\varepsilon_0 = 0$ and $g_i = 1$.)

5. From the results of Problem (4), determine the number of molecules in each of the first six levels if there are 10^{20} molecules present.

6. Evaluate the partition function for Problem (4) using as many levels as necessary to get 0.1% accuracy.

7. Find the average energy of the system in Problem (4) using the results of Problems (4) and (6).

8. The compressibility of an ideal gas $\kappa = 1/P$. Determine the density fluctuation of a mole of N_2 gas at 300 K, and 1.2 atm.

9-1.7 SELF-TEST

1. You are given a perfect four-sided die (a tetrahedron) with faces numbered 1, 2, 3, and 4. Give the probability of observing 2 ones, 3 twos, 1 three, and 1 four in seven tosses.

2. Define or discuss briefly:

   a. Microcanonical ensemble
   b. Partition function
   c. Thermodynamic probability

3. Determine the number of molecules in each energy level of the following system if 10^{12} molecules are present at 100 K.

$g_4 = 2$ ———————— $\varepsilon_4 = 13 \times 10^{-21}$ J molecule^{-1}

$g_3 = 1$ ———————— $\varepsilon_3 = 10 \times 10^{-21}$ J molecule^{-1}

$g_2 = 4$ ———— $\varepsilon_2 = 7 \times 10^{-21}$ J molecule^{-1}

$g_1 = 2$ ———— $\varepsilon_1 = 4 \times 10^{-21}$ J molecule^{-1}

$g_0 = 1$ ———— $\varepsilon_0 = 1.0 \times 10^{-21}$ J molecule^{-1}

4. Determine the partition function and the average energy of the system in Problem (3).

9-1.8 ADDITIONAL READING

1.* T. L. Hill, *Statistical Thermodynamics*, Addison-Wesley, Reading, Mass., 1960.
2.* S. M. Blinder, *Advanced Physical Chemistry*, Macmillan, London, 1969.

9-2 STATISTICAL THERMODYNAMICS

OBJECTIVES

You shall be able to derive the relationships between thermodynamic functions and the partition function and calculate these functions from appropriate data.

9-2.1 INTRODUCTION

You have just completed the development of the concept of an ensemble and the derivation of ensemble averages. The treatment of the material in Sec. 9-1 is good for calculating averages of mechanical properties such as energy and pressure. It is now time to derive relationships that allow the calculation of thermodynamic quantities from known energy states of molecules.

The first thing that needs to be done is to investigate more thoroughly the relationship between disorder and entropy. It was stated in Chap. 4 that as disorder increases, entropy increases. The equation that will be presented below relating S to the thermodynamic probability W can be made plausible by various arguments, but it should be taken as a postulate. The equation is *not derived* below. Rather, it is presented with supporting arguments. Once the connection between *statistical mechanics* and *thermodynamics* is made (or postulated), any of the thermodynamic variables can be calculated if sufficient information about the energy states of the molecules present is known.

9-2.2 ENTROPY AND PROBABILITY

Every process that occurs spontaneously within a system of given energy results in an increase in randomness. There is also an increase in entropy in the system. From that we conclude that entropy and disorder must somehow be related.

Entropy, you may recall, is additive, while you saw in Sec. 9-1 that probability is multiplicative. The total probability of a system from two subsystems designated as 1 and 2 is $W = W_1 \cdot W_2$. The corresponding entropy is $S = S_1 + S_2$. A relation that fits this requirement is a logarithmic one which in the most general form can be written as

$$S_{sm} = b \ln W + c \tag{9-2.2.1}$$

where b and c are constants that must be evaluated, and the sm subscript stands for statistical mechanics. To establish the values of b and c consider the system shown in Fig. 9-2.2.1. Assume we have, initially, N molecules of an ideal gas in $V_1 = V_A$ with V_B empty. Now assume the partition X is removed, allowing the N molecules to equilibrate in the new volume $V_2 = V_A + V_B$. The entropy change on such a free expansion for an ideal gas is (for 1 mol)

$$S_{th} = R \ln \frac{V_2}{V_1} \tag{9-2.2.2}$$

where the th subscript means thermodynamic entropy. What is the probability of finding a molecule in V_1 after X is removed? The probability is just the ratio of the volumes, V_1/V_2. Since probability is multiplicative, we see for N molecules that $W_1 = (V_1/V_2)^N$: each molecule has the same probability of being in V_1. The probability of a molecule being in V_2 is one, since every molecule is in the system somewhere. Thus, $W_2 = 1^N = 1$. The ratio of the probability of every molecule being in V_1 to the probability of every molecule being in $V_2 = V_A + V_B$ is

$$\frac{W_2}{W_1} = \frac{1^N}{(V_1/V_2)^N} = \left(\frac{V_2}{V_1}\right)^N \tag{9-2.2.3}$$

Thus, from (9-2.2.1),

$$\begin{aligned} S_{sm} = S_2 - S_1 &= b \ln W_2 + c - (b \ln W_1 + c) \\ &= b \ln \frac{W_2}{W_1} = b \ln \left(\frac{V_2}{V_1}\right)^N = N\,b \ln \frac{V_2}{V_1} \end{aligned} \tag{9-2.2.4}$$

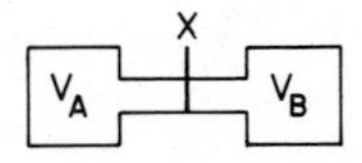

FIG. 9-2.2.1 Two connected containers enclosing N molecules of an ideal gas. X is a movable partition.

Comparison of (9-2.2.4) with (9-2.2.2) shows

$$R = Nb \tag{9-2.2.5}$$

Thus, b = k, Boltzmann's constant. We then write

$$S_{sm} = k \ln W + c$$

There remains the constant c that must be evaluated. This can be accomplished by considering a perfect crystalline material at T = 0 K. The third law states that $S_{th} = 0$ for such a case. How many ways are there to get a perfect crystal? A perfect crystal of a material can be attained only if every molecule is in the appropriate location in the crystal. There is, then, only one way to get a perfectly ordered array. Thus,

$$S_{sm} = k \ln W + c = k \ln 1 + c = c$$

$$S_{th} = 0$$

$$c = 0 \tag{9-2.2.7}$$

The final result is Boltzmann's formula

$$\boxed{S = k \ln W} \tag{9-2.2.8}$$

This exceedingly simple formula is the vital link between statistical mechanics (based on molecular model) and thermodynamics (not based on any model). We can proceed to calculate thermodynamic functions from thermodynamic probabilities derived from statistical mechanics.

9-2.3 GENERAL RELATIONSHIPS

In the remainder of this chapter we will be considering *only* systems of noninteracting particles. That is, we assume there are no interactions between the particles that make up the system. An equivalent way to state this is "the energy of the system can be expressed in terms of separate level patterns for the individual molecules." Using that assumption we can derive the general relationships between the partition function and the thermodynamic variables. The general equations will be developed, then the results given for particular types of motion.

ENERGY. The total energy of a microcanonical ensemble (an isolated system with N_t, V_t, and E_t given) is given by

$$E = n_0\varepsilon_0 + n_1\varepsilon_1 + \cdots = \sum_i n_i\varepsilon_i \tag{9-2.3.1}$$

where n_i is the number of molecules in energy level i. The ε_i are molecular energy levels. Equation (9-1.4.25) gives an expression for n_i.

$$n_i = \frac{N g_i \; e^{-\varepsilon_i/kT}}{\sum_i g_i \; e^{-\varepsilon_i/kT}} = \frac{N g_i \; e^{-\varepsilon_i/kT}}{z} \tag{9-1.4.25}$$

Thus $$E = \sum_i n_i \varepsilon_i = \frac{N \sum_i g_i \varepsilon_i \; e^{-\varepsilon_i/kT}}{z} \tag{9-2.3.2}$$

You should evaluate $(\partial/\partial T) g_i \; e^{-\varepsilon_i/kT}$ before continuing. If you do that you can show

$$kT^2 \frac{\partial}{\partial T} g_i \exp\left(-\frac{\varepsilon_i}{kT}\right) = g_i \varepsilon_i \exp\left(-\frac{\varepsilon_i}{kT}\right) \tag{9-2.3.3}$$

The numerator in Eq. (9-2.3.2) can be rewritten:

$$N \sum_i g_i \varepsilon_i \exp\left(-\frac{\varepsilon_i}{kT}\right) = N \sum_i kT^2 \frac{\partial}{\partial T} g_i \exp\left(-\frac{\varepsilon_i}{kT}\right)$$

$$= N \; kT^2 \frac{\partial}{\partial T} \sum_i g_i \exp\left(-\frac{\varepsilon_i}{kT}\right) \tag{9-2.3.4}$$

(*Note*: The above is valid only if the ε_i are constant. We will see later that this is true in a noninteracting system if the volume is constant. Thus, the derivations are taken *at constant volume*.) But,

$$\frac{\partial}{\partial T} \sum_i g_i \exp\left(-\frac{\varepsilon_i}{kT}\right) = \frac{\partial z}{\partial T} \tag{9-2.3.5}$$

and $$E = \frac{N \; kT^2 \; (\partial z/\partial T)}{z} = N \; kT^2 \frac{1}{z} \frac{\partial z}{\partial T}$$

$$= N \; kT^2 \frac{\partial \ln z}{\partial T} \tag{9-2.3.6}$$

To emphasize the V is held constant, rewrite this as

$$\boxed{E = N \; kT^2 \left(\frac{\partial \ln z}{\partial T}\right)_V} \tag{9-2.3.7}$$

Remember, z refers to the partition function for a single molecule.

ENTROPY. To find an expression for entropy, let us write an expression for the thermodynamic probability W for a microcanonical ensemble. The expression is similar to that given in Eq. (9-1.4.9), except if we include the degeneracy factors we find [see F. T. Wall, *Chemical Thermodynamics* (Sec. 9-2.10, Ref. 1).]

$$\ln W = \ln N! + \sum_i n_i \ln g_i - \sum_i \ln n_i! \qquad (9\text{-}2.3.8)$$

Applying Stirling's approximation,

$$\ln W = N \ln N - N + \sum_i n_i(\ln g_i - \ln n_i + 1) \qquad (9\text{-}2.3.9)$$

But $\ln g_i$ can be found from Eq. (9-1.4.25):

$$\ln g_i = \ln n_i - \ln N + \ln z + \frac{\varepsilon_i}{kT} \qquad (9\text{-}2.3.10)$$

Thus, $$\ln W = N \ln N - N + \sum_i n_i\left(\ln n_i - \ln N + \ln z + \frac{\varepsilon_i}{kT} - \ln n_i + 1\right)$$

$$= N \ln N - N - \sum_i n_i \ln N + \sum_i n_i \ln z + \sum_i \frac{n_i\varepsilon_i}{kT} + \sum_i n_i$$

$$= N \ln N - N - N \ln N + N \ln z + \frac{E}{kT} + N$$

or $$\ln W = N \ln z + \frac{E}{kT} \qquad (9\text{-}2.3.11)$$

And $S = k \ln W$, therefore,

$$S = Nk \ln z + \frac{E}{T} \qquad (9\text{-}2.3.12)$$

This may be written (prove it)

$$\boxed{S = Nk \left[\frac{\partial}{\partial T}(T \ln z)\right]_V} \qquad (9\text{-}2.3.13)$$

OTHER FUNCTIONS. The other functions are easy to write. C_V can be written directly from Eq. (9-2.3.7).

$$C_V = \left(\frac{\partial E}{\partial T}\right)_V$$

$$= 2NkT\left(\frac{\partial \ln z}{\partial T}\right)_V + nkT^2\left(\frac{\partial^2 \ln z}{\partial T^2}\right) \qquad (9\text{-}2.3.14)$$

Also, the *Helmholtz free energy function* is easily found.

$$A = E - TS = -NkT \ln z \qquad (9\text{-}2.3.15)$$

Pressure is found from Eq. (9-2.3.15).

$$P = -\left(\frac{\partial A}{\partial V}\right)_T = NkT\left(\frac{\partial \ln z}{\partial V}\right)_T \qquad (9\text{-}2.3.16)$$

Enthalpy is found directly from the energy expression by adding a PV term. Similarly, the *Gibbs free energy function* is obtained form the Helmholtz function be adding a PV term.

We now have the general formulas necessary for calculating various functions. The equations for specific types of motion are continued in the following. Note that the total partition function for N molecules is

$$Z_{tot} = \frac{1}{N!}(z_{tr})^N (z_{rot})^N (z_{vib})^N (z_{elec})^N \cdots \qquad (9\text{-}2.3.17)$$

Capital Z indicates the total partition function and the lowercase z an individual molecule.

Equation (9-2.3.17) assumes that each type of motion is independent of the other types. The origin of the N! is discussed in the section on translational motion.

Exercise 9-2.3.1. Assuming an ideal gas, write explicit equations for enthalpy and the Gibbs free energy.

9-2.4 TRANSLATIONAL MOTION

A problem solved in quantum mechanics in Sec. 7-1.7 is of use here. In that section we derived the energy levels for a particle in a one-dimensional potential energy well (box). The result was

$$\varepsilon_n = \frac{n^2 h^2}{8ma^2} \qquad n = 0, 1, 2, \ldots \qquad (9\text{-}2.4.1)$$

where n is a quantum number, m is the particle mass, and a is the dimension of the box. This equation is easily extended to three dimensions (assume the three dimensions are all the same length):

$$\varepsilon_{n_x,n_y,n_z} = \frac{(n_x^2 + n_y^2 + n_z^2)\, h^2}{8ma^2} \qquad (9\text{-}2.4.2)$$

where n_x, n_y, and n_z are each integers. This is all we need to find the partition function.

$$z_{tr} = \sum_{n_x,n_y,n_z=1}^{\infty} \exp\left(-\frac{(n_x^2 + n_y^2 + n_z^2)\, h^2}{8ma^2 kT}\right)$$

$$z_{tr} = \sum_{n_x=0}^{\infty} \exp\left(-\frac{n_x^2 h^2}{8ma^2kT}\right) \sum_{n_y=0}^{\infty} \exp\left(-\frac{n_y^2 h^2}{8ma^2kT}\right) \sum_{n_z=0}^{\infty} \exp\left(-\frac{n_z^2 h^2}{8ma^2kT}\right) \quad (9\text{-}2.4.3)$$

(no degeneracy factor is needed since we are summing over all levels).

For ordinary particles (molecules) in vessels of macroscopic dimensions, the energy levels predicted by Eq. (9-2.4.2) are very closely spaced. If the levels are close enough together, the $\sum$ can be replaced by $\int$. For the x component, this gives

$$z_{tr,x} = \int_0^{\infty} \exp\left(-\frac{n_x^2 h^2}{8ma^2kT}\right) dn_x = \frac{(2\pi mkT)^{1/2} a}{h} \quad (9\text{-}2.4.4)$$

All the directions are the same, so

$$z_{tr} = z_{tr,x} z_{tr,y} z_{tr,z}$$

$$= \frac{(2\pi mkT)^{3/2}}{h^3} a^3 = \frac{(2\pi mkT)^{3/2}}{h^3} V \quad (9\text{-}2.4.5)$$

From this partition function the thermodynamic function can be written immediately. However, Eq. (9-2.4.5) refers to one molecule. If there are N molecules in a given container, the total partition function is

$$(Z_{tr})_{tot} = \frac{z_1, z_2, z_3, \ldots}{N!} = \frac{(z_{tr})^N}{N!} \quad (9\text{-}2.4.6)$$

The N! takes into account the fact that the molecules are indistinguishable. (Note: For an *internal degree of freedom* such as rotation or vibration the N! term is left out, since the N! refers to the number of ways of putting N molecules into N portions of the volume of the container.) Thus for 1 mol,

$$\boxed{Z_{tr} = \left[\frac{(2\pi mkT)^{3/2}}{h^3} V\right]^N \frac{1}{N!}} \quad (9\text{-}2.4.7)$$

From the above we can find all the thermodynamic functions for translation.

ENERGY. Equation (9-2.3.7) can be used to find the energy. However, it needs to be modified slightly to take into account Eq. (9-2.4.6).

$$\bar{E} = kT^2 \left(\frac{\partial \ln Z}{\partial T}\right)_V = kT^2 \frac{\partial}{\partial T}\left(\ln \frac{z^N}{N!}\right) \quad (9\text{-}2.4.8)$$

But this reduced to (show it, since N! is a constant)

$$\bar{E} = kT^2 \left[\frac{\partial}{\partial T} (\ln z^N)\right]_V \tag{9-2.4.9}$$

For energy it does not matter whether the molecules are distinguishable. So we have

$$\bar{E}_{tr} = NkT^2 \left[\frac{\partial}{\partial T} \ln \frac{(2\pi mkT)^{3/2}}{h^3} V\right]_V$$

$$= NkT^2 \frac{[(2\pi mk)^{3/2}/h^3](3/2)(T^{1/2})(V)}{[(2\pi mkT)^{3/2}/h^3](V)}$$

$$= \frac{3}{2} NkT^2 \frac{1}{T} = \frac{3}{2} NkT \tag{9-2.4.10}$$

$$\bar{E}_{tr} = \frac{3}{2} RT \tag{9-2.4.11}$$

This is the result obtained previously (Chap. 2) on the basis of classical considerations. Do you see now that our choice of $\beta = 1/kT$ is correct?

ENTROPY. Equation (9-2.3.12) must be modified to take into account the N! term.

$$\bar{S}_{tr} = k \ln \frac{z_{tr}^N}{N!} + \frac{E}{T} \tag{9-2.4.12}$$

$$\bar{S}_{tr} = Nk \ln z_{tr} - k \ln N! + \frac{E}{T} \tag{9-2.4.13}$$

Using Stirling's approximation,

$$\bar{S}_{tr} = Nk \ln z_{tr} - Nk \ln N + Nk + \frac{E}{T}$$

$$= Nk \ln z_{tr} - Nk \ln N + Nk + \frac{3}{2} \frac{NkT}{T}$$

$$= \frac{5}{2} Nk + Nk \ln z_{tr} - Nk \ln N \tag{9-2.4.14}$$

Substituting for z_{tr}:

$$\bar{S}_{tr} = \frac{5}{2} R - R \ln N + R \ln \left[\frac{(2\pi mkT)^{3/2}}{h^3} V\right] \tag{9-2.4.15}$$

Replacing m by M/N and rearranging (do it) gives

$$\bar{S}_{tr} = \frac{5}{2} R - R \ln N + R \ln \left[\frac{(2\pi MkT)^{3/2}}{h^3 N^{3/2}} V\right] \tag{9-2.4.16}$$

and if $V = RT/P$,

$$\bar{S}_{tr} = \frac{5}{2} R + R \ln \left[\frac{(2\pi MkT)^{3/2}}{h^3 N^{5/2}} \frac{RT}{P}\right]$$

$$= R\left\{\frac{5}{2} + \ln \left[\left(\frac{2\pi k}{h^2 N^{5/3}}\right)^{3/2} R \frac{M^{3/2} T^{5/2}}{P}\right]\right\}$$

$$= R\left[\frac{5}{2} + \ln \left(8.201 \times 10^7 \frac{M^{3/2} T^{5/2}}{P}\right)\right] \tag{9-2.4.17}$$

P must be in SI units, not atmospheres. This is known as the Sacker-Tetrod equation, in honor of the scientists who first derived it.

9-2.4.1 EXAMPLE

Calculate $\bar{S}^0_{tr}$ for Kr ($M = 0.0838$ kg mol^{-1}) at 298.15 K and 1 atm $= 1.013 \times 10^5$ N m^{-2}.

Solution

$$\bar{S}_{tr}(Kr) = R\left\{\frac{5}{2} + \ln \left[\frac{8.201 \times 10^7 (0.0838)^{3/2} (298.15)^{5/2}}{1.013 \times 10^5}\right]\right\}$$

$$= 19.72 R = \underline{\underline{164.0 \text{ J mol}^{-1} \text{ K}^{-1}}}$$

The experimental value is 164.0 J mol^{-1} K^{-1}. Not all gases yield such close agreement. We will see that later when polyatomic molecules with degrees of freedom besides translation are considered.

9-2.4.2 EXAMPLE

Find the translational entropy for $H_2O(g)$ at 298.15 K and 101,325 N m^{-2}.

Solution

$$\bar{S}^0_{tr}(H_2O) = R\left\{\frac{5}{2} + \ln \left[\frac{(8.201 \times 10^7)(0.0180)^{3/2} (298.15)^{5/2}}{101,325}\right]\right\}$$

$$= R(17.4) = \underline{\underline{144.8 \text{ J mol}^{-1} \text{ K}^{-1}}}$$

HELMHOLTZ FREE ENERGY. You should derive the exact form for A for translational motion. Some of the problems will use your result. However an interesting result is found from

$$P = -\left(\frac{\partial A}{\partial V}\right)_T = kT\left(\frac{\partial \ln \frac{z^N}{N!}}{\partial V}\right)_T$$

$$= NkT\left(\frac{\partial \ln z}{\partial V}\right)_T \tag{9-2.4.18}$$

$$P = NkT\left\{\frac{\partial}{\partial V} \ln \left[\frac{(2\pi mkT)^{3/2}\ V}{h^3}\right]_T\right\}$$

$$= NkT \frac{[(2\pi mkT)^{3/2}/h^3]}{[(2\pi mkT)^{3/2}\ V/h^3]}$$

$$= \frac{NkT}{V} = \frac{RT}{V}$$

or

$$PV = RT \tag{9-2.4.19}$$

We have obtained the ideal gas law from our model of noninteracting particles confined to a given volume. Our quantization condition for the translation of particles in a given volume has led to an equation we have encountered in studies of both thermodynamics and gas properties.

If necessary, you should be able to find other thermodynamic functions from what has been presented. We will turn to other types of motion. Remember that for all types of motion besides translational

$$Z = z^N$$

Just to emphasize it, it should be again stated that these apply to noninteracting particles. Also, we assume that the energies are separable. That is $E = E_{tr} + E_{rot} + E_{vib} + \cdots$. With that assumption,

$$Z_{tot} = Z_{tr}Z_{rot}Z_{vib} \cdots$$

9-2.5 ROTATIONAL MOTION

PARTITION FUNCTION. Rotational energy levels have an energy level spacing that is intermediate between those of translation and vibration. We will consider the case of diatomic molecules here and then extend the results to others. In Sec. 8-2 we found that

$$\varepsilon_J = J(J + 1)B = J(J + 1)\frac{h^2}{8\pi^2 I} \tag{9-2.5.1}$$

For rotation, the degeneracy factor is $g_J = 2J + 1$. Thus

$$z_{rot} = \sum_{J=0}^{\infty} (2J + 1) \exp\left(- \frac{J(J + 1)B}{kT}\right) \tag{9-2.5.2}$$

This series does not sum to a closed analytical form. But, for many cases at reasonable temperature and all but very light molecules, the energy level spacings for rotation are small enough to be treated as a continuum. Under these circumstances, the Σ can be replaced by an integral.

$$z_{rot} \approx \int_0^{\infty} (2J + 1) \exp\left(- \frac{J(J + 1)B}{kT}\right) dJ \tag{9-2.5.3}$$

Evaluation of this is straightforward. You should do it to show

$$z_{rot} = \frac{kT}{B} = \frac{8\pi^2 IkT}{h^2} \tag{9-2.5.4}$$

This equation is valid as long as B/kT is small compared to 1. Also, another qualification is required. For homonuclear diatomics (and some other symmetrical polyatomic molecules considered later), a *symmetry factor* σ is included. This factor is the number of times a molecule looks the same in a 360° rotation. A molecule like O = O will assume the same configuration twice in a 360° rotation, therefore σ for O_2 is 2. The appropriate modified partition function (without justification) is

$$\boxed{z_{rot} = \frac{kT}{\sigma B} = \frac{8\pi^2 IkT}{\sigma h^2}} \tag{9-2.5.5}$$

Derivation of the partition function for symmetric and asymmetric rotors is beyond the scope of this book. The result is

$$\boxed{z_{rot} = \frac{8\pi^2 (8\pi^3 I_A I_B I_C)^{\frac{1}{2}} (kT)^{3/2}}{\sigma h^3}} \tag{9-2.5.6}$$

where I_A, I_B, and I_C are the moments of inertia. In a symmetric rotor, $I_A = I_C$ and $\sigma \neq 1$.

9-2.5.1 EXAMPLE

Calculate the partition function for $H_2O(g)$ at 298 K. $A = 8.332 \times 10^{11}\ s^{-1}$, $B = 4.347 \times 10^{11}\ s^{-1}$, and $C = 2.985 \times 10^{11}\ s^{-1}$; $\sigma = 2$.

Solution

$$I_A = \frac{h}{8\pi^2 A} \quad \text{(if A is per second); similarly for B and C}$$

$$z_{rot} = \frac{8\pi^2\,[8\pi^3(h/8\pi^2A)(h/8\pi^2B)(h/8\pi^2C)]^{1/2}\,(kT)^{3/2}}{\sigma h^3}$$

$$= \frac{8\pi^2\,\{8\pi^3[h^3/(8\pi^2)^3](1/ABC)\}^{1/2}\,(kT)^{3/2}}{\sigma h^3}$$

$$= \frac{(kT)^{3/2}\,\pi^{1/2}}{\sigma(ABC)^{1/2}\,h^{3/2}}$$

$$= \frac{[(1.3806\times10^{-23}\text{ J molecule}^{-1}\text{ K}^{-1})(298.15\text{ K})]^{3/2}\,\pi^{1/2}}{2(8.332\times4.347\times2.985\times10^{33}\text{ s}^{-3})^{1/2}\,(6.626\times10^{-34}\text{ J s})^{3/2}}$$

$$= \underline{\underline{41.73}}$$

This is a unitless quantity.

For N molecules the total rotational partition function is $Z = (z)^N$, since rotation is an *internal* motion. No factor of $1/N!$ is required.

ENERGY AND HEAT CAPACITY. The energy is readily calculated. First, for a diatomic molecule, for 1 mol,

$$\bar{E}_{rot} = kT^2\left(\frac{d\ln Z_{rot}}{dT}\right)_V$$

$$= kT^2\frac{d}{dT}\ln\left(\frac{8\pi^2IkT}{\sigma h^2}\right)^N$$

$$= NkT^2\left(\frac{d}{dT}\ln T\right) + NkT^2\frac{d}{dT}\left(\ln\frac{8\pi^2k}{\sigma h^2}\right)$$

$$= NkT^2\frac{1}{T} = NkT$$

or $$\bar{E}_{rot} = RT \quad \text{linear molecule} \tag{9-2.5.7}$$

You should use Eq. (9-2.5.6) to determine $\bar{E}_{rot}$ for a nonlinear molecule. The result is

$$E_{rot} = \frac{3}{2}RT \quad \text{nonlinear molecule} \tag{9-2.5.8}$$

Also. you should recall from our discussion of degrees of freedom that a linear molecule has two degrees of rotational freedom and a nonlinear molecule has three. Thus, for each degree of rotational freedom,

$$E_{rot}(\text{per degree of freedom}) = \frac{1}{2} RT \tag{9-2.5.9}$$

The corresponding heat capacity is

$$C_V(\text{per degree of freedom}) = \frac{1}{2} R \tag{9-2.5.10}$$

These results are only approximately true. The accuracy is determined by the accuracy of replacing the sum in Eq. (9-2.5.2) with an integral. Usually the error involved is small.

ENTROPY. Writing the entropy expression is straightforward. Equation (9-2.3.12) tells us the molar entropy is

$$\boxed{\bar{S} = Nk \ln z + \frac{\bar{E}}{T}} \tag{9-2.5.11}$$

For the linear case this becomes

$$\bar{S}_{rot}(\text{linear}) = R \ln \left(\frac{8\pi^2 IkT}{\sigma h^2}\right) + R$$

$$= R \ln \frac{kT}{\sigma B} + R \tag{9-2.5.12}$$

The nonlinear case is illustrated below.

9-2.5.2 EXAMPLE

Find the rotational entropy for $H_2O(g)$ at 298.15 K. Use the results of Example 9-2.4.2.

Solution

From Example 9-2.4.2, $z_{rot} = 41.73$. Thus

$$\bar{S}_{rot}(H_2O) = R \ln z + \frac{\bar{E}}{T} = R \ln z + \frac{3}{2} R$$

$$= 8.314 \text{ J mol}^{-1} \text{ K}^{-1}[(\ln 41.73) + 1.5]$$

$$= \underline{\underline{43.5 \text{ J mol}^{-1} \text{ K}^{-1}}}$$

9-2.6 VIBRATIONAL MOTION

PARTITION FUNCTION. In both translation and rotation we could assume that the energy level spacing was small compared to kT. This is not true for vibration. For vibration we cannot replace the sum in the partition function with an integral. As it turns out the sum is easily expressed in a relatively simple closed form. Another consideration is the choice of a zero point for our reference energy. For translation and rotation this presents no problem, since $E_0 = 0$. However, in vibrational motion (see Fig. 9-2.6.1) the lowest energy state is $\varepsilon_0 = (0 + \frac{1}{2})h\nu_0 = \frac{1}{2}h\nu_0$ above the bottom of the potential well. We usually choose ε_0 as the zero for vibrational energy. You must keep in mind that this is $\frac{1}{2}h\nu_0$ above the bottom of the potential well and D_0 below the energy of the separated atoms (see Fig. 9-2.6.1).

The partition function for a molecule is

$$z_{vib} = \sum_{v=0}^{\infty} \exp\left(-\frac{(v + \frac{1}{2})h\nu_0}{kT}\right) \tag{9-2.6.1}$$

For one-dimensional vibration the degeneracy factors are all 1. Replace $h\nu_0/kT$ by x to get

$$z_{vib} = \sum_{v=0}^{\infty} e^{-(v + \frac{1}{2})x}$$

$$= e^{-(x/2)}(1 + e^{-x} + e^{-2x} + e^{-3x} + \cdots) \tag{9-2.6.2}$$

This can be written in closed form exactly as

$$z_{vib} = \frac{e^{-(x/2)}}{1 - e^{-x}} = \frac{e^{-h\nu_0/2kT}}{1 - e^{-h\nu_0/kT}} \tag{9-2.6.3}$$

There is a term like (9-2.6.3) for each degree of vibrational freedom. Thus for the ith type of vibration

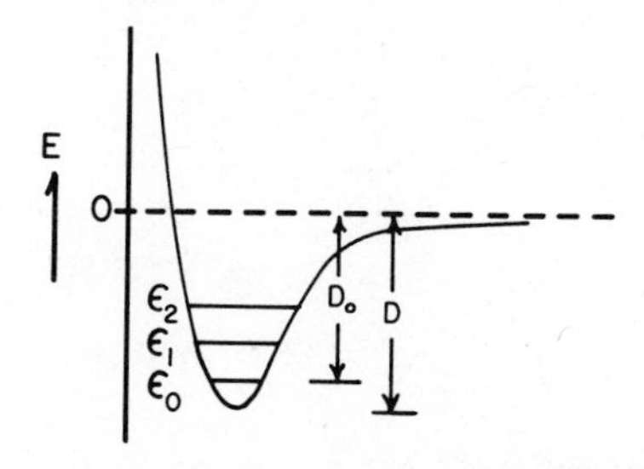

FIG. 9-2.6.1 Schematic potential energy diagram for a diatomic molecule. D_0 is the energy required to dissociate the molecule from the lowest vibrational state; D is measured from the bottom of the potential well.

$$z_{vib,i} = \frac{e^{-h\nu_i/2kT}}{1 - e^{-h\nu_i/kT}} \tag{9-2.6.4}$$

The total partition function for all the vibrations for *one* molecule is

$$z_{vib,tot} = z_{vib,1} \cdot z_{vib,2} \cdot z_{vib,3} \cdot \cdots$$

$$= \prod_{i=1}^{\substack{3n - 5 \text{ (linear)} \\ 3n - 6 \text{ (nonlinear)}}} \frac{e^{-h\nu_i/2kT}}{1 - e^{-h\nu_i/kT}} \tag{9-2.6.5}$$

where n = the number of atoms in the molecule. For N molecules, since vibration is an internal motion,

$$Z_{vib} = (z_{vib,tot})^N$$

ENERGY AND HEAT CAPACITY. Applying the now familiar formula for energy yields for *1 mol* and *one degree of vibrational freedom:*

$$\bar{E}_{vib,i} = kT^2 \left(\frac{\partial \ln Z_i}{\partial T}\right) = NkT^2 \left(\frac{\partial \ln z_i}{\partial T}\right)$$

$$= NkT^2 \frac{\partial}{\partial T} \left(\left(- \frac{h\nu_i}{2kT}\right) - \ln \left[1 - \exp\left(- \frac{h\nu_i}{kT}\right)\right]\right)$$

$$= NkT^2 \frac{h\nu_i}{2kT^2} - NkT^2 \left(- \frac{h\nu_i/kT^2}{1 - e^{-h\nu_i/kT}}\right) \exp\left(- \frac{h\nu_i}{kT}\right)$$

$$= N \frac{h\nu_i}{2} + \frac{Nh\nu_i e^{-h\nu_i/kT}}{1 - e^{-h\nu_i/kT}} \frac{e^{h\nu_i/kT}}{e^{h\nu_i/kT}}$$

$$\bar{E}_{vib,i} = \frac{Nh\nu_i}{2} + \frac{Nh\nu_i}{e^{h\nu_i/kT} - 1} \tag{9-2.6.6}$$

The total energy will be the sum of terms like that in Eq. (9-2.6.6), one for each degree of vibrational freedon

$$\bar{E}_{vib,tot} = \sum_{i=1}^{3n-5 \text{ or } 3n-6} \left(\frac{Nh\nu_i}{2} + \frac{Nh\nu_i}{e^{h\nu_i/kT} - 1} \right) \qquad (9\text{-}2.6.7)$$

Define $x_i = h\nu_i/kT$; then, by evaluating $(\partial \bar{E}_i/\partial T) = \bar{C}_{vib,i}$ show $\bar{C}_{vib,i}$ is given by

$$\bar{C}_{vib,i} = \frac{R\, x^2\, e^x}{(e^x - 1)^2} \qquad (9\text{-}2.6.8)$$

OTHER FUNCTIONS. The other functions have been left as an exercise for you. Before going on to the example given below, derive the equation for entropy to show:

$$\bar{S}_{vib,i} = -R \ln(1 - e^{-x}) + \frac{Rx}{e^x - 1} \qquad (9\text{-}2.6.9)$$

9-2.6.1 EXAMPLE

Evaluate $\bar{S}_{vib}$ for $H_2O(g)$ at 298.15 K, if $\bar{\nu}_1 = 3693.8\ cm^{-1}$, $\bar{\nu}_2 = 1614.5\ cm^{-1}$, and $\bar{\nu}_3 = 3801.7\ cm^{-1}$.

Solution

$$x_1 = \frac{h\nu_1}{kT} = \frac{hc\bar{\nu}_1}{kT}$$

$$= \frac{(6.626 \times 10^{-34}\ J\ s)(2.998 \times 10^{10}\ cm\ s^{-1})(3693.8\ cm^{-1})}{(1.38 \times 10^{-23}\ J\ K^{-1})(298.15\ K)}$$

$$x_1 = 17.8$$

$$x_2 = 7.79$$

$$x_3 = 18.4$$

$$\bar{S}_{vib,1} = -8.314\ J\ mol^{-1}\ K^{-1} \ln(1 - e^{-17.8}) + \frac{(8.314\ J\ mol^{-1}\ K^{-1})(17.8)}{e^{17.8} - 1}$$

$$\approx 0 + 0 \approx 0$$

$$\bar{S}_{vib,2} = +3.4 \times 10^{-3} + 2.68 \times 10^{-2} = 0.030\ J\ mol^{-1}\ K^{-1}$$

$$\bar{S}_{vib,3} = 0.0$$

$$\bar{S}_{vib,tot} = \sum_{i=1}^{3} \bar{S}_{vib,i} = 0.030 \text{ J mol}^{-1} \text{ K}^{-1}$$

9-2.6.2 EXAMPLE

From the examples in the last three sections find the total entropy for $H_2O(g)$ and compare to the thermodynamic value of 188.7 J mol^{-1} K^{-1}.

Solution

From Example 9-2.4.2:	$\bar{S}^0_{tr}$ =	144.8 J mol^{-1} K^{-1}
From Example 9-2.5.2:	$\bar{S}^0_{rot}$ =	43.5 J mol^{-1} K^{-1}
From Example 9-2.6.1:	$\bar{S}^0_{vib}$ =	0.030 J mol^{-1} K^{-1}
	Total =	188.3 J mol^{-1} K^{-1}

The slight difference is due to the fact that S(0) is not exactly equal to zero. The true calculated value will be $\bar{S}^0 + S(0)$. The reason that S(0) is not zero is due to some disorder existing in the H_2O crystal even at absolute zero. Ice does not form a perfect crystal.

9-2.7 ELECTRONIC ENERGY LEVELS

In general, the electronic energy levels are so widely separated that only the lowest level is populated. The partition function is easily written:

$$z_{elec} = g_0 \exp\left(-\frac{\varepsilon_0}{kT}\right) \tag{9-2.7.1}$$

where ε_0 is the energy of the ground state. The thermodynamic functions are easily written using this partition function. You can do that.

We now have all the partition function expressions we need to calculate thermodynamic functions if appropriate energy levels are known. The next section (9-3) will apply these partition functions to problems of chemical equilibrium. In Chap. 10, application will be made to determine reaction rate constants from known molecular energies.

9-2.8 PROBLEMS

1. Determine $\bar{S}^0$ and $\bar{A}^0$ for 1 mol $^{32}S^{16}O_2$ at 298.15 K. $A = 6.078 \times 10^{10}$ s^{-1}, $B = 1.032 \times 10^{10}$ s^{-1}, $C = 0.880 \times 10^{10}$ s^{-1}, $\bar{\nu}_1 = 1151.2$ cm^{-1}, $\bar{\nu}_2 = 519$ cm^{-1}, and $\bar{\nu}_3 = 1361$ cm^{-1}.

2. Determine G^0 for 1 mol NO_2 at 298 K. $I_A I_B I_C = 1.44 \times 10^{-137}$ $(\text{kg m}^2)^3$, $\bar{\nu}_1 = 1332$ cm^{-1}, $\bar{\nu}_2 = 751$ cm^{-1}, and $\bar{\nu}_3 = 1616$ cm^{-1}; $\sigma = 2$ and $g_{elec} = 2$.

9-2.9 SELF-TEST

1. Determine $\bar{E}^0$ and $\bar{S}^0$ for 1 mol of $H_2{}^{32}S$ at 298.15 K. $A = 3.163 \times 10^{11}$ s^{-1}, $B = 2.765 \times 10^{11}$ s^{-1}, $C = 1.475 \times 10^{11}$ s^{-1}, $\bar{\nu}_1 = 2610.8$ cm^{-1}, $\bar{\nu}_2 = 1290$ cm^{-1}, $\bar{\nu}_3 = 2684$ cm^{-1}, and $\sigma = 2$.

9-2.10 ADDITIONAL READING

1.* F. T. Wall, *Chemical Thermodynamics*, 3rd ed., W. H. Freeman Co., 1974, p. 244.

9-3 EQUILIBRIUM CONSTANTS

OBJECTIVES

You shall be able to calculate equilibrium constants from given energy level patterns for a set of reactants and products.

9-3.1 INTRODUCTION

So far in this chapter we have developed the equations for calculating thermodynamic variables for molecules. All that is needed is a set of energy levels for each type of motion possible. Next we will develop the equations required to calculate the equilibrium constant for a reaction. In Chap. 10, the partition function derived up to this point will be used to calculate reaction rate constants.

There are many other topics that could be included, but the two mentioned above show the wide applicability of the principles of statistical mechanics. Before turning to the equilibrium constant calculation you should be aware of a precaution. This has to do with the different kinds of statistics that are applicable to other systems besides molecules. This entire chapter has been concerned with *Boltzmann statistics*, which are applicable to distinguishable particles. In calculating the total partition function for N molecules for translation, a factor N! was arbitrarily introduced to account for the fact that molecules are, in fact, indistinguishable. This was done to make the calculated translational entropy agree with experiment. The difficulty can be avoided by considering statistics based on the symmetry of the wave functions for the different types of motion. In quantum mechanics, only wave functions that are *totally symmetric* or *totally antisymmetric* with respect to the interchange of coordinates are allowed. *Bose-Einstein statistics* apply to particles (or motion) for which the wave functions are *symmetric*. *Fermi-Dirac statistics* apply to *antisymmetric* wave functions. Bose-Einstein statistics apply to particles with integral spins (0, 1, 2, ...) and to photons. The Planck distribution law (for black body radiation) can be derived from these statistics. Fermi-Dirac statistics, on the other hand, apply to particles with spins of 1/2, 3/2, 5/2 ... . Electrons fall into this category. One of the principle uses of Fermi-Dirac statistics is in the electronic theory of metals [see Rice, Chap. 14, for more detail (Sec. 9-3.6, Ref. 1).]

9-3.2 CHEMICAL EQUILIBRIA

EQUILIBRIUM CONSTANT EQUATION: IDEAL GASES. In Chap. 5 the condition for equilibrium was given as $\Delta G = 0$. From the definition we know $G = A + PV$ which, when combined with Eq. (9-2.3.16), yields (for 1 mol)

$$\bar{G} = -NkT \ln z + P\bar{V} \tag{9-3.2.1}$$

Another definition that is found in Chap. 5 and again in Chap. 6, is that of the chemical potential. The chemical potential is defined as

$$\mu = \left(\frac{\partial G}{\partial n_i}\right)_{n_j,T,P} \tag{9-3.2.2}$$

Take a moment to read Sec. 5-3.5 and 6-1.1 before continuing.

Consider a general reaction involving gases, liquids, and/or solids.

$$aA + bB \rightleftarrows cC + dD \tag{9-3.2.3}$$

The total Gibbs free energy of this mixture is

$$G = N_A\mu_A + N_B\mu_B + N_C\mu_C + N_D\mu_D \tag{9-3.2.4}$$

where N_A, etc., are the number of molecules of each type. The next step is to assume that some infinitesimal amount of A [say $(ad\xi)$ molecules] reacts. If this occurs the equation for the reaction tells us that $bd\xi$ molecules of B also react to produce $cd\xi$ molecules of C and $dd\xi$ molecules of D. The change in G is

$$\begin{aligned} dG &= \mu_A\, dN_A + \mu_B\, dN_B + \mu_C\, dN_C + \mu_D\, dN_D \\ &= \mu_A(-ad\xi) + \mu_B(-bd\xi) + \mu_C(cd\xi) + \mu_D(dd\xi) \\ &= (c\mu_C + d\mu_D - a\mu_A - b\mu_B)\, d\xi \end{aligned} \tag{9-3.2.5}$$

or
$$dG = \Delta\mu_r\, d\xi \tag{9-3.2.6}$$

$\Delta\mu_r$ is the chemical potential change for the reaction which is given by

$$\Delta\mu_r = c\mu_C + d\mu_D - a\mu_A - b\mu_B \tag{9-3.2.7}$$

Of course, μ_A, μ_B, and so on, must be the values of the chemical potentials at the given temperature and pressure. The equation applies to liquids, gases, or solids. Relating μ to the *molar free energy* $\bar{G}$ we can rewrite Eq. (9-3.2.1) as

$$\mu_A \equiv \bar{G}_A = -N_A kT \ln z_A + P_A \bar{V}_A \tag{9-3.2.8}$$

where N_A is Avagadro's number of A molecules, as usual, and $\bar{V}_A$ is the molar volume at pressure P_A. An equation like (9-3.2.8) can be written for each different chemical species in the reaction.

Equation (9-3.2.8) is a general equation. However, all we will deal with here are ideal gases, since other systems are too complex to treat theoretically. For an *ideal gas*,

$$\mu_A = -kT \ln (z_A)^{N_A} + N_A kT \tag{9-3.2.9}$$

Let us write z_{tr} explicitly and leave all the other partition functions as $z'_A = z_{A,rot} \times z_{A,vib} \times \cdots$.

$$z_{A,tr} = \left[\frac{(2\pi m_A kT)^{3/2} V_A}{h^3}\right]^{N_A} \frac{1}{N_A!} \tag{9-3.2.10}$$

Then,
$$\mu_A = -kT \ln \left\{z'_A \left[\frac{(2\pi m_A kT)^{3/2} V_A}{h^3}\right]\right\}^{N_A} + kT \ln N_A! + N_A kT \tag{9-3.2.11}$$

Apply Stirling's approximation to show the last two terms give $kT \ln (N_A)^{N_A}$. This gives

$$\mu_A = -N_A kT \ln \left\{z'_A \left[\frac{(2\pi m_A kT)^{3/2} V_A}{h^3 N_A}\right]\right\} \tag{9-3.2.12}$$

But
$$\frac{V_A}{N_A} = \frac{kT}{P_A} = \frac{N_A kT}{N_A P_A}$$

Therefore,

$$\mu_A = -N_A kT \ln \left\{z'_A \left[\frac{(2\pi m_A kT)^{3/2}}{h^3}\right] \frac{N_A kT}{N_A P_A}\right\} \tag{9-3.2.13}$$

As mentioned previously, μ_A is a *molar free energy* so $N_A = N$, Avagadro's number. The standard state usually chosen for ideal gases is 1 atm, so the chemical potential in the standard state is

$$\mu_A^0 = -RT \ln \left\{ z_A' \left[\frac{(2\pi m_A kT)^{3/2}}{h^3} \right] \frac{RT}{N} \right\} \tag{9-3.2.14}$$

This equation and corresponding ones for B C, and D can be included in Eq. (9-3.2.9) to give

$$\Delta\mu_r^0 = c\mu_C^0 + d\mu_D^0 - a\mu_A^0 - b\mu_B^0 \tag{9-3.2 15}$$

where each chemical potential term has the form given in Eq. (9-3.2.14).

Another relation that is vital to our development is the equilibrium expression for a reaction involving ideal gases.

$$\Delta\mu_r^0 = -RT \ln Kp \tag{9-3.2.16}$$

If we define (at the standard pressure of 1 atm)

$$z_A^0 = \left\{ z_A' \left[\frac{(2\pi m_A kT)^{3/2}}{h^3} \right] \frac{RT}{N} \right\} \tag{9-3.2.17}$$

and similar terms for z_B^0, z_C^0, and z_D^0, we can write Eq. (9-3.2.15) as

$$\Delta\mu_r^0 = -RT \ln \frac{(z_C^0)^c (z_D^0)^d}{(z_A^0)^a (z_B^0)^b} \tag{9-3.2.18}$$

From which

$$Kp = \frac{(z_C^0)^c (z_D^0)^d}{(z_A^0)^a (z_B^0)^b} \tag{9-3.2.19}$$

Equation (9-3.2.19) is the one required to calculate the equilibrium constant for the reaction of ideal gases.

THE PARTITION FUNCTIONS. To evaluate Kp we must know the individual partition functions. This generates no problem for *rotation* and *translation*. We simply use the equations given before for these. For the *electronic* and *vibrational* partition functions we must take care in specifying our reference energy. Any value could be used as a reference energy, but that value must be given explicitly. The most convenient choice is to set the energy of the *dissociated atoms at infinite separation equal to zero.* In this way, reference energies for all molecules are the same.

With this choice of reference energy our equations for the *vibrational* and *electronic* partition functions must be modified. The electronic function becomes

$$z_{A,elec} = g_{A,0} \qquad (9\text{-}3.2.20)$$

where $g_{A,0}$ is the degeneracy of the ground electronic state. Equation for vibration must be modified to read

$$\boxed{z_{A,vib} = \frac{e^{-D_{A,0}/kT}}{1 - e^{-h\nu_0/kT}}} \qquad (9\text{-}3.2.21)$$

where D_0, the dissociation energy, is defined in Fig. 9-2.6.1. Kp can now be written

$$Kp = \frac{(\mathcal{C})^c\,(\mathcal{D})^d}{(\mathcal{A})^a\,(\mathcal{B})^b} \qquad (9\text{-}3.2.22)$$

where

$$(\mathcal{C})^c = \left[\frac{(2\pi m_C kT)^{3/2}}{h^3}\,\frac{RT}{N}\,g_{C,0}\,z_{C,rot}\,z_{C,vib}\right]^c$$

and similar expressions apply for $(\mathcal{D})^d$, $(\mathcal{A})^a$, and $(\mathcal{B})^b$. The form of z_{rot} will depend on whether linear or nonlinear molecules are involved. The best way to demonstrate the procedures is by an example.

9-3.2.1 *EXAMPLE*

The following data are known for the reaction of

$$I_2(g) \rightleftarrows 2I(g)$$

$g_{I,0} = 4 \qquad I_{I_2} = 7.50 \times 10^{-45}\ \text{kg m}^2$

$g_{I_2,0} = 1 \qquad \nu_{I_2,0} = 6.41 \times 10^{12}\ \text{s}^{-1}$

$\sigma_{I_2} = 2 \qquad D_{I_2,0} = 2.466 \times 10^{-19}\ \text{J}$

Find Kp at 1000°C = 1273 K.

Solution

$$Kp = \frac{\{[(2\pi m_I kT)^{3/2}/h^3](RT/N)(g_{I,0})(1)(1)\}^2}{\{[(2\pi m_{I_2} kT)^{3/2}/h^3](RT/N)(g_{I_2,0})(8\pi^2 I_{I_2} kT/\sigma h^2)[e^{D_{I_2,0}/kT}/(1 - e^{-h\nu_0/kT})]}$$

After considerable rearrangement we obtain

$$Kp = \frac{(2\pi kT)^{3/2}}{h^3} \frac{m_I^{\;3}}{m_{I_2}^{\;3/2}} \frac{RT}{N} \frac{4^2}{1} \frac{\sigma h^2}{8\pi^2 IkT} \frac{1 - e^{-h\nu_0/kT}}{e^{D_0/kT}}$$

$$= (1.26 \times 10^{71}\ J^{-(3/2)}\ s^{-3})(3.42 \times 10^{-38}\ kg^{3/2})(1.76 \times 10^{-20}\ J)(4)^2$$

$$(8.44 \times 10^{-5})(1.73 \times 10^{-7})$$

$$= 1.77 \times 10^4\ N\ m^2$$

$$= \underline{\underline{0.175\ atm}}$$

The experimental value is 0.165 atm. The agreement is better than one could hope for, since ideal gas behavior has been assumed and no allowance for anharmonicity in the vibrational levels been made. At 1273 K, enough vibrational levels may be populated to make this significant in some cases.

9-3.3 PROBLEMS

1. The following data are known for the $Na_2 \rightleftarrows 2Na$ reaction. Find Kp at 1000 K.

$\sigma_{Na_2} = 2 \qquad \sigma_{Na} = 1$

$g_{Na_2} = 1 \qquad g_{Na} = 2$

$I_{Na_2} = 1.81 \times 10^{-45}\ kg\ m^2$

$\bar{\nu}_0 = 159.23\ cm^{-1}$

$D_0 = 1.17 \times 10^{-19}\ J$

2. For the water-gas reaction, find Kp at 900 K.

$$CO_2 + H_2 \rightleftarrows CO + H_2O$$

	CO_2	H_2	CO	H_2O
g	1	1	1	1
σ	2	2	1	2
I (10^{-47} kg m^2)	71.9	0.472	14.5	1.024, 1.921, 2.947
$\bar{\nu}_i$ (10^{13} s^{-1})	3.94, 7.00, 1.96, 1.96	12.9	6.39	10.95, 4.77, 11.3

$$\Delta D_0 = D_0(H_2O) + D_0(CO) - D_0(H_2) - D_0(CO_2) = 6.70 \times 10^{-20}\ J$$

9-3.4 SELF-TEST

1. From the data given for $H_2 + D_2 \rightleftarrows 2HD$ find Kp at 670 K.

	H_2	D_2	HD
g	1	1	1
σ	2	2	1
I (10^{-47} kg m^2)	0.472	0.944	0.630
$\bar{\nu}$ (10^{14} s^{-1})	1.29	0.912	1.13

$$\Delta D_0 = 2D_0(HD) - D_0(H_2) - D_0(D_2) = 0$$

9-3.5 ADDITIONAL READING

1.* O. K. Rice, *Statistical Mechanics, Thermodynamics and Kinetics*, W. H. Freeman, 1967.

2. G. Herzberg, *Molecular Spectra and Molecular Structure*, Vol. I. *Spectra of Diatomic Molecules*, 2nd ed., 1950; Vol. II. *Electronic Spectra and Electronic Structures of Polyatomic Molecules*, 1966. D. Van Nostrand Company. Source of useful spectroscopic data.

3. For other data tabulations see Sec. 1-1.9D.

9-4 SUMMARY

Statistical mechanics can be applied in many fields of science. The equilibrium constants calculated above and reaction rate constants in Chap. 10 are only two examples. The heat capacity of solids can be predicted by assuming each atom in the solid is an oscillator. The assembly of oscillators makes up an ensemble which can be treated statistically. Planck's distribution law mentioned in Sec. 7-2.2 can be derived by applying Bose-Einstein statistics to a *photon gas*. A photon gas consists of particles each with a rest mass of zero and a velocity equal to the velocity of light. The electrons in a metal can be treated as an *electron gas* to which Fermi-Dirac statistics apply. Statistical mechanics finds considerable use in the theoretical investigation of polymer molecules. The random motion of segments of a flexible polymer chain (see Chap. 12) can be studied statistically to determine the most likely configuration of the polymer molecule. The Maxwell-Boltzmann velocity distribution function found in Chap. 2 can be derived from the translational energy distribution function. Also, diffusion and Brownian motion are amenable to treatment by the methods of statistical mechanics. There are other examples. These should, however, show the value of statistical mechanics in many areas of science. These treatments are usually rather complex, but all represent extensions of what you have studied in this chapter.

10

Chemical Kinetics

SYMBOLS AND DEFINITIONS

τ_i	Stoichiometric coefficient for a reaction
A_i	Reactant or product in a reaction
k	Boltzmann constant (1.38×10^{-23} J molecule^{-1} K^{-1})
$\underline{k}$	Reaction rate constant (units vary)

$[A_i]$	Concentration of A_i (mol dm^{-3})
n_i	Number of moles of reactant or product
$n_{i,0}$	Initial number of moles of reactant or product
ξ	Extent of reaction variable
a, b	Initial concentrations of reactants A and B (mol dm^{-3})
t	Time (s, min)
$t_{1/2}$	Half-time of a reaction (s, min)
$\tau_i x$	Amount of i that has reacted at any time (mol dm^{-3})
λ_i	Physical property of substance i
λ_t	Physical property of system of reactants and products at time t
λ_0	Physical property of system at time $t = t_0$
λ_∞	Physical property of system at time $t = \infty$
A	Constant in the Arrhenius equation
E_a	Activation energy in the Arrhenius equation (J mol^{-1})
$E_{f,0}$	Activation energy for a forward reaction (J mol^{-1})
$E_{r,0}$	Activation energy for a reverse reaction (J mol^{-1})
$\underline{k_f}$	Rate constant for a forward reaction
$\underline{k_r}$	Rate constant for a reverse reaction
ΔE_0	$E_{f,0} - E_{r,0}$ (J mol^{-1})
A_f	Pre-exponential factor for forward reaction
A_r	Pre-exponential factor for reverse reaction
A'	A_f/A_r
K_{eq}	Equilibrium constant for a reaction (units vary)
B	Temperature-independent pre-exponential factor
σ_{AB}	Collision diameter of A - B $[(\sigma_A + \sigma_B)/2]$, 10^{-10} m
σ'	$\pi\sigma_{AB}^2$, Collision cross-section (10^{-20} m^2)
μ	Reduced mass (kg $molecule^{-1}$)
c_A, c_B	Relative speeds of A and B (m s^{-1})
dZ_{AB}	Differential collision rate (s^{-1})
ε_0	Threshold kinetic energy (J $molecule^{-1}$)
c_0'	Threshold relative speed (m s^{-1})
A'	Collision theory pre-exponential factor
N_A^*, N_B^*	Molecular density (molecule m^{-3})
A''	Line of centers pre-exponential factor
p	Steric factor
$C^\ddagger$	Activated complex
$\underline{\delta}$	Distance along reaction coordinate
$\bar{v}$	Velocity across transition state
$z_\ddagger$	Partition function for the activated complex
$z_\ddagger^\delta$	Partition function for the activated complex minus the motion across the transition state
κ	Transmission coefficient
I_A, I_B, I_C	Moments of inertia (kg m^2)
ν_i	Fundamental vibrational frequency (s^{-1})

As yet we have completely ignored the time variable in chemical systems. We have assumed that all processes and phenomena are independent of time. Our study of thermodynamics has been concerned with initial and final equilibrium states without reference to how long it might take for the change to occur. In quantum mechanics we ignored any time dependence for the wave functions we found, although in many phenomena (such as transitions between energy levels) such dependence is of importance.

We now turn our attention to the study of kinetics, a study of motion or time dependence. We will want to know how long it takes for a given set of reactants to produce a set of products. Thus, kinetics is a dynamic science, whereas thermodynamics (in spite of the name) is static. The study of kinetics is not limited to rates of reaction. The *mechanism* of the reaction is very important, also, if a particular reaction is to be well understood. A reaction mechanism is a detailed account of the individual collisions and sequences of collisions that are present in the conversion of reactants to products. Detailed study of mechanisms allows classification of reactions into types (as you know from your study of organic chemistry). This allows prediction of how related compounds (that have not been studied) will react. Kinetic studies are obviously vital if one has to design a process to produce a product. The reaction rate must be investigated as a function of such variables as temperature, pressure, solvent, etc. In many cases, the mechanism of a reaction is not of immediate interest, whereas the rate must be known for that reaction to be used in a process. Later, studies of the mechanisms involved may lead to suggestions of how to increase the reaction rate or minimize unwanted products.

One area that is receiving particular attention is gas phase reaction in the atmosphere. How long do hydrocarbons, carbon monoxide, nitrogen oxides, etc., reside in the atmosphere before undergoing changes? What products result from the reactions of these materials? What role does sunlight play in the rate of the reactions and the type of products formed? These questions, and related ones, need answers before intelligent decisions can be made on what types and amounts of emission can be allowed. The answers are not easily obtained due to the complexity of the system involved. Below, the discussion will start with rates of reaction and how these can be determined experimentally. No theory is involved in this; it is completely empirical. After detailing the common types of rate expressions encountered, we turn to an investigation of simple mechanisms. Finally, you will get a chance to apply the collision theory of gases encountered in Chap. 2 and statistical mechanics found in Chap. 9 to theoretical aspects of reaction rates and mechanisms.

10-1 EXPERIMENTAL TECHNIQUES AND SIMPLE RATE LAWS

OBJECTIVES

You shall be able to

1. Discuss various experimental techniques, briefly indicating which techniques would apply to fast, slow, or intermediate reactions
2. Integrate rate expressions to find the rate law

3. Fit given data to the appropriate rate law and determine the rate constant

10-1.1 INTRODUCTION

The rate of a chemical reaction is expressed as the change in concentration of a reactant or product in unit time. The method used to investigate such a change depends on the rate. Some techniques work only for slow reactions, others for rapid ones. Illustration of these will be given below. The rate laws to be derived in the following are for simple rate expressions. If a reaction mechanism is complex, the rate law is more than likely going to be complex. In such a case, the derivation of an appropriate rate law will be difficult. This section is restricted to cases that can be readily solved. Also, interpretation of kinetic data can be very tedious if the rate law is not simple. This section will deal only with examples that fit the simple rate laws. Section 10-2 will mention some procedures for analyzing more complex reactions. A precaution before we begin: *The stoichiometry of a reaction in no way implies the mechanism of the reaction.* More detail on that later. Right now, memorize that statement.

10-1.2 EXPERIMENTAL METHODS

The fundamental variables (as you will see as the development proceeds) for reaction rates are temperature, time, and concentration of reactants. We will consider how each of these can be measured for different rates of reaction, then we will proceed with a discussion of some of the methods used to study fast or slow reactions.

TEMPERATURE. Reactants can be put into a bath accurately controlled by a thermostat to achieve a given temperature. For low or high temperatures, control may be difficult but not impossible. However, knowing the bath temperature does not always give sufficient information about the reaction vessel in the bath. For a very rapid reaction that is either exothermic or endothermic, the temperature inside the reaction vessel may be significantly different from that in the bath. For fast reactions, knowledge of the reaction time, reaction heat, and thermal conductivity of the system is necessary to estimate the temperature history of the reaction.

TIME. The major difficulty in measuring time is not in the actual measurement itself but in deciding when to start the clock. In almost any reaction that is to be studied the reactants must be mixed. This is not an instantaneous process. Also, if a given temperature is required, a finite amount of time is required to get the reactants to that temperature. For slow reactions these considerations are negligible. For very fast reactions they put an upper limit on the rates that can be studied.

CONCENTRATION. Concentration measurements are usually the most difficult of the three. As you will see below, in many rate studies, some function of concentration (such as absorption of radiation, electrical conductivity, pressure, refractive index, etc.) can be measured without disturbing the system. Such a procedure is valuable if the physical property measured can be related to the concentration of interest. That is not always easily done. Also, the measurement must be fast relative to the reaction rate. Another common technique is to intercept the reaction by taking a sample of the reaction mixture and "quenching" it. Quenching involves rapid cooling or otherwise treating of the sample to stop the reaction. The concentration is then determined. Perhaps a discussion of a few specific techniques is useful at this point.

STATIC METHODS. A static method implies a process of bringing a reaction mixture instantaneously to a given pressure, temperature, and concentration. Then, as the mixture is held at the given temperature and pressure, measurements of the concentration are made. In some cases it is convenient to allow pressure to vary and use that variable as a measure of the reaction rate. For slow reactions, this procedure works for gases, and the quenching procedure works for solutions. For fast reactions some modifications are necessary.

A common procedure for fast reactions is to perturb the system (by a flash of light, a sudden temperature or pressure change, an electrical discharge, or other means) to disturb the equilibrium. The *relaxation* of the system is then observed using a rapid response instrument. Relaxation is a term used to describe the return of a system to equilibrium after it has been disturbed. One technique is to follow the voltage drop across the solution (which is related to the electrical conductivity of the solution) on an oscilloscope. Oscilloscopes with response times on the order of nanoseconds are available, so rather rapid reaction can be studied. Another technique for observing relaxation is by absorption spectroscopy. A change in intensity of the absorption of a reactant or product with time can yield information on the concentration changes in the solution. This technique has been refined to such an extent that changes that take only a few picoseconds (10^{-12} s) can be observed. Flash photolysis is used for studying very fast reactions in which heating would take much too long or would not produce the desired reaction. For reactions that can be initiated by light, flash photolysis offers a means of very rapidly bringing about a nonequilibrium situation in a system. Measurements are then made on the systems as they return toward equilibrium (relax). Figure 10-1.2.1 gives a schematic of a simple flash photolysis system. In this type (or similar types) of apparatus, reactions of gases, liquids, and solids can be studied.

In a typical experiment, an intense beam of visible or ultraviolet radiation is used. This flash must be intense enough to significantly alter the reactants,

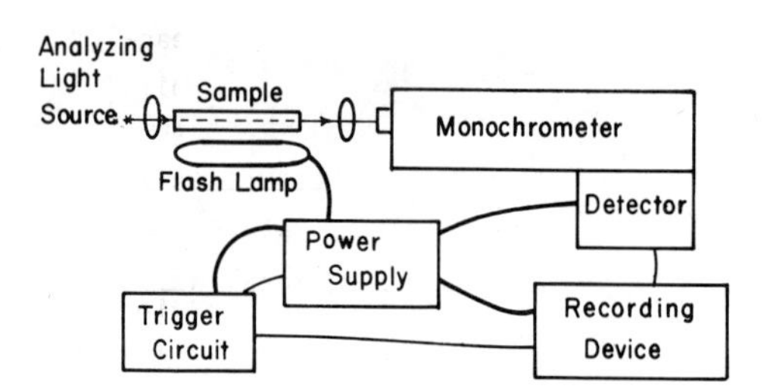

FIG. 10-1.2.1 Schematic of a simple flash photolysis apparatus.

yet be of short enough duration that it does not interfere with subsequent measurements that are to be made on the relaxing system. Normally, the more intense the flash, the longer its duration. So, as always, some compromises must be made in many experimental situations. Typical flash times are on the order of microseconds and flash powers from 20 to 20,000 J. As you can see, there is a tremendous influx of energy in a very short time. Normally, subsequent measurements on the system are made by spectroscopic means, since these can be made without further disturbing the system. Also, spectroscopic measurements themselves can be made extremely rapidly: a necessity if very fast reactions are occurring in the sample compartment. For a detailed treatment of the experimental techniques associated with flash photolysis see G. Porter, "Flash Photolysis" (see Sec. 10-1.8, Ref. 1).

The apparatus shown in Fig. 10-1.2.2 can be used either for rather slow reactions (see below under Flow Methods) or for rapid reactions. In either case, the plungers are driven in at a known rate and reactants forced into the mixing chamber. In the specially designed mixing chamber, the reactants are thoroughly and quickly mixed. The mixture flows down the exit tube at rates up to 10 m s^{-1}. In the flow method, measurements of concentrations are made at selected locations along the tube. These correspond to definite reaction times which can be readily calculated from flow velocities and tube diameter. Reactions that occur in only a few milliseconds can be studied if the observing ports are only a few cm from the mixing chamber. Measurements can be made spectroscopically, potentiometrically conductiometrically, or by other means. Any physical property that can be measured without appreciably disturbing the system and that can be made quickly relative to the reaction rate can be used.

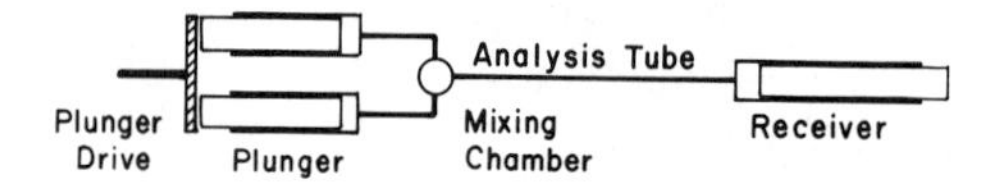

FIG. 10-1.2.2 An apparatus suitable for flow or stopped-flow methods.

In the stopped flow method that is sometimes applied to fast reactions, the reactants are forced into the mixing chamber, but flow is stopped before any measurements are made. This technique allows one to take advantage of the very rapid and complete mixing that occurs in a well-designed mixing chamber. Otherwise, the measurements are the same as for any other mixing system. Any of the methods discussed above can be used to determine concentration of reactant or product as a function of time. For an extensive review see F. J. W. Roughton and B. Chance "Rapid Reactions" (see Sec. 10-1.8, Ref. 2).

FLOW METHODS. In slow reactions, flow through a tubular reactor at a given temperature can be used. Analysis before and after the reaction zone allows calculation of the concentration changes. The calculations are difficult to carry out, however, since the flow patterns in the reactor must be known accurately. A simple stirred reactor is another method. Here two input streams feed into a well-stirred vessel. The output stream is analyzed. Residence times in the reactor must be long enough for complete mixing, so only slow reactions can be studied. The method described above (see Fig. 10-1.2.2) can also be used. For fast reactions a common method involves a shock tube, as shown in Fig. 10-1.2.3. In the shock tube a diaphragm separating a high-pressure gas from a low-pressure reaction mixture is ruptured. This generates a shock wave that travels down the tube at about the speed of sound. The sample is heated to several thousand degrees in a few microseconds by compression. An expansion or surge tank allows rapid decompression of the gas as it returns back up the tube. This lowers the temperature dramatically. Thus, either the hot, compressed gas or cooled gas can be studied. Another method involves molecular beams which are sources for streams of molecules with selected velocities (and, therefore, selected translational temperatures). Studies are made of interactions of two beams of this type.

Figure 10-1.2.4 shows a sketch of an experimental setup for studying the interaction of molecular beams. Each source generates a stream of atoms or molecules whose velocity distribution is defined by the temperature and the Maxwell-Boltzmann

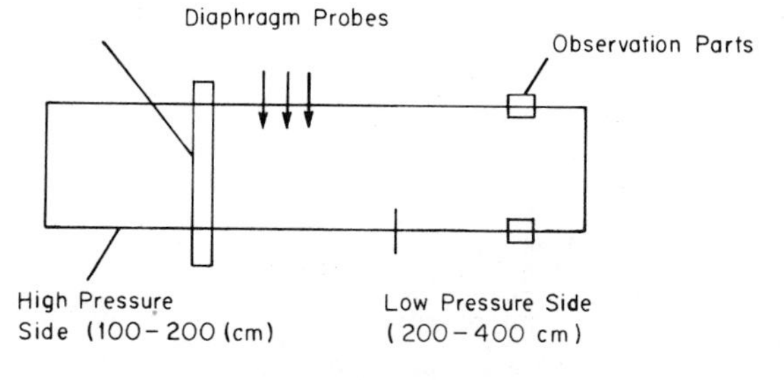

FIG. 10-1.2.3 Schematic of a shock tube.

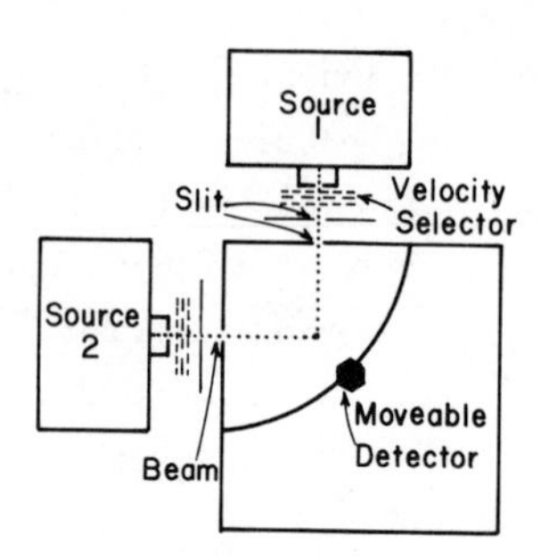

FIG. 10-1.2.4 A molecular beam apparatus for studying the interactions of two beams of atoms or molecules. Each compartment in the system is in a high vacuum.

distribution function (see Sec. 2-3). Rotating slits on either of both sources can be used to select very narrow velocity distributions. These molecules are collimated, and the beams are made to collide in the scattering chamber. The scattered beam is analyzed as a function of scattering angle. Much more sophisticated procedures can be used to choose molecules only in given rotational or other states. Also, the scattered beam can be analyzed for products, kinetic energy, electronic states, etc., all as a function of the scattering angle. As you might guess, much fundamental knowledge can be obtained about individual molecular collisions by this technique. It has been particularly useful for studying ion-molecule and atom-molecule reactions. Further detail will not be given. See R. E. Weston, Jr. and H. A. Schwarz, "Chemical Kinetics" for additional information (see Sec. 10-1.8, Ref. 3). Good references for all the above techniques (and others) are G. B. Skinner, *Introduction to Chemical Kinetics* and Vol. VIII of Weissberger (see Sec. 10-1.8, Refs. 2 and 4).

10-1.3 SOME DEFINITIONS

Some definitions are in order before working out the rate laws. Correct use of the terminology is important to help avoid confusion. The term *stoichiometry* means the number of moles of each reactant and product involved in the overall reaction. In the reaction given in Eq. (10-1.3.1), $-\tau_1$ mol A_1 react with $-\tau_2$ mol A_2 to produce τ_3 mol A_3 and τ_4 mol A_4.

$$-\tau_1 A_1 - \tau_2 A_2 \rightarrow \tau_3 A_3 + \tau_4 A_4 \tag{10-1.3.1}$$

τ_i *will be positive for products and negative for reactants.* (The notation looks a little clumsy here, but leads to simpler expressions later. With this notation and the convention that τ_i is *negative* for reactants and *positive* for products, we can write our rate laws in general form without having to worry about signs.) That bit of information tells us nothing about how the reaction occurs. It is a statement of the overall changes involved. The *order* of a reaction is determined from the *rate law* describing that reaction. For the reaction in (10-1.3.1) the rate is given by (these are always written as a rate of formation)

$$\text{rate} = \frac{1}{\tau_1}\frac{d\,[A_1]}{dt} = \frac{1}{\tau_2}\frac{d\,[A_2]}{dt}$$

$$= \frac{1}{\tau_3}\frac{d\,[A_3]}{dt} = \frac{1}{\tau_4}\frac{d\,[A_4]}{dt} \tag{10-1.3.2}$$

where the brackets imply concentration. Note that τ_1 and τ_2 will be negative while τ_3 and τ_4 are positive. Also note that if we write the equation for the rate of formation of A_1 or A_2 we get rate 1 or 2 = negative. (τ_1 and τ_2 are negative.) This means A_1 and A_2 are decreasing in concentration. A negative rate of formation is the same as a positive rate of consumption.

For the reaction in Eq. (10-1.3.1) any of the following rate laws might apply (or some other; *the stoichiometry of a reaction tells us nothing about the order of the reaction*). The actual rate law that applies must be determined experimentally.

$\text{Rate}_1 = \underline{k}_1[A_1]$	first order in A_1: overall, first order
$\text{Rate}_2 = \underline{k}_2[A_1][A_2]$	first order in A_1 and first order in A_2: overall, second order
$\text{Rate}_3 = \underline{k}_3[A_2]^2$	second order in A_2: overall, second order
$\text{Rate}_4 = \underline{k}_4[A_1]^2[A_2]$	second order in A_1 and first order in A_2: overall, third order

or some other, more complicated form. The overall order is the sum of the individual orders and $\underline{k}$ is the *rate constant*. Any of the above might apply to (10-1.3.1). We cannot tell from the stoichiometry. To emphasize that, consider the two reactions and corresponding rates in Eq. (10-1.3.3):

$$H_2(g) + I_2(g) \rightarrow 2HI(g) \qquad \text{rate} = \underline{k}[H_2][I_2] \qquad \text{second order}$$

$$H_2(g) + Br_2(g) \rightarrow 2HBr(g) \qquad \text{rate} = \frac{\underline{k}[H_2][Br_2]^{1/2}}{1 + \underline{k}'[HBr]/[Br_2]} \tag{10-1.3.3}$$

Different orders for two reactions with the same stoichiometry result from different mechanisms. (You are not yet supposed to know the source of these last two equations.)

Another term of a theoretical nature which you will see later is *molecularity*. Molecularity refers to the number of molecules involved in a simple collisional reaction process.

Unimolecular: one molecule involved (a decomposition)
Bimolecular: two molecules involved
Termolecular: three molecules involved

Molecularity is a theoretical concept derived from detailed study of the mechanism of a reaction. *Order*, on the other hand, is an empirical concept and is derived from the experimentally determined rate law. Another term you will see is *half-life* or *half-time*. This is the time required for one-half the initial amount of a given reactant to be used in a reaction.

Finally, let us define the *extent of reaction variable* ξ which is a measure of the amount of reaction that has occurred. ξ varies from 0 at zero reaction to 1 for complete conversion of reactants to products. With this definition we can write

$$n_i = n_{i,0} + \tau_i \xi \qquad (10\text{-}1.3.4)$$

where n_i is the number of moles of A_i at some time and $n_{i,0}$ is the number of moles initially present. For example, consider the reaction

$$H_2(g) + I_2(g) \rightarrow 2HI(g)$$

$\tau_{H_2} = -1$, $\tau_{I_2} = -1$ $\tau_{HI} = +2$, $n_{H_2} = n_{H_2,0} - 1\xi$ and $n_{HI} = n_{HI,0} + 2\xi$. Usually there is not 100% conversion to product, so ξ reaches a value of ξ_{eq} at equilibrium.

You can see immediately that $dn_i = \tau_i \, d\xi$. The *reaction rate can be defined as the rate of change of the extent of reaction.*

$$\text{Rate} = \frac{d\xi}{dt} = \frac{1}{\tau_i}\frac{dn_i}{dt} \qquad (10\text{-}1.3.5)$$

The value of this formula is that the reaction rate defined in this way is independent of the choice of the reactant. The rate defined by (10-1.3.5) can be related to a rate based on concentration by

$$\frac{d\xi}{dt} = \frac{V}{\tau_i}\frac{d\,[A_i]}{dt} = \frac{1}{\tau_i}\frac{dn_i}{dt} \qquad (10\text{-}1.3.6)$$

The reaction rate constant $\underline{k}$ calculated will be different for the two definitions in Eq. (10-1.3.2) and Eq. (10-1.3.5). Equation (10-1.3.2) will be used in general. The point to be noted is you cannot use the change in the number of moles alone if concentrations are being considered; the volume must be included. Also, you should be aware that the units on the rate constant will depend on what concentration units are being used. A rate constant is not complete without a set of units.

10-1.4 SIMPLE RATE LAWS

ZERO ORDER. The simplest rate law results when the reaction is zero-order in reactants (say A_1):

$$\text{Rate:} = \frac{1}{\tau_1}\frac{d\,[A_1]}{dt} = \underline{k}[A]^0 = \underline{k} \qquad (10\text{-}1.4.1)$$

Integration from a time defined as zero when $[A_1] = [A_1]_0$ to time t, when $[A_1] = [A_1]$, gives

$$\int_{[A_1]_0}^{[A_1]} d[A_1] = \underline{k}\tau_1 \int_0^t dt$$

$$\boxed{[A_1] - [A_1]_0 = \underline{k}t} \qquad (10\text{-}1.4.2)$$

Figure 10-1.4.1 shows the variations of $[A_1]$ with t for a zero-order reaction.

The half-life occurs when one-half the initial reactant is used: $[A_1] = \frac{1}{2}[A_1]_0$.

$$\tfrac{1}{2}[A_1]_0 - [A_1]_0 = -\tfrac{1}{2}[A_1]_0$$

$$= \underline{k}\tau_1 t_{1/2} \qquad (10\text{-}1.4.3)$$

or $$t_{1/2} = -\frac{[A_1]_0}{2\tau_1\underline{k}}$$

This is not a negative number. τ_1 is negative for a reactant. For a zero-order reaction, $t_{1/2}$ *is proportional to the initial concentration of the reactant.* Some examples of zero-order reactions can be found in photochemical and catalysis studies. In a photochemical reaction, if the incident radiation is completely absorbed by the reacting molecules (exciting them so they can undergo reaction), the rate of reaction will not depend on the initial concentration. Also, if a reaction takes place on a catalyst surface, the reaction will be independent of reactant concentration if the concentration is high enough to completely cover the catalyst surface.

Assume the reaction given in Eq. (10-1.3.1) is first order in A_2 and zero order in A_1:

$$\text{Rate} = \frac{1}{\tau_2}\frac{d\,[A_2]}{dt} = \underline{k}[A_2]^1 \qquad (10\text{-}1.4.4)$$

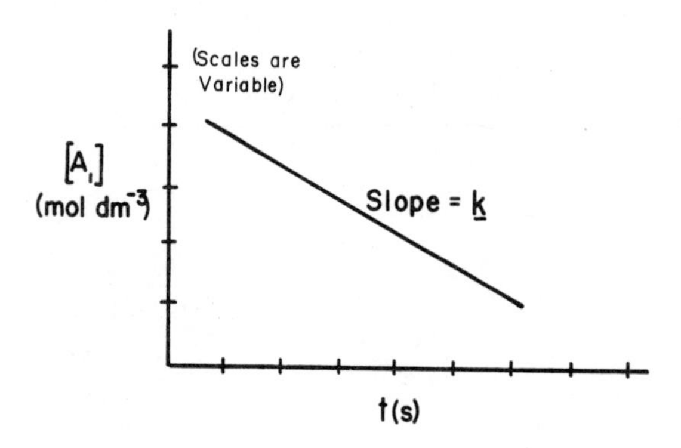

FIG. 10-1.4.1 Variation of $[A_1]$ with t for a zero-order reaction.

Again, integration from t = 0 to $t([A_2] = [A_2]_0$ to $[A_2] = [A_2])$ gives

$$\int_{[A_2]_0}^{[A_2]} \frac{d\,[A_2]}{[A_2]} = \int_0^t \underline{k}\tau_2\, dt$$

$$\ln [A_2] - \ln [A_2]_0 = \boxed{\ln \frac{[A_2]}{[A_2]_0} = \underline{k}\tau_2 t} \qquad (10\text{-}1.4.5)$$

This brings us to an important question. If a reaction is first order in A_2, what function of concentration must be plotted against time to give a straight line? Equation (10-1.4.5) provides the answer. A plot of ln $[A_2]$ vs. t should be linear if the reaction under consideration is first order in A_2 and zero order in any other reactants. (*Note*: Experimentally this can be accomplished by making the concentration of all other reactants very large with respect to $[A_2]$. Then the concentration of these other reactants will not change appreciably during the course of the reaction. The reaction will appear to be first order in A_2, if the power of $[A_2] = 1$ in the rate law. This is called a pseudo first-order reaction. You should note that pseudo first-order reactions of this type occur only if there is more than one reactant.) The plot of interest is shown in Fig. 10-1.4.2. Also shown is a plot of [B] vs. t for comparison, where B is benzenediazonium tetrafluoroborate. *Note that* ln [B] - ln $[B]_0$ *is negative. This is consistent, since* τ_2 *is negative for reactant* A_2.

The half-life expression is easily obtained by setting $[A_2] = \frac{1}{2}[A_2]_0$ for $t_{1/2}$.

$$\ln [A_2] - \ln [A_2]_0 = \ln \tfrac{1}{2} [A_2]_0 - \ln [A_2]_0$$

$$= \ln \tfrac{1}{2} = -\ln 2 = \underline{k}\tau_2 t_{1/2}$$

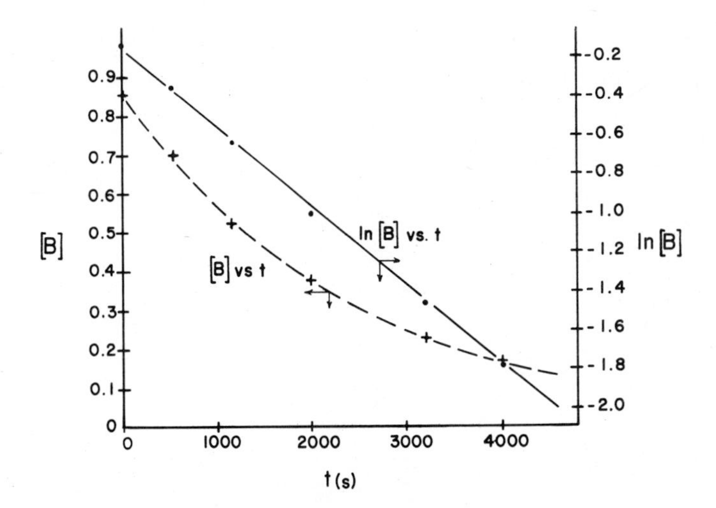

FIG. 10-1.4.2 Plot of [benzenediazonium tetrafluoroborate] and ln [benzenediazonium tetrafluoroborate] vs. t at 40°C. The linear ln [B] plot indicates first order.

$$t_{\frac{1}{2}} = -\frac{\ln 2}{\underline{k}\tau_2} \qquad (10\text{-}1.4.6)$$

You should see immediately that $t_{\frac{1}{2}}$ *is independent of the amount of material initially present for a first-order reaction.* The most well-known example of a first-order process is radioactive decay.

10-1.4.1 EXAMPLE

The isomerization reaction of p-tolyl isocyanide at 190°C to form p-tolunitrile ($CH_3C_6H_4CN$) yields the following data:

t (min)	9.0	14.0	19.0	25.0
p-tolunitrite (%)	17.5	27.0	35.0	43.0

If $\tau_{\text{isocyanide}} = -1$, show the reaction is first order and find $\underline{k}$ and $t_{\frac{1}{2}}$.

Solution

We need the concentration of isocyanide. This will be proportional to 100 - (percent tolunitrile).

[Iso] = b (100 - % tolunitrile)

where b is a constant. We then have ln [iso] = ln b + ln (100 - % tolunitrile) (the plot of these data is also shown):

t (min)	0.0	9.0	14.0	19.0	25.0
100 - % tolunitrile	100	82.5	73.0	65.0	57.0
ln [iso] - ln b	4.61	4.41	4.29	4.17	4.04

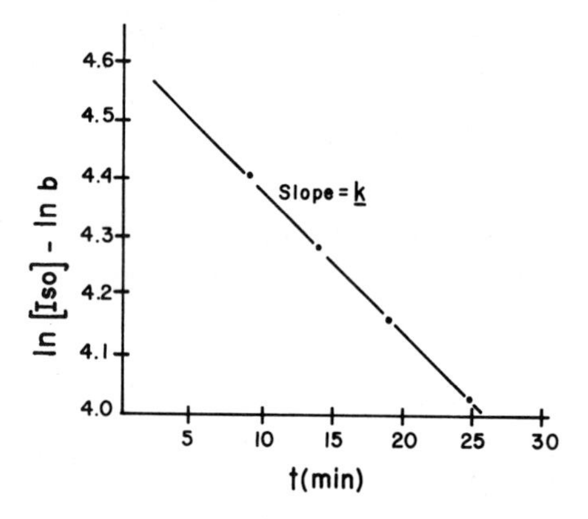

$$\text{Slope} = \frac{\ln\,[\text{iso}]_2 - \ln b - \ln\,[\text{iso}]_1 + \ln b}{t_2 - t_1}$$

$$= \frac{\ln\,[\text{iso}]_2 - \ln\,[\text{iso}]_1}{t_2 - t_1} = \underline{k}\tau_2$$

$$= -\underline{k} = 2.28 \times 10^{-2}\ \text{min}^{-1}$$

The reaction is first order (the plot is linear) with a rate constant $\underline{k}$ = 0.0228 min^{-1}.

$$t_{1/2} = \frac{-\ln 2}{-\underline{k}} = \frac{0.693}{0.0228\ \text{min}^{-1}} = 30.4\ \text{min}$$

This implies that it takes 30.4 min to use one-half the initial amount present. There is a nongraphical way to solve this problem. Rearrange Eq. (10-1.4.5) with $\tau_2 = -1$ for two times t_1 and t_2. Show that the following can be obtained.

$$\underline{k} = \frac{\ln\,[A]_2 - \ln\,[A]_1}{t_2 - t_1} = \frac{\ln\,([A]_2/[A]_1)}{t_2 - t_1}$$

Take several combinations of t_1s and t_2s to show $\underline{k}$ is constant. For example, with $t_2 = 19$ and $t_1 = 9$,

$$\underline{k} = -\frac{\ln (65/82.5)}{19 - 9} = \underline{\underline{0.0238\ \text{min}^{-1}}}$$

If several values of $\underline{k}$ agree, first order is established.

Many reactions involve gases. If this is the case, and if a reaction is carried out at constant volume, the pressure of any reactant or product is ideally proportional to its concentration. A general reaction will be given first, then a specific example. Assume ideal gas behavior. For the reaction ($\tau_A = -2$, $\tau_B = -3$, $\tau_C = 2$)

$$2A(g) + 3B(g) \rightarrow 2C(g)$$

Let us relate the total pressure at any time to the pressure of each reactant. Let $P_0(A)$ be the initial pressure of A and $P_0(B)$ the initial pressure of B. Assume $P_0(C) = 0$. The *total initial pressure* is $P_0(T) = P_0(A) + P_0(B)$. Let $2P_x$ be the amount of A that has reacted at t_x (remember at constant volume $P \propto n$, the number of moles). $2P_x$ of A reacts with $3P_x$ of B to give $2P_x$ of C. So, at time t_x, we have

$$\begin{array}{ccccc} 2A & + & 3B & \rightarrow & 2C \\ P_0(A) - 2P_x & & P_0(B) - 3P_x & & 2P_x \end{array}$$

$$P(\text{total}) = P(T) = P_0(A) - 2P_x + P_0(B) - 3P_x + 2P_x$$

$$= P_0(A) + P_0(B) - 3P_x$$

But, $P_0(T) = P_0(A) + P_0(B)$. Thus,

$$P(T) = P_0(T) - 3P_x \qquad \text{or} \qquad P_x = \frac{P_0(T) - P(T)}{3}$$

If we know the initial pressure and the pressure at time t_x we can find the amount that has reacted and the amount of A and B remaining.

$$P(A) = P_0(A) - 2P_x = P_0(A) - 2\left[\frac{P_0(T) - P(T)}{3}\right]$$

$$= P_0(A) - \frac{2}{3} P_0(T) + \frac{2}{3} P(T)$$

$$P(B) = P_0(B) - P_0(T) + P(T)$$

The pressures are then used in the rate equation that applies for that reaction.

10-1.4.2 EXAMPLE

The following reaction occurs in the gas phase

$$(CH_3)_3\text{-C-O-O-C}(CH_3)_3 \rightarrow 2(CH_3\text{-}\overset{\overset{\displaystyle O}{\|}}{C}\text{-}CH_3) + C_2H_6$$

From the following total pressure data, show the reaction is first order and find the ocrresponding rate constant. [Data from J. R. Raley, R. F. Rust, and W. E. Vaughan, *J. Am. Chem. Soc.,70*:88, (1948)].

t (min)	0	2	6	10	18	26	34	46
P_T(atm)	0.2362	0.2466	0.2613	0.2770	0.3051	0.3322	0.3569	0.3909

Solution

The reaction is of the form given above. $P_0(B) = P_0(C) = 0$.

$$\begin{array}{ccc} A & \rightarrow 2B & + \; C \\ P_0(A) - P_x & 2P_x & P_x \end{array}$$

$$P(T) = P_0(A) - P_x + 2P_x + P_x = P_0(A) + 2P_x$$

or $$P_x = \frac{P(T) - P_0(A)}{2}$$

$$P(A) = P_0(A) - P_x$$

$$= P_0(A) - \tfrac{1}{2} P(T) + \tfrac{1}{2} P_0(A)$$

$$\boxed{P(A) = \frac{3}{2} P_0(A) - \tfrac{1}{2} P(T)}$$

t (min)	0	2	6	10	18	26	34	46
P_T(atm)	0.2362	0.2466	0.2613	0.2770	0.3051	0.3322	0.3569	0.3909
P(A)	0.2362	0.2320	0.2237	0.2158	0.2018	0.1882	0.1750	0.1589
ln P(A)	-1.443	-1.461	-1.498	-1.533	-1.601	-1.670	-1.738	-1.840

Convince yourself before continuing that if $P(A) \propto [A]$, then ln P(A) vs. t has the same slope as ln [A] vs. t.

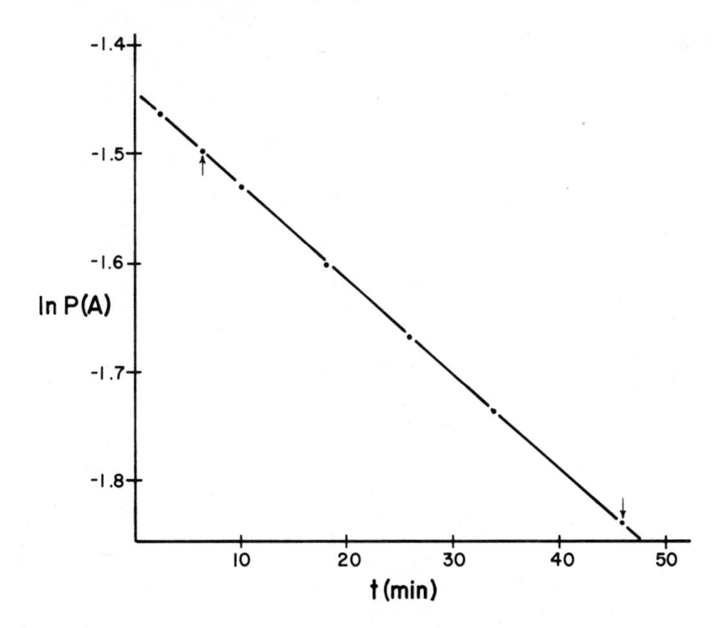

$$\text{Slope} = -\underline{k} = -8.50 \times 10^{-3}\ \text{min}^{-1}$$

$$\underline{k} = 8.50 \times 10^{-3}\ \text{min}^{-1} = \underline{\underline{1.42 \times 10^{-4}\ \text{s}^{-1}}}$$

SECOND ORDER: ONE REACTANT. Consider a reaction $-\tau_A A \rightarrow \tau_B B + \tau_C C$ which is second order in A. The rate law is

$$\text{Rate} = \frac{1}{\tau_A}\frac{d\,[A]}{dt} = \underline{k}\,[A]^2 \tag{10-1.4.7}$$

With the usual definitions,

$$\int_{[A]_0}^{[A]} \frac{d\,[A]}{[A]^2} = \int_0^t \underline{k}\tau_A\,dt = -\left.\frac{1}{[A]}\right|_{[A]_0}^{[A]}$$

$$\boxed{\frac{1}{[A]_0} - \frac{1}{[A]} = \underline{k}\tau_A t} \tag{10-1.4.8}$$

Before continuing, determine what type of plot would be required to give a straight line for a second-order reaction. Equation (10-1.4.8) gives a linear plot if $1/[A]$ is plotted against t. The slope is $-\underline{k}\tau_A$. Example 10-1.4.3 illustrates the processing of second-order data.

10-1.4.3 EXAMPLE

For the following data plot [A] and 1/[A] vs. t. Find the second-order rate constant.

Solution

t (s)	[A] (mol dm^{-3})	1/[A] (dm^3 mol^{-1})
0	1.00	1.00
10	0.95	1.05
25	0.88	1.14
40	0.82	1.22
75	0.71	1.41
100	0.65	1.54
150	0.55	1.82

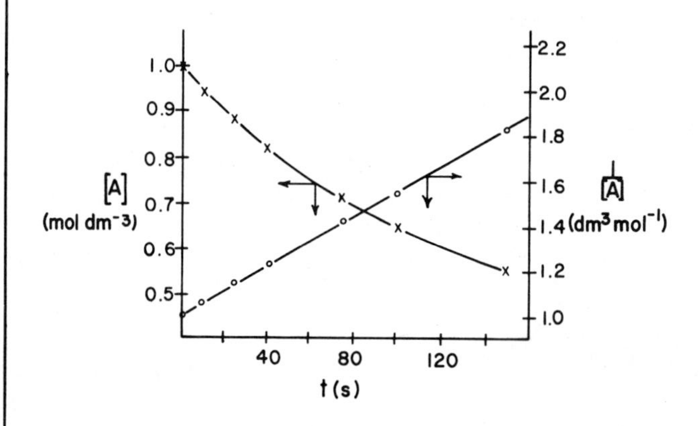

The slope of the 1/[A] vs. t plot (which is linear, indicating second order) is 0.0055 dm^3 mol^{-1} s^{-1}. According to Eq. (10-1.4.8) this should be $k\tau$.

If pressures are given, 1/P(A) can be plotted against time. The value of k will depend on what units are used in the plot. (For the first-order case, k was independent of choice of units for the concentration.) If [A] is given in *moles per liter* (moles per cubic decimeter), a common unit, k will have units of *liters per mole per second*. If pressure in *atmospheres* is used, k will be *per atmosphere per second* for a second-order reaction. The half-time for a second-order reaction in one component is

$$\frac{1}{[A]_0} - \frac{1}{\frac{1}{2}[A]_0} = \underline{k}\tau_A t_{1/2}$$

$$-\frac{1}{[A]_0} = \underline{k}\tau_A t_{1/2}$$

$$\boxed{t_{1/2} = -\frac{1}{\underline{k}\tau_A [A]_0}} \qquad (10\text{-}1.4.9)$$

The half-time is proportional to $1/[A]_0$, so as the initial concentration is increased, the half-time decreases.

SECOND ORDER: FIRST ORDER IN EACH OF TWO REACTANTS. A more complicated rate expression results when a reaction is second order overall, but first order in each of two reactants. Assume the reaction of interest is (τ_A and τ_B are negative)

$$-\tau_A A - \tau_B B \rightarrow \tau_C C$$

The rate is given by

$$\text{Rate} = \frac{1}{\tau_A}\frac{d\,[A]}{dt} = \frac{1}{\tau_B}\frac{d\,[B]}{dt}$$

$$= \underline{k}[A][B] \qquad (10\text{-}1.4.10)$$

Let $\tau_A x = \text{mol dm}^{-3}$ be the amount of A that has reacted at any time (x will be equal to ξ/V). The amount of A remaining at time t is $[A]_0 + \tau_A x$, and the amount of B is $[B]_0 + \tau_B x$. Thus the rate is

$$\frac{1}{\tau_A}\frac{d\,([A]_0 + \tau_A x)}{dt} = \underline{k}([A]_0 + \tau_A x)([B]_0 + \tau_B x)$$

$$\frac{dx}{dt} = \underline{k}([A]_0 + \tau_A x)([B]_0 + \tau_B x)$$

or

$$\frac{dx}{([A]_0 + \tau_A x)([B]_0 + \tau_B x)} = \underline{k}\,dt \qquad (10\text{-}1.4.11)$$

For ease of writing define $a \equiv [A]_0$ and $b \equiv [B]_0$. The resulting equation is

$$\int_{x=0}^{x} \frac{dx}{(a + \tau_A x)(b + \tau_B x)} = \int_0^t \underline{k}\,dt \qquad (10\text{-}1.4.12)$$

This can be solved by the *method of partial fractions*. Write the left side of Eq. (10-1.4.12) as

$$\frac{1}{(a + \tau_A x)(b + \tau_B x)} = \frac{d_1}{a + \tau_A x} + \frac{d_2}{b + \tau_B x}$$

where d_1 and d_2 are constants to be determined. Multiply both sides by $(a + \tau_A x) \times (b + \tau_B x)$.

$$1 = d_1(b + \tau_B x) + d_2(a + \tau_A x)$$

$$1 = bd_1 + ad_2 + (\tau_B d_1 + \tau_A d_2)x \tag{10-1.4.13}$$

For this to equal 1 for all values of x, the following equations apply:

$$bd_1 + ad_2 = 1 \qquad \tau_B d_1 + \tau_A d_2 = 0$$

Solving these two equations simultaneously gives (show it!)

$$d_1 = \frac{\tau_A}{b\tau_A - a\tau_B} \qquad d_2 = -\frac{\tau_B}{b\tau_A - a\tau_B} \tag{10-1.4.14}$$

Substituting these into the integral of Eq. (10-1.4.12) gives a function that can be integrated.

$$\frac{\tau_A}{b\tau_A - a\tau_B}\int_{x=0}^{x}\frac{dx}{a + \tau_A x} - \frac{\tau_B}{b\tau_A - a\tau_B}\int_0^x \frac{dx}{b + \tau_B x} = \int_0^t \underline{k}\, dt$$

$$-\frac{\tau_A}{b\tau_A - a\tau_B}\ln\frac{a + \tau_A x}{a} + \frac{\tau_B}{b\tau_A - a\tau_B}\ln\frac{b + \tau_B x}{b} = \underline{k}t \tag{10-1.4.15}$$

Finally, after rearranging,

$$\boxed{\frac{1}{(a\tau_B - b\tau_A)}\ln\frac{b^{\tau_B}(a + \tau_A x)^{\tau_A}}{a^{\tau_A}(b + \tau_B x)^{\tau_B}} = \underline{k}t} \tag{10-1.4.16}$$

Given a set of data relating x to t, Eq. (10-1.4.16) can be solved explicitly or plotted (what would you plot?) to give $\underline{k}$.

Normally we will be concerned with reactions for which $\tau_A = \tau_B = -1$. Equation (10-1.4.16) reduces to

$$\frac{1}{(a - b)}\ln\frac{b(a - x)}{a(b - x)} = \underline{k}t \tag{10-1.4.17}$$

or, in terms of initial variable definitions,

$$\boxed{\frac{1}{([A]_0 - [B]_0)}\ln\frac{[B]_0[A]}{[A]_0[B]} = \underline{k}t} \tag{10-1.4.18}$$

The example below will show how to apply this equation.

10-1.4.4 EXAMPLE

For the reaction A + B → products, prove that the given data fit a rate equation that is first order in A and first order in B.

t (s)	167	320	490	914	1190	∞ (complete reaction)
[A] ($mol\ dm^{-3}$)	0.0990	0.0906	0.0830	0.0706	0.0653	0.0424
[B] ($mol\ dm^{-3}$)	0.0566	0.0482	0.0406	0.0282	0.0229	0 0000

Solution

We cannot apply Eq. (10-1.4.18) directly, since $[A]_0$ and $[B]_0$ are not known. We do know $[A]_0 - [B]_0$ from the value of t = ∞ (or any other time). Rearranging the equation gives

$$\frac{1}{[A]_0 - [B]_0} \ln \frac{[A]}{[B]} + \frac{1}{[A]_0 - [B]_0} \ln \frac{[B]_0}{[A]_0} = \underline{k}t \qquad (10\text{-}1.4.19)$$

A plot of $1/([A]_0 - [B]_0) \ln([A]/[B])$ vs. t should give a straight line with slope $\underline{k}$ (what would the intercept be?).

t	167	320	490	914	1190
$\frac{1}{[A]_0 - [B]_0} \ln \frac{[A]}{[B]}$	13.2	14.9	16.9	21.6	24.7

From the following plot

$$\text{Slope} = \underline{k} = \underline{\underline{1.12 \times 10^{-2}\ dm^3\ mol^{-1}\ s^{-1}}}$$

If we desired values of $[A]_0$ and $[B]_0$ we could use the intercept and the fact that $[A]_0 - [B]_0 = 0.0424$.

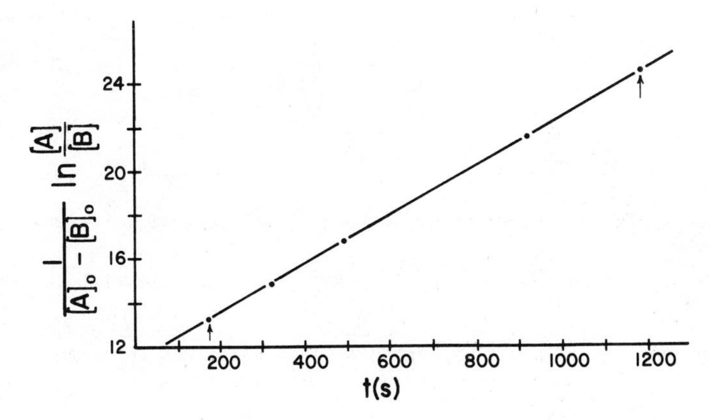

10-1.5 COMMENTS

For the simple rate laws covered in this section, it is sufficient to assume zero, first, or second order or first order in each of two reactants, then plot the data in the appropriate form for the assumed rate law. If the plot is linear, the assumed rate law is valid. The next section will concern more complicated reactions and the techniques used to analyze the data in these complicated reactions. So, enjoy the simple cases presented in the problems below; the next section will be a bit more challenging.

10-1.6 PROBLEMS

1. A reaction that follows the rate expression

$$\text{Rate} = \underline{k}[A]^{5/2}$$

is to be studied by the half-life method. Derive an appropriate expression for $t_{1/2}$. (Assume $\tau_A = -1$.)

2. A reaction is first order. After 350 s, 28% of the reactant remains. What is the rate constant? How long will it take 90% of the material to react?

3. Assume population growth can be expressed as

$$P = P_0 e^{\underline{k}t}$$

where P_0 is the population in some base year.

a. Determine $\underline{k}$ from the following data for the United States and the number of years required for the population to double.

Year	Population × 10^{-6}
1790	3.93
1800	5.31
1810	7.24
1820	9.64
1830	12.90
1840	17.10
1850	23.20
1860	31.40

b. If the population increases by 1% year^{-1}, how long will it take to double?

c. If the population increases by 2% year^{-1}, how long will it take to double?

4. The dimerization reaction of p-methoxybenzonitrile N-oxide has been studied.

At 40°C the following data were observed for an initial concentration of 0.011 mol dm^{-3}.

t (s)	0	3600	7200	19,500	33,900	56,520	72,720	91,080
Reaction (%)	0	9.1	16.7	32.7	47.3	60.9	66.6	70.3

Find the rate law and determine $\underline{k}$.

5. In a study of the decomposition of dimethyl ether $(CH_3)_2O$ in a constant volume apparatus, the following partial pressures of ether were observed:

t (s)	390	777	1195	3155	∞
P_{ether} (atm)	0.347	0.295	0.246	0.104	0

Find the order and $\underline{k}$.

6. C. N. Hinshelwood and W. K. Hutchinson [*Proc. Roy. Soc.* (London), *A111*:380 (1926)] observed the following data for the decomposition of acetaldehyde to methane and carbon monoxide ($CH_3CHO \rightarrow CH_4 + CO$) in a constant volume system

t (s)	0	42	105	190	310	480	840
P (atm)	0.474	0.518	0.571	0.624	0.676	0.729	0.795

Find the order and $\underline{k}$.

7. For the reaction $A(g) \rightarrow 2B(g) + C(g)$, determine the order and rate constant from the following data

t (min)	0	3	6	9	15	∞
P_{total} (mm Hg)	173.5	193.4	211.3	228.6	249.2	491.8

10-1.7 SELF-TEST

1. Discuss two kinetic methods that might be used to study very slow reactions.

2. A reaction is found to obey the following rate expression

$$\text{Rate} = -\frac{d[B]}{dt} = \underline{k}[B]^{3/2}$$

Integrate this and determine an equation for $t_{1/2}$.

3. A reaction that is second order in reactant A and second order overall is found

to have a rate constant of 8.32×10^{-6} dm^3 mol^{-1} s^{-1}. Find $t_{1/2}$ if $[A]_0 = 0.026$ mol dm^{-3}.

4. Y. Chia and R. E. Connick [*J. Phys. Chem.*, *63*: 1518 (1959)] have studied the reaction of I^- with ClO^- in 1 M OH^- at 298 K. The following data are based on their results.

t (s)	1	2	3	4	5	∞
I^- (mol dm^{-3})	0.00359	0.00330	0.00307	0.00290	0.00276	0.00200
ClO^- (mol dm^{-3})	0.00159	0.00130	0.00107	0.00090	0.00076	0.000

Determine the order of the reaction and the corresponding value of $\underline{k}$.

5. Find the expression required to find $P_A(t)$ for the reaction given below if $P_A(t = 0)$ is known and $P_{total}(t)$ is given for several times.

$$2P_A(g) + B(s) \rightarrow 3C(g)$$

10-1.8 ADDITIONAL READING

1.* G. Porter, "Flash Photolysis," in A. Weissberger, ed., *Techniques of Organic Chemistry*, Vol. VIII, Part II, 2nd ed., Interscience, New York, 1963, pp. 1055-1106.

2.* F. J. W. Roughton and B. Chance, "Rapid Reactions," in A. Weissberger, ed., *Techniques of Organic Chemistry*, Vol. VIII, Part II, 2nd ed., Interscience, New York, 1963, pp. 704-792.

3.* R. E. Weston, Jr., and H. A. Schwarz, *Chemical Kinetics*, Prentice-Hall, Englewood Cliffs, N. J., 1972, Sec. 3-10 and Appendix B.

4.* G. B. Skinner, *Introduction to Chemical Kinetics*, Academic, New York, 1974, Chap. 7.

5. For additional references on kinetic methods and data see Sec. 1-1.9E.

10-2 MECHANISMS AND MORE RATE LAWS

OBJECTIVES

You shall be able to

1. Define or briefly discuss: elementary reactions; unimolecular, bimolecular, and termolecular reactions; parallel and consecutive reactions; stationary (steady) state approximation
2. Determine rate laws from given experimental data and find the corresponding rate constants
3. Discuss briefly any or all mechanisms mentioned in this section and derive the appropriate rate laws

4. Apply the steady state approximation to complex mechanisms to obtain rate laws

10-2.1 INTRODUCTION

Section 10-1 was concerned with basic definitions and the simple rate laws that can be solved easily. In this section more complex reactions will be studied. The last section was strictly empirical. This one begins an introduction of some theoretical concepts, and Sec. 10-3 will be mostly theoretical. After a few examples of how to treat data that do not fit simple first-order or second-order rate equations, some discussion of mechanisms of reactions will be given. Then methods of deriving a theoretical rate law from a proposed mechanism will be presented. Such a derived rate law has no validity until it has been verified experimentally. An attempt will be made to show how experimental data can be used to test predicted mechanisms.

10-2.2 USE OF A PHYSICAL PROPERTY RELATED TO CONCENTRATION

In Sec. 10-1 we had at least one occasion to use a property rather than concentration in solving a rate problem. This property was pressure. We can use any physical property that is proportional to concentration to determine whether a given rate law is valid. Assume we have a reaction (τ_A and τ_B are negative)

$$- \tau_A A - \tau_B B \rightarrow \tau_C C \qquad (10\text{-}2.2.1)$$

which goes to completion, given sufficient time. Also, assume there is some physical property λ which varies as the reaction proceeds. This could be pressure, refractive index, dielectric constant, conductivity, etc. At any time

$$\lambda_t = \lambda_A + \lambda_B + \lambda_C + \lambda_M \qquad (10\text{-}2.2.2)$$

where the given λ values are for the reactants, products, *and* the medium in which the reaction is carried out. We have stated that this property is proportional to concentration, so

$$\lambda_A = d_A[A] \qquad \lambda_B = d_B[B]$$
$$\lambda_C = d_C[C] \qquad \lambda_M = \text{constant} \qquad (10\text{-}2.2.3)$$

where d_A, d_B, and d_C are constants. Let $\tau_A x \equiv \text{mol dm}^{-3}$ A that has reacted at any time. With this definition, you should conclude that $\tau_B x$ mol dm^{-3} B have reacted to give $\tau_C x$ mol dm^{-3} C at the given time. Then, if we define $a \equiv [A]_0$ and $b \equiv [B]_0$, the initial concentrations (τ_A and τ_B are negative):

$$[A] = [A]_0 + \tau_A x = a + \tau_A x$$

$$[B] = [B]_0 + \tau_B x = b + \tau_B x \qquad (10\text{-}2.2.4)$$

$$[C] = \tau_C x$$

From which, at the times t = t, t = 0, and t = ∞ (assume $a < b$),

$$\lambda_t = \lambda_M + d_A(a + \tau_A x) + d_B(b + \tau_B x) + d_C \tau_C x$$

$$\lambda_0 = \lambda_M + d_A a + d_B b \qquad (10\text{-}2.2.5)$$

$$\lambda_\infty = \lambda_M + d_B \left(b - \frac{\tau_B}{\tau_A} a\right) - d_C \frac{\tau_C}{\tau_A} a$$

The last equation may require some explanation. We have assumed $a < b$, so all A will be lost first. When all A has reacted, $a + \tau_A x = 0$, so $x = -a/\tau_A$ and $[C]_\infty = -\tau_C a/\tau_A$. Hence, the results shown for λ_∞.

Find $(\lambda_\infty - \lambda_0)$, $(\lambda_\infty - \lambda_t)$, and $(\lambda_t - \lambda_0)$ for later use.

$$\lambda_\infty - \lambda_0 = \lambda_M - d_C \frac{\tau_C}{\tau_A} a + d_B b - d_B \frac{\tau_B}{\tau_A} a - \lambda_M - d_A a - d_B b$$

$$= -d_C \frac{\tau_C}{\tau_A} a - d_B \frac{\tau_B}{\tau_A} a - d_A \frac{\tau_A}{\tau_A} a$$

$$= \frac{a}{\tau_A} (-\tau_C d_C - \tau_B d_B - \tau_A d_A)$$

$$= \frac{a}{\tau_A} \Delta d \qquad (10\text{-}2.2.6)$$

where Δd is defined by the equation. You should show

$$\lambda_t - \lambda_0 = -x\, \Delta d \qquad (10\text{-}2.2.7)$$

and

$$\lambda_\infty - \lambda_t = \lambda_\infty - \lambda_0 - (\lambda_t - \lambda_0)$$

$$= \left(\frac{a}{\tau_A} + x\right) \Delta d \qquad (10\text{-}2.2.8)$$

In terms of [A] we can write immediately

$$\frac{\lambda_t - \lambda_0}{\lambda_\infty - \lambda_0} = -\frac{x\, \Delta d}{(a/\tau_A)\, \Delta d} = -\frac{\tau_A x}{a}$$

$$= -\frac{\tau_A x}{[A]_0} \qquad (10\text{-}2.2.9)$$

and
$$\frac{\lambda_\infty - \lambda_0}{\lambda_\infty - \lambda_t} = \frac{(a/\tau_A)\,\Delta d}{(a/\tau_A + x)\,\Delta d}$$

$$= \frac{a}{a + \tau_A x} = \frac{[A]_0}{[A]_0 + \tau_A x}$$

$$= \frac{[A]_0}{[A]} \qquad (10\text{-}2.2.10)$$

An example will show the utility of these expressions.

10-2.2.1 EXAMPLE

Use the data from Example 10-1.4.2 and Eqs. (10-2.2.9) and (10-2.2.10) to find the first-order rate constant; $P_\infty = 0.7086$ atm.

Solution

If we assume first order in A (a good assumption since we know that is the answer), we need to evaluate Eq. (10-1.4.5) at several times to see if $\underline{k}$ is, indeed, a constant. Note that $\tau_A = -1$ for this problem using the definitions of Eq. (10-1.3.1) .

$$\underline{k} = -\frac{1}{t}\ln\frac{[A]}{[A]_0} = +\frac{1}{t}\ln\frac{[A]_0}{[A]}$$

Apply Eq. (10-2.2.10) with $\tau_A = -1$ to give

$$\underline{k} = \frac{1}{t}\ln\left(\frac{\lambda_\infty - \lambda_0}{\lambda_\infty - \lambda_t}\right) = \frac{1}{t}\ln\left(\frac{P_\infty - P_0}{P_\infty - P_t}\right)$$

Try several t; $P_0 = 0.2362$, $P_\infty = 0.7086$.

t	P_t	$\underline{k}$ (min^{-1})
6	0.2613	0.00910
18	0.3051	0.00876
26	0.3322	0.00874
34	0.3569	0.00868

The average $\underline{k}$ is 0.00882 ± 0.0002 (min^{-1}). The difference is in taking the slope in Example 10-1.4.2 at other points.

10-2.2.2 EXAMPLE

Use the following absorbance (abs) data to determine the order and the rate constant for an organic rearrangement reaction for which $\tau_A = -1$.

Solution

t (min)	0	20	60	100	140	190	∞
abs	0.438	0.496	0.604	0.690	0.770	0.854	1.280
k (× 10^3)	--	3.57	3.66	3.56	3.58	3.59	--

$$\underline{k}_{avg} = 3.59 \pm 0.04 \times 10^{-3}\ s^{-1}$$

The k values given are calculated from Eqs. (10-1.4.5) and (10-2.2.10) with $\tau_A = -1$. As you can see, the first-order equation provides a consistent set of k values. (You should plot the data in appropriate form to show that a straight line results.) This verifies first-order kinetics. If the reaction had been second order (or any other order) the values of k obtained from (10-1.4.5) and (10-2.2.10) would not have been consistent. To try a second-order fit A_0, A_t. and A_∞ would have to be combined with Eq. (10-1.4.8) into a form useful for calculating k. Such an equation is more complex than the first-order case. More practice will be gained when you work the problem section.

10-2.3 HALF-LIFE METHODS

Consider our usual reaction

$$-\tau_A A - \tau_B B \rightarrow \tau_C C$$

Let us suppose the reaction is nth order in A and zero order in B (this can be achieved by making [B] very large relative to [A]). If [B] is very large, then it is essentially constant since very little B has reacted even if all A reacts. Let $[A] = a + \tau_A x$; $n > 1$ and $[A]_0 = a$.

$$\text{Rate} = \frac{1}{\tau_A}\frac{d\,[A]}{dt} = \frac{1}{\tau_A}\frac{d(a + \tau_A x)}{dt}$$

$$= \frac{dx}{dt} = \underline{k}_n [A]^n$$

or

$$\frac{dx}{dt} = \underline{k}_n (a + \tau_A x)^n \tag{10-2.3.1}$$

Integration of Eq. (10-2.3.1) yields

$$\int_{x=0}^{x=x} \frac{dx}{(a + \tau_A x)^n} = \int_0^t \underline{k}_n \, dt$$

$$\frac{1}{\tau_A(1 - n)} \left[\frac{1}{(a + \tau_A x)^{n-1}}\right]_{x=0}^{x=x} = \frac{1}{\tau_A(1 - n)} \left[\frac{1}{(a + \tau_A x)^{n-1}} - \frac{1}{a^{n-1}}\right]$$

$$= \underline{k}_n t \qquad (10\text{-}2.3.2)$$

Choose $t = t_{1/2}$ where $a + \tau_A x = a/2$ and $\tau_A = -1$

$$\frac{1}{n-1}\left[\frac{1}{(a/2)^{n-1}} - \frac{1}{a^{n-1}}\right] = \frac{1}{n-1}\left(\frac{2^{n-1} - 1}{a^{n-1}}\right)$$

$$= \underline{k}_n t_{1/2} \qquad (10\text{-}2.3.3)$$

or

$$\boxed{t_{1/2} = \frac{2^{n-1} - 1}{\underline{k}_n (n-1) a^{n-1}}} \qquad n > 1, \ \tau_A = -1 \qquad (10\text{-}2.3.4)$$

This is the equation we can use to find the order of a reaction. Stop for a moment to see if you can figure out how to treat data giving $t_{1/2}$ for given values of a to find n. It can be done, as is shown below.

Notice that everything in the equation is constant except $t_{1/2}$ and a. So rewrite (10-2.3.4) as

$$t_{1/2} = \frac{\text{const}}{a^{n-1}} \qquad (10\text{-}2.3.5)$$

$$\ln t_{1/2} = \ln \text{const} - (n - 1) \ln a \qquad (10\text{-}2.3.6)$$

A plot of $\ln t_{1/2}$ against $\ln a$ will yield a line of slope $(1 - n)$. (a is the initial concentration.)

10-2.3.1 EXAMPLE

From the following data for half-time as a function of initial pressure of nitrous oxide, determine the order at the decomposition.

P_0 (N m^{-2})	7730	12500	18700	26900	37600
$\ln P_0$	8.95	9.43	9.83	10.2	10.5
$t_{1/2}$ (min)	13.5	9.97	7.82	6.05	5.00
$\ln t_{1/2}$	2.60	2.30	2.06	1.80	1.61

Solution

The plot of $\ln t_{1/2}$ vs. $\ln P_0$ gives a straight line with a slope of -0.64. According to Eq. (10-2.3.6) this equals -(n -1), which yields n = 1.64. A nonintegral order (something other than n = 0, 1, 2, ... ,) usually implies a rather complex mechanism. Section 10-2.5 will give you some complex mechanisms that lead to nonintegral orders.

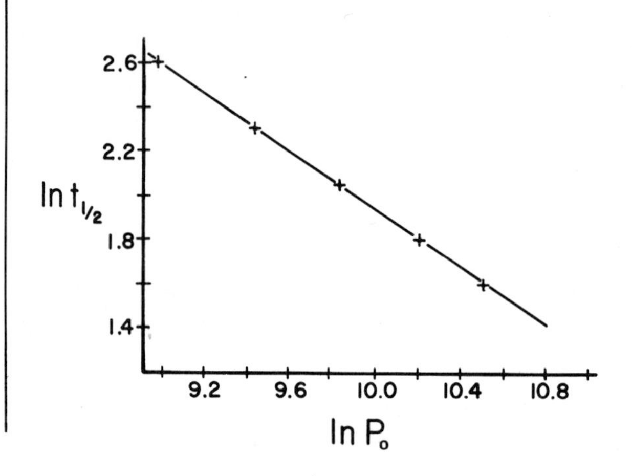

10-2.4 COMMENTS ON MECHANISMS

GENERAL. A mechanism for a reaction is a detailed description of the manner in which reactants are converted into products. A mechanism is stated in terms of a postulated sequence of *elementary* reactions. The term elementary reaction indicates a reaction which occurs in a single event. For example, the collision of two molecules to form a new species would be an *elementary* reaction. This term will become clearer below. Usually it is very difficult, if not impossible, to determine the mechanism of a reaction from kinetic data. Some possible mechanisms will be ruled out on the basis of experimental data, but not all are usually ruled out. Thus, in many cases it is impossible to choose which of two or more postulated mechanisms describes the reaction. Back to elementary reactions. If two molecules collide in an elementary reaction, the reaction is termed *bimolecular*. If only one molecule is involved, as in a dissociation, the reaction is termed *unimolecular*. There may be a few rare cases of *termolecular* reactions. However, these are unlikely due to the very low probability of three molecules colliding at precisely the same instant.

If the molecularity of a reaction is known, the rate law can be written immediately. For a unimolecular (elementary) reaction

$$A \rightarrow \text{products}$$

the rate law is simply

$$-\frac{d[A]}{dt} = \underline{k}[A]$$

similarly for a bimolecular (elementary) reaction of the type

$$2A \rightarrow \text{products}$$

the rate law is

$$-\frac{d[A]}{dt} = \underline{k}[A]^2$$

Note: We can write the rate law immediately *from the elementary reaction*. We cannot do so from the stoichiometric equation for the reaction. The $H_2 + I_2 \rightarrow 2HI$ and $H_2 + Br_2 \rightarrow 2HBr$ examples below will emphasize this point.

We need the definitions of two more terms before writing a few possible mechanisms. If a mechanism involves a reactant which can undergo two or more reactions independently and at the same time, the reactions are *parallel*. Another possibility is that the product of one reaction may undergo further reaction, so it is a reactant for a subsequent reaction. In this case the term *consecutive* reaction is used. In complex reactions both parallel and consecutive reactions may be present. Many types will be presented below.

PARALLEL FIRST-ORDER REACTIONS. Let us assume reactant A can undergo two concurrent reactions forming two different products.

$$A \xrightarrow{\underline{k}_1} B$$

$$A \xrightarrow{\underline{k}_2} C$$

The rate is easily written as a sum of the individual rates.

$$-\frac{d[A]}{dt} = \underline{k}_1[A] + \underline{k}_2[A] = (\underline{k}_1 + \underline{k}_2)[A]$$

$$= \underline{k}[A] \qquad \underline{k} = \underline{k}_1 + \underline{k}_2 \qquad (10\text{-}2.4.1)$$

The reaction is first order with respect to A.

CONSECUTIVE FIRST-ORDER REACTIONS. Consider the following elementary reactions

$$A \xrightarrow{\underline{k}_1} B$$

$$B \xrightarrow{\underline{k}_2} C$$

Such a case is common in radioactive decay. An element decays to an unstable isotope which undergoes a radioactive decay. The rate laws are easily written.

$$-\frac{d[A]}{dt} = \underline{k}_1[A] \qquad \frac{d[B]}{dt} = \underline{k}_1[A] - \underline{k}_2[B] \tag{10-2.4.2}$$

Note: Justify the signs that have been given in Eq. (10-2.4.2). You must keep signs in order if you hope to solve any problems of this type. The equation for A is easily integrated to give

$$\ln [A] - \ln [A]_0 = -\underline{k}_1 t$$

or

$$[A] = [A]_0 e^{-\underline{k}_1 t} \tag{10-2.4.3}$$

Then for B,

$$\frac{d[B]}{dt} = \underline{k}_1 [A]_0 e^{-\underline{k}_1 t} - \underline{k}_2 [B]$$

This can be integrated to give

$$[B] = \frac{[A]_0 \underline{k}_1}{\underline{k}_2 - \underline{k}_1} (e^{-\underline{k}_1 t} - e^{-\underline{k}_2 t}) \tag{10-2.4.4}$$

(The integrating factor method is used. Can you do it?)

PARALLEL SECOND-ORDER REACTIONS. Assume A and B react to form two different products.

$$A + B \xrightarrow{\underline{k}_1} C$$

$$A + B \xrightarrow{\underline{k}_2} D$$

$$-\frac{d[A]}{dt} = -\frac{d[B]}{dt} = \underline{k}_1[A][B] + \underline{k}_2[A][B]$$

$$= (\underline{k}_1 + \underline{k}_2)[A][B] \tag{10-2.4.5}$$

To integrate this equation, we must have knowledge of the relative concentrations of A and B.

OTHER REACTIONS. The number of combinations of parallel and consecutive reactions is almost limitless. As the number of steps in the mechanism increases, the complexity of the resulting rate law increases rapidly. In many cases it is not possible to integrate the resulting rate law without an assumption. This assumption is the *stationary state, or steady state, approximation*.

10-2.5 THE STEADY STATE APPROXIMATION

Consider series first-order reactions.

$$A \xrightarrow{k_1} B$$

$$B \xrightarrow{k_2} C$$

$$-\frac{d[A]}{dt} = k_1[A] \qquad \frac{d[B]}{dt} = k_1[A] - k_2[B] \qquad \frac{d[C]}{dt} = k_2[B] \tag{10-2.5.1}$$

See Eq. (10-2.4.2): note the signs again; justify them to yourself. The *steady state assumption* is that the concentration of an intermediate (B in this case) is small and constant after a short period of reaction time (the *induction period*). Notice the term steady state does not imply equilibrium in any way. Rather, a state is reached in which the concentrations of intermediates are constant, even though there is a continuing conversion of reactants to products.

According to the steady state approximation,

$$\frac{d[B]}{dt} = 0 = k_1[A] - k_2[B] \tag{10-2.5.2}$$

Thus,

$$\frac{[B]}{[A]} = \frac{k_1}{k_2} \tag{10-2.5.3}$$

You should be able to show that

$$[A] = [A]_0 e^{-k_1 t} \tag{10-2.4.3}$$

which gives

$$[B] = [A]_0 \frac{k_1}{k_2} e^{-k_1 t} \tag{10-2.5.4}$$

Note the relationship to the exact solution given in Eq. (10-2.4.4). From that solution

$$[B] = \frac{[A]_0 k_1}{(k_2 - k_1)} (e^{-k_1 t} - e^{-k_2 t}) \tag{10-2.4.4}$$

The two are equivalent if $k_2 >> k_1$ and $t >> 1/k_2$. The condition that $k_2 >> k_1$ means the intermediate is much more reactive than the initial reactant. If so, the [B] will always be small. The second condition is satisfied if t is large compared to the induction period. If these two conditions are met, then the steady state approximation is valid.

Note we assume [B] is constant. Equation (10-2.5.4) says that [B] decreases as t increases. However, $k_2 \gg k_1$, so [B] is always small and d[B]/dt will be very small (≈ 0). A couple of examples will illustrate the technique in more detail.

10-2.5.1 EXAMPLE

The proposed mechanism for the decomposition of acetaldehyde to form methane and carbon monoxide involves the formation of free radicals as shown:

$$1.\quad CH_3CHO \xrightarrow{k_1} CH_3\cdot + \cdot CHO$$

$$2.\quad CH_3CHO + CH_3\cdot \xrightarrow{k_2} CH_4 + CO + CH_3\cdot$$

$$3.\quad 2CH_3\cdot \xrightarrow{k_3} C_2H_6$$

Find an expression for the rate of formation of methane by applying the steady state approximation to $CH_3\cdot$.

Solution

$$\frac{d\,[CH_4]}{dt} = k_2[CH_3\cdot]\,[CH_3CHO]$$

$$\frac{d\,[CH_3\cdot]}{dt} = k_1[CH_3CHO] - 2k_3[CH_3\cdot]^2 = 0$$

[Reaction (2) is not included, since $CH_3\cdot$ is both a reactant and a product.] Solving the last equation:

$$[CH_3\cdot] = \left(\frac{k_1}{2k_3}\right)^{1/2} [CH_3CHO]^{1/2}$$

Substitution into the methane rate expression gives the desired result.

$$\frac{d\,[CH_4]}{dt} = k_2\left(\frac{k_1}{2k_3}\right)^{1/2} [CH_3CHO]^{3/2}$$

To test this mechanism one would need to check experimentally whether the rate of formation of methane depended on the concentration of acetaldehyde to the 3/2 power. If so, support is given to the proposed mechanism, but it is not proven. Other mechanisms might be formulated to give the same results. The next example is a little more involved.

10-2.5.2 EXAMPLE

In 1919, J. A. Christensen, K. F. Hertzfeld, and M. Polanyi proposed a mechanism for the $H_2 + Br_2 \rightarrow 2HBr$ reaction involving the following five elementary reactions.

$$Br_2 \xrightarrow{k_1} 2Br\cdot \qquad \text{initiation}$$

$$\left.\begin{aligned} Br\cdot + H_2 &\xrightarrow{k_2} HBr + H\cdot \\ H\cdot + Br_2 &\xrightarrow{k_3} HBr + Br\cdot \end{aligned}\right\} \quad \text{propagation}$$

$$H\cdot + HBr \xrightarrow{k_4} H_2 + Br\cdot \qquad \text{inhibition}$$

$$Br\cdot + Br\cdot \xrightarrow{k_5} Br_2 \qquad \text{termination}$$

You should agree that the rate of formation of HBr is (in this problem parentheses will stand for concentration and the dot on radials will not be shown)

$$\frac{d\,(HBr)}{dt} = k_2(H_2)(Br) + k_3(H)(Br_2) - k_4(H)(HBr) \qquad (10\text{-}2.5.5)$$

To eliminate (H) and (Br), the intermediates, we apply the steady state approximation to each:

$$\frac{d\,(H)}{dt} = 0 = k_2(Br)(H_2) - k_3(H)(Br_2) - k_4(H)(HBr) \qquad (10\text{-}2.5.6)$$

$$\frac{d\,(Br)}{dt} = 0 = 2k_1(Br_2) - k_2(Br)(H_2) + k_3(H)(Br_2) + k_4(H)(HBr) - 2k_5(Br)^2 \qquad (10\text{-}2.5.7)$$

Addition of Eqs. (10-2.5.6) and (10-2.5.7) gives

$$0 = 2k_1(Br_2) - 2k_5(Br)^2$$

So
$$(Br) = \left[\frac{k_1}{k_5}(Br_2)\right]^{1/2} \qquad (10\text{-}2.5.8)$$

Solution of (10-2.5.6) for (H) gives

$$(H) = \frac{k_2(Br)(H_2)}{k_3(Br_2) + k_4(HBr)} \qquad (10\text{-}2.5.9)$$

which on substitution of Eq. (10-2.5.8) for (Br) yields

$$(\mathrm{H}) = \frac{\underline{k}_2(\underline{k}_1/\underline{k}_5)^{\frac{1}{2}}(\mathrm{Br}_2)^{\frac{1}{2}}(\mathrm{H}_2)}{\underline{k}_3(\mathrm{Br}_2) + \underline{k}_4(\mathrm{HBr})} \qquad (10\text{-}2.5.10)$$

Substitute (10-2.5.8) and (10-2.5.10) into (10-2.5.5) to get the desired results.

$$\frac{d(\mathrm{HBr})}{dt} = \underline{k}_2(\mathrm{H}_2)\left(\frac{\underline{k}_1}{\underline{k}_5}\right)^{\frac{1}{2}}(\mathrm{Br}_2)^{\frac{1}{2}} + [\underline{k}_3(\mathrm{Br}_2) - \underline{k}_4(\mathrm{HBr})]\,\frac{\underline{k}_2(\underline{k}_1/\underline{k}_5)^{\frac{1}{2}}(\mathrm{Br}_2)^{\frac{1}{2}}(\mathrm{H}_2)}{\underline{k}_3(\mathrm{Br}_2) + \underline{k}_4(\mathrm{HBr})}$$

$$= \frac{\underline{k}_2\underline{k}_3(\mathrm{H}_2)(\underline{k}_1/\underline{k}_5)^{\frac{1}{2}}(\mathrm{Br}_2)^{3/2} + \underline{k}_2\underline{k}_4(\underline{k}_1/\underline{k}_5)^{\frac{1}{2}}(\mathrm{Br}_2)^{\frac{1}{2}}(\mathrm{HBr})(\mathrm{H}_2)}{\underline{k}_3(\mathrm{Br}_2) + \underline{k}_4(\mathrm{HBr})}$$

$$+ \frac{\underline{k}_2\underline{k}_3(\mathrm{H}_2)(\underline{k}_1/\underline{k}_5)^{\frac{1}{2}}(\mathrm{Br}_2)^{3/2} - \underline{k}_2\underline{k}_4(\underline{k}_1/\underline{k}_5)^{\frac{1}{2}}(\mathrm{HBr})(\mathrm{Br}_2)^{\frac{1}{2}}(\mathrm{H}_2)}{\underline{k}_3(\mathrm{Br}_2) + \underline{k}_4(\mathrm{HBr})}$$

$$= \frac{2\underline{k}_2\underline{k}_3(\underline{k}_1/\underline{k}_5)^{\frac{1}{2}}(\mathrm{H}_2)(\mathrm{Br}_2)^{3/2}}{\underline{k}_3(\mathrm{Br}_2) + \underline{k}_4(\mathrm{HBr})} \qquad (10\text{-}2.5.11)$$

This is the desired equation. About 13 years prior to the derivation of this equation, M. Bodenstein and S. C. Lind had experimentally determined the rate law to be

$$\frac{d\,(\mathrm{HBr})}{dt} = \frac{\underline{k}(\mathrm{H}_2)(\mathrm{Br}_2)^{\frac{1}{2}}}{1 + \underline{k}'(\mathrm{HBr})/(\mathrm{Br}_2)} \qquad (10\ 2.5.12)$$

Divide the numerator and denominator of the right side of Eq. (10-2.5.11) by $\underline{k}_3(\mathrm{Br}_2)$:

$$\frac{d\,(\mathrm{HBr})}{dt} = \frac{2\underline{k}_2\,(\underline{k}_1/\underline{k}_5)^{\frac{1}{2}}(\mathrm{H}_2)(\mathrm{Br}_2)^{\frac{1}{2}}}{1 + (\underline{k}_4/\underline{k}_3)(\mathrm{HBr})/(\mathrm{Br}_2)}$$

which has exactly the same form as the experimental result. This, of course, does not prove the proposed mechanism but does lend support to it.

10-2.6 CLOSING COMMENTS

The last two examples have shown rather complex reactions involving very reactive intermediates. Atoms and free radicals are not very stable, as you might guess. They react rapidly with whatever is available in the reaction vessel. In the above, the active intermediates are used up as rapidly as they are formed, so there is no build-up of these active species with time. There are some reactions which produce atoms or radicals faster than they can be used. Such a case is a *branched chain* reaction. In the foregoing only *linear chain* reactions were involved. A linear chain reaction produces no more than one active species in each step. However, in a *branched chain* reaction, a given step can form more than one active species. This may result in the formation of active species faster than they can be removed. If this continues, an explosion may occur. A linear chain reaction can lead to an explosion also, but this will likely be a *thermal explosion*. As we will see in Sec. 10-3, the rate of a reaction increases rapidly as the temperature increases. Thus, if heat generated in a reaction is not removed, the reaction vessel gets warmer, accelerating the reaction and producing more heat until explosive rates are reached. In explosions of the branched chain variety the temperature increase is not as important as the rapid build-up of active species. The reaction of H_2 and O_2 is a branched chain reaction.

It is necessary to postulate a branched chain mechanism to explain the rather strange behavior of H_2 and O_2 mixtures as a function of total pressure. Figure 10-2.6.1 shows the explosion limits for a stoichiometric mixture of the two gases. Look at the isotherm at 440°C. At 1 atm total pressure there is no perceptible reaction. As the pressure is increased at 440°C, nothing appears to happen until about 2 atm pressure is reached. At that point there is a reaction that is rapid enough to be classified as an explosion. Any pressure between about 2 and 14 atm will give an explosion at 440°C. But, if a mixture (stoichiometric) is brought to 440°C and to any pressure between about 14 and 4200 atm, no explosion occurs. Again, above about 4200 atm, explosive rates are observed. Such behavior is rather strange and a little difficult to rationalize at first. However, a branched chain mechanism can be invoked that will account for the observed behavior.

As mentioned above, a branched chain reaction will produce more than one radical from a single radical. Radicals rapidly accumulate in sufficient numbers to cause the explosive rates. Saying that does not at all explain the pressure dependence of the above reaction rate. The reaction

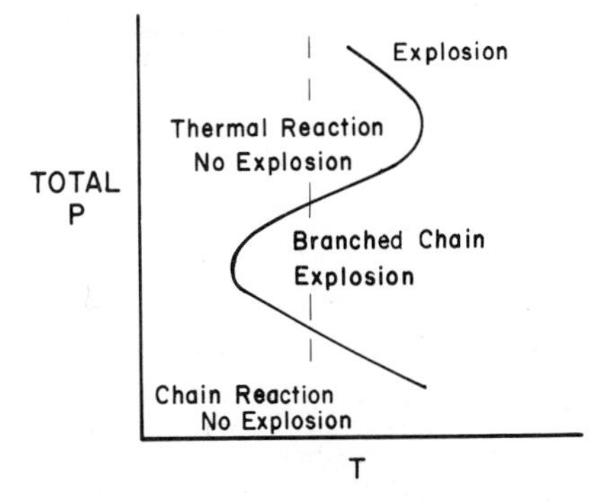

FIG. 10-2.6.1 Reaction behavior of H_2 + O_2 stoichiometric mixture as a function of T and P. (Figure not to scale.)

$$H_2 + O_2 \rightarrow H_2O + O\cdot \qquad (10\text{-}2.6.1)$$

is apparently the chain initiation step. Following the production of the O radical several reactions occur:

$$O\cdot + H_2 \rightarrow OH\cdot + H\cdot \qquad (10\text{-}2.6.2)$$

$$H\cdot + O_2 \rightarrow OH\cdot + O\cdot \qquad (10\text{-}2.6.3)$$

$$OH\cdot + H_2 \rightarrow H_2O + H\cdot \qquad (10\text{-}2.6.4)$$

As you can see, each O atom produced by reaction (10-2.6.1) leads to more than one additional radical in the sequence (10-2.6.2) through (10-2.6.4). If no other process were involved, this would always lead to an explosion. However, at low pressures it is possible for the OH and H radicals to reach the wall and be "quenched" (meaning to recombine at the wall). If this occurs to a large enough degree, the concentration of radicals will not increase without limit but will reach some steady value.

When the pressure is increased, collisions between molecules become more likely than collisions with the wall, and an excess of radicals are produced, leading to an explosion. Some supporting evidence for this explanation is that the first explosion limit occurs at a lower pressure in a larger vessel. (Stop for a moment and see if that is consistent.) If an inert gas is added to the mixture, the lower explosion limit occurs at a lower pressure. This is probably due to the interference of the inert gas with the diffusion of the radicals to the walls of the container. Finally, an inert packing material added to the reaction vessel will raise the explosion limit. (Can you explain that?)

We have taken care of the first explosion limit, but what makes the reaction slow down above a given pressure? It seems that in this region, collisions of the radicals with other gas molecules with which they do not react occurs often enough to prevent radical build-up. The proposed mechanism for this is

$$H\cdot + O_2 \rightarrow HO_2^*\cdot \qquad (10\text{-}2.6.5)$$

$$HO_2^*\cdot + M \rightarrow HO_2\cdot + M^* \qquad (10\ 2.6.6)$$

The radical H reacts with O_2 to form an excited species (indicated by the asterisk). Normally, this would decompose to form O and OH radicals which would contribute to the propogation of the reaction. At high pressures, there are so many other molecules present that a collision is likely. Such a collision takes the excess energy from the HO_2^* and stabilizes it enough for it to reach the wall and undergo recombination. The reasons for the third explosion limit are apparently not very well understood.

Throughout the above discussion, we have neglected the temperature effect by assuming constant temperature. Even in the absence of sufficient chain branching to produce an explosion, if the rate is high enough to generate a large quantity of heat due to ΔH of the reaction, a thermal explosion can occur. Thus just because you might have an O_2-H_2 mixture outside the explosion limits, there is still the possibility of a thermal explosion if heat is not removed fast enough.

10-2.7 PROBLEMS

1. The reaction $2NO + 2H_2 \rightleftarrows N_2 + 2H_2O$ was studied with equimolar amounts of NO and H_2 at various initial pressures yielding the following data:

P_0 (atm)	0.466	0.448	0.493	0.379	0.330	0.320	0.266
$t_{1/2}$ (min)	81	102	95	140	180	176	224

Determine the overall order of the reaction.

2. The reaction of n-propyl bromide with thiosulfate ion

$$C_3H_7Br + S_2O_3^{2-} \rightarrow C_3H_7SSO_3^- + Br^-$$

has been studied by T. I. Croswell and L. P. Hammett [*J. Am. Chem. Soc., 70*:3444 (1948)]. Samples (0.01002 dm^3) were withdrawn periodically from the reaction mixture. I_2 solution (0.02572 N) was used to titrate the $S_2O_3^{2-}$ remaining at that time. The following data were obtained.

t (s)	0	1110	2010	3192	5052	7380	11232	∞
I_2 (dm^3)	0.03763	0.03520	0.03363	0.03190	0.02986	0.02804	0.02601	0.02224

Determine the order and rate constant for this reaction.

3. A proposed mechanism for the reaction $NO_2Cl \rightleftarrows NO_2 + \frac{1}{2} Cl_2$ is

$$NO_2Cl \xrightarrow{k_1} NO_2 + Cl\cdot$$

$$NO_2Cl + Cl\cdot \xrightarrow{k_2} NO_2 + Cl_2$$

Apply the steady state approximation to $Cl\cdot$ to derive the rate law predicted by this mechanism.

4. In 1934, F. O. Rice and K. F. Herzfeld investigated the role of free radicals in many organic reactions. The Rice-Herzfeld mechanism for the decomposition of ethane is

$$C_2H_6 \xrightarrow{k_1} 2CH_3\cdot$$

$$CH_3\cdot + C_2H_6 \xrightarrow{k_2} CH_4 + C_2H_5\cdot$$

$$C_2H_5\cdot \xrightarrow{k_3} C_2H_4 + H\cdot$$

$$H\cdot + C_2H_6 \xrightarrow{k_4} H_2 + C_2H_5\cdot$$

$$H\cdot + C_2H_5\cdot \xrightarrow{k_5} C_2H_6$$

Determine the rate of reaction of C_2H_6 by applying the steady state approximation to CH_3, C_2H_5, and H. Assume k_1 is small relative to other rate constants. [Hint: $(k_1)^{1/2} >> k_1$]

5. If $k_2 = 4.00 \times 10^{-6}$ dm^3 mol^{-1} s^{-1} at 600 K for the decomposition of HI according to the rate expression $-d[HI]/dt = k_2[HI]^2$, how many molecules of HI decompose per second at 600 K and 1 atm when t is near t = 0?

6. At 600 K, $k_2 = 6.3 \times 10^2$ ml mol^{-1} s^{-1} for $2NO_2 \rightarrow 2NO + O_2$. At 600 K how long will it take for 1/10 of a sample of NO_2 at 0.526 atm to decompose by this reaction?

10-2.8 SELF-TEST

1. The absorption of a solution is related to its composition. D. S. Noyce et. al. [*J. Am. Chem. Soc.*, *84*: 1632 (1962)] observed the following absorbance (abs) data for the isomerization of cis-cinnamic acid to trans-cinnamic acid in a sulfuric acid solution:

t (s)	0	1200	3600	6000	8400	11400	16250	∞
abs	0.219	0.248	0.302	0.345	0.385	0.427	0.479	0.640

Determine the order and k.

2. Define or discuss briefly three of the terms listed in the Objectives section.

3. R. L. Perrine and H. S. Johnstone [*J. Chem. Phys.*, *21*:2202 (1953)] have investigated the mechanism of the reaction $2NO_2 + F_2 \rightarrow 2NO_2F$. The proposed mechanism is

$$NO_2 + F_2 \xrightarrow{k_1} NO_2F + F\cdot$$

$$NO_2F + F\cdot \xrightarrow{k_1'} NO_2 + F_2$$

$$F\cdot + NO_2 + M \xrightarrow{k_2} NO_2F + M^*$$

$$F\cdot + F\cdot + M \xrightarrow{k_2'} F_2 + M^*$$

M* is any species present which can remove energy from the collision. (a) Apply the steady state appxoximation to F and find the rate of formation of NO_2F. (Assume $k_2'[F][F][M]$ is negligible.) (b) If $k_2[NO_2][M] >> k_1'[NO_2F]$ show that the rate is first order in NO_2 and F_2 as observed experimentally.

10-2.9 ADDITIONAL READING

1. See Sec. 10-1.8 and Sec. 1-1.9E for pertinent readings.

10-3 COLLISION AND ABSOLUTE RATE THEORY

OBJECTIVES

You shall be able to

1. Determine reaction rates at various temperatures from the Arrhenius equation or, given rates at different temperatures, determine the constants in the Arrhenius or similar equations
2. Discuss the assumptions and procedures of both the collision theory and the absolute rate theory of reaction rates
3. Apply the equations derived from the collision theory and the absolute rate theory to estimate rate constants for reactions

10-3.1 INTRODUCTION

This continues (and concludes) our progression from experiment to theory. Section 10-1 was essentially all experimentally based. Section 10-2 introduced the theoretical concept of mechanisms which can be used to predict the form of a rate law for a given reaction. The rate law thus obtained must be checked experimentally. In this section, reaction rates are approached from a theoretical point of view. An empirical expression relating rate constants to temperature will be presented. The collision theory and the absolute rate theory will use the molecular model and statistical mechanics results to estimate rate constants. The only use to be made of experimental results is for comparison of the rate constants derived.

10-3.2 EFFECT OF TEMPERATURE ON REACTION RATES

The fact that temperature affects reaction rates has been known from the time of the earliest kinetic studies. Arrhenius, in 1889, first proposed a satisfactory equation to fit the observed dependence. The Arrhenius equation has a very simple form.

$$\boxed{\underline{k} = A \exp - \frac{E_a}{RT}} \qquad (10\text{-}3.2.1)$$

where A and E_a are constants. In fact, A and E_a both are probably slightly dependent on temperature, but seldom are kinetic data precise enough to show any dependence. For our purposes at present, assume A and E_a to be constant. Assume you are given a set of rate constants $\underline{k}_i$ for a set of temperatures T_i. How would you go about finding A and E_a? Think about that a moment before continuing. Of course, the first thing you must do is write Eq. (10-3.2.1) in a linear form. This can be done by taking logarithms of both sides:

$$\ln \underline{k} = \ln A - \frac{E_a}{RT} \qquad (10\text{-}3.2.2)$$

Then a graph of $\ln \underline{k}_i$ vs. $1/T_i$ will yield a straight line of slope $(-E_a/R)$ and intercept of $(\ln A)$. The following example will illustrate the procedure.

10-3.2.1 EXAMPLE

The rate constants for the cis-trans isomerization of cis-ethylene-d_2 are as follows:

$\underline{k}$ ($cm^3\ mol^{-1}\ s^{-1}$)	3.18	6.57	15.7	34.2	70.0
T (K)	561.2	578.0	600.6	620.5	639.9

Find E_a and A in Eq. (10-3.2.2).

Solution

$\ln \underline{k}$	1.16	1.88	2.75	3.53	4.24
$1/T \times 10^3$ (K^{-1})	1.78	1.73	1.67	1.61	1.56

The slope $= 13.7 \times 10^3$ K $= E_a/R$. This gives $E_a = 114$ kJ mol^{-1}. Then

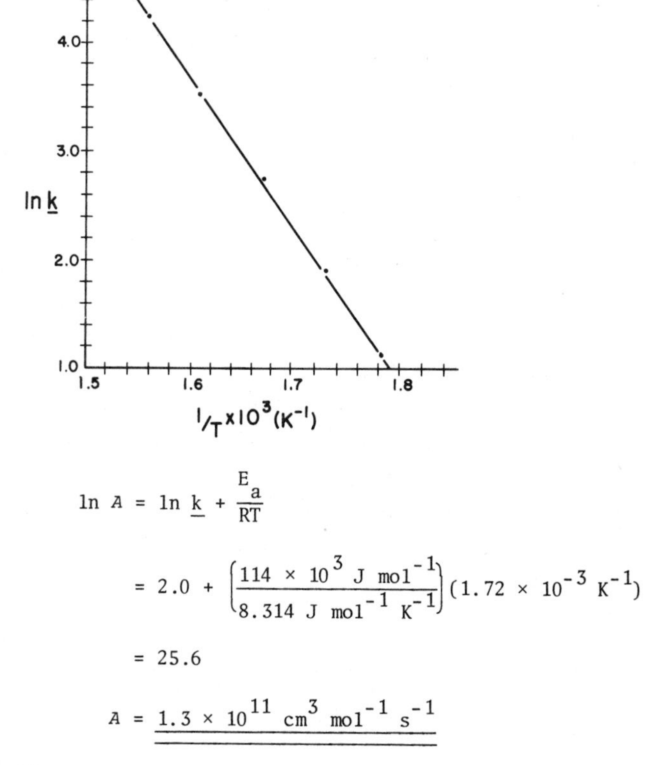

$$\ln A = \ln \underline{k} + \frac{E_a}{RT}$$

$$= 2.0 + \left(\frac{114 \times 10^3 \text{ J mol}^{-1}}{8.314 \text{ J mol}^{-1} \text{ K}^{-1}}\right)(1.72 \times 10^{-3} \text{ K}^{-1})$$

$$= 25.6$$

$$A = \underline{\underline{1.3 \times 10^{11} \text{ cm}^3 \text{ mol}^{-1} \text{ s}^{-1}}}$$

With these values of A and E_a the rate constant for the reaction can be calculated.

The energy in the Arrhenius equation is referred to as an activation energy. Arrhenius assumed this was the minimum energy a molecule or pair of molecules had to possess to react. For any elementary reaction there is a forward and a reverse reaction. There will be an activation energy associated with each direction. Figure 10-3.2.1 shows this schematically for the reaction.

$$A \underset{\underline{k}_r}{\overset{\underline{k}_f}{\rightleftarrows}} B$$

The term reaction coordinate is somewhat nebulous. Moving to the right along the reaction coordinate implies progression from reactants to products. X* represents an *activated complex*. This is some high-energy intermediate species formed when A is converting to B. This activated complex is an unstable substance in the process

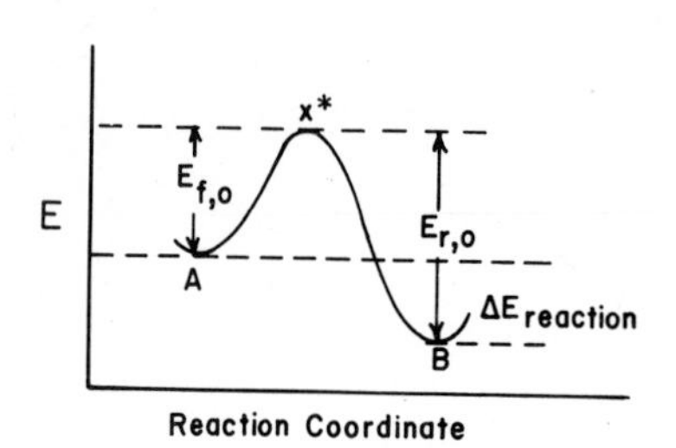

FIG. 10-3.2.1 Schematic of the activation energies for the elementary reaction A $\rightleftarrows$ B.

of converting to product or reverting to reactant. For A → B and B → A the same activated complex is involved. More will be said about properties of this activated complex in the next section and when absolute rate theory is treated (see Sec. 10-3.5). *Note*: $E_{f,0}$ in Fig. 10-3.2.1 will be the difference between zero-point energies of the reactants and the activated complex. It differs somewhat from E_a in the Arrhenius equation. That difference will be important in the final results of Sec. 10-3.3.

10-3.3 THERMODYNAMIC CONSIDERATIONS

The rate constant for the forward reaction is given by

$$\underline{k}_f = A_f \exp\left(-\frac{E_{f,0}}{RT}\right) \tag{10-3.3.1}$$

And for the reverse reaction,

$$\underline{k}_r = A_r \exp\left(-\frac{E_{r,0}}{RT}\right) \tag{10-3.3.2}$$

The equilibrium constant for the elementary reaction is

$$K_{eq} = \frac{\underline{k}_f}{\underline{k}_r} = \frac{A_f e^{-E_{f,0}/RT}}{A_r e^{-E_{r,0}/RT}}$$

$$= A' \exp\left(-\frac{E_{f,0} - E_{r,0}}{RT}\right)$$

$$= A' \exp\left(-\frac{E_0}{RT}\right) \tag{10-3.3.3}$$

A' is A_f/A_r. It tells nothing about either the forward or the reverse reaction alone, just as E_0 does not give any information about E_f or E_r alone. From our study of thermodynamics for a reaction with no change in the number of moles of gases, or for liquid and solid reactions with essentially no volume change, we can write

$$K_{eq} = \exp\left(-\frac{\Delta G^0}{RT}\right) = \exp\left(-\frac{\Delta H^0 - T\,\Delta S}{RT}\right)$$

$$K_{eq} = \exp\left(\frac{\Delta S^0}{R}\right)\exp\left(-\frac{\Delta H^0}{RT}\right) \tag{10-3.3.4}$$

Comparing the two equations gives

$$A' = \exp\left(-\frac{\Delta S^0}{R}\right) \qquad E_0 = \Delta H^0 \tag{10-3.3.5}$$

These equations are valid if ΔS^0 and ΔH^0 are *temperature independent* and there is *no volume change*.

If the number of moles of gas change in the reaction it can be shown that (assuming ideal behavior)

$$A' = \left(\frac{1}{RT}\right)^{\Delta n}\exp\left(-\frac{\Delta n + \Delta S^0}{R}\right) \qquad \Delta E_0 = \Delta H^0 - \Delta nRT \tag{10-3.3.6}$$

In this case A' is shown as a function of T. A more general form that is sometimes used in place of the Arrhenius equation is

$$\boxed{\underline{k} = BT^n \exp\left(-\frac{E_0}{RT}\right)} \tag{10-3.3.7}$$

This allows an explicit determination of the dependence of the pre-exponential factor on temperature. B is temperature independent in this equation. We will make use of these relationships a little later.

10-3.4 COLLISION THEORY OF CHEMICAL REACTIONS*

The Arrhenius formula given in Sec. 10-3.3 can be understood on the basis of a simple collision model. For a bimolecular reaction,

$$A + B \rightarrow \text{products}$$

assume A and B are spheres with *collision diameter* σ_{AB} (see Sec. 2-3.7):

---

* Based on C. E. Nordman and S. M. Blinder, *J. Chem. Educ., 51*:790 (1974) (see Sec. 10-3.9, Ref. 1).

$$\sigma_{AB} = \frac{\sigma_A + \sigma_B}{2} \tag{10-3.4.}$$

where σ_A and σ_B are the collision diameters of A and B, respectively. Define a *collision cross-section* σ':

$$\sigma' = \pi\sigma_{AB}^2 \tag{10-3.4.2}$$

The distribution of speeds for A and B (if Maxwellian, see Sec. 2-3.2) is given by (c_A and c_B are the speeds of A and B, respectively)

$$\frac{dN_c\,(A)}{N} = 4\pi c_A^2 \left(\frac{m_A}{2\pi kT}\right)^{3/2} \exp\left(-\frac{mc_A^2}{2kT}\right)$$

$$\frac{dN_c\,(B)}{N} = 4\pi c_B^2 \left(\frac{m_B}{2\pi kT}\right)^{3/2} \exp\left(-\frac{mc_B^2}{2kT}\right) \tag{10-3.4.3}$$

The distribution of *relative speeds* will be Maxwellian if the individual speeds are Thus, the distribution of relative velocities is

$$\frac{dN_c\,(AB)}{N} = 4\pi c_{AB}^2 \left(\frac{\mu}{2\pi kT}\right)^{3/2} \exp\left(-\frac{\mu c_{AB}^2}{2kT}\right) \tag{10-3.4.4}$$

where $c_{AB} = (8kT/\pi\mu)^{\frac{1}{2}}$ is the average relative speed of the molecules A and B, and μ = reduced mass $m_A m_B/(m_A + m_B)$.

By arguments that parallel those in Sec. 2-3.7 the *differential collision rate* is found to be

$$dZ_{AB} = \sigma' c_{AB} \left[\frac{dN_c\,(AB)}{N}\right] N_A^* N_B^* \, dc_{AB} \tag{10-3.4.5}$$

where N* is a molecular density (molecules per cubic meter). One would not expect every collision to lead to a reaction from what was stated in the previous section. It seems appropriate to stipulate that only collisions with a kinetic energy ε greater than some *threshold value* ε_0 can react to form products. ε_0 can be related to a *threshold relative speed* c_0 by

$$\varepsilon_0 = \tfrac{1}{2}\mu c_0^2 \tag{10-3.4.6}$$

(Drop the AB subscript on c from this point on.)

The rate of the reaction will be given by the number of collisions per second with $c \geq c_0$. Therefore

$$\text{Rate} = \int_{c_0}^{\infty} dZ_{AB} = \sigma' N_A^* N_B^* \int_{c_0}^{\infty} \frac{dN_c(AB)}{N} c\, dc$$

$$= 4\pi\sigma' N_A^* N_B^* \left(\frac{\mu}{2\pi kT}\right)^{3/2} \int_{c_0}^{\infty} c^3 \exp\left(-\frac{\mu c^2}{2kT}\right) \qquad (10\text{-}3.4.7)$$

Now substitute $\varepsilon = \frac{1}{2}\mu c^2$, so (show the steps)

$$\int_{c_0}^{\infty} c^3 \exp\left(-\frac{\mu c^2}{2kT}\right) dc = \frac{2}{\mu^2} \int_0^{\infty} \varepsilon \exp\left(-\frac{\varepsilon}{kT}\right) d\varepsilon$$

$$= \frac{2}{\mu^2} \exp\left(-\frac{\varepsilon_0}{kT}\right)(\varepsilon_0 kT + k^2T^2) \qquad (10\text{-}3.4.8)$$

This combines with Eq. (10-3.4.7) to give

$$\text{Rate} = \sigma' N_A^* N_B^* \left(\frac{8kT}{\pi\mu}\right)^{1/2} \left(1 + \frac{\varepsilon_0}{kT}\right) \exp\left(-\frac{\varepsilon_0}{kT}\right) \qquad (10\text{-}3.4.9)$$

You should recall from previous definitions

$$\text{Rate} = \underline{k} N_A^* N_B^* \qquad (10\text{-}3.4.10)$$

Define the pre-exponential factor of Eq. (10-3.4.9) as

$$A' = \sigma' \left(\frac{8kT}{\pi\mu}\right)^{1/2} \left(1 + \frac{\varepsilon_0}{kT}\right) \qquad (10\text{-}3.4.11)$$

If $\varepsilon_0 >> kT$, as is usually the case, 1 can be ignored:

$$A' \approx \sigma' \left(\frac{8kT}{\pi\mu}\right)^{1/2} \frac{\varepsilon_0}{kT}$$

$$= \sigma' \varepsilon_0 \left(\frac{8}{\pi\mu kT}\right)^{1/2} \qquad (10\text{-}3.4.12)$$

There is an implicit assumption in the above derivation that needs modification. Figure 10-3.4.1 shows a collision between A and B assuming each is a hard sphere. We have assumed that all collisions with a given c are equally effective. But, it seems reasonable that as b in Fig. 10-3.4.1 decreases the collision is more likely

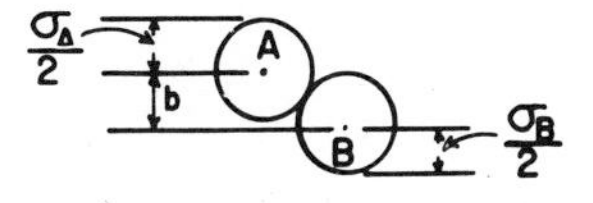

FIG. 10-3.4.1 A collision between hard-sphere molecules A and B.

to result in a reaction. [A head-on collision (b = 0) should be more effective than a glancing collision.] An analysis of this effect [see W. C. Gardiner, Jr., *Rates and Mechanisms of Chemical Reactions* (Sec. 10-3.9, Ref. 2) for a thorough discussion of this subject] results in a different form of the rate constant equation as shown in Eq. (10-3.4.13).

$$\underline{k}'(T) = \left(\frac{8kT}{\pi\mu}\right)^{\frac{1}{2}} \sigma' \exp\left(-\frac{\varepsilon_0}{kT}\right)$$

$$= A'' \exp\left(-\frac{\varepsilon_0}{kT}\right) \qquad (10\text{-}3.4.13)$$

where $\underline{k}'(T)$ is a "line-of-centers rate constant." This works out to be

$$\text{Rate} = Z_{AB} \exp\left(-\frac{E_0}{RT}\right) \qquad (10\text{-}3.4.14)$$

where E_0 is the zero-point activation energy and Z_{AB} is the collision frequency of A and B. Z_{AB} can be found with a modified form of (10-3.4.9).

Empirically it is found that n in Eq. (10-3.3.7) varies dramatically from ½ for many reactions. Thus, Eq. (10-3.4.14) is found to be rather inaccurate. It is usually "fixed" by including a factor p that has become known as the *steric factor*. This steric factor recognizes the possibility that not all orientations are equally likely to give a reaction even in head-on collisions. This is particularly true for more complex molecules. The modified equation reads

$$\text{Rate} = pZ_{AB} \exp\left(-\frac{E_0}{RT}\right) \qquad (10\text{-}3.4.15)$$

The relationship between A in the Arrhenius equation and A'' is

$$\boxed{A = e^{\frac{1}{2}} A'' p} \qquad (10\text{-}3.4.16)$$

The p is our steric factor correction term, and the $e^{\frac{1}{2}}$ term corrects for the difference between the Arrhenius activation energy E_a and the zero-point activation energy. To show how this is derived, define $BT^{\frac{1}{2}} = A''$ so B has no temperature dependence.

$$\underline{k}_r = BT^{\frac{1}{2}} \exp\left(-\frac{E_0}{RT}\right) = A \exp\left(-\frac{E_a}{RT}\right)$$

$$\ln B + \tfrac{1}{2} \ln T - \frac{E_0}{RT} = \ln A - \frac{E_a}{RT} \qquad (10\text{-}3.4.17)$$

Differentiate with respect to T:

$$\frac{1}{2T} - \frac{E_0}{RT^2} = \frac{E_a}{RT^2} \qquad \text{or} \qquad E_a = E_0 + \tfrac{1}{2} RT$$

therefore,

$$\ln B + \tfrac{1}{2} \ln T + \tfrac{1}{2} = \ln A$$

$$A = e^{\frac{1}{2}} BT^{\frac{1}{2}} = e^{\frac{1}{2}} A'' \qquad (10\text{-}3.4.18)$$

Then the steric factor is included. Table 10-3.4.1 gives appropriate parameters for a number of reactions. These are experimental values.

10-3.4.1 EXAMPLE

Calculate the pre-exponential factor in Eq. (10-3.4.13) and compare with Table 10-3.4.1 for the reaction

$$CO + O_2 \rightarrow CO_2 + O$$

Solution

$$A'' = \left(\frac{8kT}{\pi\mu}\right)^{\frac{1}{2}} \sigma'$$

$$\sigma' = \pi\sigma_{AB}^2 = \pi\left(\frac{\sigma_A + \sigma_B}{2}\right)^2$$

We can find an estimate for σ_{CO} and σ_{O_2} in the Chemical Rubber Company *Handbook of Chemistry and Physics*. An average of viscosity and van der Waal's values gives $\sigma_{CO} = 3.16 \times 10^{-10}$ m and $\sigma_{O_2} = 2.95 \times 10^{-10}$ m.

$$\sigma' = (3.1416)\left(\frac{3.16 + 2.95}{2}\right)^2 \times 10^{-20} \text{ m}^2$$

$$= \underline{\underline{29.3 \times 10^{-20} \text{ m}^2 \text{ molecule}^{-1}}}$$

$$\mu = \frac{m_{CO} m_{O_2}}{m_{CO} + m_{O_2}} = \frac{(0.028)(0.032)}{(0.028) + (0.032)} \text{ kg mol}^{-1}$$

$$= 1.49 \times 10^{-2} \text{ kg mol}^{-1} = \underline{\underline{2.48 \times 10^{-26} \text{ kg molecule}^{-1}}}$$

$$A'' = \left[\frac{(8)(1.38 \times 10^{-23}\ \text{J molecule}^{-1}\ \text{K}^{-1})(298\ \text{K})}{(3.14)(2.48 \times 10^{-26}\ \text{kg molecule}^{-1})}\right]^{1/2} \times (29.3 \times 10^{-20}\ \text{m}^2\ \text{molecule}^{-1})$$

$$= (6.50 \times 10^2\ \text{m s}^{-1})(29.3 \times 10^{-20}\ \text{m}^2\ \text{molecule}^{-1})$$

$$= 1.88 \times 10^{-16}\ \text{m}^3\ \text{s}^{-1}\ \text{molecule}^{-1}$$

$$= (1.88 \times 10^{-16})(\text{m}^3\ \text{s}^{-1}\ \text{molecule}^{-1})\left(\frac{10\ \text{dm}}{1\ \text{m}}\right)^3\left(\frac{6.023 \times 10^{23}\ \text{molecules}}{\text{mol}}\right)$$

$$= \underline{\underline{1.14 \times 10^{11}\ \text{dm}^3\ \text{mol}^{-1}\ \text{s}^{-1}}}$$

$$\ln A'' = \underline{\underline{25.5}}$$

This compares with a given value of $\ln A = 22.0$ or $A = 3.6 \times 10^9\ \text{dm}^3\ \text{mol}^{-1}\ \text{s}^{-1}$. But the value corresponding to A in the table is $pA''\,e^{1/2}$. Thus, the calculated A is

$$A = pA''\,e^{1/2}$$

$$= 0.0043(1.14 \times 10^{11}\ \text{dm}^3\ \text{mol}^{-1}\ \text{s}^{-1})(1.65)$$

$$= \underline{\underline{8.1 \times 10^8\ \text{dm}^3\ \text{mol}^{-1}\ \text{s}^{-1}}} \quad \text{and} \quad A_{table} = \underline{\underline{3.6 \times 10^9\ \text{dm}^3\ \text{mol}^{-1}\ \text{s}^{-1}}}$$

Table 10-3.4.1
Experimental Arrhenius Parameters (Including Steric Factors) for Some Bimolecular Gas Reactions

Reaction	$\ln A$[a]	E_a (kJ)	p
$H + D_2 \rightarrow HD + D$	24.61	39.3	0.088
$D + H_2 \rightarrow HD + H$	24.50	31.8	0.094
$H + HCl \rightarrow H_2 + Cl$	23.80	14.6	0.039
$H + HBr \rightarrow H_2$	25.42	15.5	0.076
$H + Cl_2 \rightarrow HCl + Cl$	25.3	22.0	0.074
$O + O_3 \rightarrow O_2 + O_2$	23.21	20.0	0.037
$N + O_2 \rightarrow NO + O$	22.8	30.0	0.034
$CO + O_2 \rightarrow CO_2 + O$	22.0	210.0	0.0043
$O_3 + NO \rightarrow NO_2 + O_2$	20.2	10.3	0.002
$O^+ + O_2 \rightarrow O_2^+ + O$	22.6	- 1.6	0.031
$O^- + NO_2 \rightarrow NO_2^- + O$	27.31	(0)	2.6
$O_3^- + NO_2 \rightarrow NO_3^- + O$	22.5	(0)	0.018

[a] A in $\text{dm}^3\ \text{mol}^{-1}\ \text{s}^{-1} = pA''\,e^{1/2}$.

All we can say is that the answer is the right order of magnitude. It is impossible to calculate p from molecular parameters, and the value of σ obtained from transport properties is probably very different from the true collision cross-section for a reaction. In many cases, all that can be obtained is an order of magnitude for the reaction rate. However, in many other cases, rather good agreement can be obtained between theory and experiment.

10-3.5 ABSOLUTE RATE THEORY

Absolute rate theory (ART) or *activated complex theory* (ACT) was first formulated about 40 years ago by Henry Eyring (1935). It was not generally accepted as a viable theory until the 1950's when comparisons of calculated and experimental rates were found to be satisfactory for a number of cases. ACT considers the behavior of reactants and an *activated complex* on a reaction potential energy surface. Potential energy surfaces are three-dimensional descriptions of the potential energy of a system of reactants, activated complex, and products for all combinations of internuclear distances. All we will be concerned with is that portion of the potential energy surface that gives the minimum energy path from reactants to products. This is just the potential energy vs. the reaction coordinate plot given in Fig. 10-3.2.1. ACT is concerned with the rate of formation of the activated complex and its rate of reaction to products. What we need to find is the concentration of activated complexes and the frequency with which these complexes pass to the product side of the reaction.

Figure 10-3.5.1 reproduces Fig. 10-3.2.1 in a little more detail for the reaction $aA + bB \rightleftarrows C^{\ddagger} \rightleftarrows$ products. The activation energy has been corrected to account for zero-point vibrational energies of the reactants and the activated complex. δ is a small distance along the reaction coordinate which will be used later.

There are several assumptions needed to proceed with the derivation. The major one is that equilibrium exists between $C^{\ddagger}$ and the reactants. This is not strictly true [see H. Eyring and E. M. Eyring, *Modern Chemical Kinetics* (Sec. 10-3.9, Ref. 3)] but causes appreciable error only in very fast reactions. Given the previous assumptions, there are at least two ways to proceed. The common method is to assume that one of the vibrational modes of the activated complex is unstable and is converted to translation along the reaction coordinate. This approach has been challenged (see B. H. Mahan, Sec. 10-3.9, Ref. 4) and replaced with a derivation based on motion across a surface in phase space of the system. Since most readers are not likely to be familiar with operations in phase space we will follow the conventional approach.

Consider Fig. 10-3.5.1 in more detail. The transition state corresponds to a small (and somewhat arbitrary) distance along the reaction coordinate. If we assume

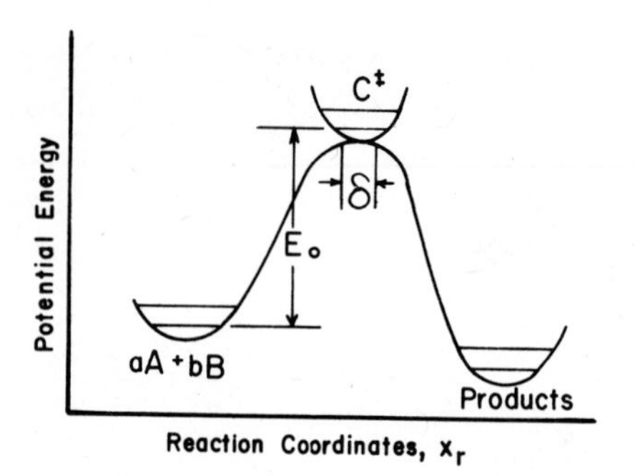

FIG. 10-3.5.1 Potential energy of reaction path aA + bB ⇄ C‡ ⇄ products. E_0 is the difference of the zero-point energies of the reactants and activated complex.

equilibrium between activated complexes and reactants we can immediately write an equilibrium constant in terms of concentrations:

$$K_{eq} = \frac{[C^{\ddagger}]}{[A]^a [B]^b} \tag{10-3.5.1}$$

or in terms of $[C^{\ddagger}]$:

$$[C^{\ddagger}] = K_{eq} [A]^a [B]^b \tag{10-3.5.2}$$

The average velocity across the transition state along the reaction coordinate x_r is (see Sec. 2-3)

$$\bar{v} = \frac{\int_0^{\infty} e^{-m_c^2 x_r^2/2kT} x_r \, dx_r}{\int_0^{\infty} e^{-m_c^2 x_r^2/2kT} dx_r} \tag{10-3.5.3}$$

where m_c is the mass of the activated complex. Evaluation of the integral (note only positive values of x_r are included since we are considering only those complexes that convert to products) gives

$$\bar{v} = \left(\frac{2kT}{\pi m_c}\right)^{1/2} \tag{10-3.5.4}$$

The average time required to traverse the distance δ along x_r is

$$\frac{\delta}{\bar{v}} = \delta \left(\frac{\pi m_c}{2kT}\right)^{1/2} \tag{10-3.5.5}$$

At equilibrium, half the complexes will be from products and half from reactants (forward and reverse reactions are equal at equilibrium). The rate of reaction will be given by the number of complexes crossing to the right per unit time per unit volume:

$$\text{Rate} = \frac{[C^{\ddagger}]}{2}\left(\frac{\delta}{\bar{v}}\right)^{-1}$$

$$= \frac{[C^{\ddagger}]}{2\delta}\left(\frac{2kT}{\pi m_c}\right)^{1/2}$$

$$= \frac{[C^{\ddagger}]}{\delta}\left(\frac{kT}{2\pi m_c}\right)^{1/2} \qquad (10\text{-}3.5.6)$$

Combining this with Eq. (10-3.5.2) yields the rate and rate constant

$$\text{Rate} = \left(\frac{K_{eq}}{\delta}\right)\left(\frac{kT}{2\pi m_c}\right)^{1/2}[A]^a[B]^b \qquad (10\text{-}3.5.7)$$

$$\underline{k}_r = \left(\frac{K_{eq}}{\delta}\right)\left(\frac{kT}{2\pi m_c}\right)^{1/2} \qquad (10\text{-}3.5.8)$$

If we express K_{eq} in terms of partition functions we will have an equation which can be solved explicitly. There is a modification that must be made to take the assumption we made into account. One degree of vibrational freedom is treated as a translation in one dimension. Applying Eq. (9-2.4.4) allows the partition function for this *special* degree of freedom to be written:

$$z^{\ddagger}_{tr} = (2\pi mkT)^{1/2}\,\frac{\delta}{h} \qquad (10\text{-}3.5.9)$$

The total partition function for the activated complex will be written as

$$z_{\ddagger} = z^{\circ}_{\ddagger}(2\pi mkT)^{1/2}\,\frac{\delta}{h} \qquad (10\text{-}3.5.10)$$

where $z^{\circ}_{\ddagger}$ includes all degrees of freedom other than the special motion along x_r. The equilibrium constant can now be written and included in Eq. (10-3.5.8). For review of the details of finding equilibrium constants from partition functions see Sec. 9-3.2.

$$K_{eq} = \frac{z^{\circ}_{\ddagger}(2\pi m_C kT)^{1/2}\,(\delta/h)}{(z^{\circ}_A)^a(z^{\circ}_B)^b}\exp\left(-\frac{E_0}{RT}\right) \qquad (10\text{-}3.5.11)$$

and

$$\underline{k}_r = \frac{(2\pi m_C kT)^{1/2}\,(\delta/h)}{\delta}\left(\frac{kT}{2\pi m_C}\right)^{1/2}\frac{z^{\circ}_{\ddagger}\,e^{E_0/RT}}{(z^{\circ}_A)^a(z^{\circ}_B)^b}$$

$$\underline{k}_r = \frac{kT}{h} \frac{z^{\circ}_{\ddagger} e^{-E_0/RT}}{(z^{\circ}_A)^a (z^{\circ}_B)^b} \qquad (10\text{-}3.5.12)$$

This last equation allows calculation of rate constants from nonkinetic data. z°_A and z°_B can be calculated from known energy level patterns. $z^{\circ}_{\ddagger}$ can also be obtained if the structure of the transition state complex is known. Usually this process involves estimating the shape and size of the complex to determine the moments of inertia. Vibrational frequencies can be determined experimentally or estimated from comparison with known molecules. Of course, this can give rise to some error, especially for cases in which low vibrational frequencies (and, therefore, vibrational partition functions greater than 1) are present. Also, there is difficulty in determining the zero-point barrier height.

A final difficulty arises from our assumption that each activated complex produced from reactants went on to form products. This may not always be true. To take that into account a *transmission coefficient* κ is introduced. The value of κ must be between 0 and 1. As you probably recognize, this gives a good "fudge factor" for forcing agreement between rate constants calculated from ACT and those found experimentally. All is not in vain, however. It is possible to make calculations and correlations between reactions that are similar. If sufficient numbers of reactions are studied, κ can likely be related to reaction type.

10-3.5.1 EXAMPLE

By applying the *statistical mechanical formulas* of Sec. 9-2 find the rate constant for the reaction at 298 K of

$$2ClO \rightarrow Cl_2 + O_2 \qquad E_0 = 0$$

The activated complex is assumed to be

```
Cl———Cl
 \   /
  O—O
```

For ^{35}ClO: $g = 2$, $\sigma = 1$, $I = 4.3 \times 10^{-46}$ kg m^2, and $\nu = 2.40 \times 10^{13}$ s^{-1}. For $(ClO)^{\ddagger}_2$: $g = 1$, $\sigma = 2$, $I_A I_B I_C = 2.20 \times 10^{-135}$ (kg m^2)^3; $\nu_1 = 4.50 \times 10^{13}$ s^{-1}, $\nu_2 = 2.1 \times 10^{13}$ s^{-1}, $\nu_3 = 2.4 \times 10^{13}$ s^{-1}, $\nu_4 = 1.80 \times 10^{13}$ s^{-1}, and $\nu_5 = 6.00 \times 10^{12}$ s^{-1}. [*Note*: $(ClO)^{\ddagger}_2$ should have 3(4) - 6 = 6 vibrational degrees of freedom. One has been converted to translation.]

Solution

It is probably better to do this in sections to ease the calculation.

^{35}ClO

$$z_{tr} = \left(\frac{2\pi mkT}{h^2}\right)^{3/2}$$

$$= \left[\frac{2(\pi)(0.051/6.02 \times 10^{23}\ \text{kg mlc}^{-1})(1.38 \times 10^{-23}\ \text{J mlc}^{-1})(298)}{(6.63 \times 10^{-34}\ \text{J s mlc}^{-1})^2}\right]^{3/2}$$

(mlc = molecule)

$$= 3.53 \times 10^{32} \left(\frac{\text{kg molecule}^{-1}\ \text{kg m}^2\ \text{s}^{-2}\ \text{molecule}^{-1}}{\text{kg}^2\ \text{m}^4\ \text{s}^{-4}\ \text{s}^2\ \text{molecule}^{-2}}\right)^{3/2}$$

$$= \underline{\underline{3.53 \times 10^{32}\ \text{m}^{-3}}}$$

$$z_{rot} = \frac{8\pi^2 IkT}{\sigma h^2}$$

$$= \frac{8\pi^2(4.3 \times 10^{-46}\ \text{kg m}^2\ \text{molecule}^{-1})(1.38 \times 10^{-23}\ \text{J molecule}^{-1})(298)}{(1)(6.63 \times 10^{-34}\ \text{J s molecule}^{-1})^2}$$

$$= 3.20 \times 10^3 \left(\frac{\text{kg m}^2\ \text{molecule}^{-1}\ \text{kg m}^2\ \text{s}^{-2}\ \text{molecule}^{-1}}{\text{kg}^2\ \text{m}^4\ \text{s}^{-4}\ \text{s}^2\ \text{molecule}^{-2}}\right)$$

$$= \underline{\underline{3.20 \times 10^2}} \quad \text{(diatomic)}$$

$$z_{vib} = \left[1 - \exp\left(-\frac{h\nu_0}{kT}\right)\right]^{-1}$$

$$= \left\{1 - \exp\left[-\frac{(6.63 \times 10^{-34}\ \text{J s molecule}^{-1})(3.40 \times 10^{13}\ \text{s})}{(1.38 \times 10^{-23}\ \text{J s molecule}^{-1}\ \text{K}^{-1})(298\ \text{K})}\right]\right\}^{-1}$$

$$= \left[1 - e^{5.48}\right]^{-1}$$

$$= \underline{\underline{1.00}}$$

$$z_{elec} = g = \underline{\underline{2}}$$

$$(z_{ClO})^2 = [(3.53 \times 10^{32}\ \text{m}^{-3})(3.2 \times 10^2)(1.00)(2)]^2 = \underline{\underline{5.10 \times 10^{70}\ \text{m}^{-6}}}$$

$(ClO)_2^{\ddagger}$

$$z_{tr} = \left(\frac{2\pi mkT}{h^2}\right)^{3/2}$$

$$= \left[\frac{2(\pi)(0.102/6.023 \times 10^{23})(1.38 \times 10^{-23})(298)}{(6.63 \times 10^{-34})^2}\right]^{3/2}$$

$$= \underline{\underline{9.93 \times 10^{32}\ \text{m}^{-3}}}$$

$$z_{rot} = \frac{8\pi^2(8\pi^3 I_A I_B I_C)^{1/2}(kT)^{3/2}}{\sigma h^3}$$

$$= \frac{8(\pi)^2(8\pi^3 \cdot 2.20 \times 10^{-135})^{1/2}(1.38 \times 10^{-23} \cdot 298)^{3/2}}{(2)(6.63 \times 10^{-34})^3}$$

$$= \underline{\underline{2.63 \times 10^4}}$$

$$z_{vib}: \quad \nu_1 = \{1 - \exp[-(4.50 \times 10^{13})(1.61 \times 10^{-13})]\}^{-1} = 1.00$$

$$\nu_2 = \quad = 1.04$$

$$\nu_3 = \quad = 1.02$$

$$\nu_4 = \quad = 1.06$$

$$\nu_5 = \quad = 1.61$$

$$z_{vib} = \prod z_{vib,i} = \underline{\underline{1.81}}$$

$$z_{elec} = 1$$

$$z_{(ClO)_2^\ddagger} = (9.93 \times 10^{32}\ m^{-3})(2.63 \times 10^4)(1.81)(1) = \underline{\underline{4.74 \times 10^{37}\ m^{-3}}}$$

$$\underline{k}_r = \frac{kT}{h} \frac{z_{(ClO)_2^\ddagger}}{\left[z_{(ClO)}\right]^2} \exp\left(-\frac{0}{RT}\right)$$

$$= \frac{(1.38 \times 10^{-23}\ J\ molecule^{-1})(298)}{6.63 \times 10^{-34}\ J\ s\ molecule^{-1}} \frac{4.74 \times 10^{37}\ m^{-3}}{5.10 \times 10^{70}\ m^{-6}}$$

$$= 5.76 \times 10^{-21}\ m^3\ s^{-1}\ molecule^{-1}$$

$$= 3.47 \times 10^3\ m^3\ s^{-1}\ mol^{-1}$$

$$= \underline{\underline{3.47 \times 10^6\ dm^3\ s^{-1}\ mol^{-1}}}$$

10-3.6 PROBLEMS

1. D. R. Herschbach, H. S. Johnston, K. S. Pitzer, and R. E. Powell, *J. Chem. Phys.*, *25*:736 (1956), presents the following data for the reaction

$$2NO_2 \rightarrow 2NO + O_2$$

through the proposed intermediate

```
O   O   N
 \ / \ / \
  N   O   O
```

$E_0 = 111\ \text{kJ mol}^{-1}$

Molecule	g	σ	Moment of inertia	$\bar{\nu}$ (cm^{-1})
NO_2	2	2	1.44×10^{-137} (kg m^2)^3	1332, 751, 1616
O–N–O–O–N–O	1	2	2.95×10^{-135} (kg m^2)^3	1700, 1700, 1300, 1300, 700, 700, 350, 250[a]

[a] These are the vibrations considered to be important for this complex.

Determine the pre-exponential factor and the rate constant for this reaction at 500 K.

2. What is the activation energy for a reaction that triples its rate for a 10°C rise in temperature?

3. The following rate constants have been determined by the cis-trans isomerization of cis-ethylene-d_6 in the presence of nitric oxide. Find the constants in the Arrhenius equation.

k (cm^3 mol^{-1} s^{-1})	3.18	6.57	15.7	34.2	70.0
T (K)	561.2	578.0	600.6	620.5	639.9

4. M. Bodenstein [*Z. Physik. Chem.*, *29*:295 (1899)] found $E_a = 184\ \text{kJ mol}^{-1}$ for the reaction

$$2HI \underset{k_1'}{\overset{k_1}{\rightleftarrows}} H_2 + I_2$$

If $\sigma = 5.0 \times 10^{-10}$ m for HI, determine k_1 from collision theory at 700 K and compare with the observed value of 1.16 cm^3 mol^{-1} s^{-1}.

10-3.7 SELF-TEST

1. The following rate constants were obtained Sullivan, [*J. Chem. Phys.*, *46*:73 (1967)] for the reaction: $2I_2(g) + H_2(g) \rightarrow 2HI(g)$.

T (K)	417.9	480.7	520.1	633.2	710.3	737.9
k (10^{-5} mol^{-2} dm^{-6} s^{-1})	1.12	2.60	3.96	9.38	16.10	18.54

Determine the constants in the Arrhenius equation.

2. D. R. Herschbach, H. S. Johnston, K. S. Pitzer, and R. E. Powell [*J. Chem. Phys.*, *25*:736 (1956)] have summarized ACT and collision theory calculations for several bimolecular reactions. They list the following data for the reaction:

$$NO + O_3 \rightarrow NO_2 + O_2$$

(The assumed activated complex is

```
O   O   O
 \ / \ /
  O   N
```

and E_0 = 10.5 kJ mol^{-1}.)

Molecule	g	σ	Moment of Inertia	$\bar{\nu}$ (cm^{-1})
NO	2	1	1.64×10^{-46} kg m^2	1904
O_3	1	2	2.77×10^{-137} $(kg\ m^2)^3$	1110, 705, 1043
$(O_3NO)^{\ddagger}$	2	1	0.922×10^{-135} $(kg\ m^2)^3$	1300, 1200, 1100, 700, 700, 350[a]

[a] These are the vibrations considered to be important for this complex.

Evaluate the pre-exponential factor and the rate constants for this reaction at 200 K.

3. Given that the bond length in O_3 is 2.0×10^{-10} m and in NO is 1.4×10^{-10} m, estimate the collision diameters of O_3 and NO. Calculate the pre-exponential factor (A'') in Eq. (10-3.4.13) and from A'' find the Arrhenius pre-exponential factor at 200 K.

10-3.8 ADDITIONAL READING

1.* C. E. Nordman and S. M. Blinder, *J. Chem. Educ.*, *51*:790 (1974).

2.* W. C. Gardiner, Jr., *Rates and Mechanisms of Chemical Reactions*, W. A. Benjamin, New York, 1969, Secs. 4-2 and 4-3.

3.* H. Eyring and E. M. Eyring, *Modern Chemical Kinetics*, Reinhold, New York, 1967, p. 49.

4.* B. H. Mahan, *J. Chem. Educ.*, *51*:709-711 (1974).

10-4 SUMMARY AND ADDITIONAL COMMENTS

This concludes your introduction to kinetics and mechanisms of chemical reactions. In this chapter you have seen how empirical rate laws are derived from experimental rate data for a few cases. You have been introduced to the concept of a reaction mechanism and have used a major simplifying assumption (the steady state approximation) to derive rate laws for proposed mechanisms. Finally, Sec. 10-3 has introduced you to some of the theoretical aspects of chemical kinetics.

There are many topics that have been essentially ignored. This is not due to any lack of importance. Rather it is due to a limited amount of space. Below, a short discussion of catalysis will be given. Also a list of some of the important topics that have been omitted or slighted will be given.

Some of the areas that have not received attention are:

Reactions in solution: solvent effects, ionic reaction mechanisms, dielectric effects, the role of diffusion, and diffusion-controlled reactions.

Surface reactions: oxidation catalysis, surface potentials, role of surface preparation.

Solid state reactions: rate processes for crystallization and grain growth in pure metals and alloys, ceramic production-particle size, and grain boundary control.

Photochemistry: the use of radiation to produce molecules in an *excited* state that can then react. This is in contrast to the reactions considered in this chapter in which *thermally active* molecules (the ones with high velocities) were seen to react. Fluorescence and phosphorescence and phenomena associated with transitions of excited molecules (the *production of cold light* or *chemiluminescence* is due to a reaction forming products in an excited state which then decays by emission of light). Photochemical studies of the interaction of sunlight with atmospheric materials is an important ongoing activity. We did investigate briefly a photochemical phenomenon in our discussion of methods of studying fast reactions. Flash photophotolysis involves the action of light on a reaction mixture (see Sec. 10-1).

There are other areas that could be mentioned, but instead of listing them, let us turn our attention to a brief discussion of catalysis. In the discussion some mention will be made of catalysis in biological systems (enzyme catalysis), as well as homogeneous and heterogeneous catalysis.

CATALYSIS. A catalyst is a substance that changes the rate of a reaction without changing the equilibrium constant. Since the equilibrium constant is not changed by the presence of a catalyst, we can readily conclude that the catalyst affects both the forward and reverse rates. An interesting dilemma would occur if the equilibrium constant were dependent on the presence or absence of the catalyst. Can you construct an imaginary perpetual motion machine using a reaction occurring in two vessels: one with a catalyst present and one without, if the catalyst did affect the position of equilibrium? It could be done. *Homogeneous catalysis* refers to catalytic activity in which the catalyst is present in the same phase as the reacting materials. A catalyst can be used to speed up a reaction (what is normally associated with catalysis) or slow down a reaction (in which case, the term *inhibitor* is used). *Heterogeneous catalysis* refers to a system in which the catalytic action is due to a substance in a different phase than the reacting materials. Finally, in biological systems, *enzyme catalysis* is of major importance. Only an example of each type will be given. No attempt will be made to give a thorough account. See the references at the end of this section for further details.

In *homogeneous catalysis*, in general, the catalyst must offer another pathway to the formation of the product. This pathway must involve a series of reactions that are faster than the original if the rate of product formation is to be increased. The acid-base catalyzed hydrolysis of esters is an example of a catalytic effect of H^+ ions, or OH^- ions. The hydrolysis of an ester proceeds according to the following stoichiometry (not mechanism):

$$R{-}C(=O){-}OR' + H_2O \rightarrow R{-}C(=O){-}OH + HOR'$$

R and R′ are organic groups. In the absence of H^+ or OH^- the rate of hydrolysis is very slow. But if either hydronium or hydroxyl ions are present, the rate is rapid, due to the ability of the small ions to take part in a new reaction sequence to form product. In this case the acid or base is consumed in the reaction. However, there are situations in which the catalyst is regenerated in one of the steps in the mechanism and can initiate a new reaction. In these cases, a small amount of catalyst can cause a large increase in the rate of product formation.

Heterogeneous catalysis involves the presence of a second phase within the system. Normally, a solid phase is the catalyst and either a gas or solution reaction occurs at the interface. Catalysis, in this case involves the adsorption of a reactant onto the surface of the catalyst, some change in the form or energy of the reactant, reaction on the surface with other materials that are in contact with it, and desorption of the product. The adsorption and desorption steps are discussed in some detail in Sec. 11-3 in which surface phenomena are treated. Some examples of heterogeneous catalyst systems that are of importance are:

1. Fe catalysts used in the Haber process for forming ammonia from nitrogen and hydrogen, one of the most important industrial processes.
2. Silica alumina gel used for cracking heavy petroleum fractions to give gasoline and other lighter products.
3. Platinum for isomerization of hydrocarbons. Also, as you saw in Sec. 6-5, Pt is essential in the operation of fuel cells that convert chemical energy into electrical energy.
4. Noble metals used for the catalytic convertors used to clean automobile exhaust.

Enzyme catalysis is important in biological systems. Enzymes are biological materials that interact with a material (the *substrate*) to form products. Normally, enzyme catalysis involves the combination of an enzyme with its substrate to form a complex. The complex can then undergo the appropriate reaction. After reaction, the enzyme is still present unaltered and can attach itself to another substrate molecule if one is present. Enzymes are very specific in their behavior. A given enzyme will react only with a given substrate. In some cases, only very minor changes in a substrate result in no reaction at all. There are cases in which an enzyme will react with a given substrate but not with the optical isomer of that substrate. Not all enzymes are that specific. In addition to enzymes in living systems, there are materials in living cells that act as *inhibitors*. As the name implies, inhibitors slow down some cell processes. In many cases, there are inhibitors present that control the behavior of enzymes to maintain a balance of necessary cell materials.

ADDITIONAL REFERENCES.

1. G. W. Castellan, *Physical Chemistry*, 2nd ed., Addison-Wesley, Reading, Mass., 1971, Chaps. 31, 32, and 33.
2. S. A. Bernard, "Homogeneous Catalysis"; and J. C. Jungers and J. C. Balaceanu, "Heterogeneous Reactions and Catalysis," in A. Weissberger, ed., *Techniques of Organic Chemistry*, Vol. VIII, Part I, Interscience, New York, 1961, Chaps. XII and XIII.

11

Colloid and Surface Chemistry

SYMBOLS AND DEFINITIONS

M_n	Number average molecular weight (kg mol^{-1})
M_w	Weight average molecular weight (kg mol^{-1})
f_g	Force due to gravity (N)
f_b	Buoyant force (N)

f_d	Net driving force (N)
f_r	Frictional force (N)
g	Acceleration due to gravity (m s^{-2})
ρ	Solution density (kg m^{-3}, kg dm^{-3})
ρ_2	Particle density (kg m^{-3}, kg dm^{-3})
v	Particle volume (m^3, dm^3)
m_{eff}	Particle effective mass (kg, g)
ϕ	Friction factor (unitless)
u	Particle velocity (m s^{-1})
r	Particle radius (m, cm)
η	Coefficient of viscosity (kg m^{-1} s^{-1})
ω	Angular velocity (radians s^{-1} or s^{-1})
x	Distance (m,cm)
$\omega^2 x$	Angular acceleration (s^{-2})
$\bar{v}$	Partial specific volume (m^3 kg^{-1}, cm^3 g^{-1})
s	Sedimentation constant (s)
D	Diffusion coefficient (m^2 s^{-1})
C	Concentration (various units)
ψ_0	Electrical potential at a surface (V = volts)
ψ_d	Electrical potential at the edge of the bound layer
ζ	Zeta potential (V)
E	Electric field (V)
L	Length (m, cm)
ε_0	Permittivity of vacuum = 8.854×10^{-12} kg^{-1} m^{-3} $s^4 A^2$ (A = amps)
θ	Dielectric constant (unitless)
$\varepsilon = \theta\varepsilon_0$	Permittivity of solvent (kg^{-1} m^{-3} s^4 A^2)
υ	Electrophoretic mobility (m^2 s^{-1} V^{-1})
R	Tube radius (m, cm)
t	Time (s)
V	Volume flow (m^3, dm^3, cm^3)
$\bar{P}$	Electroosmotic pressure (atm, N m^{-2})
P	Pressure (atm, N m^{-2})
I_s	Streaming current (A = amp)
E_s	Streaming potential (V)
k_0	Solution conductivity (ohm^{-1} m^{-1})
$\bar{\theta}$	Fraction of adsorption sites covered
b, b'	Langmuir constants
y	Amount of material adsorbed
y_m	Amount adsorbed for monolayer
k	Freundlich constant
ΔH_1	Heat of adsorption of monolayer (kJ mol^{-1})
ΔH_L	Heat of liquefaction (= $-\Delta H_{vap}$)
P_0	Normal vapor pressure
γ	Surface tension (N m^{-1}, J m^{-2})
θ	Contact angle
w_c	Work of cohesion (J m^{-2})
w_a	Work of adhesion (J m^{-2})
γ_{sa}, $\gamma_{\ell a}$, $\gamma_{s\ell}$, etc.	Interfacial surface tensions
Γ	Excess surface concentration (mol m^{-2})
A	Area (m^2)
μ	Chemical potential
E^s	Surface energy
S^s	Surface entropy
n	Number of moles
a	Activity
x_0	Effective empty layer thickness

Colloids consist of a suspension of finely divided particles in a continuous medium. Colloidal phenomena are extremely important in many areas of everyday life. Colloids are encountered in such diverse products as drilling muds used in the petroleum industry, mayonnaise, instant puddings, hair sprays, paints, etc. Emulsion polymerization (see Chap. 12) is an industrial process utilizing colloidal systems. Smoke and dust are two colloid types that are important in air pollution. The list goes on and on. Surface-active materials are important in the formulation of detergents and cleaning compounds. Surface coatings (paints) must be manufactured with proper wetting properties to insure good protection of the surfaces to which they are applied. Catalysis in many chemical reactions is a surface phenomenon. The "spreading" of an oil slick is another example of an effect of surface properties. A few of the basic phenomena associated with colloids and with surfaces will be investigated in some detail. Some topics which could be included here, such as methods of determining molecular weights, are discussed in the following chapter on polymers (which in many cases can be considered colloidal systems). Rheological properties (flow properties) are also considered in the following chapter. As you may gather, the distinction between colloid properties and polymer properties is arbitrary in many cases.

Considerable time will be devoted to a discussion of electrical phenomena in colloidal systems, since many of the unique properties of colloids (other than the properties discussed in the chapter on polymers) are the result of electrical charges on the surface of the colloid. The properties of colloids that depend on both motion and charge are called electrokinetic properties. Water purification, chemical separation, and mineral concentration can be carried out using electrokinetic phenomena. Kidney machines purify blood by using electrodialysis, an electrokinetic phenomena. Units in the section on electrical properties are, as usual, troublesome. However, most problem areas should be resolved by referring to Sec. 14-2, where a detailed unit analysis of electrical units is given. The excellent text *Introduction to Colloid Chemistry*, by K. J. Mysels (see Sec. 11-1.8, Ref. 1), is recommended for continued study. It is an introduction that will allow intelligent reading in other, more specialized, books on colloids.

11-1 DEFINITIONS AND MOLECULAR WEIGHT METHODS

OBJECTIVES

You shall be able to

1. Define or briefly discuss each of the following terms in relation to colloids: lyophobic (hydrophobic), lyophilic (hydrophilic), polydispersity, shape and flexibility, hydration (solvation), flocculation, deflocculation,

diuternal, caducous, number average and weight average molecular weight, interparticle forces

2. List and give an example of each class of colloids
3. Apply sedimentation, diffusion, and centrifugation equations

11-1.1 COMPARISON OF COLLOIDS TO OTHER CHEMICAL SYSTEMS

Perhaps the best way to introduce you to the subject of colloids is to compare many of the characteristics of colloids to those of "normal" chemical compounds. This is done in Table 11-1.1.1. Of course, the entries in the table are generalizations and will not fit every specific case. It should be obvious from the table that colloids are larger and have a much greater variability than ordinary chemical compounds (see also Fig. 11-1.1.1). We will investigate many of these properties in more detail as we go along. We will see that the charges on colloidal particles have a very marked effect on their properties. In fact, the whole area of electrokinetic phenomena (Sec. 11-2) depends on the charged surface of the colloid particles. The molecular weights of colloid particles vary considerably even within one type of system. This means we cannot define a molecular weight for a colloid (in general) as we do for a regular chemical compound which has a fixed stoichiometry. Instead we have to define some type of average as done in the following section.

11-1.2 POLYDISPERSITY AND MOLECULAR WEIGHT AVERAGES

In most colloidal systems we find there are particles with different weights. However, there are some systems in which all the particles have the same weight. These systems are called *monodisperse*. Those colloids that have a range of molecular weights are termed *polydisperse*. The concept of molecular weight offers no problem for a monodisperse colloid, but we cannot define a single molecular weight for a

Table 11-1.1.1
Comparison of the Characteristics of Colloids and "Normal" Chemical Compounds[a]

Characteristic	Colloids	"Normal" chemicals
Size	1, $10-10^3$ nm	0.1-1 nm
Weight	$10-10^3$ kg mol^{-1} or more (10^4-10^6 g mol^{-1})	0.001-1 kg mol^{-1}
Charge	Hundreds or thousands per particle	0, 1, 2, or 3, per particle
Shape	Spheres, filaments, platelets; very variable	Almost fixed; limited variability
Flexibility	Filaments very flexible; others less so	Very little
Composition	Usually variable	Fixed

[a] nm = nanometer = 10^{-9} m.

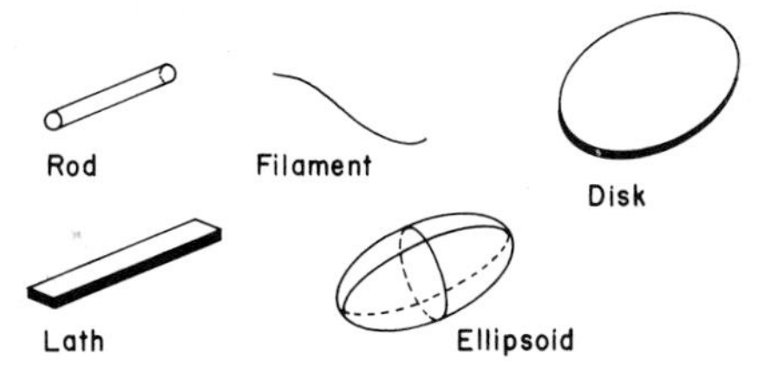

FIG. 11-1.1.1 The relative sizes and idealized shapes of some typical colloidal systems.

polydisperse system. Instead we define two *average molecular weights*. These two are the *number averages* and the *weight average*. The reason for having two different averages is that some colloidal properties depend on the number average and others depend on the weight average, as we will see below. The number average is defined as

$$M_n = \frac{\sum_i w_i}{\sum_i n_i} = \frac{\sum_i m_i n_i}{\sum_i n_i} \tag{11-1.2.1}$$

where m_i is the mass of each particle of fraction i, n_i is the number of particles in fraction i, and w_i is, then, the total mass of fraction i. The weight average is given by

$$M_w = \frac{\sum_i m_i w_i}{\sum_i w_i} = \frac{\sum_i m_i m_i n_i}{\sum_i n_i m_i} = \frac{\sum_i n_i m_i^2}{\sum_i n_i m_i} \tag{11-1.2.2}$$

The symbols have the same meaning here as in the previous equation. The major difference in these two definitions is that the weight average emphasizes the heavier particles more than the number average.

Experiments that use colloids for measuring colligative properties (see Chap. 10) produce results that are dependent on the number average molecular weight. Some other experiments especially light scattering, depend on the weight average molecular weight.

11-1.2.1 EXAMPLE

Determine the number average and the weight average of a system containing the following particles:

5 particles with m = 1

3 particles with m = 2

3 particles with m = 3

1 particle with m = 4

Solution

$$M_n = \frac{\sum_i m_i n_i}{\sum_i n_i}$$

$$= \frac{(1)(5) + (2)(3) + (3)(3) + (1)(4)}{5 + 3 + 3 + 1}$$

$$= \frac{5 + 6 + 9 + 4}{12} = \frac{24}{12} = \underline{\underline{2}}$$

$$M_w = \frac{(5)(1)^2 + (3)(2)^2 + (3)(3)^2 + (1)(4)^2}{(5)(1) + (3)(2) + (3)(3) + (1)(4)}$$

$$= \frac{5 + 12 + 27 + 16}{24} = \frac{60}{24} = \underline{\underline{2.5}}$$

As you can see the weight average is greater than the number average. This is always the case except in monodisperse systems in which two averages are the same.

11-1.3 MORE TERMS AND DEFINITIONS

As indicated in Table 11-1.1.1 and Fig. 11-1.1.1, the shapes of colloids can vary from spheres to platelets to filaments to rods. Only a few colloids can be characterized as spherical. This leads to difficulty in some experiments, such as the rate of sedimentation, in which we need to know the shape in order to interpret the results. In cases where the shape of the colloid is unknown, a shape factor (fudge factor) is included in the equations used to calculate colloid properties.

Spherical and ellipsoidal particles are assumed to be rigid. However, those colloids that form filaments, and to a lesser extent those that form rods, are far from rigid. Filamentous materials such as polymers (long-chain polymers) are very flexible. If the filament is very flexible, the colloid tends to assume a random coil arrangement due to random motion of the different segments of the filament. In many cases this random coil can be "wound up" so tightly that the filament becomes almost spherical. The point is that we can observe everything from "stringy" colloids to spheres and, in general, it is impossible to predict the shape of a colloid just by looking at its makeup. In fact, a given colloid will behave differently in different solvents. In a solvent which is compatible with the colloid, the colloid will tend to expand and "unwind" (if it is a filament), whereas in a poor solvent (one which is not attracted to the colloid) the particle will tend to form as compact a shape as possible. So a colloid might behave as a filament in one solvent and as a sphere in another solvent.

This brings us to a discussion of a few more terms. In many colloids the surface is covered with a tightly bound layer of solvent. The colloid is said to be solvated in this case. In addition to the tightly bound layer of solvent (if the solvent is water we speak of hydration) there may very well be several layers of solvent that are loosely bound to the surface. Or the solvent may merely be trapped in the crevices formed in the colloid. In a colloidal suspension, this solvent will move with the colloid and will affect the properties of the colloid. The observed volume of a colloid will be larger than if there were no solvent associated with it. This, of course, results in an observed density that is smaller than for the unsolvated colloid. (Solvation strictly applies only to the solvent bound to the colloids, not that which is merely trapped within the colloid. It is rather difficult, at times, to distinguish the two.)

There are terms that are used to indicate the degree of interaction between a solvent and a colloid. A colloid is said to be *lyophilic* ("solvent loving") if there is a tendency for the solvent to interact strongly with the colloid. Conversely, if there is no tendency for the two to interact, the colloid is said to be *lyophobic* ("solvent hating"). The terms used if the solvent is water are *hydrophilic* and *hydrophobic*. You should be aware of the fact that the distinction offered by the above two definitions is not always a sharp one. In many colloids, one portion may be lyophilic while another is lyophobic. (See the discussion of surfactants in Sec. 11-3.)

Whether a particular colloid is lyophilic or lyophobic will affect its stability in a suspension. In general, a lyophilic colloid is stable toward settling out from suspension (*flocculation*) while a lyophobic one is not. In many cases colloids can be thought of as *aggregates*. That is, they are formed by the union of smaller units to form the colloid. An *aggregate* is then defined as a particle formed from smaller units that are held together by interparticle attractions. Whether a colloid will *flocculate* depends on the forces operating between the individual colloid particles. We can distinguish several types of colloids on the basis of their tendency to flocculate. Before doing that it would be a good idea to define flocculation. *Flocculation* is the process of colloidal particles coming together, "sticking", and forming larger particles (or larger aggregates). *Deflocculation*, on the other hand, is the process of breaking up large aggregates into smaller ones which then become suspended. As we shall see a little later in this section, the degree of aggregation (or size) of a colloid affects its stability with respect to sedimentation (settling). If a colloid flocculates, it forms larger particles which will tend to settle much more rapidly than the smaller ones. This leads to the following classification. A *stable* colloid is one which does not flocculate over any finite period of time. The system will be termed *metastable* if it has a tendency to flocculate but cannot do so without some external assistance (such as stirring or heating). If a colloidal system flocculates over a period of time we call it *caducous*, and if a system appears to be stable over a period of time, then flocculates slowly, the term *diuternal* is used.

Whether a colloid is classified in a particular category will depend on a number of factors such as interparticle forces, the nature of the suspending medium, and whether there are any chemical agents present that serve to stabilize or destabilize the system. The principal forces between particles can be classified as coulombic and van der Waals. Many colloidal systems are made up of particles, all of which take on the same charge (not the same magnitude of charge, which is very variable, but the same sign of charge). Coulombic repulsion will tend to keep these particles separated, and such a system will tend to be stable with respect to flocculation. However, if the particles can get sufficiently close together the van der Waals attraction may become large enough to cause the particles to flocculate. The solvent plays an important role, also. If a colloid is in a solvent in which it can be termed lyophilic, the solvent will be attracted to the surface of the colloid and will tend to keep it from sticking to another particle, thereby preventing flocculation. Thus, in solvents in which a colloid is lyophilic, it is also stable. Conversely, a lyophobic colloid will tend to flocculate if there are no other forces (such as surface charge) to keep it from doing so. Lyophobic colloids are then classified as unstable. It is observed that some colloids can be made to flocculate by gentle stirring, whereas others require heat. Some require the addition of electrolyte to facilitate the flocculation process. And, of course, there are some that will not flocculate at all. Perhaps you can offer some explanation for these observations. Some are given if you cannot think of any.

If the forces of repulsion between two colloids are not very large, it will not take a great deal of energy to bring the particles in close enough contact for the van der Waals forces to bind the particles together. Thus, gentle stirring is sufficient to force some colloid particles together so they can stick. Heating has a similar effect. As the temperature rises, the kinetic energy of the particles increases, and collisions between particles become harder. If the repulsion forces are not too large, these collisions will be sufficient to cause flocculation. If a colloidal system consists of particles that are charged, the repulsion between particles will be large, and too much energy will be required for the previous two processes to bring the particles into contact. The addition of an electrolyte will, in many cases, neutralize the surface charge and allow the particles to approach

Table 11-1.4.1
Types of Colloidal Systems

Dispersed phase	Dispersion medium	Name	Examples
Liquid	Gas	Liquid aerosol	Fog, liquid sprays
Solid	Gas	Solid aerosol	Smoke, dust
Gas	Liquid	Foam	Foam on beer, froth on soap solutions, whipped cream
Liquid	Liquid	Emulsion	Milk, mayonnaise, paint
Solid	Liquid	Sol, colloidal suspension, or paste	AgI sol, paint, drilling mud, toothpaste
Gas	Solid	Solid foam	Expanded polystyrene and polyurethane
Liquid	Solid	Gel (solid emulsion)	Opal, pearl, jellies
Solid	Solid	Solid suspension	Pigmented plastics

each other. More will be said about this in Sec. 11-2. It may have occurred to you by now that you have not been given a good definition of a colloid. The next section remedies this oversight.

11-1.4 TYPES OF COLLOIDS AND EXAMPLES

Colloids consist of a *dispersed phase* (the colloid particle itself, if particles are present) and a *dispersion medium* (the phase in which the particles are distributed). Table 11-1.4.1 lists the common classification of colloids.

11-1.5 SEDIMENTATION, CENTRIFUGATION, AND DIFFUSION

SEDIMENTATION. Any particle in any medium will be subject to a number of forces. There is the force of gravity which tends to pull the particle to the bottom of the container. The buoyant force of the surrounding medium tends to force the particle to the top of the container. Whether a particle will settle depends on the relative magnitude of these two forces. Brownian motion (the random motion of small particles in a suspension due to the constant bombardment of the particles by molecules of the dispersing medium) is another important process which tends to keep a dispersed phase from settling out. We mentioned Brownian motion briefly in Chap. 2; this motion leads to diffusion as we saw in Chap. 2 (Sec. 2-5). Diffusion is another process that will tend to counteract any tendency of the colloidal system to settle. Diffusion will be considered in more detail shortly.

The gravitational force on a particle will be designated by f_g. And the buoyant force due to the medium will be labeled f_b. Let us define the particle volume as v and the particle density as ρ_2. Finally we will call the solution density ρ. The net driving force on a suspended particle will be

$$f_d = f_g - f_b \tag{11-1.5.1}$$

If f_d is greater than zero the particle will settle; if it is less than zero the particle will rise to the surface; and if f_d is equal to zero, there will be no tendency to rise or settle. The gravitational force will be given by

$$f_g = v\rho_2 g \tag{11-1.5.2}$$

where g is the acceleration due to gravity. The term $v\rho_2$, gives the mass of the particle. However, in a suspension we are not concerned with the mass of the particle. Rather, we need to know the effective mass which will take into account the buoying effect of the medium. The effective mass will be

$$m_{eff} = v(\rho_2 - \rho) \tag{11-1.5.3}$$

The net driving force is therefore

$$f_d = m_{eff}g = v(\rho_2 - \rho)g \tag{11-1.5.4}$$

You should be able to deduce the results (will the particle sink or rise?) if $\rho_2 > \rho$.

Useful information about the size of a colloidal particle can be gained in many cases by observing the sedimentation behavior of the particles in a particular dispersing medium. If the particle is not settling too rapidly, Stokes law will hold, and we can write the frictional force of the surrounding fluid on the particle as

$$-f_r = \phi u \tag{11-1.5.5}$$

where u is the velocity of the particle relative to the surrounding fluid and ϕ is a frictional or drag coefficient. The negative sign indicates the frictional resistance is in opposition to the direction of motion of the particle. ϕ will depend on the shape of the particle, which is something we do not know in general. We can, however, assume a spherical shape and find an effective radius of the particle as shown below. (There are methods to treat other shapes, but they are outside the scope of our discussion, so we will have to be content to examine only the spherical case.)

The terminal velocity (the constant velocity reached by a particle after it has been in a fluid for some time) will depend on the net driving force f_d and the frictional force f_r due to the medium. When these two forces are balanced, we will have a steady state system, and the particle will be rising or falling with a constant velocity. Thus, when the steady state is reached,

$$f_d = - f_r \tag{11-1.5.6}$$

For a sphere, Stokes law gives the frictional drag as

$$- f_r = 6\pi\eta r u \tag{11-1.5.7}$$

in which η is the coefficient of viscosity of the medium, r is the effective radius of the particle (assuming a spherical shape), and u is the terminal velocity.

Equating Eq. (11-1.5.4) and Eq. (11-1.5.7) yields

$$v(\rho_2 - \rho)g = 6\pi\eta r u \tag{11-1.5.8}$$

But the volume of a sphere is $4\pi r^3/3$, which gives

$$\frac{4}{3}\pi r^3(\rho_2 - \rho)g = 6\pi\eta r u \tag{11-1.5.9}$$

This may be arranged in two different ways (do it) to give

$$u = \frac{2}{9\eta} r^2 (\rho_2 - \rho)g \tag{11-1.5.10}$$

or

$$\boxed{r = \left[\frac{9\eta u}{2(\rho_2 - \rho)g}\right]^{1/2}} \tag{11-1.5.11}$$

The conditions under which we can use these last two equations are: spheres only and very low sedimentation rates. Any deviation from spherical shape will lead to a ϕ that is larger than we have used. Also, if the particle is hydrated (solvated) appreciably the value of ϕ will be larger. Even with these restrictions, the development is useful since it gives us at least an estimate of particle size that we would not otherwise have.

11-1.5.1 EXAMPLE

Find the terminal velocity of a particle with a radius of $r = 1.0 \times 10^{-5}$ m and a density of $\rho_2 = 1500$ kg m^{-3} in a fluid with a density of $\rho = 1000$ kg m^{-3}. The acceleration due to gravity can be taken as $g = 9.8$ m s^{-2}. The viscosity of the fluid is $\eta = 8.9 \times 10^{-4}$ kg m^{-1} s^{-1}.

Solution

From Eq. (11-1.5.10):

$$u = \frac{(2)(10^{-5}\ \text{m})^2(1500 - 1000\ \text{kg m}^{-3})(9.8\ \text{m s}^{-2})}{(9)(8.9 \times 10^{-4}\ \text{kg m}^{-1}\ \text{s}^{-1})}$$

$$u = 1.2 \times 10^{-4} \frac{m^2 \text{ kg m}^{-3} \text{ m s}^{-2}}{\text{kg m}^{-1} \text{ s}^{-1}}$$

$$u = \underline{\underline{1.2 \times 10^{-4} \text{ m s}^{-1}}}$$

CENTRIFUGATION. Sedimentation for colloids is usually a very slow process. The use of a centrifuge can greatly speed up the process by increasing the force on the particle far above that due to gravitation alone. Ultracentrifuges available today can produce accelerations on the order of 10^6 g. This, of course, allows the determination of effective radii of colloids much more rapidly than previously. Figure 11-1.5.1 gives a diagram of an ultracentrifuge and a concentration profile that results from centrifugation. The profiles are usually generated optically, generally using a Schlieren system. Figure 11-1.5.2 shows a schematic of the optics associated with such a system. The method gives a direct trace of the refractive index gradient in the cell. Since the refractive index depends on the concentration

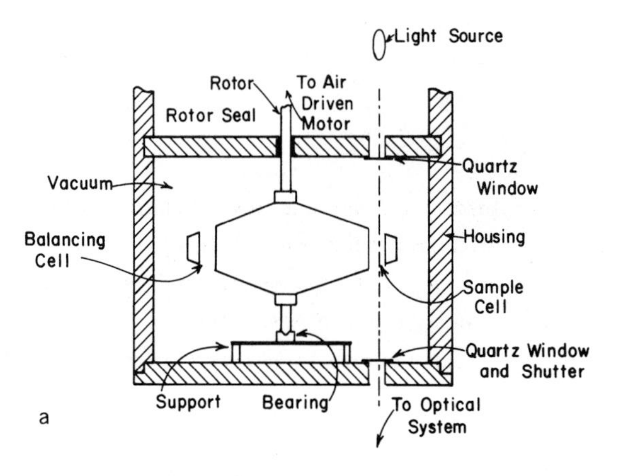

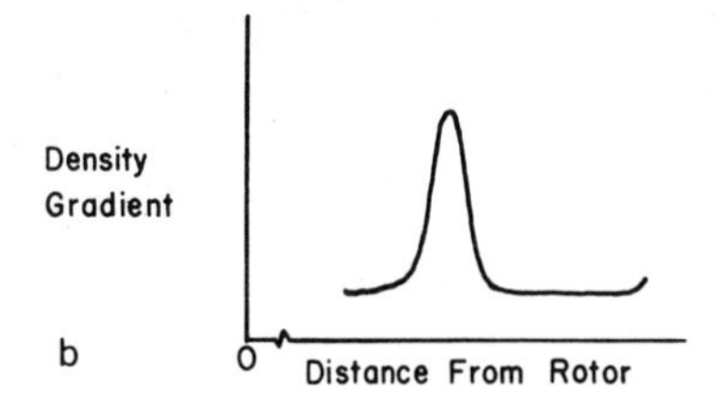

FIG. 11-1.5.1 (a) Schematic and (b) concentration profile of an ultracentrifuge. [From A. Weissberger, ed., *Techniques of Organic Chemistry*, Vol. I (*Physical Methods*), Part II, 3rd ed., Interscience, New York, 1959 (see Sec. 11-1.8, Ref. 3).]

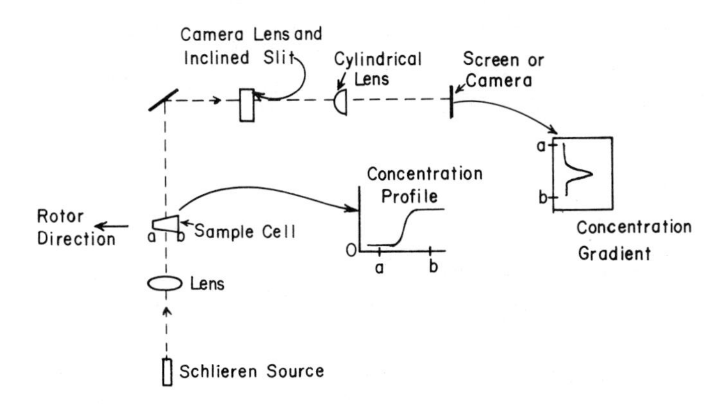

FIG. 11-1.5.2 Sketch of the optical system in an ultracentrifuge. [From A. Weissberger, ed., *Techniques of Organic Chemistry*, Vol. I (*Physical Methods*), Part II, 3rd ed., Interscience, New York, 1959, p. 1036 (see Sec. 11-1.8, Ref. 3).]

of the solution, this method allows almost instantaneous determination of the sedimenting boundary. Instrumentally, all centrifuges require vibrationless, convection-free operation and a high-grade optical system to produce sharp pictures. The highest speeds available are about 75,000 rpm. Lower speeds (up to about 20,000 rpm) are obtained with a direct drive arrangement and produce accelerations up to about 20,000 g. The pioneering work in ultracentrifuges has been done by T. Svedberg in Uppsala, Sweden. He has produced more technical improvements than any other individual. He has also pioneered the work on ultracentrifugal analysis of proteins.

To use data from ultracentrifuge measurements we merely have to replace g in our previous equations by the acceleration due to the rotation. A centrifuge with an angular velocity of ω (in radians per second) will generate an acceleration given by $\omega^2 x$, where x is the distance in the centrifuge from the center of rotation. Our sedimentation equation becomes

$$f_d = v(\rho_2 - \rho)\omega^2 x \qquad (11\text{-}1.5.12)$$

As before $f_d = -f_r = \phi u$, so we can write

$$\phi u = v(\rho_2 - \rho)\omega^2 x \qquad (11\text{-}1.5.13)$$

If we know ϕ, which we seldom do, we can find u readily.

A couple of substitutions will get this equation in a form that will be useful shortly. Define the *partial specific volume* of the particle as

$$\bar{v} = \frac{v}{m} \qquad (11\text{-}1.5.14)$$

The units on this should be obvious. In fact, $\bar{v}$ is the reciprocal of the density of the particle ρ_2. You should be able to justify that $v\rho_2 = m$, the mass of the particle. Using these definitions we can write

$$(v\rho_2 - v\rho)\omega^2 x = (m - m\bar{v}\rho)\omega^2 x$$

$$= (1 - \bar{v}\rho)m\omega^2 x \qquad (11\text{-}1.5.15)$$

Now, define a *sedimentation constant* s as follows

$$s = \frac{u}{\omega^2 x} \qquad (11\text{-}1.5.16)$$

This constant is the sedimentation rate for a unit centrifugal acceleration and is convenient for comparing different substances. We are still no better off, even though we can rewrite Eq. (11-1.5.13) in a more compact form.

$$\phi u = (1 - \bar{v}\rho)m\omega^2 x \qquad (11\text{-}1.5.17)$$

or

$$\phi = \frac{(1 - \bar{v}\rho)m}{s} \qquad (11\text{-}1.5.18)$$

We need some other relationship that will allow us to eliminate the dependence on ϕ, since we cannot determine ϕ in most cases. The other relationship we need comes from Sec. 2-5.3 in which the diffusion coefficient is defined. *You should reread that section to refresh your memory.* You may find it makes a little more sense the second time through.

DIFFUSION. Einstein, in his work on diffusion processes, developed the following relationship which is presented without proof.

$$\phi = \frac{RT}{ND} \qquad (11\text{-}1.5.19)$$

R is the gas constant, T is the absolute temperature, N is Avogadro's number, and D is the diffusion coefficient defined in Sec. 2-5.3. This gives us two equations for ϕ which allows the elimination of that quantity. Equating Eq. (11-1.5.19) and (11-1.5.18) gives the following:

$$\frac{RT}{ND} = \frac{(1 - \bar{v}\rho)m}{s} \qquad (11\text{-}1.5.20)$$

$$\boxed{M = Nm = \frac{RTs}{D(1 - \bar{v}\rho)}} \qquad (11\text{-}1.5.21)$$

where M is the molecular weight of the particles in the system. This last equation allows us to determine the molecular weights of certain systems if s, $\bar{v}$, and D are

Table 11-1.5.1
Constants for Some Proteins at 20°C (in H_2O)

Protein	$\bar{v}$ (m^3 kg^{-1}) × 10^3	s (10^{-13} s)	D (m^2 s^{-1}) × 10^{11}
Myoglobin (beef)	0.741	2.04	11.3
Hemoglobin (human)	0.749	4.48	6.9
Serum Albumin (human)	0.736	4.67	5.9
Pepsin (pig)	0.750	3.3	9.0
Insulin (beef)	0.749	3.58	7.53
Tobacco Mosaic (virus)	0.73	185	0.53

known from other experiments. They are known for several substances. Both D and s can be found experimentally, and $\bar{v}$ can be estimated. Table 11-1.5.1 lists values for some macromolecules. The following example will indicate how to use the data in Table 11-1.5.1.

11-1.5.2 EXAMPLE

Find the molecular weight of a tobacco mosaic virus from the data found in Table 11-1.5.1; ρ_{H_2O} (20°C) = 998.2 kg m^{-3}.

Solution

All we have to do for the solution to this problem is apply Eq. (11-1.5.21):

$$M = \frac{RTs}{D(1 - \bar{v}\rho)}$$

$$= \frac{(8.314 \text{ kg m}^2 \text{ s}^{-2} \text{ mol}^{-1} \text{ K}^{-1})(293 \text{ K})(185 \times 10^{-13} \text{ s})}{(0.53 \times 10^{-11} \text{ m}^2 \text{ s}^{-1})[1 - (0.73 \times 10^{-3} \text{ m}^3 \text{ kg}^{-1})(998.2 \text{ kg m}^{-3})]}$$

$$= \underline{\underline{3.1 \times 10^4 \text{ kg mol}^{-1}}}$$

There is one other very useful relationship we can write that involves the equilibrium between sedimentation (in a centrifuge) and diffusion. This equation will give another method for finding the molecular weight of a set of colloidal particles. In a centrifuge we will have an equilibrium situation when the rate of diffusion of the particles just balances the rate of sedimentation due to the centrifugal force. The sedimentation rate is given by

$$\text{Rate}_s = Cu = \frac{C\omega^2 xm(1 - \bar{v}\rho)}{\phi} \qquad (11\text{-}1.5.23)$$

where C is the concentration of particles. The rate of diffusion is [derived from equations in Sec. 2-5.3 and Eq. (11-1.5.19)]

$$\text{Rate}_d = \frac{kT}{\phi}\frac{dC}{dx} \qquad (11\text{-}1.5.23)$$

where dC/dx is a concentration gradient along a particular direction; in this case the direction is defined from the rotor of the centrifuge. Equating these two rates and rearranging allows us to eliminate the troublesome ϕ and obtain

$$\frac{dC}{dc} = \frac{C\omega^2 xm(1 - \bar{v}\rho)\,dx}{kT}$$

$$= \frac{C\omega^2 xM(1 - \bar{v}\rho)\,dx}{RT} \qquad (11\text{-}1.5.24)$$

we now integrate between two points along x (between two distances from the centrifuge rotor) to see how the concentration varies.

$$\int_{C_1}^{C_2} \frac{dC}{C} = \int_{x_1}^{x_2} \frac{\omega^2 M(1 - \bar{v}\rho)x\,dx}{RT} \qquad (11\text{-}1.5.25)$$

This is easily done to find

$$\ln\frac{C_2}{C_1} = \frac{\omega^2 M(1 - \bar{v}\rho)}{RT}\,\frac{x_2^2 - x_1^2}{\cdot 2} \qquad (11\text{-}1.5.26)$$

Finally, this can be rearranged into a form useful for finding the molecular weight.

$$\boxed{M = \frac{2RT\ln(C_2/C_1)}{(1 - \bar{v}\rho)\omega^2(x_2^2 - x_1^2)}} \qquad (11\text{-}1.5.27)$$

It has been found that Eq. (11-1.5.27) gives a weight average molecular weight for a polydisperse system.

Practical applications of the ultracentrifuge involve the determination of the molecular weights of natural and synthetic colloids (e.g., proteins and polymers) and the distribution of molecular weights in a polydisperse system. Space does not allow further detail. However, an exhaustive treatment can be found in Weissberger (see Sec. 11-1.8, Ref 3).

11-1.5.3 EXAMPLE

An aqueous suspension of a protein was centrifuged to equilibrium at 370 radians s^{-1} at 300 K. The following equilibrium concentrations were observed in the centrifuge tube:

Distance from axis of rotation (cm)	4.90	4.95	5.00	5.05	5.10	5.15
Concentration ($g\ dm^{-3}$)	1.30	1.46	1.64	1.84	2.06	2.31

The specific volume $\bar{v}$ is 0.75 $cm^3\ g^{-1}$, and the solution density can be taken as 1.00 $g\ cm^{-3}$. Calculate the molecular weight of the material.

Solution

Choose any two distances and corresponding concentrations.

$$M = \frac{2RT \ln (C_2/C_1)}{\omega^2(1 - \bar{v}\rho)(x_2^2 - x_1^2)}$$

$$= \frac{(2)(8.31\ kg\ m^2\ s^{-2}\ mol^{-1}\ K^{-1})(300\ K)\ \ln (2.06/1.46)}{(370\ s^{-1})^2[1 - (0.75\ cm^3\ g^{-1})(1\ g\ cm^{-3})][(5.1)^2 - (4.95)^2]cm^2}$$

$$= (0.0333\ kg\ m^2\ cm^{-2}\ mol^{-1})\ \frac{10^2\ cm^2}{m^2}$$

$$= \underline{\underline{3.33\ kg\ mol^{-1}}}$$

To verify this you should check some other points.

11-1.6 PROBLEMS

1. (a) What is the weight of a gold particle (ρ = 19.3 $kg\ dm^{-3}$) having a radius of 10 nm (nanometer)? (b) What is the weight of 1 mol of these particles? (c) From (a) and (b) find the degree of aggregation (the number of gold atoms per particle).
2. Assume we have a spherical particle which settles from a suspension at a rate of 0.015 $m\ s^{-1}$. What will be the rate of sedimentation for another spherical particle of the same composition but having a radius that is 10 times as large?
3. Find the weight average and the number average molecular weights for the following system of particles.

Number of particles	4	5	3	2	2
Particle weight (mass)	5	6	7	8	10

4. In an experiment to find the molecular weight for a colloid using an ultracentrifuge, the following data were observed for particles with $\bar{v} = 0.92\ dm^3\ kg^{-1}$ in water at 293 K ($\rho = 0.998\ kg\ dm^3$), for a centrifuge velocity of 10^4 radians s^{-1}. Determine the molecular weight of this colloid. $x_1 = 0.10$ m, $x_2 = 0.11$ m, $C_1 = 0.0010\ kg\ dm^{-3}$, $C_2 = 10.0\ kg\ dm^{-3}$.

5. Find the radius of a particle with $\rho_2 = 1350\ kg\ m^{-3}$ in a fluid of $\rho = 1037$ kg m^{-3} if the terminal velocity is $2.72 \times 10^{-4}\ m\ s^{-1}$ ($\eta = 1.2 \times 10^{-3}\ kg\ m^{-1}\ s^{-1}$).

6. The diffusion coefficient of a particular colloid at 291 K is $1.27 \times 10^{-13} m^2\ s^{-1}$. If $\bar{v} = 0.736 \times 10^{-3}\ m^3\ kg^{-1}$, $M = 4.21 \times 10^4\ kg\ mol^{-1}$, and $\rho = 1.071\ kg\ dm^{-3}$, determine the sedimentation constant.

11-1.7 SELF-TEST

1. Without looking at the answers, list five types of colloids, an example of each, and the dispersed phase and dispersing medium for each.

2. Write concise definitions for at least six terms listed in the objectives.

3. By referring to the discussion of intermolecular forces in Sec. 14-1 and the discussion in Sec. 11-1.3, write a one-page description of the forces that might be present in a suspension of AgI particles in an aqueous medium. Assume the AgI particles have positive surface charges. If we neutralize the surface charge in some way, will the AgI sol be more or less stable with respect to flocculation than the original charged sol?

4. Horse hemoglobin in a water suspension at 20°C is found to have a diffusion coefficient of $D = 6.3 \times 10^{-7}\ cm^2\ s^{-1}$ and a sedimentation coefficient of $s = 4.41 \times 10^{-13}$ s. If $\rho = 0.9982\ kg\ dm^{-3}$ and $\bar{v} = 0.749\ dm^3\ kg^{-1}$, find the molecular weight of the horse hemoglobin.

5. Use the results of Sec. 2-5.3 to find the mean time required for an insulin molecule to diffuse across a living cell (10^{-6} m) at 20°C if the diffusion coefficient of insulin is $8.2 \times 10^{-11}\ m^2\ s^{-1}$.

11-1.8 ADDITIONAL READING

1.* K. J. Mysels, *Introduction to Colloid Chemistry*, Interscience, New York, 1967.

2.* W. J. Moore, *Physical Chemistry*, 4th ed., Prentice-Hall, Englewood Cliffs, N. J., 1972.

3.* J. R. Nichols and E. D. Bailey, "Determinations by the Ultra Centrifuge," in A. Weissberger, ed., *Techniques of Organic Chemistry*, Vol. I (*Physical Methods*), Part II, 3rd ed., Interscience, New York, 1959, pp. 1007-1138.

11-2 ELECTROKINETIC PHENOMENA

OBJECTIVES

You shall be able to

1. Discuss three ways in which charges can be formed on a colloid surface
2. Discuss the double-layer model for colloids, including the definitions of such terms as diffuse double layer, Stern layer, counterions, and semiliions
3. Discuss in detail any of the electrokinetic phenomena mentioned in this section
4. Apply equations related to electrokinetic phenomena
5. Discuss the meaning of the terms sensitizing and desensitizing of colloids and indicate how this can be done

11-2.1 INTRODUCTION

After the introduction to some of the terms that apply to colloids in the previous section, you are ready to delve into some of the properties of colloids. Many of the unique properties of colloids come from the fact that they have a charged surface. In contrast to "ordinary" chemicals which may have up to three or four charges, a colloidal particle may have hundreds. This highly charged surface will have a great influence on the behavior of the particles in a suspension. This is especially true if an electric field is present.

Some space in this section will be devoted to detailing some of the properties of the surface of colloidal particles, then a discussion is given of some of the phenomena that are unique to colloids (electrokinetic phenomena). There will be a short section giving (with little in the way of a rigorous derivation) equations that can be used to calculate these electrokinetic properties. Finally, a few comments will be made on practical aspects of electrokinetic phenomena.

In most of the following discussion sols, dispersion of a solid in a liquid phase will be considered. The electrokinetic phenomena to be discussed can only take place if the colloid is dispersed in a phase in which it is free to move under some driving force. Emulsions could also be considered, since the dispersion medium is a liquid in which the dispersed phase can freely move. The emphasis on sols and emulsions does not in any way lessen the significance of the other types of colloids. The limitations of producing a text of finite length requires that many interesting and important topics be either omitted or touched on only very briefly. You should be able to pick up additional information necessary when the need arises, given the background provided in this and the previous section. Remember that the methods for finding molecular weights of colloids (except for the sedimentation method presented in the previous section) will not be discussed in this chapter. These are found in the following chapter on Polymers.

11-2.2 THE CHARGED SURFACE

FORMATION OF THE CHARGED SURFACE. A colloid is not usually charged when it is formed (or dispersed). Instead, it acquires a surface charge by one of the three processes described below. One method for a surface to become charged is illustrated by a AgI colloid as shown in Fig. 11-2.2.1. When colloidal AgI is placed in a solution containing HI, the AgI preferentially adsorbs the I^- ions to become negatively charged. On the other hand, if AgI is placed in a solution of $AgNO_3$, Ag^+ ions are adsorbed leading to a positive surface charge. Thus, *preferential adsorption of ions* of a particular charge from solution is one way of obtaining a charged surface. Shortly, discussion of the *double layer,* the fate of the excess ions of the opposite charge that are not adsorbed, will be discussed. Other examples are: the adsorption of H^+ ions from a solution of H_2S by an arsenic trisulfide sol; the strong affinity of colloidal gold for OH^- or Cl^- leading to negatively charged surfaces.

A second method of obtaining a charge is illustrated by some proteins which have carboxyl (-COOH) or amine ($-NH_2$) groups. If a colloidal protein has -COOH groups in its structure, these will ionize at high pH to give $-COO^-$ groups in the particle which will produce a net surface charge. In a similar manner, if $-NH_2$ groups are present at low pH where an excess of H^+ ions are present, the $-NH_2$ groups can react with the H^+ ions to form $-NH_3^+$ which will give the colloid a charge. In each case there is a chemical reaction that leads to the observed surface charge. Thus, *chemical reactions or ionization* can produce charged particles. An interesting

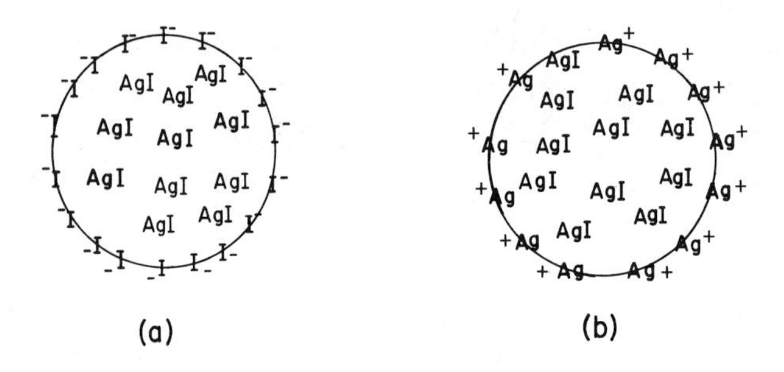

FIG. 11-2.2.1 An AgI colloid that has become (a) negatively or (b) positively charged by adsorption of ions.

example is presented by proteins that have both amine and carboxyl groups present in their structures. At low pH the $-NH_2$ groups are present as $-NH_3^+$, while the carboxyl groups remain un-ionized. At intermediate pH neither the amine nor the carboxyl are charged (there are cases in which self-ionization takes place giving $-NH_3^+$ and $-COO^-$ groups both present on the same molecule; such an ion is given the name "Zwitterion".) At high pH where an excess of OH^- ions are present the amine is uncharged and the carboxyl is present as $-COO^-$. Other important examples of the type of system that ionizes to form surface charges are the zeolites (naturally occurring silicates with SiO_2^- groups exposed) and synthetic cross-linked materials with attached ionizing groups. These two types of materials are used for ion exchange resins in which one ion in a solution in contact with the resin can be replaced by the ion already present in the resin loosely attached to the ionizing group. A further discussion of ion exchange resins and properties is beyond the scope of this section. These are immensely important materials in many industrial and pollution control processes.

The final manner in which a colloid surface can become charged is by adsorption of solvent or other polar material from the solution. For example, let us assume we have a material dispersed in an aqueous medium. Also, assume the material adsorbs some of the solvent due to an attraction for the positive end of the water molecule (this will be the hydrogen ends since they are partially positively charged due to the high electronegativity of the oxygen which is partially negatively charged). Figure 11-2.2.2 will help you visualize this process. The preferential adsorption of the positive end of the water molecule leads to a colloid that has no net charge. However, the surface appears to be negatively charged, since the oxygen atoms of the water are surrounding the surface of the colloid. In Fig. 11-2.2.2(b) we see an oil droplet that has adsorbed a long-chain organic acid. The aliphatic (carbon chain) portion of the acid is more compatible with the oil in the droplet than is the polar carboxyl group, thus the oil will tend to adsorb the carbon chain and leave the -COOH group dangling in the dispersing medium. We now have a situation similar to

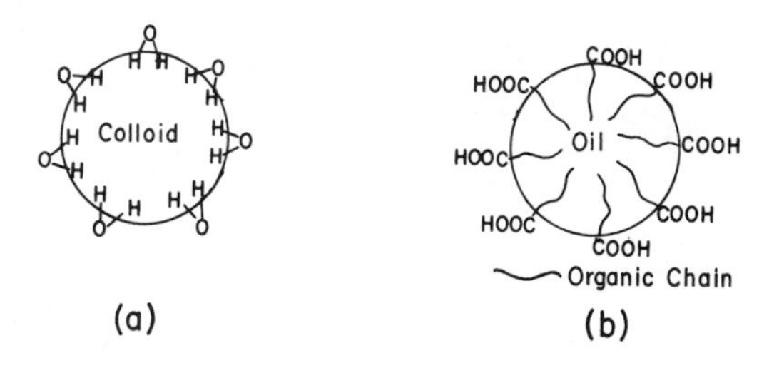

FIG. 11-2.2.2 (a) Preferential adsorption of the positive portion of the water molecule by a colloid, and (b) adsorption of the organic portion of a long-chain organic acid by an oil droplet in an emulsion.

that of the proteins described in the previous section. At high pH the carboxyl groups will ionize to form $-COO^-$, which leads to a negative charge on the surface of the colloid. This particular example is important, and you will see it again in the section on surface chemistry. Molecules that have one end soluble in nonpolar materials and the other end soluble in polar meterials are important in the preparation of cleaning agents (soaps and detergents).

THE DOUBLE LAYER. It should be clear that if we have a colloid that has gained a charged surface by some process, this charge will have an effect on the distribution of ions near that surface. If a surface is negatively charged, then there will be a tendency for the positive ions in the dispersing medium to arrange themselves close to the surface while the negative ions will tend to be repelled. This leads to the formation of the double layer. (The ions that have charges opposite the surface charge are called *counterions*. Those with the same charge as the surface are called *co-ions* or *similiions*.)

The double layer can be visualized as being made up of a layer of ions tightly bound to the surface of the colloid and a diffuse region of ions outside the bound layer. The distribution of ions in the diffuse region will be determined by a balance of electrostatic attraction for the bound ions which tends to order the layer, and random thermal motion which tends to make the layer diffuse. Figure 11-2.2.3 shows these two regions. The existence of charges on or near the colloid surface leads to an electrical potential between the particle surface and the bulk of the solution. It has proven to be a very difficult task to characterize the changes in the potential between the particle surface and the bulk of the dispersing medium mathematically. Only a very abbreviated account of the behavior of the potential will be presented below.

Several assumptions are necessary before any headway can be made in the description. The only ones listed explicitly here are:

1. The particle surface is assumed to be an infinitely extended flat plane with a uniform charge.

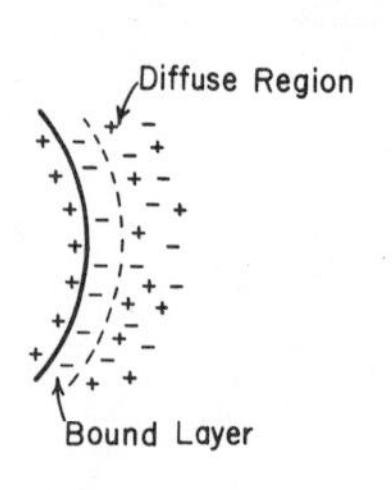

FIG. 11-2.2.3 The surface of a colloid particle with both the bound and diffuse layers shown.

2. The ions in the diffuse layer are taken as point charges which are distributed according to a Boltzmann distribution.

Figure 11-2.2.4 shows a positively charged surface with a tightly bound layer of negative ions. Immediately outside the bound layer is the diffuse layer. The potential will vary from some value designated by ψ_0 at the surface to a value of ψ_d at the edge of the bound layer. ψ_d is known as the Stern potential, and the bound layer is called the Stern layer after the man who first worked out a model for the double layer. As we continue out from the edge of the bound layer the charges become more and more random due to the thermal motion within the solution. Near the surface the electrostatic attractions are sufficient to order the charges, but as the distance from the surface increases this force is overcome by the random thermal (Brownian) motion. Thus, the potential finally decreases to zero at some point within the solution.

The potential of particular interest is the *zeta potential* ζ. *This is the potential at the plane of shear.* The plane of shear is the line just outside the bound layer that can be stripped off if the particle is forced through a solution or if solvent is forced to flow over the particle. The diffuse layer which can be stripped off by motion of the dispersing medium has a net positive charge in the case shown in Fig. 11-2.2.4. (The bound layer contains mostly negative ions leaving an excess of positive ions in the diffuse layer.) If the diffuse layer is stripped off the particle by some motion of the medium, we will get a separation of the charges which leads to the electrokinetic phenomena to be discussed later. The existence of a potential on the surface of the particles will lead to a more stable colloid. All the particles have the same sign of surface charge leading to a repulsion between

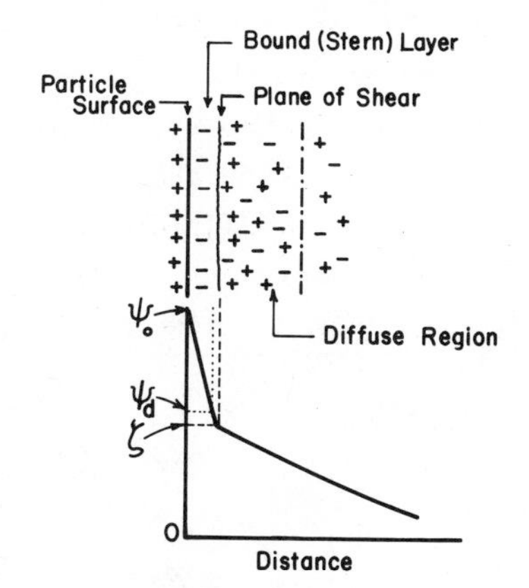

FIG. 11-2.2.4 A charged surface with associated bound layer and diffuse layer of ions. Also, the manner in which the electrical potential decreases with distance from the surface is shown.

particles that tends to prevent flocculation. More will be said about this after the discussion of electrokinetic phenomena.

11-2.3 ELECTROKINETIC PHENOMENA

The term *electrokinetic phenomenon* is applied to each of the four phenomena that have been observed when the diffuse layer of the double layer is sheared off by particle or solvent motion (or both). The four phenomena are:

1. *Electrophoresis*: the movement of a charged particle along with any ions or solvent attached to its surface through a stationary liquid due to the application of an electric field across the liquid.
2. *Electroosmosis*: the movement of a liquid relative to a stationary charged surface due to the application of an electric field. The pressure required to prevent the liquid movement is called the *electroosmotic pressure*.
3. *Streaming potential*: the electrical potential created by the flow of liquid past a stationary charged surface. This is just the opposite of electroosmosis.
4. *Sedimentation potential*: the electrical potential created by the motion of charged particles relative to a stationary liquid. This is just the opposite of electrophoresis.

Each of these will be discussed in a little greater detail, and the equations used for calculating the magnitude of the effects will be given. Some of the areas in which these phenomena are of importance will be pointed out.

ELECTROPHORESIS. Assume we have a sol or an emulsion which consists of particles large enough to be seen with the aid of a microscope. If we put this suspension in a cell such as shown in Fig. 11-2.3.1 and apply an electric field to the electrodes at the ends of the cell we will observe one of three things. The colloidal particles will move toward the positive electrode if the particles are negatively charged. They will move toward the negative electrode if they are positively charged. Or they will not move at all under the influence of the electric field if they are neutral. Thus, observation of the behavior of colloidal particles in the presence of an electric field tells us immediately what the sign of the charge on the particle is. More detailed observation of the velocity of the particles as a function of

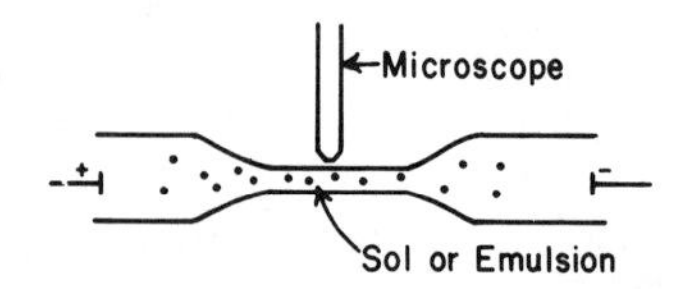

FIG. 11-2.3.1 An electrophoresis apparatus.

the applied field strength and the size of the particles will give us information as to the number of charges associated with a particular particle. We can estimate the zeta potential ζ from the above type of measurement.

Some of the applications of electrophoresis are interesting. It is possible to investigate the surface of bacteria, viruses, etc., by doing electrophoretic studies on these as a function of pH, added surface-active agents (see Sec. 11-3), added electrolyte, and added chemicals such as enzymes. Proteins can be characterized by a slightly different electrophoretic technique in which the liquid is adsorbed onto a solid substrate such as silica, which is deposited on a solid plate and an electric field imposed across the plate. A drop or two of the suspension is added to the center of the plate, and the direction and rate of movement is observed to identify the protein. A suspension containing several different proteins (or other types of colloids) can be separated into components with the proper choice of solvent and substrate.

Figure 11-2.3.2(a) shows a schematic of an electrophoresis apparatus and part (b) shows a typical separation pattern for proteins. The substrate (inert support) can be paper, agar, starch, or cellulose acetate membrane, among others. A conducting solution (buffered) is used to moisten the support, and an electric field is applied. The cover prevents loss of solvent. Since power is being dissapated across the cell some type of cooling is required to maintain constant temperature. The potential applied can be as high as 125 V cm^{-1}. The voltage is limited by electrical heating in the cell. Electrophoresis, when combined with chromatography in a two-dimensional separation, is a very valuable chemical identification tool. More detail on many different techniques can be found in Strickland (see Sec. 11-2.8, Ref. 1).

The equations relating electrophoretic mobility (the velocity of a particle under the influence of a unit electrical field) to the zeta potential are now given

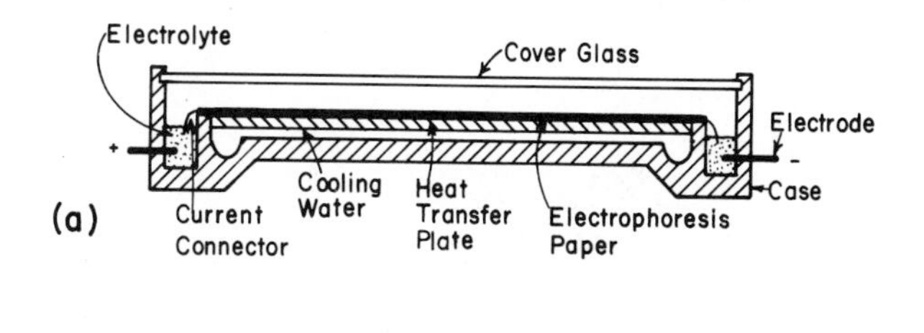

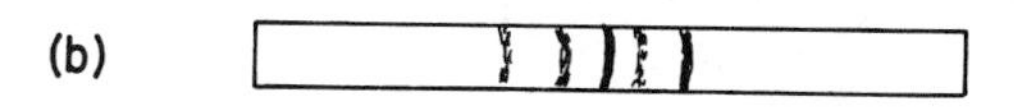

FIG. 11-2.3.2 (a) Schematic of an electrophoresis apparatus and (b) typical separations for several proteins.

without derivation or proof. A partial derivation of some of the electrokinetic equations can be found in D. J. Shaw, *Introduction to Colloid and Surface Chemistry* (see Sec. 11-2.8, Ref. 2). Several simplifying assumptions are made in the derivation of all the equations presented in this section. For example, the curvature of the particle's surface is neglected, and a flat, infinite plane is assumed, which eliminates any end effects which might be present. Also, as the particle moves, the diffuse double layer will tend to lag behind the particle instead of remaining symmetrical about it. The distortion of the diffuse double layer will have a retarding effect on the motion of the particle, since this layer is charged oppositely to the bound layer. This is called the "relaxation effect."

The simplified equation for the velocity of a charged particle in an electric field is

$$u = \frac{\varepsilon \zeta E}{L\eta} \tag{11-2.3.1}$$

in which ε is the permittivity of the solvent (which is equal to the dielectric constant θ times the permittivity of vacuum ε_0, which has a value of 8.854×10^{-12} $kg^{-1}\ m^{-3}\ s^4\ A^2$; A stands for amp). ζ is the zeta potential previously defined, L is the length of the tube (or distance between the electrodes), and η is the viscosity coefficient of the solvent. E is the electric field applied. Rewriting this equation in terms of the dielectric constant we can obtain

$$\boxed{u = \frac{\theta \varepsilon_0 E \zeta}{\eta L}} \tag{11-2.3.2}$$

Perhaps the best way to illustrate the meaning of this equation is to work through a problem. Notice why these are called electrokinetic phenomena. We have the kinetic variables L and η and the electrical variables ε_0, θ, ζ, and E. The application of an electric field E results in the movement of the colloid particles with a velocity of u. All other variables are measureable except for the zeta potential which we determine from our data. The electrophoretic mobility υ will be given by

$$\boxed{\upsilon = \frac{uL}{E} = \frac{\theta \varepsilon_0 \zeta}{\eta}} \tag{11-2.3.3}$$

11-2.3.1 EXAMPLE

Spherical particles of 0.55-mm radius, when suspended in water are observed to have an electrophoretic mobility υ of $3.0 \times 10^{-8}\ m^2\ s^{-1}\ V^{-1}$. Find an approximate value for the zeta potential if the coefficient of viscosity of water at

298 K is 8.9×10^{-4} kg m^{-1} s^{-1} and the dielectric constant is 78.5 (no units).

Solution

Rearrange Eq. (11-2.3.3) in terms of the zeta potential:

$$\zeta = \frac{\upsilon\eta}{\theta\varepsilon_0} \tag{11-2.3.4}$$

Inserting the appropriate values leads to

$$\zeta = \frac{(3.0 \times 10^{-8}\ \mathrm{m^2\ s^{-1}\ V^{-1}})(8.9 \times 10^{-4}\ \mathrm{kg\ m^{-1}\ s^{-1}})}{(78.5)(8.854 \times 10^{-12}\ \mathrm{kg^{-1}\ m^3\ s^4\ A^2})}$$

$$= 3.84 \times 10^{-2}\ \frac{\mathrm{m^2\ s^{-1}\ V^{-1}\ kg\ m^{-1}\ s^{-1}}}{\mathrm{kg^{-1}\ m^{-3}\ s^4\ A^2}}$$

$$= 3.84 \times 10^{-2}\ \frac{\mathrm{m^4\ kg^2}}{\mathrm{A^2\ s^6\ V}}$$

Refer to Sec. 14-2.2 for some useful conversion factors for electrical units. 1 A s = 1 C = 1 J V^{-1} = 1 kg m^2 s^{-2} V^{-1}. Making this set of substitutions (do it step by step) leads to

$$\zeta = 3.84 \times 10^{-2}\ \frac{\mathrm{m^4\ kg^2}}{\mathrm{(kg^2\ m^4\ s^{-4}\ V^{-2})\ s^4\ V}}$$

$$= \underline{\underline{3.84 \times 10^{-2}\ \mathrm{V}}}$$

We see there is a way to estimate zeta potentials for colloids. However, even though the above is of interest to the theoretical development of colloid chemistry, the major value of electrophoresis is the separation and identification of colloidal substances.

ELECTROOSMOSIS. Electroosmosis is the case in which the colloid is stationary and the solvent is forced to flow over or past it. Figure 11-2.3.3 shows the schematic for a cell in which electroosmosis can be studied. In this apparatus a potential is applied to the solution across the porous plug, which may be an inert material such as quartz. Along the surface of the pores in the quartz plug a double layer forms with the bound layer being negatively charged. This makes the diffuse layer in the water positively charged. (Assuming, of course, that water is the solvent.) The application of a potential will result in the movement of the diffuse (positive) portion of the double layer toward the negative electrode, as indicated in Fig. 11-2.3.4. Each positive particle will be surrounded by water which will move with

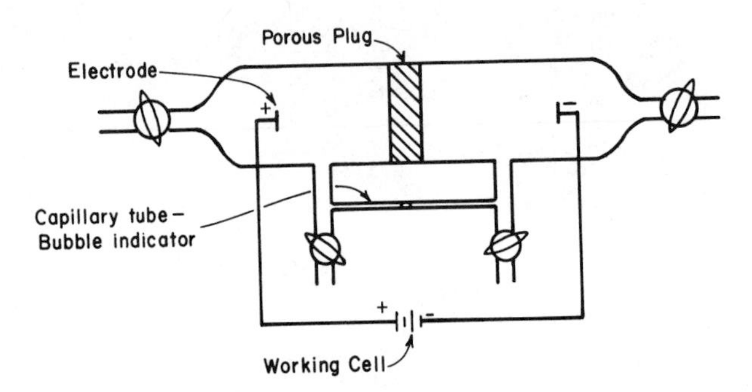

FIG. 11-2.3.3 An apparatus for studying electroosmosis.

the particle. This results in the transport of solvent toward the negative electrode along with the positive particles. The rate of movement can be measured by observing the air bubble in the return capillary. Note we can form a double layer even though there may be no ions in the solution originally. Polarization in the vicinity of the quartz surface and ions in the glass are sufficient to form the layer. The movement of solvent toward the negative electrode will generate an opposing pressure if the volume of that compartment is held constant, and no return path is available. The pressure will increase until the force due to the electroosmotic effect is just equal to the force due to the pressure (Fig. 11-2.3.3b).

Again, without derivation, the equations needed to treat this phenomenon are presented. The terminal velocity of the fluid under a potential E is given by

$$u = \frac{\zeta \varepsilon_0 \theta E}{L\eta} \tag{11-2.3.5}$$

This results in a volume flow of

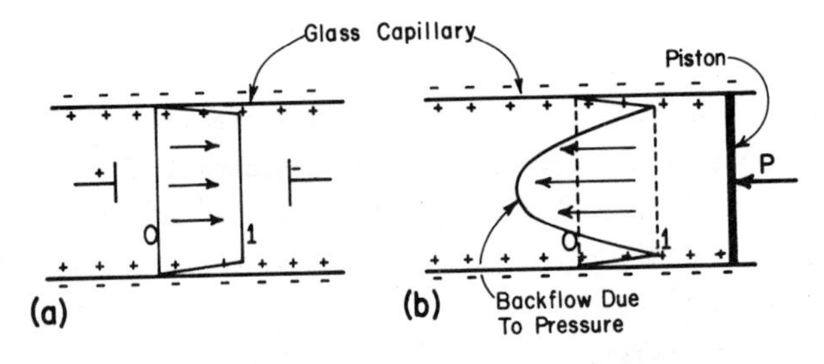

FIG. 11-2.3.4 (a) Model for electroosmotic flow and (b) opposing electroosmotic pressure. 0 indicates the position of an arbitrary layer of water at zero time, and 1 is the position after unit time. In (b) note the opposing pressure causes a backflow of water. This represents one pore in the porous plug of Fig. 11-2.3.3.

$$V = u\pi R^2 t \qquad (11\text{-}2.3.6)$$

where R is the radius of the tube and t is the time being considered. If the only return path is through the tube itself, the opposing pressure will result in a volume flow given by Poiseuille's law

$$V = \frac{\pi \bar{P} R^4 t}{8L\eta} \qquad (11\text{-}2.3.7)$$

At equilibrium, when all forces and flows are balanced, the last two volumes must be equal, from which we get the *electroosmotic pressure* (show it).

$$\boxed{\bar{P} = \frac{\zeta \varepsilon_0 \theta 8E}{R^2}} \qquad (11\text{-}2.3.8)$$

Equations (11-2.3.5) and (11-2.3.8) are the two that can be used to characterize electroosmotic phenomena. Note that we can measure every variable in these equations except for the zeta potential. Thus we have another method of finding a value of ζ. As before, these equations are only approximate.

11-2.3.2 EXAMPLE

The rate of electroosmotic flow of water through a glass capillary 10 cm long and 1 mm in diameter under the influence of a potential difference of 200 V is observed to be 6.22×10^{-5} m s^{-1}. Given that θ of H_2O at 298 K is 78.5, and η is 8.9×10^{-4} kg m^{-1} s^{-1}, find (1) the zeta potential for the glass-water interface and (2) the electroosmotic pressure that the system would exhibit at equilibrium.

Solution

1. Application of Eq. (11-2.3.5) after rearrangement gives

$$\zeta = \frac{uL\eta}{\varepsilon_0 \theta E} \qquad (11\text{-}2.3.9)$$

$$= \frac{(6.22 \times 10^{-5}\ \text{m s}^{-1})(0.10\ \text{m})(8.9 \times 10^{-4}\ \text{kg m}^{-1}\ \text{s}^{-1})}{(8.854 \times 10^{-12}\ \text{kg}^{-1}\ \text{m}^{-3}\ \text{s}^{4}\ \text{A}^{2})(78.5)(200\ \text{V})}$$

$$= 3.98 \times 10^{-2}\ \frac{\text{kg m s}^{-2}}{\text{kg}^{-1}\ \text{m}^{-3}\ \text{s}^{4}\ \text{A}^{2}\ \text{V}} = \underline{\underline{3.98 \times 10^{-2}\ \frac{\text{kg}^2\ \text{m}^4}{\text{s}^6\ \text{A}^2\ \text{V}}}}$$

As in Example 11-2.3.1, this unit converts to V. Thus

$$\zeta = 3.98 \times 10^{-2}\ \text{V} = \underline{\underline{39.8\ \text{mV}}}$$

2. We can now use (11-2.3.8) to find the pressure. It is easier, however, to combine Eqs. (11-2.3.8) and (11-2.3.9) to give

$$\bar{P} = \frac{uL\eta\varepsilon_0\theta 8E}{\varepsilon_0\theta ER^2} = \frac{8uL\eta}{R^2} \qquad (11\text{-}2.3.10)$$

Then

$$\bar{P} = \frac{(8)(6.22 \times 10^{-5}\ \mathrm{m\ s^{-1}})(0.10\ \mathrm{m})(8.9 \times 10^{-4}\ \mathrm{kg\ m^{-1}\ s^{-1}})}{(0.001\ \mathrm{m}/2)^2}$$

$$= 17.7 \times 10^{-2}\ \frac{\mathrm{m\ kg\ s^{-2}}}{\mathrm{m^2}} = 17.7 \times 10^{-2}\ \mathrm{N\ m^{-2}}$$

$$= \underline{\underline{1.7 \times 10^{-6}\ \mathrm{atm}}}$$

STREAMING POTENTIAL. A *streaming current* and *streaming potential* result when a liquid is driven through a capillary tube by an applied pressure. As pointed out in the discussion of electroosmosis, water in a glass capillary will produce a double layer with the surface of the glass being negative and the diffuse layer being positive. If pressure is applied to one end of the tube to force liquid through the capillary, the positively charged layer will be swept along with the liquid, leading to a charge separation. This results in an electrical potential which is called the *streaming potential*, and the motion of the charges in the liquid is a current called the *streaming current*. Figure 11-2.3.5 is a representation of this phenomenon for water in a glass tube.

In the figure δ is the thickness of the double layer. It can be shown that the velocity of the ions in the diffuse portion of the double layer is approximately

$$u = \frac{PR}{2L} \qquad (11\text{-}2.3.11)$$

where the variables are defined in the figure. This flow of ions results in a current which is given by

$$I_s = \frac{\varepsilon_0\theta P\pi R^2\zeta}{L\eta} \qquad (11\text{-}2.3.12)$$

where πR^2 is the cross-sectional area of the tube. Use can be made of Ohm's law in which the resistance involved will be that of the liquid itself. The resistance is related to the conductivity k_0 by

$$R_s = \frac{L}{k_0 A} \qquad (11\text{-}2.3.13)$$

where $A = \pi R^2$, the area of the tube. Thus

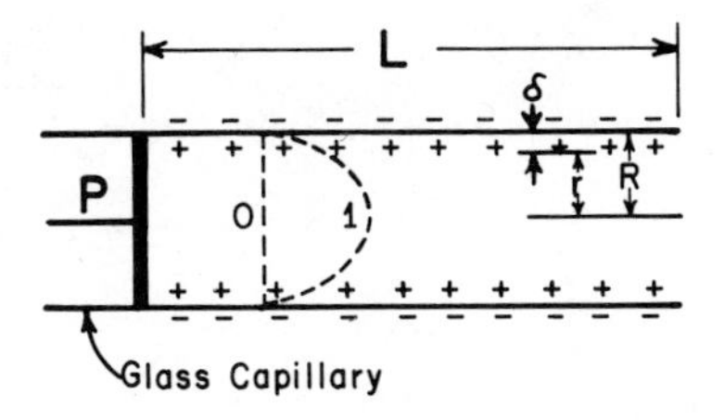

FIG. 11-2.3.5 The generation of a streaming potential. 0 is the initial position of a given layer of water, and 1 is the position after a given time of flow under the influence of a driving pressure P. (Laminar flow assumed.)

$$E_s = \frac{\varepsilon_0 \theta P \zeta}{\eta k_0} \tag{11-2.3.14}$$

Note that here R_s is assumed to be only the liquid resistance. If there are other conductors present the resistance value used in Eq. (11-2.3.13) will be the net resistance of the system. For example, the tube itself might have a low enough resistance to be important if it is made out of some other material. If we have a liquid with a very low conductivity such as a hydrocarbon (or gasoline), a very large streaming potential can develop. Sufficient, in fact, to generate sparks. Thus, it is necessary to properly ground trucks and tanks that are drained through a hose. Streaming potentials can be measured, as can every other variable in Eq. (11-2.3.14) except zeta. This gives us another method for estimating zeta potentials. As before, the values obtained are only approximate.

11-2.3.3 EXAMPLE

Given the data from Example 11-2.3.2, predict the streaming potential if a pressure of 10 atm is applied to one end of the capillary. Use the zeta potential determined in the solution to Example 11-2.3.2; $k_0 = 7.0 \times 10^{-5}$ ohm^{-1} m^{-1} (water in equilibrium with air).

Solution

We know

$\theta = 78.5$ $\qquad$ $\eta = 8.9 \times 10^{-4}$ kg m^{-1} s^{-1}

$\varepsilon_0 = 8.854 \times 10^{-12}$ kg^{-1} m^{-3} s^{4} A^{2} $\qquad$ $\zeta = 3.98 \times 10^{-2}$ V

$$P = (10 \text{ atm}) \frac{101{,}325 \text{ N m}^{-2}}{1 \text{ atm}} = 1.013 \times 10^{6} \text{ N m}^{-2} \qquad k_0 = 7.0 \times 10^{-5} \text{ ohm}^{-1} \text{ m}^{-1}$$

$$E_s = \frac{(8.85 \times 10^{-12}\ \text{kg}^{-1}\ \text{m}^{-3}\ \text{s}^4\ \text{A}^2)(78.5)(1.013 \times 10^6\ \text{kg m}^{-1}\ \text{s}^{-2})(3.98 \times 10^{-2}\ \text{V})}{(7.0 \times 10^{-5}\ \text{ohm}^{-1}\ \text{m}^{-1})(8.9 \times 10^{-4}\ \text{kg m}^{-1}\ \text{s}^{-1})}$$

$$= 4.5 \times 10^2\ \frac{\text{m}^{-4}\ \text{s}^2\ \text{V A}^2}{\text{ohm}^{-1}\ \text{m}^{-2}\ \text{kg s}^{-1}}$$

$$= 450\ \text{kg}^{-1}\ \text{m}^{-2}\ \text{s}^3\ \text{V ohm A}^2 \qquad (1\ \text{V} = 1\ \text{A ohm})$$

$$= 450\ \text{kg}^{-1}\ \text{m}^{-2}\ \text{s}^2\ \text{V}^2\ \text{A s} \qquad (1\ \text{A s V} = 1\ \text{J})$$

$$= 450\ \text{kg}^{-1}\ \text{m}^{-2}\ \text{s}^2\ \text{V J}$$

$$= 450\ \text{kg}^{-1}\ \text{m}^{-2}\ \text{s}^2\ \text{V kg m}^2\ \text{s}^{-2}$$

$$= \underline{\underline{450\ \text{V}}}$$

Notice that a material such as a hydrocarbon that has a lower conductivity will have a higher E_s. You can see perhaps why a spark could be generated in pumping such a fluid through a nonconducting hose.

SEDIMENTATION POTENTIAL. The last electrokinetic phenomenon to be considered is the *sedimentation potential*. This is the potential that results when a charged particle settles in a solvent. The solvent can be assumed stationary and the particle in motion. This is just the opposite of electrophoresis, in which an electric field is applied to force the motion of the charged particles present. The equations applicable to this problem are similar to those for electrophoresis and will not be presented here. The important point to remember is that sedimentation results for charged particles will be different than for those that are not charged. As the charged particles move through the solvent some of the counterions in the diffuse portion of the double layer are stripped off, leading to a charge separation within the system and therefore an electrical potential. The counterions that are stripped off tend to attract the particles that have lost some of their counterions (the particles that have lost some of their diffuse double layer are oppositely charged from the ions that have been stripped off, leading to an attraction between the two). This attraction acts as a retarding force which slows down the motion of the particle that is settling, leading to a sedimentation rate that is smaller than if no charges were present. (Look back to the discussion of sedimentation and see how this would affect an estimate of the radius of the particle if we mistakenly assumed no charges were present, even though there were some.)

This concludes the discussion of electrokinetic effects. However, there is one more effect that occurs due to charges on particles in the presence of a semipermeable membrane. This is the Donnan effect and is of utmost importance in biological systems.

11-2.4 INTERLUDE

DONNAN EQUILIBRIUM. Assume we have a membrane separating two solutions that contain ions at different concentrations. Also assume the membrane is permeable to solvent and to some of the ions but not to all ions present. Such a situation will lead to the development of a potential across the membrane. The following treatment is due to W. J. Moore, *Physical Chemistry*, 4th ed., (C) 1972, pp 544-546, 622-623. Reprinted by permission of Prentice-Hall, Inc., Englewood Cliffs, New Jersey (see Sec. 11-2.8, Ref. 3). (A review of osmotic pressure in Sec. 6-2.5 would be advisable before continuing in this section.) Figure 11-2.4.1 is a schematic of the system we will consider. We have K^+ and Cl^- ions and H_2O present which will diffuse through the membrane. Also present is some particle (colloidal or otherwise) which has a charge of z^+ and which will not pass through the membrane. At equilibrium the following conditions must apply (see Sec. 6-1.1)

$$\bar{\mu}_{K+} = \bar{\mu}'_{K+} \qquad \bar{\mu}_{Cl-} = \bar{\mu}'_{Cl-} \qquad \bar{\mu}_{H_2O} = \bar{\mu}'_{H_2O} \tag{11-2.4.1}$$

where $\bar{\mu}$ is the electrochemical potential defined in Eq. (11-2.4.3).

If the activities of the water are different on the two sides there will be an osmotic pressure which develops across the membrane. This pressure can be determined from

$$\Pi = \frac{RT}{\bar{V}_{H_2O}} \ln \frac{a'_{H_2O}}{a_{H_2O}} \tag{11-2.4.2}$$

If you accept the following definition of the electrochemical potential $\bar{\mu}$ we can derive an interesting result.

$$\bar{\mu}_i = \mu_i + z_i \mathcal{F} \Phi \tag{11-2.4.3}$$

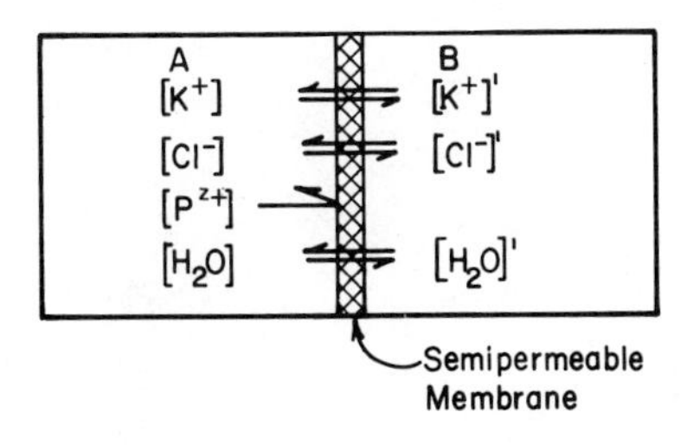

FIG. 11-2.4.1 An example of Donnan equilibrium with a multivalent cation (P^{Z+}) and KCl as the neutral salt. The variables on side B are primed, while those on side A are unprimed.

where μ_i is the chemical potential of ion of type i (see Sec. 6-1.1), z_i is the charge of the ion of type i, $\mathcal{F}$ is the Faraday (96,500 C), and Φ is the electrical potential in the solution.

The chemical potential μ_i can be written in terms of activities

$$\mu_i = \mu_i^0 + RT \ln a_i \tag{11-2.4.4}$$

Thus

$$\bar{\mu}_i = \mu_i^0 + RT \ln a_i + z_i\mathcal{F}\Phi \tag{11-2.4.5}$$

Also, since for an isothermal, reversible process

$$d\bar{\mu}_i = \bar{V}_i \, dP \tag{11-2.4.6}$$

We can write

$$\Delta\bar{\mu}_i = \bar{\mu}_i' - \bar{\mu}_i = \bar{V}_i \, \Delta P = \bar{V}_i\Pi \tag{11-2.4.7}$$

where $\bar{V}_i$ is the molar volume of the i type ion. Combining Eqs. (11-2.4.7) and (11-2.4.5) yields

$$\Delta\bar{\mu}_i = \bar{V}_i\Pi = \mu_i^{0\prime} + RT \ln a_i' + z_i\mathcal{F}\Phi' - \mu_i^0 - RT \ln a_i - z_i\mathcal{F}\Phi \tag{11-2.4.8}$$

But $\mu_i^0 = \mu_i^{0\prime}$; the standard state chemical potential is the same in each side. Thus,

$$\Pi\bar{V}_i = RT \ln \frac{a_i'}{a_i} + z_i\mathcal{F}\Delta\Phi \tag{11-2.4.9}$$

We now list the results for H_2O and each ion from (11-2.4.2) and (11-2.4.9):

$$\Pi = \frac{RT}{\bar{V}_{H_2O}} \ln \frac{a_{H_2O}'}{a_{H_2O}} \tag{11-2.4.2}$$

For K^+ (z = 1),

$$\Delta\Phi = \frac{\Pi\bar{V}_{K^+}}{\mathcal{F}} - \frac{RT}{\mathcal{F}} \ln \frac{a_{K^+}'}{a_{K^+}} \tag{11-2.4.10}$$

For Cl^- (z = - 1),

$$\Delta\Phi = -\frac{\Pi\bar{V}_{Cl^-}}{\mathcal{F}} + \frac{RT}{\mathcal{F}} \ln \frac{a_{Cl^-}'}{a_{Cl^-}} \tag{11-2.4.11}$$

Combining the last three equations gives

$$\Delta\Phi = \frac{RT\bar{V}_{K^+}}{\mathcal{F}\,\bar{V}_{H_2O}} \ln \frac{a_{H_2O}'}{a_{H_2O}} - \frac{RT}{\mathcal{F}} \ln \frac{a_{K^+}'}{a_{K^+}} \tag{11-2.4.12a}$$

Also,

$$\Delta\Phi = -\frac{RT\bar{V}_{Cl^-}}{\mathcal{F}\,\bar{V}_{H_2O}} \ln \frac{a'_{H_2O}}{a_{H_2O}} + \frac{RT}{\mathcal{F}} \ln \frac{a'_{Cl^-}}{a_{Cl^-}} \tag{11-2.4.12b}$$

Rearranging, we can see

$$\Delta\Phi = \frac{RT}{\mathcal{F}} \ln \left[\left(\frac{a'_{H_2O}}{a_{H_2O}}\right)^{\bar{V}_{K^+}/\bar{V}_{H_2O}} \left(\frac{a_{K^+}}{a'_{K^+}}\right) \right] \tag{11-2.4.13a}$$

$$= \frac{RT}{\mathcal{F}} \ln \left[\left(\frac{a_{H_2O}}{a'_{H_2O}}\right)^{\bar{V}_{Cl^-}/\bar{V}_{H_2O}} \left(\frac{a'_{Cl^-}}{a_{Cl^-}}\right) \right] \tag{11-2.4.13b}$$

If dilute, $a_{H_2O} \approx a'_{H_2O}$, thus

$$\Delta\Phi = \frac{RT}{\mathcal{F}} \ln \frac{a_{K^+}}{a'_{K^+}} = \frac{RT}{\mathcal{F}} \ln \frac{a'_{Cl^-}}{a_{Cl^-}} \tag{11-2.4.14}$$

which gives

$$\frac{a_{K^+}}{a'_{K^+}} = \frac{a'_{Cl^-}}{a_{Cl^-}} \tag{11-2.4.15}$$

or, if sufficiently dilute, $c \approx a$,

$$(c_{K^+})(c_{Cl^-}) = (c'_{K^+})(c'_{Cl^-}) = c'^2 \tag{11-2.4.16}$$

This is known as the *Donnan equilibrium* condition. However, we are not through yet.

The system must be electrically neutral on each side. Thus,

$$c_{K^+} + z^+ c_{p^{z+}} = c_{Cl^-}$$
$$c'_{K^+} = c'_{Cl^-} = c' \tag{11-2.4.17}$$

We can write c'^2 as [from (11-2.4.16) and (11-2.4.17)]

$$c'^2 = (c_{K^+})(c_{Cl^-}) = (c_{K^+})[(c_{K^+}) + z^+ c_{p^{z+}}] \tag{11-2.4.18}$$

or

$$\frac{c'}{c_{K^+}} = \left[1 + \frac{z^+ c_{p^{z+}}}{c_{K^+}}\right]^{1/2} \tag{11-2.4.19}$$

This last equation allows us to calculate the ratios of concentrations of ions on the two sides of the membrane if we know the concentration and charge of the multivalent nonpermeable ion. You will get a chance to do that in the problem section.

11-2.4.1 EXAMPLE

Find the ratio of c'/c_{K^+} and c'_{Cl^-}/c_{Cl^-} for a solution that has $c'_{K^+} = 0.0010$ mol dm^{-3} and $z^+c_{p^{z+}} = 0.0020$ mol dm^{-3}.

Solution

This is not as simple as it looks. We need c_{K^+} which appears on both sides of (11-2.4.19). Assume a c_{K^+} and iterate. The best initial guess would be $c_{K^+} = c'_{K^+}$ since the membrane is permeable to K^+ and we would not expect too large a difference across the membrane.

$$\frac{c'}{c_{K^+}} \approx \left[1 + \frac{0.002}{0.001}\right]^{\frac{1}{2}} = 1.73$$

$$c_{K^+} = \frac{c'}{1.73} = \frac{0.001}{1.73} = 0.00058$$

Now iterate:

$$\frac{c'}{c_{K^+}} \approx \left[1 + \frac{0.002}{0.00058}\right]^{\frac{1}{2}} = 2.1$$

$$c_{K^+} = \frac{0.001}{2.1} = 0.00047$$

Again,

$$\frac{c'}{c_{K^+}} \approx \left[1 + \frac{0.002}{0.00047}\right]^{\frac{1}{2}} = 2.3$$

$$c_{K^+} = \frac{0.001}{2.3} = 0.00043$$

Finally,

$$\frac{c'}{c_{K^+}} \approx \left[1 + \frac{0.002}{0.00043}\right]^{\frac{1}{2}} = 2.38$$

$$c_{K^+} = \frac{0.001}{2.38} = \underline{\underline{0.00042}}$$

and $c_{Cl^-} = c_{K^+} + z^+c_{p^{z+}} = 0.00042 + 0.002 = 0.00242$

Therefore

$$\frac{c'_{Cl^-}}{c_{Cl^-}} = \frac{0.001}{0.00242} = \underline{\underline{0.41}}$$

The above has assumed equilibrium. This is not the case for biological membranes where *active transport* occurs. Active transport refers to processes in biological systems in which material is transported into or out of a cell against a concentration or free energy gradient. This is not an equilibrium process, since some other cell reactions must furnish the driving force required to move the material against its concentration gradient. The above does show how concentration gradients across membranes can occur. The presence of active transport in many cases increases the concentration difference across the membrane.

NERVE CELLS. Another effect of biological importance is observed when we have a membrane permeable to one type of small ion but not to others. Such a situation arises in mammalian nerve cells which are permeable (in the resting state) to K^+ but nearly impermeable to Na^+ and Cl^-. A development similar to that given above yields (without proof)

$$\Delta\Phi = \frac{RT}{\mathcal{F}} \ln \frac{a_{K^+}}{a'_{K^+}} \qquad (11\text{-}2.4.20)$$

It has been observed that $K^{+\prime} \approx 20\ K^+$ in the resting nerve cell. This gives a membrane potential (at 298 K) of

$$\Delta\Phi = \frac{(8.314\ \text{J mol}^{-1}\ \text{K}^{-1})(298\ \text{K})}{96{,}500\ \text{C}} \ln \frac{1}{20}$$

$$= \underline{\underline{-77\ \text{mV}}}$$

(The inside of the cell is negative.) This is very close to the observed value. The system is obviously not this simple, since equilibrium is not attained, but the close agreement is interesting.

11-2.5 COLLOID STABILITY AND MISCELLANEOUS TOPICS

CHARGE AND STABILITY. The stability of colloids was briefly mentioned in Sec. 11-1. Here that discussion will be extended somewhat in light of what you have just learned about charges on the surface of colloids. As you encountered earlier in this section, the fact that all the colloid particles in a particular system have the same charge tends to keep the colloid in suspension. However, a lyophobic colloid is very sensitive to the addition of a small amount of an electrolyte. One explanation of the effect of added electrolyte on a charged lyophobic colloid is as follows. The electrolyte causes a compression of the diffuse layer of the double layer allowing two particles to make a closer approach to each other. If the approach is close enough the attractive forces due to van der Waals interactions can be large enough to cause flocculation. Another effect is the adsorption of ions into the Stern layer which will lower the surface potential. This allows the particles to make a closer approach to each other. From these considerations you should conclude that electrolytes with higher ionic charge will be more effective than those of lower charge in causing flocculation to occur. In fact, we find that all monovalent ions have approximately the same effect on flocculation (it takes about the same amount of Na^+ as K^+ to flocculate a negative colloid, for example). And it takes about 1/100 as much of a divalent ion of the same sign.

The state of flocculation is important in determining the extent and type of sedimentation. Large, deflocculated particles will pack into a dense sediment that is difficult to redisperse. On the other hand, flocculated material will form a loosely packed sediment. In some cases the floc will occupy the entire volume. [An example you may have observed is an $Al(OH)_3$ precipitate. $Al(OH)_3$ forms a very loose, fluffy network.] In the extreme case of strong interparticle attractions between the points of contact of the floc particles a network will be established. Such a network is called a *gel*. The rigidity and stability of the gel depends on the attraction of the points of contact.

In lyophilic systems the addition of small amounts of electrolytes has very little, if any, effect on the degree of dispersion. However, large amounts of electrolytes will lead to flocculation. Both solvation and charge determine whether a particular system of particles will remain in suspension. If particles are strongly solvated, they will have difficulty approaching each other closely enough to flocculate. Some materials such as gelatin in water have enough attraction for the solvent to remain in suspension even when all surface charges are neutralized.

PROTECTING AND SENSITIZING AGENTS. There are cases in which lyophobic colloids can be made more stable by the addition of a material which adsorbs on the surface to form a new lyophilic surface. This case was mentioned in the discussion of the formation of a charged surface. *Protective agents* can be of the following types:

1. Adsorbed material may have ionizable groups which ionize and impart a charge to the surface. (See Fig. 11-2.2.2b for an example.)
2. A protective agent may form a film around the particles which must be removed or modified before the particles can get close enough together to flocculate. (See Fig. 11-2.5.1c.)
3. Large molecules may be adsorbed on the surface which leads to steric interaction between these groups on different particles.
4. Some adsorbed materials lower the van der Waals attractions between particles.

Some materials, when added to a sol, make flocculation easier. There are some substances that act as *sensitizing agents* at very low concentrations and as protective agents at higher concentrations. Some examples are:

1. Surface-active agents (see Sec. 11-3) sometimes form an adsorbed layer with a lyophobic portion extending out into the solvent sensitizing the sol.

Higher concentrations lead to a second layer which has the lyophilic layer pointed into the solvent leading to protective action.

2. Long-chain additives tend to sensitize some sols at low concentration by acting as a bridging mechanism between particles. At high concentration protective action is again obtained, presumably due to the fact that all possible points of attachment on the surface of the particles are occupied, and bridging cannot take place.

Figure 11-2.5.1 shows how a material may behave as a protecting agent at low concentrations but become sensitizing at higher concentrations or vice-versa. A long-chain molecule that has a lyophilic and a lyophobic end can either sensitize or protect a colloid. If the lyophobic end is preferentially adsorbed by the colloid (leaving the lyophilic end projecting into the solution) the colloid will be stabilized. Part (c) of Fig. 11-2.5.1 shows this. The colloid is surrounded by a lyophilic surface. Note what happens between parts (c) and (d) when excess agent is added. After the colloid has adsorbed all the agent it can, any additional agent will tend to orient itself, as shown in part (d). The lyophilic ends of the added agent are compatible with the lyophilic ends projecting from the colloid. Thus, the added agent will tend to orient itself with the lyophobic end sticking out into the solvent. The lyophobic surface formed destabilizes the colloid. An opposite argument applies when the lyophilic end is adsorbed by the colloid. Draw diagrams to show that such a situation initially gives a sensitized colloid but leads to a stabilized system when an excess of agent is added.

OTHER TOPICS. Most of the space in this section has been devoted to a discussion of sols when specific examples were given. This is not to be interpreted as indicating that sols are the most important type of colloid. Sols are more easily

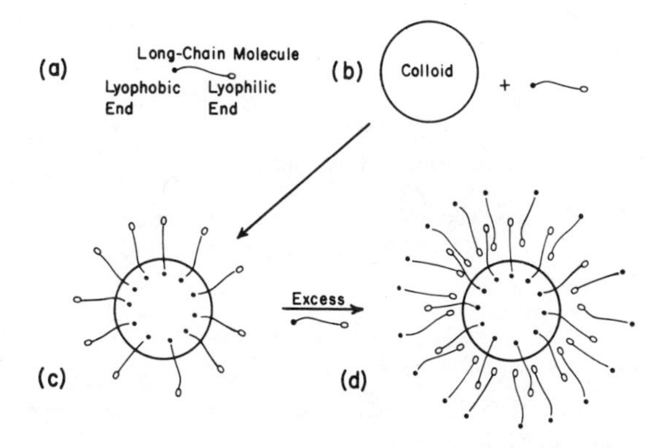

FIG. 11-2.5.1 (a) A molecule with a lyophobic and a lyophilic end (b) colloid plus protective agent, (c) stabilized colloid, and (d) sensitized colloid in the presence of excess agent.

interpreted in terms of charged surfaces and interparticle interactions than are systems such as gels. Emulsions and aerosols can be discussed in much the same way as sols have been. In the following section more will be said about emulsions in relation to surface-active agents and processes. Some properties of gels and pastes will be discussed in Chap. 12 when rheological properties are studied (Sec. 12-5).

11-2.6 PROBLEMS

1. Write a one-page summary of the double-layer model including a discussion of the methods of forming surface charge on a solloid.
2. If the conductivity of a 0.01 M solution of KCl is 1.2×10^{-1} ohm^{-1} m^{-1}, and the zeta potential is - 11.0 mV, find the streaming potential if a pressure of 3.0 atm is forcing the solution through the capillary.
3. Estimate the zeta potential of a spherical particle with a radius of 5.0×10^{-7} m if it has an electrophoretic mobility of 1.5×10^{-8} m^2 s^{-1} V^{-1}. Assume the particles are dispersed in water at 298 K.
4. A cell which is divided into two compartments by a membrane has a solution of 0.01 M NaCl on one side and a solution of 0.01 M NaX on the other. The membrane is permeable to all ions except X^-. The membrane is also permeable to H_2O. Find the concentration of ions at equilibrium on both sides of the membrane, and determine the potential developed across the membrane.
5. Determine the zeta potential of a particle that has an electrophoretic mobility of 5.6×10^{-8} m^2 s^{-1} V^{-1} in a liquid with a viscosity of 5.11×10^{-4} kg m^{-1} s^{-1} and a permittivity ε of 3.1×10^{-10} kg m^{-3} s^4 A^2.

11-2.7 SELF-TEST

1. Discuss why proteins which contain both amine and carboxyl groups are more easily precipitated at intermediate than at low or high pH values.
2. Pick two electrokinetic phenomena and discuss the origins of the phenomena and the methods used to characterize them.
3. Give a brief discussion of sensitization of colloids.
4. Determine the electroosmotic pressure developed across a 0.12-m glass tube of radius 0.0011 m if a colloid with a zeta potential of 33.2 mV suspended in water is in the tube, and the tube has an electric field of 255 V imposed across it.
5. Assume the system desrribed in Problem (4) is used to measure streaming potentials. What will be the potential developed if a pressure of 3.56×10^{-5} N m^{-2}

is applied to one end of the tube. Take the conductivity of the system to be 4.0×10^{-3} ohm^{-1} m^{-1}.

11-2.8 ADDITIONAL READING

1.* R. D. Strickland, "Electrophoresis", *Analytical Chemistry*, *42*:32R (1970).

2.* D. J. Shaw, *Introduction to Colloid and Surface Chemistry*, 2nd ed., Butterworths, London, 1970, Chap. 7.

3.* W. J. Moore, *Physical Chemistry*, 4th ed., Prentice-Hall, Englewood Cliffs, N. J., 1972.

4. D. H. Moore, "Electrophoresis", in A. Weissberger, ed., *Techniques of Organic Chemistry*, Vol. I (*Physical Methods*), Part IV, 3rd ed., Interscience, New York, 1959.

5. H. Van Olphen and K. J. Mysels, eds., *Physical Chemistry: Enriching Topics from Colloid and Surface Science*, Theorex, LaJolla, Cal., 1975.

11-3 SURFACE CHEMISTRY

OBJECTIVES

You shall be able to

1. Define or discuss all the following terms: physical adsorption, chemisorption, monolayer, surface concentration in a solution, surfactant, micelle, critical micelle concentration, film spreading, three adsorption isotherms, contact angle, wetting, detergency
2. Apply appropriate equations to calculate: constants in the Freundlich or Langmuir adsorption isotherms, surface concentrations and effective molecular areas from the Gibbs isotherm for solutions, surface tensions of solutions, whether spreading of one liquid on another occurs, contact angles

11-3.1 INTRODUCTION

Surface chemistry is important in many areas of industry as well as in everyday life. Any time you wash dishes you are taking advantage of a surface phenomenon involving the soap, water, and soil on the dish. If you have ever painted a room or house, you have used the product of countless hours of research in adjusting the properties of the paint to cover the surface with ease, with a smooth, nonsagging, nonrunning coating. Additives are put into paint to make it flow easily yet not run when the stress of brushing is removed. More detail on that particular aspect of coatings will be found in the section on rheology (Sec. 12-5). A nonstick cooking surface is an example of tailoring the surface properties of a material to do a particular job. Whether one material will spread over another or tend to form globules on the surface depends on the relative values of the surface tensions. Many important chemical processes such as the refining of petroleum depend on the availability and effectiveness of catalysts. Catalysis is, usually, a surface effect.

In this section an attempt will be made to introduce you to some of the important basic concepts related to surface chemistry. In Sec. 14-1, the concept of surface tension is introduced in relation to liquid properties. Read that section if you have not done so (Sec. 14-1.4); it will be expanded here. Definitions of different types of adsorption will be given, and equations will be presented to describe some of the observed phenomena. A little will be said about emulsions since the stability of an emulsion is very dependent on surface properties. This, as always, will not be an exhaustive (even if it is exhausting) development. What you should gain from this are the definitions and some idea of the widespread importance of surface effects. An excellent additional reference for this subject is D. J. Shaw, *Introduction to Colloid and Surface Chemistry* (see Sec. 11-3.9, Ref. 1).

11-3.2 ADSORPTION ON SOLID SURFACES

A general term used to describe the movement of a material from one phase to another is *sorption*. If sorption involves only the surface the term *adsorption* is used. If the material being sorbed is distributed throughout the phase, *absorption* is indicated. Water, when sorbed into cotton, is distributed throughout the cotton and is an example of absorption. (There could be some argument if we look on a microscopic scale. Is the water only on the surface of the fibers or is it throughout each fiber?) An example of adsorption is acetic acid on charcoal. In this case the acetic acid is found only on the surface. There are many cases in which the distinction is not clear, and in many of these cases the distinction is not important. Only *adsorption* is considered in this section.

Adsorption onto a surface is divided into two different types: *physical adsorption and chemical adsorption or chemisorption*. The major differences between the two are listed in Table 11-3.2.1. It may be apparent from Table 11-3.2.1 that physical adsorption has the characteristics of a physical process such as condensation, whereas chemical adsorption involves a chemical reaction. Note that the heats of adsorption for physical adsorption are on the order of the heats of vaporization, while the heats for chemisorption are on the order of the energies of the chemical reaction.

ADSORPTION ISOTHERMS.

General. Adsorption of a gas or liquid onto a solid surface is usually described in terms of an adsorption isotherm. An adsorption isotherm details how the

Table 11-3.2.1
Comparison of Physical and Chemical Adsorption

Physical adsorption	Chemical adsorption
Heat released on adsorption less than about 40 kJ mol^{-1}	Heat of adsorption greater than about 80 kJ mol^{-1}
Adsorption only at temperatures less than the boiling point of the adsorbate (the material being adsorbed)	Adsorption can occur at high temperatures
Multilayer adsorption occurs	At most, monolayer can form
No activation energy involved in the adsorption process	Activation energy may be involved
Extent of adsorption is mostly dependent on properties of the adsorbent (the material doing the adsorbing)	Extent of adsorption depends on both adsorbate and adsorbent

volume or weight of a gas or liquid adsorbed varies with the pressure or concentration of the adsorbate at constant temperature. An apparatus similar to that shown in Fig. 11-3.2.1 can be used to experimentally determine adsorption isotherms. Figure 11-3.2.2 illustrates two experimental isotherms. In a typical adsorption experiment, the volume of gas (absorbate) absorbed by a given weight of absorbent at a particular temperature and total pressure is measured. These data can then be used to generate the adsorption curve given in Fig. 11-3.2.2. An example of the type of isotherm shown in Fig. 11-3.2.2(a) (Type I) is ammonia on charcoal. The ammonia is adsorbed rapidly as the pressure is increased up to a limiting value which occurs when a monolayer (a layer one-molecule thick) of adsorbate has been formed. These are referred to as Langmuir-type adsorption isotherms. Chemisorption, which is restricted to monolayer adsorption, has a curve similar to this. It has also been observed that adsorption onto surfaces that possess a fine pore structure sometimes

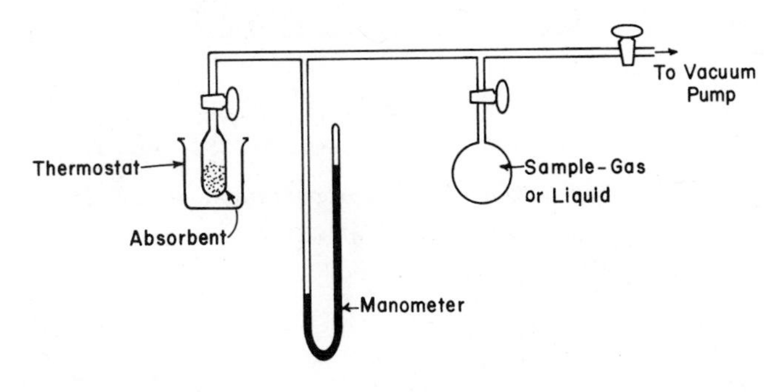

FIG. 11-3.2.1 Apparatus for measuring adsorption isotherms.

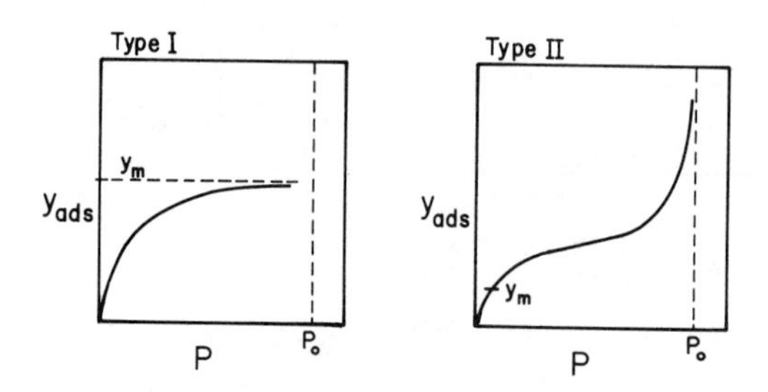

FIG. 11-3.2.2 Two common types of adsorption isotherms: (a) monolayer-formed and (b) multilayer adsorption. y_m is the volume adsorbed to form the monolayer, and P_0 is the saturation vapor pressure.

produces an isotherm of this general shape. Nitrogen adsorbed onto silica gel is an example of Type II adsorption as shown in Fig. 11-3.2.2(b). This represents multilayer adsorption onto a nonporous surface.

The Langmuir Isotherm. In 1916, Langmuir proposed a model for the adsorption process which forms only a monolayer. The proposed model is valid for most chemisorption processes and for Type I physical adsorption processes. The basic assumptions used by Langmuir are as follows.

1. There are fixed adsorption sites on the surface of the solid. At a given T and gas pressure P some fraction of these sites will be occupied by adsorbate molecules. Designate this fraction as $\bar{\theta}$.
2. Each site on the surface of the solid substrate (adsorbent) can hold one adsorbate molecule.
3. The heat of adsorption is the same for each site and is independent of $\bar{\theta}$.
4. There is no interaction between molecules on different sites.

By considering the equilibrium between the rate of evaporation and condensation on the surface, Langmuir was able to derive an expression that fits experimental data for a number of cases.

The rate of evaporation from the surface is proportional to the fraction of the surface that is covered.

$$\text{Evaporation rate} = k_d\bar{\theta} \tag{11-3.2.1}$$

where k_d is a constant. The rate of condensation will be proportional to the fraction of the surface not covered, $1 - \bar{\theta}$, and to the number of molecules striking the surface which is proportional to the pressure of the gas.

$$\text{Condensation rate} = k_a P\,(1 - \bar{\theta}) \tag{11-3.2.2}$$

where k_a is a constant. At equilibrium, these two rates have to be the same, so

$$k_d\bar{\theta} = k_a P(1 - \bar{\theta}) \tag{11-3.2.3}$$

which can be rearranged to give

$$\bar{\theta} = \frac{k_a P}{k_d + k_a P} = \frac{bP}{1 + bP} \tag{11-3.2.4}$$

where b is defined as k_a/k_d and is called the adsorption coefficient. For graphical presentation of data it is more convenient to put Eq. (11-3.2.4) into the form

$$\boxed{\frac{1}{\bar{\theta}} = 1 + \frac{1}{bP}} \tag{11-3.2.5}$$

which is a straight line form (if the assumptions are valid) when $1/\bar{\theta}$ is plotted against $1/P$.

An even more useful form is obtained if we define the fraction adsorbed in terms of the amount that can be adsorbed at monolayer. Let y_m by the amount that can be adsorbed at monolayer coverage and y the amount adsorbed at any pressure (for a unit mass of adsorbent). Then

$$\bar{\theta} = \frac{y}{y_m} \tag{11-3.2.6}$$

We can now rewrite the previous two equations as

$$y = \frac{y_m P}{\flat' + P} \tag{11-3.2.7}$$

and

$$\boxed{\frac{P}{y} = \frac{\flat'}{y_m} + \frac{P}{y_m}} \tag{11-3.2.8}$$

where $\flat'$ (= 1/b) is a constant; y_m is defined above. This last equation is the one of use in analyzing adsorption data. The amount of material adsorbed (usually in weight or volume of adsorbate per gram adsorbent) as a function of pressure is measured, and P/y is plotted against P. The slope of the line (if it is straight) will be $1/y_m$ and the intercept is $\flat'/y_m$. For chemisorptions the Langmuir development works fairly well, but it does not work for Type II physical adsorption curves. ΔH has been assumed independent of $\bar{\theta}$. This is usually not the case, since interactions between adjacent adsorbate molecules will affect the strength with which a given molecule is bound to the surface.

11-3.2.1 EXAMPLE

The following data have been observed for the adsorption of H_2 onto a Cu surface (the volume of H_2 is determined at 0°C and 1 atm). From these data determine the volume of H_2 necessary to form a monolayer. Also, use the y_m value measured and the liquid density of H_2 (the density of liquid H_2 is 0.070 kg dm^{-3}) to estimate the surface area of 1 g Cu.

P (atm)	0.05	0.10	0.15	0.20	0.25
P/y (atm dm^{-3})	42	75	115	147	175

Solution

A plot of P/y vs. P is required as shown.

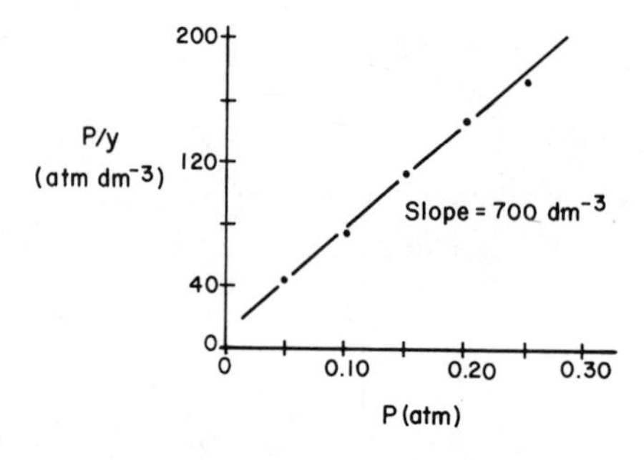

The slope is $1/y_m = 700\ dm^{-3}$, thus $y_m = 1.43 \times 10^{-3}\ dm^3$ at 273 K and 1 atm. This is the volume of H_2 gas required to produce a monolayer on the Cu surface. We can easily convert this to the number of molecules of H_2.

$$\text{No. molecules } (H_2) = \frac{(1\ \text{atm})(1.43 \times 10^{-3}\ dm^3)}{(0.0821\ dm^3\ \text{atm mol}^{-1}\ K)(273\ K)}\ 6.023 \times 10^{23}\ \frac{\text{molecules}}{\text{mol}}$$

$$= 3.84 \times 10^{19}\ \text{molecules}$$

From the density of *liquid* H_2 we can estimate the area of an individual molecule (mlc).

$$\text{Area}_{mlc} \approx (V_{mlc})^{2/3}$$

$$= \left[\frac{0.002\ \text{kg mol}^{-1}}{(0.070\ \text{kg dm}^{-3})(6.023 \times 10^{23}\ \text{molecules mol}^{-1})}\right]^{2/3}$$

$$= (4.74 \times 10^{-26}\ dm^3\ \text{molecule}^{-1})^{2/3}$$

$$= 1.31 \times 10^{-17}\ dm^2\ \text{molecule}^{-1}$$

The number of molecules times the area per molecule will give the area occupied.

Thus the area of 1 g Cu is

$$\text{Area}_{\text{Cu}} = (1.31 \times 10^{-17}\ \text{dm}^2\ \text{molecule}^{-1})(3.84 \times 10^{19}\ \text{molecule})$$

$$= 503\ \text{dm}^2\ (\text{g Cu})^{-1}$$

$$= \underline{\underline{5.03\ \text{m}^2\ (\text{g Cu})^{-1}}}$$

The Langmuir isotherm is only one of several that have been proposed to fit adsorption data. Two other types are described below.

The Freundlich Isotherm. The Freundlich isotherm is a completely empirical attempt to fit adsorption data. There is no theory behind it as there is behind the Langmuir isotherm previously considered. The Freundlich isotherm fits data that correspond to Type I adsorptions. This isotherm is used in describing the adsorption of solute from a solution as well as a gas adsorption. The form of the Freundlich isotherm is:

$$\boxed{y = kP^{(1/n)}} \qquad (11\text{-}3.2.9)$$

where k and 1/n are empirical constants and y is the amount adsorbed per unit mass of adsorbent. If we are interested in solutions we write

$$\boxed{y = kc^{(1/n)}} \qquad (11\text{-}3.2.10)$$

in which k and 1/n are again constants to be determined empirically and c is the concentration of solute in the solution. The value of c in Eq. (11-3.2.10) is the concentration of solute after equilibrium has been attained in the solution.

You may wonder of what use is such an empirical expression. The following example and discussion will, perhaps, give you an indication. In many industrial and municipal wastewater treatment facilities, one of the last purification steps is passing the water over activated carbon to remove most of the last traces of contaminant. Before some estimate of the amount of carbon that will be needed can be made, the adsorption capabilities of the carbon must be measured. An experiment like the one described below in the example would have to be run for each contaminant of interest.

11-3.2.2 EXAMPLE

The following data were obtained from measurements of the adsorption of acetic acid onto a charcoal surface. Determine the constants in the Freundlich equation.

c (mol dm^{-3})	0.018	0.031	0.062	0.126	0.268	0.471	0.882
y (mol)	0.47	0.62	0.80	1.11	1.55	2.04	2.48

ln c	-4.02	-3.47	-2.78	-2.07	-1.32	-0.75	-0.13
ln y	-0.76	-0.48	-0.22	0.104	0.438	0.713	0.908

Solution

The logarithms of the above quantities are included since we can rewrite Eq. (11-3.2.10) as

$$\ln y = \ln k + \frac{1}{n} \ln c \qquad (11\text{-}3.2.11)$$

You should be able to tell what type of plot of the data will yield a straight line if the Freundlich isotherm fits. The required plot is given below. From this plot we see the slope is 0.44 and the intercept which occurs when ln c = 0 is equal to ln k = 1.04 or k = 2.83. The Freundlich equation for this adsorption is

$$y = 2.83c^{0.44}$$

This equation is good only if the same concentration units are used throughout. Also, the equation will be different if the temperature changes.

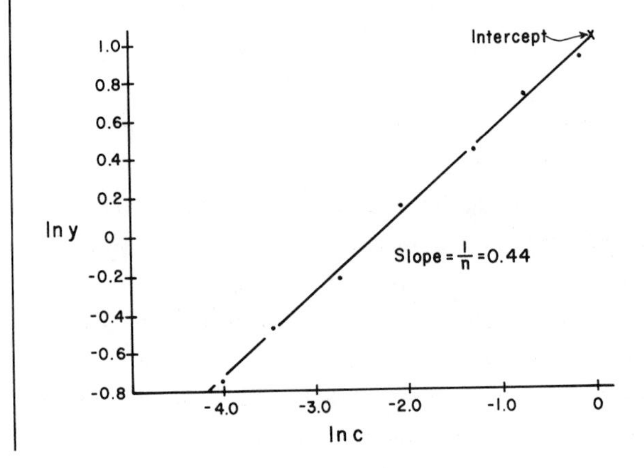

The BET Isotherm. The BET (Brunauer, Emmett, and Teller) theory is an extension of Langmuir's theory which allows for multilayer adsorption. The Langmuir theory is limited to monolayer adsorption and is therefore restricted to chemisorption and a few special cases of physical adsorption. In allowing for multilayer adsorption, the BET theory is able to describe physical adsorption processes that are of Type II. As in the derivation of the Langmuir equation, the BET equation is found by balancing the rate of evaporation from the surface with the rate of condensation on the surface. The major difference in the two derivations is the allowance made by BET for multilayer adsorption. The theory assumes that the first adsorbed layer is bound by a characteristic heat of adsorption ΔH_1. This heat of adsorption will be different than the heat of adsorption of the second and subsequent layers. The heat of adsorption of layers other than the first is assumed to be simply the heat of liquefaction ΔH_L. The form of the equation that results is

$$\frac{P}{y(P_0 - P)} = \frac{1}{y_m \varsigma'} + \frac{(\varsigma' - 1)P}{y_m \varsigma' P_0} \tag{11-3.2.12}$$

where y is the amount adsorbed per unit mass adsorbent, y_m is the amount adsorbed to form a monolayer, P is the pressure of the adsorbate, P_0 is the normal vapor pressure of the gas, and ς' is a constant given by

$$\varsigma' = \exp\left(\frac{\Delta H_L - \Delta H_1}{RT}\right) \tag{11-3.2.13}$$

The BET equation reduces to the Langmuir equation for monolayer adsorption.

11-3.3 LIQUIDS ON SOLIDS AND LIQUIDS

When a liquid comes into contact with a solid surface, the liquid will either spread over the surface or it will remain as droplets. If the liquid spreads we say the liquid has *wet* the solid, whereas if the liquid aggregates into droplets it is called *nonwetting* on that particular surface. Whether a liquid wets a surface depends on the relative values of the surface tensions of the two materials involved.

In Sec. 14-1.4 a brief discussion is given of the definition of surface tension and one method of determining that quantity for liquids. You should read that section before continuing if you have not already done so. Surface tension is defined as the energy required to produce a unit area of new surface. It has the units of Joules per square meter (or Newtons per meter) and is given the symbol γ. If more than one phase is involved we must speak of *interfacial surface tensions*. The interfacial surface tension is the relative surface tension between the two phases present. Figure 11-3.3.1 defines the *contact angle* formed when a liquid is placed on a solid surface. This angle is related to the interfacial surface tensions of the material involved. If we separate a column of liquid that has a cross-sectional area of 1 m^2, we produce 2m^2 of new surface. The *work of cohesion* necessary to do this is just twice the surface tension.

$$w_c = 2\gamma \tag{11-3.3.1}$$

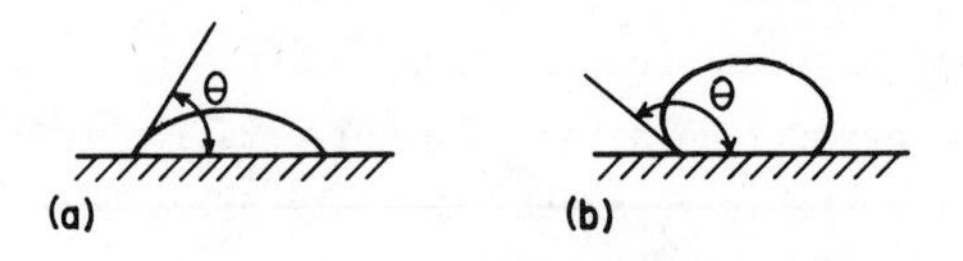

FIG. 11-3.3.1 Contact angle between a liquid and a solid. (a) $\cos\theta > 0$ (for example, water on a slightly dirty glass plate. (b) $\cos\theta < 0$ (for example, Hg on glass).

The *work of adhesion* is the work required to separate one surface from another. It is related to the interfacial surface tension. Figure 11-3.3.2 indicates the inter-

$$w_a = \gamma_{sa} + \gamma_{\ell a} - \gamma_{s\ell} \tag{11-3.3.2}$$

facial tensions of two surfaces in air. γ_{sa} is the surface tension of the solid in air, $\gamma_{\ell a}$ is the same for the liquid, and $\gamma_{s\ell}$ is the interfacial tension between the solid and the liquid. The *spreading coefficient* is a result of the balance of surface tension and cohesion forces. ($S_{\ell s}$ is the spreading coefficient of ℓ on s. Note the order given for the γs. The order given is important in determining the sign of $S_{\ell s}$.)

$$S_{\ell s} = w_a - w_c$$

$$= \gamma_{sa} - \gamma_{\ell a} - \gamma_{s\ell} \tag{11-3.3.3}$$

A positive value of $S_{\ell s}$ implies that a drop of liquid will spread across the solid surface. A similar treatment can be applied to a liquid on a liquid if the liquids are immiscible. Miscible liquids will be considered a little later. Figure 11-3.3.3 shows a liquid drop on another liquid. Assume we have an oil drop on water. An equation analogous to Eq. (11-3.3.3) can be applied.

$$S_{ow} = \gamma_{wa} - \gamma_{oa} - \gamma_{ow} \tag{11-3.3.4}$$

In each case spreading occurs when the work of adhesion is greater than the work of cohesion.

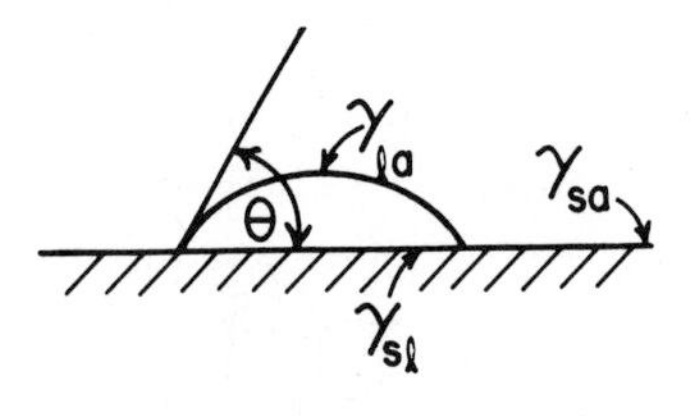

FIG. 11-3.3.2 A drop of liquid on a solid surface.

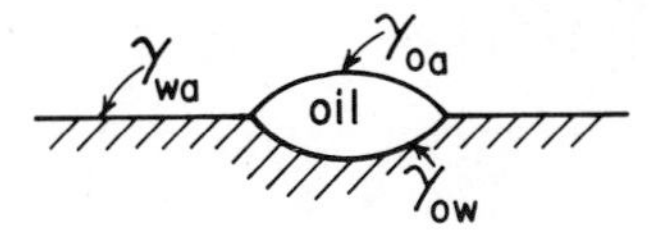

FIG. 11-3.3.3 A nonspreading liquid on the surface of another liquid.

11-3.3.1 EXAMPLE

Will n-octanol spread on a water surface if $\gamma_{oa} = 27.5 \times 10^{-3}$ N m^{-2}, $\gamma_{ow} = \gamma_{wo} = 8.5 \times 10^{-3}$ N m^{-2}, and $\gamma_{wa} = 72.8 \times 10^{-3}$ N m^{-2}.

Solution

$$S_{ow} = \gamma_{wa} - \gamma_{oa} - \gamma_{ow}$$

$$= (72.8 - 27.5 - 8.5) \times 10^{-3} \text{ N m}^{-2}$$

$$= \underline{\underline{36.8 \times 10^{-3} \text{ N m}^{-2}}}$$

Thus, n-octanol will spread on the surface of water.

Table 11-3.3.1 gives some values of interfacial surface tensions. You can use these to determine which "oils" will spread on water. These kinds of considerations are important in cleaning up oil spills. There are chemicals called "herders" which have a much greater spreading coefficient than ordinary oils. These can be placed around an oil slick and they will spread in such a manner that the oil is contained

Table 11-3.3.1
Some Values of Interfacial Surface Tensions Relative to Air and to Water ($\times 10^3$ N m^{-1})

Liquid	$\gamma_{\ell a}$	$\gamma_{\ell w}$
Water	72.75	--
Benzene	28.88	35.0
Acetic acid	27.6	--
Acetone	23.7	--
Carbon tetrachloride	26.8	45.1
Ethanol	22.3	--
n-Octanol	27.5	8.5
n-Hexane	18.4	51.1
n-Octane	21.8	50.8
Mercury	485	375

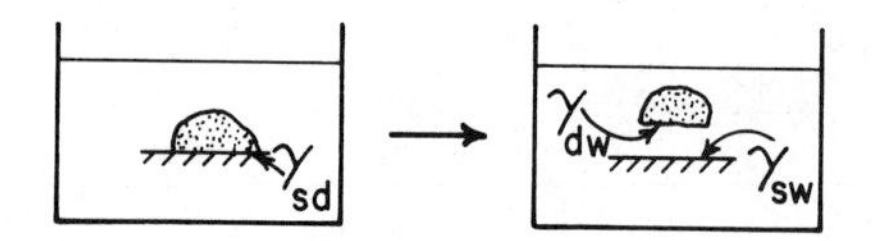

FIG. 11-3.3.4 The interfacial surface tensions present before and after detachment of a dirt particle from a surface.

within the circle formed by the added material. You should try some calculations with a few combinations of the fluids that are immiscible with water to see which will spread.

An important application of the concept of interfacial surface tension comes in the action of detergents in helping water remove dirt from a surface or fabric. The effect of detergents and soaps on oily soil will be discussed a little later. Consider the interfacial surface tensions that are depicted in Fig. 11-3.3.4. From Eq. (11-3.3.2), we can write immediately that the work of adhesion between the dirt and the surface (per unit area) will be given by

$$w_a = \gamma_{dw} + \gamma_{sw} - \gamma_{sd} \qquad (11\text{-}3.3.5)$$

Anything that will lower this work of adhesion will make the dirt easier to remove from the surface. The addition of a detergent lowers both γ_{dw} and γ_{sw}, which in turn lowers the adhesion and tends to free the dirt.

Before going into detail on surface-active agents (*surfactants*), return for a moment to Fig. 11-3.3.2 where the contact angle of a liquid drop on a solid surface is shown. Balancing the forces due to the three surface-tension components results in the following expression.

$$\gamma_{sa} = \gamma_{s\ell} + \gamma_{\ell a} \cos\theta \qquad (11\text{-}3.3.6)$$

Combining this with Eq. (11-3.3.2) leads to an expression that relates the work of adhesion between a liquid and solid surface to the contact angle formed between the two.

$$w_a = \gamma_{\ell a}(1 + \cos\theta) \qquad (11\text{-}3.3.7)$$

This equation gives a very convenient method of measuring the work of adhesion between liquids and solids. And, as you have already seen above, this is an important parameter in determining the cleaning capabilities of various solutions.

Let us see how these equations can be used to determine the interfacial tensions γ_{sa} and $\gamma_{s\ell}$. These quantities are not readily determined by direct measurement. Instead, the results of several measurements are combined to yield the values sought. If a liquid is placed on a solid, a contact angle is formed as described previously. This quantity can be measured yielding a value of $\cos\theta$. The value of $\gamma_{\ell a}$ can be found by methods described in Sec. 14-1.4. Thus, we can find cos θ and $\gamma_{\ell a}$. This allows w_a in Eq. (11-3.3.7) to be determined. In addition, w_c is known [since w_c = 2γ, Eq. (11-3.3.1)]. Using w_a and w_c we can find $S_{\ell s}$ in Eq. (11-3.3.3). Or, alternately, $S_{\ell s}$ can be measured directly. In any event, we now have $S_{\ell s} = \gamma_{sa} - \gamma_{\ell a} - \gamma_{s\ell}$, in which $S_{\ell s}$ and $\gamma_{\ell a}$ are known. Finally, from Eq. (11-3.3.6) we know $\gamma_{sa} = \gamma_{s\ell} - \gamma_{\ell a} \cos\theta$. Cos θ and $\gamma_{\ell a}$ are known. Thus, we have two equations involving the two unknowns γ_{sa} and $\gamma_{s\ell}$. This allows their evaluation. Note that this is a case in which the desired variables cannot be obtained directly from one measurement. Two or three separate experimental measurements ($\gamma_{\ell a}$, cos θ, and perhaps $S_{\ell s}$) are required.

11-3.3.2 EXAMPLE

The work of adhesion between a paraffin wax and water at 20°C is 0.054 J m^{-2}, and the spreading coefficient is - 0.0915 J m^{-2} (or N m^{-1}). Find the contact angle between the water and the wax if a drop of water is placed on the wax surface.

Solution

From Eq. (11-3.3.2) we can write

$$\cos\theta = \frac{w_a}{\gamma_{\ell a}} - 1$$

and from Table 11-3.3.1, $\gamma_{\ell a}$ for water is 0.07275 J m^{-2} (N m^{-1}), so

$$\cos\theta = \frac{0.054}{0.07275} - 1 = -0.258$$

$$= \underline{\underline{105°}}$$

11-3.4 SURFACE-ACTIVE AGENTS (SURFACTANTS): SOLUBLE MATERIALS

Two different but related phenomena will be presented in this and the following section. This section concerns the tendency of some solutes in a solution to concentrate themselves near the surface. The next section concerns the behavior of molecules that have both polar and nonpolar ends which makes them partially soluble in both polar (water, for example) and nonpolar (oil) materials. If there is a mixture of oil and water present, the substance with polar and nonpolar ends will try to dissolve in both. The polar end will prefer to remain in the water phase, and the nonpolar end tends to stay in the oil. This leads to an emulsification of the oil in the water (or of water in the oil, depending on the relative amounts of the two phases). If we have a hydrocarbon with a polar group such as a carboxyl or alcohol attached, we have a molecule that fits the above description. As the chain length of the hydrocarbon increases, it becomes more and more difficult for the polar end to pull the hydrocarbon end into aqueous solution, and the molecules will tend to orient themselves on the surface with the polar end in the water and the nonpolar end extending into the air above the solution. Even for the short-chain alcohols and acids there is some tendency to orient at the water-air interface. This leads to surface activity that will be described below.

Positive Excess Surface Concentration. First, let us consider the case in which the solute is soluble yet tends to concentrate itself at the interface between two phases. (We will consider only aqueous solutions here with the second phase being air.) Define the concentration of solute as c in the bulk of the solution. The *excess surface concentration* Γ represents the excess of solute in the vicinity of the surface. Γ can be either positive or negative (there are cases in which the solute tends to avoid the surface). We will express Γ in moles per square meter, consistent with our definition of Γ as excess surface concentration. Γ is defined mathematically as

$$\Gamma_i = \frac{n_i^s}{A}$$

where n_i^s is the amount of i in the surface phase s, and A is the area of the surface. In general the surface tension of a solution is different from that of the pure solvent. Those solutes that tend to concentrate near the surface lower the surface tension, while those that tend to concentrate away from the surface increase the surface tension relative to the pure solvent. Let us now focus our attention only on those solutes that tend to concentrate near the surface. These materials are *surface-active agents: surfactants.*

Equilibrium in such a system is attained when the free energy decrease brought about by the migration to the surface (and consequent lowering of the surface tension) is just balanced by the opposing free energy increase due to the nonuniform increase in concentration near the surface. The energy of the surface phase can be written as

$$E^s = TS^s - PV^s + \gamma A + \sum \mu_i n_i^s \tag{11-3.4.1}$$

where s stands for variables for the surface phase, and μ_i is the chemical potential for component i. Differentiating this expression and applying the general expression for dE from the first and second laws yields

$$S^s\, dT - V^s\, dP + A\, d\gamma + \sum n_i^s\, d\mu_i = 0 \tag{11-3.4.2}$$

If T and P are constant,

$$A\ d\gamma + \sum n_i^s\ d\mu_i = 0 \qquad (11\text{-}3.4.3)$$

Thus, $$d\gamma = -\sum \frac{n_i^s}{A}\ d\mu_i$$

or $$d\gamma = -\sum \Gamma_i\ d\mu_i \qquad (11\text{-}3.4.4)$$

For a two-component system (solvent A plus one solute B)

$$d\gamma = -\Gamma_A\ d\mu_A - \Gamma_B\ d\mu_B \qquad (11\text{-}3.4.5)$$

We can somewhat arbitrarily choose the boundary for our surface, so let us choose the boundary as the point where $\Gamma_A = 0$ (this is where the excess surface concentration of the solvent is zero: the surface concentration equals the bulk concentration). We then have

$$d\gamma = -\Gamma_B\ d\mu_B \qquad (11\text{-}3.4.6)$$

The chemical potential of a substance is related to its activity (see Sec. 6-1.4).

$$\mu_B = \mu_B^0 + RT\ \ln a_B \qquad (6\text{-}1.4.5)$$

and $$d\mu_B = RT\ d\ \ln a_B \qquad (11\text{-}3.4.7)$$

We can finally write (show the missing steps)

$$\Gamma_B = -\frac{1}{RT}\frac{d\gamma}{d\ \ln a_B} = -\frac{a_B}{RT}\frac{d\gamma}{da_B} \qquad (11\text{-}3.4.8)$$

If the solution is dilute $a_B \approx c_B$, then

$$\boxed{\Gamma_B = -\frac{1}{RT}\frac{d\gamma}{d\ \ln c_B} = -\frac{c_B}{RT}\frac{d\gamma}{dc_B}} \qquad (11\text{-}3.4.9)$$

This last equation is the one of interest. It relates the surface concentration to the change in surface tension with concentration. (Note: c_B is the bulk concentration of solute.) Equation (11-3.4.9) is known as the *Gibbs isotherm*.

11-3.4.1 EXAMPLE

Determine the surface concentration of n-butanol from the following surface tension data obtained at 298 K. Also, find the surface area of the butanol molecule from the surface concentration.

c (mol dm^{-3})	0.80	0.60	0.45	0.34	0.25	0.19
$\ln c$	-0.22	-0.51	-0.80	-1.08	-1.39	-1.66
γ ($\times 10^{-3}$ N m^{-1})	27.2	32.4	37.3	40.5	45.3	48.6

Solution

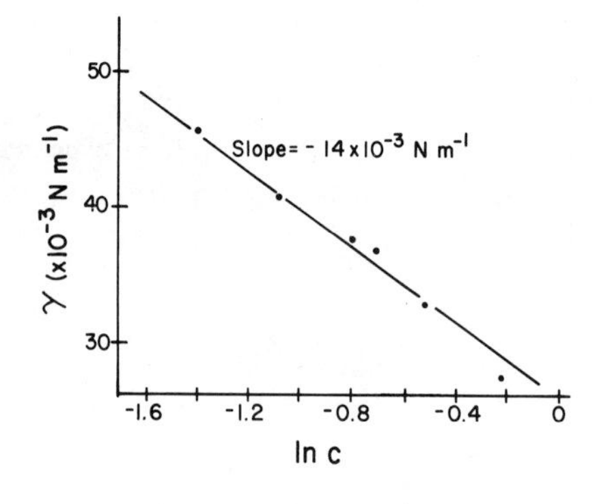

$$\Gamma = -\frac{1}{RT}\frac{d\gamma}{d \ln c}$$

$$= \frac{(-14 \times 10^{-3} \text{ N m}^{-1})}{(8.314 \text{ N m K}^{-1} \text{ mol}^{-1})(298 \text{ K})}$$

$$= \underline{\underline{5.8 \times 10^{-6} \text{ mol m}^{-2}}}$$

Convert this to molecules (mlc):

$$\Gamma_{mlc} = 3.5 \times 10^{18} \text{ molecules m}^{-2}$$

The area of a molecule = $(\Gamma_{mlc})^{-1} = 2.9 \times 10^{-19}$ m^2 molecules, or in terms of $Å^2$

$$\text{Area} = \underline{\underline{29\ Å^2 \text{ molecule}^{-1}}}$$

The significance of this area will be more apparent after you read Sec. 11-3.5.

Negative Excess Surface Concentration. Many electrolytes such as KCl and NaCl cause the surface tension to increase as the concentration of the solute increases. This is due to the tendency of these electrolytes to draw together away from the surface leaving a negative excess surface concentration. We can rewrite Eq. (11-3.4.9) in terms of concentrations as

$$\frac{\Gamma}{c} = -\frac{1}{RT}\frac{d\gamma}{dc} \qquad (11\text{-}3.4.10)$$

The left side of this equation has the units of length and is a measure of the "effective thickness" of the region at the surface where the surface concentration is significantly different than the bulk concentration. Figure 11-3.4.1 illustrates schematically how the solute concentration changes with the distance from the surface of the solution. The concentration of the solute is assumed to vary from zero at the surface to the bulk concentration c some small distance from the surface. The "effective empty layer thickness" defined by the figure is the average distance in which solute is lower in concentration than in the bulk. x_0 can be found from

$$x_0 = \frac{\Gamma}{c} = \frac{1}{RT}\frac{d\gamma}{dc} \qquad (11\text{-}3.4.11)$$

11-3.5 SURFACE-ACTIVE AGENTS: INSOLUBLE SURFACE FILMS

If a solute has a long aliphatic (hydrocarbon) chain attached to a polar group, the whole molecule cannot go into an aqueous solution. Instead, the polar end will dissolve leaving the nonpolar end extending into the air (or organic phase if one is present). Figure 11-3.5.1 illustrates this point. When enough molecules have been added to the surface, a *monolayer* will be formed. This leads to an experimental method for estimating the effective area of an individual molecule as shown in Fig. 11-3.5.2. A known amount of the film-forming material is added to a liquid surface. The film formed is compressed by the compressing float, while the force necessary for compression is observed. The force is measured by increasing torsion on the torsion wire to keep the pointers at the top of the apparatus aligned. When the area of the sample on the surface has been decreased enough to form a monolayer, the force required for the compression rises very sharply. The area can be measured at that point, and the area of each molecule can be calculated (see Fig. 11-3.5.3). It is interesting to note some representative results. Stearic acid (18-carbon chain) has an effective area of 0.20 nm^2. n-Hexatriacontanoic acid (a straight chain acid with 32 carbons) has exactly the same area. This is strong evidence in

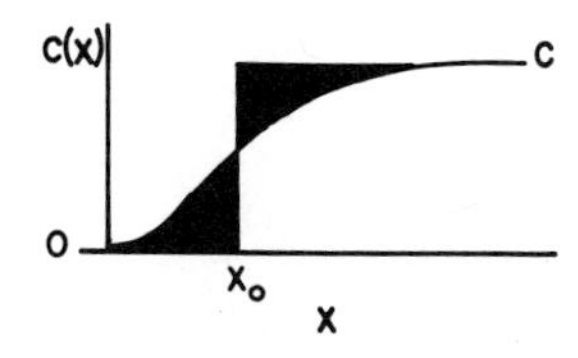

FIG. 11-3.4.1 The variation of solute concentration with distance from the surface. x_0 is the effective empty layer thickness.

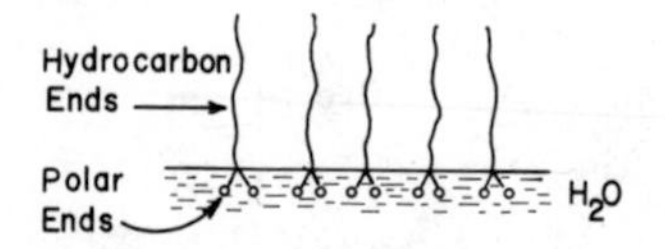

FIG. 11-3.5.1 A sketch of a group of film-forming molecules on a liquid surface. The polar end is soluble while the non-polar end is not.

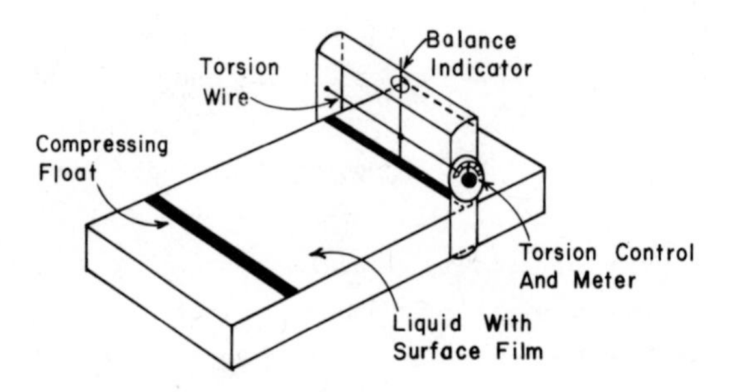

FIG. 11-3.5.2 A device for measuring surface film forces.

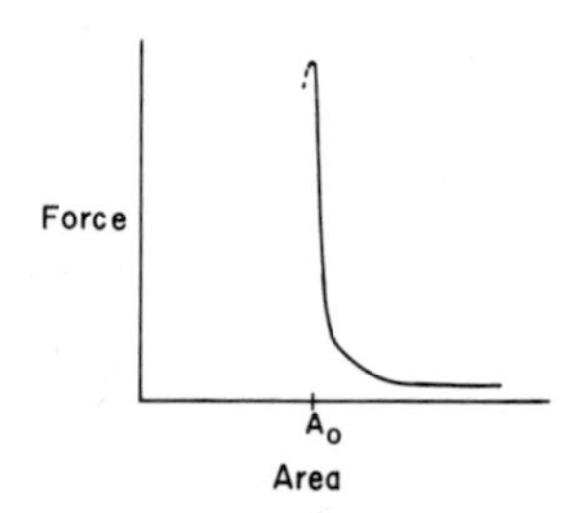

FIG. 11-3.5.3 Film force versus film area.

support of our model showing that the hydrocarbon chains extend out from the solution surface. Isostearic acid (18 carbons with a methyl side chain) has an area of 0.32 nm^2, a little larger than stearic acid due to the bulk of the methyl group hanging on the chain.

One proposed model of some cell membranes involves a bilayer of lipid molecules. The lipid molecules align themselves so that the hydrocarbon ends are facing each other with polar groups forming the outside of the bilayer. These polar groups are attached to proteins or water or other materials.

11-3.6 SURFACE-ACTIVE AGENTS: OTHER PHENOMENA

MICELLAR FORMATION. It has been observed that solutions of highly surface-active materials behave normally (for example, the surface tension of the solution varies as predicted above) until a certain concentration is reached. Beyond this concentration anomalous behavior in such properties as surface tension, osmotic pressure, and electrical conductivity is observed. This has been attributed to *micelle formation*. A micelle is an organized aggregate of the surface-active solute molecules. Apparently the lyophobic ends of the surfactant molecules attract each other with enough force to cause aggregation to take place. The lyophilic ends are pointed out from the aggregation into the solution. Figure 11-3.6.1 depicts this situation. Figure 11-3.6.2 shows how some physical properties vary near the concentration at which micelles form. The concentration at which the micelle is formed is termed the *critical micelle concentration* (*cmc*). The most popular model for these micelles is a spherical droplet of colloidal dimensions in which the lyophobic ends are pointed into the center of the drop, and the lyophilic ends form the surface of the particle. Both the size of the micelle (which is fairly constant for a given material) and the cmc depend almost entirely on the nature of the lyophobic part of the surfactant.

As you read in the previous section, a surface-active agent tends to collect at the surface of a solvent. A material such as sodium dodecyl sulfate (an anionic

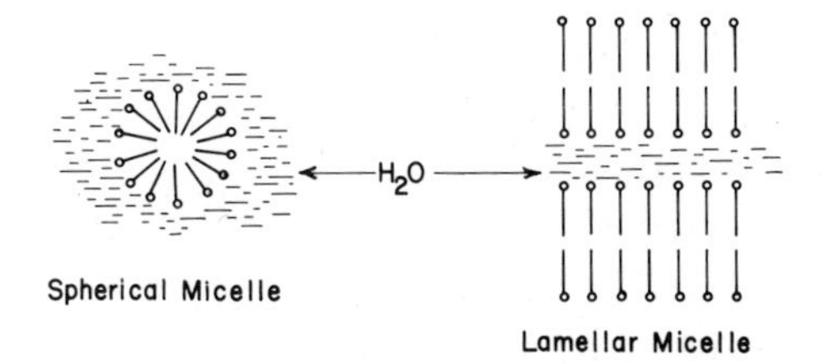

FIG. 11-3.6.1 Micelles formed from surfactants in water.

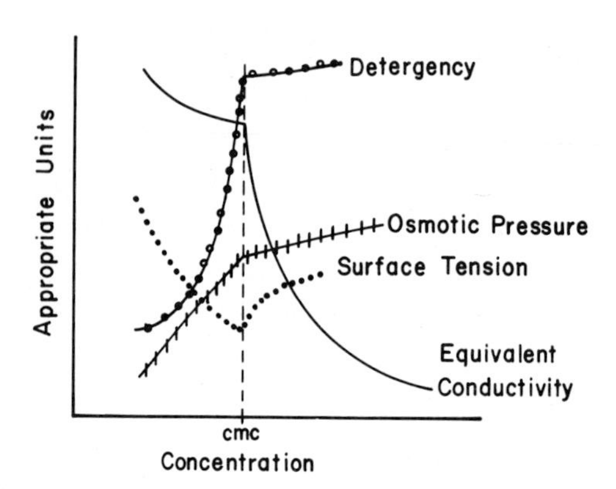

FIG. 11-3.6.2 Change in physical properties near the critical micellar concentrations.

surfactant) seems to form a surface film as evidenced by the variation of the surface tension of the solution. However, when the cmc is reached there is a discontinuity in the surface tension curve as the sodium dodecyl sulfate anions aggregate in the solution as micelles (see Fig. 11-3.6.2).

Micelles have the ability to help solubilize materials that are ordinarily not soluble in a particular solvent. For example, you know that oil is not ordinarily soluble in water. The addition of a surface-active agent does not increase the solubility below the cmc. However, above the cmc the oil can be dispersed in water. Presumably, this is due to the absorption of the oil into the interior of the micelle. As the concentration of the surfactant goes up, the solubility of the oil goes up. This may be due to the formation of more micelles into which the oil can go. A tertiary recovery process in oil production depends on this effect. It is called *micellar flooding*. This leads us to the topic of emulsification. An oil or wax can be dispersed into water giving a uniform suspension if strong stirring is used. This suspension is usually not stable, and the emulsion will "break", leading to a separation of the two phases. The addition of a surfactant will stabilize the dispersion. The model usually given has been presented briefly already. The lyophobic ends of the surfactant enter the oil droplets leaving the lyophilic ends on the surface of the drop. The presence of a lyophilic surface stabilizes the suspension.

An interesting problem of the opposite type is found in the production of petroleum. In many cases there is water associated with the oil (or oil associated with the water). Most of the water can be separated from the oil without difficulty. There is, usually, a small amount of oil that separates out with the water as an emulsion. Not only is this oil a valuable resource, but the oily water is a pollution problem. There are surface-active chemicals in the oil that help to stabilize the emulsion. Separation of the water and oil usually involves heat treating (this gives the emulsion particles enough kinetic energy to overcome repulsion forces and coalesce, the equivalent of flocculation for sols) or chemical addition to break down the micelles and allow the oil to separate.

MISCELLANEOUS. Surface films are used in many other areas, some of which you may have observed but never associated with surface phenomena. The water repellency imparted to garments is due to a surface-active chemical being added to the fabric. A molecule is chosen that has one end that can be bound to the fabric (either chemically or by physical forces). The other end of the molecule is hydrophobic. Enough of these molecules are put on the fabric surface to form a monolayer. The hydrophobic surface thus formed repels water and does not allow it to get to the fabric. Surface films of long-chain alcohols have been added to the surfaces of ponds in some areas of the country in an attempt to lower the evaporation rate. The reasons for the success are not entirely clear. There is some controversy as to whether the effect is due only to the surface coating preventing evaporation. Some have indicated that the major effect is due to the suppression of small waves caused by wind. In any event, a surface layer of alcohol does slow evaporation from water surfaces. Another whole industry dependent on surface phenomena is the adhesive industry. Adhesion of materials depends, as we have seen, on the relative surface tensions of the surfaces involved. For an adhesive to bond two surfaces, it has to be formulated with the proper surface properties. Even in soldering and welding, the fluxes used have to have the proper surface properties to wet the metal surfaces. The list goes on and on. Perhaps you can add a few examples of your own. Enough has been said to give you the basic properties of surface phenomena and point out the importance in many applications.

Catalysis is a very important surface process that is not discussed at all in this treatment. Most chemical manufacturing processes require catalysts to make the reactions involved proceed at a pace that is fast enough to be economically sound. Treatment of automobile exhausts requires catalytic surfaces in many systems. A useful reference to begin a study of catalysis is A. W. Adamson, *Physical Chemistry of Surfaces* (see Sec. 11-3.9, Ref. 2). Before such an endeavor, however, you need a basic understanding of chemical kinetics (Chap. 10).

11-3.7 PROBLEMS

1. Use the following data for the adsorption of CO on charcoal to find the constants in the Freundlich equation. The pressure is in atmospheres and the volume of gas is in cubic decimeters measured at standard temperature and pressure. The charcoal sample weighs 3.00 g.

P (atm)	0.096	0.237	0.407	0.711	1.16
y (dm^3)	7.5	16.5	25.1	38.1	52.3

2. The following data were observed for the adsorption of N_2 on a mica surface at 90 K; y is given in units of ($\times$ 10^{-6} m^3 g^{-1}) measured at 1 atm and 293 K.

P (atm)	2.8	3.4	4.0	6.0	9.4	17.1	23.5
y	12.0	13.4	15.1	17.0	23.9	28.2	30.8

Determine the constants in the Langmuir equation; find the area covered by a single N_2 molecule if the density of liquid N_2 is 0.81 kg dm^{-3}; and estimate the surface area of the mica sample.

3. Will toluene, with a surface tension of 37 $\times$ 10^{-3} N m^{-1}, spread on water?

4. In a surface balance experiment with an organic acid the area of a monolayer formed by the addition of 1.53 $\times$ 10^{-7} mol of the acid to the surface of the water was found to be 0.0293 m^2. Determine the area of a single molecule of the acid.

5. What is the work of adhesion of benzene on a plastic surface if benzene makes a contact angle of 95° with the surface?

6. Use the Freundlich equation $y = 2.83\ C^{1/n}$ and the fact that 0.62 mol of acetic acid is adsorbed on a charcoal surface in a 0.031 mol dm^{-3} solution to predict the amount of acetic acid that would be absorbed from a 0.268 mol dm^{-3} solution.

7. Calculate the Gibbs adsorption isotherm of n-butyl alcohol at m = 0.05 in an aqueous solution at 25°C from the following data. Also estimate the surface area of a n-butyl alcohol molecule.

m (mol kg^{-1})	0.00329	0.01320	0.0264	0.1050	0.2100
γ (N m^{-1})	0.07280	0.07082	0.0680	0.05631	0.04808

11-3.8 SELF-TEST

1. If the following data fit a Langmuir isotherm, determine the constants; y is in g C_2H_2 adsorbed per gram C.

P (atm)	4.0	9.7	13.4	19.0	27.1
y	0.163	0.189	0.198	0.206	0.206

2. Briefly (one-half page) discuss the major differences among the Freundlich, Langmuir, and BET theories.

3. From the data in Table 11-3.3.1 determine if H_2O will spread on a mercury surface.

4. The flotation of a mineral depends on the attachment of gas bubbles to the surface of the mineral. These bubbles bouy the mineral to the surface (the process is sometimes called froth flotation). Discuss the factors at work in this process in terms of what you have learned about interfacial surface tensions, work of adhesion, and any other concepts you feel are appropriate.

5. The effective area of a sodium dodecyl sulfate molecule in a surface tension vs. concentration experiment was found to be 30×10^{-20} m^2 at 298 K. What is the slope of the γ vs. ln c curve?

11-3.9 ADDITIONAL READING

1.* D. J. Shaw, *Introduction to Colloid and Surface Chemistry*, 2nd ed., Butterworths, London, 1970.

2.* A. W. Adamson, *Physical Chemistry of Surface*, Interscience, New York, 1960, Chap. XII.

3. See Sec. 1-1.9, Ref. A-16D.

4. R. Aveyard and D. A. Haydon, *An Introduction to the Principles of Surface Chemistry*, Cambridge University Press, 1973.

5. H. Van Olphen and K. J. Mysels, eds., *Physical Chemistry: Enriching Topics from Colloid and Surface Science*, Theorex, LaJolla, Cal., 1975.

11-4 SUMMARY

You have completed your introduction to colloid and surface chemistry. By now you should be convinced of the importance of both topics. You should also realize your knowledge is limited at this time, although you do know much basic material. In this chapter we have defined a number of terms and listed the types of colloids. We have shown how molecular weights of colloids can be measured. This aspect of colloids will be considered in more depth in Chap. 12 (Polymers). Electrokinetic phenomena have been defined and equations presented for ideal, simple cases. Some related phenomena, such as dialysis (used in artificial kidneys) have been discussed.

Another topic which was mentioned briefly is membrane properties. Dialysis, ultrafiltration, reverse osmosis, active transport in cells, etc., all involve membranes. Reverse osmosis or ultrafiltration occur when pressure is applied to a solution in contact with a semipermeable membrane (see Chap. 6). As you read in Chap. 6 (Sec. 6-2.5), solvent normally flows through a membrane from a dilute solution (or pure solvent) into a more concentrated solution due to the free energy difference. Sufficient pressure is applied to force solvent from the solution, through the membrane, into a container of solvent; the reverse of what would occur naturally. This process is very important industrially in removing impurities from water or other solvents. The distinction between reverse osmosis and ultrafiltration is not clear cut. In general, reverse osmosis is the term applied for low molecular weight solutes and ultrafiltration for higher molecular weights. Insufficient time has been spent on such topics, but membrane properties do not readily fit any one area: many physical processes are involved.

Colloid stability was mentioned in some detail. If you were manufacturing a whipped cream substitute, the stability of the foam formed would be crucial to the storageability of the material. On the other hand, some colloids are too stable to suit us. An example is the water-oil emulsion sometimes produced in oils. Many commercial products involve emulsions (mayonnaise), foams (whipped cream substitute), or gels (some toothpastes, some hair creams). These must be stabilized with additives to give them a long enough shelf life.

Some properties of surfaces have been given. As mentioned before, paint formulation requires detailed study of both the paint and the surface to which it is to be applied. In addition, the interaction between the two must be controlled in order to assure good adhesion, coverage, and appearance (no brush strokes, sags, runs, etc.). The whole cleaning products industry is continually researching new surfactant formulations to handle new jobs and to do old jobs better. Catalysis is important in numerous chemical conversion processes. Also, catalysts for fuel cells (see Sec. 6-5) is an active area of research.

Perhaps the best way to end this discussion is with a couple of additional references for further study. *Colloid Chemistry* (J. Alexander, ed., *The Chemical Catalog Company*, New York, 1928-1946) is a six-volume set containing a plethora of practical applications of colloidal phenomena. It makes for fascinating reading. A continuing series of advances in the field is *Surface and Colloid Science* (E. Matijevic, Wiley Interscience, New York). These volumes contain recent theoretical and experimental results.

12

Polymer Chemistry and Physics

SYMBOLS AND DEFINITIONS

M_v	Viscosity average molecular weight (kg mol^{-1}, g mol^{-1})
M_n	Number average molecular weight (kg mol^{-1}, g mol^{-1})
M_w	Weight average molecular weight (kg mol^{-1}, g mol^{-1})
Π	Osmotic pressure (atm, torr, mm H_2O)
m_B	Mass of solute (kg, g)
ρ	Density (kg dm^{-3}, g cm^{-3})
g	Acceleration due to gravity (m s^{-2})
h	Height of liquid level (m, cm)
ΔR	Difference in thermistor resistances in vapor phase osmometer
c_B	Concentration (various units)
k	Constant for vapor-phase osmometer
η	Solution viscosity coefficient (kg m^{-1} s^{-1}, g cm^{-1} s^{-1})
η_0	Solvent viscosity coefficient
η_r	η/η_0 = relative viscosity (viscosity ratio, no unit)
η_{sp}	$\eta_r - 1$ = specific viscosity (no unit)
η_{red}	η_{sp}/c = reduced viscosity (viscosity number)
η_{inh}	$(\ln \eta_r)/c$ = inherent viscosity (logarithm viscosity number)
$[\eta]$	$\lim_{c \to 0} (\eta_{sp}/c) = \lim_{c \to 0} (\ln \eta_r)/c$ = intrinsic viscosity (limiting viscosity number)
K, a	Constants in Staudinger's equation
τ	Turbidity
I_0	Initial light intensity
I	Light intensity at a given point in the sample
H, α, β	Constants in light scattering equations
n_0	Refractive index
λ_0	Wavelength of incident light
N	Avogadro's number
dn/dc	Variation of refractive index with concentration
T_m	Melting temperature
T_g	Glass transition temperature
ε_B	Elongation at break
ε_L	Tensile strain (strength per unit length)
ε_y	Elongation at yield
σ_L	Tensile stress (force per unit area)
σ_y	Yield stress (force)
σ_B	Ultimate strength or tensile strength
E	Young's modulus
σ_{ij}	Stress components
ε_{ij}	Strain components
L	Length
L_0	Initial length
γ	Poisson's ratio
c_1, c_2	Constants in the WLF equation
ϕ	Stress/strain
$\phi(T)$	Creep function
t	Time
Δ_T	Shift factor for time-temperature equivalence curve
F	Force
ΔL	Change in length
λ	Extension ratio
σ_A	"Nominal stress:" force/initial cross-sectional area
σ_T	"True stress:" force/actual cross-sectional area

The term *polymer* or *macromolecule* is applied to those molecules with molecular weights above about 10 kg mol^{-1} (10,000 g mol^{-1}). These materials are of colloidal dimension and have been called *molecular colloids* to distinguish them from the *association colloids* that were the subject of the previous chapter. Synthetic polymers are formed from smaller units called *monomers*. The monomers are joined together by chemical reactions to form chains which may be several thousand units long and have molecular weights in the thousands (kilograms per mole). The importance of synthetic polymers can be seen from the amounts produced and used each year. The major synthetic polymers are polyethylene, polystyrene, polyvinyl chloride, and butadiene-styrene rubber. Between one and five billion pounds each of these are used each year. This compares with the four to five billion pounds of the natural fiber, cotton, consumed each year.

You are most likely familiar with the common uses of many polymeric materials. They are found in toys, furniture, automobiles, piping, clothing, carpets, containers, shoes, flooring, synthetic tissues (arteries, veins, valves), lubricants (synthetic oils), and containers, to name a few. The forms of the polymers that are common are foams, rigid plastics, rubbers, fibers, liquids, and solutions. The principle reason for the widespread use of polymeric materials is the ease with which their properties can be modified to fit a particular application. By changing the additives (as well as controlling the molecular weight and degree of branching in the polymerization reaction), one polymer can find use in many different products. For example, polyvinyl chloride is found in sewer pipes, raincoats, plastic bottles, and phonograph records, a rather diverse group of products. There are also many natural polymers of major significance in that they are essential to life itself. These are the proteins, nucleic acids, and polysaccharides. Other natural macromolecules are cotton, silk, gums, resins, and rubber. The first synthetic polymer was synthesized in 1860. However, chemists of that era were convinced it was merely an association colloid. It was not until the pioneering work of Staudinger after 1920 that polymers were accepted as single molecules of very high molecular weight. Since that time, tremendous progress has been made in understanding the relationship between structure and properties in this class of chemicals. We will delve into this area in some detail in the following sections.

The first topic to be considered is the synthesis of polymers. The structural units will be presented, and some representative reactions will be described. Following that, descriptions of some of the methods of determining the molecular weights of these materials will be given. Then, the relationships between morphology (large-scale structure) and polymer properties will be presented. Also, development of the large-scale structure will be discussed in terms of microscopic structure. Following that, the thermal behavior of various polymers will be described, since an investigation of the thermal properties of a particular material reveals much about other properties. Finally, the dynamic-molecular properties (stress-strain behavior) and the rheological (flow) properties will be reviewed briefly. Polymers occupy more research hours than any other single branch of chemistry. No matter what field you are in, you will need a knowledge of polymers and their properties. Perhaps enough material is included here to whet your curiosity.

12-1 POLYMERIZATION REACTIONS

OBJECTIVES

You shall be able to

1. Write the structural units (monomers) for common polymers and write a few repeating units in the polymer chain that results from reaction of the monomers
2. Describe and/or define the two types of polymerization reactions; also, show several steps in the polymerization sequence for given monomers
3. Name three biologically important macromolecules (polymers) and tell what the repeating units are in each
4. Describe briefly several types of polymerization systems

12-1.1 INTRODUCTION

In this section an attempt will be made to introduce you to the chemistry of polymerization and point out the manner in which several important polymers are synthesized. The first topic will be a description of some of the more important monomers that go into the production of polymers. Some attention will be given to the manner in which these monomers bond to form polymers. This will lead to a discussion of the major types of polymerization reactions. Little will be said about the kinetics of the various polymerization reactions, even though this is a very important topic if you happen to be in the business of making polymers. Some examples will be given to illustrate the polymerization types. Finally, a few words will be devoted to the ways in which polymers are actually synthesized on a commercial scale. You will see that some of the topics discussed in the chapter on colloids are very important in the manufacture of polymers.

12-1.2 MONOMERS AND RESULTING ADDITION POLYMERS

A monomer is a small (relative to the polymer size) molecular unit that can be used in a chemical reaction or reactions to form a high molecular weight material, a polymer. Table 12-1.2.1 lists some of the more common monomers that form addition

Table 12-1.2.1
Some Common Monomers and the Addition Polymers Resulting from Them

Name	Structural formula of monomer	Repeating polymer linkage
Polyethylene	$CH_2=CH_2$	$-CH_2-CH_2-$
Polyisobutylene	$CH_2=C(CH_3)_2$	$-CH_2-C(CH_3)_2-$
Polyacrylonitrile	$CH_2=CH-CN$	$-CH_2-CH(CN)-$
Polyvinyl chloride	$CH_2=CH-Cl$	$-CH_2-CH(Cl)-$
Polystyrene	$CH_2=CH-\emptyset$	$-CH_2-CH(\emptyset)-$
Polymethyl methacrylate	$CH_2=C(CH_3)(CO_2CH_3)$	$-CH_2-C(CH_3)(CO_2CH_3)-$
Polyvinyl acetate	$CH_2=CH-OCOCH_3$	$-CH_2-CH(OCOCH_3)-$
Polyvinylidene chloride	$CH_2=CCl_2$	$-CH_2-CCl_2-$
Polytetrafluorethylene	$CF_2=CF_2$	$-CF_2-CF_2-$
Polyformaldehyde	$CH_2=O$	$-CH_2-O-$
Polyacetaldehyde	$CH(=O)CH_3$	$-CH(CH_3)-O-$
Polyisoprene (natural rubber)	$CH_2=C(CH_3)-CH=CH_2$	$-CH_2(CH_3)C=CH-CH_2-$

polymers. These monomers can react to produce polymers that have linkage shown in the table. The types of chemical reactions involved in forming polymers from monomeric units are presented in some detail in the following section.

12-1.3 POLYMERIZATION REACTION TYPES

There are two major classes of polymerization reactions. These are *addition* and *condensation* reactions of polymers. The major difference is that the polymer formed from a condensation reaction usually lacks some of the atoms present in the monomer from which it was formed. That is, in the process of forming the polymer from the monomer some of the atoms on the monomer are usually lost. (The more general term *step-reaction* is often used instead of *condensation*.) On the other hand, in an addition polymer there is no such loss. The polymer contains all the atoms initially present in the monomer. (In this case, the more general term *chain-reaction* is often used instead of addition.) Another distinction that will become clearer as we look more closely at the reactions involved is that condensation polymers usually result from the stepwise reaction of two molecules that contain active groups. Addition polymers usually result from chain reactions involving some active center, a free radical in many cases.

Let us look at these two classes of reaction in more detail after a tabulation of the major differences between the two types. Table 12-1.3.1 gives these differences. By the time you have completed this section you should be able to give the reasons for each of the statements given in the table.

CONDENSATION POLYMERIZATION. As mentioned previously, a condensation polymerization results in the elimination of some of the atoms that were present in the monomer. To illustrate this point, let us use the formation of a polyester as an example. A polyester is formed by the reaction (condensation) of an organic acid with an alcohol. Consider the organic monomer that has an alcohol group and an acid group on the same molecule. This can be made to react as shown in the following equation:

$$\mathrm{HO{-}\underset{\underset{H}{|}}{\overset{\overset{H}{|}}{C}}{-}(CH_2)_{x-1}{-}\overset{\overset{O}{\|}}{C}{-}OH} \rightarrow \mathrm{HO{-}(CH_2)_x{-}\overset{\overset{O}{\|}}{C}{-}O{-}[(CH_2)_x{-}\overset{\overset{O}{\|}}{C}{-}O]_y{-}H} + y\mathrm{H_2O}$$

The x subscript indicates there are several CH_2 units in the organic chain, and the y subscript shows that a number of the monomer units have joined together to form the polymer. For a particular system, x will be a constant and y, in general, will vary from polymer molecule to polymer molecule in the system. There are very few cases in which y is the same for all molecules formed in a polymerization reaction.

Table 12-1.3.2 lists several of the common linkages formed in condensation polymerization. You should take a minute to write out the reactions of a few of the monomers in greater detail, showing how the linkages form. In the table, the symbols R and R′ are used to designate organic groups attached to the main organic chain. You can probably see that by varying R and R′ we can form a wide variety of

Table 12-1.3.1
Distinguishing Features of Addition and Condensation Polymerizations

Addition Polymerization	Condensation Polymerization
Growth reaction adds repeating units (monomers) one at a time to the chain.	Any two molecular species present can react.
Monomer concentration decreases steadily throughout the reaction.	Monomer disappears early in the reaction.
High polymer formed quickly; the molecular weight of the polymer changes little during the course of the reaction.	Polymer molecular weight rises steadily throughout the reaction process.
Long reaction times increase the yield but affect the molecular weights very little.	Long reaction times are necessary to obtain high molecular weights.
Reaction mixture contains only monomer, high polymer, and a very small concentration of the active growing chains.	At any time during the reaction, the reaction mixture contains a distribution of molecular species from short-chain polymers to high polymers.

polymers with a particular linkage. You can see from the table that many useful polymers are formed from condensation reactions. You should be familiar with some of the terms. Most likely you are wearing clothing made from some polyester. If you happen to be wearing fabric of natural origin such as cotton, wool, or silk, you are still using the polymer that results from a condensation reaction. Much of the padding in automobiles and furniture are polyurethanes. And, much of the plastic dinnerware you have seen is based on melamine-formaldehyde monomers.

ADDITION POLYMERIZATION.

Free Radical Polymerization. Table 12-1.2.1 lists several of the more common monomers used in forming addition polymers. One thing that should strike you immediately is the fact that all these monomers have a double bond in the structure. This is necessary for the addition reaction to take place. The general reaction of these monomers is

$$n(CH_2{=}CHY) \rightarrow -(CH_2{-}CHY)_n-$$

where Y is either an organic or inorganic molecule or atom attached to the organic chain. In all these reactions, there must first be a reactive center formed. This is usually accomplished by the addition of a material that forms a free radical that can initiate the polymerization reaction. A free radical is a molecule that has an unpaired electron. The dissociation of an active additive to form the free radical is the first step, or initiation reaction. The steps in the polymerization

Table 12-1.3.2
Some Typical Condensation Polymers

Type	Characteristic linkage	Polymerization reaction
Polyamide	-CO-NH-	$H_2N\text{-}R\text{-}NH_2 + HO_2C\text{-}R'\text{-}CO_2H \rightarrow H\text{-}(\text{-}NH\text{-}R\text{-}NHCO\text{-}R'\text{-}CO\text{-})_n\text{-}OH + H_2O$
		$H_2N\text{-}R\text{-}NH_2 + ClCO\text{-}R'\text{-}COCl \rightarrow H\text{-}(\text{-}NH\text{-}R\text{-}NHCO\text{-}R'\text{-}CO\text{-})_n\text{-}Cl + HCl$
		$H_2N\text{-}R\text{-}CO_2H \rightarrow H\text{-}(\text{-}NH\text{-}R\text{-}CO\text{-})_n\text{-}OH + H_2O$
Protein, wool, silk	-CO-NH-	Naturally occurring polypeptide polymers; degradable to mixtures of different amino acids
		$H\text{-}(\text{-}NH\text{-}R\text{-}NHCO\text{-}R'\text{-}CO\text{-})_n\text{-}OH + H_2O \rightarrow H_2N\text{-}R\text{-}CO_2H + H_2N\text{-}R'\text{-}CO_2H$
Polyester	-CO-O-	$HO\text{-}R\text{-}OH + HO_2C\text{-}R'\text{-}CO_2H \rightarrow HO\text{-}(\text{-}R\text{-}OCO\text{-}R'\text{-}COO\text{-})_n\text{-}H + H_2O$
		$HO\text{-}R\text{-}OH + R''O_2C\text{-}R'\text{-}CO_2R'' \rightarrow HO\text{-}(\text{-}R\text{-}OCO\text{-}R'\text{-}COO\text{-})_n\text{-}R'' + R''OH$
		$HO\text{-}R\text{-}CO_2H \rightarrow HO\text{-}(\text{-}R\text{-}COO\text{-})_n\text{-}H + H_2O$
Polyurethane	-O-CO-NH-	$HO\text{-}R\text{-}OH + OCN\text{-}R'\text{-}NCO \rightarrow \text{-}(\text{-}O\text{-}R\text{-}OCO\text{-}NH\text{-}R'\text{-}NH\text{-}CO\text{-})_n\text{-}$
Polysiloxane	-Si-O-	$Cl\text{-}SiR_2\text{-}Cl \xrightarrow[-HCl]{H_2O} HO\text{-}SiR_2\text{-}OH \rightarrow HO\text{-}(\text{-}SiR_2\text{-}O\text{-})_n\text{-}H + H_2O$
Phenol-formaldehyde	$\text{-}Ar\text{-}CH_2\text{-}$	$C_6H_5OH + CH_2O \rightarrow [\text{-}C_6H_3(OH)\text{-}CH_2\text{-}]_n + H_2O$
Urea-formaldehyde	$\text{-}NH\text{-}CH_2\text{-}$	$H_2N\text{-}CO\text{-}NH_2 + CH_2O \rightarrow \text{-}(\text{-}HN\text{-}CO\text{-}NH\text{-}CH_2\text{-})_n\text{-} + H_2O$

Melamine-formaldehyde	$-NH-CH_2-$	$H_2N-C_3N_3(NH_2)-NH_2 + CH_2O \rightarrow -[-HN-C_3N_3(NH_2)-NH-CH_2-]_n-$
Cellulose	-O-C-	Naturally occurring; degradable to glucose $-(-C_6H_{12}O_4-)_n- + H_2O \rightarrow C_6H_{12}O_6$
Polysulfide	$-S_m-$	$Cl-R-Cl + Na_2S_m \rightarrow -(-R-S_m-)_n- + NaCl$
Polyacetal	-O-CH-O- R	$R-CHO + HO-R'-OH \rightarrow -(-O-R'-OCHR-)_n- + H_2O$

are: initiation, propagation, termination, and chain transfer. Now consider each of these steps.

Initiation. One common chemical that is used to produce free radicals is benzoyl peroxide that decomposes as follows:

$$(C_6H_5COO)_2 \rightarrow 2C_6H_5COO\cdot \rightarrow 2C_6H_5\cdot + 2CO_2$$

A free radical is designated as R·. The initiation step is

$$R\text{-}R \rightarrow 2R\cdot$$

$$R\cdot + CH_2{=}CHY \rightarrow RCH_2\overset{\displaystyle H}{\underset{\displaystyle Y}{C}}\cdot$$

The free radical has reacted with a monomer to produce a new free radical, which in turn can react with another monomer to produce a longer chain that is also a free radical, etc.

Propagation. The propagation steps are those that lengthen the chain and continue to produce free radicals as shown below.

$$R\text{-}(CH_2CHY\text{-})_x\,CH_2\overset{\displaystyle H}{\underset{\displaystyle Y}{C}}\cdot + CH_2{=}CHY \rightarrow R\text{-}(CH_2CHY\text{-})_{x+1}\,CH_2\overset{\displaystyle H}{\underset{\displaystyle Y}{C}}\cdot$$

This reaction can continue as long as there is monomer present, or until some reaction that destroys the free radical occurs. The destruction of a free radical is called a termination step, since the polymer chain ceases to grow further.

Termination. If two radicals approach each other, they have a very strong tendency to react. Such a reaction will lead to the termination of two growing polymer chains.

$$\text{-}CH_2\overset{\displaystyle H}{\underset{\displaystyle Y}{C}}\cdot + \cdot\overset{\displaystyle H}{\underset{\displaystyle Y}{C}}CH_2 \rightarrow \text{-}CH_2\overset{\displaystyle H}{\underset{\displaystyle Y}{C}}\text{-}\overset{\displaystyle H}{\underset{\displaystyle Y}{C}}CH_2\text{-}$$

Another type of termination occurs by *disproportionation.*

$$\text{-}CH_2\overset{\displaystyle H}{\underset{\displaystyle Y}{C}}\cdot + \cdot\overset{\displaystyle H}{\underset{\displaystyle Y}{C}}CH_2\text{-} \rightarrow \text{-}CH_2\overset{\displaystyle H}{\underset{\displaystyle Y}{C}}\text{-}H + \overset{\displaystyle H}{\underset{\displaystyle Y}{C}}{=}CH\text{-}$$

In this case we have a transfer of an H atom from one radical to another which eliminates two radicals. Both the above types of termination reactions are known to occur.

Chain Transfer. In a chain transfer reaction, the free radical can donate an atom to another molecule producing a free radical on the new molecule and eliminating the initial free radical. Such an occurrence leads to branching in the polymer chain. By careful control of the types of molecules added as transfer agents, the degree of branching can be controlled to produce polymers with particular properties.

Linkage in Addition Polymers. When a free radical reacts with a vinyl monomer (one with a double bond) it can do so in either of two ways.

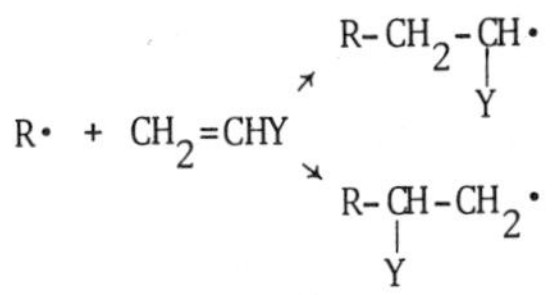

$$R\cdot + CH_2{=}CHY \nearrow R\text{-}CH_2\text{-}\underset{\displaystyle Y}{\underset{|}{CH}}\cdot \qquad (A)$$

$$R\cdot + CH_2{=}CHY \searrow R\text{-}\underset{\displaystyle Y}{\underset{|}{CH}}\text{-}CH_2\cdot \qquad (B)$$

Whether reaction A or B is favored depends on the nature of Y; the reaction leading to the most stable product will be the one that occurs most frequently. If reaction A were the only one to occur, we would obtain a *head-to-tail* polymer of the form

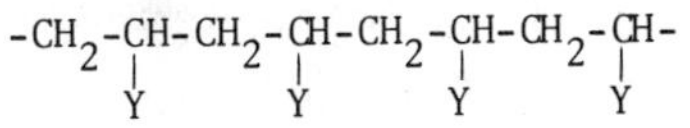

$$\text{-}CH_2\text{-}\underset{\displaystyle Y}{\underset{|}{CH}}\text{-}CH_2\text{-}\underset{\displaystyle Y}{\underset{|}{CH}}\text{-}CH_2\text{-}\underset{\displaystyle Y}{\underset{|}{CH}}\text{-}CH_2\text{-}\underset{\displaystyle Y}{\underset{|}{CH}}\text{-}$$

The polymerization reaction can result in some *head-to-head, tail-to-tail* links as follows:

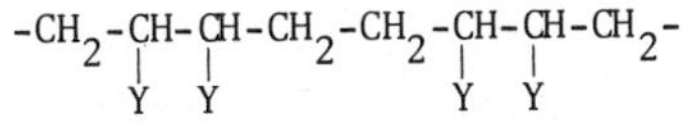

$$\text{-}CH_2\text{-}\underset{\displaystyle Y}{\underset{|}{CH}}\text{-}\underset{\displaystyle Y}{\underset{|}{CH}}\text{-}CH_2\text{-}CH_2\text{-}\underset{\displaystyle Y}{\underset{|}{CH}}\text{-}\underset{\displaystyle Y}{\underset{|}{CH}}\text{-}CH_2\text{-}$$

The final possibility is a random arrangement of the two sequences shown above.

Ionic Chain Polymerization. Addition polymerization can occur by an ionic mechanism instead of a free radical intermediate. These types of reaction are not as well understood as the previously discussed cases. The mechanisms for both anionic and cationic chain reactions can be found in standard organic textbooks and will not be considered further. You should now go back to Table 12-1.3.1 and explain why each of the statements given is true, based on what you have learned about the mechanisms involved.

12-1.4 EXAMPLES OF SOME POLYMERIZATION TECHNIQUES

This section will present a short discussion of some of the methods used to polymerize various monomers. Several different systems will be considered:

1. Homogeneous polymerization, in which the reaction all takes place in one phase. This may be in a solvent or in a reaction vessel containing only monomer.

2. Heterogeneous polymerization, in which more than one phase is involved. (You might turn to the first section in Chap. 13 for a definition of the term phase.)
3. Interfacial polymerization, in which the reaction takes place at the boundary of two different phases.

An excellent book to consult for more detail is E. A. Collins, J. Bareš, and F. W. Billmeyer, Jr., *Experiments in Polymer Science* (see Sec. 12-1.6, Ref. 1). Much of the material in this chapter is based on that text and the book by F. W. Billmeyer, Jr., *Textbook of Polymer Science* (Sec. 12-1.6, Ref. 2). Both these books are highly recommended for a more thorough introduction to polymer science. Table 12-1.4.1 compares the different polymerization systems. Glance over it now and refer to it as you read the following sections.

HOMOGENEOUS POLYMERIZATION.

Solution Polymerization. In solution polymerization, the monomer is dissolved in a suitable solvent, an initiator is added, and the reaction is allowed to proceed. Additives may be dissolved in the solution to vary the amount of branching and/or chain length or the molecular weight distribution (discussed in Sec. 12-2). Table 12-1.4.1 lists some of the advantages and disadvantages for this type of system. The term in the table, *autoacceleration*, denotes the tendency of some products of the reaction to cause the reaction to proceed at a faster rate. The faster rate produces more product that then increases the rate even more. There are cases in which autoacceleration must be controlled or the rate can reach explosive proportions.

Bulk Polymerization. For a bulk polymerization, the reaction vessel contains only monomer with small amounts of initiator and modifiers present. Both condensation and radical chain polymerizations can be carried out in a bulk reactor. However, the highly exothermic nature of radical chain reactions leads to difficulty in temperature control. Also, in this case more than in the solution system, there is a problem with viscosity. As the polymerization proceeds, the molecular weight of the polymer increases, leading to an increase in the viscosity of the reaction mixture. This leads to problems in heat transfer and mixing of reactants. This problem is more severe for free radical reactions than for condensation reactions.

HETEROGENEOUS POLYMERIZATION.

Emulsion Systems. The term heterogeneous implies that more than one phase is involved in the reaction. In the emulsion system, a surface-active material (a "soap") is added to the solution to form micelles that are dispersed throughout the solution. (Section 11-3.6 gives a short description of the formation of emulsions.

Table 12-1.4.1
Advantages and Disadvantages of Various Polymerization Systems[a]

Type	Advantages	Disadvantages
Homogeneous		
Bulk (batch type)	Low impurity level Low residual initiator	Difficult to control thermally Difficult to remove traces of monomer and initiator Autoacceleration can lead to violent or explosive reactions
Bulk (continuous)	Better thermal control than in bulk Narrower MWD[b] than in bulk	Difficult to isolate polymer Requires agitation, monomer recycling
Solution	Easy thermal control Easy mixing due to low viscosity	Difficult to remove solvent and other ingredients Cost of solvent recovery
Heterogeneous		
Suspension	Low viscosity throughout High purity product compared to emulsion systems Simple polymer isolation Easy thermal control	Reaction highly sensitive to agitation Particle size and surface characteristics difficult to control Possible contamination by stabilizers Polymer may require washing, drying, and compaction
Emulsion	Low viscosity throughout Easy thermal control Latex may be directly usable Virtually 100% conversion may be achieved High MW[c] and narrow MWD possible, at high rates	Emulsifiers, surfactants, and coagulants hard to remove High residual impurity level may degrade polymer properties Higher cost than suspension systems Polymer may require washing, drying, and compacting

[a] From E. A. Collins, J. Bareš, and F. W. Billmeyer, Jr., *Experiments in Polymer Science*, John Wiley and Sons, New York, 1973 (see Sec. 12-1.6, Ref. 1).

[b] MWD = molecular weight distribution.

[c] MW = molecular weight.

You might look over that briefly if you have not already done so.) In addition to the micelles that are formed by the soap and subsequently saturated with monomer, there will be small droplets of monomer if the monomer is not soluble in the solvent. The addition of an initiator leads to a reaction in the micelles. As the reaction proceeds these micelles increase in size due to the formation of polymer in them.

Monomer diffuses from the monomer droplets to replenish that used in the reaction. Since the polymerization sites are isolated from each other, termination reactions are less likely than in other systems, and high molecular weights can be obtained. The most common example of this type of system is the polymerization of vinyl chloride to give polyvinyl chloride.

Suspension Polymerization. The difference between this system and the previous one is that no soap is added, and the polymerization occurs in the monomer droplets that form when an insoluble monomer is added to an appropriate solvent. Some type of additive must be included to maintain the monomer in suspension. In the emulsion system, we observed that each micelle usually contains one initiator molecule. This is due to the very small size and large number of micelles formed. In the suspension system, each droplet may contain several initiators. Thus the reaction behaves much like a bulk reaction. The temperature control problems are removed, however, because of the small size of the droplets. Vinyl chloride is sometimes polymerized in this system, also.

INTERFACIAL POLYMERIZATION. Usually, interfacial polymerizations are limited to step-reaction (condensation) reactions. One of the two monomers is dissolved in one phase of a two-phase system, while the other reactant is found in the other phase. The only place the reaction can occur is at the interface between the two solutions, since this is the only place the two monomers can come into contact. Some of the nylons are prepared in this way. In fact, a very interesting demonstration can be perfommed by pulling a nylon rope out of a beaker containing sebacoyl chloride in carbon tetrachloride and hexamethylenediamine in water. [See *J. Chem. Educ., 36*:182-184 (1959)]. There are problems with solvent removal that limit the commercial application of this technique.

12-1.5 NATURAL MACROMOLECULES

Some reference to natural macromolecules is given in Table 12-1.3.2 in which the polymer linkages in cellulose, protein, wool, and silk are shown. Here, only those polymers that are of biological significance will be discussed. The three classes to be described are proteins, nucleic acids, and polysaccharides.

Proteins. The characteristic linkage for proteins is the *peptide bond*, as shown in Table 12-1.3.2. Proteins are formed from amino acids which have the generalized structure (these are called α amino acids) and reactions shown below.

$$H_2N\text{-}\underset{\displaystyle R}{\underset{|}{C}H}\text{-}\overset{\displaystyle O}{\overset{\|}{C}}\text{-}OH + H_2N\text{-}\underset{\displaystyle R'}{\underset{|}{C}H}\text{-}\overset{\displaystyle O}{\overset{\|}{C}}\text{-}OH \rightarrow H_2N\text{-}(\text{-}\underset{\displaystyle R}{\underset{|}{C}H}\text{-}\overset{\displaystyle O}{\overset{\|}{C}}\text{-}NH\text{-}\underset{\displaystyle R'}{\underset{|}{C}H}\text{-}\overset{\displaystyle O}{\overset{\|}{C}}\text{-})_x\text{-}\overset{\displaystyle O}{\overset{\|}{C}}\text{-}OH$$

Very large molecules can be formed from this type of reaction. A protein is not that simple, however. Proteins are composed of large numbers of amino acids (R and R' can vary from unit to unit in the chain) bonded together in a well-defined sequence. The sequence of amino acids in the protein structure gives the protein its characteristic properties. Not only is the sequence important, the morphology (the large-scale structure of macromolecules; see Sec. 12-3) is crucial to the function of the protein. The actual structure of the protein is determined by *intramolecular* interaction, such as hydrogen bonding.

Nucleic Acids. Nucleic acids are the materials found in the living cell that control the function of the cell and the various biochemical processes that occur. The two main types are deoxyribonucleic acid (DNA) and ribonucleic acid (RNA). These are composed of sugar residues and phosphate groups. The structures of these two nucleic acids can be found in biochemical texts. No attempt will be made to draw them here.

Polysaccharides. The final, biologically significant polymers to be considered are the polysaccharides, which are the condensation products of simple sugars. Some examples are celluloses and starches. As you are aware, cellulose molecules are threadlike, while starches are much nearer a spherical shape. The difference is in the way in which the sugar units bond when the condensation reaction takes place. There is no need to emphasize the importance of the two common cellulose examples: wood and cotton.

12-1.6 SELF-TEST

1. On the basis of the discussion in this section give a brief explanation of each of the statements in Table 12-1.3.1. Do not simply repeat the statements, explain them.
2. Write the structures and several repeating units for three monomers.
3. Use vinyl chloride ($CH_2{=}CHCl$) with some initiator (R••R) to illustrate all the steps in an addition polymerization.
4. Teraphthalic acid

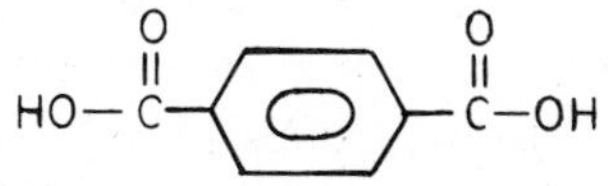

   $$HO-\overset{\overset{\displaystyle O}{\|}}{C}-C_6H_4-\overset{\overset{\displaystyle O}{\|}}{C}-OH$$

   can react with hexamethylenediamine $H_2N\text{-}(CH_2)_6\text{-}NH_2$ to produce a polyamide. What type of polymerization is this? Draw the repeating unit for the resulting polymer.

5. Discuss two different polymerization systems. List the characteristics and the advantages and disadvantages for those systems.

12-1.7 ADDITIONAL READING

1.* E. A. Collins, J. Bareš, and F. W. Billmeyer, Jr., *Experiments in Polymer Science*, 2nd ed., John Wiley and Sons, New York, 1971.

2.* F. W. Billmeyer, Jr., *Textbook of Polymer Science*, 2nd ed., John Wiley and Sons, New York, 1971.

3. G. Odian, *Principles of Polymerization*, McGraw-Hill, New York, 1970.

4. P. J. Flory, *Principles of Polymer Chemistry*, Cornell University Press, Ithaca, New York, 1953.

12-2 MOLECULAR WEIGHT METHODS

OBJECTIVES

You shall be able to

1. Define and calculate the *number average* and *weight average* molecular weights
2. Discuss in detail any of the methods for determining molecular weights presented in this section; include a description of the experimental methods and some of the problems associated with the method
3. Determine molecular weights from data given for any of the methods discussed here

12-2.1 INTRODUCTION

This section will be concerned with determination of the molecular weights of polymers. Many of the equations required for this section are developed in other portions of the book. These developments will be cross-referenced, and you are urged to review the necessary portions before continuing your study of any of the methods. The emphasis in this section is on experimental methods and limitations. The theory is found elsewhere in most cases.

In Sec. 11-1.2 you saw that there are several ways to express the molecular weight of a polydisperse system. If you recall, a *polydisperse system* is one in which the molecules have a range of molecular weights, while in a *monodisperse system* all molecules are the same weight. In the synthesis of almost all the polymeric materials, polydispersity results. Since many of the properties of a polymer depend on both the molecular weight and the molecular weight distribution, it is important to be able to determine these values. Figure 12-2.1.1 shows a typical distribution of molecular weights for a polydisperse polymer. M_n is the *number average molecular*

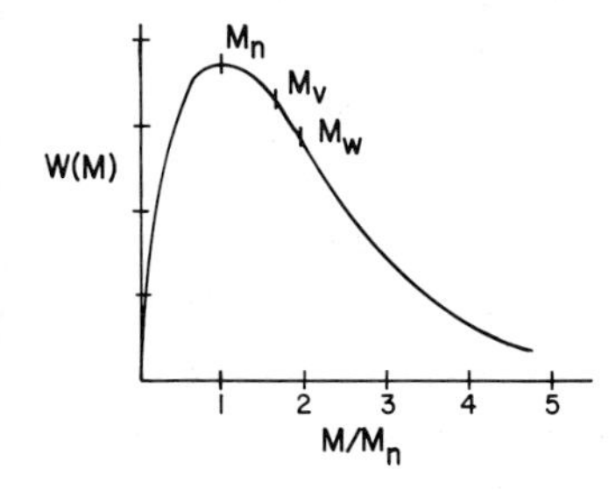

FIG. 12-2.1.1 A typical molecular weight distribution for a polydisperse polymer. (The symbols are defined in the text.)

weight defined by Eq. (11-1.2.1). M_w is the *weight average molecular weight* defined by Eq. (11-1.2.2). Example 11-1.2.1 should be reviewed at this time to show how these can be calculated. M_v is the *viscosity average molecular weight* that will be defined in Sec. 12-2.3. There are other averages, but these will not be considered. The ratio of M_w/M_n is useful in defining the general shape of the molecular weight distribution. These two averages can be obtained without a tremendous expenditure of effort. M_w is most readily obtained from the light scattering measurements very briefly mentioned below. M_n results from measurements involving colligative properties such as vapor pressure lowering or osmometry.

For completeness, let us rewrite the definitions of M_n and M_w. The number average is just the mass of the sample divided by the number of particles

$$M_n = \frac{W}{N} = \frac{\sum_i w_i}{\sum_i n_i} = \frac{\sum_i n_i m_i}{\sum_i n_i} \tag{12-2.1.1}$$

where N is the total number of particles, w_i is the mass of a particular fraction which contains n_i particles, and m_i is the mass of each particle in the i fraction. The weight average is given by

$$M_w = \frac{\sum_i m_i w_i}{\sum_i w_i} = \frac{\sum_i n_i m_i^2}{\sum_i n_i m_i} \tag{12-2.1.2}$$

M_n and M_w are the first and second moments of the molecular weight distribution.

12-2.2 METHODS FOR THE DETERMINATION OF M_n

BOILING POINT ELEVATION. In Sec. 6-2.3 you will find a detailed development of the equations required to find molecular weights from the elevation of the boiling point of a solution. You should study the appropriate equations before continuing. Boiling point elevation is very limited in polymer work due to the small changes in the boiling point associated with the formation of a solution. For example, a 0.01 kg dm^{-3} solution of a polymer with a molecular weight of 20 kg mol^{-1} results in a boiling point elevation of only 0.0013 K. This is a low molecular weight polymer. As the molecular weight of the polymer increases, the dissolution of a given weight of the material results in even less of a temperature change in the boiling point. [Prove that to yourself from Eq. (6-2.3.14).] In general this technique is limited to molecular weights less than about 30 kg mol^{-1}.

FREEZING POINT DEPRESSION. Section 6-2.4 gives the equations necessary to use freezing point depression data to find M_n. As in the case of boiling point elevation, this method is limited to values of M_n less than about 30 kg mol^{-1}. The experimental techniques used here are similar to those in the previous method. Thermistors record the freezing points of both a solvent and a solution. The difference observed is used to obtain M_n. A known substance is usually used to standardize the equipment. Supercooling (the case in which the material remains in the liquid state even after the freezing point is passed) can be a problem. Nucleating agents are sometimes added to facilitate freezing at the normal freezing point.

MEMBRANE OSMOMETRY. The theory behind the osmotic effect has been given in Sec. 6-2.5. It will not be repeated here. Turn to that section and read it before continuing your study in this chapter. In Sec. 6-2.5, osmosis is shown to be a colligative property that involves the transport of solvent across a semipermeable membrane when a concentration difference exists across that membrane. The requirement that a membrane separate the solutions of two concentrations leads to problems that limit the usefulness of this technique. Some of these problems will be discussed shortly.

The osmosis effect can yield values for M_n up to between 500 and 1000 kg mol^{-1} (500,000 to 1,000,000 g mol^{-1}). The upper limit is determined by the smallest pressure change that can be determined accurately. Assume we have a solution with a concentration of 0.001 kg solute per cubic decimeter of solvent. If the value of M_n for this polymer is 1000 kg mol^{-1} (10^6 g mol^{-1}), apply Eq. (6-2.5.12) to find the pressure difference across a membrane. Try to find the pressure in terms of an equivalent height of water before reading on. The result will be given below. The lower limit on the molecular weights that can be found by this technique is determined by the permeability of available membranes. Currently, particles with weights of less than 10 to 50 kg mol^{-1} will diffuse through the membrane enough to make the osmotic pressure measurements unreliable. If any foreign particles are present that diffuse through the membrane slowly or not at all, the osmotic pressure measured will not be the true osmotic pressure due only to the polymer.

12-2.2.1 EXAMPLE

The solution to the problem posed above is as follows. From Eq. (6-2.5.12) we find

$$\Pi = \frac{m_B/M_B}{V} RT = \frac{m_B}{V} \frac{1}{M_B} RT$$

Inserting the variables given yields

$$\Pi = (10^{-3} \text{ kg dm}^{-3}) \left(\frac{1}{10^3 \text{ kg mol}^{-1}}\right) (0.0821 \text{ dm}^3 \text{ atm mol}^{-1} \text{ K}^{-1})(298 \text{ K})$$

$$= 2.45 \times 10^{-5} \text{ atm} \frac{1.01 \times 10^5 \text{ N m}^{-2}}{1 \text{ atm}}$$

$$= \underline{\underline{2.47 \text{ N m}^{-2}}}$$

If you recall that the pressure due to a fluid head is

$$P = \rho g h$$

where ρ is the fluid density, g is the acceleration due to gravity, and h is the height of the liquid column, you can show that (for water)

$$h = \frac{\Pi}{\rho g} = \frac{2.47 \text{ kg m}^{-1} \text{ s}^{-2}}{(10^3 \text{ kg m}^{-3})(9.8 \text{ m s}^{-1})}$$

$$= 2.5 \times 10^{-4} \text{ m}$$

$$= \underline{\underline{0.25 \text{ mm } H_2O}}$$

This is a rather small pressure to try to measure accurately.

As mentioned earlier, membrane problems are a major limitation for the use of osmometry for the determination of molecular weights. Most of the materials that have been used are derivatives of cellulose. To use a membrane it must be conditioned to the solvent system in which the experiments are going to be run. Usually the membranes are received saturated with water which must be gradually replaced with the solvent of interest. This replacement is achieved through an involved process of conditioning in a series of solutions that get progressively richer in solvent until the final liquid bath is pure solvent. About the best that can be said about membrane selection and processing is that it is a very empirical process even today. Equipment is now available that allows osmotic measurements to be made quickly and accurately (within the limitations that have already been mentioned) once the membrane has been properly prepared. Old osmometers required the attainment of equilibrium by allowing solvent to diffuse through the membrane until the pressure due to the liquid column formed (see Fig. 6-2.5.1) just balanced the osmotic pressure. This required a long period of time, since an appreciable amount of solvent had to diffuse through the membrane (a slow process).

By replacing the liquid column with a piston or other device for opposing the osmotic pressure the time required is shortened significantly. No solvent is allowed to diffuse through the membrane. The pressure on the solution side is increased until the net solvent transport across the membrane is zero. The pressure necessary is just the osmotic pressure. Several modern systems have sensors that respond to pressure changes due to solvent flow through the membrane and automatically adjust the opposing pressure to yield a net solvent flow of zero. The pressure applied is digitally displayed and is the osmotic pressure of the solution. Another type of osmometer uses a strain gauge on the solvent side of the membrane and measures the pressure changes directly. Since the solvent and solution compartments are small, very little solvent must diffuse before equilibrium is attained.

VAPOR-PHASE OSMOMETRY. Vapor-phase osmometry is a misnomer since there is no membrane involved, and therefore no osmotic pressure is developed. Instead, vapor-phase osmometry refers to a method of determining the vapor pressure lowering caused by the addition of a solute to a solvent. The essentials of the method are as follows. Two thermistor beads are enclosed in a chamber saturated with vapor of the solvent being used. A drop of solvent is placed on one bead, and a drop of solution is placed on the other. There will be a small difference in the evaporation rate of the two drops since the vapor pressure of the solvent is lower in the solution drop. This results in a slight difference in temperature between the two beads that can be measured and recorded. The temperature difference is noted as a difference in the resistances of the two thermistors. A semiempirical equation that can be used to find M_n is

$$\frac{\Delta R}{c_B} = \frac{k}{M_n} \qquad (12\text{-}2.2.1)$$

where ΔR is the observed resistance difference, c_B is the concentration of the polymer solution in appropriate units (meaning the same units in which the calibration material is measured), and k is an empirical apparatus constant. The following example based on data found in the book by Collins et al. (See Sec. 12-2.8, Ref. 1) will illustrate the procedure.

12-2.2.2 EXAMPLE

In a vapor-phase osmometer, tristearin (M_n = 0.891 kg mol^{-1}) is used as a standard. A solution of 0.00441 kg dm^{-3} gives a ΔR reading of 3.41. A polyethylene sample is the same apparatus gives a ΔR value of 1.91 for a 0.00945 kg dm^{-3} solution. Find the value of M_n for polyethylene.

Solution

From (12-2.2.1) for tristearin

$$k = \frac{\Delta R}{c_B} M_n$$

$$= \frac{3.41\ (0.891\ \text{kg mol}^{-1})}{0.00441\ (\text{kg dm}^{-3})}$$

$$= \underline{689\ \text{dm}^3\ \text{mol}^{-1}}$$

For polyethylene

$$M_n = \frac{k\ c_B}{\Delta R} = \frac{(689\ \text{dm}^3\ \text{mol}^{-1})(0.00945\ \text{kg dm}^{-3})}{1.91}$$

$$= \underline{\underline{3.41\ \text{kg mol}^{-1}}}$$

Sources of difficulty in the method are: (1) presence of impurities of low molecular weight that have a relatively large effect on the vapor pressure for a given weight of impurity; (2) presence of volatile material in the solution; and (3) lack of consistency in drop size. These and other problems limit the method to $M_n < 50$ kg mol^{-1}.

12-2.3 SOLUTION VISCOSITY

The viscosity of a solution of a polymeric material can be related to the avergae molecular weight of the material. The average obtained by viscosity measurements is not the same as either M_n or M_w (see Fig. 12-2.1.1). A description of liquid viscosity and the method commonly used for its measurement is found in Sec. 14-1.5. Read that section before continuing. Before we can relate the viscosity of a solution to the molecular weight of the solute, we must define several terms. These are given below. The name given each variable is the common name for that variable. The name in parentheses is the recommended name. If we have measured the viscosity η of the solution we can define the following variables.

Relative viscosity (viscosity ratio):

$$\eta_r = \frac{\eta}{\eta_0} \tag{12-2.3.1}$$

where η_0 is the viscosity of the pure solvent.

Specific (sp) viscosity:

$$\eta_{sp} = \eta_r - 1 = \frac{\eta - \eta_0}{\eta_0} \tag{12-2.3.2}$$

Reduced (red) viscosity (viscosity number):

$$\eta_{red} = \frac{\eta_{sp}}{c} \tag{12-2.3.3}$$

where c is the concentration of the solution. Values for c are almost always listed as *grams of solute per 100 ml* solution. We will use that convention even though it is not an SI unit. This can also be written as grams per deciliter (dl) when dl = deciliter = 10^{-1} liter = 10^{-1} dm^3.

Inherent (inh) viscosity (logarithmic viscosity number):

$$\eta_{inh} = \frac{\ln \eta_r}{c} \tag{12-2.3.4}$$

Intrinsic viscosity (limiting viscosity number):

$$[\eta] = \lim_{c \to 0} \frac{\eta_{sp}}{c} = \lim_{c \to 0} \frac{\ln \eta_r}{c} \tag{12-2.3.5}$$

The details will not be given, but it can be shown that the intrinsic viscosity can be related to the viscosity average molecular weight by

$$\boxed{[\eta] = KM_v^{\ a}} \tag{12-2.3.6}$$

where K and a are empirical constants that must be evaluated for a *particular polymer* in a *particular solvent* at a *particular temperature*. To obtain the constants, known molecular weight samples must be available. The last equation is attributed to *Staudinger* and is based on the result of a statistical treatment of the shape acquired by a long, flexible molecule (see M. A. Volkenstein, "Configurational Statistics of Polymer Chains," *Interscience High Polymers*, Vol. 17, 1963).

Perhaps you can see how all the above equations can be used to get M_v once K a and a have been experimentally determined. The viscosity of the pure solvent is measured, then values of η are obtained for several concentrations of polymer at the same T. The two plots suggested by Eq. (12-2.3.5) are made, and the curves are extrapolated to c = 0 to get the intrinsic viscosity. This value is then used to obtain a value for M_v. As usual there are restrictions on the validity of Eq. (12-2.3.6). The polymer must be linear (no branching), and the aforementioned restrictions for T and solvent must be noted. There are other theories that are more exact and more general, but a discussion of these is beyond the scope of this book. The advantages in determining M_v instead of M_n or M_w are that the equipment is simple and the results are quickly available. These considerations make this a good method for following the progress of a polymerization reaction be sampling at given intervals. The following example illustrates how the equations can be used. Also, Table 12-2.3.1 lists some values for the constants K and a for several systems. *These constants yield M_v values in grams per mole.*

12-2.3.1 EXAMPLE

From the following data for the times of flow through a viscometer for polystyrene in benzene, determine the molecular weight of the polystyrene sample; $\eta_{benzene} = 0.652 \times 10^{-3}$ kg m^{-1} s^{-1}.

c (g/100 ml)	0.00	1.00	0.800	0.500	0.222
t (sec)	208.2	421.3	371.8	303.4	248.1

Solution

Equation (12-2.3.5) suggests the quantities needed: η_{sp}/c and ln η_r/c. Section 14-1.5 should be consulted to see how to handle the data. If we assume $\rho_{benzene} = \rho_{solution}$ (a good assumption for dilute solutions), Eq. (14-1.5.14) reduces to

$$\eta_r = \frac{\eta}{\eta_0} = \frac{t}{t_0}$$

c (g/100 ml)	0.00	1.00	0.800	0.500	0.222
t (s)	208.2	421.3	371.8	303.4	248.1
$\eta_r = \eta/\eta_0$	--	2.0235	1.7858	1.4573	1.1916
$(\ln \eta_r)/c$	--	0.705	0.725	0.7531	0.790
$\eta_{sp} = \eta_r - 1$	--	1.0235	0.7858	0.4573	0.1916
η_{sp}/c	--	1.024	0.982	0.915	0.863

The intrinsic viscosity can be found from the figure below. (Note how closely the two limits approach each other.)

$$[\eta] = 0.812 = KM_v^{a}$$

$K = 0.20 \times 10^{-4}$ and $a = 0.74$ (from Table 12-2.3.1).

$$M_v = \left(\frac{[\eta]}{K}\right)^{1/a} = 1.7 \times 10^{6}\ \text{g mol}^{-1}$$

$$= \underline{\underline{1700\ \text{kg mol}^{-1}}}$$

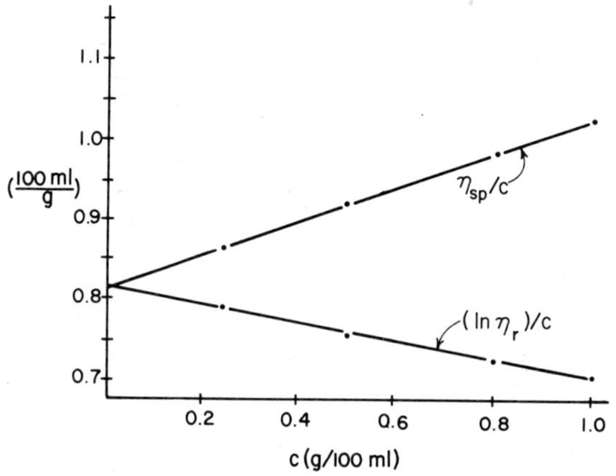

1.1
1.0
$\left(\frac{100\ \text{ml}}{\text{g}}\right)$
0.9
0.8
0.7
η_{sp}/c
$(\ln \eta_r)/c$
0.2 0.4 0.6 0.8 1.0
c (g/100 ml)

12-2.4 DETERMINATION OF M_w: LIGHT SCATTERING

The principal method of finding the weight average molecular weight is from light scattering data. Only a very brief sketch of some of the important considerations

Table 12-2.3.1
Values for K and a in Staudinger's Equation for Several Systems[a]

Polymer	Solvent	T (K)	$K \times 10^4$	a
Polystyrene	Benzene	298	0.2	0.74
	Cyclohexane	308	7.6	0.50
	Toluene	298	1.7	0.69
Polyvinyl alcohol	Water	298	2.0	0.76
Polyvinyl acetate	Acetone	298	2.1	0.68
Polymethyl methacrylate	Acetone	298	0.75	0.70
	Benzene	298	0.55	0.76
	Chloroform	293	0.60	0.79

[a] For M_v in grams per mole. [From E. A. Collins, J. Bares, and F. W. Billmeyer, Jr., *Experiments in Polymer Science*, John Wiley and Sons, New York, 1973 (see Sec. 12-2.8, Ref. 1).]

in light scattering measurements will be given. The development of the theory is rather involved and will not be included.

If light of intensity I_0 is directed toward a solution, the intensity of the beam will be diminished as it travels through the solution and is scattered. The decrease in intensity can be related to a quantity called the turbidity τ by

$$\tau = \frac{1}{\ell} \ln \frac{I_0}{I} \qquad (12\text{-}2.4.1)$$

where ℓ is the length of the cell containing the solution. Debye derived an equation relating the molecular weight of a polymer in solution to the scattering by that solution.

$$\frac{Hc}{\tau} = \frac{1}{M'} + Ac + Bc^2 + \cdots \qquad (12\text{-}2.4.2)$$

where c is the concentration in kilograms per cubic decimeter (or grams per cubic centimeter), A and B are constants. H is given by

$$H = \frac{32\pi^2 n_0^2 (dn/dc)^2}{3N\lambda_0^4} \qquad (12\text{-}2.4.3)$$

where n_0 is the refractive index of the solvent, N is Avagadro's number, dn/dc gives the variation of the refractive index with the concentration of the solution, and λ_0 is the wavelength of the incident light (in vacuum). The quantity M' in Eq. (12-2.4.2) is related to the true molecular weight of the polymer by

$$M = M'\alpha\beta \qquad (12\text{-}2.4.4)$$

Both α and β are correction terms that account for the size of the polymer molecules and their nature. Turbidities are determined at several concentrations. H is calculated from measured values of n_0, λ_0 and dn/dc. Detail on the experimental methods for finding all these variables will be omitted. A plot of Hc/τ is extrapolated to $c = 0$ to give a value for $1/M'$ as indicated by Eq. (12-2.4.2). Applying the corrections terms (also determined experimentally) allows the value of M_w to be obtained. Much more detail can be found in Billmeyer (Sec. 12-2.8, Ref. 2).

As usual, there are some problems associated with the use of the light scattering technique. Extreme care must be taken to remove dust and other extraneous material. The method is, however, applicable over a wide range of molecular weights from as low as less than 10 kg mol^{-1} to 10,000 kg mol^{-1}. This range is much larger than that available from any other method.

12-2.5 OTHER METHODS

The other methods that can be used in finding molecular weights have been detailed in Sec. 11-1. Refer to that section for more information. In Sec. 11-1, centrifugation, sedimentation, diffusion, and ultracentrifugation are described. The data obtained from the experiments described in Sec. 11-1.5, concerning the balance of diffusion and centrifugation, lead to the actual distribution of molecular weights in the sample. Of course, if this is known, either M_n or M_w can be found. The major disadvantage to this particular method is the long time period required to reach an equilibrium distribution.

12-2.6 PROBLEMS

1. From the following light scattering data, find M_w for sucrose and compare to the true value of 0.342 kg mol^{-1}; λ = 5461 Å T = 298 K, α = 1.00, and β = 0.935.

c (kg dm^{-3})	0.0352	0.0614	0.106	0.163
$Hc/\tau \times 10^3$	2.84	2.91	3.08	3.32

2. The following values of η_r were observed for polymethyl methacrylate in ethylene dichloride. Find [η] by two plots.

c (g/100 ml)	0	0.125	0.250	0.500	1.000
η_r	--	1.142	1.298	1.641	2.475

3. In an experiment to find the molecular weight for a colloid using an ultracentrifuge, the following data were observed for particles with $\bar{v}$ = 0.92 cm^3 g^{-1} in water at 293 K (ρ = 0.998 kg dm^3), for a centrifuge velocity of 10^4 radians s^{-1}. Determine the molecular weight of this colloid.

$$x_1 = 0.10 \text{ m} \qquad C_1 = 0.0010 \text{ kg dm}^{-3}$$

$$x_2 = 0.11 \text{ m} \qquad C_2 = 10.0 \text{ kg dm}^{-3}$$

4. The following osmotic pressure data were obtained for a polymer in an organic solvent at 293 K. Determine the molecular weight of the polymer.

c (kg dm^{-3})	0.00200	0.00400	0.00600	0.00800
Π (N m^{-2})	62.5	156	281	437

5. A solution of 1 g of a polymer per 100 cm^3 solution is made. If M = 10.0 kg mol^{-1}, what is the vapor pressure of the solution at 18°C. $P^0_{H_2O}$ = 21 mm Hg.

6. $[\eta] = 1.02$ for a polymer with M = 34.0 kg mol^{-1}. If a = 0.73 in Staudinger's equation, find K. What is M for a polymer with $[\eta] = 2.75$?

7. The observed (obs) molecular weight of a polymer in an organic solvent is found to vary with concentration at 293 K as follows. (a) Estimate the true value of M. (b) Explain the variation of M_{obs} with concentration.

c (kg dm^{-3})	0.0020	0.0040	0.0060	0.0080
M_{obs} (kg mol^{-1})	60.1	55.8	52.1	44.0

8. For a certain polymer in CCl_4 the following osmotic pressure data were obtained. Given that $\rho = 1.594$ kg dm^{-3} find the molecular weight of the polymer.

c (kg dm^{-3})	0.00199	0.00402	0.00600	0.00797
Δh (m of CCl_4)	0.0040	0.0100	0.0180	0.0280

12-2.7 SELF-TEST

1. Give a one-page discussion (each) of the sedimentation-diffusion equilibrium and the viscosity methods of determining molecular weight.

2. Horse hemoglobin in a water suspension at 20°C is found to have a diffusion coefficient of $D = 6.3 \times 10^{-7}$ cm^2 s^{-1} and a sedimentation coefficient of $s = 4.41 \times 10^{-13}$ s. If $\rho = 0.9982$ kg dm^{-3} and $\bar{v} = 0.749$ dm^3 kg^{-1}, find the molecular weight of the horse hemoglobin.

3. The following data were observed for polyisobutylene in diisobutylene at 20°C. Determine the constants in the Staudinger equation.

$[\eta]$	0.20	0.31	0.55	1.05	2.3
M (g mol^{-1})	2×10^4	3.1×10^4	10^5	3×10^5	10^6

12-2.8 ADDITIONAL READING

1.* E. A. Collins, J. Bareš, and F. W. Billmeyer, Jr., *Experiments in Polymer Science*, John Wiley and Sons, New York, 1973.

2.* F. W. Billmeyer, Jr., *Textbook of Polymer Science*, 2nd ed., John Wiley and Sons, New York, 1971, Sec. 3C.

3. P. J. Flory, *Principles of Polymer Chemistry*, Cornell University Press, Ithaca, N. Y., 1953.

12-3 POLYMER MORPHOLOGY

OBJECTIVES

You shall be able to

1. Discuss the methods used to characterize polymers
2. Give a detailed account of the morphology of polymers, including a discussion of the terms lamellae, spherulites, necking, and drawing

12-3.1 INTRODUCTION

This section will be a bridge between the discussion of basic properties and practical properties. In the previous two sections you have studied some of the basics of polymerization and the methods of determining molecular weight and/or the molecular weight distribution in the resulting polymer. The following sections are concerned with the behavior of polymers in given circumstances and how this behavior can be modified to give a polymer that has desirable characteristics for a particular application. We will see how a knowledge of the manner in which properties vary with certain physical parameters (such as molecular weight or degree of crystallinity, for example) allows us to tailor a polymer to fit a particular application.

The first part of this section gives a brief discussion of some of the methods used to characterize polymers. It is necessary to know some of the properties of a polymer in order for an intelligent decision to be made regarding whether further study of the polymer is justified. The simple tests outlined below give some information of this type. Following the study of analysis techniques, a fairly detailed review of polymer morphology is given. Polymer morphology is the study of the form and structure of polymers. We shall see that the amount of crystallinity and the way in which crystalline regions are arranged in a polymer sample are important in determining the properties of that polymer. The last few years have seen a tremendous amount of work done in this area in an attempt to understand how crystallization occurs in polymers and how the regions of crystallinity can be controlled. These studies are now yielding results that allow the physical properties of polymers to be controlled more closely than previously. This section will deal mainly with polymers in the solid and liquid states. However, the morphology of polymers in solution is also important. The discussion in Sec. 11-1.3 about solvent-colloid interactions is applicable to polymers in solution.

12-3.2 SIMPLE ANALYSIS TECHNIQUES

THERMAL CHARACTERIZATION. A brief study of the behavior of a polymer as its temperature is changed yields important information about the makeup of the polymer. However, a complete description of a polymer is not possible on the basis of thermal behavior alone. Whether a material is a thermoplastic or a thermoset plastic can readily be determined by observing the behavior of a sample through a heating and cooling cycle. A *thermoplastic* material is one that can be heated and cooled repeatedly without appreciably changing its nature or properties. On the other hand, a *thermosetting material* (one formed by chemical reaction at elevated temperatures) will undergo radical changes in properties when heated and will not return to its initial condition on cooling. In other words, a thermosetting material decomposes (at least partially) on heating.

In the following, two terms will be used repeatedly. These are crystalline and amorphous. A *crystalline* material is one in which the molecules in the solid are arranged in an orderly fashion. The immediate and distant environment about any given molecule is the same as about any other molecule of the same type. More detail on crystallinity can be found in Sec. 14-5. An *amorphous* material, on the other hand, lacks long-range order. There may be some ordering in the immediate vicinity of a particular molecule, but none at some distance away. Amorphous materials are random in nature, similar to liquids. Whether a material is crystalline or amorphous can readily be determined by observing whether it "melts" over a small temperature range or softens over a wide range. Of course, even crystalline polymers soften before melting, but at the melting point abrupt changes in viscosity (and other properties) occur. For amorphous polymers it is possible to observe a glass-transition temperature, T_g. *The glass transition temperature is that temperature below which the material is glassy and above which it is rubbery.* You will see more about this in Sec. 12-4 where structure-property relationships are considered.

Perhaps the easiest way of making a preliminary classification of a material into the proper class of polymers is on the basis of flammability tests. By observing such things as flame color, whether the flame propagates after removal of the ignition source, the amount and color of any smoke evolved, whether the formation of molten drops occurs, and the odor of the combustion products, we can classify the substance into a polymer type (for simple polymers). However, these tests can be hazardous. Several polymers emit toxic gases (Teflon, polyacrylonitrile, etc.) upon thermal decomposition. Thus, flame tests mast be made only after suitable precautions are taken. Not all polymers can be identified by flame tests alone, of course. Many polymeric materials are not synthesized from a single monomer. Some have two or more monomers that will cause confusing results for the flame tests. Also, the presence of fillers and modifiers will affect the results of the flame tests as well as other tests. The degree of solubility of a sample in a series of solvents allows the sample to be classified in many cases. As an example, if a polymer sample is found to be soluble in toluene and ethyl acetate, but insoluble in methanol and carbon tetrachloride, it is possible that the sample is polymethyl methacrylate. Solubility alone is not sufficient to identify a polymer conclusively. There are other such tests that are useful, but these will not be considered in detail. We will see later that differential scanning calorimetry can be used to estimate the glass transition temperature and determine if any crystallinity is present, and, perhaps, the percent crystallinity. The above is not meant to be an exhaustive treatment. In fact, you do not have enough information even to begin to identify the simplest of polymers. However, you should now be aware of some of the simple means that can be used to start an identification. More detail on the effect of temperature on polymers will be given in Sec. 12-4.

MECHANICAL PROPERTIES. Some of the mechanical properties that must be measured before a polymer can be properly characterized are outlined below. How a polymer behaves under the influence of a stress will determine its suitability for a given application. Little detail will be found in this section. Detailed descriptions of mechanical properties are in Sec. 12-5. We do need to know a few terms that will appear in the discussion of morphology in the following.

Mechanical behavior involves the deformation of a material by an applied stress. An example of this is the stretching of a bar of polymer by the application of a force to the ends. The behavior of simple polymers is sketched in Fig. 12-3.2.1. The terms used in the figure are as follows:

σ_L = tensile stress = force per unit area
ε_L = tensile strain = stretch per unit length
σ_y = yield stress
ε_y = elongation at yield
σ_B = ultimate strength or tensile strength
ε_B = ultimate elongation or elongation at break

The Young's modulus for a polymer, as for any other material that behaves elastically

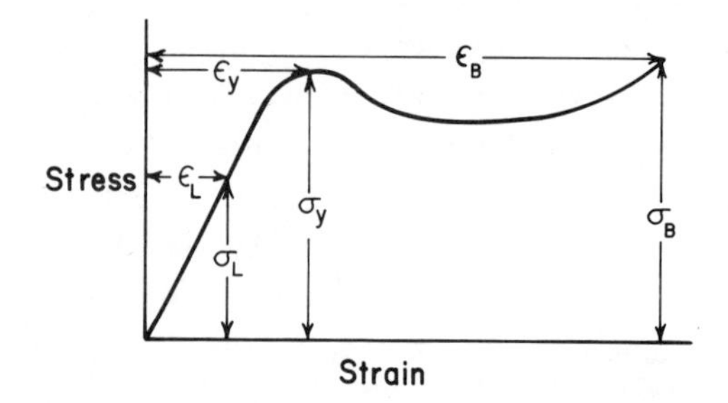

FIG. 12-3.2.1 The stress-strain behavior of a typical polymer.

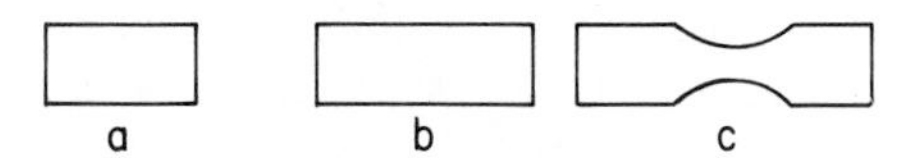

FIG. 12-3.2.2 The behavior of a bar of polymer under stress. (a) undeformed, (b) deformation in the elastic region, (c) deformation beyond the elastic limit, necking and irreversible deformation occurs.

(follows Hooke's law, the linear region in Fig. 12-3.2.1) is defined as

$$E = \frac{\sigma_L}{\varepsilon_L}$$

Most polymers follow Hooke's law for at least a small applied stress. If, however, the stress is increased, there is a point where the polymer will be irreversibly deformed. This point is called the *yield point* and is designated by σ_y and ε_y. A process called *necking* occurs after the yield point is reached. Figure 12-3.2.2 illustrates the necking process. It should be obvious that the amount of deformation a polymer undergoes under stress and the magnitude of the elastic limit are very important parameters in the determination of applications for that polymer. Section 12-5.2 gives some numerical values of some of the above quantities. Another important property is the time behavior of a sample under a constant applied stress. Figure 12-3.2.3 shows typical behavior of a material subject to *creep*. To measure this phenomenon a sample is put under stress and the elongation is measured as a function of time. Also shown in the figure is the process called *stress-relaxation* which sometimes occurs when the stress is removed (at t_1). More detail will be given in Sec. 12-5.

OTHER METHODS.

Density. The density of a polymer sample is a physical parameter that is easily measured, and it is one that is widely used. The principle use is in determining

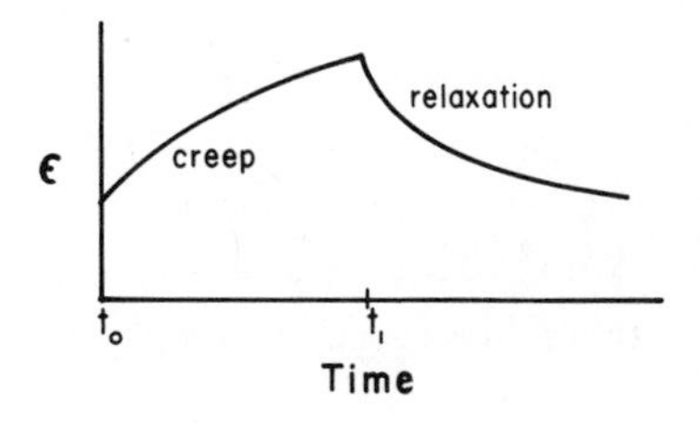

FIG. 12-3.2.3 Creep in a material under stress. A stress is applied at t_0 and removed at t_1.

the amount of crystallinity in a sample. An amorphous polymer will have a density different from one that is crystalline. Thus, a measurement of the density can be used to determine the amount of crystalline material present if purely crystalline and purely amorphous standards are available for that polymer. Such a quick check is valuable in process control.

Infrared Spectroscopy. Infrared spectroscopy is used for determining the types of monomer in a polymer. In the manufacture of many polymers, more than one monomer is used to produce a *copolymer* in an attempt to obtain a product with specific characteristics. A copolymer is one containing monomers of more than one type. There are several ways in which copolymers can form. These are discussed in Sec. 12-4.2. If each of the monomers has unique bonds somewhere in the molecule, infrared spectroscopy can be used to detect these in the polymer, and the ratio of the amounts of each monomer can be determined. For example, in a styrene-methyl methacrylate copolymer, the styrene monomer has a benzene ring that gives a distinct infrared absorption. The methacrylate has a carbonyl (-C=O) group that also gives a distinct infrared absorption. By a detailed analysis of these absorptions, it is possible to find the amount of each of the monomers incorporated into the polymer. Infrared spectroscopy can also be used as an identifying tool for simple polymers. Each polymer gives a characteristic fingerprint of absorptions in the infrared region. Comparison of a spectrum with a library of known spectra allows the identification of the unknown polymer.

X-Ray Spectroscopy. X-Ray studies of polymers are more complicated than such studies in purely crystalline materials (like those encountered in Sec. 14-5). In a polymer, as we shall see below, there are regions that are crystalline and regions of amorphous material within the same sample. An X-ray diffraction can only occur in a substance that has an ordered arrangement of the units making up the material. This ordered array must extend over several hundred angstroms (Å: 1 Å = 10^{-10} m). This can only be true in the case of a crystalline substance (almost be definition, since crystallinity implies an ordered array over long distances within the sample). It is possible from X-ray analysis to determine the type of crystals formed as well as estimate the percent of amorphous and crystalline regions. Another application of X-ray spectroscopy is in determining the gross structure (on a microscopic scale) of a polymer. The diffraction pattern from a film is very different from that of a fiber or a three-dimensional crystal. The diffraction patterns from the sample are indicative of the orientation in the polymer. (The possible orientations are: one dimensional in a fiber, two dimensional in a film, and three dimensional in a bulk sample.)

Differential Thermal Analysis. The final analytical technique to be mentioned in this section is *differential thermal analysis* (DTA). DTA is the heating of a sample at a predetermined rate with observation of the changes in heat emitted or absorbed at each temperature relative to an inert standard in the apparatus. Such transitions as the melting point (of crystalline polymers), the glass transition temperature, solid state rearrangements, and decompositions can be observed with DTA. A related technique is *thermogravimetric analysis* (TGA) in which the weight loss of a sample is measured as the temperature is raised. Further details on DTA can be found in Sec. 12-4.5.

12-3.3 MORPHOLOGY

LAMELLAE. You should have concluded by now that polymers are capable of some degree of crystallization in spite of the very large size of the individual molecules making up the polymer. You should also have deduced that there are regions in the polymer sample that are lacking long-range order. For simplicity we will call the former *crystalline* and the latter *amorphous*.

In the following discussion, polyethylene will be used as the example since it is the best characterized of all the polymers. You might consider for a moment before continuing your reading the problems associated with taking a long-chain molecule and somehow folding it into a shape that can be incorporated into a crystal. The crystal formed seems to be constructed in such a way that the building blocks making up the crystal are about 100 Å in thickness. Since polyethylene molecules are up to several thousand angstroms in length, they must somehow be folded to produce a building block that is about 100 Å thick. The thickness can vary appreciably from polymer to polymer; 100 Å is a typical value for polyethylene. In a given molecule one of the segments of the chain is incorporated into a crystal. Subsequent segments may be in the same crystallite or in another. The portions of the chains between the segments in the crystallites are variable in length and are found in a conformation different from that of the portions in the crystallites. This brings us to an important point. If a polymer consists of molecules that are not linear (that is, if there is branching from the main polymer chain) the polymer cannot

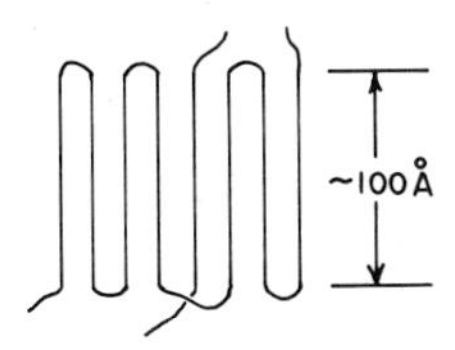

FIG. 12-3.3.1 Folding polyethylene chains to produce crystallites.

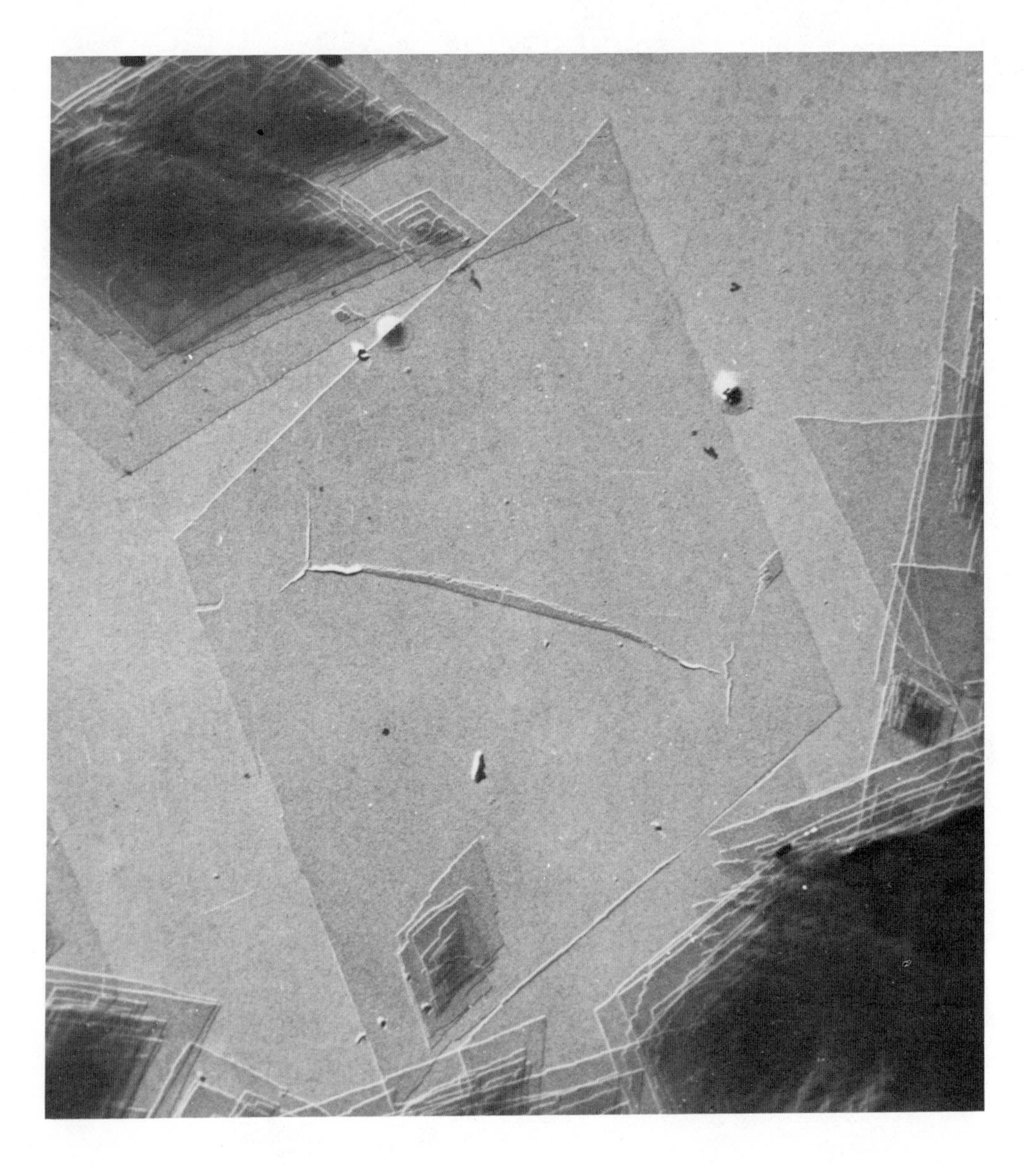

FIG. 12-3.3.2 Crystals of linear polyethylene. (Courtesy of P. H. Geil, Case Western Reserve University.)

crystallize, since it is impossible except in very special cases to arrange irregular segments in a manner that will produce a regular crystal.

Back to consideration of Fig. 12-3.3.1. If the model shown is correct, then we would expect the crystallites formed to have a layered structure with each layer being about 100 Å thick. Figure 12-3.3.2 shows that such a structure is, in fact, observed. This is an electron micrograph of crystals of linear polyethylene grown from a dilute solution. The most prominent feature in this figure is the presence of flat crystallites called *lamellae*. In this particular sample, the lamellae are found to be 104 Å thick on the average. Crystallization that results in the lamellae like those shown can occur either from solution or a melt. Crystallization begins and spreads out from individual *nucleation centers*. A nucleation center is an added "seed" crystal or an impurity already present that provides a surface on which the polymer crystallization can occur. In order to grow large crystals it is necessary to limit the number of nuclei in the system. If there are many nuclei present there will be too many growing centers in the solution or melt, and small crystals will result. Dilute solutions are normally used to grow large crystals, but there are exceptions. The most remarkable thing about the lamellae formed when a crystal is grown at a constant temperature is the degree of uniformity in their thickness. It is also observed that the thickness of the lamellae tend to increase as the temperature of growth is increased. The only way in which the crystals seem to grow in the direction normal to the surface of the lamellae is by the addition of more lamellae on top of those already there. As in the case of most crystal growth processes, the presence of screw dislocations greatly enhances the rate of growth (see Sec. 14-3.5 for a brief description of screw dislocations). Screw dislocations lead to spiral growth on the surface.

We have by no means obtained a complete description of the manner in which lamellae form and polymer crystals grow from long-chain polymer molecules. There is still a great deal of uncertainty as to what happens to the folds where the chains turn around to reenter the crystal. Also, it is observed that the lamellae are not perfectly flat as you may have thought from the above discussion. Instead, there is some distortion, such that the lamellae that appear flat in Fig. 12-3.3.2 are actually slightly pyramidal in shape due to restraints on how the folds can be incorporated in the structure. Another interesting phenomenon resulting from the fact that, as in all crystals, the melting point increases with an increase in crystal size. If we could obtain an infinite crystal of polyethylene, it would melt at about 140°C. However, crystals such as those shown previously melt at about 130°C. If we heat a crystal of polyethylene to near its melting point, a process called annealing, it will tend to become thicker. Figure 12-3.3.3 shows what happens in such a case. Portions of the crystal have become thicker at the expense of material from other portions in the crystal. Something for you to ponder is: How do the

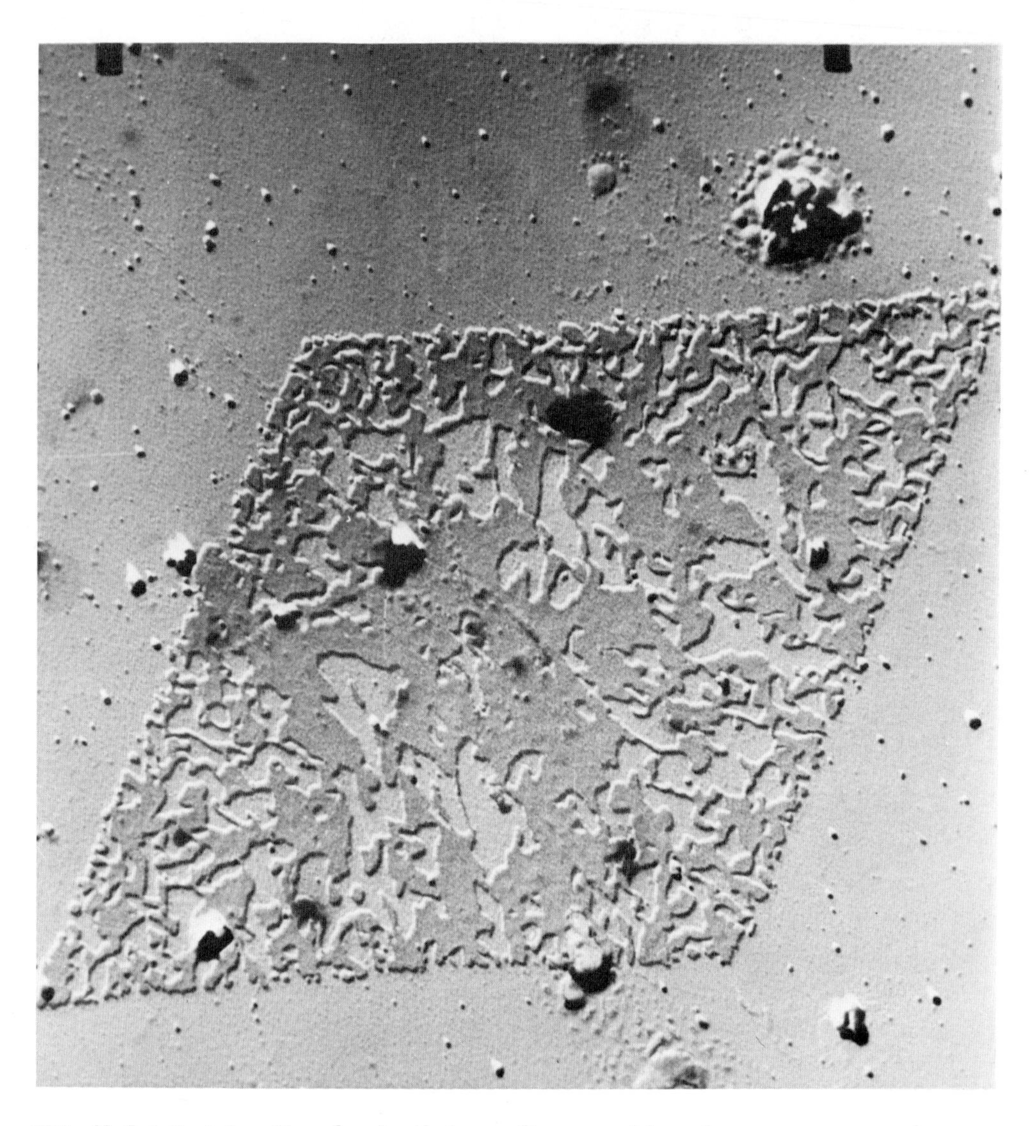

FIG. 12-3.3.3 A lamella of polyethylene after annealing at a temperature near the melting point. (Courtesy of P. H. Geil, Case Western Reserve University.)

long chains of polyethylene migrate so easily that the results shown in Fig. 12-3.3.3 can occur in a rather short period of time?

SPHERULITES. If a melt is quenched rapidly, a different type of structure appears: the *spherulite*. Figure 12-3.3.4 shows some spherulites formed by the rapid crystallization of polypropylene. (This is a cross-section of a solid sample.) As you can see, the spherulites grow from nucleation centers until they meet, filling the volume almost completely. As the crystals grow, material in the melt is incorporated. The density of the crystal is greater than the density of the melt, causing a decrease in the volume. This process continues until all the melt has been depleted. The melt is consumed before the spherulites completely fill the volume originally occupied by the melt. This leaves voids and/or weak points between the spherulites which result in a solid with a strength much lower than we would observe in the absence of the voids. One way to alleviate this problem partially is to add some nucleating agents that help produce a large number of small spherulites instead of a small number of large ones. Regardless of the method used, there will still be regions of amorphous material within the spherulites. The spherulites seem to be made up of aggregates of lamellar crystals that are radially oriented from the center. In the center of each spherulite there is a group of lamellae stacked one on top of the other. As growth proceeds from the center, the lamellae twist and turn in order to fill all the available space. In three dimensions (you are looking at a two-dimensional cross-section in Fig. 12-3.3.4) spherical symmetry is approached, whence the name spherulite.

We need now to look at how the method of growth affects the strength of the material. If well-formed lamellae are present, such as shown in Fig. 12-3.3.2, we will observe very little strength, since bending such a material will result in fractures between the lamellar sheets. However, if spherulitic growth has been achieved, (especially if a large number of small spherulites are formed instead of a small number of large ones) the material will be much more ductile. There is not total agreement on the reasons for the greater strength of the spherulitic structure. Apparently, in rapid crystallization from a melt, segments of an individual polymer chain are incorporated into more than one lamella resulting in ties between lamellae. Support for this model comes from experiments in which a film of polymer is stretched. Figure 12-3.3.5 shows a specimen of polyethylene that has been stretched. Note the cracked regions (photomicrograph 2) which have fibrous material (called *fibrils*) in them. These are probably regions of molecules that are tying lamellae together, and they are about 100 to 200 Å in diameter. As the pulling of the crystal continues, the connecting fibrils "neck down" as they are drawn off the edge of the crack. It seems there is a simple unfolding of the molecules off the edge of the crack to form the fibril. These molecules apparently

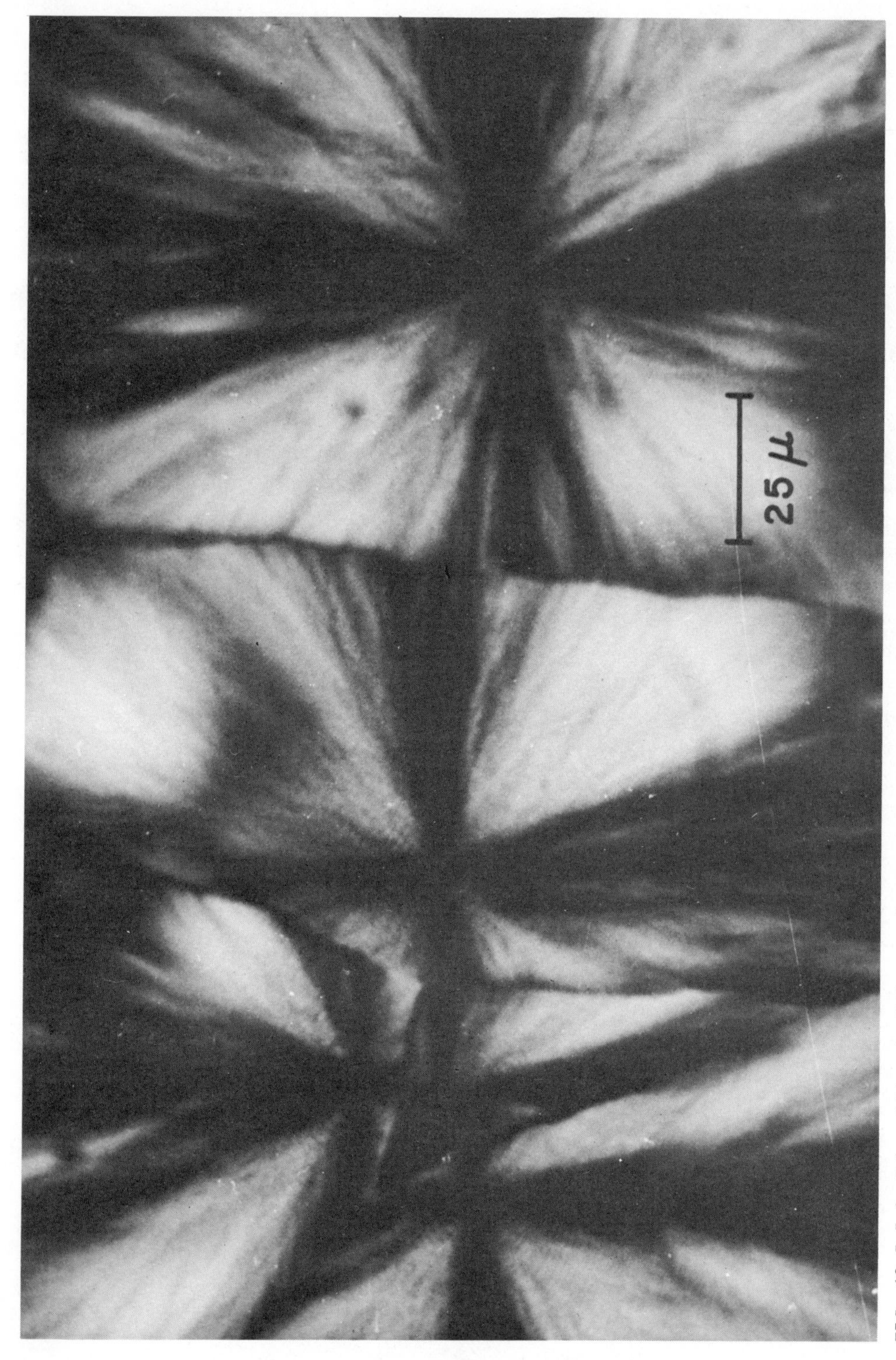

FIG. 12-3.3.4 Spherulites in Polypropylene. (Courtesy of P. H. Geil, Case Western Reserve University.)

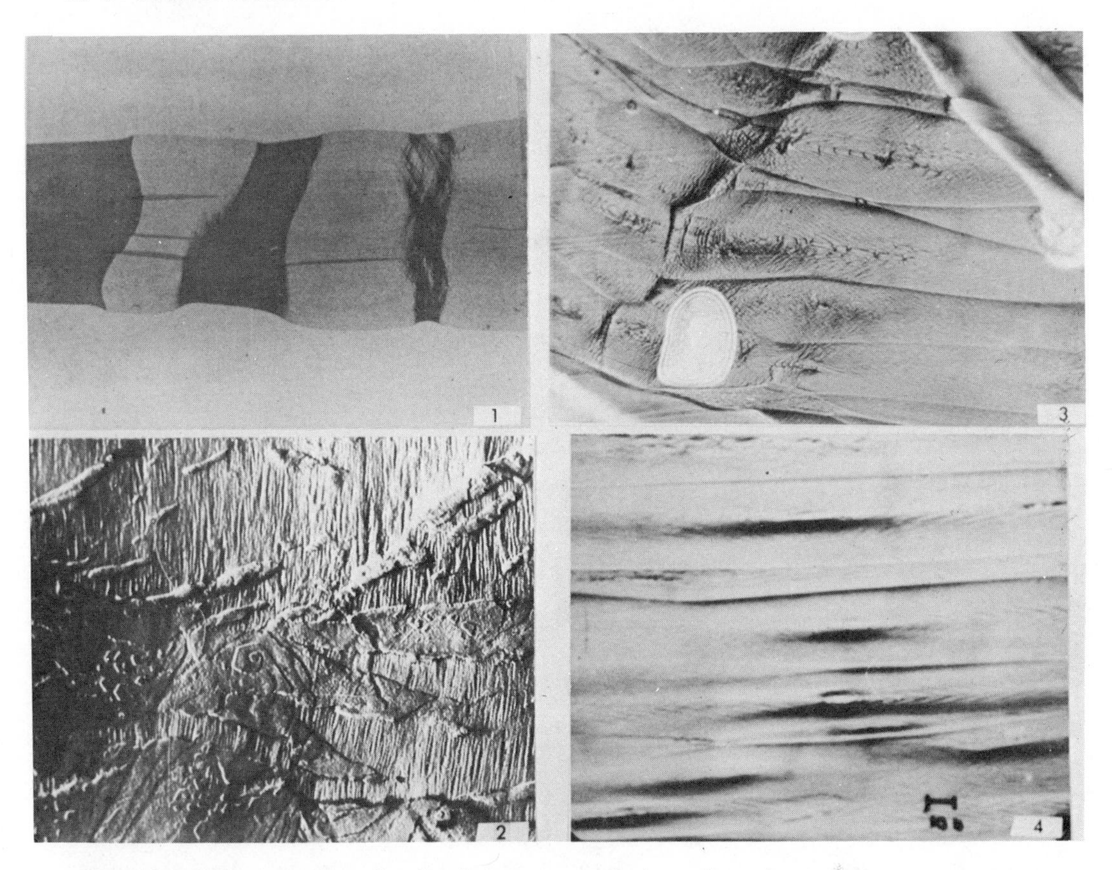

FIG. 12-3.3.5 A sample of polyethylene stretched to show the fibrils connecting the lamellae. (Courtesy of P. H. Geil, Case Western Reserve University.)

align with other molecules in the formation of the fibers. The actual mechanism of the formation of the fibrils is probably much more complicated than the procedure just described.

NECKING AND DRAWING. As mentioned briefly earlier in this section, the application of a force to the ends of a bulk polymer will result in a deformation. If the force is increased enough, we will observe the phenomenon called *necking*. Figure 12-3.2.2 gives a schematic representation of the effect. As the tension is increased on a sample of material, it will begin to form a *neck* which then yields at some point. As the tension is increased even further, the neck region travels from the yield point toward the two ends. In some cases, the neck will extend all the way to the ends before failure occurs. After the neck has traversed the full length of the sample, continued exertion of tension will result either in failure of the material or in the *drawing* of the sample into a fiber. This is, in fact, the way in which many fibers are produced from bulk polymers. Figure 12-3.3.5 (photomicrograph 1) vividly shows the region of deformation in the necking area of a sample of linear polyethylene. The mechanism is thought to be similar to that described for the formation of fibrils between spherulites when a force is applied. The molecules on either side of the neck unravel to form fibers. It is observed that the diameter of the fibers in a fully drawn sample is about 100 Å. There are still many problems associated with the topological aspects of folding and unfolding the polymer chains. So, do not interpret the above discussion as the way in which crystallization and necking occur. The above explanations are simplified descriptions of complex phenomena.

The considerable amount of research effort being expended in industry in trying to unravel (pardon the expression) the morphology of polymers is justified. If the nature of the packing in crystals and fibers can be understood, then control of these will lead to the production of much stronger polymeric materials than are currently available.

AMORPHOUS MATERIALS. Almost nothing has been said about the noncrystalline state. The reason for this is that there is no model available to describe adequately the amorphous state. It was assumed in the past that wholly amorphous solid polymers consist of randomly coiled and entangled molecular chains. More recently there has been some speculation that regions of varying order in the solid form clusters within the solid. Little is known about the effect of these ordered regions on the strength of polymers. There are other morphologies found in some polymer systems. An example is liquid crystalline polymers. These exhibit some ordering (the amount and nature of which is, in many cases, highly temperature dependent), but not as much as true crystalline polymers. (There is no Self-Test for this section, but be prepared to discuss morphology in detail.)

12-4 THERMAL PROPERTIES

OBJECTIVES

You shall be able to

1. Define the five classes of polymers considered in this section
2. Define the term glass transition temperature T_g
3. Relate the thermal properties of a polymer to its structure
4. Briefly discuss differential thermal analysis and differential scanning calorimetry

12-4.1 INTRODUCTION

A knowledge of the thermal behavior of a polymer is one of the most important pieces of information we can have about it. The conditions required for processing a polymer and fabricating products requires detailed study of the temperature dependence of that polymer's properties. The end use will be dictated by the effect of temperature on the polymer. A polymer cannot be used for a product that must retain its structural rigidity if it is subject to creep at the temperature which will be encountered by that product. Also, a material that is very brittle at some temperature will be unsuitable for many applications at that temperature. You have already been introduced to the two most important thermal parameters for a polymer. If the polymer is strictly crystalline in the solid state, the melting temperature T_m is the important temperature for that polymer. If it is amorphous, the glass transition temperature T_g is all important. Finally, it is possible to have both a glass transition and a melting temperature in the same polymer.

In this section the glass transition temperature will be defined again; then, its value will be related to some structural characteristics. To aid in relating T_g to polymer structure, five classes of polymers will be defined. Finally, two important experimental methods for determining thermal behavior will be given. Everything in these first four sections is leading up to the fifth section in which the mechanical properties of polymers are given in some detail. Obviously, it is the final mechanical properties of a material that determine whether it is suitable for use in a particular application. What we hope to do in these first four sections is to give enough basics to allow you to understand how polymer properties may be varied to fit a given need.

12-4.2 POLYMER CLASSES

Polymers will be grouped into five classes for the purposes of this discussion. Detailed treatment of each of the classes and its properties will be reserved for Sec. 12-4.4.

Class I: Polymers with Perfectly Repeating Chain Units. Examples of this type of polymer are polyethylene, nylons, natural rubber, and polyesters. These, in general, fit readily into a crystalline lattice and exhibit a well-defined melting point as discussed in Sec. 12-3. In fact, polyethylene fits into a crystalline structure so readily that it has been impossible to cool samples rapidly enough to get amorphous solids. It should be noted, however, that not all polyesters and rubbers are crystalline. If some of the $-CH_2-$ units on a polymer such as polyethylene are replaced on a regular basis by such atoms or molecules as -O-, -N-, -S-, ⌬ , etc. we will still have a Class I polymer if regularity is maintained. The function of these atoms is to serve as chain modifiers to change the properties of the polymer. Much of the discussion of thermal properties will be restricted to this class.

Class II: Random Copolymers. These polymers result from the reaction of more than one type of monomer. As the name implies, there is no order in the way in which the monomers react. Whether one monomer is adjacent to another of a particular type is a matter of chance.

Class III: Stereopolymers. This type of polymer is distinguished by the orientation of groups attached to the main carbon chain. For example, assume we start with a molecule of polyethylene and replace the hydrogen atoms on every other carbon atom with a substituent R. There are three ways we can do this. If all the R groups are on the same side of the carbon chain as shown in Fig. 12-4.2.1(a) the polymer is called *isotactic*. If the groups alternate regularly from one side to the other as shown in part (b) the structure is named *syndiotactic*. And, if a random substitution results, the name *atactic* is used. Many times, atactic polymers will be amorphous while their isotactic counterparts are crystalline.

Class IV: Block and Graft Copolymers. The best way to describe these polymers is to give an example of each. Figure 12-4.2.2(a) gives the block copolymer that results from the reaction of monomers A and B under certain conditions. In part (b) a graft copolymer is depicted. In both cases, the AAA and BBB chains can be any of the previously defined classes. Another possibility that does not fit readily in any group are alternating copolymers, in which monomer units of different types

(a)

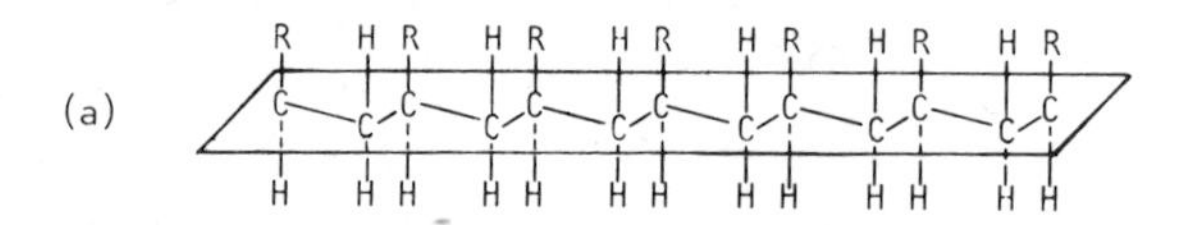

(b)

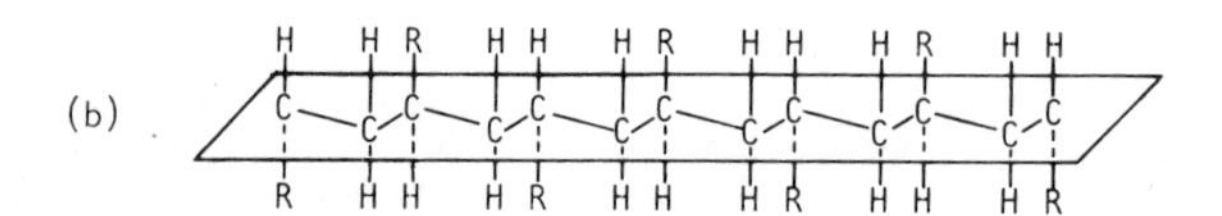

(c)

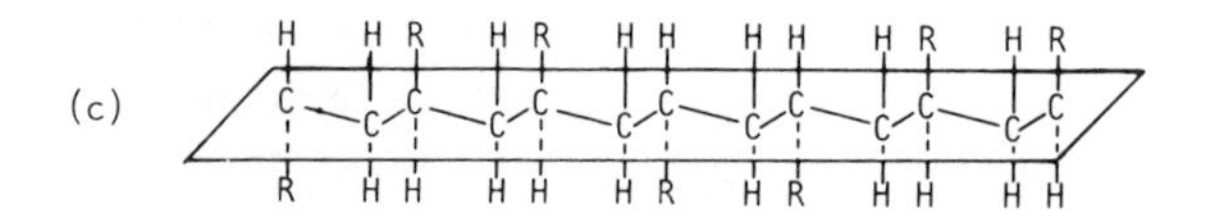

FIG. 12-4.2.1 A representation of stereopolymers: (a) isotactic, (b) syndiotactic, and (c) atactic.

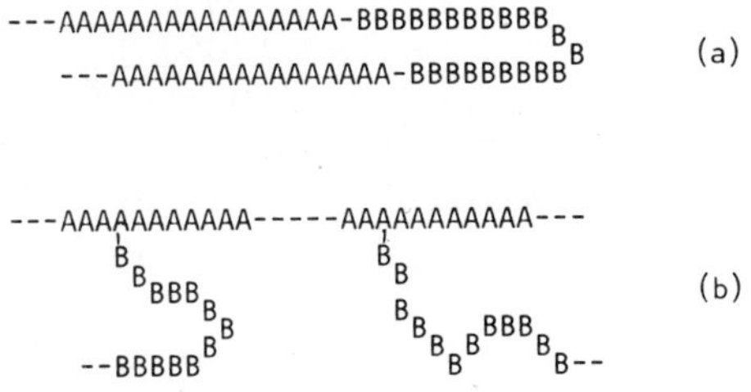

```
---AAAAAAAAAAAAAAAA-BBBBBBBBBBB
                               B
                                B       (a)
  ---AAAAAAAAAAAAAAAA-BBBBBBBBB

---AAAAAAAAAAA-----AAAAAAAAAAA---
    B               B
     BBBB            B                  (b)
         B           B
          B           B  BBB
  --BBBBB              BB   B--
```

FIG. 12-4.2.2 A representation of (a) block and (b) graft copolymers.

alternate in the chain making up the polymer.

Class V: Short-Sequence Polymers that Assume a Helical Conformation. Examples of this type are isotactic addition polymers than cannot crystallize due to interference of the substituted groups. The strain of this interference can be relieved by twisting of the molecule. If all the twists are in the same direction and of the same magnitude, the result is a helix. Many polymers in biological systems are of this type.

12-4.3 THE GLASS TRANSITION TEMPERATURE

The *glass transition temperature* T_g can be defined as that temperature above which a polymer is rubbery and below which it is brittle. (If $T > T_g$, the polymer is rubbery, and if $T < T_g$ it is brittle.) A glass transition applies strictly only to amorphous materials. If there are crystalline regions in the solid, we must consider the melting temperature T_m for those regions and the glass transition temperature T_g for the amorphous portions. As you are aware, the process of formation of a crystal is called *crystallization*. The corresponding process of forming an amorphous solid (a glass) is called *vitrification*.

On the molecular level, we can visualize the glass transition as the point at which large-scale molecular motion begins to occur. Below T_g there is some motion of the molecular chains which increases as the temperature rises. At a well-defined temperature, the glass transition temperature, motion of segments within the polymer begins. Well below T_g only very localized motion is possible, and the material is observed to be brittle. When the glass transition is passed, the segmental movement of the polymer molecules leads to flexibility and rubbery characteristics. Our observation that polystyrene is brittle and natural rubber is pliable depends on our choice of temperauure. Below T_g (- 72°C) rubber behaves like polystyrene at room temperature. Likewise, polystyrene above T_g (100°C) behaves somewhat like rubber at room temperature.

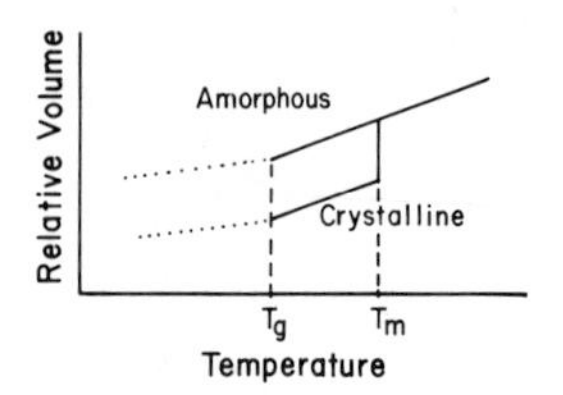

FIG. 12-4.3.1 Volume behavior of a polymer as a function of temperature. Both crystalline and amorphous states are shown.

Several theories have been advanced in attempts to explain the origin of the glass transition. One of the most successful is the *free-volume* model. This model is the result of study by T. G. Fox and P. J. Flory. Before continuing consideration of this model, look at Fig. 12-4.3.1 where the relative volume of a polymer is plotted against temperature. This figure shows a polymer that can be either crystalline or amorphous below T_m, depending on the treatment of the sample. You should note immediately that there is a T_g present even if the polymer crystallizes at T_m. This is because of the lack of 100% crystallization. Even in polyethylene, the maximum amount of crystallization is about 90%. Other polymers crystallize to a lesser extent. Thus there are amorphous regions left even after crystallization occurs. Note also that if the polymer crystallizes, there is an abrupt change in the volume. However, at T_g there is no volume change, only a change in the slope of the volume-temperature curve. The plot given in Fig. 12-4.3.1 is an equilibrium plot. In any experiment designed to measure T_g, one must be aware of kinetic effects. For example, if a sample is heated very rapidly, a higher T_g is observed than if slow cooling is used. This is due to the finite time required for rearrangements to take place in the polymer. Normally, several heating rates must be used and the T_g values obtained extrapolated to zero heating rate to obtain a "true" value of T_g.

As mentioned previously, in the glassy state there are only very localized motions, involving at most a few atoms or units. However, large-scale movements of segments of the polymer are possible at T_g. This large-scale motion requires more *free volume* than the localized motion. The theory of Fox and Flory states that below the glass transition temperature the free volume is constant. Above T_g, the free volume increases with an increase in temperature much as in an ordinary liquid due to the increasing amount of chain movement. As you might guess, anything that affects the free volume also affects the value of T_g.

12-4.4 THERMAL PROPERTIES AND STRUCTURE

GENERAL. In most cases we find there is a relation between T_g and T_m. This is not surprising, since the polymer characteristics that affect one temperature have similar effects on the other. Some of the parameters important in fixing the value of T_g (and T_m) are molecular weight, presence of low molecular weight diluents (plasticizers), intermolecular forces between polymer chains, chain stiffness, and symmetry in the chains. We will see below how some of these factors affect T_g for the different classes of polymers. Usually, T_g is found to increase as the molecular weight M_N increases. This is generally true for all classes. As you read through the following, refer to Sec. 12-4.2 for the definitions of the different classes.

CLASS I.

Structural Effects. These polymers are easily crystallized. However, they do not crystallize completely. It is observed that T_g and T_m are related to each other for this class of polymers. We will see later how T_g and T_m vary as modifications are made on Class I polymers. Polyethylene (T_g = - 115°C and T_m = 137°C) will be used as our model compound. Changes in T_g (and T_m) can be effected by replacing $-CH_2-$ units in the polyethylene polymer with other atoms or molecules.

In any organic molecule, there is restricted rotation about the carbon-carbon bond due to steric interactions of the substituents on the carbons. Any modification we make that allows for easier or more difficult rotation will change the value of the glass transition temperature. For example, the ether, ester, and sulfur linkages shown in Fig. 12-4.4.1 all allow for freer rotation than about the carbon-carbon bond itself. You have already seen that the glass transition temperature is that temperature at which segmental motion begins to occur. You should conclude that any change in the structure that makes such motion easier will reduce the value of T_g or T_m.

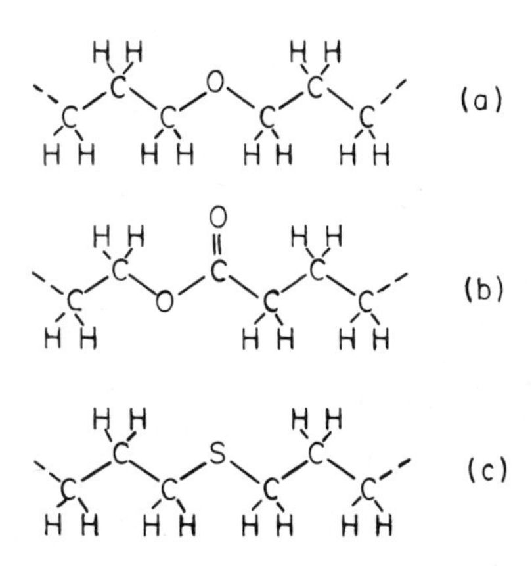

FIG. 12-4.4.1 Polymers containing (a) ether, (b) ester, and (c) sulfur linkages.

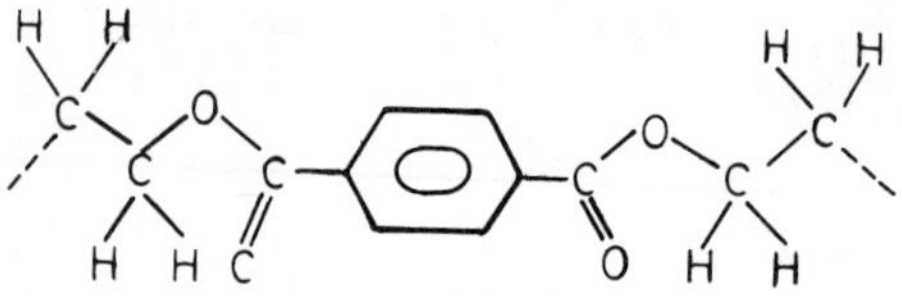

FIG. 12-4.4.2 A rigid molecule inserted into a polyethylene chain to increase T_g and T_m (T_m = 265°C).

This is indeed the case, and each of the linkages shown in Fig. 12-4.4.1 tends to lower T_g and T_m relative to polyethylene. On the other hand, a molecule such as that shown in Fig. 12-4.4.2 will "stiffen" the polymer chain and lead to an increased T_g. In this and all cases, we are assuming that everything else (such as molecular weight) is held constant for the comparison. The example given in Fig. 12-4.4.2 is polyethylene teraphthlate.

Another important factor in determining T_g and T_m is the amount of intermolecular interaction between molecular chains in the polymer. Aliphatic hydrocarbon chains (such as polyethylene) are nonpolar and have very weak interactions which are mainly due to van der Waals forces. This weak interaction favors a low T_g. In polymers such as nylon (shown in Fig. 12-4.4.3), we have interactions between polar molecules and hydrogen bonding that both favor high T_g, since both give strong intermolecular attractions. From the value of T_m given for nylon in Fig. 12-4.4.3, you can see that strong interactions lead to a high value of T_m (and T_g). Other molecules of this type are the polyamides, polyurethanes, and polyureas. (See Tables 12-1.2.1 and 12-1.3.2 for review of the structures of these and other polymers mentioned in this section.) A final possibility is the combination of chain stiffness (provided by molecules such as the teraphthlate given above) and strong intermolecular forces (given by a polyamine) as illustrated below.

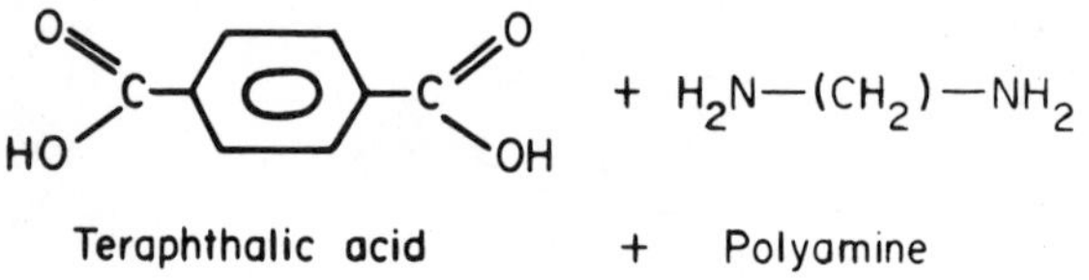

$H_2N-(CH_2)-NH_2$

Teraphthalic acid + Polyamine

This reaction yields a polyamide with a very high T_m (> 400°C). Table 12-4.4.1 gives the values of T_m and T_g for a number of polymers of *Class I*.

Effects of Monomer Length. There is a definite effect on T_g and T_m due to the number of $-CH_2-$ units between the active groups in condensation polymers. Figure 12-4.4.4 illustrates this effect when the number of $-CH_2-$ units is increased from

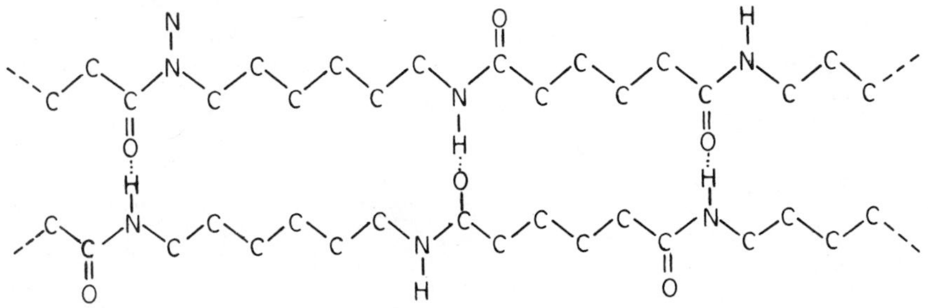

FIG. 12-4.4.3 A schematic representation of nylon 6,6 (T_m = 265°C); (- - -) denotes a hydrogen bond.

Table 12-4.4.1
Melting and Glass Transition Temperatures for Several Class I Polymers

Polymer	Repeating unit	T_g (°C)	T_m (°C)
Polydimethylsiloxane	$-OSi(CH_3)_2-$	-123	-85 to -65
Polyethylene	$-CH_2CH_2-$	-115	137
Polyoxymethylene	$-CH_2O-$	- 85	181
Natural rubber	$-CH_2C(CH_3)=CHCH_2-$	- 73	14
Polyisobutylene	$-CH_2C(CH_3)_2-$	- 73	44
Polyethylene oxide	$-CH_2CH_2O-$	- 67	66
Polypropylene	$-CH_2CH(CH_3)-$	- 20	176
Polyvinyl fluoride	$-CH_2CHF-$	- 20	200
Polyvinylidene chloride	$-CH_2CCl_2-$	- 19	190
Polyvinyl acetate	$-CH_2CH(OCOCH_3)-$	28	
Polychlorotrifluoroethylene	$-CF_2CFCl-$	45	220
Poly(ε-caprolactam)	$-(CH_2)_5CONH-$	50	223
Polyhexamethylene adipamide	$-NH(CH_2)_6NHCO(CH_2)_4CO-$	53	265
Polyethylene terephthalate	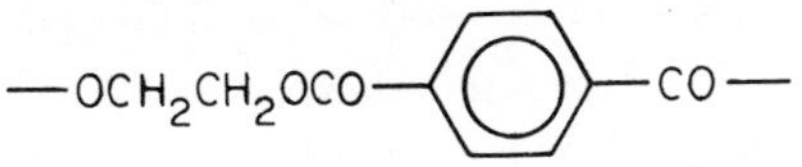 $-OCH_2CH_2OCO-C_6H_4-CO-$	69	265
Polyvinyl chloride	$-CH_2CHCl-$	81	212
Polystyrene	$-CH_2CHØ-$	100	240
Polymethyl methacrylate	$-CH_2C(CH_3)(CO_2CH_3)-$	105	200
Cellulose triacetate	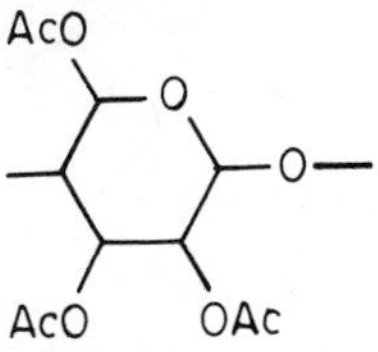 AcO, O, O, AcO, OAc (ring structure)	105	306
Polytetrafluoroethylene	$-CF_2CF_2-$	127	327

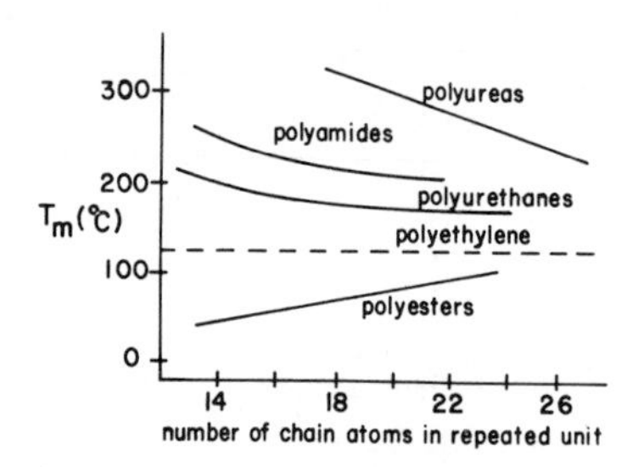

FIG. 12-4.4.4 Effect of increasing length of the repeat unit in several condensation polymers.

14 to 25 or 26. Polyethylene is sketched in as a reference. In general, increasing the length of the repeating unit leads to a lower value of T_m. The polyesters are exceptions. The figure shows how *increasing the number* of atoms in the repeat unit affects T_m. In all cases we have to assume that the *total molecular weight is constant* since an increase in molecular weight leads to higher T_m.

Comparison of T_g and T_m. As you have already read, the same features that favor high T_g also favor high T_m. A simple sketch will illustrate this point. Such a sketch is given in Fig. 12-4.4.5 where T_m is plotted against T_g. You should observe the correlation between T_m and T_g. There are exceptions, as usual, but the general correlation exists. It is possible to design a polymer of Class I with high T_g and T_m or low T_g and T_m, but the two cannot be varied independently. Additions will be made to this plot as we consider the other classes of polymers to show how such conditions as high T_g and low T_m can be obtained. (Within limits, of course: you cannot find a polymer with T_g greater than T_m. Why?) The two lines in the figure define the range of variation of T_g with T_m.

CLASS II. Random copolymers make up this class. These are not as symmetrical as the Class I polymers discussed above; therefore, they are not as easily crystallized. This class, in general has rather low values for T_m and crystallizes to only a low percent. Figure 12-4.4.6 shows a schematic of the way T_m varies with composi-

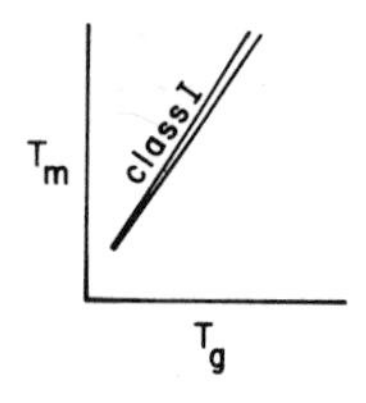

FIG. 12-4.4.5 Comparison of T_g and T_m for Class I polymers.

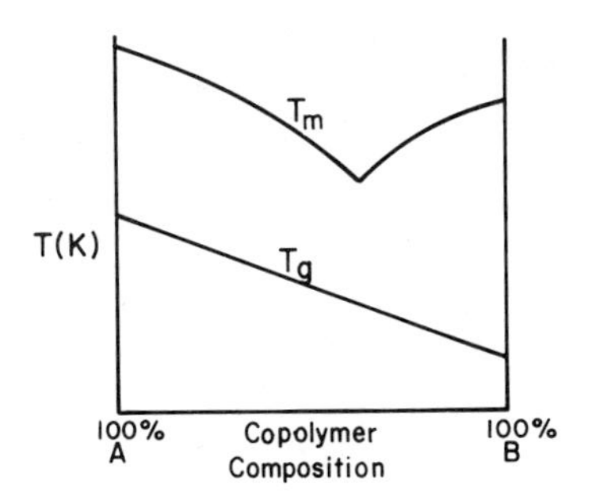

FIG. 12-4.4.6 Variation of T_m and T_g with composition in a random copolymer.

tion for a copolymer of this type. (Section 13-2 gives an explanation of the shape of the T_m curve.) Also shown is the way T_g varies with composition. T_g is not greatly affected by the copolymerization since the vitrification process does not involve the fitting of molecules into a lattice (an ordered array in the crystal). T_g varies almost linearly between the values of T_g of the two materials making up the copolymer. We can now make another entry on our T_m vs. T_g plot. The random copolymers allow some independent control of T_m and T_g not available in Class I. Figure 12-4.4.7 is the revised T_m vs. T_g plot. Lines are not used to bound Class II since there is a rather wide range over which T_g and T_m can be varied independently.

CLASS III. Class III involves the stereopolymers. These do not crystallize to any appreciable extent due to the inability of the unsymmetrical molecules to fit into a well-defined lattice. Thus, T_m does not exist. There are some monomers such as butadiene and isoprene that can polymerize in more than one configuration due to the presence of two double bonds in the molecules. This is illustrated in Fig. 12-4.4.8 for butadiene (isoprene is the same as butadiene except for a methyl group on the second carbon atom in the molecule). What this amounts to is a random polymer of these three units. It is not crystallizable, so there is no T_m. Figure

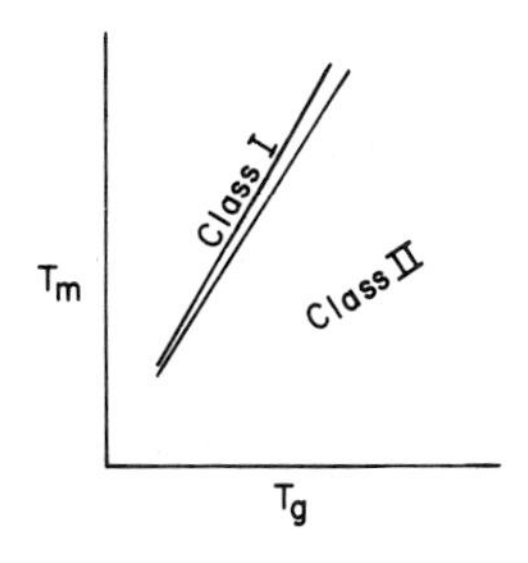

FIG. 12-4.4.7 T_m vs. T_g for two classes of polymer.

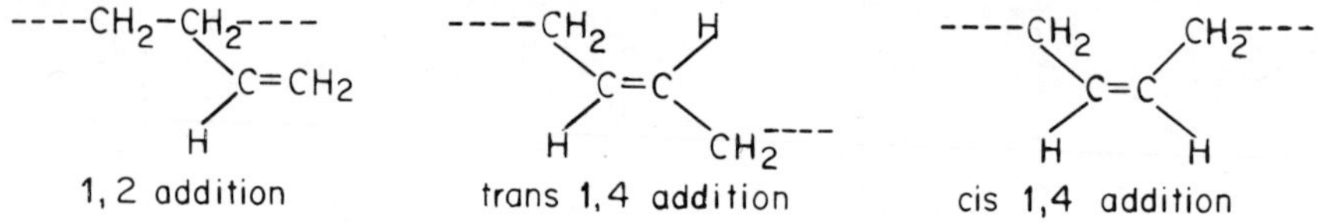

----CH_2–CH_2----, C=CH_2, H — 1,2 addition

----CH_2, H, C=C, H, CH_2---- — trans 1,4 addition

----CH_2, CH_2----, C=C, H, H — cis 1,4 addition

FIG. 12-4.4.8 Possible polymerizations of butadiene.

12-4.4.9 adds this class to the T_m vs. T_g diagram.

CLASS IV. This class offers some very interesting possibilities. By varying the composition of the A and B polymer components, a wide range of polymer properties is possible. Just a few of the possible combinations will be mentioned, then T_m and T_g will be related. First, assume that both A and B chains are Class I. If this is the case, each of the components can crystallize separately. It is unlikely that the two can both fit into the same lattice. Also, there will be certain restraints on the crystallization of the two separately since they are both connected together. We term the separation of each of the chains into different lattices *microseparation*, since it involves separation on the molecular level and is not a phase separation in the ordinary sense of the word. It is impossible to get *macroseparation* when the chains are attached to each other. For this process, we can envision regions that consist of *A domains* and other regions that are *B domains* with little or no mixing between the two. T_m for this case is depressed much less than in the random copolymer system of Class II. Other possible AB combinations are: A, high melting Class I; B, amorphous material; A, high T_g amorphous polymer; B, low T_g amorphous polymer, etc. There are a wide range of possible variations. This leads to even more independent control of T_g and T_m than is possible in the other classes alone. Figure 12-4.4.10 continues our chart of T_m vs. T_g. (Just a reminder, T_g is never larger than T_m.)

Class V will not be discussed, since helical polymers are not of major importance for the production of materials. This class is vital for life processes, but

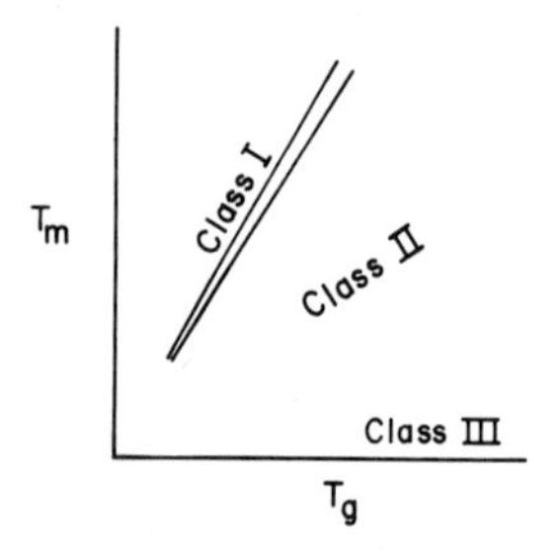

FIG. 12-4.4.9 T_m vs. T_g for three classes of polymers.

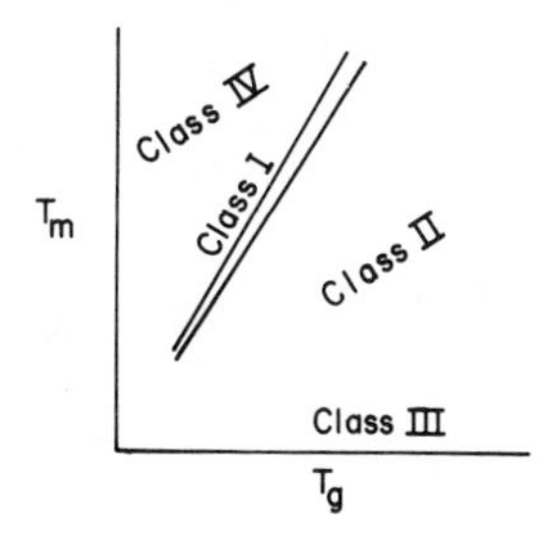

FIG. 12-4.4.10 T_m vs. T_g for four classes of polymers.

this discussion is limited to synthetic polymers. You can see from the last figure that we can independently vary T_m and T_g over a very wide range. Up to now though, we have not discussed the utility of this capability. In the next section (12-5) mechanical and rheological properties are presented. Then the variation of these properties with T_m and T_g will be given. At that time you should begin to see how a polymer system is designed from fundamental concepts such as those already presented to produce a desired set of properties.

12-4.5 DIFFERENTIAL THERMAL ANALYSIS AND DIFFERENTIAL SCANNING CALORIMETRY

Section 12-3.2 very briefly mentioned differential thermal analysis (DTA). Here a little more detail will be given, but not a complete description. The text *Experiments in Polymer Science* by E. A. Collins, et al. (see Sec. 12-4.6, Ref. 1) is an excellent place to start your study should you require more information on the subject. Also, see Weissberger (Sec. 11-2.8, Ref. 4). DTA allows a fairly complete analysis of the thermal effects that accompany physical and chemical changes in a polymer sample as the temperature is raised at a constant rate. This is done by comparing the temperature of a sample compartment to that of a reference compartment filled with some inert material as the temperature of the system is raised at a controlled rate. Any chemical or physical change that evolves or absorbs heat will show up on the instrument as a difference in heating rate of the two compartments. The instrumental aspects of DTA will not be presented, except to say that reliable results depend on careful control of a number of operating parameters such as heat transfer charactersitics of the sample and reference, sample packing, sample and reference particle size, amount of sample, and heating rate.

Figure 12-4.5.1 shows a typical plot of the temperature difference between the sample and reference compartments as the temperature is raised at a steady rate. Any positive difference means the sample is heating faster than the reference. Conversely, any negative value for ΔT is indicative of an endothermic process, since

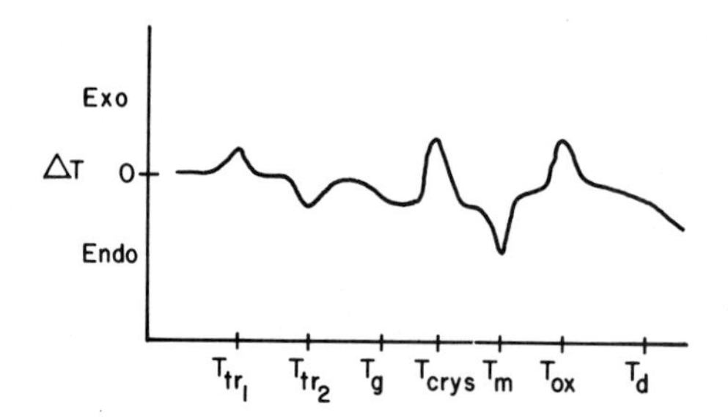

FIG. 12-4.5.1 A hypothetical DTA trace for a polymer.

the sample is heating slower than the reference. We can now interpret the peaks in Fig. 12-4.5.1. At very low temperatures only solid is present that may be amorphous, crystalline, or have aspects of both. In this case, let us assume both are present. As the temperature is raised, the first transition T_{tr_1} is observed. It is an exothermic solid phase transition which could be due to some rearrangement in the solid that leads to a lower energy (thus, energy is given off in the transition). The second transition T_{tr_2} is an endothermic solid state rearrangement. At T_g the glass transition is observed. At the glass transition there is an increase in the rate of change of the heat capacity of the sample with temperature, so the sample heats more slowly than before T_g, leading to a negative ΔT. After the glass transition temperature is passed, the amorphous material can crystallize. If it does so, the heat of crystallization (cry) is released and ΔT increases since the sample will be heating more rapidly than the reference. After T_{cry} we will observe a melting of the crystalline solid at T_m. In this case we have an endothermic process giving a negative ΔT. T_{ox} represents a reaction with air, an exothermic process of oxidation. Finally, if the temperature is raised high enough to T_d, a decomposition reaction will take place. This decomposition is an endothermic reaction. Every polymer will not have every transition shown in the figure. Also, the curves actually obtained are usually not so symmetrical. But DTA does offer a very good way to determine the major thermal characteristics of a sample quickly.

Differential scanning calorimetry (DSC) is not appreciably different. Instrumentally, changes are made that allow the plots to be used directly to determine the energy released or absorbed during any of the transitions that can occur. In DSC the area under the transition curve, any of the peaks shown in Fig. 12-4.5.1, is the energy change for that transition. The same type of information can be obtained from DTA, but it is more difficult to do so. This concludes the discussion of the thermal behavior of polymers. Some of the concepts introduced here will be referred to in the following section on mechanical and rheological properties. (There is no Self-Test section other than being able to satisfy all the points listed under the Objectives.)

12-4.6 ADDITIONAL READING

1.* E. A. Collins, J. Bareš, and F. W. Billmeyer, Jr., *Experiments in Polymer Science*, John Wiley and Sons, New York, 1973, Chap. 9.

2. A. D. Jenkins, ed., *Polymer Science: A Materials Handbook*, North Holland Publishing, Amsterdam, 1972.

3. W. J. Burlant and A. S. Hoffman, *Block and Graft Polymers*, Reinhold, New York, 1960.

12-5 MECHANICAL AND RHEOLOGICAL PROPERTIES

OBJECTIVES

You shall be able to

1. Define the terms associated with the measurement and evaluation of mechanical and rheological properties
2. Discuss in detail the phenomenon of viscoelasticity, including linear and nonlinear behavior
3. Characterize the mechanical behavior of crystalline polymers
4. Define fracturing and crazing and relate these phenomena to structural features you have already studied
5. Correlate everything given previously to determine what properties would have to be incorporated into a polymer to meet a given engineering application

12-5.1 INTRODUCTION

You have been introduced in the previous four sections to the following concepts:

Reactions and polymerization methods and some nomenclature
Characterization methods, including means for the determination of molecular weights and simple thermal tests for identification
Structure and morphology
Thermal properties

It is now time to apply these concepts to the behavior of the polymers in actual use. We must know how a polymer will deform under a continuous or oscillating load if its application requires those conditions. Interaction of a polymer with solvents and vapors that might be present is important if the application is a package of some sort. Permeability will have to be determined and controlled for materials that are to find use as films. Resistance to oxidation will have to be built in for polymers used for tires, insulation, etc. Resistance to detergents must be a property of fibers. The list goes on and on.

In every application it is impossible to optimize all the parameters of interest. It might, for example, be necessary to sacrifice a little strength in a polymer in order to gain flexibility. Or it might be necessary to reduce strength in one dimension in order to gain strength in another, as in a film. There will always be trade-offs in the design. This is no different than any other engineering task: you can never have the best for all possible properties. When you finish this section you probably will not be able to design a polymer to fit a particular application. But you should have sufficient background to indicate some of the considerations that would to into the design. You should be able to tell what the desirable characteristics are and indicate where conflicts of properties will occur.

12-5.2 TERMS AND DEFINITIONS

Some of the important terms related to mechanical properties are given in Fig. 12-3.2.1. Turn to that section for review before continuing your reading here. The terms of particular importance now are the Young's modulus and the tensile strength. Table 12-5.2.1 lists some values of these quantities for common metals and some bulk polymers. A few fibers are also listed. There are, as usual, some

Table 12-5.2.1
Comparison of Mechanical Properties

Material	E (dyn/cm^2)	Tensile strength (psi)	Tensile strength / density
Bulk materials			
Aluminum	7×10^{11}	9,000	3,300
Copper	12×10^{11}	39,000	4,300
Lead	1.5×10^{11}	2,000	176
Cast iron	9×10^{11}	15,000	1,900
Steel	22×10^{11}	60,000	7,500
Glass	6×10^{11}	10,000	4,000
Granite	3×10^{11}	19,000	7,000
Polystyrene	3.4×10^{10}	6,000	5,600
Polymethyl methacrylate	3.7×10^{10}	7,000	5,900
Nylon 66	2×10^{10}	10,000	9,100
Polyethylene (0.92 g ml^{-1})	2.4×10^{9}	2,000	2,200
Rubber	2×10^{7}	2,000	2,200
Hard phenolic	8×10^{9}		
High density polyethylene	2×10^{9}		
Low density polyethylene	8×10^{9}		
Fibers			
Steel wire		450,000	
Oriented polypropylene		120,000	
Oriented nylon		120,000	

Table 12-5.2.2
Some Conversion Factors for Mechanical Properties

To convert	to	Multiply by
psi	dyn cm^{-2}	6.895×10^4
dyn cm^{-2}	psi	1.450×10^{-5}
g $denier^{-1}$ [a]	psi	$1.28 \times 10^4 \times$ density

[a] This unit is usually reserved for fibers. It includes both the strength and the density of the material. It is related to the last column of Table 12-5.2.1.

conversion factors you may need at times, since you will find mechanical properties expressed in several different units in different sources. These are given in Table 12-5.2.2.

As a point of interest, there are some new high-modulus fibers being introduced for use that have values of the Young's modulus of about 7×10^{11}. This approaches the values for some metals, such as zinc. If you consider the much smaller density of a polymer compared to a metal, you can see that such strength for a lightweight material gives a valuable new engineering material. Another note of interest is that fiber-reinforced plastics are between hard phenolics and the metals in Young's modulus. The high strength coupled with light weight make these materials ideal for such things as boat hulls and other rigid structural parts.

Let us continue the definition of some of the important terms for mechanical behavior. We have defined stress as a force per unit area applied to a specimen. This can be applied in either a *tensile* or *shearing* fashion as shown in Fig. 12-5.2.1. The stress applied to an object can be broken down into components of tensile and shearing stress. These components are easily expressed in a tensor of the form

$$\begin{pmatrix} \sigma_{11} & \sigma_{12} & \sigma_{13} \\ \sigma_{21} & \sigma_{22} & \sigma_{23} \\ \sigma_{31} & \sigma_{32} & \sigma_{33} \end{pmatrix}$$

The diagonal elements are the tensile stresses along the three principal axes, and the off-diagonal elements are shearing stress components. (If you are unfamiliar with matrix or tensor notation, do not give up hope, since it is used only for definitions.) Figure 12-5.2.1 also shows a representation of two types of strain. Strain can also be written in tensor notation as is stress with the same interpretation of the components.

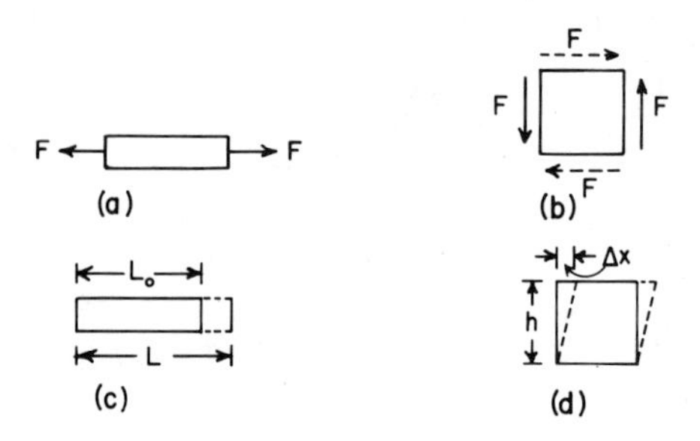

FIG. 12-5.2.1 Stress and strain types: (a) tensile stress σ, (b) shearing stress σ, (c) tensile strain $\varepsilon = (L - L_0)/L$, and (d) shearing strain $\varepsilon = \Delta x/h$.

$$\begin{pmatrix} \varepsilon_{11} & \varepsilon_{12} & \varepsilon_{13} \\ \varepsilon_{21} & \varepsilon_{22} & \varepsilon_{23} \\ \varepsilon_{31} & \varepsilon_{32} & \varepsilon_{33} \end{pmatrix}$$

If we have a uniaxial stress applied as in Fig. 12-5.2.1(a), σ_{11} will be the only stress component; the others are zero. However there is more than one strain component. The uniaxial stress will cause an elongation (ε_{11} is positive), and there will be a decrease in the cross-sectional area leading to a decrease in the values of ε_{22} and ε_{33}. All other values are zero. Using this information we can define two mechanical parameters. You have already seen the definition of Young's modulus: $E = \sigma_{11}/\varepsilon_{11}$. The Poisson's ratio γ is given by $|\varepsilon_{22}|/\varepsilon_{11}$ and indicates the cross-sectional area change for a given length change. Another quantity we will use later, especially in the discussion of nonlinear effects, is the extension ratio $\lambda = L/L_0$. This quantity is particularly important in the description of the drawing of fibers from bulk polymers. We now have the basic terms that will be used throughout the remainder of this section. Make sure you have learned the notation well. That will save having to refer to the terms as you read the following material.

12-5.3 LINEAR VISCOELASTICITY

All polymers exhibit *viscoelastic* effects to some extent. When a stress is applied to the polymer, part of the energy is stored elastically while part of it is lost as heat generated in the deformation of the sample. The permanent deformation is the *visco* portion of viscoelasticity. It is related to a flow of the sample and is a rheological property. You have already been introduced to rheology in Sec. 12-2 where viscosity of solutions is discussed. Also, in Fig. 12-3.2.3 the creep behavior of a polymer under a constant stress is given. Creep is a rheological property.

Linear viscoelastic behavior is very limited and can be thought of as an idealization of real behavior. Real behavior will be discussed in the following section. If we limit ourselves to consideration of small strains, to *isotropic* polymers (meaning polymers with properties independent of direction; a fiber is an example of a non-isotropic polymer), to homogeneous amorphous polymers, and to temperatures near or above T_g, we can treat many systems as though they were behaving linearly.

From the discussion in Sec. 12-4 you should agree that above T_g the polymer chains are in constant motion. This motion allows the polymer strands to move about slowly. So if we apply a stress to an amorphous polymer above T_g, it will slowly respond in such a way as to eliminate that stress. Figure 12-3.2.3 shows the resultant creep. On removing the stress, the energy that has been stored elastically is recovered as the polymer "rebounds" toward the original shape. The rebound is not instantaneous for the same reason the deformation is not instantaneous: there is a finite amount of time required for the polymer strands to rearrange themselves. Also, the recovery is not complete, since some of the energy has been lost in the deformation process (in a purely elastic solid, all the energy is recovered, and the sample returns to the original shape when the stress is removed).

There is a fundamental assumption (that has been proven in many cases) in the study of viscoelastic phenomena. This is the assumption of *time-temperature equivalence*. The time-temperature equivalence states that behavior at increased temperature is equivalent to the behavior expected at lower temperature but longer time. The utility of the assumption should be obvious. If we want to know how a polymer will behave under a given stress for a given length of time that is too long to observe readily (for example, we might want to know how a part deforms in a given use after several years, an experiment that takes too long to run), we study the material at elevated temperatures and convert the behavior to a corresponding time at a lower temperature. Figure 12-5.3.1 gives the basis of the equivalence assumption. Part (a) is a plot of ϕ (strain per unit stress) vs. log time. The curve obtained for a certain polymer is the *creep function* $\phi(t)$ for that polymer. Part (b) shows the temperature dependence of the creep function. Finally, part (c) gives the *master creep curve* relative to some reference temperature. The function Δ_T is called the *shift factor*. The shift factor can be calculated from the Williams, Landel, Ferry (WLF) equation:

$$\Delta_T = \frac{c_1(T - T_g)}{c_2 + T - T_g}$$

where c_1 and c_2 are constants with values of about 17.44 and 51.6, respectively. There is some variation from polymer to polymer. T_g is chosen as the reference temperature.

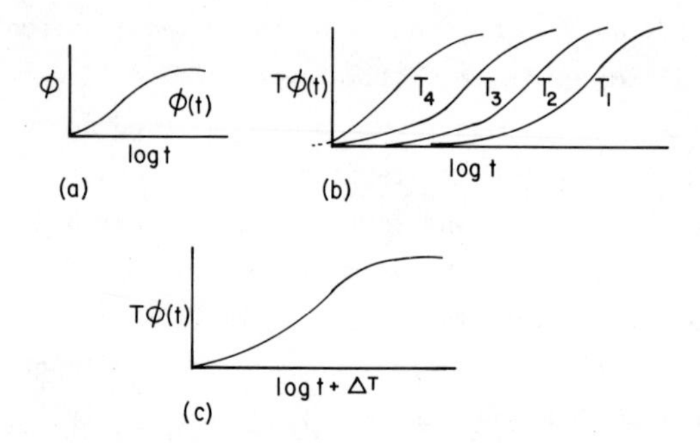

FIG. 12-5.3.1 Demonstration of the time-temperature equivalence for a polymer: (a) creep function $\phi(t)$ (ϕ = strain/stress); (b) time-temperature equivalence $T_4 > T_3 > T_2 > T_1$; and (c) master creep curve.

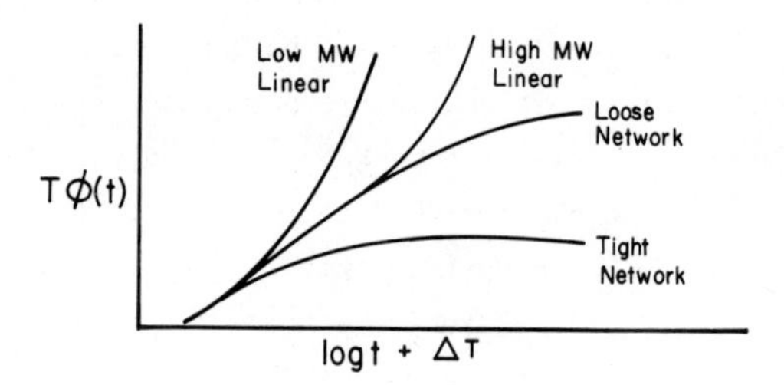

FIG. 12-5.3.2 Effect of molecular architecture on the creep curve.

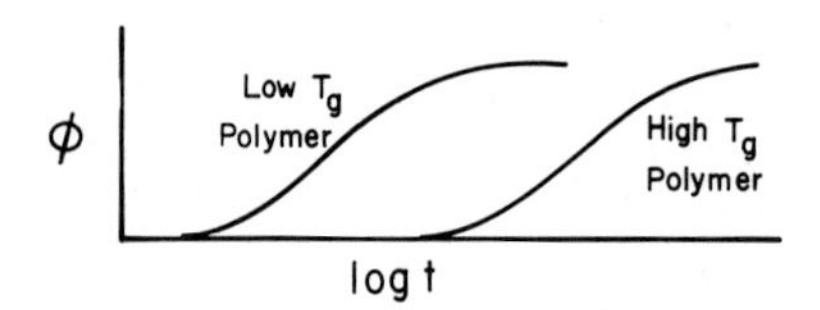

FIG. 12-5.3.3 Effect of T_g on the creep curve.

There are some structure effects on the master creep curve. Figure 12-5.3.2 shows a schematic of the effect of molecular architecture. You should notice that all types of polymers follow the same pattern at low temperature or short time, but diverge significantly at high temperature or long time. There is also an effect of chemical composition on the creep curve. Figure 12-5.3.3 shows the behavior of two polymers with different T_g values compared at the same temperature. There is a definite effect of T_g. However, if the shift factor calculated from the WLF equation is applied, the behavior of the two appears to be the same.

That concludes our short introduction to viscoelastic behavior of ideal systems. You have not been given any information about models used to calculate linear viscoelastic properties for two reasons. First, their very limited utility: The complexity of the models becomes unwieldy very quickly except in the very simplest cases. Second, your objective is to learn some of the language and basic concepts of polymer chemistry, not become an expert polymer chemist or physicist upon completion of the chapter.

12-5.4 NONLINEAR VISCOELASTICITY

The question to be answered in this section is: What happens to a polymer deformed beyond the linear response region presented previously? There is no simple answer to this. There is no theory of any practical use to predict nonlinear behavior. The models required to reproduce the observed behavior are so complicated and cumbersome they are not of use except form a theoretical standpoint. If you want to know how a polymer behaves, you must either look up the experimental data or, if it is a new polymer, measure it yourself.

The most common test for a polymer to determine its stress-strain behavior is to determine the *tensile load-elongation curve*. Figure 12-5.4.1 shows how this test is done and some of the common results observed for different types of materials. In part (a) a polymer specimen with a standard set of dimensions is placed in a

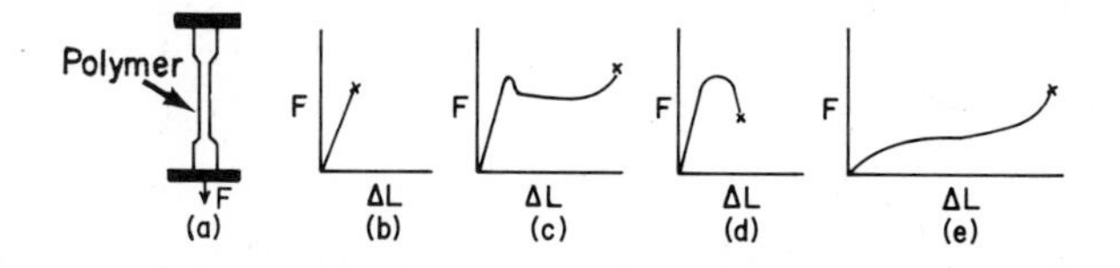

FIG. 12-5.4.1 Test apparatus and some typical force vs. elongation curves (x indicates failure): (a) test apparatus; (b) brittle fracture; (c) deformation with yield, necking, and drawing, then ductile fracture; (d) localized necking; and (e) rubber behavior.

machine that exerts a force and records the resulting elongation. Part (b) shows the results if the sample fails by brittle fracture after a small amount of deformation. More will be said about the ways in which polymers fail in Sec. 12-5.6. Part (c) shows a sample that deforms linearly until, at a given elongation, the sample yields and a neck begins to form (see Figs. 12-3.2.1 and 12-3.2.2). The neck travels along the sample until a length called the *natural draw ratio* is reached. Continued elongation beyond this length requires increased force, and failure results. In part (d) a neck forms but does not propagate. The sample fails at the place of necking. Finally, part (e) shows rubbery behavior. The elongation is reversible as long as the elasticity limit of the rubber is not exceeded. There may be a little hysteresis in the recovery curve, but almost 100% recovery is possible.

There are a number of ways to present the above data. When you have occasion to research force-elongation behavior be sure to look carefully at the manner in which the data are plotted. As you will see in Fig. 12-5.4.2 the same data can be made to look very different if a different means of plotting them is used. Many times the data are plotted as *stress vs. strain*. Stress is force/area and the strain can be expressed as an *extension ratio* $\lambda = L/L_0$. Also, the stress may be plotted as σ_A = "nominal stress" = force/area$_0$, where area$_0$ is the initial cross-sectional area. This is the most common way of expressing stress. Sometimes the more accurate "true stress" σ_T = force/area, where area is the true cross-sectional area of the sample, a quantity that changes as the specimen is elongated. The "nominal strain" is defined as $\lambda - 1$, which (you should show) $= \Delta L/L_0$. The "true strain" or "rational strain" is given as $\ln \lambda$. Finally, the "tensile strength" is usually defined by breaking force divided by area$_0$. Now look at Fig. 12-5.4.2 to see how the same results look when different plots are made.

The only purpose of the above is to point out the dependence of the appearance of data on the manner in which it is plotted. The true stress vs. extension ratio gives a very useful curve for predicting if necking will occur. Plots for three different types of material are given in Fig. 12-5.4.3. Part (a) shows a material that forms a neck, then draws to a natural draw ratio designated by λ_b. The specimen stretches uniformly until point a is reached. A neck forms, and further pulling generates a neck along the entire length until the entire specimen has $\lambda = \lambda_b$, the natural draw ratio. Further pulling results in failure. You should note the dotted lines in part (a). If a tangent from the origin can be drawn to the top of the curve, a neck will form. And if a tangent can be drawn to the bottom of the curve as shown, point b, the neck is stable and the natural draw ratio will be obtained. In case (b) a neck will form, as indicated by the tangent shown. However, no tangent can be constructed to the bottom of the curve, and the neck is unstable. Local necking that does not propagate occurs. This leads to failure in the specimen. Finally, part (c) shows a material that does not neck at all. Instead, uniform

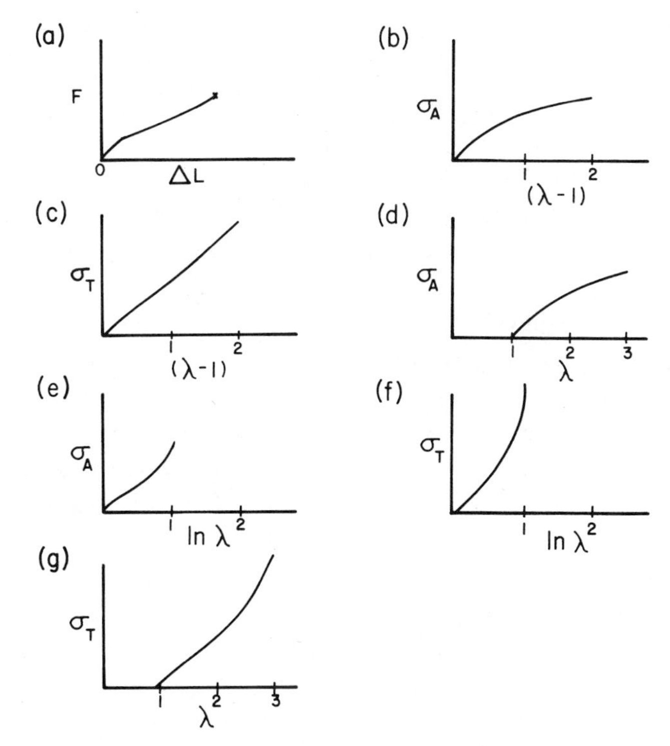

FIG. 12-5.4.2 Force-elongation data plotted in several forms: (a) raw data; (b) nominal stress vs. nominal strain; (c) true stress vs. nominal strain; (d) nominal stress vs. extension ratio; (e) nominal stress vs. true strain; (f) true stress vs. true strain; and (g) true stress vs. extension ratio.

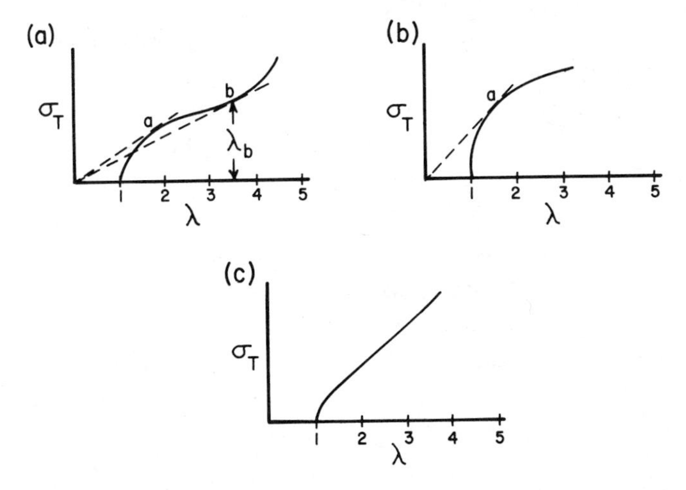

FIG. 12-5.4.3 True stress-extension ratio plots for three types of polymers: (a) a material that necks and draws; (b) a material that forms an unstable neck that fails; and (c) a material that does not neck but stretches uniformly.

stretching is observed.

You can see some of the utility of force-extension data in defining polymer characteristics. As an example, if you need a material for a fiber, you must start with something that gives a curve like Fig. 12-5.4.3(a). A fiber must be drawn to have much strength. This requires that stable necking occur. Another bit of information that can be obtained is the manner in which the material fails. Examination of the test pieces that have been pulled to failure gives valuable information about the morphology of the polymer.

12-5.5 MECHANICAL BEHAVIOR OF CRYSTALLINE POLYMERS

If you have forgotten about the morphology of crystalline polymers, you should return to Sec. 12-3 and reread that. Use will be made of the concepts developed there in this discussion. The amount of crystallinity in a polymer depends on the nature of the polymer. Polyethylene can crystallize up to about 95% for the linear type to about 30% for the highly branched variety. Other materials have less percent crystallization than linear polyethylene. In polyvinyl chloride there is only a trace of crystallization, if any at all. The presence of crystalline regions in a polymer will affect its mechanical properties. The mechanical response is nonlinear with stress load for a crystalline polymer. To properly characterize a crystalline polymer, we must do a lot more work than for a well-behaved amorphous material. There is no single function for crystalline polymers like the *master creep curve* we can draw for amorphous ones. Instead, a large number of experiments with different stress levels and different stress sequences is required to properly characterize the mechanical response of the crystalline polymer. This procedure then has to be repeated for a number of temperatures if the quantitative behavior of the substance is to be revealed. Crystalline polymers do not follow the *time-temperature equivalence* principle we encountered previously. Also, the response to a multiaxial stress cannot be predicted from uniaxial measurements. All in all, this makes for a great deal of difficulty in studying crystalline polymers.

Amorphous polymers can undergo crystallization if sufficient stress is applied. *The melting temperature of a polymer increases as the applied stress is increased.* Figure 12-5.5.1 shows how the melting temperature changes with stress. The change can be up to 100°C in some cases. Also shown in the figure are two possible temperatures for stretching the specimen. In case (a) the polymer is maintained above the melting temperature of the unstretched polymer. As the stress is increased the value of T_m increases until the two lines cross. At this point in part (a), the sample crystallizes into an oriented crystalline fiber. Now, if the stress is removed, the temperature of melting falls below the temperature of the sample, and it will melt. In case (b) the temperature of the sample is kept below T_m^0, the melting

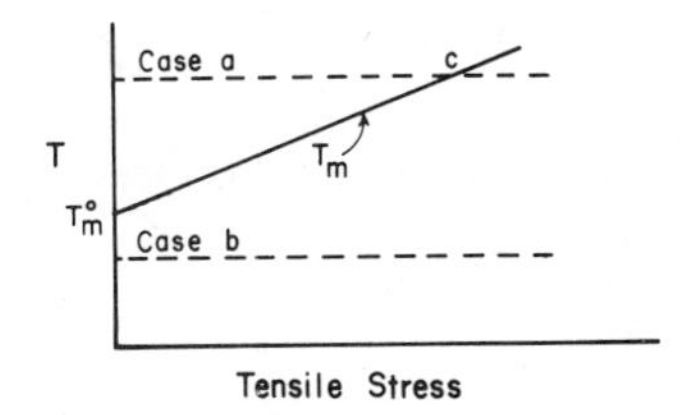

FIG. 12-5.5.1 The effect of tensile stress on the melting temperature of a polymer.

point of the unstressed polymer. Stretching the specimen leads to orientation in the sample and a crystallization. This occurs because the sample is in a *metastable, supercooled* amorphous state (see Fig. 12-4.3.1). Since the amorphous material is not stable at this temperature, once the polymer has crystallized due to the stress applied it will remain in crystalline form even when the stress is removed.

Now turn to Fig. 12-3.2.1, where the stress-strain behavior of a typical polymer is shown. What happens to a crystalline polymer before and after the yield stress σ_y has been reached? Before that point, the specimen deforms uniformly. The spherulites become distorted while still maintaining their identity. At the yield stress, a local neck forms which then grows as the stress is maintained. For this neck to propagate, there has to be a complete structural reorganization as discussed in a little detail in Sec. 12-3.3. The spherulites deform before the yield stress with the laminar ribbons turning in the draw direction. As the pulling is continued the polymer chains begin to orient in the draw direction by slipping past each other and tilting. At the yield point where necking occurs there is an abrupt jump in the extension ratio. This can be attributed to unfolding lamellae. The result is an oriented crystalline fibrous material. If drawing is continued after the neck has traveled the whole length of the polymer, the fiber is drawn to a smaller radius with more orientation (unless it fails first). Another way to get an oriented crystalline state was described above. Pulling a melt until the polymer crystallization temperature is reached will yield an oriented fiber. If the fiber is then cooled (while stretched) below the melting temperature, a stable crystalline fiber will be obtained. (Remember, if the stress is removed while the polymer is above T_m^0 it will melt.) The drawing described leads to a fiber that has a very large tensile strength in one dimension. However, the fiber will be weak in the other dimensions. In the draw direction, the polymer chains afford strength. In the other directions only intermolecular interactions keep the chains together, and this leads to much smaller strength.

12-5.6 CRAZING AND FRACTURING

It is now time to discuss what happens to a polymer if it is stressed to failure. There are several ways a polymer can fail. There can be a *brittle fracture*, *ductile fracture*, or *necking and drawing* until failure. Brittle and ductile fracture have not been defined. Perhaps you can visualize the difference from the names. Brittle and ductile fractures occur in different portions of the stress-strain curve as indicated in Fig. 12-5.4.1. Which type of failure that occurs depends on

1. Chemical composition and architecture of the polymer
2. Geometry of the stress: tensile, or shearing, or both
3. Orientation in the polymer
4. Temperature
5. Rate of application of the load

For example, unoriented polystyrene at 25°C fails in a brittle fashion under tensile stress. On the other hand, oriented (fibrous) polystyrene can withstand much greater tensile stress. The oriented material is, however, weaker in directions perpendicular to the orientation.

There is another phenomenon of importance in the study of failure. This is *crazing*. Crazing involves the formation of many *microcracks* in a material (especially glassy polymers) under tensile loading. These microcracks have a very specialized structure that distinguish them from ordinary cracks. They are filled with "craze-matter" which is highly oriented (fibrous), load-bearing material: it has some strength. The formation of crazes absorbs energy and will contribute to "toughness" in a material. A polymer that tends to form crazes is not as likely to fail in a brittle fashion. Since brittle failure is a catastropic failure that can have severe consequences in many applications (such as high-pressure pipe), it is desirable to adjust the properties of the polymer to try to eliminate that type of failure. Ductile failure is a slower and less severe type of failure, so materials are designed to fail in a ductile manner if they should be stressed beyond design limits. Crazing helps in the production of ductile instead of brittle failure.

Crazing is the precursor to fracture in glassy polymers. In tension tests, the rate of craze formation is observed to increase rapidly with temperature and with stress. Also, many liquids or vapors will accelerate crazing if they are in contact with the polymer. The reasons for this are not well understood. Many plastics are toughened by the addition of rubber particles. These particles seem to act as centers for stress concentration within the polymer, and the formation of crazes seems to be initiated at these points. What this amounts to is that a large number of crazes are formed which help relieve the stress. Some larger rubber particles appear to serve as termination points for the crazes that are formed, so they do not have the opportunity to develop into cracks that could propagate through the sample and cause failure. For many materials there is a definite temperature that separates brittle and ductile fracture. If all other properties are the same, the lower the brittle-ductile transition the better the polymer. The important point to remember from this discussion is that anything that can be done to promote the formation of crazes instead of actual cracks leads to a tougher material. The formation of crazes absorbs some of the impact energy and helps prevent catastrophic failure.

12-5.7 FABRICATION EFFECTS AND POLYMER ENGINEERING

FABRICATION METHODS. There are many ways in which polymers can be formed into useful products. Some of these are

1. Injection molding
2. Compression molding
3. Extrusion
4. Vacuum forming of sheets
5. Bottle blowing
6. Foaming
7. Cold forming (rolling, pressing, etc.)
8. Wet fabrication (spinning of viscose rayon fibers, cellophane formation)
9. Electrodeposition

Some of these operations involve pumping the hot polymer fluid through narrow channels at high temperature. These conditions can lead to degradation of the polymer, as we shall see shortly. Other operations such as sheet forming and bottle blowing are carried out at warm conditions with much less deformation: little flow is required. There are processes designed to give the finished product a certain molecular orientation. These are done near (above) T_g. Finally, some glassy polymers can be cold formed using metal-forming equipment.

EFFECTS OF FABRICATION VARIABLES ON PRODUCTS. Fabrication variables can strongly influence the final properties of a part or a product. Not only chemical reactions during the fabrication, but also molecular orientation and morphology imparted by the processing can be important. Control of these can lead to very desirable characteristics as evidenced by the biaxial strength found in polymer films that is the result of orientation during the film-forming process.

Chemical Reactions. Chemical reactions are affected by high temperature, presence of oxygen, mechanical working, and contact with metals. These reactions, in general, result in polymers with less than optimum properties, so they are usually referred to as degradation. Both high temperature and the presence of oxygen can lead to *scission reactions* and *cross-linking*. Scission reactions refer to those that break polymer bonds giving a lowered molecular weight for the polymer. Cross-linking usually leads to a stiffer, more brittle product and a higher molecular weight. Large mechanical stresses can break polymer chains to give a lower molecular weight material. The presence of oxygen can greatly accelerate the mechanical degradation. Some materials such as polyvinyl chloride (PVC) will be degraded if they are in contact with certain metals at elevated temperatures. PVC releases HCl in the above situation. This leads to two problems. First, the polymer is degraded, and second, the HCl attacks the metal surface causing corrosion.

Molecular Orientation During Fabrication. Molecular orientation usually occurs during mechanical fabrication. The orientation achieved depends on the flow patterns in the forming equipment and/or the viscoelastic deformations induced. The effects can be either favorable or unfavorable. Controlled orientation will lead to added strength in a desired dimension, whereas uncontrolled orientation usually leads to product weakness. A good example is the injection molding of a thin-walled tumbler (drinking cup). The polymer chains will preferentially orient in the direction of flow parallel to the sides of the tumbler. Such a tumbler is very strong in the flow direction (see Fig. 12-5.7.1) but has very low hoop strength (strength around the perimeter of the product). The tumbler will be fragile and will split easily in the flow direction. For example, if you squeeze such a tumbler, splits will readily form from top to bottom. Rotating the mold while the tumbler is being formed leads to orientation in several directions and strength in all directions. Another problem can develop: If there is too much rotation, the flow lines will be around the circumference, giving good hoop strength but low strength for stress applied along the tumbler axis.

There are cases in which a polymer can be made to have improved properties by imposing surface or internal stresses to it. This is similar to tempering glass, where a surface stress is imparted that tends to compress the interior of the material leading to increased strength. Internal stresses are obtained during cooling

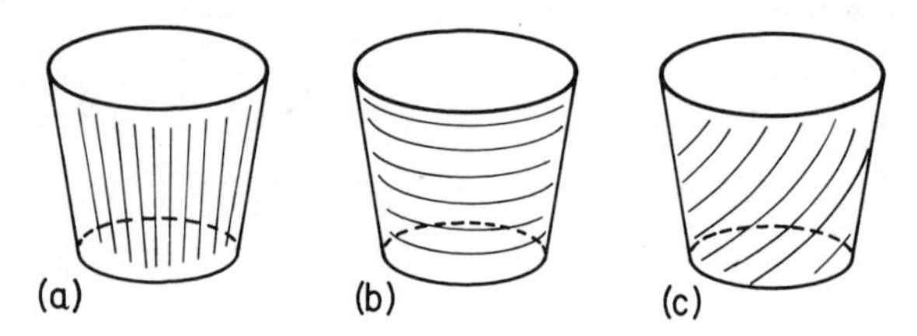

FIG. 12-5.7.1 Possible orientations of an injection-molded tumbler: (a) flow lines parallel to the sides of the cup; (b) flow lines in the hoop direction due to rapid rotation during forming; and (c) flow lines that develop with some rotation of the mold.

since there will usually be differential cooling, and therefore differential shrinkage as a part cools. Also, stresses can be developed during the curing of thermosetting resins. These stresses can be favorable or unfavorable depending on the way in which loading is imposed on the finished product.

MATERIAL SELECTION AND DESIGN. A typical problem of material selection and design will involve consideration of a wide range of information on such things as modulus, strength, creep, fatigue strength, toughness, fracture methods, etc. It is impossible to design a polymer with the most desirable of all these properties. Compromises will have to be made. Then there is the problem of selecting processing variables that will not degrade the polymer or lead to unwanted orientation effects. Perhaps an example will be the best way to illustrate these points. Assume we need a material with the following properties.

Property Requirements.

1. High modulus > 30,000 psi
2. Dimensional stability
   a. while processing
   b. under load (impact resistance, creep, deformation)
   c. thermal
   d. with time
3. Environmental stability
   a. to air oxidation
   b. to chemicals

Molecular Parameters to Meet Property.

1. The polymer must be semicrystalline or have a high T_g; high T_g is the best. Either of these gives high modulus.

2a. Shrinkage of piece occurs in mold and while cooking. Crystalline materials are worst, as crystallization means a volume loss of 5 to 15% on cooling. If this is not isotropic, the piece is distorted or retains severe strains. Best avoided by using noncrystallizing polymers.

2b. Low-T_g crystalline polymers tend to be impact resistant, as the amorphous portions can flow under load. High-T_g polymers tend to be brittle. Polystyrene and polyvinyl chloride tend to be brittle unless blended with rubber particles. These particles impart impact resistance while lowering the modulus slightly. Usually, the higher the modulus, the less creep under a given load, but the lower the impact resistance.

2c. Polymers slowly lose modulus as the temperature is raised. The loss is greater for crystalline low-T_g polymers than for amorphous, high-T_g ones. As modulus is lost, polymers can distort due to imposed load or internal strains. Low-T_g crystalline polymers (polyethylene, nylon) can be used up to 40 to 50°C below T_m. High-T_g, amorphous polymers can be used up to 20°C below T_g.

2d. High modulus and high-T_m or T_g or distortion of piece occurs during aging. To some degree, time and temperature are interdependent. A piece that distorts on heating will eventually distort on standing at room temperature.

3a. Oxidation occurs in accessible regions of polymer. Crystalline polymers are more stable, as polymer cannot be attacked in crystalline regions. Generally, antioxidant is included in most polymers. Oxidation breaks polymer chains and lowers molecular weight and mechanical strength. Also, UV radiation breaks polymer chains in many cases.

3b. Solvent resistance is also dependent on crystallinity. Crystalline polymers are not easily attacked.

In all the above, note the cases in which one property requirement is in opposition to another. For example, compare (2a) and (3a).

12-5.8 SELF-TEST

1. Define or discuss four of the variables associated with stress and strain.
2. In your own words relate how crazing helps eliminate brittle fracture.
3. Discuss as many parameters as possible that would be important in the design of a polymer film used for a food wrap. List the desired characteristics of the film and indicate briefly what property is needed to give each characteristic.

12-6 SUMMARY

You have now been introduced to a significant body of material concerning polymers. You should have some feel for the make-up, properties, and processing of polymers. You are not yet an expert, but you should be able to read polymer literature (other than research reports on theoretical developments) intelligently. When you need to know, you should be able to obtain what you need from continued reading.

13

Phase Equilibria

SYMBOLS AND DEFINITIONS

P	Number of phases
C	Number of components
S	Number of distinct chemical species in a system
R	Number of restraints applicable to a system
Φ	Number of degrees of freedom
μ	Chemical potential (J mol^{-1})
P	Pressure (atm)
T	Temperature (K)
G	Gibbs free energy (J)
T_{bp}	Boiling point temperature
T_{mp}	Melting point temperature
$\Delta H_{\ell \to v}$	Heat of transition (vaporization) from liquid to vapor (J mol^{-1})
$\Delta H_{s \to v}$	Heat of transition (sublimation) from solid to vapor (J mol^{-1})
$\Delta H_{s \to \ell}$	Heat of transition (fusion) from solid to liquid (J mol^{-1})
$\bar{V}$	Molar volume (dm^3, m^3, cm^3)
$\Delta\bar{V}$	Change in volume for phase change
α, β, γ	Solid solution phases

As yet we have considered only pure materials or mixtures of materials that form solutions. Other than a brief mention of liquid-vapor equilibria in Chap. 6, nothing has been said about systems that form more than one phase. However, not all materials are miscible. Those that are not will separate into two or more phases when they are mixed. In this chapter, an introduction to how immiscible materials behave will be given. After a group of definitions, examples of various possible phase diagrams will be presented. A phase diagram is a plot that shows the phase behavior (how many phases and their compositions) for one or more components as a function of temperature and composition or pressure and composition. In fact, after the derivation of a very important relation, the *Gibbs phase rule*, the majority of the rest of the chapter will be devoted to examples. The chapter has been divided in such a way that you can study the type of system of interest to your discipline. Chemical engineers are probably more interested in liquid-liquid and liquid-vapor equilibria. These are found in Sec. 13-3. On the other hand, geologists will have a greater interest in solid-liquid equilibria and examples of geological significance. These are given in Sec. 13-2. The last section contains three-component diagrams and more complicated two-component systems.

Phase behavior is important in numerous fields of science and engineering. Alloys such as steel (iron and carbon), stainless steel (iron, chromium, and nickel), bronze (copper and tin), brass (copper and zinc), etc., are important in structural applications. Refractories (high-melting inorganic materials) are used for firebricks, furnace linings, and crucibles. Many purification processes in chemical manufacturing involve partition of materials between immiscible solvents or distillation from one or more phases. Minerals are very complicated naturally occurring multiphase systems. Ice water is a multiphase system with which you are familiar. By the end of this chapter you should have the information necessary to interpret any of the above phenomena or systems.

13-1 DEFINITIONS AND ONE-COMPONENT SYSTEMS

OBJECTIVES

You shall be able to

1. Define and/or briefly discuss the terms component, phases, degrees of freedom
2. For a given system, determine the number of components, the number of phases, and the degrees of freedom
3. State and apply the Gibbs phase rule
4. Interpret the phase behavior of a one-component system given the phase diagram

13-1.1 DEFINITIONS

The introduction has mentioned most of the terms defined in this section. You probably have an idea as to what these terms mean. However, we need precise definitions for each.

PHASES. A phase *P* is that part of a system that is *chemically and physically* uniform throughout. This definition does not imply that the phase is continuous. For example, ice cubes in water consists of two phases even though there may be several cubes present. Each ice cube is chemically and physically the same, hence, all the cubes make up one phase. There is a limit, however, to just how finely something can be subdivided. If the subdivision approaches molecular dimensions, a solution is approached. In the case of a solution, we cannot define more than one phase. You are familiar with the designation of gases, liquids, and solids as phases. If we mechanically mix table salt and sugar, we still have a solid system, but it consists of two distinct phases. Small particles of one phase are intermingled with small particles of the other. Particles of sugar are not the same chemically as those of salt, even though they may appear the same physically. To the unaided eye no classification other than a single phase could be made. However, an X-ray analysis would definitely show characteristic patterns for both sodium chloride and sucrose. This brings us to another statement of the definition of a phase: A phase is an agglomeration of matter having distinctly identifiable properties such as a distinct refractive index, viscosity, density, X-ray pattern, etc.

*COMPONENTS**. You may have seen the term component C used many times before. Here we will use the term very precisely. To define a component we must define some other terms with which you are undoubtedly familiar but the definitions of which you may never have considered carefully. The first term is *stoichiometry*. A *stoichiometric formula* gives the overall composition of a material, not the actual molecularity of each molecule in the sample. For example, a stoichiometric formula AB implies *only* that for each A atom there is a B atom in the system. It does not imply that every molecule has *molecularity* AB. In fact, every molecule could be A_2B_2 or there could be some A_2B_3, A_3B_2, etc., present as long as overall there is one A for one B. Stoichiometry, as applied to chemical reactions, is discussed in Sec. 10-1. A *species* is any unique molecular aggregation. If the term chemical species is used, it designates a particular molecularity. If we say species A_2B_2 is present, that means the system is made up of A_2B_2 units, not A_3B_3, AB, etc. A *component is an independent chemical variable* which can be added to a system without altering the *actual quantity* of any other component already present. This definition needs some explanation. An independent chemical variable will have a thermodynamic significance. We do not have a choice in its designation. The phase rule derived below will tell us how many of these variables are allowed for a system with a given number of phases. We do have a choice as to how we designate a component. We do not have a choice as to how many there are. A component can be symbolized in many ways, so long as the proper overall stoichiometry is retained. Perhaps a digression into how to find the proper number of components will be helpful at this point.

Define the total *number* of distinct chemical species present as S. The number of equilibria possible plus any other restraints on these is designated R. The allowed number of components C is

$$C = S - R \tag{13-1.1.1}$$

For example, assume we have a solution of Na^+, Cl^-, Ag^+, NO_3^-, AgCl(s), and H_2O. There are six distinct chemical species present. How many restraints do we have on the system? To determine R you must know something of the chemistry of the system. In this case you must know that AgCl(s) is in equilibrium with Ag^+ and Cl^-.

$$AgCl(s) \rightleftarrows Ag^+ + Cl^-$$

This gives one restraint on the system. Electroneutrality of the solution requires

$$[Ag^+] + [Na^+] = [Cl^-] + [NO_3^-]$$

where the brackets indicate concentration. This is the second restraint. Thus

---

* Following the treatment given by A. Reisman (see Sec. 13-1.7, Ref. 1).

$$C = S - R$$

$$C = 6 - 2 = 4$$

There are four components for this solution. How we choose to designate these is somewhat arbitrary, as mentioned above. Either of the following choices would do: NaCl, AgCl, H_2O, $NaNO_3$, or H_2O, NaCl, $NaNO_3$, $AgNO_3$. This illustrates the arbitrariness in choosing the *designation* of the components and the lack of arbitrariness in determining the number of components.

Now what would happen if we chose to include H^+ and OH^-, from the dissociation of H_2O in the above solution? Would this affect the number of components?

$S = 8$ Na^+, Cl^-, Ag^+, NO_3^-, AgCl(s), H_2O, H^+, OH^-

$R = ?$ $[+] = [-]$

$$AgCl(s) \rightleftarrows Ag^+ + Cl^-$$

$$H_2O \rightleftarrows H^+ + OH^-$$

$[H^+] = [OH^-]$ if H_2O is the only source of H^+ and OH^-

Thus, $R = 4$.

$$C = 8 - 4 = 4$$

There is no change. *This result is important.* If we add new species that come *solely* from existing species in the system there are sufficient restraints introduced to leave the number of components unchanged. What constitutes a restraint? Any equation which can be written in which the concentrations within the system are related to each other restricts the behavior of the system and constitutes a restraint.

13-1.1.1 EXAMPLE

Find the number of components and the number of phases for a system containing the following *species*: H_2O, Na^+, Cl^-, K^+, NO_3, NH_4^+, NH_3, H^+, and OH^-.

Solution

$S = 9$

$R = ?$ $[+] = [-]$

$$NH_3 + H_2O \rightleftarrows NH_4^+ + OH^-$$

$$H_2O \rightleftarrows H^+ + OH^-$$

$R = 3$

$C = 9 - 3 = 6$

These might be designated as H_2O, NH_3, NaCl, KNO_3, $NaNO_3$, NH_4NO_3. There is only one phase; $P = 1$.

DEGREES OF FREEDOM. The number of *degrees of freedom* Φ is the number of intensive variables of the system that must be fixed (specified) in order to fix the values of all remaining intensive variables. (An intensive variable is independent of the amount of material present; see Chap. 3.) A degree of freedom can be temperature, pressure, or a composition variable. For a given system to be completely defined a certain number of these variables must be assigned values. The number to be specified can be determined from the *Gibbs phase rule*, which is derived below.

13-1.2 THE GIBBS PHASE RULE

In 1878, J. Willard Gibbs published one of his three definitive papers to thermodynamics in an obscure journal. This paper concerned the *variance* or *degrees of freedom* possible for a system with a specified number of phases P and components C. We can very quickly arrive at Gibbs' result by considering the number of variables possible in a multiphase system and the number of restraints required at equilibrium.

NUMBER OF VARIABLES[a]

Variable	Number of variables of this type per phase	Total number of variables of this type in P phases
T	1	$1 \cdot P = P$
Comment:	We are assuming T is uniform in *each* phase (one condition in the definition of phase). Thus, a single T for each phase is needed. For P phases we must know P values of T to have a complete knowledge of the system temperature.	
P	1	$1 \cdot P = P$
Comment:	Same argument as for temperature. Remember, we are finding the *maximum number* of variables in a system of C components and P phases. Some restraints will be applied below.	
x_i	$C - 1$	$P(C - 1)$
Comment:	x_i is the mole fraction of component i. Since $\sum_{i=1}^{C} x_i = 1$, from the definition of mole fraction, we have only $C - 1$ mole fractions in each phase that can vary. If we specify $C - 1$ mole fractions, the equation $\sum x_i = 1$ fixes the last one. In P phases we then have $P(C - 1)$ concentration (or mole fraction) variables.	

[a] For C components and P phases.

NUMBER OF RESTRAINTS AT EQUILIBRIUM

Restraint	Number
$T_j = T_i$	$P - 1$
Comment: At equilibrium the temperature is the same in every phase. When we specify one temperature, the remaining $P - 1$ are automatically fixed. Thus, $P - 1$ of the initial P temperature variables are fixed by assigning a temperature value to any one phase	
$P_j = P_i$	$P - 1$
Comment: Same as temperature	
$\mu_j = \mu_i$	$C(P - 1)$
Comment: By definition, the chemical potential of a given component is the same in every phase. Thus, if the chemical potential of a given component is specified, the remaining $P - 1$ values for that component are fixed. There are C components. Thus, $C(P - 1)$ gives the total number of equilibrium conditions or restraints.	

Maximum number of variables:

$$P + P + P(C - 1) = P(C + 1) \tag{13-1.2.1}$$

This is the sum of temperature, pressure, and mole fraction variables.

Total number of restraints:

$$P - 1 + P - 1 + C(P - 1) = (C + 2)(P - 1) \tag{13-1.2.2}$$

DEGREES OF FREEDOM. The *variance* or number of degrees of freedom is

$$\begin{aligned} \Phi &= \text{variables} - \text{restraints} \\ &= P(C + 1) - (C + 2)(P - 1) \\ &= C - P + 2 \end{aligned} \tag{13-1.2.3}$$

This is the *Gibbs phase rule*

$$\Phi = C - P + 2 \tag{13-1.2.4}$$

Equation (13-1.2.4) states that the least number of intensive variables that must be specified to fix all other intensive variables is equal to the number of components minus the number of phases plus 2. This is an extremely important equation in describing the phase behavior of a system. Obviously, Φ cannot be < 0. So Eq. (13-1.2.4) limits the number of phases present. For example, for one component the *minimum* Φ is 0 and the maximum P is

$$\Phi = 0 = 1 - P + 2$$

or $\quad P = 3$

For a one-component system the maximum number of phases allowed at equilibrium is *three*. Many other examples of the use of this equation will be found throughout the remainder of this chapter.

13-1.2.1 EXAMPLE

For the system described in Example 13-1.1.1 determine the number of degrees of freedom.

Solution

$C = 6$ and $P = 1$.

$$\Phi = C - P + 2 = 6 - 1 + 2$$
$$= 7$$

There are seven variables that must be specified to fix this system. These would have to be T, P, and the mole fractions of five of the six components. (The sixth would automatically be fixed.)

13-1.2.2 EXAMPLE

What is the variance (degrees of freedom) allowed for a three-component system in two phases?

Solution

$C = 3$ and $P = 2$.

$$\Phi = C - P + 2 = 3 - 2 + 2$$
$$= 3$$

These three degrees of freedom could be T, P, and one composition variable, or P and two composition variables, or some other combination of T, P, and x_i.

13-1.3 THERMODYNAMIC CONSIDERATIONS

FREE ENERGY. As you probably would suspect, the free energy of a phase relative to another phase will determine whether that phase is stable at a particular T and P. If we knew precisely how G for each possible phase in the system varied with T and P, then we could predict what phases would be stable at a given T and P. For example, look at Fig. 13-1.3.1 in which G for the gas, liquid, and solid phases is plotted as a function of temperature (pressure is held constant; this is an isobaric plot). The curves shown are idealized. However, they can be found using techniques developed in Sec. 5-2.3. In that section, equations were derived relating free

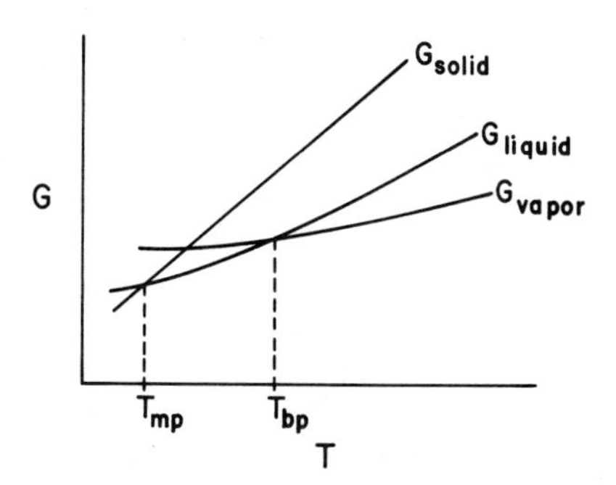

FIG. 13-1.3.1 Free energy vs. temperature for the three phases of a one-component system.

energy changes to temperature changes. Also in Sec. 5-3.2, the variation of G with pressure is given. Although we will not use any G vs. P plots, they can be calculated for each phase to determine what phase is most stable at a given pressure. With the background you have already, you could find G vs. T or G vs. P and determine the phase diagram for a system (assuming appropriate data are available).

Let us consider Fig. 13-1.3.1 further. Below T_{mp} the solid is the phase with the lowest free energy and is, thereby, the stable phase. At T_{mp}, $G_{solid} = G_{liquid}$, and solid and liquid will be in equilibrium. This is a good point to check the Gibbs phase rule. We have two phases and one component.

$$\Phi = C - P + 2 = 1 - 2 + 2 = 1$$

There is one degree of freedom and this has been taken when we assigned a pressure for the plot we have drawn. The temperature is *invariant* (cannot change) as long as the two phases are present. When all the solid has melted, the temperature can change again. Between T_{mp} and T_{bp} the liquid is lowest in free energy. At T_{bp} both liquid and gas have the same free energy. If heat is added the liquid will vaporize at constant T_{bp} ($\Phi = 1$, again with the two phases present and the pressure assumed in drawing the figure uses that degree of freedom). Above T_{bp} the gas phase is the most stable.

THE CLAUSIUS-CLAPEYRON EQUATION. In Sec. 6-2.3 the Clausius-Clapeyron equation was derived for the vapor pressure of a liquid as a function of temperature. This is given by

$$\frac{dP}{dT} = \frac{\Delta\bar{H}_{\ell\to v}\,P}{RT^2} \qquad (13\text{-}1.3.1)$$

A similar equation can be written for the vapor pressure of a solid.

$$\frac{dP}{dT} = \frac{\Delta\bar{H}_{s\to v}\, P}{RT^2} \tag{13-1.3.2}$$

For a solid to liquid transition we can write

$$\frac{dP}{dT} = \frac{\Delta\bar{H}_{s\to\ell}}{T\,\Delta\bar{V}_{s\to\ell}} \tag{13-1.3.3}$$

In both the $s \to v$ and $\ell \to v$ cases dP/dT will always be positive, since the volume of the vapor is much greater than that of the condensed phase, which means the slope of a P vs. T plot is positive for the s, v and ℓ, v equilibrium lines (see Fig. 13-1.3.2). However, the slope of dP/dT for the $s \to \ell$ transition depends on $\Delta\bar{V}_{s\to\ell} = \bar{V}_\ell - \bar{V}_s$. If $\bar{V}_\ell$ is greater than $\bar{V}_s$, the slope of the P vs. T curve will be positive as shown in Fig. 13-1.3.2. This is the normal situation. In some cases, such as the ice to water transition, $\bar{V}_\ell < \bar{V}_s$ which makes $\Delta\bar{V}_{s\to\ell}$ negative and dP/dT negative. The final possibility is $\bar{V}_\ell = \bar{V}_s$ for which dP/dT = 0. The lines drawn in Fig. 13-1.3.2 represent the equilibrium between two phases at different pressures and temperatures. On any line in the diagram the free energy of two phases are equal. At the intersection of three lines there are three phases in equilibrium at that T and P. For example, at point a $G(s, T_a, P_a) = G(\ell, T_a, P_a)$. At point b $G(s, T_b, P_b) = G(\ell, T_b, P_b) = G(g, T_b, P_b)$, or the gas, liquid, and solid are all in equilibrium at point b. Finally, at point c, $G(\ell, T_c, P_c) = G(g, T_c, P_c)$. The details of this diagram and how it is interpreted follow.

13-1.4 ONE-COMPONENT SYSTEMS

SIMPLE SYSTEMS. The diagram shown in Fig. 13-1.3.2 is for a simple one-component system. Let us look at a general diagram to see how it can be interpreted. Such a diagram is shown in Fig. 13-1.4.1. Point a is on the $s \to \ell$ transition line. This line shows how the melting point temperature changes with pressure. From the

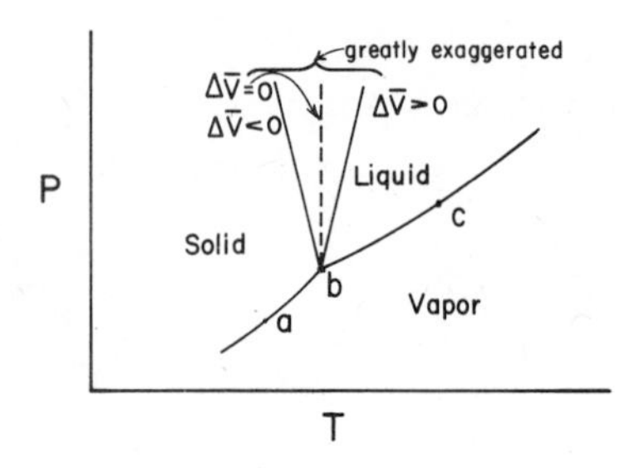

FIG. 13-1.3.2 P vs. T plot for a one-component system.

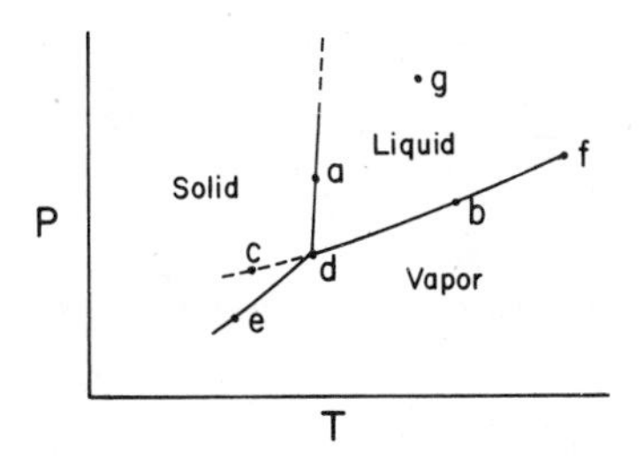

FIG. 13-1.4.1 A general one-component system.

slope we know $\Delta\bar{V} = \bar{V}_\ell - \bar{V}_s > 0$. An increase in pressure will favor the smaller volume ($\bar{V}_s$ in this case), and the temperature required to melt the solid will be higher for higher pressure. Along this line there are two phases present. The Gibbs phase rule tells us that

$$\Phi = C - P + 2 = 1 - 2 + 2 = 1$$

There is one degree of freedom. If T or P is specified, the other is fixed by the phase diagram. (There are no composition variables here since only one component is present.) Point b is on the liquid-vapor equilibrium line. It shows how the vapor pressure of the liquid varies with T. As for point a, there is only one degree of freedom for this two-phase line. Each of these two-phase lines are called *univariant* lines. Point c is a *metastable* or *supercooled* liquid. Normally, at a P and T below that specified as d, the liquid freezes to a solid. However, in many cases the liquid does not freeze. Instead it *supercools* (up to several degrees) before solidifying. If the liquid is disturbed or some foreign substance or a crystal of the solid is added to this *metastable* (*supercooled*) liquid, it will solidify (usually) almost instantly. The supercooled condition is called *metastable* becuase the system is not in its thermodynamically most stable state; the free energy of this liquid is higher than the solid, and given time and/or the proper disturbance the system will revert to the stable state (the lowest free energy). Point d is the *triple point* at which all three phases, liquid, vapor, and solid are in equilibrium.

$$\Phi_{tp} = C - P + 2 = 1 - 3 + 2 = 0$$

There is no variance at this point; T and P are fixed. For this reason triple points make excellent temperature calibration points. Point e is on the solid-vapor equilibrium line. The line shows the variation of the sublimation temperature with pressure. Again there is one degree of freedom. Point f is the critical point. To refresh your memory, the critical point is the highest temperature at which liquid and vapor can coexist in a system. Above the critical temperature there is no distinction between liquid and vapor. Point g is in a single phase region.

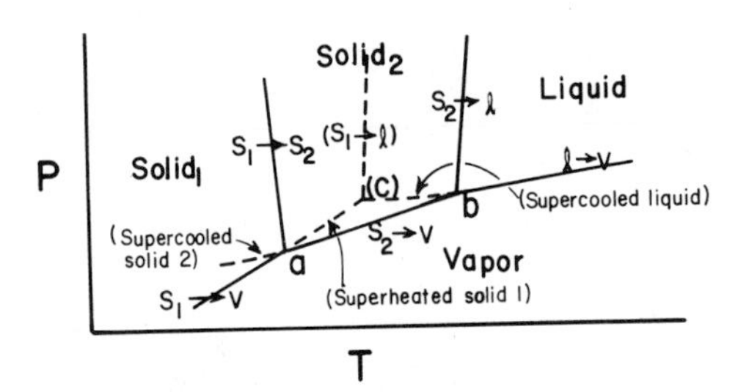

FIG. 13-1.4.2 A one-component system with a solid-solid phase transformation.

$$\Phi = C - P + 2 = 1 - 1 + 2 = 2$$

T and P can vary independently.

A MORE COMPLEX SYSTEM. Many substances can have more than one solid phase present. A typical diagram of a case in which two solid phases exist is shown in Fig. 13-1.4.2. Do not despair. This is not nearly as difficult as it looks. The solid lines are stable regions. The dashed lines and titles in parentheses represent unstable (or metastable) conditions. The labels on each line should be self-explanatory. Point a is the triple point, Solid 1-Solid 2-Vapor, and point b is the Solid 2-Liquid-Vapor triple point. The only difficulty you may have is with the *triple point (c)*. This is a metastable triple point of Solid 1-Liquid-Vapor. There are only two ways to obtain this. At the temperature of point a, Solid 1 should spontaneously convert to Solid 2. If there are kinetic limitations on this transformation (it might be a very slow process for the molecules in Solid 1 to rearrange to the order needed for Solid 2), it is possible to "superheat" Solid 1 until it melts to liquid at T_c. Another way to obtain point (c) is from the liquid. If the liquid does not form Solid 2 at T_b but supercools to T_c, then it is possible for Solid 1 to form directly from the liquid. Solid 1 and Liquid at any T between T_a and T_b are metastable with respect to Solid 2, and a spontaneous transformation to Solid 2 can occur. More than two solid phases are formed by some systems. Their phase diagrams are extensions of those we have already considered. More complicated systems will not be discussed.

13-1.5 PROBLEMS

1. How many phases, components, and degrees of freedom are there in each of the following:

   a. A bulb filled with $N_2(g)$, $Ar(g)$, and $O_2(g)$

b. A bulb half-filled with $C_6H_6(\ell)$, the other half filled with $C_6H_6(g)$ and Ne(g)

c. (sol'n) = solution

$H_2O(g) + CO_2(g)$

$H_2O(\ell)$

Ca^{2+}(sol'n) CO_3^{2-}(sol'n)

$CaCO_3(s)$

d. A vessel containing $H_2O(\ell) + H_2O(s) + H_2O(v)$

e. An aqueous solution containing H_3PO_4, $H_2PO_4^-$, HPO_4^{2-}, PO_4^{3-}, Na^+, and H^+ at 1 atm pressure.

2. For the following free energy diagram, describe the phase transitions that occur for a one-component system as the temperature is raised. Specify Φ, C, and P at several points.

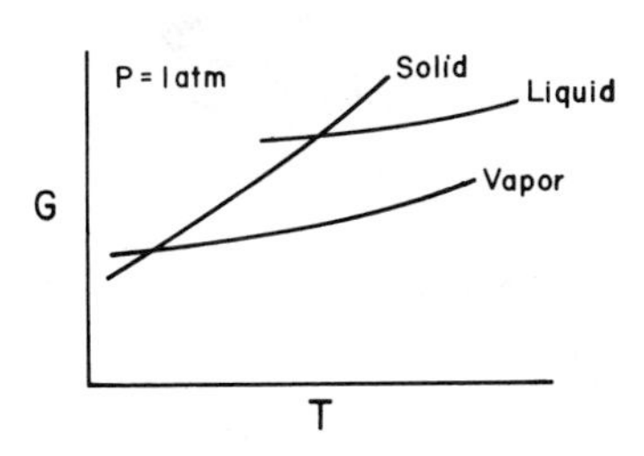

3. Given the one-component phase diagram shown below:

a. Label all areas, two-phase lines, and three-phase points, and give Φ for each

b. Describe in detail what should (thermodynamically favored) happen and what could (metastable) happen in heating from T_a to T_b

c. Repeat part (b) for cooling from T_c to T_d

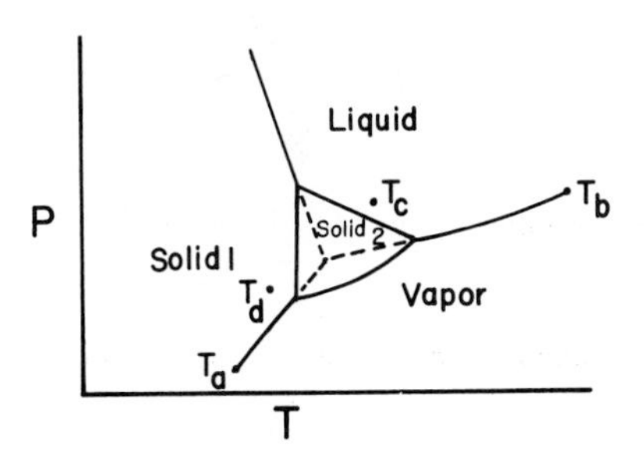

4. Assume that ΔV and ΔH for the solid → liquid transition are independent of T

and P for iron. Calculate and plot the melting point of iron up to 100,000 atm (1 atm = 1.01×10^5 N m^{-2}). $\Delta\bar{V}_{s\to\ell} = 2.7 \times 10^{-7}$ m^3 mol^{-1}, $\Delta\bar{H} = 1.55 \times 10^4$ J mol^{-1}, and T_{mp} (1 atm) = 1809 K.

5. Determine *C*, *P*, and Φ for each of the following:

   a. A closed flask containing AgCl(s), $H_2O(s)$, $H_2O(\ell)$, $H_2O(g)$, $H^+(aq)$, $Cl^-(aq)$, HCl(g), $CO_2(aq)$, $CO_2(g)$, and HCl(aq)

   b. A closed flask containing ice, distilled water, and water vapor

6. Discuss how the results of Problem 5(b) can be used in the precise calibration of a temperature sensing device.

7. Is it possible to have the following system at equilibrium? $H_2O(\ell)$, $H_2O(s)$, $H_2O(g)$, benzene(s), benzene(ℓ), and benzene(g). (Benzene and water are only slightly soluble in each other.) Explain carefully.

13-1.6 SELF-TEST

1. Determine the number of degrees of freedom in each of the following systems. (The horizontal lines indicate boundaries between liquid phases.)

(a)

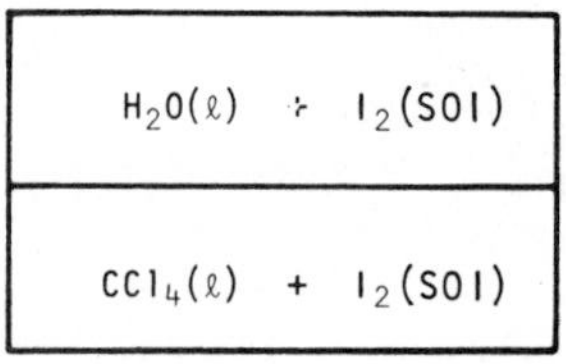

$H_2O(\ell)$ + I_2(sol)
$CCl_4(\ell)$ + I_2(sol)

(b)

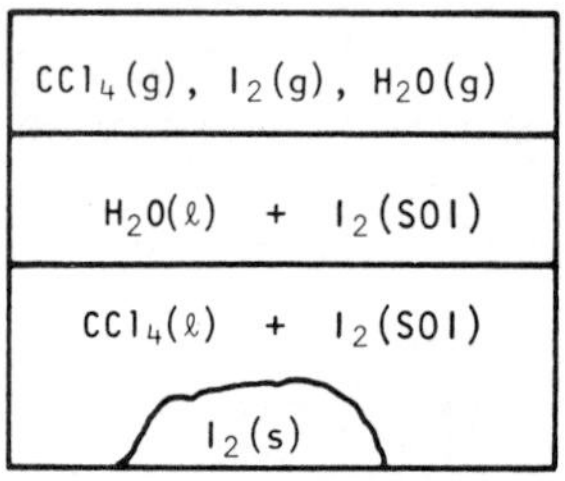

$CCl_4(g)$, $I_2(g)$, $H_2O(g)$
$H_2O(\ell)$ + I_2(sol)
$CCl_4(\ell)$ + I_2(sol)
$I_2(s)$

2. Write definitions of the terms phase, component, and degree of freedom in your own words.

3. For water, $\Delta\bar{V}_{s\to\ell} < 0$. Sketch the water phase diagram and discuss how $\Delta\bar{V}_{s\to\ell} < 0$ is important in ice skating.

4. If we have a system at point (a) in the following diagram, what will happen when we try to increase the temperature, if no new phases appear or disappear.

Explain carefully.

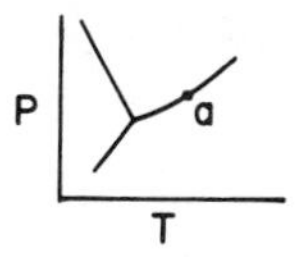

13-1.7 ADDITIONAL READING

1.* A. Reisman, *Phase Equilibria*, Academic, New York, 1970, Chap. 3.

2. F. E. W. Wetmore and D. J. LeRoy, *Principles of Phase Equilibria*, Dover, New York, 1950.

13-2 BINARY SOLID-LIQUID DIAGRAMS

OBJECTIVES

You shall be able to

1. Use cooling curve data to produce simple two-component solid-liquid phase diagrams, or vice versa
2. Label the regions in any two-component phase diagram given
3. Define the terms eutectic, congruent melting, incongruent melting, peritectic, liquidus, solidus
4. Completely describe, from a phase diagram, all changes that occur on cooling or heating a given composition of a two-component system; give all phases present at each temperature and approximate amounts of each phase; also, give the degrees of freedom at any point in the cooling or heating process

13-2.1 INTRODUCTION

This section will introduce simple two-component solid-liquid phase diagrams. More complex diagrams will be discussed in Sec. 13-4. The approach here will be by example. After introducing cooling curves, several examples will illustrate the various types of behavior that may be encountered in two-component systems. Also, examples will be used to aid in the interpretation of equilibrium cooling and heating of two-component systems. This section should be of particular interests to students of geology. Some of the examples will have some more complex phase diagrams. Liquid-liquid and liquid-vapor equilibria are treated in Sec. 13-3.

13-2.2 SIMPLE EUTECTIC SYSTEMS

A EUTECTIC DIAGRAM. Figure 13-2.2.1 gives the T-composition diagram of a simple, eutectic-forming, two-component system. The word *eutectic* means "easily melted". The eutectic mixture is the lowest melting temperature composition, point a in the figure. Above the curved two-phase lines only liquid is present. If the temperature and composition are such that a point is found between one of the two curves and the horizontal line, the point is in a two-phase region: liquid plus either solid A or solid B, depending on which side of the diagram you are considering. Below the horizontal line there are two solid phases: pure A + pure B. The curved lines are *liquidus lines* which represent the conditions at which solid first forms from the liquid. The horizontal line is the *solidus*, which is the line below which only solid is present. You might wonder what determines the shape of the curve between the liquid and solid regions in this figure. These are freezing point depression curves (see Sec. 6-2.4). Consider freezing point of pure A (about 350 K). As B is added to A, the freezing point of A falls toward point a. On the opposite side of the diagram, pure B freezes at about 230 K. As A is added to B, the freezing point of B falls toward point a.

A good method for determining what phases are present in any region is as follows.

1. Draw a horizontal line through the T and composition of interest (say, point b in this case; T = 250 K and 20 mol% B).
2. The intersection of this line with the boundaries gives the phases present.

In this case the horizontal line intersects the liquidus line at point c and the pure A line at point d. This means we have pure solid A and a liquid of composition c (about 30 mol% B) in equilibrium. There is no material in the system of composi-

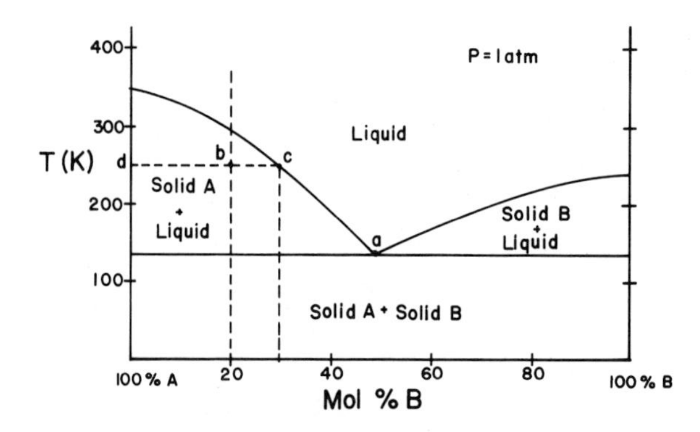

FIG. 13-2.2.1 A simple eutectic system.

tion b. There is some solid A and liquid of composition c which give an average composition of b, but *there is nothing* of composition b present. It is possible to determine the relative amounts of liquid and solid. A *lever rule* applies here. The amount of liquid present for the average composition b is related to the length of the line segment bd and the amount of solid related to the length of bc. In fact,

$$\frac{\text{Moles liquid}}{\text{Moles solid A}} = \frac{\overline{bd}}{\overline{bc}} \qquad (13\text{-}2.2.1)$$

If we had used *weight precent* instead of mole percent, the formula would read

$$\frac{\text{Weight liquid}}{\text{Weight solid A}} = \frac{\overline{bd}}{\overline{bc}} \qquad (13\text{-}2.2.2)$$

Be sure to note units on the abscissa if you wish to calculate the amounts of various phases present.

COOLING CURVES AND DESCRIPTION OF COOLING BEHAVIOR. Many solid-liquid phase diagrams are obtained by noting the cooling behavior of known compositions. Figure 13-2.2.2 shows several cooling curves for the simple eutectic system sketched in Fig. 13-2.2.1. The cooling behavior of three compositions will be described.

Pure A. Above 375 K liquid A exists, and if the surroundings are cooler than the liquid, the temperature of the liquid will fall. At 375 K and 1 atm, solid A begins to form. Here we have two phases present and one component. Thus, since P is already fixed at 1 atm (a phase diagram showing T vs. composition must be drawn at fixed pressure, which eliminates one degree of freedom):

$$\Phi = C - P + 1 = 1 - 2 + 1 = 0$$

The system is *invariant*, meaning no changes can take place in independent variables. The temperature will remain constant until all of one phase disappears: this is the reason for the horizontal portion of *line (a)* in Fig. 13-2.2.2. In this case, since we are cooling, all the liquid will disappear leaving only solid A which then cools.

$$\Phi = C - P + 1 = 1 - 1 + 1 = 1$$

(The temperature of the solid can change as shown in the figure.)

20% B. Above 290 K only liquid exists which will cool as heat is lost to the surroundings. At 290 K the liquidus curve is reached (point f). Pure solid A begins to separate from the liquid. There are now two phases and two components, so at 1 atm,

$$\Phi = C - P + 1 = 2 - 2 + 1 = 1$$

This is a *univariant* system, meaning one variable can change independently. The temperature continues to decrease, but there is a change in slope. As pure A forms

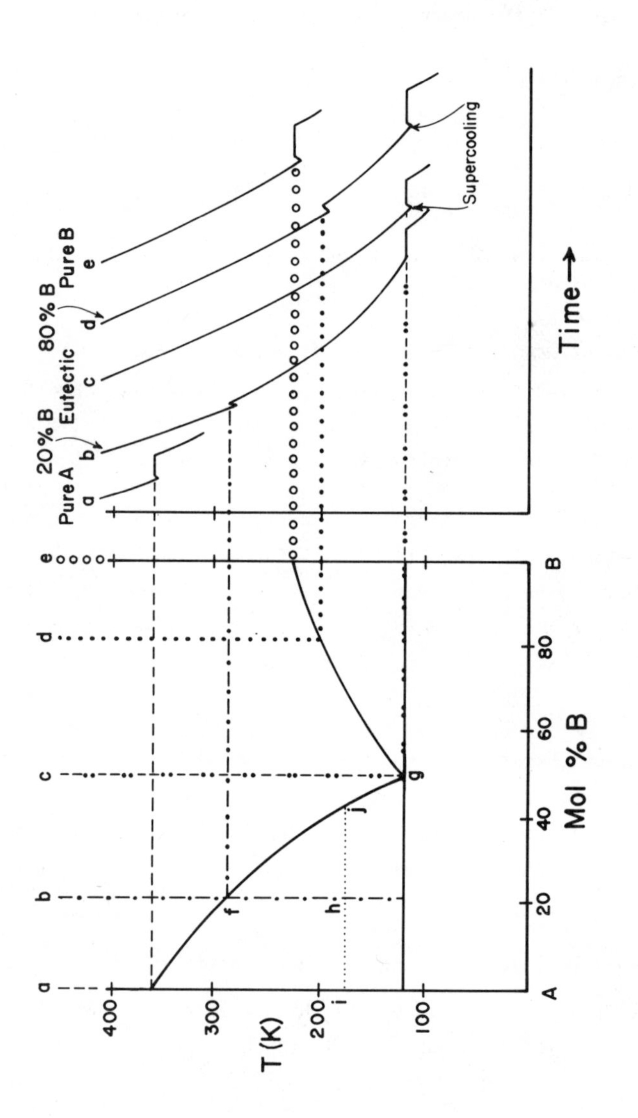

FIG. 13-2.2.2 Cooling curves for a simple eutectic system.

from the liquid, its heat of fusion is released. This tends to warm the liquid. This is opposed to the loss of heat to the surroundings and results in a slower rate of cooling. As pure A separates the liquid is losing A, so it becomes richer in B. The composition of the liquid follows the liquidus curve from f to g as the temperature drops from 290 K to 115 K. Throughout there is an equilibrium between the solid A and the liquid. At point g another change occurs. Pure solid B begins to precipitate. This gives us three phases, so

$$\Phi = C - P + 1 = 2 - 3 + 1 = 0$$

We have an invariant system. No temperature change can occur until one of the phases disappears. If we continue to remove heat, T will remain constant until all the liquid has been precipitated as pure solid A and pure solid B. The two solids will be intermingled, but they will exist as separate phases. After the liquid disappears, cooling of the two solids can continue with no further change.

Eutectic Composition. If composition c is cooled from 400 K, no changes occur until T = 115 K (point g), the eutectic temperature. At this point both A and B begin to precipitate (as two separate phases). This three-phase system (A, B, and liquid) is invariant. No temperature change can occur until all of one phase has disappeared. After the liquid has been depleted, the two solids cool with no further change.

80% B and Pure B. No new concepts are involved here. Describe the cooling behavior in your own words, then compare with the description for pure A and 20% B given above.

13-2.3 COMPOUND FORMATION: CONGRUENTLY MELTING

Figure 13-2.3.1 shows the schematic of an A-B system that forms a compound AB. The compound AB is a *congruently melting compound*. This means if we start with solid AB and heat to melting, the liquid will have the same composition as the original solid. The figure is labeled with the phases present in each region. You should have no difficulty with labeling these diagrams. A description of what happens when liquid of composition (a) is cooled from 500 K will be given. You should try to describe what happens when the composition is heated from 300 K. This type of system is, in essence, simply two eutectic systems put together. One-half the diagram is the A-AB system, the other half the AB-B system. (Remember, we can choose components as we see fit.)

Assume we have composition a (45% B) at 500 K where only liquid is present. Cooling results in no change until point b (440 K) is reached. At this temperature solid compound AB begins to precipitate. The compound is richer in B than the

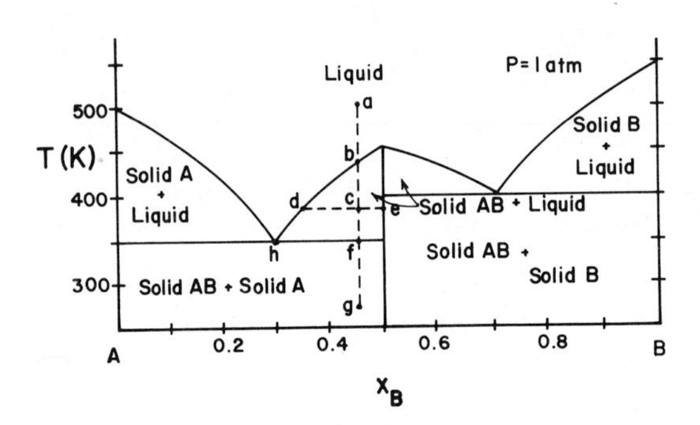

FIG. 13-2.3.1 A phase diagram for a two-component system which forms a congruently melting compound.

liquid (50% vs. 45%), so the liquid becomes depleted in B (richer in A) as solid AB separates. The liquidus curve follows the *line bdh* as the temperature is lowered ($\Phi = C - P + 1 = 2 - 2 + 1 = 1$). At point c both liquid and solid are present. We can find the amount of each by using the lever rule.

$$\frac{\text{Moles liquid}}{\text{Moles solid AB}} = \frac{e - c}{c - d} = \frac{50 - 45}{45 - 34} = 0.45$$

which states there are more moles of solid than there are moles of liquid. When 350 K is reached (point f) the liquid has a composition of $x_B = 0.30$ (point h), and the solid is AB. Solid A begins to separate from the liquid along with solid AB ($\Phi = 0$), and the temperature is constant until all the liquid is gone. Below (f) only cooling of the two solids occurs. Figure 13-2.3.2 is a geochemical example. It looks complicated, but is in fact a simple extension of what we just did. You should be able to handle its interpretation.

13-2.4 COMPOUND FORMATION: INCONGRUENTLY MELTING

A solid compound without sufficient stability to melt a liquid of the same composition is an *incongruently melting compound*. Such a compound decomposes, on heating, into a solution and a solid, each of which has a composition different from the initial compound. This is called a *peritectic* reaction or an *incongruent melting*. Figure 13-2.4.1 is another geochemical example; this one is the $Na_2O{\cdot}SiO_2$ - $CaO{\cdot}SiO_2$ system. Note that the two materials chosen as components ($Na_2O{\cdot}SiO_2$ and $CaO{\cdot}SiO_2$) are not simple materials. Also note that weight percent is being used. The diagram is labeled. The compound A, $2Na_2O{\cdot}CaO{\cdot}3SiO_2$, is an incongruently melting compound. If a solid of this composition is heated from, say, 1100°C, nothing happens until 1141°C. At this temperature, A decomposes into liquid of composition h and compound B ($Na_2O{\cdot}2CaO{\cdot}3SiO_2$). Three phases are present so the temperature is constant until

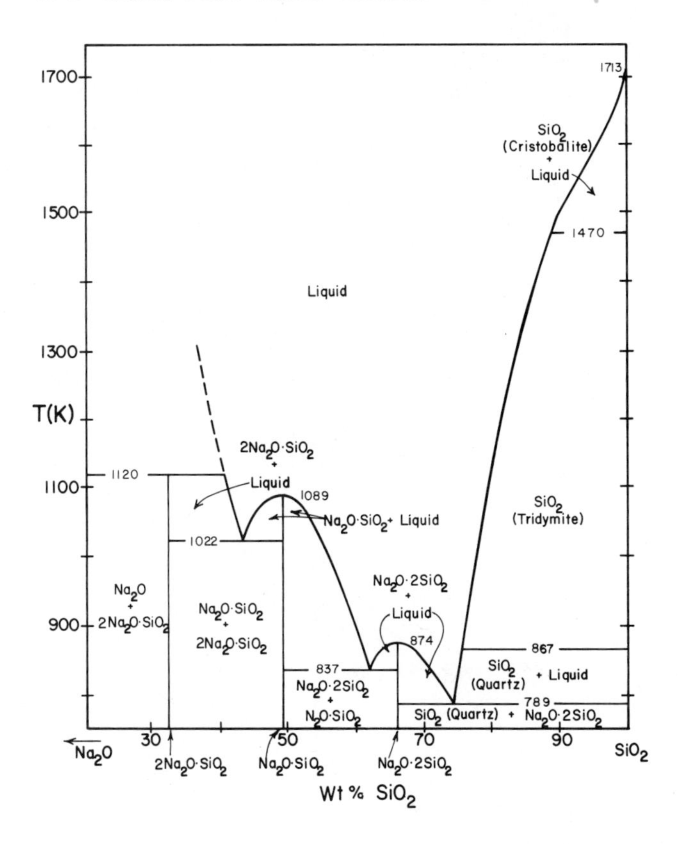

FIG. 13-2.3.2 The binary system Na_2O-SiO_2 (from the U. S. Geological Survey Professional Paper 440-L, Data of Geochemistry, 6th ed., Chap L.)

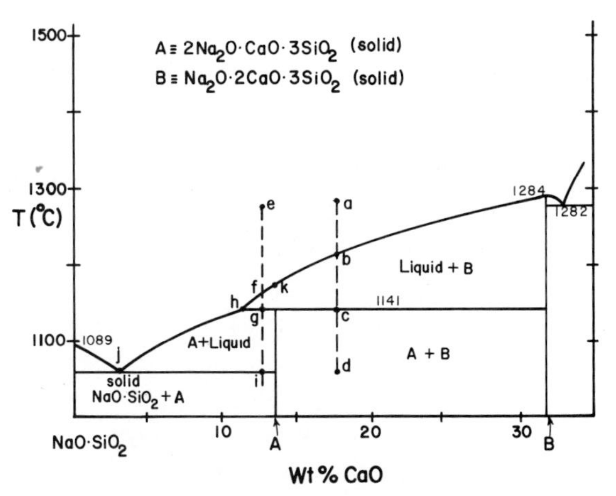

FIG. 13-2.4.1 A portion of the $NaO \cdot SiO_2$-$CaO \cdot SiO_2$ phase diagram (from the U. S. Geological Survey Professional Paper 440-L, Data of Geochemistry, 6th ed.)

all A has decomposed. If heating is continued, B will begin to melt. This increases the percent $CaO \cdot SiO_2$ in the liquid, since the B that is melting is richer in $CaO \cdot SiO_2$ than the liquid already present. So, as the temperature rises, the liquidus curve follows the line hk. At about 1170°C (point k) the last bit of B melts leaving only liquid. Now what happens on cooling from a to d (17.5% $CaO \cdot SiO_2$)? At point b (1205°C) solid B begins to separate with the liquid getting richer in $NaO \cdot SiO_2$ along the line bh. When point c (1141°C) is reached, the peritectic reaction is encountered. Liquid of composition h reacts with some solid B formed previously to give solid A. Since three phases are present the temperature is constant until all the liquid has been converted to A. With only two phases present (A and B) the temperature can again fall, resulting in the cooling of the two solid phases with no further changes. The cooling along ei is described in a similar manner; this is left as an exercise for you. Write out the steps. You need the practice.

Note: Throughout *equilibrium* cooling has been described. If sufficient time is not allowed for equilibrium to be established at each step, the above description will not be valid. We will see the importance of nonequilibrium cooling below. In any natural system (for example, geological systems) you must be aware of the possibility of nonequilibrium behavior when you are interpreting the phases present.

13-2.5 SOLID SOLUTIONS AND ALLOYS

A discussion of solid-liquid phase equilibria would not be complete without a mention of systems which form solid solutions. Metal alloys and steel are examples of materials whose properties are dependent to some extent on the partial solubility of one material in the other. Several general examples will be given. There are many different possible types of phase diagrams, but if you can tie together all the elements discussed so far, plus solubility considerations, you should be able to interpret most two-phase solid-liquid diagrams.

EUTECTIC PHASE DIAGRAM WITH LIMITED SOLID SOLUBILITY. The simplest type of phase diagram which exhibits partial solubility is shown in Fig. 13-2.5.1. The Ag-Cu system is of this type. Phase α is mostly A with some dissolved B, and phase β is mostly B with some dissolved A. Consider the *equilibrium* cooling behavior along the ag line (10% B). Cooling from 600 K results in no change until 530 K (point b) is reached. At this temperature solid α forms (A with about 3% B, point h). As the temperature continues to decrease the liquid gets richer in B as does the solid solution phase α. At point c the α phase is about 7% B (point i), and the liquid is about 22% B (point j). Cooling continues (if we continue to remove heat) until, at point d, the last of the liquid freezes to form α which is now 10% B, the same as the initial liquid. (*Note*: If this has been at equilibrium all the

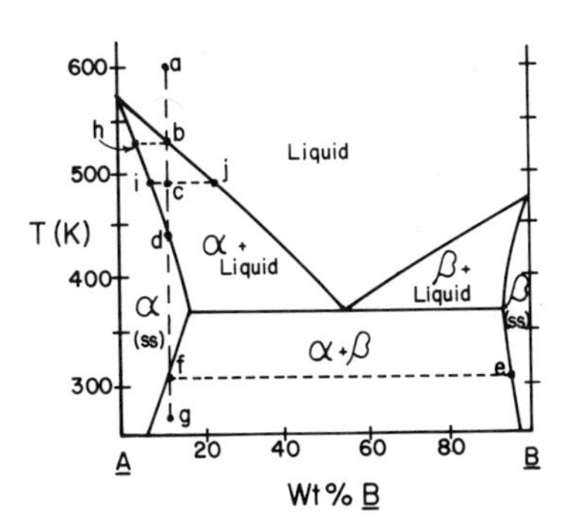

FIG. 13-2.5.1 Phase diagram for the binary system AB in which A and B show partial solubility (ss = solid solution).

way through, α will be homogeneous. If we cool too rapidly for diffusion to take place in the solid, we will find α with a range of composition from about 3 to 10%.) From 440 K (point d) to 300 K (point f) we merely cool α with no further changes. At point f solid β begins to separate from α. This is a *subsolidus* (completely in the solid phase) reaction in which diffusion and rearrangement in the solid phase must occur. β has a composition of about 95% B (point e). Below 300 K more β forms, and the composition of α and β slowly change as the temperature is lowered. You should describe the cooling (or heating) behavior of, say, 30% B between 300 K and 600 K. Hopefully, this will give you no problem.

If we quenched (cooled very rapidly) composition a from 600 K to 250 K we would "freeze in" phase α, since β would not have time to form. α at temperatures below 300 K (for the composition under discussion) is metastable, and given time, will form the two phases α-β system. This behavior is related to the "hardening" with time of some alloys. If the alloy is formed and quenched as one phase it may be soft and ductile, whereas, the stable, two-phase system is hard. Thus, as the system approaches equilibrium by forming two phases, much of the ductility can be lost. Of course, age hardening is more involved than that, but the above is a partial explanation.

LIMITED SOLUBILITY BELOW THE SOLIDUS: EUTECTIC. Figure 13-2.5.2 shows a system with complete miscibility of solids in the solid-liquid region. However, at lower temperatures a phase transformation occurs, leading to partial solid-solid miscibility. The interpretation of this is no more difficult than the previous example. The diagram itself needs a little explanation. T_A^0 is the normal melting point of A and T_B^0 that of B. $T_{\alpha-\beta}$ indicates the solid-solid phase transition temperature for $\alpha \rightarrow \beta$ or $\beta \rightarrow \alpha$ for pure A. Likewise $T_{\alpha-\gamma}$ is the temperature of the $\alpha \rightarrow \gamma$ or $\gamma \rightarrow \alpha$ transition for pure B. If composition a is cooled nothing occurs until point b,

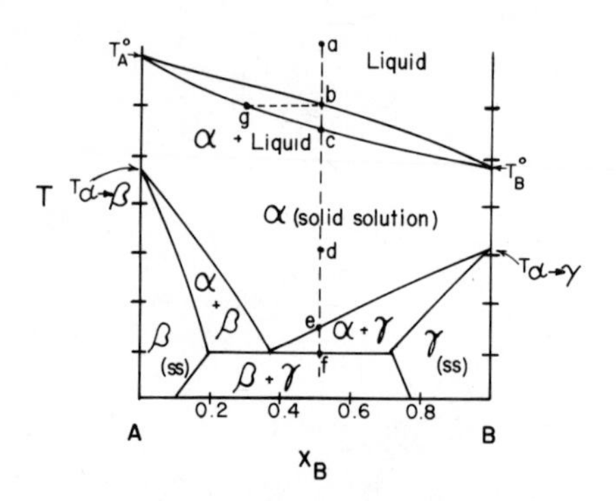

FIG. 13-2.5.2 A system in which there is complete AB solubility in the solid-liquid region but partial miscibility at lower temperatures and a subsolid eutectic interaction.

when α begins to precipitate. α has the composition of about $x_B = 0.35$. The region α is all solid solution. Within the region α, A and B are miscible in all proportions. Nothing new occurs from point c to f, except that the eutectic interaction involves three solid phases (α, β, and γ) instead of two solid and one liquid phase you have encountered previously.

LIMITED SOLUBILITY BELOW THE SOLIDUS: PERITECTIC. One more example before finishing this section. Figure 13-2.5.3 shows a system similar to that shown in Fig. 13-2.5.2 except the subsolidus interaction is of the peritectic type. Below is a list of what happens at each point. You should fill in the details

a. All liquid
b. Solid α begins to appear
c. α and liquid in equilibrium
d. All liquid disappears, only α left
e. α only
f. Solid phase γ begins to form
g. α and γ present
h. All solid α disappears
i. Only solid γ present
j. Solid β begins to form
k. β and γ both present

13-2.6 COMMENTS

If you have understood all the above, you are prepared to interpret almost any two-phase solid-liquid or solid-solid system. Many phase diagrams look impossibly dif-

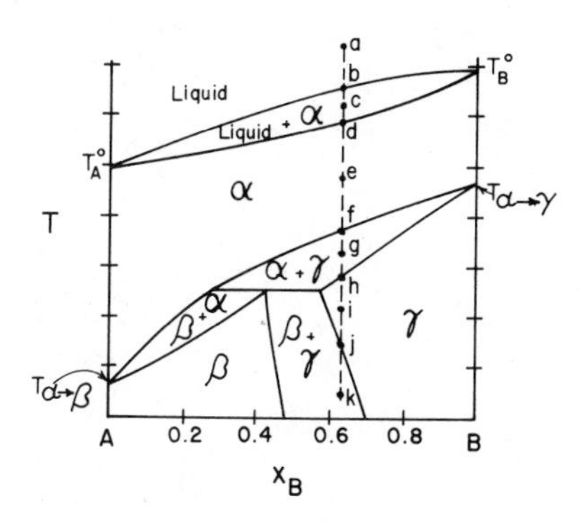

FIG. 13-2.5.3 A system in which there is complete AB solubility in the solid-liquid region but partial miscibility at lower temperatures and a subsolidus peritectic interaction.

ficult until you start breaking them down into familiar components. The problems below will give you some practice in listing the behavior of two-phase systems. Some more complex examples and problems are given in Sec. 13-4.

13-2.7 PROBLEMS

1. Draw cooling curves expected for composition (a) and (e) of Fig. 13-2.4.1.

2. For the phase diagram given below, complete the labeling and describe the heating of the two compositions marked from 1300°C to 2100°C.

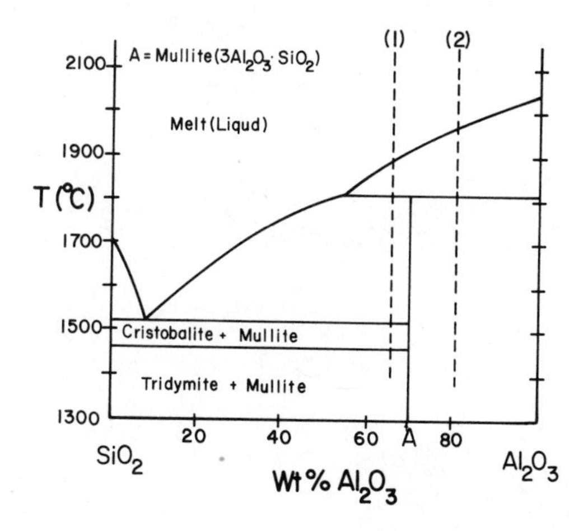

3. (a) Draw a phase diagram for the system AB which forms compounds A_2B, AB, and AB_2. Use whatever data you need from below.

(b) Label all areas in your phase diagram and discuss in detail all changes that occur when a mixture of $x_B = 0.35$ is cooled from 50 to 5°C.

(c) For each phase change in (b) find Φ, C, and P (assume pressure = 1 atm).

Substance	A	A_2B	AB	AB_2	B
Melting point	32	25	30	22	41

Eutectic composition (x_B)	20	40	60	80
Melting temperature (°C)	18	21	15	12

4. Label all areas on the following diagram, give the number of phases present at each composition (a, b, c, and d) and the approximate concentration of each phase at each point. Finally, describe the cooling behavior *completely* along the abcde line.

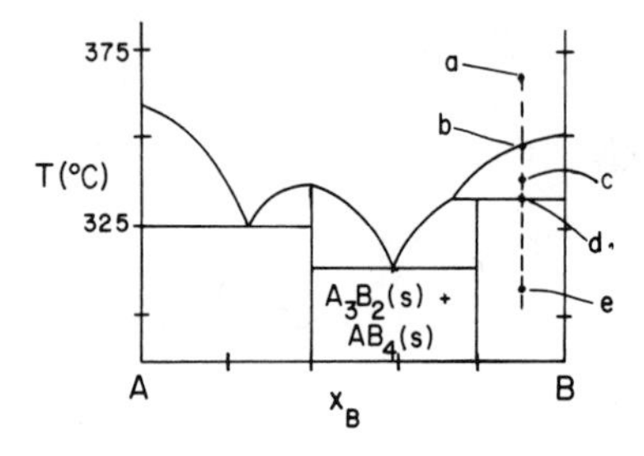

13-2.8 SELF-TEST

1. Define or briefly discuss the terms: eutectic, liquidus line, peritectic melting.

2. Label the following phase diagram completely and discuss in detail the behavior on cooling of the three indicated compositions. Also give Φ, C, and P at each phase change.

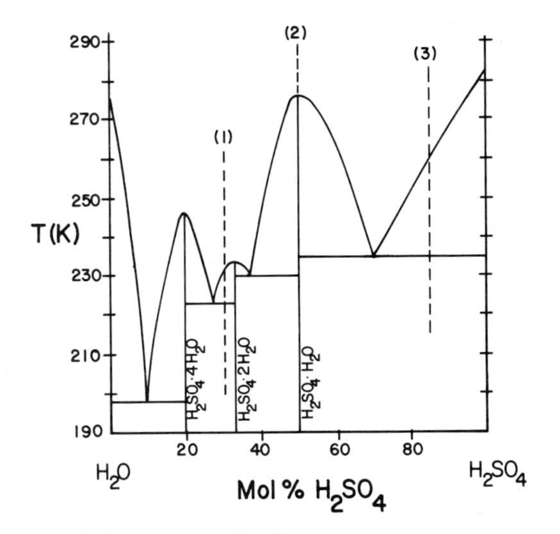

3. Given the cooling curves below, construct the corresponding phase diagram for the XY system. Label all areas. Desciibe in detail all changes that take place in cooling a melt of 65 mol% Y from 400 to $0^{\circ}C$.

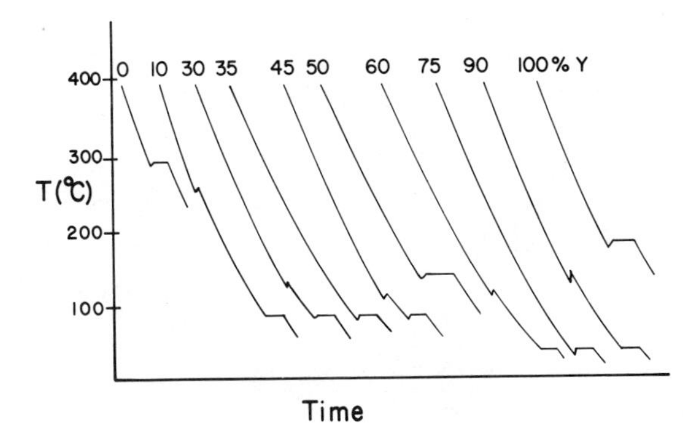

13-2.9 ADDITIONAL READING

1. L. H. Van Vlack, *Elements of Materials Science*, 2nd ed., Addison-Wesley, Reading Mass., 1964.
2. U. S. Geological Survey Professional Paper 440-L, Data of Geochemistry, 6th ed.
3. See Sec. 1-1.9.

13-3 BINARY LIQUID-LIQUID AND LIQUID-VAPOR SYSTEMS

OBJECTIVES

You shall be able to

1. Interpret fully any liquid-liquid or liquid-vapor phase diagram given by labeling the diagram; describing cooling or heating behavior, listing the phases present at any point, the compositions of these phases, and the approximate amounts of these phases
2. Describe *in detail* the distillation and fractional distillation of miscible and immiscible fluids if a phase diagram is given

13-3.1 INTRODUCTION

The major portion of this section will be concerned with liquid-vapor equilibria. Only a very short segment will discuss liquid-liquid equilibria. This is due to the relative simplicity of liquid-liquid diagrams. A more extensive treatment will be given to liquid-vapor systems. These are very important in understanding distillation and in many separation processes. Section 6-1.3, in which the vapor pressure of solutions as a function of composition is discussed, has already laid the groundwork for this section. If you have not read that section, you must do so before continuing with this. If you have forgotten the material you should review.

13-3.2 LIQUID-LIQUID EQUILIBRIA

To simplify matters, let us assume we have a binary (two-component) system which exhibits no vapor pressure. The two possible extremes for such a system: (1) The liquids could be completely miscible, in which case there would always be only one phase present, or (2) they could be completely immiscible so that two phases (each pure) are always present. We are concerned with the intermediate region of partial miscibility.

Figure 13-3.2.1 is a representation of three possible liquid-liquid diagrams. In part (a), the miscibility increases with decreasing T, while in part (b) the miscibility increases with increasing T. Any composition and temperature that falls within the boundary of the heavy line will result in two phases. The composition of the two phases will be given by the intersections of a horizontal (isothermal) line through the point and the boundary of the two-phase region. More on this shortly. In part (c) of Fig. 13-3.2.1 we find a system completely miscible at high and low temperatures but forms a two-phase system at intermediate temperature. Such a system is water-nicotine, shown in Fig. 13-3.2.2. This figure will be used for the remainder of the discussion.

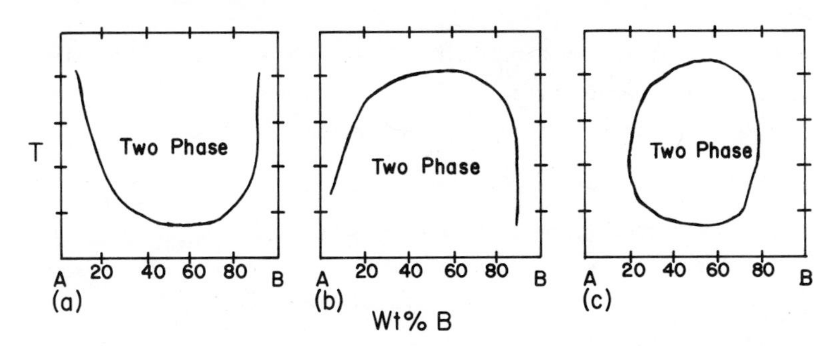

FIG. 13-3.2.1 Generalized phase diagrams for two-component systems which are partially miscible.

The region within the boundary is a two-phase region. Let us run two experiments: (1) add nicotine to pure water at 150°C and see what happens; (2) Heat a 65% nicotine solution from 50 to 225°C.

1. Start at point a and add nicotine to water. The nicotine is soluble until about 6% has been added. At this *solubility limit* (point b) a second phase of composition 80% nicotine separates (point d). Continued addition of nicotine does not change the composition of the two phases: they are fixed (at 150°C) at points b and d. However, the relative amounts of the phases do change. At an average composition given by point c (there is nothing of composition c in the system; the average of b and d gives composition c), we can find the relative amounts of the phases present by the lever rule (see Sec. 13-2.2).

$$\frac{\text{Weight of phase b}}{\text{Weight of phase d}} = \frac{d - c}{c - b} = \frac{80 - 25}{25 - 6} = 2.9 \tag{13-3.2.1}$$

This gives the ratio of weights. If we had used mole fraction or mole percent instead of weight percent we would find the ratio of phase a moles to phase b moles.

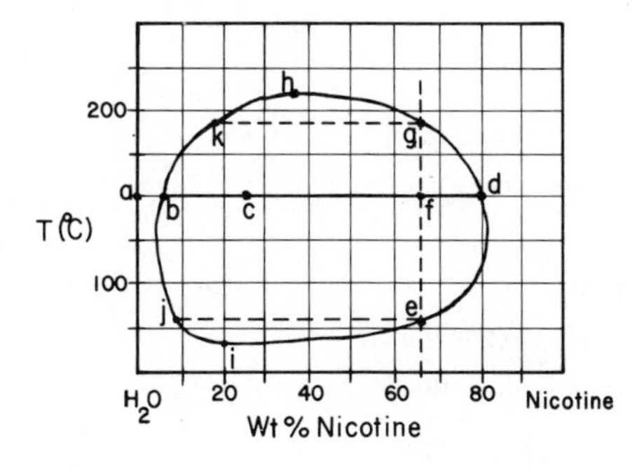

FIG. 13-3.2.2 The water-nicotine system.

If we continue adding nicotine we approach point d. (What is the ratio of weights at point f?) At point d the two-phase system reverts back to a single phase. Let us see how the Gibbs phase rule works in the two-phase region.

$$\Phi = C - P + 2 = 2 - 2 + 2 = 2$$

Along the 150° isotherm we have specified T, and when we drew the diagram we specified P, so we have used the two degrees of freedom. This means *along a given isotherm at some specified* P, *if there are two phases present the compositions of those two phases are fixed.* Adding either component merely changes the amounts of the two phases; not their compositions.

2. If we choose to heat a mixture of 65% nicotine from 50°C we find nothing changes until point e is reached. At point e, a second phase of composition j separates. Continued heating leads to changes in both the compositions and amounts of the two phases. At 150° (point f) the two phases present are d and b, and the relative amounts are

$$\frac{\text{Weight phase b}}{\text{Weight phase d}} = \frac{d - f}{f - b} = \frac{80 - 65}{65 - 6} = 0.25$$

Two phases are present until the temperature indicated by point g (185°C) is reached. At this temperature the last bit of the water-rich phase (with a composition given by point k) disappears, leaving a one-phase system. A final note on this type of diagram. Point h represents the highest temperature at which two phases can exist (at the given pressure) and is called the *critical solution, or upper consolute temperature*. Point i is the corresponding low-temperature limit for two phases and is called the *lower consolute temperature*.

13-3.3 LIQUID-VAPOR EQUILIBRIA: MISCIBLE LIQUIDS

DIAGRAMS OF NEARLY IDEAL SOLUTIONS. The vapor pressures of solutions of two miscible liquids were investigated in Sec. 6-1. You need to read or reread that section before continuing. Example 6-1.3.1 and Fig. 6-1.3.1 illustrate how the composition of the vapor in equilibrium with a solution varies. For an ideal binary mixture, the general shape of the liquid and vapor composition curves are shown in Fig. 13-3.3.1. This would hold for a system such as the ethylene dibromide-propylene dibromide system shown in Fig. 6-1.3.3.

A brief interpretation of the behavior of this system at constant temperature might be instructive. Usually, we study systems of this type at constant pressure, and we will do that next, but first, what happens as we reduce the pressure on a liquid that has $x_B = 0.7$? At point a only liquid is present. Reducing the pressure (by evacuating the flask) results in no change until P_1 (point b) is reached. At point b the solution will begin to boil, giving a vapor with the composition at

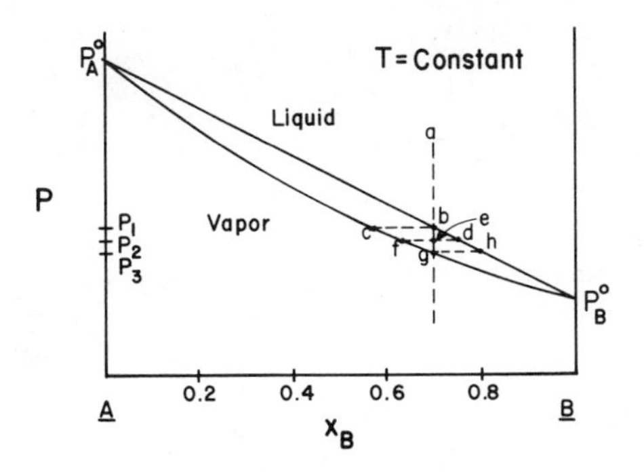

FIG. 13-3.3.1 Vapor pressure diagram for an almost ideal system. Liquid and vapor compositions are shown.

point c. (Note: the line from b to c must be horizontal since the pressure on the liquid and vapor will be the same at equilibrium.) The vapor formed is richer in A, so the liquid gets richer in B. The liquid composition follows the bdh line, and the vapor composition follows the cfg line as the pressure is lowered from P_1 to P_3. At point e (P_2) we find liquid with a composition d and vapor of composition f (there is no composition e present). The Gibbs phase rule reads

$$\Phi = C - P + 2 = 2 - 2 + 2 = 2$$

but temperature is constant, leaving $\Phi = 1$. When we specify a pressure (such as P_2), the compositions of the liquid and vapor are both fixed. We can also find the relative amounts of liquid and vapor.

$$\frac{\text{Moles vapor}}{\text{Moles liquid}} = \frac{d - e}{e - f}$$

This, plus the composition of liquid and vapor and the total moles in the starting material, gives us enough information to find the amount of each component in each phase. If we reduce the pressure to P_3 (point g) the last bit of liquid will vaporize. The last drop of liquid has composition h. The vapor now has the same composition as the original liquid ($x_B = 0.7$).

It is sometimes convenient to present data for vapor pressure on a temperature vs. composition diagram. Figure 13-3.3.2 gives such a diagram for a nearly ideal solution. These are called boiling point diagrams since they give the boiling point of a solution at constant pressure for different compositions. Note this is almost the inverse of Fig. 13-3.3.1. That should be reasonable, since high T favors a vapor while high pressure favors a liquid. What happens if we heat a liquid with $x_B = 0.7$ from a to i? At T_1 (point b) the solution begins to boil, producing vapor (composition c). This vapor is richer in A so it leaves the solution richer in B. As heating continues the liquid composition follows the bdg line, and the vapor

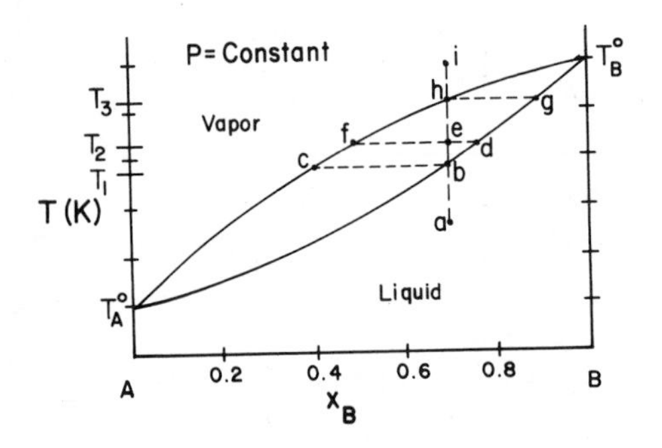

FIG. 13-3.3.2 Boiling point diagram for a nearly ideal solution.

composition follows the cfh line. (*Note*: In this and the previous example we are assuming the liquid and vapor are maintained in equilibrium with each other in the same vessel. If the vapor is removed we have distillation, which is discussed below.) At T_2 we have liquid (point d) and vapor (point f). Continued heating brings the system to point h, where the last drop of liquid (composition g) vaporizes. The vapor composition above T_3 is the same as the original liquid composition.

NEARLY IDEAL SOLUTIONS: DISTILLATION. If we heat a solution and remove the vapor we have a nonequilibrium process called *distillation*, and the above discussion does not apply. Consider Fig. 13-3.3.3 which is the same system as in Fig. 13-3.3.2. Two different processes are possible. (1) We can remove the vapor as it forms and see what happens to the liquid as the temperature rises, or (2) we can remove the vapor, condense it, revaporize, condense, etc. This process is called *fractional distillation*. Take a solution that has $x_B = 0.6$. If the temperature is increased to point a, vapor begins to form (composition c). If this vapor, which is rich in A, is removed from the system the liquid will become richer in B. After some time the liquid will reach point b which has vapor d in equilibrium with it. This vapor is still rich in A, relative to the liquid, so continued removal of vapor enriches the liquid in B. If this process is continued for long enough the liquid composition follows the abe line all the way to point e which is pure B. We can, by removing the vapor, get a liquid that is pure in the higher boiling component. As you might guess, you will not get much B since most of the original volume has been removed as vapor by the time point e is reached.

In *fractional distillation* the vapor (point c) formed from liquid (point a) is separated and condensed (along line cf; this is not an equilibrium process). We take the vapor formed from the system, place it in a separate container, and condense it to a liquid of the same composition as the vapor. The liquid formed (composition f) is then vaporized to give vapor of composition g. This vapor is separated and condensed to liquid h. The process is continued until pure A is obtained. (As above,

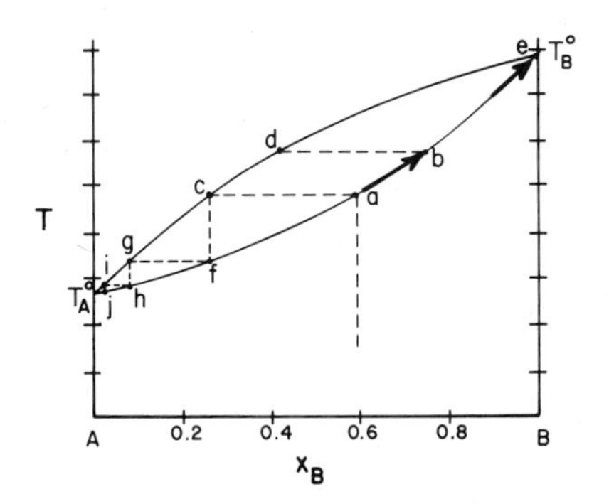

FIG. 13-3.3.3 A boiling point diagram showing distillation and fractional distillation.

you cannot get absolutely pure material except with an infinite number of steps, which leaves you with a vanishingly small volume of the pure material.) The above processes are usually carried out automatically in a *distillation column*. Each step in Fig. 13-3.3.3 (acf, cfg, etc.) is called a *theoretical plate*. The efficiency of a column is given in terms of theoretical plates. For example, if we start with a composition a in a column and find that the distillate from the top of the column at equilibrium has composition j, we say the column has *three* theoretical plates.

NONIDEAL SOLUTIONS: POSITIVE DEVIATIONS FROM RAOULT'S LAW. Figure 6-1.4.2 shows a system that exhibits positive deviation from Raoult's law. Below, Fig. 13-3.3.4 gives the liquid and vapor composition on a P vs. mole fraction plot, and Fig. 13-3.3.5 gives the same data on a T vs. x_B plot. Positive deviation from Raoult's law means that the vapor pressure above a solution is greater than predicted

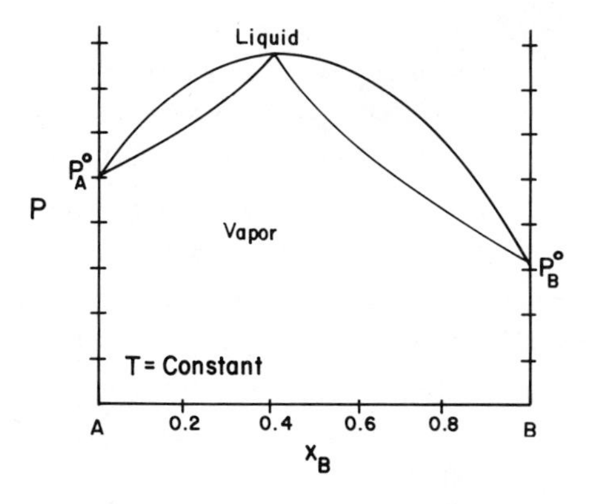

FIG. 13-3.3.4 Isothermal liquid-vapor diagram for a system showing positive deviation from Raoult's law.

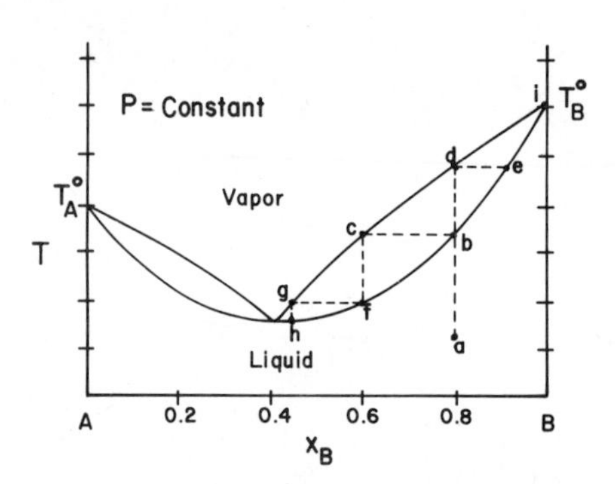

FIG. 13-3.3.5 Isobaric liquid-vapor diagram (boiling point diagram) for a system showing positive deviation from Raoult's law.

for the ideal case. Marked positive (or negative) deviation from ideal behavior leads to the formation of an *azeotrope*, or constant boiling mixture.

The following describes the behavior exhibited when a solution with $x_B = 0.8$ is heated while keeping the vapor in contact with the liquid (equilibrium). Distillation results are discussed after that. If we heat composition a to point b we observe vapor of composition c is formed in equilibrium with liquid b. This vapor is richer in A than the liquid, so as the temperature increases, the liquid becomes richer in B following the be line. Also, the vapor follows the line cd, but it is always richer in A than the liquid. The amount of liquid decreases, and the amount of vapor increases as the temperature increases from b to d. When point d is reached, the last drop of liquid (point e) vaporizes giving a vapor with $x_B = 0.8$, the same as the initial liquid. If we carry out a distillation by removing the vapor, the results are the same as for the distillation of an ideal solution. The liquid composition varies along bi, until finally at T_B^0 pure B (almost pure B) remains in the distillation flask.

A fractional distillation does not result in obtaining pure A or B. Instead, if we carry out a fractional distillation on a solution with $x_B = 0.8$ we first form vapor c which is condensed separately to liquid f. Liquid f vaporizes to vapor g, which is separated and condensed to liquid h. Instead of approaching a pure material, we come to a constant boiling mixture, the *azeotrope*. When the azeotrope composition is reached, the liquid and vapor in equilibrium have the same composition, and all the liquid will vaporize at this point with no further change in composition. We cannot obtain pure A (if we begin our distillation to the right side of the azeotrope) by fractional distillation. To get past the azeotrope we could add a third component, or perhaps change the pressure on the system. Such a complication will not be discussed.

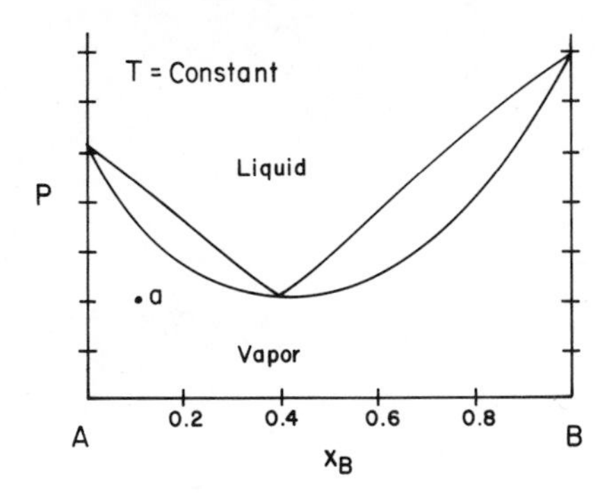

FIG. 13-3.3.6 Isothermal liquid-vapor diagram of a system with negative deviation from Raoult's law.

NONIDEAL SOLUTIONS: NEGATIVE DEVIATION FROM RAOULT'S LAW. Figures 13-3.3.6 and 13-3.3.7 show P vs. x_B and T vs. x_B diagrams for a system exhibiting a negative deviation from Raoult's law. Figure 6-1.4.3 is an example of such a system. An exercise for you is to describe the equilibrium heating, distillation, and fractional distillation behavior of this type of system.

13-3.4 IMMISCIBLE AND PARTIALLY MISCIBLE LIQUIDS

For immiscible liquids the vapor pressure of the system is simply the sum of the vapor pressures of the two components. The situation is not quite that simple for partially miscible liquids. Figure 13-3.4.1 shows a system which is a combination of those described in Secs. 13-3.2 and 13-3.3. No new problems are presented here, and you should be able to interpret the behavior of such a system.

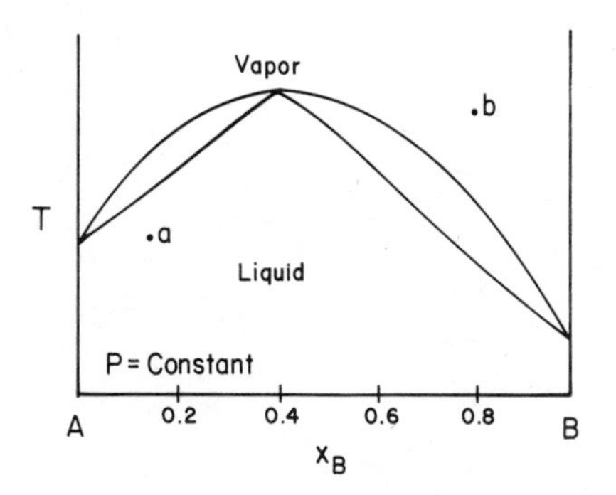

FIG. 13-3.3.7 Isobaric liquid-vapor diagram (boiling point diagram) for a system with negative deviation from Raoult's law.

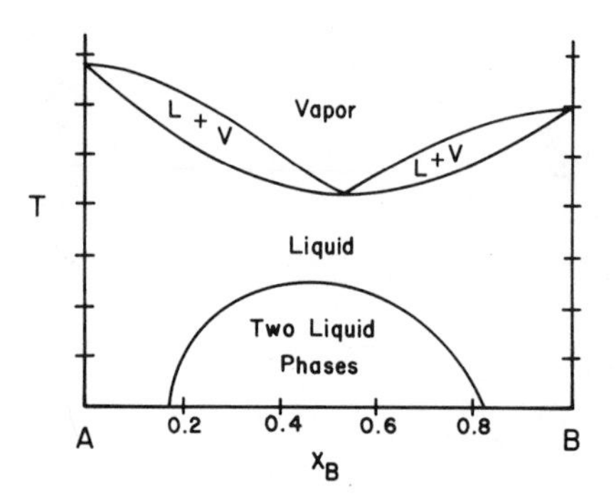

FIG. 13-3.4.1 A liquid-vapor phase diagram for two liquids that are partially miscible.

A more complex behavior is exhibited when the liquid gap between the vapor and the liquid phases disappears. The resulting diagrams are analogous to those shown in Sec. 13-2.5 for limited miscibility in the solid phase. Figure 13-3.4.2 shows a system such as water-butanol. Figure 13-3.4.2 could be the same system as Fig. 13-3.4.1 only at a lower pressure. As the pressure on the system is lowered, the temperature of vaporization is lowered. The two-liquid phase region is not affected nearly as much by a pressure change, so at some pressure the two halves will merge, and the intervening liquid region will disappear. As before, a description of the behavior of this system when it is heated in different ways will be given. First, start with liquid 1 at point a. This is a one-phase system. When this phase is heated to point b it begins to boil, giving vapor with composition e. If we keep the liquid and vapor together nothing new is encountered. The liquid follows line bd and the vapor line ec as the temperature is raised ($\Phi = C - P + 2 = 2 - 2 + 2 =$ 2. Pressure is fixed, meaning we still have one variable, either T or composition

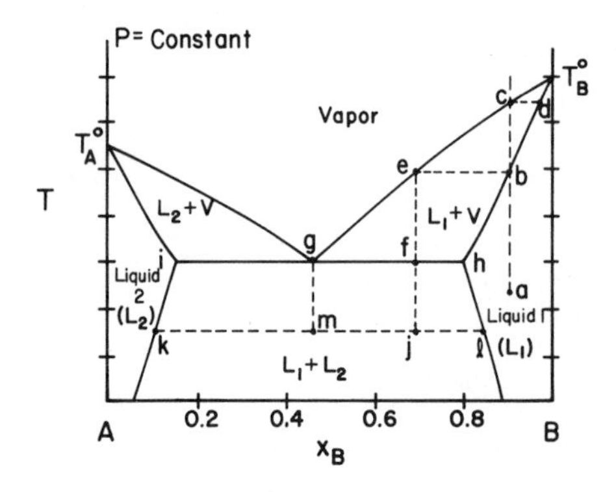

FIG. 13-3.4.2 Liquid-vapor diagram for two components exhibiting limited liquid miscibility.

of vapor, or liquid). When T_c is reached the last of the liquid (point d) vaporizes giving vapor of composition c. Distillation from point b presents us with nothing new. The liquid will approach pure B along the bd line as vapor is removed.

Fractional distillation does present a new situation. When vapor e is condensed (in a separate vessel from liquid b) to, say, the temperature of point j, we find that two liquid phases form with compositions k and ℓ. If we now heat this two-phase system nothing happens (except for small liquid composition changes along ki and ℓh) until the temperature reaches f where the system begins to boil, giving vapor of composition g. We have three phases present (i, h, and g). $\Phi = C - P + 2 = 2 - 3 + 2 = 1$, which is the specified pressure. The system is invariant, which means the temperature (and compositions) cannot change.

What would happen if we took some vapor from the three-phase system (vapor composition g) and condensed it in a separate container? We would get a liquid with average composition m (if we go back to the initial liquid temperature) with liquid 1 and liquid 2 in equilibrium. Heating this mixture brings the liquids back to h and i, at which point vapor g forms. The average composition line goes through g, which means both liquids 1 and 2 disappear at the same time if we heated long enough. If the vapor is cooled, two liquids with average composition m are again obtained. In a system such as this it is impossible to separate A and B by distillation and fractional distillation. Let us consider what happens if a liquid in the two-phase region (say, point j) is heated to form a vapor under equilibrium conditions. The ratio of moles of the two liquids just below point f is approximately

$$\frac{\text{Moles 2}}{\text{Moles 1}} \approx \frac{h - f}{f - i} < 1$$

Since there is less liquid 2 than liquid 1, the liquid depleted is liquid 2. The temperature of the system is constant until all liquid 2 vaporizes. Now only two phases are present, liquid 1 and vapor. Continued heating (with liquid and vapor retained in the same vessel) causes the liquid composition to follow the hb line and the vapor the ge line. The last drop of liquid (composition b) vaporizes to vapor of composition e.

13-3.5 PROBLEMS

1. In the water-nicotine system (Fig. 13-3.2.2), what is the maximum amount of water soluble in nicotine and the maximum amount of nicotine soluble in water at 150°C?

2. From Fig. 13-3.3.2 describe in detail the system behavior when a vapor with x_B = 0.30 is cooled until only liquid is present.

3. Give the composition of the first drop of liquid formed in Fig. 13-3.3.7 when

a. The temperature is lowered from point b

b. A is added isothermally to the vapor at composition b

4. Describe in detail all changes that occur in the equilibrium heating, distillation, and fractional distillation from point a in Fig. 13-3.3.7.

5. Draw a phase diagram for a system showing positive deviation from Raoult's law with a region of liquid-liquid immiscibility at lower temperature. Pick one composition and describe distillation and fractional distillation from that point. Pick a second composition and describe equilibrium cooling of the vapor.

6. For the figure given below (a) label all areas and (b) describe *in detail* the distillation and fractional distillation results for the composition indicated by point x. Also describe *in detail* the equilibrium cooling of the vapor from point y.

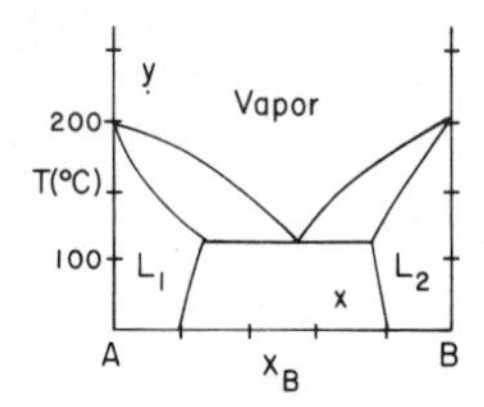

13-3.6 SELF-TEST

1. Estimate the weight ratio of the water-rich phase to the nicotine-rich phase from Fig. 13-3.2.2 at 100°C if the average solution composition is 20% nicotine.

2. From Fig. 13-3.3.5 describe in detail the system behavior when a vapor with x_B = 0.20 is cooled at equilibrium until only liquid is left.

3. For the water-isobutyl alcohol system given below, label the diagram completely and describe all changes that occur in the equilibrium heating, distillation, and fractional distillation from point a.

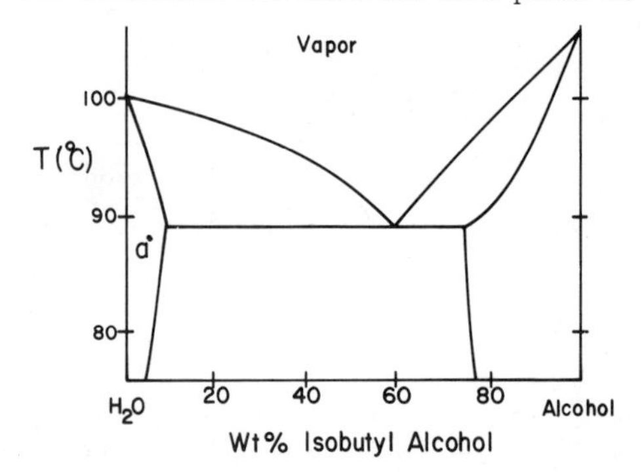

13-3.7 ADDITIONAL READING

1. F. E. W. Wetmore and D. J. LeRoy, *Principles of Phase Equilibria*, Dover, New York, 1950.
2. See Sec. 1-1.9B.

13-4 TERNARY AND MORE COMPLEX BINARY DIAGRAMS

OBJECTIVES

You shall be able to

1. Label and interpret fully simple three-component (ternary) phase diagrams
2. Interpret more complex binary and ternary phase diagrams

13-4.1 INTRODUCTION

The first part of this section is concerned with the phase behavior of three-component systems. A few simple examples will be given to aid you in understanding the meaning of ternary plots. The systems considered will be liquid-liquid and liquid-solid. In Sec. 13-4.3, more complex binary and ternary systems will be presented. Even though some of the diagrams below are fairly complicated, you should be aware that they are portions (in the geological examples) of much more complicated four-component systems. A tetrahedron is required to plot a four-component system, and interpretation of these becomes difficult. We will not discuss these systems.

13-4.2 TERNARY PHASE DIAGRAMS

GENERAL. Ternary systems are, by definition, three-component systems and must be plotted using three axes. This is accomplished by using triangular graph paper. Figure 13-4.2.1 shows the triangular plot for the three components A, B, and C. Look at this plot and try to figure out the compositions of points a and b before reading on. Notice that each apex corresponds to 100% of one of three components. The opposite base is the 0% line for that component, and lines parallel to this base correspond to varying percents of that component. Point a lies on the 30% A line, the 50% B line, and (by necessity) the 20% C line. Percents have been written on all the markers. These could be weight percent, mole percent, mole fraction or weight fraction. The reading would be done in the same manner. Point b is about 55% A, 23% B, and 22% C. Do you agree? Pick a few more points and label their percentages. If the sum of your values is 100%, you are probably correct in your reading.

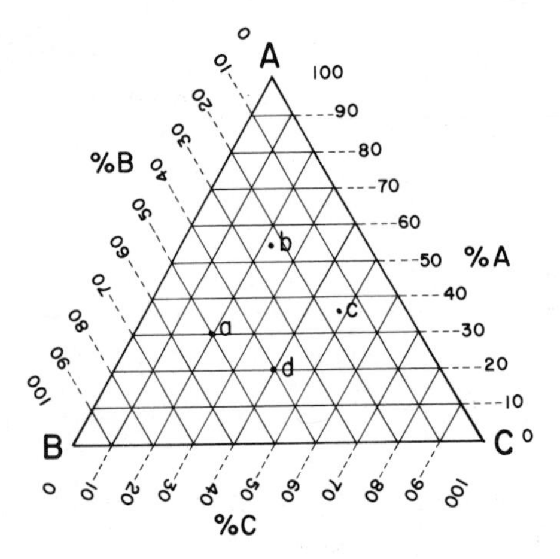

FIG. 13-4.2.1 A ternary plot on triangular graph paper for the three-component system ABC.

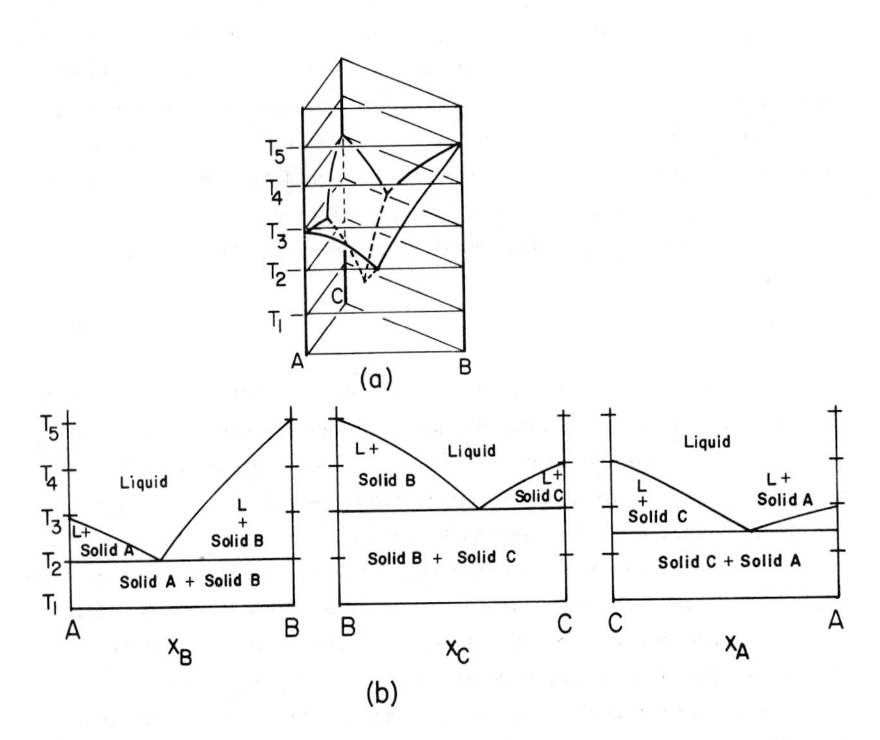

FIG. 13-4.2.2 (a) Ternary plot with temperature shown along the fourth axis. (b) Corresponding binary system from the three faces of the triangular prism.

SOLID-LIQUID. There are two ways to present data on a ternary diagram. Three-component systems actually have a fourth variable (T or P) on the vertical axis. So we have a triangular prism as shown in Fig. 13-4.2.2(a). The triangles shown are constant temperature slices which will be considered shortly. If we look at the three faces of this figure we find binary systems as shown in Fig. 13-4.2.2(b). You should be able to see how these are derived from part (a). These binary systems are labeled and can be interpreted as in Sec. 13-2. Note the three binary eutectics and the lower temperature ternary eutectic in part (a). To present ternary systems in usable form, one of two types of projections are used.

Figure 13-4.2.3(a) shows some isothermal slices through the phase diagram depicted in Fig. 13-4.2.2(a). At T_4 the temperature is above the melting point of A and C. (Can you tell from Fig. 13-4.2.2(a) that $T_A^0 = T_3$, $T_B^0 = T_5$, and $T_C^0 = T_4$?) The only solid present will be B. The shaded area in the isotherm at T_4, shows the region in which solid B will be in equilibrium with liquid. T_3 is below the melting point of both B and C but above that of A. The isothermal slice at T_3 shows the region in which solid B and solid C are in equilibrium with liquid. Finally, T_2 is below the melting temperature of all three solids. The T_2 isotherm shows the region in which solids A, B, and C are in equilibrium with liquid (the shaded areas). The central area is liquid, and the nonshaded areas along the BC and AC lines are three phase areas. These areas are labeled, and their interpretations will be described after finishing the discussion of projections.

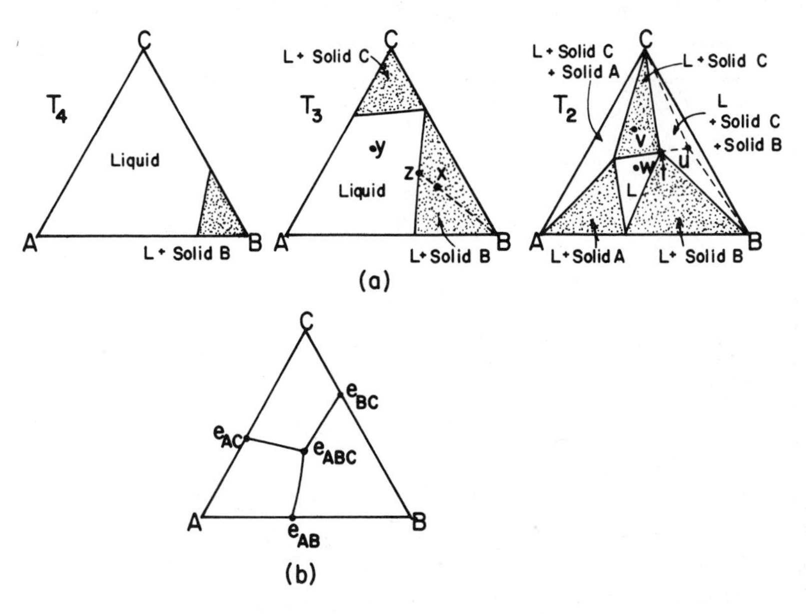

FIG. 13-4.2.3 (a) Isothermal slices through Fig. 13-4.2.2(a).
(b) Projection of Fig. 13-4.2.2(a) onto the base of the prism.

Figure 13-4.2.3(b) is a projection of Fig. 13-4.2.2 onto the prism base. It is really a projection of the liquid-solid equilibrium surface. The points e_{AB}, e_{AC}, and e_{BC} are the binary eutectic points. Point e_{ABC} is the ternary eutectic and is the lowest temperature at which liquid can be present in the system (at this pressure). [*Note*: Since P is constant, $\Phi = C - P + 1$. For three components $C = 3$ and $\Phi = 4 - P$. At each binary eutectic there are three phases (A, B, and liquid at e_{AB}, for example), so $\Phi - 4 - 3 = 1$. There is one degree of freedom. If we specify a T, the composition of the system is fixed. The univariant lines (e_{AC} to e_{ABC}, for example) reflect this fact. At the ternary eutectic, we have four phases (A, B, C, and liquid), and $\Phi = 4 - 4 = 0$; it is now an invariant system.] Usually, when the surface projection is used as shown in Fig. 13-4.2.3(b), isotherms are drawn on the projection to ease in interpretation. Figure 13-4.2.4 shows a projection with the isothermal lines indicated on the solid-liquid equilibrium surface. The dotted line shows how this projection can be used to obtain an isothermal slice. The dotted line is taken at T_6 and produces a slice similar to the third one in Fig. 13-4.2.3 (a). (See example 13-4.2.1)

You have now seen how to make various projections from ternary diagrams. How do we interpret these? Refer to Fig. 13-4.2.3(a) and the isotherm at T_3. Point y is in the all-liquid region, and we would read its composition directly from the axis if a scale such as shown in Fig. 13-4.2.1 were given. Point x is in a two-phase region with solid B in equilibrium with liquid. To get the composition of the liquid, we merely draw a *tie line* from the origin B through point x and extend to the liquid boundary. Point z is the composition of the liquid. The relative amount of liquid and solid are given by the lever rule.

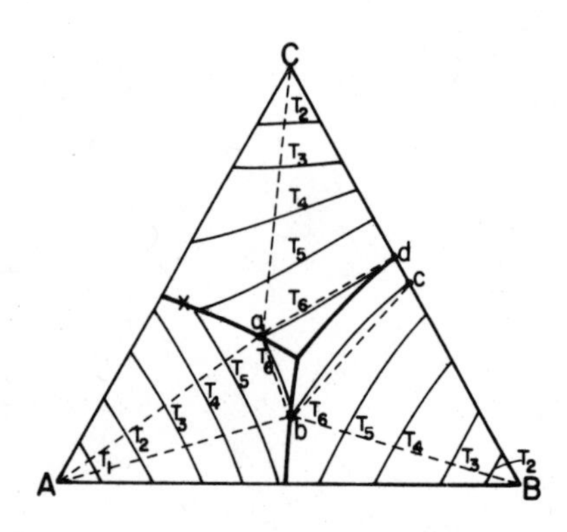

FIG. 13-4.2.4 Projection of a ternary phase diagram onto the base with isothermal lines drawn in.

$$\frac{\text{Amount of liquid}}{\text{Amount of solid}} = \frac{x - 0}{z - x}$$

Now look at the T_2 slice. Point w is in the all-liquid region. Point v has liquid and solid C in equilibrium and is interpreted in the same manner as point x above. Point u is in the three-phase region. Solid B, solid C, and liquid of composition t are present, as indicated by the dotted lines. In fact, any point within this three-phase region has the same composition. The relative amounts of the various phases will change. Perhaps the following examples will help clarify these concepts. An example for each type of projection will be given.

13-4.2.1 EXAMPLE

What changes occur if C is slowly added to solution of AB at 80% B at temperature T_6 in Fig. 13-4.2.4? See the following T_6 isothermal slice.

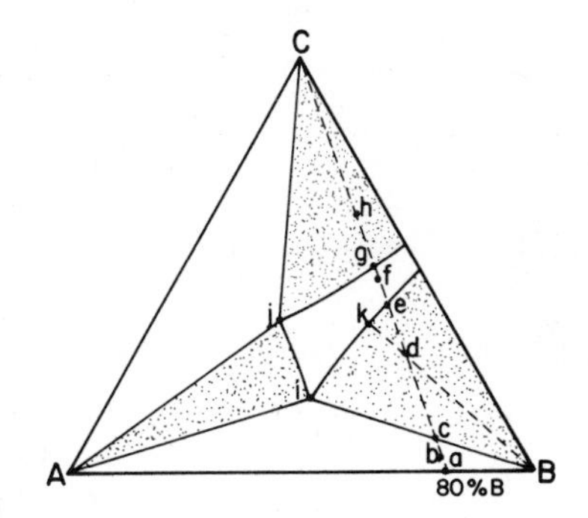

Solution

Starting at 80% B (20% A, point a) addition of C will take us along the dotted line. We start at a with solid A plus solid B (we are below the AB eutectic). Addition of a small amount of C will bring us to point b in the three-phase region: solid B, solid A, and liquid of composition i. Continued addition of C brings us to point c, at which point the last bit of solid A disappears, leaving solid B and solution i. As more C is added we maintain solid B and solution, but the solution composition changes. When the average composition is d we have solid B and solution k as indicated by the dotted tie line. At point e the last of the solid B goes into solution, and we have one phase, solution of composition e. Continued addition of C just changes the solution's composition until point g is reached. Here, solid C begins to precipitate. Any composition between g and pure C results in solid C and solution of composition g. The relative amounts of liquid and solid change along that line.

13-4.2.2 EXAMPLE

Let us use a figure similar to Fig. 13-4.2.4 to show what happens when a given composition is cooled from a temperature at which only solution exists to one at which only solid is present. Temperatures are on the diagram. Point x is the initial composition, and let us assume the initial temperature is 1500°C.

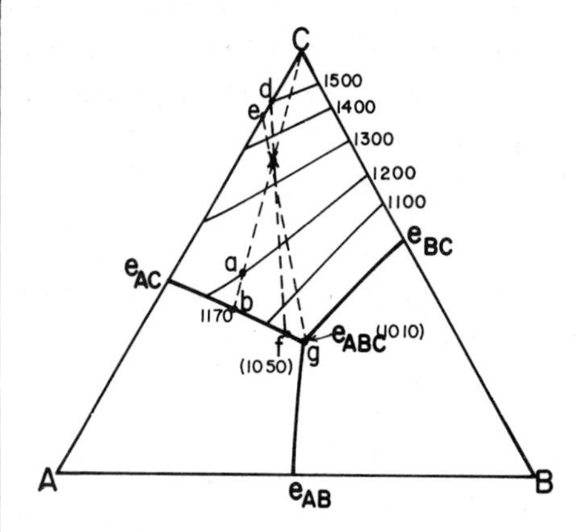

Solution

At 1500°C and composition x only liquid is present. In fact, no solid forms until the temperature reaches the surface at point x, which occurs at about the 1350°C isotherm. At 1350°C solid C begins to crystallize, and the liquid composition follows the xab line as the temperature is lowered. At 1200°C (point a) we find solid C and liquid of composition a. Finally at 1170°C, the e_{AC}-e_{ABC} line is reached at point b. Solid A begins to crystallize, along with solid C, as the temperature is lowered. The solid composition was pure C up to point b, but now A is being added to C. The average solid composition follows the AC line (Cde). Point f illustrates how to determine the solid composition. At 1050°C, liquid of composition f is in equilibrium with solid A and solid C with an average composition d, as determined by the tie line from f through x to the AC line. The liquid composition changes along the e_{AC}-e_{ABC} line until point g, the ternary eutectic, is reached. We now begin to crystallize B. Just after point g is reached we have four phases. Solids A, B, and C and liquid of composition g are present. The system is invariant. Continued removal of heat does not change the temperature. A, B, and C solids separate from the eutectic mixture until all the liquid is gone. During this time the average solid composition is progressing along the ex line until just as the last drop of liquid freezes, point x is reached. (The final solid must have the same average composition as the initial liquid.)

The above is a general description of solid-liquid ternary diagrams. For an excellent treatment in much greater detail see A. Reisman, *Phase Equilibria* (see Sec. 13-4.6, Ref. 1). This reference covers ternary systems with peritectic reactions and solid solubility, which have not been mentioned above.

LIQUID-LIQUID. With the above background, two examples of liquid-liquid systems should suffice. Figure 13-4.2.5 shows the ternary liquid system ABC at a specific T and P in which AC are partially immiscible. Then, $\Phi = C - P = 3 - P$. In the region above the solid curved line, there is a single-solution phase. Below the curved line there are two phases. In this two-phase region, $\Phi = 3 - 2 = 1$. There is only one degree of freedom. If one solution composition is given, the other is fixed by the tie line. For example, if we have point a in the two-phase region, the tie line (which must be determined experimentally since such lines are not parallel to any side) fixes the composition of the two phases in equilibrium, points b and c. Point d is a unique point. It is the *isothermal critical point* or *plait point*. At d the concentrations of the two phases in equilibrium become equal, and a single phase results.

Figure 13-4.2.6 shows a case in which all three pairs are partially immiscible. The diagram is labeled in part. The remaining labeling should be obvious. (Do it.) Tie lines are necessary in the two-phase region but not in the three-phase region ($\Phi = C - P = 0$), since at a given temperature and pressure there are no degrees of freedom. The apexes of the three-phase triangle give the composition of the phases in equilibrium at any point in the three-phase region. The dotted line will be referred to in one of the problems.

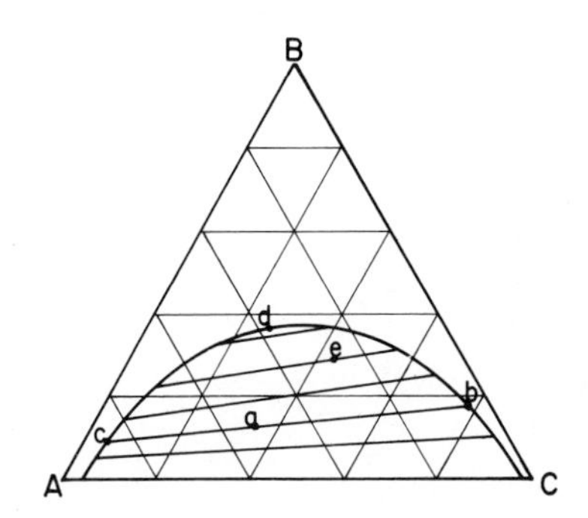

FIG. 13-4.2.5 Ternary liquid system at constant T and P in which AC are partially immiscible.

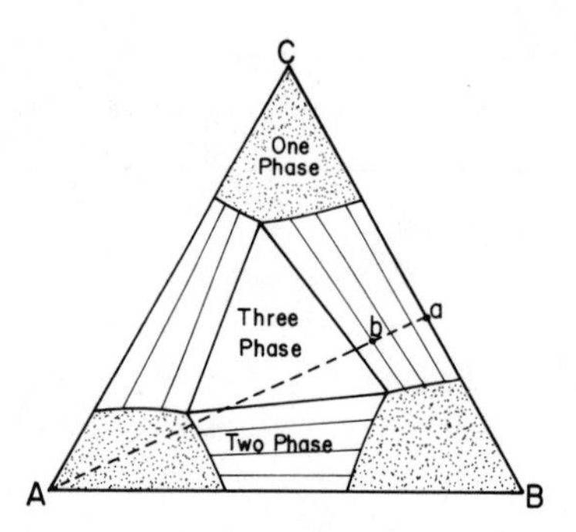

FIG. 13-4.2.6 Ternary system at a given T and P in which AB, BC, and AC pairs are all partially immiscible.

13-4.3 MORE COMPLEX BINARY AND TERNARY SYSTEMS

An example of a geological system for both a binary and ternary system will be worked through. An excellent publication for more such examples is the U. S. Geological Survey Professional Paper 440-L, Data of Geochemistry (see Sec. 13-4.6, Ref. 2).

13-4.3.1 EXAMPLE

Given the phase diagram shown in Fig. 13-4.3.1, describe all changes that occur as the composition designated by point y is cooled from 1500 to 1050°C.

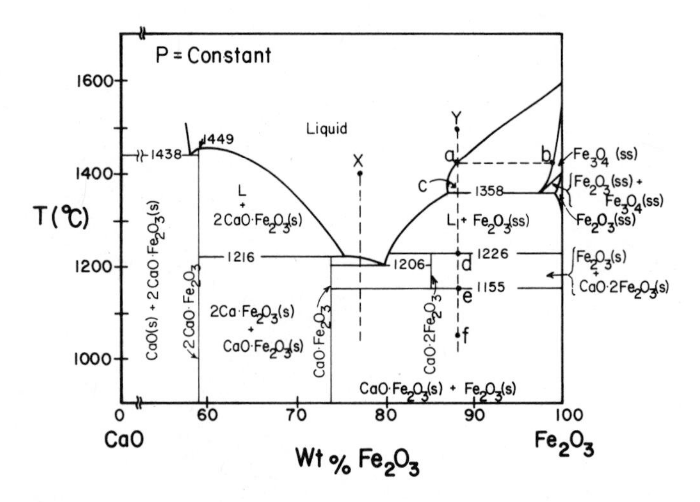

FIG. 13-4.3.1 The CaO-Fe_2O_3 system.

Solution

No changes occur on cooling until point a is reached. At point a a solid of composition b begins to precipitate. This solid is a solid-solution that contains about 98% Fe_3O_4 and about 2% CaO. Further cooling to point c produces only minor changes in the composition of both the liquid and solid phases. At point c however, a second solid phase begins to form. This phase is Fe_2O_3 solid-solution which has about 98% Fe_2O_3. We now have a three-phase system, so the temperature is invariant until one phase has been lost. As we remove heat from the system at c the Fe_3O_4 is converted to Fe_2O_3. When the last of the Fe_3O_4 is gone we have two phases and one degree of freedom, so the temperature can again change. Between points c and d the liquid becomes richer in CaO, and the solid solution becomes richer in Fe_2O_3. At point d a peritectic reaction takes place in which the liquid reacts with the solid Fe_2O_3 to form the solid compound $CaO \cdot 2Fe_2O_3$. Again the temperature is invariant until all the liquid has been used. Between d and e the two solid phases, Fe_2O_3 and $CaO \cdot 2Fe_2O_3$ are present. At 1155°C (point e) a subsolidus reaction takes place in which the $CaO \cdot 2Fe_2O_3$ is transformed to $CaO \cdot Fe_2O_3$. The temperature is invariant until the $CaO \cdot 2Fe_2O_3$ is gone. Below 1155°C no further changes occur.

13-4.3.2 EXAMPLE

A final example is a ternary liquid-solid system in which two components (AB in this case) form a congruently melting compound D. Figure 13-4.3.2 shows the projection onto the base of the prism. The isotherms are not drawn in.

Solution

When a congruently melting compound forms we can simple divide the ternary

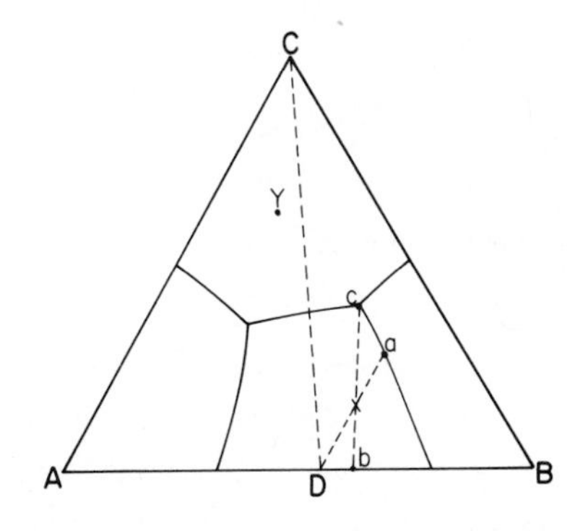

FIG. 13-4.3.2 A ternary solid-liquid system which forms a binary compound D (congruently melting).

diagram ABC into two diagrams ADC and CBD, as done above with the dashed line CD. The interpretation of cooling behavior is then very simple. For point x, start cooling from a temperature at which only liquid is present. When the solid-liquid temperature is reached, compound D begins to separate. As the temperature is lowered the liquid composition follows the xa line. At point a, solid B begins to separate also. The liquid composition goes along ac toward the ternary eutectic, and the solid (average) composition varies between D and b. At the ternary eutectic (point c) B, D, and C all continue to precipitate until all the liquid is gone. The average solid composition changes from b to x.

There are many types of systems we have not seen which lead to more complicated behavior. Such things as the formation of incongruently melting compounds and solid solutions complicate the situation considerably. These are, however, beyond the scope of this book.

13-4.4 PROBLEMS

1. Give the composition of points c and d in Fig. 13-4.2.1.
2. (a) Sketch the T_5 isotherm from Fig. 13-4.2.4. (b) Describe in detail all changes that occur when B is added (at equilibrium) to a mixture of A and C (60% C) at T_5.
3. Estimate the composition of the two phases present if the system depicted in Fig. 13-4.2.5 has an average composition of e.
4. (a) Draw the three binary diagrams from Fig. 13-4.3.2. (Assume $T_A^0 > T_D^0 > T_B^0 > T_C^0$.) (b) Describe in detail all changes that occur in cooling a liquid of composition Y until only solid remains.

13-4.5 SELF-TEST

1. Sketch the iostherm that is half-way between T_2 and T_3 in Fig. 13-4.2.2(a).
2. From Fig. 13-4.2.4 describe in detail all changes that occur when a solution (no solid) of composition x is cooled until all liquid has been crystallized.
3. (a) Describe in detail all changes that occur when A is added (at equilibrium) to a composition designated by point a in Fig. 13-4.2.6. (b) Estimate the composition and relative amounts of each phase present at point b. (c) Describe in detail all changes that occur when composition x is cooled to 1000°C in Fig. 13-4.3.1.

13-4.6 ADDITIONAL READING

1.* A. Reisman, *Phase Equilibria*, Academic, New York, 1970, Chap. 34.

2.* U. S. Geological Survey Professional Paper 440-L, Data of Geochemistry, 6th ed.

3. A. W. Frances, *Liquid-Liquid Equilibrium*, Interscience, New York, 1963. An excellent monograph with extensive tabulations of systems.

14

The Liquid and Solid States

SYMBOLS AND DEFINITIONS

$\Delta\bar{H}_{fus}$	Molar heat of fusion (kJ mol^{-1})
$\Delta\bar{H}_{vap}$	Molar heat of vaporization (kJ mol^{-1})
$\Delta\bar{H}_{sub}$	Molar heat of sublimation (kJ mol^{-1})
α	Thermal expansion coefficient (K^{-1})
β	Isothermal compression coefficient (atm^{-1})
f	Force (N)
q	Charge of particle (coulomb, C)
ε	Permittivity of a material (farad/meter = F m^{-1} = C^2 N^{-1} m^{-2})
ε_0	Permittivity of a vacuum (F m^{-1} = C^2 N^{-1} m^{-2})
$\theta = \varepsilon/\varepsilon_0$	Dielectric constant (unitless)
$\vec{r}$	Internuclear distance (m)
$\bar{E}_0$	Electric field, no dielectric (volt m^{-1} = V m^{-1})
σ	Charge density
$\bar{E}$	Electric field, with dielectric (V m^{-1})
p	Polarization due to electric field
$\vec{\mu}$	Dipole moment (Debye = D = 3.338×10^{-30} C m)
$\vec{\mu}_{ind}$	Induced moment (D)
$\underline{\alpha}$	Molar polarizability (m^3 $molecule^{-1}$)
$\bar{N}^*$	Molecules per unit volume (molecule m^{-3})
$\bar{E}_{eff}$	Effective electric field (V m^{-1})
$\mathcal{P}$	Molar polarization (m^3 mol^{-1})
k	Boltzmann's constant (1.38×10^{-23} J $molecule^{-1}$ K^{-1})
$1C^2$	8.98×10^9 J m
n_R	Refractive index
$\bar{M}$	Magnetism (Amp $meter^{-1}$ = A m^{-1})
$\bar{B}$	Magnetic flux density (tesla = T = kg s^{-2} A^{-1})
$\vec{\mathcal{H}}$	Magnetic field strength (A m^{-1})
μ_0	Magnetic permeability of vacuum (Henrys $meter^{-1}$ = H m^{-1} = kg A^{-2} s^{-2} m)
μ_m	Magnetic permeability (same as μ_0)
χ	Molecular magnetic susceptibility (unitless)
χ_M	Molar magnetic susceptibility (m^3 mol^{-1})
α_m	Magnetic polarizability (m^3 $molecule^{-1}$)
P_m	Permanent magnetic moment (A m^2)
$\mathcal{A}$	Area (m^2)

See Section 14-2 for several conversion factors.

U_{D-D}	Dipole-dipole interaction potential energy (J $molecule^{-1}$)
U_{ind}	Induced potential energy (J $molecule^{-1}$)
U_{London}	Dispersion potential energy (J $molecule^{-1}$)
w	Work (J)
x	Distance (m)
γ	Surface tension (J m^{-2}, N m^{-1})

f	Force due to surface tension (N)
g	Acceleration due to gravity (m s^{-2})
h	Height of capillary rise (m)
η	Viscosity coefficient (kg m^{-1} s^{-3} = 10^1 poise)
R	Reynolds number (unitless)
$\bar{v}$	Average fluid velocity (m s^{-1})
ℓ	Length (m)
B	Viscometer constant (appropriate units)
A	Constant
ΔE_{visc}^*	Activation energy of viscous flow (J mol^{-1})
$E_{cohesion}$	Energy of cohesion (J $molecule^{-1}$)
E_{Coul}	Coulombic energy of attraction (J mol^{-1})
M	Madelung constant
z_i	Ionic charge
N	Avagadro's number
ΔH_{cryst}	Heat of Crystallization (J mol^{-1})
D	Dissociation energy (J mol^{-1})
$\Delta\bar{H}_f^0$	Standard heat of formation (J mol^{-1})
A	Electron affinity (J mol^{-1})
I	Ionization potential (J mol^{-1})
P, I, F, C	Primitive, body-centered, face-centered, and end-centered unit cells
a, b, c	Unit cell parameters
α, β, γ	Unit cell parameters
(h k l)	Miller indices
λ	Wave length of X radiation (angstrom = Å = 1×10^{-10} m)
d_{hkl}	Distance between planes in the crystal
z	Atoms per unit cell
2θ	Recorded X-ray powder diffraction angle

This chapter will consider some of the important aspects of the two condensed phases of matter. In Chap. 2, we dealt with gases in some detail, more detail than will be involved here. This is due to the simplicity of gaseous systems relative to either liquids or solids. In solids and liquids, molecules are very close to each other, allowing strong intermolecular interactions to be present (or the molecules are close together due to strong intermolecular interactions). In either case, the interactions cannot be ignored or simply approximated as was done for gases. Thus, theoretical models for liquids and solids are very complex. The solid state is easier to study than the liquid state. The orderly (almost orderly) array of atoms and molecules in the solid crystal is more amenable to mathematical description than the limited order found in liquids. In gases, the almost complete disorder allows a rather simple treatment. The intermediate order in liquids prohibits exact mathematical description. We have previously dealt with liquids to a limited extent. Solution properties were discussed in parts of Chap. 6, and liquid-liquid and liquid-vapor equilibria were described in Chap. 13. In each of these cases, little was said about the properties of the liquids themselves. That will be remedied in this chapter. A general discussion of liquid and solid properties, a brief treatment of interactions in liquids, and descriptions of some specific liquid properties are found in Sec. 14-1. We will see that many of the equations developed in our study of thermodynamics apply to the liquid state. Section 14-2 is an introduction to electrical and magnetic properties. The final three sections deal with solid properties and structure. In all cases, the theory presented will not be very sophisticated. Only enough is given to help you understand the origin of the various phenomena being studied. The last two sections in particular are mainly descriptive accounts of symmetry and structure in the solid state.

14-1 GENERAL PROPERTIES

OBJECTIVES

You shall be able to

1. Define or briefly discuss and calculate appropriate quantities when applicable, for all the following:

   Structural differences between solids, liquids, and gases
   Heats of fusion, vaporization, and sublimation
   Vapor pressure
   Coefficients of isobaric thermal expansion and isothermal compression
   Dipole-dipole interactions, induction effects, dispersion forces, and hydrogen bonding
   Surface tension, vapor pressure, viscosity, and activation energy of flow
2. Apply the Clausius-Clapeyron equation (or the Clausius equation)
3. Find $C_p - C_V$ for a liquid given the isothermal compressibility and the coefficient of thermal expansion
4. Apply appropriate surface tension and viscosity equations

14-1.1 INTRODUCTION

This section begins with a short survey of some structural characteristics and differences of liquids and solids. Then, equations and terms developed previously that apply to condensed phases will be reviewed. Reference will be made to the section in the text where particular concepts are developed. Only results will be reproduced here for those concepts. You are urged to read appropriate preceding sections if you have not done so or have forgotten (heaven forbid!). The remainder of the section deals with liquid properties.

The relationship between vapor pressure and temperature (the Clapeyron and Clausius-Clapeyron equations) have already been derived in Sec. 6-2.3. Only the results are presented here. More detail will be given on compressibilities, thermal expansion, surface tension, and viscosity. Finally, intermolecular forces are considered. Intermolecular forces are very important in determining the behavior of a material. If there were no intermolecular forces, there would be no solids or liquids. The only phase present would be the gas phase, and all gases would be ideal. This section will be a very qualitative treatment of these intermolecular forces. Some equations will be given, but these indicate the relative magnitudes of the interactions, not absolute values. For more detail on all types of interactions described below see Pryde (Sec. 14-1.9, Ref. 1).

14-1.2 A FEW COMMENTS ON STRUCTURAL DIFFERENCES

As pointed out in the chapter introduction, liquids and solids are more difficult to describe theoretically than gases. The large distances between gas molecules result in weak interactions between molecules (or atoms), which in turn allows us to approximate the behavior of the gas with an equation of state. The short distance between molecules or atoms in solids and liquids gives rise to strong interactions which cannot be ignored. Gases have translational freedom due to the distance between molecules. However, the volume occupied by liquids and solids is approximately the same as the volume occupied by the molecules themselves. Thus, translational motion is severely restricted for condensed phases. In the case of solids, translational motion is almost completely suppressed, and the vibration of molecules, atoms, or ions about their equilibrium positions is the predominant motion involved. In liquids, the intermolecular forces are intermediate between solids and gases, and some translational motion is observed. Molecules in a liquid "flow" past each other, but the motion is restricted. There is usually some short-range ordering but no long-range ordering. Gases have high compressibilities, while liquids and solids do not. That should be easy for you to justify on the basis of what has just been said. In condensed phases, any compression usually involves deformations within the molecules themselves, since so little "free" volume is present. On the other hand, gases can be compressed, giving diminished distances between molecules but little other change (within limits, of course). As mentioned earlier, most of the energy in a gas is kinetic energy. A solid has a lower potential energy than a liquid, thus energy (heat) must be added to convert a solid to a liquid. And a liquid has less potential energy than a gas. As you know, energy must be added to convert a liquid to a gas. The nature of the forces binding liquids and solids will be discussed in various portions of the remainder of this chapter.

14-1.3 REVIEW OF EQUATIONS AND TERMS

HEATS OF FUSION, VAPORIZATION, AND SUBLIMATION. In Sec. 3-2.5 the phase changes (changes from one physical state to another) of interest in this text are listed. The heat (enthalpy) absorbed when a solid melts to a liquid (or the heat released when a liquid freezes to a solid) is called the *heat of fusion* $\Delta\bar{H}_{fus}$. The heat required to vaporize a liquid to a gas (or the heat released on condensation from gas to liquid) is the *heat of vaporization* $\Delta\bar{H}_{vap}$. As we will see below, the heat of vaporization will vary as the temperature of vaporization varies (and, by necessity, the pressure of vaporization). Finally, if a solid converts directly to a gas, the required heat is the *heat of sublimation* $\Delta\bar{H}_{sub}$. Heats of fusion are not very sensitive to pressure, whereas $\Delta\bar{H}_{vap}$ and $\Delta\bar{H}_{sub}$ are. These three quantities can be related, as shown in Fig. 14-1.3.1. Following the procedure developed in our study of thermodynamics, you should see immediately that

$$\Delta\bar{H}_{sub} = \Delta\bar{H}_{fus} + \Delta\bar{H}_{vap} \tag{14-1.3.1}$$

(This is true if all changes are measured at the same temperature.) Table 14-1.3.1 lists some $\Delta\bar{H}$ values.

VAPOR PRESSURE. The variation of the vapor pressure of a liquid with temperature is given by the Clausius-Clapeyron equation developed in Sec. 6-2.3. Read that section now. The results of interest are

$$\frac{dP}{dT} = \frac{\Delta\bar{S}_{vap}}{\Delta\bar{V}_{vap}} \tag{6-2.3.2}$$

$$\frac{dP}{dT} = \frac{\Delta\bar{H}_{vap}}{T\,\Delta\bar{V}_{vap}} \tag{6-2.3.3}$$

And after a series of approximations

$$\frac{dP}{dT} = \frac{\Delta\bar{H}_{vap}\,P}{RT^2} \tag{6-2.3.4}$$

or

$$\boxed{\frac{d\,\ln P}{dT} = \frac{\Delta\bar{H}_{vap}}{RT^2}} \tag{6-2.3.5}$$

It would be a good idea for you to review Example 6-2.3.1. Some of the problems at the end of this section will involve similar procedures. Table 14-1.3.2 lists vapor pressure as a function of temperature for several substances.

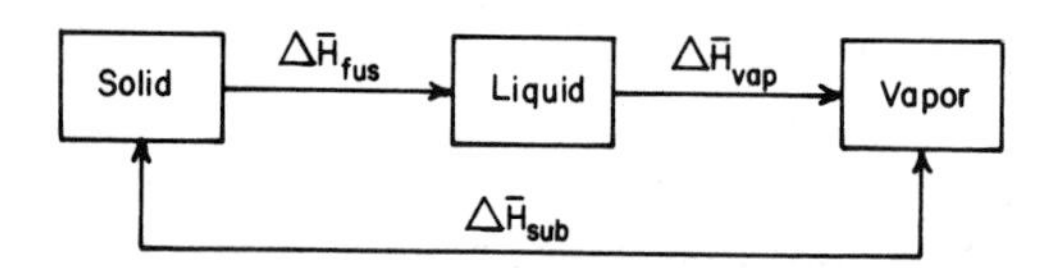

FIG. 14-1.3.1 Relationship between heats of fusion, vaporization, and sublimation.

Table 14-1.3.1
Selected Values of $\Delta\bar{H}_{vap}$ and $\Delta\bar{H}_{fus}$

Substance	T_{fus} (K)	$\Delta\bar{H}_{fus}$ (kJ mol^{-1})	T_{vap} (K)	$\Delta\bar{H}_{vap}$ (kJ mol^{-1})
Helium (He)	--	--	4.3	0.100
Hydrogen (H_2)	--	--	20.5	0.904
Nitrogen (N_2)	63.29	--	77.7	5.560
Water (H_2O)	273.16	6.0095	373.16	40.670
Sulfur dioxide (SO_2)	200.4	--	263.14	24.900
Benzene (C_6H_6)	278.7	9.832	353.3	30.760
Acetic acid (CH_3COOH)	289.8	11.7	391.4	24.390
Methanol (CH_3OH)	175.4	--	337.9	35.270
Ethanol (C_2H_5OH)	155.8	--	351.7	38.570
Cyclohexane (C_6H_{10})	279.7	2.67	353.9	30.080

COEFFICIENTS OF THERMAL EXPANSION AND ISOTHERMAL COMPRESSION. Section 2-2.2 describes the equation of state for a gas. Some of the equations derived are valid for solids and liquids as well. Below are the pertinent equations. Review Sec. 2-2.2 for background information.

The *thermal expansion coefficient* is given by Eq. (2-2.2.3):

$$\alpha = \frac{1}{V}\left(\frac{\partial V}{\partial T}\right)_P \tag{2-2.2.3}$$

and the *isothermal compression coefficient* is given by Eq. (2-2.2.6):

$$\beta = -\frac{1}{V}\left(\frac{\partial V}{\partial P}\right)_T \tag{2-2.2.6}$$

For gases, we used an equation of state to find $(\partial V/\partial T)_P$ and $(\partial V/\partial P)_T$. This is not possible for condensed phases, since we have no general equations of state. For solids and liquids, α and β are determined experimentally. Table 14-1.3.3 lists some values of α and β. Values of α can be used to determine the volume at a given temperature if volume is known at another temperature. For small temperature ranges

$$V = V_0[1 + \alpha(T - 273.15)] \tag{14-1.3.2}$$

where V_0 is the volume at T = 273.15 K. For precise calculations over a wide range of temperatures an expanded form such as Eq. (14-1.3.3) must be used:

Table 14-1.3.2
Vapor Pressure as a Function of Temperature, T_{vap}, $\Delta\bar{H}_{vap}$, and $\Delta\bar{S}_{vap}$ for Many Substances[a]

Substance	T (K)						T_{vap}	$\Delta\bar{H}_{vap}$	$\Delta\bar{S}_{vap}$
	0.0013[b]	0.013	0.0526	0.132	0.526	1	(K)	(kJ mol^{-1})	(J mol^{-1} K^{-1})
Helium (He)	1.5	1.9	2.5	2.9	3.9	4.6	4.6	0.100	22
Hydrogen (H_2)	(s)	(s)	(s)	15.3	18.7	20.7	20.7	0.904	44
Nitrogen (N_2)	(s)	(s)	(s)	63.5	72.3	77.4	77.4	5.56	72
Water (H_2O)	(s)	284.5	307.3	324.8	356.2	373.2	373.2	40.67	109
Sulfur dioxide (SO_2)	(s)	(s)	212.7	226.3	250.2	263.2	263.2	24.9	95
Benzene (C_6H_6)	(s)	(s)	280.8	299.3	333.8	353.3	353.3	30.76	87
Acetic acid (CH_3COOH)	(s)	(s)	279.9	298.7	334.0	353.9	353.9	30.08	85
Methanol (CH_3OH)	229.2	257.0	278.2	294.4	323.1	337.9	337.9	35.27	104
Ethanol (C_2H_5OH)	241.9	270.9	292.2	308.1	308.1	336.7	336.7	38.57	115
Cyclohexane (C_6H_{10})	(s)	290.7	316.2	336.2	372.2	391.3	391.3	24.39	62

[a] T is given for the vapor pressure (atm) shown. From the Chemical Rubber Company *Handbook of Chemistry and Physics*.
[b] Pressure in atm.

Table 14-1.3.3
Selected Values of Coefficients of Thermal Expansion and Compressibility at 20°C[a]

Substance	State	$\alpha \times 10^4$ (K^{-1})	$\beta \times 10^6$ (atm^{-1})
Copper	s	0.492	0.78
Graphite	s	0.24	3.0
Quartz (SiO_2)	s	0.15	2.8
Silver	s	0.583	1.0
Sodium chloride	s	1.21	4.2
Benzene (C_6H_6)	ℓ	12.4	94
Carbon tetrachloride (CCl_4)	ℓ	12.4	103
Ethanol	ℓ	11.2	110
Methanol	ℓ	12.0	120
Water (H_2O)	ℓ	2.07	45.3

[a] From G. W. Castellan, *Physical Chemistry*, 2nd ed., Addison Wesley, Reading, Mass., 1971, p. 90.

$$V = V_0(1 + at + bt^2 + \cdots) \qquad (14\text{-}1.3.3)$$

where t is in degrees centigrade and a and b are constants. The effect of pressure can be found from

$$V = V_0'[1 - \beta(P - 1)] \qquad (14\text{-}1.3.4)$$

V_0' is the volume at some temperature for P = 1 atm. P will be expressed in atmospheres. Values of β are constant over wide ranges of pressure.

14-1.3.1 EXAMPLE

Find (1) the percent and (2) the actual volume changes for copper and ethanol associated with a temperature increase from 25 to 45°C.

Solution

1. The percent volume change will be given by

$$\frac{\Delta V}{V_{25}} \times 100\% = \frac{V_{45} - V_{25}}{V_{25}} \times 100\%$$

From Eq. (14-1.3.2) this can be written as (V_0 refers to V at 273.15 K)

$$\frac{\Delta V}{V_{25}} = \frac{V_0[1 + \alpha(318.15 - 273.15)] - V_0[1 + \alpha(298.15 - 273.15)]}{V_0[1 + \alpha(298.15 - 273.15)]}$$

$$= \frac{1 + \alpha(45) - 1 - \alpha(25)}{1 + \alpha(25)} = \frac{\alpha(20)}{1 + \alpha(25)}$$

For Cu:

$$\frac{\Delta V}{V_{25}} \times 100\% = \frac{(0.492 \times 10^{-4}\ K^{-1})(20\ K)}{1 + (0.492 \times 10^{-4}\ K^{-1})(25\ K)} \times 100\%$$

$$= 9.8 \times 10^{-2} = \underline{\underline{0.098\%}}$$

For C_2H_5OH:

$$\frac{\Delta V}{V_{25}} \times 100\% = \frac{(11.2 \times 10^{-4}\ K^{-1})(20\ K)}{1 + (11.2 \times 10^{-4}\ K^{-1})(25\ K)} \times 100\%$$

$$= \underline{\underline{2.18\%}}$$

Ethanol expands about 22 times as much as Cu. Liquids, in general, do expand much more than solids on heating.

2. To get ΔV, we need $V_{45} - V_{25}$.

$$\Delta V(25 - 45) = V_{45} - V_{25}$$

$$= V_0[1 + \alpha(318.15 - 273.15)] - V_0[1 + \alpha(298.15 - 273.15)]$$

$$= V_0[1 + \alpha(45\ K) - 1 - \alpha(25\ K)]$$

$$= V_0\ \alpha(20\ K)$$

Now we need V_0, the volume at 273.15 K = 0°C. From the density values,

$$V_{25}\ (Cu) = \frac{1}{\rho_{Cu}} = 0.112\ dm^3\ kg^{-1}$$

$$V_{25}\ (ethanol) = \frac{1}{\rho_{eth}} = 1.28\ dm^3\ kg^{-1}$$

For Cu:

$$V_0 = \frac{V_{25}}{1 + \alpha(298.15 - 273.15)}$$

$$= \frac{0.112\ dm^3\ kg^{-1}}{1 + (0.492 \times 10^{-4}\ K^{-1})(25\ K)} = 0.112\ dm^3\ kg^{-1}$$

$$\Delta V\ (25 - 45) = (0.112\ dm^3\ kg^{-1})(0.492 \times 10^{-4}\ K^{-1})(20\ K)$$

$$= \underline{\underline{1.1 \times 10^{-4}\ dm^3\ kg^{-1}}}$$

For C_2H_5OH:

$$V_0 = \frac{V_{25}}{1 + \alpha(298.15 - 273.15)}$$

$$= \frac{1.28 \text{ dm}^3 \text{ kg}^{-1}}{1 + (11.2 \times 10^{-4} \text{ K}^{-1})(25 \text{ K})} = 1.25 \text{ dm}^3 \text{ kg}^{-1}$$

$$\Delta V\ (25 - 45) = (1.25 \text{ dm}^3 \text{ kg}^{-1})(11.2 \times 10^{-4} \text{ K}^{-1})(20 \text{ K})$$

$$= \underline{\underline{2.79 \times 10^{-2} \text{ dm}^3 \text{ kg}^{-1}}}$$

Note that Eq. (14-1.3.2) had to be manipulated considerably to get the quantities we wanted from it. Do not be afraid to rearrange and tamper with equations. They are not always in the proper form.

HEAT CAPACITIES. In Sec. 3-2.2 the difference of heat capacity at constant pressure and constant volume $C_p - C_v$ was shown to be

$$C_p - C_v = \left[\left(\frac{\partial E}{\partial V}\right)_T + P\right]\left(\frac{\partial V}{\partial T}\right)_P \qquad (14\text{-}1.3.5)$$

This equation offers a good opportunity to review some thermodynamic manipulations. We need to find $(\partial E/\partial V)_T$. From Eq. (5-1.5.8) we know

$$\left(\frac{\partial H}{\partial P}\right)_T = V - T\left(\frac{\partial V}{\partial T}\right)_P \qquad (5\text{-}1.5.8)$$

But, H = E + PV, by definition.

$$\left(\frac{\partial E}{\partial P}\right)_T + \left(\frac{\partial PV}{\partial P}\right)_T = V - T\left(\frac{\partial V}{\partial T}\right)_P \qquad (14\text{-}1.3.6)$$

$$\left(\frac{\partial E}{\partial P}\right)_T + P\left(\frac{\partial V}{\partial P}\right)_T + V = V - T\left(\frac{\partial V}{\partial T}\right)_P$$

or

$$\left(\frac{\partial E}{\partial P}\right)_T + P\left(\frac{\partial V}{\partial P}\right)_T = -T\left(\frac{\partial V}{\partial T}\right)_P \qquad (14\text{-}1.3.7)$$

From the definition of E = E(T, P),

$$dE = \left(\frac{\partial E}{\partial P}\right)_T dP + \left(\frac{\partial E}{\partial T}\right)_P dT \qquad (14\text{-}1.3.8)$$

We can get the following by dividing by ∂V at constant T.

$$\left(\frac{\partial E}{\partial V}\right)_T = \left(\frac{\partial E}{\partial P}\right)_T\left(\frac{\partial P}{\partial V}\right)_T + \left(\frac{\partial E}{\partial T}\right)_P\left(\frac{\partial T}{\partial V}\right)_T$$

or

$$\left(\frac{\partial E}{\partial V}\right)_T = \left(\frac{\partial E}{\partial P}\right)_T\left(\frac{\partial P}{\partial V}\right)_T \qquad (14\text{-}1.3.9)$$

This, when substituted into (14-1.3.5), gives

$$C_p - C_v = \left[\left(\frac{\partial E}{\partial P}\right)_T\left(\frac{\partial P}{\partial V}\right)_T + P\right]\left(\frac{\partial V}{\partial T}\right)_P \tag{14-1.3.10}$$

Substituting (14-1.3.7)

$$C_p - C_v = \left[-P\left(\frac{\partial V}{\partial P}\right)_T\left(\frac{\partial P}{\partial V}\right)_T - T\left(\frac{\partial V}{\partial T}\right)_P\left(\frac{\partial P}{\partial V}\right)_T + P\right]\left(\frac{\partial V}{\partial T}\right)_P$$

$$= -T\left(\frac{\partial V}{\partial T}\right)_P^2\left(\frac{\partial P}{\partial V}\right)_T$$

$$= -\frac{T(\partial V/\partial T)_P^{\ 2}}{(\partial V/\partial P)_T} \tag{14-1.3.11}$$

The definitions of α and β given in Sec. 2-2,

$$\alpha = +\frac{1}{V}\left(\frac{\partial V}{\partial T}\right)_P \tag{2-2.2.3}$$

$$\beta = -\frac{1}{V}\left(\frac{\partial V}{\partial P}\right)_T \tag{2-2.2.6}$$

are substituted to yield

$$\boxed{C_p - C_v = \frac{\alpha^2 TV}{\beta}} \tag{14-1.3.12}$$

From manipulations of thermodynamic functions, we have derived a useful equation that relates the heat capacities to compression and thermal expansion coefficients. Some data for liquids and solids are found in Table 14-1.3.3.

14-1.3.2 EXAMPLE

From data in Table 14-1.3.3 determine the heat capacity difference $C_p - C_v$ at 20°C for CCl_4

Solution

We need the molar volume $\bar{V}$ which will be

$$\bar{V} = \frac{M}{\rho} = \frac{0.1538 \text{ kg mol}^{-1}}{1594.2 \text{ kg m}^{-3}} = \underline{9.65 \times 10^{-5} \text{ m}^3 \text{ mol}^{-1}}$$

From Table 14-1.3.3,

$$\alpha = 12.4 \times 10^{-4}\ (\text{K}^{-1}) \qquad \beta = 1.03 \times 10^{-4} \text{ atm}^{-1}$$

Then, $\bar{C}_p - \bar{C}_v = \frac{\alpha^2 T\bar{V}}{\beta}$

$$= \frac{(12.4 \times 10^{-4}\ K^{-1})^2 (293.2\ K)(9.65 \times 10^{-5}\ m^3\ mol^{-1})}{1.03 \times 10^{-4}\ atm^{-1}}$$

$$= (4.22 \times 10^{-4}\ atm\ m^3\ K^{-1}\ mol^{-1})\left(\frac{10^3\ dm^3}{1\ m^3}\right)\left(\frac{8.314\ J}{0.0821\ dm^3\ atm}\right)$$

$$= \underline{\underline{42.8\ J\ mol^{-1}\ K^{-1}}}$$

If $\bar{C}_p$ or $\bar{C}_v$ is known the other can be found.

14-1.4 SURFACE TENSION

DEFINITION. There are several phenomena associated with the surfaces of solids and liquids. Of these, only surface tension will be considered here. Other surface phenomena are discussed in Chap. 11. The origin of many surface phenomena is the imbalance of forces on the molecules of the liquid. In the bulk of a liquid, each molecule is completely surrounded by other molecules, which leads to a balanced or symmetrical force about itself. This is not the case for a molecule on the surface. A surface molecule is attracted to molecules beneath and adjacent to it. However, there are no molecules above it to balance the forces beneath it. The resulting imbalance of forces results in a net inward force which causes the liquid to behave as though it were surrounded by an invisible membrane. The name of this phenomena is *surface tension*. Some observations you may have made are explained by the *surface tension* of the liquid. Some of these are

1. A "water bug" walks on the water without penetrating the surface
2. Water rises in a capillary tube, mercury falls
3. Raindrops are spherical
4. A steel needle will float on water

The thermodynamic explanation of surface tension is the tendency of a system to achieve the lowest free energy. Thus, surface tension is the result of a system trying to minimize its surface area.

Any attempt to increase the surface area of a system will require an input of work to increase the free energy of the system. The easiest way to derive the equations needed to describe surface tension is to consider a thin film of substance floating on the surface of another liquid. This is actually a problem that belongs in Chap. 11, in which surface chemistry is studied, but it will serve a useful purpose here. Consider a film of material covering area ABCD in Fig. 14-1.4.1. If the BC end of the frame is moved a distance x, to EF by the application of force f, the

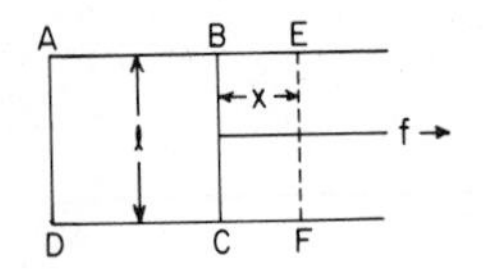

FIG. 14-1.4.1 A wire piston supporting a liquid film.

work done is given by

$$w = fx \tag{14-1.4.1}$$

This force is just balanced by the surface tension along BC. Let us designate γ as the force per meter along BC whose length is ℓ. The force is given by

$$f = 2\gamma\ell \tag{14-1.4.2}$$

The 2 is needed to account for the liquid having two surfaces. Combining these equations leads to

$$w = fx = 2\gamma\ell x \tag{14-1.4.3}$$

We can write

$$\frac{\text{Work}}{\text{Change in surface area}} = \frac{w}{\Delta \mathcal{a}} = \frac{2\gamma\ell x}{2\ell x} = \gamma \tag{14-1.4.4}$$

This shows that the *surface tension* γ is the work required to generate a unit surface area, or γ is the *energy per unit of surface area*. This energy is mechanical rather than thermal, and the tendency for the liquid to minimize its surface area is simply an attempt by the system to minimize its free energy.

MEASUREMENT. Many methods can be used to determine the surface tension of a liquid. Some of these are capillary rise, tensiometer, drop weight, and bubble pressure. The simplest and perhaps the most accurate is the capillary rise method. This method depends on the fact that most liquids will rise in a fine glass capillary tube to a level above the external surface (see Fig. 14-1.4.2). Such liquids "wet" the glass surface: that is, they adhere to it. Some liquids such as mercury do not "wet" the glass, and the liquid level in the capillary will be below the level outside the tube. If one end of a glass tube is immersed in a liquid which wets the glass surface, an area of new liquid surface is created on the inside of the glass tube. The liquid cannot rise indefinitely, however. It will rise only until the surface tension force just balances the force of gravity acting on the column of liquid.

Since γ is defined as the surface tension in newtons per meter of surface, and since the active surface is the circumference of the capillary ($2\pi r$), the force due to surface tension is

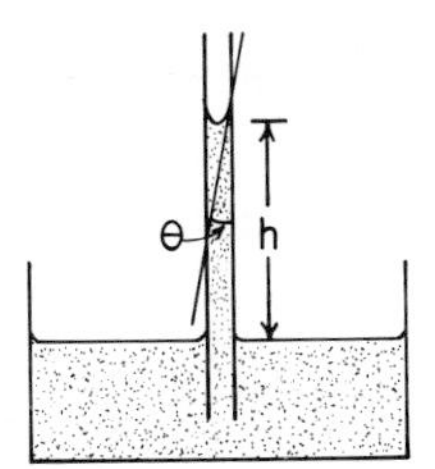

FIG. 14-1.4.2 Capillary rise in a glass tube of radius r.

$$f_\gamma = 2\pi r\gamma \qquad (14\text{-}1.4.5)$$

This must be modified slightly, since the liquid contacts the glass at some angle θ. This angle is called the *contact angle*. The actual force due to surface tension is

$$f_\gamma = 2\pi r\gamma \cos\theta \qquad (14\text{-}1.4.6)$$

The force due to gravity on the column of liquid of height h and density ρ is

$$f_g = \pi r^2 \rho g h \qquad (14\text{-}1.4.7)$$

where g is the acceleration due to gravity (9.80 m s^{-2}). Equilibrium is reached when $f_\gamma = f_g$, from which we can write (show this)

$$\gamma = \frac{r\rho g h}{2 \cos\theta} \qquad (14\text{-}1.4.8)$$

For most liquids which wet glass θ ≈ 0, so

$$\boxed{\gamma \approx \frac{r\rho g h}{2}} \qquad (14\text{-}1.4.9)$$

If you are given the capillary radius, the density of the liquid, and the height of the rise, you can quickly determine the surface tension of a liquid.

For very precise work, the volume of liquid under the meniscus must be included. Also, a correction factor for the density of the gas above the liquid must be applied. The modified equation takes the form

$$\boxed{\gamma = \frac{(h + r/3)(\rho_1 - \rho_g)rg}{2}} \qquad (14\text{-}1.4.10)$$

14-1.4.1 EXAMPLE

Water ($\gamma = 0.07275$ N m^{-1}) at 20°C is observed to rise 0.030 m in a capillary. Methanol at the same temperature an in the same capillary rises to a height of

0.012 m. What is the surface tension of methanol? $\rho^{20}_{methanol} = 791.4\ kg\ m^{-3}$ and $\rho^{20}_{H_2O} = 998.2\ kg\ m^{-3}$.

Solution

The data for water can be used to find the capillary radius. From Eq. (14-1.4.9):

$$r = \frac{2\gamma}{\rho gh} = \frac{(2)(0.07275\ N\ m^{-1})}{(998.2\ kg\ m^{-3})(9.80\ m\ s^{-2})(0.030\ m)}$$

$$= \underline{4.96 \times 10^{-4}\ m}\ (= 0.496\ mm)$$

Then, for methanol,

$$\gamma = \frac{r\rho gh}{2}$$

$$= \frac{(4.96 \times 10^{-4}\ m)(791.4\ kg\ m^{-3})(9.8\ m\ s^{-2})(0.012\ m)}{2}$$

$$= \underline{\underline{0.023\ N\ m^{-1}}}$$

Table 14-1.4.1 contains values for the surface tension of several liquids at different temperatures. (*Note*: All the values given in Table 14-1.4.1 are really interfacial surface tensions. That is, the value given is for the surface tension of the liquid in contact with either air or its own vapor. More is said about this in Chap. 11.) The surface tension decreases with increasing temperature. At the critical temperature (the temperature above which a vapor cannot be made to liquefy at any pressure), surface tension goes to zero. As mentioned earlier, most of the discussion of surface tension is found in Chap. 11, in which surface properties are considered. Further discussion can be found in that chapter.

Table 14-1.4.1
Surface Tension of Some Liquids at Various Temperatures ($N\ m^{-1}$)

	T (K)					
Liquid	273.2	293.2	313.2	333.2	353.2	373.2
Water	0.07564	0.07275	0.06956	0.06618	0.06261	0.05885
Methanol	0.0245	0.0226	0.0209	--	--	--
Ethanol	0.02405	0.02227	0.02060	0.01901	--	--
CCl_4	--	0.0268	0.0243	0.0219	--	--
Acetone	0.0262	0.0237	0.0212	0.0186	0.0162	--
Benzene	0.0316	0.0289	0.0263	0.0237	0.0213	--
Toluene	0.03074	0.02843	0.02613	0.02381	0.02153	0.01939

14-1.5 LIQUID VISCOSITY

COEFFICIENT OF VISCOSITY. Here the equations describing the flow of fluids will be derived. Gas viscosity was treated in Sec. 2-3.9, in which the molecular interpretation was given. Here the equations that describe fluid flow will be presented with little reference to the why of fluid viscosity.

Consider a cylinder of fluid within a tube subject to a pressure drop of P across the tube length ℓ (Fig. 14-1.5.1). The pressure (force over area) will cause a flow of liquid through the tube. There will be a resistance to flow due to frictional drag within the liquid, the liquid viscosity. This drag is proportional to the velocity gradient dv/dr, the area of the cylinder of fluid $\mathcal{A} = 2\pi r\ell$, and the coefficient of viscosity η.

$$f_d = \eta\mathcal{A}\left(-\frac{dv}{dr}\right) \tag{14-1.5.1}$$

The negative sign takes into account that dv/dr itself is negative. The frictional drag just balances the imposed force $f_p = P(\pi r^2)$. So,

$$f_d = f_p$$

$$\eta(2\pi r\ell)\left(-\frac{dv}{dr}\right) = P(\pi r^2) \tag{14-1.5.2}$$

or

$$dv = -\frac{P}{2\eta\ell}\, r\, dr \tag{14-1.5.3}$$

To find the fluid velocity at any radius r, we integrate (14-1.5.3) from r = R to r = r and recognize that at the tube wall (r = R) v = 0.

$$\int_{v=0}^{v} dv = -\frac{P}{2\eta\ell}\int_{r=R}^{r=r} r\, dr \tag{14-1.5.4}$$

$$v = \frac{P}{4\eta\ell}(R^2 - r^2) \tag{14-1.5.5}$$

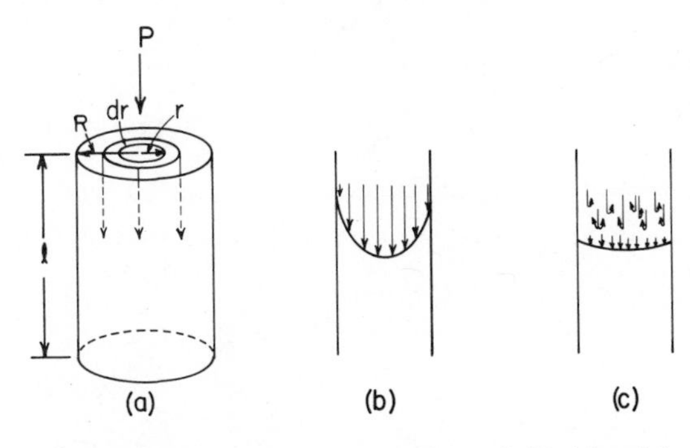

FIG. 14-1.5.1 (a) A cylinder of fluid subject to a pressure drop of P across the distance ℓ. (b) Laminar velocity profile. (c) Turbulent flow.

Figure 14-1.5.1(b) shows the velocity profile predicted by this last equation. This profile is correct if the fluid velocity is not too high. This is called *laminar flow*. For high velocities, laminar, or smooth flow, is not observed. Instead, the fluid exhibits a considerable turbulence, as shown in Fig. 14-1.5.1(c), where *turbulent flow* is indicated. A dimensionless number has been derived from fluid properties which gives an indication of whether a fluid is in turbulent or laminar flow. This number is called the *Reynolds number*.

$$R = \frac{d\bar{v}\rho}{\eta} \qquad (14\text{-}1.5.6)$$

where d is the tube diameter, $\bar{v}$ is the average velocity of the fluid, ρ is the fluid density, and η is the viscosity coefficient. It has been observed that if the Reynolds number is less than about 2100, laminar flow results. Above R = 4000, turbulent flow results. Between these two numbers is the transition region in which the type of flow depends on many system parameters and cannot be easily predicted. In any case, laminar flow exists near the wall, even if the bulk of the material is in turbulent flow.

The volume rate of flow of a liquid can be found by integrating the product of the rate of flow of a cylinder of fluid at r times the area of that cylinder.

$$\frac{dV}{dt} = \int_0^R v(2\pi r)\, dr \qquad (14\text{-}1.5.7)$$

Introducing (14-1.5.5) and integrating leads to (supply the missing steps)

$$\frac{dV}{dt} = \frac{\pi P R^4}{8\eta \ell} \qquad (14\text{-}1.5.8)$$

This can be rearranged to give the Poiseuille equation for the viscosity coefficient of a liquid.

$$\eta = \frac{\pi P R^4}{8\ell} \frac{t}{V} \qquad (14\text{-}1.5.9)$$

If the pressure is due to the fluid itself, P will be given by

$$P = \rho g h \qquad (14\text{-}1.5.10)$$

where ρ is the fluid density, g is the acceleration due to gravity, and h is the height of the liquid level above a datum level. This, upon substitution into Eq. (14-1.5.9), gives

$$\eta = \frac{\pi \rho g h R^4}{8\ell} \frac{t}{V}$$

$$= \frac{\pi g h R^4}{8\ell V} t\rho \qquad (14\text{-}1.5.11)$$

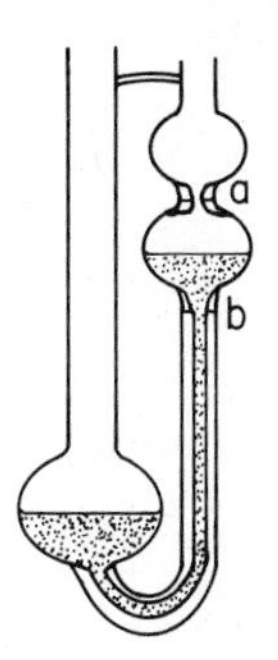

FIG. 14-1.5.2 An Ostwald viscometer.

Viscosity coefficients are usually measured in viscometers such as the Ostwald viscometer shown in Fig. 14-1.5.2. The time for a fluid to flow between points (a) and (b) is measured. Every term on the right side of (14-1.5.11) (before $t\rho$) is a constant for a given viscometer, so we can write

$$\boxed{\eta = B\rho t} \qquad B = \frac{\pi g h R^4}{8\ell V} \tag{14-1.5.12}$$

To obviate the need to measure B accurately, a difficult task, usually a material of known viscosity (η_0) is run, and B is calculated by

$$B = \frac{\eta_0}{\rho_0 t_0} \tag{14-1.5.13}$$

Then, for any other material,

$$\eta_1 = B\rho_1 t_1 = \frac{\eta_0 \rho_1 t_1}{\rho_0 t_0} \tag{14-1.5.14}$$

There are other methods of determining viscosity coefficients, such as the falling ball viscometer, in which the rate of fall of a ball of known density is measured. The fluid drag on the ball as it falls through the liquid can be related to the rate of fall. This gives an estimate of the viscosity coefficient of the fluid. The above method is the simplest in most cases, however. Table 14-1.5.1 lists some viscosity coefficients as a function of temperature. The temperature dependence is discussed following an example.

14-1.5.1 EXAMPLE

The time required for H_2O to drain in an Ostwald viscometer at 20°C is 1.52 min. An organic liquid with ρ = 800 kg m^{-3} requires 2.25 min in the same viscometer at 20°C. Find the viscosity coefficient of the organic liquid.

Solution

$\rho_{H_2O}^{20} = 998 \text{ kg m}^{-3} \qquad \eta_{H_2O}^{20} = 1.002 \times 10^{-3} \text{ kg m}^{-1} \text{ s}^{-1}$

From Eq. (14-1.5.14),

$$\eta_1 = \frac{(1.002 \times 10^{-3} \text{ kg m}^{-1} \text{ s}^{-1})(800 \text{ kg m}^{-3})(2.25 \text{ min})}{(998 \text{ kg m}^{-3})(1.52 \text{ min})}$$

$$= \underline{\underline{1.19 \times 10^{-3} \text{ kg m}^{-1} \text{ s}^{-1}}}$$

Note: The common unit for the coefficient of viscosity is the

$$\text{Centipoise} = 10^{-2} \text{ poise} = 10^{-2} \text{ g cm}^{-1} \text{ s}^{-1}$$

$$1 \text{ kg m}^{-1} \text{ s}^{-1} = 10^1 \text{ g cm}^{-1} \text{ s}^{-1} = 10^1 \text{ poise}$$

$$= 10^3 \text{ centipoise}$$

TEMPERATURE DEPENDENCE. From an investigation of the behavior of the viscosity coefficient with temperature (as is done in Example 14-1.5.2), we can conclude that

$$\eta = A \exp\left(\frac{D}{RT}\right) \tag{14-1.5.15}$$

where A and D are constants and R is the gas constant. This equation is of the form observed for activation processes (see, for example, the Arrhenius theory in Chap. 10) in which only molecules with energies greater than some threshold value can take part in a chemical reaction. In a liquid, there is a certain amount of energy required for the molecules to flow past each other. Only molecules with energy greater than this value can take part in the flow. The Boltzmann factor which was discussed in Sec. 2-3 gives the fraction of molecules with sufficient energy to mount the energy barrier. Such considerations lead to the identification of D in (14-1.5.15) with ΔE_{visc}, the activation energy of (barrier to) viscous flow. Then

Table 14-1.5.1
Viscosity Coefficients of Some Liquids at Various Temperatures (10^{-3} kg m^{-1} s^{-1})

Liquid	273.2	293.2	313.2	333.2	353.2
C_6H_6	0.912	0.652	0.503	0.392	0.329
CCl_4	1.329	0.969	0.739	0.585	0.468
$CHCl_3$	0.700	0.563	0.464	0.389	--
C_2H_5OH	1.773	1.200	0.834	0.592	--
H_2O	1.792	1.002	0.656	0.469	0.357
Hg	1.685	1.554	1.450	1.367	1.298

$$\eta = A \exp\left(\frac{\Delta E_{visc}}{RT}\right) \qquad (14\text{-}1.5.16)$$

The utility of this equation comes from its predictive capabilities. If η is known at several temperatures, then ΔE_{visc} can be determined and η can be estimated for other T. Such use is illustrated in Example 14-1.5.2.

14-1.5.2 EXAMPLE

Use the data of Table 14-1.5.1 to find ΔE_{visc} for water, then predict the viscosity coefficient at 373.2 K.

Solution

Take the logarithm of Eq. (14-1.5.16) to get a linear form for a plot to determine ΔE_{visc}.

$$\ln \eta = \ln A + \frac{\Delta E_{visc}}{R}\frac{1}{T} \qquad (14\text{-}1.5.16)$$

A plot of ln η vs. 1/T should give a straight line with slope $\Delta E_{visc}/R$ and intercept at ln A. From Table 14-1.5.1,

$\eta \times 10^3$ ($kg\ m^{-1}\ s^{-1}$)	ln η	T (K)	$1/T \times 10^3$ (K^{-1})
1.792	- 6.324	273.2	3.660
1.002	- 6.906	293.2	3.411
0.656	- 7.33	313.2	3.193
0.469	- 7.66	333.2	3.001
0.357	- 7.94	353.2	2.831

These data are plotted in Fig. 14-1.5.3. The slope is 1.80×10^3 K. Then, ΔE_{visc} = slope × R = $(1.80 \times 10^3\ K)(8.314\ J\ mol^{-1}\ K^{-1})$.

$$\Delta E_{visc} = \underline{\underline{1.50 \times 10^4\ J\ mol^{-1}}}$$

Note that the low T point has been ignored and the slope taken near the T of interest. To find ln A pick some point on the line (0)

$$\ln A = \ln \eta - \frac{\Delta E_{visc}}{R}\frac{1}{T}$$

$$= -7.2 - \frac{1.50 \times 10^4\ J\ mol^{-1}}{8.314\ J\ mol^{-1}\ K^{-1}}(3.25 \times 10^{-3}\ K^{-1})$$

$$= \underline{\underline{-13.1}}$$

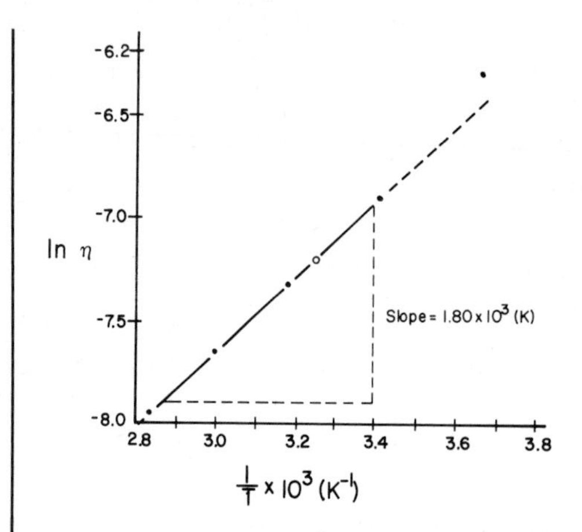

FIG. 14-1.5.3 A plot of ln η vs. 1/T for H_2O.

Thus the equation for water is

$$\ln \eta = -13.1 + \frac{1.50 \times 10^4 \text{ J mol}^{-1}}{R} \frac{1}{T}$$

Finally, solving for T = 373.2 K,

$$\ln \eta = -13.1 + \frac{1.50 \times 10^4 \text{ J mol}^{-1}}{8.314 \text{ J mol}^{-1} \text{ K}^{-1}} \frac{1}{373.2 \text{ K}}$$

$$= -8.21$$

$$\eta = \underline{\underline{2.70 \times 10^{-4} \text{ kg m}^{-1} \text{ s}^{-1}}}$$

The tabulated value is 0.284×10^{-3} kg m^{-1} s^{-1}, a 5% error, most of which is probably in the determination of the slope.

14-1.6 INTERACTIONS IN LIQUIDS

DIPOLE-DIPOLE INTERACTIONS. The dipole-dipole interaction has been called the orientation effect, since the strength of the interaction depends on the orientation of one dipole relative to another. If you are not sure you know what a dipole is, you should read Sec. 14-2.3 at this time. The interactions can range from repulsion when like ends of two dipoles are pointing toward each other, to attractions when unlike ends approach. Figure 14-1.6.1 illustrates these two possibilities.

In a liquid, molecules are free to rotate to some extent, whereas in a solid the dipoles (assuming a molecular crystal has been formed) are fixed. If molecules are rotating at random in a liquid, any pair will at times exhibit repulsion and at other times exhibit attraction. If the rotation were completely random, there would

FIG. 14-1.6.1 Two of many possible orientations of two dipoles.

be no net dipole-dipole attraction since the repulsions and attractions would cancel. However, as two dipoles rotate into the attractive position they tend to slow down each other's rotation. On the other hand, as they approach the repulsive state they tend to repel each other and speed up their rotation. This leads to a net amount of time spent in the attractive position and, therefore, a net attractive force. The potential energy of the two molecules averaged over all orientations and weighted in favor of the attractive orientations is given by

$$U_{D-D} = -\frac{2}{3}\frac{\vec{\mu}^4}{(4\pi\varepsilon_0)^2 r^6 kT} \qquad (14\text{-}1.6.1)$$

where r is the separation between the dipoles. The effect of an increase in temperature is to lower the interaction energy. This is due to the effect of thermal motion. As the temperature becomes higher, the motion of the dipoles becomes more random, leading to a decreased interaction. Table 14-1.6.1 has an estimate of some values of U_{D-D} as well as other interaction potentials.

INDUCTION EFFECTS. Another interaction involving a molecule with a dipole moment is called the *induction effect*. This may also be designated as the dipole-induced dipole effect. Here, a molecule with a dipole moment induces a moment in an adjacent molecule (whether the second molecule has a dipole is not important). Figure 14-1.6.2 illustrates this process. The induction of a dipole can take place since the electron cloud about a molecule or atom is easily distorted by nearby charges. You should conclude that the induction effect will always be attractive. It is difficult to envision inducing a repulsive interaction. If a molecule with dipole moment $\vec{\mu}$ interacts with one that has a polarizability $\underline{\alpha}$ this attractive potential is given by

$$U_{ind} = -\frac{2\underline{\alpha}\vec{\mu}^2}{4\pi\varepsilon_0 r^6} \qquad (14\text{-}1.6.2)$$

Note there is no temperature dependence in this equation. Some values are given in Table 14-1.6.1. Notice that, in general, U_{D-D} is greater than U_{ind}. The effect due to polarizability is much smaller than the effect due to a permanent dipole moment.

Table 14-1.6.1
Intermolecular Attraction Terms for Some Simple Molecules[a]

Molecule	Dipole moment (Debyes)	Polarizability ($Å^3$)	Interaction energy (J molecule^{-1}) for two molecules at r = 5 Å and T = 298 K.			
			$U_{D-D} \times 10^{22}$	$U_{ind} \times 10^{22}$	$U_{London} \times 10^{22}$	Total $\times 10^{22}$
He	0	0.2	0	0	0.005	0.5
A	0	1.6	0	0	2.9	2.9
CO	1.12	2.0	0.00021	0.0037	4.6	4.6
Xe	0	4.0	0	0	18	18
CCl_4	0	10.5	0	0	116	116
HCl	1.07	2.6	1.2	0.36	7.8	9.4
HBr	0.79	3.6	0.39	0.28	15	16
HI	0.38	5.4	0.021	1.10	33	33
H_2O	1.82	1.5	11.9	0.65	2.6	15
NH_3	1.47	2.2	5.2	0.63	5.6	11

[a] An arbitrary intermolecular distance of 5 Å has been used for the comparison. From G. M. Barrow, *Physical Chemistry*, 3rd ed., McGraw-Hill, New York, 1973, p. 516.

DISPERSION FORCES. Even in the absence of any permanent dipoles in a material, interactions exist. We "know" this must be so, or nonpolar materials such as argon or neon could not be made to liquefy. The sources of the interactions in these cases are the *induced dipole-induced dipole* interactions. These are called *dispersion forces* or *London forces* after F. London who first recognized the effect.

Dispersion forces arise from the random motion of the electrons in individual molecules. If an instantaneous dipole is generated by a temporary unsymmetrical charge distribution (more electron density found on one side of a molecule than on the other due to electron motion), a dipole can be induced in an adjacent molecule. This new, temporary dipole, in turn, induces a moment in another molecule. The original dipole is short lived, but its induced effects can be transmitted to several molecules. These interactions will always be attractive for the same reason that the induction effect is always attractive. The actual calculation of the magnitude of London forces requires detailed quantum mechanics and is not very accurate. It is found that the following dependence of the potential on the polarizability and the distance is

FIG. 14-1.6.2 Dipole inducing a dipole in an initially nonpolar molecule.

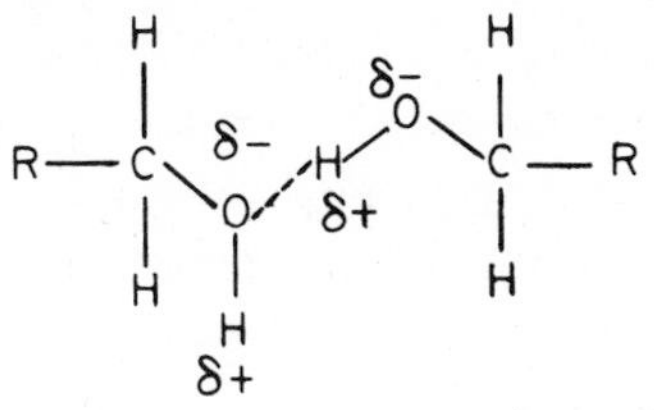

FIG. 14-1.6.3 A hydrogen bond. The R stands for an organic group.

$$U_{dispersion} \propto \frac{\alpha^2}{r^6} \qquad (14\text{-}1.6.3)$$

Dispersion forces are present whether a dipole moment is present or not. Estimates of $U_{dispersion}$ are given in Table 14-1.6.1 under the heading U_{London}.

HYDROGEN BONDING. A final interaction of major significance in biological as well as other systems is the *hydrogen bond*. If a hydrogen atom is attached to an electronegative element such as O, N, or F, the hydrogen will become slightly positive as the electronegative element withdraws electrons to itself. This partially charged hydrogen is then attracted to an electronegative (and thereby partially negatively charged) atom on an adjacent molecule, or even to other portions of itself in the case of polymer molecules. Figure 14-1.6.3 represents this effect for alcohol. The importance of hydrogen bonding in determining physical properties can be seen in Fig. 14-1.6.4 in which boiling points are plotted for the hydrides of two groups of elements. In each series you should see that when the hydrogen is attached to a very electronegative element (N or O), the boiling point is much higher than would be expected from extrapolation of the boiling points of the other hydrides in the group. This is attributed to the strong H-bonding interaction.

Hydrogen bonding accounts for unusually high melting and boiling points for such compounds as sugars, alcohols, organic acids, and other compounds containing hydrogen bonded to O, F, and N. The compound BF_3 is a gas at room temperature. In

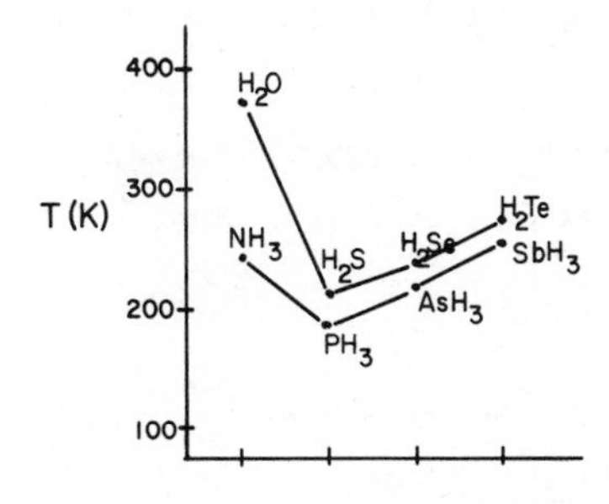

FIG. 14-1.6.4 Boiling points of some hydrides.

the absence of other data we would expect boric acid, $B(OH)_3$, with about the same weight to be a gas or low-boiling liquid. It is, however, a solid. This can be attributed to the hydrogen bonding possible. Hydrogen bonding is very important in proteins and other natural products. In many cases, the conformation (shape) of the molecule is the result of the molecule arranging itself so that hydrogen bonding can be maximized. The lower density of ice relative to water is due to the expansion of the lattice to maximize hydrogen bonding. At the melting point, thermal energy is sufficient to break up the hydrogen-bonded lattice and produce the liquid, which has a larger density.

14-1.7 PROBLEMS

1. At 20°C a rigid container is filled completely with methanol. What pressure is developed if the temperature is raised 15 K? ρ_{20} (methanol) = 0.791 kg dm^{-3}.

2. Use the Claussius-Clapeyron equation to calculate and plot the vapor pressure of cyclohexane from 0°C to its boiling point, assuming $\Delta\bar{H}_{vap}$ is constant.

3. If $\alpha = (1/V)(\partial V/\partial T)_P$ show that $\alpha = (-1/\rho)(\partial\rho/\partial T)_P$ if ρ is the density.

4. From the vapor pressure data in Table 14-1.3.2 find $\Delta\bar{H}_{vap}$ for SO_2 and compare with the value tabulated in Table 14-1.3.1.

5. The following data are known for ethanol at 20°C: $\alpha = 11.2 \times 10^{-4}\ K^{-1}$, $\beta = 1.1 \times 10^{-4}\ atm^{-1}$, $\rho = 789.3$ kg m^{-3}, and $\bar{C}_p = 114.4$ J $mol^{-1}\ K^{-1}$. What is $\bar{C}_v$ for ethanol at 20°C?

6. In a surface tension experiment using a capillary tube, the angle θ was actually 10° instead of 0°. What percent error would result if you assumed $\theta = 0$ in all your calculations?

7. At 20°C a sample of acetone is observed to rise in a capillary to a height of 2.03×10^{-2} m. Use Eq. (14-1.4.10) to find the height ethanol would rise to in the same capillary at the same temperature.

8. What will be the volume flow rate of benzene through a 0.15 m diameter pipe if there is a 1 atm pressure drop across a 100 m section at 313.2 K?

9. An Ostwald viscometer filled with C_2H_5OH drains in 313 s at 20°C. Diethyl ether, $(C_2H_5)_2O$, under the same conditions requires 67 s to drain. What is the viscosity coefficient of $(C_2H_5)_2O$? $\rho^{\circ}_{C_2H_5OH} = 789.3$ kg m^{-3} and $\rho_{(C_2H_5)_2O} =$ 714 kg m^{-3}.

10. From the data in Table 14-2.3.1 and equations in this section find U_{D-D} and U_{ind} for H_2O with $r = 5 \times 10^{-10}$ m and compare with the values in Table 14-1.6.1.

11. Perform the integral indicated in Eq. (14-1.5.7).

14-1.8 SELF-TEST

1. Define, or discuss briefly, seven terms from this section.
2. Calculate the change in volume of quartz when the pressure is increased from 1 atm to 10,000 atm at 25°C. The density of quartz at 25°C is 2.32 kg dm^{-3}.
3. The vapor pressure of CCl_4 is 1 atm at 349.9 K. If the heat of vaporization can be assumed constant and equal to 3.0×10^4 J mol^{-1}, determine the vapor of CCl_4 at 325 K.
4. To what height will a liquid rise in a capillary of radius 0.021 cm if the surface tension is found to be 0.024 N m^{-1}? ρ = 0.85 kg dm^{-3}.
5. Give a one-page discussion of the types of interaction you would expect in a solution of water and methanol, CH_3OH.

14-1.9 ADDITIONAL READING

1.* J. A. Pryde, *The Liquid State*, Hutchinson and Co., Ltd., London, 1966.

14-2 ELECTRICAL AND MAGNETIC PROPERTIES

OBJECTIVES

You shall be able to

1. Calculate the force between charges in a given medium
2. Define the dipole moment and calculate its magnitude for a given system of charges
3. Calculate molar polarization, molecular polarizability, or dipole moment if two of these quantities are given for a molecule
4. Calculate molar or molecular magnetic polarizabilities or magnetic moments of molecules if two of these quantities are given
5. Define diamagnetic and paramagnetic and discuss the origin of these phenomena

14-2.1 INTRODUCTION

Some important aspects of electrical and magnetic properties of molecules will be given in this section. The first portion will be a review of the units used throughout the chapter. Electrical and magnetic units can be somewhat confusing. Be especially careful to include units on all quantities in an equation to be certain that proper units are being used. After the unit review, electrical properties of molecules will be considered. This will involve a review of our introduction to dielectric and electrostatic equations. Dipole moments and polarizabilities will be defined and calculated here. You have seen how they can be used in one case in Sec. 14-1, in which intermolecular forces are considered. Finally, a few concepts and equations concerning magnetic properties of molecules will be given. In all cases, only the briefest of treatments will be given to the physical basis for each phenomenon.

14-2.2 BASIC ELECTROSTATICS AND UNITS OF INTEREST

FORCE. One of the basic equations of electrostatics is Coulomb's law which describes the force between charges q_1 and q_2, separated by a distance r when these charges are immersed in a medium with a *permittivity constant* of ε. This equation is

$$\boxed{f = \frac{q_1 q_2}{4\pi\varepsilon r^2}} \tag{14-2.2.1}$$

The permittivity constant of a material is related to the *permittivity of a vacuum* ε_0 and the dielectric constant θ of the material by

$$\frac{\varepsilon}{\varepsilon_0} = \theta \tag{14-2.2.2}$$

ε_0 has a value of

$$\varepsilon_0 = 8.854 \times 10^{-12}\ \text{F m}^{-1}\ \text{(farad/meter)}$$
$$= 8.854 \times 10^{-12}\ \text{C}^2\ \text{N}^{-1}\ \text{m}^{-2}\ (\text{coulomb}^2\ \text{newton}^{-1}\ \text{meter}^{-1})$$

Some useful conversions you may need are given below:

$$\text{Volt} = \text{Joule (coulomb)}^{-1}$$
$$\text{V} = \text{J C}^{-1}$$

$$\text{Coulomb} = \text{ampere} \cdot \text{sec} = \text{Joule (volt)}^{-1}$$
$$\text{C} = \text{A s} = \text{J V}^{-1}$$

$$\text{Volt} = \text{Joule (ampere)}^{-1}\ (\text{sec})^{-1}$$
$$\text{V} = \text{J A}^{-1}\ \text{s}^{-1}$$

$$\text{Farad (meter)}^{-1} = (\text{coulomb})^2\ (\text{newton})^{-1}\ (\text{meter})^{-2}$$
$$\text{F m}^{-1} = \text{C}^2\ \text{N}^{-1}\ \text{m}^{-2}$$

$$C^2 = 8.98 \times 10^9 \text{ J m}$$

$$\varepsilon_0 = (8.854 \times 10^{-12} \text{ C}^2 \text{ N}^{-1} \text{ m}^{-2})(8.98 \times 10^9 \text{ J m}/1 \text{ C}^2)$$

$$= 0.0795 = \frac{1}{4\pi}$$

$$\text{Coulomb} = \text{Volt-Farad}$$

$$C = V\ F$$

Let us now apply Eq. (14-2.2.1) to a specific problem.

14-2.2.1 EXAMPLE

Find the force exerted between two protons ($q = 0.160 \times 10^{-18}$ C) which are 10^{-9} m apart in water at 298 K. Water has a dielectric constant of 78.5.

Solution

$$f = \frac{q_1 q_2}{4\pi\varepsilon r^2} = \frac{q_1 q_2}{4\pi\varepsilon_0 \theta r^2}$$

$$= \frac{(0.160 \times 10^{-18} \text{ C})^2}{(4)(3.1416)(8.85 \times 10^{-12} \text{ C}^2 \text{ N}^{-1} \text{ m}^{-2})(78.5)(1 \times 10^{-9} \text{ m})^2}$$

$$= \frac{2.56 \times 10^{-38}}{8.73 \times 10^{-27}} \frac{\text{C}^2}{\text{C}^2 \text{ N}^{-1} \text{ m}^{-2} \text{ m}^2}$$

$$= \underline{\underline{2.93 \times 10^{-12} \text{ N}}}$$

This is the repulsive force exerted between two protons separated by 10 Å in an aqueous solution. This force is repulsive if positive, attractive if negative.

POLARIZATION. If an electrical field $\bar{E}_0$ is applied between two parallel plates separated by a material with a dielectric constant θ (or permittivity $\varepsilon = \varepsilon_0\theta$) and a distance d, a capacitor is formed. The charges that build up are shown in Fig. 14-2.2.1. The charge buildup can be either from actual charge migration or from orientation of dipoles within the medium. The *charge density* σ on the capacitor surface in the absence of a dielectric is given by Eq. (14-2.2.3).

$$\bar{E}_0 = \frac{\sigma}{\varepsilon_0} \tag{14-2.2.3}$$

The effect of a dielectric material is to reduce the field within the cavity. This is accomplished by polarization of the dielectric. This polarization compensates for part of the charge density produced (part of the charge on the face of the

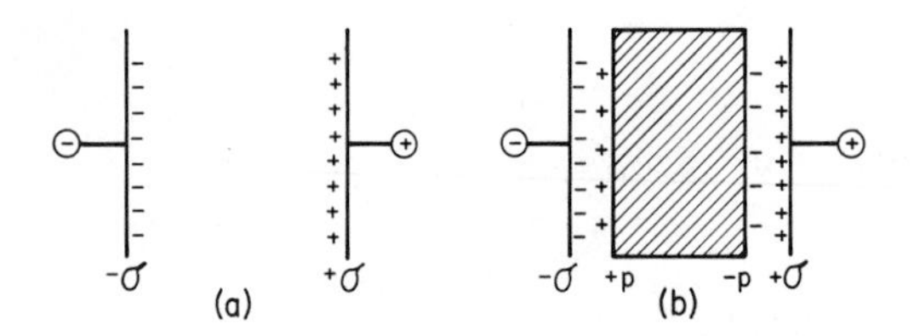

FIG. 14-2.2.1 (a) Charged capacitor, no dielectric material.
(b) Charged capacitor, dielectric present.

capacitor plate is balanced by a charge induced on the dielectric material). The *polarization* of the dielectric is given the symbol p. (The effective charge density is now σ - p.) The electric field in this *polarizable* medium is

$$\bar{E} = \frac{\sigma}{\varepsilon} = \frac{\sigma_{eff}}{\varepsilon_0} = \frac{\sigma - p}{\varepsilon_0} = \frac{\sigma}{\varepsilon_0} - \frac{p}{\varepsilon_0} \tag{14-2.2.4}$$

$$\varepsilon_0\bar{E} = \sigma - p = \varepsilon\bar{E} - p \qquad \text{since } \sigma = \varepsilon\bar{E}$$

Thus,

$$p = (\varepsilon - \varepsilon_0)\bar{E} \qquad \text{or} \qquad \frac{p}{\varepsilon_0} = (\theta - 1)\bar{E} \tag{14-2.2.5}$$

where θ is the dielectric constant $\varepsilon/\varepsilon_0$. These equations will be used shortly to find an equation relating *molar polarization* to the *dipole moment* and *molecular polarizibility*. Before doing that a definition of *dipole moment* needs to be given.

14-2.3 DIPOLE MOMENT, POLARIZATION, AND REFRACTIVE INDEX

DIPOLE MOMENT. In any molecule, if one atom has a greater tendency to attract electrons (more electronegativity) than the one to which it is bonded, an unsymmetrical charge distribution will result. The center of positive charge will not coincide with center of negative charge. This gives rise to a dipole moment for the molecule. For a system of charges, the dipole moment is defined relative to some origin by

$$\boxed{\vec{\mu} = \sum_i q_i\vec{r}_i} \tag{14-2.3.1}$$

where the arrow indicates a vector quantity. To get the dipole moment of a system of charges, add vectorially the products of each charge times its vector distance from the origin. For example, the system shown in Fig. 14-2.3.1 has a dipole moment given by

$$\vec{\mu} = q_1\vec{r}_1 + q_2\vec{r}_2 + q_3\vec{r}_3$$

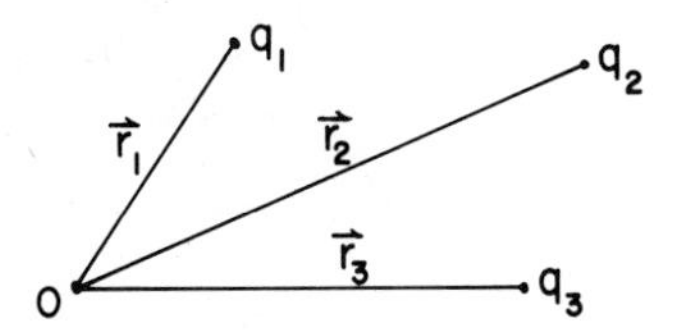

FIG. 14-2.3.1 An assembly of charges from which a dipole moment is calculated.

Assume

$$q_1 = q_2 = q_3 = +q$$

We can add these quantities vectorially as follows

$$\vec{\mu} = q_1\vec{r}_1 + q_2\vec{r}_2 + q_3\vec{r}_3$$

$$= q\vec{r}_1 + q\vec{r}_2 + q\vec{r}_3$$

$$= q(\vec{r}_1 + \vec{r}_2 + \vec{r}_3)$$

If we know angles and distances and charges we can find $\vec{\mu}$.

14-2.3.1 EXAMPLE

For the following system find the dipole moment.

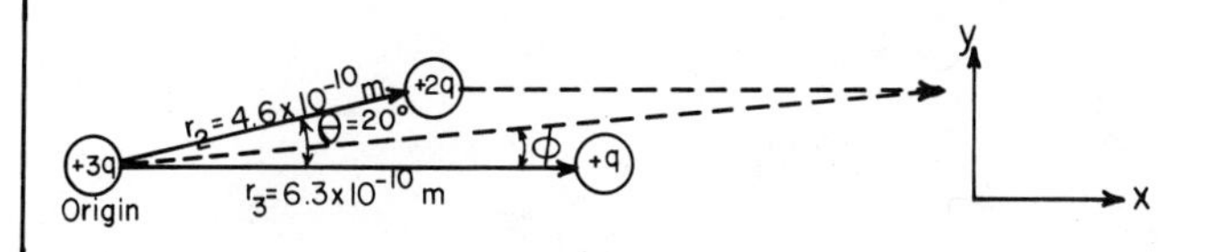

Solution

$$\vec{\mu} = \sum_i q_i\vec{r}_i$$

$$= 3q(0) + 2q(\vec{r}_2) + q(\vec{r}_3)$$

$$= 2q\vec{r}_2 + q\vec{r}_3$$

Before determining the magnitude and direction of $\vec{\mu}$, find $\vec{\mu}_x$ and $\vec{\mu}_y$, the projections onto the x and y axes.

$$\vec{\mu}_x = \vec{\mu}_{2,x} + \vec{\mu}_{3,x}$$

$$= 2q\vec{r}_2 \cos 20^\circ + q\vec{r}_3 \cos 0^\circ$$

$$= 2q(4.6 \times 10^{-10}\ \text{m})(0.940) + q(6.3 \times 10^{-10}\ \text{m})$$

$$= q(1.49 \times 10^{-9}\ \text{m})$$

$$\vec{\mu}_y = \vec{\mu}_{2,y} + \vec{\mu}_{3,y}$$

$$= 2q\vec{r}_2 \sin 20^\circ + q\vec{r}_3 \sin 0^\circ$$

$$= q(3.15 \times 10^{-10}\ \text{m})$$

The dipole moment can now be found readily.

$$\vec{\mu} = \left(\vec{\mu}_x^{\,2} + \vec{\mu}_y^{\,2}\right)^{1/2}$$

$$= \left[(2.32 \times 10^{-18})(q)^2\right]^{1/2}$$

$$= \underline{q(1.52 \times 10^{-9}\ \text{m})}$$

$$\tan \phi = \frac{\vec{\mu}_y}{\vec{\mu}_x} = \frac{q(3.14 \times 10^{-10}\ \text{m})}{q(1.49 \times 10^{-9}\ \text{m})} = 0.211$$

$$\phi = \underline{\underline{11.9^\circ}}$$

All we need is a charge to calculate the actual dipole moment. If we choose q = electronic charge = 1.602×10^{-19} C, then

$$\vec{\mu} = q(1.52 \times 10^{-9}\ \text{m})$$

$$= (1.602 \times 10^{-19}\ \text{C})(1.52 \times 10^{-9}\ \text{m})$$

$$= \underline{2.44 \times 10^{-28}\ \text{C m}}$$

Values of $\vec{\mu}$ are usually given in Debye with 1 D = 3.338×10^{-30} C m.

$$\vec{\mu} = 2.44 \times 10^{-28}\ \text{C m}\ \frac{1\ \text{D}}{3.338 \times 10^{-30}\ \text{C m}} = 73\ \text{D}$$

This is a very large number. Dipole moments are usually not that large for molecules, since bare electronic charges are not actually found in solution. Molecules have only fractional charges (due to electronegativity differences) leading to smaller moments.

Table 14-2.3.1 contains some dipole moment values. One final note, the dipole moment calculated will be dependent on your choice of origin. In molecules, the *origin is the center of mass of the molecule.*

POLARIZABILITY. If a molecule is placed in an electric field, it is *polarized*: the electron distribution is distorted. This gives rise to an *induced moment* within the molecule. The induced moment is proportional (in many cases) to the effective field acting on the molecule; we say effective because the external field imposed is screened somewhat by the material within the field, so a molecule in the system feels the effect of a field less than that imposed. The induced moment can be written

$$\vec{\mu}_{ind} = \underline{\alpha}\bar{E}_{eff} \tag{14-2.3.2}$$

where $\underline{\alpha}$ is the *molecular polarizability*. The effective field for gases and nonpolar liquids is found to be (see Slater and Frank, Sec. 14-2.7, Ref. 1)

$$\bar{E}_{eff} = \bar{E} + \frac{p}{3\varepsilon_0} \tag{14-2.3.3}$$

where the last term is due to polarization about an imaginary spherical cavity in the medium in which the molecule is contained.

The polarization p is just equal to the *induced moment times the number of molecules per unit volume* N*. Thus

$$p = N^*\vec{\mu}_{ind} = N^*\underline{\alpha}\bar{E}_{eff} \tag{14-2.3.4}$$

Combined with (14-2.3.3),

$$p = N^*\underline{\alpha}\left(\bar{E} + \frac{p}{3\varepsilon_0}\right) \tag{14-2.3.5}$$

But, from (14-2.2.5)

$$p = \varepsilon_0(\theta - 1)\bar{E} \tag{14-2.2.5}$$

Substitution of p from (14-2.2.5) for p in (14-2.3.5) upon considerable simplification gives

$$\frac{1}{3}\frac{N^*\underline{\alpha}}{\varepsilon_0} = \frac{\theta - 1}{\theta + 2} \tag{14-2.3.6}$$

$\theta = \varepsilon/\varepsilon_0$ is the dielectric constant for the material.

We can express this in terms of a *molar polarization* $\mathcal{P}$. With $N^* = (\rho/M)N$ where ρ is the density, M is the molecular weight, and N is Avagadro's number, we define

$$\boxed{\mathcal{P} = \frac{1}{3}\frac{N\underline{\alpha}}{\varepsilon_0} = \frac{4\pi N\underline{\alpha}}{3} = \frac{\theta - 1}{\theta + 2}\frac{M}{\rho}} \tag{14-2.3.7}$$

Table 14-2.3.1
Values of Dielectric Quantities for Several Molecules

Molecule	Dipole moment Debye[a]	Polarizability $m^3\ mlc^{-1} \times 10^{30}$[b]	Refractive index	Dielectric constant
HF	1.8	0.80	1.90 (gas)	84 (0°C)
HCl	1.07	--	--	4.6 (28°C)
H_2O	1.82	1.44	1.333	80.37 (20°C) 78.54 (25°C)
NH_3	1.47	2.34	1.325 (16.5°C)	17.8 (15°C)
PH_3	0.55	--	1.317 (liquid)	--
$CHCl_3$	1.94	--	1.4433 (25°C)	4.806 (20°C)
CH_4	0	2.60	--	1.70 (- 173°C)
CCl_4	0	10.5	1.4664 (20°C)	2.238 (20°C)
C_6H_6	0	25.1	1.5011 (20°C)	2.284 (20°C)
He	0	0.20	--	1.054 (3.9 K)
N_2	0	1.73	--	1.454 (- 203°C)
CH_3OH	1.70	3.0	1.3288 (20°C)	32.63 (25°C)
C_2H_5OH	1.70	5.2	1.3611 (20°C)	24.3 (25°C)

[a] 1 Debye = 3.338×10^{-30} C m.
[b] mlc = molecule.

This is the Clausius-Mosotti equation which relates the molecular polarizability or molar polarization to the dielectric constant for a material. *This equation applies only to molecules with no permanent dipole moment.*

If a molecule has a permanent dipole moment, there will be an additional contribution to the polarizability. The contribution of the dipole will depend on its orientation in the electric field. We can obtain an average effect by averaging over each orientation (an integral) weighted by the probabilities of that orientation (the weighting factor is the energy of that orientation). The average dipole moment in the direction of the field is found to be

$$\vec{\mu}_{av} = \frac{\vec{\mu}^2}{3kT} \bar{E}_{eff} \tag{14-2.3.8}$$

The total effective $\vec{\mu}$ is $\vec{\mu}_{ind} + \vec{\mu}_{av}$ [see Eq. (14-2.3.2)], and the total polarizability is given by

$$\mathcal{P} = \frac{1}{3}\frac{N}{\varepsilon_0}\left(\underline{\alpha} + \frac{\vec{\mu}^2}{3kT}\right) = \frac{4\pi N}{3}\left(\underline{\alpha} + \frac{\vec{\mu}^2}{3kT}\right)$$

$$\mathcal{P} = \frac{4\pi N}{3}\left(\underline{\alpha} + \frac{\vec{\mu}^2}{3kT}\right) = \frac{\theta - 1}{\theta + 2}\frac{M}{\rho} \qquad (14\text{-}2.3.9)$$

The units on this can be a problem. The example below may help clarify them.

14-2.3.2 EXAMPLE

For H_2O, $\vec{\mu} = 1.82$ D $= 6.08 \times 10^{-30}$ C m molecule^{-1}, and $\underline{\alpha} = 1.44 \times 10^{-30}$ m^3 molecule^{-1}. Find the molar polarization at 298 K. (mlc = molecule.)

Solution

$$\mathcal{P} = \frac{N}{3\varepsilon_0}\left[(1.44 \times 10^{-30}\ \text{m}^3\ \text{mlc}^{-1}) + \frac{(6.08 \times 10^{-30}\ \text{C m mlc}^{-1})^2}{(3)(1.38 \times 10^{-23}\ \text{J mlc}^{-1}\ \text{K}^{-1})(298\ \text{K})}\right]$$

$$= \frac{N}{3\varepsilon_0}(1.44 \times 10^{-30}\ \text{m}^3\ \text{molecule}^{-1} + 2.99 \times 10^{-39}\ \text{C}^2\ \text{m}^2\ \text{J}^{-1}\ \text{molecule}^{-1})$$

The needed conversion factor is

$$1\ \text{C}^2 = 8.98 \times 10^9\ \text{J m}$$

$$\mathcal{P} = \frac{N}{3\varepsilon_0}\left(1.44 \times 10^{-30}\ \text{m}^3\ \text{mlc}^{-1} + 2.99 \times 10^{-39}\ \text{C}^2\ \text{m}^2\ \text{J}^{-1}\ \text{mlc}^{-1}\ \frac{8.98 \times 10^9\ \text{J m}}{\text{C}^2}\right)$$

$$= \frac{1}{3}\frac{N}{\varepsilon_0}(2.83 \times 10^{-29}\ \text{m}^3\ \text{molecule}^{-1})$$

$$= \frac{1}{3\varepsilon_0}(6.023 \times 10^{23}\ \text{molecules mol}^{-1})(2.83 \times 10^{-29}\ \text{m}^3\ \text{molecule}^{-1})$$

$$= \frac{5.68 \times 10^{-6}\ \text{m}^3\ \text{mol}^{-1}}{\varepsilon_0} = 4\pi(5.68 \times 10^{-6}\ \text{m}^3\ \text{mol}^{-1})$$

$$= \underline{\underline{7.14 \times 10^{-5}\ \text{m}^3\ \text{mol}^{-1}}}$$

$\varepsilon_0 \approx 1/4\pi$: see conversion factors following Eq. (14-2.2.2).

The form of Eq. (14-2.3.9) should give you a hint as to how it can be used. If we measure the molar polarization at different temperatures we can get values for $\underline{\alpha}$ and $\vec{\mu}$.

$$\mathcal{P} = \frac{N\underline{\alpha}}{3\varepsilon_0} + \frac{N\vec{\mu}^2}{9\varepsilon_0 k}\frac{1}{T} \qquad (14\text{-}2.3.10)$$

A plot of $\mathcal{P}$ vs. 1/T should yield a straight line with slope $(N\vec{\mu}^2/9\varepsilon_0 k)$ and intercept $(N\underline{\alpha}/3\varepsilon_0)$.

REFRACTIVE INDEX. The velocity of light in a dielectric medium is less than in a vacuum. This is due to interaction of the oscillating electromagnetic field with the polarizable molecules in the dielectric. The electromagnetic field of visible light oscillates so rapidly ($\approx 10^{15}$ Hz) that the dipoles cannot respond. The material behaves as though the only polarization is due to the electron polarizability described by $\underline{\alpha}$. Clerk Maxwell has shown that the permittivity of a material is related to the index of refraction n_R by

$$\frac{\varepsilon}{\varepsilon_0} = n_R^2 \tag{14-2.3.11}$$

if no dipole is present. Even if a dipole moment is present, it cannot respond to the oscillating field at high frequencies, so Eq. (14-2.3.11) still applies. (This will not be true if lower frequencies are used to measure n_R.) It follows directly from this and Eq. (14-2.3.7),

$$\left(\frac{n_R^2 - 1}{n_R^2 + 2}\right)\frac{M}{\rho} = \frac{1}{3}\frac{N\underline{\alpha}}{\varepsilon_0} \tag{14-2.3.12}$$

Measurement of refractive indices provides a method for finding values for $\underline{\alpha}$. Table 14-2.3.1 lists some values of dipole moments, polarizabilities, and other related data. You might like to use some of these data to calculate other entries in the table for cross-checking to see how well the equations work. (Whether you like it or not, it would be good practice.) Note that Eq. (14-2.3.12) *does not hold well* for materials in which *hydrogen bonding* may occur. Which equation holds better for predicting $\underline{\alpha}$, Eq. (14-2.3.9) or Eq. (14-2.3.12)?

14-2.3.3 EXAMPLE

The index of refraction for CCl_4 (carbon tetrachloride) is 1.4664 at 20°C. If $\rho_{20} = 1594.2$ kg m^{-3} and M = 0.1538 kg mol^{-1} find the polarizability $\underline{\alpha}$.

Solution

$$\underline{\alpha} = \frac{3\varepsilon_0}{N}\left(\frac{n_R^2 - 1}{n_R^2 + 2}\right)\frac{M}{\rho}$$

$$= \frac{3}{4\pi N}\left(\frac{n_R^2 - 1}{n_R^2 + 2}\right)\frac{M}{\rho}$$

Substitution of given values yields

$$\underline{\alpha} = \frac{3}{4\pi(6.023 \times 10^{23}\ \text{molecules mol}^{-1})} \frac{(1.4664)^2 - 1}{(1.4664)^2 + 2} \frac{0.1538\ \text{kg mol}^{-1}}{1594.2\ \text{kg m}^{-3}}$$

$$= \underline{\underline{1.06 \times 10^{-29}\ \text{m}^3\ \text{molecule}^{-1}}}$$

The accepted value is 1.05×10^{-29} m^3 $molecule^{-1}$.

14-2.4 MAGNETIC PROPERTIES

MAGNETIC MOMENTS AND SUSCEPTIBILITIES. In this section only the very basics of magnetic properties of molecules will be treated. Nothing will be said about uses of magnetic phenomena and little about the causes of magnetism. Only two types of magnetism, diamagnetism and paramagnetism will be defined. You should be aware that there are many types (such as ferromagnetism, ferrimagnetism, antiferromagnetism, etc.) that are important if the complete magnetic behavior of a material is to be described. As in the case of dielectric properties, a molecule can have a permanent magnetic moment and an induced moment. The treatment is similar to that given for dielectric properties.

A magnetic field can be described by $\bar{\mathcal{H}}$, the magnetic field strength and $\bar{B}$, the magnetic flux density. The units on $\bar{B}$ are kg s^{-2} A^{-1} and are called tesla (T); (1 T = 10^4 gauss in electromagnetic units). A material in a magnetic field responds to the magnetic flux density and gains a magnetism $\bar{M}$ which is the magnetic moment per unit volume.

$$\bar{M} = \frac{\bar{B}}{\mu_0} - \bar{\mathcal{H}} \qquad (14\text{-}2.4.1)$$

μ_0 is the *magnetic permeability of vacuum* and is the analog to ε_0 of Sec. 14-2.3. Introducing the *magnetic permeability* μ_m of a substance we can write a magnetic equation analogous to the first part of Eq. (14-2.2.4)

$$\bar{B} = \mu_m \bar{\mathcal{H}} \qquad (14\text{-}2.4.2)$$

which gives

$$\bar{M} = \bar{\mathcal{H}} \left(\frac{\mu_m}{\mu_{0,m}} - 1 \right)$$

Defining the *molecular magnetic susceptibility* χ as

$$\chi = \frac{\bar{M}}{\bar{\mathcal{H}}} = \frac{\mu_m}{\mu_{0,m}} - 1 \qquad (14\text{-}2.4.3)$$

The corresponding *molar magnetic susceptibility* is given by

$$\chi_M = \frac{M\chi}{\rho} \tag{14-2.4.4}$$

where, as before, M = molecular weight and ρ = density. An equation similar to (14-2.3.9) can be written

$$\boxed{\chi_M = N\left(\alpha_m + \frac{\mu_{0,m}P_m^2}{3kT}\right)} \tag{14-2.4.5}$$

where P_m is the permanent magnetic moment in the molecule, and α_m is the magnetic polarizability induced by the magnetic field.

If χ_M is positive, the material is called *paramagnetic*. Such a material is attracted by a magnetic field and is pulled toward the field. Or, the magnetic lines are drawn into the sample. Paramagnetism arises from molecules having unpaired electrons. These unpaired electrons, when spinning, behave as though they were little magnets which are attracted to any external magnetic field. If no unpaired electrons are present, only an induced moment is possible (an induced moment is present even in molecules with a permanent magnetic moment). The induced moment aligns in opposition to an applied field and will be repelled by the field. Or, the magnetic field is reduced within such a sample. A *diamagnetic* material has only an induced moment, and the *susceptibility is negative*. Also, χ_M for a diamagnetic material is smaller by a factor of about 10^2 to 10^3 than for a paramagnetic one.

EXPERIMENTAL DETERMINATIONS. Magnetic moments are usually measured by suspending a sample in a magnetic field. A sensitive balance (called a Gouy balance) records the increase or decrease in force on the sample in the presence of a magnetic field. Knowledge of the force exerted, the magnetic strength, the sample area, and the density is sufficient to find the susceptibility. The following equation applies

$$\boxed{f = \frac{1}{2}\mu_{0,m}(\chi - \chi_0)\bar{\mathcal{H}}^2 \mathcal{A}} \tag{14-2.4.6}$$

where f is the force, χ is the susceptibility of the sample, χ_0 is the susceptibility of air, $\mathcal{A}$ is the cross-sectional area of the sample exposed to $\bar{\mathcal{H}}$, and $\bar{\mathcal{H}}$ is the magnetic field strength. This equation can be used to find χ for the samples, then Eq. (14-2.4.4) can be applied to get the molar susceptibility. The following, taken from G. Pass and H. Sutcliffe, *J. Chem. Educ.*, *48*:180-181 (1971), should clarify the units for this section.

UNITS.

Eq. (14-2.4.1):

$$\bar{M} = \frac{\bar{B}}{\mu_0} - \bar{\mathcal{H}}$$

where $\bar{B}$ = magnetic flux density = webers per meter^2 (Wb m^{-2})

$\bar{\mathcal{H}}$ = magnetic field strength = amperes per meter (A m^{-1})

μ_0 = permeability of vacuum = $4\pi \times 10^{-7}$ Henrys per meter (H m^{-1} = Wb m^{-1} A^{-1} = kg A^{-2} s^{-2} m)

$\bar{M}$ = magnetization (A m^{-1})

Eq. (14-2.4.3):

$$\chi = \frac{\bar{M}}{\bar{\mathcal{H}}} \quad \text{(unitless)}$$

Eq. (14-2.4.4):

$$\chi_M = \frac{M\chi}{\rho} \ (\text{m}^3 \ \text{mol}^{-1})$$

Some conversion factors (SI units underlined)

1 gauss (G) = 10^{-4} $\underline{\text{Wb m}^{-2}}$ = 10^{-4} $\underline{\text{kg s}^{-2}\ \text{A}^{-1}}$

1 tesla (T) = 10^4 gauss = 1 $\underline{\text{Wb m}^{-2}}$ = 1 $\underline{\text{kg s}^{-2}\ \text{A}^{-1}}$

1 Oersted = $\frac{1}{4\pi}\ 10^3$ $\underline{\text{A m}^{-1}}$

$\chi_M = \frac{M\chi}{\rho}$ (m^3 mol^{-1}) = $4\pi \times 10^{-6}$ χ_M (cm^3 mol^{-1})

Eq. (14-2.4.5):

$$\chi_M = N\left(\alpha_m + \frac{\mu_{0,m} P_m^2}{3kT}\right)$$

where N = Avagadro's number

α_m = magnetic polarizability (m^3 molecule^{-1})

$\mu_{0,m}$ = see above

P_m = magnetic moment (A m^2)

k = Boltzmann constant (1.38×10^{-23} J molecule^{-1} K^{-1})

T = absolute temperature

$$\frac{\mu_{0,m} P_m^2}{3kT} = \frac{(\mathrm{H\ m^{-1}})(\mathrm{A\ m^2})^2}{(\mathrm{J\ molecule^{-1}\ K^{-1}})(\mathrm{K})} \times \frac{(\mathrm{Wb\ m^{-1}\ A^{-1}})(\mathrm{m^{-1}})}{(1\ \mathrm{H\ m^{-1}})(\mathrm{m^{-1}})}$$

$$= \frac{\mathrm{A^2\ m^4\ (Wb\ m^{-2})\ A^{-1}}}{\mathrm{J\ molecule^{-1}\ m^{-1}}} \times \frac{1\ \mathrm{kg\ s^{-2}\ A^{-1}}}{1\ \mathrm{Wb\ m^{-2}}}$$

$$= \frac{\mathrm{A^2\ m^5\ A^{-2}\ kg\ s^{-2}}}{\mathrm{J\ molecule}} \times \frac{1\ \mathrm{J}}{\mathrm{kg\ m^2\ s^{-2}}}$$

$$= \underline{\underline{\mathrm{m^3\ molecule^{-1}}}} \quad \text{which is the proper unit}$$

Eq. (14-2.4.6):

$$f = \frac{1}{2}\,\mu_{0,m}(\chi - \chi_0)\bar{\mathcal{H}}^2\,\mathcal{A}$$

where f = force (N)

χ = magnetic susceptibility (unitless)

χ_0 = magnetic susceptibility of air $(4\pi \times 0.029 \times 10^{-6})$

$\mathcal{A}$ = area $(\mathrm{m^2})$

$\bar{\mathcal{H}}$ = magnetic field strength; Oersted = $[\mathrm{A\ m^{-1}}/(4\pi \times 10^{-3})]$

$\mu_{0,m}$ = see above

$$f = \frac{1}{2}\,(\mathrm{H\ m^{-1}})(\mathrm{A\ m^{-1}})^2\,\frac{\mathrm{kg\ A^{-2}\ s^{-2}\ m}}{\mathrm{H\ m^{-1}}}$$

$$= \mathrm{A^2\ m^{-2}\ m^2\ kg\ A^{-2}\ s^{-2}\ m}$$

$$= \mathrm{kg\ m\ s^{-2}} = \text{Newton (N)} \quad \text{which is the proper force unit}$$

14-2.4.1 EXAMPLE

A sample of an organometallic compound was placed in a susceptibility balance. The following data were obtained. Find the susceptibility of the compound.

$f = 2.47 \times 10^{-4}$ N $\quad \bar{\mathcal{H}} = 4000$ Oersted

$\mathcal{A} = 1.8 \times 10^{-5}\ \mathrm{m^2} \quad \chi_0 = 4\pi \times 0.029 \times 10^{-6}$

Solution

From Eq. (14-2.4.6):

$$f = \frac{1}{2}\,\mu_{0,m}(\chi - \chi_0)\bar{\mathcal{H}}^2\,\mathcal{A}$$

Substituting the experimental data gives

$$2.47 \times 10^{-4}\ \text{N} = \frac{1}{2}\ (4\pi \times 10^{-7}\ \text{H m}^{-1})\left(\frac{1\ \text{kg A}^{-2}\ \text{s}^{-2}\ \text{m}}{1\ \text{H m}^{-1}}\right)(\chi - \chi_0)(4000\ \text{Oersted})^2$$

$$\times \left(\frac{1\ \text{A m}^{-1}}{4\pi \times 10^{-3}\ \text{Oersted}}\right)^2 (1.8 \times 10^{-5}\ \text{m}^2)$$

$$2.47 \times 10^{-4}\ \frac{\text{kg m}}{\text{s}^2} = \frac{1}{2}\ (4\pi \times 10^{-7})(4000)^2\left(\frac{1}{4\pi \times 10^{-3}}\right)^2 (1.8 \times 10^{-5})(\chi - \chi_0)\text{kg m s}^{-2}$$

$$2.155 \times 10^{-4} = \chi - \chi_0$$

$$\chi = 2.155 \times 10^{-4} + 4\pi \times 0.029 \times 10^{-6}$$

$$= \underline{2.16 \times 10^{-4}}$$

MOLECULAR INTERPRETATION. The magnetic susceptibility of a diamagnetic substance is independent of temperature and is in the range of - 10 to - 100 × 10^{-12} $m^3\ mol^{-1}$. Diamagnetism arises from the action of a magnetic field upon the electrons in a material. Application of a magnetic field changes the velocity of the electrons. A magnetic field is induced in opposition to the applied magnetic field. χ is always negative for a diamagnetic material.

Paramagnetism is related to the orbital angular momentum and the spins of the electrons in a material. The electron in orbit behaves as a current in a wire giving rise to a magnetic moment. Also, the spinning electron behaves as a small magnet giving rise to a magnetic moment. In most molecules the angular momentum of the electrons are directed due to bonding and cannot respond to an external magnetic field. So the moment measured for a molecule in a field is due almost exclusively to the spins of the electrons. The magnetic moment measured for a molecule leads directly to an estimate of the number of unpaired spins. With that simplified explanation, we will leave the introduction to magnetic behavior.

14-2.5 PROBLEMS

1. Calculate the force between two + 1 C charges separated by 1 m in a solvent with a dielectric constant θ of 2.

2. Calculate the dipole moment of the following system if $q = 1.602 \times 10^{-19}$ C. Also, find $\vec{\mu}_x$ and $\vec{\mu}_y$ if $\phi = 42°$; $R = 1 \times 10^{-9}$ m.

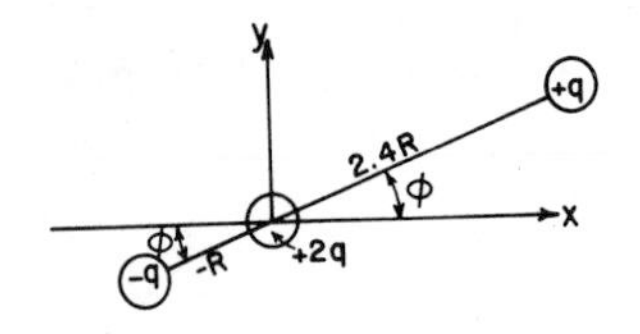

3. From data in Table 14-2.3.1 calculate the molar polarization of N_2 from $\underline{\alpha}$ and from θ. Compare the two values. ρ_{N_2} at - 196°C = 0.81 (kg dm^{-3}).

4. From the following values of the dielectric constant for methanol as a function of temperature find $\mathcal{P}$ for each T. From a suitable plot evaluate $\underline{\alpha}$ and μ for Eq. (14-2.3.9) and compare to the values given in Table 14-2.3.1.

θ	32.63	40	54	64
T (K)	298	253	193	160

5. From the data in Table 14-2.3.1 predict the molar polarization of NH_3 at 298.2 K K.

6. Would you expect NO to be paramagnetic or diamagnetic? Explain.

7. For the compound given in Example 14-2.4.1 find the molar magnetic susceptibility if the molecular weight of the compound is 0.3519 kg mol^{-1} and its density is 0.564×10^3 kg m^{-3}.

8. χ for a sample at 275 K is found to be 4.26×10^{-4}. If the magnetic moment $\mathcal{P}_m$ = 1.91×10^{-23} m^2 A $molecule^{-1}$, the molecular weight is 0.055 kg mol^{-1}, and the density is 0.971 kg dm^{-3}, estimate the value of the magnetic polarizability, α_m.

9. What is the ratio of the force between two particles with charges q_1 and q_2 in a vacuum and in a solvent with a dielectric constant of 8.61.

10. Derive the units on the quantity $\alpha_m k/\mu_0 P_m^2$. Obtain the simplest units you can.

14-2.6 SELF-TEST

1. The molar polarization of HCl at 333 K is 27.5×10^{-6} m^3 mol^{-1}. Given that the dipole moment of HCl is 1.070 Debye, find the molecular polarizability.

2. Discuss briefly the origin of diamagnetic and paramagnetic behavior.

3. In the following system $\vec{\mu}_x$ is found to be 9.8 Debye and $\vec{\mu}_y$ is 6.850. If r = 2.4×10^{-10} m and $q_1 = 1.6 \times 10^{-16}$ C find q_2 and ϕ.

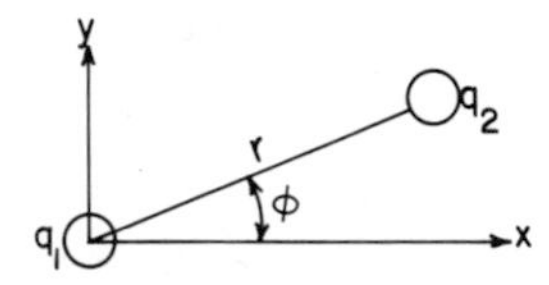

4. Supply the missing steps in obtaining Eq. (14-2.3.6) from (14-2.2.5) and (14-2.3.5).

5. A material with a molar magnetic susceptibility of $\chi_m = 0.220 \times 10^{-6}$ m^3 mol^{-1}, a molecular weight of 0.275 kg mol^{-1}, and a density of 1.15×10^3 kg m^{-3} is placed in a susceptibility apparatus which has a magnetic field strength of 4300 Oersted. The sample area is 3.0×10^{-7} m^2. Find the force exerted on the sample.

14-2.7 ADDITIONAL READING

1.* J. C. Slater and N. H. Frank, *Introduction to Theoretical Physics*, McGraw-Hill, New York, 1933, p. 278.

2. G. Pass and H. Sutcliffe, *J. Chem. Educ.*, *48*:180-181 (1971).

14-3 BONDING IN SOLIDS

OBJECTIVES

You shall be able to

1. Calculate the cohesive energy of a crystal given appropriate data
2. Apply the Born-Haber cycle
3. Discuss the bond and band models for solids
4. Briefly discuss the types of defects in crystals

14-3.1 PRELIMINARY REMARKS

This section will present the basics of solid state theory and introduce several important concepts related to bonding in the solid state. These concepts will not be developed rigorously. They are given to familiarize you with terminology. Chapter 2 gives a fairly detailed investigation of the gaseous state. The first two sections of this chapter are concerned with general properties of condensed phases. If you have not read those it would be useful for you to do so now. The last two sections (14-4 and 14-5) are concerned with identification of and structures of solids. This section will be devoted exclusively to the types of forces holding crystals together and to models for the solid state. It is not necessary to have an understanding of this section to study the next two sections, but such an understanding will be an aid to you.

The types of forces holding solid structures together are ionic, van der Waals, covalent, hydrogen bonding, and metallic. Section 14-1 gives an adequate discussion of hydrogen bonding and van der Waals interactions (induction effects). Covalent bonding, as the force holding the solid together, is important in only a few types of solid substances such as diamond, graphite, silicon carbide, and silica (SiO_2). In polymers (see Chap. 12), covalent bonding is important in the formation of polymer chains. However, except for steric restrictions caused by covalent bonding within the molecules, other types of interaction determine the shape of the crystal and its stability. The covalent bond as related to structure in the solid state will not be considered further. Ionic forces and metallic bonding will receive more detailed treatment below. A good book for further study is C. Kittel, *Introduction to Solid State Physics*, (see Sec. 14-3.8, Ref. 1).

14-3.2 COHESIVE ENERGY: THE BOND MODEL

In ionic crystals the cohesive energy can be calculated satisfactorily for simple ionic substances from Coulomb's law. For two ions (see Sec. 14-2 for comments on units) with charges q_1 and q_2 separated by a distance r_{12}, the attractive energy is given by

$$E_{12}^{att} = \frac{q_1 q_2}{4\pi\varepsilon_0 r_{12}} \qquad (14\text{-}3.2.1)$$

There is also a repulsion term which can be written in the form

$$E_{12}^{rep} = b\ e^{-cr_{12}} \qquad (14\text{-}3.2.2)$$

where b and c are constants. In a crystal we must include the energy terms for all pairs of ions in the crystal. To do that we sum over all pairs and multiply by ½ to keep from counting each pair twice ($E_{ij} = E_{ji}$). This gives

$$E_{cohesion} = \frac{1}{2} \sum_{ij}{}' \frac{q_1 q_2}{4\pi\varepsilon_0 r_{ij}} + \frac{1}{2} b \sum_{ij}{}' \exp(-\ cr_{ij}) \qquad (14\text{-}3.2.3)$$

The prime on the Σ indicates the i = j term has been eliminated. The first sum gives all the attractive interactions a given ion experiences for all other ions in the crystal. Sums of this type were calculated by E. Madelung in 1918. Madelung calculated the sum given in the first part of Eq. (14-3.2.3) and tabulated it as the *Madelung constant M* for different structural types. (See Sec. 14-4 for a description of the structural types.)

The *Coulombic energy of attraction* can be written using the Madelung constant as shown in Eq. (14-3.2.4).

$$E_{coul} = -\frac{Nz_i z_j e^2 M}{4\pi\varepsilon_0 r_0} \qquad (14\text{-}3.2.4)$$

Here the charges on the ions are now designated by $z_i e$ and $z_j e$, z_i and z_j are integers, r_0 is the distance between the nearest oppositely charged ions, N is Avagadro's number, and e is the electronic charge. Table 14-3.2.1 gives some values of the Madelung constant for some important crystal types. The values given are for use in Eq. (14-3.2.4). Some tabulations use equations of slightly different form. Note the form of the equation being used should you ever have to use other tabulations. Eq. (14-3.2.3) can be written now as

$$E_{coh} = -\frac{Ne^2}{4\pi\varepsilon_0 r} M + B\exp\left(-\frac{r}{\rho}\right) \qquad (14\text{-}3.2.5)$$

where B and ρ are empirical constants related to the repulsion constants of Eq. (14-3.2.3). By noting that $\partial E/\partial r = 0$ at equilibrium and by considering the compressibility of the crystal, which is controlled mainly by repulsion between ions, we can evaluate B and ρ. We will not do that. As a rough guess for cohesive energies we can use only the coulombic terms of (14-3.2.5). This usually gives a value about 10% too high. It can be shown that the enthalpy of crystallization is given by:

$$\Delta H_{cryst} = \frac{N e^2 M}{4\pi\varepsilon_0 r_0}\left(1 - \frac{\rho}{r_0}\right) + 2RT \qquad (14\text{-}3.2.6)$$

where $\rho \approx 0.34 \times 10^{-10}$ for alkali halide crystals. Some representative values are given in Table 14-3.3.1 in the next section.

Table 14-3.2.1
Madelung Constants for Several Crystal Types

Structure	Stoichiometry	Coordination	Madelung constant M (no unit)
Rock salt	AB	6-6	1.7476
Cesium chloride	AB	8-8	1.7627
Zincblende	AB	4-4	1.6380
Wurtzite	AB	4-4	1.6413
Fluorite	AB_2		2.5194
Rutile	AB_2		2.3850

14-3.3 THE BORN-HABER CYCLE

The Born-Haber cycle is a thermodynamic cycle that allows us to calculate the *lattice energy* of a crystal. The lattice energy is the energy required for the process

$$AB(cryst) \rightarrow A^+(g) + B^-(g) \tag{14-3.3.1}$$

This value cannot be determined directly, but by outlining an appropriate thermodynamic cycle, we can determine it. LiCl will be used as an example to illustrate the method. The cycle needed to find ΔH_{cryst} is given below.

ΔH_f^0 = standard heat of formation of LiCl

ΔH_{sub} = heat of sublimation of Li

D = dissociation energy of Cl_2 (RT is the PV term, assuming ideal behavior, to convert to enthalpy)

I = ionization potential of Li

A = electron affinity of Cl

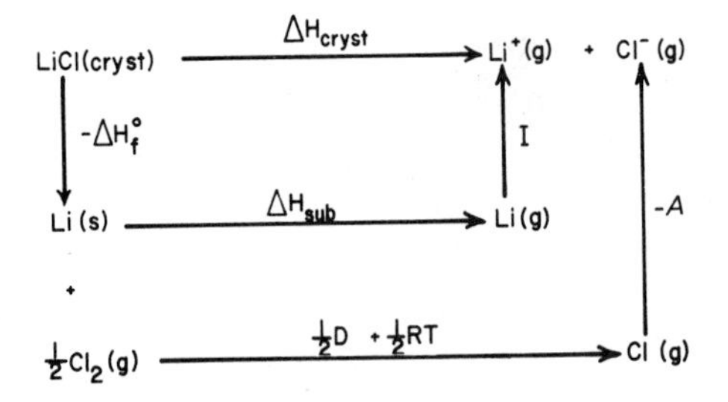

Following normal procedures for such cycles you should be able to show:

$$\Delta H_{cryst} = -\Delta H_f^0 + \Delta H_{sub} + (\tfrac{1}{2}D + \tfrac{1}{2}RT) + I - A \tag{14-3.3.2}$$

Taking values for LiCl from Table 14-3.3.1 we find (all values in kilojoules per mole)

$$\Delta H_{cryst} = 409 + 161 + 122 + 520 - 350$$
$$= \underline{\underline{862 \text{ kJ mol}^{-1}}}$$

The corresponding value calculated theoretically from Eq. (14-3.2.6) is 819 kJ mol^{-1}, which is only 5% below the experimental value. Table 14-3.3.1 contains relevant data and results for the alkali halides. The agreement between theory and experiment is very good for the alkali halides. It is not as good for other systems,

Table 14-3.3.1
Crystal Enthalpies and Related Data for Born-Haber Cycle Calculations[ab]

Crystal	ΔH_f^0	ΔH_{sub} (metal)	I	½(D + RT)	*A*	ΔH_{B-H}	ΔH	r_0 (10^{-10} m)
LiF	- 612	161	520	80	333	1040	1003	2.01
NaF	- 569	108	496	80	333	920	896	2.31
LiCl	- 409	161	520	122	350	862	819	2.57
NaCl	- 411	108	496	122	350	787	759	2.81
KCl	- 436	89	419	122	350	716	689	3.14
NaBr	- 376[c]	108	496	97	330	747	724	2.97
KBr	- 408[c]	89	419	97	330	683	661	3.29
KI	- 359[c]	89	419	77	300	644	621	3.53

[a] This table is taken from G. M. Barrow, *Physical Chemistry*, 3rd ed., McGraw-Hill, New York, 1973, p. 495 (see Sec. 14-3.8, Ref. 2).

[b] All values in kilojoules per mole.

[c] Calculated for gas-phase Br_2 and I_2.

especially in cases where polyatomic anions or cations are present. In these cases the simple point-charge model we have used in inadequate. Even for the alkali halide, the theory is not good enough to predict which crystal structure the material will assume. The energy between two different possible structures is usually small. Uncertainties in the calculated energies of the two different structures is usually greater than that energy difference.

14-3.4 THE BAND MODEL

We have considered one model for solids in the above: the *bond model*. In Sec. 14-3.3 we pictured the solid as made up of regularly spaced ions which were bonded to neighboring ions through coulombic forces. If the solid were made up of molecular units, then we would suppose van der Waals interactions, or hydrogen bonding, or both were holding the crystal together. In each case, the crystal is visualized as being made up of discrete atomic or molecular units held together by one or more forces. There is another approach to bonding in solids. This is the *band model* in which the nuclei are at fixed positions within the crystal. The electrons are then poured into this nuclear matrix and occupy available energy "bands" within the crystal.

In isolated atoms, the electronic energy levels are well defined and can be characterized by a single energy value (see Chap. 7). However, if the nuclei are packed into an ordered solid array, the influence of adjacent nuclei affect the energies of the electronic levels, causing them to "smear out" into a range of

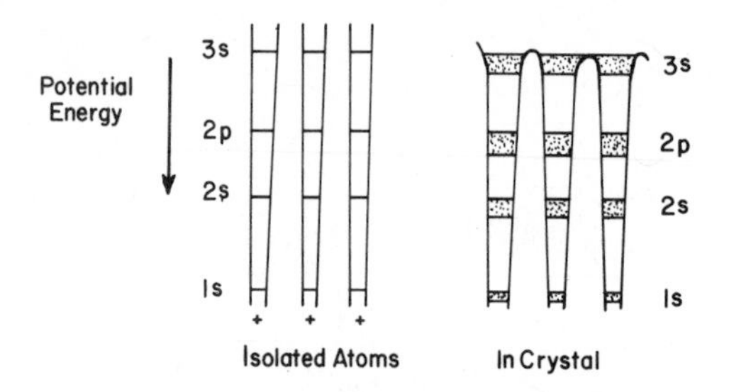

FIG. 14-3.4.1 Energy levels in sodium showing the bands of energies formed in the crystal from the energy levels of the isolated atom. (From W. J. Moore, *Physical Chemistry*, 4th ed., © 1972, p. 837. Reprinted by permission of Prentice-Hall, Inc., Englewood Cliffs, New Jersey.)

possible energies, a *band*. Figure 14-3.4.1 illustrates this for sodium. The detailed treatment of the extent (energy range) of the bands formed requires a quantum mechanical treatment beyond the scope of this chapter. However, the results can be presented in a qualitative fashion. The presentation of the results of the band model is justified, since this model has been very successful in treating such phenomena as conduction and nonconduction as well as semiconduction behavior in solids.

Figure 14-3.4.2 shows how the energy bands in copper change as the internuclear distance in the solid is decreased. At large internuclear distances, discrete energy levels exist. However, as the nuclei approach each other, the discrete levels broaden. This means the allowed ranges of energy for a given electron in a given level increase as the distances between nuclei decreases. Only the upper two energy levels of copper are shown in the figure. Note the overlapping of the bands at the observed internuclear distance. This overlapping of energy bands gives rise to the conduction properties of metals. The uppermost occupied band in a crystal is called the valence band, and the band in which electrons can be moved through the crystal easily is called the conduction band. Figure 14-3.4.3 should show the difference between conducting and nonconducting materials. You should note that the conduction band for Na is partially occupied by electrons. The electrons are readily transported through the metal in this conduction band, making metals such as sodium good electrical (and heat) conductors. In the case of Ne, the conduction band is vacant (meaning the next lowest band is fully occupied). There is no path for electrons to move through the solid. The only way conduction can take place is for an electron to somehow be promoted across the forbidden zone into the conduction band. In nonconductors this requires too much energy and does not occur to an appreciable degree. On the other hand, semiconductors do undergo such a process. As indicated in Fig. 14-3.4.4 for intrinsic semiconductors, the forbidden energy region is not as large as in nonconductors. Thermal energy is sufficient (at some reason-

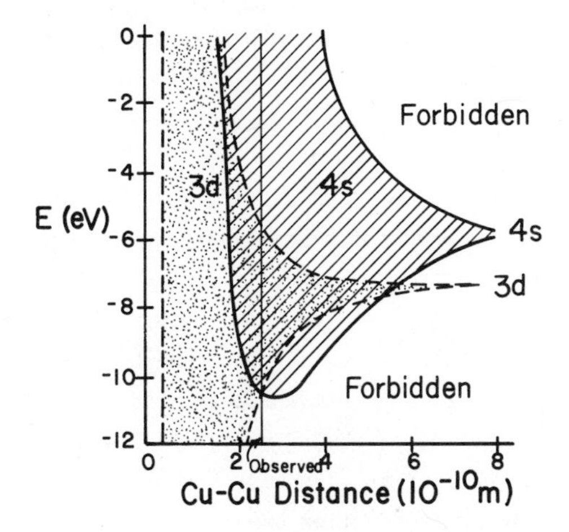

FIG. 14-3.4.2 Energy bands in solid copper. [From H. M. Krutter, *Phys. Rev.*, *48*:664 (1935).]

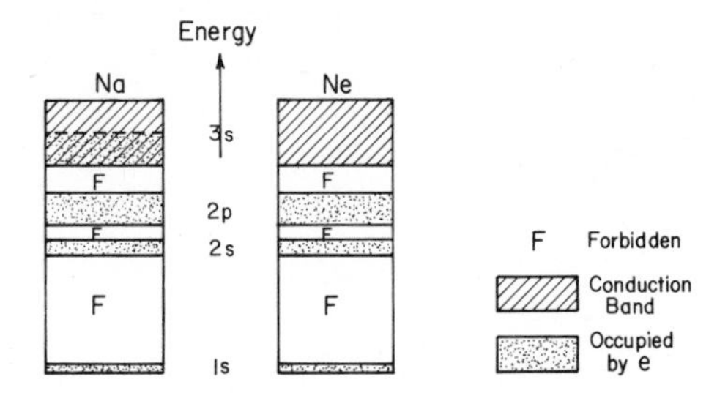

FIG. 14-3.4.3 Energy bands for a conductor and a nonconductor.

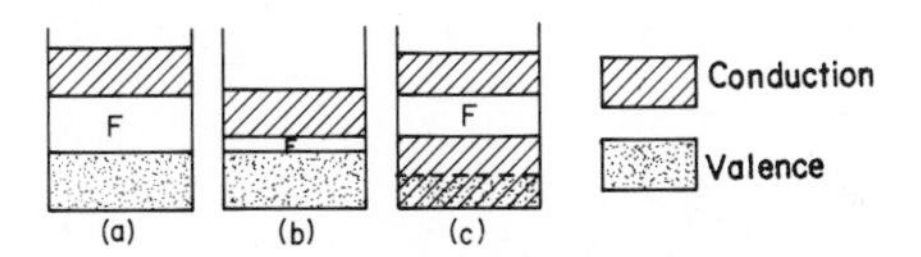

FIG. 14-3.4.4 Schematic band models for solids of different electronic types.

able temperature) to promote some valence electrons into the conduction band in the intrinsic semiconductors. This allows some conduction, but not as much as in metals.

There is one more type of system we must consider. It is of tremendous importance, since a vast array of solid state electronic devices depend on it. This is the *impurity semiconductor*. An insulator (nonconductor) such as ultrapure silicon (Si) or germanium (Ge) can be converted to a semiconductor by the addition of controlled amounts of elements with similar sizes and chemical characteristics. For example, the addition of a very small amount of phosphorus (P) to Si results in the incorporation of the P into the Si lattice. The P has an extra electron which cannot be accomodated by the Si valence band structure. The extra electron establishes an energy band slightly below the Si conduction band, as shown in Fig. 14-3.4.5(a). This is a negative (*n-type*) charge carrier semiconductor. The addition of P results in a system in which the P electron is close to the conduction band, and thermal energy is sufficient to excite some electrons into the conduction band. Another type of semiconductor results when an impurity such as B or Al which has fewer electrons than Si are added to Si. B or Al has one less electron from the Si. If an Si donates an electron to a hole, another hole is created on that Si. In fact, these holes can migrate just like electrons, except in the opposite direction. This type of semiconductor is called a *p-type* (positive holes carry the charge) semiconductor. Figure 14-3.4.5(b) shows a typical case in which an acceptor impurity (one with holes that can accept electrons from the main element in the crystal) is added. In both cases (n-type and p-type), the forbidden zone is no longer forbidden because of the available impurity levels.

14-3.5 DEFECTS IN SOLIDS

Thus far, the assumption has been made that a crystal is a perfectly ordered array of atomic or molecular units. This is not true in real crystals. The possibility of impurities was mentioned in the previous section. Also previously it was pointed

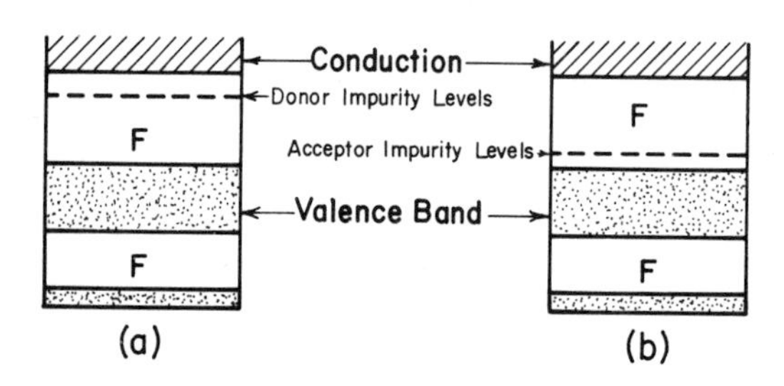

FIG. 14-3.4.5 Electron bands for (a) n-type and (b) p-type semiconductors.

out that many compounds are nonstoichiometric. This situation usually arises when there is an impurity of a different charge from the bulk and a corresponding *vacancy* in the crystal to make up for charge change. For example, in Fe_2O_3, if there are two Fe^{2+} ions in the lattice (in place of the normal Fe^{3+}) then an O^{2-} ion must be missing somewhere within the crystal to compensate for the charge difference and to keep the whole crystal electrically neutral. There are other types of defects also possible. These are point defects and linear defects.

POINT DEFECTS. The two principal point defects are the Frenkel defect and the Schottky defect. The Frenkel defect occurs when an ion (usually the cation due to its smaller size) leaves its lattice site and takes up an interstitial position (interstitial means between points normally occupied by ions). The Schottky defect occurs when there is both an anion and a cation vacancy. (See Sec. 14-4.3 for a description of lattice types in crystals.) In both cases, the overall stoichiometry of the crystal is the same as for the perfect case. These defects help to account for the ease with which atoms can move about in some crystals. It is much easier for an atom to move into a vacancy (thereby creating another vacancy) than it is for two atoms to change places as they would have to do in a perfect array.

LINEAR DEFECTS: DISLOCATIONS. It is observed that metals have an elastic limit (the applied stress at which permanent deformation occurs) that is 10^{-2} to 10^{-4} times what would be predicted for the perfect crystal. This lack of strength is explained on the basis of imperfections in the crystal. Two types of dislocations have been identified: the edge dislocation and the screw dislocation shown in Fig. 14-3.5.1. As is apparent in the edge dislocation, one plane of atoms is terminated at some point, and the other planes converge around the one that is terminated. This defect allows the crystal to be readily deformed under a shear stress. A shear stress applied in the horizontal direction in Fig. 14-3.5.1(a) would readily cause a displacement of the top half of the crystal with respect to the lower half. The screw dislocation occurs when one group of atoms slips relative to another as shown in Fig. 14-3.5.1(b). This is called a screw dislocation because addition of atoms

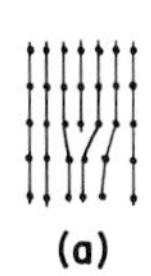

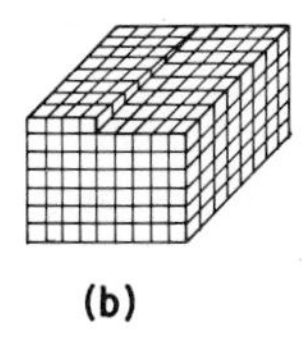

FIG. 14-3.5.1 Schematic of (a) an edge dislocation and (b) a screw dislocation. [Each cube in (b) represents an atom or molecule.]

to this surface during crystal growth leads to the formation of a spiral. (Can you visualize how this occurs?) Screw dislocations are extremely important in the growth of crystals. In the absence of an imperfection, the probability of an atom depositing on the crystal surface is very low. The screw dislocation offers a "parking" place for atoms to become fixed on the crystal. If you have ever grown crystals from solution, you probably have noted the spiral growth patterns that result from screw dislocations. The effect of the addition of a small amount of an alloying substance to a metal in many cases greatly enhances the strength of the metal. This can be explained on the basis of dislocations. When a metal deforms under stress it does so by the migration (and multiplication) of dislocations. A foreign atom tends to take up a position at a dislocation and stabilize it. Defects are also important in the chemical reactivity of a crystal. Crystal growth, as pointed out, tends to occur at dislocations. Likewise, chemical reactivity is greater at a dislocation on the surface of the crystal.

14-3.6 PROBLEMS

1. Assume $\rho = 0.34 \times 10^{-10}$ for all alkali halides. Calculate ΔH_{cryst} for NaCl using the value of r_0 given in Table 14-3.3.1 and compare with ΔH_{cryst} determined from the Born-Haber cycle.
2. For KF find ΔH_{cryst} given that $\Delta H_f^0 = -563$ kJ mol^{-1} and whatever additional data you need from Table 14-3.3.1.
3. Write a summary of the *band* model for solids. Include a discussion of conduction, nonconduction, and semiconduction.
4. Determine ΔH_f^0 for NaCl given the following Born-Haber data (in kJ mol^{-1}): $\Delta H_{sub} = 108$, $I = 496$, $\frac{1}{2}(D + RT) = 122$, $A = 350$, and $\Delta H_{cryst} = 787$.

14-3.7 SELF-TEST

1. Given ΔH_{cryst} of NaI is 700 kJ mol^{-1} and appropriate data from Table 14-3.3.1, find ΔH_f^0 for NaI from gas-phase I_2.
2. Write a one-page discussion of the types of interaction that can exist in the solid state (the *bond* model).
3. Briefly discuss the types of defects present in solids and their importance.

14-3.8 ADDITIONAL READING

1.* C. Kittel, *Introduction to Solid State Physics*, John Wiley and Sons, New York, 1966.

2.* G. Barrow, *Physical Chemistry*, 3rd ed., McGraw-Hill, New York, 1973, p. 495.
3. See Sec. 1-1.9, Ref. A-4, A-5, and A-6.
4. B. Henderson, *Defects in Crystalline Solids*, Crane, Russak and Company, New York, 1972.
5. W. J. Moore, *Seven Solid States*, W. A. Benjamin, New York, 1967.
6. L. H. Van Vlack, *Elements of Materials Science*, 2nd ed., Addison-Wesley, Reading, Mass., 1964.
7. J. S. Blakemore, *Solid State Physics*, W. B. Saunders, Philadelphia, 1969.

14-4 SYMMETRY AND CRYSTAL LATTICES

OBJECTIVES

You shall be able to

1. List the symmetry elements of a given figure, or given a list of symmetry elements, draw a figure with those elements
2. Name and draw the Bravais lattices
3. Define Miller indices, unit cell, and lattice
4. Determine the Miller indices for lines and planes in two- and three-dimensional figures, or given Miller indices draw planes corresponding to these in appropriate figures

14-4.1 INTRODUCTION

Symmetry is a very important property of materials (and mathematical functions) that finds application in many aspects of chemistry. Quantum mechanical equations can many times be simplified by consideration of the symmetry of the molecule being described. The determination of the structure of solids from single crystal X-ray diffraction work is changed from a task that is impossible to one that is merely difficult through the use of symmetry considerations. The interpretation of the electronic spectra of molecules is made tractable because of symmetry relationships. In this section, we will consider only *point symmetry*, defined as all symmetry operations that leave at least one point in the crystal (figure) unchanged. We will not discuss symmetry of mathematical operations, a study called *group theory* which is essential for a thorough study of molecular structure and spectra. In this section, symmetry elements of two- and three-dimensional figures will be considered. Then the notation related to the *unit cell* in crystals, the basic building block of a crystal lattice will be introduced. Also, Miller indices will be defined. These are used in the following section (14-5) in which X-ray diffraction from *planes* within a lattice is discussed.

14-4.2 SYMMETRY

The first thing that must be done is to list the symmetry elements. After that we will consider several examples showing what these elements mean. Table 14-4.2.1 gives the list. An operation gives a valid symmetry element if the figure after the operation is indistinguishable from that before the operation. For example, consider the following figure in which there is a two-fold rotation axis, as designated. A *two-fold* rotation ($180^\circ = 360^\circ/2$) about this designated axis gives (b) from (a). (b) and (a) are identical (the numbers are imaginary; they are included

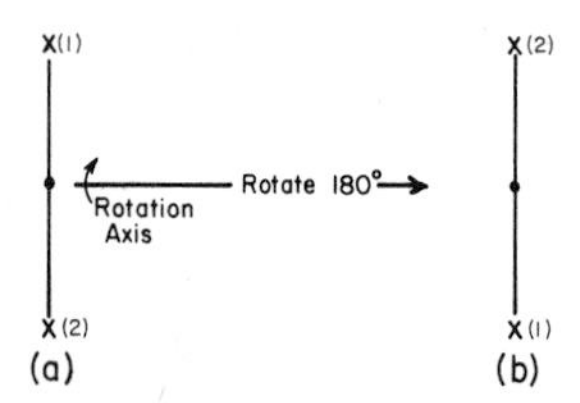

to help you visualize the operation). An n-fold rotation is a rotation of $360^\circ/n$ as you will see below. Now for a few examples.

Identity, 1. Everything has an identity element; rotation of 360° about any axis gives the figure back again.

Three-fold Rotation (axis perpendicular to paper), 3. The 120° rotation leaves the figure unchanged. Note there are 3 two-fold axes present, from the central point through each x. (All x are identical in this and the following figures.)

Center of Symmetry, $\bar{1}$. This operation is a little difficult to imagine at first but after a little thought it should give you no trouble. It is merely the reflection of every point of the figure through an imaginary mirror at the origin. Notice that this, like all the point symmetry elements, leaves one point (at least) unchanged. Every point of the figure is reflected through the origin to a point on the opposite side. The dashed lines in the following figure show how the reflection is achieved.

Table 14-4.2.1
Symmetry Elements[a]

Element	Operation	Symbol[b]
No symmetry (identity)	360° rotation about an axis	1
Center of symmetry	Invert through central point	$\bar{1}$
Two-fold rotation	180° rotation about an axis	2
Three-fold rotation	120° rotation about an axis	3
Three-fold rotation inversion	120° rotation about an axis, then inversion through a central point	$\bar{3}$
Four-fold rotation	90° rotation about an axis	4
Four-fold rotation inversion	90° rotation about an axis, then inversion through a central point	$\bar{4}$
Six-fold rotation	60° rotation about an axis	6
Six-fold rotation inversion	60° rotation about an axis, then inversion through a central point	$\bar{6}$
Perpendicular mirror	Reflection through a mirror perpendicular to a given axis	/m
Parallel mirror	Reflection through a mirror parallel to a given axis	m

[a] Note the absence of five-fold and seven-fold. Such figures cannot be used to fill space,
[b] Hermann-Mauquin symbols.

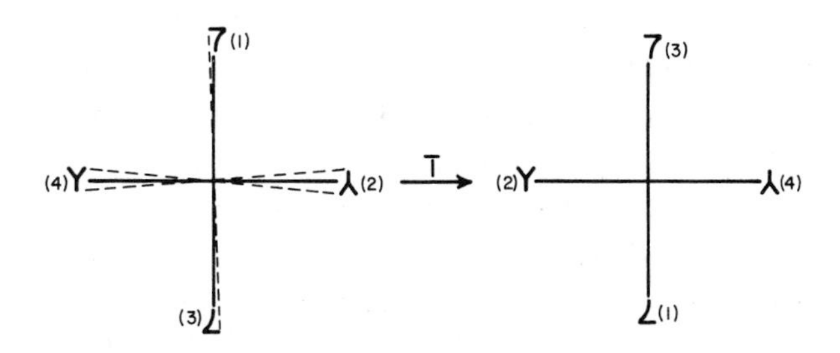

Six-fold Rotation Inversion, $\bar{6}$. This is a two-step operation but is considered as one element. Rotation-inversion is the only two-step operation allowed. X is in the plane, the axis is perpendicular to the paper.

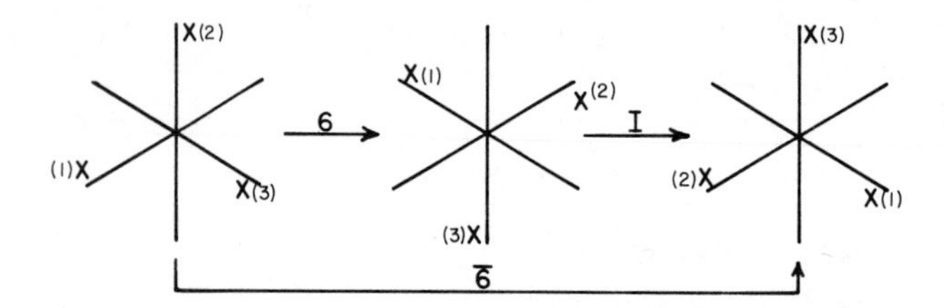

Three-fold with Mirror. So far you have seen only two-dimensional figures. Look at the next one for a three-dimensional example. The rotation axis is shown in (a).

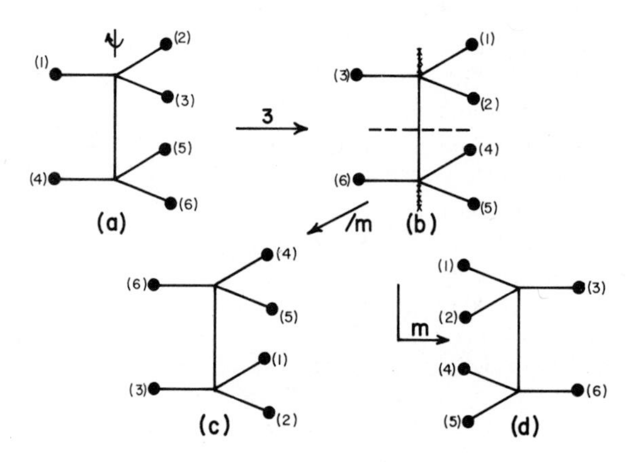

Between (b) and (c) a mirror perpendicular to the three-fold axis has been indicated by the dashed lines. The top half of the figure is reflected through the mirror to the bottom, and the bottom is reflected to the top. (c) is the same as (a) and (b). (Ignore the labeling numbers.) So /m is valid. The xxx along the rotation axis indicates a mirror parallel to the axis and perpendicular to points (3) and (6). This parallel mirror (parallel to the rotation axis) reflects the left side to the right and the right to the left. The figure in (d) results. It is not the same as (a) and (b), so m is not a valid element here. (Prove to yourself that this figure is also $\bar{6}$. Representing the figure as

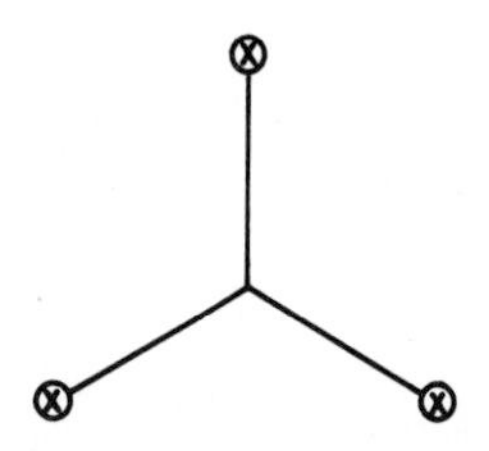

where the x indicates a point above the plane and 0 is a point below the plane might help you visualize this.)

Two-fold with Parallel and Perpendicular Mirrors. The interpretation is left to you.

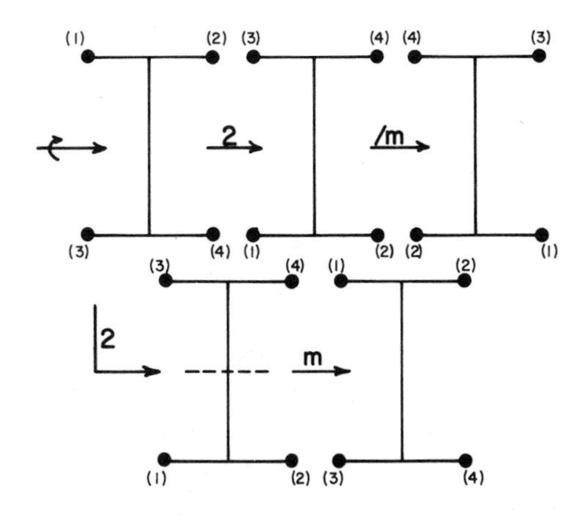

Symmetry of a Cube. All the symmetry elements of a cube are listed below. Try to visualize these with and without a model. If you have trouble see W. J. Moore, *Physical Chemistry* (see Sec. 14-4.7, Ref. 1). The symmetry elements are:

Three 4-fold axes with parallel and perpendicular mirrors
Six 2-fold axes with parallel and perpendicular mirrors
Four 3-fold axes, $\bar{1}$.

14-4.3 POINT GROUPS, CRYSTAL SYSTEMS, AND CLASSES

The operations presented above form the basis for defining possible crystal types. The *unit cell* of a crystal, a collection of atoms or molecules which, by translation along the three principal axes, will reproduce the whole crystal, must be constructed such that a combination of unit cells fills space. This restricts unit cells to a small number of the many possible ways of arranging points. A lattice is an idealization of a crystal in which each group of atoms is visualized as a point. Each molecule or atom at a *lattice* position is reduced to a single point to ease visualization of the positions. The *lattice* of the crystal is broken down into *unit cells*. In 1848, A. Bravais showed there are only 14 possible lattices that can be drawn in three dimensions. Unit cells of these lattices can be defined and are shown in Fig. 14-4.3.1. The unit cells of a lattice are somewhat arbitrary, since in some cases it is possible to choose a simpler unit cell. The more complex unit cell in these cases is kept because of its high symmetry.

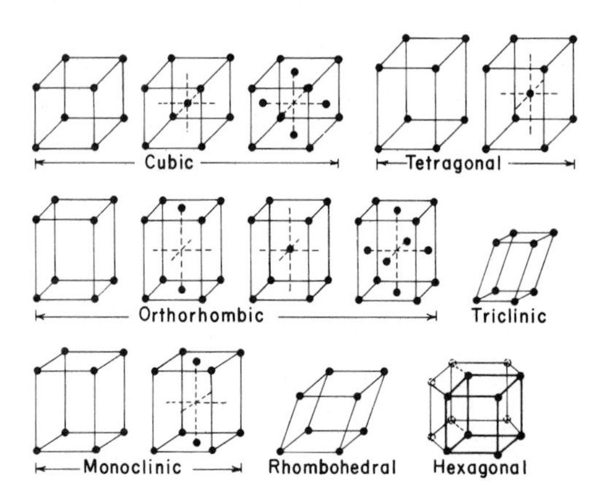

FIG. 14-4.3.1 The 14 Bravais lattices.

A *primitive* P unit cell has points only at the corners. There are eight points in the unit cell, but each point is shared by eight cells, leaving a net number of points of *one* (8 × 1/8) per unit cell. A *face centered* F unit cell has a point at each corner and one on each face. The points at the corners furnish *one* (8 × 1/8) point per unit cell. Each face is shared by two cells, and there are six faces, so there are *three* (6 × 1/2) points per cell from the faces. This gives a total of *four* points per unit cell. In a *body-centered* I unit cell there is a point on each corner and one in the center. This gives a total of *two* points in each unit cell. *End-centered* C unit cells have points at all corners and one on two opposing faces. Each point on the face is shared by two cells, so the end centered points give *one* (2 × 1/2) point to the cell, resulting in *two* points total per unit cell. As can be seen in Fig. 14-4.3.1 crystals can be divided into *seven crystal systems*. These are listed in Table 14-4.3.1. Figure 14-4.3.2 shows how the unit cell parameters are defined. Exactly 32 crystallographic point groups can be formed from symmetry elements that leave at least one point of the crystal invariant, the elements listed in Table 14-4.2.1. This leads to *32 crystal classes* from the *seven crystal systems*. If we now add two additional operations which *translate* the unit cell leaving no points invariant, we have *space operations*. The two new operations combine with the point operations (rotations and reflections) to give *glide planes* and *screw axes*. These will not be defined here. However, it is found that combining point operations and space operations defines *230 space groups*. Every crystal *must* belong to one of these 230 groups. This is a large number of possibilities, but certainly much smaller than the infinite number that might be possible in the absence of symmetry. X-Ray diffraction (Sec. 14-5) is a method of determining the inner structure of crystals. The symmetry of the crystal is extremely important in simplifying the interpretation of X-ray diffraction data.

Table 14-4.3.1
Unit Cell Parameters for the Seven Crystal Systems

System	Number of lattice types	Axes	Angles	Minimum symmetry
Cubic	3	$a = b = c$	$\alpha = \beta = \gamma = 90^\circ$	Four 3-fold axes
Tetragonal	2	$a = b \neq c$	$\alpha = \beta = \gamma = 90^\circ$	One 4-fold axis
Orthorhombic	4	$a \neq b \neq c$	$\alpha = \beta = \gamma = 90^\circ$	Three 2-fold axes
Monoclinic	2	$a \neq b \neq c$	$\alpha = \gamma = 90^\circ;\ \beta \neq 90^\circ$	One 2-fold axis
Rhombohedral	1	$a = b = c$	$\alpha = \beta = \gamma \neq 90^\circ$	One 3-fold axis
Hexagonal	1	$a = b \neq c$	$\alpha = \beta = 90^\circ;\ \gamma = 120^\circ$	One 6-fold axis
Triclinic	1	$a \neq b \neq c$	$\alpha \neq \beta \neq \gamma \neq 90^\circ$	None

14-4.4 CRYSTAL PLANES: MILLER INDICES

In X-ray diffraction, the interaction of the X radiation with a crystal can be characterized by diffraction from imaginary *planes* within the crystal. These planes are defined by groups of atoms or molecules (the lattice points). It is necessary for us to have some way of labeling the planes in the crystal. This is done by the Miller indices. Figure 14-4.4.1 will serve to define the Miller indices. The Miller indices are defined in terms of the unit cell parameters a, b, and c.

$$h = \frac{\text{unit length } a}{\text{intercept on } a} = \frac{a}{OA}$$

$$k = \frac{\text{unit length } b}{\text{intercept on } b} = \frac{b}{OB} \qquad (14\text{-}4.4.1)$$

$$l = \frac{\text{unit length } c}{\text{intercept on } c} = \frac{c}{OC}$$

Some examples in two dimensions, then some more in three dimensions should be sufficient to clarify this concept.

TWO DIMENSIONS. Figure 14-4.4.2 gives a two-dimensional lattice with some lines drawn in. In any extended lattice, there will be an infinite number of lines

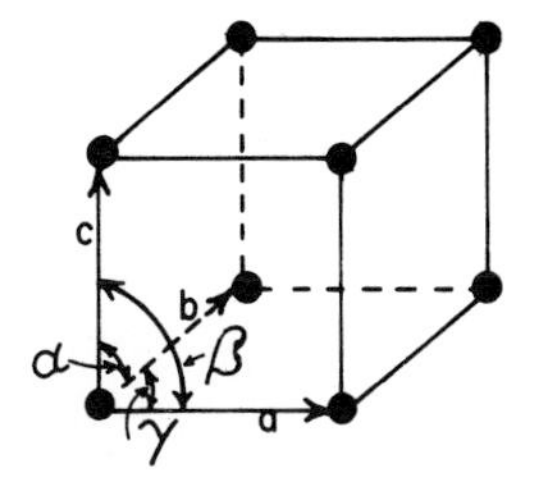

FIG. 14-4.3.2 Definition of unit cell parameters.

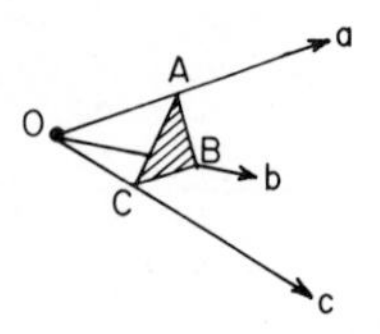

FIG. 14-4.4.1 A plane (ABC) in a unit cell with unit lengths of a, b, and c.

of each type. Only two or three have been drawn for simplicity. The arrows indicate positive a and b directions and the unit lengths of these. Now let us give the Miller indices of each set of lines.

Set A: Origin X(A). The a intercept of the closest line is at a. We must find the b intercept of the *same* line. This line is parallel to b so it intercepts at ∞ b.

$$h = \frac{a}{a \text{ intercept}} = \frac{a}{a} = 1$$

$$k = \frac{b}{b \text{ intercept}} = \frac{b}{\infty b} = 0$$

The Miller indices are (1 0).

Set B: Origin at X(B). The a intercept of the nearest line is a. That same line intercepts the b axis at b.

$$h = \frac{a}{a} = 1$$

$$k = \frac{b}{b} = 1$$

Miller indices (1 1).

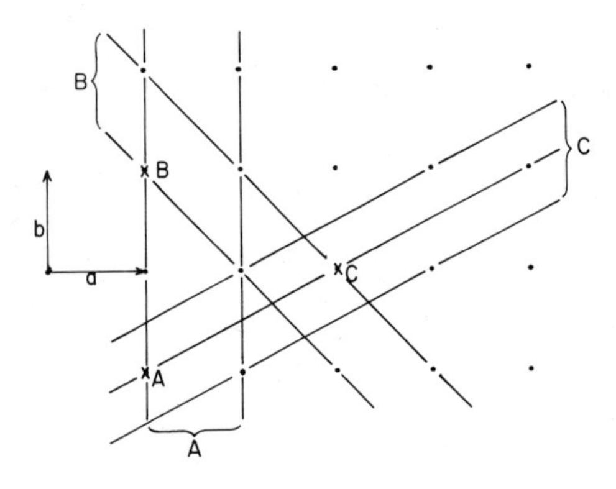

FIG. 14-4.4.2 A two-dimensional lattice with some lines represented.

Set C: Origin at X(C). The a intercept is a. The same line intercepts b at $-\frac{1}{2}$ b.

$$h = \frac{a}{a} = 1$$

$$k = -\frac{b}{b/2} = -2 = \bar{2}$$

Miller indices $(1\ \bar{2})$.

We could just as easily choose b intercept = ½ b, and then the a intercept for the same line would be - a giving $(\bar{1}\ 2)$. Thus $(1\ \bar{2})$ and $(\bar{1}\ 2)$ are equivalent representations of the same set of lines.

THREE DIMENSIONS. Figure 14-4.4.3 shows some planes in a three-dimensional lattice. Assume each plane represents the one of the set nearest the indicated origin. (There are an infinite number of planes parallel to each one shown. Only one of each set has been drawn in the figure.)

Set	Origin	Intercept on a	Intercept on b	Intercept on c	$\frac{a}{\text{a int.}}$ [a]	$\frac{b}{\text{b int.}}$	$\frac{c}{\text{c int.}}$	Miller indices (h k l)
A	X	a	b	c	1	1	1	(1 1 1)
B	X	a	b		1	1	0	(1 1 0)
C	X	2/5 a	1/2 b	1/5 c	5/2	2	5	(5 4 10)[b]
D	Y	1/2 a	1/2 b	- 1/4 c	2	2	- 4	$(2\ 2\ \bar{4})$[c]

[a] int. = intercept.

[b] The smallest set of whole numbers is chosen: (5/2 2 5) becomes (5 4 10).

[c] $(2\ 2\ \bar{4})$ is not the same as $(1\ 1\ \bar{2})$. There are twice as many $(2\ 2\ \bar{4})$ planes as there are $(1\ 1\ \bar{2})$ planes. Prove that for yourself.

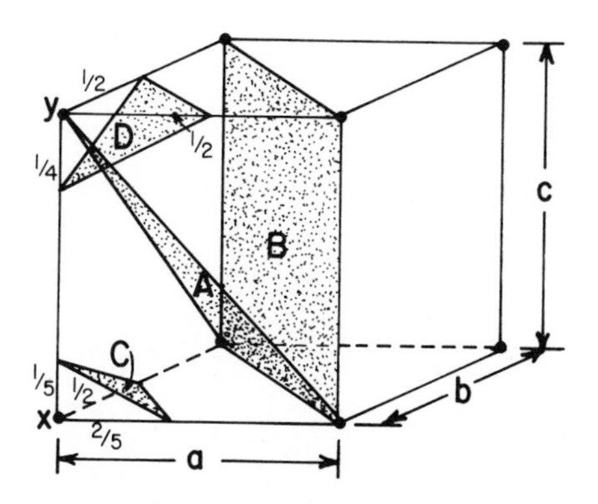

FIG. 14-4.4.3 Some planes in a three-dimensional lattice. X is the origin for planes A, B, and C; Y is the origin for D. The arrows indicate positive direction for a, b, and c.

14-4.5 PROBLEMS

1. List the symmetry elements of the following 3-dimensional figure. An O means a point above the plane of the paper and X means a point the same distance below the plane of the paper.

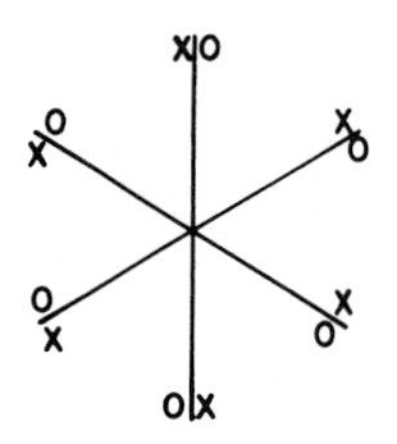

2. Draw a portion of a three-dimensional lattice and sketch in the following planes $(1\ \bar{1}\ 0)$, $(1\ 2\ \bar{2})$, $(2\ 2\ 2)$, $(3\ 2\ 1)$, and $(\bar{1}\ \bar{1}\ 2)$.

3. Determine the Miller indices of the following lines. Choose your own origins.

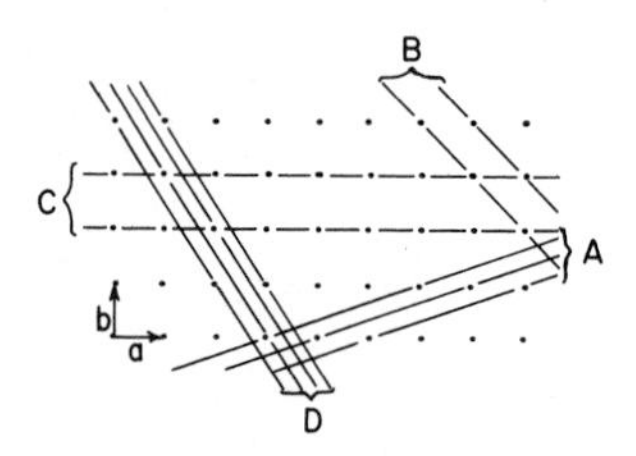

4. List the symmetry elements of the figure given. (Angles are 60°.)

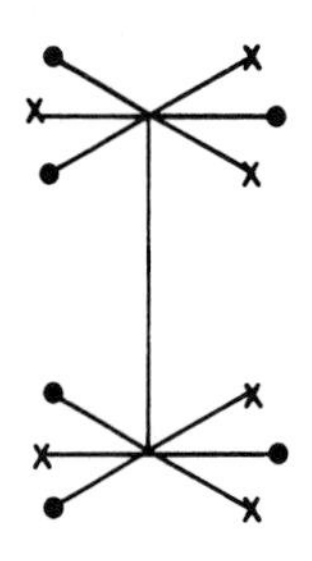

5. Name and draw five Bravais lattices. You must have at least four crystal classes represented.

6. Sketch a three-dimensional figure and draw the $(2\ \bar{2}\ 2)$, $(3\ 3\ 3)$, $(\bar{1}\ \bar{2}\ 2)$, and $(2\ 0\ \bar{2})$ planes.

14-4.6 SELF-TEST

1. List all of the symmetry elements of the following figure. (Angles are 90°.)

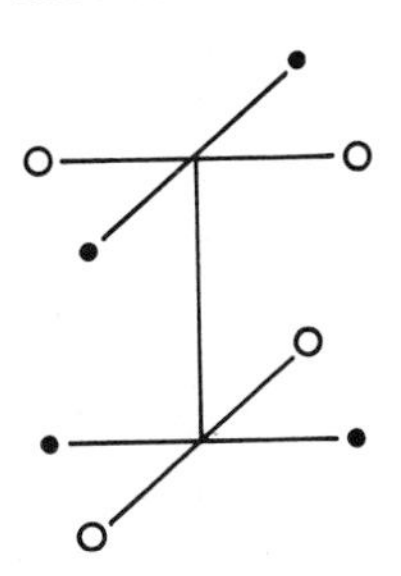

2. Draw the following planes in a two-dimensional lattice: (0 1), (0 $\bar{2}$), (2 $\bar{1}$), (3 2), and ($\bar{2}$ 2).

3. Determine the Miller indices of each plane in the following figure.

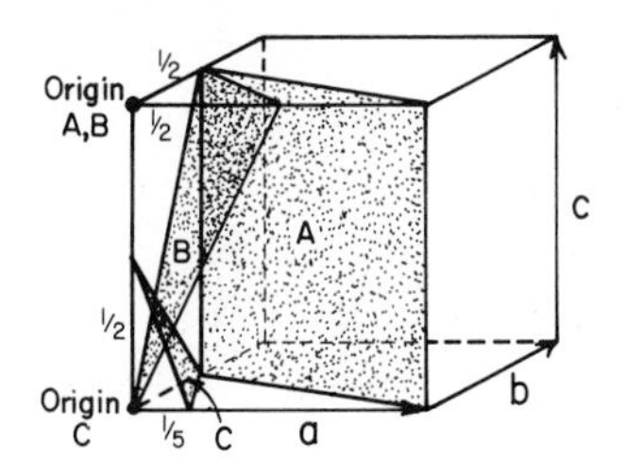

14-4.7 ADDITIONAL READING

1.* W. J. Moore, *Physical Chemistry*, 4th ed., Prentice-Hall, Englewood Cliffs, New Jersey, 1972.

2. W. J. Moore, *Seven Solid States*, W. A. Benjamin, New York, 1967.

14-5 CRYSTAL STRUCTURE AND X-RAY DIFFRACTION

OBJECTIVES

You shall be able to

1. Discuss packing in crystalline solids briefly
2. Draw the common crystal structures or given a structure, name the crystal type
3. Apply the Bragg equation to predict the diffraction angle of a given set of planes in a crystal, or given diffraction angles determine interplanar spacings

4. Given a set of diffraction angles and other appropriate data find the unit cell parameters for a cubic system and the number of molecular units per cell
5. Discuss the procedures for using X-ray diffraction from powdered samples for identification, and/or given diffraction data determine the materials present by using the powder diffraction file.

14-5.1 INTRODUCTION

This section will introduce some basic ideas concerning X-ray diffraction and its use in elucidating the internal structure of crystalline substances. In Secs. 14-3 and 14-4 you have seen some solid state theory (what holds a solid together) and how the ordered arrangement of atoms can be described by a lattice of points. This lattice then defines the planes whose spacings are of interest in X-ray diffraction. We will make use of geometrical considerations to help establish the equations that relate the angle of diffraction of an X-ray to the interplanar spacing as defined by the crystalline lattice. An extremely important tool in identifying crystalline materials will be introduced. A powdered sample of a crystalline material will diffract an X-ray beam at definite angles. (Crystalline is specified, since an amorphous material, one with no long-range order, will not produce sharp diffraction patterns. In some materials, both crystalline and amorphous regions are present. The following discussion is restricted to the crystalline regions.) The angles of diffraction for a given specimen serve as a very useful "fingerprint" for identification purposes. We will not discuss single crystal X-ray diffraction, even though it is the most powerful method available for determining the molecular arrangement in crystals. The technique is now refined to such an extent that proteins and other natural products can be structurally characterized. The method is, however, not simple. It requires painstaking work and large computing capacity to solve the structure of a complex molecule. An introduction can be found in Moore, Chap. 18, Secs. 14 and 15 (see Sec. 14-5.9, Ref. 1).

14-5.2 CLOSE-PACKED SPHERES AND EXAMPLES OF COMMON STRUCTURES

CLOSE PACKING. Many of the structures in the solid state, especially the metal oxides, are based on the close-packed structure. This is the structure that results if spheres of identical size are stacked in layers. There are two ways in which identical spheres can be packed. These are shown in Fig. 14-5.2.1. After the spheres in the first layer are placed in contact with each other, there is only one way to add the second layer. So, the first two layers in Fig. 14-5.2.1(a) and (b) are the same. The difference between the two figures is in the manner in which the third layer is added. In part (a) the spheres of the third layer are directly over

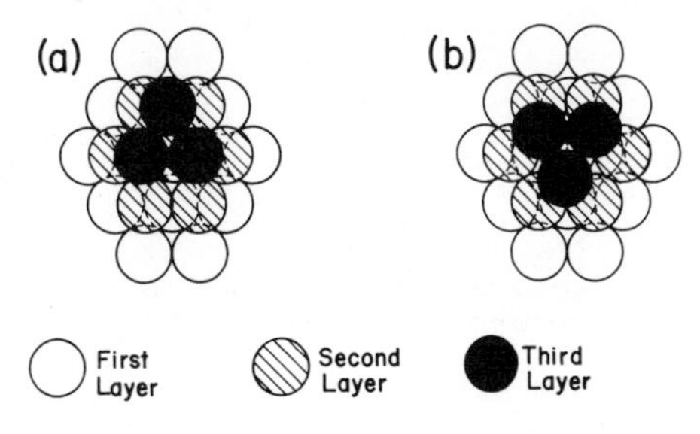

FIG. 14-5.2.1 Hexagonal close packing (a) and cubic close packing (b) of identical spheres.

the spheres of the first layer, giving a packing sequence of A B A B A B .... (The A layers are directly above and below each other. The B layers are directly above and below each other but offset from the spheres in the A layers.) In part (b) the sequence is A B C A B C A B C .... The third layer of spheres is above the "holes" between spheres in the first layer. The fourth layer spheres are directly above the spheres of the first layer. You might try to see the difference by stacking coins or marbles. All the coins must be the same size, of course.

An atom in the close packed structure has 12 nearest neighbors (atoms closest to it, all the same distance away). Six are in the same plane, three above and three below. A face-centered cubic (fcc) unit cell is a cubic close-packed structure as you can tell if you view a fcc structure along the body diagonal. (For a more complete exposition of structure and packing see W. F. Sheehan, *Physical Chemistry*.) (See Sec. 14-5.9, Ref. 2.) Many elements and, as mentioned, many metal oxides, crystallize in the close-packed structure. In both cases such a result is reasonable. Pure elements can be visualized as small spheres which can easily pack in the close-packed structure. For metal oxides, the size of the O^{2-} ion, in general, is much greater than the associated metal cations. The large, spherical O^{2-} ions tend to form close packed structures. The small cations distribute themselves in the various vacancies left after the spheres are packed as tightly as possible. About 74% of available volume is actually occupied by spheres in an ideal close-packed structure. In fact, for spheres of any size, only ≈ 74% of the volume can be occupied. The remaining 26% is void space between the spheres. The vacancies are of three types:

1. Trigonal: surrounded by three oxide ions (or other ions). The radius of a cation that will fit into this vacancy is 0.1547 times the radius of the close-packed ion. (Try your geometry to show this result is correct.)
2. Tetrahedral: surrounded by four spheres, three forming the base, and the fourth forming the apex of a tetrahedron. Can you identify such a vacancy in Fig. 14-5.2.1? The largest cation that will fit in such a void is 0.2247 times the radius of the close-packed spheres.

3. Octahedral: surrounded by six spheres, four in a plane, one above and one below. This is difficult to visualize from Fig. 14-5.2.1. The voids are between the layers with three spheres above and three below. The largest ion that will fit in the octahedral hole is 0.4142 times the size of the packing spheres.

In all cases, somewhat larger cations can be accomodated due to the ability of the anion structure to be "expanded" a little by the pressure of a cation that is a little too large for the void.

The presence of these voids into which the cations can be "slipped" to balance the charge on the anions leads to a marked *nonstoichiometry* in the metal oxides (and sulfides and others). For example, the compound FeO is never found as FeO. It usually occurs as $Fe_{0.95}O$. (This occurs because a few Fe^{2+} ions are replaced by Fe^{3+} ions.) In Fe_2O_3 the Fe^{3+} is distributed in the octahedral holes. In Fe_3O_4 $[Fe(II)Fe(III)_2O_4]$ the Fe^{2+} ions are found in the octahedral holes, and Fe^{3+} ions are distributed equally in the octahedral and tetrahedral holes. Both these last two compounds are not exactly of the composition given (Fe_2O_3, Fe_3O_4). Instead, the stoichiometry will vary somewhat depending on the method of preparation. The point to be remembered is that *nonstoichiometry is more likely to be the rule than the exception* in solids.

COMMON STRUCTURAL TYPES. Below structures for the common crystal types will be shown. These will be divided into groups labeled AX, AX_2, and Perovskite. AX indicates a cation to anion ratio of 1:1. AX_2 means a ratio of cations to anions of 1:2. You will see later what Perovskite indicates. Very little explanation will be given with the following figures. The intent here is to familarize you with some common solid state structures.

AX Structures. Figure 14-5.2.2 shows the rock salt structure type of which NaCl is the most common example. In this structure, each anion and cation has a coordination number of *six*, meaning each is surrounded by six nearest neighbors (in an octahedral arrangement) of the opposite charge. More than 100 compounds crystallize into this structure. The anions can be visualized as cubic close-packed with

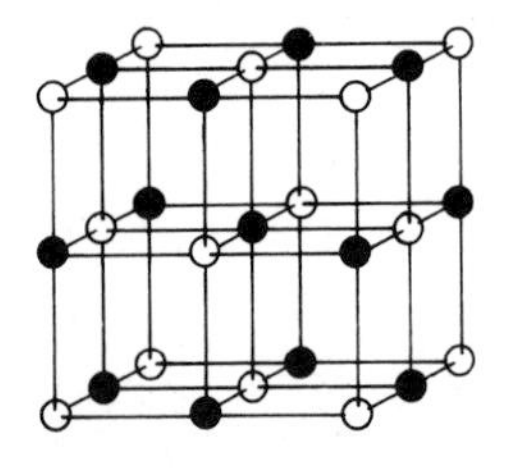

FIG. 14-5.2.2 The rock-salt structure; ● indicates Na^+ and O indicates Cl^-.

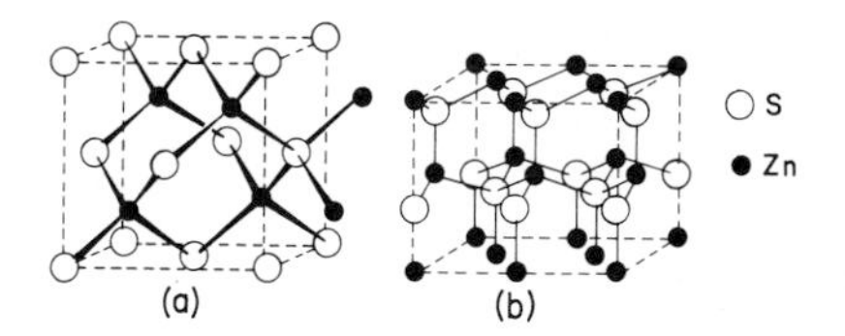

FIG. 14-5.2.3 The crystal structure of (a) zincblende, and (b) wurtzite.

cations in octahedral voids. Another structure with cubic close-packed anions is the ZnS (zincblende or sphalerite) structure shown in Fig. 14-5.2.3(a). The Zn^{2+} ions occupy half the tetrahedral voids formed by the close-packed S^{2-}. About 40 compounds crystallize in this structure. ZnS can also crystallize in a hexagonal close-packed array as shown in Fig. 14-5.2.3(b). About 25 compounds form structures of this type. In each of these two structures, the cation fills some of the tetrahedral voids and has, therefore, a coordination number of *four*. The final AX type we will consider is the CsCl structure shown in Fig. 14-5.2.4. The Cs^+ ion is located at the center of the unit cell with a chloride at each corner. Hence the Cs^+ has a coordination number of *eight*.

AX_2 Structures. The most common AX_2 structure is silica, shown in Fig. 14-5.2.5. In this structure, the oxide ions are close-packed with the silicon ion located in tetrahedral positions. Appropriate substitutions and rearrangements lead to a vast array of silicate minerals. Silicon has a coordination number of *four*. Rutile, TiO_2 is shown in Fig. 14-5.2.6. Each Ti is octahedrally coordinated, and each oxygen is surrounded by a trigonal-planar group of three Ti ions. The third common AX_2 type is fluorite CaF_2 shown in Fig. 14-5.2.7. Each Ca^{2+} is surrounded by eight F^-, and each F^- by four Ca^{2+} ions. The Ca^{2+} has, therefore, a coordination number of *eight*. Note the similarities and differences between this and the CsCl structure.

PEROVSKITE AND OTHER CLOSE-PACKED STRUCTURES. *Perovskite* has the general formula ABX_3 in which A and B are two different cations. X is usually O^{2-}. Figure 14-5.2.8 shows the ideal perovskite structure (B is much smaller than A). In this

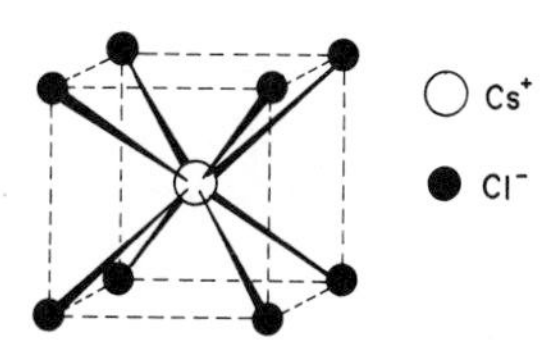

FIG. 14-5.2.4 The CsCl structure.

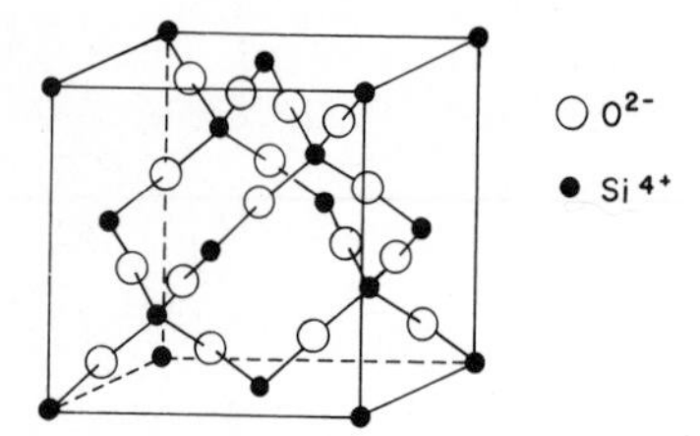

FIG. 14-5.2.5 Structure of silica (β, cristobalite).

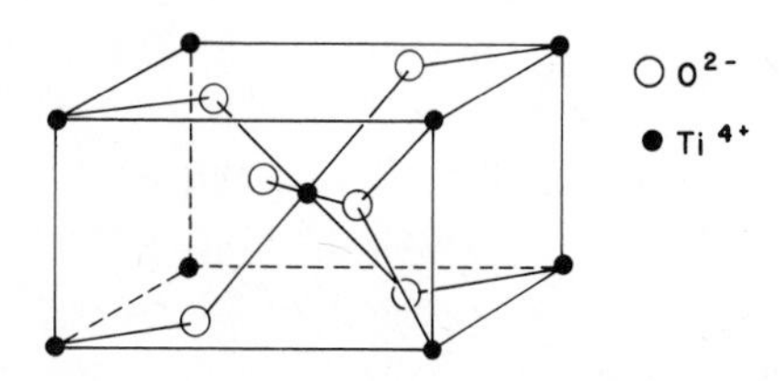

FIG. 14-5.2.6 The rutile structure.

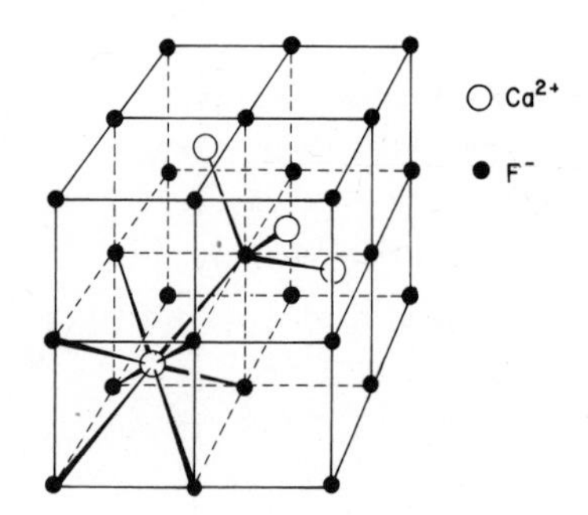

FIG. 14-5.2.7 The fluorite structure.

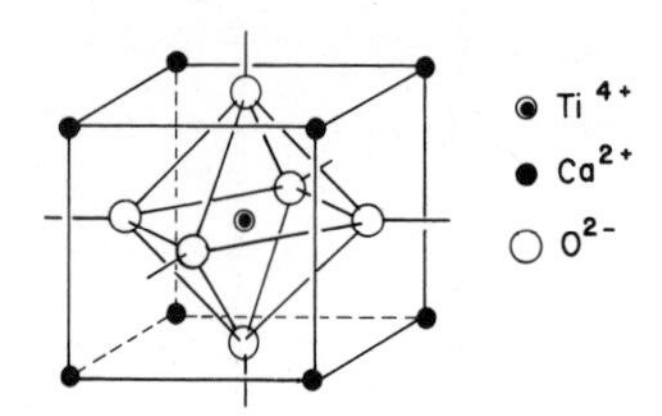

FIG. 14-5.2.8 Ideal perovskite structure.

case the CaO_3 units are cubic close packed with Ti^{4+} in the octahedral holes. Perhaps the most important example of a perovskite structure is $BaTiO_3$. Barium titanate is a ferroelectric material which finds extensive use in capacitors and other electronic devices. The remarkable ferroelectric properties of $BaTiO_3$ is due to the very large size of Ba^{2+} and O^{2-} relative to Ti^{4+}. The octahedral hole is large enough for the Ti to "rattle around" in it. In fact, the Ti is displaced a few hundredths of an angstrom from the center of the octahedral cavity leading to a material whose center of electric charge does not coincide with its center of symmetry. $BaTiO_3$, as a result, is electrically polarized. Imposing an electric field leads to even larger polarizations, and thereby the capacitor properties of $BaTiO_3$.

Two other close-packed structures are *spinel* and *ilmenite*. Fe_3O_4, discussed previously, is a spinel. The general formula of a spinel is AB_2O_4 with ccp O^{2-}. A ions are found in the tetrahedral voids, and B ions in the octahedral voids. Finally, *ilmenite* has the formula ABO_3 (as does perovskite). Ilmenite is a slightly distorted hcp O^{2-} structure with A and B ions distributed in the octahedral voids. Examples are $FeTiO_3$, $CoTiO_3$, and α-Fe_2O_3 (discussed previously; in this case A = B).

14-5.3 BRAGG'S X-RAY DIFFRACTION LAW

We now begin the discussion of how X-ray diffraction from solids can be used to help identify and characterize crystalline materials. Geometrical arguments will be used to derive the basic diffraction law that applies to crystalline materials. (If you would prefer a mathematical proof of Bragg's law, see W. J. Moore, *Physical Chemistry* (Sec. 14-5.9, Ref. 1)). We have already idealized a crystal as a lattice of points. These points, as we have seen, form planes within the structure which we can identify by Miller indices. X-Rays, which are directed toward a crystal, interact with the electrons of the atoms in the lattice and behave as though they were being reflected from the planes of atoms within the lattice. Using this model, we can easily formulate the Bragg diffraction law.

Consider a set of planes defined by the lattice points shown in Fig. 14-5.3.1. Some X-rays impinging on the crystal are reflected by the first plane, and others penetrate to the second, third, etc., planes and are diffracted. Look now at only the first two planes. X-Rays coming into the crystal are in phase. The X-rays which are reflected will interfere with each other either constructively or destructively depending on their relative phases. If the X-rays from each of the planes are in phase, constructive interference will result, and X-rays can be detected coming out of the crystal. If the X-rays from the planes are out of phase, destructive interference results and no X-rays are observed coming out. For the X-rays shown in Fig. 14-5.3.1 to be in phase, the extra distance traveled by those X-rays diffracted from the second plane must equal an integral number of wavelengths (X-rays

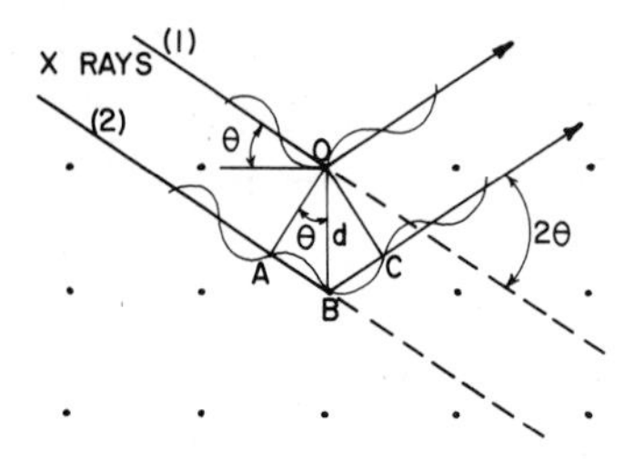

FIG. 14-5.3.1 Diffraction of X-rays from a lattice of atoms.

at points O and A are in phase. To be in phase at O and C, the distance AB + BC must equal $n\lambda$ where n is an integer and λ is the wavelength of the X-ray.). We have then

$$n\lambda = AB + BC \tag{14-5.3.1}$$

But AB = BC = $d \sin \theta$ from the geometry of the system (θ is the angle of incidence and also the angle between OA and OB). This leads to the result

$$n\lambda = 2d \sin \theta \tag{14-5.3.2}$$

which is the *Bragg diffraction law*. The integer n is associated with the order of the diffraction. If we limit ourselves to the first-order diffractions (n = 1) and associate d with the spacing between planes with Miller indices (h k 1), we can write

$$\boxed{\lambda = 2d_{hk1} \sin \theta} \tag{14-5.3.3}$$

Now, if we can find a relationship between (h k 1) and d_{hk1} we will have a valuable equation. Given a set of (h k 1) planes and X-ray of wavelength λ (fixed by the X-ray source chosen), we can find d_{hk1} by observing values for θ and calculating from Eq. (14-5.3.3). This is done in the next section.

14-5.4 LATTICE TYPES AND UNIT CELL DIMENSIONS FROM X-RAY DIFFRACTION

If a powdered specimen of a crystalline material is placed in a diffraction apparatus (see Sec. 14-5.5 for details of such an apparatus), X-ray diffraction maxima will be observed for various values of the diffraction angle θ. We can determine d_{hk1} directly from (14-5.3.3) if we know which set of planes is involved for a particular diffraction angle. Usually it is very difficult to assign a set of θ values to a set of (h k 1) planes. If we can make the assignment (see Sec. 14-5.5 and Example 14-5.5.1 for the procedures for making an assignment), we can determine the unit cell parameters (values of the unit cell dimensions). It can be shown, from geometrical arguments, that (if all angles are 90°)

$$\frac{1}{d_{hkl}} = \left(\frac{h^2}{a^2} + \frac{k^2}{b^2} + \frac{l^2}{c^2}\right)^{\frac{1}{2}} \tag{14-5.4.1}$$

If we know several values of Θ, we can find values of d_{hkl} for several (h k l) planes from (14-5.3.3). This allows calculation of a, b, and c from (14-5.4.1). For a cubic system Eq. (14-5.4.1) reduces to a simpler form, since a = b = c.

$$\frac{1}{d_{hkl}} = \frac{1}{a}(h^2 + k^2 + l^2)^{\frac{1}{2}} \tag{14-5.4.2}$$

or

$$d_{hkl} = \frac{a}{(h^2 + k^2 + l^2)^{\frac{1}{2}}} \tag{14-5.4.3}$$

All we need for a cubic system is one known diffraction angle for one known set of (h k l) planes.

14-5.4.1 EXAMPLE

Silver is known to form a cubic close-packed structure. The smallest observed diffraction angle is at $\theta = 19.076°$. This diffraction is associated with the (1 1 1) plane. Find the cubic unit cell parameter a. A copper X-ray target was used giving X-rays with $\lambda = 1.5418$ Å $= 1.5418 \times 10^{-10}$ m.

Solution

From (14-5.3.3),

$$d_{111} = \frac{\lambda}{2 \sin \theta} = \frac{1.5418 \text{ Å}}{2 \sin 19.076}$$

$$= 2.359 \text{ Å}$$

[*Note*: Angstrom units (Å) (10^{-10} m) will be used throughout this and the next section. Almost all X-ray diffraction data are in this unit.] Now applying (14-5.4.3):

$$d_{111} = 2.359 \text{ Å} = \frac{a}{(h^2 + k^2 + l^2)^{\frac{1}{2}}}$$

$$= \frac{a}{(1 + 1 + 1)^{\frac{1}{2}}} = \frac{a}{(3)^{\frac{1}{2}}}$$

$$a = \underline{\underline{4.086 \text{ Å}}}$$

14-5.5 ATOMS PER UNIT CELL AND ALLOWED DIFFRACTIONS

ATOMS PER UNIT CELL. Given the dimensions of a unit cell and the density of a material we can find the number of atoms in the unit cell. Let us use the silver example just solved. Given $\rho_{Ag} = 10500 \text{ kg m}^{-3}$ and $M = 0.10787 \text{ kg mol}^{-1}$ we can determine the effective volume of each Ag atom from the molecular weight and density.

$$V_{Ag} = \frac{M}{\rho} = \frac{0.10787 \text{ kg mol}^{-1}}{10500 \text{ kg m}^{-3}} \frac{1}{6.023 \times 10^{23} \text{ atom mol}^{-1}}$$

$$= (1.706 \times 10^{-29} \text{ m}^3 \text{ atom}^{-1}) \left(\frac{10^{10} \text{ Å}}{1 \text{ m}}\right)^3$$

$$= \underline{1.706 \times 10^1 \text{ Å}^3 \text{ atom}^{-1}}$$

The volume of a cubic unit cell is a^3.

$$V_{cell} = a^3 = (4.086 \text{ Å})^3$$

$$= \underline{6.82 \times 10^1 \text{ Å}^3 \text{ (unit cell)}^{-1}}$$

The number of atoms per unit cell Z is

$$Z = \frac{V_{cell}}{V_{Ag}} = \frac{68.2 \text{ Å}^3 \text{ (unit cell)}^{-1}}{17.06 \text{ Å}^3 \text{ (atom)}^{-1}}$$

$$= \underline{4.0 \text{ atom (unit cell)}^{-1}}$$

There are four Ag atoms per unit cell, which tells us Ag is f c c. If we do not know the assignment of (h k l) for observed values of θ, we have to do a little trial and error work. Due to interference effects of atoms not included in some planes, we find that some (h k l) planes do not give diffractions. Table 14-5.5.1 gives the allowed reflections for several (h k l) indices for a cubic cell. An example will show how θ values can be used to find an assignment for a simple system.

14-5.5.1 EXAMPLE

Assume MgO is known to form a cubic unit cell, but the type of cell is not known. Given the following data for the first five observed θ values and $\rho = 3580 \text{ kg m}^{-3}$ and $M = 0.04031 \text{ kg mol}^{-1}$, find the type of cubic cell, the unit cell dimension, and the number of molecules per unit cell ($\lambda = 1.5418$): θ = 18.5°, 21.5°, 31.2°, 37.4°, 39.4°.

Solution

Find the corresponding d_{hkl} values from the Bragg equation.

Table 14-5.5.1
Indices for Cubic Crystals

Index (h k l)	$(h^2 + k^2 + l^2)$[a]	Cubic lattice type that allows reflection[b]
1 0 0	1	P
1 1 0	2	P I
1 1 1	3	P F
2 0 0	4	P F I
2 1 0	5	P
2 1 1	6	P I
2 2 0	8	P F I
3 0 0, 2 2 1	9	P
3 1 0	10	P I
3 1 1	11	P F
2 2 2	12	P F I
3 2 0	13	P

[a] Note absence of $(h^2 + k^2 + l^2) = 7$

[b] P = primitive, Z = 1; I = body centered, Z = 2; F = face centered, Z = 4.

For $\theta = 18.5°$

$$d_{hkl} = \frac{1.5418}{2 \sin 18.5} = \underline{2.43 \text{ Å}}$$

The other values are tabulated below. Now we start our guessing.

1. If the unit cell is primitive, the first allowed (h k l) is 1 0 0 (See Table 14-5.5.1). Solve (14-5.4.3):

$$a_{100} = d_{100} (h^2 + k^2 + l^2)^{1/2}$$

$$= d_{100} (1 + 0 + 0)^{1/2} = 2.43 \text{ Å}$$

The other values are tabulated.

2. If the unit cell is body-centered, the first allowed (h k l) is 1 1 0.

$$a_{110} = d_{110} (1^2 + 1^2 + 0^2)^{1/2}$$

$$= \sqrt{2}\, d_{110} = 3.44 \text{ Å}$$

See below for other allowed (h k l).

3. Finally, if the unit cell is face centered, the first (h k l) is 1 1 1.

$$a_{111} = d_{111} (1^2 + 1^2 + 1^2)^{1/2}$$

$$= \sqrt{3}\, d_{111} = 4.21 \text{ Å}$$

The calculated a values for the first five allowed (h k l) combinations for each type of lattice are given below.

		If P		If I		If F	
θ	d_{hkl} (Å)	(h k l)	a (Å)	(h k l)	a (Å)	(h k l)	a (Å)
18.5°	2.43	1 0 0	2.43	1 1 0	3.44	1 1 1	4.21
21.5°	2.10	1 1 0	2.97	2 0 0	4.21	2 0 0	4.21
31.2°	1.49	1 1 1	2.58	2 1 1	3.65	2 2 0	4.21
37.4°	1.27	2 0 0	2.54	2 2 0	3.59	3 1 1	4.21
39.4°	1.21	2 1 0	2.71	3 1 0	3.84	2 2 2	4.21

It should be obvious that calculated a values are consistent only for the face-centered case. So,

$$a = \underline{\underline{4.21\ \text{Å}}}$$

Since it is face centered we know there are four molecules per unit cell, but let us prove it.

$$V_{MgO} = \frac{M}{\rho} = \frac{(0.04031\ \text{kg mol}^{-1})(10^{30}\ \text{Å}^3\ \text{m}^{-3})}{(3580\ \text{kg m}^{-3})(6.023 \times 10^{23}\ \text{molecules mol}^{-1})}$$

$$= 1.87 \times 10^{1}\ \text{Å}^3\ \text{molecule}^{-1}$$

$$V_{cell} = a^3 = (4.21\ \text{Å})^3 = 74.6\ \text{Å}^3\ (\text{cell})^{-1}$$

$$\frac{V_{cell}}{V_{MgO}} = \frac{74.6\ \text{Å}^3\ \text{cell}^{-1}}{18.7\ \text{Å}^3\ \text{molecule}^{-1}} = \underline{\underline{3.99\ \text{molecules cell}^{-1}}}$$

So we find 4.0 molecules per unit cell, as expected for a f c c cell.

14-5.6 POWDER DIFFRACTION: CRYSTAL "FINGERPRINTS"

In the previous section we saw how to find unit cell parameters from the X-ray diffraction angles from a powder. Except for cubic systems, the assignment of (h k l) values to the observed diffractions is usually too difficult to be practical. Thus, the major use of X-ray diffraction data is for the identification of samples. If diffraction angles (and d_{hkl}) are tabulated for known materials, an unknown can be identified by comparison. After a description of typical X-ray diffraction equipment, an example will be worked through indicating how a *powder pattern* can be used to identify an unknown powdered sample.

DIFFRACTION APPARATUS. Figure 14-5.6.1 shows a schematic of a diffraction

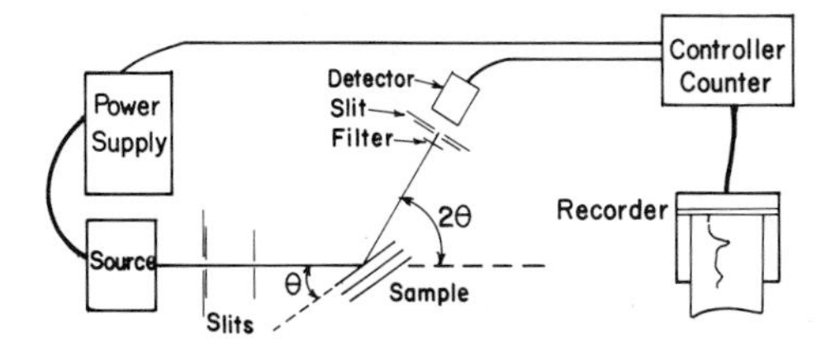

FIG. 14-5.6.1 Schematic of a simple X-ray diffraction apparatus.

apparatus. A source produces a beam of X-rays whose wavelength depends on the target material in the source. Copper is a common target and produces X-rays with λ = 1.5418 Å. Slits collimate the beam and direct it to a powdered specimen mounted on a rotating sample holder. Diffracted X-rays are detected by the detector that rotates synchronously with the sample. The counter converts the detected X-rays into a signal that can be presented on a recorder or other output device. The sample holder and detector rotate together, so that as the sample rotates through an angle θ with respect to the X-ray beam, the detector rotates 2θ. The recorder output is then given as *intensity* I of the X-rays detected vs. 2θ. Only specific orientations of the sample with respect to the beam (those orientations that fulfill the Bragg condition) will diffract the X-ray beam to the detector. The angle of the diffraction is related directly to the d spacing of some set of planes within the crystal. Once we have the *intensity* (I) vs. 2θ values we can convert these to corresponding d spacings. (This is necessary so data from different sources can be compared. For a given d spacing, the 2θ value observed depends on the wavelength of the X-rays used. If we use a copper target in our apparatus, and someone else uses an iron target, the 2θ values observed for the same sample will be different. However, the calculated d spacings will be the same.) The following example will indicate the procedure required to identify an unknown sample given I vs. 2θ data determined experimentally. Figure 14-5.6.2 gives a typical recorder output for a scan.

14-5.6.1 EXAMPLE

The following data were taken from a recorder output for a powdered sample. The X-ray source was a copper target (λ = 1.5418 Å). The intensities listed are relative intensities with the largest diffraction peak assigned a value of 100. Values for d are calculated from the Bragg law:

$$d = \frac{\lambda}{2 \sin \theta}$$

Several weak peaks are not listed. Identify the sample.

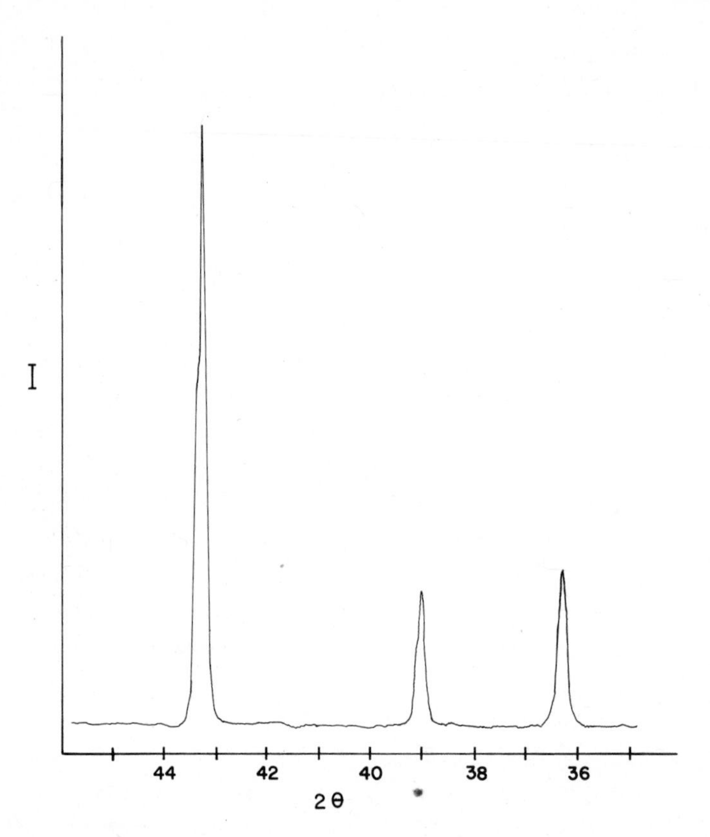

FIG. 14-5.6.2 A portion of a typical recorder output for an X-ray diffraction scan (for Zn powder; the 2θ angles are for Cu Kα radiation.)

2θ	d (Å)	I
19.8	4.48	50
25.7	3.46	10
29.5	3.03	40
31.4	2.86	15
38.8	2.32	100
39.0	2.31	30
47.9	1.90	20
55.7	1.65	60
70.2	1.34	15

Solution

Rearrange these in decreasing order of intensity.

I	100	60	50	40	30	20	15	15	10
d	2.32	1.65	4.48	3.03	2.31	1.90	2.85	1.34	3.47

The next step is to turn to the Powder Diffraction File Index (published by the Joint Committee on Powder Diffraction Standards, 1601 Park Lane, Swarthmore, Pennsylvania) to the d-spacing range 2.30 to 2.34, which includes the most intense peak observed. A portion of a page from the index is shown below.

	d	d	d	d	d	d	d	d	Compound	File No.	Index
	2.30x	1.73_9	2.80_8	1.53_8	5.49_7	1.55_7	1.39_7	1.26_7	$(Fe_{.73}Mg_{.22}Mn_{.05})(OH)_2$	15- 125	I- 64-E 8
*	2.33_9	1.72_8	2.55x	2.85_6	2.01_4	1.52_4	1.09_2	1.01_2	NiAsS	12- 705	I- 50-B12
	2.33x	1.72x	2.55x	2.85_9	1.52_9	1.58_8	1.10_7	2.02_7	(SiP_2) 12C	19-1132	I-113-B10
o	2.29x	1.72x	2.09_8	1.24_8	1.13_8	0.96_6	1.95_3	1.78_3	$AsCo_2$	9- 93	I- 30-F12
	2.34x	1.71_9	1.09_9	2.63_8	1.81_6	2.06_5	1.60_5	0.0_1	Ti_3S_4	9- 294	I- 32-D10
	2.35x	1.70_3	1.65_3	2.85_2	5.80_1	2.03_1	1.46_1	1.90_1	NaOH	1-1173	I- 6-D 8
	2.32_7	1.70_6	2.21x	1.31_6	2.50_5	1.44_5	0.81_5	1.23_4	$(AlTi_2)$ 6H	9- 98	I- 31-B 4
*	2.30_4	1.68_2	2.72x	1.39_2	2.47_2	1.47_2	1.63_1	1.09_1	In	5- 642	I- 19-B 4
	2.29x	1.68_9	1.17_9	3.51_7	1.92_7	1.43_7	0.88_7	4.27_6	Al_3Ta	2-1128	I- 9-E 3
	2.35x	1.67x	2.72x	1.42x	1.06x	0.80x	5.39_8	2.83_8	$NaYbO_2$	19-1260	I-114-C12
	2.35_9	1.67x	2.71_9	1.43_9	0.80_9	0.79_8	1.06_7	0.97_7	$NaTlO_2$	17- 828	I- 87-C10
	2.35_8	1.67_8	2.71x	1.42_8	1.36_5	1.08_5	1.05_5	0.96_5	$HfH_{1.70}$	5- 641	I- 19-B 4
	2.35_4	1.66x	3.85_6	2.01_4	1.42_4	1.39_4	1.72_2	2.11_1	Fe_2O_3	19- 615	I-108-D 5
	2.35_4	1.66x	2.83_6	1.05_6	1.80_4	1.09_4	0.99_4	1.22_2	Co_3S_4	19- 367	I-106-C 4
*	2.35_9	1.66_4	2.71x	1.42_3	1.35_1	1.05_1	0.96_1	1.08_1	CdO	5- 640	I- 19-B 3
	2.35_8	1.66_5	2.71x	1.41_5	0.79_3	1.05_2	0.78_2	1.35_2	(ZrC) 8F	19-1487	I-116-D 3
	2.35_8	1.66_4	2.71x	1.42_3	1.36_1	1.05_1	1.08_1	0.96_1	Ca_2Si	3- 798	I- 13-D 2
*	2.35_4	1.66_2	2.57x	0.79_3	1.59_2	0.78_2	1.53_2	1.25_2	$CaPd_3O_4$	15- 52	I- 64-B 6
o	2.35x	1.66x	1.48x	3.21_8	1.99_8	2.26_6	2.20_6	1.96_6	Bi_4PbTe_7	15- 167	I- 65-B 7
	2.35x	1.66_6	1.44_6	1.36x	1.01x	1.23_8	1.00_8	1.41_6	(Ag_5Cd_8) 52B	14- 4	I- 56-C 3
	2.34_7	1.66_6	2.71x	0.78_8	0.79_7	1.41_6	1.05_3	0.96_3	$Sc_{0.3-0.5}C$	20-1028	I-126-D12
	2.34_8	1.66x	2.70_8	1.41_8	1.05_8	0.95_8	0.90_8	1.08_7	Na_2PbO_3	8- 245	I- 28-D 9
	2.32x	1.66_6	2.87_5	1.27x	1.20x	1.35_8	1.16_8	1.63_5	(Nb_2AsC) H	20-1288	I-128-F11
	2.31x	1.66_8	1.62_8	2.64_5	1.39_5	1.32_5	5.38_3	1.79_3	Na_2ZrO_3	8- 242	I- 28-D 7
	2.33x	1.65_5	2.68_4	1.41_1	1.07_1	1.04_1	1.34_1	1.17_1	Na_2ZrO_3	19-1218	I-113-F12
	2.29_4	1.65_4	3.11x	1.30_4	1.21_4	2.11_3	1.97_3	1.92_3	Bi_2Te_2S	9- 447	I- 33-F 3
	2.33_7	1.64x	4.00_9	2.42_7	2.13_5	1.60_5	1.37_4	1.48_3	Mn-Mn-O-(OH)	17- 510	I- 84-D 4
	2.33_9	1.64_8	2.61x	2.53_7	1.86_7	1.06_5	2.42_4	1.26_4	FeAs	18- 635	I- 94-B10
	2.32x	1.64_8	2.81_4	1.34x	1.04x	1.16_8	1.81_4	3.84_1	(Al, Ta) x	14- 482	I- 60-C11
	2.32x	1.64_6	1.34_2	1.04_1	0.95_1	1.16_1	2.68_1	0.82_1	NaF	4- 793	I- 16-F 2
	2.33x	1.63_7	2.55x	1.49x	1.25x	1.26_9	1.60_8	1.23_8	(AlCoU) 9H	19- 7	I-103-B 6
	2.32x	1.63_9	2.67_8	1.39_5	1.03_4	0.94_4	1.33_3	1.06_3	$NaInO_2$	21-1124	I-141-B 9
	2.32_7	1.63_6	2.53x	1.32_6	1.31_6	2.20_5	1.96_5	1.86_5	MgB_4	15- 299	I- 66-C 3
	2.31_8	1.63_8	2.84x	1.07x	1.47_8	1.32_8	1.20_8	1.11_8	$(AuPb_2)$ 12U	8- 419	I- 29-E 5
	2.31x	1.63_6	2.67x	1.39_4	1.03_2	0.94_2	1.33_2	1.06_1	ZrO	20- 684	I-123-D 4
	2.30x	1.63_8	2.68x	1.39_6	1.03_4	1.06_4	2.51_3	0.0_1	Hf-B-N	14- 110	I- 57-B11
	2.35_6	1.62_6	2.66x	1.41_6	1.35_6	1.67_4	2.21_3	1.33_3	(Pu) 2U	9- 316	I- 32-E 8
*	2.34x	1.62_9	3.18_7	2.56_7	2.74_5	1.69_5	1.44_5	2.12_2	$RbScO_2$	18-1126	I- 98-E 1
o	2.31x	1.62_3	2.34x	1.26_3	2.58_2	1.32_2	3.94_1	2.21_1	(Ag, Ga, In)	15- 135	I- 64-F 1
	2.29_8	1.62_8	2.98x	3.87_7	2.74_7	1.93_7	1.77_7	1.48_7	$BaBi_4Ti_4O_{15}$	8- 261	I- 28-E 3
	2.29_8	1.62x	2.80_9	1.55_8	1.19_8	1.16_8	3.06_7	2.86_7	(CeSi) 8O	18- 320	I- 91-C 5
	2.29x	1.62_8	2.64x	1.38_7	1.02_5	0.93_5	1.32_4	1.05_4	ZrN	2- 956	I- 9-B 8
	2.34x	1.60_7	2.55x	1.48x	1.25x	1.21_8	1.47_7	1.26_7	(AlFeU) 9H	19- 23	I-103-C 2
	2.35x	1.59x	2.75x	1.50x	1.04x	1.01x	0.92x	0.90x	(Al_2Tm) 24F	15- 492	I- 67-F 6
	2.33_6	1.59x	3.10_8	1.93_6	4.10_6	1.64_6	2.48_5	1.75_4	$TaBO_4$	7- 131	I- 24-E11

Note the second column is arranged in decreasing order of d spacing. Find the first two d values 2.32, 1.65 ($\pm$ 0.02) if possible. Remember, these d spacings may be in error 0.01 to 0.02 Å if the diffraction unit is not calibrated. There are several entries that fit 2.32, 1.65 (indicated by the brace). Now look for the third entry, 4.48. You will not find one that fits, so perhaps the 4.48 belongs to another compound. Try to find 3.03 in column 3: no luck; same for 2.31 and 1.90. The value of 2.85 fits the d spacing but not the intensity. [x As a subscript indicates the most intense line. So if the

$(AuPb_2)$12U entry were correct, the observed intensity of the 2.85 peak should be greater than 2.31 and 1.63. Also the 2.31 and 1.63 peaks should have the same intensity, which is not observed.] Finally, try the 1.44 peak which does fit. NaF fits the first three peaks. To verify that it is NaF we need more information, although three peaks matching is a good indicator that we are right. The complete card from the Powder Diffraction File #4-793 is reproduced below.

4-0793 MAJOR CORRECTION

d 4-0794	2.32	1.64	1.34	2.68	NaF
I/I_1 4-0793	100	60	17	3	Sodium Fluoride (Villiaumite)

Rad. CuKα λ 1.5405 Filter Ni
Dia. Cut off Coll.
I/I_1 G. C. Diffractometer d corr. abs.?
Ref. Swanson and Tatge, JC Fel. Reports, NBS 1949

Sys. Cubic S.G. O_H^5 - Fm3m
a_0 4.64 b_0 c_0 A C
α β γ Z 4
Ref. Ibid.

ε α nωβ 1.3270 (Na) εγ Sign
2V D_x 2.802 mp Color
Ref. Ibid.

Spec. anal. Si-V, weak others only as traces.
At 26°C
To replace 1-1181, 1-1184, 2-1115

d Å	I/I_1	hkl	d Å	I/I_1	hkl
2.680	3	111			
2.319	100	200			
1.639	60	220			
1.399	2	311			
1.338	17	222			
1.1588	7	400			
1.0633	<1	331			
1.0363	12	420			
0.9458	8	422			
.8920	1	511			
.8192	3	440			

We would need to observe the peak at 1.036 Å ($2\theta = 96°$) to make a definite assignment (or observe one of the very weak peaks, which might be difficult). Scratch off all peaks from the original list that have now been accounted for. We are left with the following. I′ are rescaled intensities.

I	50	40	30	20	15	10
d	4.48	3.03	2.31	1.90	2.85	3.46
I′	100	80	60	40	30	20

Turn to the section of the Powder Diffraction Index that covers 4.30 to 4.59 and look for 4.48 and 3.03. (See below.) Again, the brace indicates the range of values possible. There is an entry in the third column of 2.33, one in the fourth at 1.90, etc. This fits $CaCl_2$ very closely. The card reproduced below from the file (#1-338) gives the other data and shows that we can account for all diffraction peaks if the sample is NaF + $CaCl_2$. If peaks were left, we would repeat the process for a third compound. A couple of notes: (1) you do not have to rearrange observed peaks in decreasing intensity order. All permutations of the three most intense diffraction peaks are listed in the index,

*	4.39$_7$	3.06$_6$	4.33x	3.90$_4$	5.22$_3$	3.14$_3$	3.12$_3$	3.00$_3$	$(NH_4)_2SO_4$	10- 343	I- 36-E11
o	4.37x	3.06x	2.74x	5.50$_8$	2.90$_8$	2.83$_8$	2.63$_8$	2.49$_8$	$Na_3VO_4.1/4NaOH.12H_2O$	19-1256	I-114-C10
	4.33x	3.06x	2.50x	2.16x	1.53x	1.30x	1.93$_6$	1.76$_6$	$(YAl_3)4C$	20- 70	I-117-E 7
*	4.32$_5$	3.06x	3.15$_7$	2.03$_4$	3.91$_3$	3.25$_3$	2.83$_3$	1.93$_3$	HgSbBr	18- 825	I- 95-F 5
	4.62$_8$	3.05x	2.77x	4.17$_8$	4.04$_8$	3.75$_8$	3.41$_8$	3.24$_8$	$Na_2B_8O_{13}$	16- 306	I- 74-C 1
	4.54$_3$	3.05x	2.96$_5$	2.52$_3$	6.54$_3$	2.56$_3$	1.65$_2$	1.63$_2$	$NaScSi_2O_6$	21-1369	I-143-D 3
	4.49x	3.05$_8$	2.33$_6$	1.90$_4$	2.85$_3$	3.46$_2$	2.24$_2$	2.09$_2$	$CaCl_2$	1- 338	I- 3-C 5
	4.47$_3$	3.05x	3.95$_7$	5.81$_3$	4.36$_3$	3.69$_2$	3.43$_2$	2.34$_1$	$Na_2H_2(PO_3)_4$	9- 100	I- 31-B 5
	4.39$_6$	3.05x	4.96$_6$	2.58$_6$	4.25$_5$	2.61$_3$	1.96$_3$	3.00$_3$	$CsBO_2.4H_2O$	21- 201	I-132-F 1
	4.30x	3.05x	1.93x	2.81$_7$	2.37$_7$	2.16$_7$	5.25$_4$	2.48$_4$	$CaCO_3.H_2O$	17- 528	I- 84-E 3
	4.64x	3.04$_9$	5.06$_8$	1.98$_7$	4.97$_5$	3.91$_5$	3.80$_5$	3.32$_5$	$(NH_4,K)NaSO_4.2H_2O$	15- 370	I- 66-F 3
	4.54$_5$	3.04x	3.85$_7$	2.98$_5$	2.79$_4$	5.30$_3$	4.78$_3$	3.44$_3$	$PbZn_2B_2O_6$	19- 709	I-109-C 9
	4.42$_7$	3.04$_7$	2.84x	6.05$_6$	3.51$_6$	1.87$_6$	2.17$_5$	2.13$_5$	$PrPO_4.H_2O$	20- 966	I-126-B 4
	4.31$_5$	3.04x	4.98$_8$	2.84$_5$	2.16$_5$	1.76$_5$	4.04$_3$	2.60$_3$	K_3NbO_4	14- 283	I- 58-E 4
	4.27x	3.04x	3.24$_7$	2.53$_5$	1.78$_4$	1.75$_4$	5.10$_3$	2.69$_3$	$Ca(IO_3)_2$	1- 386	I- 3-D 2
	4.26x	3.04x	5.37x	4.82$_4$	3.89$_4$	2.96$_4$	2.85$_4$	2.48$_4$	$Al(PO_4).2H_2O$	15- 281	I- 66-B 6
	4.26$_6$	3.04x	3.60$_7$	2.53$_5$	2.79$_4$	3.56$_3$	2.93$_3$	2.56$_3$	$SmBO_3$	13- 489	I- 55-C 4
	4.26$_5$	3.04x	3.60$_6$	2.91$_4$	2.77$_4$	3.55$_4$	0.0 $_1$	0.0 $_1$	$EuBO_3$	13- 487	I- 55-C 3
	4.42x	3.03$_9$	2.85$_8$	6.06$_7$	2.16$_6$	1.87$_5$	3.49$_5$	3.07$_5$	$BiPO_4$	15- 766	I- 70-C 9
	4.41x	3.03x	2.84x	6.09$_6$	3.50$_6$	2.15$_6$	1.86$_6$	1.93$_5$	$YbSO_4$	17- 774	I- 86-F 5
	4.29$_3$	3.03x	3.02x	2.47$_3$	2.14$_2$	1.91$_2$	2.48$_1$	1.92$_1$	$RbNO_3$	17- 516	I- 84-D 7
	4.28$_8$	3.03x	2.47$_9$	3.04$_8$	1.75$_5$	1.92$_4$	1.92$_3$	1.76$_3$	$TlNO_3$	14- 39	I- 56-D 8
	4.49x	3.02$_9$	3.98$_8$	2.50$_7$	3.33$_5$	1.85$_5$	3.72$_3$	1.78$_3$	$Al_2(SO_4)_3.18H_2O$	14- 559	I- 61-B 4
	4.45x	3.02x	15.10x	2.60x	2.49x	1.65x	1.49x	5.03$_8$	Al-Na-Al-Si-O-OH	3- 13	I- 11-B 8
	4.28x	3.02$_9$	10.50$_6$	2.81$_6$	3.17$_5$	3.16$_4$	1.91$_4$	1.88$_4$	$7(Ca,Na)O.LaO.11BO.7H_2O$	21- 158	I-132-D 3
	4.61$_2$	3.01x	1.87$_3$	2.51$_2$	1.53$_2$	2.10$_2$	1.66$_2$	2.80$_1$	$HoVO_4$	16- 482	I- 75-E 9
	4.53$_6$	3.01x	3.67$_6$	4.10$_6$	2.73$_6$	2.53$_5$	2.81$_4$	2.30$_4$	$AgNO_3$	6- 363	I- 21-C 1
	4.43$_8$	3.01x	4.90$_8$	3.48$_8$	2.96$_6$	2.59$_6$	2.42$_6$	2.34$_6$	$Tl_2Al_2Si_3O_{10}.2H_2O$	2- 636	I- 8-C 1
	4.42$_6$	3.01x	2.58$_8$	2.12$_5$	2.03$_5$	3.37$_4$	1.94$_4$	3.20$_4$	$KF.2H_2O$	1- 854	I- 5-C 1
	4.32$_9$	3.01x	2.53x	3.51$_9$	3.92$_6$	1.75$_6$	1.63$_6$	5.17$_5$	$(Li,Mn,Fe)PO_4$	13- 338	I- 53-F 8

1-0338 MAJOR CORRECTION

0781 d 1-0349	4.49	3.05	2.33	4.49	$CaCl_2$
I/I_1 1-0338	100	80	60	100	Calcium Chloride (Hydrophilite)

Rad. MoKα λ 0.709 Filter ZrO_2
Dia. 16 inches Cut off Coll.
I/I_1 Calibrated strips d corr. abs.? No
Ref. H

Sys. Orthorhombic S.G. D_{2h}^{12} - Pnnm
a_0 6.25 b_0 6.44 c_0 4.21 A 0.971 C 0.644
α β γ Z 2 D_x 2.17
Ref. D_7 (Artif.)

εα 1.600 nωβ 1.605 εγ 1.613 Sign +
2V mod. D 2.512 mp 772 Color
Ref. C.C., Wa., HCP

B.P. > 1600

d Å	I/I_1	hkl	d Å	I/I_1	hkl
4.49	100	110			
3.46	16	101			
3.05	80	111			
2.85	32	120			
2.33	60	211			
2.24	16	220			
2.09	16	002			
1.90	36	112,031			
1.79	8	311			
1.68	12	212			
1.56	4	400,140			
1.51	8	410			
1.49	4	Indexed			
1.33	12	by LGB			
1.24	12				
1.21	12				
1.17	4				

so each compound is listed three times. (2) There is much trial and error involved, and only a lot of practice will make you skilled at identifying compounds. (3) Many substances, especially hydrates, may be difficult due to structural changes resulting from the method of sample synthesis. (4) Powder diffraction will not solve all identification problems, but it is an extremely valuable tool. (5) Another index, the Fink Index, lists the first eight diffractions for each compound. Thus, each compound is listed eight times. In many cases this index is more convenient to use than the one described above. The Fink Index is also published by the Joint Committee on Powder Diffraction Standards.

14-5.7 PROBLEMS

1. The following d spacings for NaOH are given. For X-rays with λ = 1.542 Å (= 0.1542×10^{-9} m = 0.1542 nm) find the corresponding values of θ; d = 2.35, 1.70, 1.65, 2.85, and 5.80 Å.

2. A cubic crystalline solid is observed to have d spacings of 2.68, 2.32, 1.64, 1.40, 1.34, and 1.16. Determine the type of cubic cell, the unit cell parameters, and the density of the crystal if the material has a molecular weight of 0.042 kg mol^{-1}.

3. Geometrically determine the largest sphere that will fit in the tetrahedral void formed by four spheres. (Hint: A tetrahedral arrangement can be obtained by

placing one sphere at every other corner of a cube with the spheres touching along the face diagonals.)

4. The orthorhombic unit cell parameters for the mineral, topaz are a_0 = 8.39 Å, b_0 = 8.79 Å, and c_0 = 4.65 Å. Determine the X-ray diffraction angles from the (1 0 1), (1 1 0), (1 1 1), and (2 2 2) planes.

5. Sketch the unit cells of two of the following common crystal structures: zinc-blende; perovskite; fluorite.

6. The X-ray diffraction from the (1 1 1) plane in a cubic material is observed 19.1°. Determine the diffraction angles for the (1 0 1), (0 1 1), and (2 2 0) planes.

14-5.8 SELF-TEST

1. (a) Find the values of d_{hkl} and corresponding values of θ (X-ray wavelength of 0.710 Å) for an orthorhombic system with a = 7.36, b = 8.04, and c = 7.34 Å for the planes (1 1 0), (1 0 1), (2 1 0), and (2 2 1). (b) If the molecular weight of the substance is 0.09413 kg mol^{-1}, and there are four molecules per unit cell, determine the crystal density.

2. Find the radius of the largest sphere that will just fit into an octahedral void formed by six identical spheres.

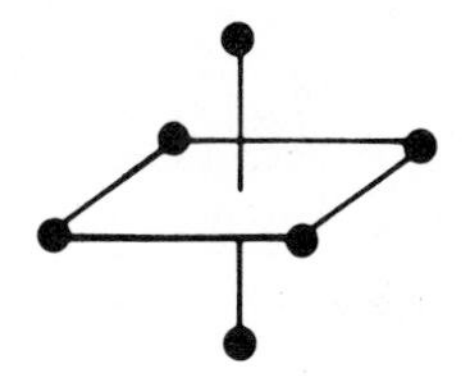

(Spheres touch along edges.)

3. The following diffraction data were observed for a powdered sample. Use the Powder Diffraction File and its index to identify the materials present.

d	7.02	2.97	2.53	3.51	1.61	1.71	1.75	3.46	2.34	1.48	2.95	2.34	2.10	2.31
I	20	70	100	100	90	60	30	20	30	90	20	30	70	20

14-5.9 ADDITIONAL READING

1.* W. J. Moore, *Physical Chemistry*, 4th ed., Prentice-Hall, Englewood Cliffs, N. J., 1972.

2.* W. F. Sheehan, *Physical Chemistry*, 2nd ed., Allyn and Bacon, Boston, 1970, Secs. 16.12 and 16.17.

3. F. A. Cotton and G. Wilkenson, *Advanced Inorganic Chemistry*, Interscience, New York, 1962, Sec. 2-5.
4. "The Powder Diffraction File Index" and "The Fink Index," Joint Committee on Powder Diffraction Standards, 1601 Park Lane, Swarthmore, Pennsylvania.
5. For information on actual molecular structures see Sec. 1-1.9, Ref. D-1.

14-6 SUMMARY

This completes an introduction to properties of the liquid and solid states. It has been a rather long chapter. In spite of its length, the chapter has only scratched the surface of many topics and ignored many others. The first section reviewed the structural differences between liquids and solids and gathered together terms and equations from other chapters applicable to this study. In addition, specific liquid properties were given some attention. Of the sections found in this chapter, Sec. 14-2 will probably be one of the most useful due to the detailed treatment of electrical and magnetic units. You will probably make use of many of the conversions given in other courses. Both Secs. 14-2 and 11-2 are good references, should you ever have to work with electrical and magnetic equations. The final three sections dealt with solid state properties. Symmetry and its relationship to crystal structure was covered in some detail. Finally, a practical application, the use of X-ray powder diffraction to identify unknown solid samples, was given. In all the above, the theory has been minimal. As mentioned before, this is due to the very complex nature of the liquid and solid states relative to a gas. To adequately give the theory would take us far beyond the realm of an introductory course. You should have developed some understanding of at least the theoretical concepts used to describe liquid and solid properties.

Appendix A
Notes to the Instructor

It was pointed out in the preface that this text can be used by instructors in individualizing their courses to fit the needs of different disciplines. In the past, all students have received the same introductory course regardless of their individual backgrounds or interests. By selecting topics related to individual fields of study, you, as the instructor can do much to alleviate that problem. Before suggesting topics for different disciplines, I should mention how my course operates. For those who have never tried a self-paced course, an example might be useful. These procedures are not meant as rules. Each person will want to develop procedures to fit his or her own situation.

I have approximately 120 students, including the following disciplines: petroleum engineering, chemical engineering, chemistry, physics, geology, and life science. No lectures are given other than the first day of class. Each student is given handouts describing course procedures, and verbal instructions are given. Students are issued a record card showing which objectives must be completed for a C grade and which for a B grade. I have found that about 15 objectives are sufficient for a C grade with about five more needed for a B grade. In my course, an A can be obtained only by passing a comprehensive final exam. I require the final exam to ensure that the student can interrelate the material in the objectives. The material in the course is covered rather piecemeal. The final makes sure the A grade represents comprehension of the course as a whole. Some instructors might prefer to require additional objectives for an A grade.

Objective exams are given during several scheduled times (including the "lecture" hour). When a student wants an exam, he or she presents a completed Self-Test, which is checked by me or a grader. After clearing up any doubtful answers on the Self-Test, the student is issued an exam and sheet of paper. The date and exam number are noted on the record card (which the student keeps). The student is sent to a designated area in the room to complete the exam. Only the exam card (4" × 6"), blank sheet of paper, pencil, calculator, and formula sheet are allowed. The formula sheets give general formulas for the whole course (one 8-1/2" × 11" sheet, both sides, is sufficient), and each student has his or her own copy on which no extraneous marks are allowed. After completion, the student brings the exam solution back for scoring. If only minor errors are present, one hint is given, after which the answers are either right or wrong. If all is correct, the record card is marked "Pass". If the answer is partially correct (say, one problem of three is worked correctly) a notation to that effect is made on the grade card. Only the portion of the exam that was missed is required when the student returns to take another exam on the same Objective. Some instructors may wish to have the whole exam repeated. That is a matter of personal taste. In any event, the grader and the student go through the solution step-by-step to ensure that no misconceptions remain.

I use five or six graders during each exam period. This is the bare minimum for the number of students in the course (120). I would recommend a ratio of one grader per 15 students (not including yourself). I pay my graders, but some universities offer course credit instead of pay. Others use students enrolled in the course who are ahead of the average in the class as graders. These are just a few options. A couple of very good references that should be consulted before attempting an individualized instruction format are: J. G. Sherman, ed., *PSI-41 Germinal Papers*, W. A. Benjamin, Menlo Park, Calif., 1974; J. E. Stice, *Expansion of Keller Plan Instruction in Engineering and Selected Other Disciplines: A Final Report*, The University of Texas at Austin, 1975. One additional note: As an incentive to complete the course early, I offer a final exam (required for an A) about three weeks before the end of the term for those who have completed all 20 Objectives. About 25% of the class is normally eligible for this early final.

Normally, petroleum engineering, earth science, and life science students take only one semester of physical chemistry. I have tried to include topics of interest to these students in that one semester. It seems that the order of topics is not very important (after completion of the thermodynamics), and I allow students to complete Objectives in any order they wish. This has created no major problems. Below, I list the 20 Objectives I require for a B for different disciplines. (The assignments change from year to year as laboratory experiments are changed.) Also, I give a brief statement as to why these topics are included while others are omitted. I have not indicated the B and C Objectives. How they are assigned is a matter of personal choice. There are no A Objectives listed. I require satisfactory completion of a comprehensive final examination to gain an A. These are just some possible combinations. The ones you choose will depend on your preference and the background of the students in your class.

CHEMISTRY, First Semester

Objective	Comments
1	Errors, units, error analysis, data treatment
2-1, 2-2, 2-3	Gas laws
3-1, 3-2, 3-3, 4-1, 4-2, 4-3, 5-1, 5-2, 5-3	A thorough introduction to thermodynamics
6-1, 6-2, 6-3	Introduction to the treatment of thermodynamics of solution
13-1, 13-2, 13-3, 13-4	Introduction to phase equilibria (associated with a laboratory exercise)

The strange combination is dictated by requirements for some of the laboratory experiments.

CHEMISTRY, Second Semester

Objective	Comments
6-4, 6-5	Nonideal solution; electromotive force
7-1, 7-3, 7-4	Atomic structure and properties; the detailed solution to the hydrogen atom is omitted
8-1, 8-2, 8-3, 8-4	Rotation-vibration in molecules, molecular structure, spectroscopy (section on rotation-vibration is a laboratory exercise)
9-1, 9-2	Introduction to statistical mechanics
10-1, 10-2	Introduction to kinetics; statistical treatment is omitted
11-1, 11-2, 11-3	Colloid and surface properties
12-2, 12-3	Introduction to polymer properties
14-1	Liquid properties
14-2	Electrostatic and magnetic properties of molecules

(*Note*: More quantum mechanics, statistical mechanics, and theoretical kinetics could be included, but only at the expense of some descriptive material in polymers and colloids.)

CHEMICAL ENGINEERING, First Semester

Objective	Comments
1	Treatment of data
2-1, 2-2, 2-3	Gas laws
3-1, 3-2, 3-3, 4-2, 4-3, 5-1, 5-2, 5-3	Chemical engineering students already have a thermodynamics background before they take physical chemistry but need additional study to prepare them for thermodynamics of solutions as well as to give a theoretical background to the thermodynamics they have had
6-1, 6-2, 6-3, 6-4	Solution thermodynamics
13-1, 13-3	Phase equilibria (associated with a laboratory exercise)
11-3	Surface properties (associated with a laboratory exercise
12-2	Polymer molecular weight methods, associated with a laboratory experiment

CHEMICAL ENGINEERING, Second Semester

Objective	Comments
2-5	Diffusion
6-5	Electromotive force (associated with a laboratory exercise)
7-1, 7-3	Postulates of quantum mechanics, atomic structure, periodic properties
8-2, 8-4	Molecular structure, spectroscopy
9-1, 9-2	Introduction to statistical thermodynamics
10-1, 10-2	Empirical kinetics
12-3, 12-4, 12-5	Polymers
11-1, 11-2	Colloid chemistry
13-2, 13-4	Phase equilibria
14-1, 14-2, 14-3	Solid and liquid properties; electrostatics

PETROLEUM ENGINEERING, One Semester Only

Objective	Comments
1	Treatment of data
2-1, 2-2, 2-3	Gas laws
3-1, 3-2, 4-2, 5-2	Introduction to terms and symbols used in thermodynamics in the course; these students already have a background in thermodynamics
6-1, 6-2, 6-3	Ideal solutions, colligative properties
11-1, 11-2, 11-3	Colloids and surface chemistry (very important to a petroleum engineer)
12-2, 12-3, 12-4	Polymers (also, very important)
13-1, 13-2	Phase equilibria
14-1	Liquid properties

GEOLOGY, One Semester Only

Objective	Comments
1	Treatment of data
2-1, 2-2, 2-3	Gas laws
3-1, 3-2, 3-3, 4-1, 4-2, 5-1, 5-2, 5-3	Strong thermodynamics; geology students, typically, are weak in this area
11-1, 11-3	Colloid properties, surface properties, minus electrokinetics
13-1, 13-2, 13-4	Solid-liquid phase equilibria
14-2, 14-4, 14-5	Electrostatics, symmetry, solid state structure

LIFE SCIENCE, One Semester Only

Objective	Comments
1	Treatment of data
2-1, 2-3, 2-5	Gas laws, diffusion
3-1, 3-2, 3-3, 4-1, 4-2, 5-1, 5-2	Thermodynamics; normally, life science students are weak in thermodynamics
6-1, 6-2, 6-3	Ideal solutions, colligative properties, electrolytes (important for life science majors)
11-1, 11-2, 11-3	Surface activity and colloids (very important)
12-1, 12-2, 12-3	Polymer properties

Perhaps the most important aspect of a course taught in a "Self-Paced" or "Individualized" format is the sense of community established in the class. Competition is eliminated, since all students can reach the "A" or "B" level if they have the ability and perserverence. Each person's achievement is independent of that of any other student. Students sense and respond to the concern shown by the instructor and tutors. Overall, there is a positive response to such a course. The students leave the course with at least as positive an attitude as when they entered.

Appendix B
Physical Constants and Conversion Factors

Table B-1
Some Physical Constants

Constant	Symbol	Value	Units
Gas constant	R	8.3143	$J\ K^{-1}\ mol^{-1}$
		1.987	$cal\ K^{-1}\ mol^{-1}$
		82.06	$cm^3\ atm\ K^{-1}\ mol^{-1}$
		0.08206	$dm^3\ atm\ K^{-1}\ mol^{-1}$
Avagadro's number	N	6.023×10^{23}	$molecule\ mol^{-1}$
Faraday's constant	$\mathcal{F}$	9.6487×10^{3}	C (coulomb) mol^{-1}
Planck's constant	h	6.6256×10^{-34}	J s
Boltzmann's constant	k	1.3805×10^{-23}	$J\ molecule^{-1}\ K^{-1}$
Velocity of light	c	2.9979×10^{8}	$m\ s^{-1}$
Charge of electron	e	1.602×10^{-19}	C
Mass of electron	m_e	9.109×10^{-31}	kg
Permittivity of vacuum	ε_0	8.8542×10^{-12}	$kg^{-1}\ m^{-3}\ s^4\ A^2$

Table B-2
Some Common Conversion Factors

Length

$$1 \text{ meter (m)} = 10^2 \text{ centimeters (cm)}$$
$$= 10 \text{ decimeters (dm)}$$
$$= 10^{10} \text{ angstrom (Å)}$$

Area

$$1 \text{ m}^2 = 10^4 \text{ cm}^2 = 10^2 \text{ dm}^2 = 10^{20} \text{ Å}^2$$

Volume

$$1 \text{ m}^3 = 10^6 \text{ cm}^3 = 10^3 \text{ dm}^3 = 10^3 \text{ liter}$$

Force

$$1 \text{ Newton (N = kg m s}^{-2}) = 1 \times 10^5 \text{ dyn (g cm s}^{-2})$$

Power

$$1 \text{ watt} = 1 \text{ J s}^{-1} = 0.0569 \text{ BTU (min)}^{-1}$$
$$= 1.341 \times 10^{-3} \text{ horsepower (HP)}$$
$$= 0.01433 \text{ kilocalorie per min (kcal min}^{-1})$$

Work and Energy

$$1 \text{ J (N m}^2 \text{ s}^{-2}) = 1 \text{ N m} = 1 \times 10^7 \text{ erg (dyn cm = g cm}^2 \text{ s}^{-2})$$
$$= 1 \text{ Volt Coulomb (V C)}$$
$$= 2.389 \times 10^{-4} \text{ kcal} = 9.48 \times 10^{-4} \text{ BTU}$$
$$= 2.778 \times 10^{-7} \text{ kilowatt hours (kw·hr)}$$

Density

$$1 \text{ kg m}^{-3} = 10^{-3} \text{ kg dm}^{-3} = 10^{-6} \text{ kg cm}^{-3}$$
$$= 10^{-3} \text{ g cm}^{-3} = 10^{-3} \text{ kg liter}^{-1}$$

Pressure

$$1 \text{ atmosphere (atm)} = 1.01325 \times 10^5 \text{ N m}^{-2}$$
$$= 1.0132 \times 10^6 \text{ dyn cm}^{-2}$$
$$= 760 \text{ mm Hg}$$

Appendix C
Problem Solutions

1-1.9 PROBLEMS

1. $y = 4.35x^3 - 6.55x^2z + 1.22xz^2 + z^3$

$$\left(\frac{\partial y}{\partial x}\right) = 3(4.35)x^2 - 2(6.55)xz + 1.22z^2; \quad \left(\frac{\partial y}{\partial z}\right) = -6.55x^2 + 2.44xz + 3z^2$$

$$\Delta y = \left(\frac{\partial y}{\partial x}\right)\Delta x + \left(\frac{\partial y}{\partial z}\right)\Delta z = 145.39(0.07) + |-92.00|(0.02)$$

$$= 10.18 + 1.84 = \underline{\underline{12}} \text{ Absolute error}$$

$\underline{\underline{y = 154.40}}$

$$\text{Relative error} = \frac{\Delta y}{y} = \frac{12.02}{154.4} = 0.0779 = \underline{\underline{8.0\%}}$$

2. $\ell n\ y = 2\ \ell n\ u + 3\ \ell n\ v - \frac{1}{2}\ \ell n\ w - \frac{1}{2}\ \ell n\ x - \frac{1}{4}\ \ell n\ z$

Take absolutes

$$\ell n\ y = 2\ \ell n\ u + 3\ \ell n\ v + \frac{1}{2}\ \ell n\ w + \frac{1}{2}\ \ell n\ x + \frac{1}{4}\ \ell n\ z$$

$$\frac{\Delta y}{y} = \frac{2\Delta u}{u} + \frac{3\Delta v}{v} + \frac{\frac{1}{2}\Delta w}{w} + \frac{\frac{1}{2}\Delta x}{x} + \frac{\frac{1}{4}\Delta z}{z}$$

$$= 2(0.13) + 3(0.006) + \frac{1}{2}(0.022) + \frac{1}{2}(0.0033) + \frac{1}{4}(0.0102)$$

$$\frac{\Delta y}{y} = \underline{\underline{0.0592}} = \underline{\underline{6\%}}$$

3. $$\left(\frac{\Delta y}{y}\right)^2 = \left(\frac{2\Delta u}{u}\right)^2 + \left(\frac{3\Delta v}{v}\right)^2 + \left(\frac{1}{2}\frac{\Delta w}{w}\right)^2 + \left(\frac{1}{2}\frac{\Delta x}{x}\right)^2 + \left(\frac{1}{4}\frac{\Delta z}{z}\right)^2$$

$$= 0.00113$$

$$\frac{\Delta y}{y} = \underline{\underline{0.0336}} = \underline{\underline{3\%}}$$

This is smaller than in problem 2, as predicted.

4. $\frac{\Delta y}{y} = 5\left(2\,\frac{\Delta u}{u}\right) = 5\left(3\,\frac{\Delta v}{v}\right) = 5\left(\frac{1}{2}\,\frac{\Delta w}{w}\right) = 5\left(\frac{1}{2}\,\frac{\Delta x}{x}\right) = 5\left(\frac{1}{4}\,\frac{\Delta z}{z}\right)$

$= 0.0049$

$10\,\frac{\Delta u}{u} = 0.0049;\ \frac{\Delta u}{u} = 0.049\%\ :\ 2.5\,\frac{\Delta x}{x} = 0.0049;\ \frac{\Delta x}{x} = 0.196\%$

$15\,\frac{\Delta v}{v} = 0.0049;\ \frac{\Delta v}{v} = 0.033\%\ :\ \frac{5}{4}\,\frac{\Delta z}{z} = 0.0049;\ \frac{\Delta z}{z} = 0.392\%$

$2.5\,\frac{\Delta w}{w} = 0.0049;\ \frac{\Delta w}{w} = 0.196\%$

5. $y = \frac{y_m P}{b + P};\quad yb + yP = y_m P;\quad \frac{yb}{y_m} + \frac{yP}{y_m} = P;\quad \frac{b}{y_m} + \frac{P}{y_m} = \frac{P}{y}$

or

$$\frac{P}{y} = \frac{P}{y_m} + \frac{b}{y_m}$$

A plot of P/y vs. P will yield a straight line with slope = $1/y_m$. The intercept (P = 0) is

$$\left(\frac{P}{y}\right)_{P=0} = \frac{b}{y_m}$$

To get b multiply intercept by 1/slope.

$$\left(\frac{b}{y_m} \cdot \frac{1}{1/y_m}\right) = b$$

6. 45010 - 4; 54.900 - 5; 0.000034 - 2; 5.004 - 4;
780 - 2; 33.1 - 3.

7. $\ln \eta = \ln m + \ln \bar{c} - \ln 2 - \frac{1}{2}\ln 2 - \ln \pi - 2\ln \sigma$

$\therefore\ \frac{\Delta\eta}{\eta} = \frac{\Delta m}{m} + \frac{\Delta\bar{c}}{\bar{c}} + 0 + 0 + 0 + 2\,\frac{\Delta\sigma}{\sigma}$

$= 3.2\% + 1.23\% + 2(4.81\%) = \underline{\underline{14.05\%}}$

$= \underline{\underline{14\%}}$ = Relative error in η

8. $\bar{x} = 2.19114;\quad \sigma = \left[\frac{1}{7 - 1}\sum_i (x_i - 2.19114)^2\right]^{1/2} = 0.08725$

$\bar{x} = \underline{\underline{2.191}};\quad \sigma = \underline{\underline{0.087}}$

9. $\ln \Delta T_f = \ln R + 2\ln T_f + \ln m - \ln \Delta H_f - \ln n_A$

$\therefore\ \frac{\Delta(\Delta T_f)}{\Delta T_f} = 0 + \frac{2\Delta T_f}{T_f} + \frac{\Delta m}{m} + \frac{\Delta(\Delta H_f)}{\Delta H_f} + \frac{\Delta n_A}{n_A}$

$= 2(1.03\%) + 1.45\% + 1.03\% + 1.45\%$

$= \underline{\underline{5.99\%}}$ = Relative error in ΔT_f

2-1.5 PROBLEMS

1. $$V = \frac{nRT}{P} = \frac{\left(\dfrac{0.0163\ \text{kg}}{0.0320\ \text{kg mol}^{-1}}\right)(0.08206\ \text{dm}^3\ \text{atm mol}^{-1}\ \text{K}^{-1})(296.4\ \text{K})}{(0.966\ \text{atm})}$$

$$= \underline{\underline{12.8\ \text{dm}^3\ \text{(liters)}}}$$

2. (a) $$n = \frac{PV}{RT} = \frac{(1.00\ \text{atm})V}{(0.08206\ \text{dm}^3\ \text{atm mol}^{-1}\ \text{K}^{-1})(290\ \text{K})} = 0.0420\ V\ \text{mol dm}^{-3}$$

$$\therefore\ \frac{n}{V} = \underline{\underline{0.0420\ \text{mol dm}^{-3}}}$$

$$\frac{Nn}{V} = \frac{(6.023 \times 10^{23}\ \text{mlc mol}^{-1})(0.0420\ \text{dm}^{-3}\ \text{mol})}{V}$$

$$= (2.53 \times 10^{22}\ \text{mlc dm}^{-3})\left(\frac{1\ \text{dm}}{10\ \text{cm}}\right)^3$$

$$= \underline{\underline{2.53 \times 10^{19}\ \text{mlc cm}^{-3}}}$$

(b) $$\frac{n}{V} = \underline{\underline{4.20 \times 10^{-10}\ \text{mol dm}^{-3}}}\ :\ \frac{Nn}{V} = \underline{\underline{2.53 \times 10^{11}\ \text{mlc cm}^{-3}}}$$

$$\frac{V}{nN} = \frac{1}{2.53 \times 10^{11}} = \underline{\underline{3.96 \times 10^{-12}\ \text{cm}^3\ \text{mlc}^{-1}}}$$

3. $$n = \frac{w}{M} = \frac{PV}{RT};\quad \frac{w}{V} = \frac{PM}{RT}:\quad \rho = \frac{w}{V} = \frac{PM}{RT} = \frac{Pw}{nRT}$$

4. $$\rho = \frac{PM}{RT} = \frac{(1.00\ \text{atm})(0.0320\ \text{kg mol}^{-1})(101{,}325\ \text{N m}^{-2}\ \text{atm}^{-1})(1000\ \text{g kg}^{-1})}{(8.3143\ \text{kg m}^2\ \text{s}^{-1}\ \text{mol}^{-1}\ \text{K}^{-1})(273.15\ \text{K})(100\ \text{cm m}^{-1})^3}$$

$$\underline{\underline{1.43 \times 10^{-3}\ \text{g cm}^{-3}}}$$

or

$$\rho = \frac{(1.00\ \text{atm})(32.00\ \text{g mol}^{-1})}{(0.08206\ \text{dm}^3\ \text{atm mol}^{-1}\ \text{K}^{-1})(273.15\ \text{K})} = 1.43\ \text{g dm}^{-3}$$

$$= 1.43 \times 10^{-3}\ \text{g cm}^{-3}$$

5. $$M = \frac{\rho RT}{P} = \frac{(1.63 \times 10^{-6}\ \text{kg cm}^{-3})(82.058\ \text{cm}^3\ \text{atm mol}^{-1}\ \text{K}^{-1})(298.15\ \text{K})}{1.00\ \text{atm}}$$

$$= 0.0399\ \text{kg mol}^{-1} = \underline{\underline{39.9\ \text{g mol}^{-1}}} = \text{Ar}$$

$$R = \frac{P_1M}{\rho_1T_1} = \frac{P_2M}{\rho_2T_2}\quad \therefore\ \rho_2 = \rho_1\left(\frac{T_1}{T_2}\right)\left(\frac{P_2}{P_1}\right) = (1.63\ \text{g dm}^{-3})\left(\frac{298.15\ \text{K}}{273.15\ \text{K}}\right)\left(\frac{1.00\ \text{atm}}{1.00\ \text{atm}}\right)$$

$$\rho_2 = \underline{\underline{1.78\ \text{g dm}^{-3}}}$$

6. $n = \dfrac{0.0567 \text{ kg}}{0.065 \text{ kg mol}^{-1}} = \underline{0.872 \text{ mol}}$

$$P = \frac{nRT}{V} = \frac{0.872 \text{ mol}}{20.456 \text{ dm}^3}(0.0821 \text{ dm}^3 \text{ atm mol}^{-1} \text{ K}^{-1})(304.5 \text{ K}) = 1.07 \text{ atm}$$

$\underline{\underline{P = 1.1 \text{ atm}}}$

7. $\rho = \left(\dfrac{0.872 \text{ mol}}{20.456 \text{ dm}^3}\right)\left(\dfrac{0.065 \text{ kg}}{1 \text{ mol}}\right) = 0.00277 \text{ kg dm}^{-3} = \underline{\underline{2.8 \text{ g dm}^{-3}}} = \rho$

8. $\dfrac{n}{V} = \dfrac{P}{RT};\quad P = \dfrac{n}{V}RT = \left(\dfrac{2.034 \times 10^{22} \text{ mlc}}{\text{dm}^3}\right)\left(\dfrac{1 \text{ mol}}{6.02 \times 10^{23} \text{ mlc}}\right)$

$$\times \left(0.0821 \frac{\text{dm}^3 \text{ atm}}{\text{mol K}}\right)(304.5 \text{ K})$$

$\underline{\underline{P = 0.84 \text{ atm}}}$

2-2.10 PROBLEMS

1. For H_2O, $T_c = 647.15$ K $\quad \therefore T_R = \dfrac{T}{T_c} = \dfrac{776}{647.15} = \underline{1.20} \qquad z = 0.60$

From Fig. 2-2.3.3

$$P_R = 1.9 \quad \therefore \quad P = P_c \cdot P_R = (1.9)(217.7 \text{ atm})$$

$$= 413.6 \text{ atm} = \underline{\underline{410 \text{ atm}}}$$

2. $PV = nRT + nRTbP \rightarrow b = \dfrac{PV}{nRTP} - \dfrac{nRT}{nRTP} = \dfrac{V}{nRT} - \dfrac{1}{P}$

$$= \frac{27.0 \text{ dm}^3}{(1.00 \text{ mol})(0.08206 \text{ dm}^3 \text{ mol}^{-1} \text{ K}^{-1} \text{ atm})(273 \text{ K})} - \frac{1}{\text{atm}}$$

$b = 1.205 \text{ atm}^{-1} - 1.00 \text{ atm}^{-1} = \underline{\underline{0.205 \text{ atm}^{-1}}}$

$$V = \frac{nRT(1 + bP)}{P} = \frac{(1.00 \text{ mol})(0.08206 \text{ dm}^3 \text{ atm mol}^{-1} \text{ K}^{-1})(295 \text{ K})}{2.00 \text{ atm}}$$

$$\times \frac{(1 + 0.205 \text{ atm}^{-1} \cdot 2.00 \text{ atm})}{2.00 \text{ atm}} = \underline{\underline{17.1 \text{ dm}^3}}$$

3. $T = \dfrac{\left(P + \dfrac{an^2}{V^2}\right)(V - nb)}{nR}$

$$= \frac{[1.00 \text{ atm} + 1.35 \text{ dm}^6 \text{ mol}^{-2} \text{ atm})(2.0 \text{ mol})^2][51.5 \text{ dm}^3 - 2 \text{ mol}(0.0322)]}{(2.0 \text{ mol})(0.08206 \text{ dm}^3 \text{ atm mol}^{-1} \text{ K}^{-1})}$$

$\underline{\underline{T = 314 \text{ K}}}$

4. *Ideal* $P = \frac{nRT}{V} = \frac{(10\ \text{mol})(0.08206\ \text{dm}^3\ \text{atm mol}^{-1}\ \text{K}^{-1})(300\ \text{K})}{4.86\ \text{dm}^3}$

$= \underline{\underline{50.7\ \text{atm}}} = P_I$

Beattie-Bridgeman:

$$P = \frac{RT}{\bar{V}^2}\left(1 - \frac{c}{\bar{V}T^3}\right)\left(\bar{V} + B_o - \frac{bB_o}{\bar{V}}\right) - \frac{A_o}{\bar{V}^2}\left(1 - \frac{a}{\bar{V}}\right) \quad : \quad \underline{\underline{\bar{V} = 0.486\ \text{dm}^3\ \text{mol}^{-1}}}$$

$$P = \frac{(0.08206)(300)}{(0.486)^2}\left[1 - \frac{90.0 \times 10^4}{(0.486)(300)^3}\right]\left[(0.486) + 0.0940 - \frac{(0.01915)(0.0940)}{(0.486)}\right] - \frac{5.880}{(0.486)^2}\left[1 - \frac{0.05861}{0.486}\right]$$

$= 104.2[0.9314][0.576] - [21.89] = 55.93 - 21.89$

$= \underline{\underline{34\ \text{atm}}} = P_{BB}$

Redlich-Kwong:

$$P = \frac{RT}{\bar{V} - b} - \frac{a}{T^{1/2}\bar{V}(\bar{V} + b)}$$

$a = 0.4278\ R^2T_c^{2.5}/P_c \qquad T_c = 305.4\ \text{K}$

$b = 0.0867\ RT_c/P_c \qquad P_c = 48.2\ \text{atm}$

$a = 0.4278(0.08206)^2(305.4)^{2.5}/(48.2) = \underline{\underline{97.42 = a}}$

$b = 0.0867(0.08206)(305.4)/(48.2) = \underline{\underline{0.04508 = b}}$

$$P = \frac{(0.08206)(300)}{(0.486 - 0.04508)} - \frac{(97.42)}{(300)^{1/2}(0.486)(0.486 + 0.04508)}$$

$= 55.83 - 21.79 = \underline{\underline{34.0\ \text{atm}}} = P_{RK}$

Both BB and RK give very good results. The ideal gas law is very poor as expected for high pressure.

5. $$P = \frac{RT}{\bar{V}^2}\left(1 - \frac{c}{\bar{V}T^3}\right)\left(\bar{V} + B_o - \frac{bB_o}{\bar{V}}\right) - \frac{A_o}{\bar{V}^2}\left(1 - \frac{a}{\bar{V}}\right)$$

$$= \frac{(0.0821)(275)}{(11.2)^2}\left[1 - \frac{4.2 \times 10^4}{(11.2)(275)^3}\right]\left[11.2 + 0.05046 - \frac{(-0.00691)(0.05046)}{11.2}\right] - \frac{1.3445}{(11.2)^2}\left(1 - \frac{0.02617}{11.2}\right)$$

$= 0.18[1 - 0.00018][11.25] - [0.0107][0.9977] = 2.01\ \text{atm}$

$P = \underline{\underline{2.01\ \text{atm}}}$

6. $a = 0.4278\ R^2T_c^{2.5}/P_c = (0.4278)(0.0821)^2(405.6)^{2.5}/(112.2) = 85.15$

$b = 0.0867\ RT_c/P_c = (0.0867)(0.0821)(405.6)/(112.2) = 0.0257$

Solve RK for T:

$$T_{RK} \approx \frac{P(\bar{V} - b)}{R} + \frac{a(\bar{V} - b)}{T^{1/2}\bar{V}(\bar{V} + b)R}$$

Find $T_{ideal} = 935K$. To substitute in right side of equation

$$T_{RK} \approx \frac{(10.1)(7.6 - 0.0257)}{(0.0821)} + \frac{(85.15)(7.6 - 0.0257)}{(935)^{1/2}(7.6)(7.6 + 0.0257)(0.0821)}$$

$$= 931.8 + 4.4 = \underline{\underline{936K = T_{RK}}}$$ No need to iterate.

2-3.11. PROBLEMS

1. $\eta = 9.0 \times 10^{-6}\ kg\ m^{-1}\ s^{-1}$

$$N^* = \frac{NP}{RT} = \frac{(6.023 \times 10^{23}\ mlc\ mol^{-1})(1.00\ atm)}{(0.08206\ dm^3\ atm\ mol^{-1}\ K^{-1})(298.15K)} = \underline{\underline{2.46 \times 10^{22}\ mlc\ dm^{-3}}}$$

$$\bar{c} = [8RT/\pi M]^{1/2} = \left[\frac{(8)(8.314\ J\ mol^{-1}\ K^{-1})(298.15K)}{(3.1416)(0.002016\ kg\ mol^{-1})}\right]^{1/2}$$

$$= \underline{\underline{1.77 \times 10^3\ m\ s^{-1} = \bar{c}}}$$

$$\lambda = \frac{2\eta}{m\bar{c}N^*} = \frac{(2)(9.0 \times 10^{-6}\ kg\ m^{-1}\ s^{-1})}{\left(\frac{0.002016\ kg\ mol^{-1}}{6.023 \times 10^{23}\ mlc\ mol^{-1}}\right)(1.77 \times 10^3\ m\ s^{-2})(2.46\quad 10^{22}\ mlc\ dm^{-3})(10^3\ dm^{+3}\ m^{-3})}$$

$$\lambda = \underline{\underline{1.24 \times 10^{-7}\ m}}$$

$$\sigma = \left[\frac{1}{\sqrt{2}\pi\lambda N^*}\right]^{1/2}$$

$$= \left[\frac{1}{(1.414)(3.1416)(1.24 \times 10^{-7}\ m)(2.46 \times 10^{22}\ mlc\ dm^{-3})(10^3\ dm^{+3}\ m^{-3})}\right]^{1/2}$$

$$= \underline{\underline{2.72 \times 10^{-10}\ m = \sigma}}$$

$$Z_1 = \pi\sigma^2\bar{c}N^*\sqrt{2}$$

$$= (3.1416)(2.72 \times 10^{-10}\ m)^2(1.77 \times 10^3\ m\ s^{-1})(2.46 \times 10^{25}\ mlc\ m^{-3})\sqrt{2}$$

$$= \underline{\underline{1.43 \times 10^{10}\ s^{-1}}} = Z_1$$

$$Z_{11} = \frac{1}{2} Z_1 N^* = \frac{1}{2}(1.43 \times 10^{10}\ s^{-1})(2.46 \times 10^{25}\ mlc\ m^{-3})$$

$$= \underline{\underline{1.76 \times 10^{35}\ mlc\ m^{-3}\ s^{-1}}} = Z_{11}$$

2. Find λ and compare to sun-moon distance.

$$\lambda = \frac{1}{\sqrt{2}\pi\sigma N^*}$$

$$\lambda = \frac{1}{(1.414)(3.1416)(1 \times 10^{-10}\ m)(4\ m^{-3})}$$

$$= \underline{\underline{5.63 \times 10^{18}\ m >> 1.56 \times 10^{8}\ m}}$$

∴ No collision between H atoms.

3. $Z_1 = \pi\bar{c}\sigma^2 N^*$ (see #1 for units)

Also, $P_{O_2} = 0.19$ atm, $P_{N_2} = 0.76$ atm (20% O_2, 80% N_2 in atmosphere)

For O_2:

$$\bar{c} = \left[\frac{8RT}{\pi M}\right]^{1/2} = \left[\frac{(8)(8.314)(273.15)}{(3.1416)(0.032)}\right]^{1/2} = \underline{\underline{425\ m\ s^{-1} = \bar{c}}}$$

$$N^* = \frac{NP}{RT} = \frac{(6.023 \times 10^{23})(0.19)}{(0.08206)(273.15)}\left(\frac{10^3\ dm^3}{1\ m^3}\right) = \underline{\underline{5.10 \times 10^{24}\ mlc\ m^{-3} = N^*}}$$

$$Z_{1,O_2} = \sqrt{2}\pi(3.57 \times 10^{-10}\ m)^2(425\ m\ s^{-1})(5.10 \times 10^{24}\ mlc\ m^{-3})$$

$$= \underline{\underline{1.23 \times 10^{9}\ s^{-1}}} = Z_{1,O_2}$$

For N_2:

$$\bar{c} = 455\ m\ s^{-1};\quad N^* = 2.04 \times 10^{25}\ mlc\ m^{-3};\quad Z_{1,N_2} = \underline{\underline{5.76 \times 10^{9}\ s^{-1}}}$$

4. $$\lambda = 10^5\ m;\quad N^* = \frac{1}{\sqrt{2}\pi\sigma^2\lambda} = \frac{1}{\sqrt{2}\pi(2.18 \times 10^{-10})^2(10^5\ m)}$$

$$= \underline{\underline{4.73 \times 10^{13}\ mlc\ m^{-3}}}$$

$$P = \frac{N^*RT}{N} = \frac{(4.73 \times 10^{13}\ mlc\ m^{-3})(8.206 \times 10^{-5}\ m^3\ atm\ mol^{-1}\ K^{-1})(298.15\ K)}{(6.023 \times 10^{23}\ mlc\ mol^{-1})}$$

$$= \underline{\underline{1.92 \times 10^{-12}\ atm}}$$

5. $$\bar{c} = \left(\frac{8RT}{\pi M}\right)^{1/2} = \left[\frac{(8)(8.314\ \text{J mol}^{-1}\ \text{K}^{-1})(325\ \text{K})}{\pi(0.044\ \text{kg mol}^{-1})}\right]^{1/2} = \underline{\underline{395\ \text{m s}^{-1} = \bar{c}}}$$

$$v_{rms} = \left(\frac{RT}{M}\right)^{1/2} = \left[\frac{(8.314)(325)}{(0.044)}\right]^{1/2} = \underline{\underline{248\ \text{m s}^{-1}}} = v_{rms}$$

6. $\rho = \frac{PM}{RT}$ We need T. Calculate from $v_{rms}^2 = \frac{RT}{M}$ or $T = \frac{Mv_{rms}^2}{R}$

$$\therefore\ T = \frac{(340\ \text{m s}^{-1})^2(0.060\ \text{kg mol}^{-1})}{(8.314\ \text{kg m}^2\ \text{mol}^{-1}\ \text{K}^{-1}\ \text{s}^{-2})} = \underline{\underline{834\ \text{K} = T}}$$

$$\rho = \frac{(0.80\ \text{atm})(0.060\ \text{kg mol}^{-1})}{(0.08206\ \text{dm}^3\ \text{atm mol}^{-1}\ \text{K}^{-1})(834\ \text{K})} = \underline{\underline{7.01 \times 10^{-4}\ \text{kg dm}^{-3} = \rho}}$$

7. $$\lambda = \frac{1}{\sqrt{2}\pi\sigma^2 N^*};\quad Z_1 = \sqrt{2}\pi\sigma^2\bar{c}N^* \quad \text{or} \quad \sqrt{2}\pi\sigma^2 = \frac{Z_1}{\bar{c}N^*}$$

$$\lambda = \frac{1}{\frac{Z_1}{\bar{c}N^*} N^*} = \frac{Z_1}{\bar{c}} \quad \text{or} \quad Z_1 = \frac{\bar{c}}{\lambda} = \frac{380\ \text{m s}^{-1}}{7.31 \times 10^{-8}\ \text{m}} = \underline{\underline{5.20 \times 10^9\ \text{s}^{-1}}} = Z_1$$

8. $$N^* = \frac{NP}{RT} = \frac{(6.02 \times 10^{23}\ \text{mlc mol}^{-1})(2.2 \times 10^{-4}\ \text{atm})}{(0.0821\ \text{dm}^3\ \text{atm mol}^{-1}\ \text{K}^{-1})(100\ \text{K})} \cdot \frac{10^3\ \text{dm}^3}{1\ \text{m}^3}$$

$$= \underline{\underline{1.61 \times 10^{22}\ \text{mlc m}^{-3}}}$$

$$\lambda = \frac{1}{\sqrt{2}\pi\sigma^2 N^*} = \frac{1}{\sqrt{2}\pi(3.74 \times 10^{-10}\ \text{m})^2(1.61 \times 10^{22}\ \text{mlc m}^{-3})}$$

$$\lambda = \underline{\underline{1.0 \times 10^{-4}\ \text{m}}}$$

9. $$\eta = \frac{m\bar{c}}{2\sqrt{2}\pi\sigma^2};\quad m = \frac{M}{N} = \frac{0.128\ \text{kg mol}^{-1}}{6.023 \times 10^{23}\ \text{mlc mol}^{-1}} = \underline{2.125 \times 10^{-25}\ \text{kg mlc}^{-1} = m}$$

$$\bar{c} = \left(\frac{8RT}{\pi M}\right)^{1/2} = \left[\frac{8(8.314\ \text{J mol}^{-1}\ \text{K}^{-1})(298\ \text{K})}{\pi(0.128\ \text{kg mol}^{-1})}\right]^{1/2} = \underline{222\ \text{m s}^{-1} = \bar{c}}$$

$$\therefore\ \eta = \frac{(2.125 \times 10^{-25}\ \text{kg mlc}^{-1})(222\ \text{m s}^{-1})}{2\sqrt{2}\pi(5.55 \times 10^{-10}\ \text{m})^2} = \underline{\underline{1.72 \times 10^{-5}\ \text{kg m}^{-1}\ \text{s}^{-1} = \eta}}$$

10. $$Z_{11}(T_2) = \frac{\sqrt{2}}{2}\pi\sigma^2\bar{c}_2(N_2^*)^2;\quad Z_{11}(T_1) = \frac{\sqrt{2}}{2}\pi\sigma^2\bar{c}_1(N_1^*)^2$$

$$\therefore\ \frac{Z_{11}(T_2)}{Z_{11}(T_1)} = \frac{\bar{c}_2(N_2^*)^2}{\bar{c}_1(N_1^*)^2} = \frac{[8RT_2/(\pi M)]^{1/2}[NP/(RT_2)]^2}{[8RT_1/(\pi M)]^{1/2}[NP/(RT_1)]^2} = \frac{T_1^{3/2}}{T_2^{3/2}}$$

$$Z_{11}(600) = Z_{11}(300)\left(\frac{300}{600}\right)^{3/2} = (9.54 \times 10^{34}\ m^{-3}\ s^{-1})\left(\frac{1}{2}\right)^{3/2}$$

$$= 3.37 \times 10^{34}\ m^{-3}\ s^{-1} = Z_{11}(600)$$

2-4.1 PROBLEMS

1. $$\overline{u^2} = \int_{-\infty}^{\infty} u^2 \frac{dN_u}{N}\, du = c' \int_{-\infty}^{\infty} u^2 e^{-3mu^2/2kT} du$$

We need to evaluate c'.

$$c' \int_{-\infty}^{\infty} e^{-3mu^2/2kT} du = 1$$

or

$$c' = \frac{1}{\int_{-\infty}^{\infty} e^{-3mu^2/2kT} du} = \frac{1}{\int_{-\infty}^{\infty} e^{-au^2} du} : a = \frac{3m}{2kT}$$

From integrals in Appendix D, we find

$$c' = \left(\frac{3m}{2\pi kT}\right)^{1/2}$$

Then

$$\overline{u^2} = \left(\frac{3m}{2\pi kT}\right)^{1/2} \int_{-\infty}^{\infty} u^2 e^{-3mu^2/2kT} du = \left(\frac{3m}{2\pi kT}\right)^{1/2} \int_{-\infty}^{\infty} u^2 e^{-au^2} du$$

From Appendix D:

$$\overline{u^2} = \left(\frac{3m}{2\pi kT}\right)^{1/2}\left(\frac{1}{2a}\right)\left(\frac{\pi}{a}\right)^{1/2} = \left(\frac{3m}{2\pi kT}\right)^{1/2}\left(\frac{2kT}{2 \cdot 3m}\right)\left(\frac{2kT\pi}{3m}\right)^{1/2}$$

$$= \frac{kT}{3m} = \frac{RT}{3M} = \overline{u^2} \qquad u_{rms} = (\overline{u^2})^{1/2}$$

$$\overline{u^2} = \frac{(8.314\ J\ mol^{-1}\ K^{-1})(300\ K)}{(3)(0.036\ kg\ mol^{-1})} = 2.31 \times 10^4\ m^2\ s^{-2} : u_{rms} = 152\ m\ s^{-1}$$

2-5.4 PROBLEMS

1. $$R_{CH_4} = \left(\frac{M_{HE}}{M_{CH_4}}\right)^{1/2} R_{HE} = \left(\frac{0.004}{0.016}\right)^{1/2}(.55\ dm^3\ min^{-1}) = 0.28\ dm^3\ min^{-1}$$

$$= 396\ dm^3\ day^{-1}$$

$$(396\ dm^3\ day^{-1})\left(\frac{1\ ft^3}{28\ dm^3}\right)\left(\frac{\$1.25}{1000\ ft^3}\right) = 0.018\ \$/day = \$6.45/year$$

2. $t = \frac{\ell^2}{2D} = \frac{1\ \text{cm}^2}{(2)(6.1 \times 10^{-7}\ \text{cm}^2\ \text{s}^{-1})} = \underline{\underline{8.25 \times 10^5\ \text{s} = 9.5\ \text{day}}}$

3. $Q = -DA_{eff}\left(\frac{\partial c}{\partial x}\right)t : t = \frac{-Q}{DA_{eff}\left(\frac{\partial c}{\partial x}\right)} : \underline{\underline{\frac{\partial c}{\partial x}}} \approx \frac{\Delta c}{\Delta x} = \frac{-100\ \text{g dm}^{-3}}{(3\ \text{mm})(10^{-2}\ \text{dm mm}^{-1})}$

$= \underline{\underline{-3330\ \text{g dm}^{-4}}}$

D(HANDBOOK) = $\underline{\underline{0.521 \times 10^{-5}\ \text{cm}^2\ \text{s}^{-1}}}$

$\underline{\underline{A_{eff}}} = 0.33\pi r^2 = (0.33)\pi(1\ \text{cm})^2 = \underline{\underline{1.04\ \text{cm}^2}}$

$t = \frac{1}{(0.521 \times 10^{-5}\ \text{cm}^2\ \text{s}^{-1})(1.04\ \text{cm}^2)(-3330\ \text{g dm}^{-4})(.1\ \text{dm cm}^{-1})^4}$

$= \underline{\underline{5.54 \times 10^5\ \text{s} = t}}$

4. RATE $\alpha \left(\frac{1}{\rho}\right)^{1/2}$; $\rho = \frac{m}{V} = \frac{nM}{V} = \frac{MP}{RT}$

$\therefore \frac{\text{RATE}_1}{\text{RATE}_2} = \left[\frac{MP_2/(RT_2)}{MP_1/(RT_1)}\right]^{1/2} = \left(\frac{T_1}{T_2}\right)^{1/2}$ since P is constant.

Or

$\text{RATE}_2 = \text{RATE}_1\left(\frac{T_2}{T_1}\right)^{1/2} = (1.45 \times 10^{-10}\ \text{mol min}^{-1})\left(\frac{203}{406}\right)^{1/2}$

$\underline{\underline{\text{RATE}_2}} = \underline{\underline{1.03 \times 10^{-10}\ \text{mol min}^{-1}}}$

5. $Q = AJt = -DA\left(\frac{\partial c}{\partial x}\right)t$; for unit area, unit time Q = J.

The number given is actually the flux,

$J = -D\left(\frac{\partial c}{\partial x}\right)$

$\therefore \left(\frac{\partial c}{\partial x}\right) = -\frac{J}{D} = \frac{-2.5 \times 10^{-5}\ \text{mol m}^{-2}\ \text{s}^{-1}}{8.2 \times 10^{-11}\ \text{m}^2\ \text{s}^{-1}} = \underline{\underline{-3.0 \times 10^5\ \text{mol m}^{-4}}}$

3-1.6 PROBLEMS

1. (a) $dH = d(E + PV) = dE + P\,dV + V\,dP = \underline{\underline{dE + V\,dP}}$ (V = constant)

(b) $dE = n\bar{C}_V\,dT + \left[T\left(\frac{\partial P}{\partial T}\right)_V - P\right]dV = n\bar{C}_V\,dT$

(if V is constant or if ideal gas)

$\therefore \underline{\underline{\Delta E}} = \int n\bar{C}_V\,dT = \underline{\underline{\int C_V\,dT}}$

2. If gas is ideal, $\Delta E = \Delta H = 0$. Then $Q = -W$.

$$W = -\int P\,dV = -\int \frac{nRT\,dV}{V} \qquad \text{(gas ideal)}$$

$$W = -nRT \int_{V_1}^{V_2} \frac{dV}{V}$$

(Isothermal, otherwise T cannot be taken out of integral)

$$W = -nRT\,\ell n \frac{V_2}{V_1} = -(3\text{ moles})(8.314\text{ J mol}^{-1}\text{ K}^{-1})(300\text{ K})\,\ell n\left(\frac{2.5\text{ dm}^3}{100\text{ dm}^3}\right)$$

$$= \underline{\underline{27.6\text{ kJ} = W = -Q}}$$

3. Since T is constant, $\Delta H = \Delta E = 0$. Then $Q = -W$.

$\underline{\underline{W = -475\text{J};\ Q = 475\text{J}}}$

4. $n = \dfrac{10\text{ g}}{40\text{ g mol}^{-1}} = 0.25\text{ mol}$; $\quad P_1 = P_2 = 0.75\text{ atm}$

If ideal gas and $\bar{C}_p$ is constant:

$$\Delta\bar{H} = \int_{T_1}^{T_2} n\bar{C}_p\,dT = n\bar{C}_p \int_{T_1}^{T_2} dT = n\bar{C}_p(T_2 - T_1)$$

$$= (0.25\text{ mol})(\tfrac{5}{2})(8.314\text{ J mol}^{-1}\text{ K}^{-1})(330\text{K} - 105\text{K})$$

$$= \underline{\underline{1.17\text{ kJ} = \Delta H}}$$

5. If ideal and $\bar{C}_v$ is constant:

$$\Delta E = \int_{T_1}^{T_2} n\bar{C}_v\,dT = n\bar{C}_v\,\Delta T = (0.25)(\tfrac{3}{2})(8.314)(330 - 105)$$

$$= \underline{\underline{701\text{ J} = \Delta E}}$$

6. $W = -\int P dV = -\int \frac{nRT}{V}\,dV$ (If Rev. and Ideal)

Cannot integrate since T, V both change. However, the problem states the process is isobaric.

$$\therefore\ W = -\int P_{opp} dV = -P_{opp}\int dV = -P_{opp}(V_2 - V_1) = \underline{-P_1(V_2 - V_1) = W}$$

$$P_1 = nRT_1/V_1 = (1.72\text{ mol})(0.0821\text{ dm}^3\text{ mol}^{-1}\text{ K}^{-1}\text{ atm})(307\text{K})/(33.2\text{ dm}^3)$$

$$= \underline{1.31\text{ atm} = P_1}$$

$$\therefore\ W = 720\text{ J} = 7.13\text{ dm}^3\text{ atm} = -1.31\text{ atm}(V_2 - V_1)$$

$$\underline{V_2 - V_1 = -5.46\text{ dm}^3}$$

$$V_2 = V_1 - 5.46 = 33.2\ dm^3 - 5.46\ dm^3 = \underline{27.7\ dm^3 = V_2}$$

$$T_2 = \frac{P_2V_2}{nR} = \frac{(1.31\ atm)(27.7\ dm^3)}{(1.72\ mol)(0.0821\ dm^3\ atm\ mol^{-1}\ K^{-1})} = \underline{\underline{257\ K = T_2}}$$

7. If ideal, isothermal, $\Delta E = \Delta H = 0$; $Q = W$.

$$W = -\int PdV = -P_{ext}\int dV = -P_{ext}(V_2 - V_1)$$

$$V_1 = nRT/P_1 = 6.43\ dm^3;\quad V_2 = 13.43\ dm^3$$

$$W = -(1.05\ atm)(13.43 - 6.43)dm^3 = -7.36\ dm^3\ atm$$

$\underline{\underline{W = -743\ J}}$; $\underline{\underline{Q = +743\ J}}$

8. $$\left[T\left(\frac{\partial P}{\partial T}\right)_V - P\right] = \left[T\,\frac{nR}{V} - \frac{nRT}{V}\right] = 0 \quad \text{for ideal gas}$$

$$\therefore\ \Delta E = \int n\bar{C}_V\, dT = n\bar{C}_V\, \Delta T$$

$$\Delta H = \Delta E + \Delta(PV) = \Delta E + \Delta(nRT) = n\bar{C}_V\, \Delta T + nR\, \Delta T = n\bar{C}_p\, \Delta T$$

$\therefore$ $\underline{\underline{\Delta H = n\bar{C}_p\, \Delta T}}$ for ideal gas

3-2.6 PROBLEMS

1. Ideal, constant volume $\Delta E = n\bar{C}_V\, \Delta T$; $\bar{C}_V = \bar{C}_p - R$; $\Delta H = n\bar{C}_p\, \Delta T$

$$\Delta H = (5\ mol)(\tfrac{7}{2})(8.314\ J\ mol^{-1}\ K^{-1})(240K - 120K)$$

$= \underline{\underline{17.5\ kJ = \Delta H}}$: $\underline{\underline{\Delta E}} = 5(\tfrac{5}{2})(8.314)(\Delta T) = \underline{\underline{12.5\ kJ}}$

At constant volume, $Q = \Delta E$ since $W = 0$

$\therefore$ $\underline{\underline{Q = 12.5\ kJ}}$

2. (a) We need ΔT to apply $\Delta E = \int n\bar{C}_V\, dT = n\bar{C}_V\, \Delta T$ for ideal gas; C_V constant. Equation (3-2.4.11) gives

$$\left(\frac{P_2}{P_1}\right) = \left(\frac{T_2}{T_1}\right)^{\bar{C}_p/R} = \left(\frac{T_2}{T_1}\right)^{5/2}$$

since $\bar{C}_p = \bar{C}_V + R = \frac{5}{2}R$

$$\left(\frac{T_2}{T_1}\right) = \left(\frac{P_2}{P_1}\right)^{2/5} = \left(\frac{1.0}{7.0}\right)^{0.40} = 0.459$$

$\therefore$ $T_2 = 0.459\ T_1 = (0.459)(600K) = \underline{\underline{275K = T_2}}$

$\Delta E = (1\ mol)(\tfrac{3}{2})(8.314\ J\ mol^{-1}\ K^{-1})(275 - 600)K = \underline{\underline{-4.05\ kJ = \Delta E}}$

$\Delta H = (1)(\tfrac{5}{2})(8.314)(-325) = \underline{\underline{-6.76\ kJ = \Delta H}}$

(b) $Q = 0;\quad W = -\int P\,dV = -P_{opp}(V_2 - V_1)$ (P_{opp} is constant at 1 atm)

$$= -1\text{ atm}\left(\frac{RT_2}{P_2} - \frac{RT_1}{P_1}\right)$$ (P_1, P_2 are internal pressures)

Thus

$$\Delta E = W = \bar{C}_v(T_2 - T_1) = -1\text{ atm}\left(\frac{RT_2}{P_2} - \frac{RT_1}{P_1}\right)$$

Then

$$\frac{3}{2}\cancel{R}T_2 - \frac{3}{2}\cancel{R}T_1 = -\frac{\cancel{R}T_2}{P_2} + \frac{\cancel{R}T_1}{P_1}$$

$$\frac{3}{2}T_2 + \frac{T_2}{1} = \frac{3}{2}T_1 + \frac{T_1}{7};\quad T_2 = \frac{2}{5}(1.64)T_1 = 0.657\ T_1$$

$$\underline{\underline{T_2 = 394\ \text{K}}}$$

$$\Delta E = n\bar{C}_v\ \Delta T = (1\text{ mol})(\tfrac{3}{2})(8.314)(394 - 600\text{K})$$

$$= \underline{\underline{-2.57\ \text{kJ} = \Delta E}}$$

$$\underline{\underline{\Delta H}} = (1)(\tfrac{5}{2})(R)(\Delta T) = \underline{\underline{-4.28\ \text{kJ}}}$$

3. $\Delta H = \int n\bar{C}_p\,dT = (2\text{ mol})\int_{298}^{423}(22.63 + 6.28\times 10^{-3}\,T)\,dT\,(\text{J mol}^{-1})$

$$= 2\left[22.63(T_2 - T_1) + \frac{6.28\times 10^{-3}}{2}\left(T_2^2 - T_1^2\right)\right]$$

$$= 2\left[22.63(423 - 298) + \frac{6.28\times 10^{-3}}{2}(423^2 - 298^2)\right]\text{J}$$

$$= 2[2829 + 283]\text{J} = \underline{\underline{6.22\ \text{kJ} = \Delta H}}$$

$\Delta E \approx \Delta H$ since ΔV is very small. Would need thermal expansion coefficient to get W.

4. $\dfrac{37.02\ \text{g}}{18.02\ \text{g mol}^{-1}} = 2.06\text{ mol } H_2O;\quad \Delta\bar{H}_{vap} = 40.66\text{ kJ mol}^{-1}$

$\therefore\ \underline{\underline{\Delta\bar{H}}} = (2.06\text{ mol})(40.66\text{ kJ mol}^{-1}) = \underline{\underline{83.8\ \text{kJ}}}$

5. From Beattie-Bridgeman:

$$P = \left(\frac{RT}{\bar{V}^2} - \frac{cRT}{\bar{V}^3T^3}\right)(\bar{V} + B_0 - bB_0/\bar{V}) - \frac{A_0}{\bar{V}^2}\left(1 - \frac{a}{\bar{V}}\right)$$

$$-W = \int P dV = \int\left(\frac{RT}{\bar{V}} - \frac{cRT}{\bar{V}^2T^3} + \frac{B_0RT}{\bar{V}^2} - \frac{cB_0RT}{\bar{V}^3T^3} - \frac{bB_0RT}{\bar{V}^3} + \frac{cRTbB_0}{\bar{V}^4T^3} - \frac{A_0}{\bar{V}^2} + \frac{aA_0}{\bar{V}^3}\right)dV$$

$$= \int\left[\frac{RT}{\bar{V}} + \frac{1}{\bar{V}^2}\left(RTB_0 - \frac{cR}{T^2} - A_0\right) + \frac{1}{\bar{V}^3}\left(-RTbB_0 - \frac{cB_0R}{T^2} + aA_0\right) + \frac{1}{\bar{V}^4}\left(\frac{cbB_0R}{T^2}\right)\right]dV$$

This can be integrated (you should do it). After collecting terms:

$$-W = RT \ln \frac{\bar{V}_2}{\bar{V}_1} - \left(\frac{1}{\bar{V}_2} - \frac{1}{\bar{V}_1}\right)\left(-\frac{cR}{T^2} + B_0RT - A_0\right)$$

$$- \frac{1}{2}\left(\frac{1}{\bar{V}_2^2} - \frac{1}{\bar{V}_1^2}\right)\left(-\frac{cB_0R}{T^2} - bB_0RT + aA_0\right) - \frac{1}{3}\left(\frac{1}{\bar{V}_2^3} - \frac{1}{\bar{V}_1^3}\right)\left(\frac{cRbB_0}{T^2}\right)$$

$A_0 = 1.3445$; $a = 0.02617$; $B_o = 0.05046$; $b = -0.00691$;

$c = 4.20 \times 10^4$

$$-W = (0.0821)(310) \ln\left(\frac{30}{10}\right) - \left(\frac{1}{30} - \frac{1}{10}\right)\left[-\frac{4.20 \times 10^4(.0821)}{310^2}\right.$$

$$\left. + (0.0506)(0.0821)(310) - 1.3445\right]$$

$$+ \frac{1}{2}\left(\frac{1}{30^2} - \frac{1}{10^2}\right)\left[\frac{(4.20 \times 10^4)(0.05046)(0.0821)}{310^2}\right.$$

$$\left. + (-0.00691)(0.05046)(0.0821)(310) - (1.3445)(0.02617)\right]$$

$$- \frac{1}{3}\left(\frac{1}{30^3} - \frac{1}{10^3}\right)\left[\frac{(4.20 \times 10^4)(0.0821)(-0.00691)(0.05046)}{310^2}\right]$$

$$= 27.96 - 6.4 \times 10^{-3} + 1.87 \times 10^{-4} - 4 \times 10^{-9}$$

$$= 27.95 \text{ dm}^3 \text{ atm mol}^{-1} = \underline{\underline{2.82 \times 10^3 \text{ J mol}^{-1} = -W}}$$

Note: The ideal gas law gives about the same value in this case.

6. $\Delta H_{fusion} = \Delta E_{fus} + \Delta(PV)$ so $\Delta H_{fus} - \Delta E_{fus} = P\Delta V = P(V_\ell - V_s)$

(if P constant)

$$\Delta H_{fus} - \Delta E_{fus} = P\left(\frac{M}{\rho_\ell} - \frac{M}{\rho_s}\right) = (1 \text{ atm})(0.018 \text{ kg mol}^{-1})\left(\frac{1}{0.9998} - \frac{1}{0.917}\right)\frac{\text{dm}^3}{\text{kg}}$$

$$= -1.62 \times 10^{-3} \text{ dm}^3 \text{ atm mol}^{-1} = \underline{\underline{-0.164 \text{ J mol}^{-1} = \Delta H_{fus} - \Delta E_{fus}}}$$

The volume decreases in the transition from $s \to \ell$ so the system has work done on it.

$$\Delta H_{vap} - \Delta E_{vap} = P(V_g - V_\ell) = (1 \text{ atm})(0.018 \text{ kg mol}^{-1})$$

$$\times \left(\frac{1}{5.96 \times 10^{-4}} - \frac{1}{0.9584}\right)\frac{\text{dm}^3}{\text{kg}}$$

$$= 30.2 \text{ dm}^3 \text{ atm mol}^{-1} = \underline{\underline{3.05 \text{ kJ mol}^{-1} = \Delta H_{vap} - \Delta E_{vap}}}$$

7. (a) $\left(\frac{P_2}{P_1}\right) = \left(\frac{T_2}{T_1}\right)^{C_p/R}$: $T_2 = T_1\left(\frac{P_2}{P_1}\right)^{R/C_p} = \left(\frac{0.1}{1.0}\right)^{R/\frac{7}{2}R}$

$$T_1 = \left(\frac{0.1}{1.0}\right)^{2/7}(400\ \text{K}) = (0.518)(400\ \text{K}) = \underline{\underline{207\ \text{K} = T_2}}$$

$$\Delta E = \int nC_v\, dT = (3\ \text{moles})(\tfrac{5}{2})(8.314\ \text{J mol}^{-1}\ \text{K}^{-1})(207 - 400)\text{K}$$

$$= \underline{\underline{-12.0\ \text{kJ} = \Delta E}}$$

(b) From equation (3-2.4.13)

$$\left(\frac{T_2}{T_1}\right)^{\bar{C}_v/R} = \left(\frac{V_1 - nb}{V_2 - nb}\right)$$

assuming $\bar{C}_v$ is constant. If not, the procedure of Example 3-2.4.4 must be used.

$$\left(\frac{T_2}{T_1}\right)^{\bar{C}_v/R} = \frac{(P_2 + an^2/V_2^2)(T_1)}{(P_1 + an^2/V_1^2)(T_2)}$$

Thus

$$\left(\frac{T_2}{T_1}\right)^{\frac{\bar{C}_v}{R} + 1} = \left(\frac{P_2 + an^2/V_2^2}{P_1 + an^2/V_1^2}\right)$$

or

$$T_2 = T_1\left(\frac{P_2 + an^2/V_2^2}{P_1 + an^2/V_1^2}\right)^{\frac{1}{\bar{C}_v/R + 1}}$$

We need V_1 and V_2.

$$a = 1.46\ \text{dm}^6\ \text{atm mol}^{-2};\quad b = 0.0392\ \text{dm}^3\ \text{mol}^{-1} : n = 3\ \text{mol}$$

$$V_1 \approx \frac{nRT}{(P + an^2/V_{ideal}^2)} - nb : V_{ideal} = \frac{nRT_1}{P_1}$$

$$= (3)(0.0821)(400)/(1) = \underline{\underline{98.52 = V_{ideal}}}$$

$$V_1 = \frac{(3)(0.0821)(400)}{[1 + (1.46)(3)^2/(98.5)^2]} - 3(0.0392) = 98.3\ \text{dm}^3$$

Iterate → $\underline{\underline{98.3\ \text{dm}^3 = V_1}}$

Check V_2 using $T_2 = 207$ K as found in (a).

$$V_{ideal} = 510;\quad V_{vdw} = 509.5 = \underline{\underline{510\ \text{dm}^3 = V_2}}$$

$$T_2 = 400\left[\frac{0.1 + 1.46(3)^2/(510)^2}{1 + 1.46(3)^2/(98.3)^2}\right]^{2/7} = 400(0.518) = \underline{\underline{207\ K = T_2}}$$

Same as (a). So the assumption in finding V_2 was good. Note that, if T_2 were not 207, we would have to recalculate V_2, then find T_2 again and iterate until a constant set is obtained. We assumed $\bar{C}_V$ a constant which is probably a good assumption there.

8. From equation (3-2.3.6)

$$\mu = -\frac{1}{C_p}\left[V - T\left(\frac{\partial V}{\partial T}\right)_p\right]$$

(a) $V = T\left(\frac{\partial V}{\partial T}\right)_p$ for ideal gas. $\therefore\ \mu = 0$

(b) For CO_2, $b = 0.0428\ dm^3\ mol^{-1}$; $a = 3.60\ dm^6\ atm^2\ mol^{-2}$

$$C_p(300K) = 44.22 + 8.79 \times 10^{-3}(300) - 8.62 \times 10^5(300)^{-2}$$

$$= 37.28\ J\ mol^{-1}\ K^{-1} = 0.369\ \frac{dm^3\ atm}{mol\ K}$$

$$\mu = \left(\frac{2a}{RT} - b\right)/\bar{C}_p = \left[\frac{2(3.60)}{(0.0821)(300)} - 0.0428\right]/0.369$$

$$= \underline{\underline{0.68\ K\ atm^{-1}}} = \underline{\underline{\mu_{vdw}}}$$

μ can be found from the Beattie-Bridgeman equation.

$$\mu = \frac{1}{\bar{C}_p}\left\{-B_0 + \frac{2A_0}{RT} + \frac{4c}{T^3} + \left[\frac{2B_0 b}{RT} - \frac{3A_0 a}{(RT)^2} + \frac{5B_0 c}{(RT^4)}\right]P\right\}$$

$$A_0 = 5.0065,\ a = 0.07132,\ B_0 = 0.10476,\ b = 0.07235,\ c = 66 \times 10^4$$

$$\mu = \frac{1}{0.369}\{-0.105 + 0.407 + 0.0978 + [0.00062 - 0.00177 + 0.00052]10$$

$$= \frac{0.393}{0.369} = \underline{\underline{1.07 = \mu_{BB}}}$$

The figure gives

$$\mu \approx \underline{\underline{1.0\ K\ atm^{-1}}}$$

Therefore μ_{BB} is much better than van der Waals

9. Redlich-Kwong:

$$P = \frac{RT}{\bar{V} - b} - \frac{a}{T^{1/2}\bar{V}(\bar{V} + b)}$$

$$a = 0.4278\ R^2T_c^{2.5}/P_c; \quad b = 0.0867\ RT_c/P_c$$

$$a = (0.4278)(0.08206)^2(405.6)^{2.5}/(112.2) = \underline{\underline{85.1 = a}}$$

$$b = (0.0867)(0.08206)(405.6)/(112.2) = \underline{\underline{0.0257 = b}}$$

(a) $$dE = C_V\ dT + \left[T\left(\frac{\partial P}{\partial T}\right)_V - P\right]dV; \quad \left(\frac{\partial P}{\partial T}\right)_V = \frac{R}{\bar{V} - b} + \frac{a}{2T^{3/2}\bar{V}(\bar{V} + b)}$$

$$\therefore \quad dE = C_V\ dT + \left[\frac{RT}{\bar{V} - b} + \frac{a}{2T^{1/2}\bar{V}(\bar{V} + b)} - \frac{RT}{\bar{V} - b} + \frac{a}{T^{1/2}\bar{V}(\bar{V} + b)}\right]dV$$

$$= \underline{\underline{C_V\ dT + \frac{3a\ dV}{2T^{1/2}\bar{V}(\bar{V} + b)} = dE}}$$

The last term can not be integrated readily since it involves both T and V and these both change.

(b) T = constant in this part. So

$$d\bar{E} = \frac{3a\ dV}{2T^{1/2}\bar{V}(\bar{V} + b)} \quad \text{or} \quad \Delta\bar{E} = \frac{3a}{2T^{1/2}} \int \frac{dV}{\bar{V}(\bar{V} + b)}$$

From an integral table:

$$\Delta\bar{E} = -\frac{1}{b}\frac{3a}{2T^{1/2}}\ \ell n \left.\frac{\bar{V} + b}{\bar{V}}\right|_{\bar{V}_1}^{\bar{V}_2}$$

$$\Delta\bar{E} = -\frac{1}{0.0257}\frac{(3)(85.1)}{2(280K)^{1/2}}\ \ell n \left\{\left[\frac{(30.6 + 0.0257)}{(30.6)}\right]\left[\frac{40.1}{(40.1 + 0.0257)}\right]\right\}$$

$$= -5.9 \times 10^{-2}\ dm^3\ atm\ mol^{-1}$$

27 g = 1.59 mol $\quad \therefore$ $\underline{\underline{\Delta E = -0.094\ dm^3\ atm}}$ (a very small change)

Return to part (a) and assume the dV term contributes only a negligible amount, as found in part (b).

$$\Delta E = (1.59\ mol) \int_{T_1}^{T_2} \bar{C}_p\ dT = 1.59 \int_{295}^{376} [29.75 + 25.1 \times 10^{-3}\ T - 1.55 \times 10^5\ T^{-2}]dT$$

$$= (1.59)\left[29.75(376 - 295) + \frac{25.10 \times 10^{-3}}{2}(376^2 - 295^2) + 1.55 \times 10^5\left(\frac{1}{376} - \frac{1}{295}\right)\right]$$

$$= 1.59[2410 + 682 - 113] = 1.59(2980) = 4.74 \times 10^3\ J$$

$\underline{\underline{\Delta E = 46.9\ dm^3\ atm}}$

This is much larger than the dV contribution, so we can safely forget the dV term in this case.

10. $Q_p = \int n\bar{C}_p \, dT = (1 \text{ mol})(75.5 \text{ J mol}^{-1} \text{ K}^{-1})(99°C - 0°C) \quad (1 \text{ C}° = 1 \text{ K}°)$

$$\underline{\underline{Q_p = \Delta H = 7470 \text{ J}}}$$

$$\Delta E = \Delta H - \Delta(PV) = \Delta H - P\,\Delta V = \Delta H - P(V_2 - V_1) = \Delta H - P\left(\frac{M}{\rho_2} - \frac{M}{\rho_1}\right)$$

$$= 7470 \text{ J} = 1 \text{ atm}\left(\frac{0.018 \text{ kg mol}^{-1}}{0.9584 \text{ kg dm}^{-3}} - \frac{0.018 \text{ kg mol}^{-1}}{0.9998 \text{ kg dm}^{-3}}\right)(1 \text{ mol})\left(\frac{101 \text{ J}}{1 \text{ dm}^3 \text{ atm}}\right)$$

$$= 7470 \text{ J} - 0.08 \text{ J} = \underline{\underline{7470 \text{ J} = \Delta E}}$$

11. $\dfrac{V_1}{V_2} = \left(\dfrac{T_2}{T_1}\right)^{C_V/R}$ or $T_2 = T_1\left(\dfrac{V_1}{V_2}\right)^{R/C_V} = (290)\left(\dfrac{32.72}{41.6}\right)^{8.314/25.7}$

$$= 290 \ (0.925) = \underline{\underline{268 \text{ K} = T_2}}$$

$$P_2 = nRT_2/V_2 = (0.76 \text{ mol})(0.0821 \text{ dm}^3 \text{ mol}^{-1} \text{ K}^{-1})(268\text{K})/(41.6 \text{ dm}^3)$$

$$= \underline{\underline{0.402 \text{ atm} = P_2}}$$

$$\Delta E = \int n\bar{C}_V \, dT + \int\left[T\left(\frac{\partial P}{\partial T}\right)_V - P\right]dV$$

Can evaluate dT term. Need an equation of state to find the dV terms.

3-3.8 PROBLEMS

1. The reactions are:

$C_2H_2 + \frac{5}{2} O_2 \rightarrow 2CO_2 + H_2O \qquad \Delta H_{comb} = -1300$ kJ (for the reactions as written)

$2(H_2 + \frac{1}{2} O_2 \rightarrow H_2O \qquad \Delta H_{comb} = -285.9$ kJ)

$-(C_2H_6 + \frac{7}{2} O_2 \rightarrow 2CO_2 + 3H_2O \qquad \Delta H_{comb} = -1560$ kJ)

$$C_2H_2 + \frac{5}{2} O_2 + 2H_2 + O_2 - C_2H_6 - \frac{7}{2} O_2 \rightarrow 2CO_2 + H_2O + 2H_2O - 2CO_2 - 3H_2O$$

$$C_2H_2 + 2H_2 - C_2H_6 \rightarrow 0 \qquad \Delta H_{reac} = -1300 + 2(-285.9) - (-1560)$$

$$C_2H_2 + 2H_2 \rightarrow C_2H_6 \qquad \underline{\underline{\Delta H_{reac} = -311.8 \text{ kJ}}}$$

2. $\Delta H_{reac} = \Delta H^\circ_f(C_2H_6) - 2\Delta H^\circ_f(H_2) - \Delta H^\circ_f(C_2H_2) = -84.7 \text{ kJ} - 2(0) - (226.75)\text{kJ}$

$$= \underline{\underline{-311.4 \text{ kJ} = \Delta H_{reac}}}$$

3. Apply combustion data.

$$C_2H_5OH + 3O_2 \rightarrow 2CO_2 + 3H_2O \qquad \Delta H_{comb} = -1367\ kJ$$

$$-(CH_3COOH + 2O_2 \rightarrow 2CO_2 + 2H_2O \quad \Delta H_{comb} = -872.4\ kJ)$$

$$C_2H_5OH + 3O_2 - CH_3COOH - 2O_2 \rightarrow 2CO_2 + 3H_2O - 2CO_2 - 2H_2O$$

$$C_2H_5OH + O_2 - CH_3COOH \rightarrow H_2O \qquad \Delta H_{reac} = (-1367) - (-872.4)$$

$$C_2H_5OH + O_2 \rightarrow CH_3COOH + H_2O \qquad \underline{\underline{\Delta H_{reac} = -494.6\ kJ}}$$

4. $S(g) + O_2(g) \rightarrow SO_2(g)$

$$\underline{\underline{\Delta H^\circ_R}} = \Delta H^\circ_f(SO_2) - \Delta H^\circ_f(S) - \Delta H^\circ_f(O_2) = -296.9 - (222.8) - 0 = \underline{\underline{-519.7\ kJ}}$$

	Item	S(s)	$O_2(g)$	$SO_2(g)$	Δ(Item)
kJ	ΔH°_f	222.8	0	-296.9	-519.7
JK^{-1}	a	22	29.96	38.60	-13.36
	$b \times 10^3$	-0.418	4.18	9.25	5.49
	$c \times 10^5$	1.50	-1.67	5.02	5.19

$$C_p = a + bT + cT^{-2}$$

$$\Delta H_R(600) = \Delta H^\circ_R(298) + \int_{298}^{600} (\Delta a + \Delta bT + \Delta cT^{-2})dT$$

$$= -519{,}700\ J + \Delta a(T_2 - T_1) + \frac{\Delta b}{2}(T_2^2 - T_1^2) - \Delta c\left(\frac{1}{T_2} - \frac{1}{T_1}\right)$$

$$= -519{,}700\ J - 13.36(600 - 298) + \frac{5.49 \times 10^{-3}}{2}(600^2 - 298^2)$$

$$- 5.19 \times 10^{-5}\left(\frac{1}{600} - \frac{1}{298}\right)$$

$$= -519{,}700 - 4035 + 1489 - (-877) = -5.214 \times 10^5\ J$$

$$\underline{\underline{\Delta H_R(600) = -5.21 \times 10^5\ J}}$$

5. The reactions are:

$$CO_2(g) + H_2O(\ell) \rightarrow CO_3^=(aq) + 2H^+(aq)$$

$$Ca^{2+}(aq) + CO_3^=(aq) \rightarrow CaCO_3(s)$$

$$Ca^{2+}(aq) + CO_2(g) + H_2O(\ell) \rightarrow CaCO_3(s) + 2H^+(aq)$$

$$\Delta H_{reac} = 2\Delta H_f^\circ(H^+) + \Delta H_f^\circ(CaCO_3(s)) - \Delta H_f^\circ(Ca^{2+}(aq)) - \Delta H_f^\circ(CO_2(g)) - \Delta H_f^\circ(H_2O(\ell))$$

$$= 2(0) - 1206.0 - (-542.96) - (-393.51) - (-285.84)\ \text{kJ}$$

$$\Delta H^\circ_{reaction} = 16.3\ \text{kJ}$$

6. $H_2(g) + I_2(g) \rightarrow 2HI(g)$

	Item	$H_2(g)$	$I_2(g)$	2HI (g)	Δ(Item)
kJ	ΔH_f°	0	62.24	2(25.9)	-10.44
JK^{-1}	a	27.28	37.40	2(29.2)	-6.28
	$b \times 10^3$	3.26	0.59	0	-3.85
	$c \times 10^{-5}$	0.50	-0.71	0	0.21

$$\Delta H_{reac} = \Delta H_R^\circ + \int_{298}^{520} \Delta C_p\, dT$$

$$= \Delta H_R^\circ + \Delta a(T_2 - T_1) + \frac{\Delta b}{2}\left(T_2^2 - T_1^2\right) - \Delta c\left(\frac{1}{T_2} - \frac{1}{T_1}\right)$$

$$= -10{,}440 - 6.28(520 - 298) - \frac{3.85 \times 10^{-3}}{2}(520^2 - 298^2) - 0.21 \times 10^5\left(\frac{1}{520} - \frac{1}{298}\right)$$

$$= -10{,}440 - 1394 - 350 - (-30) = -12{,}150\ \text{J}$$

$$\Delta H_{reac} = -12.2\ \text{kJ}$$

7. $4[C(g) + O_2(g) \rightarrow CO_2(g)]$

$$4(\Delta H_r = \Delta H_{comb}(C) = -393.51) = -1574.04\ \text{kJ}$$

$$5[H_2(g) + \tfrac{1}{2}\,O_2(g) \rightarrow H_2O(g)]$$

$$5(\Delta H_r = \Delta H_{comb}(H_2) = -285.84) = -1429.20\ \text{kJ}$$

$$4CO_2(g) + 5H_2O(g) \rightarrow \tfrac{13}{2}\,O_2(g) + C_4H_{10}(g)$$

$$\Delta H_r = -\Delta H_{comb}(C_4H_{10}) = -(-2877.1) = 2877.1\ \text{kJ}$$

$$4C(s) + 5H_2(g) \rightarrow C_4H_{10}(g) \qquad \Delta H = \Delta H_f^\circ = -126.1\ \text{kJ(mol } C_4H_{10})^{-1}$$

8. $H_2(g) + S(g) \rightarrow H_2S(g)$

$$\Delta C_p = C_p(H_2S) - C_p(S) - C_p(H_2) = -16.8 + 9.54 \times 10^{-3}\ T$$

$$\Delta H_f^\circ(300) = \Delta H_f^\circ(450) + \int_{450}^{300} \Delta C_p dT = \Delta H_f^\circ(450) + \int_{450}^{300} (-16.8 + 9.54 \times 10^{-3}\ T)dT$$

$$= -22{,}131 + (-16.8)(300 - 450) + \frac{9.54 \times 10^{-3}}{2}(300^2 - 450^2)$$

$$= -22{,}131 + 2520 - 537 = -20{,}148\ J$$

$$\underline{\underline{\Delta H_f^\circ(300) = -20.148\ kJ(mol\ H_2S)^{-1}}}$$

9. $CH_4(g) + 2O_2(g) \rightarrow CO_2(g) + 2H_2O(\ell)$

$$\Delta H_r^\circ = 2\Delta H_f^\circ(H_2O) + \Delta H_f^\circ(CO_2) - 2\Delta H_f^\circ(O_2) - \Delta H_f(CH_4)$$

$$= 2(-285.8) + (-393.51) - 0 - (-74.81) = \underline{\underline{-890.3\ kJ = \Delta H_r^\circ}}$$

10. $H_2(g) + Br_2(g) \rightleftharpoons 2HBr(g)$

$$\Delta H_r^\circ = 2\Delta H_f^\circ(HBr) - \Delta H_f^\circ(H_2) - \Delta H_f^\circ(Br_2, g)$$

$$\Delta H_r^\circ = 2(-36.2) - 0 - (30.7)kJ = \underline{\underline{-103.1\ kJ = \Delta H_r^\circ}}$$

4-1.8 PROBLEMS

1. The system is initially at equilibrium at T_{int}. If T is increased by dT the energy of the system will increase by $\Delta E = \int C_v\, dT$. Assume we want q = 0, so no heat is dissipated to the surroundings. To restore equilibrium the energy must be changed somehow. This can be done if work is performed.

$$q = 0 \qquad \therefore \quad dE = w$$

The gas in the cylinder can expand, performing work on the surroundings until the external energy has fallen to its equilibrium value.

If T is decreased, the reverse of the above process occurs.

2. Apply equations (4-1.4.8) and (4-1.4.13) to find the volumes that are unknown.

$$\underline{\underline{V_3 = 1.2\ dm^3;\ V_2 = ?}} \qquad \underline{\underline{V_1 = 0.10\ dm^3;\ V_4 = ?}}$$

$$\frac{\bar{C}_V}{R} \ell n \frac{T_\ell}{T_h} = - \ell n \frac{V_3 - nb}{V_2 - nb} :$$

$$\frac{3}{2}\frac{R}{R} \ell n \frac{300}{1100} = - \ell n \frac{1.2 - (2 \text{ mol})(0.0237 \text{ dm}^3 \text{ mol}^{-1})}{V_2 - (2 \text{ mol})(0.0237 \text{ dm}^3 \text{ mol}^{-1})}$$

$$7.02 = (1.2 - 0.0474)/(V_2 - 0.0474) : \underline{\underline{V_2 = 0.21 \text{ dm}^3}}$$

$$\frac{\bar{C}_V}{R} \ell n \frac{T_h}{T_\ell} = - \ell n \frac{V_1 - nb}{V_4 - nb} : 0.142 = \frac{0.10 - 0.0474}{V_4 - 0.0474} : \underline{\underline{V_4 = 0.418 \text{ dm}^3}}$$

Step 1

$$W_1 = -nRT_h \ell n \frac{V_2 - nb}{V_1 - nb} - an^2\left(\frac{1}{V_2} - \frac{1}{V_1}\right)$$

$$= -2(8.314 \text{ J})(1100) \ell n \frac{0.21 - 0.0474}{0.10 - 0.0474}$$

$$- 0.0341(2)^2\left(\frac{1}{0.21} - \frac{1}{0.1}\right)\text{dm}^3 \text{ atm}$$

$$= -2.1 \times 10^4 \text{ J} + 0.71 \text{ dm}^3 \text{ atm}$$

$$= -2.1 \times 10^4 \text{ J} + 72 \text{ J} = \underline{\underline{-2.1 \times 10^4 \text{ J} = W_1}}$$

$$\Delta E_1 = -an^2\left(\frac{1}{V_2} - \frac{1}{V_1}\right) = 0.71 \text{ dm}^3 \text{ atm} = \underline{\underline{72 \text{ J} = \Delta E_1}} :$$

$$Q_1 = \Delta E_1 - W_1 = \underline{\underline{+2.1 \times 10^4 \text{ J} = Q_1}}$$

$$\Delta S_1 = Q_1/T_h = \underline{\underline{19 \text{ JK}^{-1} = \Delta S_1}}$$

Step 2

$$\underline{\underline{Q_2 = 0}} \quad \underline{\underline{\Delta S_2 = 0}} : \Delta E_2 = nC_v(T_\ell - T_h) - an^2\left(\frac{1}{V_3} - \frac{1}{V_2}\right) = W_2$$

$$\Delta E_2 = (2 \text{ mol})(\tfrac{3}{2})(8.314 \text{ J mol}^{-1} \text{ K}^{-1})(300 - 1100)\text{K}$$

$$- (0.0341)(2)^2\left(\frac{1}{1.2} - \frac{1}{0.21}\right)\text{dm}^3 \text{ atm}$$

$$= -2.0 \times 10^4 \text{ J} + 0.54 \text{ dm}^3 \text{ atm} = \underline{\underline{-2.0 \times 10^4 \text{ J} = \Delta E_2 = W_2}}$$

Step 3

$$W_3 = -nRT_\ell \ell n \frac{V_4 - nb}{V_3 - nb} - an^2\left(\frac{1}{V_4} - \frac{1}{V_3}\right)$$

$$= -(-5.6 \times 10^3 \text{ J}) - (0.21 \text{ dm}^3 \text{ atm}) = \underline{\underline{5.6 \times 10^3 \text{ J} = W_3}}$$

$$\Delta E_3 = -an^2\left(\frac{1}{V_4} - \frac{1}{V_3}\right) = -0.21 \text{ dm}^3 \text{ atm} = \underline{\underline{-21 \text{ J} = \Delta E_3}}$$

$$\underline{\underline{Q_3 \approx W_3 = -5.6 \times 10^3 \text{ J}}} : \Delta S_3 = Q_3/T_\ell = \underline{\underline{-19 \text{ JK}^{-1} = \Delta S_3}}$$

Step 4

$$\underline{\underline{Q_4 = 0;\ \Delta S_4 = 0}}$$

$$\Delta E_4 = W_4 = n\bar{C}_v(T_h - T_\ell) - an^2\left(\frac{1}{V_1} - \frac{1}{V_4}\right)$$

$$= 2.0 \times 10^4 \text{ J} - 1 \text{ dm}^3 \text{ atm} = \underline{\underline{2.0 \times 10^4 \text{ J} = \Delta E_4 = W_4}}$$

$$\Delta E_{total} = \sum_i \Delta E_i = 72 \text{ J} - 2.0 \times 10^4 - 21 \text{ J} + 2.0 \times 10^4 \text{ J} = 0$$

within our accuracy

$$\Delta S_{total} = \sum_i \Delta S_i = 19 + 0 - 19 + 0 = 0$$

3. See text.

4. $\varepsilon = \dfrac{T_h - T_\ell}{T_h} = \dfrac{673 \text{ K} - 288 \text{ K}}{673 \text{ K}} = 57\%;$ $\underline{\underline{43\% \text{ wasted}}}$

4-2.3 PROBLEMS

1. For ideal gas

$$\bar{C}_p = \bar{C}_v + R$$

Thus,

$$\bar{C}_v = 20.8 - 8.3 = \underline{\underline{12.5 \text{ J mol}^{-1} \text{ K}^{-1}}}$$

$$dS = \frac{n\bar{C}_v}{T} dT + \left(\frac{\partial P}{\partial T}\right)_V dV$$

For ideal $\left(\frac{\partial P}{\partial T}\right)_V = \frac{nR}{V}$

$$\therefore \Delta S = \int dS = \int_{300}^{400} \frac{n\bar{C}_v}{T} dT + \int_{.0462}^{0.0574} \frac{nR}{V} dV$$

$$= (2 \text{ mol})(12.5 \text{ J mol}^{-1} \text{ K}^{-1}) \ln \frac{400}{300}$$

$$+ (2 \text{ mol})(8.3 \text{ J mol}^{-1} \text{ K}^{-1}) \ln\left(\frac{0.0574}{0.0462}\right) = \underline{\underline{10.8 \text{ JK}^{-1} = \Delta S}}$$

2. For van der Waals

$$P = \frac{nR'T}{V - nb} - \frac{an^2}{V^2} \quad \left(\frac{\partial P}{\partial T}\right)_V = \frac{nR'}{V - nb}$$

$$\therefore\ \Delta S = \int \frac{n\bar{C}_v}{T}\,dT + \int \frac{nR'}{V - nb}\,dV \simeq \int \frac{n(\bar{C}_p - R')}{T}\,dT + \int \frac{nR'}{V - nb}\,dV$$

$$= (2\ \text{mol})\left[\int \frac{28.58 + 3.76 \times 10^{-3}\ T - 6.5}{T}\,dT + \int \frac{R'dV}{V - nb}\right]$$

$$= (2\ \text{mol})\left\{22.08\ \ln \frac{400}{300} + 3.76 \times 10^{-3}(400 - 300)\right.$$

$$\left. + 6.5\ \ln\left[\frac{0.0574 - 2(3.0 \times 10^{-5})}{0.0462 - 2(3.0 \times 10^{-5})}\right]\right\}$$

$$= 2(6.35 + 0.38 + 1.41) = \underline{\underline{16.3\ JK^{-1} = \Delta S}}$$

No comparison can be made. C_p is different.

3. The T will rise. Initially, the room and refrigerator are at the same T. When the refrigerator is plugged in, the inside will absorb heat from the room. This heat is rejected to the room from the refrigerator's condensor. The two heats would be the same if no losses occurred in the compressor but no cycle is 100% efficient, so more heat is rejected than is absorbed.

4. Assume ideal gas.

$$\left(\frac{\partial V}{\partial T}\right)_p = \frac{nR}{P}$$

$$\Delta S = \int \frac{n\bar{C}_p\ dT}{T} - \int \left(\frac{\partial V}{\partial T}\right)_p dP$$

$$\Delta S = n \int \frac{27.28 + 3.26 \times 10^{-3}\ T + 0.50 \times 10^5\ T^{-2}}{T}\,dT - n \int \frac{RdP}{P}$$

$$= (3\ \text{mol})\left\{27.28\ \ln \frac{1000}{300} + 3.26 \times 10^{-3}(1000 - 300)\right.$$

$$\left. - \frac{0.50 \times 10^5}{2}\left(\frac{1}{1000^2} - \frac{1}{300^2}\right) - 8.314\ \ln\left(\frac{3}{1}\right)\right\}$$

$$= 3[32.8 + 2.28 - (-0.25) - 9.13] = 3(26.2) = \underline{\underline{78.6\ J\ K^{-1} = \Delta S}}$$

5. $$\Delta S_p = \int \frac{C_p}{T}\,dT = \int_{100}^{225} \frac{1.265 + 14.008 \times 10^{-3}\ T - 103.31 \times 10^{-7}\ T^2}{T}\,dT$$

$$= 1.265\ \ln\left(\frac{225}{100}\right) + 14.008 \times 10^{-3}(225 - 100)$$

$$- \frac{103.31 \times 10^{-7}}{2}(225^2 - 100^2)$$

$$= 1.03 + 1.75 - 0.21 = \underline{\underline{2.57\ J\ K^{-1}\ mol^{-1}}} = \Delta S_p$$

$$\Delta S_V = \int \frac{C_V\, dT}{T} = \int \frac{C_p - 0.002R}{T}\, dT = \Delta S_p - 0.002R \int \frac{dT}{T}$$

$$= 2.57 \text{ J mol}^{-1}\text{ K}^{-1} - 0.002(8.314 \text{ J mol}^{-1}\text{ K}^{-1})\, \ell n \frac{225}{100}$$

$$= 2.57 - 0.01 = \underline{\underline{2.56 \text{ J mol}^{-1}\text{ K}^{-1}}} = \Delta S_V$$

Small difference due to heating of the solid which causes a small expansion.

6. $$\Delta S = \int \frac{n\bar{C}_V}{T}\, dT + \int \left(\frac{\partial P}{\partial T}\right)_V dP = \int \frac{n\bar{C}_V}{T}\, dT + \int \frac{nR}{V}\, dV; \quad C_V = 3.3\, R;$$

$$n = 0.051/0.044 = 1.16 \text{ mol}$$

$$= 1.16 \text{ mol}\left[3.3(8.314 \text{ J mol}^{-1}\text{ K}^{-1})\, \ell n \frac{350}{400} + (8.314 \text{ J mol}^{-1}\text{ K}^{-1})\, \ell n \left(\frac{93.0}{71.0}\right)\right]$$

$$= 1.16(-3.66 + 2.24)\text{JK}^{-1} = \underline{\underline{-1.65 \text{ JK}^{-1} = \Delta S}}$$

7. $$\left(\frac{\partial P}{\partial T}\right)_V = \frac{R}{\bar{V} - b} - \frac{a}{\bar{V}^2}\left(-\frac{1}{T^2}\right) = \frac{R}{\bar{V} - b} + \frac{a}{\bar{V}^2 T^2}$$

$$\Delta S = \int \frac{n\bar{C}_V}{T}\, dT + \int_{26.2}^{54.3} \left(\frac{R}{\bar{V} - b} + \frac{a}{\bar{V}^2 T^2}\right) dV = R\, \ell n \frac{\bar{V}_2 - b}{\bar{V}_1 - b} - \frac{a}{T^2}\left(\frac{1}{\bar{V}_2} - \frac{1}{\bar{V}_1}\right)$$

(the dT term is struck through and marked 0)

$$= (8.314 \text{ J K}^{-1}\text{ mol}^{-1})\, \ell n \left(\frac{54.3 - 0.019}{26.2 - 0.019}\right) - \frac{3.82 \times 10^4}{(225)^2}\left[\frac{1}{54.3} - \frac{1}{26.2}\right] \text{dm}^3 \text{ atm mol}^{-1}\text{ K}^{-1}$$

$$= 6.06 \text{ J mol}^{-1}\text{ K}^{-1} - (-0.0149 \text{ dm}^3 \text{ atm mol}^{-1}\text{ K}^{-1})\left(\frac{101 \text{ J}}{\text{dm}^3 \text{ atm}}\right)$$

$$= 6.06 + 1.51 = \underline{\underline{7.56 \text{ J mol}^{-1}\text{ K}^{-1} = \Delta S}}$$

8. $$\Delta S = \int \frac{n\bar{C}_V dT}{T} + \int \left(\frac{\partial P}{\partial T}\right)_V dV$$

$$P = \frac{RT}{\bar{V}^2}\left(1 - \frac{c}{\bar{V}T^3}\right)\left(\bar{V} + B_o - \frac{bB_o}{\bar{V}}\right) - \frac{A_o}{\bar{V}^2}\left(1 - \frac{a}{\bar{V}}\right); \quad \left(\frac{\partial P}{\partial T}\right)_V = \frac{R}{\bar{V}^2}\left(\bar{V} + B_o - \frac{bB_o}{\bar{V}}\right) + \frac{2Rc}{\bar{V}^3 T^3}\left[\bar{V} + B_o - \frac{bB_o}{\bar{V}}\right]$$

$$\Delta S = \int_{T_1}^{T_2} \frac{nC_V}{T}\, dT + nR \int_{V_1}^{V_2} \left(\frac{1}{\bar{V}} + \frac{B_o}{\bar{V}^2} - \frac{bB_o}{\bar{V}^3}\right) dV + \int \left[\frac{2Rc}{\bar{V}^3 T^3}\left(\bar{V} + B_o - \frac{bB_o}{\bar{V}}\right)\right] dV$$

The last term can not be integrated.

$$\Delta S = n\bar{C}_V\, \ell n \frac{T_2}{T_1} + nR\, \ell n \frac{\bar{V}_2}{\bar{V}_1} - nRB_o\left(\frac{1}{\bar{V}_2} - \frac{1}{\bar{V}_1}\right) + \frac{1}{2}\, nRbB_o\left(\frac{1}{\bar{V}_2^2} - \frac{1}{\bar{V}_1^2}\right) + \int \left[\frac{2Rc}{\bar{V}^3 T^3}\left(\bar{V} + B_o - \frac{bB_o}{\bar{V}}\right)\right] dV$$

4-3.4 PROBLEMS

1. $N_2(g) + 3H_2(g) \rightarrow 2NH_3(g)$

$$\Delta S° = 2S°_{NH_3} - 3S°_{H_2} - S°_{N_2} = 2(192.5) - 3(130.59) - (191.50)$$

$$\underline{\underline{\Delta S° = -198.27\ JK^{-1}}}$$

2. $C(graphite) + 3H_2(g) + \frac{1}{2} O_2(g) \rightarrow CH_3OH(\ell)$

$$\Delta S° = S°_{CH_3OH} - 2S°_{H_2} - \frac{1}{2} S°_{O_2} - S°_{C} = 127 - 2(130.59)$$

$$- \frac{1}{2}(205.1) - 5.69 = \underline{\underline{-242\ JK^{-1} = \Delta S°}}$$

3. Plot C_p vs ℓn T or C_p/T vs T. Graphically integrate. Accepted value for Cd(s) is

$$\underline{\underline{S°_{Cd(s)} = 102.0\ J\ mol^{-1}\ K^{-1}}}$$

4. $H_2O(\ell) \rightarrow H_2O(g)$

$$\Delta S°_{vap} = S°_{H_2O(g)} - S°_{H_2O(\ell)} = 188.72 - 70.00 = \underline{\underline{118.72\ J\ mol^{-1}\ K^{-1} = \Delta S°_{vap}}}$$

$S_{H_2O(s)} > S_{H_2O(\ell)}$ due to greater disorder in the gas phase.

5.

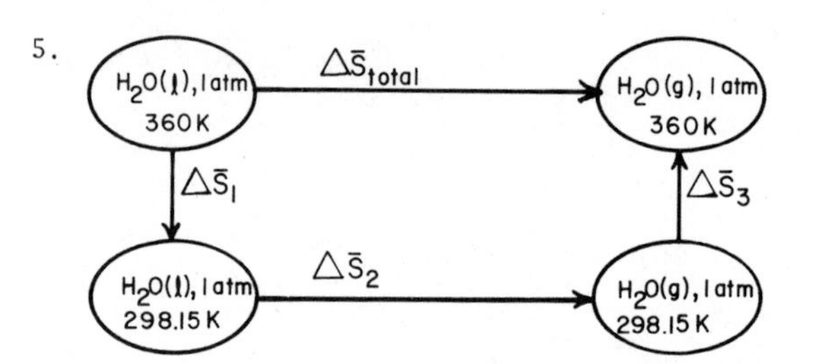

$$\Delta\bar{S}_{total} = \Delta\bar{S}_1 + \Delta\bar{S}_2 + \Delta S_3$$

$$\Delta\bar{S}_1 = \int_{360}^{298.15} \frac{75.88}{T}\, dT = 75.88\ \ell n\ \frac{298.15}{360} = \underline{\underline{-14.30\ J\ mol^{-1}\ K^{-1} = \Delta S_1}}$$

$\underline{\underline{\Delta\bar{S}_2 = 118.72\ J\ mol^{-1}\ K^{-1}}}$ (see #4)

$$\Delta\bar{S}_3 = \int_{298.15}^{360} \frac{30.54 + 10.29 \times 10^{-3}\ T}{T}\, dT = 30.54\ \ell n\ \frac{360}{298.15}$$

$$+ 10.29 \times 10^{-3}(360 - 298.15) = \underline{\underline{6.39\ J\ mol^{-1}\ K^{-1} = \Delta\bar{S}_3}}$$

$$\Delta\bar{S}_{total} = -14.30 + 118.72 + 6.39 = \underline{\underline{110.8\ J\ mol^{-1}\ K^{-1} = \Delta\bar{S}_{total}}}$$

6. At constant T

$$\Delta S = \frac{\Delta H}{T} = \frac{\Delta H_f^\circ(g) - \Delta H_f^\circ(\ell)}{T}$$

and at equilibrium:

$$= \frac{-100 \text{ kJ mol}^{-1} - (-131.8 \text{ kJ mol}^{-1})}{334 \text{ K}} = \frac{31800}{334} \text{ J mol}^{-1} \text{ K}^{-1}$$

$$\underline{\underline{\Delta S = 95.2 \text{ J mol}^{-1} \text{ K}^{-1}}}$$

7. $\Delta S^\circ(298) = 89.26 \text{ J mol}^{-1} \text{ K}^{-1}$

$$\Delta C_p = C_p(O) - \frac{1}{2} C_p(O_2) - C_p(C) = -3.43 - 2.76 \times 10^{-3} T + 8.92 \times 10^5 T^{-2}$$

$$\Delta S^\circ(500) = \Delta S^\circ(298) + \int \frac{-3.43}{T} dT + \int \frac{-2.76 \times 10^{-3} T}{T} dT + \int \frac{8.92 \times 10^5}{T^3} dT$$

$$= 89.26 - 3.43 \ln\left(\frac{500}{298}\right) - 2.76 \times 10^{-3}(500 - 298) - \frac{1}{2}(8.92 \times 10^5)\left(\frac{1}{500^2} - \frac{1}{298^2}\right)$$

$$= 89.26 - 1.78 - 0.56 - (-3.24)$$

$$= \underline{\underline{90.17 \text{ J mol}^{-1} \text{ K}^{-1} = \Delta S^\circ(500)}}$$

5-1.6 PROBLEMS

1. $\Delta G = \Delta H - T\,\Delta S$ (for isothermal)

$= -102{,}000 \text{ J} - (302)(-330)\text{J} = \underline{\underline{-2.34 \text{ kJ} = \Delta G}}$

∴ Process tends to be spontaneous.

2. (a) $K = H + TS = K(T, S)$

$$dK = dH + T\,dS + S\,dT = d(E + PV) + T\,dS + S\,dT$$

$$dK = dE + P\,dV + V\,dP + T\,dS + S\,dT$$

$$= q + w + P\,dV + V\,dP + T\,dS + S\,dT$$

$$= q - \cancel{P\,dV} + \cancel{P\,dV} + V\,dP + T\,dS + S\,dT$$

$$\therefore \underline{\underline{dK = q + V\,dP + T\,dS + S\,dT}}$$

(b) If reversible, $q = T\,dS$. If isobaric, $dP = 0$.

$\therefore\ dK = T\,dS + T\,dS + S\,dT = \underline{\underline{2T\,dS + S\,dT = dK}}$

But $dK = \left(\frac{\partial K}{\partial T}\right)_S dT + \left(\frac{\partial K}{\partial S}\right)_T dS$. Equate coefficients of dT, dS

$$\therefore\ \underline{\underline{\left(\frac{\partial K}{\partial T}\right)_S = S}} \quad \text{and} \quad \underline{\underline{\left(\frac{\partial K}{\partial S}\right)_T = 2T}}$$

3. (a) $dG = d(H - TS) = d(E + PV - TS)$. Proceed as in #2 to get:

$$dG = q - \cancel{P\,dV} + \cancel{P\,dV} + V\,dP - T\,dS - S\,dT$$

If reversible

$$q = T\,dS \text{ and } dG = V\,dP - S\,dT = P\,dV - S\,dT$$

$\therefore$ <u>False</u>

(b) $dA = d(E - TS) = q - P\,dV - T\,dS - S\,dT$

If reversible, $q = T\,dS$

$$dA = -P\,dV - S\,dT$$

$\therefore$ <u>True</u> for reversible process involving only PV work.

(c) $G = H - TS$. By definition. <u>True</u> for all processes.

4. See text.

5. $\left(\frac{\partial T}{\partial P}\right)_S = \left(\frac{\partial V}{\partial S}\right)_P$

6. (a) $A = E - TS$

(b) $A = A(T, V)$; $dA = dE - T\,dS - S\,dT = q + w - T\,dS - S\,dT$

If reversible, $q = T\,dS$

$$dA = \cancel{q} + w - \cancel{T\,dS} - S\,dT = \underline{-P\,dV - S\,dT} \text{ (if PV work only)}$$

From $A = A(T, V)$

$$\underline{dA = \left(\frac{\partial A}{\partial T}\right)_V dT + \left(\frac{\partial A}{\partial V}\right)_T dV}$$

Equating coefficients of dT and dV yields

$$\underline{\underline{\left(\frac{\partial A}{\partial T}\right)_V = -S; \quad \left(\frac{\partial A}{\partial V}\right)_T = -P}}$$

(c) $dA = -P\,dV - S\,dT + w_{other}$

7. By Maxwell's relations

$$dB = L\,dT + J\,dP$$

5-2.4 PROBLEMS

1. Find ΔH°_R, ΔS°_R from tables in Chapters 3 and 4. Assume T = 298.15 K. Then evaluate $\Delta G^\circ_R = \Delta H^\circ_R - T\,\Delta S^\circ_R$ and compare with result from Table 5-2.2.1

$$\Delta G^\circ_R = \Delta G^\circ_f(CO) - \frac{1}{2}\Delta G^\circ_f(O_2) - \Delta G^\circ_f(C)$$

$$= -137.27 - 0 - 0 = \underline{\underline{-137.3\ \text{kJ}}} = \Delta G^\circ_R$$

$$\Delta H^\circ_R = \Delta H^\circ_f(CO) - \frac{1}{2}\Delta H^\circ_f(O_2) - \Delta H^\circ_f(C) = -110.52 - 0 - 0$$

$$= \underline{\underline{-110.52\ \text{kJ}}} = \Delta H^\circ_R$$

$$\Delta S^\circ_R = 197.5 - \frac{1}{2}(205.1) - 5.695 = \underline{\underline{89.26\ \text{J K}^{-1}}} = \Delta S^\circ_R$$

Then $\Delta G^\circ_R = \Delta H^\circ_R - 298.15\ \Delta S^\circ_R = \underline{\underline{-137.1\ \text{kJ}}} = \Delta G^\circ_R$

The agreement is good to the fourth place; the limit of accuracy of the S° data.

2. From Eq. (5-2.3.1)

$$\int dG = -\int S\,dT;\ \text{but } S = S^\circ + \int_{T_o}^{T} \frac{C_p}{T}\,dT$$

$$\therefore\ \Delta G = -\int_{T_o}^{T_1}\left\{S^\circ + \int_{T_o}^{T} \frac{C_p}{T'}\,dT'\right\}dT$$

We need C_p as a function of T. However, to simplify the calculation let's use $\bar{C}_p$ (average) over the temperature range. Using Table 3-2.2.1, we find $\bar{C}_p(306) = 34.20$ J mol^{-1} K^{-1}; $\bar{C}_p(750) = 38.04$ J mol^{-1} K^{-1}; $\bar{C}_p(\text{avg}) = 36.1$ J mol^{-1} K^{-1}. This procedure introduces some error. For precise work, C_p as a function of T would have to be inserted and integrated.

$$\Delta G = -\int_{T_o}^{T}\left\{S^\circ + \int_{306}^{T}\frac{36.1}{T'}\,dT'\right\}dT = -\int_{T_o}^{T}\left\{S^\circ + 36.1\ \ell n\ T'\Big|_{306}^{T}\right\}dT$$

$$= -\int_{306}^{750}\left\{(S^\circ - 207) + 36.1\ \ell n\ T\right\}dT \qquad \left[\int \ell n\ x\,dx = x\ \ell n\ x - x\right]$$

$$= -\left\{(S^\circ - 207)(750 - 306) + 36.1[T\ \ell n\ T - T]\Big|_{306}^{750}\right\}$$

$$= -\{(S^\circ - 207)(444) + 36.1[750\ \ell n\ 750 - 750 - 306\ \ell n\ 306 + 306]\}$$

$$S^\circ = S_{298} + \int_{298}^{T_o}\frac{C_p}{T}\,dT.\quad \text{This gives } \underline{\underline{S^\circ = 224\ \text{J mol}^{-1}\ \text{K}^{-1}}}$$

$$\therefore\ \Delta G = -[(224 - 207)(444) + 100{,}000] = -107{,}500\ \text{J mol}^{-1}.$$

For 3 mol,

$\Delta G = -322.5$ kJ

Note: Actually inserting C_p and doing the integrals (try it) gives $\Delta G = -321.8$ kJ. Thus our assumption was good.

3. $C(s) + \frac{1}{2} O_2(g) \rightarrow CO(g)$

	Item	C(s)	$\frac{1}{2} O_2(g)$	CO(g)	Δ(Item)
kJ	ΔH_f°	0	0	-110.52	-110.52
$J\ K^{-1}$	S°	5.694	$\frac{1}{2}(205.1)$	197.5	89.26
	a	16.86	$\frac{1}{2}(29.96)$	28.41	-3.43
	$b \times 10^{+3}$	4.77	$\frac{1}{2}(4.18)$	4.10	-2.76
	$c \times 10^{-5}$	-8.54	$\frac{1}{2}(-1.67)$	-0.46	8.92

$$\Delta H_{1100} = \Delta H_{298} + \int_{298}^{1100} (\Delta C_p) dT$$

$$= -110{,}520 + \int_{298}^{1100} (-3.43 - 2.76 \times 10^{-3}\ T + 8.92 \times 10^5\ T^{-2}) dT$$

$$= -110{,}520 - 3.43(1100 - 298) - \frac{2.76 \times 10^{-3}}{2}(1100^2 - 298^2)$$

$$- 8.92 \times 10^5 \left(\frac{1}{1100} - \frac{1}{298}\right) = -112{,}600\ \text{J} = \Delta H_{1100}$$

$$\Delta S_{1100} = \Delta S_o + \int_{298}^{1100} \frac{\Delta C_p}{T} dT = 89.26 - 3.43\ \ln \frac{1100}{298}$$

$$- 2.76 \times 10^{-3}(1100 - 298) - \frac{8.92 \times 10^5}{2}\left(\frac{1}{1100^2} - \frac{1}{298^2}\right)$$

$$\Delta S_{1100} = 87.2\ \text{J K}^{-1}$$

$$\Delta G_{1100} = 112{,}600\ \text{J} - 1100(87.2)\ \text{J} = -209\ \text{kJ}$$

By Method II:

$$\frac{\Delta G_{1100}}{1100} = \frac{\Delta G_{298}}{298} - \int_{298}^{1100} \frac{\Delta H(T)}{T^2} dT$$

But

$$\Delta H(T) = \Delta H_{298} + \int_{298}^{T} \Delta C_p\ dT'$$

$$\frac{\Delta G_{1100}}{1100} = \frac{-137{,}100}{298} - \int_{298}^{1100}$$

$$\cdot \left\{ \frac{-110{,}520 + \left[-3.43(T - 298) - 1.35 \times 10^{-3}(T^2 - 298^2) - 8.92 \times 10^5\left(\frac{1}{T} - \frac{1}{298}\right)\right]}{T^2} \right\} dT$$

$$= -460 - \int_{298}^{1100} \left\{ -\frac{106{,}385}{T^2} - \frac{3.43}{T} - 1.35 \times 10^{-3} - \frac{8.92 \times 10^{+5}}{T^3} \right\} dT$$

$$= -460 - \left\{ 106{,}385\left(\frac{1}{1100} - \frac{1}{298}\right) - 3.43 \ \ln \frac{1100}{298} \right.$$

$$\left. - 1.35 \times 10^{-3}(802) + \frac{8.92 \times 10^5}{2}\left(\frac{1}{1100^2} - \frac{1}{298^2}\right) \right\}$$

$$= -460 - (-270.5) = \underline{\underline{-189.5 \text{ J} = \frac{\Delta G_{1100}}{1100}}}$$

$\therefore$ $\underline{\underline{\Delta G_{1100}}} = 1100(-189.5) = \underline{\underline{-208 \text{ kJ}}}$

These two agree within our accuracy.

4. (a) $\Delta G^\circ_R = \Delta G^\circ_f(BaCl_2) - 2\Delta G^\circ_f(Cl^-) - \Delta G^\circ_f(B_A^{2+})$

$= -810.9 - 2(-131.17) - (-560.7) = \underline{\underline{12.14 \text{ kJ}}}$

(b) $\Delta G^\circ_R = -1353 - (-741.99) - (-560.7) = \underline{\underline{-50.31 \text{ kJ}}}$

(c) $\Delta G^\circ_R = 2(-16.64) - 3(0) - 2(0) = \underline{\underline{-33.28 \text{ kJ}}}$

5. $\Delta G^\circ_{298} = \Delta G^\circ_f(H_2O,g) - \Delta G^\circ_f(H_2O,\ell) = -228.596 - (-237.192)$

$= \underline{\underline{+8.60 \text{ kJ}}} = \Delta G^\circ_{298}$

Positive sign indicates the reverse reaction is favored. Water is a liquid at 25°C, 1 atm as you know.

$$\Delta H_{373} = \Delta H_{298} + \int_{298}^{373} \Delta C_p \, dT = 44{,}010 + \left[-44.94(373 - 298)\right.$$

$$\left. + \frac{0.01029}{2}(373^2 - 298^2)\right]$$

$$= \underline{\underline{40{,}898 \text{ J} = \Delta H_{373}}}$$

$$\Delta S_{373} = \Delta S_{298} + \int_{298}^{373} \frac{\Delta C_p}{T} \, dT = 118.72$$

$$+ \left[-44.94 \ \ln \frac{373}{298} + 0.01029(373 - 298)\right]$$

$$= \underline{\underline{109.4 \text{ J K}^{-1} = \Delta S_{373}}}$$

$$\Delta G_{373} = \Delta H_{373} - 373\ \Delta S_{373} = 40{,}898 - 373(109.4)$$

$$= 40{,}898 - 40{,}807 = \underline{\underline{91\ J = \Delta G_{373}}}$$

This should be zero and it is within the accuracy of our calculation since 373 and 298 were used instead of 373.15 and 298.15. At 373.15 K, the process is at equilibrium.

5-3.6 PROBLEMS

1. From 5-2.4

$$\Delta G^\circ_{298} = -137.27\ \text{kJ};\quad \Delta G^\circ_{1100} = -208\ \text{kJ}$$

$$\Delta G^\circ = -RT\ \ln K_p \qquad \therefore\ \ln K_p(298) = \frac{-137{,}270\ \text{J}}{(8.314)(298.15)\text{J}} = 55.4$$

$$\underline{\underline{K_p(298) = 1.12 \times 10^{24}}}$$

At 1100 K

$$\ln K_p = \frac{-208{,}000}{(8.314)(1100)} = 22.7$$

$$\therefore\quad \underline{\underline{K_p(1100) = 7.54 \times 10^{9}}}$$

Assume ΔH° is constant (probably a poor assumption over a wide T range). Equation (5-3.4.5) gives:

$$\ln \frac{K_p(1100)}{K_p(298)} = -\frac{\Delta H^\circ}{R}\left(\frac{1}{1100} - \frac{1}{298}\right)\ :$$

$$\Delta H^\circ = \frac{-(8.314\ \text{J K}^{-1})\ \ln\left(\dfrac{7.54 \times 10^{9}}{1.12 \times 10^{24}}\right)}{\left(\dfrac{1}{1100} - \dfrac{1}{298}\right)\text{K}^{-1}} = \underline{\underline{-111\ \text{kJ} = \Delta H^\circ}}$$

This is an average value for the whole T range. From the tables (Chapter 3)

$$\underline{\underline{\Delta H^\circ_{298} = -110.5\ \text{kJ}}}$$

The agreement is better than we might expect.

2. $H_2O(\ell) \rightarrow H_2O(g)$

$$\Delta G = \Delta G^\circ + RT\ \ln P^\circ_{H_2O}\ \text{(ideal)}$$

At equilibrium $\Delta G = 0$, therefore

$$\Delta G^\circ = -RT\ \ln P^\circ_{H_2O} = [-228{,}596 - (-237{,}192)] = 8600\ \text{J mol}^{-1}$$

$$\ln P^{\circ}_{H_2O} = \frac{-8600\ \text{J mol}^{-1}}{(8.314)(298)\text{J mol}^{-1}} = -3.47$$

$$\therefore \quad P^{\circ}_{H_2O} = \underline{\underline{0.0311\ \text{atm at } 298.15\ \text{K}}}$$

The handbook value is $\underline{\underline{0.0313\ \text{atm}}}$. The vapor is nearly ideal.

3. $\Delta G = \int V\, dP$

$$P_{vdw} = \frac{nRT}{V - nb} - \frac{an^2}{V^2}$$

At constant T,

$$dP = \left[-\frac{nRT}{(V - nb)^2} + \frac{2an^2}{V^3}\right] dV$$

$$\therefore \quad \Delta G = \int \left[\frac{-nRTV}{(V - nb)^2} + \frac{2an^2}{V^2}\right] dV$$

A table of integrals is required to do this integral (see #4).

4. (a) $0.015\ \text{kg}\ CCl_4 = 0.10\ \text{mol}\ CCl_4$

$$\underline{\underline{\text{Ideal } V_1 = 12.6\ \text{dm}^3; \quad V_2 = 1.26\ \text{dm}^3}}$$

$$\Delta G(\text{ideal}) = \int V\, dP = \int \frac{nRT}{P}\, dP = nRT \ln \frac{P_2}{P_1}$$

since T is constant.

$$\Delta G(\text{ideal}) = (0.1\ \text{mol})(8.314\ \text{J mol}^{-1}\ \text{K}^{-1})(306\ \text{K}) \ln\left(\frac{2}{0.2}\right)$$

$$= \underline{\underline{590\ \text{J} = \Delta G(\text{ideal})}}$$

(b) From Problem 3

$$\Delta G(\text{vdw}) = \int \left[\frac{-nRTV}{(V - nb)^2} + \frac{2an^2}{V^2}\right] dV$$

From integral tables:

$$\Delta G(\text{vdw}) = -nRT\left[\ln(V - nb) - \frac{nb}{V - nb}\right]_{V_1}^{V_2} - 2an^2\left(\frac{1}{V}\right)\Bigg|_{V_1}^{V_2}$$

$$= -nRT\left\{\ln\left(\frac{V_2 - nb}{V_1 - nb}\right) - nb\left(\frac{1}{V_2 - nb} - \frac{1}{V_1 - nb}\right)\right\}$$

$$- 2an^2\left(\frac{1}{V_2} - \frac{1}{V_1}\right)$$

$$\underline{\underline{V_1(\text{ideal}) = 12.6\ \text{dm}^3; \quad V_2 = 1.26\ \text{dm}^3}}$$

$$V_1(\text{vdw}) \simeq \frac{nRT}{P + an^2/V_{\text{ideal}}^2} + nb = \frac{(0.1)(0.0821)(306)}{0.2 + (19.6)(0.1)^2/(12.6)^2}$$

$$+ (0.1)(0.127) = \underline{\underline{12.5 \text{ dm}^3 = V_1(\text{vdw})}}$$

No further iteration needed.

$$V_2(\text{vdw}) = \frac{(0.1)(0.0821)(306)}{(2) + (19.6)(0.1)^2/(1.26)^2} + (0.1)(0.127)$$

$$= 1.20 \text{ dm}^3 \qquad \text{Iterate} = \underline{\underline{1.19 \text{ dm}^3 = V_2(\text{vdw})}}$$

Then

$$\Delta G(\text{vdw}) = -(0.1)(0.0821)(306)\ \ln\left(\frac{1.19 - 0.0127}{12.5 - 0.0127}\right)$$

$$+ (0.1)(0.0821)(306)(.0127)$$

$$\times \left[\frac{1}{1.19 - 0.0127} - \frac{1}{12.5 - 0.0127}\right]$$

$$- 2(19.6)(0.1)^2\left(\frac{1}{1.19} - \frac{1}{12.5}\right) = 5.66 \text{ dm}^3 \text{ atm}$$

$\therefore$ $\underline{\underline{\Delta G(\text{vdw}) = 573 \text{ J}}}$

5. $\Delta G_R^\circ = \Delta G_f^\circ(C_2H_4) - \Delta G_f^\circ(C_2H_2) - \Delta G_f^\circ(H_2) = 68.124 - 209 - 0$

$$= \underline{\underline{-140.9 \text{ kJ}}} = \Delta G_R^\circ$$

$\Delta G = 0$ at equilibrium.

$$\therefore \Delta G_R^\circ = -RT \ln K_p = -RT \ln \frac{P_{C_2H_4}}{P_{H_2} P_{C_2H_2}} \quad \text{(ideal)}$$

$$\ln \frac{P_{C_2H_4}}{P_{H_2} P_{C_2H_2}} = -\frac{\Delta G_R^\circ}{RT} = -\frac{(-140{,}900)}{(8.314)(298)} = 56.8 = \ln \frac{P_{C_2H_4}}{10^{-4}}$$

$$\therefore P_{C_2H_4} = 10^{-4}(e^{56.8}) = 10^{-4}(5 \times 10^{24}) = \underline{\underline{5 \times 10^{20} \text{ atm}}} = P_{C_2H_4}$$

This is an impossible pressure. This tells us that we can not maintain P_{H_2} or $P_{C_2H_2} = 10^{-2}$ atm in this reaction.

6. $d \ln K_p = \dfrac{\Delta H^\circ}{RT^2}\, dT$

$$\ln \frac{K_{p_2}}{K_{p_1}} = \int \frac{\Delta H^\circ(T)}{RT^2}\, dT = \int_{T_1}^{T_2} \frac{\Delta H^\circ(T_1) + \int_{T_1}^{T} \Delta C_p\, dT'}{RT^2}\, dT$$

From Table 3-2.2.1 for $H_2(g) + \frac{1}{2} O_2(g) \rightarrow H_2O(g)$

$$\Delta C_p = -11.72 + 4.94 \times 10^{-3}\, T + 0.335 \times 10^5\, T^{-2};$$

$\Delta H^\circ_{298} = -241{,}830$ J

Then

$$\ln \frac{K_{p_2}}{K_{p_1}} =$$

$$\int \left[\frac{-241{,}830 + (-11.72)(T - 298) + \frac{4.94 \times 10^{-3}}{2}(T^2 - 298^2) - 0.335 \times 10^5 \left(\frac{1}{T} - \frac{1}{298}\right)}{RT^2}\right] dT$$

$$= \frac{1}{R} \int_{298}^{1000} \left[-238{,}440\left(\frac{1}{T^2}\right) - \frac{11.72}{T} + 2.47 \times 10^{-3} - 0.335 \times 10^5 \left(\frac{1}{T^3}\right)\right] dT$$

$$= \frac{1}{R}\left[238{,}440\left(\frac{1}{1000} - \frac{1}{298}\right) - 11.72 \ln \frac{1000}{298} + 2.47 \times 10^{-3}(1000 - 298) + \frac{1}{2}(0.335 \times 10^5)\left(\frac{1}{1000^2} - \frac{1}{298^2}\right)\right]$$

$$= \frac{1}{8.314}[-561.7 - 14.2 + 1.73 + 0.17] = \underline{69.0 \text{ (no units)}} = \ln(K_{p_2}/K_{p_1})$$

$$\ln K_{p_1} = -\frac{\Delta G^\circ_{298}}{RT}$$

ΔG°_{298} from Table 5-2.2.1 = $\underline{-228{,}596 \text{ J mol}^{-1}} = \Delta G^\circ_{298}$

$\ln K_{p_1} = 92.3$; $\quad K_{p_1} = \underline{1.21 \times 10^{40}}$

$\underline{K_{p_2}/K_{p_1}} = e^{69.0} = \underline{1.08 \times 10^{-30}}$

$\therefore\ K_{p_2} = (1.21 \times 10^{40})(1.08 \times 10^{-30})$

$K_{p_2} = \underline{1.31 \times 10^{10}}$

K_p decreases with T as predicted by Le Chatelier's principle since $\Delta H < 0$. The reaction shifts to the left as T increases.

7. From example 5-3.2.2

$$RT \ln f = P\bar{V} - RT - RT \ln \frac{\bar{V} - b}{RT} - \frac{a}{bT^{1/2}} \ln \frac{\bar{V} + b}{\bar{V}}$$

$$T = T_c T_R = 527 \text{ K}; \quad P = P_c P_R = 224 \text{ K}$$

Find $\bar{V}$ first:

$$\text{(i)} \quad P = \frac{RT}{\bar{V} - b} - \frac{a}{T^{1/2}\bar{V}(\bar{V} + b)}$$

$$\text{(ii)} \quad \bar{V} - b = \frac{RT}{P} - \frac{a(\bar{V} - b)}{T^{1/2}\bar{V}(\bar{V} + b)P}$$

or

$$\bar{V} = \frac{RT}{P} + b - \frac{a(\bar{V} - b)}{T^{1/2}\bar{V}(\bar{V} + b)P}$$

Find $\bar{V}$(ideal) then insert in (ii) and iterate.

$$a = 0.4278\ R^2T_c^{2.5}/P_c = (0.4278)(0.08206)^2(405.6)^{2.5}/112.2$$

$$= \underline{\underline{85.1 = a}}$$

$$b = 0.0867\ RT_c/P_c = (0.0867)(0.08206)(405.6)/112.2$$

$$= \underline{\underline{0.0257 = b}}$$

$$\bar{V}(\text{ideal}) = RT/P = (0.08206)(527)/(224) = 0.193 \text{ dm}^3 \text{ mol}^{-1}$$

$$\bar{V}(\text{RK})_1 \approx 0.193 + 0.0257 - \frac{85.1(0.193 - 0.0257)}{(527)^{1/2}(0.193)(224)(0.193 + 0.0257)}$$

$$= 0.153$$

Iterate

$V(\text{RK})_2$ (using 0.153) → 0.142

$V(\text{RK})_3$ (using 0.142) → 0.138

Does converge to 0.136

As a check evaluate equation (i) for several values of $\bar{V}$ (trial and error)

$\bar{V}$	0.0349	0.050	0.090	0.15	0.14	0.135
P	2947	800	316	207	218	225

Close enough. Now evaluate f.

Note: A useful technique many times is to make a plot of P vs $\bar{V}$, then read $\bar{V}$ at the required pressure from the plot.

$$RT\ \ell n\ f = (224)(0.135) - (0.08206)(527)$$

$$- (0.08206)(527)\ \ell n \frac{(0.135 - 0.0257)}{(0.08206)(527)}$$

$$- \frac{85.1}{(0.0257)(527)^{1/2}}\ \ell n \frac{0.135 + 0.0257}{(0.135)}$$

$$RT\ \ell n\ f = 220.5$$

$$\ell n\ f = 5.099$$

$\therefore$ $\underline{\underline{f = 164 \text{ atm}}}$ compared to $\underline{\underline{P = 224 \text{ atm}}}$

8. $$\ell n\ K_p^\circ = \frac{-\Delta G_r^\circ}{RT} = \frac{-(-50{,}790 \text{ J mol}^{-1})}{(8.314 \text{ J mol}^{-1} \text{ K}^{-1})(298 \text{ K})} = 20.5;\ K_p^\circ = 8.00 \times 10^8$$

$$\ell n \frac{K'_p}{K_p^\circ} = -\frac{\Delta H_{avg}}{R}\left(\frac{1}{T'} - \frac{1}{T^\circ}\right):$$

$$\Delta H_{avg} = \frac{-R\ \ell n\left(\frac{K'_p}{K_p^\circ}\right)}{\left(\frac{1}{T'} - \frac{1}{T^\circ}\right)} = \frac{-(8.314 \text{ J mol}^{-1} \text{ K}^{-1})\ \ell n \left(\frac{4.8 \times 10^{17}}{8.0 \times 10^8}\right)}{\left(\frac{1}{900} - \frac{1}{298}\right) \text{K}^{-1}}$$

$$= 74{,}800 \text{ J} \qquad \underline{\underline{\Delta H_{avg} = 75 \text{ kJ}}}$$

9. (a) $$K_p = \frac{P_{CH_4}}{(P_{H_2})^2}$$

(b) Reaction should shift toward the side of the reaction that has the least number of moles of gas if the pressure is increased by compressing the system to a smaller volume. Therefore, shift to right. This gives more methane, increases K_p.

However, if an inert gas is used to increase the pressure at constant volume, no shift in equilibrium. Justify that conclusion!

10. $$\left(\frac{\partial P}{\partial V}\right)_T = -\frac{RT}{(\bar{V} - b)^2} + \frac{2a}{T}\left(\frac{1}{\bar{V}^3}\right)$$

$$\therefore\ \Delta G = \int V\ dP = \int\left[-\frac{RT}{(\bar{V} - b)^2} + \frac{2a}{T}\frac{1}{\bar{V}^3}\right]\bar{V}\ d\bar{V}$$

$$\Delta G = \int\left[-\frac{RT\bar{V}}{(\bar{V} - b)^2} + \frac{2a}{T}\frac{1}{\bar{V}^2}\right]dV$$

Need integral table to evaluate.

6-1.6 PROBLEMS

1. At equilibrium

$$\mu_1(A) = \mu_1(B) = \mu_1(C)$$

and

$$\mu_2(A) = \mu_2(B) = \mu_2(C)$$

by definition. Thus, if

$$\mu_1(A) = 0.85\ \mu_2(B)$$

then

$$\mu_1(B) = 0.85\ \mu_2(C)$$

2. $$n_s = \frac{0.01\ \text{kg}}{0.60\ \text{kg mol}^{-1}} = 0.01667\ \text{mol s}$$

$$n_w = \frac{0.090\ \text{kg}}{0.018\ \text{kg mol}^{-1}} = 5.00\ \text{mol w}$$

$$x_s = \frac{n_s}{n_s + n_w} = 0.00332; \quad x_w = 1 - x_s = 0.9968$$

$$\Delta\bar{G}_{mix} = x_s\ RT\ \ell n\ x_s + x_w\ RT\ \ell n\ x_w = RT(x_s\ \ell n\ x_s + x_w\ \ell n\ x_w)$$

$$= (8.31\ \text{J mol}^{-1}\ \text{K}^{-1})(373\ \text{K})[(0.00332)\ \ell n\ (0.00332)$$

$$+ (0.9968)\ \ell n\ (0.9968)] = \underline{\underline{-68.6\ \text{J(mol solution)}^{-1}}}$$

$$\Delta\bar{H}_{mix}(\text{ideal}) = 0;$$

$$\Delta\bar{S}_{mix} = -\frac{\Delta G_{mix}}{T} = -\frac{-68.6}{373} = +\ \underline{\underline{0.18\ \text{J mol}^{-1}\ \text{K}^{-1}}} = \Delta S_{mix}$$

$$P_w(\text{ideal}) = x_w P_w^\circ = (0.9968)(1\ \text{atm}) = \underline{\underline{0.9968\ \text{atm}}} = P_w(\text{ideal})$$

$$(P_w^\circ = 1\ \text{atm at the normal boiling point})$$

$$a_w = \frac{P_w}{P_w^\circ} = \frac{0.87}{1.00} = \underline{\underline{0.87}} = a_w; \quad \gamma_w = \frac{a_w}{x_w^\ell} = \frac{0.87}{0.9968} = \underline{\underline{0.873}} = \gamma_w$$

(Also, $\gamma_w = P_w(\text{real})/P_w(\text{ideal})$)

3. If ideal,

$$\Delta\bar{V}_{mix} = \Delta\bar{H}_{mix} = 0$$

$$x_A = \frac{0.61\ \text{mol A}}{1.94\ \text{total mol}} = \underline{\underline{0.31}}; \quad x_B = 1 - x_A = \underline{\underline{0.69}}$$

$$\Delta\bar{G}_{mix} = RT(x_A \ell n\ x_A + x_B \ell n\ x_B) = (8.314)(298)[(0.31)\ \ell n\ (0.31) + (0.69)\ \ell n\ (0.69)]$$

$$\underline{\underline{\Delta\bar{G}_{mix} = -1540 \text{ J (mol solution)}^{-1}}}$$

$$\Delta\bar{S}_{mix} = -\Delta G_{mix}/T = -(-1540 \text{ J mol}^{-1})/298 \text{ K}$$

$$= \underline{\underline{5.2 \text{ J (mol solution)}^{-1} \text{ K}^{-1}}} = \underline{\underline{\Delta\bar{S}_{mix}}}$$

4. From Eq. (6-1.4.5)

$$\mu_A = \mu_A^\circ + RT\ \ell n\ a_A = \mu_A^\circ + RT\ \ell n\ x_A \quad \text{(if ideal)}$$

Moles solute = 0.1; moles H_2O = (0.120 kg)/(0.018 kg mol^{-1}) = 6.67

$$x_w = \frac{6.67}{6.77} = \underline{\underline{0.985}} = x_w$$

$$\mu_{H_2O} = -285{,}840 \text{ J mol}^{-1} + (8.314 \text{ J mol}^{-1} \text{ K}^{-1})(298.15 \text{ K})\ \ell n\ (0.985)$$

$$\mu_{H_2O} = -285{,}840 - 37 = \underline{\underline{-285.88 \text{ kJ mol}^{-1}}} = \mu_{H_2O}$$

5. $P_{acet}\text{(ideal)} = x^{\ell}_{acet} P^\circ_{acet} = (0.40)(0.336) = \underline{\underline{0.134 \text{ atm}}};$

$$P_{CHCl_3}\text{(ideal)} = (0.60)(0.586) = \underline{\underline{0.352 \text{ atm}}}$$

$$a_A = \frac{P_A}{P_A^\circ} = \frac{0.261}{0.336} = \underline{\underline{0.777}}; \qquad a_C = \frac{P_C}{P_C^\circ} = \frac{0.553}{0.586} = \underline{\underline{0.944}}$$

$$\gamma_A = a_A/x_A = (0.777)/(0.40) = \underline{\underline{1.94}} = \gamma_A$$

$$\gamma_C = a_C/x_C = (0.944)/(0.60) = \underline{\underline{1.57}} = \gamma_C$$

or

$$\gamma_A = P_A/P_A\text{(ideal)} = (0.261)/(0.134) = \underline{\underline{1.94}} = \gamma_A$$

$$\gamma_C = (0.553)/(0.352) = \underline{\underline{1.57}} = \gamma_C$$

6. $\Delta\bar{G}_{mix}\text{(real)} = RT(x_A\ \ell n\ a_A + x_C\ \ell n\ a_C)$

$$= (8.31)(330)[(0.44)\ \ell n\ (0.777) + (0.60)\ \ell n\ (0.944)]$$

$$\underline{\underline{\Delta\bar{G}_{mix}\text{(real)} = -372 \text{ J mol}^{-1} \text{ K}^{-1}}}$$

$$\Delta\bar{G}_{mix}\text{(ideal)} = -RT(x_A\ \ell n\ x_A + x_C\ \ell n\ x_C)$$

$$= \underline{\underline{-1840 \text{ J mol}^{-1} \text{ K}^{-1} = \Delta\bar{G}_{mix}\text{(ideal)}}}$$

$\Delta\bar{G}_{mix}$(real) > $\bar{G}_{mix}$(ideal) indicates positive deviation.

But $P^{(CHCl_3)}_{actual} < P_{ideal}$ indicates negative deviation, in agreement with the discussion in the text. There is an inconsistency in the data given. P° values incorrect.

7. (a) Solution I: $x_A^I = 0.75, \quad x_B^I = 0.25$

Solution II: $x_B^{II} = 0.50, \quad x_B^{II} = 0.50$

(b) $P_{total} = x_A P_A^\circ + x_B P_B^\circ$

$$\therefore \left.\begin{aligned} 400 \text{ mm Hg} &= 0.75P_A^\circ + 0.25P_B^\circ \\ 460 \text{ mm Hg} &= 0.50P_A^\circ + 0.50P_B^\circ \end{aligned}\right\} P_A^\circ = 340 \text{ mm Hg}; \; P_B^\circ = 580 \text{ mm Hg}$$

(c) $x_A^{III} = 0.333, \quad x_B^{III} = 0.667$

$$\therefore \; P_{total}^{III} = (0.333)(340) + (0.667)(580) = 500 \text{ mm Hg} = P_{total}^{III}$$

(d) $\Delta G_{mix}^I = RT(x_A \ln x_A + x_B \ln x_B)$

$= 2660[0.75 \ln (0.75) + 0.25 \ln (0.25)]$

$\Delta G_{mix}^I = -1500 \text{ J(mol solution)}^{-1}$

$\Delta G_{mix}^{II} = -1800 \text{ J(mol solution)}^{-1}$

(e) $x_B^V = P_B/P_T = (0.25)(580)/(400) = 0.36$

8. (a) $P_{O_2} = K_{O_2} x_{O_2}; \quad x_{O_2} = P_{O_2}/K_{O_2} = \dfrac{2.5 \times 10^3 \text{ atm}}{3.88 \times 10^4 \text{ atm}} = 0.0644 = x_{O_2}$

$$x_{O_2} = \frac{\text{mole } O_2}{\text{mole } O_2 + \text{mole } H_2O}$$

Thus there are 0.0644 mole O_2 per 0.936 mole H_2O.

$$[O_2] = \frac{0.0644 \text{ mole } O_2}{(0.936 \text{ mole } H_2O)\left(\dfrac{0.018 \text{ kg } H_2O}{1 \text{ mole } H_2O}\right)}$$

$$= \frac{3.83 \text{ mole } O_2}{1 \text{ kg } H_2O} \times \frac{0.997 \text{ kg } H_2O}{1 \text{ dm}^3 \; H_2O}$$

$[O_2] = 3.81 \text{ mol dm}^{-3}$

(b) $P_{H_2O} = x_{H_2O}P^\circ_{H_2O} = (0.936)(20) = \underline{\underline{18.7 \text{ Torr}}} = P_{H_2O}$ (ideal)

(c) $$a_{H_2O} = \frac{P_{H_2O}(\text{real})}{P^o_{H_2O}(\text{ideal})} = \frac{18.36}{20} = \underline{\underline{0.918}} = a_{H_2O}$$

$$\gamma = \frac{a}{x} = \frac{0.918}{0.936} = \underline{\underline{0.98}} = \gamma_{H_2O}$$

9. (a) $n_B = 0.407$ mol; $n_A = 2.57$ $\therefore$ $x_A = \underline{\underline{0.863}}$; $x_B = \underline{\underline{0.137}}$

(b) $P^\circ_A > P^\circ_B$ $\therefore$ Boiling Point (A) < Boiling Point (B)

(c) If ideal, $\Delta H = 0$, $\Delta G = -T\,\Delta S$

$$\therefore \Delta S = -\frac{\Delta G}{T} = -R(x_A \ln x_A + x_B \ln x_B)$$

$$\Delta S = (-8.314 \text{ J mol}^{-1}\text{ K}^{-1})[0.863 \ln (0.863) + (0.137) \ln (0.137)]$$

$$= \underline{\underline{3.32 \text{ J K}^{-1}(\text{mol solution})^{-1}}}$$

(d) $P_T = x_A P^\circ_A + x_B P^\circ_A = (0.863)(400) + (0.137)(300) = \underline{\underline{386 \text{ mm Hg}}}$

6-2.6 PROBLEMS

1. $$\text{Moles Sucrose} = \frac{0.010 \text{ kg}}{0.342 \text{ kg mol}^{-1}} = 0.0292 \text{ mol};$$

$$\text{Moles Water} = \frac{(0.350 \text{ dm}^3)(0.997 \text{ kg dm}^{-3})}{(0.018 \text{ kg mol}^{-1})}$$

$n_s = \underline{\underline{0.0292 \text{ mol}}}$; $n_w = \underline{\underline{19.38 \text{ mol}}}$

(a) $$x_w = \frac{19.38}{19.41} = \underline{\underline{0.998}} = x_w$$

$P_w(\text{ideal}) = x_w P^\circ_w = (0.998)(0.03126 \text{ atm}) = \underline{\underline{0.0312 \text{ atm}}} = P_w$

(b) $$\gamma_w = \frac{P_w(\text{real})}{P_w(\text{ideal})} = \frac{0.0296}{0.0312} = \underline{\underline{0.949}} = \gamma_w$$

(c) Assuming ideal:

$$\pi = \frac{n_s}{V_w} RT = \frac{(0.0292 \text{ mol})}{(0.350 \text{ dm}^3)}(0.0821 \text{ dm}^3 \text{ atm mol}^{-1}\text{ K}^{-1})(298 \text{ K})$$

$\underline{\underline{\pi = 2.04 \text{ atm}}}$

$$\Delta T_B = K_B m: \quad m = \frac{\text{moles solute}}{\text{kg solvent}} = \frac{(0.0292 \text{ moles s})}{(0.350 \text{ dm}^3)(0.997 \text{ kg dm}^{-3})}$$

$= 0.0837$ molal

$$\therefore\ \Delta T_B = (0.512\ \text{K molal}^{-1})(0.0837\ \text{molal}) = \underline{\underline{0.04\ \text{K}}} = \Delta T_B$$

$$T_B = 373.15 + 0.04;\quad T_B = \underline{\underline{373.19\ \text{K}}}$$

$$\Delta T_f = -K_f m = -(1.86\ \text{K molal}^{-1})(0.0837\ \text{molal})$$

$$\Delta T_f = \underline{\underline{0.16\ \text{K}}};\quad T_f = 273.15 - 0.16 = \underline{\underline{273.0\ \text{K}}} = T_f$$

2. From Eq. (6-2.2.2)

$$\frac{\Delta P_A}{P_A^\circ} = x_B = \frac{0.582 - 0.560}{0.582} = \underline{\underline{0.0378\ \text{mol solute/mol solution}}}$$

We need mol solute/kg solvent. Remember 1 mol solution contains 0.0378 moles solute plus 0.962 moles solvent. Thus

$$\frac{0.0378\ \text{moles solute}}{(0.962\ \text{mol solvent})\left(\frac{0.074\ \text{kg ether}}{1\ \text{mol ether}}\right)} = \underline{\underline{0.531\ \frac{\text{mol solute}}{\text{kg solvent}} = \text{molality}}}$$

But, the solution has

$$\frac{0.01\ \text{kg solute}}{0.10\ \text{kg ether}} = \underline{\underline{0.10\ \text{kg solute/kg solvent}}}$$

Equate these two quantities:

$$0.531\ \text{mol} = 0.10\ \text{kg}\qquad \therefore\ \underline{\underline{M_{\text{solute}} = 0.188\ \text{kg mol}^{-1}}}$$

3. We need the total concentration in the solution:

$$3\%\ \text{urea} = \frac{0.03\ \text{kg urea}}{1\ \text{kg solution}} = \frac{0.03\ \text{kg urea}}{0.93\ \text{kg water}}$$

(3% urea, 4% glycerol, 93% water)

$$= \frac{(0.03\ \text{kg})\left(\frac{1\ \text{mol}}{0.060\ \text{kg}}\right)}{(0.93\ \text{kg water})} = \underline{\underline{\frac{0.538\ \text{mol U}}{\text{kg w}} = \text{molality}_U}}$$

$$4\%\ \text{glycerol} = \frac{(0.04\ \text{kg})\left(\frac{1\ \text{mol}}{0.092\ \text{kg}}\right)}{(0.93\ \text{kg w})} = \underline{\underline{\frac{0.468\ \text{mol gly}}{\text{kg w}} = \text{molality}_G}}$$

$$\text{Total molality} = \underline{\underline{1.005 = m}}$$

$$\Delta T_f = -K_f m = -(1.86\ \text{K m}^{-1})(1.00\ \text{m}) = -1.86\ \text{K};\quad T_f = \underline{\underline{271.3\ \text{K}}}$$

$$\Delta T_B = +K_B m = (0.512\ \text{K m}^{-1})(1.00\ \text{m}) = +0.512\ \text{K};\ T_B = \underline{\underline{373.66\ \text{K}}}$$

4. $\Delta T_B = 354.7\ \text{K} - 353.3\ \text{K} = 1.40\ \text{K} = K_B m\ :\ m = \dfrac{1.40\ \text{K}}{253\ \text{K m}^{-1}} = 0.553\ \text{molal}$

(a) Thus we have

$$\frac{0.553 \text{ mole solute}}{(1 \text{ kg solvent})\left(\frac{1 \text{ mole B}}{0.078 \text{ kg B}}\right)} = \frac{0.553 \text{ mol solute}}{12.82 \text{ mol B}}$$

$$x_B^{(\ell)} = \frac{12.82 \text{ mol}}{13.373 \text{ mol}} = \underline{\underline{0.959 = x_B^{(\ell)}}} :$$

$$P_B(\text{ideal}) = x_B^{\ell} P_B^{\circ} = (0.959)(0.131) = \underline{\underline{0.126 \text{ atm}}} = P_B(\text{ideal})$$

(b) $$\Pi = \frac{\text{mol s}}{V_B} RT = \frac{0.553 \text{ mol s}}{(1 \text{ kg B})\left(\frac{1 \text{ dm}^3 \text{ B}}{0.875 \text{ kg B}}\right)} (0.0821 \text{ dm}^3 \text{ atm mol}^{-1} \text{ K}^{-1})(299.3 \text{ K})$$

$$= \underline{\underline{11.9 \text{ atm} = \Pi}}$$

(c) $$\Delta T_f = -K_f m = -(5.12 \text{ K m}^{-1})(0.553 \text{ m}) = -2.83 \text{ K}$$

$$T_f = 278.7 - 2.83 = \underline{\underline{275.9 \text{ K}}} = T_f$$

5. Equation (6-2.3.5) indicates we need a plot of ℓn P vs $\frac{1}{T}$.

T(K)	236.45	261.65	280.75	299.25	333.75	353.75
$\frac{1}{T} \times 10^3 (K^{-1})$	4.229	3.822	3.562	3.342	2.996	2.827
$P \times 10^3$(atm)	1.31	13.1	52.6	131.6	526.3	1000
ℓn P	-6.64	-4.34	-2.95	-2.03	-0.642	0

You should plot ℓn P vs 1/T. Slope = $-\Delta\bar{H}_{vap}/R$. Such a plot yields the results given below.

At 333 K, $\Delta\bar{H}_{vap} = -R \times \text{Slope} = -(8.31 \text{ J mol}^{-1} \text{ K}^{-1})(-3.8 \times 10^{+3} \text{ K})$

$= \underline{\underline{3.2 \times 10^4 \text{ J mol}^{-1}}}$

At 278 K, $\Delta\bar{H}_{vap} = -(8.31)(-4.5 \times 10^{+3}) = \underline{\underline{3.7 \times 10^4 \text{ J mol}^{-1}}}$

6. Slope at the boiling point ≈ slope at 333 K (B.P. = 354 K)

$\therefore \Delta\bar{H}_{vap} \approx 3.2 \times 10^4 \text{ J mol}^{-1}$

n_A = mol solvent/kg solvent = 1 kg/0.078 kg mol^{-1}

= 12.8 mol. (Per kg solvent) = $\underline{\underline{12.8 \text{ molal}}}$

$$K_{bp} = \frac{RT_{bp}^2}{\Delta\bar{H}_{vap}\, n_A} = \frac{(8.314 \text{ J mol}^{-1} \text{ K}^{-1})(353.3 \text{ K})^2}{(3.2 \times 10^4 \text{ J mol}^{-1})(12.8 \text{ molal})} = \underline{\underline{2.5 \text{ K molal}^{-1}}} = K_{bp}$$

This agrees with the table exactly. Such good agreement is mostly luck. Our slope was not very good; it was only an estimate.

7. $\ln \frac{P_2}{P_1} = -\frac{\Delta H}{R}\left(\frac{1}{T_2} - \frac{1}{T_1}\right)$

$$\ln \frac{P_2}{3.95 \times 10^4} = \frac{-2.91 \times 10^4 \text{ J mol}^{-1}}{8.314 \text{ J mol}^{-1} \text{ K}^{-1}}\left(\frac{1}{275 \text{ K}} - \frac{1}{283 \text{ K}}\right)$$

$$\ln \frac{P_2}{3.95 \times 10^4} = -0.359$$

$$P_2 = (3.95 \times 10^4 \text{ N m}^{-2})(0.698) = \underline{\underline{2.76 \times 10^4 \text{ N m}^{-2}}} = P_2$$

8. (a) $\frac{2.58 \text{ g P}/M_p}{0.100 \text{ kg Bromo}} = m = \frac{\Delta T_f}{K_f} = \frac{280.95 - 278.58}{14.4}$

$$= 0.165 \text{ mol P(kg Bromo)}^{-1}$$

$$\therefore \quad M_p = \frac{2.58 \text{ g P}}{(0.100)(0.165)\text{mol P}} = \underline{\underline{156 \text{ g mol}^{-1}}}$$

(Greater than 94 due to association)

(b) $m = \frac{(2.21 \text{ g})/(156 \text{ g mol}^{-1})}{(0.090 \text{ kg Bromo})} = 0.157 \text{ molal} = \underline{\underline{0.16 \text{ molal} = m}}$

$$\Delta T_{bp} = K_{bp} m = (3.63 \text{ K molal}^{-1})(0.157 \text{ molal}) = 0.568 \text{ K}$$

$$T = T_{bp} + \Delta T = 334.5 + 0.57 = 335.07 \text{ K} = \underline{\underline{335.1 \text{ K} = T}}$$

6-3.6 PROBLEMS

1. $\Lambda^\circ_{NH_4OH} = \lambda^\circ_{NH_4^+} + \lambda^\circ_{OH^-} = \Lambda^\circ_{NH_4Cl} + \Lambda^\circ_{NaOH} - \Lambda^\circ_{NaCl}$

$$= \lambda^\circ_{NH_4^+} + \lambda^\circ_{Cl^-} + \lambda^\circ_{Na^+} + \lambda^\circ_{OH^-} - \lambda^\circ_{Na^+} - \lambda^\circ_{OH^-}$$

$$= 0.01497 + 0.02478 - 0.012645$$

$$= \underline{\underline{0.02710 \text{ ohm}^{-1} \text{ m}^2 \text{ equiv}^{-1}}} = \Lambda^\circ_{NH_4OH}$$

$c = 0.001$ $\qquad NH_4OH \rightleftharpoons NH_4^+ + OH^-$

$\qquad\qquad\qquad c(1 - \alpha) \quad c\alpha \quad c\alpha$

$$K_{diss} = \frac{(c\alpha)(c\alpha)}{c(1 - \alpha)} = \frac{c\alpha^2}{(1 - \alpha)} \; : \; \alpha = \frac{\Lambda}{\Lambda^\circ}$$

$$\alpha = \frac{0.0034}{0.0271} = 0.125; \quad K_{diss} = \frac{(10^{-3})(0.125)^2}{(1 - 0.125)} = \underline{\underline{1.8 \times 10^{-5}}} = K_{diss}$$

$c = 0.01$

$$\alpha = \frac{0.00113}{0.0271} = 0.0417; \quad \underline{\underline{K_{diss} = 1.81 \times 10^{-5}}}$$

$c = 0.100$

$$\alpha = \frac{0.0036}{0.0271} = 0.0133; \qquad \underline{\underline{K_{diss} = 1.79 \times 10^{-5}}}$$

2. Plot Λ vs $\sqrt{c}$ to give the intercept at $c = 0$

$$\underline{\underline{\Lambda^\circ = 0.04296 \text{ ohm}^{-1} \text{ m}^2 \text{ equiv}^{-1}}}$$

3. $$\Lambda^\circ_{H_2SO_4} = \lambda^\circ_{H^+} + \lambda^\circ_{SO_4^-} = 0.034982 + 0.00798$$

$$= \underline{\underline{0.04296 \text{ ohm}^{-1} \text{ m}^2 \text{ equiv}^{-1}}} = \Lambda^\circ_{H_2SO_4}$$

4. (a) $L = \kappa(\text{const})$:

$$\text{const} = \frac{L}{\kappa} = \frac{0.120 \text{ ohm}^{-1}}{0.14087 \text{ ohm}^{-1} \text{ m}^{-1}} = \underline{\underline{0.852 \text{ m} = \text{const}}}$$

(b) From Eq. (6-3.4.2):

$$\Lambda_{0.005} = 0.01435 \text{ (see Table 6-3.4.2)}$$

$$\kappa = \frac{\Lambda c}{10^{-3}} = \frac{(0.01435)(0.005)}{(10^{-3})} = \underline{\underline{0.07175 \text{ ohm}^1 \text{ m}^{-1}}} = \kappa$$

$$L = \kappa(\text{const}) = (0.07175 \text{ ohm}^{-1} \text{ m}^{-1})(0.852 \text{ m}) = \underline{\underline{0.06113 \text{ ohm}^{-1}}} = L$$

5. $$\underline{i} = \frac{\Delta T_f(\text{real})}{\Delta T_f(\text{ideal})} = \frac{-273.15 + 272.94}{-K_f m} = \frac{-0.21 \text{ K}}{-(1.86 \text{ K molal}^{-1})(0.10 \text{ molal})}$$

$$= \underline{\underline{1.1}} = \underline{i}$$

$$\Delta T_{bp} = \underline{i} K_{bp} m = (1.1)(0.512 \text{ K molal}^{-1})(0.10 \text{ molal}) = 0.06 \text{ K}$$

$$\underline{\underline{T_{bp} = 373.16 + 0.06 = 373.2 \text{ K}}}$$

6. $$\underline{i} = \frac{\Delta T_{bp}(\text{real})}{\Delta T_{bp}(\text{ideal})} = \frac{(373.62 - 373.15)\text{K}}{(0.512 \text{ K molal}^{-1})(0.50 \text{ molal})} = \underline{\underline{1.8}} = \underline{i}$$

$$\alpha = \frac{\underline{i} - 1}{\nu - 1} = \frac{1.8 - 1}{2 - 1} = \underline{\underline{0.8 = \alpha}}$$

7. For Problem 5

(Assume $\underline{i}$ is independent of T. A poor assumption.)

$$\Pi = \underline{i}MRT = (1.1)\left(\frac{0.10 \text{ mol}}{1 \text{ dm}^3}\right)(0.0821 \text{ dm}^3 \text{ atm mol}^{-1} \text{ K}^{-1})(298 \text{ K})$$

$$= \underline{\underline{2.7 \text{ atm} = \Pi}}$$

$M \approx m$ in this case. Why?

Problem 6

$$\Pi = \underline{i}MRT = \underline{i}mRT = (1.8)(0.5)(0.0821)(298) = \underline{\underline{22 \text{ atm} = \Pi}}$$

8. $CuSO_4(aq) \rightleftharpoons Cu^{2+} + SO_4^{2-}$

$c(1 - \alpha) \quad c\alpha \quad c\alpha$

$$K_{diss} = \frac{c\alpha^2}{1 - \alpha}; \qquad \alpha = \frac{\underline{i} - 1}{\nu - 1} = \frac{1.1 - 1}{2 - 1} = 0.1$$

$$K = \frac{(0.1)(0.1)^2}{(1 - 0.1)} = \underline{\underline{1 \times 10^{-3} = K_{diss}}}$$

9. Plot $\frac{\Delta T_f}{m}$ vs m. The $\lim_{c \to 0} \left(\frac{\Delta T_f}{m}\right) = i_o K_f = 3.60$. Thus,

$$\underline{i}_o = \frac{3.60}{1.86} = \underline{\underline{1.94 = \underline{i}_o}}$$

NH_4Cl is almost completely dissociated at m = 0.

From the plot,

$$\text{at } 0.189, \ \underline{i} = \frac{3.371}{1.86} = 1.812; \ \alpha = \frac{1.812 - 1}{2 - 1} = \underline{\underline{0.81}} = \alpha_{0.189}$$

$$\text{at } 0.778, \ \underline{i} = \frac{3.31}{1.86} = 1.78; \quad \alpha = \frac{1.78 - 1}{2 - 1} = \underline{\underline{0.78}} = \alpha_{0.778}$$

10. (a) $\Lambda^\circ_{NaOH} = \Lambda^\circ_{NaCl} + \Lambda^\circ_{Ca(OH)_2} - \Lambda^\circ_{CaCl_2} = 0.02486 \text{ ohm}^{-1} \text{ m}^2 \text{ equiv}^{-1}$

$$\alpha = \frac{\Lambda}{\Lambda^\circ} = \frac{0.0238}{0.02486} = \underline{\underline{0.958 = \alpha}}; \quad K_{diss} = \frac{(0.01)^2\alpha^2}{(0.01)(1 - \alpha)}$$

$$= \underline{\underline{0.22 = K_{diss}}}$$

(b) $\Delta T_f = -\underline{i}K_f m$

$$i = \alpha(\nu - 1) + 1 = 0.958(2 - 1) + 1 = 1.958$$

$$\Delta T_f = -(1.958)(1.86)(0.01) = -0.036 = -0.04 \text{ K} = \Delta T_f$$

$$T_f = \underline{\underline{273.11 \text{ K}}}$$

6-4.6 PROBLEMS

1. $A_2B_3 \rightleftharpoons 2A^{+3} + 3B^{-2}$

$$\gamma_\pm^5 = \gamma_+^2\gamma_-^3 \qquad \text{or} \qquad \gamma_\pm = \left(\gamma_+^2\gamma_-^3\right)^{1/5}$$

2. $0.005\ m\ NaCl \rightarrow 0.005\ m\ Na^+ + 0.005\ m\ Cl^-$

$$\mu = \frac{1}{2}\sum_i C_i Z_i^2 = \frac{1}{2}[(0.005)(1)^2 + (0.005)(-1)^2] = \underline{\underline{0.005\ m = \mu}}$$

$$\log_{10}\gamma_\pm = (0.5091)(+1)(-1)\sqrt{(0.005)} = -3.60 \times 10^{-2}$$

$$\gamma_\pm = \underline{\underline{0.920}}$$

$$a_\pm = (\gamma_\pm^2 C_+ C_-)^{1/2} = \gamma_\pm C_\pm = (0.920)(0.005) = \underline{\underline{4.60 \times 10^{-3}}} = a_\pm$$

3. $K_3Fe(CN)_6 \rightarrow 3K^+ + Fe(CN)_6^{3-}$
   0.03 m 0.01 m

$$m_\pm^4 = m_+^3 m_- = (0.03)^3(0.01) = 2.7 \times 10^{-7}; \quad \underline{\underline{m_\pm = 0.023}}$$

$$a_\pm = \gamma_\pm m_\pm = (0.571)(0.023) = \underline{\underline{0.013 = a_\pm}}$$

4. $K_3Fe(CN)_6 \rightarrow 3K^+ + Fe(CN)_6^{3-}$
   0.003 m 0.001 m

$$\mu = \frac{1}{2}[(0.003)(1)^2 + (0.001)(3)^2] = \underline{\underline{0.0060 = \mu}}$$

$$\log_{10}\gamma_\pm = (0.5091)(-3)(+1)\sqrt{0.006} = -0.118; \quad \underline{\underline{0.762}} = \gamma_\pm$$

This compares with an observed value of 0.808. The theory is very limited for multivalent ions.

5. $C_{Na^+} = 0.003; \quad C_{Cl} = 0.003; \quad C_{Zn^{2+}} = 0.002; \quad C_{SO_4^{2-}} = 0.002$

$$\mu = \frac{1}{2}[(0.003)(1)^2 + (0.003)(-1)^2 + (0.002)(2)^2 + (0.002)(-2)^2]$$
$$= \underline{\underline{0.011 = \mu}}$$

$$\log\gamma_{SO_4^{2-}} = -(0.5091)(-2)^2\sqrt{0.011} = -0.213 \qquad \gamma_{SO_4^{2-}} = \underline{\underline{0.611}}$$

6. (a) $\text{mol } FeSO_4 = (0.050\ dm^3)(0.003\ mol\ dm^{-3}) = 0.00015\ mol$

$\text{mol } FeCl_3 = (0.100\ dm^3)(0.002\ mol\ dm^{-3}) = 0.00020\ mol$

Total vol = $0.150\ dm^3$

$$C_{FeSO_4} = (0.00015\ mol)/(0.150\ dm^3) = 1.00 \times 10^{-3}\ mol\ dm^{-3}$$

$$C_{FeCl_3} = 0.00020/0.150 = 1.33 \times 10^{-3}\ mol\ dm^{-3}$$

$$C_{Fe^{2+}} = C_{SO_4^{2-}} = \underline{\underline{1.0 \times 10^{-3}\ M}}$$

$$C_{Fe^{3+}} = \underline{\underline{1.33 \times 10^{-3}\ M}}$$

$$C_{Cl^-} = 3C_{Fe^{3+}} = \underline{\underline{4.0 \times 10^{-3}\ M}}$$

(b) $\mu = \frac{1}{2}[(1 \times 10^{-3})(2)^2 + (1 \times 10^{-3})(-2)^2 + (1.33 \times 10^{-3})(3)^2$

$+ (4 \times 10^{-3})(-1)^2] = \underline{\underline{1.20 \times 10^{-2} = \mu}}$

(c) $\log \gamma_{FeSO_4} = (0.5091)(-2)(+2)\sqrt{0.012} = -0.223; \quad \gamma_{FeSO_4} = \underline{\underline{0.598}}$

$\log \gamma_{FeCl_3} = (0.5091)(+3)(-1)\sqrt{0.012} = -0.167; \quad \gamma_{FeCl_3} = \underline{\underline{0.680}}$

(d) $a^2_{FeSO_4} = a_{Fe^{2+}} a_{SO_4^{2-}} = \gamma_{Fe^{2+}} C_{Fe^{2+}} \gamma_{SO_4^{2-}} C_{SO_4^{2-}}$

$= \gamma^2_{FeSO_4} C_{Fe^{2+}} C_{SO_4^{2-}} = \gamma^2_{FeSO_4} C^2_{FeSO_4}.$

$a_{FeSO_4} = \gamma_{FeSO_4} C_{FeSO_4} = (0.598)(1.0 \times 10^{-3})$

$= \underline{\underline{5.98 \times 10^{-4}}} = a_{FeSO_4}$

$a^4_{FeCl_3} = a_{Fe^{3+}} a_{Cl^-} = \gamma^4_{FeCl_3} C_{Fe^{3+}} C^3_{Cl^-}$

$= (0.680)^4 (1.33 \times 10^{-3})(4 \times 10^{-3})^3 = 1.82 \times 10^{-11}$

$a_{FeCl_3} = \underline{\underline{2.07 \times 10^{-3}}}$

7. (a) $K = \frac{c\alpha^2}{1 - \alpha}$

Assume $\alpha \ll 1$. Then

$$\alpha^2 \approx \frac{K(1 - \alpha)}{c} \approx \frac{K}{c} = \frac{1.34 \times 10^{-5}}{0.01} = 1.34 \times 10^{-3}$$

$\alpha = 0.0367$

Iterate

$$\alpha^2 = \frac{(1.34 \times 10^{-5})(1 - 0.0367)}{(0.01)} = 1.29 \times 10^{-3}$$

$\underline{\underline{\alpha = 0.0359}}$

(b) $C_{H^+} = 0.01\alpha = C_{A^-}$

$\mu = \frac{1}{2}[(0.01\alpha)(1)^2 + (0.01)\alpha(-1)^2] = 0.01\alpha$

$= \underline{\underline{3.59 \times 10^{-4} = \mu}}$

(c) $\log_{10} \gamma_\pm = (0.5091)(+1)(-1)(3.59 \times 10^{-4})^{1/2} = -9.65 \times 10^{-3}$

$\gamma_\pm = \underline{\underline{0.978}}$

$$a_{\pm}^2 = \gamma_{\pm}^2 C_+ C_- = (0.978)^2(3.59 \times 10^{-4})(3.59 \times 10^{-4}) = 1.23 \times 10^{-7}$$

$$a_{\pm} = \underline{\underline{3.51 \times 10^{-4}}}$$

(d) $$K_{th} = \frac{a_{H^+} a_{A^-}}{a_{HA}} = \frac{a_{\pm}^2}{a_{HA}} \simeq \frac{a_{\pm}^2}{C_{HA}}$$

$$C_{HA} = C_o(1 - \alpha) = (0.01)(1 - 0.0359) = 9.64 \times 10^{-3}$$

$$K_{th} = \frac{(3.51 \times 10^{-4})^2}{(9.64 \times 10^{-3})} = \underline{\underline{1.27 \times 10^{-5}}}$$

Show that this gives the same result as $K_{th} = \gamma_{\pm}^2 K_{diss}$

8. $BaCl_2 \rightarrow Ba^{2+} + 2Cl^-$

$$\mu = 0.015 = \frac{1}{2}[C(2)^2 + 2C(1)^2] = 3C \qquad \therefore \quad C = 0.005$$

$$[Cl^-] = 2C = \underline{\underline{0.010 \text{ molar}}}$$

$$\log \gamma_- = -0.5091(-1)^2\sqrt{0.015} = -0.0624 \qquad \underline{\underline{\gamma_- = 0.87}}$$

9. $CaCl_2 \rightarrow Ca^{2+} + 2Cl^-$; $\quad Ca_3(PO_4)_2 \rightarrow 3Ca^{2+} + 2PO_4^{3-}$

$C_1 \quad 2C_1 \qquad\qquad 3C_2 = 0.0015; \quad 2C_2 = 0.001$

$$\mu = \frac{1}{2}[C_1(2)^2 + 2C_1(1)^2 + 3C_2(2)^2 + 2C_2(3)^2]$$

$$= \frac{1}{2}[4C_1 + 2C_1 + (0.0015)(4) + (0.0010)(9)]$$

$$= \frac{1}{2}[6C_1 + 0.015] = 3C_1 + 0.0075 = \mu = 0.0825$$

$$C_1 = \frac{0.0825 - 0.0075}{3} = 0.025$$

$$[Cl^-] = 2C_1 = \underline{\underline{0.050 \text{ molar}}}$$

6-5.10 PROBLEMS

1. $Br_2(\ell) + 2e^- \rightarrow 2Br^-$. Reduction $Br^-(m)|Br_2(\ell)|Pt$

$\varepsilon° = 1.06$ V

2. Red. $\quad e^- + AgBr(s) \rightarrow Ag(s) + Br^-(m) \qquad \varepsilon° = 0.0713$

Oxid. $\quad Br^-(m) \rightarrow \frac{1}{2} Br_2(\ell) + e^- \qquad \varepsilon° = -1.065$

$Pt|Br_2(\ell)|Br^-(m)|AgBr|Ag$

$$AgBr(s) \longrightarrow Ag(s) + \frac{1}{2} Br_2(\ell) \quad \varepsilon° = -0.9937$$

3. (a) Red. $2e^- + Hg_2Cl_2(s) \rightarrow 2Hg(\ell) + 2Cl^-(1N)$ $\varepsilon° = 0.280$

Oxid. $Cd(s) \rightarrow Cd^{2+}(a = 1) + 2e^-$ $\varepsilon° = -(-0.4026)$

$Hg_2Cl_2(s) + Cd(s) \rightarrow Cd^{2+}(a = 1) + 2Hg(\ell) + 2Cl^-(1N)$ $\varepsilon° = 0.6826V$

(b) Red. $2e^- + Br_2(\ell) \rightarrow 2Br^-$ $\varepsilon° = 1.065$

Oxid. $2Cl^- \rightarrow Cl_2(g) + 2e^-$ $\varepsilon° = -1.3585$

$2Cl^- + Br_2(\ell) \rightarrow Cl_2(g) + 2Br^-$ $\varepsilon° = -0.2935V$

(c) Red. $2(Cu^{2+} + 1e^- \rightarrow Cu^+)$ $\varepsilon° = 0.158$

Oxid. $Co \rightarrow Co^{2+} + 2e^-$ $\varepsilon° = -0.28$

$2Cu^{2+} + Co(s) \rightarrow Co^{2+} + 2Cu^+$ $\varepsilon° = 0.438V$

(d) Red. $2(Cu^{2+} + 1e^- \rightarrow Cu^+)$ $\varepsilon° = 0.158$

Oxid. $Cu \rightarrow Cu^{2+} + 2e^-$ $\varepsilon° = -(0.342)$

$Cu(s) + Cu^{2+} \rightarrow 2Cu^+$ $\varepsilon° = -0.184V$

(e) Red. $Zn^{2+} + 2e^- \rightarrow Zn$ $\varepsilon° = -0.7628$

Oxid. $Pb \rightarrow Pb^{2+} + 2e^-$ $\varepsilon° = -(-0.1263)$

$Zn^{2+} + Pb(s) \rightarrow Zn(s) + Pb^{2+}$ $\varepsilon° = -0.6365$

4. (a) $\Delta G° = -nF\varepsilon° = -2(96{,}500)(0.6826) = \underline{-1.32 \times 10^5 \text{ J}}$:

$\ell n\, K_{eq} = -\frac{\Delta G°}{RT} = +53.2$ $K_{eq} = \underline{1.27 \times 10^{23}}$

(b) $\Delta G° = -2(96{,}500)(-0.2935) = \underline{5.66 \times 10^4 \text{ J}}$:

$\ell n\, K_{eq} = -22.9$ $K_{eq} = \underline{1.13 \times 10^{-10}}$

(c) $\Delta G° = -2(96{,}500)(0.438) = \underline{-8.45 \times 10^4 \text{ J}}$:

$\ell n\, K_{eq} = 34.1$ $K_{eq} = \underline{6.45 \times 10^{14}}$

(d) $\Delta G° = -2(96{,}500)(-0.184) = \underline{3.55 \times 10^4 \text{ J}}$:

$\ell n\, K_{eq} = -14.1$ $K_{eq} = \underline{6.16 \times 10^{-7}}$

(e) $\Delta G° = -2(96{,}500)(-0.6365) = \underline{1.23 \times 10^5 \text{ J}}$:

$\ell n\, K_{eq} = -49.6$ $K_{eq} = \underline{2.93 \times 10^{-22}}$

5. We need the reaction $PbSO_4(s) \rightarrow Pb^{2+} + SO_4^{2-}$

$$PbSO_4(s) + 2e^- \rightarrow Pb(s) + SO_4^{2-} \qquad \varepsilon^\circ = -0.3588$$

$$Pb(s) \rightarrow Pb^{2+} + 2e^- \qquad \varepsilon^\circ = -(-0.1263)$$

$$PbSO_4(s) \rightarrow Pb^{2+} + SO_4^{2-} \qquad \varepsilon^\circ = -0.2325V$$

$$\ell n\ K_{sp} = \frac{nF\varepsilon^\circ}{RT} = \frac{(2)(96,500)(-0.2325)}{(8.314)(298)} = -18.1$$

$$\underline{\underline{K_{sp} = 1.36 \times 10^{-8}}}$$

6. The net reaction is $Zn(Hg)(a = x) \rightarrow Zn(Hg)(a = y) \quad \varepsilon^\circ = 0$

$$\varepsilon = -\frac{RT}{nF}\ \ell n\ \frac{a_{Zn}(y)}{a_{Zn}(x)} = -\frac{RT}{nF}\ \ell n\ \frac{y}{x}$$

or

$$\ell n\ \frac{y}{x} = -\frac{(0.062)(2)(96.500)}{(8.314)(312)} = -4.61 :$$

$$\frac{y}{x} = 9.9 \times 10^{-3} = \underline{\underline{0.01 = y/x}}$$

7. The reactions are:

$$\text{Red.}\quad AgCl(s) + 1e^- \rightarrow Ag(s) + Cl^-(m = 0.0171) \qquad \varepsilon^\circ = 0.2225$$

$$\text{Oxid.}\quad \frac{1}{2}H_2(g) \rightarrow H^+(m = 0.0171) + 1e^- \qquad \varepsilon^\circ = 0.00$$

$$AgCl(s) + \frac{1}{2}H_2(g) \rightarrow Ag(s) + HCl(0.0171\ m) \qquad \varepsilon^\circ = 0.2225$$

$$\varepsilon = \varepsilon^\circ - \frac{RT}{nF}\ \ell n\ \frac{a_{H^+}a_{Cl^-}a_{Ag}}{a_{AgCl}a_{H_2}^{1/2}} = \varepsilon^\circ - \frac{RT}{nF}\ \ell n\ a_{H^+}a_{Cl^-}$$

$$= \varepsilon^\circ - \frac{RT}{nF}\ \ell n\ \gamma_\pm^2 C_{H^+}C_{Cl^-}$$

$(a_{solid} \equiv 1;\ a_{H_2} \approx 1$ at low pressures)

$$-\ \ell n\ \gamma_\pm^2 = \frac{\varepsilon - \varepsilon^\circ}{RT}(nF) + \ell n\ C_{HCl}^2 = \frac{(0.43783 - 0.2225)(1)(96,500)}{(8.314)(298.15)}$$

$$-\ \ell n(0.0171)^2$$

$$-\ell n\ \gamma_\pm^2 = 0.245 : \quad \gamma_\pm^2 = 0.783 : \quad \underline{\underline{\gamma_\pm = 0.885}}$$

8. $HClO + H^+ + 2e^- = Cl^- + H_2O \qquad \varepsilon^\circ = 1.49V$

$$Cl^- + 2OH^- = H_2O + ClO^- + 2e^- \qquad = -0.90$$

$$2H_2O + 2e^- = H_2 + 2OH^- \qquad = -0.828$$

$$H_2 = 2e^- + 2H^+ \qquad = 0$$

$$HClO = H^+ + ClO^- \qquad \varepsilon^\circ = -0.24V$$

$$\Delta G° = -nF\varepsilon° = -(2)(96{,}500)(-0.24) = \underline{\underline{46\ \text{kJ} = \Delta G°}}$$

$$\ell n\ K_{diss} = \frac{-46000}{(8.314)(298)} = -18.6 \qquad \underline{\underline{K_{diss} = 8.4 \times 10^{-9}}}$$

7-1.9 PROBLEMS

1. $$\bar{\nu} = 109{,}700\ \text{cm}^{-1}\left(\frac{1}{k^2} - \frac{1}{n^2}\right) = 109{,}700\ \text{cm}^{-1}\left(\frac{1}{k^2} - \frac{1}{\infty^2}\right) = 109{,}700\ \text{cm}^{-1}\left(\frac{1}{k^2}\right)$$

$$k = 1 \quad \bar{\nu} = 109{,}700\ \text{cm}^{-1}\left(\frac{1}{1^2}\right) = 109{,}700\ \text{cm}^{-1} = 3.29 \times 10^{15}\ \text{s}^{-1}$$

$$= 2.18 \times 10^{-18}\ \text{J mlc}^{-1}$$

$$k = 2 \quad \bar{\nu} = 109{,}700\ \text{cm}^{-1}\left(\frac{1}{2^2}\right) = 27{,}400\ \text{cm}^{-1} = 8.23 \times 10^{14}\ \text{s}^{-1}$$

$$= 5.45 \times 10^{-19}\ \text{J mlc}^{-1}$$

$$k = 6 \quad \bar{\nu} = 109{,}700\ \text{cm}^{-1}\left(\frac{1}{6^2}\right) = 3050\ \text{cm}^{-1} = 9.14 \times 10^{13}\ \text{s}^{-1}$$

$$= 6.06 \times 10^{-20}\ \text{J mlc}^{-1}$$

2. $$\lambda = \frac{h}{mv} = \frac{6.626 \times 10^{-34}\ \text{J} \cdot \text{s}}{(5 \times 10^{-2}\ \text{kg})(35\ \text{m s}^{-1})} = 3.8 \times 10^{-34}\ \text{m}$$

3. $$E_t = \frac{1}{2} m_e v^2 + \omega = 2.1 \times 10^{-19}\ \text{J} + 2.8 \times 10^{-19} = 4.9 \times 10^{-19}\ \text{J} = h\nu$$

$$\nu = \frac{4.90 \times 10^{-19}\ \text{J}}{6.63 \times 10^{-34}\ \text{J} \cdot \text{s}} = \underline{7.39 \times 10^{14}\ \text{s}^{-1}} :$$

$$\lambda = c\nu^{-1} = \frac{3 \times 10^{8}\ \text{m s}^{-1}}{7.39 \times 10^{14}\ \text{s}^{-1}} = \underline{4.06 \times 10^{-7}\ \text{m}}$$

4. $$E_0 = \frac{0h^2}{8ma^2} = 0$$

$$E_1 = \frac{(1)^2(6.626 \times 10^{-34}\ \text{J s})^2}{(8)(0.028\ \text{kg mol}^{-1})\left(\dfrac{1\ \text{mol}}{6.023 \times 10^{23}\ \text{mlc}}\right)(0.1\ \text{m})} = \underline{\underline{1.18 \times 10^{-41}\ \text{J mlc}^{-1}}}$$

$$E_2 = (2)^2 E_1 = \underline{\underline{4.72 \times 10^{-41}\ \text{J mlc}^{-1}}}$$

$$RT = (8.314\ \text{J mol}^{-1}\ \text{K}^{-1})(298\ \text{K})\left(\frac{1\ \text{mol}}{6.023 \times 10^{23}\ \text{mlc}}\right)$$

$$= \underline{\underline{4.11 \times 10^{-21}\ \text{J mlc}^{-1}}}$$

$\therefore$ E_1, E_2 are negligible compared to RT.

5. Probability between a/4 and 3a/4

$$= \int_{a/4}^{3a/4} \Psi_2\Psi_2\, dx = \left(\frac{2}{a}\right) \int_{a/4}^{3a/4} \sin^2\left(\frac{2\pi x}{a}\right) dx$$

$$= \frac{2}{a}\left[\frac{1}{2}x + \frac{a}{2\pi} \cdot \frac{1}{4}\sin\left(2 \cdot \frac{2\pi x}{a}\right)\right]_{a/4}^{3a/4}$$

$$= \frac{2}{a}\left[\frac{3a}{8} - \frac{a}{8} + \frac{a}{2\pi}\left(\frac{1}{4}\right)\sin 2\left(\frac{2\pi \cdot 3a}{4a}\right) - \frac{a}{2\pi}\left[\frac{1}{4}\sin 2\left(\frac{2\pi a}{4a}\right)\right]\right]$$

$$= \frac{2}{a}\left[\frac{2a}{8} + \frac{a}{8\pi}\sin 3\pi - \frac{a}{8\pi}\sin\pi\right] = \frac{1}{2} = \text{Probability}$$

(the terms $\frac{a}{8\pi}\sin 3\pi$ and $\frac{a}{8\pi}\sin\pi$ are struck through → 0)

If you plot Ψ_2 as requested in the text, you will see that, indeed, half the time the particle is between a/4 and 3a/4.

6. $$\langle x\rangle = \int_0^a \left(\frac{2}{a}\right)^{1/2} \sin\frac{3\pi x}{a}(x)\left(\frac{2}{a}\right)^{1/2} \sin\frac{3\pi x}{a}\, dx = \frac{2}{a}\int_0^a x \sin^2\frac{3\pi x}{a}\, dx$$

Let $y = \frac{3\pi x}{a}$; then $dx = \frac{a}{3\pi}dy$ and $y = 0$ when $x = 0$; $y = 3\pi$ when $x = a$.

$$\langle x\rangle = \frac{2}{a}\int_0^{3\pi} \frac{a}{3\pi} y \sin^2 y \frac{a}{3\pi}\, dy = \frac{2}{a}\left(\frac{a}{3\pi}\right)^2 \int_0^{3\pi} y \sin^2 y\, dy$$

$$= \frac{2}{a}\left(\frac{a}{3\pi}\right)^2 \left[\frac{y^2}{4} - \frac{y \sin 2y}{4} - \frac{\cos 2y}{8}\right]_0^{3\pi}$$

$$= \frac{2}{a}\left(\frac{a}{3\pi}\right)^2 \left[\frac{(3\pi)^2}{4} - 0 - \frac{3\pi \sin 2(3\pi)}{4} + 0 - \frac{\cos 6\pi}{8} + \frac{\cos 0}{8}\right]$$

(the term $\frac{3\pi \sin 2(3\pi)}{4}$ is struck through → 0)

$$= \frac{2}{a}\left(\frac{a}{3\pi}\right)^2 \left[\frac{3\pi^2}{4} - \frac{1}{8} + \frac{1}{8}\right] = \frac{2}{a}a^2\frac{1}{4} = \frac{1}{2}a = \langle x\rangle_{n=3}$$

7. $\bar{\nu} = 2.06 \times 10^6\ m^{-1} = 2.06 \times 10^4\ cm^{-1}$

$\nu = c\bar{\nu} = (3 \times 10^8\ m\ s^{-1})(2.06 \times 10^6\ m) = 6.18 \times 10^{14}\ s^{-1} = \nu$

$\Delta\varepsilon = h\nu = (6.626 \times 10^{-34}\ J \cdot s)(6.18 \times 10^{14}\ s^{-1}) = 4.10 \times 10^{-19}\ J = \Delta\varepsilon$

$\bar{\nu} = 2.06 \times 10^4\ cm^{-1} = 109{,}700\ cm^{-1}\left(\frac{1}{2^2} - \frac{1}{K_2^2}\right)$

$$\frac{109{,}700}{k_2^2} = \frac{109{,}700}{4} - 2.06 \times 10^4 = 6825$$

$$k_2 = \left(\frac{109{,}700}{6825}\right)^{1/2} = (16)^{1/2} = 4 = k_2$$

8. $\lambda = \dfrac{h}{mv}; \quad m = \dfrac{h}{\lambda v} = \dfrac{6.626 \times 10^{-34}\ \text{J s}}{(1.22 \times 10^{-7}\ \text{m})(3 \times 10^{8}\ \text{m s}^{-1})}$

$$= \underline{\underline{1.81 \times 10^{-35}\ \text{kg} = m}}$$

7-2.6 PROBLEMS

1. From the figure and comparison with Example 7-2.3.3

$$x_1 = x + r_1 \sin\theta\cos\phi = x + \frac{m_2}{M} R \sin\theta\cos\phi$$

$$y_1 = y + \frac{m_2}{M} R \sin\theta\sin\phi$$

$$z_1 = z + \frac{m_2}{M} R \cos\theta$$

$$x_2 = x - r_2 \sin\theta\cos\phi = x - \frac{m_1}{M} R \sin\theta\cos\phi$$

$$y_2 = y - \frac{m_1}{M} R \sin\theta\sin\phi$$

$$z_2 = z - \frac{m_1}{M} R \cos\theta$$

$$\dot{x}_1 = \dot{x} + \frac{m_2}{M} R \cos\theta\cos\phi\,\dot\theta - \frac{m_2}{M} R \sin\theta\sin\phi\,\dot\phi$$

$$\dot{x}_2 = \dot{x} - \frac{m_1}{M} R \cos\theta\cos\phi\,\dot\theta + \frac{m_1}{M} R \sin\theta\sin\phi\,\dot\phi$$

$$\dot{y}_1 = \dot{y} + \frac{m_2}{M} R \cos\theta\sin\phi\,\dot\theta + \frac{m_2}{M} R \sin\theta\cos\phi\,\dot\phi$$

$$\dot{y}_2 = \dot{y} - \frac{m_1}{M} R \cos\theta\sin\phi\,\dot\theta - \frac{m_1}{M} R \sin\theta\cos\phi\,\dot\phi$$

$$\dot{z}_1 = \dot{z} - \frac{m_2}{M} R \sin\theta\,\dot\theta; \qquad \dot{z}_2 = \dot{z} + \frac{m_1}{M} R \sin\theta\,\dot\theta$$

$$\dot{x}_1^2 = \dot{x}^2 + 2\frac{m_2}{M} R\dot{x}\dot\theta\cos\theta\cos\phi - 2\frac{m_2}{M} R\dot{x}\dot\phi\sin\theta\sin\phi$$
$$+ \left(\frac{m_2}{M} R\right)^2 \dot\theta^2\cos^2\theta\cos^2\phi - 2\left(\frac{m_2 R}{M}\right)^2 \dot\theta\dot\phi\cos\theta\sin\theta\cos\phi\sin\phi$$
$$+ \left(\frac{m_2 R}{M}\right)^2 \dot\phi^2\sin^2\theta\sin^2\phi$$

$$\dot{y}_1^2 = \dot{y}^2 + 2\frac{m_2 R}{M}\dot{y}\dot\theta\cos\theta\sin\phi + 2\frac{m_2 R}{M}\dot{y}\dot\phi\sin\theta\cos\phi$$
$$+ \left(\frac{m_2 R}{M}\right)^2 \dot\theta^2\cos^2\theta\sin^2\phi + 2\left(\frac{m_2 R}{M}\right)^2 \dot\phi\dot\theta\cos\theta\sin\theta\cos\phi\sin\phi$$
$$+ \left(\frac{m_2 R}{M}\right)^2 \dot\phi^2\sin^2\theta\cos^2\phi$$

$$\dot{z}_1^2 = \dot{z}^2 - \frac{2m_2R}{M}\,\dot{z}\dot{\theta}\,\sin\theta + \left(\frac{m_2R}{M}\right)\dot{\theta}^2\,\sin^2\theta$$

$$\dot{x}_2^2 = \dot{x}^2 - 2\,\frac{m_1R}{M}\,\dot{x}\dot{\theta}\,\cos\theta\,\cos\phi + 2\,\frac{m_1R}{M}\,\dot{x}\dot{\phi}\,\sin\theta\,\sin\phi$$
$$+ \left(\frac{m_1R}{M}\right)^2\dot{\theta}^2\,\cos^2\theta\,\cos^2\phi - 2\left(\frac{m_1R}{M}\right)^2\dot{\theta}\dot{\phi}\,\sin\theta\;\;\sin\phi\,\cos\phi\,\cos\theta$$
$$+ \left(\frac{m_1R}{M}\right)^2\dot{\phi}^2\,\sin^2\theta\,\sin^2\phi$$

$$\dot{y}_2^2 = \dot{y}^2 - 2\left(\frac{m_1R}{M}\right)\dot{y}\dot{\theta}\,\cos\theta\,\sin\phi - 2\left(\frac{m_2R}{M}\right)\dot{y}\dot{\phi}\,\sin\theta\,\cos\phi$$
$$+ \left(\frac{m_1R}{M}\right)^2\dot{\theta}^2\,\cos^2\theta\,\sin^2\phi + 2\left(\frac{m_1R}{M}\right)^2\dot{\theta}\dot{\phi}\,\cos\theta\,\sin\phi\,\sin\theta\,\cos\phi$$
$$+ \left(\frac{m_1R}{M}\right)^2\dot{\phi}^2\,\sin^2\theta\,\cos^2\phi$$

$$\dot{z}_2^2 = \dot{z}^2 + 2\dot{\theta}\dot{z}\left(\frac{m_1R}{M}\right)\sin\theta + \left(\frac{m_1R}{M}\right)^2\dot{\theta}^2\,\sin^2\theta$$

$$T = \frac{1}{2}\,m_1(\dot{x}_1^2 + \dot{y}_1^2 + \dot{z}_1^2) + \frac{1}{2}\,m_2(\dot{x}_2^2 + \dot{y}_2^2 + \dot{z}_2^2)$$

Make substitutions and simplify.

$$T = \frac{1}{2}(m_1 + m_2)(\dot{x}^2 + \dot{y}^2 + \dot{z}^2) + \frac{1}{2}\left[m_1\left(\frac{m_2R}{M}\right)^2 + m_2\left(\frac{m_1R}{M}\right)^2\right]\dot{\theta}^2\,\cos^2\theta$$
$$+ \frac{1}{2}\left[m_1\left(\frac{m_2R}{M}\right)^2 + m_2\left(\frac{m_1R}{M}\right)^2\right]\dot{\phi}^2\,\sin^2\theta + \frac{1}{2}\left[m_1\left(\frac{m_2R}{M}\right)^2\right.$$
$$\left. + m_2\left(\frac{m_1R}{M}\right)^2\right]\dot{\theta}^2\,\sin^2\theta$$

But

$$\frac{m_1m_2^2R^2}{(m_1 + m_2)^2} + \frac{m_2m_1R^2}{(m_1 + m_2)^2} = \frac{R^2m_1m_2}{(m_1 + m_2)^2}\,[m_1 + m_2]$$

$$= R^2\,\frac{m_1m_2}{m_1 + m_2} = \mu R^2 = I$$

Then

$$T = \frac{M}{2}(\dot{x}^2 + \dot{y}^2 + \dot{z}^2) + \frac{I}{2}[\dot{\theta}^2 + \dot{\phi}^2\,\sin^2\theta]$$

$V = 0 \quad \therefore \quad L = T - V = T = H.$ Below we will prove $T = H$.

$$P_x = \frac{\partial L}{\partial \dot{x}} = M\dot{x};\quad P_y = M\dot{y};\quad P_z = M\dot{z};\quad P_\theta = I\dot{\theta};\quad P_\phi = I\,\sin^2\theta\,\dot{\phi}$$

$$H = \Sigma p_i\dot{q}_i - L = M(\dot{x}^2 + \dot{y}^2 + \dot{z}^2) + I\dot{\theta}^2 + I\,\sin^2\theta\,\dot{\phi}^2$$
$$- \frac{1}{2}\,M(\dot{x}^2 + \dot{y}^2 + \dot{z}^2) - \frac{1}{2}\,I\dot{\theta}^2 - \frac{1}{2}\,I\,\sin^2\theta\,\dot{\phi}^2$$

$$\therefore \quad H = \frac{M}{2}(\dot{x}^2 + \dot{y}^2 + \dot{z}^2) + \frac{I}{2}[\dot{\theta}^2 + \dot{\phi}^2 \sin^2\theta]$$

$$= \frac{1}{2M}[P_x^2 + P_y^2 + P_z^2] + \frac{1}{2I}\left[P_\theta^2 + \frac{P_\phi^2}{\sin^2\theta}\right]$$

2. $$\langle r\rangle_{10} = \frac{\int_0^\infty R_{10}^* r R_{10} r^2 dr}{\int_0^\infty R_{10}^* R_{10} r^2 dr}$$ (Θ, Φ integrate to 1 since they are normalized)

$$= \int_0^\infty R_{10}^* r R r^2 dr = 4\left(\frac{1}{a_o}\right)^3 \int_0^\infty r^3 e^{-2r/a_o} dr$$

$$= 4\left(\frac{1}{a_o}\right)^3 \left[\frac{3!}{(2/a_o)^4}\right] = \frac{4}{a_o^3}\frac{3 \cdot 2a_o^4}{2 \cdot 2 \cdot 2 \cdot 2} = \frac{3}{2}a_o = \langle r\rangle_{10}$$

But $a_o = 0.0529$ n m $\therefore$ $\langle r\rangle_{10} = 0.080$ n m

See Fig. 7-3.4.2 to check this value.

$$\langle r\rangle_{21} = \frac{1}{3}\left(\frac{1}{2a_o}\right)^3\left(\frac{1}{a_o}\right)^2 \int_0^\infty r^3 e^{-r/a_o} r^2 dr$$

$$= \frac{1}{3}\left(\frac{1}{2a_o}\right)^3\left(\frac{1}{a_o}\right)^2 \int_0^\infty r^5 e^{-r/a_o} dr$$

$$= \left(\frac{1}{3}\right)\left(\frac{1}{8a_o^5}\right)\left[\frac{5!}{(1/a_o)^6}\right] = \left(\frac{1}{3}\right)\left(\frac{1}{8a_o^5}\right)\left[\frac{5 \cdot 4 \cdot 3 \cdot 2 \cdot 1}{1} a_o^6\right] = 5a_o$$

$\langle r\rangle_{21} = 5a_o = 0.26$ n m See Fig. 7-3.4.2 to check this value.

3. $$\Theta(1,0) = \left(\frac{3}{2}\right)^{1/2} \cos\theta$$

$$\Theta(2,0) = \left(\frac{5}{8}\right)^{1/2} (3\cos^2\theta - 1) \qquad d\tau = \sin\theta\, d\theta$$

$$\int_0^\pi \Theta^2(1,0)\sin\theta\, d\theta = \frac{3}{2}\int_0^\pi \cos^2\theta \sin\theta\, d\theta$$

$$= -\frac{3}{2}\left(\frac{1}{3}\right)\cos^3\theta\Big|_0^\pi = -\frac{1}{2}[\cos^3\pi - \cos^3 0]$$

$$= -\frac{1}{2}[(-1)^3 - (1)^3] = -\frac{1}{2}(-2) = 1$$

$\therefore$ $\Theta(1,0)$ is normalized.

$$\int_0^\pi \Theta^2(2,0)\sin\theta\, d\theta = \left(\frac{5}{8}\right)\int_0^\pi (9\cos^4\theta\sin\theta - 6\cos^2\theta\sin\theta + \sin\theta)d\theta$$

$$= \left(\frac{5}{8}\right)\left[-\frac{9}{5}\cos^5\theta + \frac{6}{3}\cos^3\theta - \cos\theta\right]_0^\pi$$

$$= \frac{5}{8}\left[-\frac{9}{5}(-2) + 2(-2) - (-2)\right]$$

$$= \frac{5}{8}\left[\frac{18}{5} - \frac{20}{5} + \frac{10}{5}\right] = \frac{5}{8}\left(\frac{8}{5}\right) = \underline{\underline{1}}$$

$\therefore$ $\underline{\underline{\Theta(2,0)\text{ is normalized}}}$.

$$\int_0^\pi \Theta(1,0)\Theta(2,0)\sin\theta\, d\theta = \left(\frac{5}{8}\right)^{1/2}\left(\frac{3}{2}\right)^{1/2}$$

$$\cdot \int_0^\pi (3\cos^3\theta\sin\theta - \cos\theta\sin\theta)d\theta$$

$$= \left(\frac{5}{8}\right)^{1/2}\left(\frac{3}{2}\right)^{1/2}\left[-\frac{3}{4}\cos^4\theta + \frac{1}{2}\cos^2\theta\right]_0^\pi$$

$$= \left(\frac{5}{8}\right)^{1/2}\left(\frac{3}{2}\right)^{1/2}\left[\left(-\frac{3}{4}\right)\left\{\overbrace{(-1)^2 - (1)^2}^{0}\right\} + \frac{1}{2}\left\{\overbrace{(-1)^2 - (1)^2}^{0}\right\}\right]$$

$= \underline{\underline{0}}$ $\quad\therefore$ $\underline{\underline{\Theta(1,0)\text{ and }\Theta(2,0)\text{ are orthogonal}}}$.

4. $$\alpha^2 = \left(\frac{4\pi^2 e^2\mu}{nh^2}\right)^2 = \left(\frac{16\pi^4 e^4\mu}{n^2h^2}\right):\ E_n = \frac{-h^2\alpha^2}{8\pi^2\mu}$$

$$\therefore\ E_n = \frac{-h^2}{8\pi^2\mu}\cdot\frac{16\pi^4e^4\mu^2}{n^2h^4} = -\frac{2\pi^2e^4\mu}{n^2h^2} = E_n$$

5. $$R(2,1) = \left(\frac{1}{3}\right)^{1/2}\left(\frac{1}{2a_o}\right)^{3/2}\left(\frac{r}{a_o}\right)e^{-r/2a_o}$$

Ignore all multiplying constants. They divide out later.

$H\psi - E\psi = 0$ $\quad$ or for R, $HR - ER \overset{?}{=} 0$ $\quad$ $E = -e^2/2a_o;\ n = 2$

$$\frac{1}{r^2}\frac{d}{dr}\left[r^2\frac{d}{dr}re^{-r/2a_o}\right] - \frac{2r}{r^2}e^{-r/2a_o} + \frac{2}{a_oe^2}\left[-\frac{e^2}{2n^2a_o} + \frac{e^2}{r}\right]re^{-r/2a_o} \overset{?}{=} 0$$

$$\frac{1}{r^2}\frac{d}{dr}\left[r^2\left(1 - \frac{r}{2a_o}\right)\right]e^{-r/2a_o} - \frac{2}{r}e^{-r/2a_o} - \left[\frac{r}{4a_o^2} - \frac{2r}{ra_o}\right]e^{-r/2a_o} \overset{?}{=} 0$$

$$\frac{1}{r^2}\left\{2r - \frac{3r^2}{2a_o} - \frac{r^2}{2a_o} + \frac{r^3}{4a_o^2}\right\} e^{-r/2a_o} - \frac{2}{r} e^{-r/2a_o} - \left(\frac{r}{4a_o^2} - \frac{2}{a_o}\right) e^{-r/2a_o} \stackrel{?}{=} 0$$

$$\div\ e^{-r/2a_o} : \frac{2}{r} - \frac{3}{2a_o} - \frac{1}{2a_o} + \frac{r}{4a_o^2} - \frac{2}{r} - \frac{r}{4a_o^2} + \frac{2}{a_o} \stackrel{?}{=} 0 : -\frac{4}{2a_o} + \frac{2}{a_o} = 0$$

∴ R(2,1) is a solution to the Schrödinger equation.

6. $E = \frac{-\text{constant}}{n^2}$

Let the constant equal G

n =	1	2	3	4	5	6	7	8	··· ∞
-E =	$\frac{G}{1}$	$\frac{G}{4}$	$\frac{G}{9}$	$\frac{G}{16}$	$\frac{G}{25}$	$\frac{G}{36}$	$\frac{G}{49}$	$\frac{G}{64}$	0

n = 1 2 3 456

E = $-G$ $\quad -\frac{3G}{4} \quad -\frac{G}{2} \quad -\frac{G}{4} \quad 0$

7. $\frac{\partial}{\partial r}(r^2R^2) = 0;\quad R(1,0) = 2\left(\frac{1}{a_o}\right)^{3/2} e^{-r/a_o}$

$$r^2R^2 = \frac{4}{a_o^3} r^2 e^{-2r/a_o}$$

$$\frac{\partial}{\partial r}(r^2R^2) = 0 = \frac{4}{a_o^3}\left(2r - \frac{2r^2}{a_o}\right) e^{-2r/a_o} = 0 \rightarrow \left(1 - \frac{r}{a_o}\right) = 0$$

$(r_{max})_{10} = a_o$ (This is the Bohr orbit.)

$$R(2,1) = \left(\frac{1}{3}\right)^{1/2}\left(\frac{1}{2a_o}\right)^{3/2} \frac{r}{a_o} e^{-r/2a_o}$$

$$r^2R^2 = \left(\frac{1}{3}\right)\left(\frac{1}{2a_o}\right)^3\left(\frac{1}{a_o}\right)^2 r^4 e^{-r/a_o}$$

$$\frac{\partial}{\partial r}(r^2R^2) = 0 = \left(\frac{1}{3}\right)\left(\frac{1}{8a_o^5}\right)\left(4r^3 - \frac{r^4}{a_o}\right) e^{-r/a_o} = 4r^3 - \frac{r^4}{a_o} = 0$$

∴ $4 - \frac{r}{a_o} = 0$ $\qquad (r_{max})_{2,1} = 4a_o = 0.21$ n m

7-3.7 PROBLEMS

1. $$\mu_B = \frac{eh}{4\pi mc} = \frac{(0.1602 \times 10^{-18}\ c)(6.63 \times 10^{-34}\ J\ s)}{4\pi(9.11 \times 10^{-31}\ kg)(3 \times 10^{8}\ m\ s^{-1})}$$

$$= 3.09 \times 10^{-32}\ \frac{C\ J\ s}{kg\ m\ s^{-1}}$$

$$1C = 1\ \text{amp} \cdot \text{sec} = 1A \cdot s$$

$$= 3.09 \times 10^{-32}\ \frac{A \cdot s^2 \cdot J}{kg\ m\ s^{-1}}\left(\frac{1\ kg\ s^{-2}\ A^{-1}}{1T}\right)$$

$$= \left(3.09 \times 10^{-32}\ \frac{J \cdot s}{m\ T}\right)(3 \times 10^{8}\ m\ s^{-1}) = \underline{\underline{9.27 \times 10^{-24}\ J \cdot T^{-1}}}$$

2. From Eq. 7-3.5.12 and Fig. 7-3.5.1

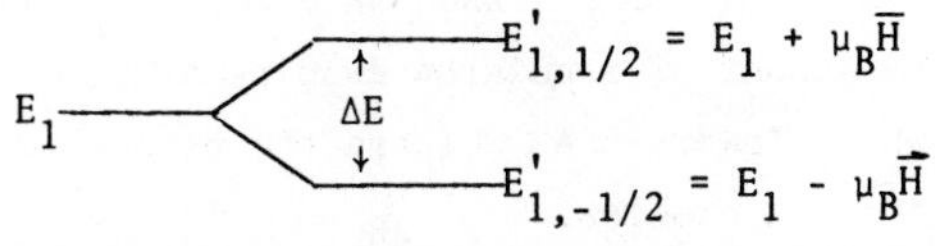

$$E_1 \to E'_{1,1/2} = E_1 + \mu_B \bar{H}$$

$$\Delta E$$

$$E_1 \to E'_{1,-1/2} = E_1 - \mu_B \vec{H}$$

$$\Delta E = E'_{1,1/2} - E'_{1,-1/2} = 2\mu_B\vec{H} = 2(9.27 \times 10^{-24}\ J\ T^{-1})(2T)$$

$$= \underline{3.70 \times 10^{-23}\ J = \Delta E = \text{Splitting}}$$

3. ^{22}Ti; ^{31}Ga; ^{40}Zr

4. Each successive atom adds 1 proton + 1 electron. The electron is not perfectly effective in shielding the extra proton charge so Z_{eff} increases in general across the d orbitals leading to a smaller radius.

5. $\Delta E_H = 109{,}700\left(\frac{1}{k^2} - \frac{1}{n^2}\right)cm^{-1}$. For ionization, n = ∞

$$\therefore\ \Delta E_H(\text{Ionization}) = 109{,}700\left(\frac{1}{k^2}\right).$$

For other multi-electron atoms:

$$\Delta E(\text{Ionization}) = 109{,}700\left(\frac{Z_{eff}}{k}\right)^2\ cm^{-1}$$

$$= 2.18 \times 10^{-18}\left(\frac{Z_{eff}}{k}\right)^2\ J$$

Boron I.E. #	1	2	3	4	5
k	2	2	2	1	1
Value ($\times 10^{-18}$ J)	1.33	4.03	6.07	41.5	54.2
Z_{eff}	1.56	2.72	3.34	4.37	4.99

$$Z_{eff} = \left(\frac{k^2 \Delta E}{2.18 \times 10^{-18}}\right)^{1/2} \qquad \text{Actual } Z = 5$$

Z_{eff} increases as electrons are removed. This is due to the loss of screening of one electron by others as the total number of electrons decrease.

6. F is much lower than would be predicted for the trend. This is likely due to the small size of F. When an electron is added to F to form F^- a large amount of repulsion results among electrons about the small F which lowers E.A. Cl and others are large enough that this is not a problem.

7. $\Phi = \frac{1}{\sqrt{2\pi}} e^{im\phi}$

$$\hat{L}_\phi \Phi = \frac{ih}{2\pi} \frac{\partial}{\partial \phi}\left(\frac{1}{\sqrt{2\pi}}\right)(e^{im\phi}) = \frac{ih}{2\pi}\left(\frac{1}{\sqrt{2\pi}}\right)(im)e^{im\phi}$$

$$= \text{constant} \cdot e^{im\phi} \quad \therefore \text{All } \Phi \text{ are eigenfunctions of } \hat{L}_\phi.$$

8. See text.

9. (a) P; (b) Mn; (c) Ge; (d) Rb.

10. (a) $[Ne]3s^2 3p^6$ (b) $[Ar]4s^2 3d^3$
    (c) $[Kr]5s^2$ (d) $[Ne]3s^2 3p^2$

7-4.6 PROBLEMS

1. $1s^2 2p^1$

$$\Sigma \ell_i = (0 + 0 + 1) = 1 = L \rightarrow P$$

$$\Sigma s_i = (+\tfrac{1}{2} - \tfrac{1}{2} + \tfrac{1}{2}) = \tfrac{1}{2} = S$$

$$L + S = 1 + \tfrac{1}{2} = \tfrac{3}{2} : L - S = 1 - \tfrac{1}{2} = \tfrac{1}{2}$$

$$\therefore \ {}^2P_{1/2};\ {}^2P_{3/2}$$

2. From equation 7-4.2.16, we see that

$$E_1' = \int \Psi_{1s} \hat{H}' \Psi_{1s} \, d\tau$$

$$\hat{H}' = e\vec{E}z = e\vec{E}r \cos\theta$$

$$\Psi_{1s} = \frac{1}{\sqrt{\pi}}\left(\frac{1}{a_o}\right)^{3/2} e^{-r/a_o}$$

Then

$$E_1' = \int_o^{\pi}\int_o^{2\pi}\int_o^{\infty} \frac{1}{\sqrt{\pi}}\left(\frac{1}{a_o}\right)^{3/2} e^{-r/a_o}(e\vec{E}r\cos\theta)\frac{1}{\sqrt{\pi}}\left(\frac{1}{a_o}\right)^{3/2}$$

$$\cdot\, e^{-r/a_o} r^2 \sin\theta \, dr\, d\theta\, d\phi$$

$$= e\vec{E}\,\frac{1}{\pi}\left(\frac{1}{a_o}\right)^3 \int_o^{\pi} \cos\theta \sin\theta \, d\theta \int_o^{2\pi} d\phi \int_o^{\infty} e^{-2r/a_o} r^2 \, dr$$

But

$$\int_o^{\pi} \cos\theta \sin\theta \, d\theta = \left.\frac{\sin^2\theta}{2}\right|_o^{\pi} = 0$$

$$\therefore \; E_1' = 0$$

There is no first order effect. The small effect must be calculated from second order perturbation theory.

3. See a basic text on atomic absorption spectroscopy.

8-1.10 PROBLEMS

1. (MO energy level diagram, from top: two unlabeled empty levels; σ_p ↑↓; π_p ↑↓ ↑↓ π_p; ↑↓ σ^*_{2s}; ↑↓ σ_{2s}; ↑↓ σ^*_{1s}; ↑↓ σ_{1s})

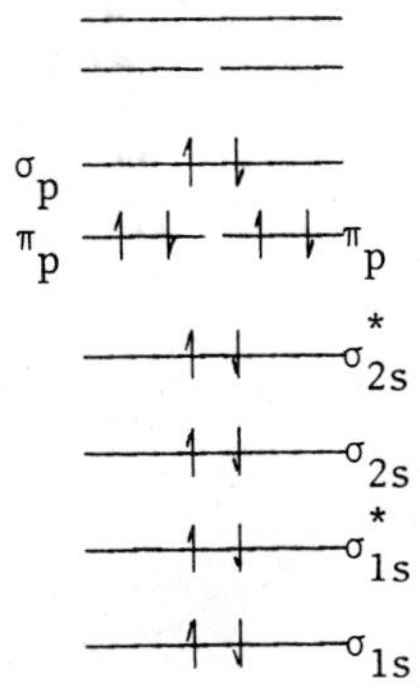

CN^- has fourteen electrons

Bond order $= \frac{10 - 4}{2} = 3$; a triple bond.

Zero unpaired electrons

Highest <u>two</u> occupied orbitals are σ_p and π_p.

See text for shapes.

2. $$\psi_1 = \frac{1}{\sqrt{3}}\phi_s + c_{1x}\phi_x + c_{1y}\phi_y$$

$$\psi_2 = \frac{1}{\sqrt{3}}\phi_s + c_{2x}\phi_x + c_{2y}\phi_y$$

$$\psi_3 = \frac{1}{\sqrt{3}}\phi_s + c_{3x}\phi_x + c_{3y}\phi_y$$

$$c_{1x} = c_{1y} \quad \therefore \; \frac{1}{3} + c_{1x}^2 + c_{1y}^2 = 1$$

$$2c_{1x}^2 = \frac{2}{3}$$

$$c_{1x} = \frac{1}{\sqrt{3}} = c_{1y}$$

$$\psi_1 = \frac{1}{\sqrt{3}}\phi_s + \frac{1}{\sqrt{3}}\phi_x + \frac{1}{\sqrt{3}}\phi_y \qquad c_{2y} = c_{3x}$$

$$\psi_2 = \frac{1}{\sqrt{3}}\phi_s + c_{2x}\phi_x + c_{2y}\phi_y \qquad c_{3y} = c_{2x} \qquad \tan 15° = \frac{c_{2x}}{-c_{2y}} = 0.268$$

$$\psi_3 = \frac{1}{\sqrt{3}}\phi_s + c_{3x}\phi_x + c_{3y}\phi_y \qquad \therefore\ \underline{\underline{c_{2x} = -0.268\ c_{2y}}}$$

$$\frac{1}{3} + (0.268\ c_{2y})^2 + c_{2y}^2 = 1 \rightarrow 1.072\ c_{2y}^2 = 2/3$$

(Note signs)

$$\underline{\underline{c_{2y} = -0.789;}} \qquad \underline{\underline{c_{2x} = -0.268\ c_{2y} = 0.211}}$$

Thus:

$$\underline{\underline{c_{3x} = -0.789 \text{ and } c_{3y} = 0.211}}$$

Finally:

$$\psi_1 = \frac{1}{\sqrt{3}}\phi_s + \frac{1}{\sqrt{3}}\phi_x + \frac{1}{\sqrt{3}}\phi_y$$

$$\psi_2 = \frac{1}{\sqrt{3}}\phi_s + 0.211\ \phi_x - 0.789\ \phi_y$$

$$\psi_3 = \frac{1}{\sqrt{3}}\phi_s - 0.789\ \phi_x + 0.211\ \phi_y$$

3. $c_1 - c_2 - c_3 - c_4$ $\quad 4e^-$ in π system. $E_{loc} = 4(\alpha + \beta) = \underline{\underline{4\alpha + 4\beta}}$

$$\begin{vmatrix} \alpha-\varepsilon & \beta & 0 & 0 \\ \beta & \alpha-\varepsilon & \beta & 0 \\ 0 & \beta & \alpha-\varepsilon & \beta \\ 0 & 0 & \beta & \alpha-\varepsilon \end{vmatrix} = \begin{vmatrix} x & 1 & 0 & 0 \\ 1 & x & 1 & 0 \\ 0 & 1 & x & 1 \\ 0 & 0 & 1 & x \end{vmatrix} = 0$$

Solve: $x^4 - 3x^2 + 1 = 0$. Let $y = x^2 \rightarrow y^2 - 3y + 1 = 0$.

Then $y = 2.62;\ 0.382$ and $\underline{\underline{x = \pm 1.62;\ \pm 0.62.}}$

But $x = (\alpha - \varepsilon)/\beta \quad \therefore\ \varepsilon_1 = \alpha + 1.62\beta$

$$\varepsilon_2 = \alpha + 0.62\beta$$

$$\varepsilon_3 = \alpha - 0.62\beta$$

$$\varepsilon_4 = \alpha - 1.62\beta$$

———— ε_4

———— ε_3

——↑↓—— ε_2

——↑↓—— ε_1

$$E_{mo} = 2(\alpha + 1.62\ \beta) + 2(\alpha + 0.62\ \beta)$$

$$E_{mo} = \underline{\underline{4\alpha + 4.48\beta}}$$

$$E_{res} = \underline{\underline{E_{loc} - E_{mo} = 0.48\beta}}$$

4. 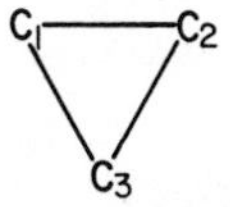

Localized $2e^-$ in π system, 1 in σ

$$E_{loc} = 2(\alpha + \beta) + 1(\alpha) = \underline{\underline{3\alpha + 2\beta = E_{loc}}}$$

$$\begin{vmatrix} x & 1 & 1 \\ 1 & x & 1 \\ 1 & 1 & x \end{vmatrix} = 0 = x^3 - 3x + 2$$

By synthetic division

$$x = 1,\ x = 1,\ x = -2$$

and

$$x = (\alpha - \varepsilon)/\beta$$

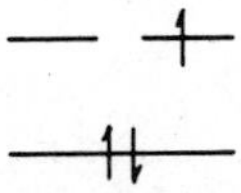

$$\varepsilon_2 = \varepsilon_3 = \alpha - \beta$$

$$\varepsilon_1 = \alpha + 2\beta$$

$$E_{mo} = 2(\alpha + 2\beta) + (\alpha - \beta) = \underline{\underline{3\alpha + 3\beta}} = E_{mo}$$

$$E_{res} = E_{loc} - E_{mo} = \underline{\underline{\beta = E_{res}}}$$

5. 1 - Unpaired e^-; bond order - 2.5

$$(\sigma_{1s})^2(\sigma^*_{1s})^2(\sigma_{2s})^2(\pi_{2p})^2(\pi_{2p})^2(\sigma_{2p})^2(\pi^*_{2p})^1$$

NO has a bond order of 2.5. NO^+ loses an electron from (π^*_{2p}). Therefore, NO^+ has a bond order of 3. It is more stable than NO.

6. $\psi_1 = a_{1s}s + a_{1x}P_x + a_{1y}P_y$

$\psi_2 = a_{2s}s + a_{2x}P_x + a_{2y}P_y$

$\psi_3 = a_{3x}P_x + a_{3y}P_y$

$\psi_4 = P_z$

$$a_{1s} = a_{2s} = \frac{1}{\sqrt{2}};\quad -a_{3x} = a_{3y} = \frac{1}{\sqrt{2}};\quad a_{1x} = a_{1y}$$

$$\therefore\ a_{1x}^2 + a_{1y}^2 + a_{1s}^2 = 1;\quad \therefore\ a_{1x} = a_{1y} = \frac{1}{2}$$

By analogy, $a_{2x} = a_{2y} = -\frac{1}{2}$

$$\psi_1 = \frac{1}{\sqrt{2}} s + \frac{1}{2} P_x + \frac{1}{2} P_y$$

$$\psi_2 = \frac{1}{\sqrt{2}} s - \frac{1}{2} P_x - \frac{1}{2} P_y$$

$$\psi_3 = \quad - \frac{1}{\sqrt{2}} P_x + \frac{1}{\sqrt{2}} P_y$$

8-2.8 PROBLEMS

1. $B = \dfrac{h}{8\pi^2 I}$

$$I = \mu r^2 = \frac{m_1 m_2}{m_1 + m_2} r^2 = \frac{(0.023)(0.035)}{(0.023 + 0.035)} (2.36 \times 10^{-10})^2 \left(\frac{1 \text{ kg m}^2}{6.02 \times 10^{23}}\right)$$

$$I = 1.28 \times 10^{-45} \text{ kg m}^2$$

$$B = \frac{6.626 \times 10^{-34} \text{ J} \cdot \text{s}}{8\pi^2 (1.28 \times 10^{-45} \text{ kg m}^2)} = 6.54 \times 10^9 \text{ s}^{-1}$$

$$\Delta\varepsilon = 2(J + 1)B$$

$J \rightarrow J + 1$	$\Delta\varepsilon(s^{-1})$	$\Delta\varepsilon(cm^{-1})$	$\Delta\varepsilon(J)$
$0 \rightarrow 1$	1.31×10^{10}	0.437	8.66×10^{-24}
$1 \rightarrow 2$	2.62×10^{10}	0.872	1.74×10^{-23}
$2 \rightarrow 3$	3.92×10^{10}	1.31	2.60×10^{-23}
$3 \rightarrow 4$	5.23×10^{10}	1.74	3.47×10^{-23}

2. $\nu_o = 1285.1 \text{ cm}^{-1}$

$$\varepsilon_{v=0} = \frac{1}{2} h\nu_o = 1.277 \times 10^{-20} \text{ J}$$

$$\varepsilon_{v=1} = \frac{3}{2} h\nu_o = 3.832 \times 10^{-20} \text{ J}$$

$$\varepsilon_{v=2} = \frac{5}{2} h\nu_o = 6.386 \times 10^{-20} \text{ J}$$

$$kT = (1.38 \times 10^{-23} \text{ J} \cdot \text{mlc}^{-1} \text{ K}^{-1})(298 \text{ K}) = 4.11 \times 10^{-21} \text{ J}$$

The following values are shifted from v = 0, v = 1, v = 2:

$$\varepsilon_{J=0} = J(J + 1)B = 0 \times 24.58435 \times 10^9 \text{ s}^{-1} = 0$$

$$\varepsilon_{J=1} = 2B = 3.25 \times 10^{-23} \text{ J} = 0.0032 \times 10^{-20} \text{ J}$$

$$\varepsilon_{J=2} = 6B \qquad = 0.0098 \times 10^{-20} \text{ J}$$

$$\vdots \qquad \qquad \vdots$$

$$\varepsilon_{J=10} = 110B \qquad = 0.179 \times 10^{-20} \text{ J}$$

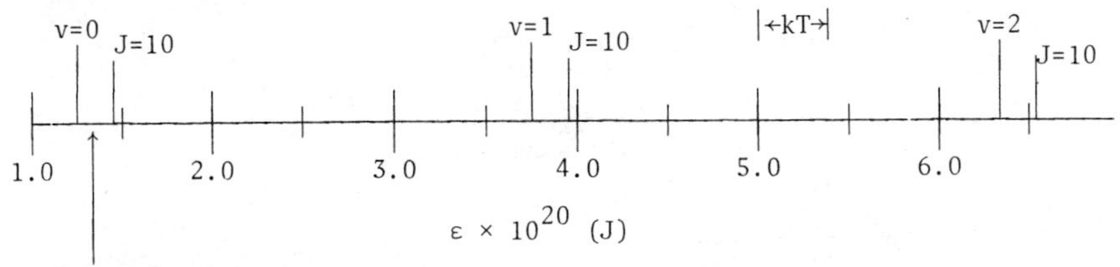

3. $\nu_o = \frac{1}{2\pi}\sqrt{\bar{k}/\mu}$: $\mu = \frac{(0.012)(0.016)}{(0.012 + 0.016)} \frac{1}{(6.023 \times 10^{23})}$ kg mlc^{-1}

$$= \underline{\underline{1.14 \times 10^{-26}\ \text{kg mlc}^{-1}}}$$

$$\bar{k} = 4\pi^2\mu\nu_o^2 = 4\pi^2(1.14 \times 10^{-26}\ \text{kg mlc}^{-1})(2170\ \text{cm}^{-1})^2(3 \times 10^{10}\ \text{cm s}^{-2})^2$$

$$= 1.90 \times 10^3\ \text{kg s}^{-2} = 1.90 \times 10^3\ \text{kg m s}^{-2}\ \text{m}^{-1}$$

$$= \underline{\underline{1.90 \times 10^3\ \text{N m}^{-1}}}$$

Table 8-2.6.1 gives 18.7×10^2 N m^{-1}.

4. $B = \frac{h}{8\pi^2\mu r^2}$: $\Delta\varepsilon_{J=1\to2} = 2(J + 1)B = 2(2)B = 4B$: $I = \mu r^2$

$$42.6228 \times 10^9\ \text{s}^{-1} = 4B \quad \therefore \quad \underline{\underline{B = 1.066 \times 10^{10}\ \text{s}^{-1}}}$$

$$I = \frac{h}{8\pi^2 B} = \frac{6.626 \times 10^{-34}\ \text{J} \cdot \text{s}}{8\pi^2(1.066 \times 10^{10}\ \text{s}^{-1})} = \underline{\underline{7.876 \times 10^{-46}\ \text{kg m}^2 = I}}$$

$$\mu = \frac{(0.081)(0.019)}{(0.081 + 0.019)} \frac{1}{6.023 \times 10^{23}}\ \text{kg mlc}^{-1} = \underline{\underline{2.555 \times 10^{-26}\ \text{kg mlc}^{-1} = \mu}}$$

$$r = \left(\frac{I}{\mu}\right)^{1/2} = 1.756 \times 10^{-10}\ \text{m} = r$$

$$\Delta\varepsilon_{1\to2} = 4B = 4(1.066 \times 10^{10}\ \text{s}^{-1})(6.626 \times 10^{-34}\ \text{J s}) = \underline{\underline{2.825 \times 10^{-23}\ \text{J}}} = \Delta\varepsilon_{1\to2}$$

$$\Delta\varepsilon_{2\to3} = 6B = \underline{\underline{4.238 \times 10^{-23}\ \text{J}}} = \Delta\varepsilon_{2\to3}$$

$$\Delta\varepsilon_{4\to5} = 10B = \underline{\underline{7.063 \times 10^{-23}\ \text{J}}} = \Delta\varepsilon_{4\to5}$$

5. From Example 8-2.2.2

$$I = \frac{m_N m_N r_{NN}^2 + m_N m_O r_{NO}^2 + m_N m_O (r_{NN} + r_{NO})^2}{(m_N + m_N + m_O)N}$$

$$I = \frac{(0.014)^2(1.126 \times 10^{-10}\text{m}) + (0.014)(0.016)\text{kg}^2(1.191 \times 10^{-10})^2 + (0.014)(0.016)\text{kg}^2(2.317 \times 10^{-10})^2}{(0.014 + 0.014 + 0.016)(6.023 \times 10^{23})\text{kg}}$$

$\underline{\underline{I = 6.674 \times 10^{-46} \text{ kg m}^2}}$

$$B = \frac{h^2}{8\pi^2 I} = \frac{(6.626 \times 10^{-34} \text{ J} \cdot \text{s})^2}{8\pi^2(6.674 \times 10^{-46} \text{ kg m}^2)}$$

$\underline{\underline{B = 8.331 \times 10^{-24} \text{ J}}}$

$$\Delta\varepsilon_{1\to 2} = 2(J + 1)B = 2(2)B = \underline{\underline{3.332 \times 10^{-23} \text{ J}}}$$

6. Assume that Br vibrates against the CN unit and N vibrates against the BrC unit.

$$\mu(\text{Br - CN}) = \frac{(0.079)(0.012 + 0.014)}{(0.079 + 0.026)(6.023 \times 10^{23})} \text{ kg mlc}^{-1}$$

$$= 3.25 \times 10^{-26} \text{ kg mlc}^{-1}$$

$$\mu(\text{N - BrC}) = \frac{(0.014)(0.079 + 0.012)}{(0.014 + 0.091)(6.023 \times 10^{23})} \text{ kg mlc}^{-1}$$

$$= 2.014 \times 10^{-26} \text{ kg mlc}^{-1}$$

The higher wave number should belong to the C ≡ N stretch.

$$\bar{k}_1 = 4\pi^2\mu(\text{Br - CN})\bar{\nu}_1^2 = 4\pi^2(3.25 \times 10^{-26} \text{ kg})(580 \text{ cm}^{-1})^2 \cdot (3 \times 10^{10} \text{ cm s}^{-1})^2$$

$$\bar{k}_1 = \underline{\underline{388.2 \text{ N m}^{-1} = 390 \text{ N m}^{-1}}}$$

$$\bar{k}_2 = 4\pi^2\mu(\text{N - CBr})\bar{\nu}_2^2 = 4\pi^2(2.014 \times 10^{-26} \text{ kg})(2187 \text{ cm}^{-1})^2(3 \times 10^{10} \text{ cm s}^{-1})^2$$

$$\underline{\underline{\bar{k}_2 = 3400 \text{ N m}^{-1}}}$$

8-3.5 PROBLEMS

1. $v = 0 \to v = 1$

R Branch:

$$\Delta\varepsilon_{\substack{J\to J+1 \\ v=0\to 1}} = \bar{\nu}_e - 2\overline{\nu_e x_e} + 2(J + 1)\bar{B}_e - (J + 3)(J + 1)\bar{\alpha}_e$$

$$\bar{B}_e = 1.9314 \text{ cm}^{-1}$$

$$\bar{\alpha}_e = 0.01748 \text{ cm}^{-1}$$

$$\overline{\nu_e x_e} = 13.461 \text{ cm}^{-1}$$

$$\bar{\nu}_e = 2170.21 \text{ cm}^{-1}$$

$$\Delta\varepsilon_{0\to1} = \bar{\nu}_e - 2\overline{\nu_e x_e} + 2\bar{B}_e - 3\bar{\alpha}_e$$

$$= 2170.21 - 2(13.461) + 2(1.9314) - 3(0.01748)$$

$$= \underline{\underline{2147.10\ \text{cm}^{-1}}} = \Delta\varepsilon_{0\to1}$$

$$\Delta\varepsilon_{1\to2} = \bar{\nu}_e - 2\overline{\nu_e x_e} + 4\bar{B}_e - 8\bar{\alpha}_e = \underline{\underline{2150.87\ \text{cm}^{-1}}} = \Delta\varepsilon_{J=1\to2}$$

P Branch:

$$\Delta\varepsilon_{\substack{J\to J-1 \\ v=0\to1}} = \bar{\nu}_e - 2\overline{\nu_e x_e} - 2J\bar{B}_e - J(J-2)\bar{\alpha}_e$$

$$\Delta\varepsilon_{1\to0} = \bar{\nu}_e - 2(\overline{\nu_e x_e}) - 2\bar{B}_e + \bar{\alpha}_e = \underline{\underline{2139.44\ \text{cm}^{-1}}} = \Delta\varepsilon_{J=1\to0}$$

$$\Delta\varepsilon_{2\to1} = \bar{\nu}_e - 2\overline{\nu_e x_e} - 4\bar{B}_e - 0 = \underline{\underline{2135.56\ \text{cm}^{-1}}} = \Delta\varepsilon_{J=2\to1}$$

2. $\Delta\varepsilon_{J\to J+1} = 2(J+1)B_v - 4D(J+1)^3$

$$B_{v=0} = B_e - \frac{1}{2}\alpha_e$$

$$B_e = B_o + \frac{1}{2}\alpha_e = 127.3582 \times 10^9\ \text{s}^{-1} + \frac{1.258 \times 10^9}{2}\ \text{s}^{-1}$$

$$B_e = \underline{\underline{127.987 \times 10^9\ \text{s}^{-1}}}$$

(a) $\Delta\varepsilon_{3\to4} = 8B_o - 4D(4^3) = 8(127.3582 \times 10^9\ \text{s}^{-1})$

$$- 4(4)^3(2.8 \times 10^6\ \text{s}^{-1}) = \underline{\underline{1.01815 \times 10^{12}\ \text{s}^{-1}}}$$

$$\Delta\varepsilon'_{3\to4}(\text{without D}) = 8(127.3582 \times 10^9\ \text{s}^{-1})$$

$$= \underline{\underline{1.01886 \times 10^{12}\ \text{s}^{-1}}} \quad \underline{\underline{0.07\%\ \text{error}}}$$

(b) $\Delta\varepsilon_{10\to11} = 22B_o - 4D(11)^3 = \underline{\underline{2.78697 \times 10^{12}\ \text{s}^{-1}}}$

$$\Delta\varepsilon'_{10\to11} = \underline{\underline{2.80188 \times 10^{12}\ \text{s}^{-1}}} \qquad \underline{\underline{0.53\%\ \text{error}}}$$

(c) $\Delta\varepsilon_{20\to21} = 42B_o - 4D(21)^3 = \underline{\underline{5.2717 \times 10^{12}\ \text{s}^{-1}}}$

$$\Delta\varepsilon'_{20\to21} = 42B_o = \underline{\underline{5.34904 \times 10^{12}\ \text{s}^{-1}}} \qquad \underline{\underline{2.0\%\ \text{error}}}$$

3. $\Delta\varepsilon_{0\to v'} = \bar{\nu}_e v' - \overline{\nu_e x_e}[v'(v'+1)]$

$$\Delta\varepsilon_{0\to1} = \bar{\nu}_e - 2\,\overline{\nu_e x_e} = 3019\ \text{cm}^{-1}$$

$$\Delta\varepsilon_{0\to4} = 4\bar{\nu}_e - 20\overline{\nu_e x_e} = 11315\ \text{cm}^{-1}$$

$$-4\Delta\varepsilon_{0\to 1} = -4\bar{\nu}_e + 8\overline{\nu_e x_e} = -4(3019) = -12076\ \text{cm}^{-1}$$

$$+\Delta\varepsilon_{0\to 4} = 4\bar{\nu}_e - 20\overline{\nu_e x_e} = 11315 = 11315\ \text{cm}^{-1}$$

$$-12\overline{\nu_e x_e} = -761 \qquad \underline{\underline{\overline{\nu_e x_e} = 63.4\ \text{cm}^{-1}}}$$

$$\bar{\nu}_e = 3019 + 2(63.4) = \underline{\underline{3145.8\ \text{cm}^{-1}}}$$

Check: $\Delta\varepsilon_{0\to 3} = 8700 \overset{?}{=} 3\bar{\nu}_e - 12\overline{\nu_e x_e}$

$$= 3(3145.8) - 12(63.4) = \underline{\underline{8677\ \text{cm}^{-1}}}$$

An error of 0.3%

Other combinations give slightly different results. The best procedure would be a least squares fit.

4. $$\varepsilon_{J+1,v+1} = (J+1)(J+2)B_e - \alpha_e(v+1+\tfrac{1}{2})(J+1)(J+2) - D(J+1)^2(J+2)^2 + \bar{\nu}_e(v+\tfrac{3}{2}) - \overline{\nu_e x_e}(v+\tfrac{3}{2})^2$$

$$\varepsilon_{J,v} = (J)(J+1)B_e - \alpha_e(v+\tfrac{1}{2})(J)(J+1) - D(J)^2(J+1)^2 + \bar{\nu}_e(v+\tfrac{1}{2}) - \overline{\nu_e x_e}(v+\tfrac{1}{2})^2$$

$$\Delta\varepsilon_{J,v\to J+1,v+1} = \varepsilon_{J+1,v+1} - \varepsilon_{J,v}$$

$$\Delta\varepsilon_{J,v\to J+1,v+1} = 2(J+1)B_e - \alpha_e(J+1)(J+2v+3) - 4D(J+1)^3 + \bar{\nu}_e - \overline{\nu_e x_e}(2v+2)$$

$$\begin{aligned}\Delta\varepsilon_{10,1\to 11,2} &= 2(11)B_e - \alpha_e(11)[10+2+3] - 4(11)^3 D + \bar{\nu}_e - \overline{\nu_e x_e}(4)\\ &= 22(5.78975\times 10^{10}\ \text{s}^{-1}) - (11)(15)(5.24\times 10^{8}\ \text{s}^{-1})\\ &\quad - 4(1331)(1.834\times 10^{5}\ \text{s}^{-1})\\ &\quad + (2170.21\ \text{cm}^{-1})(3\times 10^{10}\ \text{cm s}^{-1})\\ &\quad - 4(13.5\ \text{cm}^{-1})(3\times 10^{10}\ \text{cm s}^{-1})\\ &= \underline{\underline{6.467\times 10^{13}\ \text{s}^{-1} = 4.285\times 10^{-20}\ \text{J}}}\\ &= \underline{\underline{\Delta\varepsilon_{10,1\to 11,2}}}\end{aligned}$$

8-4.7 PROBLEMS

1. $\lambda = 4358.35\times 10^{-10}$ m

$$\nu = \frac{c}{\lambda} = \frac{3.0\times 10^{8}\ \text{m s}^{-1}}{4358.35\times 10^{-10}\ \text{m}} = 6.883\times 10^{14}\ \text{s}^{-1}$$

$$\nu_1'' = \nu \pm \nu_1' = 6.883 \times 10^{14}\ \text{s}^{-1} \pm (1320\ \text{cm}^{-1})(3 \times 10^{10}\ \text{cm s}^{-1})$$

$$= 7.2793 \times 10^{14}\ \text{s}^{-1} \text{ or } 6.487 \times 10^{14}\ \text{s}^{-1}$$

$$\nu_2'' = \nu \pm \nu_2' = 6.883 \times 10^{14}\ \text{s}^{-1} \pm (668\ \text{cm}^{-1})(3 \times 10^{10}\ \text{cm s}^{-1})$$

$$= 7.0837 \times 10^{14}\ \text{s}^{-1} \text{ or } 6.6829 \times 10^{14}\ \text{s}^{-1}$$

$$\nu_3'' = \nu \pm \nu_3' = 6.883 \times 10^{14}\ \text{s}^{-1} \pm (2350\ \text{cm}^{-1})(3 \times 10^{10}\ \text{cm s}^{-1})$$

$$= 7.5883 \times 10^{14}\ \text{s}^{-1} \text{ or } 6.1783 \times 10^{14}\ \text{s}^{-1}$$

2. (a) CH_3COOH

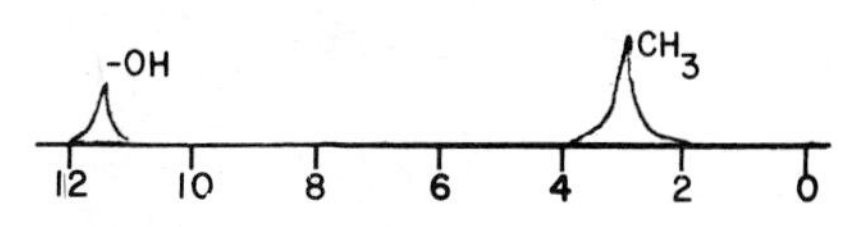

(b) $CH_3 - CH_2 - \overset{O}{\overset{\|}{C}} - CH_2CH_3$

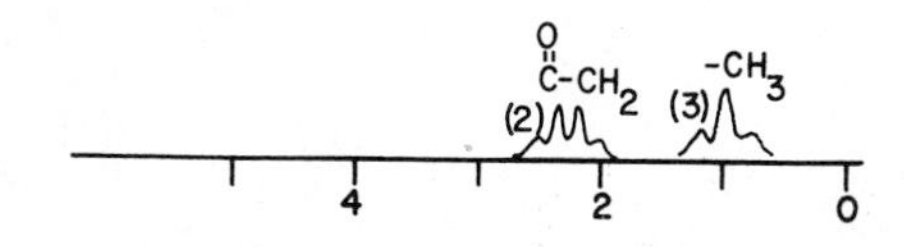

(c) $CH_2 = CH - CH_2 - \overset{O}{\overset{\|}{C}} - H$

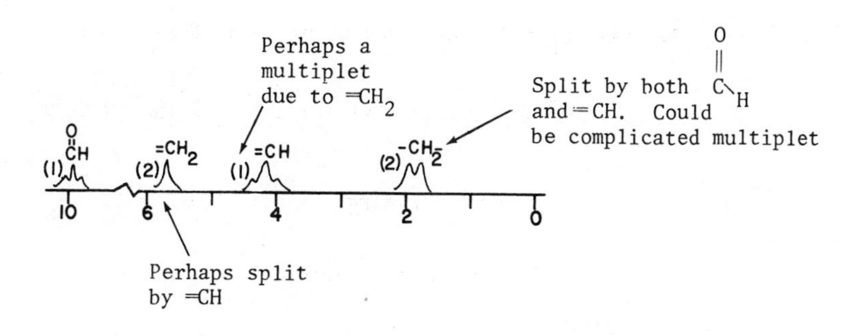

(d) $CH_3-\underset{CH_3}{\overset{OH}{C}}-CH_3$

Variable position

-OH (1)

(9) -CH3

6 4 2 0

(e) $CH_3-\overset{O}{\overset{\|}{C}}-CH_2-CH_2-CH_3$

$\overset{O}{\overset{\|}{C}}CH_2$ (2); $\overset{O}{\overset{\|}{C}}CH_3$ (3); $-CH_2-$ (2); $-CH_3$ (3)

6 4 2 0

Split by both $-CH_3$ and $-CH_2-$. A complicated multiplet.

3. (a) $CH_3-\overset{O}{\overset{\|}{C}}-CH_2-CH_3$

(b) $C_6H_5CH_2OH$ (benzene ring with H, H, H, H, H and CH_2OH)

9-1.6 PROBLEMS

1. (a) $\Sigma P_i = 1$

$$P_6 = 1 - \frac{1}{6} - \frac{1}{5} - \frac{1}{7} - \frac{1}{12} - \frac{1}{3} = 1 - 0.926 = \underline{\underline{0.074}} = P_6$$

(b) $$Q^8_{2,2,0,0,1,3} = (\tfrac{1}{6})^2(\tfrac{1}{5})^2(\tfrac{1}{7})^0(\tfrac{1}{12})^0(\tfrac{1}{3})^1(0.074)^3$$

$$\times \frac{8!}{2!2!0!0!1!3!} = (1.50 \times 10^{-7})(1680) = \underline{\underline{2.52 \times 10^{-4}}}$$

2. (a) $P_1 = \frac{1}{6}$ independent of when it is thrown.

(b) $$Q^4_{0,0,0,4,0,0} = (\tfrac{1}{6})^0(\tfrac{1}{5})^0(\tfrac{1}{7})^0(\tfrac{1}{12})^4(\tfrac{1}{3})^0(0.074)^0$$

$$\times \frac{4!}{0!0!0!4!0!0!} = (\tfrac{1}{12})^4 = \underline{\underline{4.8 \times 10^{-5}}}$$

3. Only a few of numerous possibilities are given.

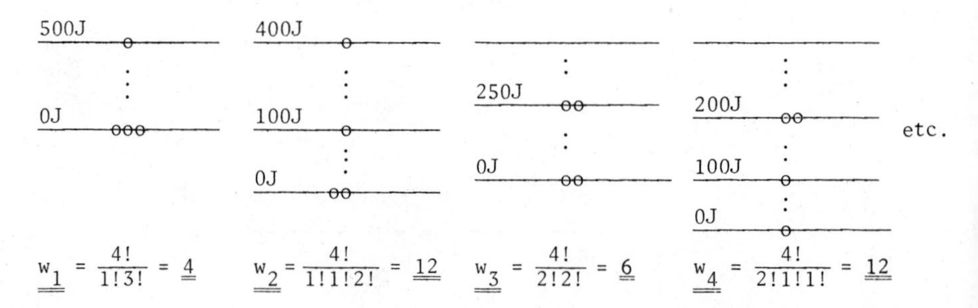

$\underline{\underline{w_1}} = \frac{4!}{1!3!} = \underline{\underline{4}}$ $\underline{\underline{w_2}} = \frac{4!}{1!1!2!} = \underline{\underline{12}}$ $\underline{\underline{w_3}} = \frac{4!}{2!2!} = \underline{\underline{6}}$ $\underline{\underline{w_4}} = \frac{4!}{2!1!1!} = \underline{\underline{12}}$

4. $\frac{N_1}{N_0} = e^{-(\varepsilon_1 - \varepsilon_0)/kT} = e^{-4 \times 10^{-12}/(1.38 \times 10^{-23})(298)} = 0.378$

$\frac{N_2}{N_0} = e^{-(\varepsilon_2 - \varepsilon_0)/kT} = e^{-8 \times 10^{-12}/(1.38 \times 10^{-23})(298)} = 0.143$

$\frac{N_3}{N_0} = 0.054; \quad \frac{N_4}{N_0} = 0.020; \quad \frac{N_5}{N_0} = 0.008$

5. Must assume all 10^{20} molecules are in the first six energy levels or use the partition function. Best to use partition function. See Problem 6 for z.

$$n_j = \frac{N\, e^{-\varepsilon_j/kT}}{z}; \quad n_0 = \frac{(10^{20})(1.0)}{1.6077} = 6.220 \times 10^{19};$$

$$n_1 = \frac{(10^{20})(0.378)}{1.6077} = 2.351 \times 10^{19}$$

$$n_2 = 8.894 \times 10^{18}; \quad n_3 = 3.36 \times 10^{18}; \quad n_4 = 1.27 \times 10^{18};$$

$$n_5 = 4.8 \times 10^{17}$$

$\sum_{i=0}^{5} n_i = \underline{\underline{9.97 \times 10^{19}}}$ molecules in first six levels = 99.7%

6. $z = \sum_i g_i e^{-\varepsilon_i/kT}$

To get 0.1% accuracy, retain all terms > 0.1% of the total.

$$z = 1.0 + 0.3781 + 0.1429 + 0.0540 + 0.0204 + 0.0077 + 0.0029 + 0.0011 + 0.00042 + 0.00016$$

$\underline{\underline{z = 1.6077 \text{ (better than 0.1\%)}}}$

7. $$\bar{E} = \sum \frac{\varepsilon_j e^{-\varepsilon_j/kT}}{z} = 0 + \frac{(4 \times 10^{-21})(2.351 \times 10^{19})}{10^{20}}$$

$$+ \frac{(8 \times 10^{-21})(8.894 \times 10^{18})}{10^{20}} + \frac{(12 \times 10^{-21})(3.36 \times 10^{18})}{10^{20}}$$

$$+ \frac{(16 \times 10^{-21})(1.27 \times 10^{18})}{10^{20}} + \frac{(20 \times 10^{-21})(4.8 \times 10^{17})}{10^{20}}$$

$= \underline{\underline{2.35 \times 10^{-21} \text{ J mlc}^{-1} = \bar{E}}}$

Note that $\frac{e^{-\varepsilon_j/kT}}{z} = \frac{n_j}{N}$

However, this is not the true value of $\bar{E}$. If level 7 (ε_6) is included, $\bar{E} = 2.39 \times 10^{-21}$ J mlc^{-1}. Several more levels must be included to get the true value.

8. $$\frac{\overline{\delta\rho^2}}{\rho^{-2}} = \frac{kT\kappa}{V} = \frac{kT}{VP} = \frac{kT}{RT} = \frac{kT}{NkT} = \frac{1}{N} = \underline{\underline{1.66 \times 10^{-24}}} \quad \text{a miniscule value}$$

9-2.8 PROBLEMS

1. $$S_{tr} = R\left[\frac{5}{2} + \ln(8.201 \times 10^7 \frac{M^{3/2}T^{5/2}}{P}\right]$$

$$= 8.314 \text{ J mol}^{-1} \text{ K}^{-1}\left[\frac{5}{2} + \ln 8.201 \times 10^{-7} \frac{(0.064)^{1.5}(298.15)^{2.5}}{(1.013 \times 10^5)}\right]$$

$$= (8.314)(2.5 + 16.82) = \underline{\underline{160.6 \text{ J mol}^{-1} \text{ K}^{-1}}} = S_{tr}$$

$$z_{rot} = \frac{8\pi^2(8\pi^3 I_A I_B I_C)^{1/2}(kT)^{3/2}}{\sigma h^3} = \frac{(kT)^{3/2}\pi^{1/2}}{\sigma(ABC)^{1/2}h^{3/2}}$$

$$= \frac{[(1.38 \times 10^{-23} \cdot 298.15)^{3/2}(3.1416)^{1/2}]}{2[(6.078)(1.032)(0.880) \times 10^{+30}]^{1/2}(6.626 \times 10^{-34})^{3/2}}$$

$$z_{rot} = \underline{\underline{5.84 \times 10^3}} : S_{rot} = R \ln z + \frac{\bar{E}}{T} = R \ln (5.84 \times 10^3) + \frac{3}{2} R$$

$$= R(8.67 + 1.5) = \underline{\underline{84.6 \text{ J mol}^{-1} \text{ K}^{-1}}} = S_{rot}$$

$$x_1 = h\nu_1/kT = \frac{(6.626 \times 10^{-34})(1151.2 \text{ cm}^{-1})(3 \times 10^{10} \text{ cm s}^{-1})}{(1.38 \times 10^{-23})(298.15)}$$

$$= \underline{\underline{5.56 = x_1; \quad x_2 = 2.51; \quad x_3 = 6.58}}$$

$$S_{vib,1} = -R \ln(1 - e^{-x}) + \frac{Rx}{e^x - 1} = -R \ln (1 - e^{-5.56}) + \frac{R(5.56)}{e^{5.56} + 1}$$

$$= R[0.0039 + 0.021] = 0.207 \text{ J mol}^{-1} \text{ K}^{-1}$$

$$S_{vib,1} = 0.207; \quad S_{vib,2} = 2.27; \quad S_{vib,3} = 0.088$$

$$\underline{\underline{S_{vib,tot} = S_{v,1} + S_{v,2} + S_{v,3} = 2.57 \text{ J mol}^{-1} \text{ K}^{-1}}}$$

$$S^\circ_{total} = S_{tr} + S_{rot} + S_{vib} = 160.6 + 84.6 + 2.57$$

$$= \underline{\underline{247.8 \text{ J mol}^{-1} \text{ K}^{-1}}} = S^\circ_{total}$$

To find A:

$$Z_{tr} = \left[\frac{(2\pi mkT)^{3/2}V}{h^3}\right]^N \frac{1}{N!} : \ell n\, z_{tr}^N = N\,\ell n\left[\frac{(2\pi mkT)^{3/2}V}{h^3}\right] - \ell n\, N!$$

$$\ell n\, z_{tr}^N = N\,\ell n\left[\frac{(2\pi mkT)^{3/2}}{h^3}\frac{RT}{P}\right] - (N\,\ell n\, N - N)$$

$$= N\,\ell n\left[\frac{(2\pi mkT)^{3/2}RT}{h^3NP}\right] + N$$

$$A_{tr} = -\,kT\,\ell n\, z_{tr}^N = -kT\left[N\,\ell n\left\{\frac{(2\pi mkT)^{3/2}RT}{h^3NP}\right\} + N\right]$$

$$= -\,RT\left\{\ell n\,\frac{(2\pi mkT)^{3/2}RT}{h^3NP}\right\} - RT = -(8.314)(298.15)$$

$$\times\left\{\ell n\left[\frac{\{(1.38\times 10^{-23})2\pi(0.064/6.02\times 10^{23})(298.15)\}^{1.5}(8.314)(298.15)}{(6.626\times 10^{-34})^3(1.013\times 10^5)(6.023\times 10^{23})}\right] + 1\right\}$$

(All SI units)

$$= -2.479\times 10^3\left\{\ell n[2.00\times 10^7] + 1\right\} = \underline{\underline{-44{,}200\ \text{J mol}^{-1}}} = A_{tr}$$

$$A_{rot} = -RT\ \ell n\, z_{rot} = -\,RT\,\ell n(5.84\times 10^3) = \underline{\underline{-21500\ \text{J mol}^{-1}}} = A_{rot}$$

$$A_{vib,1} = -\,RT\,\frac{e^{-x_1/2}}{1-e^{-x_1}} = -\,RT(0.0623) = \underline{\underline{-154\ \text{J mol}^{-1}}};$$

$$\underline{\underline{A_{vib,2}}} = \underline{\underline{-769\ \text{J mol}^{-1}}};\qquad \underline{\underline{A_{vib,3}}} = \underline{\underline{-92\ \text{J mol}^{-1}}}$$

$$\bar{A}_{total} = A_{tr} + A_{rot} + \Sigma A_{vib,i} = \underline{\underline{-66{,}700\ \text{J mol}^{-1} = \bar{A}_{total}}}$$

2. $G° = H° - TS = E° + PV - TS° = E° + RT - TS°$

$$E° = E_{tr} + E_{rot} + E_{vib}$$

$$\underline{E_{tr} = \frac{3}{2}RT};\quad \underline{E_{rot} = \frac{3}{2}RT};\quad E_{vib_i} = \frac{Nhc\bar{\nu}_i}{2} + \frac{Nhc\bar{\nu}_i}{e^{hc\bar{\nu}_i/kT} - 1}$$

$E_1 = 7972 + 26 = 7998$

$E_2 = 4495 + 246 = 4741$

$E_3 = 9672 + 8 = 9680$

$\underline{E_{total}(\text{vib}) = 22{,}400\ \text{J mol}^{-1}}$

$E^\circ_{total} = 22{,}400 + 3RT$

$H^\circ_{total} = 22{,}400 + 4RT = \underline{\underline{32{,}330 \text{ J mol}^{-1}}} = H^\circ$

$$S^\circ_{tr} = \frac{5}{2}R + R\,\ell n\left[\frac{(2\pi MkT)^{3/2}}{h^3N^{5/2}}\frac{RT}{P}\right] = \frac{5}{2}R + R\,\ell n(1.22 \times 10^7)$$

$$= R(18.82) = \underline{\underline{156}} = S^\circ_{tr}$$

$$z_{rot} = \frac{8\pi^2(8\pi^3 I_A I_B I_C)^{1/2}(kT)^{3/2}}{\sigma h^3}$$

$$= \frac{8\pi^2[8\pi^3(1.44 \times 10^{-137})]^{1/2}[(1.38 \times 10^{-23})(298)]^{3/2}}{2(6.626 \times 10^{-34})^3} = 2139$$

$$S^\circ_{rot} = R\,\ell n\, z + 1.5\,R = R[\ell n(2139) + 1.5] = \underline{\underline{76.2}} = S^\circ_{rot}$$

$$S^\circ_{vib} = -R\,\ell n(1 - e^{-x}) + \frac{Rx}{e^x - 1};$$

$$x_i = hc\bar{\nu}_i/kT = 4.83 \times 10^{-3}\,\bar{\nu}_i$$

$x_1 = 6.43 \qquad S_1 = 0.0133 + 0.0863 = 0.0996$

$x_2 = 3.63 \qquad S_2 = 0.224 + 0.822 = 1.05$

$x_3 = 7.81 \qquad S_3 = 0.003 + 0.026 = 0.029$

$S^\circ_{vib} = \underline{1.17 \text{ J mol}^{-1}\text{ K}^{-1}}$

$z_{elc} = g_{elc}; \qquad S_{elc} = R\ln z = R\ln 2 = \underline{\underline{5.76 \text{ J mol}^{-1}\text{ K}^{-1}}}$

$\underline{\underline{S^\circ_{total} = 239 \text{ J mol}^{-1}\text{ K}^{-1}}}$

$G^\circ = H^\circ - TS^\circ = 32{,}330 - 298(239) = \underline{\underline{-38{,}900 \text{ J mol}^{-1}}} = G^\circ$

9-3.3 PROBLEMS

1. $Na_2 \rightleftharpoons 2Na \qquad K_p = \frac{[Na]^2}{[Na_2]}$

See text for following equation forms.

$$K_p = \left(\frac{2\pi kT}{h^2}\right)^{3/2}\frac{m_{Na}^3}{m_{Na_2}^{3/2}}\;\frac{RT}{N}\;\frac{2^2}{1}\;\frac{\sigma_{Na_2}h^2}{8\pi^2 IkT}\left(1 - e^{-h\nu_o/kT}\right)\frac{1}{e^{-D_o/kT}}$$

$$= (8.78 \times 10^{70}\text{ J}^{-3/2}\text{ s}^{-3})(2.64 \times 10^{-39}\text{ kg}^{3/2})(1.38 \times 10^{-20}\text{ J})$$

$$\times (4)(4.45 \times 10^{-4})(0.205)(2.08 \times 10^{-4})$$

$$= 2.42 \times 10^5 \text{ N m}^{-2} = \underline{\underline{2.39 \text{ atm}}} = K_p$$

2. $CO_2 + H_2 \rightarrow CO + H_2O$

$$K_p = \frac{z'_{CO} z'_{H_2O}}{z'_{CO_2} z'_{H_2}} e^{-\Delta D_o/kT}$$

z' is total partition function minus electronic.

$$z'_{CO} = \frac{(2\pi mkT)^{3/2}}{h^3} \frac{RT}{N} \cdot 1 \cdot \frac{8\pi^2 I_{CO} kT}{\sigma h^2} \cdot \frac{1}{1 - e^{-h\nu_o/kT}}$$

$$= (7.53 \times 10^{32})(1.24 \times 10^{-20})(1)(3.24 \times 10^2)(1.034)$$

$$= \underline{\underline{3.13 \times 10^{15}}} = z'_{CO}$$

$$z'_{H_2O} = \frac{(2\pi mkT)^{3/2}}{h^3} \frac{RT}{N} \quad 1 \quad \frac{8\pi^2 (8\pi^3 I_A I_B I_C)^{1/2} (kT)^{3/2}}{2h^3} \prod_{i=1}^{3} \frac{1}{1 - e^{-h\nu_i/kT}}$$

$$= (3.87 \times 10^{32})(1.24 \times 10^{-20})(1)(225)(1.003)(1.085)(1.002)$$

$$= \underline{\underline{1.18 \times 10^{15}}} = z'_{H_2O}$$

$$z'_{H_2} = (1.43 \times 10^{31})(1.24 \times 10^{-20})(5.27)(1.001) = \underline{\underline{9.35 \times 10^{11}}} = z'_{H_2}$$

$$z'_{CO_2} = (1.48 \times 10^{33})(1.24 \times 10^{-20})(803)(1.139)(1.024)(1.542)(1.542)$$

$$= \underline{\underline{4.09 \times 10^{16}}} = z'_{CO_2}$$

$$e^{-\Delta D_o/kT} = e^{-5.39} = 0.00454$$

$$K_p = \frac{(3.13 \times 10^{15})(1.18 \times 10^{15})}{(9.35 \times 10^{11})(4.09 \times 10^{16})} (0.00454) = \underline{\underline{0.44 = K_p \quad \text{No units}}}$$

10-1.6 PROBLEMS

1. $$\frac{d[A]}{dt} = -k[A]^{5/2} \rightarrow \int_{[A]_o}^{[A]} \frac{d[A]}{[A]} = -k\,dt \rightarrow$$

$$-\frac{2}{3}\left\{[A]^{-3/2} - [A_o]^{-3/2}\right\} = -kt$$

At $t_{1/2}$,

$$[A] = \frac{1}{2}[A]_o$$

$$\therefore -\frac{2}{3}\left\{\frac{1}{(\frac{1}{2}[A]_o)^{3/2}} - \frac{1}{[A]_o^{3/2}}\right\} = -\frac{2}{3}\left\{\frac{2\sqrt{2}}{[A]_o^{3/2}} - \frac{1}{[A]_o^{3/2}}\right\} = -kt_{1/2}$$

$$kt_{1/2} = \frac{1.22}{[A]_o^{3/2}}$$

2. $\ln \frac{[A]}{[A]_o} = -kt$

But $[A] = 0.28[A]_o$ when $t = 350$ sec.

$$k = -\frac{1}{t} \ln \frac{0.28[A]_o}{[A]_o} = -\frac{1}{350} \ln 0.28 = 3.64 \times 10^{-3}\ s^{-1} = k$$

What is t when $[A] = 0.10[A]_o$?

$$t = \frac{1}{k} \ln \frac{0.1[A]_o}{[A]_o} = -\frac{1}{3.64 \times 10^{-3}\ s^{-1}} \ln 0.1 = 633\ s = t$$

3. (a) $\ln P - \ln P_o = \ln \frac{P}{P_o} = kt$

You should plot $\ln P/P_o$ vs. t to show a straight line result. The slope = k.

t	0	·10	20	30	40	50	60	70
$P \times 10^{-6}$	3.93	5.31	7.24	9.64	12.9	17.1	23.2	31.4
$\ln P/P_o$	0	0.301	0.611	0.897	1.19	1.47	1.78	2.08

From the plot, slope $= k = \frac{2.08 - 0.301}{70 - 10} = 0.030\ \text{year}^{-1} = k$

Time for $P = 2P_o$:

$$t = \frac{1}{0.030\ yr^{-1}} \ln \frac{2P_o}{P_o} = 23\ \text{year} = t$$

(b) For 1% population growth, $P = 1.01\ P_o$ for t = 1 year.

$$k = \frac{1}{t} \ln \frac{P}{P_o} = \frac{1}{1\ yr} \ln \frac{1.01\ P_o}{P_o} = 0.00995\ yr^{-1}$$

Time to double: $P = 2P_o$

$$t = \frac{1}{0.00995\ yr^{-1}} \ln \frac{2P_o}{P_o} = 69.7\ yr$$

(c) For 2%, $P = 1.02\ P_o$ for $t = 1$ year.

$$k = \frac{1}{1\ yr}\ \ell n\ \frac{1.02\ P_o}{P_o} = 0.0198\ yr^{-1}$$

To double

$$t = \frac{1}{0.0198}\ \ell n\ \frac{2P_o}{P_o} = \underline{\underline{35\ yr}}$$

4. If first order, plot of $\ell n\ A/A_o$ would be linear; if second order, plot of $1/[A]$ vs. t would be linear. You should make the two plots. $\ell n\ \frac{\%A}{100}$ is definitely nonlinear. Plot of $\frac{1}{[\%A]}$ is linear.

$A \rightarrow B$ $\quad$ %A = 100 - % reaction

t (s)	0	3600	7200	19500	33900	56520	72720	91080
%A	100	90.9	83.3	67.3	52.7	39.1	33.4	29.7
$\ell n\ \frac{\%A}{100}$	0	-0.0954	-0.183	-0.396	-0.641	-0.939	-1.097	-1.214
$\frac{1}{\%A}$	0.010	0.0110	0.0120	0.0149	0.0190	0.0256	0.0299	0.0337

Slope of $\frac{1}{[\%A]}$ vs. t plot is $\frac{0.034 - 0.010}{85000 - 0.00}\ \%^{-1}\ s^{-1}$

$\underline{\underline{k = 2.82 \times 10^{-7}\ \%^{-1}\ s^{-1}}}$

5. Plot of $\ell n\ P_E$ vs. t (do it) is linear. $\therefore$ first order.

$$\text{Slope} = -k = \frac{-2.204 - (-1.058)}{3155 - 390}$$

t (s)	390	777	1195	3155	∞
P_{ether} (atm)	0.347	0.295	0.246	0.104	0
$\ell n\ P_E$	-1.058	-1.221	-1.402	-2.263	-

$\underline{\underline{k = 4.3 \times 10^{-4}\ s^{-1}}}$

6. $CH_3CHO \rightarrow CH_4 + CO$

$P_o - P_x \quad P_x \quad P_x$

$$P_{total} = P_o - P_x + P_x + P_x = P_o + P_x$$

Thus

$$P_x = P_{total} - P_o$$

$$P_{CH_3CHO} = P_o - P_x = P_o - (P_{total} - P_o) = \underline{\underline{2P_o - P_{total}}} = P_{CH_3CHO}$$

Plot of $\ln P_{CH_3CHO}$ vs. t is nonlinear. Plot of $\frac{1}{P_{CH_3CHO}}$ vs. t (do it) is linear.

t(s)	0	42	105	190	310	480	840
P_{total}(atm)	0.474	0.518	0.571	0.624	0.676	0.729	0.795
P_{CH_3CHO}(atm)	0.474	0.430	0.377	0.324	0.272	0.219	0.153
$\ln P_{CH_3CHO}$	-0.747	-0.844	-0.976	-1.13	-1.302	-1.52	-1.88
$1/P_{CH_3CHO}$	2.11	2.33	2.65	3.09	3.68	4.57	6.54

Slope of $\frac{1}{P_{CH_3CHO}}$ vs. t plot $= k = \frac{6.54 - 2.11}{840 - 0}$ atm^{-1} s^{-1}

$$= \underline{\underline{5.27 \times 10^{-3} \text{ atm}^{-1} \text{ s}^{-1} = k}}$$

7. $A(g) \longrightarrow 2B(g) + C(g)$

$P_A = P_A^\circ - P_x \qquad 2P_x \qquad P_x$

We are given P_{total}; however we need P_A.

$$P_T = P_A + P_B + P_C = P_A^\circ - P_x + 2P_x + P_x = P_A^\circ + 2P_x$$

Thus

$$P_x = \frac{P_T - P_A^\circ}{2}$$

Finally,

$$P_A = P_A^\circ - P_x = P_A^\circ - \frac{P_T - P_A^\circ}{2} = \frac{3P_A^\circ - P_T}{2}$$

t	0	3	6	9	15	∞
P_T	173.5	193.4	211.3	228.6	249.2	491.8
P_A	173.5	163.6	154.6	145.95	135.65	14.35
$\ln P_A$	5.156	5.097	5.041	4.983	4.910	2.66
$\frac{1}{P_A} \times 10^3$	5.76	6.11	6.47	6.85	7.37	69.7

Plot $\ln P_A$ vs. t; if linear, first order.

Plot $1/P_A$ vs. t; if linear, second order.

Difficult to determine, in this case, which is most nearly linear.

$\underline{\underline{k_1 = 3.2 \times 10^{-4} \text{ s}^{-1}}}$; $\quad \underline{\underline{k_2 = 2.02 \times 10^{-6} (\text{mm Hg})^{-1} \text{ s}^{-1}}}$

10-2.7 PROBLEMS

1. $\text{Rate} = -k[H_2]^{n_1}[NO]^{n_2}$

But

$$[H_2] = [NO]$$

or

$$P_{H_2} = P_{NO}$$

$$\text{Rate} = -k[H_2]^{n_1+n_2} = -kP^n$$

$n = n_1 + n_2$ = overall rate equation order

Plot $\ell n\ t_{1/2}$ vs. $\ell n\ P_o$ (do it). Slope = -(n - 1).

$\ell n\ t_{1/2}$	4.39	4.63	4.55	4.94	5.19	5.17	5.41
$\ell n\ P_o$	-0.764	-0.803	-0.707	-0.970	-1.11	-1.14	-1.32

Data points show considerable scatter. However, a straight line is defined. Slope = -1.51 = -(n - 1), therefore

n = 2.5 = overall reaction order.

Do not know individual orders (n_1 or n_2).

2. $a = [S_2O_3^{-2}]_0 = \dfrac{\text{Equivalents } I_2 \text{ consumed}}{V_{S_2O_3^{2-}}} = \dfrac{N_{I_2}V_{I_2}}{V_{S_2O_3^{2-}}}$

$$= \frac{(0.03763)(0.02572)}{(0.01002)} = 0.0966N$$

$$b = [PrBr]_0 = [S_2O_3^{2-}]_0 - [S_2O_3^{2-}]_\infty = \frac{(0.03763 - 0.02224)(0.02572)}{(0.01002)}$$

$$= 0.0395N = b$$

(This assumes all PrBr reacts. There is an excess of $S_2O_3^{2-}$.)

a - b = 0.0571

t (s)	0	1110	2010	3192	5052	7380	11232
$[S_2O_3^{2-}]$	0.0966	0.0904	0.0863	0.0819	0.0767	0.07197	0.0668
PrBr	0.0395	0.0333	0.0292	0.0248	0.0196	0.0149	0.00968
k*	--	0.00165	0.00165	0.00165	0.00163	0.00162	

$$*k = \frac{1}{t}\left(\frac{1}{a - b}\right) \ell n \left(\frac{b}{a}\right)\left(\frac{a - x}{b - x}\right)$$ Units are $dm^3\ mol^{-1}\ s^{-1}$.

$k_{avg} = 0.00162 \pm 0.00001\ dm^3\ mole^{-1}\ s^{-1}$

Reaction is first order in each reactant. You should plot $\ell n\left(\frac{a - x}{b - x}\right)$ vs. t to verify the results given.

3. $$\frac{d(Cl)}{dt} = 0 = k_1(NO_2Cl) - k_2(NO_2Cl)(Cl)$$

$\therefore$ $\underline{\underline{(Cl) = k_1/k_2}}$ Note: [ ] ≡ ( )

$$\frac{d(NO_2Cl)}{dt} = -k_1(NO_2Cl) - k_2(NO_2Cl)(Cl) = -k_1(NO_2Cl) - k_2(NO_2Cl)\left(\frac{k_1}{k_2}\right)$$

$$\therefore \frac{d(NO_2Cl)}{dt} = \underline{\underline{-2k_1(NO_2Cl)}}$$

4. (a) $$\frac{d(CH_3)}{dt} = 0 = 2k_1(C_2H_6) - k_2(CH_3)(C_2H_6)$$

$$\underline{\underline{(CH_3) = \frac{2k_1(C_2H_6)}{k_2(C_2H_6)} = 2k_1/k_2}}$$

(b) $$\frac{d(C_2H_5)}{dt} = 0 = k_2(CH_3)(C_2H_6) - k_3(C_2H_5) + k_4(H)(C_2H_6) - k_5(H)(C_2H_5)$$

(c) $$\frac{d(H)}{dt} = 0 = k_3(C_2H_5) - k_4(H)(C_2H_6) - k_5(H)(C_2H_5)$$

Add (b) + (c) $$0 = k_2(CH_3)(C_2H_6) - 2k_5(H)(C_2H_5)$$

From (a)

$$\therefore (H) = \frac{k_2(CH_3)(C_2H_6)}{2k_5(C_2H_5)} = \frac{k_2(2k_1/k_2)(C_2H_6)}{2k_5(C_2H_5)} = \underline{\underline{\frac{k_1(C_2H_6)}{k_5(C_2H_5)} = (H)}}$$

Substitute (H) into (c):

$$0 = k_3(C_2H_5) - \frac{k_4k_1(C_2H_6)^2}{k_5(C_2H_5)} - k_5\frac{k_1(C_2H_6)(C_2H_5)}{k_5(C_2H_5)}$$

$$(C_2H_5)^2 - \frac{k_1}{k_3}(C_2H_6)(C_2H_5) - \frac{k_1k_4}{k_3k_5}(C_2H_6)^2 = 0$$

Thus

$$(C_2H_5) = \left[\frac{k_1}{2k_3} \pm \frac{\sqrt{\left(\frac{k_1}{k_3}\right)^2 + 4\left(\frac{k_1k_4}{k_3k_5}\right)}}{2}\right](C_2H_6) \equiv \mathbb{D}(C_2H_6) = (C_2H_5)$$

$\mathbb{D}$ is defined by the equation.

Then

$$(H) = \frac{k_1}{k_5\mathbb{D}}; \qquad (CH_3) = \frac{2k_1}{k_2}$$

(d) $$-\frac{d(C_2H_6)}{dt} = k_1(C_2H_6) + k_2(CH_3)(C_2H_6) + k_4(H)(C_2H_6) - k_5(H)(C_2H_5)$$

$$= k_1(C_2H_6) + 2k_1(C_2H_6) + \frac{k_4k_1}{k_5\mathbb{D}}(C_2H_6) - \frac{k_1k_5}{k_5\mathbb{D}}\mathbb{D}(C_2H_6)$$

$$= 2k_1(C_2H_6) + \frac{k_4k_1}{k_5\mathbb{D}}(C_2H_6) = -\frac{d(C_2H_6)}{dt}$$

If k_1 is small, higher order terms in k_1 are negligible.

$\mathbb{D}$ becomes $\left(\frac{k_1k_4}{k_3k_5}\right)^{1/2}(C_2H_6)$; i.e., $k_1^2 << k_1$ and $k_1 << k_1^{1/2}$

Then

$$-\frac{d(C_2H_6)}{dt} = 2k_1(C_2H_6) + \frac{k_4k_1k_3^{1/2}k_5^{1/2}}{k_5k_1^{1/2}k_4^{1/2}}(C_2H_6)$$

$$= 2k_1(C_2H_6) + \frac{k_4^{1/2}k_1^{1/2}k_3^{1/2}}{k_5^{1/2}}(C_2H_6)$$

But $k_1 << k_1^{1/2}$

$$\therefore \quad -\frac{d(C_2H_6)}{dt} \approx \left(\frac{k_4k_1k_3}{k}\right)^{1/2}(C_2H_6);$$ first order in C_2H_6

5. The rate is so slow that for a one second period near t = 0, we can assume [HI] = constant. Therefore,

$$-\frac{d[HI]}{dt} = -k_2[HI]^2$$

may be written as:

$$\frac{\Delta[HI]}{\Delta t} = k_2[HI]^2$$

From ideal gas law,

$$[HI] = 2.03 \times 10^{-2} \text{ mol dm}^{-3}$$

Therefore

$$\frac{\Delta[HI]}{\Delta t} = (4.0 \times 10^{-6} \text{ dm}^3 \text{ mol}^{-1} \text{ s}^{-1})(2.03 \times 10^{-2} \text{ mol dm}^{-3})^2$$

$$\times (6.023 \times 10^{23} \text{ mlc mol}^{-1})$$

$$= \underline{\underline{9.93 \times 10^{14} \text{ mlc dm}^{-3} \text{ s}^{-1}}}$$

6. $$\frac{d[NO_2]}{dt} = -2k_2[NO_2]^2$$

$$\int \frac{d[NO_2]}{[NO_2]^2} = -\int 2k_2 \, dt$$

$$-\frac{1}{[NO_2]} + \frac{1}{[NO_2]_0} = -2K_2(t - 0)$$

$$\frac{1}{[NO_2]} - \frac{1}{[NO_2]_0} = +2k_2t$$

Need $[NO_2]_0 = \frac{1}{\bar{V}_o} = \frac{P}{RT} = \frac{0.526 \text{ atm}}{(82.1 \text{ ml atm mol}^{-1} \text{ k}^{-1})(600\text{K})}$

$$= 1.068 \times 10^{-5} \text{ mol ml}^{-1}$$

$$[NO_2] = 0.9[NO_2]_0 = 0.9(1.068 \times 10^{-5}) = 9.61 \times 10^{-6} \text{ mol ml}^{-1}$$

Thus the time required for one tenth to be consumed is

$$t = +\frac{1}{2k_2}\left(\frac{1}{0.9[NO_2]_0} - \frac{1}{[NO_2]_0}\right)$$

$$= \frac{1}{2(6.3 \times 10^2 \text{ ml mol}^{-1} \text{ s}^{-1})}\left(\frac{1}{9.61 \times 10^{-6}} - \frac{1}{1.068 \times 10^{-5}}\right) \text{ ml mol}^{-1}$$

$$\underline{\underline{t = 8.3 \text{ s}}}$$

10-3.6 PROBLEMS

1. Follow Example 10-3.5.1; all units kept in SI.

<u>For NO_2:</u>

$$z_{tr} = \left(\frac{2\pi mkT}{h^2}\right)^{3/2} = \left\{\frac{(2)(\pi)\left(\frac{0.046}{6.023\times 10^{23}}\right)(1.38\times 10^{-23})(500)}{(6.626\times 10^{-34})^2}\right\}^{3/2}$$

$$= \underline{\underline{6.55\times 10^{-32}\ m^{-3}\ mlc}} = z_{tr}(NO_2)$$

$$z_{rot} = \frac{8\pi^2(8\pi^3 I_A I_B I_C)^{1/2}(kT)^{3/2}}{\sigma h^3}$$

$$= \frac{8\pi^2\{8\pi^3(1.44\times 10^{-137})^{1/2}[(1.38\times 10^{-23})(500)]^{3/2}\}}{2(6.626\times 10^{-34})^3}$$

$$= \underline{\underline{4.64\times 10^3}} = z_{rot}(NO_2)$$

$$z_{vib,1} = \left(1 - e^{-h\nu_1/kT}\right)^{-1} = \left[1 - e^{-3.84}\right]^{-1} = 1.02$$

$$z_{vib,2} = (1 - e^{-2.16})^{-1} = 1.13$$

$$z_{vib,3} = (1 - e^{-4.66})^{-1} = 1.01$$

$$\underline{\underline{z_{vib,total}(NO_2) = 1.17}}$$

$$z_{elec} = 2$$

$$\therefore (z_{NO_2})^2 = [(6.55\times 10^{+32}\ m^{-3})(4.64\times 10^3)(1.17)(2)]^2$$

$$= \underline{\underline{5.03\times 10^{73}\ m^{-6}\ mlc^2}}$$

For $(N_2O_4)^{\ddagger}$:

$$z^{\ddagger}_{tr} = \left\{\frac{(2)(\pi)\left(\frac{0.092}{6.02\times 10^{23}}\right)(1.38\times 10^{-23})(500)}{(6.626\times 10^{-34})^2}\right\}^{3/2}$$

$$= \underline{\underline{1.85\times 10^{33}\ m^{-3}\ mlc}} = z^{\ddagger}_{tr}$$

$$z^{\ddagger}_{rot} = \frac{8\pi^2[8\pi^3(2.95\times 10^{-135})^{1/2}][(1.38\times 10^{-23})(500)]^{3/2}}{2(6.626\times 10^{-34})^3}$$

$$= \underline{\underline{6.65\times 10^4 = z^{\ddagger}_{rot}}}$$

$$z^{\ddagger}_{vib,1} = 1.008;\quad z^{\ddagger}_{vib,2} = 1.008;\quad z^{\ddagger}_{vib,3} = 1.024;$$

$z^{\ddagger}_{vib,4} = 1.024$; $z^{\ddagger}_{vib,5} = 1.154$; $z^{\ddagger}_{vib,6} = 1.154$;

$z^{\ddagger}_{vib,7} = 1.574$; $z^{\ddagger}_{vib,8} = 1.948$; $\underline{z^{\ddagger}_{vib,total} = 4.35}$

$z^{\ddagger}_{elec} = 1$

$\underline{z^{\ddagger}_{total} = 5.35 \times 10^{38}\ m^{-3}\ mlc}$

$$A = \frac{kT}{h}\frac{z^{\ddagger}_{N_2O_4}}{z^2_{NO_2}} = \frac{(1.38 \times 10^{-23}\ J\ mlc\ K^{-1})(500K)}{(6.626 \times 10^{-34}\ J\ s\ mlc^{-1})}\ \frac{5.35 \times 10^{38}\ m^{-3}\ mlc}{5.03 \times 10^{73}\ m^{-6}\ mlc^2}$$

$$= 1.099 \times 10^{-22}\ m^3\ mlc^{-1}\ s^{-1}$$

$$= 6.62 \times 10^{1}\ m^3\ mol^{-1}\ s^{-1}$$

$$\underline{A = 6.62 \times 10^4\ dm^3\ mol^{-1}\ s^{-1}}$$

$$k_r = Ae^{-E/RT} = (6.62 \times 10^4\ dm^3\ mol^{-1}\ s^{-1})e^{-111{,}000/(8.314)(500)}$$

$$= \underline{1.68 \times 10^{-7}\ dm^3\ mol^{-1}\ s^{-1}} = k_r$$

2. $$\frac{k_2}{k_1} = \frac{Ae^{-E_A/RT_2}}{Ae^{-E_A/RT_1}} = e^{-(E_A/R)(\frac{1}{T_1} - \frac{1}{T_2})} \ :\ \ell n\frac{k_2}{k_1} = -\frac{E_A}{R}\left(\frac{1}{T_2} - \frac{1}{T_1}\right)$$

$k_2 = 3k_1$ when $T_2 = T_1 + 10°$. Choose $T_1 = 300K$.

$$E_A = \left(-R\ \ell n\frac{k_2}{k_1}\right)\Bigg/\left(\frac{1}{300} - \frac{1}{310}\right) = -(8.314)(\ell n\ 3)/(-1.075 \times 10^{-4})$$

$$= \underline{8.5 \times 10^4\ J\ mol^{-1} = E_A}$$

The value obtained depends to some extent on the choice of T_1.

3. A plot of $\ell n\ k$ vs. $1/T$ is necessary. Slope of plot $= -E_A/R$.

Check a couple of data sets:

$T_1 = 561.2$; $T_2 = 620.5$; $k_1 = 3.18$; $k_2 = 34.2$

$$E_A = \frac{-R\ \ell n\frac{k_2}{k_1}}{\left(\frac{1}{T_2} - \frac{1}{T_1}\right)} = \frac{-(8.314\ J\ mol^{-1}\ K^{-1})\ \ell n\left(\frac{34.2}{3.18}\right)}{\left(\frac{1}{620.5} - \frac{1}{561.2}\right)K^{-1}}$$

$$= \underline{1.16 \times 10^5\ J\ mol^{-1} = E_A}$$

For $T_1 = 578$, $T_2 = 639.9$; $k_1 = 6.57$; $k_2 = 70$

$$E_A = \underline{\underline{1.18 \times 10^5 \text{ J mol}^{-1}}} = E_A$$

Now pick a point to calculate A. Should use a point on the line in your plot.

$$T = 578,\ K = 6.57,\ k = Ae^{-E_A/(RT)}$$

$$A = \frac{k}{e^{-E_A/RT}} = \frac{6.57}{e^{-1.17 \times 10^5/(8.314)(578)}}$$

$$= \underline{\underline{2.46 \times 10^{11} \text{ cm}^3 \text{ mol}^{-1} \text{ s}^{-1} = A}}$$

(Verify the units.)

4. $$A'' = \left(\frac{8kT}{\pi\mu}\right)^{1/2} \pi\sigma^2 = \left[\frac{(8)(1.38 \times 10^{-23} \text{ J mlc}^{-1} \text{ K}^{-1})(700\text{K})}{\pi \frac{(0.127)(0.127)}{(0.127 + 0.127)}\left(\frac{1}{6.023 \times 10^{23}}\right) \text{kg mlc}^{-1}}\right]^{1/2}$$

$$\times \pi(5.0 \times 10^{-10} \text{ m})^2$$

$$= (4.83 \times 10^2)(\pi)(25 \times 10^{-20})\text{m}^3 \text{ mlc}^{-1} \text{ s}^{-1}$$

$$= \underline{\underline{3.79 \times 10^{-16} \text{ m}^3 \text{ mlc}^{-1} \text{ s}^{-1} = A''}}$$

$$A = A''e^{1/2} = \underline{\underline{6.25 \times 10^{-16} \text{ m}^3 \text{ mlc}^{-1} \text{ s}^{-1}}}$$

$$k = Ae^{-E_A/(RT)} = (6.25 \times 10^{-16} \text{ m}^3 \text{ mlc}^{-1} \text{ s}^{-1})(e^{-184{,}000/(8.314)(700)})$$

$$= (6.25 \times 10^{-16})(1.86 \times 10^{-14})$$

$$= 1.16 \times 10^{-29} \text{ m}^3 \text{ mlc}^{-1} \text{ s}^{-1}$$

$$= 1.16 \times 10^{-23} \text{ cm}^3 \text{ mlc}^{-1} \text{ s}^{-1}$$

$$\underline{\underline{k = 7.0 \text{ cm}^3 \text{ mol}^{-1} \text{ s}^{-1}}}$$

This k does not include any steric factor which would lower the value.

11-1.6 PROBLEMS

1. (a) $$V = \frac{4}{3}\pi r^3 = \frac{4}{3}\pi(10 \times 10^{-9})^3$$

$$= (4.19 \times 10^{-24} \text{ m}^3)\left(\frac{10^3 \text{ dm}^3}{1 \text{ m}^3}\right) = \underline{\underline{4.19 \times 10^{-21} \text{ dm}^3 = V}}$$

$$W = \rho V = (19.3\ \text{kg dm}^{-3})(4.19 \times 10^{-21}\ \text{dm}^3)$$

$$= \underline{\underline{8.08 \times 10^{-20}\ \text{kg (per particle)}}}$$

(b) $$M = (8.08 \times 10^{-20}\ \text{kg})(6.023 \times 10^{23}\ \text{mol}^{-1})$$

$$= \underline{\underline{4.9 \times 10^{4}\ \text{kg mol}^{-1} = M}}$$

(c) $$\frac{4.9 \times 10^4\ \text{kg(mol particles)}^{-1}}{0.197\ \text{kg(mol Au atoms)}^{-1}} = \underline{\underline{2.47 \times 10^5\ \frac{\text{Au atoms}}{\text{particle}}}}$$

2. Equation (11-1.5.10) gives

$$u_1 = \frac{2}{9\eta} r_1^2(\rho_2 - \rho)g; \quad u_2 = \frac{2}{9\eta} r_2^2(\rho_2 - \rho)g$$

$$\therefore \quad \frac{u_2}{u_1} = \frac{r_2^2}{r_1^2}; \quad r_2 = 10\ r_1$$

$$\therefore \quad u_2 = \frac{(10\ r_1)^2}{r_1^2} u_1 = 100\ u_1 = \underline{\underline{1.5\ \text{ms}^{-1} = u_2}}$$

3. $$M_N = \frac{\Sigma m_i n_i}{\Sigma n_i} = \frac{(4)(5) + (5)(6) + (3)(7) + (2)(8) + (2)(10)}{4 + 5 + 3 + 2 + 2}$$

$$= \frac{107}{16} = \underline{\underline{6.6\dot{9} = M_N}}$$

$$M_W = \frac{\Sigma n_i m_i^2}{\Sigma m_i n_i} = \frac{(4)(5)^2 + (5)(6)^2 + (3)(7)^2 + (2)(8)^2 + (2)(10)^2}{107}$$

$$= \frac{755}{107} = \underline{\underline{7.06 = M_W}}$$

4. Apply equation (11-1.5.27)

$$M = \frac{2RT\ \ell n\ (C_2/C_1)}{(1 - \bar{v}\rho)\omega^2(x_2^2 - x_1^2)}$$

(Assume value given in radians per second)

$$M = \frac{(2)(8.314\ \text{J mol}^{-1}\ \text{K}^{-1})(298\text{K})\ell n(10/0.001)}{[1 - (0.998)(0.92)](10^4\ \text{s}^{-1})^2(0.11^2 - 0.1^2)\text{m}^2}$$

$$= 2.6\ \frac{\text{kg m}^2\ \text{s}^{-2}\ \text{mol}^{-1}\ \text{K}^{-1}\ \text{K}}{\text{s}^{-2}\ \text{m}^2}$$

$$\underline{\underline{M = 2.6\ \text{kg mol}^{-1} = 2600\ \text{g mol}^{-1}}}$$

5. $$r = \left[\frac{9\eta u}{2(\rho_2 - \rho)g}\right]^{1/2} = \left[\frac{9(1.2\times10^{-3}\ \text{kg m}^{-1}\ \text{s}^{-1})(2.72\times10^{-4}\ \text{m s}^{-1})}{2(1350 - 1037)\text{kg m}^{-3}(9.8\ \text{m s}^{-2})}\right]^{1/2}$$

$$= [4.8\times10^{-10}\ \text{m}^2]^{1/2} = \underline{\underline{2.2\times10^{-5}\ \text{m} = r}}$$

6. $$D = \frac{RTs}{M(1 - \bar{v}\rho)} \qquad \text{or} \qquad s = \frac{DM(1 - \bar{v}\rho)}{RT}$$

Units must be consistent ↓

$$s = \frac{(1.27\times10^{-13}\ \text{m}^2\ \text{s}^{-1})(4.21\times10^4\ \text{kg mol}^{-1})[1 - (0.736)(1.071)]}{(8.31\ \text{kg m}^2\ \text{s}^{-2}\ \text{mol}^{-1}\ \text{K}^{-1})(291\text{K})}$$

$$\underline{\underline{s = 4.68\times10^{-13}\ \text{seconds}}}$$

11-2.6 PROBLEMS

1. See text.

2. $$E_s = \frac{\varepsilon_o \Theta P\zeta}{\eta\kappa_o}$$

$$= \frac{(8.854\times10^{-12}\ \text{kg}^{-1}\ \text{m}^{-3}\ \text{s}^4\ \text{A}^2)(78.5)(3\ \text{atm})\left(\frac{1.01\times10^5\ \text{N m}^{-2}}{1\ \text{atm}}\right)(-0.011\text{V})}{(8.9\times10^{-4}\ \text{kg m}^{-1}\ \text{s}^{-1})(1.2\times10^{-1}\ \text{ohm}^{-1}\ \text{m}^{-1})}$$

$$= -2.17\times10^{-2}\ \frac{\text{kg}^{-1}\ \text{m}^{-3}\text{s}^4\ \text{A}^2\ \text{kg m s}^{-2}\ \text{m}^{-2}\ \text{V}}{\text{kg m}^{-1}\ \text{s}^{-1}\ \text{ohm}^{-1}\ \text{m}^{-1}}$$

$$= \underline{\underline{-2.17\times10^{-2}\ \text{V}}} = E_s$$

(See example 11-2.3.3 for units.)

3. $$\zeta = \frac{v\eta}{\theta\varepsilon_o} = \frac{(1.5\times10^{-8}\ \text{m}^2\ \text{V}^{-1}\ \text{s}^{-1})(8.9\times10^{-4}\ \text{kg m}^{-1}\ \text{s}^{-1})}{(78.5)(8.854\times10^{-12}\ \text{kg}^{-1}\ \text{m}^{-3}\ \text{s}^4\ \text{A}^2)}$$

$$= \underline{\underline{1.92\times10^{-2}\text{V} = \zeta}}$$

(See example 11-2.3.1 for units.)

4. Assume y Cl^-, Na^+ diffuses to left compartment.

Initial

$Na^+ = 0.01$	$Na^{+\prime} = 0.01$
$X^- = 0.01$	$Cl^{-\prime} = 0.01$

Final

$Na^+ = 0.01 + y$	$Na^{+\prime} = 0.01 - y$
$X^- = 0.01$	$Cl^{-\prime} = 0.01 - y$
$Cl^- = y$	

Equation (11-2.4.19) reads $\dfrac{C'}{C_{Cl^-}} = \left[1 + \dfrac{C_{X^-}}{C_{Cl^-}}\right]^{1/2}$

Does not converge as did Example 11-2.4.1.

Use equation (11-2.4.16)

$$C_{Na^+}C_{Cl^-} = C'_{Na^+}C'_{Cl^-}$$

or

$$(0.01 + y)(y) = (0.01 - y)(0.01 - y) \rightarrow y = 0.00333$$

$\therefore\ C_{Na^+} = 0.01333;\quad C_{Cl^-} = 0.00333;\quad C_{X^-} = 0.01;\quad C'_{Na^+} = 0.00666$

$C'_{Cl^-} = 0.00666. \qquad \dfrac{C'}{C_{Cl^-}} = 2.0$

5. $$\zeta = \frac{v\eta}{\varepsilon} = \frac{(5.6 \times 10^{-8}\ m^2\ s^{-1}\ V^{-1})(5.11 \times 10^{-4}\ kg\ m^{-1}\ s^{-1})}{(3.1 \times 10^{-10}\ kg^{-1}\ m^{-3}\ s^4\ A^2)}$$

$$= 0.092\ \frac{m^5\ kg^2}{V\ m\ s^6\ A^2}\left(\frac{1\ A^2\ s^2}{kg^2\ m^4\ s^{-4}\ V^{-2}}\right) = \underline{\underline{0.092V = \zeta}}$$

11-3.7 PROBLEMS

1. Plot $\ell n\ y$ vs. $\ell n\ P$ (do it). Slope $= \dfrac{3.64 - 3.22}{0.34 - 0.90} = 0.75 = \dfrac{1}{n}$.

The intercept ($\ell n\ P = 0$) gives $\ell n\ K = 3.95$. $K = 51.9$.

P (atm)	0.096	0.237	0.407	0.711	1.16
ℓn P	-2.34	-1.44	-0.90	-0.34	0.15
y (dm^3)	7.5	16.5	25.1	38.1	52.3
ℓn y	2.02	2.80	3.22	3.64	3.96

Thus $\underline{\underline{y = 51.9\ P^{0.75}}}$

2. (a)

P/y ($\times 10^6$ atm m^{-3} g)	0.233	0.254	0.265	0.353	0.393	0.606	0.763
P (atm)	2.8	3.4	4.0	6.0	9.4	17.1	23.5

Plot P/y vs. P (do it). Slope $= 0.0253 \times 10^6\ m^{-3}$ g.

$\therefore\ \underline{\underline{y_m = 3.95 \times 10^{-5}\ m^3\ g^{-1}}}$

Intercept $= 0.17 \times 10^6$ atm m^{-3} g $= b/y_m$

$\underline{\underline{b = 6.71 \text{ atm}}}$

(b) $\rho_{N_2} = 0.81\ \text{kg dm}^{-3}$

$$\bar{V} = \frac{M}{\rho} = \frac{0.028\ \text{kg mol}^{-1}}{0.81\ \text{kg dm}^{-3}} = (0.0346\ \text{dm}^3\ \text{mol}^{-1})\left(\frac{1\ \text{mol}}{6.023\times 10^{23}\ \text{mlc}}\right)$$

$$\underline{\underline{\bar{V} = 5.74 \times 10^{-26}\ \text{dm}^3\ \text{mlc}^{-1}}}$$

$$A \approx \bar{V}^{2/3} = (1.49 \times 10^{-17}\ \text{dm}^2\ \text{mlc}^{-1})\left(\frac{10^{-2}\ \text{m}^2}{1\ \text{dm}^2}\right)$$

$$= \underline{\underline{1.49 \times 10^{-19}\ \text{m}^2\ \text{mlc}^{-1} = \text{Area per molecule}}}$$

(c) From $y_m = 3.95 \times 10^{-5}\ \text{m}^3\ \text{g}^{-1}$ (at 293K, 1 atm), the number of molecules per gram of solid is

$$Nn = N\frac{PV}{RT} = \frac{NP}{RT}y_m$$

$$= \frac{(6.023\times 10^{23}\ \text{mlc mol}^{-1})(1\ \text{atm})(3.95\times 10^{-5}\ \text{m}^3\ \text{g}^{-1})(1\ \text{dm}^3)}{(0.0821\ \text{dm}^3\ \text{atm mol}^{-1}\ \text{K}^{-1})(293\text{K})(10^{-3}\ \text{m}^3)}$$

$$= \underline{\underline{9.9 \times 10^{+20}\ \text{mlc g}^{-1}}} = \text{number of molecules per gram solid}$$

Area of 1 g = Number molecules × Area per molecule

$$= (9.9 \times 10^{-20}\ \text{mlc g}^{-1})(1.49 \times 10^{-19}\ \text{m}^2\ \text{mlc}^{-1})$$

$$= \underline{\underline{147\ \text{m}^2\ \text{g}^{-1}}} = \text{Area of solid}$$

3. $\gamma_{TA} = 37 \times 10^{-3}\ \text{N m}^{-1}$

$$S_{TW} = \gamma_{WA} - \gamma_{TA} - \gamma_{TW} = ?$$

We do not know γ_{TW} so we cannot solve the problem.

4. $(1.53 \times 10^{-7}\ \text{mol})\left(\frac{6.023 \times 10^{23}\ \text{mlc}}{1\ \text{mol}}\right) = 9.22 \times 10^{16}\ \text{mlc}$

Area per mlc $= (0.0293\ \text{m}^2)/(9.22 \times 10^{16}\ \text{mlc})$

$$= 3.19 \times 10^{-19}\ \text{m}^2\ \text{mlc}^{-1} = \underline{\underline{32\ \text{Å}^2\ \text{mlc}^{-1}}}$$

5. $W_A = \gamma_{\ell A}(1 + \cos\theta) = (28.8 \times 10^{-3}\ \text{N m}^{-1})(1 + \cos 95°)$

$$= \underline{\underline{2.63 \times 10^{-2}\ \text{N m}^{-1} = W_A}}$$

6. $y = 2.83C^{1/n}$. We know $y = 0.62$ when $C = 0.031\ \text{mol dm}^{-3}$.

$\therefore\ \ln(0.62) = \ln(2.83) + \frac{1}{n}\ln(0.031)$. This gives $\frac{1}{n} = 0.44$.

Thus $y = 2.83C^{0.44}$.

When C = 0.268 mol dm^{-3} (units must be same as above),

$$y = 2.83(0.268)^{0.44} = \underline{\underline{1.6 \text{ mol acetic acid adsorbed}}}$$

7. Plot of γ vs. $\ln m$ yields a curve. The slope at m = 0.05 is about 0.0084 N m^{-1}. Thus

$$\Gamma = \frac{1}{(8.314 \text{ N m mol}^{-1} \text{ K}^{-1})(298\text{K})}(-0.0084 \text{ N m}^{-1})$$

$$= \underline{3.39 \times 10^{-6} \text{ mol m}^{-2}}$$

$$\text{Area} = \frac{1}{\Gamma N} = \frac{1}{(3.39 \times 10^{-6} \text{ mol m}^{-2})(6.023 \times 10^{23} \text{ mlc mol}^{-1})}$$

$$= 4.9 \times 10^{-19} \text{ m}^2 \text{ mlc}^{-1}$$

$$\underline{\underline{\text{Area} = 4.9 \times 10^{-19} \text{ m}^2 \text{ mlc}^{-1} = 49 \text{ Å}^2 \text{ mlc}^{-1}}}$$

12-1.5 PROBLEMS

See text for solutions to these problems.

12-2.6 PROBLEMS

1. $$\frac{Hc}{\tau} = \frac{1}{M'} + Ac + Bc^2 + \cdots$$

Plot Hc/τ vs. c. Intercept is $1/M'$ = 2.69.

$\underline{\underline{M' = 0.371}}$

But $M_w = M'\alpha\beta$ = (0.371)(1.00)(0.935)

$\underline{\underline{M_w = 0.346 \text{ kg mol}^{-1}}}$

$$\% \text{ Error} = \frac{0.346 - 0.342}{0.342} \times 100\% = \underline{\underline{1.17\%}}$$

2. Plot η_{sp}/c vs. c and $\ln \eta_r/c$. Two lines intersect at c = 0 giving $[\eta] = \lim_{c\to 0} \eta_{sp}/c = \lim_{c\to 0} (\ln \eta_r)/c = 1.10 = [\eta]$.

c (g/100 ml)	0.125	0.250	0.500	1.000
η_r	1.142	1.298	1.641	2.475
$\ln \eta_r/c$	1.062	1.04	0.991	0.906
η_{sp}	0.142	0.298	0.641	1.475
η_{sp}/c	1.14	1.19	1.28	1.475

3. $M = 2RT \ln(c_2/c_1)/[(1 - \bar{v}\rho)\omega^2(x_2^2 - x_1^2)]$

$$= \frac{2(8.31 \text{ J mol}^{-1} \text{ K}^{-1})(293\text{K}) \ln\left(\frac{10}{0.001}\right)}{[1 - (0.92 \text{ cm}^3 \text{ g}^{-1})(0.998 \text{ g cm}^{-3})](10^4 \text{ radians} \cdot \text{s}^{-1})^2[0.11^2 - 0.1^2]\text{m}^2}$$

$M = 2.61 \text{ kg mol}^{-1}$

4.

Π (atm)	6.17×10^{-4}	1.54×10^{-3}	2.77×10^{-3}	4.31×10^{-3}
$\frac{c}{\Pi}\left(\frac{\text{kg}}{\text{dm}^3 \text{ atm}}\right)$	3.23	2.60	2.17	1.86

Plot $\frac{c}{\Pi}$ vs. c. Take $\lim_{c\to 0} \frac{c}{\Pi} = 3.63 \text{ kg dm}^{-3} \text{ atm}^{-1}$. This should be an ideal solution value since $c \to 0$.

$$M = RT \lim_{c\to 0} \frac{c}{\Pi} = (0.08206 \text{ dm}^3 \text{ atm mol}^{-1} \text{ K}^{-1})(293\text{K})(3.63 \text{ kg dm}^{-3} \text{ atm}^{-1})$$

$M = 87.3 \text{ kg mol}^{-1}$

5. $n_{H_2O} \approx (100 \text{ cm}^3 \text{ sol'n})\left(\frac{1 \text{ g } H_2O}{1 \text{ g sol'n}}\right)\left(\frac{1 \text{ mole } H_2O}{18 \text{ g } H_2O}\right) = 5.56 \text{ mol } H_2O$

$$n_{polymer} = (1 \text{ g polymer})\left(\frac{1 \text{ mole polymer}}{10 \times 10^3 \text{ g}}\right) = 1 \times 10^{-4} \text{ mole polymer}$$

$$x_{H_2O} = \frac{n_{H_2O}}{n_{total}} = \frac{5.56}{5.56 + 1 \times 10^{-4}} = 1.000$$

$$P_{H_2O} = x_{H_2O} P^\circ_{H_2O} = (1)(21 \text{ mmHg}) = 21 \text{ mmHg} = P_{H_2O}$$

(No change in vapor pressure)

6. $[\eta] = KM_v^a$ or $K = \frac{[\eta]}{M_v^a} = \frac{1.02}{(34{,}000 \text{ g mol}^{-1})^{0.73}} = 5.0 \times 10^{-4} = K$

$$M_v = \left[\frac{[\eta]}{K}\right]^{1/a} = \left[\frac{2.75}{5.0 \times 10^{-4}}\right]^{1/0.73} = 130{,}000 \text{ g mol}^{-1} = M_v$$

7. Plot M_{obs} vs. c. Extrapolate to c = 0 to find

$$M_{c=0} \approx 62.5 \text{ kg mol}^{-1}$$

The variation with concentration is likely due to nonideality. The equations we use to find M assume ideal (no interactions in solution) behavior.

8. We know from 10-2 that

$$\Pi = \frac{n_B}{V} RT = \frac{m_B}{M_B}\frac{RT}{V} = \frac{c_B RT}{M_B}$$

$$\therefore \quad \frac{\Pi}{c_B} = \frac{RT}{M_B}$$

Determine Π/c as $c \to 0$; then solve for M_B.

From the data $\Pi = \rho g h$

$c\ (\mathrm{kg\ m^{-3}})$	1.99	4.02	6.00	7.97
$\Pi\ (\mathrm{N\ m^{-2}})$	62.5	156	281	437
$\Pi/c\ (\mathrm{m^2\ s^{-2}})$	31.4	38.8	46.8	54.8

Plot Π/c vs. c. Extrapolate to c = 0. This gives $\left(\frac{\Pi}{c}\right)_{c=0} \approx 25$

$$M_B = \frac{RT}{(\Pi/c)} = \frac{(8.314\ \mathrm{kg\ m^2\ s^{-2}\ mol^{-1}\ K^{-1}})(298\mathrm{K})}{(25\ \mathrm{m^2\ s^{-2}})} = \underline{\underline{99\ \mathrm{kg\ mol^{-1}}}} = M_B$$

13-1.5 PROBLEMS

1. (a) Three gases, ∴ 1 phase; 3 species; 0 restraints;

$\therefore\ \underline{C} = 3,\ \underline{P} = 1$

$$\Phi = \underline{C} - \underline{P} + 2 = 3 - 1 + 2 = \underline{\underline{4 = \Phi}}$$

(b) $\underline{P} = 2$ (gas + liquid); $\underline{C} = S - R = 3 - 1 = 2$

$R : [C_6H_6(g) \rightleftharpoons C_6H_6(\ell)]$

$$\therefore\ \Phi = \underline{C} - \underline{P} + 2 = 2 - 2 + 2 = \underline{\underline{2 = \Phi}}$$

(c) $\underline{P} = 3$; $S = 6$; $R: CaCO_3(s) \rightleftharpoons Ca^{2+} + CO_3^{2-}$; $H_2O(\ell) \rightleftharpoons H_2O(g)$; [+] = [-]

$\underline{C} = S - R = 3$; $R = 3$

$$\Phi = \underline{C} - \underline{P} + 2 = 3 - 3 + 2 = \underline{\underline{2 = \Phi}}$$

(d) $\underline{P} - 3$; $S = 3$; $R = 2: H_2O(\ell) \rightleftharpoons H_2O(g)$; $H_2O(\ell) \rightleftharpoons H_2O(s)$

$\underline{C} = 3 - 2 = 1$

$$\Phi = \underline{C} - \underline{P} + 2 = 1 - 3 + 2 = 0 = \Phi$$

(e) Pressure = 1 atm, therefore

$\Phi = \underline{C} - \underline{P} + 1$; $S = 7$ (including H_2O)

$P = 1$

$$R = 4: H_3PO_4 \rightleftharpoons H_2PO_4^- + H^+$$
$$H_2PO_4^- \rightleftharpoons HPO_4^{2-} + H^+$$
$$HPO_4^{2-} \rightleftharpoons PO_4^{3-} + H^+$$
$$[+] = [-]$$

$\underline{C} = 3$

$\Phi = \underline{C} - \underline{P} + 1 = 3 - 1 + 1 = \underline{\underline{3 = \Phi}}$

2. Start heating solid at low T. At the intersection of the S, V lines the material changes to a vapor. Continued heating does not produce a liquid. Liquid is unstable at 1 atm. Will convert to a vapor or solid spontaneously.

3. (a) Areas are labeled:

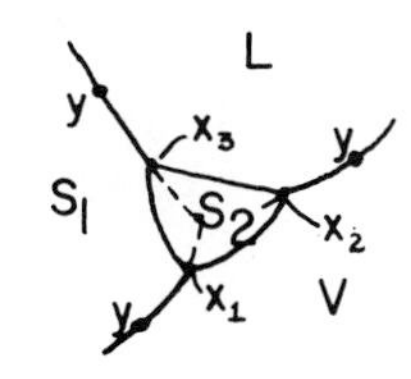

x = 3 phase points (another three phase point is at the intersection of the dotted lines)

$\Phi = 1 - 3 + 2 = 0$ (Invariant)

y = 2 phase lines

$\Phi = 1 - 2 + 2 = 1$

$\Phi = 2$ in other areas

(b) Assume we allow the pressure to change such that the equilibrium lines are followed. Heating from T_a, we have solid_1 and vapor in equilibrium until x_1 is reached where $S_1 \rightarrow S_2$. Between x_1 and x_2, S_2 + vapor in equilibrium.

At x_2, $S_2 \rightarrow \ell$ (three phase point). From x_2 to T_b, liquid and vapor in equilibrium.

(It is possible for the solid_1 to superheat past x_1. If this occurs, solid_1 will convert to a liquid at the triple point S_1, ℓ, V. This point is unstable with respect to S_2.)

(c) Assume cooling occurs along a straight line connecting c to d. Liquid will convert to solid at the intersection with the Solid_2-Liquid line. At a lower temperature at the intersection with Solid_2-Solid_1 line, Solid_2 converts to Solid_1.

(It is possible for the liquid to supercool until it freezes to solid_1 directly.)

4. $\int_1^{P_f} dP = \int \frac{\Delta\bar{H}}{\Delta\bar{V}} \frac{dT}{T} \approx \frac{\Delta\bar{H}}{\Delta\bar{V}} \int_{T_{MP}}^{T_f} \frac{dT}{T} = \frac{\Delta\bar{H}}{\Delta\bar{V}} \ln T \Big|_{T_{MP}}^{T_f}$: $P_f - 1 = \frac{\Delta\bar{H}}{\Delta\bar{V}} \ln \frac{T_f}{T_{MP}}$

$$T_f = T_{MP} \exp\left[(P_f - 1)\frac{\Delta\bar{V}}{\Delta\bar{H}}\right]$$

$$= 1809 \exp\left[(P_f - 1)\frac{(2.7 \times 10^{-7}\ m^3\ mol^{-1})(101\ J)}{(1.55 \times 10^4\ J\ mol^{-1})(1\ dm^3\ atm)}\frac{(10^3\ dm^3)}{(1\ m^3)}\right]$$

$$T_f = 1809 \exp[(P_f - 1)(1.76 \times 10^{-6})].$$

You should plot the results.

P_f (atm)	1	10^2	10^3	10^4	10^5
T_f (K)	1809	1809.3	1812	1841	2157

5. (a) S = 10

R = 6

∴ C = 4

P = 4 (two solid phases)

R : $H_2O(g) \rightleftharpoons H_2O(\ell)$

$H_2O(\ell) \rightleftharpoons H_2O(s)$

$HCl(g) \rightleftharpoons HCl(aq)$

$HCl(aq) \rightleftharpoons H^+(aq) + Cl^-(aq)$

$[H^+] = [Cl^-]$

$CO_2(g) \rightleftharpoons CO_2(aq)$

$\Phi = C - P + 2 = 4 - 4 + 2 = 2 = \Phi$

(b) C = 1, P = 3

$\Phi = C - P + 2 = 0$

6. There are no degrees of freedom. Thus, as long as three phases are present, the temperature can not vary. Such a situation gives an excellent calibration point.

7. S = 6

R = 4

C = 2

P = 6

R : $H_2O(\ell) \rightleftharpoons H_2O(g)$

$H_2O(s) \rightleftharpoons H_2O(\ell)$

Benzene(s) ⇌ Benzene(ℓ)

Benzene(ℓ) ⇌ Benzene(g)

∴ $\Phi = C - P + 2 = 2 - 6 + 2 = -2 = \Phi$ an impossible condition.

13-2.7 PROBLEMS

1.

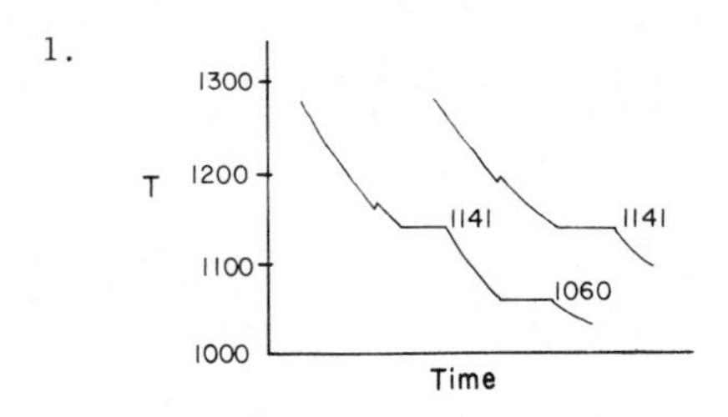

2. SEE FIGURE BELOW

Composition 1: Heating the two phases (tridymite + mullite) from below 1400 leads to no changes until 1470 (point e) at which the tridymite (SiO_2) converts to cristobalite (SiO_2). During this conversion three phases are present (tridymite, cristobalite and mullite) so $\Phi = \underline{C} - \underline{P} + 1 = 2 - 3 + 1 = 0$ (pressure is fixed already). Between e and d, cristobalite and mullite coexist. At point d (at 1515) liquid of composition x begins to form yielding three phases. $\Phi = 0$ until all cristobalite disappears leaving mullite + liquid. Mullite continues to melt and the liquid gets richer in Al_2O_3 as the temperature rises ($\Phi = 1$). At point c another 3 phase region occurs. $\Phi = 0$ (liquid, mullite, corundum in equilibrium). T constant until all mullite is converted to corundum (point c).

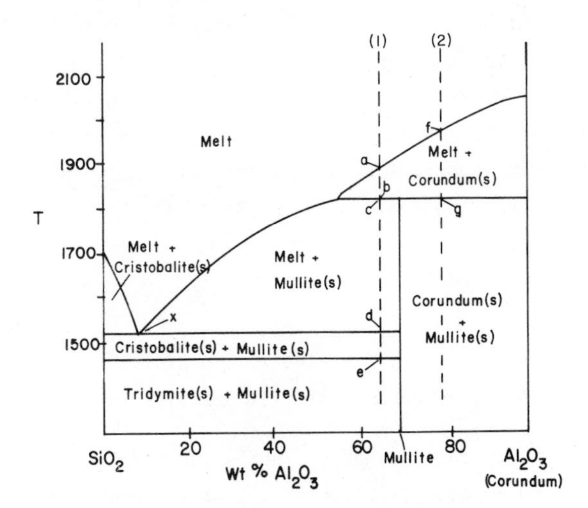

Increasing T leads to melting of corundum until at point a all solid is gone.

Composition 2: Below 1800 corundum and mullite in equilibrium. At point g solids begin to melt giving three phases. $\Phi = 0$, T = constant until all mullite disappears. Above g corundum continues to melt until all has gone at point f.

3. (a)

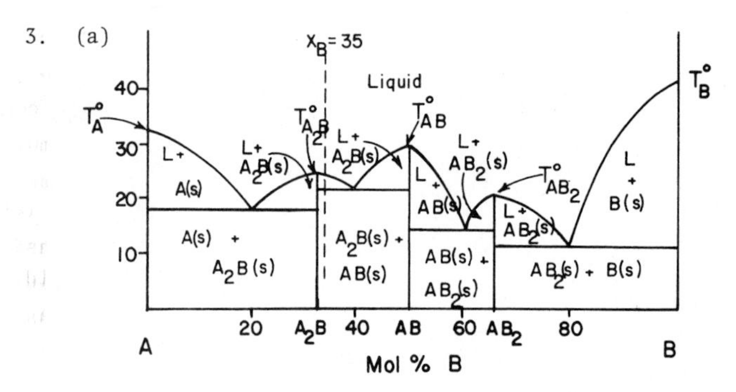

(b, c) Only liquid ($\Phi = 2 - 1 + 1 = 2$) until about 24° at which point (a), solid A_2B begins to separate ($\Phi = 2 - 2 + 1 = 1$). Continued cooling freezes out more A_2B. Liquid approaches eutectic at 40%. At 21° solid AB begins to form yielding three phases ($\Phi = 2 - 3 + 1 = 0$). Temperature invariant until all liquid freezes to AB(s) + AB_2(s). Below eutectic, no further changes.

4. From point a, cooling results in no change until point b is reached. At point b, solid B begins to separate. Continued cooling results in more B(s) separating with the liquid getting richer in A. Point c is in a two phase region, consisting of solid B and a solution (melt) with $x_B \approx 0.77$. At point d, a three phase line is reached. Melt (composition of $x_B \approx 0.75$) combines with solid B to produce a new solid AB_4. T is invariant until all the melt is consumed. Below point d, we have two solid phases with no additional change in composition.

13-3.5 PROBLEMS

1. Point b: About 6% (by weight) nicotine will dissolve in water.
   Point d: About 20% (by weight) water will dissolve in nicotine.

2. No changes occur until the two-phase line is reached. At this point, a liquid of about 0.61 mole fraction B separates. Continued cooling results in more liquid with both liquid and vapor getting progressively richer in A. The last "drop" of vapor has a composition of $x_B = 0.1$.

3. $x_B^{(\ell)} \approx 0.60$

   Liquid begins to form when $x_B^{(v)} \approx 0.60$. $x_B^{(\ell)}$ that forms is $\approx$ 0.44.

4. Equilibrium Vaporization Point (a), ($x_B^{(\ell)} = 0.16$). Heating of liquid results in a vapor with $x_B^{(v)} = 0.07$ when the two-phase temperature is reached. As heating is continued the vapor composition changes from $x_B^{(v)} = 0.07 \rightarrow x_B^{(v)} = 0.16$. The liquid composition varies from $x_B^{(\ell)} = 0.16 \rightarrow x_B^{(\ell)} = 0.27$. The last drop of liquid that vaporizes has $x_B^{(\ell)} = 0.27$.

   Distillation If vapors are removed, $x_B^{(\ell)}$ continues to rise until at $x_B^{(\ell)} \approx 0.40$ where the azeotrope is reached and all liquid will boil away at that composition.

   Fractional Distillation

Liquid composition	Vapor composition	Condensed liquid composition
$x_B^{(\ell)} = 0.16$	$x_B^{(v)} = 0.07$	$x_B^{\prime(\ell)} = 0.07$
$x_B^{\prime(\ell)} = 0.07$	$x_B^{(v)} = 0.03$	$x^{\prime\prime(\ell)} = 0.03$
$x_B^{\prime\prime(\ell)} = 0.03$	$x_B^{(v)} \approx 0.01$	$x_B^{\prime\prime\prime(\ell)} \approx 0.01$

   Not quite 0 but it is close.

5. Consider the phase diagram shown.

   Distillation from a: At a there are two liquid phases present (b, c) with $x_B = 0.22$ and 0.75, respectively. As the temperature is increased, mutual solubility increases until at about 140° (point d) the two phases form one phase. Continued heating results in no change until 240° (point e) at which point vapor of composition f ($x_B = 0.42$) is formed. If this vapor is removed and boiling continued, the composition of the liquid approaches pure A as the temperature rises to about 350°.

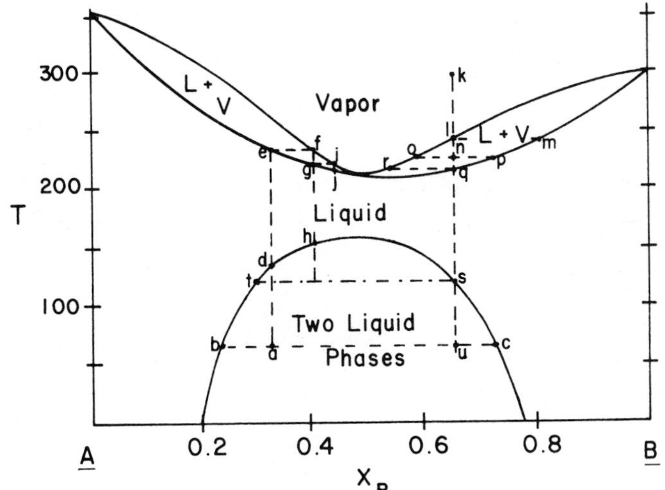

Fractional Distillation from a: If vapor f is separated and condensed, a liquid of composition $x_B = 0.42$ is formed (two liquid phases will form if this liquid is cooled below point h). When this liquid is reheated to 220° (point g), a vapor of composition i ($x_B = 0.45$) results. When this is condensed and reboiled, a vapor with the azeotropic composition ($x_B = 0.48$) is approached. Pure A or B can not be obtained by fractional distillation.

Equilibrium Cooling from k: Only vapor exists until about 245° where point l is reached. A liquid (m) with composition $x_B = 0.81$ is formed. Continued cooling results in the formation of more liquid and both liquid and vapor get richer in A. At point n, the vapor composition (o) is $x_B = 0.58$ and the liquid composition is $x_B = 0.73$. At 215° point q is reached and the last bit of vapor point r) with $x_B = 0.54$ condenses to liquid. Continued cooling results in no changes until point s is reached (at 120°). A second liquid phase (point t) with $x_B = 0.3$ separates. Small composition changes occur as T is lowered until at point u, the two phases (point b, c) with compositions $x_B = 0.22$ and $x_B = 0.75$ result.

6. Distillation from x: At point x there are two liquid phases present (a, b) with $x_B = 0.23$ and $x_B = 0.79$, respectively. Heating to point c results in minor liquid composition changes until point c is reached. At point c, there are three phases present; two liquid (d, $x_B = 0.77$ and e, $x_B = 0.25$) and a vapor f ($x_B = 0.56$). T is invariant until all of liquid 1 is vaporized. If the vapor is removed as heating is continued, the temperature rises and the liquid composition approaches pure B.

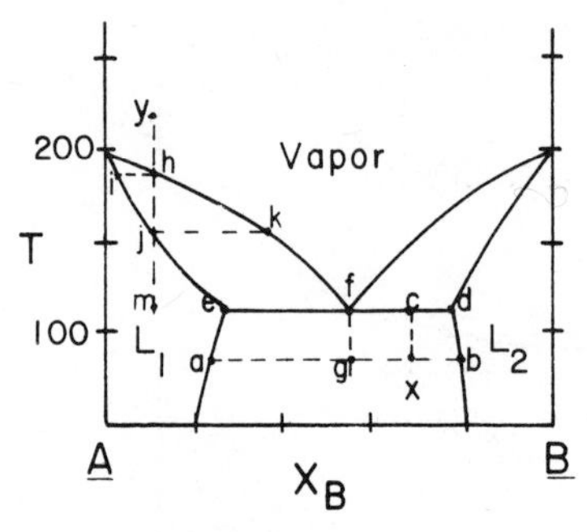

Fractional Distillation from x: If the vapor f is separated and condensed, two liquid phases form. If this mixture is heated to boiling (along gf), vapor f is again formed. Mixture cannot be separated by fraction distillation.

Equilibrium Cooling from y: Only vapor is present until the temperature falls to point h where a liquid (i) with x_B = 0.03 forms. Continued cooling of the two phase (L + V) system results in the conversion of more vapor into liquid with both liquid and vapor getting richer in B. Finally, at point j, the last bit of vapor (k) with x_B = 0.38 condenses. Below point j, only a single liquid phase is present.

13-4.4 PROBLEMS

1. Point c: 36% A, 16% B, 48% C Point d: 20% A, 40% B, 40% C
2. (a) (b)

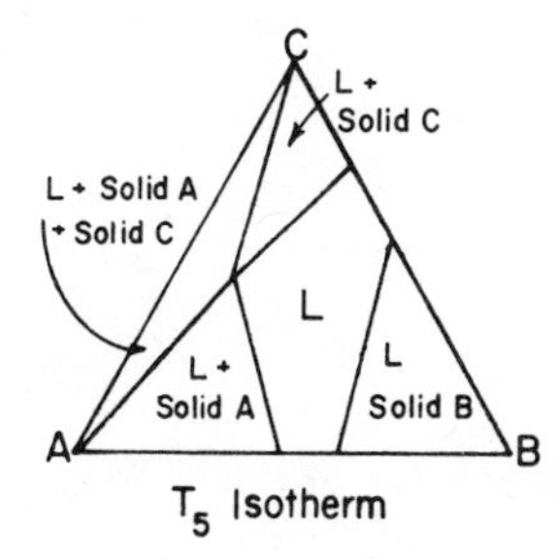

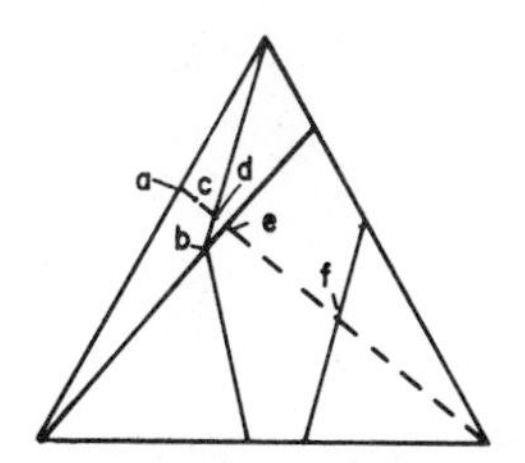

Starting at point a(60% C, 40% A), there are two phases (solid A + solid C) as B is added, the three phase region is entered, such as at point c where solid A + solid C + L of composition b are present. Continued addition of B leads to d where the last of solid A disappears and two phases (solid C + L of comp. b) are in equilibrium. As more B is added these phases remain, but the L changes composition along the line be since the liquid is getting richer in C and B. At point e, the last of solid C disappears and we are left with liquid of composition e. Continued addition of B causes the liquid composition to follow the line ef, until point f is reached. At point f, solid B begins to crystallize. Here there are two phases (solid B + L of composition f). The relative amounts are given by the lever rule along fg.

3. (a) Phases: 70% A; 22% B; 8% C and 12% A; 33% B, 55% C

4. (a)

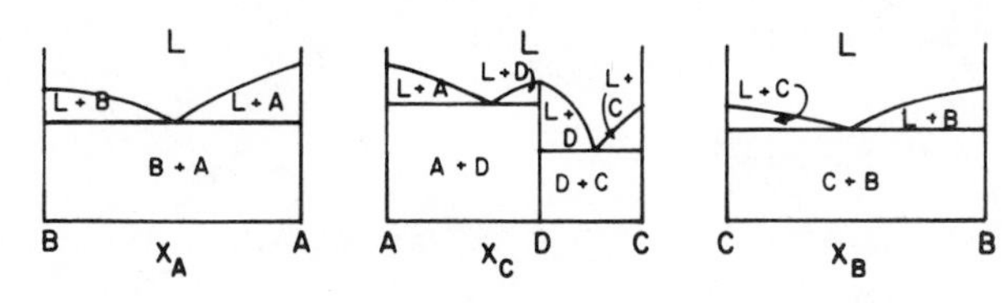

(b) In cooling liquid of composition y, we are concerned only with the ABD part of the diagram. At first only liquid is present until the solid-liquid temperature of B is reached at which time solid B + L exist. As cooling continues and B precipitates, the L composition follows the ya line. At a we reach the BD eutectic line and have three phases: solid B, solid D, L. As cooling continues, the L composition follows the ab line and the solid composition follows the BD line as B and D precipitate. At b, the ternary eutectic is reached. Three solid phases, A, B, and D, are in equilibrium with liquid of composition b. As cooling continues and the solids A, B, D are formed, the average solid composition follows cy and arrives at point y when all liquid is used. No further changes occur upon cooling.

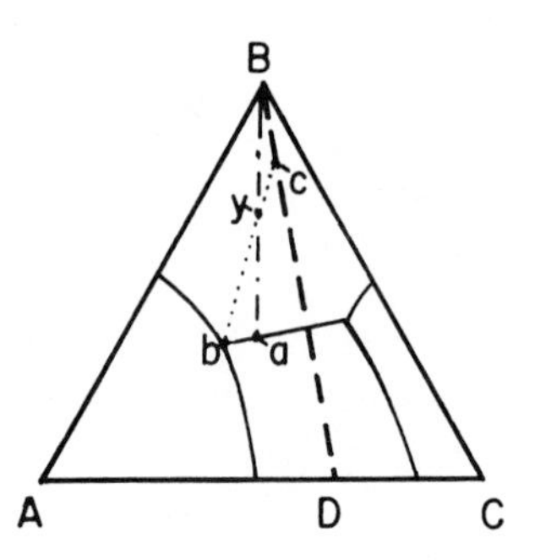

14-1.7 PROBLEMS

1. From integrating (2-2.2.3), assuming α, V are constant: $\Delta V_1 = \alpha V \Delta T$. This gives the small volume change for given temperature change. From equation (2-2.2.6), $\Delta V_2 = -\beta V \Delta P$. We want $\Delta V_2 = -\Delta V_1$. That is, the pressure is increased to bring V back to the original value. Therefore

$$\beta V \Delta P = \alpha V \Delta T$$

or

$$\underline{\underline{\Delta P}} = \frac{\alpha}{\beta} \Delta T = \frac{12 \times 10^{-4} \ K^{-1}}{120 \times 10^{-6} \ atm^{-1}} (15K) = \underline{\underline{150 \ atm}}$$

2. Apply equation (6-2.3.5)

$$d \ \ell n \ P = \frac{\Delta \bar{H}_{vap}}{RT^2} dT$$

Assume $\Delta\bar{H}_{vap}$ = const. Then

$$\ln P_2 - \ln P_1 = -\frac{\Delta\bar{H}_{vap}}{R}\left(\frac{1}{T_2} - \frac{1}{T_1}\right)$$

P_1 = 1 atm at T_{bp} = 353.9K

$\Delta\bar{H}_{vap}$ = 30,080 J mol^{-1}

Calculate P at different T's

P_2 (atm)	1.00	0.743	0.542	0.388	0.272	0.185	0.124	0.081
T_2 (K)	353.9	343.9	333.9	323.9	313.9	303.9	293.9	283.9

Plot these. Handbook gives P = 0.131 atm at 298.6. Plot of above shows P = 0.151 at 298.6K. Not too bad since we assumed ΔH was constant.

3. $\rho = \frac{m}{V}$ or $V = \frac{m}{\rho}$

$$\therefore \quad \alpha = \frac{1}{\left(\frac{m}{\rho}\right)}\left[\frac{\partial\left(\frac{m}{\rho}\right)}{\partial T}\right]_P = \frac{\rho m}{m}\left(\frac{\partial\left(\frac{1}{\rho}\right)}{\partial T}\right)_P = \rho\left[-\frac{1}{\rho^2}\frac{\partial\rho}{\partial T}\right]_P = -\frac{1}{\rho}\left(\frac{\partial\rho}{\partial T}\right)_P = \alpha$$

4. Plot $\ln P$ vs. $\frac{1}{T}$ as suggested by the equation:

$$d \ln P = -\frac{\Delta H}{R} d\left(\frac{1}{T}\right)$$

Make such a plot and take the slope near the boiling point. For this solution assume a straight line relationship and use two values near the boiling temperature.

P_1 = 0.526; T_1 = 250.2K; P_2 = 1.00; T_2 = 263.2K

$$\ln \frac{P_2}{P_1} = \frac{\Delta\bar{H}_{vap}}{R}\left(\frac{1}{T_2} - \frac{1}{T_1}\right) \quad \text{Assuming } \Delta H_{vap} \text{ is constant.}$$

$$\Delta H_{vap} = -R \ln \frac{P_2}{P_1} \Big/ \left(\frac{1}{T_2} - \frac{1}{T_1}\right) = -\frac{(8.314)\left(\ln \frac{1.00}{0.526}\right)}{\left(\frac{1}{263.2} - \frac{1}{250.2}\right)} \text{ J mol}^{-1}$$

$$= 27.1 \text{ kJ mol}^{-1} = \Delta\bar{H}_{vap}$$

Table value gives 24.9 kJ mol^{-1}. A plot should yield a better value.

5. $\bar{C}_p = \bar{C}_v + \dfrac{\alpha^2 T\bar{V}}{\beta}$

$$\bar{V} = \frac{m}{\rho} = \frac{0.046\ \text{kg mol}^{-1}}{0.789\ \text{kg dm}^{-3}} = 0.0583\ \text{dm}^3\ \text{mol}^{-1}$$

$$\bar{C}_v = \bar{C}_p + \frac{\alpha^2 T\bar{V}}{\beta} = 114.4\ \text{J mol}^{-1}\ \text{K}^{-1} - \frac{(11.2\times 10^{-4}\ \text{K}^{-1})^2 (293\text{K})(0.0583\ \text{dm}^3\ \text{mol}^{-1})}{(1.1\times 10^{-4}\ \text{atm})}$$

$$= 114.4\ \text{J mol}^{-1}\ \text{K}^{-1}$$

$$- (1.94\times 10^{-1}\ \text{dm}^3\ \text{atm mol}^{-1}\ \text{K}^{-1})\left(\frac{101\ \text{J}}{\text{dm}^3\ \text{atm}}\right)$$

$$= \underline{\underline{94.7\ \text{J mol}^{-1}\ \text{K}^{-1} = \bar{C}_v}}$$

6. $\gamma = \dfrac{r\rho gh}{2\cos\theta}$

From Chapter 1:

$$\ln\gamma = \ln r + \ln\rho + \ln g + \ln h - \ln 2 - \ln\cos\theta$$

$$\frac{\Delta\gamma}{\gamma} = \frac{\Delta r}{r} + \frac{\Delta\rho}{\rho} + \frac{\Delta g}{g} + \frac{\Delta h}{h} + \frac{\Delta 2}{2}\overset{\nearrow 0}{} + \frac{\Delta\cos\theta}{\cos\theta}$$

(Assume the only error is in θ.) Then

$$\frac{\Delta\gamma}{\gamma} = \frac{\Delta\cos\theta}{\cos\theta}$$

$\cos 0 = 1.00$; $\cos 10° = 0.984$, so $\Delta\cos\theta = 0.015$

$$\therefore\ \frac{\Delta\gamma}{\gamma} = \frac{\Delta\cos\theta}{\cos\theta} = \frac{0.015}{1.00} = 0.015 \rightarrow \underline{\underline{1.5\%\ \text{error}}}$$

7. $\gamma_{\text{acetone}} = 0.0237\ \text{N m}^{-1}$

$$= \frac{(2.03\times 10^{-2}\ \text{m} + r/3)(0.791 - \overset{*}{0.0012})(\text{kg dm}^{-3})r(9.8\ \text{m s}^{-2})}{2}$$

$* = \rho_{\text{air}}$

First trial, ignore r/3:

$$0.0237\ \text{kg m m}^{-1}\ \text{s}^{-2} = 7.86\times 10^{-3}\ \text{kg m}^2\ \text{s}^{-2}\ \text{dm}^{-3}\left(\frac{10^3\ \text{dm}^3}{1\ \text{m}^3}\right)r$$

$$r \approx \frac{0.0237}{7.86\times 10^1}\ \text{m} \approx \underline{\underline{3.01\times 10^{-4}\ \text{m} \approx r}}$$

Substitute this in r/3 term

$$r = \frac{2(0.0237)}{\left(2.03 \times 10^{-2} + \frac{3.01 \times 10^{-4}}{3}\right)(790)(9.8)}$$

$$= \underline{\underline{3.00 \times 10^{-4} \text{ m} = r}}$$

$\gamma_{ethanol} = 0.02227 \text{ N m}^{-1}$ $\rho_{ethanol} = 789 \text{ kg m}^3$

$$h = \frac{2\gamma}{(\rho_1 - \rho_{air})(rg)} - \frac{r}{3} = \frac{2(0.02227)}{(789 - 1.2)(3 \times 10^{-4})(9.8)} - \frac{3.00 \times 10^{-4}}{3}$$

$$= \underline{\underline{0.0192 \text{ m}}} = h_{ethanol}$$

8. $$\frac{dV}{dt} = \frac{\pi PR^4}{8\eta\ell} = \frac{\pi(1 \text{ atm})\left(1.01 \times 10^5 \frac{\text{N m}^{-2}}{\text{atm}}\right)(0.075 \text{ m})^4}{8(0.503 \times 10^{-3} \text{ kg m}^{-1} \text{ s}^{-1})(100 \text{ m})}$$

$$= 25 \frac{\text{N m}^2}{\text{kg s}^{-1}} \frac{(\text{m s}^{-1})}{(\text{m s}^{-1})} = \underline{\underline{25 \text{ m}^3 \text{ s}^{-1}}}$$

The value is unreasonably large.

9. $$\eta_1 = \frac{\eta_o \rho_1 t_1}{\rho_o t_o} = \frac{(0.834 \text{ kg m}^{-1} \text{ s}^{-1} \times 10^{-3})(714)(67)}{(789.3)(313)}$$

$$= \underline{\underline{0.161 \times 10^{-3} \text{ kg m}^{-1} \text{ s}^{-1} = \eta_1}}$$

10. $$U_{DD} = -\frac{2}{3} \frac{\mu^4}{(4\pi\varepsilon_o)^2 r^6 kT} = -\frac{2}{3} \frac{\mu^4}{r^6 kT} \quad \text{since } \varepsilon_o = \frac{1}{4\pi}$$

$$\mu_{H_2O} = 1.82 \text{ D} = 6.07 \times 10^{-30} \text{ C m}$$

$$\alpha = 1.44 \times 10^{-30} \text{ m}^3 \text{ mlc}^{-1}$$

$$U_{DD} = -\frac{2}{3} \frac{(6.07 \times 10^{-30} \text{ C m})^4}{(5 \times 10^{-10} \text{ m})^6 (1.38 \times 10^{-23} \text{ J K}^{-1})(298\text{K})}$$

$$= -(1.14 \times 10^{79} \times 10^{-120} \text{ C}^4 \text{ m}^4 \text{ m}^{-6} \text{ J})\left(\frac{8.98 \times 10^9 \text{ J m}}{1 \text{ C}^2}\right)^2$$

$U_{DD} = \underline{\underline{-11.4 \times 10^{-22} \text{ J}}}$ Table gives 11.9×10^{-22} J

$$U_{IND} = \frac{2\alpha\mu^2}{4\pi\varepsilon_o r^6} = -\frac{2\alpha\mu^2}{r^6}$$

$$= \frac{-2(1.44 \times 10^{-30} \text{ m}^3 \text{ mlc}^{-1})(6.07 \times 10^{-30} \text{ C m})^2}{(5 \times 10^{-10} \text{ m})^6}\left(\frac{8.98 \times 10^9 \text{ J m}}{1 \text{ C}^2}\right)$$

$= 0.61 \times 10^{-22}$ J. Table gives 0.65×10^{-22} J.

14-2.5 PROBLEMS

1. $f = (q_1q_2)/(4\pi\varepsilon r^2) = (q_1q_2)/(4\pi\varepsilon_o\theta r^2)$. But $\varepsilon^\circ = \frac{1}{4\pi}$,

$$\therefore\quad f = (q_1q_2)/(\theta r^2) = \frac{(1C)(1C)}{2(1\ m)^2} = \left(\frac{1}{2}\ \frac{C^2}{m^2}\right)\left(\frac{8.98 \times 10^9\ J\ m}{1\ C^2}\right)$$

$$= \frac{1}{2}(8.98 \times 10^9)N = \underline{\underline{4.49 \times 10^9\ N = f}}$$

2. $\mu = \sum_i q_i\vec{r}_i$

$$\mu_x = \sum_i q_i\vec{r}_{ix} = (+q)(2.4R)(\cos\phi) + (-q)(-R)(\cos\phi)$$

$$= q(\cos\phi)[3.4R] = 3.4(1.602 \times 10^{-19}\ C)(1 \times 10^{-9}\ m)(\cos 42°)$$

$$= 4.05 \times 10^{-28}\ C\ m = \underline{\underline{121D = \mu_x}}$$

$$\mu_y = 3.4qR \sin\phi = 3.64 \times 10^{-28}\ C\ m = 109D = \mu_y$$

$$\mu = \left(\mu_x^2 + \mu_y^2\right)^{1/2} = \underline{\underline{162D = \mu}}$$

3. $P = \frac{1}{3}\frac{N}{\varepsilon_o}\left(\alpha + \frac{\mu^2}{3kT}\right) = \frac{4\pi N}{3}\left(\alpha + \frac{\mu^2}{3kT}\right)$; $\mu = 0$; $\alpha = 1.73 \times 10^{-30}\ m^3\ mlc^{-1}$

$$= \frac{4\pi(6.023 \times 10^{23}\ mlc\ mol^{-1})(1.73 \times 10^{-30}\ m^3\ mlc^{-1}}{3}$$

$$= \underline{\underline{4.36 \times 10^{-6}\ m^3\ mol^{-1} = P}}$$

$$P = \frac{\theta - 1}{\theta + 2}\frac{M}{\rho} = \frac{1.454 - 1}{1.454 + 2}\left(\frac{0.028\ kg\ mol^{-1}}{0.81\ kg\ dm^{-3}}\right)\left(\frac{10^{-3}\ m^{-3}}{1\ dm^3}\right)$$

$$= \underline{\underline{4.54 \times 10^{-6}\ m^3\ mol^{-1} = P}}$$

4. From equation (14-2.3.10)

$$P = \frac{N\alpha}{3\varepsilon_o} + \frac{N\mu^2}{9\varepsilon_o k}\frac{1}{T}$$

Plot P vs. $1/T$.

Slope $= \frac{N\mu^2}{9\varepsilon_o k}$; Intercept $= \frac{N\alpha}{3\varepsilon_o}$; $P = \frac{\theta - 1}{\theta + 2}\frac{M}{\rho}$

	θ	32.63	40	54	64	
	T	298	253	193	160	(K)
Assume constant→	ρ	810	810	810	810	($kg\ m^{-3}$)
	P	36.1	36.7	37.4	37.7	($m^3\ mol^{-1} \times 10^6$)
	$\frac{1}{T}$	3.36	3.95	5.18	6.25	($K^{-1} \times 10^3$)

Slope of plot = $6.95 \times 10^{-4}\ m^3\ mol^{-1}\ K^1$

$$\mu^2 = \frac{9\varepsilon_o k}{N} \cdot \text{Slope} = \frac{9k}{4\pi N} \cdot \text{Slope}$$

$$= \frac{(9)(1.38 \times 10^{-23}\ J\ K^{-1})(6.95 \times 10^{-4}\ m^3\ mol^{-1}\ K^{-1})}{(4)(\pi)(6.02 \times 10^{23}\ mlc\ mol^{-1})}$$

$$\mu^2 = \left(1.14 \times 10^{-50}\ \frac{J\ m^3}{mlc}\right)\left(\frac{1\ C^2}{8.98 \times 10^9\ J\ m}\right) = 1.27 \times 10^{-60}\ C^2\ m^2$$

$\mu = 1.13 \times 10^{-30}$ C m = $0.34D = \mu$ Table Value = 1.70D

$$\frac{N\alpha}{3\varepsilon_o} = \frac{4\pi N\alpha}{3} = \text{Intercept} = 33.8 \times 10^{-6}\ m^3\ mlc^{-1}$$

$$\alpha = \frac{(3)(33.8 \times 10^{-6}\ m^3\ mol^{-1})}{(4)(\pi)(6.023 \times 10^{23}\ mlc\ mol^{-1})}$$

$$= 1.34 \times 10^{-29}\ m^3\ mlc^{-1} = \alpha$$

Table value: $\alpha = 3.0 \times 10^{-30}\ m^3\ mlc^{-1}$

Neither result is good, due to hydrogen bonding in the solvent as mentioned in the text. You should try a non-hydrogen bonding solvent to see how well it works.

5. $\mu = 1.47D$; $\alpha = 2.34 \times 10^{-30}\ m^3\ mlc^{-1}$; $P = \frac{4\pi N}{3}\left(\alpha + \frac{\mu^2}{3kT}\right)$

$$P = \frac{4\pi(6.023 \times 10^{23}\ mlc\ mol^{-1})}{3}\left[2.34 \times 10^{-30}\ m^3\ mlc^{-1} + \frac{(4.91 \times 10^{-30}\ C\ m)^2(8.98 \times 10^9\ \frac{J\ m}{C^2})}{3(1.38 \times 10^{-23}\ J\ mlc^{-1}\ K^{-1})(298.2\ K)}\right]$$

$$= 2.52 \times 10^{24}\ mlc\ mol^{-1}(2.34 \times 10^{-30}\ m^3\ mlc^{-1} + 1.75 \times 10^{-29}\ m^3\ mlc^{-1}) = 50.1 \times 10^{-6}\ m^3\ mol^{-1}$$

6. NO has $15e^-$. Must have an unpaired e^-; $\therefore$ paramagnetic.

7. $$\chi_m = \frac{M\chi}{\rho} = \frac{(0.3519\ kg\ mol^{-1})}{(0.564 \times 10^3\ kg\ m^{-3})}(2.16 \times 10^{-4})$$

$$= 1.35 \times 10^{-7}\ m^3\ mol^{-1} = \chi_m$$

8. $$\chi_m = \frac{M\chi}{\rho} = \frac{(0.055 \text{ kg mol}^{-1})(4.26 \times 10^{-4})}{(0.971 \text{ kg dm}^{-3})} \frac{10^{-3} \text{ m}^3}{1 \text{ dm}^3}$$

$$= 2.41 \times 10^{-8} \text{ m}^3 \text{ mol}^{-1}$$

$$\chi_m = N\left(\alpha_m + \frac{\mu_o P_m^2}{3kT}\right) \quad \text{or} \quad \alpha_m = \frac{\chi_m}{N} - \frac{\mu_o P_m^2}{3kT}$$

$$\alpha_m = \frac{2.41 \times 10^{-8} \text{ m}^3 \text{ mol}^{-1}}{6.023 \times 10^{23} \text{ mlc mol}^{-1}}$$

$$- \frac{(4\pi \times 10^{-7} \text{ kg A}^{-2} \text{ s}^{-2} \text{ m})(1.91 \times 10^{-23} \text{ m}^2 \text{ A mlc}^{-1})^2}{3(1.38 \times 10^{-23} \text{ kg m}^2 \text{ s}^{-2} \text{ mlc}^{-1} \text{ K}^{-1})(275\text{K})}$$

$$= 4.00 \times 10^{-32} \text{ m}^3 \text{ mlc}^{-1} - 4.03 \times 10^{-32} \text{ m}^3 \text{ mlc}^{-1} \approx 0 \approx \alpha_m$$

The molecule has a very small susceptibility.

9. $$\frac{f_{vac.}}{f_{sol}} = \frac{q_1q_2/(4\pi\varepsilon_o r^2)}{q_1q_2/(4\pi\varepsilon_o \theta r^2)} = \theta \neq 8.61 \text{ if r is constant.}$$

10. $$\frac{\alpha_m \kappa}{\mu_o P_m^2} = \frac{(\text{m}^3 \text{ mlc}^{-1})(\text{kg m}^2 \text{ s}^{-2} \text{ mlc}^{-1} \text{ K}^{-1})}{\text{kg A}^{-2} \text{ s}^{-2} \text{ m}(\text{A m}^2)^2 \text{ mlc}^{-2}} = \text{K}^{-1}$$

Note that P_m is actually per molecule.

14-3.6 PROBLEMS

1. $$\Delta H_{cryst} = \frac{NMe^2}{4\pi\varepsilon_o r_o}\left(1 - \frac{\rho}{r_o}\right) + 2RT$$

$$\rho = 0.34 \times 10^{-10} \text{ m}$$

$$r_o = 2.81 \times 10^{-10} \text{ m}$$

$$e = 1.62 \times 10^{-19} \text{ C}$$

$$\varepsilon_o = 8.85 \times 10^{-12} \text{ kg}^{-1} \text{ m}^{-3} \text{ s}^4 \text{ A}^2$$

M (from table) = 1.7476

$$\Delta H_{cryst} = \frac{(6.023 \times 10^{23} \text{ mlc}^{-1} \text{ mol}^{-1})(1.7476)(1.602 \times 10^{-19} \text{ C})^2}{(4)(3.1416)(8.85 \times 10^{-12} \text{ kg}^{-1} \text{ m}^3 \text{ s}^4 \text{ A}^2)(2.81 \times 10^{-10} \text{ m})}$$

$$\cdot \left(1 - \frac{0.34 \times 10^{-10}}{2.81 \times 10^{-10}}\right) + 2(8.314)(298.15)\text{J mol}^{-1}$$

$$= 7.60 \times 10^5 \frac{mol^{-1} \cancel{C^2}}{kg^{-1}\ m^{-3} \cancel{(A\ s)^2}\ s^2\ m} + 4.96 \times 10^3\ J\ mol^{-1}$$

$$= 7.60 \times 10^5\ J\ mol + 4.96 \times 10^3\ J\ mol^{-1}$$

$$= \underline{\underline{765\ kJ\ mol^{-1} = \Delta H_{cryst}}}$$

Born-Haber cycle value is 787 kJ mol^{-1}

2. $\Delta H_{cryst} = -\Delta H_f^\circ + \Delta H_{sub}(K) + \frac{1}{2}(D + RT)(F_2) + I(K) - A(F)$

$$= 563 + 89 + 80 + 419 - 333 = \underline{\underline{818\ kJ\ mol^{-1} = \Delta H_{cryst}}}$$

3. See text.

4. $\Delta H_{cryst} = -\Delta H_f^\circ + \Delta H_{sub} + \frac{1}{2}(D + RT) + I - A$

$$- 787 + 108 + 122 + 496 - 350 = -411\ kJ\ mol^{-1}$$

14-4.5 PROBLEMS

1. Axis ⊥ paper: 6-fold rotation; no mirrors
   Axis || paper: 2-fold along A line; no mirrors
   Axis || paper: 2-fold bisecting two lines; no mirrors

2.

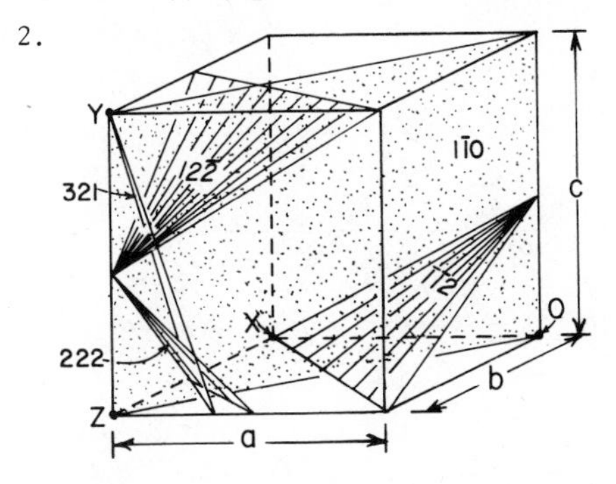

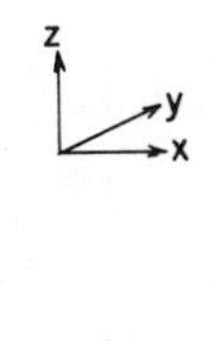

Origin	X ,	Y ,	Z ,	O
Indicies	$(1\bar{1}0)$	$(12\bar{2})$	(222),(321)	$(\bar{1}\bar{1}2)$

3. A - $(1\bar{3})$; B - (11); C - (01); D- (3,2)

4. 3m/m, $\bar{6}$; 2 fold ⊥ to the 3-fold axis.

5. See text.

6.

	a-intercept	b-intercept	c-intercept
$2\bar{2}2$	1/2	-1/2	1/2
333	1/3	1/3	1/3
$\bar{1}\bar{2}2$	-1	-1/2	1/2
$20\bar{2}$	1/2	∞	-1/2

Draw a figure with planes with the intercepts given.

14-5.7 PROBLEMS

1. $n\lambda = 2d \sin \theta$

$\theta = \arc \sin \frac{\lambda}{2d}$

d	2.35	1.70	1.65	2.85	5.80
θ	19.2	27.0	27.9	15.7	7.64

2.

	If P		If I		If F	
d_{hkl}	(hkl)	a(Å)	(hkl)	a(Å)	(hkl)	a(Å)
2.68	100	2.68	110	3.79	111	4.64
2.32	110	3.28	200	4.64	200	4.64
1.64	111	2.84	211	4.01	220	4.64
1.40	200	•	220	•	311	4.64
1.34	210	•	310	•	222	4.64
1.16	211	•	222	•	400	4.64

∴ Face centered cubic

a = 4.64 Å, four molecules per unit cell

$V_{cell} = (4.64 \times 10^{-10}\ m)^3 = 9.99 \times 10^{-29}\ m^3$

$$\text{Mass in each cell} = (4\ mlc)\left(\frac{0.042\ kg}{1\ mol}\right)\left(\frac{1\ mol}{6.023 \times 10^{23}\ mlc}\right)$$

$$= \underline{\underline{2.79 \times 10^{-25}\ kg}}$$

$\rho = m/v = \underline{\underline{2.79 \times 10^3\ kg\ m^3}}$

3. Large spheres touch on face diagonal = 2R (R = radius of large sphere)

The body diagonal = 2(R + r) (r = radius of small sphere)

Body diagonal = $\sqrt{3}$ a; face diagonal = $\sqrt{2}$ a (a = cell edge)

∴ $\sqrt{3}\,a = 2R + 2r$ and $2R = \sqrt{2}\,a \rightarrow a = \frac{2R + 2r}{\sqrt{3}} = \frac{2R}{\sqrt{2}}$

Thus, r = 0.225R

4. $$\sin\theta = \frac{\lambda}{2d} = \frac{\lambda}{2}\left[\frac{h^2}{a^2} + \frac{k^2}{b^2} + \frac{\ell^2}{c^2}\right]^{1/2} = 0.771\left[\frac{h^2}{a^2} + \frac{k^2}{b^2} + \frac{\ell^2}{c^2}\right]^{1/2}$$

$$a = 8.39\ \text{Å},\quad b = 8.79\ \text{Å},\quad c = 4.65\ \text{Å}$$

$$\sin\theta(101) = 0.771\left[\frac{1}{8.39^2} + \frac{1}{4.65^2}\right]^{1/2} = 0.190;$$

$$\theta(101) = 10.9°$$

$$\sin\theta(111) = 0.771\left[\frac{1}{8.39^2} + \frac{1}{8.79^2} + \frac{1}{4.65^2}\right]^{1/2} = 0.209;$$

$$\theta(111) = 12.1°$$

$$\theta(110) = 7.3°; \qquad \theta(222) = 24.7°$$

5. See text.

6. For cubic

$$\frac{1}{d_{hk\ell}} = \frac{1}{a}[h^2 + k^2 + \ell^2]^{1/2}$$

$$\sin\theta = \frac{\lambda}{2d} = \frac{\lambda}{2a}[h^2 + k^2 + \ell^2]^{1/2}$$

$$\therefore\quad a = \frac{\lambda}{2\sin\theta}[h^2 + k^2 + \ell^2]^{1/2} = \frac{1.542}{2\sin(19.1)}[1 + 1 + 1]^{1/2}$$

$$= \underline{\underline{4.08\ \text{Å} = a}}$$

$$\sin\theta(101) = \sin\theta(011) = 0.267; \quad \underline{\underline{\theta(101) = \theta(011) = 15.5°}}$$

$$\sin\theta(220) = 0.534; \quad \underline{\underline{\theta(220) = 32.3°}}$$

Appendix D
Some Common Integrals

$$\int_{-\infty}^{\infty} e^{-ax^2}\, dx = \sqrt{\pi/a}$$

$$\int_{-\infty}^{\infty} x^2\, e^{-ax^2}\, dx = \frac{1}{2a}\sqrt{\pi/a}$$

$$\int_{-\infty}^{\infty} x^4\, e^{-ax^2}\, dx = \frac{3}{4a^2}\sqrt{\pi/a}$$

$$\int_{-\infty}^{\infty} x^n\, e^{-ax^2}\, dx = \frac{n-1}{2a}\int_{0}^{\infty} x^{n-2}\, e^{-ax^2}\, dx$$

$$\int \frac{x\, dx}{a + bx} = \frac{x}{b} - \frac{a}{b^2}\ln\,(a + bx)$$

$$\int \frac{dx}{x(a + bx)} = -\frac{1}{a}\ln\frac{a + bx}{x}$$

$$\int_{0}^{\infty} x\, e^{-ax^2}\, dx = \frac{1}{2a}$$

$$\int_{0}^{\infty} x^3\, e^{-ax^2}\, dx = \frac{1}{2a^2}$$

Index

Index of Examples